Stochastic Modelling and Applied Probability

(Formerly:
Applications of Mathematics)

36

D0882108

Marek Musiela   Marek Rutkowski

Martingale Methods in Financial Modelling

Second Edition

 Springer

Authors

Marek Musiela
BNP Paribas
10 Harewood Avenue
London NW1 6AA
UK
marek.musiela@bnpparibas.com

Marek Rutkowski
Technical University Warszawa
Inst. Mathematics
Pl. Politechniki 1
00-661 Warszawa
Poland
markrut@alpha.mini.pw.edu.pl

Managing Editors

B. Rozovskiĭ
Division of Applied Mathematics
Brown University
182 George St
Providence, RI 02912
USA
rozovsky@dam.brown.edu

G. Grimmett
Centre for Mathematical Sciences
University of Cambridge
Wilberforce Road
Cambridge CB3 0WB
UK
g.r.grimmett@statslab.cam.ac.uk

Cover illustration: Cover pattern courtesy of Rick Durrett,
Cornell University, Ithaca, New York.

ISBN 978-3-642-05898-1   e-ISBN 978-3-540-26653-2
DOI 10.1007/978-3-540-26653-2

Stochastic Modelling and Applied Probability ISSN 0172-4568

Mathematics Subject Classification (2000): 60Hxx, 62P05, 90A09

2nd ed. 2005. Corr. 3rd printing 2009

© 2010 Springer-Verlag Berlin Heidelberg

Cover design: WMXDesign GmbH, Heidelberg

Printed on acid-free paper

9 8 7 6 5 4 3 2 1

springer.com

Preface to the Second Edition

During the seven years that elapsed between the first and second editions of the present book, considerable progress was achieved in the area of financial modelling and pricing of derivatives. Needless to say, it was our intention to incorporate into the second edition at least the most relevant and commonly accepted of these developments. Since at the same time we had the strong intention not to expand the book to an unbearable size, we decided to leave out from the first edition of this book some portions of material of lesser practical importance.

Let us stress that we have only taken out few sections that, in our opinion, were of marginal importance for the understanding of the fundamental principles of financial modelling of arbitrage valuation of derivatives. In view of the abundance of new results in the area, it would be in any case unimaginable to cover all existing approaches to pricing and hedging financial derivatives (not to mention all important results) in a single book, no matter how voluminous it were. Hence, several intensively studied areas, such as: mean-variance hedging, utility-based pricing, entropy-based approach, financial models with frictions (e.g., short-selling constraints, bid-ask spreads, transaction costs, etc.) either remain unmentioned in this text, or are presented very succinctly. Although the issue of market incompleteness is not totally neglected, it is examined primarily in the framework of models of stochastic (or uncertain) volatility. Luckily enough, the afore-mentioned approaches and results are covered exhaustively in several excellent monographs written in recent years by our distinguished colleagues, and thus it is our pleasure to be able to refer the interested reader to these texts.

Let us comment briefly on the content of the second edition and the differences with respect to the first edition.

Part I was modified to a lesser extent and thus is not very dissimilar to Part I in the first edition. However, since, as was mentioned already, some sections from the first edition were deliberately taken out, we decided for the sake of better readability to merge some chapters. Also, we included in Part I a new chapter entirely devoted to volatility risk and related modelling issues. As a consequence, the issues of hedging of plain-vanilla options and valuation of exotic options are no longer limited to the classical Black-Scholes framework with constant volatility. The theme of stochastic volatility also reappears systematically in the second part of the book.

Part II has been substantially revised and thus its new version constitutes a major improvement of the present edition with respect to the first one. We present there alternative interest rate models, and we provide the reader with an analysis of each of them, which is very much more detailed than in the first edition. Although we did not even try to appraise the efficiency of real-life implementations of each approach, we have stressed on each occasion that, when dealing with derivatives pricing models, one should always have in mind a specific practical perspective. Put another way, we advocate the opinion, put forward by many researchers, that the choice of model should be tied to observed real features of a particular sector of the financial market or even a product class. Consequently, a necessary first step in modelling is a detailed study of functioning of a given market we wish to model. The goal of this preliminary stage is to become familiar with existing liquid primary and derivative assets (together with their sometimes complex specifications), and to identify sources of risks associated with trading in these instruments.

It was our hope that by concentrating on the most pertinent and widely accepted modelling approaches, we will be able to provide the reader with a text focused on practical aspects of financial modelling, rather than theoretical ones. We leave it, of course, to the reader to assess whether we have succeeded achieving this goal to a satisfactory level.

Marek Rutkowski expresses his gratitude to Marek Musiela and the members of the Fixed Income Research and Strategies Team at BNP Paribas for their hospitality during his numerous visits to London.

Marek Rutkowski gratefully acknowledges partial support received from the Polish State Committee for Scientific Research under grant PBZ-KBN-016/P03/1999.

We would like to express our gratitude to the staff of Springer-Verlag. We thank Catriona Byrne for her encouragement and invaluable editorial supervision, as well as Susanne Denskus for her invaluable technical assistance.

London and Sydney   Marek Musiela
September 2004   Marek Rutkowski

Note on the Second Printing

The second printing of the second edition of this book expands and clarifies further its contents exposition. Several proofs previously left to the reader are now included. The presentation of LIBOR and swap market models is expanded to include the joint dynamics of the underlying processes under the relevant probability measures. The appendix in completed with several frequently used theoretical results making the book even more self-contained. The bibliographical references are brought up to date as far as possible.

This printing corrects also numerous typographical errors and mistakes. We would like the express our gratitude to Alan Bain and Imanuel Costigan who uncovered many of them.

London and Sydney   Marek Musiela
August 2006   Marek Rutkowski

Preface to the First Edition

The origin of this book can be traced to courses on financial mathematics taught by us at the University of New South Wales in Sydney, Warsaw University of Technology (Politechnika Warszawska) and Institut National Polytechnique de Grenoble. Our initial aim was to write a short text around the material used in two one-semester graduate courses attended by students with diverse disciplinary backgrounds (mathematics, physics, computer science, engineering, economics and commerce). The anticipated diversity of potential readers explains the somewhat unusual way in which the book is written. It starts at a very elementary mathematical level and does not assume any prior knowledge of financial markets. Later, it develops into a text which requires some familiarity with concepts of stochastic calculus (the basic relevant notions and results are collected in the appendix). Over time, what was meant to be a short text acquired a life of its own and started to grow. The final version can be used as a textbook for three one-semester courses – one at undergraduate level, the other two as graduate courses.

The first part of the book deals with the more classical concepts and results of arbitrage pricing theory, developed over the last thirty years and currently widely applied in financial markets. The second part, devoted to interest rate modelling is more subjective and thus less standard. A concise survey of short-term interest rate models is presented. However, the special emphasis is put on recently developed models built upon market interest rates.

We are grateful to the Australian Research Council for providing partial financial support throughout the development of this book. We would like to thank Alan Brace, Ben Goldys, Dieter Sondermann, Erik Schlögl, Lutz Schlögl, Alexander Mürmann, and Alexander Zilberman, who offered useful comments on the first draft, and Barry Gordon, who helped with editing.

Our hope is that this book will help to bring the mathematical and financial communities closer together, by introducing mathematicians to some important problems arising in the theory and practice of financial markets, and by providing finance professionals with a set of useful mathematical tools in a comprehensive and self-contained manner.

Sydney   Marek Musiela
March 1997   Marek Rutkowski

Contents

Part I

Spot and Futures Markets

1

An Introduction to Financial Derivatives

We shall first review briefly the most important kinds of financial contracts, traded either on exchanges or over-the-counter (OTC), between financial institutions and their clients. For a detailed account of the fundamental features of *spot* (i.e., *cash*) and *futures* financial markets the reader is referred, for instance, to Duffie (1989), Kolb (1991), Redhead (1996), or Hull (1997).

1.1 Options

Options are standard examples of *derivative securities* – that is, securities whose value depends on the prices of other more basic securities (referred to as *primary securities* or *underlying assets*) such as *stocks* or *bonds*. By *stocks* we mean *common stocks* – that is, shares in the net asset value not bearing fixed interest. They give the right to dividends according to profits, after payments on *preferred stocks* (the preferred stocks give some special rights to the stockholder, typically a guaranteed fixed dividend). A *bond* is a certificate issued by a government or a public company promising to repay borrowed money at a fixed rate of interest at a specified time. Generally speaking, a *call option* (respectively, a *put option*) is the right to buy (respectively, to sell) the option's underlying asset at some future date for a predetermined price. Options (in particular, warrants[1]) have been traded for centuries in many countries. Unprecedented expansion of the options market started, however, quite recently with the introduction in 1973 of listed stock options on the Chicago Board Options Exchange (CBOE). Incidentally, in the same year Black and Scholes and, independently, Merton have published the seminal papers, in which the fundamental principles of arbitrage pricing of options were elaborated. During the last thirty years, trading in derivative securities have undergone a tremendous development, and nowadays options, futures, and other financial derivatives are traded in large numbers all over the world.

---

[1] A *warrant* is a call option issued by a company or a financial institution.

We shall now describe, following Hull (1997), the basic features of traditional stock and options markets, as opposed to computerized online trading. The most common system for trading stocks is a *specialist system*. Under this system, an individual known as the specialist is responsible for being a market maker and for keeping a record of limit orders – that is, orders that can only be executed at the specified price or a more favorable price. Options usually trade under a *market maker system*. A market maker for a given option is an individual who will quote both a bid and an ask price on the option whenever he is asked to do so. The bid price is the price at which the market maker is prepared to buy and the ask price is the price at which he is prepared to sell. At the time the bid and ask prices are quoted, the market maker does not know whether the trader who asked for the quotes wants to buy or sell the option. The amount by which the ask exceeds the bid is referred to as the *bid-ask spread*. To enhance the efficiency of trading, the exchange may set upper limits for the bid-ask spread.

The existence of the market maker ensures that buy and sell orders can always be executed at some price without delay. The market makers themselves make their profits from the bid-ask spread. When an investor writes options, he is required to maintain funds in a margin account. The size of the margin depends on the circumstances, e.g., whether the option is *covered* or *naked* – that is, whether the option writer does possess the underlying shares or not. Let us finally mention that one contract gives the holder the right to buy or sell 100 shares; this is convenient since the shares themselves are usually traded in lots of 100.

It is worth noting that most of the traded options are of *American style* (or shortly, *American options*) – that is, the holder has the right to exercise an option at any instant before the option's expiry. Otherwise, that is, when an option can be exercised only at its expiry date, it is known as an option of *European style* (a *European option*, for short).

Let us now focus on exercising of an option of American style. The record of all outstanding long and short positions in options is held by the Options Clearing Corporation (OCC). The OCC guarantees that the option writer will fulfil obligations under the terms of the option contract. The OCC has a number of the so-called *members*, and all option trades must be cleared through a member. When an investor notifies his broker of the intention to exercise an option, the broker in turn notifies the OCC member who clears the investor's trade. This member then places an exercise order with the OCC. The OCC randomly selects a member with an outstanding short position in the same option. The chosen member, in turn, selects a particular investor who has written the option (such an investor is said to be *assigned*). If the option is a call, this investor is required to sell stock at the so-called *strike price* or *exercise price* (if it is a put, he is required to buy stock at the strike price). When the option is exercised, the *open interest* (that is, the number of options outstanding) goes down by one.

In addition to options on particular stocks, a large variety of other option contracts are traded nowadays on exchanges: foreign currency options, index options (e.g., those on S&P100 and S&P500 traded on the CBOE), and futures options (e.g., the Treasury bond futures option traded on the Chicago Board of Trade (CBOT)).

Derivative financial instruments involving options are also widely traded outside the exchanges by financial institutions and their clients. Let us mention here such widely popular interest-rate sensitive contracts as *caps* and *floors*. They are, basically, portfolios of call and put options on a prespecified interest rate respectively. Another important class of interest rate options are *swaptions* – that is, options on an *interest rate swap*. A swaption can be equivalently seen as an option on the *swap rate*. Finally, options are implicit in several financial instruments, for example in some bond or stock issues (*callable bonds, savings bonds* or *convertible bonds,* to mention a few).

One of the most appealing features of options (apart from the obvious chance of making extraordinary returns) is the possibility of easy speculation on the future behavior of a stock price. Usually this is done by means of so-called *combinations* – that is, combined positions in several options, and possibly the underlying asset. For instance, a *bull spread* is a portfolio created by buying a call option on a stock with a certain strike price and selling a call option on the same stock with a higher strike price (both options have the same expiry date). Equivalently, bull spreads can be created by buying a put with a low strike price and selling a put with a high strike price. An investor entering a bull spread is hoping that the stock price will increase. Like a bull spread, a *bear spread* can be created by buying a call with one strike price and selling a call with another strike price. The strike price of the option purchased is now greater than the strike price of the option sold, however. An investor who enters a bear spread is hoping that the stock price will decline.

A *butterfly spread* involves positions in options with three different strike prices. It can be created by buying a call option with a relatively low strike price, buying another call option with a relatively high strike price, and selling two call options with a strike price halfway between the other two strike prices. The butterfly spread leads to a profit if the stock price stays close to the strike price of the call options sold, but gives rise to a small loss if there is a significant stock price move in either direction. A portfolio created by selling a call option with a certain strike price and buying a longer-maturity call option with the same strike price is commonly known as a *calendar spread*. A *straddle* involves buying a call and put with the same strike price and expiry date. If the stock price is close to this strike price at expiry of the option, the straddle leads to a loss. A straddle is appropriate when an investor is expecting a large move in stock price but does not know in which direction the move will be. Related types of trading strategies are commonly known as *strips, straps* and *strangles*.

1.2 Futures Contracts and Options

Another important class of exchange-traded derivative securities comprises *futures contracts*, and options on futures contracts, commonly known as *futures options*. Futures contracts apply to a wide range of commodities (e.g., sugar, wool, gold) and financial assets (e.g., currencies, bonds, stock indices); the largest exchanges on which futures contracts are traded are the Chicago Board of Trade and the Chicago Mercantile Exchange (CME). In what follows, we restrict our attention to financial futures (as opposed to commodity futures). To make trading possible, the exchange specifies certain standardized features of the contract. Futures prices are regularly reported in the financial press. They are determined on the floor in the same way as other prices – that is, by the law of supply and demand. If more investors want to go long than to go short, the price goes up; if the reverse is true, the price falls. Positions in futures contracts are governed by a specific daily settlement procedure commonly referred to as *marking to market*. An investor's initial deposit, known as the *initial margin,* is adjusted daily to reflect the gains or losses that are due to the futures price movements. Let us consider, for instance, a party assuming a long position (the party who agreed to buy). When there is a decrease in the futures price, her margin account is reduced by an appropriate amount of money, her broker has to pay this sum to the exchange and the exchange passes the money on to the broker of the party who assumes the short position. Similarly, when the futures price rises, brokers for parties with short positions pay money to the exchange, and brokers of parties with long positions receive money from the exchange. This way, the trade is marked to market at the close of each trading day. Finally, if the delivery period is reached and delivery is made by a party with a short position, the price received is generally the futures price at the time the contract was last marked to market.

In a *futures option*, the underlying asset is a futures contract. The futures contract normally matures shortly after the expiry of the option. When the holder of a futures call option exercises the option, she acquires from the writer a long position in the underlying futures contract plus a cash amount equal to the excess of the current futures price over the option's strike price. Since futures contracts have zero value and can be closed out immediately, the payoff from a futures option is the same as the payoff from a stock option, with the stock price replaced by the futures price. Futures options are now available for most of the instruments on which futures contracts are traded. The most actively traded futures option is the Treasury bond futures option traded on the Chicago Board of Trade. On some markets (for instance, on the Australian market), futures options have the same features as futures contracts themselves – that is, they are not paid up-front as classical options, but are traded at the margin. Unless otherwise stated, by a futures option we mean here a standard option written on a futures contract.

1.3 Forward Contracts

A *forward contract* is an agreement to buy or sell an asset at a certain future time for a certain price. One of the parties to a forward contract assumes a long position and agrees to buy the underlying asset on a certain specified future date for a *delivery price*; the other party assumes a short position and agrees to sell the asset on the same date for the same price. At the time the contract is entered into, the delivery price is determined so that the value of the forward contract to both parties is zero. Thus it is clear that some features of forward contracts resemble those of futures contracts. However, unlike futures contracts, forward contracts do not trade on exchanges. Also, a forward contract is settled only once, at the maturity date. The holder of the short position delivers the asset to the holder of the long position in return for a cash amount equal to the delivery price. The following list (cf. Sutcliffe (1993)) summarizes the main differences between forward and futures contracts. A more detailed description of the functioning of futures markets can be found, for instance, in Duffie (1989), Kolb (1991), or Sutcliffe (1993).

1. Contract specification and delivery

Futures contracts. The contract precisely specifies the underlying instrument and price. Delivery dates and delivery procedures are standardized to a limited number of specific dates per year, at approved locations. Delivery is not, however, the objective of the transaction, and less than 2% are delivered.

Forward contracts. There is an almost unlimited range of instruments, with individually negotiated prices. Delivery can take place on any individual negotiated date and location. Delivery is the object of the transaction, with over 90% of forward contracts settled by delivery.

2. Prices

Futures contracts. The price is the same for all participants, regardless of transaction size. Typically, there is a daily price limit (although, for instance, on the FT-SE 100 index, futures prices are unlimited). Trading is usually by open outcry auction on the trading floor of the exchange. Prices are disseminated publicly. Each transaction is conducted at the best price available at the time.

Forward contracts. The price varies with the size of the transaction, the credit risk, etc. There are no daily price limits. Trading takes place between individual buyers and sellers. Prices are not disseminated publicly. Hence, there is no guarantee that the price is the best available.

3. Marketplace and trading hours

Futures contracts. Trading is centralized on the exchange floor, with worldwide communications, during hours fixed by the exchange.

Forward contracts. Trading takes place through direct negotiations between individual buyers and sellers. Trading is over-the-counter world-wide, 24 hours per day.

4. Security deposit and margin

Futures contracts. The exchange rules require an initial margin and the daily settlement of variation margins. A central clearing house is associated with each exchange to handle the daily revaluation of open positions, cash payments and delivery procedures. The clearing house assumes the credit risk.

Forward contracts. The collateral level is negotiable, with no adjustment for daily price fluctuations. There is no separate clearing house function. Thus, the market participant bears the risk of the counter-party defaulting.

5. Volume and market liquidity

Futures contracts. Volume (and open interest) information is published. There is very high liquidity and ease of offset with any other market participant due to standardized contracts.

Forward contracts. Volume information is not available. The limited liquidity and offset is due to the variable contract terms. Offset is usually with the original counter-party.

1.4 Call and Put Spot Options

Let us first describe briefly the set of general assumptions imposed on our models of financial markets. We consider throughout, unless explicitly stated otherwise, the case of a so-called *frictionless market*, meaning that: all investors are price-takers, all parties have the same access to the relevant information, there are no transaction costs or commissions andall assets are assumed to be perfectly divisible and liquid. There is no restriction whatsoever on the size of a bank credit, and the lending and borrowing rates are equal. Finally, individuals are allowed to sell short any security and receive full use of the proceeds (of course, restitution is required for payoffs made to securities held short). Unless otherwise specified, by an *option* we shall mean throughout a European option, giving the right to exercise the option only at the expiry date. In mathematical terms, the problem of pricing of American options is closely related to *optimal stopping* problems. Unfortunately, closed-form expressions for the prices of American options are rarely available; for instance, no closed-form solution is available for the price of an American put option in the classical framework of the Black-Scholes option pricing model.

A *European call option* written on a common stock[2] is a financial security that gives its holder the right (but not the obligation) to buy the underlying stock on at some given date and for a predetermined price. The predetermined fixed price, say K, is termed the *strike* or *exercise* price; the terminal date, denoted by T in what follows, is called the *expiry date* or *maturity*. The act of making this transaction is referred to as *exercising* the option. If an option is not exercised, we say it is *abandoned*.

Another class of options comprises so-called *American options*. These may be exercised at any time on or before the expiry date. Let us emphasize that an option gives the holder the right to do something; however, the holder is not obliged to exercise this right. In order to purchase an option contract, an investor needs to pay an option's price (or *premium*) to a second party at the initial date when the contract is entered into.

Let us denote by S_T the stock price at the terminal date T. It is natural to assume that S_T is not known at time 0, hence S_T gives rise to uncertainty in our model. We argue that from the perspective of the option holder, the payoff g at expiry date T from a European call option is given by the formula

$$g(S_T) = (S_T - K)^+ \stackrel{\text{def}}{=} \max\{S_T - K, 0\}, \tag{1.1}$$

that is to say

---

[2] Unless explicitly stated otherwise, we assume throughout that the underlying stock pays no dividends during the option's lifetime.

$$g(S_T) = \begin{cases} S_T - K, & \text{if } S_T > K \text{ (option is exercised)}, \\ 0, & \text{if } S_T \le K \text{ (option is abandoned)}. \end{cases}$$

In fact, if at the expiry date T the stock price is lower than the strike price, the holder of the call option can purchase an underlying stock directly on a spot (i.e., cash) market, paying less than K. In other words, it would be irrational to exercise the option, at least for an investor who prefers more wealth to less. On the other hand, if at the expiry date the stock price is greater than K, an investor should exercise his right to buy the underlying stock at the strike price K. Indeed, by selling the stock immediately at the spot market, the holder of the call option is able to realize an instantaneous net profit $S_T - K$ (note that transaction costs and/or commissions are ignored here). In contrast to a call option, a *put option* gives its holder the right to sell the underlying asset by a certain date for a predetermined price. Using the same notation as above, we arrive at the following expression for the payoff h at maturity T from a European put option

$$h(S_T) = (K - S_T)^+ \stackrel{\text{def}}{=} \max\{K - S_T, 0\}, \tag{1.2}$$

or more explicitly

$$h(S_T) = \begin{cases} 0, & \text{if } S_T \ge K \text{ (option is abandoned)}, \\ K - S_T, & \text{if } S_T < K \text{ (option is exercised)}. \end{cases}$$

It follows immediately that the payoffs of call and put options satisfy the following simple but useful equality

$$g(S_T) - h(S_T) = (S_T - K)^+ - (K - S_T)^+ = S_T - K. \tag{1.3}$$

The last equality can be used, in particular, to derive the so-called *put-call parity* relationship for option prices. Basically, put-call parity means that the price of a European put option is determined by the price of a European call option with the same strike and expiry date, the current price of the underlying asset, and the properly discounted value of the strike price.

1.4.1 One-period Spot Market

Let us start by considering an elementary example of an option contract.

Example 1.4.1 Assume that the current stock price is $280, and after three months the stock price may either rise to $320 or decline to $260. We shall find the rational price of a 3-month European call option with strike price $K = \$280$, provided that the simple risk-free interest rate r for 3-month deposits and loans[3] is $r = 5\%$.

Suppose that the subjective probability of the price rise is 0.2, and that of the fall is 0.8; these assumptions correspond, loosely, to a so-called *bear market*. Note that the word *subjective* means that we take the point of view of a particular individual.

---
[3] We shall usually assume that the borrowing and lending rates are equal.

Generally speaking, the two parties involved in an option contract may have (and usually do have) differing assessments of these probabilities. To model a *bull market* one may assume, for example, that the first probability is 0.8, so that the second is 0.2. The *subjective probability* $\mathbb{P}$ is also referred to as the *actual probability, real-world probability*, or *statistical probability* in various texts.

Let us focus first on the bear market case. The terminal stock price S_T may be seen as a random variable on a probability space $\Omega = \{\omega_1, \omega_2\}$ with a probability measure $\mathbb{P}$ given by

$$\mathbb{P}\{\omega_1\} = 0.2 = 1 - \mathbb{P}\{\omega_2\}.$$

Formally, S_T is a function $S_T : \Omega \to R_+$ given by the following formula

$$S_T(\omega) = \begin{cases} S^u = 320, & \text{if } \omega = \omega_1, \\ S^d = 260, & \text{if } \omega = \omega_2. \end{cases}$$

Consequently, the terminal option's payoff $X = C_T = (S_T - K)^+$ satisfies

$$C_T(\omega) = \begin{cases} C^u = 40, & \text{if } \omega = \omega_1, \\ C^d = 0, & \text{if } \omega = \omega_2. \end{cases}$$

Note that the expected value under $\mathbb{P}$ of the discounted option's payoff equals

$$\mathbb{E}_\mathbb{P}\left((1+r)^{-1}C_T\right) = 0.2 \times 40 \times (1.05)^{-1} = 7.62.$$

It is clear that the above expectation depends on the choice of the probability measure $\mathbb{P}$; that is, it depends on the investor's assessment of the market. For a call option, the expectation corresponding to the case of a bull market would be greater than that which assumes a bear market. In our example, the expected value of the discounted payoff from the option under the bull market hypothesis equals 30.48. Still, to construct a reliable model of a financial market, one has to guarantee the uniqueness of the price of any derivative security. This can be done by applying the concept of the so-called replicating portfolio, which we will now introduce.

1.4.2 Replicating Portfolios

The two-state option pricing model presented below was developed independently by Sharpe (1978) and Rendleman and Bartter (1979) (a point worth mentioning is that the ground-breaking papers of Black and Scholes (1973) and Merton (1973), who examined the arbitrage pricing of options in a continuous-time framework, were published much earlier). The idea is to construct a portfolio at time 0 that replicates exactly the option's terminal payoff at time T. Let $\phi = \phi_0 = (\alpha_0, \beta_0) \in \mathbb{R}^2$ denote a portfolio of an investor with a short position in one call option. More precisely, let α_0 stand for the number of shares of stock held at time 0, and β_0 be the amount of money deposited on a bank account or borrowed from a bank. By $V_t(\phi)$ we denote the wealth of this portfolio at dates $t = 0$ and $t = T$; that is, the payoff from the portfolio ϕ at given dates. It should be emphasized that once the portfolio is set up at time 0, it remains fixed until the terminal date T. For its wealth process $V(\phi)$, we therefore have

$$V_0(\phi) = a_0 S_0 + \beta_0 \quad \text{and} \quad V_T(\phi) = a_0 S_T + \beta_0(1 + r). \tag{1.4}$$

We say that a portfolio ϕ *replicates* the option's terminal payoff whenever $V_T(\phi) = C_T$, that is, if

$$V_T(\phi)(\omega) = \begin{cases} V^u(\phi) = a_0 S^u + (1+r)\beta_0 = C^u, & \text{if } \omega = \omega_1, \\ V^d(\phi) = a_0 S^d + (1+r)\beta_0 = C^d, & \text{if } \omega = \omega_2. \end{cases}$$

For the data of Example 1.4.1, the portfolio ϕ is determined by the following system of linear equations

$$\begin{cases} 320\,\alpha_0 + 1.05\,\beta_0 = 40, \\ 260\,\alpha_0 + 1.05\,\beta_0 = 0, \end{cases}$$

with unique solution $\alpha_0 = 2/3$ and $\beta_0 = -165.08$. Observe that for every call we are short, we hold α_0 of stock[4] and the dollar amount β_0 in risk-free bonds in the hedging portfolio. Put another way, by purchasing shares and borrowing against them in the right proportion, we are able to replicate an option position. (Actually, one can easily check that this property holds for any *contingent claim* X that settles at time T.) It is natural to define the *manufacturing cost* C_0 of a call option as the initial investment needed to construct a replicating portfolio, i.e.,

$$C_0 = V_0(\phi) = a_0 S_0 + \beta_0 = (2/3) \times 280 - 165.08 = 21.59.$$

Note that in order to determine the manufacturing cost of a call we did not need to know the probability of the rise or fall of the stock price.

In other words, it appears that the manufacturing cost is invariant with respect to individual assessments of market behavior. In particular, it is identical under the bull and bear market hypotheses. To determine the *rational* price of a call we have used the option's strike price, the current value of the stock price, the range of fluctuations in the stock price (that is, the future levels of the stock price), and the risk-free rate of interest. The investor's transactions and the corresponding cash flows may be summarized by the following two exhibits for any $t \in [0, T]$

$$\text{at time } t = 0 \quad \begin{cases} \text{one written call option} & C_0, \\ \alpha_0 \text{ shares purchased} & -\alpha_0 S_0, \\ \text{amount of cash borrowed} & \beta_0, \end{cases}$$

and

$$\text{at time } t = T \quad \begin{cases} \text{payoff from the call option} & -C_T, \\ \alpha_0 \text{ shares sold} & \alpha_0 S_T, \\ \text{loan paid back} & -\hat{r}\beta_0, \end{cases}$$

---

[4] We shall refer to the number of shares held for each call sold as the *hedge ratio*. Basically, to *hedge* means to reduce risk by making transactions that reduce exposure to market fluctuations.

where $\hat{r} = 1 + r$. Observe that no net initial investment is needed to establish the above portfolio; that is, the portfolio is costless. On the other hand, for each possible level of stock price at time T, the hedge exactly breaks even on the option's expiry date. Also, it is easy to verify that if the call were not priced at \$21.59, it would be possible for a sure profit to be gained, either by the option's writer (if the option's price were greater than its manufacturing cost) or by its buyer (in the opposite case). Still, the manufacturing cost cannot be seen as a fair price of a claim X, unless the market model is arbitrage-free, in a sense examined below. Indeed, it may happen that the manufacturing cost of a nonnegative claim is a strictly negative number. Such a phenomenon contradicts the usual assumption that it is not possible to make risk-free profits.

1.4.3 Martingale Measure for a Spot Market

Although, as shown above, subjective (or actual) probabilities are useless when pricing an option, probabilistic methods play an important role in contingent claims valuation. They rely on the notion of a *martingale*, which is, intuitively, a probabilistic model of a fair game. In order to apply the so-called *martingale method* of derivative pricing, one has to find first a probability measure $\mathbb{P}^*$ equivalent to $\mathbb{P}$, and such that the *discounted* (or *relative*) stock price process S^*, defined by the formula

$$S_0^* = S_0, \quad S_T^* = (1+r)^{-1} S_T,$$

follows a $\mathbb{P}^*$-martingale; that is, the equality $S_0^* = \mathbb{E}_{\mathbb{P}^*}(S_T^*)$ holds. Such a probability measure $\mathbb{P}^*$ is called a *martingale measure* (or a *martingale probability*) for the discounted stock price process S^*.

In the case of a two-state model, the probability measure $\mathbb{P}^*$ is easily seen to be uniquely determined (provided it exists) by the following linear equation

$$S_0 = (1+r)^{-1}(p_* S^u + (1 - p_*) S^d), \tag{1.5}$$

where $p_* = \mathbb{P}^*\{\omega_1\}$ and $1 - p_* = \mathbb{P}^*\{\omega_2\}$. Solving this equation for p_*

$$\mathbb{P}^*\{\omega_1\} = \frac{(1+r)S_0 - S^d}{S^u - S^d}, \quad \mathbb{P}^*\{\omega_2\} = \frac{S^u - (1+r)S_0}{S^u - S^d}. \tag{1.6}$$

Let us now check that the price C_0 coincides with C_0^*, where we denote by C_0^* the expected value under $\mathbb{P}^*$ of an option's discounted terminal payoff – that is

$$C_0^* \overset{\text{def}}{=} \mathbb{E}_{\mathbb{P}^*}\big((1+r)^{-1} C_T\big) = \mathbb{E}_{\mathbb{P}^*}\big((1+r)^{-1}(S_T - K)^+\big).$$

Indeed, using the data of Example 1.4.1 we find $p_* = 17/3$, so that

$$C_0^* = (1+r)^{-1}\big(p_* C^u + (1 - p_*) C^d\big) = 21.59 = C_0.$$

Remarks. Observe that since the process S^* follows a $\mathbb{P}^*$-martingale, we may say that the discounted stock price process may be seen as a fair game model in a *risk-neutral* *economy* – that is, in the stochastic economy in which the probabilities of future stock price fluctuations are determined by the martingale measure $\mathbb{P}^*$. For this reason, $\mathbb{P}^*$ is also known as the *risk-neutral probability*. It should be stressed, however, that the fundamental idea of arbitrage pricing is based exclusively on the existence of a portfolio that hedges perfectly the risk exposure related to uncertain future prices of risky securities. Thus, the probabilistic properties of the model are not essential. In particular, we do not assume that the real-world economy is actually risk-neutral. On the contrary, the notion of a risk-neutral economy should be seen rather as a technical tool. The aim of introducing the martingale measure is twofold: firstly, it simplifies the explicit evaluation of arbitrage prices of derivative securities; secondly, it describes the arbitrage-free property of a given pricing model for primary securities in terms of the behavior of relative prices. This approach is frequently referred to as the *partial equilibrium approach*, as opposed to the *general equilibrium approach*. Let us stress that in the latter theory the investors' preferences, usually described in stochastic models by means of their (expected) utility functions, play an important role.

To summarize, the notion of an arbitrage price for a derivative security does not depend on the choice of a probability measure in a particular pricing model for primary securities. More precisely, using standard probabilistic terminology, this means that the arbitrage price depends on the support of an actual probability measure $\mathbb{P}$, but is invariant with respect to the choice of a particular probability measure from the class of mutually equivalent probability measures. In financial terminology, this can be restated as follows: all investors agree on the range of future price fluctuations of primary securities; they may have different assessments of the corresponding subjective probabilities, however.

1.4.4 Absence of Arbitrage

Let us consider a simple two-state, one-period, two-security market model defined on a probability space $\Omega = \{\omega_1, \omega_2\}$ equipped with the σ-fields $\mathcal{F}_0 = \{\emptyset, \Omega\}$, $\mathcal{F}_T = 2^\Omega$ (i.e., $\mathcal{F}_T$ contains all subsets of Ω), and a probability measure $\mathbb{P}$ on $(\Omega, \mathcal{F}_T)$ such that $\mathbb{P}\{\omega_1\}$ and $\mathbb{P}\{\omega_2\}$ are strictly positive numbers. The first primary security is a stock whose price is modelled as a strictly positive discrete-time process $S = (S_t)_{t \in \{0, T\}}$. We assume that the process S is adapted to the *filtration* $\mathbb{F} = \{\mathcal{F}_0, \mathcal{F}_T\}$, meaning that the random variable S_t is $\mathcal{F}_t$-measurable for $t = 0, T$. This implies that S_0 is a real number, and

$$S_T(\omega) = \begin{cases} S^u, & \text{if } \omega = \omega_1, \\ S^d, & \text{if } \omega = \omega_2, \end{cases}$$

where, without loss of generality, $S^u > S^d$. The second primary security is a risk-free bond whose price process is $B_0 = 1$, $B_T = 1+r$ for some real $r \geq 0$. Let Φ stand for the linear space of all stock-bond portfolios $\phi = \phi_0 = (\alpha_0, \beta_0)$, where α_0 and β_0 are real numbers (clearly, the class Φ may be thus identified with $\mathbb{R}^2$). We shall consider the pricing of contingent claims in a security market model $\mathcal{M} = (S, B, \Phi)$. We shall now check that an arbitrary *European contingent claim* X that settles at time T (i.e., any $\mathcal{F}_T$-measurable real-valued random variable) admits a unique replicating portfolio in our market model. In other words, an arbitrary contingent claim X is *attainable* in the market model $\mathcal{M}$. Indeed, if

$$X(\omega) = \begin{cases} X^u, & \text{if } \omega = \omega_1, \\ X^d, & \text{if } \omega = \omega_2, \end{cases}$$

then the replicating portfolio ϕ is determined by a linear system of two equations in two unknowns, namely

$$\begin{cases} \alpha_0 S^u + (1+r)\beta_0 = X^u, \\ \alpha_0 S^d + (1+r)\beta_0 = X^d, \end{cases} \tag{1.7}$$

that admits a unique solution

$$\alpha_0 = \frac{X^u - X^d}{S^u - S^d}, \quad \beta_0 = \frac{X^d S^u - X^u S^d}{(1+r)(S^u - S^d)}, \tag{1.8}$$

for arbitrary values of X^u and X^d. Consequently, an arbitrary contingent claim X admits a unique *manufacturing cost* $\pi_0(X)$ in $\mathcal{M}$ which is given by the formula

$$\pi_0(X) \stackrel{\text{def}}{=} V_0(\phi) = \alpha_0 S_0 + \beta_0 = \frac{X^u - X^d}{S^u - S^d} S_0 + \frac{X^d S^u - X^u S^d}{(1+r)(S^u - S^d)}. \tag{1.9}$$

As was already mentioned, the manufacturing cost of a strictly positive contingent claim may appear to be a negative number, in general.

If this were the case, there would be a profitable risk-free trading strategy (so-called *arbitrage opportunity*) involving only the stock and risk-free borrowing and lending. To exclude such situations, which are clearly inconsistent with any broad notion of a rational market equilibrium (as it is common to assume that investors are *non-satiated*, meaning that they prefer more wealth to less), we have to impose further essential restrictions on our simple market model.

Definition 1.4.1 We say that a security pricing model $\mathcal{M}$ is *arbitrage-free* if there is no portfolio $\phi \in \Phi$ for which

$$V_0(\phi) = 0, \quad V_T(\phi) \geq 0 \quad \text{and} \quad \mathbb{P}\{V_T(\phi) > 0\} > 0. \tag{1.10}$$

A portfolio ϕ for which the set (1.10) of conditions is satisfied is called an *arbitrage opportunity*. A *strong arbitrage opportunity* is a portfolio ϕ for which

$$V_0(\phi) < 0 \quad \text{and} \quad V_T(\phi) \geq 0. \tag{1.11}$$

It is customary to take either (1.10) or (1.11) as the definition of an arbitrage opportunity. Note, however, that both notions are not necessarily equivalent. We are in a position to introduce the notion of an arbitrage price; that is, the price derived using the no-arbitrage arguments.

Definition 1.4.2 Suppose that the security market $\mathcal{M}$ is arbitrage-free. Then the manufacturing cost $\pi_0(X)$ is called the *arbitrage price* of X at time 0 in security market $\mathcal{M}$.

As the next result shows, under the absence of arbitrage in a market model, the manufacturing cost may be seen as the unique rational price of a given contingent claim – that is, the unique price compatible with any rational market equilibrium. Since it is easy to create an arbitrage opportunity if the no-arbitrage condition $H_0 = \pi_0(X)$ is violated, the proof is left to the reader.

Proposition 1.4.1 *Suppose that the spot market model $\mathcal{M} = (S, B, \Phi)$ is arbitrage-free. Let H stand for the rational price process of some attainable contingent claim X; more explicitly, $H_0 \in \mathbb{R}$ and $H_T = X$. Let us denote by Φ_H the class of all portfolios in stock, bond and derivative security H. The extended market model (S, B, H, Φ_H) is arbitrage-free if and only if $H_0 = \pi_0(X)$.*

1.4.5 Optimality of Replication

Let us show that replication is, in a sense, an optimal way of hedging. Firstly, we say that a portfolio ϕ *perfectly hedges* against X if $V_T(\phi) \geq X$, that is, whenever

$$\begin{cases} \alpha_0 S^u + (1+r)\beta_0 \geq X^u, \\ \alpha_0 S^d + (1+r)\beta_0 \geq X^d. \end{cases} \tag{1.12}$$

The minimal initial cost of a perfect hedging portfolio against X is called the *seller's price* of X, and it is denoted by $\pi_0^s(X)$. Let us check that the equality $\pi_0^s(X) = \pi_0(X)$ holds. By denoting $c = V_0(\phi)$, we may rewrite (1.12) as follows

$$\begin{cases} \alpha_0(S^u - S_0(1+r)) + c(1+r) \geq X^u, \\ \alpha_0(S^d - S_0(1+r)) + c(1+r) \geq X^d. \end{cases} \tag{1.13}$$

It is trivial to check that the minimal $c \in \mathbb{R}$ for which (1.13) holds is actually that value of c for which inequalities in (1.13) become equalities. This means that the replication appears to be the least expensive way of perfect hedging for the seller of X. Let us now consider the other party of the contract, i.e., the buyer of X. Since the buyer of X can be seen as the seller of $-X$, the associated problem is to minimize $c \in \mathbb{R}$, subject to the following constraints

$$\begin{cases} \alpha_0(S^u - S_0(1+r)) + c(1+r) \geq -X^u, \\ \alpha_0(S^d - S_0(1+r)) + c(1+r) \geq -X^d. \end{cases}$$

It is clear that the solution to this problem is $\pi^s(-X) = -\pi(X) = \pi(-X)$, so that replication appears to be optimal for the buyer also. We conclude that the least price the seller is ready to accept for X equals the maximal amount the buyer is ready to pay for it. If we define the *buyer's price* of X, denoted by $\pi_0^b(X)$, by setting $\pi_0^b(X) = -\pi_0^s(-X)$, then

$$\pi_0^s(X) = \pi_0^b(X) = \pi_0(X);$$

that is, all prices coincide. This shows that in a two-state, arbitrage-free model, the arbitrage price of any contingent claim can be defined using the optimality criterion. It appears that such an approach to arbitrage pricing carries over to more general models models; we prefer, however, to define the arbitrage price as that value of the price that rules out arbitrage opportunities. Indeed, the fact that observed market prices are close to arbitrage prices predicted by a suitable stochastic model is due to the presence of the traders known as *arbitrageurs*[5] on financial markets, rather than to the rational investment decisions of most market participants.

The next proposition explains the role of the so-called *risk-neutral* economy in arbitrage pricing of derivative securities. Observe that the important role of risk preferences in classical equilibrium asset pricing theory is left aside in the present context. Notice, however, that the use of a martingale measure $\mathbb{P}^*$ in arbitrage pricing corresponds to the assumption that all investors are risk-neutral, meaning that they do not differentiate between all risk-free and risky investments with the same expected rate of return. The arbitrage valuation of derivative securities is thus done as if an economy actually were *risk-neutral*. Formula (1.14) shows that the arbitrage price of a contingent claim X can be found by first modifying the model so that the stock earns at the risk-free rate, and then computing the expected value of the discounted claim. To the best of our knowledge, this method of computing the price was discovered by Cox and Ross (1976b).

Proposition 1.4.2 *The spot market* $\mathcal{M} = (S, B, \Phi)$ *is arbitrage-free if and only if the discounted stock price process* S^* *admits a martingale measure* $\mathbb{P}^*$ *equivalent to* $\mathbb{P}$. *In this case, the arbitrage price at time 0 of any contingent claim* X *that settles at time* T *is given by the risk-neutral valuation formula*

$$\pi_0(X) = \mathbb{E}_{\mathbb{P}^*}\big((1+r)^{-1}X\big), \tag{1.14}$$

or explicitly

$$\pi_0(X) = \frac{S_0(1+r) - S^d}{S^u - S^d}\,\frac{X^u}{1+r} + \frac{S^u - S_0(1+r)}{S^u - S^d}\,\frac{X^d}{1+r}. \tag{1.15}$$

Proof. We know already that the martingale measure for S^* equivalent to $\mathbb{P}$ exists if and only if the unique solution p_* of equation (1.5) satisfies $0 < p_* < 1$. Suppose there is no equivalent martingale measure for S^*; for instance, assume that $p_* \geq 1$. Our aim is to construct explicitly an arbitrage opportunity in the market model (S, B, Φ). To this end, observe that the inequality $p_* \geq 1$ is equivalent to $(1+r)S_0 \geq S^u$ (recall that S^u is always greater than S^d). The portfolio $\phi = (-1, S_0)$ satisfies $V_0(\phi) = 0$ and

$$V_T(\phi) = \begin{cases} -S^u + (1+r)S_0 \geq 0, & \text{if } \omega = \omega_1, \\ -S^d + (1+r)S_0 > 0, & \text{if } \omega = \omega_2, \end{cases}$$

so that ϕ is indeed an arbitrage opportunity. On the other hand, if $p_* \leq 0$, then the inequality $S^d \geq (1+r)S_0$ holds, and it is easily seen that in this case the portfolio

---

[5] An *arbitrageur* is that market participant who consistently uses the price discrepancies to make (almost) risk-free profits.

$\psi = (1, -S_0) = -\phi$ is an arbitrage opportunity. Finally, if $0 < p_* < 1$ then for any portfolio ϕ satisfying $V_0(\phi) = 0$, in view (1.9) and (1.6), we get

$$p_* V^u(\phi) + (1 - p_*)V^d(\phi) = 0,$$

so that $V^d(\phi) < 0$ when $V^u(\phi) > 0$, and $V^d(\phi) > 0$ if $V^u(\phi) < 0$. This shows that there are no arbitrage opportunities in $\mathcal{M}$ when $0 < p_* < 1$. To prove formula (1.14) it is enough to compare it with (1.9). Alternatively, we may observe that for the unique portfolio $\phi = (\alpha_0, \beta_0)$ that replicates the claim X we have

$$\begin{aligned}
\mathbb{E}_{\mathbb{P}^*}\big((1+r)^{-1}X\big) &= \mathbb{E}_{\mathbb{P}^*}\big((1+r)^{-1}V_T(\phi)\big) = \mathbb{E}_{\mathbb{P}^*}(\alpha_0 S_T^* + \beta_0) \\
&= \alpha_0 S_0^* + \beta_0 = V_0(\phi) = \pi_0(X),
\end{aligned}$$

so that we are done. □

1.4.6 Change of a Numeraire

The choice of the bond price process as a discount factor (or, as a *numeraire*) is not essential. Suppose, on the contrary, that we have chosen the stock price S as a numeraire. In other words, we now consider the bond price B discounted by the stock price S. We shall write $\bar{B}_t = B_t/S_t$ for $t \in \{0, T\}$. The martingale measure $\bar{\mathbb{P}}$ for the process $\bar{B}$ is uniquely determined by the equality $\bar{B}_0 = \mathbb{E}_{\bar{\mathbb{P}}}(\bar{B}_T)$, or explicitly

$$\bar{p}\,\frac{1+r}{S^u} + \bar{q}\,\frac{1+r}{S^d} = \frac{1}{S_0},$$

where $\bar{q} = 1 - \bar{p}$. One finds that

$$\bar{\mathbb{P}}\{\omega_1\} = \bar{p} = \left(\frac{1}{S^d} - \frac{1}{(1+r)S_0}\right)\frac{S^u S^d}{S^u - S^d} = \frac{p_* S^u}{(1+r)S_0} \tag{1.16}$$

and

$$\bar{\mathbb{P}}\{\omega_2\} = \bar{q} = \left(\frac{1}{(1+r)S_0} - \frac{1}{S^u}\right)\frac{S^u S^d}{S^u - S^d} = \frac{(1-p_*)S^d}{(1+r)S_0}. \tag{1.17}$$

We will now show that the properly modified version of the risk-neutral valuation formula holds under $\bar{\mathbb{P}}$.

Corollary 1.4.1 *Under the assumptions of Proposition 1.4.2, for any contingent claim X, which settles at time T, we have that*

$$\pi_0(X) = S_0 \,\mathbb{E}_{\bar{\mathbb{P}}}(S_T^{-1}X). \tag{1.18}$$

Proof. We note that the Radon-Nikodým density of $\bar{\mathbb{P}}$ with respect to $\mathbb{P}^*$ equals

$$\eta_T(\omega) := \frac{d\bar{\mathbb{P}}}{d\mathbb{P}^*}(\omega) = \begin{cases} \frac{\bar{p}}{p_*} = \frac{S^u}{(1+r)S_0}, & \text{if } \omega = \omega_1, \\ \frac{1-\bar{p}}{1-p_*} = \frac{S^d}{(1+r)S_0}, & \text{if } \omega = \omega_2. \end{cases}$$

This also means that the random variable η_T can be represented as follows

$$\eta_T = \frac{S_T}{(1+r)S_0} = \frac{B_0 S_T}{S_0 B_T}.$$

It is worth noting that $\eta_T > 0$ and $\mathbb{E}_{\mathbb{P}^*}(\eta_T) = 1$, so that the probability measures $\mathbb{P}^*$ and $\bar{\mathbb{P}}$ are equivalent. Finally,

$$S_0 \, \mathbb{E}_{\bar{\mathbb{P}}}\left(S_T^{-1} X\right) = S_0 \, \mathbb{E}_{\mathbb{P}^*}\left(\eta_T S_T^{-1} X\right) = B_0 \mathbb{E}_{\mathbb{P}^*}\left(B_T^{-1} X\right) = \mathbb{E}_{\mathbb{P}^*}\left((1+r)^{-1} X\right),$$

and thus (1.18) follows immediately from (1.14).   □

Let us apply this approach to the call option of Example 1.4.1. One finds easily that $\bar{p} = 0.62$, and thus formula (1.18) gives

$$C_0 = S_0 \, \mathbb{E}_{\bar{\mathbb{P}}}\left(S_T^{-1} (S_T - K)^+\right) = 21.59,$$

as expected.

It appears that in some circumstances the choice of the stock price as a numeraire is more convenient than that of the savings account. For instance, for the call option we denote $D = \{S_T > K\}$ and we note that the payoff can be decomposed as follows

$$C_T = (S_T - K)^+ = S_T \mathbb{1}_D - K \mathbb{1}_D = X_1 - X_2.$$

Consequently, using (1.14) and (1.18), we obtain the following convenient representation for the call price

$$C_0 = \pi_0(X_1) - \pi_0(X_2) = S_0 \bar{\mathbb{P}} \{S_T > K\} - K(1+r)^{-1} \mathbb{P}^* \{S_T > K\}.$$

1.4.7 Put Option

We refer once again to Example 1.4.1. However, we shall now focus on a European put option instead of a call option. Since the buyer of a put option has the right to sell a stock at a given date T, the terminal payoff from the option is now $P_T = (K - S_T)^+$, i.e.,

$$P_T(\omega) = \begin{cases} P^u = 0, & \text{if } \omega = \omega_1, \\ P^d = 20, & \text{if } \omega = \omega_2, \end{cases}$$

where we have taken, as before, $K = \$280$. The portfolio $\phi = (\alpha_0, \beta_0)$ replicating the European put option is thus determined by the following system of linear equations

$$\begin{cases} 320\,\alpha_0 + 1.05\,\beta_0 = 0, \\ 260\,\alpha_0 + 1.05\,\beta_0 = 20, \end{cases}$$

so that $\alpha_0 = -1/3$ and $\beta_0 = 101.59$. Consequently, the arbitrage price P_0 of the European put option equals

$$P_0 = -(1/3) \times 280 + 101.59 = 8.25.$$

Note that the number of shares in a replicating portfolio is negative. This means that an option writer who wishes to hedge risk exposure should sell short at time 0 the number $-\alpha_0 = 1/3$ shares of stock for each sold put option. The proceeds from the short-selling of shares, as well as the option's premium, are invested in an interest-earning account. To find the arbitrage price of the put option we may alternatively apply Proposition 1.4.2. By virtue of (1.14), with $X = P_T$, we get

$$P_0 = \mathbb{E}_{\mathbb{P}^*}\big((1+r)^{-1}P_T\big) = 8.25.$$

Finally, the put option value can also be found by applying the following relationship between the prices of call and put options.

Corollary 1.4.2 *The following put-call parity relationship is valid*

$$C_0 - P_0 = S_0 - (1+r)^{-1}K. \tag{1.19}$$

Proof. The formula is an immediate consequence of equality (1.3) and the pricing formula (1.14) applied to the claim $S_T - K$.   □

It is worth mentioning that relationship (1.19) is universal – that is, it does not depend on the choice of the model (the only assumption we need to make is the additivity of the price). Using the put-call parity, we can calculate once again the arbitrage price of the put option. Formula (1.19) yields immediately

$$P_0 = C_0 - S_0 + (1+r)^{-1}K = 8.25.$$

Let us write down explicit formulas for the call and put price in the one-period, two-state model. We assume, as usual, that $S^u > K > S^d$. Then

$$C_0 = \frac{S_0(1+r) - S^d}{S^u - S^d} \frac{S^u - K}{1+r}, \tag{1.20}$$

and

$$P_0 = \frac{S^u - S_0(1+r)}{S^u - S^d} \frac{K - S^d}{1+r}. \tag{1.21}$$

1.5 Forward Contracts

A *forward contract* is an agreement, signed at the initial date 0, to buy or sell an asset at a certain future time T (called *delivery date* or *maturity* in what follows) for a predetermined price K, referred to as the *delivery price*. In contrast to stock options and futures contracts, forward contracts are not traded on exchanges. By convention, the party who agrees to buy the underlying asset at time T for the delivery price K is said to assume a *long position* in a given contract. Consequently, the other party, who is obliged to sell the asset at the same date for the price K, is said to assume a *short position*. Since a forward contract is settled at maturity and a party in a long position is obliged to buy an asset worth S_T at maturity for K, it is clear that the payoff from the long position (respectively, from the short position) in a given forward contract with a stock S being the underlying asset corresponds to the time T contingent claim X (respectively, $-X$), where

$$X = S_T - K. \tag{1.22}$$

Let us emphasize that there is no cash flow at the time the forward contract is entered into. In other words, the price (or value) of a forward contract at its initiation is zero. Notice, however, that for $t > 0$, the value of a forward contract may be negative or positive. As we shall now see, a forward contract is worthless at time 0 provided that a judicious choice of the delivery price K is made.

1.5.1 Forward Price

We shall now find the rational delivery price for a forward contract. To this end, let us introduce first the following definition which is, of course, consistent with typical features of a forward contract. Recall that, typically, there is no cash flow at the initiation of a forward contract.

Definition 1.5.1 The delivery price K that makes a forward contract worthless at initiation is called the *forward price* of an underlying financial asset S for the settlement date T.

Note that we use here the adjective *financial* in order to emphasize that the storage costs that have to be taken into account when studying forward contracts on commodities, are neglected. In the case of a dividend-paying stock, in the calculation of the forward price, it is enough to substitute $S_0 - \hat{I}_0$ for S_0, where $\hat{I}_0$ is the present value of all future dividend payments during the contract's lifetime (cf. Sect. 3.2).

Proposition 1.5.1 *Assume that the one-period, two-state security market model (S, B, Φ) is arbitrage-free. Then the forward price at time 0 for the settlement date T of one share of stock S equals $F_S(0, T) = (1 + r)S_0$.*

Proof. We shall apply the martingale method of Proposition 1.4.2. By applying formulas (1.14) and (1.22), we get

$$\pi_0(X) = \mathbb{E}_{\mathbb{P}^*}\left(\hat{r}^{-1}X\right) = \mathbb{E}_{\mathbb{P}^*}(S_T^*) - \hat{r}^{-1}K = S_0 - \hat{r}^{-1}K = 0, \tag{1.23}$$

where $\hat{r} = 1 + r$. It is now apparent that $F_S(0, T) = (1 + r)S_0$. □

By combining Corollary 1.6.1 with the above proposition, we conclude that in a one-period model of a spot market, the futures and forward prices of financial assets for the same settlement date are equal.

Remarks. It seems instructive to consider a slightly more general model of a one-period market. Assume that S_0 is a given real number and S_T stands for an arbitrary random variable defined on some probability space $(\Omega, \mathcal{F}, \mathbb{P})$. Also let $B_0 = 1$ and $B_T = 1 + r$ for some real $r \geq 0$. Fix a real number K, and assume that an investor may enter into a forward contract on a financial asset whose price follows the process S with the settlement date T and delivery price K. Put another way, we extend the market by considering an additional "security" whose price process, denoted by G, is: $G_0 = 0$, $G_T = S_T - K$. Let Φ_G be the class of all trading strategies in the extended market. It may be identified with the set of all vectors $\phi_0 = (\alpha_0, \beta_0, \gamma_0) \in \mathbb{R}^3$, where α_0, β_0 have the same interpretation as in Sect. 1.4.1, and γ_0 stands for the number of forward contracts entered into at the initial date 0. As before, a security market (S, B, G, Φ_G) is said to be arbitrage-free whenever there are no arbitrage opportunities in Φ_G. We shall show that the considered market is arbitrage-free if and only if the equality $K = (1 + r)S_0$ holds. Assume, on the contrary, that $K \neq (1 + r)S_0$. If $K < (1 + r)S_0$, we consider a trading strategy $\psi = (-1, S_0, 1)$. Its wealth at time 0 equals $V_0(\psi) = -S_0 + S_0 = 0$. On the other hand, at time T we have

$$V_T(\psi) = -S_T + S_0(1 + r) + S_T - K = S_0(1 + r) - K > 0,$$

so that the portfolio $\psi \in \Phi_G$ guarantees a risk-free profit. Similarly, if the inequality $K > (1 + r)S_0$ is satisfied, the trading strategy $-\psi$ constitutes an arbitrage opportunity.

1.6 Futures Call and Put Options

We will first describe very succinctly the main features of futures contracts, which are reflected in stochastic models of futures markets to be developed later. As in the previous section, we will focus mainly on the arbitrage pricing of European call and put options; clearly, instead of the spot price of the underlying asset, we will now consider its futures price. The model of futures prices we adopt here is quite similar to the one used to describe spot prices. Still, due to the specific features of futures contracts used to set up a replicating strategy, one has to modify significantly the way in which the payoff from a portfolio is defined.

1.6.1 Futures Contracts and Futures Prices

A *futures contract* is an agreement to buy or sell an asset at a certain date in the future for a certain price. The important feature of these contracts is that they are traded on exchanges. Consequently, the authorities need to define precisely all the characteristics of each futures contract in order to make trading possible. More importantly, the *futures price* – the price at which a given futures contract is entered into – is determined on a given futures exchange by the usual law of demand and supply (in a similar way as for spot prices of listed stocks). Futures prices are therefore settled daily and the quotations are reported in the financial press. A futures contract is referred to by its delivery month, however an exchange specifies the period within that month when delivery must be made. The exchange specifies the amount of the asset to be delivered for one contract, as well as some additional details when necessary (e.g., the quality of a given commodity or the maturity of a bond). From our perspective, the most fundamental feature of a futures contract is the way the contract is settled. The procedure of daily settlement of futures contracts is called *marking to market*. A futures contract is worth zero when it is entered into; however, each investor is required to deposit funds into a *margin account*. The amount that should be deposited when the contract is entered into is known as the *initial margin*. At the end of each trading day, the balance of the investor's margin account is adjusted in a way that reflects daily movements of futures prices. To be more specific, if an investor assumes a long position, and on a given day the futures price rises, the balance of the margin account will also increase. Conversely, the balance of the margin account of any party with a short position in this futures contract will be properly reduced. Intuitively, it is thus possible to argue that futures contracts are actually closed out after each trading day, and then start afresh the next trading day. Obviously, to offset a position in a futures contract, an investor enters into the opposite trade to the original one. Finally, if the delivery period is reached, the delivery is made by the party with a short position.

1.6.2 One-period Futures Market

It will be convenient to start this section with a simple example which, in fact, is a straightforward modification of Example 1.4.1 to the case of a futures market.

Example 1.6.1 Let $f_t = f_S(t, T^*)$ be a one-period process that models the futures price of a certain asset S, for the settlement date $T^* \geq T$. We assume that $f_0 = 280$, and

$$f_T(\omega) = \begin{cases} f^u = 320, & \text{if } \omega = \omega_1, \\ f^d = 260, & \text{if } \omega = \omega_2, \end{cases}$$

where $T = 3$ months.[6] We consider a 3-month European futures call option with strike price $K = \$280$. As before, we assume that the simple risk-free interest rate for 3-month deposits and loans is $r = 5\%$.

The payoff from the futures call option $C_T^f = (f_T - K)^+$ equals

$$C_T^f(\omega) = \begin{cases} C^{fu} = 40, & \text{if } \omega = \omega_1, \\ C^{fd} = 0, & \text{if } \omega = \omega_2. \end{cases}$$

A portfolio ϕ replicating the option is composed of α_0 futures contracts and β_0 units of cash invested in risk-free bonds (or borrowed). The wealth process $V_t^f(\phi)$, $t \in \{0, T\}$, of this portfolio equals $V_0^f(\phi) = \beta_0$, since futures contracts are worthless when they are first entered into. Furthermore, the terminal wealth of ϕ is

$$V_T^f(\phi) = \alpha_0 (f_T - f_0) + (1 + r)\beta_0, \tag{1.24}$$

---

[6] Notice that in the present context, the knowledge of the settlement date T^* of a futures contract is not essential. It is implicitly assumed that $T^* \geq T$, however.

where the first term on the right-hand side represents gains (or losses) from the futures contract, and the second corresponds to a savings account (or loan). Note that (1.24) reflects the fact that futures contracts are marked to market daily (that is, after each period in our model).

A portfolio $\phi = (\alpha_0, \beta_0)$ is said to replicate the option when $V_T^f = C_T^f$, or more explicitly, if the equalities

$$V_T^f(\omega) = \begin{cases} \alpha_0(f^u - f_0) + (1+r)\beta_0 = C^{fu}, & \text{if } \omega = \omega_1, \\ \alpha_0(f^d - f_0) + (1+r)\beta_0 = C^{fd}, & \text{if } \omega = \omega_2, \end{cases}$$

are satisfied. For Example 1.6.1, this gives the following system of linear equations

$$\begin{cases} 40\,\alpha_0 + 1.05\,\beta_0 = 40, \\ -20\,\alpha_0 + 1.05\,\beta_0 = 0, \end{cases}$$

yielding $\alpha_0 = 2/3$ and $\beta_0 = 12.70$. The manufacturing cost of a futures call option is thus $C_0^f = V_0^f(\phi) = \beta_0 = 12.70$. Similarly, the unique portfolio replicating a sold put option is determined by the following conditions

$$\begin{cases} 40\,\alpha_0 + 1.05\,\beta_0 = 0, \\ -20\,\alpha_0 + 1.05\,\beta_0 = 20, \end{cases}$$

so that $\alpha_0 = -1/3$ and $\beta_0 = 12.70$ in this case. Consequently, the manufacturing costs of futures put and call options are equal in our example.

As we shall see soon, this is not a pure coincidence; in fact, by virtue of formula (1.30) below, the prices of call and put futures options are equal when the option's strike price coincides with the initial futures price of the underlying asset. The above considerations may be summarized by means of the following exhibits (note that β_0 is a positive number)

$$\text{at time } t = 0 \begin{cases} \text{one sold futures option} & C_0^f, \\ \text{futures contracts} & 0, \\ \text{cash deposited in a bank} & -\beta_0 = -C_0^f, \end{cases}$$

and

$$\text{at time } t = T \begin{cases} \text{option's payoff} & -C_T^f, \\ \text{profits/losses from futures} & \alpha_0\,(f_T - f_0), \\ \text{cash withdrawal} & \hat{r}\beta_0, \end{cases}$$

where, as before, $\hat{r} = 1 + r$.

1.6.3 Martingale Measure for a Futures Market

We are looking now for a probability measure $\tilde{\mathbb{P}}$ that makes the futures price process (with no discounting) follow a $\tilde{\mathbb{P}}$-martingale. A probability $\tilde{\mathbb{P}}$, if it exists, is thus determined by the equality

$$f_0 = \mathbb{E}_{\tilde{\mathbb{P}}}(f_T) = \tilde{p}\, f^u + (1 - \tilde{p})\, f^d. \tag{1.25}$$

It is easily seen that

$$\tilde{\mathbb{P}}\{\omega_1\} = \tilde{p} = \frac{f_0 - f^d}{f^u - f^d}, \quad \tilde{\mathbb{P}}\{\omega_2\} = 1 - \tilde{p} = \frac{f^u - f_0}{f^u - f^d}. \tag{1.26}$$

Using the data of Example 1.6.1, one finds easily that $\tilde{p} = 1/3$. Consequently, the expected value under the probability $\tilde{\mathbb{P}}$ of the discounted payoff from the futures call option equals

$$\tilde{C}_0^f = \mathbb{E}_{\tilde{\mathbb{P}}}\big((1 + r)^{-1}(f_T - K)^+\big) = 12.70 = C_0^f.$$

This illustrates the fact that the martingale approach may be used also in the case of futures markets, with a suitable modification of the notion of a martingale measure.

Using the traditional terminology of mathematical finance, we may conclude that the risk-neutral futures economy is characterized by the fair-game property of the process of a futures price. Remember that the risk-neutral spot economy is the one in which the discounted stock price (as opposed to the stock price itself) models a fair game.

1.6.4 Absence of Arbitrage

In this subsection, we shall study a general two-state, one-period model of a futures price. We consider the filtered probability space $(\Omega, (\mathcal{F}_t)_{t \in \{0,T\}}, \mathbb{P})$ introduced in Sect. 1.4.4. The first process, which intends to model the dynamics of the futures price of a certain asset, for the fixed settlement date $T^* \geq T$, is an adapted and strictly positive process $f_t = f_S(t, T^*)$, $t = 0, T$. More specifically, f_0 is assumed to be a real number, and f_T is the following random variable

$$f_T(\omega) = \begin{cases} f^u, & \text{if } \omega = \omega_1, \\ f^d, & \text{if } \omega = \omega_2, \end{cases}$$

where, by convention, $f^u > f^d$. The second security is, as in the case of a spot market, a risk-free bond whose price process is $B_0 = 1$, $B_T = 1 + r$ for some real $r \geq 0$. Let Φ^f stand for the linear space of all futures contracts-bonds portfolios $\phi = \phi_0 = (\alpha_0, \beta_0)$; it may be, of course, identified with the linear space $\mathbb{R}^2$. The wealth process $V^f(\phi)$ of any portfolio equals

$$V_0(\phi) = \beta_0, \quad \text{and} \quad V_T^f(\phi) = \alpha_0(f_T - f_0) + (1 + r)\beta_0 \tag{1.27}$$

(it is useful to compare these formulas with (1.4)). We shall study the valuation of derivatives in the futures market model $\mathcal{M}^f = (f, B, \Phi^f)$. It is easily seen that an arbitrary contingent claim X that settles at time T admits a unique replicating portfolio $\phi \in \Phi$. Put another way, all contingent claims that settle at time T are *attainable* in the market model $\mathcal{M}^f$. In fact, if X is given by the formula

$$X(\omega) = \begin{cases} X^u, & \text{if } \omega = \omega_1, \\ X^d, & \text{if } \omega = \omega_2, \end{cases}$$

then its replicating portfolio $\phi \in \Phi^f$ may be found by solving the following system of linear equations

$$\begin{cases} \alpha_0(f^u - f_0) + (1+r)\beta_0 = X^u, \\ \alpha_0(f^d - f_0) + (1+r)\beta_0 = X^d. \end{cases} \tag{1.28}$$

The unique solution of (1.28) is

$$\alpha_0 = \frac{X^u - X^d}{f^u - f^d}, \quad \beta_0 = \frac{X^u(f_0 - f^d) + X^d(f^u - f_0)}{(1+r)(f^u - f^d)}. \tag{1.29}$$

Consequently, the *manufacturing cost* $\pi_0^f(X)$ in $\mathcal{M}^f$ equals

$$\pi_0^f(X) \stackrel{\text{def}}{=} V_0^f(\phi) = \beta_0 = \frac{X^u(f_0 - f^d) + X^d(f^u - f_0)}{(1+r)(f^u - f^d)}. \tag{1.30}$$

We say that a model $\mathcal{M}^f$ of the futures market is *arbitrage-free* if there are no arbitrage opportunities in the class Φ^f of trading strategies. The following simple result provides necessary and sufficient conditions for the arbitrage-free property of $\mathcal{M}^f$.

Proposition 1.6.1 *The futures market* $\mathcal{M}^f = (f, B, \Phi^f)$ *is arbitrage-free if and only if the process* f *that models the futures price admits a (unique) martingale measure* $\tilde{\mathbb{P}}$ *equivalent to* $\mathbb{P}$. *In this case, the arbitrage price at time 0 of any contingent claim* X *that settles at time* T *equals*

$$\pi_0^f(X) = \mathbb{E}_{\tilde{\mathbb{P}}}\big((1+r)^{-1}X\big), \tag{1.31}$$

or explicitly

$$\pi_0^f(X) = \frac{f_0 - f^d}{f^u - f^d}\frac{X^u}{1+r} + \frac{f^u - f_0}{f^u - f^d}\frac{X^d}{1+r}. \tag{1.32}$$

Proof. If there is no martingale measure for f, equivalent to $\mathbb{P}$, we have either $\tilde{p} \geq 1$ or $\tilde{p} \leq 0$. In the first case, we have $f_0 - f^d \geq f^u - f^d$ and thus $f_0 \geq f^u > f^d$. Consequently, a portfolio $\phi = (-1, 0)$ is an arbitrage opportunity. Similarly, when $\tilde{p} \leq 0$ the inequalities $f_0 \leq f^d < f^u$ are valid. Hence, the portfolio $\phi = (1, 0)$ is an arbitrage opportunity. Finally, if $0 < \tilde{p} < 1$ and for some $\phi \in \Phi^f$ we have $V_0^f(\phi) = 0$, then it follows from (1.30) that

$$\frac{f_0 - f^d}{f^u - f^d}V^{fu} + \frac{f^u - f_0}{f^u - f^d}V^{fd} = 0,$$

so that $V^{fd} < 0$ if $V^{fu} > 0$, and $V^{fu} < 0$ when $V^{fd} > 0$. This shows that the market model $\mathcal{M}^f$ is arbitrage-free if and only if the process f admits a martingale measure equivalent to $\mathbb{P}$. The valuation formula (1.31) now follows by (1.26)–(1.30). $\square$

When the price of the futures call option is already known, in order to find the price of the corresponding put option one may use the following relation, which is an immediate consequence of equality (1.3) and the pricing formula (1.31)

$$C_0^f - P_0^f = (1+r)^{-1}(f_0 - K). \tag{1.33}$$

It is now obvious that the equality $C_0^f = P_0^f$ is valid if and only if $f_0 = K$; that is, when the current futures price and the strike price of the option are equal. Equality (1.33) is referred to as the *put-call parity relationship* for futures options.

1.6.5 One-period Spot/Futures Market

Consider an arbitrage-free, one-period spot market (S, B, Φ) described in Sect. 1.4. Moreover, let $f_t = f_S(t, T)$, $t \in \{0, T\}$, be the process of futures prices with the underlying asset S and for the maturity date T. In order to preserve consistency with the financial interpretation of the futures price, we have to assume that $f_T = S_T$. Our aim is to find the right value f_0 of the futures price at time 0; that is, that level of the price f_0 that excludes arbitrage opportunities in the combined spot/futures market. In such a market, trading in stocks, bonds, as well as entering into futures contracts is allowed.

Corollary 1.6.1 *The futures price at time 0 for the delivery date T of the underlying asset S that makes the spot/futures market arbitrage-free equals $f_0 = (1+r)S_0$.*

Proof. Suppose an investor enters at time 0 into one futures contract. The payoff of his position at time T corresponds to a time T contingent claim $X = f_T - f_0 = S_T - f_0$. Since it costs nothing to enter a futures contract we should have

$$\pi_0(X) = \pi_0(S_T - f_0) = 0,$$

or equivalently,

$$\pi_0(X) = S_t - (1+r)^{-1}f_0 = 0.$$

This proves the asserted formula. Alternatively, one can check that if the futures price f_0 were different from $(1+r)S_0$, this would lead to arbitrage opportunities in the spot/futures market.   □

1.7 Options of American Style

An *option of American style* (or briefly, an *American option*) is an option contract in which not only the decision whether to exercise the option or not, but also the choice of the exercise time, is at the discretion of the option's holder. The exercise time cannot be chosen after the option's expiry date T. Hence, in our simple one-period model, the exercise date can either coincide with the initial date 0, or with the terminal date T.

Notice that the value (or the price) at the terminal date of the American call or put option written on any asset equals the value of the corresponding European option with the same strike price K. Thus, the only unknown quantity is the price of the American option at time 0. In view of the early exercise feature of the American option, the concept of perfect replication of the terminal option's payoff is not adequate for valuation purposes. To determine this value, we shall make use of the general rule of absence of arbitrage in the market model. By definition, the arbitrage price at time 0 of the American option should be set in such a way that trading in American options would not destroy the arbitrage-free feature the market.

Assume, as before, that $r \geq 0$ so that $\hat{r} \geq 1$. We will first show that the American call written on a stock that pays no dividends during the option's lifetime is always equivalent to the European call; that is, that both options necessarily have identical prices at time 0. As we shall see in what follows, such a property is not always true in the case of American put options; that is, American and European puts are not necessarily equivalent. For similar reasons, American and European calls are no longer equivalent, in general, if the underlying stock pays dividends during the option's lifetime (or when $-1 < r < 0$, so that $0 < \hat{r} < 1$).

We place ourselves once again within the framework of a one-period spot market $\mathcal{M} = (S, B, \Phi)$, as specified in Sect. 1.4.1. It will be convenient to assume that European options are traded securities in our market. This causes no loss of generality, since the price process of any European option can be mimicked by the wealth process of a suitable trading strategy. For $t = 0, T$, let us denote by C_t^a and P_t^a the arbitrage price at time t of the American call and put respectively. It is obvious that $C_T^a = C_T$ and $P_T^a = P_T$. As was mentioned earlier, both arbitrage prices C_0^a and P_0^a will be determined using the following property: if the market $\mathcal{M} = (S, B, \Phi)$ is arbitrage-free, then the market with trading in stocks, bonds and American options should remain arbitrage-free. It should be noted that it is not evident a priori that the last property determines in a unique way the values of C_0^a and P_0^a.

We assume throughout that the inequalities $S^d < S_0(1 + r) < S^u$ hold and the strike price satisfies $S^d < K < S^u$. Otherwise, either the market model would not be arbitrage-free, or valuation of the option would be a trivial matter. The first result establishes the equivalence of the European and American call written on a non-dividend-paying stock.

Proposition 1.7.1 *Assume that the risk-free interest rate r is a nonnegative real number. Then the arbitrage price C_0^a of an American call option in the arbitrage-free market model $\mathcal{M} = (S, B, \Phi)$ coincides with the price C_0 of the European call option with the same strike price K.*

Proof. Assume, on the contrary, that $C_0^a \neq C_0$. Suppose first that $C_0^a > C_0$. Note that the arbitrage price C_0 satisfies

$$C_0 = p_* \frac{S^u - K}{1 + r} = \frac{(1 + r)S_0 - S^d}{S^u - S^d} \frac{S^u - K}{1 + r} > S_0 - K, \qquad (1.34)$$

if $r \geq 0$. It is now straightforward to check that there exists an arbitrage opportunity in the market. In fact, to create a risk-free profit, it is sufficient to sell the American call option at C_0^a, and simultaneously buy the European call option at C_0. If European options are not traded, one may, of course, create a replicating portfolio for the European call at initial investment C_0. The above portfolio is easily seen to lead to a risk-free profit, independently from the decision regarding the exercise time made by the holder of the American call. If, on the contrary, the price C_0^a were strictly smaller than C_0, then by selling European calls and buying American calls, one would be able to create a profitable risk-free portfolio.   □

Before continuing our analysis of American options in an arbitrage-free one-period model, let us recall *Jensen's inequality*. We consider here only a special case of a finite probability space (for the general case, see Lemma A.1.3).

Lemma 1.7.1 *Let $h : \mathbb{R} \to \mathbb{R}$ be a convex function, and ξ a random variable on a finite probability space $(\Omega, \mathcal{F}, \mathbb{P})$. Then Jensen's inequality holds, that is,* $g(\mathbb{E}_\mathbb{P}(\xi)) \leq \mathbb{E}_\mathbb{P}(g(\xi))$.

Proof. Under the present assumptions we have $\Omega = \{\omega_1, \ldots, \omega_n\}$, and thus

$$g(\mathbb{E}_\mathbb{P}(\xi)) = g\left(\sum_{i=1}^{n} p_i \xi(\omega_i)\right)$$

where $p_i = \mathbb{P}\{\omega_i\}$ for $i = 1, \ldots, n$. Using the convexity of g, we obtain

$$g\left(\sum_{i=1}^{n} p_i \xi(\omega_i)\right) \leq \sum_{i=1}^{n} p_i g(\xi(\omega_i)) = \mathbb{E}_\mathbb{P}(g(\xi)),$$

which is the desired inequality.   □

American call option. It is worthwhile to observe that inequality (1.34) is valid in a more general set-up. Indeed, if $r \geq 0$, $S_0 > K$, and S_T is a $\mathbb{P}^*$-integrable random variable, then we have always

$$\mathbb{E}_{\mathbb{P}^*}((1+r)^{-1}(S_T - K)^+) \geq \left(\mathbb{E}_{\mathbb{P}^*}((1+r)^{-1}S_T) - (1+r)^{-1}K\right)^+$$
$$= (S_0 - (1+r)^{-1}K)^+ \geq S_0 - K,$$

where the first inequality follows by Jensen's inequality (we use Lemma 1.7.1 if Ω is finite, and Lemma A.1.3 otherwise). We conclude that American and European call options are equivalent. If, on the contrary, $-1 < r < 0$ then the second inequality is not valid, in general. Consequently, the equivalence of American and European calls may fail to hold in this case.

American put option. It is interesting to notice that in the case of the put option the situation is different. First, we have

$$\mathbb{E}_{\mathbb{P}^*}((1+r)^{-1}(K - S_T)^+) \geq \left(\mathbb{E}_{\mathbb{P}^*}((1+r)^{-1}K - (1+r)^{-1}S_T)\right)^+$$
$$= ((1+r)^{-1}K - S_0)^+ > K - S_0,$$

where the last inequality holds provided that $-1 < r < 0$. If $r = 0$, we obtain

$$\mathbb{E}_{\mathbb{P}^*}((1+r)^{-1}(K - S_T)^+) \geq K - S_0.$$

Finally, if $r > 0$, no obvious relationship between P_0 and $S_0 - K$ is available. This feature suggests that for strictly positive interest rate the counterpart of Proposition 1.7.1 – the case of American put – should be more interesting. In fact, we have the following result.

Proposition 1.7.2 *Assume that $r > 0$. Then $P_0^a = P_0$ if and only if the inequality*

$$K - S_0 \leq \frac{S^u - (1+r)S_0}{S^u - S^d} \frac{K - S^d}{1 + r} \tag{1.35}$$

is valid. Otherwise, $P_0^a = K - S_0 > P_0$. If $r = 0$, then invariably $P_0^a = P_0$.

Proof. In view of (1.21), it is clear that inequality (1.35) is equivalent to $P_0 \geq K - S_0$. Suppose first that the last inequality holds. If, in addition, $P_0^a > P_0$ (respectively, $P_0^a < P_0$), by selling the American put and buying the European put (respectively, by buying the American put and selling the European put) one creates a profitable risk-free strategy. Hence $P_0^a = P_0$ in this case.[7] Suppose now that (1.35) fails to hold – that is, $P_0 < K - S_0$, and assume that $P_0^a \neq K - S_0$. We wish to show that P_0^a should be set to be $K - S_0$, otherwise arbitrage opportunities arise. Actually, if P_0^a were strictly greater that $K - S_0$, the seller of an American put would be able to lock in a profit by perfectly hedging exposure using the European put acquired at a strictly lower cost P_0. If, on the contrary, inequality $P_0^a < K - S_0$ were true, it would be profitable to buy the American put and exercise it immediately. Summarizing, if (1.35) fails to hold, the arbitrage price of the American put is strictly greater than the price of the European put. Finally, one verifies easily that if the holder of the American put fails to exercise it at time 0, the option's writer is still able to lock in a profit.

If (1.35) fails to hold, the American put should be exercised immediately, otherwise arbitrage opportunities would arise in the market. For the last statement, observe that if $r = 0$, then inequality (1.35), which now reads

$$K - S_0 \leq \frac{S^u - S_0}{S^u - S^d}(K - S^d),$$

is easily seen to be valid (it is enough to take $K = S^d$ and $K = S^u$). $\quad\square$

The above results suggest the following general "rational" exercise rule in a discrete-time framework: at any time t before the option's expiry, find the maximal expected payoff over all admissible exercise rules and compare the outcome with the payoff obtained by exercising the option immediately. If the latter value is greater,

---

[7] To be formal, we need to check that no arbitrage opportunities are present if $P_0^a = P_0$ and (1.35) holds. It is sufficient to examine an arbitrary zero net investment portfolio built from stocks, bonds and American puts.

exercise the option immediately, otherwise go one step further. In fact, one checks easily that the price at time 0 of an American call or put option may be computed as the maximum expected value of the payoff over all exercises, provided that the expectation in question is taken under the martingale probability measure. The last feature distinguishes arbitrage pricing of American options from the typical optimal stopping problems, in which maximization of expected payoffs takes place under a subjective (or actual) probability measure rather than under an artificial martingale measure. We conclude that a simple argument that the rational option's holder will always try to maximize the expected payoff of the option at exercise is not sufficient to determine arbitrage prices of American claims. A more precise statement would read: the American put option should be exercised by its holder at the same date as it is exercised by a risk-neutral individual[8] whose objective is to maximize the discounted expected payoff of the option; otherwise arbitrage opportunities would arise in the market.

For our further purposes, it will be useful to formalize the concept of an *American contingent claim*.

Definition 1.7.1 A *contingent claim of American style* X^a (or shortly, an *American claim*) is a pair (X_0, X_T), where X_0 is a real number and X_T is a random variable. We interpret X_0 and X_T as the payoffs received by the holder of the American claim X^a if he chooses to exercise it at time 0 and at time T respectively.

Observe that in our present set-up, the only admissible *exercise times*[9] are the initial date and the expiry date, say $\tau_0 = 0$ and $\tau_1 = T$. We assume also, for notational convenience, that $T = 1$. Then we may formulate the following corollary to Propositions 1.7.1–1.7.2, whose proof is left as exercise.

Corollary 1.7.1 *The arbitrage prices of an American call and an American put option in the arbitrage-free market model* $\mathcal{M} = (S, B, \Phi)$ *are given by*

$$C_0^a = \max_{\tau \in \mathcal{T}} \mathbb{E}_{\mathbb{P}^*}\big((1+r)^{-\tau}(S_\tau - K)^+\big)$$

and

$$P_0^a = \max_{\tau \in \mathcal{T}} \mathbb{E}_{\mathbb{P}^*}\big((1+r)^{-\tau}(K - S_\tau)^+\big)$$

respectively, where $\mathcal{T}$ *denotes the class of all exercise times. More generally, if* X^a *is an arbitrary contingent claim of American style, then its arbitrage price* $\pi(X^a) = (X_0^a, X_T^a)$ *in* $\mathcal{M} = (S, B, \Phi)$ *equals*

$$\pi_0(X^a) = X_0^a = \max_{\tau \in \mathcal{T}} \mathbb{E}_{\mathbb{P}^*}\big((1+r)^{-\tau}X_\tau\big), \quad \pi_T(X^a) = X_T^a = X_T.$$

---

[8] Let us recall that a risk-neutral individual is one whose subjective assessments of the market correspond to the martingale probability measure $\mathbb{P}^*$.

[9] By convention, we say that an option is exercised at expiry date T if it is not exercised prior to that date, even when its terminal payoff equals zero (so that in fact the option is abandoned). Let us also mention that in a general setting, the exercise time is assumed to be the so-called *stopping time*. The only stopping times in a one-period model are $\tau_0 = 0$ and $\tau_1 = T$, however.

Remarks. Let us observe that pricing formulas of Corollary 1.7.1 are quite general. More precisely, they can be extended to the case of a multi-period discrete-time market model, as well as to a continuous-time set-up.

American options on a dividend-paying stock. We make the standard assumption that $r \geq 0$. Suppose that the underlying stock pays at time T a dividend whose value is known in advance. It is common to assume that the stock price declines on the ex-dividend day (that is, at time T) by an amount equal to the dividend payment. Thus, the dividend payment does not reduce the wealth of a portfolio (if one sells borrowed shares, he is obliged not only to give back shares, but also to make restitution for the dividend payments). On the other hand, however, it affects the payoff from the option. In fact, the payoffs at expiry are $C_T^\kappa = (S_T - \kappa - K)^+$ and $P_T^\kappa = (K + \kappa - S_T)^+$ for the call and put option respectively, where $\kappa > 0$ represents the dividend amount.

For European options, the dividend payment lowers the value of the call and increases the value of the put. To find a proper modification of the option price it is sufficient to replace the strike price K by $K + \kappa$ in the risk-neutral valuation formula. If options are of American style, dividend payments have important qualitative consequences, in general. Indeed, the American call written on a dividend-paying stock is not necessarily equivalent to the European call. Generally speaking, it may be optimal for a risk-neutral holder of an American call to exercise the option before expiry. For instance, in a one-period model, a holder of an American call option should exercise it immediately whenever the inequality

$$S_0 - K > (1 + r)^{-1} \mathbb{E}_{\mathbb{P}^*}\left((S_T - \kappa - K)^+\right)$$

holds; otherwise, her inaction would create an arbitrage opportunity in the market. It is also important to note that – in contrast to the case of the American put – the optimal exercise rule for the American call is always restricted to the set of ex-dividend dates only. Let us finally mention that the dividend payment increases the probability of early exercise of an American put option.

1.8 Universal No-arbitrage Inequalities

We will now derive universal inequalities that are necessary for absence of arbitrage in the market. In contrast to the situation studied up to now, we no longer assume that the price of the underlying asset admits only two terminal values. Furthermore, trading may occur continuously over time, hence a specific *self-financing* property needs to be imposed on trading strategies. At the intuitive level, a strategy is self-financing if no infusion of funds or withdrawals of cash are allowed; in particular, intertemporal consumption is excluded. In other words, the terminal wealth associated with a dynamic portfolio comes exclusively from the initial investment and the capital gains generated by the trading process. We do not need to give here a more formal definition of self-financing property as it is clear that the following property is valid in any discrete-or continuous-time, arbitrage-free market.

Price monotonicity rule. In any model of an arbitrage-free market, if X_T and Y_T are two European contingent claims, where $X_T \geq Y_T$, then $\pi_t(X_T) \geq \pi_t(Y_T)$ for every $t \in [0, T]$, where we denote by $\pi_t(X_T)$ and $\pi_t(Y_T)$ the arbitrage prices at time t of X_T and Y_T respectively. Moreover, if $X_T > Y_T$, then $\pi_t(X_T) > \pi_t(Y_T)$ for every $t \in [0, T]$.

For the sake of notational convenience, a constant rate $r \geq 0$ will now be interpreted as a continuously compounded rate of interest. Hence the price at time t of one dollar to be received at time $T \geq t$ equals $e^{-r(T-t)}$; in other words, the savings account process equals $B_t = e^{rt}$ for every $t \in [0, T]$. This means that we place ourselves here in a continuous-time setting. Discrete-time counterparts of relations (1.36)–(1.40) are, of course, equally easy to obtain.

Proposition 1.8.1 *Let C_t and P_t (respectively, C_t^a and P_t^a) stand for the arbitrage prices at time t of European (respectively, American) call and put options, with strike price K and expiry date T. Then the following inequalities are valid for every $t \in [0, T]$*

$$(S_t - Ke^{-r(T-t)})^+ \leq C_t = C_t^a \leq S_t, \tag{1.36}$$

$$(Ke^{-r(T-t)} - S_t)^+ \leq P_t \leq K, \tag{1.37}$$

and

$$(K - S_t)^+ \leq P_t^a \leq K. \tag{1.38}$$

The put-call parity relationship, which in the case of European options reads

$$C_t - P_t = S_t - Ke^{-r(T-t)}, \tag{1.39}$$

takes, in the case of American options, the form of the following inequalities

$$S_t - K \leq C_t^a - P_t^a \leq S_t - Ke^{-r(T-t)}. \tag{1.40}$$

Proof. All inequalities may be derived by constructing appropriate portfolios at time t and holding them to the terminal date. Let us derive, for instance, the first one. Consider the following (static) portfolios, A and B, established at time t. Portfolio A consists of one European call option and $Ke^{-r(T-t)}$ of cash; portfolio B contains only one share of stock. The value of the first portfolio at time T equals

$$C_T + K = (S_T - K)^+ + K = \max\{S_T, K\} \geq S_T,$$

while the value of portfolio B is exactly S_T. Hence the arbitrage price of portfolio A at time t dominates the price of portfolio B – that is, we have

$$C_t + Ke^{-r(T-t)} \geq S_t, \quad \forall t \in [0, T].$$

Since the price of the option is nonnegative, this proves the first inequality in (1.36). All remaining inequalities in (1.36)–(1.38) may be verified by means of similar arguments. To check that $C_t^a = C_t$, we consider the following portfolios: portfolio A – one American call option and $Ke^{-r(T-t)}$ of cash; and portfolio B – one share of stock. If the call option is exercised at some date $t^* \in [t, T]$, then the value of portfolio A at time t^* equals

$$S_{t^*} - K + K e^{-r(T-t^*)} < S_{t^*},$$

while the value of B is S_{t^*}. On the other hand, the value of portfolio A at the terminal date T is $\max\{S_T, K\}$. Hence it dominates the value of portfolio B, which is S_T. This means that early exercise of the call option would contradict our general price monotonicity rule. A justification of relationship (1.39) is straightforward, as $C_T - P_T = S_T - K$. To justify the second inequality in (1.40), notice that in view of (1.39) and the obvious inequality $P_t^a \geq P_t$, we get

$$P_t^a \geq P_t = C_t^a + K e^{-r(T-t)} - S_t, \quad \forall t \in [0, T].$$

To prove the first inequality in (1.40), let us take the two following portfolios: portfolio A – one American call and K units of cash; and portfolio B – one American put and one share of stock. If the put option is exercised at time $t^* \in [t, T]$ then the value of portfolio B at time t^* is K. On the other hand, the value of portfolio A at this date equals

$$C_t + K e^{r(t^*-t)} \geq K.$$

Portfolio A is therefore more valuable at time t than portfolio B; that is

$$C_t^a + K \geq P_t^a + S_t$$

for every $t \in [0, T]$.   □

We shall now examine general dependencies of option prices on the time to expiry and the strike price. Let $C(S_0, T, K)$ (respectively, $C^a(S_0, T, K)$) stand for the price of the European (respectively, American) call option with the expiration date T and the strike price K.

The following relationships are easy to derive

$$C^a(S_0, T_1, K) \leq C^a(S_0, T_2, K)$$

where $T_1 \leq T_2$. It is also rather clear that

$$C(S_0, T, K_2) \leq C(S_0, T, K_1), \quad C^a(S_0, T, K_2) \leq C^a(S_0, T, K_1)$$

provided that $K_1 \leq K_2$.

Proposition 1.8.2 *Assume that $K_1 < K_2$. The following inequalities are valid*

$$e^{-rT}(K_1 - K_2) \leq C(S_0, T, K_2) - C(S_0, T, K_1) \leq 0$$

and

$$K_1 - K_2 \leq C^a(S_0, T, K_2) - C^a(S_0, T, K_1) \leq 0.$$

Proof. Let us consider, for instance, the case of European call options. Take the two following portfolios at time 0: portfolio A – one European call with exercise price K_2 and $e^{-rT}(K_2 - K_1)$ units of cash; and portfolio B – one European call with exercise price K_1. The value of portfolio A at time T is

$$(S_T - K_2)^+ + (K_2 - K_1) \geq (S_T - K_1)^+$$

and the value of portfolio B at time T equals $(S_T - K_1)^+$. Consequently

$$C(S_0, T, K_2) + e^{-rT}(K_2 - K_1) \geq C(S_0, T, K_1),$$

as expected.   □

Proposition 1.8.3 *The price of a European (or American) call (or put) option is a convex function of the exercise price K.*

Proof. Let us consider, for instance, the case of a European put option. We denote its price at time 0 by $P(S_0, T, K)$. Assume that $K_1 < K_2$ and put $K_3 = \gamma K_1 + (1 - \gamma)K_2$, where $\gamma \in [0, 1]$ is a constant. We consider the following portfolios: portfolio A consisting of γ European put options with exercise price K_1 and $1 - \gamma$ European put options with exercise price K_2; and portfolio B that consists of one European put option with exercise price K_3. At maturity T we have

$$\gamma(K_1 - S_T)^+ + (1 - \gamma)(K_2 - S_T)^+ \geq \Big((\gamma K_1 + (1 - \gamma)K_2) - S_T\Big)^+$$

since the payoff function $h(x) = (K - x)^+$ is convex in K.   □

An interesting (and more difficult) issue is the convexity of the option price as a function of the initial stock price. As we shall see in what follows, such a property is valid in several commonly used financial models, that is, in most models the option price inherits the convexity (concavity) property from the terminal payoff.

2

Discrete-time Security Markets

This chapter deals mostly with *finite markets* – that is, discrete-time models of financial markets in which all relevant quantities take a finite number of values. Essentially, we follow here the approach of Harrison and Pliska (1981); a more exhaustive analysis of finite markets can be found in Taqqu and Willinger (1987). An excellent introduction to discrete-time financial mathematics is given by Pliska (1997) and Shreve (2004). A monograph by Föllmer and Schied (2000) is the most comprehensive source in the area.

The detailed treatment of finite models of financial markets presented below is not motivated by their practical importance (except for binomial or multinomial models). The main motivation comes rather from the fact that the most important ideas and results of arbitrage pricing can be presented in a more transparent way by working first in a finite-dimensional framework.

Since the number of dates is finite, there is no loss of generality if we take the set of dates $T = \{0, 1, \ldots, T^*\}$. Let Ω be an arbitrary finite set, $\Omega = \{\omega_1, \omega_2, \ldots, \omega_d\}$, and let $\mathcal{F} = \mathcal{F}_{T^*}$ be the σ-field of all subsets of Ω, i.e., $\mathcal{F} = 2^\Omega$. We consider a filtered probability space $(\Omega, \mathbb{F}, \mathbb{P})$ equipped with a filtration $\mathbb{F} = (\mathcal{F}_t)_{t=0}^{T^*}$, where $\mathbb{P}$ is an arbitrary probability measure on $(\Omega, \mathcal{F}_{T^*})$ such that $\mathbb{P}\{\omega_j\} > 0$ for every $j = 1, 2, \ldots, d$. We assume throughout that the σ-field $\mathcal{F}_0$ is trivial; that is, $\mathcal{F}_0 = \{\emptyset, \Omega\}$. Since T and Ω are both finite sets, all random variables and all stochastic processes are necessarily bounded. Thus they are integrable with respect to any probability measure $\mathbb{P}$ considered in what follows. A vector of prices of k *primary securities* is modelled by means of an $\mathbb{F}$-adapted $\mathbb{R}^k$-valued, nonnegative stochastic process $Z = (Z^1, Z^2, \ldots, Z^k)$. A k-dimensional process $Z = (Z^1, Z^2, \ldots, Z^k)$ is said to be $\mathbb{F}$-*adapted*, if, for any $i = 1, 2, \ldots, k$ and any $t \leq T^*$, the random variable Z_t^i is $\mathcal{F}_t$-measurable. For brevity, we shall say that a given process is adapted, instead of $\mathbb{F}$-adapted, if no confusion may arise. We assume throughout that $\mathcal{F}_t = \mathcal{F}_t^Z = \sigma(Z_0, Z_1, \ldots, Z_t)$; that is, the filtration $\mathbb{F}$ is generated by the observations of the price process Z. A *trading strategy* (a *dynamic portfolio*) is an $\mathbb{R}^k$-valued $\mathbb{F}$-adapted process $\phi = (\phi^1, \phi^2, \ldots, \phi^k)$. At any date t, the ith component, ϕ_t^i, of a portfolio ϕ determines the number of units of the ith asset that are held in the portfolio at this date.

2.1 The Cox-Ross-Rubinstein Model

Before we start the analysis of a general finite market, we present a particular model, which is a direct extension of a one-period two-state model with two securities to the multi-period set-up. Let T by a positive integer, interpreted as the time to maturity of a derivative contract, expressed in some convenient units of time. A European call option written on one share of a stock S paying no dividends during the option's lifetime, is formally equivalent to the claim X whose payoff at time T is contingent on the stock price S_T, and equals

$$X = (S_T - K)^+ \stackrel{\text{def}}{=} \max\{S_T - K, 0\}. \tag{2.1}$$

The call option value (or price) at the expiry date T equals $C_T = (S_T - K)^+$. Our first aim is to put the price on the option at any instant $t = 0, \ldots, T$, when the price of a risky asset (a stock) is modelled by the Cox et al. (1979a) multiplicative binomial lattice, commonly known as the Cox-Ross-Rubinstein (CRR, for short) model of a stock price. Since, for good reasons, models of this form are by far the most popular discrete-time financial models, it seems legitimate to refer to the CRR model as the *benchmark* discrete-time model.

2.1.1 Binomial Lattice for the Stock Price

We consider a discrete-time model of a financial market with the set of dates $0, 1, \ldots, T^*$, and with two primary traded securities: a risky asset, referred to as a *stock*, and a risk-free investment, called a *savings account* (or a *bond*).

Let us first describe the savings account. We assume that it a constant rate of return $r > -1$ over each time period $[t, t + 1]$, meaning that its price process $B = (B_t)_{t=0}^{T^*}$ equals (by convention $B_0 = 1$)

$$B_t = (1 + r)^t = \hat{r}^t, \quad \forall t = 0, 1, \ldots, T^*, \tag{2.2}$$

where we set $\hat{r} = 1 + r$. We postulate that the stock price $S = (S_t)_{t=0}^{T^*}$ satisfies

$$S_{t+1}/S_t \in \{u, d\} \tag{2.3}$$

for $t = 0, 1, \ldots, T^* - 1$, where $0 < d < u$ are real numbers and S_0 is a strictly positive constant. It will be essential to assume that, given the level S_t of the stock price at time t, both possible future states at time $t + 1$, that is, $S_{t+1}^u = u S_t$ and $S_{t+1}^d = d S_t$, have strictly positive probabilities of occurrence, also when we take into account all past observations $S_0, S_1, \ldots, S_t$ of the price process S. Formally, we postulate that

$$\mathbb{P}\{S_{t+1} = u S_t \mid S_0, S_1, \ldots, S_t\} > 0, \ \mathbb{P}\{S_{t+1} = d S_t \mid S_0, S_1, \ldots, S_t\} > 0. \tag{2.4}$$

To put it more intuitively, given the full observation of the sample path of the stock price S up to time t, the investors should never be in a position to tell with certainty whether the stock price will reach the upper level uS_t or the lower level dS_t at the end of the next time period. Since (2.4) states that at any date t the conditional distribution of S_{t+1} is non-degenerate under the actual (i.e., the real-world) probability $\mathbb{P}$, we shall refer to (2.4) as the *non-degeneracy condition*. We shall argue that the replicating strategies (and thus also arbitrage prices) of derivative assets are independent of the choice of the actual probability measure $\mathbb{P}$ on the underlying probability space, provided that the non-degeneracy property (2.4) is valid under $\mathbb{P}$.

As a consequence, it is sufficient to focus on the simplest probabilistic model of the stock price with the desired features. To this end, we pick an arbitrary number $p \in (0, 1)$ and we introduce a sequence $\xi_t,\ t = 1, 2, \ldots, T^*$ of mutually independent random variables on a common probability space $(\Omega, \mathcal{F}, \mathbb{P})$, with identical probability distribution

$$\mathbb{P}\{\xi_t = u\} = p = 1 - \mathbb{P}\{\xi_t = d\}, \quad \forall t = 1, 2, \ldots, T^*.$$

We now formally define the stock price process S by setting

$$S_t = S_0 \prod_{j=1}^{t} \xi_j, \quad \forall t = 0, 1, \ldots, T^*. \tag{2.5}$$

The sequence of independent and identically distributed random variables $\xi_t,\ t = 1, 2, \ldots, T^*$ plays the role of a *driving noise* in the stochastic dynamics of the stock price, which can be seen here as a *geometric random walk*. It is apparent that under the present assumptions we have

$$\mathbb{P}\{S_{t+1} = uS_t \mid S_0, S_1, \ldots, S_t\} = \mathbb{P}\{\xi_{t+1} = u\} = p > 0,$$
$$\mathbb{P}\{S_{t+1} = dS_t \mid S_0, S_1, \ldots, S_t\} = \mathbb{P}\{\xi_{t+1} = d\} = 1 - p > 0,$$

and thus (2.4) holds. Equivalently, the value of the stock price at time t is given as

$$S_t = S_0 \exp\left(\sum_{j=1}^{t} \zeta_j \right), \quad \forall t = 0, 1, \ldots, T^*, \tag{2.6}$$

where $\zeta_1, \zeta_2, \ldots, \zeta_{T^*}$ are independent, identically distributed random variables:

$$\mathbb{P}\{\zeta_t = \ln u\} = p = 1 - \mathbb{P}\{\zeta_t = \ln d\}, \quad \forall t = 1, 2, \ldots, T^*.$$

Due to representation (2.6), the process S given by (2.5) is frequently referred to as an *exponential random walk*. It is also clear that $\ln S_t = \ln S_{t-1} + \zeta_t$ for $t = 1, 2, \ldots, T^*$. This means that the logarithm of the stock price (the *log-price*) follows an *arithmetic random walk*. Finally, let us introduce the sequence of logarithmic returns (the *log-returns*) $\mu_1, \mu_2, \ldots, \mu_{T^*}$ on the stock by setting

$$\mu_t \stackrel{\text{def}}{=} \ln\left(\frac{S_t}{S_{t-1}} \right) = \zeta_t.$$

Obviously, the log-returns $\mu_1, \mu_2, \ldots, \mu_{T^*}$ are independent and identically distributed under $\mathbb{P}$.

In view of (2.5), it is clear that a typical sample path of the stock price can be represented as a sequence $(s_0, s_1, \ldots, s_{T^*})$ of real numbers such that $s_0 = S_0$ and the ratio s_{i+1}/s_i equals either u or d. If we take, for example, $T^* = 5$, then a sample path of S may look, for instance, as follows:

$$(S_0, u S_0, u^2 S_0, u^2 d S_0, u^2 d^2 S_0, u^3 d^2 S_0)$$

or

$$(S_0, d S_0, u d S_0, u d^2 S_0, u d^3 S_0, u^2 d^3 S_0).$$

For mathematical convenience, we will later formally identify each sample path of the process S with a sequence of length T^* of zeroes and ones, by associating the number 1 to u, and the number 0 to d. Manifestly, for the two sample paths given above, the corresponding sequences of zeroes and ones are $(1, 1, 0, 0, 1)$ and $(0, 1, 0, 0, 1)$ respectively. The collection of all sequences of zeroes and ones will also play the role of the underlying probability space Ω, so that we formally set $\Omega = 2^{T^*}$. This also means that in the CRR model with T^* periods the space $\Omega = \{\omega_1, \omega_2, \ldots, \omega_d\}$ comprises 2^{T^*} elementary events, each elementary event ω_k representing in fact a particular sample path of the stock price.

2.1.2 Recursive Pricing Procedure

Let us re-emphasize that the standing assumption that the random variables ξ_t, $t = 1, 2, \ldots, T^*$ are mutually independent and identically distributed under the actual probability $\mathbb{P}$ is not essential for our further purposes; we make this assumption, without loss of generality, for mathematical convenience.

As we will see in what follows, the arbitrage price of any European or American contingent claim in the binomial model of a financial market is independent of the choice of the probability of upward and downward movements of the stock price at any particular node. Indeed, it is uniquely determined by the assumed values of the stock price – that is, by the postulated form of sample paths of the stock price process.

Formally, the arbitrage price depends only on the specification of the payoff, the initial stock price S_0 and the value of parameters u, d and r. Hence the valuation results that we are going to derive will appear to be distribution-free, meaning that they do not depend on the specification of the actual probability measure $\mathbb{P}$.

Let us introduce some notation. For any $t \leq T$, we write α_t to denote the number of shares held during the period $[t, t + 1)$, while β_t stands for the dollar investment in the savings account during this period.

To determine the arbitrage price of a European call option we shall show, using backward induction, that by adjusting his or her dynamic portfolio $\phi_t = (\alpha_t, \beta_t)$, $t = 0, 1, \ldots, T - 1$, at the beginning of each period, an investor is able to mimic the payoff of an option at time T for every state. We shall refer to this fact by saying that the contingent claim $X = (S_T - K)^+$ admits a unique, dynamic, replicating, self-financing strategy. Replication and valuation of a European put option can be reduced, through the put-call parity relationship, to that of a European call.

For concreteness, we shall now focus on a European call option. It will soon become clear, however, that the backward induction method presented below can be directly applied to any European contingent claim X of the form $X = g(S_T)$. It is common to refer to such claims as *path-independent*, as opposed to *path-dependent* claims that have the form $X = h(S_0, S_1, \ldots, S_T)$; that is, they depend on the whole sample path. In particular, we shall see that for an arbitrary path-independent European claim we need to compute only $t + 1$ values at any date $t = 0, 1, \ldots, T^*$. By contrast, in the case of a path-dependent claim one has to deal with (at most) 2^t values at any date $t = 0, 1, \ldots, T$. The backward induction method (as well as the risk-neutral valuation method presented in Sect. 2.2.2) can be applied to both cases with no essential changes, however.

First step. Given a fixed maturity date $1 \leq T \leq T^*$, we start our analysis by considering the last period before the expiry date, $[T - 1, T]$. We assume that a portfolio replicating the terminal payoff of a call option is established at time $T - 1$, and remains fixed until the expiry date T. We thus need to find a portfolio $\phi_{T-1} = (\alpha_{T-1}, \beta_{T-1})$ at the beginning of the last period for which the terminal wealth $V_T(\phi)$, which equals

$$V_T(\phi) = \alpha_{T-1} S_T + \beta_{T-1} \hat{r}, \tag{2.7}$$

replicates the option payoff C_T; that is, we have $V_T(\phi) = C_T$. Combining (2.1) with (2.7), we get the following equality

$$\alpha_{T-1} S_T + \beta_{T-1} \hat{r} = (S_T - K)^+. \tag{2.8}$$

By virtue of our assumptions, we have $S_T = S_{T-1} \xi_{T-1}$; hence we may rewrite (2.8) in a more explicit form

$$\begin{cases} \alpha_{T-1} u S_{T-1} + \beta_{T-1} \hat{r} = (u S_{T-1} - K)^+, \\ \alpha_{T-1} d S_{T-1} + \beta_{T-1} \hat{r} = (d S_{T-1} - K)^+. \end{cases}$$

This simple system of two linear equations can be solved easily, yielding

$$\alpha_{T-1} = \frac{(u S_{T-1} - K)^+ - (d S_{T-1} - K)^+}{S_{T-1}(u - d)}, \tag{2.9}$$

and

$$\beta_{T-1} = \frac{u(d S_{T-1} - K)^+ - d(u S_{T-1} - K)^+}{\hat{r}(u - d)}. \tag{2.10}$$

Furthermore, the wealth $V_{T-1}(\phi)$ of this portfolio at time $T - 1$ equals

$$V_{T-1}(\phi) = \alpha_{T-1} S_{T-1} + \beta_{T-1}$$
$$= \hat{r}^{-1} \Big(p_*(u S_{T-1} - K)^+ + (1 - p_*)(d S_{T-1} - K)^+ \Big),$$

where we write $p_* = (\hat{r} - d)/(u - d) = (1 + r - d)/(u - d)$.

As we shall show later in this section, the number p_* specifies the probability of the rise of the stock price over each period $[t, t+1]$ under the martingale measure for the process $S^* = S/B$. We assume from now on that $d < 1 + r < u$ so that the number p_* belongs to the open interval $(0, 1)$. Assuming the absence of arbitrage in the market model,[1] the wealth $V_{T-1}(\phi)$ agrees with the value (that is, the *arbitrage price*) of a call option at time $T - 1$. Put another way, the equality $C_{T-1} = V_{T-1}(\phi)$ is valid.

Second step. We continue our analysis by considering the preceding period, $[T - 2, T - 1]$. In this step, we seek a portfolio $\phi_{T-2} = (\alpha_{T-2}, \beta_{T-2})$ created at time $T - 2$ in such a way that its wealth at time $T - 1$ replicates option value C_{T-1}; that is

$$\alpha_{T-2}S_{T-1} + \beta_{T-2}\hat{r} = C_{T-1}. \tag{2.11}$$

Note that since $C_{T-1} = V_{T-1}(\phi)$, the dynamic trading strategy ϕ constructed in this way will possess the *self-financing property* at time $T - 1$

$$\alpha_{T-2}S_{T-1} + \beta_{T-2}\hat{r} = \alpha_{T-1}S_{T-1} + \beta_{T-1}. \tag{2.12}$$

Basically, the self-financing feature means that the portfolio is adjusted at time $T - 1$ (and more generally, at any trading date) in such a way that no withdrawals or inputs of funds take place. Since $S_{T-1} = S_{T-2}\xi_{T-2}$ and $\xi_{T-2} \in \{u, d\}$, we get the following equivalent form of equality (2.11)

$$\begin{cases} \alpha_{T-2}u S_{T-2} + \beta_{T-2}\hat{r} = C^u_{T-1}, \\ \alpha_{T-2}d S_{T-2} + \beta_{T-2}\hat{r} = C^d_{T-1}, \end{cases} \tag{2.13}$$

where we set

$$C^u_{T-1} = \frac{1}{\hat{r}}\left(p_*(u^2 S_{T-2} - K)^+ + (1 - p_*)(ud S_{T-2} - K)^+\right)$$

and

$$C^d_{T-1} = \frac{1}{\hat{r}}\left(p_*(ud S_{T-2} - K)^+ + (1 - p_*)(d^2 S_{T-2} - K)^+\right).$$

In view of (2.13), it is evident that

$$\alpha_{T-2} = \frac{C^u_{T-1} - C^d_{T-1}}{S_{T-2}(u - d)}, \quad \beta_{T-2} = \frac{uC^d_{T-1} - dC^u_{T-1}}{\hat{r}(u - d)}.$$

Consequently, the wealth $V_{T-2}(\phi)$ of the portfolio $\phi_{T-2} = (\alpha_{T-2}, \beta_{T-2})$ at time $T - 2$ equals

$$V_{T-2}(\phi) = \alpha_{T-2}S_{T-2} + \beta_{T-2} = \frac{1}{\hat{r}}\left(p_*C^u_{T-1} + (1 - p_*)C^d_{T-1}\right)$$

$$= \frac{1}{\hat{r}^2}\left(p_*^2(u^2 S_{T-2} - K)^+ + 2p_*q_*(ud S_{T-2} - K)^+ + q_*^2(d^2 S_{T-2} - K)^+\right).$$

---

[1] We will return to this point later in Sect. 2.2. Let us only mention here that the necessary and sufficient condition for the absence of arbitrage has the same form as in the case of the one-period model; that is, it is exactly the condition $d < 1 + r < u$ we have just imposed.

Using the same arbitrage arguments as in the first step, we argue that the wealth $V_{T-2}(\phi)$ of the portfolio ϕ at time $T-2$ gives the arbitrage price at time $T-2$, i.e., $C_{T-2} = V_{T-2}(\phi)$.

General induction step. It is evident that by repeating the above procedure, one can completely determine the option price at any date $t \leq T$, as well as the (unique) trading strategy ϕ that replicates the option. Summarizing, the above reasoning provides a recursive procedure for finding the value of a call with any number of periods to go. It is worth noting that in order to value the option at a given date t and for a given level of the current stock price S_t, it is enough to consider a sub-lattice of the CRR binomial lattice starting from S_t and ranging over $T-t$ periods.

Path-independent European claims. As already mentioned, the recursive valuation procedure applies with virtually no changes to any European claim of the form $X = g(S_T)$. Indeed, a specific formula for the option's payoff was used in the first step only. If $X = g(S_T)$, we replace (2.8) by

$$\alpha_{T-1}S_T + \beta_{T-1}\hat{r} = g(S_T), \tag{2.14}$$

so that we now need to solve the following pair of linear equations:

$$\begin{cases} \alpha_{T-1}uS_{T-1} + \beta_{T-1}\hat{r} = g(uS_{T-1}), \\ \alpha_{T-1}dS_{T-1} + \beta_{T-1}\hat{r} = g(dS_{T-1}). \end{cases}$$

The unique solution α_{T-1}, β_{T-1} depends on S_{T-1}, but not on the values of the stock price at times $0, 1, \ldots, T-2$. Hence the replicating portfolio can be written as $\alpha_{T-1} = f_{T-1}(S_{T-1})$ and $\beta_{T-1} = h_{T-1}(S_{T-1})$ for some functions $f_{T-1}, h_{T-1} : \mathbb{R}_+ \to \mathbb{R}$. Consequently, the *arbitrage price* $\pi_{T-1}(X) = V_{T-1}(\phi)$ of X at time $T-1$ can be represented as $\pi_{T-1}(X) = g_{T-1}(S_{T-1})$ for some function $g_{T-1} : \mathbb{R}_+ \to \mathbb{R}$.

In the next step, we consider the period $[T-2, T-1]$ and a path-independent claim $g_{T-1}(S_{T-1})$ that settles at time $T-1$. A similar reasoning as above shows that $\pi_{T-2}(X) = g_{T-2}(S_{T-2})$ for some function $g_{T-2} : \mathbb{R}_+ \to \mathbb{R}$.

We conclude that there exists a sequence of functions $g_t : \mathbb{R}_+ \to \mathbb{R}$ (with $g_T = g$) such that the arbitrage price process $\pi(X)$ satisfies $\pi_t(X) = g_t(S_t)$ for $t = 0, 1, \ldots, T$. As we shall see in what follows, this *Markovian feature* of the replicating strategy and the arbitrage price of a path-independent European contingent claim can also be deduced from the Markov property of the stock price under the martingale measure.

Path-dependent European claims. In the case of a general (that is, possibly path-dependent) European contingent claim, formula (2.14) becomes

$$\alpha_{T-1}S_T + \beta_{T-1}\hat{r} = g(S_0, S_1, \ldots, S_T). \tag{2.15}$$

It is essential to make clear that α_{T-1} and β_{T-1} depend not only on the level of the stock price at time $T-1$ (i.e., on the choice of the node at time $T-1$), but on the entire sample path that connects the initial state $S_0 = s_0$ with a given node at time $T-1$.

Let $(s_0, s_1, \ldots, s_{T-1})$ be a particular sample path of the stock price process S that connects s_0 with a generic value s_{T-1} at time $T-1$. In order to find the replicating portfolio for this particular sample path, we need to solve for the unknowns $\alpha_{T-1}(s_0, s_1, \ldots, s_{T-1})$ and $\beta_{T-1}(s_0, s_1, \ldots, s_{T-1})$ the following pair of linear equations

$$\begin{cases} \alpha_{T-1}(s_0, s_1, \ldots, s_{T-1})u s_{T-1} + \beta_{T-1}(s_0, s_1, \ldots, s_{T-1})\hat{r} = g^u, \\ \alpha_{T-1}(s_0, s_1, \ldots, s_{T-1})d s_{T-1} + \beta_{T-1}(s_0, s_1, \ldots, s_{T-1})\hat{r} = g^d, \end{cases}$$

where $g^u = g(s_0, s_1, \ldots, s_{T-1}, u s_{T-1})$ and $g^d = g(s_0, s_1, \ldots, s_{T-1}, d s_{T-1})$.

Of course, the number of sample paths connecting s_0 with the current level of the stock price at time $T-1$ depends on the choice of a node. It is also clear that the total number of sample paths of length $T-1$ is 2^{T-1}. Thus the arbitrage price of X at time $T-1$ will have at most 2^{T-1} different values. As soon as the arbitrage price

$$\pi_{T-1}(X) = V_{T-1}(\phi) = \alpha_{T-1}(s_0, s_1, \ldots, s_{T-1})s_{T-1} + \beta_{T-1}(s_0, s_1, \ldots, s_{T-1})$$

is found at each node s_{T-1} and for each sample path connecting s_0 with this node, we are in a position to apply the same procedure as above to the (path-dependent) claim $\pi_{T-1}(X)$ settling at $T-1$. In this way, we are able to find the sequence of prices $\pi_{T-2}(X), \pi_{T-3}(X), \ldots, \pi_1(X), \pi_0(X)$, as well as replicating strategy $\phi_t = (\alpha_t, \beta_t)$, $t = 0, 1, \ldots, T$ for a claim X. Since this procedure has a unique solution for any X, it is fairly clear that any European contingent claim can be replicated, and thus it is *attainable* in the CRR model; we refer to this property as the *completeness* of the CRR model.

Let $\mathcal{F}_t^S$ stand for the σ-field of all events of $\mathcal{F}$ generated by the observations of the stock price S up to the date t; formally $\mathcal{F}_t^S = \sigma(S_0, S_1, \ldots, S_t)$, where $\sigma(S_0, S_1, \ldots, S_t)$ denotes the least σ-field with respect to which the random variables $S_0, S_1, \ldots, S_t$ are measurable. We write $\mathbb{F}^S$ to denote the filtration[2] generated by the stock price, so that $\mathbb{F}^S = (\mathcal{F}_t^S)_{t \leq T^*}$.

By construction of a replicating strategy, it is evident that, for any fixed t, the random variables α_t, β_t defining the portfolio at time t, as well as the wealth $V_t(\phi)$ of this portfolio, are measurable with respect to the σ-field $\mathcal{F}_t^S$. Hence the processes $\phi = (\alpha, \beta)$ and $V(\phi)$ are *adapted* to the filtration $\mathbb{F}^S$ generated by the stock price, or briefly, $\mathbb{F}^S$-adapted.

We conclude that the unique replicating strategy for an arbitrary European contingent claim in the CRR model follows an $\mathbb{F}^S$-adapted stochastic process and is *self-financing*, in the sense that the equality

$$\alpha_{t-1} S_t + \beta_{t-1}\hat{r} = \alpha_t S_t + \beta_t \tag{2.16}$$

holds for every $t = 1, \ldots, T-1$. From the considerations above, we conclude that the following result is valid within the set-up of the CRR model.

---

[2] Recall that a *filtration* is simply an increasing family of σ-fields.

Proposition 2.1.1 *Any European claim X is attainable in the CRR model. Its replicating strategy ϕ and arbitrage price $\pi(X)$ are $\mathbb{F}^S$-adapted processes. If, in addition, a claim X is path-independent, the arbitrage price at time $t = 0, 1, \ldots, T$ is a function of the current level S_t of the stock price, so that, for any $t = 0, 1, \ldots, T$, we have $\pi_t(X) = g_t(S_t)$ for some function $g_t : \mathbb{R}_+ \to \mathbb{R}$.*

2.1.3 CRR Option Pricing Formula

It appears that the recursive pricing procedure leads to an explicit formula for the arbitrage price of a European call (or put) option in the CRR model. Before we state this result, we find it convenient to introduce some notation.

For any fixed natural number m, let the function $a_m : \mathbb{R}_+ \to \mathbb{N}^*$ be given by the formula ($\mathbb{N}^*$ stands hereafter for the set of all nonnegative integers)

$$a_m(x) = \inf\{\, j \in \mathbb{N}^* \mid xu^j d^{m-j} > K \,\},$$

where, by convention, $\inf \emptyset = \infty$.

Let us set $a^d = a_m(dx)$ and $a^u = a_m(ux)$. It is not difficult to check that for any $x > 0$, we have either $a^d = a^u$ or $a^d = a^u + 1$. For ease of notation, we write

$$\Delta_m(x, j) = \binom{m}{j} p_*^j (1 - p_*)^{m-j} (u^j d^{m-j} x - K).$$

Proposition 2.1.2 *For every $m = 1, 2, \ldots, T$, the arbitrage price of a European call option at time $t = T - m$ is given by the Cox-Ross-Rubinstein valuation formula*

$$C_{T-m} = S_{T-m} \sum_{j=a}^{m} \binom{m}{j} \bar{p}^j (1 - \bar{p})^{m-j} - \frac{K}{\hat{r}^m} \sum_{j=a}^{m} \binom{m}{j} p_*^j (1 - p_*)^{m-j}, \quad (2.17)$$

where $a = a_m(S_{T-m})$, $p_* = (\hat{r} - d)/(u - d)$ and $\bar{p} = p_* u/\hat{r}$. At time $t = T - m - 1$, the unique replicating strategy $\phi_{T-m-1} = (\alpha_{T-m-1}, \beta_{T-m-1})$ equals

$$\alpha_{T-m-1} = \sum_{j=a^d}^{m} \binom{m}{j} \bar{p}^j (1 - \bar{p})^{m-j} + \frac{\delta \Delta_m(u S_{T-m-1}, a^u)}{S_{T-m-1}(u - d)},$$

$$\beta_{T-m-1} = -\frac{K}{\hat{r}^{m+1}} \sum_{j=a^d}^{m} \binom{m}{j} p_*^j (1 - p_*)^{m-j} - \frac{\delta d \Delta_m(u S_{T-m-1}, a^u)}{\hat{r}(u - d)},$$

where $a^d = a_m(d S_{T-m-1})$, $a^u = a_m(u S_{T-m-1})$ and $\delta = a^d - a^u$.

Proof. Straightforward calculations yield $1 - \bar{p} = d(1 - p_*)/\hat{r}$, and thus

$$\bar{p}^j (1 - \bar{p})^{m-j} = p_*^j (1 - p_*)^{m-j} u^j d^{m-j}/\hat{r}^m.$$

Formula (2.17) is therefore equivalent to the following

$$C_{T-m} = \frac{1}{\hat{r}^m} \sum_{j=a}^{m} \binom{m}{j} p_*^j (1 - p_*)^{m-j} \left(u^j d^{m-j} S_{T-m} - K \right)$$

$$= \frac{1}{\hat{r}^m} \sum_{j=0}^{m} \binom{m}{j} p_*^j (1 - p_*)^{m-j} \left(u^j d^{m-j} S_{T-m} - K \right)^+.$$

We will now proceed by induction with respect to m. For $m = 0$, the above formula is manifestly true, since it reduces to $C_T = (S_T - K)^+$. Assume now that C_{T-m} is the arbitrage price of a European call option at time $T - m$. We have to select a portfolio $\phi_{T-m-1} = (\alpha_{T-m-1}, \beta_{T-m-1})$ for the period $[T-m-1, T-m]$ (that is, established at time $T-m-1$ at each node of the binomial lattice) in such a way that the portfolio's wealth at time $T - m$ replicates the value C_{T-m} of the option. Formally, the wealth of the portfolio $(\alpha_{T-m-1}, \beta_{T-m-1})$ needs to satisfy the relationship

$$\alpha_{T-m-1} S_{T-m} + \beta_{T-m-1} \hat{r} = C_{T-m}, \tag{2.18}$$

which in turn is equivalent to the following pair of equations

$$\begin{cases} \alpha_{T-m-1} u S_{T-m-1} + \beta_{T-m-1} \hat{r} = C_{T-m}^u, \\ \alpha_{T-m-1} d S_{T-m-1} + \beta_{T-m-1} \hat{r} = C_{T-m}^d, \end{cases}$$

where

$$C_{T-m}^u = \frac{1}{\hat{r}^m} \sum_{j=0}^{m} \binom{m}{j} p_*^j (1 - p_*)^{m-j} \left(u^{j+1} d^{m-j} S_{T-m-1} - K \right)^+$$

$$= \frac{1}{\hat{r}^m} \sum_{j=a^u}^{m} \binom{m}{j} p_*^j (1 - p_*)^{m-j} \left(u^{j+1} d^{m-j} S_{T-m-1} - K \right)$$

$$C_{T-m}^d = \frac{1}{\hat{r}^m} \sum_{j=0}^{m} \binom{m}{j} p_*^j (1 - p_*)^{m-j} \left(u^j d^{m-j+1} S_{T-m-1} - K \right)^+$$

$$= \frac{1}{\hat{r}^m} \sum_{j=a^d}^{m} \binom{m}{j} p_*^j (1 - p_*)^{m-j} \left(u^j d^{m-j+1} S_{T-m-1} - K \right).$$

Consequently, we have (we write $q_* = 1 - p_*$)

$$\alpha_{T-m-1} = \frac{C_{T-m}^u - C_{T-m}^d}{S_{T-m-1}(u - d)}$$

$$= \frac{1}{\hat{r}^m (u - d)} \sum_{j=a^d}^{m} \binom{m}{j} p_*^j q_*^{m-j} \left(u^{j+1} d^{m-j} - u^j d^{m-j+1} \right)$$

$$+ \frac{\delta \Delta_m (u S_{T-m-1}, a^u)}{S_{T-m-1}(u - d)}$$

$$= \sum_{j=a^d}^{m} \binom{m}{j} \bar{p}^j (1 - \bar{p})^{m-j} + \frac{\delta \Delta_m (u S_{T-m-1}, a^u)}{S_{T-m-1}(u - d)}.$$

Similarly,

$$\beta_{T-m-1} = \frac{uC_{T-m}^d - dC_{T-m}^u}{\hat{r}(u-d)}$$

$$= \frac{1}{\hat{r}^{m+1}(u-d)} \sum_{j=a^d}^{m} \binom{m}{j} p_*^j (1-p_*)^{m-j}(dK - uK)$$

$$- \frac{\delta d \, \Delta_m(uS_{T-m-1}, a^u)}{\hat{r}(u-d)}$$

$$= -\frac{K}{\hat{r}^{m+1}} \sum_{j=a^d}^{m} \binom{m}{j} p_*^j (1-p_*)^{m-j} - \frac{\delta d \, \Delta_m(uS_{T-m-1}, a^u)}{\hat{r}(u-d)}.$$

The wealth of this portfolio at time $T-m-1$ equals (note that just established explicit formulas for the replicating portfolio are not employed here)

$$C_{T-m-1} = \alpha_{T-m-1}S_{T-m-1} + \beta_{T-m-1}$$

$$= (u-d)^{-1}\left(C_{T-m}^u - C_{T-m}^d + \hat{r}^{-1}(uC_{T-m}^d - dC_{T-m}^u)\right)$$

$$= \hat{r}^{-1}\left(p_*C_{T-m}^u + (1-p_*)C_{T-m}^d\right)$$

$$= \frac{1}{\hat{r}^{m+1}}\left\{ \sum_{j=0}^{m} \binom{m}{j} p_*^{j+1} q_*^{m-j} \left(u^{j+1}d^{m-j}S_{T-m-1} - K\right)^+ \right.$$

$$\left. + \sum_{j=0}^{m} \binom{m}{j} p_*^j q_*^{m+1-j} \left(u^j d^{m+1-j}S_{T-m-1} - K\right)^+ \right\}$$

$$= \frac{1}{\hat{r}^{m+1}}\left\{ \sum_{j=1}^{m+1} \binom{m}{j-1} p_*^j q_*^{m+1-j} \left(u^j d^{m+1-j}S_{T-m-1} - K\right)^+ \right.$$

$$\left. + \sum_{j=0}^{m} \binom{m}{j} p_*^j q_*^{m+1-j} \left(u^j d^{m+1-j}S_{T-m-1} - K\right)^+ \right\}.$$

Using the last formula and the equality $\binom{m}{j-1} + \binom{m}{j} = \binom{m+1}{j}$, we obtain

$$C_{T-m-1} = \frac{1}{\hat{r}^{m+1}}\left\{ p_*^{m+1}\left(u^{m+1}S_{T-m-1} - K\right)^+ \right.$$

$$+ \sum_{j=1}^{m} \binom{m}{j} p_*^j q_*^{m+1-j} \left(u^j d^{m+1-j}S_{T-m-1} - K\right)^+$$

$$+ \sum_{j=1}^{m} \binom{m}{j-1} p_*^j q_*^{m+1-j} \left(u^j d^{m+1-j}S_{T-m-1} - K\right)^+$$

$$\left. + q_*^{m+1}\left(d^{m+1}S_{T-m-1} - K\right)^+ \right\}$$

$$= \frac{1}{\hat{r}^{m+1}} \sum_{j=0}^{m+1} \binom{m+1}{j} p_*^j q_*^{m+1-j} \left(u^j d^{m+1-j}S_{T-m-1} - K\right)^+. \qquad \square$$

Note that the CRR valuation formula (2.17) makes no reference to the subjective (or actual) probability p. Intuitively, the pricing formula does not depend on the investor's attitudes toward risk. The only assumption made with regard to the behavior of an individual is that all investors prefer more wealth to less wealth, and thus have an incentive to take advantage of risk-free profitable investments. Consequently, if arbitrage opportunities were present in the market, no market equilibrium would be possible. This feature of arbitrage-free markets explains the term *partial equilibrium approach,* frequently used in economic literature in relation to arbitrage pricing of derivative securities.

2.2 Martingale Properties of the CRR Model

Our next goal is to analyze the no-arbitrage features of the CRR model. As mentioned already, it is convenient to work on Ω related to the *canonical space* of the process $S = (S_t)_{t=0}^{T^*}$. We start by introducing a finite probability space Ω; namely, for a fixed natural number T^*, we consider

$$\Omega = \{\omega = (a_1, a_2, \ldots, a_{T^*}) \mid a_j = 1 \text{ or } a_j = 0\}. \tag{2.19}$$

In the present context, it will be sufficient to consider a specific class $\mathcal{P}$ of probability measures on the measurable space $(\Omega, \mathcal{F})$, where $\mathcal{F}$ is the σ-field of all subsets of Ω, i.e., $\mathcal{F} = 2^{\Omega}$. For a generic elementary event $\omega = (a_1, a_2, \ldots, a_{T^*})$, we define its probability $\mathbb{P}\{\omega\}$ by setting, for a fixed $p \in (0, 1)$,

$$\mathbb{P}\{\omega\} = p^{\sum_{j=1}^{T^*} a_j} (1 - p)^{T^* - \sum_{j=1}^{T^*} a_j}. \tag{2.20}$$

Definition 2.2.1 We denote by $\mathcal{P}$ the class of all probability measures on the canonical space $(\Omega, \mathcal{F})$ given by (2.19) that have the form (2.20).

It is clear that any element $\mathbb{P} \in \mathcal{P}$ is uniquely determined by the value of the parameter p. For any $j = 1, 2, \ldots, T^*$, we denote by A_j the event $A_j = \{\omega \in \Omega \mid a_j = 1\}$. It is not difficult to check that the events $A_j, j = 1, 2, \ldots, T^*$ are mutually independent; moreover, $\mathbb{P}\{A_j\} = p$ for every $j = 1, 2, \ldots, T^*$. We are now in a position to define a sequence of random variables $\xi_j, j = 1, 2, \ldots, T^*$ by setting

$$\xi_j(\omega) = u a_j + d(1 - a_j), \quad \forall \, \omega \in \Omega. \tag{2.21}$$

The random variables ξ_j are easily seen to be independent and identically distributed, with the following probability distribution under $\mathbb{P}$

$$\mathbb{P}\{\xi_j = u\} = p = 1 - \mathbb{P}\{\xi_j = d\}, \quad \forall \, t \leq T^*. \tag{2.22}$$

We shall show that the unique martingale measure for the process $S^* = S/B$ belongs to the class $\mathcal{P}$. The assumption that the real-world probability is selected from this class is not essential, though. It is enough to assume that $\mathbb{P}$ is such that any potential sample path of the price process S has a strictly positive probability, or equivalently, that the non-degeneracy condition (2.4) is satisfied.

2.2.1 Martingale Measures

Let us return to the multiplicative binomial lattice modelling the stock price. In the present framework, the process S is determined by the initial stock price S_0 and the sequence ξ_j, $j = 1, 2, \ldots, T^*$ of independent random variables given by (2.21). More precisely, the sequence $S = (S_t)_{t=0}^{T^*}$ is defined on the probability space $(\Omega, \mathcal{F}, \mathbb{P})$ by means of (2.5), or equivalently, by the relation

$$S_{t+1} = \xi_{t+1} S_t, \quad \forall t < T^*, \tag{2.23}$$

with $S_0 > 0$. Let us introduce the process S^* of the *discounted* stock price by setting

$$S_t^* = S_t/B_t = S_t/\hat{r}^t, \quad \forall t \le T^*. \tag{2.24}$$

Let $\mathcal{D}_t^S$ be the family of decompositions of Ω generated by random variables S_u, $u \le t$; that is, $\mathcal{D}_t^S = \mathcal{D}(S_0, S_1, \ldots, S_t)$ for every $t \le T^*$. It is clear that the family $\mathcal{D}_t^S$, $t \le T^*$, of decompositions is an increasing family of σ-fields, meaning that $\mathcal{D}_t^S \subset \mathcal{D}_{t+1}^S$ for every $t \le T^* - 1$. Note that the family $\mathcal{D}_t^S$ is also generated by the family $\xi_1, \xi_2, \ldots, \xi_{T^*}$ of random variables, more precisely,

$$\mathcal{D}_t^S = \mathcal{D}(\xi_0, \xi_1, \ldots, \xi_t), \quad \forall t \le T^*,$$

where by convention $\xi_0 = 1$. The family $\mathcal{D}_t^S$, $t \le T^*$, models a discrete-time flow of information generated by the observations of stock prices. In financial interpretation, the decomposition $\mathcal{D}_t^S$ represents the market information available to all investors at time t. Let us set $\mathcal{F}_t^S = \sigma(\mathcal{D}_t^S)$ for every $t \le T^*$, where $\sigma(\mathcal{D}_t^S)$ is the σ-field generated by the decomposition $\mathcal{D}_t^S$. It is clear that for $t \le T^*$

$$\mathcal{F}_t^S = \sigma(S_0, S_1, \ldots, S_t) = \sigma(S_0^*, S_1^*, \ldots, S_t^*) = \mathcal{F}_t^{S^*}.$$

Recall that we write $\mathbb{F}^S = (\mathcal{F}_t^S)_{t \le T^*}$ to denote the family of natural σ-fields of the process S, or briefly, the *natural filtration* of the process S.

Let $\mathbb{P}$ be a probability measure satisfying (2.4) – that is, such that $\mathbb{P}\{\omega\} > 0$ for any elementary event $\omega = (a_1, a_2, \ldots, a_{T^*})$. Note that all probability measures satisfying (2.4) are mutually equivalent.

Definition 2.2.2 A probability measure $\mathbb{P}^*$ equivalent to $\mathbb{P}$ is called a *martingale measure* for the discounted stock price process $S^* = (S_t^*)_{t=0}^{T^*}$ if

$$\mathbb{E}_{\mathbb{P}^*}(S_{t+1}^* \mid \mathcal{F}_t^S) = S_t^*, \quad \forall t \le T^* - 1, \tag{2.25}$$

that is, if the process S^* follows a *martingale* under $\mathbb{P}^*$ with respect to the filtration $\mathbb{F}^S$. In this case, we say that the discounted stock price S^* is a $(\mathbb{P}^*, \mathbb{F}^S)$-*martingale*, or briefly, a $\mathbb{P}^*$-*martingale*, if no confusion may arise.

Since a martingale measure $\mathbb{P}^*$ corresponds to the choice of the savings account as a numeraire asset, it is also referred to as a *spot martingale measure*. The problem of existence and uniqueness of a spot martingale measure in the CRR model can be solved completely, as the following result shows.

Proposition 2.2.1 *A martingale measure* $\mathbb{P}^*$ *for the discounted stock price* S^* *exists if and only if* $d < 1 + r < u$. *In this case, the martingale measure* $\mathbb{P}^*$ *is the unique element from the class* $\mathcal{P}$ *that corresponds to* $p = p_* = (1 + r - d)/(u - d)$.

Proof. Using (2.23)–(2.24), we may re-express equality (2.25) in the following way

$$\mathbb{E}_{\mathbb{P}^*}(\hat{r}^{-(t+1)}\xi_{t+1}S_t \mid \mathcal{F}_t^S) = \hat{r}^{-t}S_t, \quad \forall t \leq T^* - 1, \tag{2.26}$$

or equivalently

$$\hat{r}^{-(t+1)}S_t \mathbb{E}_{\mathbb{P}^*}(\xi_{t+1} \mid \mathcal{F}_t^S) = \hat{r}^{-t}S_t, \quad \forall t \leq T^* - 1.$$

Since $S_t > 0$, for the last equality to hold it is necessary and sufficient that

$$\mathbb{E}_{\mathbb{P}^*}(\xi_{t+1} \mid \mathcal{F}_t^S) = 1 + r. \tag{2.27}$$

Since the right-hand side in (2.27) is non-random, we conclude that

$$\mathbb{E}_{\mathbb{P}^*}(\xi_{t+1} \mid \mathcal{F}_t^S) = \mathbb{E}_{\mathbb{P}^*}(\xi_{t+1}).$$

In view of Lemma 2.2.1 below, this means that ξ_{t+1} is independent of $\mathcal{F}_t^S$ under $\mathbb{P}^*$, and thus independent of the random variables $\xi_1, \xi_2, \ldots, \xi_t$ under $\mathbb{P}^*$. Since (2.27) holds for any t, it is easy to deduce by induction that the random variables $\xi_1, \ldots, \xi_{T^*}$ are necessarily independent under $\mathbb{P}^*$. It thus suffices to find the distribution of ξ_t under $\mathbb{P}^*$. We have, for any $t = 1, 2, \ldots, T^*$

$$\mathbb{E}_{\mathbb{P}^*}(\xi_t) = u\mathbb{P}^*\{\xi_t = u\} + d(1 - \mathbb{P}^*\{\xi_t = u\}) = 1 + r.$$

By solving this equation, we find that $\mathbb{P}^*\{\xi_t = u\} = p_* = (1 + r - d)/(u - d)$ for every t. Note that the last equality defines a probability measure $\mathbb{P}^*$ if and only if $d \leq 1 + r \leq u$. Moreover, $\mathbb{P}^*$ belongs to the class $\mathcal{P}$ and is equivalent to $\mathbb{P}$ if and only if $d < 1 + r < u$. $\qquad\square$

Lemma 2.2.1 *Let* ξ *be a random variable on a probability space* $(\Omega, \mathcal{F}, \mathbb{P})$ *such that* $\mathbb{P}\{\xi = a\} + \mathbb{P}\{\xi = b\} = 1$ *for some real numbers* a, b. *Let* $\mathcal{G}$ *be a sub-σ-field of* $\mathcal{F}$. *If* $\mathbb{E}_{\mathbb{P}}(\xi \mid \mathcal{G}) = \mathbb{E}_{\mathbb{P}}\xi$ *then* ξ *is independent of* $\mathcal{G}$.

Proof. Equality $\mathbb{E}_{\mathbb{P}}(\xi \mid \mathcal{G}) = \mathbb{E}_{\mathbb{P}}\xi$ implies that for any event $B \in \mathcal{G}$ we have

$$\int_B \xi \, d\mathbb{P} = \int_B \mathbb{E}_{\mathbb{P}}\xi \, d\mathbb{P}.$$

Let us write $A = \{\xi = a\}$. Then we obtain (we denote by A^c the complement of A in Ω, so that $A^c = \Omega \setminus A$), for any event $B \in \mathcal{G}$,

$$\int_B (a\mathbb{1}_A + b\mathbb{1}_{A^c}) \, d\mathbb{P} = a\mathbb{P}\{A \cap B\} + b\mathbb{P}\{A^c \cap B\} = \mathbb{P}\{B\}(a\mathbb{P}\{A\} + b\mathbb{P}\{A^c\}).$$

The last equality implies that $\mathbb{P}\{A \cap B\} = \mathbb{P}\{A\}\mathbb{P}\{B\}$, as expected. $\qquad\square$

Note that the stock price follows under the unique martingale measure $\mathbb{P}^*$, an exponential random walk (cf. formula (2.6)), with the probability of upward movement equal to p_*. This feature explains why it was possible, with no loss of generality, to restrict attention to the special class of probability measures on the underlying canonical space Ω. It is important to point out that the martingale measure $\mathbb{P}^*$ is not exogenously introduced in order to model the observed real-world fluctuations of stock prices. Unlike the actual probability $\mathbb{P}$, the martingale measure $\mathbb{P}^*$ should be seen as a technical tool that proves to be useful in the arbitrage valuation of derivative securities.

In view of relationship (2.23) is is clear that we have $S^*_{t+1} = \hat{r}^{-1}\xi_{t+1}S^*_t$. The following corollary is thus an immediate consequence of the independence of the random variables $\xi_1, \xi_2, \ldots, \xi_{T^*}$ under $\mathbb{P}^*$.

Corollary 2.2.1 *The stock price S and the discounted stock price S^* are Markov processes under the spot martingale measure $\mathbb{P}^*$ with respect to the filtration $\mathbb{F}^S = \mathbb{F}^{S^*}$.*

Let Φ stand for the class of all self-financing (see (2.16)) trading strategies in the CRR model. As in the previous chapter, by an *arbitrage opportunity* we mean a strategy $\phi \in \Phi$ with zero initial wealth, nonnegative terminal wealth, and such that the terminal wealth is strictly positive with positive probability (the formal definition is given in Sect. 2.6.3 below). We say that a model is *arbitrage-free* if such a trading strategy does not exist.

Corollary 2.2.2 *The CRR binomial model (S, B, Φ) is arbitrage-free if and only if $d < 1 + r < u$.*

Proof. Suppose first that the condition $d < 1 + r < u$ does not hold. Then is it possible to produce an example of an arbitrage opportunity by considering, for instance, the first time-period and proceeding as in the case of a single-period model (buy stock if $1 + r \leq d$ and short stock if $1 + r \geq u$). We conclude that the condition $d < 1 + r < u$ is necessary for the arbitrage-free property of the model. The proof of sufficiency relies on the existence of a martingale measure established in Proposition 2.2.1. It is enough to show that the discounted wealth process of any self-financing strategy follows a martingale under $\mathbb{P}^*$. We do not go into details here, since this part of the proposition follows immediately from Proposition 2.6.2. □

The notion of a martingale measure (or a *risk-neutral probability*) is not universal, in the sense that it is relative to a specific choice of a numeraire asset. It can be verified, for instance, that the unique martingale measure for the relative bond price $\bar{B} = B/S$ is the unique element $\bar{\mathbb{P}}$ from the class $\mathcal{P}$ that corresponds to the following value of p (cf. (1.16))

$$p = \bar{p} = \left(\frac{1}{d} - \frac{1}{\hat{r}}\right)\frac{ud}{u-d} = \frac{p_* u}{\hat{r}}.$$

It is easily seen that $\bar{p}$ is in the open interval $(0, 1)$ if and only if $d < 1 + r < u$ (that is, if and only if p_* belongs to this interval). Hence the following result.

Proposition 2.2.2 *A martingale measure $\bar{\mathbb{P}}$ for the process $\bar{B} = B/S$ exists if and only if $d < 1 + r < u$. In this case, the martingale measure $\bar{\mathbb{P}}$ is the unique element from the class $\mathcal{P}$ that corresponds to $p = \bar{p}$.*

2.2.2 Risk-neutral Valuation Formula

We shall show that the CRR option pricing formula of Proposition 2.1.2 may be alternatively derived by the direct evaluation of the conditional expectation, under the martingale measure $\mathbb{P}^*$, of the discounted option's payoff.

Proposition 2.2.3 *Consider a European call option, with expiry date T and strike price K, written on one share of a common stock whose price S is assumed to follow the CRR multiplicative binomial process (2.5). Then for any $m = 0, 1, \ldots, T$ the arbitrage price C_{T-m}, given by formula (2.17), coincides with the conditional expectation*

$$C^*_{T-m} = \mathbb{E}_{\mathbb{P}^*}\left(\hat{r}^{-m}(S_T - K)^+ \,\big|\, \mathcal{F}^S_{T-m}\right). \tag{2.28}$$

Proof. It is enough to find the conditional expectation in (2.28) explicitly. Recall that we have

$$S_T = S_{T-m}\,\xi_{T-m+1}\xi_{T-m+2}\cdots\xi_T = S_{T-m}\eta_m,$$

where we write $\eta_m = \xi_{T-m+1}\xi_{T-m+2}\cdots\xi_T$. Note that the stock price S_{T-m} is manifestly an $\mathcal{F}^S_{T-m}$-measurable random variable, whereas the random variable η_m is independent of the σ-field $\mathcal{F}^S_{T-m}$.

Hence, by applying the well-known property of conditional expectations (see Lemma A.1.1) to the random variables $\psi = S_{T-m}$, $\eta = \eta_m$ and to the function $h(x, y) = \hat{r}^{-m}(xy - K)^+$, one finds that

$$C^*_{T-m} = \mathbb{E}_{\mathbb{P}^*}\left(\hat{r}^{-m}(S_T - K)^+ \,\big|\, \mathcal{F}^S_{T-m}\right) = H(S_{T-m}),$$

where the function $H : \mathbb{R} \to \mathbb{R}$ equals

$$H(x) = \mathbb{E}_{\mathbb{P}^*}\left(h(x, \eta_m)\right) = \mathbb{E}_{\mathbb{P}^*}\left(\hat{r}^{-m}(x\eta_m - K)^+\right), \quad \forall\, x \in \mathbb{R}.$$

The random variables $\xi_{T-m+1}, \xi_{T-m+2}, \ldots, \xi_T$ are also mutually independent and identically distributed under $\mathbb{P}^*$, with $\mathbb{P}^*\{\xi_j = u\} = p_* = 1 - \mathbb{P}^*\{\xi_j = d\}$. It is thus clear that

$$H(x) = \hat{r}^{-m}\sum_{j=0}^{m}\binom{m}{j}p_*^j(1 - p_*)^{m-j}\left(xu^j d^{m-j} - K\right)^+.$$

Using equalities $\bar{p} = p_* u/\hat{r}$ and $1 - \bar{p} = (1 - p_*)d/\hat{r}$, we conclude that

$$C^*_{T-m} = \sum_{j=a}^{m}\binom{m}{j}\left(S_{T-m}\bar{p}^j(1 - \bar{p})^{m-j} - K\hat{r}^{-m}p_*^j(1 - p_*)^{m-j}\right),$$

where $a = a_m(S_{T-m}) = \inf\{j \in \mathbb{N}^* \mid S_{T-m}u^j d^{m-j} > K\}$. $\qquad\square$

One might wonder if the risk-neutral valuation formula (2.28) remains in force for a larger class of financial models and European contingent claims. Generally speaking, the answer to this question is positive, even if the interest rate is assumed to follow a stochastic process, that is, for any claim X maturing at T and every $t = 0, 1, \ldots, T$, the arbitrage price $\pi_t(X)$ equals

$$\pi_t(X) = B_t \, \mathbb{E}_{\mathbb{P}^*}\big(B_T^{-1} X \,\big|\, \mathcal{F}_t^S\big), \quad \forall\, t \leq T. \tag{2.29}$$

For the CRR model, the last formula follows, for instance, from the recursive pricing procedure described in Sect. 2.1.2. A general result for a discrete-time finite model is established in Sect. 2.6.5.

The risk-neutral valuation formula (2.29) makes it clear, that the valuation is a linear map from the space of contingent claim that settle at time T (that is, the linear space of all $\mathcal{F}_T$-measurable random variables) to $\mathbb{R}$. The *put-call parity* relationship

$$C_t - P_t = S_t - K \hat{r}^{-(T-t)}$$

can now be easily established by applying the risk-neutral valuation formula to the payoff $X = C_T - P_T = S_T - K$.

2.2.3 Change of a Numeraire

The following variant of the risk-neutral valuation formula, based on the martingale measure $\bar{\mathbb{P}}$ of Proposition 2.2.2, is also valid for any contingent claim X

$$\pi_t(X) = S_t \, \mathbb{E}_{\bar{\mathbb{P}}}(S_T^{-1} X \,|\, \mathcal{F}_t^S), \quad \forall\, t \leq T. \tag{2.30}$$

The last formula can be proved using (2.29), the Bayes formula, and the expression for the Radon-Nikodým derivative of $\bar{\mathbb{P}}$ with respect to $\mathbb{P}^*$, which is given by the following result.

Proposition 2.2.4 *The Radon-Nikodým derivative process η of $\bar{\mathbb{P}}$ with respect to $\mathbb{P}^*$ equals, for every $t = 0, 1, \ldots, T^*$,*

$$\eta_t = \frac{d\bar{\mathbb{P}}}{d\mathbb{P}^*}\bigg|_{\mathcal{F}_t^S} = \frac{B_0 S_t}{S_0 B_t}. \tag{2.31}$$

Proof. It is clear that $\mathbb{P}^*$ and $\bar{\mathbb{P}}$ are equivalent on $(\Omega, \mathcal{F}_{T^*})$ and thus the Radon-Nikodým derivative $\eta_{T^*} = d\bar{\mathbb{P}}/d\mathbb{P}^*$ exists. Since $S^* = S/B$ is a martingale under $\mathbb{P}^*$ and, by virtue of general properties of the Radon-Nikodým derivative, we have that $\eta_t = \mathbb{E}_{\mathbb{P}^*}(\eta_{T^*} \,|\, \mathcal{F}_t^S)$, it suffices to show that

$$\eta_{T^*} = \frac{d\bar{\mathbb{P}}}{d\mathbb{P}^*} = \frac{B_0 S_{T^*}}{S_0 B_{T^*}}. \tag{2.32}$$

It is easily seen that the last equality defines an equivalent probability measure $\bar{\mathbb{P}}$ on $(\Omega, \mathcal{F}_{T^*})$ since $\eta_{T^*} > 0$ and $\mathbb{E}_{\mathbb{P}^*}\eta_{T^*} = 1$. It suffices to verify that the martingale property of the process $S^* = S/B$ under $\mathbb{P}^*$ implies that the process B/S is a martingale under $\bar{\mathbb{P}}$. We thus assume that, for $u \leq t$,

$$\mathbb{E}_{\mathbb{P}^*}(S_t/B_t \mid \mathcal{F}_u^S) = S_u/B_u.$$

Using the Bayes formula (see Lemma A.1.4), we obtain

$$
\begin{aligned}
\mathbb{E}_{\bar{\mathbb{P}}}(B_t/S_t \mid \mathcal{F}_u^S) &= \frac{\mathbb{E}_{\mathbb{P}^*}(\eta_{T^*} B_t/S_t \mid \mathcal{F}_u^S)}{\mathbb{E}_{\mathbb{P}^*}(\eta_{T^*} \mid \mathcal{F}_u^S)} \\
&= \frac{\mathbb{E}_{\mathbb{P}^*}(\mathbb{E}_{\mathbb{P}^*}(\eta_{T^*} \mid \mathcal{F}_t^S) B_t/S_t \mid \mathcal{F}_u^S)}{\mathbb{E}_{\mathbb{P}^*}(\eta_{T^*} \mid \mathcal{F}_u^S)} \\
&= \frac{\mathbb{E}_{\mathbb{P}^*}(\eta_t B_t/S_t \mid \mathcal{F}_u^S)}{\eta_u} \\
&= \frac{\mathbb{E}_{\mathbb{P}^*}((S_t/B_t)(B_t/S_t) \mid \mathcal{F}_u^S)}{S_u/B_u} \\
&= B_u/S_u.
\end{aligned}
$$

We conclude that the probability measure $\bar{\mathbb{P}}$ given by (2.32) is indeed the unique martingale measure for the process $\bar{B} = B/S$. □

Let us check that the right-hand sides in (2.29) and (2.30) coincide for any contingent claim X. Since X and S_T are $\mathcal{F}_T^S$-measurable random variables, the Bayes formula yields

$$
\begin{aligned}
S_t \, \mathbb{E}_{\bar{\mathbb{P}}}(S_T^{-1} X \mid \mathcal{F}_t^S) &= S_t \frac{\mathbb{E}_{\mathbb{P}^*}(\eta_T S_T^{-1} X \mid \mathcal{F}_t^S)}{\mathbb{E}_{\mathbb{P}^*}(\eta_T \mid \mathcal{F}_t^S)} \\
&= S_t \frac{\mathbb{E}_{\mathbb{P}^*}(\eta_T S_T^{-1} X \mid \mathcal{F}_t^S)}{\eta_t} \\
&= S_t \frac{\mathbb{E}_{\mathbb{P}^*}(S_T B_T^{-1} S_T^{-1} X \mid \mathcal{F}_t^S)}{S_t B_t^{-1}} \\
&= B_t \, \mathbb{E}_{\mathbb{P}^*}(B_T^{-1} X \mid \mathcal{F}_t^S).
\end{aligned}
$$

We are in a position to state the following corollary.

Corollary 2.2.3 *For any European contingent claim X settling at time T the arbitrage price satisfies, for every $t = 0, 1, \ldots, T$,*

$$\pi_t(X) = B_t \, \mathbb{E}_{\mathbb{P}^*}(B_T^{-1} X \mid \mathcal{F}_t^S) = S_t \, \mathbb{E}_{\bar{\mathbb{P}}}(S_T^{-1} X \mid \mathcal{F}_t^S). \tag{2.33}$$

The following result furnishes an alternative representation for the arbitrage price of European call and put options.

Proposition 2.2.5 *The arbitrage price at time t of a European call option settling at time T equals*

$$C_t = S_t \, \bar{\mathbb{P}}\{S_T > K \mid \mathcal{F}_t^S\} - K \hat{r}^{-(T-t)} \mathbb{P}^*\{S_T > K \mid \mathcal{F}_t^S\}, \tag{2.34}$$

where $\bar{\mathbb{P}}$ and $\mathbb{P}^*$ are the unique martingale measures for the processes $\bar{B} = B/S$ and $S^* = S/B$ respectively. The arbitrage price at time t of a European put option is given by

$$C_t = K\hat{r}^{-(T-t)}\mathbb{P}^*\{S_T < K \mid \mathcal{F}_t^S\} - S_t\,\bar{\mathbb{P}}\{S_T < K \mid \mathcal{F}_t^S\}. \tag{2.35}$$

Proof. To derive (2.34), let us denote $D = \{S_T > K\}$ and let us observe that

$$C_T = (S_T - K)^+ = S_T \mathbb{1}_D - K\mathbb{1}_D.$$

Consequently, using the linearity of the arbitrage price and Corollary 2.2.3, we obtain

$$C_t = \pi_t(S_T\mathbb{1}_D) - \pi_t(K\mathbb{1}_D) = S_t\,\mathbb{E}_{\bar{\mathbb{P}}}\big(S_T^{-1}S_T\mathbb{1}_D \mid \mathcal{F}_t^S\big) - KB_t\,\mathbb{E}_{\mathbb{P}^*}\big(B_T^{-1}\mathbb{1}_D \mid \mathcal{F}_t^S\big).$$

It is easily seen that the last formula yields (2.34). The proof of (2.35) is similar. □

2.3 The Black-Scholes Option Pricing Formula

We will now show that the classical Black-Scholes option valuation formula (2.40) can be obtained from the CRR option valuation result (2.17) by an asymptotic procedure, using a properly chosen sequence of binomial models. To this end, we need to examine the asymptotic properties of the CRR model when the number of steps goes to infinity and, simultaneously, the size of time and space steps tends to zero in an appropriate way.

In contrast to the previous section, $T > 0$ is a fixed, but arbitrary, real number. For any n of the form $n = 2^k$, we divide the interval $[0, T]$ into n equal subintervals I_j of length $\Delta_n = T/n$, namely, $I_j = [j\Delta_n, (j+1)\Delta_n]$ for $j = 0, 1, \ldots, n-1$. Note that n corresponds, in a sense, to the natural number T^* in the preceding section.

Let us first introduce the modified accumulation factor. We write r_n to denote the risk-free rate of return over each interval $I_j = [j\Delta_n, (j+1)\Delta_n]$. Hence the price B^n of the risk-free asset equals

$$B^n_{j\Delta_n} = (1 + r_n)^j, \quad \forall\, j = 0, 1, \ldots, n.$$

We thus deal here with a sequence B^n of savings accounts. The same remark applies to the sequence S^n of binomial lattices describing the evolution of the stock price.

For any n, we assume that the stock price S^n can appreciate over the period I_j by u_n or decline by d_n. Specifically, we set $S^n_{(j+1)\Delta_n} = \xi^n_{j+1}S^n_{j\Delta_n}$ for $j = 0, 1, \ldots, n - 1$, where for any fixed n and j, ξ^n_j is a random variable with values in the two-element set $\{u_n, d_n\}$.

In view of Proposition 2.1.2, we may assume, without loss of generality, that for any n the random variables ξ^n_j, $j = 1, 2, \ldots, n$ are defined on a common probability space $(\Omega_n, \mathcal{F}_n, \mathbb{P}_n)$, are mutually independent, and

$$\mathbb{P}_n\{\xi^n_j = u_n\} = p_n = 1 - \mathbb{P}_n\{\xi^n_j = d_n\}, \quad \forall\, j = 1, \ldots, n,$$

for some number $p_n \in (0, 1)$. Recall that the choice of the parameter $p_n \in (0, 1)$ is not relevant from the viewpoint of arbitrage pricing.

To ensure the convergence of the CRR option valuation formula to the Black-Scholes one, we need to impose, in addition, specific restrictions on the asymptotic behavior of the quantities r_n, u_n and d_n. Let us put

$$1 + r_n = e^{r\Delta_n}, \quad u_n = e^{\sigma\sqrt{\Delta_n}}, \quad d_n = u_n^{-1}, \tag{2.36}$$

where $r \geq 0$ and $\sigma > 0$ are given real numbers. It is worth noting that for every $n > r^2\sigma^{-2}T$ we have $d_n = u_n^{-1} < 1 + r_n < u_n$ and thus the CRR model (S^n, B^n, Φ^n) is arbitrage-free for n sufficiently large. Also, using elementary arguments, we obtain

$$\lim_{n \to +\infty} p_{*,n} = \lim_{n \to +\infty} \frac{1 + r_n - d_n}{u_n - d_n} = \lim_{n \to +\infty} \frac{e^{r\Delta_n} - e^{-\sigma\sqrt{\Delta_n}}}{e^{\sigma\sqrt{\Delta_n}} - e^{-\sigma\sqrt{\Delta_n}}} = 1/2$$

and

$$\lim_{n \to +\infty} \bar{p}_n = \lim_{n \to +\infty} \frac{p_{*,n} u_n}{1 + r_n} = 1/2.$$

As was mentioned earlier, we seek the asymptotic value of the call option price when the number of time periods tends to infinity. Assume that $t = jT/2^k$ for some natural j and k; that is, t is an arbitrary dyadic number from the interval $[0, T]$. Given any such number, we introduce the sequence $m_n(t)$ by setting

$$m_n(t) = n(T - t)/T, \quad \forall n \in \mathbb{N}. \tag{2.37}$$

Then the sequence $m_n(t)$ has natural values for $n = 2^l$ sufficiently large. Furthermore, $T - t = m_n(t)\Delta_n$ so that $m_n(t)$ represents the number of trading periods in the interval $[t, T]$ (at least for large l). Note also that

$$\lim_{n \to +\infty} (1 + r_n)^{-m_n(t)} = \lim_{n \to +\infty} e^{-r\Delta_n m_n(t)} = e^{-r(T-t)}. \tag{2.38}$$

For a generic value of stock price at time t, $S_t = S_{T-m_n(t)\Delta_n} = s$, we define

$$b_n(t) = \inf\{ j \in \mathbb{N}^* \mid s u_n^j d_n^{m_n(t)-j} > K\}. \tag{2.39}$$

The next proposition provides the derivation of the classical Black-Scholes option valuation formula by means of an asymptotic procedure. A direct analysis of the continuous-time Black-Scholes option pricing model, based on the Itô stochastic calculus, is presented in Sect. 3.1.

Proposition 2.3.1 *For any dyadic $t \in [0, T]$, the following convergence holds*

$$\lim_{n \to +\infty} \sum_{j=b_n(t)}^{m_n(t)} \binom{m_n(t)}{j} \{S_t \bar{p}_n^j \bar{q}_n^{m_n(t)-j} - K\hat{r}_n^{-m_n(t)} p_{*,n}^j q_{*,n}^{m_n(t)-j}\} = C_t,$$

where $q_{*,n} = 1 - p_{*,n}$, $\bar{q}_n = 1 - \bar{p}_n$, and C_t is given by the Black-Scholes formula

$$C_t = S_t N(d_1(S_t, T - t)) - K e^{-r(T-t)} N(d_2(S_t, T - t)), \qquad (2.40)$$

where

$$d_1(s, t) = \frac{\ln(s/K) + (r + \frac{1}{2}\sigma^2)t}{\sigma\sqrt{t}}, \qquad (2.41)$$

$$d_2(s, t) = d_1(s, t) - \sigma\sqrt{t} = \frac{\ln(s/K) + (r - \frac{1}{2}\sigma^2)t}{\sigma\sqrt{t}}, \qquad (2.42)$$

and N stands for the standard Gaussian cumulative distribution function,

$$N(x) = \frac{1}{\sqrt{2\pi}} \int_{-\infty}^{x} e^{-u^2/2}\,du, \quad \forall x \in \mathbb{R}.$$

Before proceeding to the proof of the proposition, let us make a few comments. We wish to make clear that the limit of the CRR option price depends essentially on the choice of sequences u_n and d_n. For the choice of u_ns and d_ns that we have made here, the asymptotic dynamic of the stock price is that of the *geometric Brownian motion* (known also as the *geometric Wiener process*). This means, in particular, that the asymptotic evolution of the stock price under the martingale measure may be described by a stochastic process whose sample paths almost all follow continuous functions; furthermore, the risk-neutral probability distribution of the continuous-time stock price at any time t is lognormal.

Proof of Proposition 2.3.1. Let $S_t = s$ be the generic value of the stock price at time t. Our first goal is to check that

$$\lim_{n \to +\infty} \sum_{j=b_n(t)}^{m_n(t)} \binom{m_n(t)}{j} \bar{p}_n^j (1 - \bar{p}_n)^{m_n(t)-j} = N(d_1(s, T - t)). \qquad (2.43)$$

It is essential to observe that

$$\sum_{j=b_n(t)}^{m_n(t)} \binom{m_n(t)}{j} \bar{p}_n^j (1 - \bar{p}_n)^{m_n(t)-j} = \mathbb{Q}\{b_n(t) \le \gamma_n \le m_n(t)\},$$

where for every n the random variable γ_n has, under $\mathbb{Q}$, the binomial distribution with parameters $m_n(t)$ and $\bar{p}_n$.

Let us write $\sigma_n = \sqrt{\bar{p}_n(1 - \bar{p}_n)}$ and let us introduce the normalized sequence $\tilde{\gamma}_n$ by setting

$$\tilde{\gamma}_n = \frac{\gamma_n - \mathbb{E}_{\mathbb{Q}}(\gamma_n)}{\sqrt{\mathrm{Var}_{\mathbb{Q}}(\gamma_n)}} = \frac{\gamma_n - m_n(t)\bar{p}_n}{\sqrt{m_n(t)\bar{p}_n(1 - \bar{p}_n)}} = \frac{\sum_{j=1}^{m_n(t)}(\zeta_j^n - \bar{p}_n)}{\sigma_n\sqrt{m_n(t)}},$$

where ζ_j^n, $j = 1, \ldots, n$ are independent, identically distributed random variables with Bernoulli distribution and the parameter $\bar{p}_n$

$$\mathbb{Q}\{\zeta_j^n = 1\} = \bar{p}_n = 1 - \mathbb{Q}\{\zeta_j^n = 0\}.$$

We wish first to check that the sequence $\tilde{\gamma}_n$ of random variables converges in distribution to the standard Gaussian distribution. To this end, let us denote by $\tilde{f}_n$ the characteristic function of the random variable $\tilde{\zeta}_j^n = \zeta_j^n - \bar{p}_n$. We have

$$\tilde{f}_n(z) = \mathbb{E}_{\mathbb{Q}}(e^{iz\tilde{\zeta}_j^n}) = e^{-iz\bar{p}_n}\,\mathbb{E}_{\mathbb{Q}}(e^{iz\zeta_j^n}) = e^{-iz\bar{p}_n}\big(\bar{p}_n e^{iz} + 1 - \bar{p}_n\big).$$

It is not difficult to check that

$$\tilde{f}_n(z) = 1 - \frac{\sigma_n^2 z^2}{2} + o(z^2),$$

where the symbol $o(z^2)$ represents a quantity satisfying $\lim_{z\to 0} o(z^2)/z^2 = 0$. Consequently, by virtue of independence of the random variables ζ_j^n, $j = 1, \dots, n$, the characteristic function $g_n(z) = \mathbb{E}_{\mathbb{Q}}(e^{iz\tilde{\gamma}_n})$ of the random variable $\tilde{\gamma}_n$ can be represented as follows:

$$g_n(z) = \left\{\tilde{f}_n\left(\frac{z}{\sigma_n\sqrt{m_n(t)}}\right)\right\}^{m_n(t)} = \left\{1 - \frac{z^2}{2m_n(t)} + o\left(\frac{z^2}{m_n(t)}\right)\right\}^{m_n(t)}.$$

From the last expression, it is thus apparent that the following point-wise convergence is valid:

$$\lim_{n\to+\infty} g_n(z) = e^{-z^2/2}, \quad \forall z \in \mathbb{R}. \tag{2.44}$$

Recall the well-known fact that the function $e^{-z^2/2}$ is the characteristic function of the standard Gaussian distribution $N(0, 1)$.

The convergence in distribution of the normalized sequence $\tilde{\gamma}_n$ to the standard Gaussian distribution now follows from the point-wise convergence (2.44) of the corresponding sequence g_n of characteristic functions to the characteristic function of the standard Gaussian distribution (see, for instance, Theorem III.3.1 in Shiryaev (1984)).

Furthermore, it is clear that

$$\lim_{n\to+\infty} \frac{m_n(t) - m_n(t)\bar{p}_n}{\sqrt{m_n(t)\bar{p}_n(1 - \bar{p}_n)}} = \lim_{n\to+\infty} \sqrt{\bar{m}_n(t)\bar{p}_n^{-1}(1 - \bar{p}_n)} = +\infty.$$

Hence, using (2.39), we obtain

$$\lim_{n\to+\infty} \frac{b_n(t) - m_n(t)\bar{p}_n}{\sqrt{m_n(t)\bar{p}_n(1 - \bar{p}_n)}}$$

$$= \lim_{n\to+\infty} \frac{\frac{\ln(K/s) + m_n(t)\sigma\sqrt{\Delta_n}}{2\sigma\sqrt{\Delta_n}} - m_n(t)\bar{p}_n}{\sqrt{m_n(t)\bar{p}_n(1 - \bar{p}_n)}}$$

$$= \lim_{n\to+\infty} \frac{\ln(K/s) + \sigma m_n(t)\sqrt{\Delta_n}(1 - 2\bar{p}_n)}{2\sigma\sqrt{m_n(t)\Delta_n\bar{p}_n(1 - \bar{p}_n)}}$$

$$= \frac{\ln(K/s) - (r + \frac{1}{2}\sigma^2)(T - t)}{\sigma\sqrt{T - t}} = -d_1(s, T - t)$$

since

$$\lim_{n\to+\infty} m_n(t)\sqrt{\Delta_n}(1 - 2\bar{p}_n) = -(T - t)\Big(\frac{r}{\sigma} + \frac{\sigma}{2}\Big).$$

The convergence in distribution of the sequence γ_n to the standard Gaussian distribution $N(0, 1)$ and the equality $1 - N(x) = N(-x)$ for $x \in \mathbb{R}$ yield

$$\lim_{n\to+\infty} \mathbb{Q}\{b_n(t) \le \gamma_n \le m_n(t)\} = 1 - N\big(-d_1(s, T - t)\big) = N\big(d_1(s, T - t)\big).$$

This completes the proof of equality (2.43). Reasoning in a similar manner, we will now check that

$$\lim_{n\to+\infty} \sum_{j=b_n(t)}^{m_n(t)} \binom{m_n(t)}{j} p_{*,n}^j (1 - p_{*,n})^{m_n(t)-j} = N(d_2(s, T - t)). \tag{2.45}$$

The sum in the left-hand side of formula (2.45) is equal to the probability $\mathbb{Q}\{b_n(t) \le \gamma_n^* \le m_n(t)\}$, where for every n the random variable γ_n^* has, under $\mathbb{Q}$, the binomial distribution with parameters $m_n(t)$ and $p_{*,n}$. Moreover, we have

$$\lim_{n\to+\infty} \frac{b_n(t) - m_n(t)p_{*,n}}{\sqrt{m_n(t)p_{*,n}(1 - p_{*,n})}}$$

$$= \lim_{n\to+\infty} \frac{\frac{\ln(K/s)+m_n(t)\sigma\sqrt{\Delta_n}}{2\sigma\sqrt{\Delta_n}} - m_n(t)p_{*,n}}{\sqrt{m_n(t)p_{*,n}(1 - p_{*,n})}}$$

$$= \lim_{n\to+\infty} \frac{\ln(K/s) + \sigma m_n(t)\sqrt{\Delta_n}(1 - 2p_{*,n})}{2\sigma\sqrt{m_n(t)\Delta_n p_{*,n}(1 - p_{*,n})}}$$

$$= \frac{\ln(K/s) - (r - \frac{1}{2}\sigma^2)(T - t)}{\sigma\sqrt{T - t}} = -d_2(s, T - t)$$

since

$$\lim_{n\to+\infty} m_n(t)\sqrt{\Delta_n}(1 - 2p_{*,n}) = (T - t)\Big(\frac{\sigma}{2} - \frac{r}{\sigma}\Big).$$

Therefore

$$\lim_{n\to+\infty} \mathbb{Q}\{b_n(t) \le \gamma_n^* \le m_n(t)\} = 1 - N\big(-d_2(s, T - t)\big) = N\big(d_2(s, T - t)\big).$$

The proposition now follows by combining (2.38), (2.43), and (2.45).   □

For a different choice of the sequences u_n and d_n, the stock price may asymptotically follow a stochastic process with discontinuous sample paths. For instance, if we put $u_n = u$ and $d_n = e^{ct/n}$ then the stock process will follow asymptotically a log-Poisson process, examined by Cox and Ross (1975). This noticeable feature is related to the fact that we deal with a triangular array of random variables and thus the class of asymptotic probability distributions is larger than in the case of the classical central limit theorem. More advanced problems related to the convergence of discrete-time financial models to continuous-time counterparts were studied by, among others, Madan et al. (1989), He (1990), Amin (1991), Willinger and Taqqu (1991), Duffie and Protter (1992), Cutland et al. (1993b), Amin and Khanna (1994), Rachev and Rüschendorf (1994), Hubalek and Schachermayer (1998), Leisen (1998), Broadie et al. (1998), Lesne et al. (2000) and Lesne and Prigent (2001). The recent monograph by Prigent (2003) is an excellent source of expertise in the area of weak convergence of financial models.

2.4 Valuation of American Options

In this section, we are concerned with the arbitrage valuation of American options written on a stock S within the framework of the CRR binomial model of a stock price. Due to the possibility of an early exercise of an American option, the problem of pricing and hedging of such claims cannot be reduced to a simple replication of the terminal payoff. Nevertheless, the valuation of American options will still be based on the no-arbitrage arguments.

2.4.1 American Call Options

Let us first consider the case of the American call option – that is, the option to buy a specified number of shares, which may be exercised at any time before the option expiry date T, or on that date. The exercise policy of the option holder is necessarily based on the information accumulated to date and not on the future prices of the stock. As in the previous chapter, we will write C_t^a to denote the arbitrage price at time t of an American call option written on one share of a stock. By arbitrage price of the American call we mean such price process C_t^a, $t \leq T$, that an extended financial market model – that is, a market with trading in risk-free bonds, stocks and American call options – remains arbitrage-free. Our first goal is to show that the price of an American call option in the CRR arbitrage-free market model coincides with the arbitrage price of a European call option with the same expiry date and strike price. For this purpose, it is sufficient to show that the American call option should never be exercised before maturity, since otherwise the option writer would be able to make risk-free profit. It is worth noting that, by convention, when an American call is exercised at time t, its payoff equals $(S_t - K)^+$, rather than $S_t - K$ (a similar convention applies to an American put). Due to this convention, we may postulate that an American call or put option should be exercised by its holder either prior to or at maturity date.

The argument hinges on the following simple inequality

$$C_t \geq (S_t - K)^+, \quad \forall\, t \leq T, \tag{2.46}$$

which can be justified in several ways. For instance, one may use the explicit formula (2.17), or apply the risk-neutral valuation formula (2.28). In the latter method, the argument is based on Jensen's conditional inequality applied to the convex function $f(x) = (x - K)^+$. In fact, we have (recall that $t = T - m$)

$$\mathbb{E}_{\mathbb{P}^*}\left(\hat{r}^{-m}(S_T - K)^+ \mid \mathcal{F}_t^S\right) \geq \left(\mathbb{E}_{\mathbb{P}^*}(\hat{r}^{-m} S_T \mid \mathcal{F}_t^S) - \hat{r}^{-m} K\right)^+ \geq (S_t - K)^+,$$

where the first inequality is the Jensen conditional inequality, and the second follows from the trivial inequality $K/\hat{r}^m \leq K$ (the assumption that $r \geq 0$ is essential here).

A more intuitive way of deriving (2.46) is based on no-arbitrage arguments. Note that since the option's price C_t is always nonnegative, it is sufficient to consider the case when the current stock price is greater than the exercise price – that is, when $S_t - K > 0$. Suppose, on the contrary, that $C_t < S_t - K$ for some t, i.e., $S_t - C_t > K$. Then it would be possible, with zero net initial investment, to buy at time t a call option, short a stock, and invest the sum $S_t - C_t$ in a savings account. By holding this portfolio unchanged up to the maturity date T, we would be able to lock in a risk-free profit. Indeed, the value of our portfolio at time T would satisfy (recall that $r \geq 0$)

$$\hat{r}^{T-t}(S_t - C_t) + C_T - S_T > \hat{r}^{T-t} K + (S_T - K)^+ - S_T \geq 0.$$

We conclude once again that inequality (2.46) is necessary for the absence of arbitrage opportunities.

Taking (2.46) for granted, we may deduce the property $C_t^a = C_t$ by simple no-arbitrage arguments. Suppose, on the contrary, that the writer of an American call is able to sell the option at time 0 at the price $C_0^a > C_0$ (it is evident that, at any time, an American option is worth at least as much as a European option with the same contractual features; in particular, $C_0^a \geq C_0$). In order to profit from this transaction, the option writer establishes a dynamic portfolio replicating the value process of the European call, and invests the remaining funds in risk-free bonds.

Suppose that the holder of the option decides to exercise it at instant t before the expiry date T. Then the option's writer locks in a risk-free profit, since the value of portfolio satisfies

$$C_t - (S_t - K)^+ + \hat{r}^t(C_0^a - C_0) > 0, \quad \forall t \leq T.$$

The above reasoning implies that the European and American call options are equivalent from the point of view of arbitrage pricing theory; that is, both options have the same price, and an American call should never be exercised by its holder before expiry.

The last statement means also that a risk-neutral investor who is long an American call should be indifferent between selling it before, and holding it to, the option's expiry date (provided that the market is efficient – that is, options are neither underpriced nor overpriced).

Let us show by still another intuitive reasoning that a holder of an American call should never exercise the option before its expiry date. Consider an investor who contemplates exercising an American call option at a certain date $t < T$. A better solution is to short one share of stock and to hold the option until its expiry date. Actually, exercising the option yields $S_t - K$ of cash at time t – that is, $\hat{r}^{T-t}(S_t - K)$ of cash at the option's expiry. The second trading strategy gives the payoff $\hat{r}^{T-t} S_t -$

$S_T + (S_T - K)^+$ at time T; that is, either $\hat{r}^{T-t} S_t - K$ (if $S_T \geq K$) or $\hat{r}^{T-t} S_t - S_T$ (if $S_T < K$).

It is thus evident that in all circumstances the second portfolio outperforms the first if $r \geq 0$. It is interesting to observe that this argument can be easily extended to the case of uncertain future interest rates.

2.4.2 American Put Options

Since the early exercise feature of American put options was examined in Sect. 1.7, we will focus on the justification of the valuation formula. Let us denote by $\mathcal{T}$ the class of all *stopping times* defined on the filtered probability space $(\Omega, \mathbb{F}, \mathbb{P})$, where $\mathcal{F}_t = \mathcal{F}_t^S$ for every $t = 0, 1, \ldots, T^*$.

By a *stopping time* we mean an arbitrary mapping $\tau : \Omega \to \{0, 1, \ldots, T^*\}$ such that for any $t = 0, 1, \ldots, T^*$, a random event $\{\omega \in \Omega \mid \tau(\omega) = t\}$ belongs to the σ-field $\mathcal{F}_t$. Intuitively, this property means that the decision whether to stop a process at time t (that is, whether to exercise an option at time t or not) depends on the stock price fluctuations up to time t only. Also, let $\mathcal{T}_{[t,T]}$ stand for the subclass stopping times τ satisfying $t \leq \tau \leq T$. Recall that, by convention, if an American put is exercised at time t, the payoff to its holder equals $(K - S_t)^+$, rather than $K - S_t$.

Corollary 1.7.1 and the preceding discussion suggest the following result.

Proposition 2.4.1 *The arbitrage price P_t^a of an American put option equals*

$$P_t^a = \max_{\tau \in \mathcal{T}_{[t,T]}} \mathbb{E}_{\mathbb{P}^*} \left(\hat{r}^{-(\tau-t)} (K - S_\tau)^+ \mid \mathcal{F}_t \right), \quad \forall t \leq T. \tag{2.47}$$

Moreover, for any $t \leq T$ the stopping time τ_t^* that realizes the maximum in (2.47) is given by the expression (by convention $\min \emptyset = T$)

$$\tau_t^* = \min \left\{ u \in \{t, t+1, \ldots, T\} \mid (K - S_u)^+ \geq P_u^a \right\}. \tag{2.48}$$

Proof. The proof of the proposition is left to the reader as an exercise. Alternatively, we refer to Sect. 2.8 for a detailed study of valuation and hedging of American claims in a finite market model, which covers also the case of the CRR model. □

The stopping time τ_t^* will be referred to as the *rational exercise time* of an American put option that is still alive at time t. Let us emphasize that the stopping time τ_t^* does not solve the optimal stopping problem for any individual, but only for those investors who are risk-neutral. Also, we do not claim that the rational stopping time after t is unique, in general. In fact, the stopping time τ_t^* given by (2.48) is the minimal rational exercise time after t. For more information on this issue, see Sect. 2.8.2.

Recall that the stock price S is Markovian under $\mathbb{P}^*$ (see Corollary 2.2.1). An application of the *Bellman principle*[3] reduces the optimal stopping problem (2.47) to an explicit recursive procedure that allows us to find the value function V^p. These observations lead to the following corollary to Proposition 2.4.1.

---

[3] For an exposition of the stochastic optimal control, see, e.g., Bertsekas and Shreve (1978) or Zabczyk (1996).

Corollary 2.4.1 *Let the nonnegative adapted process U_t, $t \leq T$, be defined recursively by setting $U_T = (K - S_T)^+$ and, for $t \leq T - 1$,*

$$U_t = \max \left\{ K - S_t, \; \mathbb{E}_{\mathbb{P}^*} \left(\hat{r}^{-1} U_{t+1} \mid \mathcal{F}_t \right) \right\}. \tag{2.49}$$

Then the arbitrage price P_t^a of an American put option at time t equals U_t. Moreover, the rational exercise time as of time t equals

$$\tau_t^* = \min \left\{ u \in \{t, t+1, \ldots, T\} \mid K - S_u \geq U_u \right\}. \tag{2.50}$$

It is also possible to go the other way around – that is, to first show directly that the price P_t^a needs to satisfy the recursive relation, for $t \leq T - 1$,

$$P_t^a = \max \left\{ K - S_t, \; \mathbb{E}_{\mathbb{P}^*} \left(\hat{r}^{-1} P_{t+1}^a \mid \mathcal{F}_t \right) \right\} \tag{2.51}$$

subject to the terminal condition $P_T^a = (K - S_T)^+$, and subsequently derive the equivalent representation (2.47). In the case of the CRR model (indeed, in the case of any discrete-time security pricing model), the latter approach appears to be the simplest way to value American options. The main reason for this is that an apparently difficult valuation problem is thus reduced to the simple one-period case. To show this we shall argue, as usual, by contradiction. To start with, we assume that (2.51) fails to hold for $t = T - 1$. If this is the case, by reasoning along the same lines as in Sect. 1.7, one may easily construct at time $T - 1$ a portfolio producing risk-free profit at time T. We thus conclude that necessarily

$$P_{T-1}^a = \max \left\{ K - S_{T-1}, \; \mathbb{E}_{\mathbb{P}^*} \left(\hat{r}^{-1} (K - S_T)^+ \mid \mathcal{F}_T \right) \right\}.$$

The next step is to consider the time period $[T - 2, T - 1]$, with $T - 1$ now playing the role of the terminal date, and P_{T-1}^a being the terminal payoff. This procedure may be repeated as many times as needed.

Summarizing, in the case of the CRR model, the arbitrage pricing of an American put reduces to the following simple recursive recipe, which is valid for every $t = 0, 1, \ldots, T - 1$,

$$P_t^a = \max \left\{ K - S_t, \; \hat{r}^{-1} \left(p_* P_{t+1}^{au} + (1 - p_*) P_{t+1}^{ad} \right) \right\} \tag{2.52}$$

with $P_T^a = (K - S_T)^+$. Note that P_{t+1}^{au} and P_{t+1}^{ad} represent the value of the American put in the next step corresponding to the upward and downward movement of the stock price starting from a given node on the lattice – that is, to the values $u S_t$ and $d S_t$ of the stock price at time $t + 1$, respectively.

2.4.3 American Claims

The above results may be easily extended to the case of an arbitrary claim of American style. We shall assume that an American claim does not produce any payoff unless it is exercised, so that it is not the most general definition one may envisage.

Definition 2.4.1 An *American contingent claim* $X^a = (X, \mathcal{T}_{[0,T]})$ expiring at time T consists of a sequence of payoffs $(X_t)_{t=0}^T$, where X_t is an $\mathcal{F}_t$-measurable random variable for $t = 0, 1, \ldots, T$, and of the set $\mathcal{T}_{[0,T]}$ of admissible exercise policies.

We interpret X_t as the payoff received by the holder of the claim X^a upon exercising it at time t. Note that the set of admissible exercise policies is restricted to the class $\mathcal{T}_{[0,T]}$ of all stopping times of the filtration $\mathbb{F}^S$ with values in $\{0, 1, \ldots, T\}$. Let $h : \mathbb{R} \times \{0, 1, \ldots, T\} \to \mathbb{R}$ be an arbitrary function. We say that X^a is an American contingent claim is associated with the *reward function* h if the equality $X_t = h(S_t, t)$ holds for every $t = 0, 1, \ldots, T$.

An American contingent claim is said to be *path-independent* when its generic payoffs X_t do not depend on the whole sample path up to time t, but only on the current value, S_t, of the stock price. It is clear the a claim is path-independent if and only if it is associated with a certain reward function h.

Arbitrage valuation of any American claim in a discrete-time model is based on a simple recursive procedure. In order to price a path-independent American claim in the case of the CRR model, it is sufficient to move backward in time along the binomial lattice. If an American contingent claim is path-dependent, such a simple recipe is no longer applicable (for examples of efficient numerical procedures for valuing path-dependent options, we refer to Hull and White (1993c)). We have, however, the following result, whose proof is omitted. Once again, we refer to Sect. 2.8 for a detailed study of arbitrage pricing of American claims in a finite market model.

Proposition 2.4.2 *For every* $t \leq T$, *the arbitrage price* $\pi_t(X^a)$ *of an arbitrary American claim* X^a *in the CRR model equals*

$$\pi_t(X^a) = \max_{\tau \in \mathcal{T}_{[t,T]}} \mathbb{E}_{\mathbb{P}^*}\left(\hat{r}^{-(\tau-t)} X_\tau \mid \mathcal{F}_t\right).$$

The price process $\pi(X^a)$ *can be determined using the following recurrence relation, for every* $t \leq T - 1$,

$$\pi_t(X^a) = \max\left\{X_t, \ \mathbb{E}_{\mathbb{P}^*}\left(\hat{r}^{-1}\pi_{t+1}(X^a) \mid \mathcal{F}_t\right)\right\} \tag{2.53}$$

subject to the terminal condition $\pi_T(X^a) = X_T$. *In the case of a path-independent American claim* X^a *with the reward function* h *we have that, for every* $t \leq T - 1$,

$$\pi_t(X^a) = \max\left\{h(S_t, t), \ \hat{r}^{-1}\left(p_*\pi_{t+1}^u(X^a) + (1 - p_*)\pi_{t+1}^d(X^a)\right)\right\}, \tag{2.54}$$

where, for a generic stock price value S_t *at time* t, *we write* $\pi_{t+1}^u(X^a)$ *and* $\pi_{t+1}^d(X^a)$ *to denote the values of the price process* $\pi(X^a)$ *at time* $t + 1$ *in the nodes that correspond to the upward and downward movements of the stock price during the time-period* $[t, t + 1]$ – *that is, for the values* uS_t *and* dS_t *of the stock price at time* $t + 1$, *respectively.*

By a slight abuse of notation, we will sometimes write $X_t^a = \pi_t(X^a)$ to denote the arbitrage price at time t of an American claim X^a. Hence (2.53) becomes (cf. (2.51))

$$X_t^a = \max\left\{X_t,\ \mathbb{E}_{\mathbb{P}*}\left(\hat{r}^{-1}X_{t+1}^a \mid \mathcal{F}_t\right)\right\} \tag{2.55}$$

with the terminal condition $X_T^a = X_T$. Similarly, formula (2.54) takes a more concise form (cf. (2.52))

$$X_t^a = \max\left\{h(S_t, t),\ \hat{r}^{-1}\left(p_* X_{t+1}^{au} + (1 - p_*)X_{t+1}^{ad}\right)\right\}. \tag{2.56}$$

2.5 Options on a Dividend-paying Stock

So far we have assumed that a stock pays no dividend during an option's lifetime. Suppose now that the stock pays dividends, and the dividend policy is of the following specific form: the stock maintains a constant yield, κ, on each ex-dividend date. We shall restrict ourselves to the last period before the option's expiry. However, the analysis we present below carries over to the more general case of multi-period trading. We assume that the option's expiry date T is an ex-dividend date. This means that the shareholder will receive at that time a dividend payment d_T that amounts to $\kappa u S_{T-1}$ or $\kappa d S_{T-1}$, according to the actual stock price fluctuation. On the other hand, we postulate that the ex-dividend stock price at the end of the period will be either $u(1 - \kappa)S_{T-1}$ or $d(1 - \kappa)S_{T-1}$. This corresponds to the traditional assumption that the stock price declines on the ex-dividend date by exactly the dividend amount. Therefore, the option's payoff C_T^κ at expiry is either

$$C_T^u = \left(u(1 - \kappa)S_{T-1} - K\right)^+ \quad \text{or} \quad C_T^d = \left(d(1 - \kappa)S_{T-1} - K\right)^+,$$

depending on the stock price fluctuation during the last period. If someone is long a stock, he or she receives the dividend at the end of the period; a party in a short position has to make restitution for the dividend to the party from whom the stock was borrowed. Under these assumptions, the replicating strategy of a call option is determined by the following system of equations (independently of the sign of α_{T-1}; that is, whether the position is long or short)

$$\begin{cases} \alpha_{T-1}u S_{T-1} + \beta_{T-1}\hat{r} = \left(u(1 - \kappa)S_{T-1} - K\right)^+, \\ \alpha_{T-1}d S_{T-1} + \beta_{T-1}\hat{r} = \left(d(1 - \kappa)S_{T-1} - K\right)^+. \end{cases}$$

Note that, in contrast to the option payoff, the terminal value of the portfolio $(\alpha_{T-1}, \beta_{T-1})$ is not influenced by the fact that T is the ex-dividend date. This nice feature of portfolio's wealth depends essentially on our assumption that the ex-dividend drop of the stock price coincides with the dividend payment. Solving the above equations for α_{T-1} and β_{T-1}, we find

$$\alpha_{T-1} = \frac{(u_\kappa S_{T-1} - K)^+ - (d_\kappa S_{T-1} - K)^+}{S_{T-1}(u - d)} = \frac{C_T^u - C_T^d}{S_T^u - S_T^d}$$

and

$$\beta_{T-1} = \frac{u(d_\kappa S_{T-1} - K)^+ - d(u_\kappa S_{T-1} - K)^+}{\hat{r}(u - d)} = \frac{uC_T^d - dC_T^u}{\hat{r}(u - d)},$$

where $u_\kappa = (1 - \kappa)u$ and $d_\kappa = (1 - \kappa)d$. By standard arguments, we conclude that the price C_{T-1}^κ of the option at the beginning of the period equals

$$C_{T-1}^\kappa = \alpha_{T-1} S_{T-1} + \beta_{T-1} = \hat{r}^{-1}\big(p_* C_T^u + (1 - p_*)C_T^d\big),$$

or explicitly

$$C_{T-1}^\kappa = \hat{r}^{-1}\Big(p_*(u_\kappa S_{T-1} - K)^+ + (1 - p_*)(d_\kappa S_{T-1} - K)^+\Big), \qquad (2.57)$$

where $p_* = (\hat{r} - d)/(u - d)$. Working backwards in time from the expiry date, one finds the general formula for the arbitrage price of a European call option, provided that the ex-dividend dates and the dividend ratio $\kappa \in (0, 1)$ are known in advance. If we price a put option, the corresponding hedging portfolio at time $T - 1$ satisfies

$$\begin{cases} \alpha_{T-1} u S_{T-1} + \beta_{T-1}\hat{r} = \big(K - u_\kappa S_{T-1}\big)^+, \\ \alpha_{T-1} d S_{T-1} + \beta_{T-1}\hat{r} = \big(K - d_\kappa S_{T-1}\big)^+. \end{cases}$$

This yields the following expression for the arbitrage price of a put option at time $T - 1$

$$P_{T-1}^\kappa = \hat{r}^{-1}\big(p_* P_T^u + (1 - p_*)P_T^d\big).$$

Once again, for any set of ex-dividend dates known in advance, the price of a European put option at time t can be derived easily by backward induction. Generally speaking, it is clear that the price of a call option is a decreasing function of the dividend yield κ (cf. formula (2.57)). Similarly, the price of a put option increases when κ increases. Both above relationships are rather intuitive, as the dividend payments during the option's lifetime make the underlying stock less valuable at an option's expiry than it would be if no dividends were paid. Also, one can easily extend the above analysis to include dividend policies in which the amount paid on any ex-dividend date depends on the stock price at that time in a more general way (we refer to Sect. 3.2 for more details).

Before we end this section, let us summarize the basic features of American options. We have argued that in the CRR model of a financial market, European and American call options on a stock paying no dividends during the option's lifetime are equivalent (this holds, indeed, in any arbitrage-free market model). This means, in particular, that an American call option should never be exercised before its expiry date. If the underlying stock pays dividends during the option's lifetime, it may be rational to exercise an American call before expiry (but only on a pre-dividend day – that is, one period before the next dividend payment). It is important to notice that the arbitrage valuation of an American call option written on a dividend-paying stock can be done, as usual, by means of backward induction.

On the other hand, we know that the properties of American and European put options with the same contractual features are distinct, in general, as in some circumstances the holder of an American put written on a non-dividend-paying stock should exercise her right to sell the stock before the option's expiry date. If the underlying stock pays dividends during a put option's lifetime, the probability of early exercise declines, and thus the arbitrage price of an American put becomes closer to the price of the otherwise identical put option of European style.

2.6 Security Markets in Discrete Time

This section deals mostly with *finite markets* – that is, discrete-time models of financial markets in which all relevant quantities take a finite number of values. The case of discrete-time models with infinite state space is treated very briefly. Essentially, we follow here the approach of Harrison and Pliska (1981); a more exhaustive analysis of finite markets can be found in Taqqu and Willinger (1987). An excellent introduction to discrete-time financial mathematics is given in a monograph by Pliska (1997). A monograph by Föllmer and Schied (2000) is the most comprehensive source in the area.

The detailed treatment of finite models of financial markets presented below is not motivated by their practical importance (except for binomial or multinomial models). The main motivation comes rather from the fact that the most important ideas and results of arbitrage pricing can be presented in a more transparent way by working first in a finite-dimensional framework.

We need first to introduce some notation. Since the number of dates is assumed to be a finite ordered set, there is no loss of generality if we take the set of dates $\mathcal{T} = \{0, 1, \ldots, T^*\}$. Let Ω be an arbitrary finite set, $\Omega = \{\omega_1, \omega_2, \ldots, \omega_d\}$ say, and let $\mathcal{F} = \mathcal{F}_{T^*}$ be the σ-field of all subsets of Ω, i.e., $\mathcal{F} = 2^{\Omega}$. We consider a filtered probability space $(\Omega, \mathbb{F}, \mathbb{P})$ equipped with a filtration $\mathbb{F} = (\mathcal{F}_t)_{t \leq T^*}$, where $\mathbb{P}$ is an arbitrary probability measure on $(\Omega, \mathcal{F}_{T^*})$, such that $\mathbb{P}\{\omega_j\} > 0$ for every $j = 1, 2, \ldots, d$. We assume throughout that the σ-field $\mathcal{F}_0$ is trivial; that is, $\mathcal{F}_0 = \{\emptyset, \Omega\}$. A vector of prices of k primary securities is modelled by means of an $\mathbb{F}$-adapted, $\mathbb{R}^k$-valued, nonnegative stochastic process $Z = (Z^1, Z^2, \ldots, Z^k)$. Recall that a k-dimensional process $Z = (Z^1, Z^2, \ldots, Z^k)$ is said to be $\mathbb{F}$-*adapted*, if for any $i = 1, 2, \ldots, k$, and any $t \leq T^*$, the random variable Z_t^i is $\mathcal{F}_t$-measurable.

Since the underlying probability space and the set of dates are both finite sets, all random variables and all stochastic processes considered in finite markets are necessarily bounded. For brevity, we shall say that a given process is adapted, instead of $\mathbb{F}$-adapted, if no confusion may arise. A *trading strategy* (also called a *dynamic portfolio*) is an arbitrary $\mathbb{F}$-adapted, $\mathbb{R}^k$-valued process $\phi = (\phi^1, \phi^2, \ldots, \phi^k)$. At any date t, the ith component, ϕ_t^i, of a portfolio ϕ determines the number of units of the ith asset that are held in the portfolio at this date.

We do not postulate, in general, that $\mathcal{F}_t = \mathcal{F}_t^Z = \sigma(Z_0, Z_1, \ldots, Z_t)$, that is, that the underlying filtration $\mathbb{F}$ is generated by the observations of the price process Z.

2.6.1 Finite Spot Markets

Unless explicitly stated otherwise, we assume throughout that all assets are perfectly divisible and the market is frictionless, i.e., no restrictions on the short-selling of assets, nor transaction costs or taxes, are present.

In this section, the security prices $Z^1, Z^2, \ldots, Z^k$ are interpreted as *spot prices* (or *cash prices*) of certain financial assets. In order to avoid any confusion with the case of *futures markets,* which will be studied in the subsequent section, we shall denote hereafter the price process Z by $S = (S^1, S^2, \ldots, S^k)$. In some places, it will be essential to assume that the price process of at least one asset follows a strictly positive process (such a process will play the role of a *numeraire* in what follows). Therefore, we assume, without loss of generality, that $S_t^k > 0$ for every $t \le T^*$. To emphasize the special role of this particular asset, we will sometimes write B instead of S^k.

As mentioned above, the component ϕ_t^i of a trading strategy ϕ stands for the number of units of the ith security held by an investor at time t. This implies that $\phi_t^i S_t^i$ represents the amount of funds invested in the ith security at time t. The term "funds" is used here for the sake of terminological convenience only. In fact, we assume only that the prices of all primary securities are expressed in units of a certain common asset, which is thus used as a benchmark. The benchmark asset should have monotone appeal, meaning that either (a) all individuals prefer more units of this asset to less, or (b) all individuals prefer less units of this asset to more (we prefer to assume that (a) holds). Thus, the value of any contingent claim will also be expressed in units of the benchmark asset.

In our further development, we will sometimes express the original prices of all traded assets in terms of a fixed primary security; referred to as a *numeraire*. The modified processes will be referred to as *relative prices* (or *discounted prices,* if the numeraire corresponds to a bond price). The original prices of primary securities may be seen as relative prices with respect to the benchmark asset, which is not explicitly specified, however.

2.6.2 Self-financing Trading Strategies

In view of our conventions, the following definition of the wealth of a spot trading strategy ϕ is self-explanatory.

Definition 2.6.1 The *wealth process* $V(\phi)$ of a spot trading strategy ϕ is given by the equality (the dot "$\cdot$" stands for the usual inner product in $\mathbb{R}^k$)

$$V_t(\phi) = \phi_t \cdot S_t = \sum_{i=1}^{k} \phi_t^i S_t^i, \quad \forall\, t \le T^*.$$

The initial wealth $V_0(\phi) = \phi_0 S_0$ is also referred to as the *initial investment* of the trading strategy ϕ. Since both S_0 and ϕ_0 are $\mathcal{F}_0$-measurable random variables, they may be identified with some vectors in $\mathbb{R}^k$, therefore the initial wealth $V_0(\phi)$

of any portfolio is a real number. Subsequently, at any instant $t = 1, 2, \ldots, T^*$, the portfolio ϕ may be rebalanced in such a way that there are no infusions of external funds, and no funds are withdrawn (in particular, the definition of a self-financing strategy assumes that no intertemporal consumption takes place). In the discrete-time spot market setup, these natural assumptions are easily formalized by means of the following definition.

Definition 2.6.2 A spot trading strategy ϕ is said to be *self-financing* if it satisfies the following condition, for every $t = 0, 1, \ldots, T^* - 1$,

$$\phi_t \cdot S_{t+1} = \phi_{t+1} \cdot S_{t+1}. \tag{2.58}$$

Intuitively, after a portfolio ϕ_0 is set up at time 0, its revisions are allowed at times $1, 2, \ldots, T^*$ only. In other words, it is held fixed over each time period $(t, t + 1)$ for $t = 0, 1, \ldots, T^* - 1$. Notice that the rebalancing of a portfolio ϕ at the terminal date T^* is also allowed. If a trading strategy ϕ is self-financing, its revision at time T^* does not affect the terminal wealth $V_{T^*}(\phi)$, however. In fact, by virtue of (2.58), the terminal wealth $V_{T^*}(\phi)$ is uniquely determined by the form ϕ_{T^*-1} of the portfolio at time $T^* - 1$ and the vector S_{T^*} of terminal prices of primary securities. Summarizing, when dealing with replication of contingent claims, we may assume that $T^* - 1$ is the last date when a portfolio may be rebalanced. No wonder that the notion of a gains process $G(\phi)$, which is assumed to represent the *capital gains* earned by the holder of the dynamic portfolio ϕ, does not take into account the random variable ϕ_{T^*}.

We denote by Φ the class of all self-financing spot trading strategies. Let us observe that the class Φ is a vector space. Indeed, it is easily seen that for every $\phi, \psi \in \Phi$ and arbitrary real numbers c, d, the linear combination $c\phi + d\psi$ also represents a self-financing trading strategy.

Definition 2.6.3 The *gains process* $G(\phi)$ of an arbitrary spot trading strategy ϕ equals, for every $t = 0, 1, \ldots, T^*$,

$$G_t(\phi) = \sum_{u=0}^{t-1} \phi_u \cdot (S_{u+1} - S_u). \tag{2.59}$$

In view of (2.59), it is clear that we consider here primary securities which do not pay intertemporal cash flows to their holders (such as dividends earned by a stockholder, or coupons received by a bond-holder from the issuer of a bond). The following useful lemma relates the gains process $G(\phi)$ of a self-financing strategy ϕ to its wealth process $V(\phi)$.

Lemma 2.6.1 *A spot trading strategy ϕ is self-financing if and only if we have that, for every $t = 0, 1, \ldots, T^*$,*

$$V_t(\phi) = V_0(\phi) + G_t(\phi). \tag{2.60}$$

Proof. Assume first that ϕ is self-financing. Then, taking into account formulas (2.58)–(2.59), we obtain

$$V_t(\phi) = \phi_0 \cdot S_0 + \sum_{u=0}^{t-1}(\phi_{u+1} \cdot S_{u+1} - \phi_u \cdot S_u)$$

$$= \phi_0 \cdot S_0 + \sum_{u=0}^{t-1}\phi_u \cdot (S_{u+1} - S_u) = V_0(\phi) + G_t(\phi),$$

so that (2.60) holds. The inverse implication is also easy to establish.   □

All definitions and results above can be easily extended to the case of a trading strategy ϕ over the time set $\{0, 1, \ldots, T\}$ for some $T < T^*$. Also, any self-financing trading strategy on $\{0, 1, \ldots, T\}$ can be extended to a self-financing trading strategy on $\{0, 1, \ldots, T^*\}$ by postulating that the total wealth is invested at time T in the last asset, B. In that case, the terminal wealth of ϕ at time T^* satisfies $V_{T^*}(\phi) = B_{T^*}B_T^{-1}V_T(\phi)$. We will use this convention in what follows, without explicit mentioning.

2.6.3 Replication and Arbitrage Opportunities

We pursue an analysis of the spot market model $\mathcal{M} = (S, \Phi)$, where S is a k-dimensional, $\mathbb{F}$-adapted stochastic process and Φ stands for the class of all self-financing (spot) trading strategies.

By a *European contingent claim* X which settles at time T we mean an arbitrary $\mathcal{F}_T$-measurable random variable. Unless explicitly stated otherwise, we shall deal with European contingent claims, and we shall refer to them as *contingent claims* or simply *claims*. Since the space Ω is assumed to be a finite set with d elements, any claim X has the representation $X = (X(\omega_1), X(\omega_2), \ldots, X(\omega_d)) \in \mathbb{R}^d$, and thus the class $\mathcal{X}$ of all contingent claims which settle at some date $T \le T^*$ may be identified with the linear space $\mathbb{R}^d$.

Definition 2.6.4 A *replicating strategy* for the contingent claim X, which settles at time T, is a self-financing trading strategy ϕ such that $V_T(\phi) = X$. Given a claim X, we denote by Φ_X the class of all trading strategies which replicate X.

The wealth process $V_t(\phi)$, $t \le T$, of an arbitrary strategy ϕ from Φ_X is called a *replicating process* of X in $\mathcal{M}$. Finally, we say that a claim X is *attainable* in $\mathcal{M}$ if it admits at least one replicating strategy. We denote by $\mathcal{A}$ the class of all attainable claims.

Definition 2.6.5 A market model $\mathcal{M}$ is called *complete* if every claim $X \in \mathcal{X}$ is attainable in $\mathcal{M}$, or equivalently, if for every $\mathcal{F}_T$-measurable random variable X there exists at least one trading strategy $\phi \in \Phi$ such that $V_T(\phi) = X$. In other words, a market model $\mathcal{M}$ is complete whenever $\mathcal{X} = \mathcal{A}$.

Generally speaking, the completeness of a particular model of a financial market is a highly desirable property. We will show that under market completeness, any European contingent claim can be priced by arbitrage, and its price process can be mimicked by means of a self-financing dynamic portfolio. We need first to examine the arbitrage-free property of $\mathcal{M}$, however.

Definition 2.6.6 A trading strategy $\phi \in \Phi$ is called an *arbitrage opportunity* if $V_0(\phi) = 0$ and the terminal wealth of ϕ satisfies

$$\mathbb{P}\{V_{T^*}(\phi) \geq 0\} = 1 \quad \text{and} \quad \mathbb{P}\{V_{T^*}(\phi) > 0\} > 0.$$

We say that a spot market $\mathcal{M} = (S, \Phi)$ is *arbitrage-free* if there are no arbitrage opportunities in the class Φ of all self-financing trading strategies.

2.6.4 Arbitrage Price

In this section, X is an arbitrary attainable claim which settles at time T.

Definition 2.6.7 We say that X is *uniquely replicated* in $\mathcal{M}$ if it admits a unique replicating process in $\mathcal{M}$; that is, if the equality

$$V_t(\phi) = V_t(\psi), \quad \forall t \leq T,$$

holds for arbitrary trading strategies ϕ, ψ belonging to Φ_X. In this case, the process $V(\phi)$ is termed the *wealth process* of X in $\mathcal{M}$.

Proposition 2.6.1 *Suppose that the market model $\mathcal{M}$ is arbitrage-free. Then any attainable contingent claim X is uniquely replicated in $\mathcal{M}$.*

Proof. Suppose, on the contrary, that there exists a time T attainable contingent claim X which admits two replicating strategies, say ϕ and ψ, such that for some $t < T$ we have: $V_u(\phi) = V_u(\psi)$ for every $u < t$, and $V_t(\phi) \neq V_t(\psi)$.

Assume first that $t = 0$ so that $V_0(\phi) > V_0(\psi)$ for some replicating strategies ϕ and ψ. Consider a strategy ζ which equals (recall that $B = S^k$)

$$\zeta_u = \psi_u - \phi_u + (0, \dots, 0, v_0 B_0^{-1}) \mathbb{1}_A,$$

where $v_0 = V_0(\phi) - V_0(\psi) > 0$. Then $V_0(\zeta) = 0$ and $V_{T^*}(\zeta) = v_0 B_0^{-1} B_{T^*} > 0$ for every ω, so that ζ is an arbitrage opportunity.

Let us now consider the case $t > 0$. We may assume, without loss of generality, that $\mathbb{P}\{A\} > 0$, where A stands for the event $\{V_t(\phi) > V_t(\psi)\}$. Denote by ξ the random variable $v_t = V_t(\phi) - V_t(\psi)$, and consider the following trading strategy η

$$\eta_u = \phi_u - \psi_u, \quad \forall u < t,$$

and

$$\eta_u = (\phi_u - \psi_u) \mathbb{1}_{A^c} + (0, \dots, 0, v_t B_t^{-1}) \mathbb{1}_A, \quad \forall u \geq t,$$

where A^c denotes the complement of the set A (i.e., if the event A occurs, both portfolio's are liquidated and the proceeds are invested in the kth asset). It is apparent that the strategy η is self-financing and $V_0(\eta) = 0$. Moreover, its terminal wealth $V_{T^*}(\eta)$ equals

$$V_{T^*}(\eta) = v_t B_t^{-1} B_{T^*} \mathbb{1}_A.$$

Hence it is clear that $V_{T^*}(\eta) \geq 0$ and $\mathbb{P}\{V_{T^*}(\eta) > 0\} = \mathbb{P}\{A\} > 0$. We conclude that η is an arbitrage opportunity. This contradicts our assumption that the market $\mathcal{M}$ is arbitrage-free. $\square$

The converse implication is not valid; that is, the uniqueness of the wealth process of any attainable contingent claim does not imply the arbitrage-free property of a market, in general. Therefore, the existence and uniqueness of the wealth process associated with any attainable claim is insufficient to justify the term *arbitrage price*. Indeed, it is trivial to construct a finite market in which all claims are uniquely replicated, but there exists a strictly positive claim, say Y, which admits a replicating strategy with negative initial investment (with negative manufacturing cost, using the terminology of Chap. 1). Suppose now that for every claim X, its price at time 0, $\pi_0(X)$, is defined as the initial investment of a strategy which replicates X. It is important to point out that the price functional π_0, on the space $\mathcal{X}$ of contingent claims, would not be supported by any kind of intertemporal equilibrium. In fact, any individual would tend to take an infinite position in any such claim Y (recall that we assume that all individuals are assumed to prefer more wealth to less).

We thus find it natural to formally introduce the notion of an arbitrage price in the following way.

Definition 2.6.8 Suppose that the market model $\mathcal{M}$ is arbitrage-free. Then the wealth process of an attainable claim X is called the *arbitrage price process* (or simply, the *arbitrage price*) of X in $\mathcal{M}$. We denote it by $\pi_t(X)$, $t \leq T$.

2.6.5 Risk-neutral Valuation Formula

As mentioned earlier, the martingale approach to arbitrage pricing was first elaborated by Cox and Ross (1976) (although the idea of "risk-neutral" probabilities goes back to Arrow (1964, 1970)). In financial terminology, they showed that in a world with one stock and one bond, it is possible to construct preferences through a risk-neutral individual who gives the value of those claims which are priced by arbitrage. In this regard, let us mention that the *martingale measures*, which we are now going to introduce, are sometimes referred to as *risk-neutral probabilities*.

For the sake of notational simplicity, we write as usual $S^k = B$. This convention does not imply, however, that S^k should necessarily be interpreted as the price process of a risk-free bond. Recall, however, that we have assumed that S^k follows a strictly positive process. Let us denote by S^* the process of *relative* (or *discounted*) prices, which equals, for every $t = 0, 1, \ldots, T^*$,

$$S_t^* = (S_t^1 B_t^{-1}, \ S_t^2 B_t^{-1}, \ldots, \ S_t^k B_t^{-1}) = (S_t^{*1}, S_t^{*2}, \ldots, S_t^{*(k-1)}, 1),$$

where we denote $S^{*i} = S^i B^{-1}$. Recall that the probability measures $\mathbb{P}$ and $\mathbb{Q}$ on $(\Omega, \mathcal{F})$ are said to be *equivalent* if, for any event $A \in \mathcal{F}$, the equality $\mathbb{P}\{A\} = 0$ holds if and only if $\mathbb{Q}\{A\} = 0$. Similarly, $\mathbb{Q}$ is said to be *absolutely continuous* with respect to $\mathbb{P}$ if, for any event $A \in \mathcal{F}$, the equality $\mathbb{P}\{A\} = 0$ implies that $\mathbb{Q}\{A\} = 0$.

Definition 2.6.9 A probability measure $\mathbb{P}^*$ on $(\Omega, \mathcal{F}_{T^*})$ equivalent to $\mathbb{P}$ (absolutely continuous with respect to $\mathbb{P}$, respectively) is called a *martingale measure for S^** (a *generalized martingale measure* for S^*, respectively) if the relative price S^* is a $\mathbb{P}^*$-martingale with respect to the filtration $\mathbb{F}$.

Recall that a k-dimensional process $S^* = (S^{*1}, S^{*2}, \ldots, S^{*k})$ is a $\mathbb{P}^*$-martingale with respect to a filtration $\mathbb{F}$ if $\mathbb{E}_{\mathbb{P}^*}|S_t^{*i}| < \infty$ for every i and $t = 0, 1, \ldots, T^*$, and the equality

$$\mathbb{E}_{\mathbb{P}^*}(S_{t+1}^{*i} \mid \mathcal{F}_t) = S_t^{*i}$$

is valid for every i and $t = 0, 1, \ldots, T^* - 1$. Of course, in the case of a finite space Ω, the integrability condition $\mathbb{E}_{\mathbb{P}^*}|S_t^{*i}| < \infty$ is always satisfied. The martingale property of S^* will be simply represented as $\mathbb{E}_{\mathbb{P}^*}(S_{t+1}^* \mid \mathcal{F}_t) = S_t^*$ for $t = 0, 1, \ldots, T^* - 1$.

We denote by $\mathcal{P}(S^*)$ and $\bar{\mathcal{P}}(S^*)$ the class of all martingale measures for S^* and the class of all generalized martingale measures for S^*, respectively. It is easily see that the inclusion $\mathcal{P}(S^*) \subseteq \bar{\mathcal{P}}(S^*)$ is valid. Moreover, it is not difficult to provide an example in which the class $\mathcal{P}(S^*)$ is empty, whereas the class $\bar{\mathcal{P}}(S^*)$ is not. Observe also that the notion of a martingale measure essentially depend on the choice of the numeraire – recall that we have chosen $S^k = B$ as a numeraire throughout. In the next step, we introduce the concept of a martingale measure for a market model $\mathcal{M}$.

Definition 2.6.10 A probability measure $\mathbb{P}^*$ on $(\Omega, \mathcal{F}_{T^*})$ equivalent to $\mathbb{P}$ (absolutely continuous with respect to $\mathbb{P}$, respectively) is called a *martingale measure for* $\mathcal{M} = (S, \Phi)$ (a *generalized martingale measure for* $\mathcal{M} = (S, \Phi)$, respectively) if for every trading strategy $\phi \in \Phi$ the relative wealth process $V^*(\phi) = V(\phi)B^{-1}$ follows a $\mathbb{P}^*$-martingale with respect to the filtration $\mathbb{F}$.

We write $\mathcal{P}(\mathcal{M})$ ($\bar{\mathcal{P}}(\mathcal{M})$ respectively) to denote the class of all martingale measures (of all generalized martingale measures, respectively) for $\mathcal{M}$. Our goal is now to show that $\mathcal{P}(S^*) = \mathcal{P}(\mathcal{M})$ and $\bar{\mathcal{P}}(S^*) = \bar{\mathcal{P}}(\mathcal{M})$.

Lemma 2.6.2 *A trading strategy ϕ is self-financing if and only if the relative wealth process $V^*(\phi) = V(\phi)B^{-1}$ satisfies, for every $t = 0, 1, \ldots, T^*$,*

$$V_t^*(\phi) = V_0^*(\phi) + \sum_{u=0}^{t-1} \phi_u \cdot \Delta_u S^*, \qquad (2.61)$$

where $\Delta_u S^* = S_{u+1}^* - S_u^*$. If ϕ belongs to Φ then, for any (generalized) martingale measure $\mathbb{P}^*$, the relative wealth $V^*(\phi)$ is a $\mathbb{P}^*$-martingale with respect to the filtration $\mathbb{F}$.

Proof. Let us write $V = V(\phi)$ and $V^* = V^*(\phi)$. Then (2.61) is equivalent to

$$\Delta_t V^* = V_{t+1}^* - V_t^* = \phi_t \cdot \Delta_t S^*, \quad \forall\, t \leq T^* - 1. \qquad (2.62)$$

But, on the one hand, we have that

$$V_{t+1}^* - V_t^* = \frac{V_{t+1}}{B_{t+1}} - \frac{V_t}{B_t} = \frac{\phi_{t+1} \cdot S_{t+1}}{B_{t+1}} - \frac{\phi_t \cdot S_t}{B_t},$$

and, on the other hand, we also see that

$$\phi_t \cdot \Delta_t S^* = \frac{\phi_t \cdot S_{t+1}}{B_{t+1}} - \frac{\phi_t \cdot S_t}{B_t}.$$

We thus conclude that condition (2.62) is indeed equivalent to the self-financing property: $\phi_t \cdot S_{t+1} = \phi_{t+1} \cdot S_{t+1}$ for $t = 0, 1, \ldots, T^* - 1$.

For the second statement, it is enough to check that, for every $t = 0, 1, \ldots, T^*-1$,

$$\mathbb{E}_{\mathbb{P}^*}(V_{t+1}^* - V_t^* \mid \mathcal{F}_t) = 0.$$

Using (2.62), we obtain

$$\mathbb{E}_{\mathbb{P}^*}(V_{t+1}^* - V_t^* \mid \mathcal{F}_t) = \mathbb{E}_{\mathbb{P}^*}(\phi_t \cdot (S_{t+1}^* - S_t^*) \mid \mathcal{F}_t) = \phi_t \cdot \mathbb{E}_{\mathbb{P}^*}(S_{t+1}^* - S_t^* \mid \mathcal{F}_t) = 0,$$

where the last equality follows from the martingale property of the relative price process S^* under $\mathbb{P}^*$.   □

In view of the last lemma, the following corollary is easy to prove.

Corollary 2.6.1 *A probability measure $\mathbb{P}^*$ on $(\Omega, \mathcal{F}_{T^*})$ is a (generalized) martingale measure for the spot market model $\mathcal{M}$ if and only if it is a (generalized) martingale measure for the relative price process S^*, i.e., $\mathcal{P}(S^*) = \mathcal{P}(\mathcal{M})$ and $\bar{\mathcal{P}}(S^*) = \bar{\mathcal{P}}(\mathcal{M})$.*

The next result shows that the existence of a martingale measure for $\mathcal{M}$ is sufficient for the no-arbitrage property of $\mathcal{M}$. Recall that trivially $\mathcal{P}(\mathcal{M}) \subseteq \bar{\mathcal{P}}(\mathcal{M})$ so that the class $\bar{\mathcal{P}}(\mathcal{M})$ is manifestly non-empty if $\mathcal{P}(\mathcal{M})$ is so.

Proposition 2.6.2 *Assume that the class $\mathcal{P}(\mathcal{M})$ is non-empty. Then the spot market $\mathcal{M}$ is arbitrage-free. Moreover, the arbitrage price process of any attainable contingent claim X, which settles at time T, is given by the risk-neutral valuation formula*

$$\pi_t(X) = B_t \mathbb{E}_{\mathbb{P}^*}(X B_T^{-1} \mid \mathcal{F}_t), \quad \forall t \le T, \tag{2.63}$$

where $\mathbb{P}^*$ is any (generalized) martingale measure for the market model $\mathcal{M}$ associated with the choice of B as a numeraire.

Proof. Let $\mathbb{P}^*$ be some martingale measure for $\mathcal{M}$. We know already that the relative wealth process $V^*(\phi)$ of any strategy $\phi \in \Phi$ follows a $\mathbb{P}^*$-martingale, and thus

$$V_t(\phi) = B_t V_t^*(\phi) = B_t \mathbb{E}_{\mathbb{P}^*}(V_T^*(\phi) \mid \mathcal{F}_t) = B_t \mathbb{E}_{\mathbb{P}^*}(V_T(\phi) B_T^{-1} \mid \mathcal{F}_t)$$

for every t. Since $\mathbb{P}^*$ is equivalent to $\mathbb{P}$, it is clear that there are no arbitrage opportunities in the class Φ of self-financing trading strategies, hence the market $\mathcal{M}$ is arbitrage-free.

Moreover, for any attainable contingent claim X which settles at time T, and any trading strategy $\phi \in \Phi_X$, we have

$$\pi_t(X) = V_t(\phi) = B_t \mathbb{E}_{\mathbb{P}^*}(V_T^*(\phi) \mid \mathcal{F}_t) = B_t \mathbb{E}_{\mathbb{P}^*}(X B_T^{-1} \mid \mathcal{F}_t). \tag{2.64}$$

This completes the proof in the case of a martingale measure $\mathbb{P}^*$. When $\mathbb{P}^*$ is a generalized martingale measure for $\mathcal{M}$, all equalities in (2.64) remain valid (let us stress that the existence of a generalized martingale measure does not imply the absence of arbitrage, however).   □

Remarks. In a more general setting (e.g., in a continuous-time framework), a generalized martingale measure no longer plays the role of a pricing measure – that is, equality (2.63) may fail to hold, in general, if a martingale measure $\mathbb{P}^*$ is merely absolutely continuous with respect to an underlying probability measure $\mathbb{P}$. The reason is that the Itô stochastic integral (as opposed to a finite sum) is not invariant with respect to an absolutely continuous change of a probability measure.

2.6.6 Existence of a Martingale Measure

This section addresses a basic question: is the existence of a martingale measure a necessary condition for absence of arbitrage in a finite model of a financial market? Results of this type are sometimes referred to as fundamental theorems of asset pricing. In the case of a finite market model, it was proved by Harrison and Pliska (1981); for a purely probabilistic approach we refer to Taqqu and Willinger (1987), who examine the case of a finite market, and to the papers of Dalang et al. (1990) and Schachermayer (1992), who deal with a discrete-time model with infinite state space (see also Harrison and Kreps (1979) for related results).

Recall that since $\Omega = \{\omega_1, \omega_2, \ldots, \omega_d\}$, the space $\mathcal{X}$ of all contingent claims that settle at time T^* may be identified with the finite-dimensional linear space $\mathbb{R}^d$. For any claim $X \in \mathcal{X}$, we write $X = (X(\omega_1), X(\omega_2), \ldots, X(\omega_d)) = (x^1, x^2, \ldots, x^d) \in \mathbb{R}^d$.

From Proposition 2.6.2, we know already that if the set of martingale measures is non-empty, then the market model $\mathcal{M}$ is arbitrage-free. Our next goal is to show that this condition is also necessary for the no-arbitrage property of the market model $\mathcal{M}$.

Let us first recall the following definition.

Definition 2.6.11 A random variable $\tau : \Omega \to \{0, 1, \ldots, T^*\}$ is said to be a *stopping time* with respect to the filtration $\mathbb{F}$ if the event $\{\tau = t\}$ belongs to $\mathcal{F}_t$ for every $t = 0, 1, \ldots, T^*$.

Proposition 2.6.3 *Suppose that the spot market model $\mathcal{M}$ is arbitrage-free. Then the class $\mathcal{P}(\mathcal{M})$ of martingale measures for $\mathcal{M}$ is non-empty.*

Proof. We consider the following closed and bounded (i.e., compact) and convex subset of $\mathbb{R}^d$

$$\mathcal{X}_+ = \{X \in \mathbb{R}^d \mid X \geq 0 \text{ and } \mathbb{E}_{\mathbb{P}} X = 1\}.$$

Also, let $\mathcal{A}_0^*$ stand for the class of all discounted claims that are attainable by trading strategies with zero initial wealth, that is,

$$\mathcal{A}_0^* = \{Y \in \mathbb{R}^d \mid Y = V_{T^*}^*(\phi) \text{ for some } \phi \in \Phi \text{ with } V_0(\phi) = 0\}.$$

Equivalently, noting that $V_{T^*}(\phi) = G_{T^*}(\phi)$ if and only if $V_0(\phi) = 0$,

$$\mathcal{A}_0^* = \{Y \in \mathbb{R}^d \mid Y = G_{T^*}^*(\phi) \text{ for some } \phi \in \Phi\}.$$

It is easily seen that $\mathcal{A}_0^*$ is a linear subspace of $\mathbb{R}^d$. The crucial observation is that the assumption that $\mathcal{M}$ is arbitrage-free implies that $\mathcal{X}_+ \cap \mathcal{A}_0^* = \emptyset$.

From a *separating hyperplane theorem* (see Corollary 2.6.3), we deduce that there exists a vector $Z = (z^1, z^2, \ldots, z^d)$ such that $Z \cdot X > 0$ for every $X \in \mathcal{X}_+$ and $Z \cdot Y = \sum_{j=1}^d z^j y^j = 0$ for every $Y \in \mathcal{A}_0^*$. Let us denote $p_j = \mathbb{P}\{\omega_j\}$. Since the vector $\tilde{e}_j = (0, \ldots, p_j^{-1}, \ldots, 0) \in \mathbb{R}^d$ belongs to $\mathcal{X}_+$ for any $j = 1, 2, \ldots, d$, we see that necessarily $z^j > 0$ for every $j = 1, 2, \ldots, d$.

We define the probability measure $\mathbb{P}^*$ on $(\Omega, \mathcal{F}_{T^*})$ by setting, for every $j = 1, 2, \ldots, d$,

$$\mathbb{P}^*\{\omega_j\} = \frac{z^j}{\sum_{m=1}^d z^m} = cz^j.$$

Since $z^j > 0$ for every $j = 1, 2, \ldots, d$, it is clear that $\mathbb{P}^*$ is equivalent to $\mathbb{P}$. In view of Corollary 2.6.1, it thus remains to show that the relative price process S^* follows a $\mathbb{P}^*$-martingale. To this end, we observe that, by construction, the equality $\mathbb{E}_{\mathbb{P}^*} Y = cZ \cdot Y = 0$ holds for every $Y \in \mathcal{A}_0^*$.

Let τ be an arbitrary stopping time with respect to the filtration $\mathbb{F}$ (see Definition 2.6.11). For any fixed $l \in \{1, 2, \ldots, k-1\}$, we consider a trading strategy $\phi \in \Phi$ defined by the formula

$$\phi_t^i = \begin{cases} 0 & \text{for } i \neq l, \ i \neq k, \\ \mathbb{1}_{[0,\tau[}(t) & \text{for } i = l, \\ -S_0^l + S_\tau^l B_\tau^{-1} \mathbb{1}_{[\tau, T^*]}(t) & \text{for } i = k. \end{cases}$$

Note that ϕ is self-financing and $V_0(\phi) = 0$. Hence the discounted terminal wealth $V_{T^*}^*(\phi)$ belongs to the space $\mathcal{A}_0^*$, and thus

$$\mathbb{E}_{\mathbb{P}^*}(V_{T^*}^*(\phi)) = cZ \cdot V_{T^*}^*(\phi) = 0.$$

Since manifestly $V_{T^*}^*(\phi) = -S_0^l + S_\tau^l B_\tau^{-1}$, the formula above yields

$$\mathbb{E}_{\mathbb{P}^*}(S_\tau^l B_\tau^{-1}) = S_0^l B_0^{-1}.$$

We conclude that the equality $\mathbb{E}_{\mathbb{P}^*} S_\tau^{*l} = S_0^{*l}$ holds for an arbitrary stopping time τ. By considering stopping times of the form $\tau = t \mathbb{1}_A + T^* \mathbb{1}_{A^c}$, where A is an arbitrary event from $\mathcal{F}_t$, one can easily deduce that the process S^{*l} is a $\mathbb{P}^*$-martingale. Since l was an arbitrary number from the set $\{1, 2, \ldots, k-1\}$, we conclude that the probability measure $\mathbb{P}^*$ is a martingale measure for the k-dimensional process S^*. $\square$

The following result, which is an immediate consequence of Propositions 2.6.2 and 2.6.3, is a version of the so-called *First Fundamental Theorem of Asset Pricing*.

Theorem 2.6.1 *A finite spot market $\mathcal{M}$ is arbitrage-free if and only if the class $\mathcal{P}(\mathcal{M})$ is non-empty.*

2.6.7 Completeness of a Finite Market

The next result links the completeness of a finite market to the uniqueness of a martingale measure. Any result of this kind is commonly referred to as the *Second Fundamental Theorem of Asset Pricing*.

Theorem 2.6.2 *Assume that a spot market model $\mathcal{M}$ is arbitrage-free so that the class $\mathcal{P}(\mathcal{M})$ is non-empty. Then $\mathcal{M}$ is complete if and only if the uniqueness of a martingale measure for $\mathcal{M}$ holds.*

Proof. ($\Rightarrow$) Assume that $\mathcal{M}$ is an arbitrage-free complete market. Then for every $\mathcal{F}_{T^*}$-measurable random variable X there exists at least one trading strategy $\phi \in \Phi$ such that $V_{T^*}(\phi) = X$. By virtue of Propositions 2.6.2 and 2.6.3, we have, for any $\mathcal{F}_{T^*}$-measurable random variable X,

$$\pi_0(X) = B_0\, \mathbb{E}_{\mathbb{P}^*}(X B_{T^*}^{-1}), \quad \forall\, X \in \mathcal{X},$$

where $\mathbb{P}^*$ is any martingale measure from the non-empty class $\mathcal{P}(\mathcal{M})$. Therefore, for any two martingale measures $\mathbb{P}_1^*$ and $\mathbb{P}_2^*$ from the class $\mathcal{P}(\mathcal{M})$, we have that, for any $\mathcal{F}_{T^*}$-measurable random variable X,

$$\mathbb{E}_{\mathbb{P}_1^*}(X B_{T^*}^{-1}) = \mathbb{E}_{\mathbb{P}_2^*}(X B_{T^*}^{-1}). \tag{2.65}$$

Let $A \in \mathcal{F}_{T^*}$ be any event and let us take $X = B_{T^*}\mathbb{1}_A$. Then (2.65) yields the equality $\mathbb{P}_1^*\{A\} = \mathbb{P}_2^*\{A\}$, and thus we conclude that $\mathbb{P}_1^*$ and $\mathbb{P}_2^*$ coincide. This ends the proof of the "only if " clause.

($\Leftarrow$) Assume now that there exists a unique element $\mathbb{P}^*$ in $\mathcal{P}(\mathcal{M})$. Clearly, the market $\mathcal{M}$ is then arbitrage-free; it rests to establish its completeness. Suppose, on the contrary, that the model $\mathcal{M}$ is not complete, so that there exists a contingent claim which is not attainable in $\mathcal{M}$, that is, $\mathcal{A} \neq \mathcal{X}$.

We will show that in that case uniqueness of a martingale measure does not hold; specifically, we will construct a martingale measure $\mathbb{Q}$ for $\mathcal{M}$ different from $\mathbb{P}^*$. We denote by $\mathcal{A}^*$ the class of all discounted claims attainable by some trading strategy, that is,

$$\mathcal{A}^* = \{\, Y \in \mathbb{R}^d \mid Y = V_0(\phi) + G_T^*(\phi) \text{ for some } \phi \in \Phi \,\}.$$

Let us define the inner product in $\mathbb{R}^d$ by the formula

$$\langle X, Y \rangle_{\mathbb{P}^*} := \mathbb{E}_{\mathbb{P}^*}(XY) = \sum_{j=1}^{d} x^j y^j \mathbb{P}^*\{\omega_j\}. \tag{2.66}$$

Observe that $\mathcal{A}^*$ is a linear subspace of $\mathbb{R}^d$ and $\mathcal{A}^* \neq \mathcal{X} = \mathbb{R}^d$. Hence there exists a non-zero random variable Z orthogonal to $\mathcal{A}^*$, in the sense that $\mathbb{E}_{\mathbb{P}^*}(ZY) = 0$ for every $Y \in \mathcal{A}^*$. Moreover, since the constant random variable $Y = 1$ belongs to $\mathcal{A}^*$ (indeed, it simply corresponds to the buy-and-hold strategy in one unit of the asset $B = S^k$), we obtain that $\mathbb{E}_{\mathbb{P}^*}Z = 0$.

Let us define, for every $\omega_j \in \Omega$,

$$\mathbb{Q}\{\omega_j\} = \left(1 + \frac{Z(\omega_j)}{2\,||Z||_\infty}\right)\mathbb{P}^*\{\omega_j\},$$

where $||Z||_\infty = \max_{m=1,2,\dots,d} |Z(\omega_m)|$. We will show that $\mathbb{Q}$ is a martingale measure, that is, $\mathbb{Q} \in \mathcal{P}(\mathcal{M})$. First, it is rather obvious that $\mathbb{Q}$ is a probability measure equivalent to $\mathbb{P}^*$ and thus it is equivalent to $\mathbb{P}$ as well. Second, we will show that for any $\mathbb{F}$-adapted process ϕ we have that

$$\mathbb{E}_\mathbb{Q}\left(\sum_{u=0}^{T-1} \phi_u \cdot \Delta_u S^*\right) = 0. \tag{2.67}$$

We now argue that last equality implies that S^* is an $\mathbb{F}$-martingale under $\mathbb{Q}$. To check this claim, we fix t and i, and we take $\phi_t^i = \mathbb{1}_A$ for an arbitrary (but fixed) $\mathcal{F}_t$-measurable event A, and $\phi_u^i = 0$ for $t \neq u$ as well as $\phi_u^k = 0$ for $k \neq i$. We obtain the required equality $\mathbb{E}_\mathbb{Q}(S_{t+1}^{*i} - S_t^{*i} \mid \mathcal{F}_t) = 0$ for any $t = 0, 1, \dots, T-1$ and every $i = 1, \dots, k-1$ (for $i = k$, this condition is trivially satisfied, since $S^{*k} = 1$). We conclude that $\mathbb{Q}$ belongs to $\mathcal{P}(S^*)$ and thus it belongs to $\mathcal{P}(\mathcal{M})$ as well. It also rather obvious that $\mathbb{Q} \neq \mathbb{P}^*$.

To complete the proof, it remains to show that (2.67) is valid. To this end, we observe that

$$\mathbb{E}_\mathbb{Q}\left(\sum_{u=0}^{T-1} \phi_u \cdot \Delta_u S^*\right) = \mathbb{E}_{\mathbb{P}^*}\left(\sum_{u=0}^{T-1} \phi_u \cdot \Delta_u S^*\right)$$
$$+ \frac{1}{2\,||Z||_\infty}\,\mathbb{E}_{\mathbb{P}^*}\left(Z\sum_{u=0}^{T-1} \phi_u \cdot \Delta_u S^*\right) =: J_1 + J_2.$$

We first note that $J_1 = 0$ since the probability measure $\mathbb{P}^*$ belongs to $\mathcal{P}(\mathcal{M})$. Furthermore, equality $J_2 = 0$ also holds, since it follows from the assumption that Z is orthogonal to $\mathcal{A}^*$ and the observation that the sum $\sum_{u=0}^{T-1} \phi_u \cdot \Delta_u S^*$ belongs to $\mathcal{A}^*$. We conclude that equality (2.67) is indeed satisfied.

In that way, we have established the existence of two distinct martingale measures under non-completeness of a model $\mathcal{M}$. Therefore, the proof of the "if" clause is completed. □

Corollary 2.6.2 *A contingent claim $X \in \mathcal{X}$ is attainable if and only if the map $\mathbb{P}^* \mapsto \mathbb{E}_{\mathbb{P}^*}(X B_T^{-1})$ from $\mathcal{P}(\mathcal{M})$ to $\mathbb{R}$ is constant.*

Proof. We consider the case of $T = T^*$; the arguments for any $T \leq T^*$ are analogous. Suppose first that X is attainable. Then its arbitrage price at time 0 is well defined and it is unique. The risk-neutral valuation formula thus shows that the map $\mathbb{P}^* \mapsto \mathbb{E}_{\mathbb{P}^*}(X B_T^{-1})$ is constant.

Suppose now that the map is constant for a given non-zero claim X (of course, the zero claim is always attainable). We wish to show that X is attainable. Assume, on the contrary, that X is not attainable and denote $X^* = XB_T^{-1}$. Let $\mathbb{P}^*$ be some martingale measure from $\mathcal{P}(\mathcal{M})$ and let $Z = X^* - X^\perp$ where $X^\perp$ is the orthogonal projection under $\mathbb{P}^*$ of X^* on the class $\mathcal{A}^*$, in the sense of the inner product introduced in the proof of Theorem 2.6.2 (see formula (2.66)).

Since X^* does not belong to $\mathcal{A}^*$, the vector Z is non-zero and we may define the associated martingale measure $\mathbb{Q} \in \mathcal{P}(\mathcal{M})$ (once again, see the proof of Theorem 2.6.2). Finally, we note that $\mathbb{E}_{\mathbb{P}^*}(X^*) \neq \mathbb{E}_{\mathbb{Q}}(X^*)$, since

$$\mathbb{E}_{\mathbb{Q}}(X^*) = \mathbb{E}_{\mathbb{P}^*}(X^*) + \frac{1}{2\,||Z||_\infty}\,\mathbb{E}_{\mathbb{P}^*}(ZX^*)$$

and

$$\mathbb{E}_{\mathbb{P}^*}(ZX^*) = \mathbb{E}_{\mathbb{P}^*}((X^* - X^\perp)X^*) = \mathbb{E}_{\mathbb{P}^*}(X^*)^2 \neq 0.$$

This completes the proof.

$\square$

2.6.8 Separating Hyperplane Theorem

We take for granted the following version of a *strong separating hyperplane theorem* (see, e.g., Rockafellar (1970) or Luenberger (1984)).

Theorem 2.6.3 *Let U be a closed, convex subset of $\mathbb{R}^d$ such that $0 \notin U$. Then there exists a linear functional $y \in \mathbb{R}^d$ and a real number $\alpha > 0$ such that $y \cdot u \geq \alpha$ for every $u \in U$.*

In the proof of Proposition 2.6.3, we made use of the following corollary to Theorem 2.6.3.

Corollary 2.6.3 *Let W be a linear subspace of $\mathbb{R}^d$ and let K be a compact, convex subset of $\mathbb{R}^d$ such that $K \cap W = \emptyset$. Then there exists a linear functional $y \in \mathbb{R}^d$ and a real number $\alpha > 0$ such that $y \cdot w = 0$ for every $w \in W$ and $y \cdot k \geq \alpha$ for every $k \in K$.*

Proof. Let U stand for the following set

$$U = K - W = \{x \in \mathbb{R}^d \mid x = k - w, \ k \in K, \ w \in W\}.$$

It is easy to check that U is a convex, closed set. Since K and W are disjoint, we manifestly have that $0 \notin U$. From Theorem 2.6.3, we deduce the existence of $y \in \mathbb{R}^d$ and $\alpha > 0$, such that $y \cdot u \geq \alpha$ for every $u \in U$. This in turn implies that, for every $k \in K$ and $w \in W$,

$$y \cdot k - y \cdot w \geq \alpha. \tag{2.68}$$

Let us fix $k \in K$ and let us take λw instead of w, where λ is an arbitrary real number (recall that W is a linear space). Since λ is arbitrary and (2.68) holds, it is clear that $y \cdot w = 0$ for every $w \in W$. We conclude that $y \cdot k \geq \alpha$ for every $k \in K$. $\square$

2.6.9 Change of a Numeraire

Assume that the price of the lth asset is also strictly positive for some $l < k$ (recall the standing assumption that $S_t^k > 0$ for every t). Then it is possible to us S^l, rather than $B = S^k$, to express prices of all assets – that is, as a *numeraire asset*. It is clear that all general results established in the previous sections will remain valid if we use S^l instead of S^k. For instance, the model is arbitrage-free if and only if there exists a martingale measure, denoted by $\bar{\mathbb{P}}$, associated with the choice of S^l as a numeraire. Moreover, the uniqueness of a martingale measure $\bar{\mathbb{P}}$ associated with S^l is equivalent to the market completeness. Assume that a model is arbitrage-free and complete. By modifying the notation in the statement and the proof of Proposition 2.6.2, we deduce that the arbitrage price process of any European contingent claim X settling at time T is given by the risk-neutral valuation formula under $\bar{\mathbb{P}}$

$$\pi_t(X) = S_t^l \, \mathbb{E}_{\bar{\mathbb{P}}}((S_T^l)^{-1} X \mid \mathcal{F}_t), \quad \forall\, t \leq T. \tag{2.69}$$

The last formula can also be proved using directly (2.63) and the expression for the Radon-Nikodým derivative of $\bar{\mathbb{P}}$ with respect to $\mathbb{P}^*$, which is given by the following result, which extends Proposition 2.2.4.

Proposition 2.6.4 *The Radon-Nikodým derivative process η of $\bar{\mathbb{P}}$ with respect to $\mathbb{P}^*$ equals, for every $t = 0, 1, \ldots, T^*$,*

$$\eta_t = \frac{d\bar{\mathbb{P}}}{d\mathbb{P}^*}\Big|_{\mathcal{F}_t} = \frac{S_0^k S_t^l}{S_0^l S_t^k}. \tag{2.70}$$

Proof. It is clear that $\mathbb{P}^*$ and $\bar{\mathbb{P}}$ are equivalent on $(\Omega, \mathcal{F}_{T^*})$ and thus the Radon-Nikodým derivative $\eta_{T^*} = d\bar{\mathbb{P}}/d\mathbb{P}^*$ exists. Note that S^l/S^k is a martingale under $\mathbb{P}^*$ and, by virtue of general properties of the Radon-Nikodým derivative, we have that $\eta_t = \mathbb{E}_{\mathbb{P}^*}(\eta_{T^*} \mid \mathcal{F}_t)$. It thus suffices to show that

$$\eta_{T^*} = \frac{d\bar{\mathbb{P}}}{d\mathbb{P}^*} = \frac{S_0^k S_{T^*}^l}{S_0^l S_{T^*}^k}. \tag{2.71}$$

It is easily seen that the last equality defines an equivalent probability measure $\bar{\mathbb{P}}$ on $(\Omega, \mathcal{F}_{T^*})$ since $\eta_{T^*} > 0$ and $\mathbb{E}_{\mathbb{P}^*}\eta_{T^*} = 1$. Hence it remains to verify that if a process S^i/S^k is a martingale under $\mathbb{P}^*$ then the process S^i/S^l is a martingale under $\bar{\mathbb{P}}$. We thus assume that, for every $u \leq t$,

$$\mathbb{E}_{\mathbb{P}^*}(S_t^i/S_t^k \mid \mathcal{F}_u) = S_u^i/S_u^k.$$

Using the Bayes formula (see Lemma A.1.4), we obtain

$$
\begin{aligned}
\mathbb{E}_{\bar{\mathbb{P}}}(S_t^i/S_t^l \mid \mathcal{F}_u) &= \frac{\mathbb{E}_{\mathbb{P}^*}(\eta_{T^*} S_t^i/S_t^l \mid \mathcal{F}_u)}{\mathbb{E}_{\mathbb{P}^*}(\eta_{T^*} \mid \mathcal{F}_u)} \\
&= \frac{\mathbb{E}_{\mathbb{P}^*}(\mathbb{E}_{\mathbb{P}^*}(\eta_{T^*} \mid \mathcal{F}_t) S_t^i/S_t^l \mid \mathcal{F}_u)}{\mathbb{E}_{\mathbb{P}^*}(\eta_{T^*} \mid \mathcal{F}_u)} \\
&= \frac{\mathbb{E}_{\mathbb{P}^*}(\eta_t S_t^i/S_t^l \mid \mathcal{F}_u)}{\eta_u} \\
&= \frac{\mathbb{E}_{\mathbb{P}^*}((S_t^l/S_t^k)(S_t^i/S_t^l) \mid \mathcal{F}_u)}{S_u^l/S_u^k} \\
&= \frac{\mathbb{E}_{\mathbb{P}^*}(S_t^i/S_t^k \mid \mathcal{F}_u)}{S_u^l/S_u^k} \\
&= \frac{S_u^i/S_u^k}{S_u^l/S_u^k} = S_u^i/S_u^l.
\end{aligned}
$$

We conclude that the probability measure $\bar{\mathbb{P}}$ given by (2.71) is indeed the unique martingale measure associated with the numeraire asset S^l. $\square$

2.6.10 Discrete-time Models with Infinite State Space

In this short subsection, we relax the standing assumption that the underlying probability space is finite. We assume instead that we are given a finite family of real-valued, discrete-time, stochastic processes $S^1, S^2, \ldots, S^k$, defined on a filtered probability space $(\Omega, \mathbb{F}, \mathbb{P})$, equipped with a filtration $\mathbb{F} = (\mathcal{F}_t)_{t \leq T^*}$. We note that all definitions and results of Sect. 2.6.2-2.6.5 remain valid within the present set-up. Moreover, as in the finite case, it is possible to give a probabilistic characterization of discrete-time models of financial markets. The following result, which is a version of the First Fundamental Theorem of Asset Pricing, was first established by Dalang et al. (1990), and later re-examined by other authors.

Theorem 2.6.4 *A discrete-time spot market model $\mathcal{M} = (S, \Phi)$ is arbitrage-free if and only if there exists a martingale measure $\mathbb{P}^*$ for S^*; that is, the class $\mathcal{P}(S^*)$ is non-empty.*

To proof of sufficiency is based on similar arguments as in Sect. 2.6.5. One needs, however, to introduce the notion of a *generalized martingale* (see Jacod and Shiryaev (1998)), and thus we do not present the details here. The proof of necessity is more complicated. It is based, among others, on the following lemma (for its proof, see Rogers (1994)) and the *optional decomposition* of a generalized supermartingale (see Föllmer and Kabanov (1998)).

Lemma 2.6.3 *Let X_t, $t \leq T^*$, be a sequence of $\mathbb{R}^k$-valued random variables, defined on a filtered probability space $(\Omega, \mathbb{F}, \mathbb{P})$, such that X_t is $\mathcal{F}_t$-measurable for* *every t, and let γ_t, $t \le T^*$, be a sequence of $\mathbb{R}^k$-valued random variables such that γ_t is $\mathcal{F}_{t-1}$-measurable for every t and $|\gamma_t| < c$ for some constant c. Suppose that we have, for $t = 1, 2, \ldots, T^*$,*

$$\mathbb{P}\{\gamma_t \cdot X_t > 0 \,|\, \mathcal{F}_{t-1}\} > 0$$

and

$$\mathbb{P}\{\gamma_t \cdot X_t < 0 \,|\, \mathcal{F}_{t-1}\} > 0.$$

Then there exist a probability measure $\mathbb{Q}$ equivalent to $\mathbb{P}$ on $(\Omega, \mathcal{F}_{T^*})$ and such that the process X_t, $t \le T^*$, is a $\mathbb{Q}$-martingale.

For the proof of the following version of the Second Fundamental Theorem of Asset Pricing, we refer to the paper by Jacod and Shiryaev (1998).

Theorem 2.6.5 *An arbitrage-free discrete-time spot market model $\mathcal{M} = (S, \Phi)$ is complete if and only if a martingale measure $\mathbb{P}^*$ is unique.*

For further results on hedging in discrete-time incomplete models, one may consult the papers by Föllmer and Leukert (1999) (quantile hedging), Föllmer and Leukert (2000) (shortfall hedging), Frittelli (2000) (entropy-based approach), Rouge and El Karoui (2000) (utility-based approach), Carassus et al. (2002) (super-replication). Let us mention that results obtained in these papers are rarely explicit; in most cases only the existence of respective hedging strategies is established.

2.7 Finite Futures Markets

We continue working under the assumption that the underlying probability space is finite. Throughout this section, the coordinates $Z^1, Z^2, \ldots, Z^{k-1}$ of an $\mathbb{R}^k$-valued process Z are interpreted as *futures prices* of some either traded or non-traded assets. More precisely, $Z_t^i = f_{S^i}(t, T^i)$ represents the value of the futures price at time t of the ith underlying security, in a futures contract with expiry date T^i. We assume here that $T^i \ge T^*$ for every $i = 1, 2, \ldots, k - 1$. For the ease of notation, we shall write briefly $f_t^i = f_{S^i}(t, T^i)$. Also, we write f to represent the random vector $(f^1, f^2, \ldots, f^{k-1})$.

Furthermore, the last coordinate, $Z^k = S^k$, is assumed to model the spot price of a certain security. The process S^k may be taken to model the price of an arbitrary risky asset; however, we shall typically assume that S^k represents the price process of a risk-free asset – that is, of a savings account. The rationale for this assumption lies in the fact that we prefer to express spot prices of derivative securities, such as futures options, in terms of a spot security. In other words, the price process S^k will play the role of a benchmark for the prices of derivative securities. It is convenient to assume that the price process S^k is strictly positive.

2.7.1 Self-financing Futures Strategies

Our first goal is to introduce a class of self-financing futures strategies, along the similar lines as in Sect. 2.6.1. By a *futures trading strategy* we mean an arbitrary $\mathbb{R}^k$-valued, $\mathbb{F}$-adapted process ϕ_t, $t \le T^*$. The coordinates ϕ_t^i, $i = 1, 2, \ldots, k - 1$, represent the number of long or short positions in a given futures contract assumed by an individual at time t, whereas $\phi_t^k S_t^k$ stands, as usual, for the cash investment in the spot security S^k at time t. As in the case of spot markets, we write $B = S^k$. In view of specific features of futures contracts, namely, the fact that it cost nothing to enter such a contract (cf. Sect. 1.2), the definition of the wealth process of a trading strategy is adjusted as follows.

Definition 2.7.1 The *wealth process* $V^f(\phi)$ of a futures trading strategy ϕ is an adapted stochastic process given by the equality

$$V_t^f(\phi) = \phi_t^k B_t, \quad \forall\, t \le T^*. \tag{2.72}$$

In particular, the *initial investment* $V_0^f(\phi)$ of any futures portfolio ϕ equals $V_0^f(\phi) = \phi_0^k B_0$. Let us denote by $\phi^f = (\phi^1, \phi^2, \ldots, \phi^{k-1})$ the futures position. Though ϕ^f is not present in formula (2.72), which defines the wealth process, it appears explicitly in the self-financing condition (2.73) as well as in expression (2.74) which describes the gains process (this in turn reflects the marking to market feature of a futures contract).

Definition 2.7.2 A futures trading strategy ϕ is said to be *self-financing* if and only if the condition

$$\phi_{t+1}^f \cdot (f_{t+1} - f_t) + \phi_t^k B_{t+1} = \phi_{t+1}^k B_{t+1} \tag{2.73}$$

is satisfied for every $t = 0, 1, \ldots, T^* - 1$.

The *gains process* $G^f(\phi)$ of any futures trading strategy ϕ is given by the equality, for every $t = 0, 1, \ldots, T^*$,

$$G_t^f(\phi) = \sum_{u=0}^{t-1} \phi_u^f \cdot (f_{u+1} - f_u) + \sum_{u=0}^{t-1} \phi_u^k (B_{u+1} - B_u). \tag{2.74}$$

Let us denote by Φ^f the vector space of all self-financing futures trading strategies. The following result is a counterpart of Lemma 2.6.1.

Lemma 2.7.1 *A futures trading strategy ϕ is self-financing if and only if we have that, for every $t = 0, 1, \ldots, T^*$,*

$$V_t^f(\phi) = V_0^f(\phi) + G_t^f(\phi).$$

Proof. Taking into account (2.73)–(2.74), for any self-financing futures strategy ϕ we obtain

$$V_t^f(\phi) = V_0^f(\phi) + \sum_{u=0}^{t-1}(V_{u+1}^f - V_u^f)$$

$$= V_0^f(\phi) + \sum_{u=0}^{t-1}(\phi_{u+1}^k B_{u+1} - \phi_u^k B_u)$$

$$= V_0^f(\phi) + \sum_{u=0}^{t-1}\left(\phi_u^f \cdot (f_{u+1} - f_u) + \phi_u^k B_{u+1} - \phi_u^k B_u\right)$$

$$= V_0^f(\phi) + \sum_{u=0}^{t-1}\phi_u^f \cdot (f_{u+1} - f_u) + \sum_{u=0}^{t-1}\phi_u^k(B_{u+1} - B_u)$$

$$= V_0^f(\phi) + G_t^f(\phi)$$

for every $t = 0, 1, \ldots, T^*$. This proves the "only if" clause. The proof of the "if" clause is left to the reader. $\qquad\square$

We say that a futures trading strategy $\phi \in \Phi^f$ is an *arbitrage opportunity* if $\mathbb{P}\{V_0^f(\phi) = 0\} = 1$, and the terminal wealth of ϕ satisfies

$$V_{T^*}^f(\phi) \geq 0 \quad \text{and} \quad \mathbb{P}\{V_{T^*}^f(\phi) > 0\} > 0.$$

We say that a futures market $\mathcal{M}^f = (f, B, \Phi^f)$ is *arbitrage-free* if there are no arbitrage opportunities in the class Φ^f of all futures trading strategies. The notions of a contingent claim, replication and completeness, as well as of a wealth process of an attainable contingent claim, remain the same, with obvious terminological modifications. For instance, we say that a claim X which settles at time T is *attainable* in $\mathcal{M}^f$ if there exists a self-financing futures trading strategy ϕ such that $V_T^f(\phi) = X$. The following result can be proved along the same lines as Proposition 2.6.1.

Proposition 2.7.1 *Suppose that the market $\mathcal{M}^f$ is arbitrage-free. Then any attainable contingent claim X is uniquely replicated in $\mathcal{M}^f$.*

The next definition, which introduces the arbitrage price in a futures market, is merely a reformulation of Definition 2.6.8.

Definition 2.7.3 Suppose that the futures market $\mathcal{M}^f$ is arbitrage-free. Then the wealth process of an attainable contingent claim X which settles at time T is called the *arbitrage price process* of X in the market model $\mathcal{M}^f$. We denote it by $\pi_t^f(X)$, $t \leq T$.

2.7.2 Martingale Measures for a Futures Market

The next step is to examine the arbitrage-free property of a futures market model. Recall that a probability measure $\tilde{\mathbb{P}}$ on $(\Omega, \mathcal{F}_{T^*})$, equivalent to $\mathbb{P}$, is called a martingale measure for f if the process f follows a $\tilde{\mathbb{P}}$-martingale with respect to the filtration

F. Note that we take here simply the futures prices, as opposed to the case of a spot market in which we dealt with relative prices. We denote by $\mathcal{P}(f)$ the class of all martingale measures for f, in the sense of the following definition (cf. Definition 2.6.10).

Definition 2.7.4 A probability measure $\tilde{\mathbb{P}}$ on $(\Omega, \mathcal{F}_{T^*})$ equivalent to $\mathbb{P}$ is called a *martingale measure for* $\mathcal{M}^f = (f, B, \Phi^f)$ if the relative wealth process $\tilde{V}^f(\phi) = V_t^f(\phi)B_t^{-1}$ of any self-financing futures trading strategy ϕ is a $\tilde{\mathbb{P}}$-martingale with respect to the filtration $\mathbb{F}$. The class of all martingale measures for $\mathcal{M}^f$ is denoted by $\mathcal{P}(\mathcal{M}^f)$.

In the next result, it is essential to assume that the discrete-time process B is *predictable* with respect to the filtration $\mathbb{F}$, meaning that for every $t = 0, 1, \ldots, T^* - 1$, the random variable B_{t+1} is measurable with respect to the σ-field $\mathcal{F}_t$. Intuitively, predictability of B means that the future value of B_{t+1} is known already at time t.

Remarks. Such a specific property of a *savings account* B may arise naturally in a discrete-time model with an uncertain rate of interest. Indeed, it is common to assume that at any date t, the rate of interest r_t that prevails over the next time period $[t, t+1]$ is known already at the beginning of this period. For instance, if the σ-field $\mathcal{F}_0$ is trivial, the rate r_0 is a real number, so that the value B_1 of a savings account at time 1 is also deterministic. Then, at time 1, at any state the rate r_1 is known, so that B_2 is a $\mathcal{F}_1$-measurable random variable and so forth. In this way, we construct a savings account process B which is predictable with respect to the filtration $\mathbb{F}$ (for brevity, we shall simply say that B is predictable).

Lemma 2.7.2 *A futures trading strategy* ϕ *is self-financing if and only if the relative wealth process* $\tilde{V}^f(\phi)$ *admits the following representation, for every* $t = 0, 1, \ldots, T^*$,

$$\tilde{V}_t^f(\phi) = \tilde{V}_0^f(\phi) + \sum_{u=0}^{t-1} B_{u+1}^{-1} \phi_u^f \cdot \Delta_u f,$$

where $\Delta_u f = f_{u+1} - f_u$. *Consequently, for any martingale measure* $\tilde{\mathbb{P}} \in \mathcal{P}(f)$, *the relative wealth process* $\tilde{V}^f(\phi)$ *of any self-financing futures trading strategy* ϕ *is a martingale under* $\tilde{\mathbb{P}}$.

Proof. Let us denote $V^f = V^f(\phi)$ and $\tilde{V}^f = \tilde{V}^f(\phi)$. For the first statement, it is sufficient to check that, for every $t = 0, 1, \ldots, T^* - 1$,

$$\tilde{V}_{t+1}^f - \tilde{V}_t^f = B_{t+1}^{-1} \phi_t^f \cdot (f_{t+1} - f_t). \tag{2.75}$$

To this end, it suffices to observe that

$$\tilde{V}_{t+1}^f - \tilde{V}_t^f = V_{t+1}^f B_{t+1}^{-1} - V_t^f B_t^{-1} = \phi_{t+1}^k - \phi_t^k,$$

since $V_t^f = \phi_t^k B_t$ for every t. It is thus clear that (2.75) is equivalent to (2.73), as expected.

For the second assertion, we need to show that for any self-financing futures trading strategy ϕ we have, for every $t = 0, 1, \ldots, T^* - 1$,

$$\mathbb{E}_{\tilde{\mathbb{P}}}(\tilde{V}^f_{t+1} - \tilde{V}^f_t \mid \mathcal{F}_t) = 0.$$

But

$$\mathbb{E}_{\tilde{\mathbb{P}}}\big(\tilde{V}^f_{t+1} - \tilde{V}^f_t \mid \mathcal{F}_t\big) = \mathbb{E}_{\tilde{\mathbb{P}}}(B^{-1}_{t+1}\phi^f_t \cdot (f_{t+1} - f_t) \mid \mathcal{F}_t)$$
$$= B^{-1}_{t+1}\phi^f_t \cdot \mathbb{E}_{\tilde{\mathbb{P}}}(f_{t+1} - f_t \mid \mathcal{F}_t) = 0,$$

as the random variable $B^{-1}_{t+1}\phi^f_t$ is $\mathcal{F}_t$-measurable (this holds since B is $\mathbb{F}$-predictable and ϕ^f is $\mathbb{F}$-adapted), and the price process f is a martingale under the probability measure $\tilde{\mathbb{P}}$.   □

The following corollary can be easily established.

Corollary 2.7.1 *A probability measure $\tilde{\mathbb{P}}$ on $(\Omega, \mathcal{F}_{T^*})$ is a martingale measure for the futures market model $\mathcal{M}^f$ if and only if $\tilde{\mathbb{P}}$ is a martingale measure for the futures price process f; that is, the equality $\mathcal{P}(f) = \mathcal{P}(\mathcal{M}^f)$ holds.*

2.7.3 Risk-neutral Valuation Formula

The next result shows that the existence of a martingale measure is a sufficient condition for the absence of arbitrage in the futures market model $\mathcal{M}^f$. In addition, the risk-neutral valuation formula is valid. Note that a claim X is interpreted as an ordinary "spot" contingent claim which settles at time T. In other words, X is a $\mathcal{F}_T$-measurable random payoff denominated in units of spot security B. Of course, X may explicitly depend on the futures prices. For instance, in the case of a path-independent European claim that settles at time T we have $X = h(f^1_T, f^2_T, \ldots, f^{k-1}_T, S^k_T)$ for some function $h : \mathbb{R}^k \to \mathbb{R}$.

Proposition 2.7.2 *Assume that the class $\mathcal{P}(\mathcal{M}^f)$ of futures martingale measures is non-empty. Then the futures market model $\mathcal{M}^f$ is arbitrage-free. Moreover, the arbitrage price in $\mathcal{M}^f$ of any attainable contingent claim X which settles at time T is given by the risk-neutral valuation formula*

$$\pi^f_t(X) = B_t \, \mathbb{E}_{\tilde{\mathbb{P}}}(X B^{-1}_T \mid \mathcal{F}_t), \quad \forall t \leq T, \qquad (2.76)$$

where $\tilde{\mathbb{P}}$ is any martingale measure from the class $\mathcal{P}(\mathcal{M}^f)$.

Proof. It is easy to check the absence of arbitrage opportunities in $\mathcal{M}^f$. Also, equality (2.76) is a straightforward consequence of Lemma 2.7.2. Indeed, it follows immediately from the martingale property of the discounted wealth of a strategy that replicates X.   □

The last result of this section corresponds to Theorems 2.6.1 and 2.6.2.

Theorem 2.7.1 *The following statements are true.*
(i) *A finite futures market* $\mathcal{M}^f$ *is arbitrage-free if and only if the class* $\mathcal{P}(\mathcal{M}^f)$ *of martingale measures is non-empty.*
(ii) *An arbitrage-free futures market* $\mathcal{M}^f$ *is complete if and only if the uniqueness of a martingale measure* $\tilde{\mathbb{P}}$ *for* $\mathcal{M}^f$ *holds.*

Proof. Both statements can be proved by means of the same arguments as those used in the case of a spot market. $\square$

2.7.4 Futures Prices Versus Forward Prices

In the preceding section, the evolution of a futures price was not derived but postulated. Here, we start with an arbitrage-free model of a spot market and we examine the forward price of a spot asset. Subsequently, we investigate the relation between forward and futures prices of such an asset. As before, we write S^i, $i \leq k$, to denote spot prices of primary assets. Also, we assume the strictly positive process $B = S^k$ plays the role of a numeraire asset – intuitively, a savings account.

It is convenient to introduce an auxiliary family of derivative spot securities, referred to as (default-free) *zero-coupon bonds* (or *discount bonds*). For any $t \leq T \leq T^*$, we denote by $B(t, T)$ the time t value of a security which pays to its holder one unit of cash at time T (no intermediate cash flows are paid before this date). We refer to $B(t, T)$ as the spot price at time t of a zero-coupon bond of maturity T, or briefly, the price of a T-maturity bond. Assume that the spot market model is arbitrage-free so that the class $\mathcal{P}(\mathcal{M})$ of spot martingale measures is non-empty. Taking for granted the attainability of the European claim $X = 1$ which settles at time T (and which represents the bond's payoff), we obtain

$$B(t, T) = B_t \, \mathbb{E}_{\mathbb{P}^*}(B_T^{-1} \,|\, \mathcal{F}_t), \quad \forall t \leq T.$$

Let us fix i and let us denote $S^i = S$ (as usual, we shall refer to S as the stock price process). We know already that a forward contract (with the settlement date T) is represented by the contingent claim $X = S_T - K$ that settles at time T. Recall that by definition, the forward price $F_S(t, T)$ at time $t \leq T$ is defined as that level of the delivery price K, which makes the forward contract worthless at time t equal to 0. Since the forward price $F_S(t, T)$ is determined at time t, it is necessarily a $\mathcal{F}_t$-measurable random variable. It appears that the forward price of a (non-dividend-paying) asset can be expressed in terms of its spot price at time t and the price at time t of a zero-coupon bond of maturity T.

Proposition 2.7.3 *The forward price at time* $t \leq T$ *of a stock* S *for the settlement date* T *equals*

$$F_S(t, T) = \frac{S_t}{B(t, T)}, \quad \forall t \leq T. \tag{2.77}$$

Proof. In view of (2.63), we obtain

$$\pi_t(X) = B_t \, \mathbb{E}_{\mathbb{P}^*}\big(B_T^{-1}(S_T - F_S(t, T)) \,\big|\, \mathcal{F}_t\big) = 0.$$

On the other hand, since by assumption $F_S(t, T)$ is $\mathcal{F}_t$-measurable, we get

$$\pi_t(X) = B_t \, \mathbb{E}_{\mathbb{P}^*}(S_T^* \mid \mathcal{F}_t) - B_t \, F_S(t, T) \mathbb{E}_{\mathbb{P}^*}(B_T^{-1} \mid \mathcal{F}_t) = S_t - B(t, T) F_S(t, T),$$

since $\mathbb{E}_{\mathbb{P}^*}(S_T^* \mid \mathcal{F}_t) = S_t^*$.   □

Though we have formally derived (2.77) using risk-neutral valuation approach, it is clear that equality (2.77) can also be easily established using standard no-arbitrage arguments. In such an approach, it is enough to assume that the underlying asset and the T-maturity bond are among traded securities (the existence of a savings account is not required). On the other hand, if the reference asset S pays dividends (coupons, etc.) during the lifetime of a forward contract, formula (2.77) should be appropriately modified. For instance, in the case of a single (non-random) dividend D to be received by the owner of S at time U, where $t < U < T$, equality (2.77) becomes

$$F_S(t, T) = \frac{S_t - D B(t, U)}{B(t, T)}, \quad \forall t \leq T. \tag{2.78}$$

Let us denote by $f_S(t, T)$ the futures price of the stock S – that is, the price at which a futures contract written on S with the settlement date T is entered into at time t (in particular, $f_S(T, T) = S_T$). We change slightly our setting, namely, instead on focusing on a savings account, we assume that we are given the price process of the T-maturity bond. We make a rather strong assumption that $B(t, T)$ follows a predictable process. In other words, for any $t \leq T - 1$ the random variable $B(t + 1, T)$ is $\mathcal{F}_t$-measurable. Intuitively, this means that on each date we know the bond price which will prevail on the next date (though hardly a realistic assumption, it is nevertheless trivially satisfied in any security market model which assumes a deterministic[4] savings account).

Proposition 2.7.4 *Let the bond price $B(t, T)$ follow a predictable process. The combined spot-futures market is arbitrage-free if and only if the futures and forward prices agree; that is, $f_S(t, T) = F_S(t, T)$ for every $t \leq T$.*

Proof. We first aim show that the asserted equality is necessary for the absence of arbitrage in the combined spot-futures market. As mentioned above, the T-maturity bond is considered as the basic spot asset; in particular, all proceeds from futures contracts are immediately reinvested in this bond. Let us consider a self-financing futures strategy ψ for which $V_0^f(\psi) = 0$. It is not difficult to check that the terminal wealth of a strategy ψ satisfies

$$V_T^f(\psi) = \sum_{t=0}^{T-1} \psi_t \, B^{-1}(t + 1, T) \, (f_{t+1} - f_t).$$

For instance, the gains/losses $\psi_0(f_1 - f_0)$ incurred at time 1 are used to purchase $B^{-1}(1, T)$ units of the bond that matures at T. This investment results in $\psi_0(f_1 - f_0)B^{-1}(1, T)$ units of cash when the bond expires at time T.

---

[4] It was observed by Lutz Schlögl that if all bond prices with different maturities are predictable then the bond prices (and thus also the savings account) are deterministic.

Let us now consider a specific futures trading strategy; namely, we take $\psi_t = -B(t+1, T)$ for every $t \leq T - 1$ (note that the bond coordinate of this strategy is uniquely determined by the self-financing condition). It is easy to see that in this case we get simply $V_T^f(\psi) = f_0 - S_T$. In addition, we shall employ the following spot trading strategy ϕ: buy-and-hold the stock S, using the proceeds from the sale of T-maturity bonds. It is clear that in order to purchase one share of stock at the price S_0, one needs to sell $S_0 B^{-1}(0, T)$ units of the bond at the price $B(0, T)$. Therefore, the initial wealth of ϕ is zero, and the terminal wealth equals

$$V_T(\phi) = S_T - S_0 B^{-1}(0, T).$$

Combining these two strategies, we obtain the spot-futures strategy with zero initial wealth, and terminal wealth that equals

$$V_T(\phi) + V_T^f(\psi) = f_0 - S_0 B^{-1}(0, T).$$

Since arbitrage opportunities in the combined spot-futures market were ruled out, we conclude that $f_0 = S_0 B^{-1}(0, T)$; that is, the futures price at time 0 coincides with the forward price $F_S(0, T)$. Similar reasoning leads to the general equality. The proof of the converse implication is left to the reader as an exercise.   □

2.8 American Contingent Claims

We will now address the issue of valuation and hedging of American contingent claims within the framework of a finite market in which k primary assets are traded at spot prices $S^1, S^2, \ldots, S^k$. As usual, we denote $B = S^k$ and we postulate that $B > 0$. Let us first recall the concept of a stopping time.

Definition 2.8.1 A *stopping time with respect to a filtration* $\mathbb{F}$ (an $\mathbb{F}$-*stopping time* or, simply, a *stopping time*) is an arbitrary function $\tau : \Omega \to \{0, 1, \ldots, T^*\}$ such that for any $t = 0, 1, \ldots, T^*$ the random event $\{\tau = t\}$ belongs to the σ-field $\mathcal{F}_t$. We denote by $\mathcal{T}_{[0, T^*]}$ the class of all *stopping times* with values in $\{0, 1, \ldots, T^*\}$ defined on the filtered probability space $(\Omega, \mathbb{F}, \mathbb{P})$.

Any stopping time τ is manifestly a random variable taking values in $\{0, 1, \ldots, T^*\}$, but the converse is not true, in general. Intuitively, the property $\{\tau = t\} \in \mathcal{F}_t$ means that the decision whether to stop a given process at time t or to continue (in our case, whether to exercise an American claim at time t or to hold it alive at least till the next date, $t + 1$) is made exclusively on the basis of information available at time t. It is worth mentioning that the property that $\{\tau = t\} \in \mathcal{F}_t$ for any $t = 0, 1, \ldots, T^*$ is equivalent to the condition that $\{\tau \leq t\} \in \mathcal{F}_t$ for any $t = 0, 1, \ldots, T^*$. Therefore, for any stopping time τ, the event $\{\tau > t\} = \{\tau \geq t + 1\}$ belongs to $\mathcal{F}_t$ for any $t = 0, 1, \ldots, T^*$.

The following lemma is easy to established and thus its proof is left to the reader as an exercise.

Lemma 2.8.1 *Let τ and σ be two stopping times. Then the random variables $\tau \wedge \sigma = \min(\tau, \sigma)$ and $\tau \vee \sigma = \max(\tau, \sigma)$ are stopping times as well.*

For convenience, we define the following subclass of stopping times.

Definition 2.8.2 We write $\mathcal{T}_{[t,T]}$ to denote the subclass of those stopping times τ with respect to $\mathbb{F}$ that satisfy the inequalities $t \le \tau \le T$, that is, take values in the set $\{t, t+1, \dots, T\}$.

Let us recall the definition of an American contingent claim (see Definition 2.4.1).

Definition 2.8.3 An *American contingent claim* $X^a = (X, \mathcal{T}_{[0,T]})$ expiring at time T consists of an $\mathbb{F}$-adapted payoff process $X = (X_t)_{t=0}^{T}$ and of the set $\mathcal{T}_{[0,T]}$ of admissible exercise times for the holder of an American claim.

For any stopping time $\tau \in \mathcal{T}_{[0,T]}$, the random variable X_τ, defined as

$$X_\tau = \sum_{t=0}^{T} X_t \mathbb{1}_{\{\tau=t\}},$$

represents the payoff received at time τ by the holder of an American claim X^a, if she decides to exercise the claim at time τ.

Note that we will adopt a convention that stipulates that an American claim is exercised at its expiry date T if it was not exercised prior to that date. We will frequently assume that a given American claim is *alive* at time t, meaning that, by assumption, it was not exercised at times $0, 1, \dots, t-1$ (notwithstanding whether is should have been exercised at some date $u \le t-1$ according to some rationality principle) and we will examine the valuation of the claim at this date, as well as its hedging from time t onwards.

Let us thus suppose that an American claim is alive at time t. Then, the set of exercise times available to the holder an American claim from time t onwards is represented by the class $\mathcal{T}_{[t,T]}$ of stopping times τ taking values in the set $\{t, t+1, \dots, T\}$. In other words, the set of admissible exercise times from time t onwards equals $\mathcal{T}_{[t,T]}$.

Assume that the finite market model $\mathcal{M} = (S, \Phi)$ is arbitrage-free, but not necessarily complete. Then the concept of an arbitrage price in $\mathcal{M}$ of an American claim is introduced through the following definition.

Definition 2.8.4 By an *arbitrage price* in $\mathcal{M}$ of an American claim X^a we mean any price process $\pi_t(X^a)$, $t \le T$, such that the extended market model – that is, the market model with trading in assets $S^1, S^2, \dots, S^k$ and an American claim – remains arbitrage-free.

Note that an American claim is traded only up to its exercise time, which is chosen at discretion of its holder, so it is essentially any stopping time from the class $\mathcal{T}_{[0,T]}$. If market completeness is not postulated, we cannot expect an arbitrage price of an American claim to be unique. Indeed, in that case, the uniqueness of an arbitrage price is not ensured even for European claims. Therefore, to simplify the analysis we will work hereafter under the standing assumption that the market model is not only arbitrage-free, but also complete, in the sense discussed in Sect. 2.6.1 (see, in particular, Definition 2.6.5).

In the case of an arbitrage-free and complete finite model, we have the following result, which can be seen as an extension of Corollary 1.7.1, in which we examined the valuation of American claims for a single-period model (see also Proposition 2.4.1 for the CRR model case).

Proposition 2.8.1 *Assume that the finite market model* $\mathcal{M} = (S, \Phi)$ *is arbitrage-free and complete. Let* $\mathbb{P}^*$ *be the unique martingale measure for the process* $S^* = S/B$. *Then the arbitrage price* $\pi(X^a)$ *in* $\mathcal{M}$ *of an American claim* X^a, *with the payoff process* X *and expiry date* T *is unique and it is given by the formula, for every* $t = 0, 1, \ldots, T$,

$$\pi_t(X^a) = \max_{\tau \in \mathcal{T}_{[t,T]}} B_t \, \mathbb{E}_{\mathbb{P}^*}\big(B_\tau^{-1} X_\tau \mid \mathcal{F}_t\big). \tag{2.79}$$

In particular, the arbitrage prices of American call and put options, written on the asset S^l *for some* $l \neq k$, *are given by the following expressions, for every* $t = 0, 1, \ldots, T$,

$$C_t^a = \max_{\tau \in \mathcal{T}_{[t,T]}} B_t \, \mathbb{E}_{\mathbb{P}^*}\big(B_\tau^{-1}(S_\tau^l - K)^+ \mid \mathcal{F}_t\big)$$

and

$$P_t^a = \max_{\tau \in \mathcal{T}_{[t,T]}} B_t \, \mathbb{E}_{\mathbb{P}^*}\big(B_\tau^{-1}(K - S_\tau^l)^+ \mid \mathcal{F}_t\big).$$

The proof of Proposition 2.8.1 is postponed to Sect. 2.8.2 (see, in particular, Corollary 2.8.3), since it requires several preliminary results regarding the so-called *optimal stopping problems*. The valuation and hedging of an American claim is clearly a more involved issue than just solving some optimal stopping problem, so let us first make some comments.

On the one hand, we note that formula (2.79) is intuitively plausible from the perspective of the holder. Indeed, the holder of an American claim may choose freely her exercise time τ, and thus it is natural to expect that she will try select a stopping time τ that would maximize the price of a European claim with the terminal payoff X_τ to be received at a random maturity τ. We will thus refer to any stopping time that realizes the maximum in (2.79) for a given t as a *rational exercise time* as of time t. Note that we do not claim that this time is unique, even when the date t is fixed.

On the other hand, it is not evident whether the price level given by (2.79) will be also acceptable by the seller of an American claim. Let us make clear that the seller's strategy should not rely on the assumption that the claim will necessarily be exercised by its holder at a rational exercise time. Therefore, starting from the initial wealth $\pi_0(X^a)$, the seller should be able to construct a trading strategy, such that the wealth process will cover his potential liabilities due to her short position in an American claim at any time t, and not only at a rational exercise time.

2.8.1 Optimal Stopping Problems

The goal of this section is to establish the most fundamental results concerning the optimal stopping problem in an abstract setup with a finite horizon. Let thus $Z = (Z_t)_{t=0}^n$ be a sequence of random variable on some probability space $(\Omega, \mathcal{F}, \mathbb{P})$ and let $\mathbb{F} = (\mathcal{F}_t)_{t=0}^n$ be some filtration such that the *payoff process* Z is $\mathbb{F}$-adapted. We assume, as usual, that $\mathcal{F}_0 = \{\emptyset, \Omega\}$ and we denote by $\mathcal{T}_{[t,n]}$ the class of all $\mathbb{F}$-stopping times taking values in the set $\{t, t+1, \ldots, n\}$.

The assumption that the underlying probability space has a finite number of elementary events is not crucial here, and thus it can be relaxed. Indeed, it suffices to make the technical assumption that $\mathbb{E}_\mathbb{P}|Z_t| < \infty$ for every $t = 1, 2, \ldots, n$, so that for any stopping time $\tau \in \mathcal{T}_{[t,n]}$, we have that

$$\mathbb{E}_\mathbb{P}|Z_\tau| < n \max_{t=0,1,\ldots,n} \mathbb{E}_\mathbb{P}|Z_t| < \infty.$$

Of course, this condition is satisfied by any process Z when Ω is finite, since in that case the random variables $|Z_0|, |Z_1|, \ldots, |Z_n|$ are all bounded by a common constant.

We will examine the following *optimal stopping problems*.

Problem (A) For $t = 0$, we wish to maximize the expected value $\mathbb{E}_\mathbb{P}(Z_\tau)$ over the class $\mathcal{T}_{[0,n]}$ of stopping times τ. In other words, we search for an *optimal stopping time* $\tau_0^* \in \mathcal{T}_{[0,n]}$ such that

$$\mathbb{E}_\mathbb{P}(Z_{\tau_0^*}) \geq \mathbb{E}_\mathbb{P}(Z_\tau), \quad \forall \tau \in \mathcal{T}_{[0,n]},$$

and for the corresponding expectation $\mathbb{E}_\mathbb{P}(Z_{\tau_0^*})$.

Problem (B) For any $t = 0, 1, \ldots, n$, we wish to maximize the conditional expectation $\mathbb{E}_\mathbb{P}(Z_\tau \mid \mathcal{F}_t)$ over the class $\mathcal{T}_{[t,n]}$ of stopping times. We thus search for a collection of *optimal stopping times* $\tau_t^* \in \mathcal{T}_{[t,n]}$ that satisfy, for each $t = 0, 1, \ldots, n$,

$$\mathbb{E}_\mathbb{P}(Z_{\tau_t^*} \mid \mathcal{F}_t) \geq \mathbb{E}_\mathbb{P}(Z_\tau \mid \mathcal{F}_t), \quad \forall \tau \in \mathcal{T}_{[t,n]},$$

and for the corresponding conditional expectations $\bar{Y}_t = \mathbb{E}_\mathbb{P}(Z_{\tau_t^*} \mid \mathcal{F}_t)$.

Remarks. It is clear that the latter problem is a direct extension of the former, and thus for $t = 0$ it suffices to examine problem (A). Note also that the solution to problem (B) for $t = n$ is manifestly $\tau_n^* = n$, since the stopping time $\tau = n$ is the unique element of the class $\mathcal{T}_{[n,n]}$. Therefore, we have that $\bar{Y}_n = \mathbb{E}_\mathbb{P}(Z_n \mid \mathcal{F}_n) = Z_n$.

Definition 2.8.5 Suppose that for every $t = 0, 1, \ldots, n$, the problem (B) admits a solution τ_t^*. Then the $\mathbb{F}$-adapted process $\bar{Y}$, given by the formula $\bar{Y}_t = \mathbb{E}_\mathbb{P}(Z_{\tau_t^*} \mid \mathcal{F}_t)$ for $t = 0, 1, \ldots, n$, is called the *value process* for the optimal stopping problem (B).

As we shall see in what follows, an optimal stopping time τ_t^* is not unique, in general. It is clear, however, that the value process is uniquely defined.

The main tool in solving the optimal stopping problem (B) is the concept of the *Snell envelope* of the payoff process Z.

Definition 2.8.6 A process $Y = (Y_t)_{t=0}^{n}$ is the *Snell envelope* of an $\mathbb{F}$-adapted process $Z = (Z_t)_{t=0}^{n}$ whenever the following conditions are satisfied:
(i) Y is a supermartingale with respect to the filtration $\mathbb{F}$,
(ii) Y dominates Z, that is, the inequality $Y_t \geq Z_t$ is valid for any $t = 0, 1, \ldots, n$,
(iii) for any other supermartingale $\tilde{Y}$ dominating Z, we have that $\tilde{Y}_t \geq Y_t$ for every $t = 0, 1, \ldots, n$.

Conditions (i)–(iii) of Definition 2.8.6 can be summarized as follows: the Snell envelope of Z is the minimal supermartingale dominating Z. In particular, property (iii) implies that the Snell envelope, if it exists, is unique.

To address the issue of existence of the Snell envelope of Z, we define the process $Y = (Y_t)_{t=0}^{n}$ by setting, for every $t = 1, 2, \ldots, n$,

$$Y_n = Z_n, \quad Y_{t-1} = \max \left\{ Z_{t-1}, \, \mathbb{E}_{\mathbb{P}}(Y_t \mid \mathcal{F}_{t-1}) \right\}. \tag{2.80}$$

Lemma 2.8.2 *The process Y given by formula* (2.80) *is the Snell envelope of the payoff process Z.*

Proof. First, we note that Y is an $\mathbb{F}$-adapted process and that (2.80) implies immediately that $Y_{t-1} \geq \mathbb{E}_{\mathbb{P}}(Y_t \mid \mathcal{F}_{t-1})$ for $t = 1, 2, \ldots, n$. This shows that Y is an $\mathbb{F}$-supermartingale. Formula (2.80) implies also that Y dominates Z. It remains to check that condition (iii) in Definition 2.8.6 is satisfied as well. Let thus $\tilde{Y}$ be any supermartingale that dominates Z. Of course, condition (iii) holds for $t = n$, since obviously $\tilde{Y}_n \geq Z_n = Y_n$. We now proceed by induction. Assume that $\tilde{Y}_t \geq Y_t$ for some $t = 1, 2, \ldots, n$. Since $\tilde{Y}$ is a supermartingale dominating Z, we have that $\tilde{Y}_{t-1} \geq \mathbb{E}_{\mathbb{P}}(\tilde{Y}_t \mid \mathcal{F}_{t-1})$ and $\tilde{Y}_{t-1} \geq Z_{t-1}$. Therefore,

$$\tilde{Y}_t \geq \max \left\{ Z_{t-1}, \, \mathbb{E}_{\mathbb{P}}(\tilde{Y}_t \mid \mathcal{F}_{t-1}) \right\} \geq \max \left\{ Z_{t-1}, \, \mathbb{E}_{\mathbb{P}}(Y_t \mid \mathcal{F}_{t-1}) \right\} = Y_{t-1},$$

since the assumed inequality $\tilde{Y}_t \geq Y_t$ and monotonicity of the conditional expectation imply that

$$\mathbb{E}_{\mathbb{P}}(\tilde{Y}_t \mid \mathcal{F}_{t-1}) \geq \mathbb{E}_{\mathbb{P}}(Y_t \mid \mathcal{F}_{t-1}).$$

We thus conclude that Y given by (2.80) is the minimal supermartingale dominating Z. $\qquad\square$

We claim that the optimal stopping time is the first moment when the Snell envelope Y hits the payoff process Z and that Y is in fact the value process of an optimal stopping problem. Formally, for any fixed $t = 0, 1, \ldots, n$, we define the stopping time τ_t^* belonging to the class $\mathcal{T}_{[t,n]}$ by setting

$$\tau_t^* = \min \left\{ u \in \{t, t+1, \ldots, n\} \mid Y_u = Z_u \right\}. \tag{2.81}$$

It is worth noting that the process Y satisfies, for any $t = 0, 1, \ldots, n-1$,

$$\mathbb{1}_{\{\tau_t^* > t\}} Y_t = \mathbb{1}_{\{\tau_t^* > t\}} \mathbb{E}_{\mathbb{P}}(Y_{t+1} \mid \mathcal{F}_t)$$

since clearly $\{\tau_t^* > t\} = \{Y_t > Z_t\}$ and the inequality $Y_t > Z_t$ implies in turn that $Y_t = \mathbb{E}_{\mathbb{P}}(Y_{t+1} \mid \mathcal{F}_t)$.

Proposition 2.8.2 *For any $t = 0, 1, \ldots, n$, the stopping time τ_t^* is a solution to problem (B), that is, τ_t^* is an optimal stopping time as of time t. Moreover, for any $t = 0, 1, \ldots, n$,*

$$Y_t = \mathbb{E}_{\mathbb{P}}(Z_{\tau_t^*} \mid \mathcal{F}_t) = \bar{Y}_t. \tag{2.82}$$

In other words, the Snell envelope Y of Z is equal to the value process $\bar{Y}$ for the optimal stopping problem (B).

First proof of Proposition 2.8.2. We first present a demonstration based on the induction with respect to t. Of course, the statement is trivially satisfied for $t = n$, since we know that $Y_n = Z_n$ and that the stopping time $\tau_n^* = n$ is optimal.

Let us now assume that the assertions are valid for some t. Our goal is to show that they also hold for time $t - 1$.

We first check that (2.82) holds for $t - 1$ if it is satisfied for t. We start by observing that the stopping time τ_{t-1}^* can be represented as follows

$$\tau_{t-1}^* = (t-1)\mathbb{1}_{\{\tau_{t-1}^* = t-1\}} + \tau_t^* \mathbb{1}_{\{\tau_{t-1}^* \geq t\}}.$$

Indeed,

$$\begin{aligned}
\tau_{t-1}^* &= \min\left\{ u \in \{t-1, t, \ldots, n\} \mid Y_u = Z_u \right\} \\
&= (t-1)\mathbb{1}_{\{Y_{t-1} = Z_{t-1}\}} + \min\{u \geq t \mid Y_u = Z_u\}\mathbb{1}_{\{Y_{t-1} > Z_{t-1}\}} \\
&= (t-1)\mathbb{1}_{\{\tau_{t-1}^* = t-1\}} + \tau_t^* \mathbb{1}_{\{\tau_{t-1}^* \geq t\}}.
\end{aligned}$$

It is important to observe that the events

$$\{\tau_{t-1}^* = t-1\} = \{Z_{t-1} \geq \mathbb{E}_{\mathbb{P}}(Y_t \mid \mathcal{F}_{t-1})\}$$

and

$$\{\tau_{t-1}^* > t-1\} = \{\tau_{t-1}^* \geq t\} = \{Z_{t-1} < \mathbb{E}_{\mathbb{P}}(Y_t \mid \mathcal{F}_{t-1})\}$$

belong to the σ-field $\mathcal{F}_{t-1}$. Consequently, we have that

$$\begin{aligned}
\mathbb{E}_{\mathbb{P}}(Z_{\tau_{t-1}^*} \mid \mathcal{F}_{t-1}) &= \mathbb{E}_{\mathbb{P}}(\mathbb{1}_{\{\tau_{t-1}^* = t-1\}} Z_{t-1} + \mathbb{1}_{\{\tau_{t-1}^* \geq t\}} Z_{\tau_t^*} \mid \mathcal{F}_{t-1}) \\
&= \mathbb{1}_{\{\tau_{t-1}^* = t-1\}} Z_{t-1} + \mathbb{1}_{\{\tau_{t-1}^* \geq t\}} \mathbb{E}_{\mathbb{P}}(Z_{\tau_t^*} \mid \mathcal{F}_{t-1}) \\
&= \mathbb{1}_{\{\tau_{t-1}^* = t-1\}} Z_{t-1} + \mathbb{1}_{\{\tau_{t-1}^* \geq t\}} \mathbb{E}_{\mathbb{P}}(\mathbb{E}_{\mathbb{P}}(Z_{\tau_t^*} \mid \mathcal{F}_t) \mid \mathcal{F}_{t-1}) \\
&= \mathbb{1}_{\{Z_{t-1} \geq \mathbb{E}_{\mathbb{P}}(Y_t \mid \mathcal{F}_{t-1})\}} Z_{t-1} + \mathbb{1}_{\{Z_{t-1} < \mathbb{E}_{\mathbb{P}}(Y_t \mid \mathcal{F}_{t-1})\}} \mathbb{E}_{\mathbb{P}}(Y_t \mid \mathcal{F}_{t-1}) \\
&= \max\left(Z_{t-1}, \mathbb{E}_{\mathbb{P}}(Y_t \mid \mathcal{F}_{t-1}) \right) = Y_{t-1},
\end{aligned}$$

where we have used, in particular, the assumption that $\mathbb{E}_{\mathbb{P}}(Z_{\tau_t^*} \mid \mathcal{F}_t) = Y_t$ and formula (2.80).

We will now show that the optimality of τ_t^* implies that τ_{t-1}^* is optimal as well. For any stopping time $\tau \in \mathcal{T}_{[t-1,n]}$, we have that

$$\tau = (t-1)\mathbb{1}_{\{\tau = t-1\}} + \tau\mathbb{1}_{\{\tau \geq t\}},$$

where the events $\{\tau = t - 1\}$ and $\{\tau \geq t\} = \{\tau > t - 1\}$ manifestly belong to the σ-field $\mathcal{F}_{t-1}$. Let us set $\tilde{\tau} = \tau \vee t = \max(\tau, t)$. It is easy to check that $\tilde{\tau}$ is a stopping time belonging to the class $\mathcal{T}_{[t,n]}$. We now obtain

$$
\begin{aligned}
\mathbb{E}_{\mathbb{P}}(Z_{\tau^*_{t-1}} \mid \mathcal{F}_{t-1}) &= \max\left(Z_{t-1}, \mathbb{E}_{\mathbb{P}}(Y_t \mid \mathcal{F}_{t-1})\right) \\
&\geq \mathbb{1}_{\{\tau=t-1\}} Z_{t-1} + \mathbb{1}_{\{\tau \geq t\}} \mathbb{E}_{\mathbb{P}}(Y_t \mid \mathcal{F}_{t-1}) \\
&= \mathbb{1}_{\{\tau=t-1\}} Z_{t-1} + \mathbb{1}_{\{\tau \geq t\}} \mathbb{E}_{\mathbb{P}}(\mathbb{E}_{\mathbb{P}}(Z_{\tau^*_t} \mid \mathcal{F}_t) \mid \mathcal{F}_{t-1}) \\
&\geq \mathbb{1}_{\{\tau=t-1\}} Z_{t-1} + \mathbb{1}_{\{\tau \geq t\}} \mathbb{E}_{\mathbb{P}}(\mathbb{E}_{\mathbb{P}}(Z_{\tilde{\tau}} \mid \mathcal{F}_t) \mid \mathcal{F}_{t-1}) \\
&= \mathbb{1}_{\{\tau=t-1\}} Z_{t-1} + \mathbb{1}_{\{\tau \geq t\}} \mathbb{E}_{\mathbb{P}}(Z_{\tilde{\tau}} \mid \mathcal{F}_{t-1}) \\
&= \mathbb{1}_{\{\tau=t-1\}} Z_{t-1} + \mathbb{E}_{\mathbb{P}}(\mathbb{1}_{\{\tau \geq t\}} Z_\tau \mid \mathcal{F}_{t-1}) \\
&= \mathbb{E}_{\mathbb{P}}\left(\mathbb{1}_{\{\tau=t-1\}} Z_{t-1} + \mathbb{1}_{\{\tau \geq t\}} Z_\tau \mid \mathcal{F}_{t-1}\right) = \mathbb{E}_{\mathbb{P}}(Z_\tau \mid \mathcal{F}_{t-1}),
\end{aligned}
$$

where the second inequality holds since $\tilde{\tau}$ belongs to $\mathcal{T}_{[t,n]}$ and thus, by assumption, we have that

$$
\mathbb{E}_{\mathbb{P}}(Z_{\tau^*_t} \mid \mathcal{F}_t) \geq \mathbb{E}_{\mathbb{P}}(Z_{\tilde{\tau}} \mid \mathcal{F}_t).
$$

We conclude that the stopping time τ^*_{t-1} is optimal for problem (B) as of time $t - 1$, as was required to demonstrate. $\qquad\square$

An alternative derivation of Proposition 2.8.2 will hinge on the following lemma, which is of independent interest.

Lemma 2.8.3 *The stopped process $Y^{\tau^*_0}$ is an $\mathbb{F}$-martingale.*

Proof. Let us denote $M = Y^{\tau^*_0}$. Then we have, for any $t = 0, 1, \ldots, n - 1$,

$$
\begin{aligned}
\mathbb{E}_{\mathbb{P}}(M_{t+1} \mid \mathcal{F}_t) &= \mathbb{E}_{\mathbb{P}}(\mathbb{1}_{\{\tau^*_0 > t\}} M_{t+1} + \mathbb{1}_{\{\tau^*_0 \leq t\}} M_{t+1} \mid \mathcal{F}_t) \\
&= \mathbb{1}_{\{\tau^*_0 > t\}} \mathbb{E}_{\mathbb{P}}(M_{t+1} \mid \mathcal{F}_t) + \mathbb{E}_{\mathbb{P}}(\mathbb{1}_{\{\tau^*_0 \leq t\}} M_t \mid \mathcal{F}_t) \\
&= \mathbb{1}_{\{\tau^*_0 > t\}} M_t + \mathbb{1}_{\{\tau^*_0 \leq t\}} M_t = M_t
\end{aligned}
$$

since manifestly $M_{t+1} = Y_{\tau^*_0} = M_t$ on the event $\{\tau^*_0 \leq t\} \in \mathcal{F}_t$, and

$$
M_t = Y_t = \mathbb{E}_{\mathbb{P}}(Y_{t+1} \mid \mathcal{F}_t) = \mathbb{E}_{\mathbb{P}}(Y^{\tau^*_0}_{t+1} \mid \mathcal{F}_t) = \mathbb{E}_{\mathbb{P}}(M_{t+1} \mid \mathcal{F}_t)
$$

on the event $\{\tau^*_0 > t\} = \{\tau^*_0 \geq t + 1\} \in \mathcal{F}_t$. The second equality above follows from the inclusion $\{\tau^*_0 > t\} \subset \{Y_t > Z_t\}$ and the observation that the inequality $Y_t > Z_t$ in (2.80) implies that $Y_t = \mathbb{E}_{\mathbb{P}}(Y_{t+1} \mid \mathcal{F}_t)$. $\qquad\square$

We are ready to present an alternative proof of Proposition 2.8.2.

Second proof of Proposition 2.8.2. For simplicity, let us focus on the case $t = 0$. From Lemma 2.8.3, the stopped process $Y^{\tau^*_0}$ is a martingale, whereas for any stopping time $\tau \in \mathcal{T}_{[0,T]}$ the stopped process Y^τ is a supermartingale (this follows from the fact that Y is a supermartingale). Since $Y \geq Z$, we thus obtain, for any stopping time $\tau \in \mathcal{T}_{[0,T]}$,

$$
\mathbb{E}_{\mathbb{P}}(Z_\tau) \leq \mathbb{E}_{\mathbb{P}}(Y_\tau) = \mathbb{E}_{\mathbb{P}}(Y^\tau_\tau) \leq \mathbb{E}_{\mathbb{P}}(Y_0) = \mathbb{E}_{\mathbb{P}}(Y^{\tau^*_0}_{\tau^*_0}) = \mathbb{E}_{\mathbb{P}}(Z_{\tau^*_0}),
$$

where the last equality follows from the observation that $Y_{\tau_0^*} = Z_{\tau_0^*}$. This proves that τ_0^* is a solution to the optimal stopping problem (A). As we shall see in what follows (cf. Corollary 2.8.1(ii)), the properties that the stopped process $Y^{\tilde{\tau}}$ is a martingale and $Y_{\tilde{\tau}} = Z_{\tilde{\tau}}$ characterize in fact any optimal stopping time $\tilde{\tau}$ for problem (A).

The case of any date t can be dealt with in an analogous way. It suffices to extend slightly Lemma 2.8.3 by checking that, for any fixed t, the process $(Y_u^{\tau_t^*})_{u=t}^n$ is an $\mathbb{F}$-martingale. The details are left to the reader. $\qquad\square$

We have shown in Proposition 2.8.2 that, for any fixed t, the stopping time τ_t^* is a solution to problem (B). This does not mean, of course, that the uniqueness of an optimal stopping problem holds, in general. We will now address this issue and we will show that τ_t^* is the *minimal solution* to problem (B), in the sense that for any fixed t and any other optimal stopping time $\tilde{\tau}_t$ for problem (B) we have that $\tilde{\tau}_t \geq \tau_t^*$. Let us observe that a stopping time $\tilde{\tau}_t$ is optimal for problem (B) at time t if $\tilde{\tau}_t$ belongs to $\mathcal{T}_{[t,n]}$ and

$$\mathbb{E}_{\mathbb{P}}(Z_{\tilde{\tau}_t} \mid \mathcal{F}_t) = \mathbb{E}_{\mathbb{P}}(Z_{\tau_t^*} \mid \mathcal{F}_t).$$

We will first find the maximal solution to the optimal stopping problem. To this end, recall that the value process Y is a supermartingale and denote by $Y = M - A$ its Doob-Meyer decomposition. This means, in particular, that M is a martingale, whereas A is an $\mathbb{F}$-predictable process, so that A_{t+1} is $\mathcal{F}_t$-measurable for any $t = 0, 1, \ldots, n - 1$. As usual, we postulate that $M_0 = Y_0$ and $A_0 = 0$ to ensure the uniqueness of the Doob-Meyer decomposition of Y.

Since M is a martingale, it follows that, for every $t = 0, 1, \ldots, n - 1$,

$$A_{t+1} - A_t = Y_t - \mathbb{E}_{\mathbb{P}}(Y_{t+1} \mid \mathcal{F}_t) \geq 0. \qquad (2.83)$$

We first consider problem (A). Let the stopping time $\hat{\tau}_0$ be given by the formula (recall that $A_0 = 0$)

$$\hat{\tau}_0 = \min\{t \in \{0, 1, \ldots, n - 1\} \mid A_{t+1} > 0\},$$

where, by convention, we set $\inf \emptyset = n$, so that $\hat{\tau}_0 = n$ if the set in the right-hand side of the last formula is empty.

Lemma 2.8.4 *The stopping time $\hat{\tau}_0$ is a solution to the optimal stopping problem (A).*

Proof. In view of (2.83), the inequality $A_{t+1} - A_t > 0$ holds if and only if $Y_t > \mathbb{E}_{\mathbb{P}}(Y_{t+1} \mid \mathcal{F}_t)$, which in turn implies that $Y_t = Z_t$. Hence the equality $Y_{\hat{\tau}_0} = Z_{\hat{\tau}_0}$. It is clear that the stopped process $Y^{\hat{\tau}_0}$ satisfies the equality $Y^{\hat{\tau}_0} = M^{\hat{\tau}_0}$, and thus the supermartingale Y stopped at $\hat{\tau}_0$ is a martingale. This implies that

$$\mathbb{E}_{\mathbb{P}}(Z_{\hat{\tau}_0}) = \mathbb{E}_{\mathbb{P}}(Y_{\hat{\tau}_0}) = \mathbb{E}_{\mathbb{P}}(Y_{\hat{\tau}_0}^{\hat{\tau}_0}) = Y_0,$$

and thus $\hat{\tau}_0$ is an optimal stopping time for problem (A). Note also that $\mathbb{E}_{\mathbb{P}}(M_{\hat{\tau}_0}) = Y_0$. $\qquad\square$

Let us observe that $\tau_0^* \le \hat{\tau}_0$. Indeed, for every $t = 0, 1, \ldots, n-1$, on the event $\{\tau_0^* > t\}$ we have that $Y_t > Z_t$ and thus $Y_t = \mathbb{E}_{\mathbb{P}}(Y_{t+1} \mid \mathcal{F}_t)$, which manifestly implies that $A_{t+1} = 0$, and thus the inequality $\hat{\tau}_0 > t$ holds on this event. Since $\tau_0^* \le \hat{\tau}_0$, it is clear that the supermartingale Y stopped at τ_0^* is a martingale.

Lemma 2.8.5 (i) *The stopping time τ_0^* is the minimal solution to problem (A), specifically, if τ is any stopping time such that $\mathbb{P}\{\tau < \tau_0^*\} > 0$ then $Y_0 > \mathbb{E}_{\mathbb{P}}(Z_\tau)$.*
(ii) *The stopping time $\hat{\tau}_0$ is the maximal solution to problem (A), meaning that if τ is any stopping time such that $\mathbb{P}\{\tau > \hat{\tau}_0\} > 0$ then $Y_0 > \mathbb{E}_{\mathbb{P}}(Z_\tau)$.*

Proof. For part (i), we note that on the event $\{\tau < \tau_0^*\}$ we have $Y_\tau > Z_\tau$ (by the definition of τ_0^*). It is thus easy to check that the equalities

$$Y_0 = \mathbb{E}_{\mathbb{P}}(Z_{\tau_0^*}) = \mathbb{E}_{\mathbb{P}}(Z_\tau)$$

would imply that for the stopping time $\tilde{\tau} = \tau \vee \tau_0^*$ we would have $\mathbb{E}_{\mathbb{P}}(Z_{\tilde{\tau}}) > \mathbb{E}_{\mathbb{P}}(Z_{\tau_0^*})$. This would contradict the optimality of τ_0^*.

To prove part (ii), we note that since M is a martingale, for stopping times τ and $\hat{\tau}_0$, we obtain

$$\mathbb{E}_{\mathbb{P}}(M_\tau) = \mathbb{E}_{\mathbb{P}}(M_{\hat{\tau}_0}) = Y_0.$$

However, since $\mathbb{P}\{\tau > \hat{\tau}_0\} > 0$, we obtain

$$\mathbb{E}_{\mathbb{P}}(Z_\tau) \le \mathbb{E}_{\mathbb{P}}(Y_\tau) < \mathbb{E}_{\mathbb{P}}(M_\tau) = Y_0,$$

since on the event $\{\tau > \hat{\tau}_0\}$ we have $A > 0$, and thus $Y = M - A < M$. Hence τ is not optimal. □

Let $T_{[0,n]}^*$ stand for the class of all *optimal stopping times* for problem (A), so that $\tilde{\tau} \in T_{[0,n]}^*$ whenever

$$\mathbb{E}_{\mathbb{P}}(Z_{\tilde{\tau}}) = \max_{\tau \in T_{[0,n]}} \mathbb{E}_{\mathbb{P}}(Z_\tau).$$

Note that on the stochastic interval $[\tau_0^*, \hat{\tau}_0]$ we merely have that $Y_t \ge Z_t$. Therefore, it is not true, in general, that any stopping time τ such that $\tau_0^* \le \tau \le \hat{\tau}_0$ belongs to the class $T_{[0,n]}^*$. The following result provides some useful characterizations of the class $T_{[0,n]}^*$.

Corollary 2.8.1 (i) *A stopping time $\tau \in T_{[0,n]}$ belongs to $T_{[0,n]}^*$ whenever $\tau_0^* \le \tau \le \hat{\tau}_0$ and $Y_\tau = Z_\tau$.*
(ii) *A stopping time $\tau \in T_{[0,n]}$ belongs to $T_{[0,n]}^*$ whenever the stopped process Y^τ is a martingale and $Y_\tau = Z_\tau$.*

Proof. For part (i), it suffices to note that the stopped process Y^τ is a martingale and thus

$$Y_0 = \mathbb{E}_{\mathbb{P}}(Y_\tau) = \mathbb{E}_{\mathbb{P}}(Z_\tau),$$

where the last equality holds if and only if $Y_\tau = Z_\tau$ (recall that $Y \ge Z$).

For part (ii), it suffices to observe that the stopped process Y^τ is always a su-permartingale, so that $Y_0 \geq \mathbb{E}_{\mathbb{P}}(Y_\tau)$. If a stopping time τ is optimal then $Y_0 = \mathbb{E}_{\mathbb{P}}(Z_\tau) \leq \mathbb{E}_{\mathbb{P}}(Y_\tau)$ and thus $Y_0 = \mathbb{E}_{\mathbb{P}}(Y_\tau)$. This implies that Y^τ is a martingale. The converse implication is also clear. □

It is sometimes convenient to focus on the minimal solution only and to formally define the optimal stopping time as the *first* moment when the decision to stop is optimal. This convention would ensure the uniqueness of a solution to optimal stopping problem (A), but it would hide an important observation that it is sometimes possible to hold an American claim after time τ_0^*, and still exercise it later at some optimal time.

The analysis above can be extended without difficulty to the case of any $t = 1, 2, \ldots, n$. For any fixed t, we define the stopping time $\hat{\tau}_t \in \mathcal{T}_{[t,n]}$ by setting

$$\hat{\tau}_t = \min \{u \in \{t, t+1, \ldots, n-1\} \mid A_{u+1} - A_u > 0\}.$$

For any fixed $t = 0, 1, \ldots, n$, the stopping time $\hat{\tau}_t$ is an optimal stopping time for problem (B) and it satisfies the inequality $\hat{\tau}_t \geq \tau_t^*$. For any fixed $t = 0, 1, \ldots, n$, we write $\mathcal{T}_{[t,n]}^*$ to denote the class of all optimal stopping times for problem (B). Let us formulate a counterpart of Corollary 2.8.1.

Corollary 2.8.2 *For a fixed* $t = 0, 1, \ldots, n$, *a stopping time* τ *belongs to* $\mathcal{T}_{[t,n]}^*$ *whenever* $\tau_t^* \leq \tau \leq \hat{\tau}_t$ *and* $Y_\tau = Z_\tau$.

For any stopping time $\tau \in \mathcal{T}_{[t,n]}$ that does not belong to $\mathcal{T}_{[t,n]}^*$ the optimality of τ_t^* and $\hat{\tau}_t$ implies immediately that

$$\mathbb{P}\{\mathbb{E}_{\mathbb{P}}(Z_{\tau_t^*} \mid \mathcal{F}_t) > \mathbb{E}_{\mathbb{P}}(Z_\tau \mid \mathcal{F}_t)\} > 0$$

and

$$\mathbb{P}\{\mathbb{E}_{\mathbb{P}}(Z_{\hat{\tau}_t} \mid \mathcal{F}_t) > \mathbb{E}_{\mathbb{P}}(Z_\tau \mid \mathcal{F}_t)\} > 0.$$

For a fixed t, let us consider problem (B) under various assumptions about the payoff process Z. The following consequences of results established in this subsection are worth stating.

(i) First, a rather trivial example of non-uniqueness of an optimal stopping time is furnished by a payoff process Z that follows an $\mathbb{F}$-martingale under $\mathbb{P}$. In that case, we manifestly have that $\mathcal{T}_{[t,n]}^* = \mathcal{T}_{[t,n]}$, that is, any stopping time is an optimal so-lution to problem (B). In particular, we have that $\tau_t^* = t$ and $\hat{\tau}_t = n$ for every $t = 0, 1, \ldots, n$.

(ii) Second, if a payoff process Z is a strict submartingale, that is,

$$Z_{t-1} < \mathbb{E}_{\mathbb{P}}(Z_t \mid \mathcal{F}_{t-1}), \quad \forall t = 0, 1, \ldots, n,$$

then $\tau_t^* = \hat{\tau}_t = n$, so that $\mathcal{T}_{[t,n]}^* = \{n\}$. If Z is a strict supermartingale, so that

$$Z_{t-1} > \mathbb{E}_{\mathbb{P}}(Z_t \mid \mathcal{F}_{t-1}), \quad \forall t = 0, 1, \ldots, n,$$

we have that $\tau_t^* = \hat{\tau}_t = t$ and thus $\mathcal{T}_{[t,n]}^* = \{t\}$.

(iii) Finally, if payoff process Z is a (not necessarily strict) submartingale then the stopping time $\hat{\tau}_t = n$ is optimal. If Z is a (not necessarily strict) supermartingale then the stopping time $\tau_t^* = t$ is optimal.

2.8.2 Valuation and Hedging of American Claims

We assume that the market model $\mathcal{M} = (S, \Phi)$ is arbitrage-free and complete. Let $\mathbb{P}^*$ be the unique martingale measure for the process $S^* = S/B$. We will first consider the concept of hedging of an American claim from the seller's perspective.

Definition 2.8.7 We say that a self-financing trading strategy $\phi \in \Phi$ is the (seller's) *super-hedging strategy* for an American claim X^a expiring at T whenever the inequality $V_t(\phi) \geq X_t$ holds for every $t = 0, 1, \ldots, T$. In other words, $\phi \in \Phi$ *super-hedges* X^a for the seller if the wealth process $V(\phi)$ dominates the payoff process X of an American claim X^a.

Of course, there is no reason to expect that a super-hedging strategy for X^a is unique. Hence it is natural to define the seller's price of X^a through minimization of the initial cost of super-hedging.

Definition 2.8.8 The *seller's price* of an American claim X^a is the minimal cost of a super-hedging strategy for X^a. Let $\Phi(X^a)$ denote the class of all super-hedging strategies for X^a. Similarly, let $\Phi_t(X^a)$ be the class of all super-hedging strategies for X^a from time t onwards. Then the seller's price of X^a equals, for $t = 0$,

$$\pi_0^s(X^a) = \inf \{ V_0(\phi) \mid \phi \in \Phi(X^a) \}$$

and, for every $t = 1, 2, \ldots, T$,

$$\pi_t^s(X^a) = \inf \{ V_t(\phi) \mid \phi \in \Phi_t(X^a) \}.$$

Let us define the process U by setting $U_T = X_T$ and, for every $t = 1, 2, \ldots, T$,

$$U_{t-1} = \max \left\{ X_{t-1}, \ B_{t-1} \, \mathbb{E}_{\mathbb{P}^*}(U_t B_t^{-1} \mid \mathcal{F}_{t-1}) \right\}.$$

Also, let $U^* = B^{-1}U$ and $X^* = B^{-1}X$ stand for the relative processes. Then, clearly, $U_T^* = X_T^*$ and, for every $t = 1, 2, \ldots, T$,

$$U_{t-1}^* = \max \left\{ X_{t-1}^*, \ \mathbb{E}_{\mathbb{P}^*}(U_t^* \mid \mathcal{F}_{t-1}) \right\}. \tag{2.84}$$

It follows immediately from Lemma 2.8.2 that the process U^* is the Snell envelope of the discounted payoff process X^*. Furthermore, from Proposition 2.8.2, we obtain

$$U_t^* = \max_{\tau \in \mathcal{T}_{[t,T]}} \mathbb{E}_{\mathbb{P}^*}(X_\tau^* \mid \mathcal{F}_t)$$

or, equivalently,

$$U_t = \max_{\tau \in \mathcal{T}_{[t,T]}} B_t \, \mathbb{E}_{\mathbb{P}^*}\left(B_\tau^{-1} X_\tau \mid \mathcal{F}_t\right). \tag{2.85}$$

Proposition 2.8.3 *For any fixed $t = 0, 1, \ldots, T$, there exists a trading strategy $\phi^* \in \Phi_t(X^a)$ such that $V_t(\phi^*) = U_t$. Hence $\pi_t^s(X^a) \leq U_t$ for every $t = 0, 1, \ldots, T$. Moreover, for any super-hedging strategy $\phi \in \Phi_t(X^a)$ for X^a we have that $V_t(\phi) = U_t$ and thus $\pi_t^s(X^a) \geq U_t$ for every $t = 0, 1, \ldots, T$. Consequently, the seller's price $\pi^s(X^a)$ coincides with the process U.*

Proof. Let us first examine the case $t = 0$. Let $U^* = M - A$ be the Doob-Meyer decomposition of the (bounded) supermartingale U^*. Here, M is a martingale with $M_0 = U_0^*$ and A an $\mathbb{F}$-predictable, increasing process with $A_0 = 0$. In particular, A_t is $\mathcal{F}_{t-1}$-measurable (recall that, by convention, $\mathcal{F}_{-1} = \mathcal{F}_0$) and, for every $t = 1, 2, \ldots, T$,

$$A_t - A_{t-1} = U_{t-1}^* - \mathbb{E}_{\mathbb{P}^*}(U_t^* \mid \mathcal{F}_{t-1}).$$

Let us set $Z = M_T B_T$ and let us consider a replicating strategy $\phi^* \in \Phi$ for Z. From Lemma 2.6.2, the relative wealth process $V^*(\phi^*)$ is a martingale. Hence

$$V_t^*(\phi^*) = \mathbb{E}_{\mathbb{P}^*}(V_T^*(\phi^*) \mid \mathcal{F}_t) = \mathbb{E}_{\mathbb{P}^*}(V_T(\phi^*) B_T^{-1} \mid \mathcal{F}_t) = \mathbb{E}_{\mathbb{P}^*}(M_T \mid \mathcal{F}_t) = M_t.$$

Consequently, for every $t = 0, 1, \ldots, T$,

$$V_t^*(\phi^*) = M_t = U_t^* + A_t^* \geq U_t^* \geq X_t^*.$$

We conclude that the wealth process of ϕ^* satisfies, for every $t = 0, 1, \ldots, T$,

$$V_t(\phi^*) \geq U_t \geq X_t$$

so that it super-hedges an American claim X^a. Also, $V_0^*(\phi^*) = M_0 = U_0^* + A_0 = U_0^*$ and thus $V_0(\phi^*) = U_0$. This shows that $U_0 \geq \pi_0^s(X^a)$.

To establish the desired inequality $U_t \geq \pi_t^s(X^a)$ for $t = 1, 2, \ldots, T$, it is enough to consider the Doob-Meyer decomposition of the supermartingale U^* restricted to the time set $\{t, t+1, \ldots, T\}$ and to introduce the corresponding trading strategy ϕ^* (note that the strategy ϕ^* will depend here on the initial date t).

Finally, since U^* is the smallest supermartingale dominating X^* and the relative wealth process of any super-hedging strategy ϕ for X^a is a martingale dominating X^*, we see that the inequality $V^*(\phi) \geq U^*$ for any super-hedging strategy ϕ for X^a. Since U^* is given by the recurrence relationship (2.84), this argument can be applied to any start date t of a super-hedging strategy for X^a. We thus conclude that $\pi_t^s(X^a) \geq U_t$ for every $t = 0, 1, \ldots, T$.   □

The next goal is to examine an American claim from the perspective of the buyer. We shall argue that, for any fixed $t = 0, 1, \ldots, T$, any stopping time that solves the optimization problem

$$\max_{\tau \in \mathcal{T}_{[t,T]}} B_t \, \mathbb{E}_{\mathbb{P}^*}\left(B_\tau^{-1} X_\tau \mid \mathcal{F}_t\right)$$

is the *rational exercise time* for the buyer of an American claim, in the sense that if the buyer purchases the claim at time t for $\pi_0^s(X^a)$ and fails to exercise it at some stopping time from the class $\mathcal{T}_{[t,T]}^*$ of optimal stopping times, then an arbitrage opportunity for the seller of the claim will arise.

Let the stopping time τ_t^* be given by

$$\tau_t^* = \min\left\{u \in \{t, t+1, \ldots, T\} \mid U_u = X_u\right\}$$

and let the stopping time $\hat{\tau}_t$ satisfy

$$\hat{\tau}_t = \min\left\{u \in \{t, t+1, \ldots, T-1\} \mid A_{u+1} - A_u > 0\right\}$$

where, by convention, $\min \emptyset = T$.

We know already that τ_t^* and $\hat{\tau}_t$ are optimal stopping times for the problem: for a fixed $t = 0, 1, \ldots, n$, maximize the conditional expectation

$$B_t \,\mathbb{E}_{\mathbb{P}^*}\left(B_\tau^{-1} X_\tau \mid \mathcal{F}_t\right).$$

Moreover, a stopping time τ such that $\tau_t^* \leq \tau \leq \hat{\tau}_t$ is optimal if and only if $U_\tau = X_\tau$. Another characterization of an optimal stopping time reads: a stopping time τ is optimal if and only if the process U^* stopped at τ is a martingale and $U_\tau = X_\tau$.

Hereafter, we will refer to any optimal stopping time as a *rational exercise time* for the buyer of an American claim and we will focus on the case $t = 0$.

Proposition 2.8.4 *Assume that an American claim X^a was sold at time 0 at the initial price $\pi_0^s(X^a)$. Then there exists an arbitrage opportunity for the seller of the claim if and only if the holder does not exercise the claim at some rational exercise time.*

Proof. Suppose first that the claim is exercised by its holder at some rational exercise time τ. Then the wealth process $V(\phi^*) = U$ of the optimal superhedging strategy ϕ^* for the seller matches exactly the payoff process X at time τ.

If, on the contrary, the claim is exercised some stopping time τ, which is not a rational exercise time then either of the following three cases is valid:
(a) $\tau_0^* \leq \tau \leq \hat{\tau}_0$ and $\mathbb{P}\{U_\tau > X_\tau\} > 0$ (equivalently, $\mathbb{P}^*\{U_\tau > X_\tau\} > 0$),
(b) $\mathbb{P}\{\tau < \tau_0^*\} > 0$ and, by the definition of τ_0^*, we have $U_\tau > X_\tau$ on the event $\{\tau < \tau_0^*\}$,
(c) $\mathbb{P}\{\tau > \hat{\tau}_0\} > 0$ and, by the definition of $\hat{\tau}_0$, we have, on the event $\{\tau > \hat{\tau}_0\}$,

$$V_\tau^*(\phi^*) = M_\tau > U_\tau^* \geq X_\tau^*,$$

so that $U_\tau > X_\tau$ on this event.

Recall that $Y \geq X$. Hence in either case the strategy ϕ^* is an arbitrage opportunity for the seller.

$\square$

Let us finally observe that no arbitrage opportunity exists for the buyer who purchases at time 0 an American claim at the seller's price $\pi_0^s(X^a)$. This is a consequence of the following lemma.

Lemma 2.8.6 *Suppose that a trading strategy $\phi \in \Phi$ is such that $V_0(\phi) = -\pi_0^s(X^a)$ and for some stopping time $\tau \in \mathcal{T}_{[0,T]}$ we have $V_\tau(\phi) + X_\tau \geq 0$. Then $V_\tau(\phi) + X_\tau = 0$.*

Proof. Assume, on the contrary, that the inequality $\mathbb{P}\{V_\tau(\phi) + X_\tau > 0\} > 0$ is valid or, equivalently, $\mathbb{P}^*\{V_\tau^*(\phi) + X_\tau^* > 0\} > 0$. Then the martingale property of the relative wealth $V^*(\phi)$ under $\mathbb{P}^*$ yields

$$\mathbb{E}_{\mathbb{P}*}(V_\tau^*(\phi) + X_\tau^*) = -\pi_0^s(X^a)B_0^{-1} + \mathbb{E}_{\mathbb{P}*}(X_\tau^*) > 0.$$

Consequently,

$$B_0 \mathbb{E}_{\mathbb{P}*}(B_\tau^{-1}X_\tau) > \pi_0^s(X^a) = U_0.$$

This clearly contradicts (2.85).   □

However, if either the seller is able to sell an American claim at the price higher than $\pi_0^s(X^a)$, or the buyer is able to buy at the price lower than $\pi_0^s(X^a)$, then an arbitrage opportunity arises. Hence the following result, which establishes also Proposition 2.8.1.

Corollary 2.8.3 *The unique arbitrage price of an American claim X^a is given by the following expression, for every $t = 0, 1, \ldots, T$,*

$$\pi_t(X^a) = \max_{\tau \in \mathcal{T}_{[t,T]}} B_t \, \mathbb{E}_{\mathbb{P}*}\left(B_\tau^{-1}X_\tau \mid \mathcal{F}_t\right) = B_t \, \mathbb{E}_{\mathbb{P}*}\left(B_{\tilde{\tau}_t}^{-1}X_{\tilde{\tau}_t} \mid \mathcal{F}_t\right),$$

where $\tilde{\tau}_t$ is any stopping time from $\mathcal{T}_{[t,T]}^*$. Equivalently, the relative price $\pi^*(X^a) = B^{-1}\pi(X^a)$ satisfies $\pi_T^*(X^a) = X_T^*$ and, for every $t = 1, 2, \ldots, T$,

$$\pi_{t-1}^*(X^a) = \max\left\{X_{t-1}^*, \, \mathbb{E}_{\mathbb{P}*}(\pi_t^*(X^a) \mid \mathcal{F}_{t-1})\right\}. \tag{2.86}$$

The last result can also be supported by introducing the concept of super-hedging strategy for the buyer (this proves useful when dealing with American claims in an incomplete market model).

Definition 2.8.9 We say that a self-financing trading strategy $\phi \in \Phi$ is a *super-hedging strategy for the buyer* of an American claim X^a whenever there exists a stopping time $\tau \in \mathcal{T}_{[0,T]}$ such that the inequality $V_\tau(\phi) \geq -X_\tau$ holds.

The definition above can be easily extended to any date $t = 0, 1, \ldots, T$.

Definition 2.8.10 For a fixed $t = 0, 1, \ldots, T$, let $V_t(\phi^*)$ be the minimal cost of a super-hedging strategy for the buyer from time t onwards. The *buyer's price* at time t of an American claim X^a is defined as $\pi_t^b(X^a) = -V_t(\phi^*)$.

The proof of the following result is left to the reader.

Corollary 2.8.4 *For an American claim X^a the equality $\pi_t^s(X^a) = \pi_t^b(X^a)$ holds for every $t = 0, 1, \ldots, T$. Hence $\pi(X^a) = \pi^s(X^a) = \pi^b(X^a)$.*

It is worth recalling that we work in this section under the standing assumption that the underlying finite market model is arbitrage-free and complete. In the case of a market incompleteness, the seller's price is typically strictly greater than the buyer's price and thus the bid-ask spread arises.

2.8.3 American Call and Put

Let us fix some $l < k$, and let us consider an *American call option* with the payoff $X_t = (S_t^l - K)^+$ for every $t = 0, 1, \ldots, T$. Assume that the process $B = S^k$ is increasing, so that $B_{t+1} \geq B_t$ for $t = 0, 1, \ldots, T-1$. When B represents the savings account, this corresponds to the assumption that interest rates are non-negative.

Using the conditional Jensen's inequality (see Lemma A.1.3), one can show that the discounted payoff

$$X_t^* = B_t^{-1} X_t = B_t^{-1}(S_t^l - K)^+ = (S_t^{l*} - B_t^{-1} K)^+$$

is a submartingale under $\mathbb{P}^*$ and thus $\hat{\tau}_0 = T$ is a rational exercise time (not unique, in general) for an American call. Hence an American call option is equivalent to a European call with expiry date T, in the sense that $\pi_t(X^a) = C_t$, where C is the price process of a European call with terminal payoff $C_T = (S_T^l - K)^+$.

This property is not shared by an *American put option* with the payoff $X_t = (K - S_t^l)^+$ for every $t = 0, 1, \ldots, T$, unless the process $B = S^k$ is decreasing, that is, unless the inequality $B_{t+1} \leq B_t$ holds for every $t = 0, 1, \ldots, T-1$. In that case, the discounted payoff

$$X_t^* = B_t^{-1} X_t = B_t^{-1}(K - S_t^l)^+ = (B_t^{-1} K - S_t^{l*})^+$$

is a submartingale under $\mathbb{P}^*$ and thus American and European put options are equivalent, in the sense described above.

Note that if the process B is intended to model the savings account, then the assumption that B is decreasing is equivalent to the property that (possibly random) interest rates are non-positive.

2.9 Game Contingent Claims

The concept of a game contingent claim was introduced by Kiefer (2000), as a natural extension of the notion of an American contingent claim to a situation where both parties to a contract may exercise it prior to its maturity date. The payoff of a game contingent claim depends not only on the moment when it is exercised, but also on which party takes the decision to exercise. In order to make a clear distinction between the exercise policies of the two parties, we will refer to the decision of the seller (also termed the issuer) as the *cancelation policy*, whereas the decision of the buyer (also referred to as the holder) is called the *exercise policy*.

Definition 2.9.1 A *game contingent claim* $X^g = (L, H, \mathcal{T}^e, \mathcal{T}^c)$ expiring at time T consists of $\mathbb{F}$-adapted payoff processes L and H, a set $\mathcal{T}^e$ admissible exercise times, and a set $\mathcal{T}^c$ of admissible cancelation times.

Remarks. A game contingent claim is also known in the literature as a *game option* or an *Israeli option*. In practical applications, the terminology for exercise times is frequently modified to reflect the feature of a particular class of derivatives that fall within the general scope of game contingent claims. For instance, in the case of a *convertible bond*, the cancelation and exercise times are referred to as the *call time* and the *put/conversion time*, respectively (see Example 2.9.1).

It is assumed throughout that $L \leq H$, meaning that the $\mathcal{F}_t$-measurable random variables L_t and H_t satisfy the inequality $L_t \leq H_t$ for every $t = 0, 1, \ldots, T$.

Unless explicitly otherwise stated, the sets of admissible exercise and cancelation times are restricted to the class $\mathcal{T}_{[0,T]}$ of all stopping times of the filtration $\mathbb{F}^S$ with values in $\{0, 1, \ldots, T\}$ – that is, it is postulated that $\mathcal{T}^e = \mathcal{T}^c = \mathcal{T}_{[0,T]}$.

We interpret L_t as the payoff received by the holder upon exercising a game contingent claim at time t. The random variable H_t represents the payoff received by the holder if a claim is canceled (i.e., exercised by its issuer) at time t. More formally, if a game contingent claim is exercised by either of the two parties at some date $t \leq T$, that is, on the event $\{\tau = t\} \cup \{\sigma = t\}$, the random payoff equals

$$X_t = \mathbb{1}_{\{\tau = t \leq \sigma\}} L_t + \mathbb{1}_{\{\sigma = t < \tau\}} H_t, \tag{2.87}$$

where τ and σ are the exercise and cancelation times, respectively. Note that both the holder and the issuer may choose freely their exercise and cancelation times. If they decide to exercise their right at the same moment, we adopt the convention that a game contingent claim is exercised, rather than canceled, so that the payoff is given by the process L. This feature is already reflected in the payoff formula (2.87).

Remarks. A comparison of Definitions 2.8.3 and 2.9.1 shows that if the class of all admissible cancelation times is assumed to be $\mathcal{T}^c = \{T\}$ then a game contingent claim X^g becomes an American claim X^a with the payoff process $X = L$. Therefore, Definition 2.9.1 of a game contingent claim covers as a special case the notion of an American contingent claim. If we postulate, in addition, that $\mathcal{T}^e = \{T\}$ then a game contingent claim reduces to a European claim L_T maturing at time T.

From the previous section, we know that valuation and hedging of American claims is related to optimal stopping problems. Game contingent claims are associated with the so-called *Dynkin games*, which in turn can be seen as natural extensions of optimal stopping problems. Therefore, before analyzing in some detail the game contingent claims, we first present a brief survey of basic results concerning Dynkin games.

2.9.1 Dynkin Games

In this section, we assume that $t = 0, 1, \ldots, n$ and we investigate the *Dynkin game* (also known as the *optimal stopping game*) associated with the payoff

$$Z(\sigma, \tau) = \mathbb{1}_{\{\tau \leq \sigma\}} L_\tau + \mathbb{1}_{\{\sigma < \tau\}} H_\sigma,$$

where $L \leq H$ are $\mathbb{F}$-adapted stochastic processes defined on a finite probability space $(\Omega, \mathcal{F}, \mathbb{P})$ endowed with a filtration $\mathbb{F}$. Stochastic games, as described by the foregoing definition, were first studied in the classic papers by Dynkin (1969) and Kiefer (1971).

Definition 2.9.2 For any fixed date $t = 0, 1, \ldots, n$, by the *Dynkin game* started at time t and associated with the payoff $Z(\sigma, \tau)$, we mean a stochastic game in which the *min-player*, who controls a stopping time $\sigma \in T_{[t,n]}$, wishes to minimize the conditional expectation

$$\mathbb{E}_{\mathbb{P}}(Z(\sigma, \tau) \mid \mathcal{F}_t), \tag{2.88}$$

while the *max-player*, who controls a stopping time $\tau \in T_{[t,n]}$, wishes to maximize the conditional expectation (2.88).

Let us fix some date t. Since the stopping times σ and τ are assumed to belong to the class $T_{[t,n]}$, formula (2.87) yields

$$\mathbb{E}_{\mathbb{P}}(Z(\sigma, \tau) \mid \mathcal{F}_t) = \mathbb{E}_{\mathbb{P}}\Big(\sum_{u=t}^{n} \big(\mathbb{1}_{\{\tau = u \leq \sigma\}} L_u + \mathbb{1}_{\{\sigma = u < \tau\}} H_u \big) \Big| \mathcal{F}_t \Big). \tag{2.89}$$

We are interested in finding the so-called *value process* of a Dynkin game and the corresponding *optimal stopping times*. We start by formulating the definition of the *upper* and *lower* value processes.

Definition 2.9.3 The $\mathbb{F}$-adapted process $\bar{Y}^u$ given by the formula

$$\bar{Y}_t^u = \min_{\sigma \in T_{[t,n]}} \max_{\tau \in T_{[t,n]}} \mathbb{E}_{\mathbb{P}}(Z(\sigma, \tau) \mid \mathcal{F}_t)$$

is called the *upper value process*. The *lower value process* $\bar{Y}^l$ is an $\mathbb{F}$-adapted process given by the formula

$$\bar{Y}_t^l = \max_{\tau \in T_{[t,n]}} \min_{\sigma \in T_{[t,n]}} \mathbb{E}_{\mathbb{P}}(Z(\sigma, \tau) \mid \mathcal{F}_t).$$

Lemma 2.9.1 *Let $L^0(\Omega, \mathcal{F}, \mathbb{P})$ stand for the class of all random variables defined on a probability space $(\Omega, \mathcal{F}, \mathbb{P})$. Let A and B be two finite sets and let $g : A \times B \to L^0(\Omega, \mathcal{F}, \mathbb{P})$ be an arbitrary map. Then*

$$\min_{a \in A} \max_{b \in B} g(a, b) \geq \max_{b \in B} \min_{a \in A} g(a, b).$$

Proof. It is clear that, for any $a_0 \in A$ and $b_0 \in B$,

$$G(a_0) := \max_{b \in B} g(a_0, b) \geq g(a_0, b_0) \geq \min_{a \in A} g(a, b_0) =: H(b_0),$$

and thus $\min_{a_0 \in A} G(a_0) \geq H(b_0)$ for every $b_0 \in B$. Consequently,

$$\min_{a \in A} \max_{b \in B} g(a, b) = \min_{a_0 \in A} G(a_0) \geq \max_{b_0 \in B} H(b_0) = \max_{b \in B} \min_{a \in A} g(a, b),$$

and thus we have established the required result. $\square$

It follows immediately from Definition 2.9.3 and Lemma 2.9.1 that the upper value process $\bar{Y}^u$ always dominates the lower value process $\bar{Y}^l$, in the sense that the inequality $\bar{Y}_t^u \geq \bar{Y}_t^l$ is valid for every $t = 0, 1, \ldots, n$.

Definition 2.9.4 If the equality $\bar{Y}^u = \bar{Y}^l$ is satisfied, we say that the *Stackelberg equilibrium* holds for a Dynkin game. Then the process $\bar{Y} = \bar{Y}^u = \bar{Y}^l$ is called the *value process*.

The next definition is based on the concept of a *saddle point* of a Dynkin game.

Definition 2.9.5 We say that the *Nash equilibrium* holds for a Dynkin game if for any t there exist stopping times $\sigma_t^*, \tau_t^* \in \mathcal{T}_{[t,n]}$, such that the inequalities

$$\mathbb{E}_{\mathbb{P}}\big(Z(\sigma_t^*, \tau) \,|\, \mathcal{F}_t\big) \leq \mathbb{E}_{\mathbb{P}}(Z(\sigma_t^*, \tau_t^*) \,|\, \mathcal{F}_t) \leq \mathbb{E}_{\mathbb{P}}\big(Z(\sigma, \tau_t^*) \,|\, \mathcal{F}_t\big) \tag{2.90}$$

are satisfied for arbitrary stopping times $\tau, \sigma \in \mathcal{T}_{[t,n]}$, that is, the pair (σ_t^*, τ_t^*) is a *saddle point* of a Dynkin game.

The next result shows that the Nash equilibrium for a Dynkin game implies the Stackelberg equilibrium.

Lemma 2.9.2 *Assume that the Nash equilibrium holds. Then the Stackelberg equilibrium holds and*

$$\bar{Y}_t = \mathbb{E}_{\mathbb{P}}(Z(\sigma_t^*, \tau_t^*) \,|\, \mathcal{F}_t),$$

so that σ_t^* and τ_t^* are optimal stopping times as of time t.

Proof. From (2.90), we obtain

$$\max_{\tau \in \mathcal{T}_{[t,n]}} \mathbb{E}_{\mathbb{P}}\big(Z(\sigma_t^*, \tau) \,|\, \mathcal{F}_t\big) \leq \mathbb{E}_{\mathbb{P}}(Z(\sigma_t^*, \tau_t^*) \,|\, \mathcal{F}_t) \leq \min_{\sigma \in \mathcal{T}_{[t,n]}} \mathbb{E}_{\mathbb{P}}\big(Z(\sigma, \tau_t^*) \,|\, \mathcal{F}_t\big).$$

Consequently,

$$\begin{aligned}
\bar{Y}_t^u &= \min_{\sigma \in \mathcal{T}_{[t,n]}} \max_{\tau \in \mathcal{T}_{[t,n]}} \mathbb{E}_{\mathbb{P}}(Z(\sigma, \tau) \,|\, \mathcal{F}_t) \leq \mathbb{E}_{\mathbb{P}}(Z(\sigma_t^*, \tau_t^*) \,|\, \mathcal{F}_t) \\
&\leq \max_{\tau \in \mathcal{T}_{[t,n]}} \min_{\sigma \in \mathcal{T}_{[t,n]}} \mathbb{E}_{\mathbb{P}}(Z(\sigma, \tau) \,|\, \mathcal{F}_t) = \bar{Y}_t^l.
\end{aligned}$$

Since the inequality $\bar{Y}_t^u \geq \bar{Y}_t^l$ is known to be always satisfies, we conclude that the value process $\bar{Y}$ is well defined and satisfies $\bar{Y}_t = \mathbb{E}_{\mathbb{P}}(Z(\sigma_t^*, \tau_t^*) \,|\, \mathcal{F}_t)$ for any $t = 0, 1, \ldots, n$. □

The following definition introduces a plausible candidate for the value process of a Dynkin game.

Definition 2.9.6 The process Y is defined by setting $Y_n = L_n$ and, for any $t = 0, 1, \ldots, n-1$,

$$Y_t = \min\Big\{H_t, \, \max\big\{L_t, \mathbb{E}_{\mathbb{P}}(Y_{t+1} \,|\, \mathcal{F}_t)\big\}\Big\}. \tag{2.91}$$

Remarks. The assumption that $L \leq H$ ensures that, for any $t = 0, 1, \ldots, n-1$,

$$Y_t = \min \Big\{ H_t, \, \max \big\{ L_t, \mathbb{E}_{\mathbb{P}}(Y_{t+1} \mid \mathcal{F}_t) \big\} \Big\}$$
$$= \max \Big\{ L_t, \, \min \big\{ H_t, \mathbb{E}_{\mathbb{P}}(Y_{t+1} \mid \mathcal{F}_t) \big\} \Big\}.$$

It is also clear from (2.91) that $L_t \leq Y_t \leq H_t$ for $t = 0, 1, \ldots, n$. In particular, $Y_t = L_t = H_t$ if the equality $L_t = H_t$ holds.

Lemma 2.9.3 *Let us set, for any fixed $t = 0, 1, \ldots, n$,*

$$\sigma_t^* = \min \big\{ u \in \{t, t+1, \ldots, n\} \mid Y_u = H_u \big\} \wedge n \qquad (2.92)$$

and

$$\tau_t^* = \min \big\{ u \in \{t, t+1, \ldots, n\} \mid Y_u = L_u \big\}. \qquad (2.93)$$

Then the following representations are valid

$$\sigma_{t-1}^* = (t-1)\mathbb{1}_{\{\sigma_{t-1}^* = t-1\}} + \sigma_t^* \mathbb{1}_{\{\sigma_{t-1}^* \geq t\}}$$

and

$$\tau_{t-1}^* = (t-1)\mathbb{1}_{\{\tau_{t-1}^* = t-1\}} + \tau_t^* \mathbb{1}_{\{\tau_{t-1}^* \geq t\}}.$$

Proof. We have

$$\sigma_{t-1}^* = \min \big\{ u \in \{t-1, t, \ldots, n\} \mid Y_u = H_u \big\} \wedge n$$
$$= (t-1)\mathbb{1}_{\{Y_{t-1}=H_{t-1}\}} + (\min \{u \geq t \mid Y_u = H_u\} \wedge n)\mathbb{1}_{\{Y_{t-1}<H_{t-1}\}}$$
$$= (t-1)\mathbb{1}_{\{\sigma_{t-1}^* = t-1\}} + \sigma_t^* \mathbb{1}_{\{\sigma_{t-1}^* \geq t\}}.$$

Similarly,

$$\tau_{t-1}^* = \min \big\{ u \in \{t-1, t, \ldots, n\} \mid Y_u = L_u \big\}$$
$$= (t-1)\mathbb{1}_{\{Y_{t-1}=L_{t-1}\}} + \min \{u \geq t \mid Y_u = L_u\}\mathbb{1}_{\{Y_{t-1}>L_{t-1}\}}$$
$$= (t-1)\mathbb{1}_{\{\tau_{t-1}^* = t-1\}} + \tau_t^* \mathbb{1}_{\{\tau_{t-1}^* \geq t\}}.$$

This ends the proof. □

The next result examines some auxiliary properties of the process Y.

Lemma 2.9.4 *For every $t = 0, 1, \ldots, n-1$, we have that $Y_t \leq \mathbb{E}_{\mathbb{P}}(Y_{t+1} \mid \mathcal{F}_t)$ on the event $\{\tau_t^* > t\}$ and $Y_t \geq \mathbb{E}_{\mathbb{P}}(Y_{t+1} \mid \mathcal{F}_t)$ on the event $\{\sigma_t^* > t\}$. Therefore, the equality $Y_t = \mathbb{E}_{\mathbb{P}}(Y_{t+1} \mid \mathcal{F}_t)$ holds on the event $\{\sigma_t^* \wedge \tau_t^* > t\} = \{\sigma_t^* > t\} \cap \{\sigma_t^* > t\}$.*

Proof. Let us fix t and let us write $a = L_t$, $b = H_t$ and $c = \mathbb{E}_{\mathbb{P}}(Y_{t+1} \mid \mathcal{F}_t)$. We wish to examine the relationship between the value of c and $h(c) = \min(b, \max(a, c))$ for arbitrary a, b such that $a \leq b$. It is clear that $Y_t > L_t$ on the event $\{\tau_t^* > t\}$. This corresponds to the inequality $h(c) > a$. But this in turn implies that $h(c) \leq c$, which means that the inequality $Y_t \leq \mathbb{E}_{\mathbb{P}}(Y_{t+1} \mid \mathcal{F}_t)$ holds. Similarly, the inequality $Y_t < H_t$ is easily seen to hold on the event $\{\sigma_t^* > t\}$. This corresponds to $h(c) < b$, which entails that $h(c) \geq c$. We have thus established that on the event $\{\sigma_t^* > t\}$ we have $Y_t \geq \mathbb{E}_{\mathbb{P}}(Y_{t+1} \mid \mathcal{F}_t)$. □

By arguing as in the proof of Lemma 2.8.3, one can also establish the following result.

Lemma 2.9.5 *The stopped process* $Y^{\sigma_0^* \wedge \tau_0^*}$ *is an* $\mathbb{F}$-*martingale.*

We are in a position to show that the process Y is equal to the value process $\bar{Y}$ of a Dynkin game.

Theorem 2.9.1 (i) *Let the stopping times* σ_t^*, τ_t^* *be given by* (2.92)–(2.93). *Then we have, for arbitrary stopping times* $\tau, \sigma \in \mathcal{T}_{[t,n]}$,

$$\mathbb{E}_{\mathbb{P}}\big(Z(\sigma_t^*, \tau) \mid \mathcal{F}_t\big) \leq Y_t \leq \mathbb{E}_{\mathbb{P}}\big(Z(\sigma, \tau_t^*) \mid \mathcal{F}_t\big) \tag{2.94}$$

and thus also

$$\mathbb{E}_{\mathbb{P}}\big(Z(\sigma_t^*, \tau) \mid \mathcal{F}_t\big) \leq \mathbb{E}_{\mathbb{P}}\big(Z(\sigma_t^*, \tau_t^*) \mid \mathcal{F}_t\big) \leq \mathbb{E}_{\mathbb{P}}\big(Z(\sigma, \tau_t^*) \mid \mathcal{F}_t\big) \tag{2.95}$$

so that the Nash equilibrium holds.
(ii) *The process* Y *is the value process of a Dynkin game, that is, for every* $t = 0, 1, \ldots, n$,

$$Y_t = \min_{\sigma \in \mathcal{T}_{[t,n]}} \max_{\tau \in \mathcal{T}_{[t,n]}} \mathbb{E}_{\mathbb{P}}\big(Z(\sigma, \tau) \mid \mathcal{F}_t\big) = \mathbb{E}_{\mathbb{P}}\big(Z(\sigma_t^*, \tau_t^*) \mid \mathcal{F}_t\big) \tag{2.96}$$

$$= \max_{\tau \in \mathcal{T}_{[t,n]}} \min_{\sigma \in \mathcal{T}_{[t,n]}} \mathbb{E}_{\mathbb{P}}\big(Z(\sigma, \tau) \mid \mathcal{F}_t\big) = \bar{Y}_t,$$

and thus the stopping times σ_t^* *and* τ_t^* *are optimal as of time* t.

Proof. We will first show that the inequality holds for $t = n$. Of course, if $\tau, \sigma \in \mathcal{T}_{[t,n]}$ then $\tau = \sigma = n$. As

$$\sigma_n^* = \min\{u \geq n \mid Y_n = H_n\} \wedge n = n = \min\{u \geq n \mid Y_n = L_n\} = \tau_n^*,$$

it follows from Definition 2.9.6 that

$$\mathbb{E}_{\mathbb{P}}\big(Z(\sigma_n^*, n) \mid \mathcal{F}_n\big) = \mathbb{E}_{\mathbb{P}}\big(Z(n, \tau_n^*) \mid \mathcal{F}_n\big) = \mathbb{E}_{\mathbb{P}}\big(Z(n, n) \mid \mathcal{F}_n\big)$$
$$= \mathbb{E}_{\mathbb{P}}(L_n \mid \mathcal{F}_n) = L_n = Y_n.$$

Evidently, for $t = n$, the upper and lower bounds hold as equalities in (2.94).
We now assume that (2.94) holds for some t, so that

$$\mathbb{E}_{\mathbb{P}}\big(Z(\sigma_t^*, \tilde{\tau}) \mid \mathcal{F}_t\big) \leq Y_t \leq \mathbb{E}_{\mathbb{P}}\big(Z(\tilde{\sigma}, \tau_t^*) \mid \mathcal{F}_t\big) \tag{2.97}$$

for arbitrary stopping times $\tilde{\tau}, \tilde{\sigma} \in \mathcal{T}_{[t,n]}$. We wish to prove that (2.94) holds for $t - 1$, that is, for arbitrary stopping times $\tau, \sigma \in \mathcal{T}_{[t-1,n]}$,

$$\mathbb{E}_{\mathbb{P}}\big(Z(\sigma_{t-1}^*, \tau) \mid \mathcal{F}_{t-1}\big) \leq Y_{t-1} \leq \mathbb{E}_{\mathbb{P}}\big(Z(\sigma, \tau_{t-1}^*) \mid \mathcal{F}_{t-1}\big). \tag{2.98}$$

Let us first establish the upper bound in (2.98). For any $\sigma \in \mathcal{T}_{[t-1,n]}$, let us define $\tilde{\sigma} = \sigma \vee t = \max(\sigma, t)$. It is clear that the stopping time $\tilde{\sigma}$ belongs to $\mathcal{T}_{[t,n]}$. We thus have that

$$\mathbb{E}_{\mathbb{P}}(Z(\sigma, \tau_{t-1}^*) \mid \mathcal{F}_{t-1}) = \mathbb{E}_{\mathbb{P}}(\mathbb{1}_{\{\sigma=t-1\}} Z(t-1, \tau_{t-1}^*) \mid \mathcal{F}_{t-1})$$

$$+ \, \mathbb{E}_{\mathbb{P}}(\mathbb{1}_{\{\sigma \geq t\}} \, Z(\tilde{\sigma}, \tau_{t-1}^*) \mid \mathcal{F}_{t-1})$$

$$= \mathbb{1}_{\{\sigma=t-1\}} \, \mathbb{E}_{\mathbb{P}}(Z(t-1, \tau_{t-1}^*) \mid \mathcal{F}_{t-1}) + \mathbb{1}_{\{\sigma \geq t\}} \, \mathbb{E}_{\mathbb{P}}(Z(\tilde{\sigma}, \tau_{t-1}^*) \mid \mathcal{F}_{t-1})$$

$$= \mathbb{1}_{\{\sigma=t-1\}} \, \mathbb{E}_{\mathbb{P}}(\mathbb{1}_{\{\tau_{t-1}^*=t-1\}} \, L_{t-1} + \mathbb{1}_{\{\tau_{t-1}^* \geq t\}} \, H_{t-1} \mid \mathcal{F}_{t-1})$$

$$+ \, \mathbb{1}_{\{\sigma \geq t\}} \, \mathbb{E}_{\mathbb{P}}(\mathbb{1}_{\{\tau_{t-1}^*=t-1\}} \, Z(\tilde{\sigma}, t-1) + \mathbb{1}_{\{\tau_{t-1}^* \geq t\}} \, Z(\tilde{\sigma}, \tau_t^*) \mid \mathcal{F}_{t-1})$$

$$= \mathbb{1}_{\{\sigma=t-1=\tau_{t-1}^*\}} \, L_{t-1} + \mathbb{1}_{\{\sigma=t-1<\tau_{t-1}^*\}} \, H_{t-1}$$

$$+ \, \mathbb{1}_{\{\tau_{t-1}^*=t-1<\sigma\}} \, L_{t-1} + \mathbb{1}_{\{\tau_{t-1}^* \geq t, \, \sigma \geq t\}} \, \mathbb{E}_{\mathbb{P}}(Z(\tilde{\sigma}, \tau_t^*) \mid \mathcal{F}_{t-1})$$

$$= \mathbb{1}_{\{\sigma=t-1=\tau_{t-1}^*\}} \, L_{t-1} + \mathbb{1}_{\{\sigma=t-1<\tau_{t-1}^*\}} \, H_{t-1} + \mathbb{1}_{\{\tau_{t-1}^*=t-1<\sigma\}} \, L_{t-1}$$

$$+ \, \mathbb{1}_{\{\tau_{t-1}^* \geq t, \, \sigma \geq t\}} \, \mathbb{E}_{\mathbb{P}}(\mathbb{E}_{\mathbb{P}}(Z(\tilde{\sigma}, \tau_t^*) \mid \mathcal{F}_t) \mid \mathcal{F}_{t-1})$$

$$\geq \mathbb{1}_{\{\tau_{t-1}^*=t-1\}} \, L_t + \mathbb{1}_{\{\sigma=t-1<\tau_{t-1}^*\}} \, H_{t-1} + \mathbb{1}_{\{\tau_{t-1}^* \geq t, \, \sigma \geq t\}} \, \mathbb{E}_{\mathbb{P}}(Y_t \mid \mathcal{F}_{t-1})$$

$$\geq Y_{t-1}$$

where we have used the right-hand side of inequality (2.97) to establish the penultimate inequality. Finally, the last inequality follows from Lemma 2.9.4, the equality

$$\mathbb{1}_{\{\tau_{t-1}^*=t-1\}} L_{t-1} = \mathbb{1}_{\{\tau_{t-1}^*=t-1\}} Y_{t-1},$$

and the inequality $H_{t-1} \geq Y_{t-1}$. We have thus established the upper bound of (2.98) and now set about establishing the lower bound in (2.98).

For any $\tau \in \mathcal{T}_{[t-1,n]}$, we set $\tilde{\tau} = \tau \vee t = \max(\tau, t)$ so that $\tilde{\tau}$ is in $\mathcal{T}_{[t,n]}$. Then

$$\mathbb{E}_{\mathbb{P}}(Z(\sigma_{t-1}^*, \tau) \mid \mathcal{F}_{t-1}) = \mathbb{E}_{\mathbb{P}}(\mathbb{1}_{\{\tau=t-1\}} Z(\sigma_{t-1}^*, t-1) \mid \mathcal{F}_{t-1})$$

$$+ \, \mathbb{E}_{\mathbb{P}}(\mathbb{1}_{\{\tau \geq t\}} \, Z(\sigma_{t-1}^*, \tilde{\tau}) \mid \mathcal{F}_{t-1})$$

$$= \mathbb{1}_{\{\tau=t-1\}} \, \mathbb{E}_{\mathbb{P}}(Z(\sigma_{t-1}^*, t-1) \mid \mathcal{F}_{t-1}) + \mathbb{1}_{\{\tau \geq t\}} \, \mathbb{E}_{\mathbb{P}}(Z(\sigma_{t-1}^*, \tilde{\tau}) \mid \mathcal{F}_{t-1})$$

$$= \mathbb{1}_{\{\tau=t-1\}} \, \mathbb{E}_{\mathbb{P}}(\mathbb{1}_{\{\sigma_{t-1}^*=t-1\}} \, H_{t-1} + \mathbb{1}_{\{\sigma_{t-1}^*>t\}} \, L_{t-1} \mid \mathcal{F}_{t-1})$$

$$+ \, \mathbb{1}_{\{\tau \geq t\}} \, \mathbb{E}_{\mathbb{P}}(\mathbb{1}_{\{\sigma_{t-1}^*=t-1\}} Z(t-1, \tilde{\tau}) + \mathbb{1}_{\{\sigma_{t-1}^* \geq t\}} \, Z(\sigma_t^*, \tilde{\tau}) \mid \mathcal{F}_{t-1})$$

$$= \mathbb{1}_{\{\tau=t-1=\sigma_{t-1}^*\}} \, H_{t-1} + \mathbb{1}_{\{\tau=t-1<\sigma_{t-1}^*\}} \, L_{t-1}$$

$$+ \, \mathbb{1}_{\{\sigma_{t-1}^*=t-1<\tau\}} \, H_{t-1} + \mathbb{1}_{\{\tau_{t-1}^* \geq t, \, \tau \geq t\}} \, \mathbb{E}_{\mathbb{P}}(Z(\sigma_t^*, \tilde{\tau}) \mid \mathcal{F}_{t-1})$$

$$= \mathbb{1}_{\{\tau=t-1=\sigma_{t-1}^*\}} \, H_{t-1} + \mathbb{1}_{\{\tau=t-1<\sigma_{t-1}^*\}} \, L_{t-1} + \mathbb{1}_{\{\sigma_{t-1}^*=t-1<\tau\}} \, H_{t-1}$$

$$+ \, \mathbb{1}_{\{\sigma^* \geq t, \, \tau \geq t\}} \, \mathbb{E}_{\mathbb{P}}(\mathbb{E}_{\mathbb{P}}(Z(\sigma_t^*, \tilde{\tau}) \mid \mathcal{F}_t) \mid \mathcal{F}_{t-1})$$

$$\leq \mathbb{1}_{\{\sigma_{t-1}^*=t-1\}} \, H_{t-1} + \mathbb{1}_{\{\tau=t-1<\sigma_{t-1}^*\}} \, L_{t-1} + \mathbb{1}_{\{\sigma_{t-1}^* \geq t, \, \tau \geq t\}} \, \mathbb{E}_{\mathbb{P}}(Y_t \mid \mathcal{F}_{t-1})$$

$$\leq \mathbb{1}_{\{\sigma_{t-1}^*=t-1\}} Y_{t-1} + \mathbb{1}_{\{\tau=t-1<\sigma_{t-1}^*\}} Y_{t-1} + \mathbb{1}_{\{\sigma_{t-1}^* \geq t, \, \tau \geq t\}} Y_{t-1}$$

$$= Y_{t-1}$$

where the penultimate inequality follows from the left-hand side inequality in (2.97) and by noting that

$$\{\tau = t-1 = \sigma_{t-1}^*\} \cup \{\sigma_{t-1}^* = t-1 < \tau\} = \{\sigma_{t-1}^* = t-1\}.$$

The last inequality follows from Lemma 2.9.4 and using the fact that

$$\mathbb{1}_{\{\sigma^*_{t-1}=t-1\}} H_{t-1} = \mathbb{1}_{\{\sigma^*_{t-1}=t-1\}} Y_{t-1}, \quad L_{t-1} \le Y_{t-1}.$$

This proves the lower bound in (2.98) and thus completes the proof of part (i). Part (ii) is an immediate consequence of part (i) and Lemma 2.9.2.   □

Let us assume that $L_0 = H_0$ and let us consider the Dynkin game started at time 0. As was already observed, the equality $L_0 = H_0$ implies that $Y_0 = L_0 = H_0$. Therefore, the stopping times $\sigma^*_0 = \tau^*_0 = 0$ are optimal and $\bar{Y}_0 = L_0 = H_0$. Note that we only assume here that $L_t \le H_t$ for $t = 1, 2, \dots, n$.

2.9.2 Valuation and Hedging of Game Contingent Claims

In this section, we make the standing assumption that the finite market model is complete or, equivalently, that there exists a unique martingale measure $\mathbb{P}^*$ for relative prices $S^* = (S^{*1}, S^{*2}, \dots, S^{*(k-1)})$. Recall that such a $\mathbb{P}^*$ is also a martingale measure for the discounted wealth process $U^*(\phi)$ of any self-financing trading strategy $\phi \in \Phi$. Our aim is to now show that the value process of the Dynkin game associated with a game contingent claim can be interpreted as its arbitrage price.

To this end, we first define the discounted payoffs $L^* = B^{-1}L$ and $H^* = B^{-1}H$. We will investigate the Dynkin game with the payoff given by the expression

$$Z^*(\sigma, \tau) = \mathbb{1}_{\{\tau \le \sigma\}} L^*_\tau + \mathbb{1}_{\{\sigma < \tau\}} H^*_\sigma.$$

In view of our financial interpretation, the max-player and the min-player will be referred to as the seller (issuer) and the buyer (holder), respectively.

Let us write $U = BU^*$, where U^* is the value process of the Dynkin game, that is,

$$U^*_t = \min_{\sigma \in \mathcal{T}_{[t,T]}} \max_{\tau \in \mathcal{T}_{[t,T]}} \mathbb{E}_{\mathbb{P}}(Z^*(\sigma, \tau) \mid \mathcal{F}_t) \tag{2.99}$$

$$= \max_{\tau \in \mathcal{T}_{[t,T]}} \min_{\sigma \in \mathcal{T}_{[t,T]}} \mathbb{E}_{\mathbb{P}}(Z^*(\sigma, \tau) \mid \mathcal{F}_t).$$

In view of Theorem 2.9.1, this also means that

$$U^*_t = \min \left\{ H^*_t, \max \left\{ L^*_t, \mathbb{E}_{\mathbb{P}}(U^*_{t+1} \mid \mathcal{F}_t) \right\} \right\} \tag{2.100}$$

or, equivalently,

$$U_t = \min \left\{ H_t, \max \left\{ L_t, B_t \mathbb{E}_{\mathbb{P}}(B^{-1}_{t+1} U_{t+1} \mid \mathcal{F}_t) \right\} \right\}. \tag{2.101}$$

By a slight abuse of language, we will refer to U (rather than U^*) as the value process of the Dynkin game associated with a game contingent claim $X^g = (L, H, \mathcal{T}^e, \mathcal{T}^c)$. Let us set, for any fixed $t = 0, 1, \dots, T$,

$$\sigma^*_t = \min \left\{ u \in \{t, t+1, \dots, T\} \mid U_u = H_u \right\} \wedge T$$

and

$$\tau^*_t = \min \left\{ u \in \{t, t+1, \dots, T\} \mid U_u = L_u \right\}.$$

We first state the following immediate corollary to results of Sect. 2.8.2.

Proposition 2.9.1 *For any stopping time* $\sigma \in \mathcal{T}_{[t,T]}$, *there exists a minimal seller's super-hedging strategy for the American contingent claim* $X^{a,\sigma}$ *with the payoff* $X_t^{\sigma} = Z(\sigma, t)$ *for* $t = 0, 1, \ldots, T$. *Moreover, the seller's price of this claim,* $\pi^s(X^{a,\sigma})$, *is the solution to the following maximization problem*

$$\pi_t^s(X^{a,\sigma}) = \max_{\tau \in \mathcal{T}_{[t,T]}} B_t \, \mathbb{E}_{\mathbb{P}^*}(Z^*(\sigma, \tau) \mid \mathcal{F}_t). \tag{2.102}$$

The following simple result, which is easy to establish by induction, will be useful in the proof of Theorem 2.9.2.

Lemma 2.9.6 *Let* $(\Omega, \mathbb{F}, \mathbb{P})$ *be a finite filtered probability space and let* M *be an* $\mathbb{F}$-martingale. *Then for any* $t = 0, 1, \ldots, T$ *and any stopping time* $\tau \in \mathcal{T}_{[t,T]}$

$$\mathbb{E}_{\mathbb{P}}(M_\tau \mid \mathcal{F}_t) = M_t.$$

In the next result, we take the perspective of the seller.

Theorem 2.9.2 *The seller's price* $\pi_t^s(X^g)$ *at time* t *of a game contingent claim is equal to the seller's price* $\pi^s(X^{a,\sigma_t^*})$ *of an American claim* X^{a,σ_t^*} *with the payoff process* $X^{\sigma_t^*}$ *given by the formula* $X_t^{\sigma_t^*} = Z(\sigma_t^*, t)$ *for every* $t = 0, 1, \ldots, T$. *Consequently,* $\pi_t^s(X^g) = U_t$ *for every* $t = 0, 1, \ldots, T$, *and thus* $\pi_t^s(X^g)$ *is the solution to the problem*

$$\pi_t^s(X^g) = \max_{\tau \in \mathcal{T}_{[t,T]}} B_t \, \mathbb{E}_{\mathbb{P}^*}(Z^*(\sigma_t^*, \tau) \mid \mathcal{F}_t) = \max_{\tau \in \mathcal{T}_{[t,T]}} \min_{\sigma \in \mathcal{T}_{[t,T]}} B_t \, \mathbb{E}_{\mathbb{P}^*}(Z^*(\sigma, \tau) \mid \mathcal{F}_t).$$

This also means that

$$\pi_t^s(X^g) = B_t \, \mathbb{E}_{\mathbb{P}^*}\big(Z^*(\sigma_t^*, \tau_t^*) \mid \mathcal{F}_t\big).$$

Proof. Let us fix some date $t = 0, 1, \ldots, T$. The seller of a game contingent claim is set to choose her cancelation time $\sigma \in \mathcal{T}_{[t,T]}$ as well as a trading strategy ϕ. We will first show that for any seller's super-hedging strategy (σ, ϕ) for a game contingent claim from time t onwards, the inequality $V_t(\phi) \geq U_t$ is satisfied.

First, as a pair (σ, ϕ) is a super-hedging strategy for the seller, we necessarily have that $V_u(\phi) \geq Z(\sigma, u)$ for every $u = t, t+1, \ldots, T$. Furthermore, the discounted wealth of a self-financing strategy is an $\mathbb{F}$-martingale under $\mathbb{P}^*$, and thus we obtain, for $\tau = \tau_t^*$,

$$V_t^*(\phi) = \mathbb{E}_{\mathbb{P}^*}(V_{\sigma \wedge \tau_t^*}^*(\phi) \mid \mathcal{F}_t) \geq \mathbb{E}_{\mathbb{P}^*}(Z^*(\sigma, \tau_t^*) \mid \mathcal{F}_t) \geq U_t^*,$$

where the equality follows from Lemma 2.9.6, the first inequality follows from the fact that ϕ is a trading strategy that super-hedges the payoff $Z(\sigma, \tau)$, and the last inequality is a consequence of (2.94) applied to Z^* and U^*.

Let us now check that there exists a trading strategy ϕ for a game contingent claim such that the wealth process $V_t(\phi) = U_t$. To this end, it suffices to take $\sigma = \sigma_t^*$ in Proposition 2.9.1. Indeed, the seller's price of the associated American claim X^{a,σ_t^*} with the payoff process $X_t^{\sigma_t^*} = Z(\sigma_t^*, t)$ satisfies

$$\pi_t^s(X^{a,\sigma_t^*}) = \max_{\tau \in \mathcal{T}_{[t,T]}} B_t \, \mathbb{E}_{\mathbb{P}^*}(Z^*(\sigma_t^*, \tau) \,|\, \mathcal{F}_t) = B_t \, \mathbb{E}_{\mathbb{P}^*}(Z^*(\sigma_t^*, \tau_t^*) \,|\, \mathcal{F}_t) = U_t,$$

where the first equality follows from (2.102) and where the second equality is a consequence of part (iii) in Theorem 2.9.1. □

It remains to examine the buyer's price of a game contingent claim.

Proposition 2.9.2 *The buyer's price $\pi^b(X^g)$ of a game contingent claim X^g is equal to the seller's price $\pi^s(X^g)$, that is, $\pi_t^b(X^g) = U_t$ for every $t = 0, 1, \ldots, T$. Hence the arbitrage price $\pi(X^g)$ of a game contingent claim is unique and it is equal to the value process U of the associated Dynkin game.*

Proof. Since the super-hedging problems for the seller and the buyer of a game contingent claim are essentially symmetric, the equality $\pi^b(X^g) = U$ can be established using the same arguments as those employed in the proof of Theorem 2.9.2. We thus obtain the asserted equality $\pi_t(X^g) = U_t$ for every $t = 0, 1, \ldots, T$. The details are left to the reader. □

Similarly as in the case of an American claim, it is possible to argue that the arbitrage price is a fair value of a game contingent claim, in the sense that if the seller and the buyer of a game contingent claim decide to exercise it at their respective rational exercise times, σ_t^* and τ_t^*, then the claim's payoff at the random time $\tau_t^* \wedge \sigma_t^* = \min(\tau_t^*, \sigma_t^*)$ will match perfectly the wealth of the minimal super-hedging strategy for the seller.

Let us note that the latter property hinges on the standing assumption that the underlying finite market model is arbitrage-free and complete. If the market completeness is not postulated, a strictly positive bid-ask spread $\pi_t^s(X^g) - \pi_t^b(X^g)$ is likely to arise when valuing a game contingent claim. For more information on pricing of American and game options in incomplete models, we refer to Kallsen and Kühn (2004) and Dolinsky and Kifer (2007) and the references therein.

Example 2.9.1 As a real life example of a game contingent claim, let us describe briefly a *convertible bond*, that is, a coupon-bearing bond issued by a company in which both the issuer and the holder have an option to exercise the contract before its maturity date T.

First, the issuer has the right to cancel the contract by *calling* the bond; in that case, the payoff at cancelation time is given by the pre-determined *call price* $\bar{C}$. Second, the holder of the bond is has typically the right to either put the bond at the pre-determined *put price* $\bar{P}$ (which is set to be lower than the call price), or to convert the bond into a pre-determined number $\kappa > 0$ (the so-called *conversion rate*) of shares of the issuer's equity S. We thus deal in fact with a callable and putable convertible bond.

Suppose momentarily that we deal with a zero-coupon convertible bond with the nominal value N. In that case, the bond can be formally defined as a game contingent claim given by the expression

$$Z(\sigma, \tau) = (\bar{P} \vee \kappa S_\tau)\mathbb{1}_{\{\tau \le \sigma < T\}} + \bar{C}\mathbb{1}_{\{\sigma < \tau\}} + N\mathbb{1}_{\{\tau = \sigma = T\}},$$

where σ stands for the *call time* and τ represents the *put/conversion time*. Hence the lower payoff satisfies $L_t = \bar{P} \vee \kappa S_t$ for $t < T$ and $L_T = N$, whereas the upper payoff equals $H_t = \bar{C}$ for $t < T$ and $H_T = N$. In fact, to ensure that $L \leq H$, we should consider the adjusted lower payoff process $L \wedge \bar{C}$, rather than L.

Of course, to analyze the actual convertible bond, one has to specify, in addition, the coupon amounts and the coupon payment dates. This leads to a suitable modification of the last formula. Let us conclude by observing that the natural way of dealing with a convertible bond is to combine a model for the equity value with some model of the term structure of interest rates. Issues related to the modeling of the term structure of interest rates are extensively studied in the second part of this book.

3

Benchmark Models in Continuous Time

The option pricing model developed in a groundbreaking paper by Black and Scholes (1973), formalized and extended in the same year by Merton (1973), enjoys great popularity. It is computationally simple and, like all arbitrage-based derivative pricing models, does not require the knowledge of an investor's risk preferences. Option valuation within the Black-Scholes framework is based on the already familiar concept of perfect replication of contingent claims. More specifically, we will show that an investor can replicate an option's return stream by continuously rebalancing a self-financing portfolio involving stocks and risk-free bonds. At any date t, the current wealth of a replicating portfolio determines the arbitrage price of an option.

Our main goal is to derive closed-form expressions for both the option's price and the replicating strategy in the Black-Scholes setting, and in other benchmark continuous-time models. To do this in a strict way, we need first to construct an arbitrage-free market model with continuous trading. This can be done relatively easily if we take for granted certain results from from the *Itô stochastic calculus*, presented in several monographs, e.g., Elliott (1982), Durrett (1996), Karatzas and Shreve (1998a), Mikosch (1999), Revuz and Yor (1999), Steele (2000), Protter (2003), and Jeanblanc et al. (2006).

In financial literature, it is not uncommon to derive the Black-Scholes formula by introducing a continuously rebalanced *risk-free portfolio* containing an option and underlying stocks. In the absence of arbitrage, the return from such a portfolio needs to match the returns on risk-free bonds. This property leads to the Black-Scholes partial differential equation satisfied by the arbitrage price of an option. It appears, however, that the risk-free portfolio does not satisfy the formal definition of a self-financing strategy, and thus this way of deriving the Black-Scholes formula is flawed.

We assume throughout that trading takes place continuously in time, and unrestricted borrowing and lending of funds is possible at the same constant interest rate. Furthermore, the market is frictionless, meaning that there are no transaction costs or taxes, and no discrimination against the short sale. In other words, we shall deal with the case of a *perfect market*.

3.1 The Black-Scholes Model

We start by describing processes modelling the prices of primary securities, a savings account (equivalently, a risk-free bond) and a common stock.

3.1.1 Risk-free Bond

The first primary security represents in our model an accumulation factor corresponding to a *savings account* (a *money market account*). We assume that the *short-term interest rate* r is constant (but not necessarily nonnegative) over the trading interval $[0, T^*]$. The risk-free security is assumed to continuously compound in value at the rate r; that is, $dB_t = rB_t\,dt$. We adopt the usual convention that $B_0 = 1$, so that its price equals $B_t = e^{rt}$ for every $t \in [0, T^*]$. When dealing with the Black-Scholes model, we may equally well replace the savings account by the *risk-free bond*. A unit zero-coupon bond maturing at time T is a security paying to its holder 1 unit of cash at a predetermined date T in the future, known as the bond's *maturity date*. Let $B(t, T)$ stand for the price at time $t \in [0, T]$ of a bond maturing at time T. It is easily seen that to replicate the payoff 1 at time T it suffices to invest B_t/B_T units of cash at time t in the savings account B. This shows that, in the absence of arbitrage opportunities, the price of the bond satisfies

$$B(t, T) = e^{-r(T-t)}, \quad \forall\, t \in [0, T].$$

Note that for any fixed T, the bond price solves the ordinary differential equation

$$dB(t, T) = rB(t, T)\,dt, \quad B(0, T) = e^{-rT}.$$

We consider here a *risk-free* bond, meaning that its issuer will not default on his obligation to pay to the bondholder the face value at maturity date.

3.1.2 Stock Price

Following Samuelson (1965) and, of course, Black and Scholes (1973), we take a *geometric* (or *exponential*) *Brownian motion* as a stochastic process modelling the stock price.

Let us fix the horizon date $T^* > 0$. The continuous-time evolution of the stock price process $(S_t)_{t \in [0, T^*]}$ is described by the following linear *stochastic differential equation* (SDE)

$$dS_t = \mu S_t\,dt + \sigma S_t\,dW_t, \quad S_0 > 0, \tag{3.1}$$

where $\mu \in \mathbb{R}$ is a constant *appreciation rate* of the stock price, $\sigma > 0$ is a constant *volatility*, and S_0 is the initial value of the stock price. As we shall see in what follows, the restrictive condition that μ and σ are constant can be weakened substantially.

The driving noise W_t, $t \in [0, T^*]$, is a one-dimensional standard Brownian motion defined on a filtered probability space $(\Omega, \mathbb{F}, \mathbb{P})$. This means that W is a process with continuous sample paths, W has independent increments with respect to the reference filtration $\mathbb{F}$, and for every $t > u$ the probability distribution of the increment $W_t - W_u$ is Gaussian with expected value 0 and variance $t - u$. Finally, by convention we set $W_0 = 0$. The formal definition of a standard Brownian motion is given in the appendix (see Definition A.4.1).

Since the sample paths of a Brownian motion are known to be almost everywhere non-differentiable functions, with probability 1, equation (3.1) cannot be interpreted as a family of ordinary differential equations (one equation for each fixed elementary event $\omega \in \Omega$). In fact, almost all sample paths of a Brownian motion have infinite variation on each time interval $[u, t]$ for $t > u$, so that we are not in a position to apply the Lebesgue-Stieltjes integration theory in order to define an integral of the form $\int_0^t X_u \, dW_u$, where X is an $\mathbb{F}$-adapted stochastic process

We refer to (3.1) as an SDE, however, expression (3.1) is merely a shorthand notation for the following *stochastic integral equation* (see Sect. A.12)

$$S_t = S_0 + \int_0^t \mu S_u \, du + \int_0^t \sigma S_u \, dW_u, \quad \forall\, t \in [0, T^*]. \tag{3.2}$$

In other words, by a stochastic differential equation of the form (3.1) we mean in fact the stochastic integral equation (3.2). The definition of the second integral in (3.2), interpreted here as the *Itô stochastic integral*, can be found in Sect. A.6. Let us only mention here that if X is a continuous, $\mathbb{F}$-adapted process that satisfies suitable integrability conditions then

$$\int_0^t X_u \, dW_u = \lim_{n \to +\infty} \sum_{j=0}^{n-1} X_{t_j^n} (W_{t_{j+1}^n} - W_{t_j^n}),$$

where the limit is in fact taken over an arbitrary sequence of finite partitions $\{t_0^n = 0 < t_1^n < \cdots < t_n^n = t\}$ of the interval $[0, t]$ such that $\lim_{n \to \infty} \max_{k=0,\dots,n-1} (t_{k+1}^n - t_k^n) = 0$.

It is worth stressing that the Itô integral $\int_0^t X_u \, dW_u$ follows a martingale (or at least a local martingale) under $\mathbb{P}$ with respect to the reference filtration $\mathbb{F}$. Other kinds of stochastic integrals, such as the *Fisk-Stratonovich integral*, do not possess this property, and thus are commonly perceived as less appropriate in financial modelling oriented toward arbitrage pricing.

It appears that (3.1) specifies the stock price process uniquely, as the following result shows.

Proposition 3.1.1 *The process S given by the formula*

$$S_t = S_0 \exp\left(\sigma W_t + \left(\mu - \tfrac{1}{2}\sigma^2\right)t\right), \quad \forall\, t \in [0, T^*], \tag{3.3}$$

is the unique solution of the stochastic differential equation (3.1), or equivalently, the stochastic integral equation (3.2).

Proof. We will first check that the process given by (3.3) satisfies (3.1). To this end, we will use the Itô formula (see Theorem A.9.1). As customary, we use the differential notation, rather than the integral notation. Let us introduce the auxiliary process X by setting

$$dX_t = \left(\mu - \tfrac{1}{2}\sigma^2\right) dt + \sigma \, dW_t$$

with $X_0 = 0$. Then the process S given by (3.3) satisfies $S_t = g(X_t)$ for $t \in [0, T^*]$, where the function $g : \mathbb{R} \to \mathbb{R}$ is defined as $g(x) = S_0 e^x$. Obviously, $g(x) = g'(x) = g''(x)$ and thus Itô's formula yields

$$dS_t = dg(X_t) = g'(X_t)\left(\mu - \tfrac{1}{2}\sigma^2\right) dt + g'(X_t)\sigma \, dW_t + \tfrac{1}{2}g''(X_t)\sigma^2 \, dt$$
$$= g(X_t)(\mu \, dt + \sigma \, dW_t) = S_t(\mu \, dt + \sigma \, dW_t),$$

as expected. Uniqueness of solutions to (3.2) is an immediate consequence of the general result, due to Itô, stating that a SDE with Lipschitz continuous coefficients has a unique solution.

Alternatively, one can prove the uniqueness by applying the Itô formula to the ratio of two solutions to (3.2). Let us assume that Y is another solution to (3.1) with the initial value $Y_0 = S_0$. We wish to show that the Itô differential of the ratio Y/S is equal to zero. To this end, we will first compute the Itô differential of the process $1/S$. Using again Itô's formula of Theorem A.9.1, we obtain (note that now we take $g(s) = 1/s$)

$$d\left(\frac{1}{S_t}\right) = -\frac{1}{S_t^2} \, dS_t + \frac{1}{S_t^3} \, d\langle S\rangle_t = -\frac{1}{S_t^2}\left(\mu S_t \, dt + \sigma S_t \, dW_t\right) + \frac{1}{S_t^3}\sigma^2 S_t^2 \, dt$$
$$= -\frac{1}{S_t}\left((\mu - \sigma^2) \, dt + \sigma \, dW_t\right),$$

where we used the fact that the quadratic variation of S equals

$$\langle S\rangle_t = \int_0^t \sigma^2 S_u^2 \, d\langle W\rangle_u = \int_0^t \sigma^2 S_u^2 \, du.$$

By applying Itô's integration by parts formula (see Corollary A.9.1) to processes Y and $1/S$, we get

$$d\left(\frac{Y_t}{S_t}\right) = \frac{1}{S_t} \, dY_t + Y_t \, d\left(\frac{1}{S_t}\right) + d\langle Y, 1/S\rangle_t$$
$$= \frac{Y_t}{S_t}\left(\mu \, dt + \sigma \, dW_t\right) - \frac{Y_t}{S_t}\left((\mu - \sigma^2) \, dt + \sigma \, dW_t\right) - \frac{Y_t}{S_t}\sigma^2 \, dt = 0,$$

since the cross-variation of Y and $1/S$ equals (recall that $\langle W\rangle_t = t$)

$$\langle Y, 1/S\rangle_t = -\int_0^t \sigma^2 \frac{Y_u}{S_u} \, d\langle W\rangle_u = -\int_0^t \sigma^2 \frac{Y_u}{S_u} \, du.$$

Since the Itô differential of the ratio Y/S is zero, this process is constant and thus it equals its initial value, which is 1. We conclude that $Y_t = S_t$ for every $t \in [0, T^*]$, and thus the uniqueness of a solution to (3.2) is established. □

We make the standing assumption that the underlying filtration $\mathbb{F} = (\mathcal{F}_t)_{t \in [0,T^*]}$ is the natural filtration $\mathbb{F}^W$ of the driving Brownian motion. More precisely, $\mathbb{F}$ is the standard augmentation of the natural filtration of W, but it will still be denoted by $\mathbb{F}^W$. For any fixed $t \in [0, T^*]$, we have $S_t = f(W_t)$ for some function $f : \mathbb{R} \to \mathbb{R}_+$ that determines the one-to-one correspondence between the values S_t of the stock price and the values W_t of the Brownian motion. Specifically, the strictly increasing function f is given by the formula

$$f(x) = S_0 \exp\left(\sigma x + \left(\mu - \tfrac{1}{2}\sigma^2\right)t\right), \quad \forall x \in \mathbb{R}.$$

Consequently, for any natural n, any dates $0 \le t_1 < t_2 < \cdots < t_n \le T^*$, and any real numbers $x_1, x_2, \ldots, x_n$, we have

$$\{\omega \in \Omega \mid S_{t_1} < x_1, \ldots, S_{t_n} < x_n\} = \{\omega \in \Omega \mid W_{t_1} < y_1, \ldots, W_{t_n} < y_n\},$$

where we set $y_i = f^{-1}(x_i)$. It is thus clear that we have, for every $t \in [0, T^*]$,

$$\mathcal{F}_t^W = \sigma\{W_u \mid u \le t\} = \sigma\{S_u \mid u \le t\} = \mathcal{F}_t^S.$$

We conclude that the filtration generated by the stock price coincides with the natural filtration of the driving noise W, and thus $\mathbb{F}^S = \mathbb{F}^W = \mathbb{F}$. This means that the information structure of the model is based on observations of the stock price process only.

It is easy to check that the Brownian motion W follows a martingale with respect to the underlying filtration $\mathbb{F}$. Indeed, for any t the random variable W_t is integrable with respect to $\mathbb{P}$, and we have, for any $u \le t$,

$$\mathbb{E}_{\mathbb{P}}(W_t \mid \mathcal{F}_u) = \mathbb{E}_{\mathbb{P}}(W_t - W_u \mid \mathcal{F}_u) + \mathbb{E}_{\mathbb{P}}(W_u \mid \mathcal{F}_u) = W_u. \tag{3.4}$$

Let us define an auxiliary process M by setting

$$M_t = \exp\left(\sigma W_t - \tfrac{1}{2}\sigma^2 t\right), \quad \forall t \in [0, T^*]. \tag{3.5}$$

Lemma 3.1.1 *The process M is a martingale under $\mathbb{P}$ with respect to the filtration $\mathbb{F}$.*

Proof. In order to show that M is a martingale under $\mathbb{P}$ with respect to $\mathbb{F}$, it suffices to note that

$$\mathbb{E}_{\mathbb{P}}\left(e^{\sigma(W_t - W_u)} \mid \mathcal{F}_u\right) = \mathbb{E}_{\mathbb{P}}\left(e^{\sigma(W_t - W_u)}\right) = \exp\left(\tfrac{1}{2}\sigma^2(t - u)\right),$$

where the first equality is an consequence of the independence of the increment $W_t - W_u$ of the σ-field $\mathcal{F}_u$ and the second follows from the fact that $W_t - W_u$ has the Gaussian distribution $N(0, t - u)$. □

Since $S_t = e^{\mu t} M_t$, we have, for every $u \le t \le T^*$,

$$\mathbb{E}_{\mathbb{P}}(S_t \mid \mathcal{F}_u) = \mathbb{E}_{\mathbb{P}}(e^{\mu t} M_t \mid \mathcal{F}_u) = e^{\mu t} M_u = S_u\, e^{\mu(t-u)}.$$

Corollary 3.1.1 *The stock price S follows a martingale under the actual probability* $\mathbb{P}$ *with respect to the filtration* $\mathbb{F}$ *if and only if its appreciation rate* $\mu = 0$. *If* $\mu > 0$ *(respectively* $\mu < 0$) *then* $\mathbb{E}_{\mathbb{P}}(S_t \mid \mathcal{F}_u) > S_u$ *(respectively* $\mathbb{E}_{\mathbb{P}}(S_t \mid \mathcal{F}_u) < S_u$) *for every* $t > u$ *and thus the stock price S is a submartingale (respectively a supermartingale) under the actual probability* $\mathbb{P}$.

It is apparent from (3.3) that the logarithmic stock returns are Gaussian, meaning that the random variable $\ln(S_t/S_u)$ has under $\mathbb{P}$ a Gaussian probability distribution for any choice of dates $u < t \leq T^*$. Moreover, S is a *time-homogeneous Markov process* under $\mathbb{P}$ with respect to the filtration $\mathbb{F}$ with the transition probability density function, for $x, y > 0$ and $t > u$,

$$p_S(u, x, t, y) = \frac{1}{\sqrt{2\pi(t-u)}\,\sigma y} \exp\left\{ -\frac{\left(\ln(y/x) - \mu(t-u) + \frac{1}{2}\sigma^2(t-u)\right)^2}{2\sigma^2(t-u)} \right\}$$

where formally $p_S(u, x, t, y) = \mathbb{P}\{S_t = y \mid S_u = x\}$. It is important to observe that even if the appreciation rate μ is replaced by an adapted stochastic process μ_t and the stock price S is no longer Markovian under the actual probability $\mathbb{P}$, it will always be a time-homogeneous Markov process under the martingale measure $\mathbb{P}^*$ for the discounted stock price (see Sect. 3.1.4).

As shown above, the stock price process given by (3.1) enjoys a lot of mathematically convenient (but not necessarily realistic) properties. It should be made clear that we are not concerned here with the question of whether the maintained model is the "correct" model of asset price fluctuations. This problem has been an ongoing subject of empirical and theoretical studies over the past forty years (see Mandelbrot (1960, 1963, 1967), Fama (1965, 1981, 1990), Fama and MacBeth (1973), Greene and Fielitz (1977, 1979), Merton (1980), French (1983), Bookstaber and Clarke (1984), Kon (1984), Keim and Stambaugh (1986), French et al. (1987), Lo and MacKinlay (1988), West (1988), Scheinkman and LeBaron (1989), Nelson (1990), Madan and Seneta (1990), Mittnik and Rachev (1993), Richardson and Smith (1993), Taylor (1994), Eberlein and Keller (1995), Barndorff-Nielsen et al. (1996), Jackwerth and Rubinstein (1996), or Willinger et al. (1999)). Let us agree that (3.1) can hardly be seen as a realistic model of the real-world behavior of a stock price. The option prices obtained within the Black-Scholes framework are reasonably close to those observed on options exchanges, however.

3.1.3 Self-financing Trading Strategies

By a *trading strategy* we mean a pair $\phi = (\phi^1, \phi^2)$ of $\mathbb{F}$-*progressively measurable* (see Sect. A.5) stochastic processes on the underlying probability space $(\Omega, \mathbb{F}, \mathbb{P})$. The concept of a self-financing trading strategy in the Black-Scholes market is formally based on the notion of the Itô integral. Intuitively, this choice of stochastic integral is supported by the fact that, in the case of the Itô integral (as opposed, for instance, to the Fisk-Stratonovich integral), the underlying process is integrated in a predictable way, meaning that we take its values on the left-hand end of each (infinitesimal) time interval.

Unless explicitly stated otherwise, we will assume that a stock S does not pay dividends (at least during the option's lifetime).

Definition 3.1.1 We say that a trading strategy $\phi = (\phi^1, \phi^2)$ over the time interval $[0, T]$ is *self-financing* if its wealth process $V(\phi)$, which is set to equal

$$V_t(\phi) = \phi_t^1 S_t + \phi_t^2 B_t, \quad \forall t \in [0, T], \tag{3.6}$$

satisfies the following condition

$$V_t(\phi) = V_0(\phi) + \int_0^t \phi_u^1 \, dS_u + \int_0^t \phi_u^2 \, dB_u, \quad \forall t \in [0, T], \tag{3.7}$$

where the first integral is understood in the Itô sense and the second it the pathwise Riemann (or Lebesgue) integral.

It is, of course, implicitly assumed that both integrals on the right-hand side of (3.7) are well-defined. It is well known that a sufficient condition for this is that[1]

$$\mathbb{P}\left\{ \int_0^T (\phi_u^1)^2 \, du < \infty \right\} = 1 \quad \text{and} \quad \mathbb{P}\left\{ \int_0^T |\phi_u^2| \, du < \infty \right\} = 1. \tag{3.8}$$

We denote by Φ the class of all self-financing trading strategies. It follows from the example below that arbitrage opportunities are not excluded a priori from the class of self-financing trading strategies.

Example 3.1.1 The following example of a suicide strategy is borrowed from Harrison and Pliska (1981). It can be modified easily to provide an example of an arbitrage opportunity in an unconstrained Black-Scholes setting. For simplicity, we take $r = 0$, $T = 1$, and $S_0 = 1$. For a strictly positive constant $b > 0$, we consider the following trading strategy

$$\phi_t^1 = \begin{cases} -b, & 0 \le t \le \tau(b), \\ 0, & \text{otherwise,} \end{cases} \qquad \phi_t^2 = \begin{cases} 1 + b, & 0 \le t \le \tau(b), \\ 0, & \text{otherwise,} \end{cases}$$

where

$$\tau(b) = \inf\{t : S_t = 1 + b^{-1}\} = \inf\{t : V_t(\phi) = 0\},$$

and $V_t(\phi) = 1 + b - bS_t$.

In financial interpretation, an investor starts with one dollar of wealth, sells b shares of stock short, and buys $1 + b$ bonds. Then he holds the portfolio until the terminal date $T = 1$, or goes bankrupt, whichever comes first. The probability of bankruptcy under this strategy is equal to $p(b) = \mathbb{P}\{\tau(b) < 1\}$, so that it increases from zero to one as b increases from zero to infinity. By selling short a very large amount of stock, the investor makes his failure almost certain, but he will probably make a great deal of money if he survives. The chance of survival can be completely

---

[1] Note that condition (3.8) is invariant with respect to an equivalent change of a probability measure.

eliminated, however, by escalating the amount of stock sold short in the following way.

To show this, we shall modify the strategy as follows. On the interval $[0, 1/2]$, we follow the strategy above with $b = 1$. The probability of bankruptcy during $[0, 1/2]$ thus equals $p = \mathbb{P}\{\tau(1) \le 1/2\}$. If $\tau(1) > 1/2$, the amount of stock sold short is adjusted to a new level b_1 at time $1/2$. Simultaneously, the number of bonds held is revised in a self-financing fashion. The number b_1 is chosen so as to make the conditional probability of ruin during the time interval $(1/2, 3/4]$, given that we have survived up to time $1/2$, equal to p again.

In general, if at any time $t_n = 1 - (1/2)^n$ we still have positive wealth, then we readjust (typically increase) the amount of stock sold short, so that the conditional probability of bankruptcy during (t_n, t_{n+1}) is always p. To keep the strategy self-financing, the amount of bonds held must be adjusted at each time t_n as well. The probability of survival until time t_n is then $(1-p)^n$, and it vanishes as n tends to 0 (so that t_n tends to 1). We thus have an example of a piecewise constant, self-financing strategy, $(\hat{\phi}^1, \hat{\phi}^2)$ say, with $V_0(\hat{\phi}) = \hat{\phi}_0^1 S_0 + \hat{\phi}_0^2 = 1$,

$$V_t(\hat{\phi}) = \hat{\phi}_t^1 S_t + \hat{\phi}_t^2 \ge 0, \quad \forall t \in [0, 1),$$

and $V_1(\hat{\phi}) = 0$. To get a reliable model of a security market we need, of course, to exclude such examples of doubling strategies from the market model.

3.1.4 Martingale Measure for the Black-Scholes Model

We find it convenient to introduce the concept of *admissibility* of a trading strategy directly in terms of a martingale measure (we refer to Chap. 8 for an alternative approach based on the concept of a *tame strategy*). For this reason, our next goal is to study the concept of a martingale measure within the Black-Scholes set-up.

As usual, we write $S^* = S/B$ to denote the discounted stock price. Using (3.3) and the fact that $B_t = e^{rt}$, we obtain, for every $t \in [0, T^*]$,

$$S_t^* = S_0^* \exp\left(\sigma W_t + \left(\mu - r - \tfrac{1}{2}\sigma^2\right) t\right).$$

Put equivalently, the process S^* is the unique solution to the SDE

$$dS_t^* = (\mu - r)S_t^* \, dt + \sigma S_t^* \, dW_t, \quad S_0^* = S_0. \tag{3.9}$$

The martingale property of S^* under $\mathbb{P}$ can thus be analyzed in exactly the same way as for the stock price S.

Corollary 3.1.2 *The discounted stock price S^* is a martingale under the actual probability $\mathbb{P}$ with respect to the filtration $\mathbb{F}$ if and only if $\mu = r$. If $\mu > r$ (respectively $\mu < r$) then $\mathbb{E}_{\mathbb{P}}(S_t^* \mid \mathcal{F}_u) > S_u^*$ (respectively $\mathbb{E}_{\mathbb{P}}(S_t^* \mid \mathcal{F}_u) < S_u^*$) for every $t > u$, so that S^* is a submartingale (respectively a supermartingale) under $\mathbb{P}$.*

The next definition refers directly to the martingale property of the discounted stock price S^*. Essentially, we seek a probability measure under which the drift term in (3.9) disappears.

Definition 3.1.2 A probability measure $\mathbb{Q}$ on $(\Omega, \mathcal{F}_{T^*})$, equivalent to $\mathbb{P}$, is called a *martingale measure for S^** if S^* is a local martingale under $\mathbb{Q}$.

In the study of the no-arbitrage property of the model and the derivation of the risk-neutral valuation formula, it is convenient to focus directly on the martingale property of the discounted wealth process.

Definition 3.1.3 A probability measure $\mathbb{P}^*$ on $(\Omega, \mathcal{F}_{T^*})$, equivalent to $\mathbb{P}$, is called a *spot martingale measure* if the discounted wealth $V_t^*(\phi) = V_t(\phi)/B_t$ of any self-financing trading strategy ϕ is a local martingale under $\mathbb{P}^*$.

The following result shows that within the present set-up both these notions co-incide. Let us stress that this is true only under the assumption that a stock S does not pay dividends during the time period $[0, T^*]$. The case of a dividend-paying stock is analyzed in Sect. 3.2.

Lemma 3.1.2 *A probability measure is a spot martingale measure if and only if it is a martingale measure for the discounted stock price S^*.*

Proof. Let ϕ be a self-financing strategy and let $V^* = V^*(\phi)$. Using Itô's integration by parts formula, we get (note that $dB_t^{-1} = -rB_t^{-1}\,dt$)

$$
\begin{aligned}
dV_t^* = d(V_t B_t^{-1}) &= V_t\,dB_t^{-1} + B_t^{-1}dV_t \\
&= (\phi_t^1 S_t + \phi_t^2 B_t)\,dB_t^{-1} + B_t^{-1}(\phi_t^1\,dS_t + \phi_t^2\,dB_t) \\
&= \phi_t^1(B_t^{-1}\,dS_t + S_t\,dB_t^{-1}) = \phi_t^1\,dS_t^*.
\end{aligned}
$$

We thus obtain the following equality

$$
V_t^*(\phi) = V_0^*(\phi) + \int_0^t \phi_u^1\,dS_u^*, \quad \forall t \in [0, T^*].
$$

It suffices to make use of the local martingale property of the Itô integral (see Sect. A.6). $\qquad\square$

In the Black-Scholes setting, the martingale measure for the discounted stock price is unique, and it is known explicitly, as the following result shows. The proof of this lemma relies on the Girsanov theorem (see Theorem A.15.1 and Proposition A.15.3 with $d = 1$).

Lemma 3.1.3 (i) *The unique martingale measure $\mathbb{Q}$ for the discounted stock price process S^* is given by the Radon-Nikodým derivative*

$$
\frac{d\mathbb{Q}}{d\mathbb{P}} = \exp\left(\frac{r - \mu}{\sigma} W_{T^*} - \frac{1}{2}\frac{(r - \mu)^2}{\sigma^2} T^* \right), \quad \mathbb{P}\text{-a.s.}
$$

(ii) *The discounted stock price S^* satisfies, under the martingale measure $\mathbb{Q}$,*

$$
dS_t^* = \sigma S_t^*\,dW_t^*, \tag{3.10}
$$

and the continuous, $\mathbb{F}$-adapted process W^* given by the formula

$$W_t^* = W_t - \frac{r - \mu}{\sigma} t, \quad \forall t \in [0, T^*],$$

is a standard Brownian motion on the probability space $(\Omega, \mathbb{F}, \mathbb{Q})$.

Proof. Let us first prove part (i). Recall that we assumed that the filtration $\mathbb{F}$ is the $\mathbb{P}$-augmentation of the natural filtration of W, that is, $\mathbb{F} = \mathbb{F}^W$ (see Sect. A.2). We thus know from Proposition A.15.3 that for any probability measure $\mathbb{Q}$ on $(\Omega, \mathcal{F}_{T^*})$ equivalent to $\mathbb{P}$ there exists an $\mathbb{F}^W$-progressively measurable process γ such that the Radon-Nikodým derivative of $\mathbb{Q}$ with respect to $\mathbb{P}$ equals

$$\frac{d\mathbb{Q}}{d\mathbb{P}}\Big|_{\mathcal{F}_t} = \mathcal{E}_t\left(\int_0^{\cdot} \gamma_u \cdot dW_u\right), \quad \mathbb{P}\text{-a.s.}$$

Moreover, by virtue of Theorem A.15.1, the process W^* given as

$$W_t^* = W_t - \int_0^t \gamma_u \, du, \quad \forall t \in [0, T^*],$$

is a standard Brownian motion on the probability space $(\Omega, \mathbb{F}, \mathbb{Q})$.

In order to identify a yet unknown process γ, we substitute the differential $dW_t = dW_t^* + \gamma_t \, dt$ into (3.9). In that way, we obtain the following SDE satisfied by S^* under $\mathbb{Q}$

$$dS_t^* = (\mu - r + \gamma_t \sigma) S_t^* \, dt + \sigma S_t^* \, dW_t^*, \quad S_0^* = S_0.$$

Hence the process S^* is a martingale under $\mathbb{Q}$ if and only if the drift term in the last formula vanishes (see Corollary A.8.1). This in turn holds whenever the equality $\int_0^t (\mu - r + \gamma_u \sigma) \, du = 0$ is satisfied for $t \in [0, T^*]$.

In view of Theorem A.15.1, it is easy to check that for the constant (and thus bounded) process $\gamma_t = (r - \mu)\sigma^{-1}$, $t \in [0, T^*]$, the probability measure $\mathbb{Q}$ is indeed well defined and equivalent to $\mathbb{P}$. To this end, it suffices to note that W_{T^*} has the Gaussian probability distribution $N(0, T^*)$ under $\mathbb{P}$ so that

$$\mathbb{E}_{\mathbb{P}}\left(\exp\left(\frac{r - \mu}{\sigma} W_{T^*} - \frac{1}{2}\frac{(r - \mu)^2}{\sigma^2} T^*\right)\right) = 1.$$

We have thus established the existence and uniqueness of a martingale measure for the discounted stock price S^*. Part (ii) is an immediate consequence of Theorem A.15.1.   □

Combining the last two lemmas, we conclude that the unique spot martingale measure $\mathbb{P}^*$ for the Black-Scholes model is given on $(\Omega, \mathcal{F}_{T^*})$ by means of the Radon-Nikodým derivative

$$\frac{d\mathbb{P}^*}{d\mathbb{P}} = \exp\left(\frac{r - \mu}{\sigma} W_{T^*} - \frac{1}{2}\frac{(r - \mu)^2}{\sigma^2} T^*\right), \quad \mathbb{P}\text{-a.s.} \tag{3.11}$$

The discounted stock price S^* follows a strictly positive martingale under $\mathbb{P}^*$ since we have, for every $t \in [0, T^*]$,

$$S_t^* = S_0^* \exp\left(\sigma W_t^* - \tfrac{1}{2}\sigma^2 t\right) = S_0^* M_t^*, \tag{3.12}$$

where M^* is given by (3.5) with W replaced by W^*. The stock price at time t equals

$$S_t = S_t^* B_t = S_0 \exp\left(\sigma W_t^* + (r - \tfrac{1}{2}\sigma^2)t\right), \tag{3.13}$$

and thus the dynamics of the stock price S under $\mathbb{P}^*$ are

$$dS_t = r S_t \, dt + \sigma S_t \, dW_t^*, \quad S_0 > 0. \tag{3.14}$$

Note that S is a time-homogeneous Markov process under $\mathbb{P}^*$ with respect to the filtration $\mathbb{F}$ with the transition probability density function given as, for $x, y > 0$ and $t > u$,

$$p_S^*(u, x, t, y) = \frac{1}{\sqrt{2\pi(t-u)}\sigma y} \exp\left\{-\frac{\left(\ln(y/x) - r(t-u) + \tfrac{1}{2}\sigma^2(t-u)\right)^2}{2\sigma^2(t-u)}\right\}$$

where we denote $p_S^*(u, x, t, y) = \mathbb{P}^*\{S_t = y \mid S_u = x\}$. Finally, it is useful to observe that all filtrations involved in the model coincide; that is, $\mathbb{F} = \mathbb{F}^W = \mathbb{F}^{W^*} = \mathbb{F}^S = \mathbb{F}^{S^*}$ (this can be checked similarly as the equality $\mathbb{F}^W = \mathbb{F}^S$).

We are in a position to introduce the class of *admissible* trading strategies. In view of Example 3.1.1, the necessity of imposing some restriction on a class of trading strategies is obvious. In an unconstrained Black-Scholes model, arbitrage opportunities are not precluded and thus reliable valuation of derivative instruments is not possible.

Definition 3.1.4 A trading strategy $\phi \in \Phi$ is called $\mathbb{P}^*$-*admissible* if the discounted wealth process

$$V_t^*(\phi) = B_t^{-1} V_t(\phi), \quad \forall t \in [0, T],$$

follows a martingale under $\mathbb{P}^*$. We write $\Phi(\mathbb{P}^*)$ to denote the class of all $\mathbb{P}^*$-admissible trading strategies. The triple $\mathcal{M}_{BS} = (S, B, \Phi(\mathbb{P}^*))$ is called the arbitrage-free Black-Scholes model of a financial market, or briefly, the *Black-Scholes model*.

For brevity, we shall say that ϕ is *admissible* rather than $\mathbb{P}^*$-admissible if no confusion may arise. It easy to check that by restricting our attention to the class of all admissible strategies, we have ensured the absence of arbitrage opportunities in the Black-Scholes model.

We say that a European contingent claim X that settles at time $T \leq T^*$ is *attainable* in the Black-Scholes model if it can be replicated by means of an admissible strategy. Given an attainable European contingent claim, we can uniquely define its arbitrage price in the Black-Scholes model.

Definition 3.1.5 Let X be an attainable European contingent claim that settles at time $T \leq T^*$. Then its *arbitrage price process* in the Black-Scholes model $\mathcal{M}_{BS}$, denoted as $\pi_t(X)$, $t \in [0, T]$, is defined as the wealth process $V_t(\phi)$, $t \in [0, T]$, of any admissible trading strategy ϕ that replicates X, that is, satisfies the equality $V_T(\phi) = X$.

If no replicating admissible strategy exists,[2] the arbitrage price of such a claim is not defined. Conforming with the definition of an arbitrage price, to value a derivative security we will usually search first for its replicating strategy. The risk-neutral valuation approach (cf. Sect. 2.6.5) to the pricing problem is also possible, as the following simple result shows.

Corollary 3.1.3 *Let X be an attainable European contingent claim that settles at time T. Then the arbitrage price $\pi_t(X)$ at time $t \in [0, T]$ in $\mathcal{M}_{BS}$ is given by the risk-neutral valuation formula*

$$\pi_t(X) = B_t \, \mathbb{E}_{\mathbb{P}^*}(B_T^{-1} X \mid \mathcal{F}_t), \quad \forall \, t \in [0, T]. \tag{3.15}$$

Proof. Let ϕ be an admissible strategy replicating X. We have

$$\frac{\pi_t(X)}{B_t} = V_t^*(\phi) = \mathbb{E}_{\mathbb{P}^*}(V_T^*(\phi) \mid \mathcal{F}_t) = \mathbb{E}_{\mathbb{P}^*}(B_T^{-1} X \mid \mathcal{F}_t).$$

where the first and the last equalities are consequences of the definition of the arbitrage price, and the middle equality follows from Lemma 3.1.2.   □

Remarks. (a) The assumption that $\mathbb{F} = \mathbb{F}^W$ is not essential if our aim is to value European and American contingent claims with the stock S being the underlying asset. If this condition is relaxed, the uniqueness of the martingale measure, and thus also the completeness of the market, may fail to hold, in general. But, more importantly, this does not affect the arbitrage valuation of standard options written on the stock S and other claims having S as the underlying asset.
(b) We could have assumed that the interest rate r and the stock price volatility σ are deterministic functions of time (see Sect. 3.1.6 below). Also, the appreciation rate μ could be an adapted stochastic process, satisfying mild regularity conditions. Although we do not discuss such generalizations here, it is rather clear that extensions of most results presented in this chapter to such a case are rather straightforward. Replicating strategies and arbitrage prices of attainable claims with any expiration date $T \le T^*$ will be exactly the same if μ is a constant and in the case of a random appreciation rate.
    In particular, Theorem 3.1.1 and its proof remain valid with no changes if we replace a constant μ by an $\mathbb{F}$-adapted process. We need, however, to assume that the process μ_t, $t \in [0, T^*]$, is such that the formula

$$\frac{d\mathbb{P}^*}{d\mathbb{P}} = \exp\left(\int_0^{T^*} \frac{r - \mu_u}{\sigma} \, dW_u - \frac{1}{2} \int_0^{T^*} \frac{(r - \mu_u)^2}{\sigma^2} \, du \right) \stackrel{\text{def}}{=} \eta_{T^*} \tag{3.16}$$

defines a probability measure $\mathbb{P}^*$ equivalent to $\mathbb{P}$ on $(\Omega, \mathcal{F}_{T^*})$. This holds if and only if $\mathbb{E}_{\mathbb{P}}(\eta_{T^*}) = 1$. In general, it may happen that the random variable η_{T^*} given by (3.16) is well-defined, but $\mathbb{E}_{\mathbb{P}}(\eta_{T^*}) < 1$ so that $\mathbb{P}^*\{\Omega\} < 1$.

---

[2] This happens only if a claim is not integrable under $\mathbb{P}^*$ (see Sect. 8.2).

3.1.5 Black-Scholes Option Pricing Formula

In Black and Scholes (1973), two alternative justifications of the option valuation formula are provided. The first relies on the fact that the risk-free return can be replicated by holding a continuously adjusted position in the underlying stock and an option. In other words, if an option is not priced in accordance with the Black-Scholes formula, there is a sure profit to be made by some combination of either short or long sales of the option and the underlying asset. The second method of derivation is based on equilibrium arguments requiring, in particular, that the option earns an expected rate of return commensurate with the risk involved in holding the option as an asset. The first approach is usually referred to as the *risk-free portfolio method,* while the second is known as the *equilibrium derivation* of the Black-Scholes formula. An analysis of economic assumptions that support these two methodologies can be found, for instance, in Gilster and Lee (1984) and McDonald and Siegel (1984).

The *replication approach* presented below is based on the observation that in the Black-Scholes setting the option value can be mimicked by holding a continuously rebalanced position in the underlying stock and risk-free bonds. Recall that we assume throughout that the financial market we are dealing with is perfect (partially, this was already implicit in the definition of a self-financing trading strategy).

We shall first consider a European call option written on a stock S, with expiry date T and strike price K. Hence the payoff (and the price) at time T equals $(S_T - K)^+$. Let the function $c : \mathbb{R}_+ \times (0, T] \to \mathbb{R}$ be given by the formula

$$c(s, t) = s N\big(d_1(s, t)\big) - K e^{-rt} N\big(d_2(s, t)\big), \tag{3.17}$$

where

$$d_1(s, t) = \frac{\ln(s/K) + (r + \frac{1}{2}\sigma^2)t}{\sigma\sqrt{t}} \tag{3.18}$$

and $d_2(s, t) = d_1(s, t) - \sigma\sqrt{t}$, or explicitly

$$d_2(s, t) = \frac{\ln(s/K) + (r - \frac{1}{2}\sigma^2)t}{\sigma\sqrt{t}}. \tag{3.19}$$

Furthermore, let N stand for the standard Gaussian cumulative distribution function

$$N(x) = \frac{1}{\sqrt{2\pi}} \int_{-\infty}^{x} e^{-z^2/2}\, dz, \quad \forall x \in \mathbb{R}.$$

We adopt the following notational convention

$$d_{1,2}(s, t) = \frac{\ln(s/K) + (r \pm \frac{1}{2}\sigma^2)t}{\sigma\sqrt{t}}.$$

Let us denote by C_t the arbitrage price at time t of a European call option in the Black-Scholes model. We are in a position to state the main result of this chapter.

Theorem 3.1.1 *The arbitrage price at time $t < T$ of the European call option with expiry date T and strike price K in the Black-Scholes model is given by the formula*

$$C_t = c(S_t, T - t), \quad \forall t \in [0, T), \tag{3.20}$$

where the function $c : \mathbb{R}_+ \times (0, T] \to \mathbb{R}$ is given by (3.17). More explicitly, we have, for any $t < T$

$$C_t = S_t N\big(d_1(S_t, T - t)\big) - K B(t, T) N\big(d_2(S_t, T - t)\big),$$

where d_1, d_2 are given by (3.18)–(3.19). Moreover, the unique admissible replicating strategy $\phi = (\phi^1, \phi^2)$ of the call option satisfies, for every $t \in [0, T]$,

$$\phi_t^1 = \frac{\partial c}{\partial s}(S_t, T - t) = N\big(d_1(S_t, T - t)\big)$$

and

$$\phi_t^2 = e^{-rt} \big(c(S_t, T - t) - \phi_t^1 S_t\big).$$

Proof. We provide two alternative proofs of the Black-Scholes result. The first relies on the direct determination of the replicating strategy. It thus gives not only the valuation formula (this, however, requires solving the Black-Scholes PDE (3.26)), but also explicit formulas for the replicating strategy.

The second method makes direct use of the risk-neutral valuation formula (3.15) of Corollary 3.1.3. It focuses on the explicit computation of the arbitrage price, rather than on the derivation of the hedging strategy. However, as soon as the option's price is found, it is also possible to derive the form of the replicating strategy by taking the partial derivative of the pricing function c with respect to the price of the underlying asset, that is, the stock.

First method. We start by assuming that the option price, C_t, satisfies the equality $C_t = v(S_t, t)$ for some function $v : \mathbb{R}_+ \times [0, T] \to \mathbb{R}$. We may thus assume that the replicating strategy ϕ we are looking for has the following form

$$\phi_t = (\phi_t^1, \phi_t^2) = (g(S_t, t), h(S_t, t)) \tag{3.21}$$

for $t \in [0, T]$, where $g, h : \mathbb{R}_+ \times [0, T] \to \mathbb{R}$ are unknown functions.

Since ϕ is assumed to be self-financing, its wealth process $V(\phi)$, given by

$$V_t(\phi) = g(S_t, t)S_t + h(S_t, t)B_t = v(S_t, t), \tag{3.22}$$

needs to satisfy the following: $dV_t(\phi) = g(S_t, t)\, dS_t + h(S_t, t)\, dB_t$. Under the present assumptions, the last equality can be given the following form

$$dV_t(\phi) = (\mu - r)S_t g(S_t, t)\, dt + \sigma S_t g(S_t, t)\, dW_t + r v(S_t, t)dt, \tag{3.23}$$

since from the second equality in (3.22) we obtain

$$\phi_t^2 = h(S_t, t) = B_t^{-1}\big(v(S_t, t) - g(S_t, t)S_t\big).$$

We shall search for the wealth function v in the class of smooth functions on the open domain $\mathcal{D} = (0, +\infty) \times (0, T)$; more exactly, we assume that $v \in C^{2,1}(\mathcal{D})$. An application of Itô's formula yields[3]

$$dv(S_t, t) = \left(v_t(S_t, t) + \mu S_t\, v_s(S_t, t) + \tfrac{1}{2}\sigma^2 S_t^2\, v_{ss}(S_t, t)\right) dt + \sigma S_t\, v_s(S_t, t)\, dW_t.$$

Combining the expression above with (3.23), we arrive at the following expression for the Itô differential of the process Y that equals $Y_t = v(S_t, t) - V_t(\phi)$

$$dY_t = \left(v_t(S_t, t) + \mu S_t\, v_s(S_t, t) + \tfrac{1}{2}\sigma^2 S_t^2\, v_{ss}(S_t, t)\right) dt + \sigma S_t\, v_s(S_t, t)\, dW_t$$
$$+ (r - \mu)S_t g(S_t, t)\, dt - \sigma S_t g(S_t, t)\, dW_t - rv(S_t, t)\, dt.$$

On the other hand, in view of (3.22), the process Y vanishes identically (i.e., is evanescent), and thus $dY_t = 0$. By virtue of the uniqueness of canonical decomposition of continuous semimartingales, the diffusion term in the above decomposition of Y vanishes. In our case, this means that we have, for every $t \in [0, T]$,

$$\int_0^t \sigma S_u \left(g(S_u, u) - v_s(S_u, u)\right) dW_u = 0,$$

where, as usual, the equality of random variables is meant to hold with probability 1, that is, $\mathbb{P}$-almost surely. In view of the properties of the Itô integral, this is equivalent to:

$$\int_0^T S_u^2 \left(g(S_u, u) - v_s(S_u, u)\right)^2 du = 0. \tag{3.24}$$

For (3.24) to hold, it is sufficient and necessary that the function g satisfies

$$g(s, t) = v_s(s, t), \quad \forall\, (s, t) \in \mathbb{R}_+ \times [0, T]. \tag{3.25}$$

Strictly speaking, equality (3.25) should hold $\mathbb{P} \otimes \lambda$-almost everywhere, where λ is the Lebesgue measure. We shall assume from now on that (3.25) holds. Then, using (3.25), we get still another representation for Y, namely

$$Y_t = \int_0^t \left\{ v_t(S_u, u) + \tfrac{1}{2}\sigma^2 S_u^2 v_{ss}(S_u, u) + r S_u v_s(S_u, u) - rv(S_u, u)\right\} du.$$

It is thus apparent that Y is evanescent whenever v satisfies the following partial differential equation (PDE), referred to as the *Black-Scholes PDE*

$$v_t(s, t) + \tfrac{1}{2}\sigma^2 s^2 v_{ss}(s, t) + rs v_s(s, t) - rv(s, t) = 0. \tag{3.26}$$

Since $C_T = v(S_T, T) = (S_T - K)^+$, we need to impose also the terminal condition $v(s, T) = (s - K)^+$ for $s \in \mathbb{R}_+$. It is not hard to check by direct computation that the function $v(s, t) = c(s, T - t)$, where c is given by (3.17)–(3.19), actually solves

---

[3] Subscripts on v denote partial derivatives with respect to the corresponding variables.

this problem.[4] It thus remains to check that the replicating strategy ϕ, given by the formula

$$\phi_t^1 = g(S_t, t) = v_s(S_t, t), \quad \phi_t^2 = h(S_t, t) = B_t^{-1}(v(S_t, t) - g(S_t, t)S_t),$$

is admissible. Let us first check that ϕ is indeed self-financing. Though this property is here almost trivial by the construction of ϕ, it is nevertheless always preferable to check directly the self-financing property of a given strategy (cf. Sect. 3.1.10). We need to check that

$$dV_t(\phi) = \phi_t^1 \, dS_t + \phi_t^2 \, dB_t.$$

Since $V_t(\phi) = \phi_t^1 S_t + \phi_t^2 B_t = v(S_t, t)$, by applying Itô's formula, we get

$$dV_t(\phi) = v_s(S_t, t) \, dS_t + \tfrac{1}{2}\sigma^2 S_t^2 \, v_{ss}(S_t, t) \, dt + v_t(S_t, t) \, dt.$$

In view of (3.26), the last equality can also be given the following form

$$dV_t(\phi) = v_s(S_t, t) \, dS_t + rv(S_t, t) \, dt - r S_t \, v_s(S_t, t) \, dt,$$

and thus

$$dV_t(\phi) = \phi_t^1 \, dS_t + r B_t \frac{v(S_t, t) - S_t \phi_t^1}{B_t} \, dt = \phi_t^1 \, dS_t + \phi_t^2 \, dB_t.$$

This ends the verification of the self-financing property. In view of our definition of admissibility of trading strategies, we need to verify that the discounted wealth process $V^*(\phi)$, which satisfies

$$V_t^*(\phi) = V_0^*(\phi) + \int_0^t v_s(S_u, u) \, dS_u^*, \tag{3.27}$$

follows a martingale under the martingale measure $\mathbb{P}^*$. By direct computation we obtain $v_s(s, t) = N(d_1(s, T - t))$ for every $(s, t) \in \mathbb{R}_+ \times [0, T]$, and thus, using also (3.10), we find that

$$V_t^*(\phi) = V_0^*(\phi) + \int_0^t \sigma S_u N(d_1(S_u, T - u)) \, dW_u^* = V_0^*(\phi) + \int_0^t \zeta_u \, dW_u^*,$$

where $\zeta_u = \sigma S_u N(d_1(S_u, T - u))$.

The existence of the stochastic integral is an immediate consequence of the sample path continuity of the process ζ. From the general properties of the Itô integral, it is thus clear that the discounted wealth $V^*(\phi)$ follows a local martingale under $\mathbb{P}^*$. To show that $V^*(\phi)$ is a genuine martingale (in fact, a square-integrable martingale), it is enough to observe that (for the first equality, see Sect. A.6)

---

[4] More precisely, the function v solves the final value problem for the backward PDE (3.26), while the function c solves the associated initial value problem for the forward PDE. To get the forward PDE, it is enough to replace $v_t(s, t)$ by $-v_t(s, t)$ in (3.26). Both of these PDEs are of parabolic type on $\mathcal{D}$.

$$\mathbb{E}_{\mathbb{P}*}\left(\int_0^T \zeta_u \, dW_u^*\right)^2 = \mathbb{E}_{\mathbb{P}*}\left(\int_0^T \zeta_u^2 \, du\right) \leq \sigma^2 \int_0^T \mathbb{E}_{\mathbb{P}*}\left(S_u^2\right) du < \infty,$$

where the last inequality follows from the existence of the exponential moments of a Gaussian random variable. Equality $c_s(S_t, T - t) = N(d_1(S_t, T - t))$ can be checked by the direct computation of the partial derivative of the pricing function c with respect to s. □

Second method. The second method of the proof puts more emphasis on the explicit evaluation of the price function c. The form of the replicating strategy will not be examined here. Since we wish to apply Corollary 3.1.3, we need to check first that the contingent claim $X = (S_T - K)^+$ is attainable in the Black-Scholes market model, however. This follows easily from the general results – more specifically, from the predictable representation property (see Theorem A.11.2) combined with the square-integrability of the random variable $X^* = B_T^{-1}(S_T - K)^+$ under the martingale measure $\mathbb{P}^*$. We conclude that there exists a predictable process θ such that the stochastic integral

$$V_t^* = V_0^* + \int_0^t \theta_u \, dW_u^*, \quad \forall t \in [0, T], \tag{3.28}$$

follows a (square-integrable) continuous martingale under $\mathbb{P}^*$, and

$$X^* = B_T^{-1}(S_T - K)^+ = \mathbb{E}_{\mathbb{P}*} X^* + \int_0^T \theta_u \, dW_u^* = \mathbb{E}_{\mathbb{P}*} X^* + \int_0^T h_u dS_u^*, \tag{3.29}$$

where we have put $h_t = \theta_t/(\sigma S_t^*)$. Let us set $V_t = B_t V_t^*$ and let us consider a trading strategy ϕ that is given by

$$\phi_t^1 = h_t, \quad \phi_t^2 = V_t^* - h_t S_t^* = B_t^{-1}(V_t - h_t S_t). \tag{3.30}$$

We shall check that the strategy ϕ is self-financing. The wealth process $V(\phi)$ agrees with V, and thus

$$\begin{aligned} dV_t(\phi) = d(B_t V_t^*) &= B_t \, dV_t^* + r V_t^* B_t \, dt \\ &= B_t h_t \, dS_t^* + r V_t \, dt = B_t h_t (B_t^{-1} \, dS_t - r B_t^{-1} S_t \, dt) + r V_t \, dt \\ &= h_t \, dS_t + r(V_t - h_t S_t) \, dt = \phi_t^1 \, dS_t + \phi_t^2 \, dB_t, \end{aligned}$$

as expected. It is also clear that $V_T(\phi) = V_T = (S_T - K)^+$, so that ϕ is in fact an admissible replicating strategy for X. So far, we have shown that the call option is represented by a contingent claim that is attainable in the Black-Scholes model $\mathcal{M}_{BS}$.

Our goal is now to evaluate the arbitrage price of X using the risk-neutral valuation formula. Since $\mathcal{F}_t^W = \mathcal{F}_t^S$ for every $t \in [0, T]$, the risk-neutral valuation formula (3.15) can be rewritten as follows

$$C_t = B_t \mathbb{E}_{\mathbb{P}*}\left((S_T - K)^+ B_T^{-1} \mid \mathcal{F}_t^S\right) = c(S_t, T - t) \tag{3.31}$$

for some function $c : \mathbb{R}_+ \times [0, T] \to \mathbb{R}$. The second equality in (3.31) can be inferred, for instance, from the Markovian property of S (recall that S follows a time-homogeneous Markov process under $\mathbb{P}^*$). Alternatively, we can make use of equality (3.13). The increment $W_T^* - W_t^*$ of the Brownian motion is independent of the σ-field $\mathcal{F}_t^W = \mathcal{F}_t^{W^*}$; on the other hand, the stock price S_t is manifestly $\mathcal{F}_t^W$-measurable.

By virtue of the well-known properties of conditional expectations (see Lemma A.1.1), we get

$$\mathbb{E}_{\mathbb{P}^*}\big((S_T - K)^+ \,|\, \mathcal{F}_t^S\big) = H(S_t, T - t), \tag{3.32}$$

where the function $H(s, T - t)$ is defined as follows

$$H(s, T - t) = \mathbb{E}_{\mathbb{P}^*}\left\{ \left(s\, e^{\sigma(W_T^* - W_t^*) + (r - \sigma^2/2)(T - t)} - K \right)^+ \right\}$$

for any $(s, t) \in \mathbb{R}_+ \times [0, T]$. It is therefore enough to focus on the case $t = 0$ and to evaluate the unconditional expectation

$$\mathbb{E}_{\mathbb{P}^*}\big((S_T - K)^+ B_T^{-1}\big) = \mathbb{E}_{\mathbb{P}^*}\big(S_T B_T^{-1} \mathbb{1}_D\big) - \mathbb{E}_{\mathbb{P}^*}\big(K B_T^{-1} \mathbb{1}_D\big) = J_1 - J_2,$$

where D stands for the set $\{S_T > K\}$. For J_2, we have

$$\begin{aligned}
J_2 &= e^{-rT} K \,\mathbb{P}^*\{S_T > K\} \\
&= e^{-rT} K \,\mathbb{P}^*\left\{ S_0 \exp\big(\sigma W_T^* + (r - \tfrac{1}{2}\sigma^2)T\big) > K \right\} \\
&= e^{-rT} K \,\mathbb{P}^*\left\{ -\sigma W_T^* < \ln(S_0/K) + (r - \tfrac{1}{2}\sigma^2)T \right\} \\
&= e^{-rT} K \,\mathbb{P}^*\left\{ \xi < \frac{\ln(S_0/K) + (r - \tfrac{1}{2}\sigma^2)T}{\sigma\sqrt{T}} \right\} \\
&= e^{-rT} K N(d_2(S_0, T)),
\end{aligned}$$

since the random variable $\xi = -W_T/\sqrt{T}$ has a standard Gaussian distribution $N(0, 1)$ under the martingale measure $\mathbb{P}^*$. For the other integral, note first that

$$J_1 = \mathbb{E}_{\mathbb{P}^*}(S_T B_T^{-1} \mathbb{1}_D) = \mathbb{E}_{\mathbb{P}^*}(S_T^* \mathbb{1}_D). \tag{3.33}$$

In order to reduce the computational burden, it is convenient to introduce an auxiliary probability measure $\bar{\mathbb{P}}$ on $(\Omega, \mathcal{F}_T)$ by setting

$$\frac{d\bar{\mathbb{P}}}{d\mathbb{P}^*} = \exp\big(\sigma W_T^* - \tfrac{1}{2}\sigma^2 T\big), \quad \mathbb{P}^*\text{-a.s.}$$

By virtue of Proposition A.15.1, the process $\bar{W}_t = W_t^* - \sigma t,\ t \in [0, T]$ is a standard Brownian motion on the space $(\Omega, \mathbb{F}, \bar{\mathbb{P}})$. Using (3.12), we obtain

$$S_T^* = S_0 \exp\big(\sigma \bar{W}_T + \tfrac{1}{2}\sigma^2 T\big). \tag{3.34}$$

Combining (3.33) with (3.34), we find that

$$J_1 = S_0 \bar{\mathbb{P}} \{ S_T^* > K B_T^{-1} \}$$
$$= S_0 \bar{\mathbb{P}} \left\{ S_0 \exp \left(\sigma \bar{W}_T + \tfrac{1}{2} \sigma^2 T \right) > K e^{-rT} \right\}$$
$$= S_0 \bar{\mathbb{P}} \left\{ - \sigma \bar{W}_T < \ln(S_0/K) + (r + \tfrac{1}{2} \sigma^2) T \right\}.$$

Using similar arguments as for J_2, we conclude that $J_1 = S_0 N(d_1(S_0, T))$. Summarizing, we have shown that the price at time 0 of a call option equals

$$C_0 = c(S_0, T) = S_0 N\big(d_1(S_0, T) \big) - K e^{-rT} N\big(d_2(S_0, T) \big),$$

where

$$d_{1,2}(S_0, T) = \frac{\ln(S_0/K) + (r \pm \tfrac{1}{2} \sigma^2) T}{\sigma \sqrt{T}}.$$

This ends the proof for the special case of $t = 0$. The valuation formula for any date $0 < t < T$ can now be deduced easily from (3.32).   □

Remarks. (a) One can check that the probability measure $\bar{\mathbb{P}}$ is the martingale measure corresponding to the choice of the stock price as a numeraire asset, that is, the unique probability measure, equivalent to $\mathbb{P}$, under which the process $\bar{B} = B/S$ is a martingale. In the second proof, we have in fact shown that (cf. formula (2.34))

$$C_t = S_t \bar{\mathbb{P}} \{ S_T > K \mid \mathcal{F}_t \} - K B(t, T) \mathbb{P}^* \{ S_T > K \mid \mathcal{F}_t \}.$$

(b) Undoubtedly, the most striking feature of the Black-Scholes result is the fact that the appreciation rate μ does not enter the valuation formula. This is not surprising, however, as expression (3.14) describing the evolution of the stock price under the martingale measure $\mathbb{P}^*$ does not involve μ. More generally, we could have assumed that the appreciation rate is not constant, but is varying in time, or even follows a stochastic process ($\mathbb{F}$-adapted for some underlying filtration $\mathbb{F}$).
(c) Let us stress that we have worked within a fully continuous-time set-up – that is, with continuously rebalanced portfolios. For obvious reasons, such an assumption is not justified from the practical viewpoint. It is thus interesting to note that the Black-Scholes result can be derived in a discrete-time set-up, by making use of the general equilibrium arguments (see Rubinstein (1976), Brennan (1979)). In such an approach, it is common to assume that the stock price has the jointly lognormal dynamics with the aggregate consumption, that is, with the market. Also, it is essential to impose specific restrictions on investors' risk preferences. Since the general equilibrium methodology is beyond the scope of this text, the interested reader is referred to the monograph by Huang and Litzenberger (1988) for details.

3.1.6 Case of Time-dependent Coefficients

Assume that the stock price process is governed by the stochastic differential equation

$$dS_t = \mu(t, S_t) S_t \, dt + \sigma(t) S_t \, dW_t, \quad S_0 > 0, \tag{3.35}$$

where $\mu : [0, T^*] \times \mathbb{R} \to \mathbb{R}$ is a deterministic function and $\sigma : [0, T^*] \to \mathbb{R}$ is also deterministic, with $\sigma(t) > \epsilon > 0$ for some constant ϵ. We assume that μ and σ are sufficiently regular, so that the SDE (3.35) admits a unique strong solution S, which is a continuous, strictly positive process. For instance, it suffices to assume that the function $\mu(t, s)$ is Lipschitz continuous in the second variable, and both μ and σ are bounded functions.

We introduce the accumulation factor B by setting

$$B_t = \exp\left(\int_0^t r(u)\, du\right), \quad \forall\, t \in [0, T], \tag{3.36}$$

for a deterministic function $r : [0, T^*] \to \mathbb{R}$. In view of (3.36), we have that

$$dB_t = r(t) B_t\, dt, \quad B_0 = 1,$$

so that $r(t)$ represents the instantaneous, continuously compounded interest rate prevailing at the market at time t. It is easy to check that the price of a zero-coupon bond satisfies

$$B(t, T) = e^{-\int_t^T r(u)\, du}, \quad \forall\, t \in [0, T].$$

Under the present hypotheses, the martingale measure $\mathbb{P}^*$ is unique, and the risk-neutral valuation formula (3.15) is still valid. Hence the price of a European call option satisfies, for every $t \in [0, T]$,

$$C_t = B(t, T)\, \mathbb{E}_{\mathbb{P}^*}\big((S_T - K)^+ \,\big|\, \mathcal{F}_t\big). \tag{3.37}$$

Note that under the martingale measure $\mathbb{P}^*$ we have

$$dS_t = r(t) S_t\, dt + \sigma(t) S_t\, dW_t^*. \tag{3.38}$$

If r and σ are, for instance, bounded functions, the unique solution to (3.38) is known to be

$$S_t = S_0 \exp\left(\int_0^t \sigma(u)\, dW_u^* + \int_0^t \big(r(u) - \tfrac{1}{2}\sigma^2(u)\big)\, du\right).$$

It is now an easy task to derive a suitable generalization of the Black-Scholes formula using (3.37). Indeed, it appears that it is enough to replace the quantities $r(T - t)$ and $\sigma^2(T - t)$ in the standard Black-Scholes formula by

$$r(t, T) = \int_t^T r(u)\, du \quad \text{and} \quad v^2(t, T) = \int_t^T \sigma^2(u)\, du, \tag{3.39}$$

respectively. The pricing function obtained in this way solves the Black-Scholes PDE with time-dependent coefficients.

3.1.7 Merton's Model

In this paragraph, the assumption that the interest rate is deterministic will be relaxed. Valuation of options under stochastic interest rates is examined at some length in the second part of this book. Here we consider only a very special model, which was put forward by Merton (1973). We assume that the stock price is given by (3.35). On the other hand, the evolution of the price $B(t, T)$ of a unit *zero-coupon bond*[5] is determined by the following expression

$$dB(t, T) = B(t, T)\big(a(t, T)\, dt + b(t, T)\, d\tilde{W}_t\big), \tag{3.40}$$

where $\tilde{W}$ is another standard Brownian motion, also defined on the underlying probability space $(\Omega, \mathbb{F}, \mathbb{P})$. For a fixed maturity T, the process $B(t, T)$, $t \in [0, T]$, is assumed to be adapted and strictly positive, also $B(T, T) = 1$ (a judicious choice of the coefficient a and b is essential here). In addition, we assume that $d\langle W, \tilde{W}\rangle_t = \rho\, dt$, where ρ is a real number. Finally, for a fixed maturity T, the volatility $b(t, T)$ of the bond price is assumed to be a deterministic function of t. Using the no-arbitrage arguments, Merton showed that the price of a European call option, with strike price K and expiry date T, equals

$$C_t = S_t N\big(h_1(S_t, t, T)\big) - K B(t, T) N\big(h_2(S_t, t, T)\big), \tag{3.41}$$

where

$$h_{1,2}(s, t, T) = \frac{\ln(s/K) - \ln B(t, T) \pm \frac{1}{2} v^2(t, T)}{v(t, T)}$$

and

$$v^2(t, T) = \int_t^T \big(\sigma^2(u) - 2\rho\sigma(u)b(u, T) + b^2(u, T)\big)\, du.$$

It is interesting to note that equality (3.41) follows easily from a much more general valuation result, which is established in Sect. 11.3.2.

Let us return to the case of a deterministic interest rate. In the present context, it corresponds to the assumption that the volatility of the bond price vanishes identically – that is, $b(t, T) = 0$ for every $t \le T$ and any maturity T. In the absence of arbitrage, this can be shown to imply also that the coefficients $a(t, T)$ are deterministic, and all zero-coupon bonds with differing maturities have identical instantaneous rates of return – that is, $a(t, T) = a(t, U)$ for every $t \le \min\{T, U\}$. We may thus define the short-term interest rate r by setting $r(t) = a(t, T)$ for every $t \le T$. Since r is a deterministic function, the bond price $B(t, T)$ equals

$$B(t, T) = \exp\left(-\int_t^T r(u)\, du\right), \quad \forall t \in [0, T]. \tag{3.42}$$

On the other hand, the price of a European call option is now given by (3.41), with

$$h_{1,2}(s, t, T) = \frac{\ln(s/K) + r(t, T) \pm \frac{1}{2} v^2(t, T)}{v(t, T)}$$

with $r(t, T)$ and $v^2(t, T)$ given by (3.39) (note that in that case we have $r(t, T) = -\ln B(t, T)$). If, in addition, the short-term rate and the volatility are constant, the pricing result above reduces, as it should, to the standard Black-Scholes formula.

---

[5] Recall that a unit zero-coupon bond maturing at time T is a security paying to its holder 1 unit of cash at a predetermined date T in the future.

3.1.8 Put-Call Parity for Spot Options

We shall show that if no arbitrage opportunities exist, puts and calls with the same strike and maturity must at all times during their lives obey the put-call parity relationship. Let us mention in this regard that empirical studies of the put-call parity are reported in Gould and Galai (1974) and Klemkosky and Resnick (1979).

A point worth stressing is that equality (3.43) does not rely on specific assumptions imposed on the stock price model. Indeed, it is satisfied in any arbitrage-free, continuous-time model of a security market, provided that the savings account B is given by the formula $B_t = e^{rt}$.

Proposition 3.1.2 *The arbitrage prices of European call and put options with the same expiry date T and strike price K satisfy the put-call parity relationship*

$$C_t - P_t = S_t - K e^{-r(T-t)} \qquad (3.43)$$

for every $t \in [0, T]$.

Proof. It is sufficient to observe that the payoffs of the call and put options at expiry satisfy the equality

$$(S_T - K)^+ - (K - S_T)^+ = S_T - K.$$

Relationship (3.43) now follows from the risk-neutral valuation formula. Alternatively, one may derive (3.43) using simple no-arbitrage arguments. □

The put-call parity can be used to derive a closed-form expression for the arbitrage price of a European put option. Let us denote by $p : \mathbb{R}_+ \times (0, T] \to \mathbb{R}$ the function

$$p(s, t) = K e^{-rt} N\big(-d_2(s, t)\big) - s N\big(-d_1(s, t)\big), \qquad (3.44)$$

with $d_1(s, t)$ and $d_2(s, t)$ given by (3.18)–(3.19). The following result is an immediate consequence of Proposition 3.1.2 combined with Theorem 3.1.1.

Corollary 3.1.4 *The Black-Scholes price at time $t < T$ of a European put option with strike price K equals $P_t = p(S_t, T - t)$, where the function $p : \mathbb{R}_+ \times [0, T] \to \mathbb{R}$ is given by (3.44). More explicitly, the price at time $t < T$ of a European put option equals*

$$P_t = K B(t, T) N\big(-d_2(S_t, T - t)\big) - S_t N\big(-d_1(S_t, T - t)\big).$$

3.1.9 Black-Scholes PDE

Our purpose is now to extend the valuation procedure of Theorem 3.1.1 to any contingent claim attainable in $\mathcal{M}_{BS}$, and whose values depend only on the terminal value of the stock price. Random payoffs of such a simple form are termed *path-independent* claims, as opposed to *path-dependent* contingent claims – that is, payoffs depending on the fluctuation of the stock price over a prespecified period of time before the settlement date. Path-dependent payoffs correspond to the various kinds of OTC options, known under the generic name of *exotic options.* For instance, a *lookback* call option, whose terminal payoff equals $(S_T - \min_{t \in [0,T]} S_t)^+$, can serve as an example of a path-dependent contingent claim. Another example of a path-dependent option is an *Asian option,* whose payoff at the expiry date is determined on the basis of the arithmetic average of the stock price during a predetermined time interval.

We start with an auxiliary result, dealing with a special case of the classical Feynman-Kac formula (see Sect. A.17). Basically, the Feynman-Kac formula expresses the solution of a parabolic PDE as the expected value of a certain functional of a Brownian motion. We refer to Sect. 4.3–4.4 in Karatzas and Shreve (1998a) for a thorough discussion of technical assumptions.

Lemma 3.1.4 *Let W be a one dimensional Brownian motion defined on a probability space $(\Omega, \mathcal{F}, \mathbb{P})$. For a Borel-measurable function $h : \mathbb{R} \to \mathbb{R}$, we define the function $u : \mathbb{R} \times [0, T] \to \mathbb{R}$ by setting*

$$u(x, t) = \mathbb{E}_{\mathbb{P}}\big(e^{-r(T-t)} h(W_T) \,\big|\, W_t = x\big) \quad \forall (x, t) \in \mathbb{R} \times [0, T]. \tag{3.45}$$

Suppose that

$$\int_{-\infty}^{+\infty} e^{-ax^2} |h(x)| \, dx < \infty \tag{3.46}$$

for some $a > 0$. Then the function u is defined for $0 < T - t < 1/2a$ and $x \in \mathbb{R}$, and has derivatives of all orders. In particular, it belongs to the class $C^{2,1}(\mathbb{R} \times (0, T))$ and satisfies the following PDE

$$-\frac{\partial u}{\partial t}(x, t) = \frac{1}{2} \frac{\partial^2 u}{\partial x^2}(x, t) - r u(x, t), \quad \forall (x, t) \in \mathbb{R} \times (0, T), \tag{3.47}$$

with the terminal condition $u(x, T) = h(x)$ for $x \in \mathbb{R}$.

Proof. From the fundamental properties of the Brownian motion, it is clear that u is given by the expression

$$u(x, t) = \frac{1}{\sqrt{2\pi(T-t)}} \int_{-\infty}^{+\infty} e^{-r(T-t)} h(y) e^{-\frac{(y-x)^2}{2(T-t)}} \, dy$$

for every $(x, y, t) \in \mathbb{R}^2 \times [0, T]$. Hence, all statements easily follow by straightforward computations (the reader is referred to Karatzas and Shreve (1998a) for details). □

Suppose we are given a Borel-measurable function $g : \mathbb{R} \to \mathbb{R}$. Then we have the following result, which generalizes Theorem 3.1.1. Let us observe that the problem of attainability of any $\mathbb{P}^*$-integrable European contingent claim can be resolved by invoking the predictable representation property of a Brownian motion (see Sect. A.11). Completeness of the multidimensional Black-Scholes model is examined in Sect. 8.2.1.

Corollary 3.1.5 *Let* $g : \mathbb{R} \to \mathbb{R}$ *be a Borel-measurable function, such that the random variable* $X = g(S_T)$ *is integrable under* $\mathbb{P}^*$. *Then the arbitrage price in* $\mathcal{M}_{BS}$ *of the claim* X *which settles at time* T *is given by the equality* $\pi_t(X) = v(S_t, t)$, *where the function* $v : \mathbb{R}_+ \times [0, T] \to \mathbb{R}$ *solves the Black-Scholes partial differential equation*

$$\frac{\partial v}{\partial t} + \frac{1}{2}\sigma^2 s^2 \frac{\partial^2 v}{\partial s^2} + rs\frac{\partial v}{\partial s} - rv = 0, \quad \forall\, (s, t) \in (0, \infty) \times (0, T), \tag{3.48}$$

subject to the terminal condition $v(s, T) = g(s)$.

Proof. We shall focus on the straightforward derivation of (3.48) from the risk-neutral valuation formula. By reasoning along similar lines to the second proof of Theorem 3.1.1, we find that the price $\pi_t(X)$ satisfies

$$\pi_t(X) = \mathbb{E}_{\mathbb{P}^*}\left(e^{-r(T-t)}\, g(S_T)\,\big|\,\mathcal{F}_t^S\right) = v(S_t, t) \tag{3.49}$$

for some function $v : \mathbb{R}_+ \times [0, T] \to \mathbb{R}$. Furthermore,

$$\pi_t(X) = \mathbb{E}_{\mathbb{P}^*}\left(e^{-r(T-t)}\, g(f(W_T^*, T))\,\big|\,\mathcal{F}_t^{W^*}\right),$$

where $f : \mathbb{R} \times [0, T] \to \mathbb{R}$ is a strictly positive function given by the formula

$$f(x, t) = S_0 e^{\sigma x + (r - \sigma^2/2)t} \quad \forall\, x \in \mathbb{R}. \tag{3.50}$$

Let us denote

$$u(x, t) = \mathbb{E}_{\mathbb{P}^*}\left(e^{-r(T-t)}\, g(f(W_T^*, T))\,\big|\, W_t^* = x\right).$$

By virtue of Lemma 3.1.4, the function $u(x, t)$ satisfies

$$-u_t(x, t) = \tfrac{1}{2} u_{xx}(x, t) - ru(x, t), \quad \forall\, (x, t) \in \mathbb{R} \times (0, T), \tag{3.51}$$

subject to the terminal condition $u(x, T) = g(f(x, T))$. A comparison of (3.49) and (3.45) yields the following relationship between $v(s, t)$ and $u(x, t)$

$$u(x, t) = v(f(x, t), t), \quad \forall\, (x, t) \in \mathbb{R} \times [0, T].$$

Therefore,

$$u_t(x, t) = v_s(s, t) f_t(x, t) + v_t(s, t),$$

where we set $s = f(x, t)$ so that $s \in (0, +\infty)$. Furthermore,

$$u_x(x, t) = v_s(s, t) f_x(x, t),$$

and thus

$$u_{xx}(x, t) = v_{ss}(s, t) f_x^2(x, t) + v_s(s, t)\, f_{xx}(x, t).$$

On the other hand, it follows from (3.50) that

$$f_x(x, t) = \sigma f(t, x), \quad f_{xx}(x, t) = \sigma^2 f(t, x),$$

and

$$f_t(x, t) = \left(r - \tfrac{1}{2}\sigma^2\right) f(x, t).$$

We conclude that

$$u_t(x, t) = s\left(r - \tfrac{1}{2}\sigma^2\right) v_s(s, t) + v_t(s, t)$$

and

$$u_{xx}(x, t) = \sigma^2 s^2 v_{ss}(s, t) + \sigma^2 s v_s(s, t).$$

Substitution into (3.51) gives

$$s\left(\tfrac{1}{2}\sigma^2 - r\right) v_s(s, t) - v_t(s, t) = \tfrac{1}{2}\sigma^2 s^2 v_{ss}(s, t) + \tfrac{1}{2}\sigma^2 s v_s(s, t) - rv(s, t).$$

Simplified, this yields (3.48). Both the terminal and boundary conditions are also evident. □

3.1.10 A Riskless Portfolio Method

A *risk-free portfolio* is a trading strategy created by taking positions in an option and a stock in such a way that the portfolio's wealth follows a process of finite variation. It can be verified that the wealth of a risk-free portfolio should appreciate at the risk-free rate (otherwise, it would be possible to create profitable risk-free strategies).

Let $v = v(s, t)$ be a smooth function $v : \mathbb{R}_+ \times [0, T] \to \mathbb{R}$ such that $v(s, T) = (s - K)^+$. As in the proof of Theorem 3.1.1, we assume a priori that the arbitrage price of a call option at time $t < T$ equals $C_t = v(S_t, t)$. The same reasoning can be applied to a put option on a stock S.

Our goal is to examine a specific trading strategy that involves a short position in one call option, and a long position in the underlying stock. Formally, we consider a dynamic portfolio $\phi_t = (\phi_t^1, \phi_t^2)$, where ϕ_t^1 and ϕ_t^2 stand for the number of shares of stock and the number of call options held at instant t respectively. More specifically, we assume that for every $t \in [0, T]$

$$\phi_t = \left(v_s(S_t, t), -1\right). \tag{3.52}$$

The wealth at time t of this strategy equals

$$V_t = V_t(\phi) = \phi_t^1 S_t + \phi_t^2 C_t = v_s(S_t, t) S_t - v(S_t, t). \tag{3.53}$$

Suppose that the trading strategy ϕ is self-financing. Then V satisfies

$$dV_t = \phi_t^1 dS_t + \phi_t^2 dC_t = v_s(S_t, t) dS_t - dv(S_t, t). \tag{3.54}$$

We wish to re-derive the Black-Scholes PDE by taking (3.54) as a starting point. Note that Itô's formula gives

$$dv(S_t, t) = \left(\mu S_t v_s(S_t, t) + \tfrac{1}{2}\sigma^2 S_t^2 v_{ss}(S_t, t) + v_t(S_t, t)\right) dt + \sigma S_t v_s(S_t, t) dW_t.$$

Substituting this into (3.54), we obtain

$$dV_t = -\left(v_t(S_t, t) + \tfrac{1}{2}\sigma^2 S_t^2 v_{ss}(S_t, t)\right) dt, \tag{3.55}$$

so that the strategy ϕ appears to be *instantaneously risk-free*, meaning that its wealth process is a continuous, adapted process of finite variation. Since under the martingale measure the discounted wealth process $V_t^* = B_t^{-1} V_t$ of any self-financing trading strategy is known to follow a local martingale, the last equality implies that the continuous process of finite variation V^* is constant.[6] Put another way, we have shown that $dV_t^* = 0$, or equivalently, that $dV_t = rV_t \, dt$. On the other hand, in view of (3.55), the differential dV_t^* satisfies also

$$dV_t^* = B_t^{-1} dV_t - r B_t^{-1} V_t \, dt,$$

or more explicitly

$$dV_t^* = -B_t^{-1}\left(v_t(S_t, t) + r S_t v_s(S_t, t) + \tfrac{1}{2}\sigma^2 S_t^2 v_{ss}(S_t, t) - rv(S_t, t)\right) dt.$$

The equality $dV_t^* = 0$ for all $t \in [0, T]$ is therefore satisfied if and only if we have, for any $t \in [0, T]$,

$$\int_0^t \left\{ v_t(S_u, u) + r S_u v_s(S_u, u) + \tfrac{1}{2}\sigma^2 S_u^2 v_{ss}(S_u, u) - rv(S_u, u) \right\} du = 0.$$

This in turn holds if and only if the function v solves the Black-Scholes PDE. This shows that the Black-Scholes PDE, and thus also the Black-Scholes option valuation result, can be obtained via the risk-free portfolio approach.

So far we have implicitly assumed that the portfolio ϕ given by (3.52) is self-financing. However, to completely justify the above proof of the Black-Scholes formula, we still need to verify whether the strategy ϕ is self-financing when v solves the Black-Scholes PDE. Unfortunately, this is not the case, as Proposition 3.1.3 shows. Thus, the rather frequent derivation of the Black-Scholes formula in existing financial literature through the risk-free portfolio approach is mathematically unsatisfactory (to our knowledge, this fact was first noted by Bergman (1982)). A point worth stressing is that the risk-free portfolio approach is unquestionable in a discrete-time setting.

Proposition 3.1.3 *Suppose that the trading strategy ϕ is given by (3.52) with the function $v(s, t) = c(s, T - t)$, where the function c is given by the Black-Scholes formula (3.17). Then the strategy ϕ is not self-financing; that is, condition (3.54) fails to hold.*

Proof. To check whether the strategy ϕ is self-financing, it is sufficient to evaluate the stochastic differential dV_t, using directly expression (3.53), and to compare the result with the self-financing condition (3.54).

---

[6] It is well known that all (local) martingales with respect to the natural filtration of a Brownian motion have sample paths that are almost all of infinite variation, unless constant.

In order to make direct calculations of dV_t we need, in particular, to find the Itô differential of the product $v_s(S_t, t)S_t$ of two Itô processes. For ease of notation, we shall write U to denote the process $U_t = v_s(S_t, t)$ for $t \in [0, T]$. An application of Itô's formula yields

$$dU_t = \left(\mu S_t \, v_{ss}(S_t, t) + \tfrac{1}{2}\sigma^2 S_t^2 \, v_{sss}(S_t, t) + v_{ts}(S_t, t)\right) dt + \sigma S_t \, v_{ss}(S_t, t) \, dW_t.$$

Consequently, once again by the Itô formula, we get

$$dV_t = d(U_t \, S_t) - dv(S_t, t) = S_t \, dU_t + U_t \, dS_t + d\langle U, S\rangle_t - dv(S_t, t)$$

and thus

$$\begin{aligned}
dV_t &= S_t\left(\mu S_t \, v_{ss}(S_t, t) + \tfrac{1}{2}\sigma^2 S_t^2 \, v_{sss}(S_t, t) + v_{ts}(S_t, t)\right) dt \\
&\quad + \sigma S_t^2 \, v_{ss}(S_t, t) \, dW_t + v_s(S_t, t) \, dS_t + \sigma^2 S_t^2 \, v_{ss}(S_t, t) \, dt - dv(S_t, t).
\end{aligned}$$

Suppose that the trading strategy ϕ is self-financing. In the present context, this means that the following equality holds (cf. (3.53))

$$dV_t = v_s(S_t, t) \, dS_t - dv(S_t, t).$$

For the last equality to be satisfied, it is necessary that (for convenience, the argument (S_t, t) is suppressed)

$$S_t\left(\mu S_t \, v_{ss} + \tfrac{1}{2}\sigma^2 S_t^2 \, v_{sss} + \sigma^2 S_t \, v_{ss} + v_{ts}\right) dt + \sigma S_t^2 \, v_{ss} \, dW_t = 0.$$

Using the Black-Scholes PDE, differentiated with respect to the variable s, we may rewrite the last equality in the following way

$$\sigma S_t^2 v_{ss}(S_t, t)\left(dW_t + \sigma^{-1}(\mu - r) \, dt\right) = \sigma S_t^2 v_{ss}(S_t, t) \, dW_t^* = 0,$$

where W^* is the Brownian motion under the martingale measure $\mathbb{P}^*$. It is evident that the last equality is not valid, since the second derivative of the option price with respect to s does not vanish, either for the call or for the put option, that is, when $v(s, t) = c(s, T - t)$ or $v(s, t) = p(s, T - t)$. An explicit expression for these derivatives is given in the next section,[7] in which we deal with the sensitivities of the option price with respect to underlying variables and model parameters.

We conclude that the trading strategy ϕ introduced in the statement of the proposition is not self-financing. Notice, however, that the (discounted) additional cost associated with ϕ up to time T is easily seen to be represented by a random variable with zero mean under the martingale measure $\mathbb{P}^*$. This peculiar feature of ϕ explains why the risk-free portfolio approach leads to the correct form of the option valuation formula.   □

---

[7] The put-call parity relationship makes it obvious that they have the same value.

3.1.11 Black-Scholes Sensitivities

We will now examine the basic features of trading portfolios involving options. Let us start by introducing the terminology widely used in relation to option contracts. We say that at a given instant t before or at expiry, a call option is *in-the-money* and *out-of-the-money* if $S_t > K$ and $S_t < K$ respectively. Similarly, a put option is said to be in-the-money and out-of-the-money at time t when $S_t < K$ and $S_t > K$ respectively. Finally, when $S_t = K$ both options are said to be *at-the-money*. The *intrinsic values* of a call and a put options are defined by the formulas

$$I_t^C = (S_t - K)^+, \quad I_t^P = (K - S_t)^+,$$

respectively, and the *time values* equal

$$J_t^C = C_t - (S_t - K)^+, \quad J_t^P = P_t - (K - S_t)^+,$$

for $t \in [0, T]$. It is thus evident that an option is in-the-money if and only if its intrinsic value is strictly positive. A short position in a call option is referred to as a *covered call* if the writer of the option hedges his or her risk exposure by holding the underlying stock; in the opposite case, the position is known as a *naked call*. Writing *covered call* options is a popular practice among portfolio managers, as such a strategy seems to offer obvious advantages. Let us consider, for example, a call option with the strike price K above the current stock price S_t. If the stock appreciates and is called away, the portfolio still gains the option price and the difference $K - S_t$ (also, it receives dividends for the period before exercise). If the stock price rises, but the option is out-of-the money at expiry (and thus is abandoned), the portfolio will gain the option price and the stock appreciation up to the strike price – that is, $S_T - S_t$. Finally, if the stock declines in price, the loss $S_t - S_T$ is buffered by the option price, so that the total loss from the position will be less than if no option were written. Summarizing, writing covered call options significantly reduces the risk exposure of a stockholder, traditionally measured by means of the variance of the return distribution. It should be observed, however, that most of the drop in variance occurs in that part of the return distribution where high variance is desirable – that is, on the right-hand side. Still, it is possible (at least theoretically) for a covered call strategy to give a higher expected return and a lower variance than the underlying stock.

When an investor who holds a stock also purchases a put option on this stock as a protection against stock price decline, the position is referred to as a *protective put*. While writing covered calls truncates, roughly speaking, the right-hand side of the return distribution and simultaneously shifts it to the right, buying protective puts truncates the left-hand side of the return distribution and at the same time shifts the distribution to the left. The last effect is due to the fact that the cost of a put increases the initial investment of a portfolio.

Note that the classical mean-variance analysis, pioneered by Markowitz (1952),[8] does not provide an appropriate performance measure for portfolios with options

---

[8] The more recent literature include Markowitz (1987), Huang and Litzenberger (1988), and Elton and Gruber (1995).

because of the skewness that may be introduced into portfolio returns, and the fact the volatility is not constant (see Chap. 7). For more details on the effectiveness of option portfolio management, the interested reader may consult Leland (1980), Bookstaber and Clarke (1984, 1985), and Sheedy and Trevor (1996, 1999).

To measure quantitatively the influence of an option's position on a given portfolio of financial assets, we will now examine the dependence of its price on the fluctuations of the current stock price, time to expiry, strike price, and other relevant parameters. For a fixed expiry date T and arbitrary $t \leq T$, we denote by τ the time to option expiry – that is, we put $\tau = T - t$. We write $p(S_t, \tau, K, r, \sigma)$ and $c(S_t, \tau, K, r, \sigma)$ to denote the price of a call and a put option respectively. The functions c and p are thus given by the formulas

$$c(s, \tau, K, r, \sigma) = sN(d_1) - Ke^{-r\tau}N(d_2) \tag{3.56}$$

and

$$p(s, \tau, K, r, \sigma) = Ke^{-r\tau}N(-d_2) - sN(-d_1) \tag{3.57}$$

with

$$d_{1,2} = d_{1,2}(s, \tau, K, \sigma, r) = \frac{\ln(s/K) + (r \pm \frac{1}{2}\sigma^2)\tau}{\sigma\sqrt{\tau}}. \tag{3.58}$$

Recall that at any time $t \in [0, T]$, the replicating portfolio of a call option involves α_t shares of stock and β_t units of borrowed funds, where

$$\alpha_t = c_s(S_t, \tau) = N(d_1(S_t, \tau)), \quad \beta_t = c(S_t, \tau) - \alpha_t S_t.$$

The strictly positive number α_t that determines the number of shares in the replicating portfolio is termed the *hedge ratio* or, briefly, the *delta* of the option. It is not hard to verify by straightforward calculations that

$$c_s = N(d_1) = \delta > 0,$$

$$c_{ss} = \frac{n(d_1)}{s\sigma\sqrt{\tau}} = \gamma > 0,$$

$$c_\tau = \frac{s\sigma}{2\sqrt{\tau}}n(d_1) + Kre^{-r\tau}N(d_2) = \theta > 0,$$

$$c_\sigma = s\sqrt{\tau}n(d_1) = \lambda > 0,$$

$$c_r = \tau Ke^{-r\tau}N(d_2) = \rho > 0,$$

$$c_K = -e^{-r\tau}N(d_2) < 0,$$

where n stands for the standard Gaussian probability density function:

$$n(x) = \frac{1}{\sqrt{2\pi}}e^{-x^2/2}, \quad \forall x \in \mathbb{R}.$$

Similarly, in the case of a put option we get

$$p_s = N(d_1) - 1 = -N(-d_1) = \delta < 0,$$
$$p_{ss} = \frac{n(d_1)}{s\sigma\sqrt{\tau}} = \gamma > 0,$$
$$p_\tau = \frac{s\sigma}{2\sqrt{\tau}} n(d_1) + Kre^{-r\tau}(N(d_2) - 1) = \theta,$$
$$p_\sigma = s\sqrt{\tau}n(d_1) = \lambda > 0,$$
$$p_r = \tau Ke^{-r\tau}(N(d_2) - 1) = \rho < 0,$$
$$p_K = e^{-r\tau}(1 - N(d_2)) > 0.$$

Consequently, the delta of a long position in a put option is a strictly negative number (equivalently, the price of a put option is a strictly decreasing function of a stock price). Generally speaking, the price of a put moves in the same direction as a short position in the asset. In particular, in order to hedge a written put option, an investor needs to short a certain number of shares of the underlying stock.

Another useful coefficient measuring the relative change of an option's price as the stock price moves is the *elasticity*. For any date $t \le T$, the elasticity of a call option is given by the equality

$$\eta_t^c \stackrel{\text{def}}{=} c_s(S_t, \tau)S_t/C_t = N(d_1(S_t, \tau))S_t/C_t,$$

and for a put option it equals

$$\eta_t^p \stackrel{\text{def}}{=} p_s(S_t, \tau)S_t/P_t = -N(-d_1(S_t, \tau))S_t/P_t.$$

Let us check that the elasticity of a call option price is always greater than 1. Indeed, for every $t \in [0, T]$, we have

$$\eta_t^c = 1 + e^{-r\tau}KC_t^{-1}N(d_2(S_t, \tau)) > 1.$$

This implies also that $C_t - c_s(S_t, \tau)S_t < 0$, and thus the replicating portfolio of a call option always involves the borrowing of funds. Similarly, the elasticity of a put option satisfies

$$\eta_t^p = 1 - Ke^{-r\tau}P_t^{-1}N(-d_2(S_t, \tau)) < 1.$$

This in turn implies that $P_t - S_t p_s(S_t, \tau) > 0$ (this inequality is obvious anyway) and thus the replicating portfolio of a short put option generates funds which are invested in risk-free bonds. These properties of replicating portfolios have special consequences when the assumption that the borrowing and lending rates coincide is relaxed. It is instructive to determine the dynamics of the option price C under the martingale measure $\mathbb{P}^*$. Using Itô's formula, one finds easily that under the martingale measure $\mathbb{P}^*$ we have

$$dC_t = C_t(r\,dt + \sigma\eta_t^c\,dW_t^*).$$

The last formula shows that the appreciation rate of the option price in a risk-neutral economy equals the risk-free rate r. The volatility coefficient equals $\sigma \eta_t^c$ so that, in contrast to the stock price volatility, the volatility of the option price follows a stochastic process (note also that $\sigma \eta_t^c > \sigma$). This feature makes the influence of an option's position on the performance of a portfolio of financial assets rather difficult to quantify. Some of the qualitative properties of the arbitrage price inherited from the corresponding properties of the terminal payoff (such as monotonicity, convexity, the sign of the hedge ratio, etc.), which in the Black-Scholes setting can be easily deduced from the knowledge of the partial derivatives, can be shown to hold also in a more general setting, when the stock price is assumed to obey a one-dimensional diffusion. Interesting results in this vein can be found in Jagannathan (1984), Grundy (1991), Bergman et al. (1996), and El Karoui et al. (1998).

The *position delta* is obtained by multiplying the *face value*[9] of the option position by its delta. Clearly, the position delta of a long call option (or a short put option) is positive; on the contrary, the position delta of a short call option (and of a long put option) is a negative number. The position delta of a portfolio is obtained by summing up the position deltas of its components. In this context, let us make the trivial observation that the position delta of a long stock equals 1, and that of a short stock is -1. It is clear that the option's (or the option portfolio's) position delta measures only the market exposure at the current price levels of underlying assets. More precisely, it gives the first order approximation of the change in option price, and thus it is sufficiently accurate only for a small move in the underlying asset price.

To measure the change in the option delta as the underlying asset price moves, one should use the second derivative with respect to s of the option's price – that is, the option's *gamma*. The *gamma effect* means that position deltas also move as asset prices fluctuate, so that predictions of revaluation profit and loss based on position deltas are not sufficiently accurate, except for small moves. It is easily seen that bought options have positive gammas, while sold options have negative gammas. A portfolio's gamma is the weighted sum of its options' gammas, and the resulting gamma is determined by the dominant options in the portfolio. In this regard, options close to the money with a short time to expiry have a dominant influence on the portfolio's gamma. Generally speaking, a portfolio with a positive gamma is more attractive than a negative gamma portfolio. Recall that by *theta* we have denoted the derivative of the option price with respect to time to expiry.

As a rule, a portfolio dominated by bought options will have a negative theta, meaning that the portfolio will lose value as time passes (other variables held constant). In contrast, short options generally have positive thetas. Finally, options close to the money will have more influence on the position theta than options far from the money.

The derivative of the option price with respect to volatility is known as the *vega* of an option. A positive vega position will result in profits from increases in volatility; similarly, a negative vega means a strategy will profit from falling volatility. To create

---

[9] The *face value* equals the number of underlying assets, e.g., the face value of an option on a lot of 100 shares of stock equals 100.

a positive vega, a trader needs to dominate the portfolio with bought options, bearing in mind that the vega will be dominated by those options that are close to the money and have significant remaining time to expiry.

Example 3.1.2 Consider a call option on a stock S, with strike price $30 and with 3 months to expiry. Suppose, in addition, that the current stock price equals $31, the stock price volatility is $\sigma = 10\%$ per annum, and the risk-free interest rate is $r = 5\%$ per annum with continuous compounding. We may assume, without loss of generality, that $t = 0$ and $T = 0.25$. Using (3.18), we obtain (approximately) $d_1(S_0, T) = 0.93$, and thus $d_2(S_0, T) = d_1(S_0, T) - \sigma\sqrt{T} = 0.88$. Consequently, using formula (3.17) and the following values of the standard Gaussian probability distribution function: $N(0.93) = 0.8238$ and $N(0.88) = 0.8106$, we find that (approximately) $C_0 = 1.52$, $\phi_0^1 = 0.82$ and $\phi_0^2 = -23.9$. This means that to hedge a short position in the call option sold at the arbitrage price $C_0 = \$1.52$, an investor needs to purchase at time 0 the number $\delta = 0.82$ shares of stock (this transaction requires an additional borrowing of 23.9 units of cash). The elasticity at time 0 of the call option price with respect to the stock price equals

$$\eta_0^c = \frac{N\big(d_1(S_0, T)\big) S_0}{C_0} = 16.72.$$

Suppose that the stock price rises immediately from $31 to $31.2, yielding a return rate of 0.65% flat. Then the option price will move by approximately 16.5 cents from $1.52 to $1.685, giving a return rate of 10.86% flat. Roughly speaking, the option has nearly 17 times the return rate of the stock; of course, this also means that it will drop 17 times as fast. If an investor's portfolio involves 5 long call options (each on a round lot of 100 shares of stock), the position delta equals $500 \times 0.82 = 410$, so that it is the same as for a portfolio involving 410 shares of the underlying stock. Let us now assume that an option is a put. The price of a put option at time 0 equals (alternatively, P_0 can be found from the put-call parity (3.43))

$$P_0 = 30 \, e^{-0.05/4} \, N(-0.88) - 31 \, N(-0.93) = 0.15.$$

The *hedge ratio* corresponding to a short position in the put option equals approximately $\delta = -0.18$ (since $N(-0.93) = 0.18$), therefore to hedge the exposure, using the Black-Scholes recipe, an investor needs to short 0.18 shares of stock for one put option. The proceeds from the option and share-selling transactions, which amount to $5.73, should be invested in risk-free bonds. Note that the elasticity of the put option is several times larger than the elasticity of the call option. If the stock price rises immediately from $31 to $31.2, the price of the put option will drop to less than 12 cents.

3.1.12 Market Imperfections

Let us summarize once again the basic assumptions of the Black-Scholes model that allowed us to value European options by arbitrage reasoning:

(a) the market for the stock, the option and cash is perfect;
(b) present and future interest rates are known with certainty;
(c) the lending and borrowing interest rates are equal;
(d) the stock price has a known constant volatility;
(e) there are no transaction costs or taxes;
(f) there are no margin requirements.

Let us comment briefly on the last assumption. Market regulations usually impose restrictions upon the amount of funds one can borrow to purchase securities. For instance, if there is a 50% *margin requirement,* 50% at most of the stock's value can be borrowed at the time of purchase. If the stock price declines after initiation, the borrowings may rise by up to 75% of the stock's value. On the other hand, when a party shorts a stock, a margin account must be established with a balance of at least 50% of stock's initial value. If the stock price rises, it remains at least 30% of the subsequent stock's value. The proceeds from the short sale are held by the broker and usually do not earn interest (on the contrary, the margin account earns interest). When purchasing naked calls, no margin is required, but when selling a naked call, a margin must be maintained, just as if it were a short sale. Alternatively, the underlying security can be placed with the broker, resulting in a covered call. When the assumption (e) is relaxed, the frequency of transactions may mean that the exact formulas are very sensitive to the imposition of even small transaction costs. In essence, positive transaction costs impose some risk on neutral hedgers who must adopt finite holding periods. If these costs are not too high, and the hedgers are not too risk-averse, the exact formulas established in Sect. 3.1 will still prove useful.

In practice, even if the portfolio adjustment is based on the continuous-time Black-Scholes model, the rebalancing is made periodically. An important problem is therefore to study the efficiency of discretely adjusted option hedges. The bias (systematic risk) of discretely rebalanced option hedges were examined by, among others, Boyle and Emanuel (1980), Cox and Rubinstein (1985), Gilster (1990), and Mercurio and Vorst (1996). Leland (1985), Gilster and Lee (1984), Henrotte (1991), Grannan and Swindle (1996), Toft (1996), Kabanov and Safarian (1997), and Whalley and Wilmott (1997) extend this methodology by taking transaction costs into account. Johnson and Stulz (1987), Hull and White (1995), Jarrow and Turnbull (1995), and Klein (1996) study the effect of credit risk on option prices.

Empirical tests of the Black-Scholes model and its alternatives were the subject of several studies; to mention just a few: Black and Scholes (1972), MacBeth and Merville (1979, 1980), Rubinstein (1985), Lauterbach and Schultz (1990), and Ncube and Satchell (1997).

3.1.13 Numerical Methods

Since a closed-form expression for the arbitrage price of a claim is not always available, an important issue is the study of numerical methods that yield approximations of arbitrage prices and hedging strategies. In the Black-Scholes framework, these methods include: the use of multinomial lattices (trees) to approximate continuous-time models of security prices; procedures based on the Monte Carlo simulation of random variables; and finite-difference methods of solving the associated partial differential equations. Since a detailed study of various techniques is far beyond the scope of this text, we shall only survey briefly the relevant literature. The interested reader may consult recent monographs and textbooks by Clewlow and Strickland (1998), Jäckel (2002), Seydel (2002), and Glasserman (2003).

Binomial (or, more generally, multinomial) models of security prices were studied by Cox et al. (1979a) and Boyle (1986), who proposed an approximation of the stock price by means of the trinomial tree. Generally speaking, the main goal of these, and several other, studies lies in improving the accuracy and/or speed of numerical pricing procedures. Let us observe in this regard that the market practice frequently requires the ability to perform a re-evaluation of a large number of derivative securities in a reasonably short period of time.

Let us return to the classical CRR binomial approximation of the exponential Brownian motion, which was examined in Sect. 2.3. If a continuous-time framework is taken as a benchmark, an increase in accuracy is gained by assuming that the first two moments of an approximating binomial process coincide with the corresponding moments of the exponential Brownian motion used to model the stock price. The corresponding modification of the CRR model of Chap. 2.1 runs as follows. For a fixed $T > 0$ and arbitrary $n \in \mathbb{N}$, we write $\Delta_n = T/n$. Notice that for every $j = 0, \dots, n-1$ we have

$$S_{(j+1)\Delta_n} = S_{j\Delta_n} \exp\left(\sigma\left(W^*_{(j+1)\Delta_n} - W^*_{j\Delta_n}\right) + \left(r - \tfrac{1}{2}\sigma^2\right)\Delta_n\right)$$

under the martingale measure $\mathbb{P}^*$. Hence the expected value of the ratio $S_{(j+1)\Delta_n}/S_{j\Delta_n}$ equals

$$m_1(n) = \mathbb{E}_{\mathbb{P}^*}(S_{(j+1)\Delta_n}/S_{j\Delta_n}) = \exp(r\Delta_n)$$

for every j, and the second moment of this ratio is

$$m_2(n) = \mathbb{E}_{\mathbb{P}^*}(S_{(j+1)\Delta_n}/S_{j\Delta_n})^2 = \exp(2r\Delta_n + \sigma^2\Delta_n).$$

The corresponding values of the parameters u_n, d_n and p_n can be found by solving the following equations

$$\begin{cases} p_n u_n + (1-p_n)d_n = m_1(n), \\ p_n u_n^2 + (1-p_n)d_n^2 = m_2(n), \end{cases}$$

with $d_n = u_n^{-1}$.

It is worth pointing out that the convergence result of Sect. 2.3 remains valid for the above modification of the CRR binomial model. More accurate approximation results were obtained by considering a trinomial lattice, as proposed by Boyle (1986). In a trinomial lattice, there are three possible future states for every node. For a fixed j, let us denote by $p_{jk}(i)$ the transition probability from the state $s_j(i)$ at time $j\Delta_n$ to the state $s_{j+1}(k)$ at time $(j+1)\Delta_n$. The transition probabilities $p_{jk}(i)$ and the state values $s_j(i)$ must be chosen in such a way that the lattice accurately approximates the behavior of the stock price S under the martingale probability. Once the lattice is already constructed, the valuation of European (and American) contingent claims is done by the standard backward induction method.

In order to evaluate the prices of options depending on two underlying assets, Boyle (1988) extends this technique to the case of a bivariate multinomial model. For further developments of the binomial approach, which take into account the presence of the volatility smile effect, we refer to Rubinstein (1994) and Derman et al. (1996a, 1996b). Related empirical studies were reported in a recent paper by Jackwerth and Rubinstein (1996).

Another widely popular numerical technique in option pricing is the Monte Carlo method. The efficient valuation of derivative securities (with the special emphasis on exotic options) using the Monte Carlo simulation was studied by, among others, Boyle (1977), Johnson and Shanno (1987), Duffie and Glynn (1995), Boyle et al. (1997), Broadie and Glasserman (1997a), Fournié et al. (1997c, 1999), Lehoczky (1997), and Glasserman et al. (1999). In particular, the Monte Carlo method was applied by Hull and White (1987a) and Fournié et al. (1997b) in the context of stochastic volatility models.

Finite-difference methods of solving partial differential equations are examined in the context of option valuation in Schwartz (1977), Brennan and Schwartz (1978a), and Courtadon (1982b). Since a presentation of these methods is beyond the scope of this book, let us mention only that one can use both explicit and implicit finite-difference schemes, as well as the Crank-Nicolson method. For a detailed analysis of finite-difference methods in option valuation, we refer the reader to the monograph by Wilmott et al. (1993), in which the authors successfully apply the PDE approach to numerous types of standard and exotic options (also accounting for the presence of (small) transaction costs). Hull and White (1988a) apply the control variate technique in order to improve the efficiency of the finite-difference method when valuing American options on a dividend-paying stock (let us mention that the use of the control variate technique in conjunction with the Monte Carlo method goes back to Boyle (1977)).

For more advanced studies related to the computation of continuous-time option prices using discrete-time approximations, we refer the reader also to Geske and Shastri (1985a), Omberg (1987b), Nelson and Ramaswamy (1990), Amin (1991), Leisen and Reimer (1996), Rogers and Stapleton (1998), and Broadie et al. (1999).

3.2 A Dividend-paying Stock

Up to now, we have studied arbitrage pricing within the Black-Scholes framework under the assumption that the stock upon which an option is written pays no dividends during option's lifetime. In this section, we assume that the dividends (or the dividend rate) that will be paid to the shareholders during an option's lifetime can be predicted with certainty.

3.2.1 Case of a Constant Dividend Yield

We will first examine the case when the dividend yield, rather than the dividend pay-offs, is assumed to be known. More specifically, we assume that the stock S continuously pays dividends at some fixed rate κ. Following Samuelson (1965), we assume that the effective dividend rate is proportional to the level of the stock price, that is, the dividend payment over an infinitesimally short time interval $[t, t + dt]$ equals $\kappa S_t\, dt$ per share. Although this is rather impractical as a realistic dividend policy associated with a particular stock, Samuelson's model fits the case of a stock index option reasonably well. The dividend payments should be used in full, either to purchase additional shares of stock, or to invest in risk-free bonds (however, intertemporal consumption or infusion of funds is not allowed). The following definition is a natural extension of Definition 3.1.1.

Definition 3.2.1 A trading strategy $\phi = (\phi^1, \phi^2)$ is said to be *self-financing* when its wealth process $V(\phi)$, which equals $V_t(\phi) = \phi_t^1 S_t + \phi_t^2 B_t$, satisfies

$$dV_t(\phi) = \phi_t^1\, dS_t + \kappa \phi_t^1 S_t\, dt + \phi_t^2\, dB_t. \tag{3.59}$$

In view of the postulated dynamics of the stock price, formula (3.59) can also be represented as follows

$$dV_t(\phi) = \phi_t^1 (\mu + \kappa) S_t\, dt + \phi_t^1 \sigma S_t\, dW_t + \phi_t^2\, dB_t.$$

To simplify the analysis, we find it convenient to introduce an auxiliary process $\tilde{S}_t = e^{\kappa t} S_t$, whose dynamics are given by the stochastic differential equation

$$d\tilde{S}_t = e^{\kappa t} (dS_t + \kappa S_t\, dt) = \mu_\kappa \tilde{S}_t\, dt + \sigma \tilde{S}_t\, dW_t,$$

where we write $\mu_\kappa = \mu + \kappa$. In terms of the process $\tilde{S}$, we obtain

$$V_t(\phi) = \phi_t^1 e^{-\kappa t}\, \tilde{S}_t + \phi_t^2 B_t$$

and

$$dV_t(\phi) = \phi_t^1 e^{-\kappa t} d\tilde{S}_t + \phi_t^2\, dB_t.$$

Also, it is not difficult to check that the discounted wealth $V^*(\phi)$ satisfies

$$dV_t^*(\phi) = \phi_t^1 e^{-\kappa t} d\tilde{S}_t^*,$$

where we denote $\tilde{S}_t^* = \tilde{S}_t B_t^{-1}$. More explicitly, we have that

$$dV_t^*(\phi) = \sigma \phi_t^1 e^{-\kappa t} \tilde{S}_t^* \left(dW_t + \sigma^{-1} (\mu_\kappa - r)\, dt \right).$$

This leads to the following lemma. Recall (cf. Definition 3.1.3) that a probability measure $\mathbb{P}^*$ on $(\Omega, \mathcal{F}_{T^*})$, equivalent to $\mathbb{P}$, is a spot martingale measure if the discounted wealth $V_t^*(\phi) = V_t(\phi)/B_t$ of any self-financing trading strategy ϕ is a local martingale under $\mathbb{P}^*$.

Lemma 3.2.1 *The unique spot martingale measure $\mathbb{P}^*$ for the Black-Scholes model in which the stock S continuously pays dividends at some fixed rate κ is given by formula (3.11) with μ replaced by μ_κ. For any self-financing trading strategy ϕ, the dynamics of $V^*(\phi)$ under $\mathbb{P}^*$ are given by the expression*

$$dV_t^*(\phi) = \sigma \phi_t^1 e^{-\kappa t} \tilde{S}_t^* \, d\tilde{W}_t = \phi_t^1 e^{-\kappa t} d\tilde{S}_t^*,$$

where in turn the dynamics of the process $\tilde{S}_t^* = e^{(\kappa - r)t} S_t$ are

$$d\tilde{S}_t^* = \sigma \tilde{S}_t^* \, d\tilde{W}_t, \tag{3.60}$$

and the process $\tilde{W}_t = W_t - (r - \mu_\kappa)t/\sigma$ is a standard Brownian motion on the probability space $(\Omega, \mathbb{F}, \mathbb{P}^*)$.

It is clear that $\mathbb{P}^*$ is also the unique martingale measure for the process $\tilde{S}^*$, but not for the discounted stock price S^*, unless $\kappa = 0$. Using Lemma 3.2.1, it is possible to construct, by defining in a standard way the class of admissible trading strategies, an arbitrage-free market in which a risk-free bond and a dividend-paying stock are primary securities. Assuming that this is done, the valuation of stock-dependent contingent claims is now standard. In particular, we have the following result.

Proposition 3.2.1 *The arbitrage price at time $t < T$ of a call option on a stock which pays dividends at a constant rate κ during the option's lifetime is given by the risk-neutral formula*

$$C_t^\kappa = B_t \, \mathbb{E}_{\mathbb{P}^*}\big(B_T^{-1}(S_T - K)^+ \,\big|\, \mathcal{F}_t\big), \tag{3.61}$$

or explicitly

$$C_t^\kappa = \bar{S}_t N\big(d_1(\bar{S}_t, T - t)\big) - Ke^{-r(T-t)} N\big(d_2(\bar{S}_t, T - t)\big), \tag{3.62}$$

where $\bar{S}_t = S_t e^{-\kappa(T-t)}$, and the functions d_1, d_2 are given by (3.18)–(3.19). Equivalently,

$$C_t^\kappa = e^{-\kappa(T-t)} \Big(S_t N\big(\hat{d}_1(S_t, T - t)\big) - Ke^{-(r-\kappa)(T-t)} N\big(\hat{d}_2(S_t, T - t)\big)\Big),$$

where the functions $\hat{d}_1$ and $\hat{d}_2$ are given by

$$\hat{d}_{1,2}(s, t) = \frac{\ln(s/K) + (r - \kappa \pm \frac{1}{2}\sigma^2)t}{\sigma \sqrt{t}}. \tag{3.63}$$

Proof. The first equality is obvious. For the second, note first that we may rewrite (3.61) as follows

$$C_t^\kappa = e^{-r(T-t)} \, \mathbb{E}_{\mathbb{P}^*}\big((S_T - K)^+ \,\big|\, \mathcal{F}_t\big) = e^{-\kappa T} e^{-r(T-t)} \, \mathbb{E}_{\mathbb{P}^*}\big((\tilde{S}_T - e^{\kappa T}K)^+ \,\big|\, \mathcal{F}_t\big).$$

Using (3.60), and proceeding along the same lines as in the proof of Theorem 3.1.1, we find that

$$C_t^\kappa = e^{-\kappa T} c \left(\tilde{S}_t, T - t, e^{\kappa T} K\right),$$

where c is the standard Black-Scholes call option valuation function. Put another way, $C_t^\kappa = c^\kappa(S_t, T - t)$, where

$$c^\kappa(s, t) = se^{-\kappa t} N\left(\hat{d}_1(s, t)\right) - Ke^{-rt} N\left(\hat{d}_2(s, t)\right)$$

and the functions $\hat{d}_1$ and $\hat{d}_2$ are given by (3.63).   □

Alternatively, to derive the valuation formula for a call option (or for any European claim of the form $X = g(S_T)$), we may first show that its arbitrage price equals $v(t, S_t)$, where v solves the following backward PDE

$$\frac{\partial v}{\partial t} + \frac{1}{2}\sigma^2 s^2 \frac{\partial^2 v}{\partial s^2} + (r - \kappa)s \frac{\partial v}{\partial s} - rv = 0 \qquad (3.64)$$

on $(0, \infty) \times (0, T)$, subject to the terminal condition $v(s, T) = g(s)$.

Under the assumptions of Proposition 3.2.1, one can show that the value at time t of the forward contract with expiry date T and delivery price K is given by the equality

$$V_t(K) = B_t \, \mathbb{E}_{\mathbb{P}^*}\left(B_T^{-1}(S_T - K) \,\big|\, \mathcal{F}_t\right) = e^{-\kappa(T-t)} S_t - e^{-r(T-t)} K. \qquad (3.65)$$

To establish this equality, it suffices to note that the process $\tilde{S}^*$ is a martingale under $\mathbb{P}^*$ (cf. (3.60)). In view of (3.65), the forward price at time $t \le T$ of the dividend-paying stock S, for settlement at date T, equals

$$F_S^\kappa(t, T) = e^{(r-\kappa)(T-t)} S_t. \qquad (3.66)$$

Since a call option does not pay any dividends prior to its expiry date, the forward price of a call option is given by the usual formula

$$F_{C^\kappa}(t, T) = e^{r(T-t)} C_t^\kappa.$$

In terms of forward prices, the call option pricing formula (3.62) can be rewritten as follows

$$F_{C^\kappa}(t, T) = F_S^\kappa(t, T) N\left(\tilde{d}_1(F_S^\kappa(t, T), T - t)\right) - K N\left(\tilde{d}_2(F_S^\kappa(t, T), T - t)\right),$$

where the functions $\tilde{d}_1$ and $\tilde{d}_2$ are given by

$$\tilde{d}_{1,2}(F, t) = \frac{\ln(F/K) \pm \frac{1}{2}\sigma^2 t}{\sigma\sqrt{t}}.$$

Finally, it is easy to see that the following version of the put-call parity relationship is valid in the present setup

$$c^\kappa(S_t, T - t) - p^\kappa(S_t, T - t) = e^{-\kappa(T-t)} S_t - e^{-r(T-t)} K, \qquad (3.67)$$

where $p^K(S_t, T-t)$ stands for the arbitrage price at time t of the European put option with maturity date T and strike price K. In particular, if the exercise price equals the forward price of the underlying stock, then

$$c^K(S_t, T-t) - p^K(S_t, T-t) = 0.$$

As already mentioned, formula (3.62) is commonly used by market practitioners when valuing stock index options. For this purpose, one needs to assume that the stock index is a lognormal process – actually, a geometric Brownian motion.[10] The dividend yield κ, which can be estimated from the historical data, slowly varies on a monthly (or quarterly) basis. Therefore, for options with a relatively short maturity, it is reasonable to assume that the dividend yield is constant.

3.2.2 Case of Known Dividends

We will now focus on the valuation of European options on a stock with declared dividends during the option's lifetime. We assume that the amount of dividends to be paid before option's expiry and the dates of their payments, are known in advance. Our aim is to show that the Black-Scholes formula remains in force, provided that the current stock price is reduced by the discounted value of all the dividends to be paid during the life of the option.

As usual, it is sufficient to consider the case $t = 0$. Assume that the dividends $\kappa_1, \ldots, \kappa_m$ are to be paid at the dates $0 < T_1 < \cdots < T_m < T$. We assume, in addition, that dividend payments are known in advance – that is, $\kappa_1, \ldots, \kappa_m$ are real numbers. The present value of all future dividend payments equals

$$\tilde{I}_t = \sum_{j=1}^m \kappa_j e^{-r(T_j-t)} \mathbb{1}_{[t,T]}(T_j), \quad \forall\, t \in [0, T], \tag{3.68}$$

and the value of all the dividends paid after time t and compounded at the risk-free rate to the option's expiry date T is given by the expression

$$I_t = \sum_{j=1}^m \kappa_j e^{r(T-T_j)} \mathbb{1}_{[t,T]}(T_j), \quad \forall\, t \in [0, T]. \tag{3.69}$$

Following Heath and Jarrow (1988), in modelling the stock price we will separate the capital gains process from the impact of dividend payments. The capital gains process G is assumed to follow the usual stochastic differential equation

$$dG_t = \mu G_t\, dt + \sigma G_t\, dW_t, \quad G_0 > 0, \tag{3.70}$$

---

[10] This does not follow from the assumption that each underlying stock is a lognormal process, unless the stock index is calculated on the basis of the geometric average, as opposed to the more commonly used arithmetic average.

hence G follows a geometric Brownian motion. The process S, representing the price of a stock that pays dividends $\kappa_1, \ldots, \kappa_m$ at times $T_1, \ldots, T_m$ may be introduced by setting

$$S_t = G_t - \sum_{j=1}^{m} \kappa_j e^{r(t-T_j)} \mathbb{1}_{[T_j,T]}(t) = G_t - D_t, \qquad (3.71)$$

where the process D, given by the equality

$$D_t = \sum_{j=1}^{m} \kappa_j e^{r(t-T_j)} \mathbb{1}_{[T_j,T]}(t),$$

accounts for the impact of ex-dividend stock price decline. Alternatively, one may put, for $t \in [0, T]$,

$$S_t = G_t + \sum_{j=1}^{m} \kappa_j e^{-r(T_j-t)} \mathbb{1}_{[0,T_j]}(t) = G_t + \tilde{D}_t, \qquad (3.72)$$

where $\tilde{D}$ now satisfies

$$\tilde{D}_t = \sum_{j=1}^{m} \kappa_j e^{-r(T_j-t)} \mathbb{1}_{[0,T_j]}(t).$$

Note that in the first case we have $G_0 = S_0$ and $G_T = S_T + D_T = S_T + I_0$, while in the second case $G_0 = S_0 - \tilde{D}_0 = S_0 - \tilde{I}_0$ and $G_T = S_T$.

Not surprisingly, the option valuation formulas corresponding to various specifications of the stock price (that is, to (3.71) and (3.72)) are not identical.

As usual, the first step in option valuation is to introduce the notion of a self-financing trading strategy. Under the present assumptions, it is natural to assume that if $\phi = (\phi^1, \phi^2)$ is a self-financing trading strategy, then all dividends received during the option's lifetime are immediately reinvested in stocks or bonds. Hence the wealth process of any trading strategy ϕ, which equals $V_t(\phi) = \phi_t^1 S_t + \phi_t^2 B_t$, satisfies, as usual,

$$dV_t(\phi) = \phi_t^1 dG_t + \phi_t^2 dB_t. \qquad (3.73)$$

Intuitively, since the decline of the stock price equals the dividend, the wealth of a portfolio is not influenced by dividend payments. It turns out that the resulting option valuation formula will not agree with the standard Black-Scholes result, however. In view of (3.73), the martingale measure $\mathbb{P}^*$ for the security market model corresponds to the unique martingale measure for the discounted capital gains process $G_t^* = G_t B_t^{-1}$. Hence $\mathbb{P}^*$ can be found in exactly the same way as in the proof of Theorem 3.1.1. We can also price options using the standard risk-neutral valuation approach.

Proposition 3.2.2 *Consider a European call option with strike price K and expiry date T, written on a stock S that pays deterministic dividends $\kappa_1, \ldots, \kappa_m$ at times*

$T_1, \ldots, T_m$. *Assume that the stock price S satisfies (3.70)–(3.71). Then the arbitrage price at time t of this option equals*

$$C_t^\kappa = S_t N\big(h_1(S_t, T - t)\big) - e^{-r(T-t)}(K + I_t)N\big(h_2(S_t, T - t)\big),$$

where I_t is given by expression (3.69), and

$$h_{1,2}(S_t, T - t) = \frac{\ln S_t - \ln(K + I_t) + (r \pm \tfrac{1}{2}\sigma^2)(T - t)}{\sigma\sqrt{T - t}}.$$

Proof. It is sufficient to consider the case of $t = 0$. An application of the risk-neutral valuation formula yields

$$C_0^\kappa = e^{-rT} \mathbb{E}_{\mathbb{P}^*}\big((S_T - K)^+\big) = \mathbb{E}_{\mathbb{P}^*}\big((G_T^* - e^{-rT}(I_0 + K))^+\big).$$

The dynamics of the capital gains process G under $\mathbb{P}^*$ are

$$dG_t = rG_t\,dt + \sigma G_t\,dW_t, \quad G_0 = S_0 > 0.$$

We conclude that the price of the European call option is given by the standard Black-Scholes formula, with strike price K replaced by $I_0 + K$. □

By virtue of Proposition 3.2.2, if the stock price behavior is described by (3.70)–(3.71), then the pricing of European options corresponds to use of the Black-Scholes formula with the strike price increased by the value of the dividends compounded to time T at the risk-free rate. The next result corresponds to the second specification of the stock price, given by (3.72). The reader should be advised that this approach appears to be rather inconvenient when dividend payments are modelled as random variables.

Proposition 3.2.3 *Consider a European call option with strike price K and expiry date T, written on a stock S paying deterministic dividends $\kappa_1, \ldots, \kappa_m$ at times $T_1, \ldots, T_m$. Assume that the stock price S satisfies (3.70) and (3.72). Then the arbitrage price at time t of this option equals*

$$\tilde{C}_t^\kappa = (S_t - \tilde{I}_t)\,N\big(d_1(S_t - \tilde{I}_t, T - t)\big) - e^{-r(T-t)}KN\big(d_2(S_t - \tilde{I}_t, T - t)\big),$$

where $\tilde{I}_t$ is given by equality (3.68), and the functions d_1, d_2 are given by (3.18)–(3.19).

Proof. Once again we consider the case of $t = 0$. We need to find

$$\tilde{C}_0^\kappa = e^{-rT} \mathbb{E}_{\mathbb{P}^*}\big((S_T - K)^+\big) = e^{-rT}\mathbb{E}_{\mathbb{P}^*}\big((G_T - K)^+\big).$$

But now under $\mathbb{P}^*$ we have

$$dG_t = G_t(r\,dt + \sigma\,dW_t), \quad G_0 = S_0 - \tilde{I}_0 > 0.$$

The assertion is now obvious. □

Remarks. The last result is consistent with the market practice of using the Black-Scholes formula with the stock price reduced by the present value of the dividends (see Rubinstein (1983), Hull (1997)). In terms of binomial approximations of the stock price, the approach first presented above (i.e., based on formulas (3.70)–(3.71)) corresponds to the following specification of the discrete-time tree

$$S_0 u_n^j d_n^{j-i} - \sum_{j=1}^m \kappa_j e^{r(i\Delta_n - T_j)} \mathbb{1}_{[i\Delta_n, T]}(T_j)$$

for $i = 0, \dots, n$ and $j = 0, \dots, i$. On the other hand, the second approach corresponds to the process

$$(S_0 - \tilde{I}_0) u_n^j d_n^{j-i} + \sum_{j=1}^m \kappa_j e^{-r(T_j - i\Delta_n)} \mathbb{1}_{[0, i\Delta_n]}(T_j)$$

for $j = 0, \dots, i$. In both cases, valuation of an option can be made either by means of the standard backward induction method, or by using the risk-neutral valuation formula. We have assumed throughout that the dividend payments are deterministic, and the ex-dividend decline of the stock price equals the dividend. The ex-dividend behavior of stock prices is examined by several authors; see, for instance, Kalay (1982, 1984), Lakonishok and Vermaelen (1983, 1986), Barone-Adesi and Whaley (1986), and Kaplanis (1986). It is commonly agreed that due to higher personal taxes on dividends than on capital gains, the equilibrium stock price decline on the ex-dividend day should be less than the dividend. However, empirical studies show that although the stock price drops by an amount which is usually less then the dividend, it is sometimes more.

3.3 Bachelier Model

In his doctoral dissertation published in 1900, Louis Bachelier not only examined the valuation of bond options, but in fact initiated the study of continuous-time processes and introduced the Wiener process. Despite the notoriety of more recent Black-Scholes model, in some markets the Bachelier model appears to be a better tool to handle derivatives (especially short-lived options). According to the contemporary version of the Bachelier model, the stock price is given by the equality

$$S_t = S_0 + \mu t + \sigma W_t, \quad \forall t \in \mathbb{R}_+,$$

and the interest rate r is constant. The process W is a one-dimensional standard Brownian motion under $\mathbb{P}$. We postulate, in addition, that the information flow is modelled by the filtration $\mathbb{F}$, which is the natural filtration of W. Note the similarity to the Black-Scholes model in that $\mathbb{F}^W = \mathbb{F}^S$.

Self-financing trading strategies are defined as in Sect. 3.1.3; in particular, Lemma 3.1.2 is still valid within the present set-up. An application of Itô's formula yields

$$dS_t^* = B_t^{-1}\big(\mu\, dt + \sigma\, dW_t - r S_t\, dt\big).$$

For any fixed $T^* > 0$, therefore, the unique spot martingale measure for the Bachelier model coincides with the unique probability measure, equivalent to $\mathbb{P}$ on $(\Omega, \mathcal{F}_{T^*})$, and such that the process $S_T^* = S_t/B_t$ is a $\mathbb{P}^*$-martingale. As a straightforward consequence of Girsanov's theorem, we get (cf. (3.11))

$$\frac{d\mathbb{P}^*}{d\mathbb{P}} = \exp\left(\int_0^{T^*} \frac{r S_t - \mu}{\sigma}\, dW_t - \frac{1}{2}\int_0^{T^*} \frac{(r S_t - \mu)^2}{\sigma^2}\, dt\right), \quad \mathbb{P}\text{-a.s.}$$

Thus, the process W_t^*, $t \in [0, T^*]$, given by the formula

$$W_t^* = W_t - \int_0^t \frac{r S_u - \mu}{\sigma}\, du$$

is a one-dimensional standard Brownian motion under $\mathbb{P}^*$ with respect to $\mathbb{F}$. Moreover, under the martingale measure $\mathbb{P}^*$ we have

$$S_t^* = S_0^* + \int_0^t \sigma B_u^{-1}\, dW_u^*, \tag{3.74}$$

so that, obviously, S^* is a martingale under $\mathbb{P}^*$. Equivalently,

$$dS_t = r S_t\, dt + \sigma\, dW_t^*. \tag{3.75}$$

It is common to consider the Bachelier model for $r = 0$. In this case, the last formula simplifies to:

$$S_t = S_0 + \sigma W_t^*. \tag{3.76}$$

We shall adopt this convention in what follows, leaving the general case as an exercise. Of course, the stock price given by (3.76) is a time-homogeneous Gaussian process with independent increments (and thus a Markov process) under the martingale measure $\mathbb{P}^*$.

3.3.1 Bachelier Option Pricing Formula

Consider a European call option written on a stock S, with expiry date T and strike price K. Let the function $c : \mathbb{R}_+ \times (0, T] \to \mathbb{R}$ be given by the formula

$$c(s, t) = \sigma\sqrt{t}\, n\big(d(s, t)\big) + (s - K)N\big(d(s, t)\big), \tag{3.77}$$

where

$$d(s, t) = \frac{s - K}{\sigma\sqrt{t}}. \tag{3.78}$$

As before, N (respectively, n) stands for the standard Gaussian cumulative distribution function (respectively, the standard Gaussian density function). As usual, we shall restrict our attention to the class of admissible trading strategies.

Theorem 3.3.1 *Assume that* $r = 0$. *Then the arbitrage price at time* $t < T$ *of the European call option with expiry date* T *and strike price* K *in the Bachelier market is given by the formula*

$$C_t = c(S_t, T - t), \quad \forall t \in [0, T),$$

where the function $c : \mathbb{R}_+ \times (0, T] \to \mathbb{R}$ *is given by (3.77)–(3.78). The unique admissible replicating strategy* ϕ *of the call option satisfies*

$$\phi_t^1 = \frac{\partial c}{\partial s}(S_t, T - t), \quad \phi_t^2 = c(S_t, T - t) - \phi_t^1 S_t.$$

Proof. We shall only derive the arbitrage price, using the risk-neutral valuation formula. For $t = 0$, we obtain

$$C_0 = \mathbb{E}_{\mathbb{P}^*}\big((S_T - K)^+\big) = \mathbb{E}_{\mathbb{P}^*}\big(S_T \mathbb{1}_D\big) - K\mathbb{P}^*\{D\},$$

where D stands for the event $\{S_T > K\}$. Consequently,

$$C_0 = \mathbb{E}_{\mathbb{P}^*}\big((S_0 + \sigma W_T^*)\mathbb{1}_D\big) - K\mathbb{P}^*\{D\}$$

$$= \sigma\mathbb{E}_{\mathbb{P}^*}\big(W_T^*\mathbb{1}_D\big) + (S_0 - K)\mathbb{P}^*\{D\}.$$

The random variable $\xi = W_T^*/\sqrt{T}$ has a standard Gaussian distribution $N(0, 1)$. Thus

$$\mathbb{E}_{\mathbb{P}^*}\big(W_T^*\mathbb{1}_{\{S_0 + \sigma W_T^* > K\}}\big) = \sqrt{T}\,\mathbb{E}_{\mathbb{P}^*}\big(\xi\mathbb{1}_{\{\xi > d(S_0, T)\}}\big) = \sqrt{T}\,n\big(d(S_0, T)\big),$$

where we have used the equality $n'(x) = -xn(x)$, and

$$\mathbb{P}^*\{D\} = \mathbb{P}^*\{W_T^* > \sigma^{-1}(K - S_0)\} = \mathbb{P}^*\{\xi > -d(S_0, T)\} = N\big(d(S_0, T)\big).$$

For $0 < t < T$, we have $S_T = S_t + \sigma(W_T^* - W_t^*)$. To evaluate

$$C_t = \mathbb{E}_{\mathbb{P}^*}\big((S_T - K)^+ \,\big|\, \mathcal{F}_t\big) = \mathbb{E}_{\mathbb{P}^*}\big(S_T \mathbb{1}_D \,\big|\, \mathcal{F}_t\big) - K\mathbb{P}^*\{D \,\big|\, \mathcal{F}_t\},$$

we observe that the random variable S_t is $\mathcal{F}_t$-measurable, and the increment $W_T^* - W_t^*$ is independent of the σ-field $\mathcal{F}_t$ under $\mathbb{P}^*$. □

Notice that the Bachelier formula (3.77) can be rewritten as follows

$$c(s, t) = \sigma\sqrt{t}\,\tilde{N}\big(d(s, t)\big),$$

where we denote

$$\tilde{N}(x) = \int_{-\infty}^x N(u)\,du = n(x) - xN(x).$$

Since the put-call parity (3.43) remains valid, the price of a put option can be found from Theorem 3.3.1 as well. Since $C_t - P_t = S_t - K$, we obtain

$$P_t = \sigma\sqrt{T-t}\, n\big(d(S_t, T-t)\big) + (S_t - K)N\big(-d(S_t, T-t)\big).$$

For a detailed comparison of the Bachelier and Black-Scholes pricing formulas, the interested reader is referred to Schachermayer and Teichmann (2008).

Non-zero interest rate. For a non-zero interest rate r, we have

$$C_0 = \mathbb{E}_{\mathbb{P}^*}\big(B_T^{-1}(S_T - K)^+\big) = \mathbb{E}_{\mathbb{P}^*}\big((S_T^* - K^*)^+\big),$$

where we write $K^* = B_T^{-1}K$. Thus, in view of (3.74), in order to derive the pricing formula, it suffices to replace the constant σ by the function $\tilde{\sigma}(t) = \sigma e^{-rt}$ in the proof of Theorem 3.3.1. As a result, we obtain

$$C_0 = v(0, T)\, n\big(d(S_0, T)\big) - (S_0 - K)N\big(d(S_0, T)\big)$$

where

$$d(S_0, T) = \frac{S_0 - K}{v(0, T)}$$

and

$$v^2(0, T) = \int_0^T \tilde{\sigma}^2(t)\, dt = \sigma^2 \int_0^T e^{-2rt}\, dt = \frac{\sigma^2}{2r}\big(1 - e^{-2rT}\big).$$

Remarks. A different expression for the price is obtained if we postulate that the stock price is given by the equality $S_t = e^{rt}(S_0 + \sigma W_t^*)$, rather than by equation (3.75). We now have $S_t^* = S_0 + \sigma W_t^*$, and the formula for the arbitrage price of a call option becomes

$$\hat{C}_t = e^{rt}\sigma\sqrt{T-t}\, n\big(\hat{d}(S_t, t, T)\big) + \big(S_t - Ke^{-r(T-t)}\big)N\big(\hat{d}(S_t, t, T)\big),$$

where we denote

$$\hat{d}(S_t, t, T) = \frac{e^{r(T-t)}S_t - K}{e^{rT}\sigma\sqrt{T-t}}.$$

This follows from the fact that, under the present assumptions, the conditional distribution of $S_T - K$ given the σ-field $\mathcal{F}_t$ is Gaussian, specifically,

$$S_T - K \overset{d}{=} N\big(e^{r(T-t)}S_t - K,\ e^{2rT}\sigma^2(T-t)\big).$$

3.3.2 Bachelier's PDE

Let us return to the general case, as specified by equation (3.75). To solve this equation explicitly with the initial condition S_0, it suffices to note that (3.75) is equivalent to (3.74). It is easily seen that S_t satisfies

$$S_t = B_t S_0 + B_t \int_0^t \sigma B_u^{-1}\, dW_u^*,$$

or more explicitly

$$S_t = e^{rt} S_0 + \int_0^t \sigma e^{r(t-u)} \, dW_u^*.$$

Thus the random variable S_t has a Gaussian distribution under the martingale measure $\mathbb{P}^*$.

Since we have, for arbitrary $s \leq t$,

$$S_t = e^{r(t-s)} S_s + \int_s^t \sigma e^{r(t-u)} \, dW_u^*,$$

it is also clear that S is a Markov process under $\mathbb{P}^*$. We conclude that we deal here with a time-homogeneous diffusion process, and thus the PDE approach can also be applied. Let $g : \mathbb{R} \to \mathbb{R}$ be a Borel-measurable function. Mimicking the proof of Corollary 3.1.5, we check that the arbitrage price in the Bachelier model of the claim $X = g(S_T)$ settling at time T is given by the equality $\pi_t(X) = v(S_t, t)$, where $v : \mathbb{R} \times [0, T] \to \mathbb{R}$ solves the Bachelier PDE

$$\frac{\partial v}{\partial t} + \frac{1}{2} \sigma^2 \frac{\partial^2 v}{\partial s^2} + rs \frac{\partial v}{\partial s} - rv = 0, \quad \forall \, (s, t) \in \mathbb{R} \times (0, T),$$

with the terminal condition $v(s, T) = g(s)$.

3.3.3 Bachelier Sensitivities

For a fixed date T and any $t \leq T$, we denote by τ the time to option expiry – that is, we put $\tau = T - t$. We consider the case when $r = 0$, and we write $p(s, \tau, K, \sigma)$ and $c(s, \tau, K, \sigma)$ to denote the pricing function of a call and a put option respectively. The functions c and p are thus given by the formulas

$$c(s, \tau, K, \sigma) = \sigma \sqrt{\tau} \, n(d) + (s - K) N(d)$$

and

$$p(s, \tau, K, \sigma) = \sigma \sqrt{\tau} \, n(d) + (s - K) N(-d)$$

with $d = d(s, \tau)$ given by (3.78). Straightforward calculations show that

$$c_s = N(d) = \delta > 0,$$
$$c_{ss} = \frac{n(d)}{\sigma \sqrt{\tau}} = \gamma > 0,$$
$$c_\tau = \frac{\sigma n(d)}{2\sqrt{\tau}} = \theta > 0,$$
$$c_\sigma = \sqrt{\tau} \, n(d) = \lambda > 0,$$
$$c_K = -c_s = -N(d) < 0.$$

Similarly, for a put option we obtain

$$p_s = -N(-d) = \delta < 0,$$

$$p_{ss} = \frac{n(d)}{\sigma\sqrt{\tau}} = \gamma > 0,$$

$$p_\tau = \frac{\sigma n(d)}{2\sqrt{\tau}} = \theta > 0,$$

$$p_\sigma = \sqrt{\tau}\, n(d) = \lambda > 0,$$

$$p_K = -p_s = N(-d) > 0.$$

The signs of sensitivities are identical in Bachelier and Black-Scholes models.

3.4 Black Model

The goal of this section is to present another classical continuous-time model, put forward in a paper by Black (1976a), who examined the valuation of options on futures contracts. Although, for the sake of expositional simplicity, we refer in this section to a futures contract on a stock, suitable variants of Black's formula will later be applied to a large variety of options.

Let $f_S(t, T^*)$, $t \in [0, T^*]$, stand for the futures price of a certain stock S for the date T^*. The evolution of futures prices $f_t = f_S(t, T^*)$ is given by the familiar expression

$$df_t = \mu_f f_t\, dt + \sigma_f f_t\, dW_t, \quad f_0 > 0, \tag{3.79}$$

where μ_f and $\sigma_f > 0$ are real numbers, and W_t, $t \in [0, T^*]$, stands for a one-dimensional standard Brownian motion, defined on a probability space $(\Omega, \mathbb{F}, \mathbb{P})$, where $\mathbb{F} = \mathbb{F}^W$. The unique solution of SDE (3.79) equals

$$f_t = f_0 \exp\left(\sigma_f W_t + (\mu_f - \tfrac{1}{2}\sigma_f^2)\, t\right).$$

The price of the second security is given, as before, by $B_t = e^{rt}$.

Remarks. In the Black-Scholes set-up, the futures price dynamics of a stock S can be found by combining (3.1) with the following chain of equalities

$$f_t = f_S(t, T^*) = F_S(t, T^*) = S_t e^{r(T^*-t)}, \tag{3.80}$$

where, as usual, $F_S(t, T^*)$ stands for the forward price of the stock for the settlement date T^*. The last equality in (3.80) can be derived easily from the absence of arbitrage in the spot/forward market; the second is a consequence of the assumption that the interest rate is deterministic. If the stock price S is governed by the SDE (3.1), then Itô's formula yields

$$df_t = (\mu - r) f_t\, dt + \sigma f_t\, dW_t, \quad f_0 = S_0 e^{rT^*},$$

and thus f satisfies (3.79) with $\mu_f = \mu - r$ and $\sigma_f = \sigma$. It follows from (3.80) that

$$f_S(t, T^*) = F_S(t, T^*) = \mathbb{E}_{\mathbb{P}^*}(S_{T^*} | \mathcal{F}_t), \tag{3.81}$$

but also

$$f_S(t, T^*) = F_S(t, T^*) = S_t / B(t, T^*), \tag{3.82}$$

where $B(t, T^*)$ is the price at time t of the zero-coupon bond maturing at T^*. It appears that under uncertainty of interest rates, the right-hand sides of (3.81) and (3.82) characterize the futures and the forward price of S respectively (see Sect. 9.6).

Since futures contracts are not necessarily associated with a physical underlying security, such as a stock or a bond, we prefer to study the case of futures options abstractly. This means that we consider (3.79) as the exogenously given dynamics of the futures price f. For ease of notation, we write $\mu = \mu_f$ and $\sigma = \sigma_f$ in what follows.

3.4.1 Self-financing Futures Strategies

By a *futures strategy* we mean a pair $\phi_t = (\phi_t^1, \phi_t^2)$ of real-valued adapted stochastic processes, defined on the probability space $(\Omega, \mathbb{F}, \mathbb{P})$. Since it costs nothing to take a long or short position in a futures contract, the wealth process $V^f(\phi)$ of a futures strategy ϕ equals

$$V_t^f(\phi) = \phi_t^2 B_t, \quad \forall t \in [0, T]. \tag{3.83}$$

We say that a futures strategy $\phi = (\phi^1, \phi^2)$ is *self-financing* if its wealth process $V^f(\phi)$ satisfies, for every $t \in [0, T]$,

$$V_t^f(\phi) = V_0^f(\phi) + \int_0^t \phi_u^1 \, df_u + \int_0^t \phi_u^2 \, dB_u.$$

We denote by Φ^f the class of all self-financing futures strategies.

3.4.2 Martingale Measure for the Futures Market

A probability measure $\tilde{\mathbb{P}}$ equivalent to $\mathbb{P}$ is called the *futures martingale measure* if the discounted wealth $\tilde{V}^f(\phi) = V^f(\phi)/B$ of any strategy $\phi \in \Phi^f$ follows a local martingale under $\tilde{\mathbb{P}}$.

Lemma 3.4.1 *Let $\tilde{\mathbb{P}}$ be a probability measure on $(\Omega, \mathcal{F}_T)$ equivalent to $\mathbb{P}$. Then $\tilde{\mathbb{P}}$ is a futures martingale measure if and only if the futures price f is a local martingale under $\tilde{\mathbb{P}}$.*

Proof. The discounted wealth $\tilde{V}^f$ for any trading strategy $\phi \in \Phi^f$ satisfies

$$d\tilde{V}_t^f(\phi) = B_t^{-1}(\phi_t^1 \, df_t + \phi_t^2 \, dB_t) - rB_t^{-1} V_t^f(\phi) \, dt = \phi_t^1 B_t^{-1} \, df_t,$$

as (3.83) yields the equality

$$B_t^{-1}(\phi_t^2 \, dB_t - rV_t^f(\phi) \, dt) = B_t^{-1}(B_t^{-1} V_t^f(\phi) \, dB_t - rV_t^f(\phi) \, dt) = 0.$$

The statement of the lemma now follows easily. $\qquad\square$

The next result is an immediate consequence of Girsanov's theorem.

Proposition 3.4.1 *The unique martingale measure $\tilde{\mathbb{P}}$ on $(\Omega, \mathcal{F}_T)$ for the process f is given by the Radon-Nikodým derivative*

$$\frac{d\tilde{\mathbb{P}}}{d\mathbb{P}} = \exp\left(-\frac{\mu}{\sigma}\, W_T - \frac{1}{2}\frac{\mu^2}{\sigma^2}\, T\right), \quad \mathbb{P}\text{-a.s.}$$

The dynamics of the futures price f under $\tilde{\mathbb{P}}$ are

$$df_t = \sigma f_t\, d\tilde{W}_t, \tag{3.84}$$

and the process $\tilde{W}_t = W_t + \mu\sigma^{-1}t$, $t \in [0, T]$, is a standard Brownian motion on the probability space $(\Omega, \mathbb{F}, \tilde{\mathbb{P}})$.

It is clear from (3.84) that

$$f_t = f_0 \exp\left(\sigma \tilde{W}_t - \tfrac{1}{2}\sigma^2 t\right), \quad \forall\, t \in [0, T], \tag{3.85}$$

so that f follows a strictly positive martingale under $\tilde{\mathbb{P}}$. As expected, we say that a futures strategy $\phi \in \Phi^f$ is $\tilde{\mathbb{P}}$-*admissible* if the discounted wealth $\tilde{V}^f(\phi)$ follows a martingale under $\tilde{\mathbb{P}}$. We shall study an arbitrage-free futures market $\mathcal{M}^f = (f, B, \Phi^f(\tilde{\mathbb{P}}))$, where $\Phi^f(\tilde{\mathbb{P}})$ is the class of all $\tilde{\mathbb{P}}$-admissible futures trading strategies. The futures market model $\mathcal{M}^f$ is referred to as the *Black futures model* in what follows. The notion of an arbitrage price is defined in a similar way to the case of the Black-Scholes market.

3.4.3 Black's Futures Option Formula

We will now derive the valuation formula for futures options, first established by Black (1976b). Let the function $c^f : \mathbb{R}_+ \times [0, T] \to \mathbb{R}$ be given by *Black's futures formula*

$$c^f(f, t) = e^{-rt}\left(f N(\tilde{d}_1(f, t)) - K N(\tilde{d}_2(f, t))\right), \tag{3.86}$$

where

$$\tilde{d}_{1,2}(f, t) = \frac{\ln(f/K) \pm \tfrac{1}{2}\sigma^2 t}{\sigma\sqrt{t}} \tag{3.87}$$

and N denotes the standard Gaussian cumulative distribution function.

Before we formulate the main result of this section, let us consider once again the futures contract written on a stock whose price dynamics satisfy (3.1). If $T = T^*$, then the futures option valuation result (3.86)–(3.87) can be found directly from the Black-Scholes formula by setting $S_t = f_t e^{-r(T-t)}$ (this applies also to the replicating strategy). Intuitively, this follows from the simple observation that in this case we have $f_T = S_T$ at the option's expiry, and thus the payoffs from both options agree. In practice, the expiry date of a futures option usually precedes the settlement date of the underlying futures contract – that is, $T < T^*$. In such a case we have

$$C_T^f = \left(f_S(T, T^*) - K \right)^+,$$

or equivalently,

$$C_T^f = e^{r(T^*-T)} \left(S_T - K e^{-r(T^*-T)} \right)^+,$$

and we may still value the futures option as if it were the spot option. Such considerations rely on the equality $f_S(t, T^*) = F_S(t, T^*)$, which in turn hinges on the assumption that the interest rate is a deterministic function. They thus cannot be easily carried over to the case of stochastic interest rates. For this reason, we prefer to give below a straightforward derivation of the Black futures formula.

Theorem 3.4.1 *The arbitrage price C^f in the arbitrage-free futures market $\mathcal{M}^f$ of a European futures call option, with expiry date T and strike price K, is given by the equality $C_t^f = c^f(f_t, T - t)$. More explicitly,*

$$C_t^f = B(t, T) \Big(f_t N\big(\tilde{d}_1(f_t, T - t)\big) - K N\big(\tilde{d}_2(f_t, T - t)\big) \Big).$$

The futures strategy $\phi \in \Phi^f$ ($\tilde{\mathbb{P}}$) that replicates a European futures call option equals for every $t \in [0, T]$,

$$\phi_t^1 = \frac{\partial c^f}{\partial f}(f_t, T - t), \quad \phi_t^2 = e^{-rt} c^f(f_t, T - t). \tag{3.88}$$

Proof. We will follow rather closely the proof of Theorem 3.1.1. We shall therefore focus mainly on the derivation of (3.86)–(3.88). Some technical details, such as integrability of random variables or admissibility of trading portfolios, are left aside.

First method. Assume that the price process C_t^f is of the form $C_t^f = v(f_t, t)$ for some function $v : \mathbb{R}_+ \times [0, T] \to \mathbb{R}$, and consider a futures strategy $\phi \in \Phi^f$ of the form $\phi_t = (g(f_t, t), h(f_t, t))$ for some functions $g, h : \mathbb{R}_+ \times [0, T] \to \mathbb{R}$. Since the replicating portfolio ϕ is assumed to be self-financing, the wealth process $V^f(\phi)$, which equals

$$V_t^f(\phi) = h(S_t, t) B_t = v(f_t, t), \tag{3.89}$$

satisfies

$$dV_t^f(\phi) = g(f_t, t)\, df_t + h(f_t, t)\, dB_t,$$

or more explicitly

$$dV_t^f(\phi) = f_t \mu g(f_t, t)\, dt + f_t \sigma g(f_t, t)\, dW_t + r v(f_t, t)\, dt. \tag{3.90}$$

On the other hand, assuming that the function v is sufficiently smooth, we find that

$$dv(f_t, t) = \left(v_t(f_t, t) + \mu f_t v_f(f_t, t) + \tfrac{1}{2}\sigma^2 f_t^2 v_{ff}(f_t, t) \right) dt + \sigma f_t\, v_f(f_t, t)\, dW_t.$$

Combining the last equality with (3.90), we get the following expression for the Itô differential of the process $Y_t = v(f_t, t) - V_t^f(\phi)$

$$dY_t = \left(v_t(f_t, t) + \mu f_t v_f(f_t, t) + \tfrac{1}{2}\sigma^2 f_t^2 v_{ff}(f_t, t)\right) dt + \sigma f_t v_f(f_t, t) dW_t$$
$$- \mu f_t g(f_t, t) dt - \sigma f_t g(f_t, t) dW_t - rv(f_t, t) dt = 0.$$

Arguing along similar lines to the proof of Theorem 3.1.1, we infer that

$$g(f, t) = v_f(f, t), \quad \forall (f, t) \in \mathbb{R}_+ \times [0, T], \tag{3.91}$$

and thus also

$$Y_t = \int_0^t \left\{ v_t(f_u, u) + \tfrac{1}{2}\sigma^2 f_u^2 v_{ff}(f_u, u) - rv(f_u, u) \right\} du = 0,$$

where the last equality follows from the definition of Y.

To guarantee the last equality we assume that v satisfies the following partial differential equation (referred to as the *Black PDE*)

$$v_t + \tfrac{1}{2}\sigma^2 f^2 v_{ff} - rv = 0$$

on $(0, \infty) \times (0, T)$, with the terminal condition $v(f, T) = (f - K)^+$. Since the function $v(f, t) = c^f(f, T - t)$, where c^f is given by (3.86)–(3.88), is easily seen to solve this problem, to complete the proof it is sufficient to note that, by virtue of (3.91) and (3.89), the unique $\tilde{\mathbb{P}}$-admissible strategy ϕ that replicates the option satisfies

$$\phi_t^1 = g(f_t, t) = v_f(f_t, t), \quad \phi_t^2 = h(f_t, t) = B_t^{-1} v_f(f_t, t).$$

Details are left to the reader.

Second method. Since the random variable

$$X^* = B_T^{-1}(f_T - K)^+$$

is easily seen to be integrable with respect to the martingale measure $\tilde{\mathbb{P}}$, it is enough to evaluate the conditional expectation

$$C_t^f = B_t \, \mathbb{E}_{\tilde{\mathbb{P}}}\big((f_T - K)^+ B_T^{-1} | \mathcal{F}_t^f\big) = B_t \, \mathbb{E}_{\tilde{\mathbb{P}}}\big((f_T - K)^+ B_T^{-1} | f_t\big).$$

This means that, in particular for $t = 0$ we need to find the expectation

$$\mathbb{E}_{\tilde{\mathbb{P}}}\big((f_T - K)^+ B_T^{-1}\big) = \mathbb{E}_{\tilde{\mathbb{P}}}(f_T B_T^{-1} \mathbb{1}_D) - \mathbb{E}_{\tilde{\mathbb{P}}}(K B_T^{-1} \mathbb{1}_D) = I_1 - I_2,$$

where D denotes the set $\{f_T > K\}$. For I_2, we have

$$I_2 = e^{-rT} K \, \tilde{\mathbb{P}}\{f_T > K\} = e^{-rT} K \, \tilde{\mathbb{P}} \left\{ f_0 \exp\left(\sigma \tilde{W}_T - \tfrac{1}{2}\sigma^2 T\right) > K \right\},$$

and thus

$$I_2 = e^{-rT} K \, \tilde{\mathbb{P}} \left\{ -\sigma \tilde{W}_T < \ln(f_0/K) - \tfrac{1}{2}\sigma^2 T \right\}$$

$$= e^{-rT} K \, \tilde{\mathbb{P}} \left\{ \xi < \frac{\ln(f_0/K) - \tfrac{1}{2}\sigma^2 T}{\sigma\sqrt{T}} \right\}$$

$$= e^{-rT} K N\big(\tilde{d}_2(f_0, T)\big),$$

since the random variable $\xi = -\tilde{W}_T/\sqrt{T}$ has under $\tilde{\mathbb{P}}$ the standard Gaussian distribution. To evaluate I_1, we define an auxiliary probability measure $\hat{\mathbb{P}}$ on $(\Omega, \mathcal{F}_T)$ by setting

$$\frac{d\hat{\mathbb{P}}}{d\tilde{\mathbb{P}}} = \exp\left(\sigma \tilde{W}_T - \tfrac{1}{2}\sigma^2 T\right), \quad \tilde{\mathbb{P}}\text{-a.s.,}$$

and thus (cf. (3.85))

$$I_1 = \tilde{\mathbb{P}}\{f_T B_T^{-1} \mathbb{1}_D\} = e^{-rT} f_0 \hat{\mathbb{P}}\{f_T > K\}.$$

Moreover, the process $\hat{W}_t = \tilde{W}_t - \sigma t$ is a standard Brownian motion on the filtered probability space $(\Omega, \mathbb{F}, \hat{\mathbb{P}})$, and

$$f_T = f_0 \exp\left(\sigma \hat{W}_T + \tfrac{1}{2}\sigma^2 T\right).$$

Consequently,

$$\begin{aligned} I_1 &= e^{-rT} f_0 \hat{\mathbb{P}}\{f_T > K\} \\ &= e^{-rT} f_0 \hat{\mathbb{P}}\left\{ f_0 \exp\left(\sigma \hat{W}_T + \tfrac{1}{2}\sigma^2 T\right) > K\right\} \\ &= e^{-rT} f_0 \hat{\mathbb{P}}\left\{ -\sigma \hat{W}_T < \ln(f_0/K) + \tfrac{1}{2}\sigma^2 T\right\} \\ &= e^{-rT} f_0 N\left(\tilde{d}_1(f_0, T)\right). \end{aligned}$$

The general valuation result for any date t is a fairly straightforward consequence of the Markov property of f. $\square$

The method of arbitrage pricing in the Black futures model can be easily extended to any path-independent claim contingent on the futures price. In fact, the following corollary follows easily from the first proof of Theorem 3.4.1. As was mentioned already, in financial literature, the partial differential equation of Corollary 3.4.1 is commonly referred to as the *Black PDE*.

Corollary 3.4.1 *The arbitrage price in $\mathcal{M}^f$ of any attainable contingent claim $X = g(f_T)$ settling at time T is given by $\pi_t^f(X) = v(f_t, t)$, where the function $v : \mathbb{R}_+ \times [0, T] \to \mathbb{R}$ is a solution of the following partial differential equation*

$$\frac{\partial v}{\partial t} + \frac{1}{2}\sigma^2 f^2 \frac{\partial^2 v}{\partial f^2} - rv = 0, \quad \forall (f, t) \in (0, \infty) \times (0, T),$$

subject to the terminal condition $v(f, T) = g(f)$.

Let us denote by $P_t^f = p^f(f_t, T - t)$ the price of a futures put option with strike price K and $T - t$ to its expiry date, provided that the current futures price is f_t. To find the price of a futures put option, we can use the following result, whose easy proof is left to the reader.

Corollary 3.4.2 *The following relationship, known as the put-call parity for futures options, holds for every $t \in [0, T]$*

$$C_t^f - P_t^f = c^f(f_t, T - t) - p^f(f_t, T - t) = e^{-r(T-t)}(f_t - K).$$

Consequently, the price of a futures put option equals

$$P_t^f = e^{-r(T-t)} \left(KN\left(- \tilde{d}_2(f_t, T - t) \right) - f_t N\left(- \tilde{d}_1(f_t, T - t) \right) \right),$$

where $\tilde{d}_1(f, t)$ and $\tilde{d}_2(f, t)$ are given by (3.87).

Example 3.4.1 Suppose that the call option considered in Example 3.1.2 is a futures option. This means, in particular, that the price is now interpreted as the futures price. Using (3.86), one finds that the arbitrage price of a futures call option equals (approximately) $C_0^f = 1.22$. Moreover, the portfolio that replicates the option is composed at time 0 of ϕ_0^1 futures contracts and ϕ_0^2 invested in risk-free bonds, where $\phi_0^1 = 0.75$ and $\phi_0^2 = 1.22$. Since the number ϕ_0^1 is positive, it is clear that an investor who assumes a short option position needs to enter ϕ_0^1 (long) futures contracts. Such a position, commonly referred to as the *long hedge,* is also a generally accepted practical strategy for a party who expects to purchase a given asset at some future date. To find the arbitrage price of the corresponding futures put option, we make use of the put-call parity relationship. We find that $P_0^f = 0.23$; moreover, for the replicating portfolio of the put option we have $\phi_0^1 = -0.25$ and $\phi_0^2 = 0.23$. Since now $\phi_0^1 < 0$, we deal here with the *short hedge* – a strategy typical for an investor who expects to sell a given asset at some future date.

3.4.4 Options on Forward Contracts

We return to the classical Black-Scholes framework of Sect. 3.1. We will consider a forward contract with delivery date $T^* > 0$ written on a non-dividend-paying stock S. Recall that the forward price at time t of a stock S for the settlement date T^* equals

$$F_S(t, T^*) = S_t e^{r(T^*-t)}, \quad \forall t \in [0, T^*].$$

This means that the forward contract, established at time t, in which the delivery price is set to be equal to $F_S(t, T^*)$ is worthless at time t. Of course, the value of such a contract at time $u \in (t, T^*]$ is no longer zero, in general. It is intuitively clear that the value $V^F(t, u, T^*)$ of such a contract at time u equals the discounted value of the difference between the current forward price of S at time u and its value at time t, that is

$$V^F(t, u, T^*) = e^{-r(T^*-u)} \left(S_u e^{r(T^*-u)} - S_t e^{r(T^*-t)} \right) = S_u - S_t e^{r(u-t)}$$

for every $u \in [t, T^*]$. The last equality can also be derived by applying directly the risk-neutral valuation formula to the claim $X = S_{T^*} - F_S(t, T^*)$ that settles at time T^*. Indeed, we have

$$V^F(t, u, T^*) = B_u \, \mathbb{E}_{\mathbb{P}^*} \left(B_{T^*}^{-1} S_{T^*} - B_{T^*}^{-1} S_t e^{r(T^*-t)} \mid \mathcal{F}_u \right)$$

$$= B_u \, \mathbb{E}_{\mathbb{P}^*}(S_{T^*}^* \mid \mathcal{F}_u) - S_t e^{r(T^*-t)} e^{-r(T^*-u)}$$

$$= S_u - S_t e^{r(u-t)} = S_u - F_S(t, u),$$

since the random variable S_t is $\mathcal{F}_u$-measurable. It is worthwhile to observe that the value $V^F(t, u, T^*)$ is in fact independent of the settlement date T^*. Therefore, we may and do write $V^F(t, u, T^*) = V^F(t, u)$ in what follows.

By definition,[11] a call option written at time t on a forward contract with the expiry date $t < T < T^*$ is simply a call option with zero strike price, written on the value of the underlying forward contract, so that the payoff at time T equals

$$C_T^F = \left(V^F(t, T) \right)^+ = \left(S_T - S_t e^{r(T-t)} \right)^+.$$

It is clear that the call option on the forward contract purchased at time t gives the right to enter at time T into the forward contract on the stock S with delivery date T^* and delivery price $F_S(t, T^*)$. If the forward price at time T is less than it was at time t, the option is abandoned. In the opposite case, the holder exercises the option, and either enters, at no additional cost, into a forward contract under more favorable conditions than those prevailing at time T, or simply takes the payoff of the option. Assume now that the option was written at time 0, so that the terminal payoff equals

$$C_T^F = (V^F(0, T))^+ = (S_T - S_0 e^{rT})^+.$$

To value such an option at time $t \leq T$, we can make use of the Black-Scholes formula with the (fixed) strike price $K = S_0 e^{rT}$. After simple manipulations, we find that the option's value at time t is

$$C_t^F = S_t N\big(d_1(S_t, t)\big) - S_0 e^{rt} N\big(d_2(S_t, t)\big), \tag{3.92}$$

where

$$d_{1,2}(S_t, t) = \frac{\ln S_t - \ln(S_0 e^{rt}) \pm \frac{1}{2}\sigma^2(T - t)}{\sigma\sqrt{T - t}}.$$

Remarks. Alternatively, in order to derive (3.92), we can make use of Black's futures formula. Since the futures price $f_S(T, T)$ equals S_T, we have

$$C_T^F = (f_S(T, T) - S_0 e^{rT})^+.$$

An application of Black's formula yields

$$C_t^F = e^{-r(T-t)} \left(f_t \, N\big(\tilde{d}_1(f_t, t)\big) - S_0 e^{rT} \, N\big(\tilde{d}_2(f_t, t)\big) \right), \tag{3.93}$$

where $f_t = f_S(t, T)$ and

---

[11] Since options on forward contracts are not traded on exchanges, the definition of an option that has a forward contract as an underlying asset is largely a matter of convention.

$$\tilde{d}_{1,2}(f_t, t) = \frac{\ln f_t - \ln(S_0 e^{rT}) \pm \frac{1}{2}\sigma^2(T - t)}{\sigma\sqrt{T - t}}.$$

Since in the Black-Scholes setting the relationship $f_S(t, T) = S_t e^{r(T-t)}$ is satisfied, it is apparent that expressions (3.92) and (3.93) are equivalent. It is rather obvious that this equivalence hinges on the postulated non-random character of interest rates in the Black-Scholes model.

3.4.5 Forward and Futures Prices

Simple relation $F_S(t, T) = f_S(t, T)$ is a consequence of very restrictive assumptions (such as: a perfect divisibility of futures contracts, no allowance for transaction costs and taxes, equality of the lending and borrowing rates, etc.) imposed on the considered model of a financial market. It is worth pointing out that this equality is not necessarily valid in the case of models of financial markets with stochastic interest rates. For instance, in Sect. 11.5, a non-trivial relationship between a forward and a futures prices in the Gaussian HJM set-up is established. Generally speaking, the spread between the forward and futures price of an asset depends on the correlation between the asset price and the savings account (or the zero-coupon bond).

No-arbitrage relationships between the forward and futures prices of commodities, stocks, stock indices, currencies, bonds, etc. are analyzed by several authors (see, for instance, Cox et al. (1981b), Levy (1989), or Flesaker (1991)). Polakoff and Dizz (1992) demonstrate that when futures are not perfectly divisible, even if interest rates are constant, futures and forward prices will differ. The impact of different income and capital gains taxes on stock index futures and on the no-arbitrage condition was examined by Cornell and French (1983a, 1983b), who allowed also for stochastic interest rates. Finally, when transaction costs are taken into account, the non-existence of arbitrage opportunities leads to a determination of a band (instead of a specific value) for the futures price (see Klemkosky and Lee (1991)).

Theoretical relationships between forward and futures prices predicted by the no-arbitrage condition, and the effects of marking to market, were also objects of exhaustive statistical studies. In particular, such tests were performed for commodity futures markets by French (1983); for stock index futures by Park and Chen (1985), Park and Sears (1985), Ng (1987), Morse (1988), MacKinlay and Ramaswamy (1988), Chung (1991), and Stoll and Whaley (1994); for foreign exchange markets by Cornell and Reinganum (1981); and for interest rate derivatives by Sundaresan (1991) and Meulbroek (1992).

There is some evidence that the actual futures prices frequently depart from the values predicted theoretically (although some researchers of futures prices conclude that the no-arbitrage condition is supported by the empirical evidence). Put another way, futures contracts appear to be over- or underpriced (one says, for instance, that a given futures contract is *overpriced* when the actual futures price exceeds the theoretical futures price predicted by the no-arbitrage condition). In consequence, after allowing for marking to market and transaction costs, profitable arbitrage opportunities do sometimes arise between the real-world futures market and the corresponding spot market for a given financial asset. It may be shown, however, that allowance for trading lags leads to clear reductions in the average arbitrage profit; moreover, these profits become risky. It is frequently argued that the observed deviations from theoretical prices are due to particular regulatory restrictions and to an inadequate allowance for transaction costs and taxes.

3.5 Robustness of the Black-Scholes Approach

The crucial postulate of the Black-Scholes approach is the constancy of the volatility parameter (or at least its deterministic and predictable character). In this section, we present a few results related to the impact of an uncertain volatility of the underlying asset on the pricing and hedging of a derivative contract. By an *uncertain volatility* we mean any stochastic process (or an unknown deterministic function) which determines the real-life dynamics of the underlying asset. Since we postulate that the real-life dynamics of the underlying asset are given by some roughly described stochastic process, and in practical implementations we use a well-specified volatility process, this is also referred to as the issue of *mis-specified volatility*.

Let us stress that we reserve the term *stochastic volatility* to an exogenously given volatility process occurring in a postulated model of the financial market. In Chap. 7, we shall analyze modelling of stochastic volatility, and the implications of the volatility specification on pricing and hedging of derivatives. Since we do not examine here questions related to the concept of implied volatility, the results of this section are of a minor practical importance.

3.5.1 Uncertain Volatility

In this subsection, we assume that the stock price dynamics under the real-life probability $\mathbb{P}$ are

$$dS_t = S_t\big(\mu_t\,dt + \bar{\sigma}_t\,dW_t\big), \quad S_0 > 0, \tag{3.94}$$

where W is a Brownian motion with respect to some filtration $\mathbb{F}$, and μ and $\bar{\sigma} \geq 0$ are certain $\mathbb{F}$-adapted stochastic processes, such that the stock price process S is well defined. This means that S is given by the formula

$$S_t = S_0 \exp\Big(\int_0^t \bar{\sigma}_u\,dW_u + \int_0^t \big(\mu_u - \tfrac{1}{2}\bar{\sigma}_u^2\big)\,du\Big),$$

where the first integral is the Itô integral. Equation (3.94) will be referred to as *true dynamics* of the stock price. It should be made clear that there is no assumption here that the processes μ and $\bar{\sigma}$ are known explicitly. Put differently, we only know that the true dynamics fall within a certain subclass of strictly positive Itô processes. Also, there is no need to postulate that the filtration $\mathbb{F}$ is generated either by the stock price, or by the process W. We only assume that W is a Wiener process with respect to $\mathbb{F}$ under $\mathbb{P}$. Let us stress that equation (3.94) will not be used in the derivation of the arbitrage price and/or the replicating strategy for a given contingent claim. We are only going to employ this equation by taking it as a true model of the real-life behavior of the stock price. Our goal is to examine the robustness of the hedging strategy derived within a particular model with respect to an uncertain real-life volatility of the stock price.

Remarks. Let us stress that the process of uncertain volatility cannot be observed. At best, a trader can observe the so-called realized volatility along a single sample path of the process S. Formally, for a given sample path $S.(\omega)$, the associated sample path $\bar{\sigma}.(\omega)$ of the uncertain volatility is called the *realized volatility* of the stock price, and the integral

$$\bar{v}^2(t, T)(\omega) = \int_t^T \bar{\sigma}_u^2(\omega)\, du$$

is informally termed the *realized variance* of the stock price over the time interval $[t, T]$. Some authors examine the issue of uncertain volatility from the pathwise perspective by focusing on the so-called *worst-case scenarios*. For results in this vein we refer to Avellaneda et al. (1995), who consider pricing and hedging derivative securities in markets with uncertain volatilities, and to Bick (1995), who studies quadratic-variation-based trading strategies.

3.5.2 European Call and Put Options

We make a crucial assumption that the nonnegative uncertain volatility process $\bar{\sigma}_t$, $t \in [0, T]$, is bounded by a constant: $\bar{\sigma}_t \leq \sigma_{max} < \infty$, where

$$\sigma_{max} = \text{ess sup}_{\omega \in \Omega}\, \sup_{t \in [0,T]} \bar{\sigma}_t.$$

Following El Karoui et al. (1998), Davis (2000) and others, we shall consider the following situation: a trader sells a call (or put) option at the Black-Scholes price $c(S_0, T) = \bar{c}(S_0, 0)$ evaluated for some constant volatility parameter $\sigma \geq \sigma_{max}$. Recall that, in accordance with our notation, the Black-Scholes price of the call option equals $C_t = c(S_t, T - t) = \bar{c}(S_t, t)$, where the function c is given by (3.17) and where we set $\bar{c}(s, t) = c(s, T - t)$.

Starting from the initial endowment $V_0(\phi) \geq c(S_0, T)$, a trader consistently hedges his short position in a call option using the Black-Scholes delta, so that $\phi_t^1 = N(d_1(S_t, t)) = \bar{c}_s(S_t, t)$ for every $t \in [0, T]$. Our goal is to show that the wealth process $V_t(\phi) = \phi_t^1 S_t + \phi_t^2 B_t$ of the self-financing trading strategy $\phi = (\phi^1, \phi^2)$ satisfies

$$V_T(\phi) = V_0(\phi) + \int_0^T \phi_u^1\, dS_u + \int_0^T \phi_u^2\, dB_u \geq (S_T - K)^+, \quad \mathbb{P}\text{-a.s.}$$

Since the strategy is self-financing, this can be rewritten using the discounted values, namely,

$$V_T(\phi) B_T^{-1} = V_0(\phi) + \int_0^T \phi_u^1\, dS_u^* \geq B_T^{-1}(S_T - K)^+, \quad \mathbb{P}\text{-a.s.}$$

Proposition 3.5.1 *Assume that the short-term interest rate r is constant. Suppose that a constant $\sigma > 0$ is such that the inequality $\bar{\sigma}_t \leq \sigma$ holds for every $t \in [0, T]$. Let the initial endowment satisfy $V_0(\phi) \geq c(S_0, T)$. Then for the self-financing strategy $\phi = (\phi^1, \phi^2)$, where $\phi_t^1 = N(d_1(S_t, t)) = \bar{c}_s(S_t, t)$ for every $t \in [0, T]$, we have $V_T(\phi) \geq (S_T - K)^+$.*

Proof. Since the true dynamics of the stock price are given by (3.94), an application of Itô's formula to the value function $\bar{c}$ yields

$$d\bar{c}(S_t, t) = \bar{c}_t(S_t, t)\, dt + \bar{c}_s(S_t, t)\, dS_t + \tfrac{1}{2}\bar{c}_{ss}(S_t, t)\bar{\sigma}_t^2 S_t^2 dt. \qquad (3.95)$$

It should be stressed that $\bar{c}(S_t, t)$ is simply an auxiliary process; it represents neither the market price, nor the model's price of the option. It can be seen as the wrongly calculated price of the option by a trader who uses the Black-Scholes model with constant volatility σ, while in reality the stock price evolves according to (3.94).[12] However, it is clear that for $t = T$ we get $\bar{c}(S_T, T) = (S_T - K)^+$, so that at the option's expiration date the value of the process $\bar{c}(S_t, t)$ matches the option's payoff. Recall that the function $\bar{c}(s, t)$ satisfies the Black-Scholes PDE (cf. (3.26))

$$\bar{c}_t(s, t) + \tfrac{1}{2}\sigma^2 s^2 \bar{c}_{ss}(s, t) + rs\bar{c}_s(s, t) - r\bar{c}(s, t) = 0. \qquad (3.96)$$

Combining (3.95) with (3.96), and using the assumption $\phi_t^1 = \bar{c}_s(S_t, t)$, we arrive at the following expression

$$d\bar{c}(S_t, t) = \tfrac{1}{2}\bar{c}_{ss}(S_t, t)(\bar{\sigma}_t^2 - \sigma^2)S_t^2\, dt + \phi_t^1(dS_t - rS_t\, dt) + r\bar{c}(S_t, t)\, dt.$$

Consequently,

$$d\bar{c}(S_t, t) = \tfrac{1}{2}\bar{c}_{ss}(S_t, t)(\bar{\sigma}_t^2 - \sigma^2)S_t^2\, dt + dV_t(\phi) - rV_t(\phi)\, dt + r\bar{c}(S_t, t)\, dt,$$

or equivalently,

$$d(\bar{c}(S_t, t)B_t^{-1}) = \tfrac{1}{2}\bar{c}_{ss}(S_t, t)(\bar{\sigma}_t^2 - \sigma^2)B_t^{-1}S_t^2\, dt + dV_t^*(\phi).$$

Upon integration of the last equality, we obtain

$$V_T^*(\phi) - V_0(\phi) = B_T^{-1}(S_T - K)^+ - \bar{c}(S_0, 0) + \int_0^T \tfrac{1}{2}\bar{c}_{ss}(S_t, t)(\sigma^2 - \bar{\sigma}_t^2)B_t^{-1}S_t^2\, dt.$$

We know already that the price of the call option in the Black-Scholes model is a strictly convex function of s, specifically, $\bar{c}_{ss}(s, t) > 0$ for every $t > 0$. Consequently, we have that

$$V_T^*(\phi) \geq V_0(\phi) - \bar{c}(S_0, 0) + B_T^{-1}(S_T - K)^+.$$

The conclusion of the proposition follows easily. $\square$

From the theoretical perspective, Proposition 3.5.1 is an interesting result, since it shows the robustness of the Black-Scholes pricing and hedging recipe with respect

---

[12] Of course, the price $\bar{c}(S_t, t)$ is correct if $\bar{\sigma}_t$ happens to coincide with σ.

to a model mis-specification. It seems unlikely, however, that any result of this kind could have a strong practical appeal. First, to the best of our knowledge, no robust statistical method allowing for the estimation of the instantaneous volatility process $\bar{\sigma}_t$ is available. A reliable prediction of the future behavior of this process is even more questionable.

Second, the concept of an uncertain volatility, formally represented by an unspecified stochastic process, seems to be inconsistent with the postulated dynamics (3.94). Indeed, if we have no hints on how to model the behavior of $\bar{\sigma}_t$, what makes us believe that an adequate description of real-life dynamics of the underlying asset has the form (3.94).

Third, assuming that a trader is lucky enough to know in advance the upper bound σ_{max}, we need to make an unrealistic assumption that he can sell an option for the price corresponding to the maximal level of the forecasted volatility. The real issue in the over-the-counter trading is, rather, how to find a competitive price of a contract. It is rather clear that a more precise assessment of volatility risk is needed for this purpose.

Finally, from the viewpoint of hedging, it is not at all clear whether the use of the Black-Scholes hedge ratio could be the best hedging strategy for a trader needing to address the issue of an uncertain volatility.

Let us return to the theoretical study. The following result is an easy consequence of the proof of Proposition 3.5.1. Note that we now postulate that the inequality $\bar{\sigma}_t \geq \sigma > 0$ is valid; that is, the uncertain volatility is assumed to be bounded from below by a positive constant.

Corollary 3.5.1 *Assume that the short-term interest rate r is constant and a constant $\sigma > 0$ is such that the inequality $\bar{\sigma}_t \geq \sigma$ holds for every $t \in [0, T]$. If the initial endowment satisfies $V_0(\phi) \geq -c(S_0, T)$, then for the self-financing trading strategy $\phi = (\phi^1, \phi^2)$, where ϕ^1 equals*

$$\phi_t^1 = N(d_1(S_t, t)) = -\bar{c}_s(S_t, t), \quad \forall t \in [0, T],$$

we have $V_T(\phi) \geq -(S_T - K)^+$.

Proposition 3.5.1 and Corollary 3.5.1 allow us to examine the seller's price $\pi_t^s(C_T)$ and the buyer's price $\pi_t^b(C_T) = -\pi_t^s(-C_T)$ of a call (or put) option. Recall that we now work under the standing assumption that the real-world dynamics of the stock are specified by (3.94).

Seller's price. We shall first consider the seller's price of an option. We stress that $\pi_t^s(C_T)$ stands for the seller's price of the call option in the real-world model, that is, under (3.94). It is easy to deduce from Proposition 3.5.1 that this seller's price satisfies $\pi_t^s(C_T) \leq c_{max}(S_t, T - t)$, where $c_{max}(S_t, T - t)$ is the Black-Scholes price of the call option evaluated for the constant volatility $\sigma_{max}(t)$ given by

$$\sigma_{max}(t) = \operatorname{ess\,sup}_{\omega \in \Omega} \sup_{u \in [t, T]} \bar{\sigma}_u.$$

It is also clear that the same result is valid for a European put option written on S, that is, $\pi_t^s(P_T) \leq p_{max}(S_t, T - t)$, where $p_{max}(S_t, T - t)$ the Black-Scholes price of the put option evaluated for the volatility $\sigma_{max}(t)$.

Buyer's price. Let us now consider the buyer's price $\pi_t^b(C_T)$ in the real-world model. Using Corollary 3.5.1, we obtain $\pi_t^s(-C_T) \leq -c_{min}(S_t, T - t)$, where $c_{min}(S_t, T - t)$ is the Black-Scholes price of the call option evaluated for the volatility $\sigma_{min}(t)$, where

$$\sigma_{min}(t) = \operatorname{ess\,inf}_{\omega \in \Omega} \inf_{u \in [t, T]} \bar{\sigma}_u,$$

provided, of course, that $\sigma_{min}(t) > 0$. Consequently, $\pi_t^b(C_T) \geq c_{min}(S_t, T - t)$ and, by similar arguments, $\pi_t^b(P_T) \geq p_{min}(S_t, T - t)$,

No-arbitrage bounds. We conclude that the following inequalities are valid:

$$c_{min}(S_t, T - t) \leq \pi_t^b(C_T) \leq \pi_t^s(C_T) \leq c_{max}(S_t, T - t)$$

and

$$p_{min}(S_t, T - t) \leq \pi_t^b(P_T) \leq \pi_t^s(P_T) \leq p_{max}(S_t, T - t).$$

Of course, since the real-life model is not completely specified, there is no chance of being more precise. In particular, we are not in a position to find the exact values for no-arbitrage bounds that are $\pi_t^b(C_T)$ and $\pi_t^s(C_T)$ in the case of a call option ($\pi_t^b(P_T)$ and $\pi_t^s(P_T)$ for a put).

3.5.3 Convex Path-independent European Claims

We refer the reader to El Karoui and Jeanblanc (1990), Lyons (1995), El Karoui et al. (1998), Frey and Sin (1999), Frey (2000), Romagnoli and Vargiolu (2000), and Ekström et al. (2004) for further results on robustness of the Black-Scholes formula, and to Bergman et al. (1996), Bergman (1998) or Kijima (2002) for more information on the general properties of option prices. We will not analyze the general properties of option prices in diffusion-type models, but rather we shall focus on the impact of uncertain volatility on prices of contingent claims and the respective hedging strategies.

We shall now extend the result of the previous subsection to the case of a convex (concave) path-independent European claim and a general diffusion-type model of the stock price. As before, given the (true) volatility process $\bar{\sigma}_t$, we have the following dynamics of the stock price under the real-world probability $\mathbb{P}$

$$dS_t = S_t(\mu_t \, dt + \bar{\sigma}_t \, dW_t), \quad S_0 > 0, \tag{3.97}$$

for some $\mathbb{F}$-adapted processes μ and $\bar{\sigma} \geq 0$. The process W is a Brownian motion under $\mathbb{P}$, but we do not assume that the filtration $\mathbb{F}$ is generated by W. As was mentioned before, we shall restrict our attention to a specific class of models. Let $\tilde{S}$ stand for the process describing the stock price in the model. We assume that the volatility $\tilde{\sigma}$ in the model has the form $\tilde{\sigma}_t = \tilde{\sigma}(\tilde{S}_t, t)$, where $\tilde{\sigma} : \mathbb{R}_+ \times \mathbb{R}_+ \to \mathbb{R}_+$ is a deterministic function. We shall refer to a volatility process of this form as the *path-independent asset-based volatility* or, more informally, *local volatility*. We model the stock price dynamics under the martingale measure $\tilde{\mathbb{P}}$ as follows

$$d\tilde{S}_t = \tilde{S}_t(r(t) \, dt + \tilde{\sigma}(\tilde{S}_t, t) \, d\tilde{W}_t), \quad \tilde{S}_0 > 0, \tag{3.98}$$

where the process $\tilde{W}$ is a Brownian motion under $\tilde{\mathbb{P}}$.

We assume that $r : \mathbb{R}_+ \to \mathbb{R}_+$ is a deterministic function, so that

$$B_t = \exp\left(\int_0^t r(u)\, du\right).$$

We postulate that $\tilde{S}_0 = S_0$ and the function $\tilde{\sigma}(s, t)$ is sufficiently regular to ensure the existence and uniqueness of a nonnegative solution of (3.98).

The model (3.98) will later be used to price a given contingent claim X, through the risk-neutral valuation formula under $\tilde{\mathbb{P}}$. We therefore find it convenient to assume that this model is complete, so that $\tilde{\mathbb{P}}$ is the unique martingale measure. Let us stress that there is no relationship between the two probabilities $\mathbb{P}$ and $\tilde{\mathbb{P}}$ (in general, they are defined on different probability spaces). The only instance when the probability measure $\tilde{\mathbb{P}}$ intervenes is the calculation of the arbitrage price of a given contingent claim within the model given by (3.98).

One of our goals is to examine the monotonicity of a call (put) option price with respect to the level of the volatility. We shall check that in the case of the stock price given by (3.98) the option's price is an increasing function of the volatility in the following sense: if the two volatility functions satisfy $\tilde{\sigma}(s, t) \geq \bar{\sigma}(s, t)$ for every (s, t) then the price a call (put) option assuming the volatility function $\tilde{\sigma}$ is never less than the price evaluated for a model with the same interest rate, but with volatility function $\bar{\sigma}$. This property extends to a larger class of convex path-independent European contingent claims. It is no longer valid, in general, when the underlying stock pays dividends (see Kijima (2002)) or when the volatility coefficient is path-dependent, as opposed to a state-dependent volatility $\sigma(s, t)$ (see El Karoui et al. (1998)).

Technical assumptions. We make the following standing assumptions regarding the regularity of the volatility function $\tilde{\sigma}$ and the payoff function h:
(i) $\tilde{\sigma} : [0, T] \times (0, \infty) \to \mathbb{R}$ is continuous and bounded,
(ii) $(\partial/\partial s)(s\tilde{\sigma}(s, t))$ is continuous in (s, t) and Lipschitz continuous and bounded in $s \in (0, \infty)$, uniformly in $t \in [0, T]$,
(iii) h is a convex (or concave) function on $(0, \infty)$ with bounded one-side derivatives: $|h'(x\pm)| < \infty$ for every $x > 0$.

The first issue that we are going to address is the preservation of the convexity (or concavity) property of the terminal payoff. The following result is borrowed from El Karoui et al. (1998).

Lemma 3.5.1 *Assume that the dynamics of the stock price under $\tilde{\mathbb{P}}$ are given by (3.98) with deterministic interest rate r.[13] If the payoff of a path-independent European claim $X = h(\tilde{S}_T)$ is represented by a convex (concave) function h, then for any fixed $t \in [0, T]$ the arbitrage price*

$$\tilde{\pi}_t(X) = B_t\, \mathbb{E}_{\tilde{\mathbb{P}}}\big(h(\tilde{S}_T)B_T^{-1} \mid \tilde{S}_t\big) = \tilde{v}(\tilde{S}_t, t) \tag{3.99}$$

is a convex (concave) function of the variable s.

---

[13] In fact, this last restriction can be relaxed (see El Karoui et al. (1998)).

For the proof of Lemma 3.5.1, the reader is referred to the original paper. Similar convexity preserving properties were established in Bergman et al. (1996), Bergman (1998), Hobson (1998b), Martini (1999), and Kijima (2002).

We shall assume from now on that the function $\tilde{v} : (0, \infty) \times [0, T] \to \mathbb{R}$ is continuous, of class $C^{2,1}((0, \infty) \times [0, T])$, with bounded partial derivative $\tilde{v}_s(s, t)$ (we refer to El Karoui et al. (1998) for technical conditions that are sufficient for these properties to hold). Under the present assumptions, the function $\tilde{v}(s, t)$ satisfies the following extension of the Black-Scholes PDE

$$\tilde{v}_t(s, t) + \tfrac{1}{2}\tilde{\sigma}^2(s, t)s^2\tilde{v}_{ss}(s, t) + r(t)s\tilde{v}_s(s, t) - r(t)\tilde{v}(s, t) = 0. \qquad (3.100)$$

By virtue of Lemma 3.5.1, in order to establish the monotonicity property of the arbitrage price, we may use in the present context similar techniques as in the proof of Proposition 3.5.1. Suppose that we have sold the claim X at time 0 at the price $\tilde{\pi}_0(X)$ that was calculated in accordance with our model, as specified by (3.98). We define a hedging strategy of the short position by setting $\phi_t^1 = \tilde{v}_s(S_t, t)$ for every $t \in [0, T]$ (note that the process ϕ^1 is bounded). Assume that true dynamics of S are given by (3.97). Our goal is to show that under some circumstances the wealth process of the self-financing trading strategy $\phi = (\phi^1, \phi^2)$ satisfies $V_T(\phi) \geq h(S_T)$.

Proposition 3.5.2 *Let the short-term interest rate $r(t)$ be deterministic. Assume that $\tilde{\sigma}(S_t, t) \geq \bar{\sigma}_t$ for every $t \in [0, T]$, where the process S is given by (3.97). Consider a European claim $X = h(\tilde{S}_T)$ represented by a convex function h. Let $\phi = (\phi^1, \phi^2)$ be the self-financing trading strategy such that $\phi_t^1 = \tilde{v}_s(S_t, t)$, where $\tilde{v}(s, t)$ is the function given by (3.99). If $V_0(\phi) \geq \tilde{\pi}_0(X)$, then $V_T(\phi) \geq h(S_T)$, $\mathbb{P}$-a.s.*

Proof. The proof is based on arguments similar to those of the proof of Proposition 3.5.1. We shall focus on the case of a convex payoff function h. First, Itô's formula yields

$$d\tilde{v}(S_t, t) = \tilde{v}_t(S_t, t)\, dt + \tilde{v}_s(S_t, t)\, dS_t + \tfrac{1}{2}\tilde{v}_{ss}(S_t, t)\bar{\sigma}_t^2 S_t^2 dt.$$

Using (3.100) and the assumption that $\phi_t^1 = \tilde{v}_s(S_t, t)$, we obtain

$$d\tilde{v}(S_t, t) = \tfrac{1}{2}\tilde{v}_{ss}(S_t, t)(\bar{\sigma}_t^2 - \tilde{\sigma}^2(S_t, t))S_t^2 dt + \phi_t^1(dS_t - r(t)S_t dt) + r(t)\tilde{v}(S_t, t)dt$$

and thus

$$d\tilde{v}(S_t, t) = \tfrac{1}{2}\tilde{v}_{ss}(S_t, t)(\bar{\sigma}_t^2 - \tilde{\sigma}^2(S_t, t))S_t^2 dt + dV_t(\phi) - r(t)V_t(\phi)dt + r(t)\tilde{v}(S_t, t)dt.$$

Finally,

$$d(\tilde{v}(S_t, t)B_t^{-1}) = \tfrac{1}{2}\tilde{v}_{ss}(S_t, t)(\bar{\sigma}_t^2 - \tilde{\sigma}^2(S_t, t))B_t^{-1} S_t^2\, dt + dV_t^*(\phi).$$

Integrating the last equality, we obtain

$$V_T^*(\phi) = V_0(\phi) - \tilde{\pi}_0(X) + B_T^{-1}h(S_T) + \tfrac{1}{2}\tilde{v}_{ss}(S_t, t)(\tilde{\sigma}^2(S_t, t) - \bar{\sigma}_t^2)B_t^{-1} S_t^2\, dt.$$

Since the convexity of the pricing function $\tilde{v}$ yields $\tilde{v}_{ss}(s, t) \geq 0$ and by assumption $\tilde{\sigma}^2(S_t, t) \geq \bar{\sigma}_t^2$, we conclude that

$$V_T(\phi) \geq B_T(V_0(\phi) - \tilde{\pi}_0(X)) + h(S_T).$$

This ends the proof of the proposition. $\qquad\qquad\qquad\qquad\qquad\qquad\quad$ $\square$

Remarks. Under assumptions of Proposition 3.5.2, if $\bar{\sigma}_t \geq \tilde{\sigma}(S_t, t)$ then

$$V_T(\phi) \leq B_T(V_0(\phi) - \tilde{\pi}_0(X)) + h(S_T).$$

Suppose now that the claim $X = h(S_T)$, where $-h$ satisfies the technical assumptions of Proposition 3.5.2 (in particular, h is concave). Then it follows easily from the proof of Proposition 3.5.2 that for the strategy $\phi = (\phi^1, \phi^2)$ with $\phi_t^1 = \tilde{v}_s(S_t, t)$ and $V_0(\phi) \geq \tilde{\pi}_0(X)$ the inequality $V_T(\phi) \geq h(S_T)$ will hold, if we assume that $\bar{\sigma}_t \geq \tilde{\sigma}(S_t, t)$ for every $t \in [0, T]$.

Corollary 3.5.2 *Suppose that instead of dynamics (3.97), we postulate that S is governed under some probability $\bar{\mathbb{P}}$ by the following SDE*

$$dS_t = S_t(r(t)\, dt + \bar{\sigma}_t\, d\bar{W}_t), \tag{3.101}$$

where $\bar{W}$ is a Brownian motion under $\bar{\mathbb{P}}$. Then under the assumptions of Proposition 3.5.2, we have

$$\tilde{\pi}_0(X) = B_0 \, \mathbb{E}_{\bar{\mathbb{P}}}\big(h(\tilde{S}_T)B_T^{-1}\big) \geq B_0 \, \mathbb{E}_{\bar{\mathbb{P}}}\big(h(S_T)B_T^{-1}\big). \tag{3.102}$$

Before proceeding to the proof, let us observe that since $\bar{\sigma}$ is an uncertain volatility in risk-neutral dynamics (3.101), the expected value occurring in the right-hand side of (3.102) is, practically speaking, not available. If instead we interpret (3.101) as another model of the stock price (so that the process $\bar{\sigma}$ is known), then (3.102) may be used to compare arbitrage prices yielded by two different models; this idea is exploited in Corollary 3.5.3.

Proof of Corollary 3.5.2. It suffices to observe that Proposition 3.5.2 yields, for $V_0(\phi) = \tilde{\pi}_0(X)$,

$$V_T(\phi)B_T^{-1} = \tilde{\pi}_0(X) + \int_0^T \phi_u^1\, dS_u^* \geq B_T^{-1}h(S_T), \quad \bar{\mathbb{P}}\text{-a.s.}$$

Since the process ϕ^1 is bounded and $S^* = S_t B_t^{-1}$ is a martingale under $\bar{\mathbb{P}}$, we obtain

$$\tilde{\pi}_0(X) \geq B_0 \, \mathbb{E}_{\bar{\mathbb{P}}}\big(h(S_T)B_T^{-1}\big),$$

as desired. $\qquad\qquad\qquad\qquad\qquad\qquad\qquad\qquad\qquad\qquad\qquad\qquad\qquad$ $\square$

Let us consider another diffusion-type local volatility model

$$d\bar{S}_t = \bar{S}_t(r(t)\, dt + \bar{\sigma}(\bar{S}_t, t)\, d\bar{W}_t), \tag{3.103}$$

where $\bar{\sigma} \geq 0$ is such that the SDE has a unique solution under $\bar{\mathbb{P}}$. We assume that a given contingent claim X represented by the payoffs $h(\tilde{S}_T)$ and $h(\bar{S}_T)$ can be replicated in corresponding models, as given by SDEs (3.98) and (3.103) respectively.

Corollary 3.5.3 *Assume that $\tilde{S}_0 = \bar{S}_0$ and the inequality $\tilde{\sigma}(s, t) \geq \bar{\sigma}(s, t)$ holds for every (s, t). Then for any path-independent European contingent claim X represented by a convex function h, we have $\tilde{\pi}_0(X) \geq \bar{\pi}_0(X)$. Similarly, if X is given by a concave function h, we have $\tilde{\pi}_0(X) \leq \bar{\pi}_0(X)$.*

Proof. It suffices to make use of Corollary 3.5.2.   □

In the general case, when X is not attainable in either of the two models, we claim that the inequality $\tilde{\pi}_0^b(X) \geq \bar{\pi}_0^s(X)$ is valid. To check this, it suffices to observe that (3.102) yields

$$\inf_{\tilde{\mathbb{P}} \in \tilde{\mathcal{M}}} B_0\, \mathbb{E}_{\tilde{\mathbb{P}}}\big(h(\tilde{S}_T)B_T^{-1}\big) \geq \sup_{\bar{\mathbb{P}} \in \bar{\mathcal{M}}} B_0\, \mathbb{E}_{\bar{\mathbb{P}}}\big(h(\bar{S}_T)B_T^{-1}\big). \qquad (3.104)$$

where $\tilde{\mathcal{M}}$ (respectively, $\bar{\mathcal{M}}$) stands for the class of all martingale measures for the first (respectively, second) model. We take for granted the following equalities (see El Karoui and Quenez (1995) or Cvitanić and Karatzas (1992, 1993)

$$\pi_0^s(X) = \sup_{\bar{\mathbb{P}} \in \bar{\mathcal{M}}} B_0\, \mathbb{E}_{\bar{\mathbb{P}}}\big(h(\bar{S}_T)B_T^{-1}\big) \qquad (3.105)$$

and

$$\pi_0^b(X) = \inf_{\tilde{\mathbb{P}} \in \tilde{\mathcal{M}}} B_0\, \mathbb{E}_{\tilde{\mathbb{P}}}\big(h(\tilde{S}_T)B_T^{-1}\big). \qquad (3.106)$$

Thus, inequality (3.104) immediately implies that $\tilde{\pi}_0^b(X) \geq \bar{\pi}_0^s(X)$. A more straightforward proof of this inequality is based on the following reasoning: we define the function $\tilde{v}$ by setting

$$\tilde{v}(\tilde{S}_t, t) = B_t\, \mathbb{E}_{\tilde{\mathbb{P}}}\big(h(\tilde{S}_T)B_T^{-1} \,\big|\, \tilde{S}_t\big),$$

and we check (using Lemma 3.5.1 and mimicking the proof of Proposition 3.5.2) that the strategy $\phi_t^1 = \tilde{v}_s(\bar{S}_t, t)$, with $V_0(\phi) = \tilde{v}(\bar{S}_0, 0)$, satisfies $V_T(\phi) \geq h(\bar{S}_T)$, $\bar{\mathbb{P}}$-a.s. This shows that $\bar{\pi}_0^s(X) \leq \tilde{v}(\bar{S}_0, 0)$ for any choice of the martingale probability $\bar{\mathbb{P}}$. Consequently, we have $\tilde{\pi}_0^b(X) \geq \bar{\pi}_0^s(X)$.

Corollary 3.5.3 yields a desirable, and intuitively plausible, monotonicity property of arbitrage price of European call and put options with respect to a state-dependent volatility coefficient. As was mentioned already, when path-dependent volatility processes are allowed, it is possible to give an example in which the monotonicity property of Corollary 3.5.3 is violated, even for a standard European call option. El Karoui et al. (1998) construct an explicit example[14], of two volatility processes depending on the stock price only (so that we may formally write $\tilde{\sigma}_t = f(t, S_{[0,t]})$ and $\bar{\sigma}_t = g(t, S_{[0,t]})$) such that $\tilde{\sigma}_t \geq \bar{\sigma}_t$ for every $t \in [0, T]$, but the price at time 0 of a European call option with expiration date T evaluated for $\tilde{\sigma}$ is strictly below the price yielded by a model with a higher volatility $\bar{\sigma}$. Finally, let us mention that in the case of a path-dependent claim, there is no reason to expect that the price will be an increasing function of a volatility, even in the classical Black-Scholes set-up.

---

[14] This counter-example does not invalidate Proposition 3.5.2, since it relies on a comparison of two models with path-dependent stochastic volatilities.

3.5.4 General Path-independent European Claims

Avellaneda et al. (1995) (see also, Avellaneda and Parás (1996), and Gozzi and Vargiolu (2002)) develop a PDE approach to *super-replication* of contingent claims for the generalized Black-Scholes model, in which the uncertain volatility $\bar{\sigma}$ is assumed to follow a stochastic process that is bounded from below and from above by strictly positive deterministic constants denoted, as before, by σ_{min} and σ_{max}. Consider a model of the form

$$dS_t = S_t(r\,dt + \bar{\sigma}_t\,dW_t^*). \tag{3.107}$$

The process W^* is a one-dimensional Brownian motion on a filtered probability space $(\Omega, \mathbb{F}, \mathbb{P}^*)$, and $\bar{\sigma}_t$ stands for an arbitrary adapted stochastic process on this space, with a priori known lower and upper bounds. No restrictions on the underlying filtration $\mathbb{F}$ are imposed, and thus the foregoing results cover also the case of stochastic volatility models, which are examined in Chap. 7.

It is worth stressing that Avellaneda et al. (1995) consider a fairly general class of path-independent European contingent claims of the form $X = h(S_T)$, without imposing either the convexity or the concavity condition on the terminal payoff h (this should be contrasted with results of the two previous subsections).

Let us denote by $S_u^{s,t,\bar{\sigma}}$, $u \in [t, T]$, the process given by (3.107) and starting at time t from the level s; that is, with the initial condition $S_t^{s,t,\bar{\sigma}} = s$. Also, for any fixed $(s, t) \in \mathbb{R}_+ \times [0, T]$, let $\mathcal{Q}^{s,t}$ be the family of all (infinite-dimensional) probability distributions of such processes, when $\bar{\sigma}$ ranges over all $\mathbb{F}$-adapted stochastic processes that satisfy the constraint

$$\sigma_{min} \leq \bar{\sigma}_u \leq \sigma_{min}, \quad \forall\, u \in [t, T].$$

Let us write $I = [\sigma_{min}, \sigma_{max}]$, and let us denote by $\mathcal{A}_t(I)$ the class of all such processes $\bar{\sigma}$. In view of (3.105), in order to examine the seller's price, it is natural to introduce the function $\bar{v} : \mathbb{R}_+ \times [0, T] \to \mathbb{R}$ by setting

$$\bar{v}(s, t) = \sup_{\bar{\sigma} \in \mathcal{A}_t(I)} \mathbb{E}_{\mathbb{P}^*}\big(e^{-r(T-t)}h(S_T^{s,t,\bar{\sigma}})\big),$$

or equivalently,

$$\bar{v}(s, t) = \sup_{\mathbb{Q} \in \mathcal{Q}^{s,t}} \mathbb{E}_{\mathbb{Q}}\big(e^{-r(T-t)}h(S_T)\big),$$

where, of course, $\mathbb{Q}\{S_t = s\} = 1$ for any $\mathbb{Q} \in \mathcal{Q}^{s,t}$. Similarly, we define the function $\underline{v}$ by the formula

$$\underline{v}(s, t) = \inf_{\bar{\sigma} \in \mathcal{A}_t(I)} \mathbb{E}_{\mathbb{P}^*}\big(e^{-r(T-t)}h(S_T^{s,t,\bar{\sigma}})\big),$$

or equivalently,

$$\underline{v}(s, t) = \inf_{\mathbb{Q} \in \mathcal{Q}^{s,t}} \mathbb{E}_{\mathbb{Q}}\big(e^{-r(T-t)}h(S_T)\big).$$

In the present set-up, it is natural to define the seller's price as the minimal initial endowment at time t given the level s of the stock price, such that for any choice of $\bar{\sigma} \in \mathcal{A}_t(I)$ it is possible to construct a trading strategy ϕ which *super-replicates* a given contingent claim, in the sense that the inequality $V_T(\phi) \geq h(S_T^{s,t})$ holds. Formally, we set

$$\pi_t^s(X) = \inf\{V_t(\phi) \mid \forall\, \bar{\sigma} \in \mathcal{A}_t(I)\, \exists\, \phi \in \Phi_{t,T} : V_T(\phi) \geq h(S_T^{s,t})\},$$

where $\Phi_{t,T}$ is the class of self-financing strategies over the time interval $[t, T]$. Let us first observe that $V_t(\phi)$ will in fact depend on s and t only. Notice also that for any choice of a volatility process $\bar{\sigma} \in \mathcal{A}_t(I)$, the inequality $V_T(\phi) \geq X$ should hold for some strategy ϕ. Hence the choice of the strategy may depend on a given form of the uncertain volatility process $\bar{\sigma}$. For any date $t \in [0, T]$, the seller's price can be defined in an analogous way.

Avellaneda et al. (1995) show that $\pi_t^s(X) = v(S_t, t)$, where the function $v :$ $\mathbb{R}_+ \times [0, T] \to \mathbb{R}$ satisfies the following non-linear PDE,[15] which is termed the *Black-Scholes-Barenblatt equation* (BSB equation, for short)

$$\frac{\partial \bar{v}}{\partial t} + \frac{1}{2} s^2 \bar{w} \left(\frac{\partial^2 \bar{v}}{\partial s^2}\right) \frac{\partial^2 \bar{v}}{\partial s^2} + rs \frac{\partial \bar{v}}{\partial s} - r\bar{v} = 0 \tag{3.108}$$

with the obvious terminal condition $\bar{v}(s, T) = h(s)$, where the auxiliary function $\bar{w} : \mathbb{R} \to \{\sigma_{min}^2, \sigma_{max}^2\}$ equals

$$\bar{w}(x) = \begin{cases} \sigma_{min}^2 & \text{if } x < 0, \\ \sigma_{max}^2 & \text{if } x \geq 0. \end{cases}$$

Notice that equation (3.108) can also be rewritten as follows

$$\frac{\partial \bar{v}}{\partial t} + \frac{1}{2} s^2 \sup_{\sigma_{min} \leq \sigma \leq \sigma_{max}} \left(\sigma^2 \frac{\partial^2 \bar{v}}{\partial s^2}\right) + rs \frac{\partial \bar{v}}{\partial s} - r\bar{v} = 0.$$

This emphasizes the fact that the volatility parameter is chosen dynamically, according to the local convexity (concavity) of the value function $\bar{v}(s, t)$.

The buyer's pricing function $\underline{v}$ solves an analogous equation, with the function $\bar{w}$ replaced by the function $\underline{w}$, which is given by the formula

$$\underline{w}(x) = \begin{cases} \sigma_{max}^2 & \text{if } x < 0, \\ \sigma_{min}^2 & \text{if } x \geq 0, \end{cases}$$

so that the corresponding BSB equation is

$$\frac{\partial \underline{v}}{\partial t} + \frac{1}{2} s^2 \inf_{\sigma_{min} \leq \sigma \leq \sigma_{max}} \left(\sigma^2 \frac{\partial^2 \underline{v}}{\partial s^2}\right) + rs \frac{\partial \underline{v}}{\partial s} - r\underline{v} = 0$$

with the terminal condition $\underline{v}(s, T) = h(s)$.

---

[15] This PDE is in fact the Hamilton-Jacobi-Bellman (HJB) equation associated with a specific exit problem.

It is clear that the properties of the BSB equation depend in an essential manner on the convexity (concavity) of the value function $\bar{v}$. As we already know from Lemma 3.5.1, the convexity (concavity) of the terminal payoff is generally preserved by arbitrage pricing for diffusion-type models, in the sense that the value function has this property as well. Thus it is by no means surprising that in the case of standard call and put options, we obtain simply the Black-Scholes prices that correspond to the extreme values of the volatility parameter. To be more specific, the seller's price (respectively, the buyer's price) of a call (and put) option in the sense of Avellaneda et al. (1995) is given by the Black-Scholes formula with the volatility parameter equal to σ_{max} (respectively, σ_{min}). Of course, the situation becomes much more complicated when we deal with contracts with non-convex terminal payoffs, with multi-asset derivatives, or with non-diffusion-type models.

We have, however, the following result extending Proposition 3.5.1.

Proposition 3.5.3 *Assume that the short-term interest rate r is constant. Suppose that the inequality $\sigma_{min} \leq \bar{\sigma}_u \leq \sigma_{min}$ holds for every $t \in [0, T]$, and the initial endowment satisfies $V_0(\phi) \geq \bar{v}(S_0, T)$. Assume that there exists a classical solution $\bar{v}$ to the BSB equation (3.108). Then for the self-financing trading strategy $\phi = (\phi^1, \phi^2)$, where $\phi_t^1 = \bar{v}_s(S_t, t)$ for every $t \in [0, T]$, we have $V_T(\phi) \geq h(S_T)$.*

Proof. The proof goes along similar lines to the proof of Proposition 3.5.1. In particular, we obtain the following equality

$$V_T^*(\phi) - V_0(\phi) = B_T^{-1} h(S_T) - \bar{v}(S_0, 0)$$

$$+ \frac{1}{2} \int_0^T \left(\sup_{\sigma_{min} \leq \sigma \leq \sigma_{max}} \left(\sigma^2 \bar{v}_{ss}(S_t, t) \right) - \bar{\sigma}_t^2 \bar{c}_{ss}(S_t, t) \right) B_t^{-1} S_t^2 \, dt.$$

This shows that

$$V_T^*(\phi) - V_0(\phi) \geq B_T^{-1} h(S_T) - \bar{v}(S_0, 0),$$

which is the desired result.   □

Proposition 3.5.3 shows that a super-replication of (path-independent) European contingent claims is feasible under the following assumptions. First, our initial endowment should be at least equal to the seller's price $\bar{v}(S_0, T)$. Second, we should be able to solve the corresponding BSB equation and to find a bound on the partial derivative $\bar{v}_s$. Recall once again that we postulate here that actual dynamics of the stock price are given by an uncertain volatility process that is bounded from below and from above by strictly positive constants.

Option prices and hedging strategies based on the maximization of the exponential utility function in a model with mis-specified volatility were studied by Ahn et al. (1997) and Huang and Davison (2002). Robustness of Gaussian hedges under mis-specification was examined by Dudenhausen et al. (1999).

4

Foreign Market Derivatives

In this chapter, an arbitrage-free model of the domestic security market is extended by assuming that trading in foreign assets, such as foreign risk-free bonds and foreign stocks (and their derivatives), is allowed. We will work within the classical Black-Scholes framework. More specifically, both domestic and foreign risk-free interest rates are assumed throughout to be nonnegative constants, and the foreign stock price and the exchange rate are modelled by means of geometric Brownian motions. This implies that the foreign stock price, as well as the price in domestic currency of one unit of foreign currency (i.e., the exchange rate) will have lognormal probability distributions at future times. Notice, however, that in order to avoid perfect correlation between these two processes, the underlying noise process should be modelled by means of a multidimensional, rather than a one-dimensional, Brownian motion. Our main goal is to establish explicit valuation formulas for various kinds of currency and foreign equity options. Also, we will provide some indications concerning the form of the corresponding hedging strategies. It is clear that foreign market contracts of certain kinds should be hedged both against exchange rate movements and against the fluctuations of relevant foreign equities.

4.1 Cross-currency Market Model

All processes considered in what follows are defined on a common filtered probability space $(\Omega, \mathbb{F}, \mathbb{P})$, where the filtration $\mathbb{F}$ is assumed to be the $\mathbb{P}$-augmentation of the natural filtration generated by a d-dimensional Brownian motion $W = (W^1, \ldots, W^d)$. The domestic and foreign interest rates, r_d and r_f, are assumed to be given real numbers. Consequently, the domestic and foreign savings accounts, denoted by B^d and B^f respectively, satisfy

$$B_t^d = \exp(r_d t), \quad B_t^f = \exp(r_f t), \quad \forall t \in [0, T^*].$$

It is worth stressing that B_t^d and B_t^f are denominated in units of domestic and foreign currency respectively.

We now introduce the exchange rate process Q that will play an important role as a tool used to convert foreign market cash flows into units of domestic currency. Moreover, it can also play the role of an option's underlying "asset". We adopt the convention that the exchange rate is denominated in units of domestic currency per unit of foreign currency. This means that Q_t represents the domestic price at time t of one unit of the foreign currency. Note, however, that the exchange rate process Q cannot be treated on an equal basis with the price processes of domestic assets; put another way, the foreign currency cannot be seen as just an additional traded security in the domestic market model, unless the impact of the foreign interest rate is taken into account.

We assume throughout that the exchange rate process Q is governed under the actual probability $\mathbb{P}$ by the following stochastic differential equation

$$dQ_t = Q_t \left(\mu_Q \, dt + \sigma_Q \cdot dW_t \right), \quad Q_0 > 0, \tag{4.1}$$

where $\mu_Q \in \mathbb{R}$ is a constant drift coefficient and $\sigma_Q \in \mathbb{R}^d$ denotes a constant volatility vector. As usual, the dot " $\cdot$ " stands for the Euclidean inner product in $\mathbb{R}^d$, for instance,

$$\sigma_Q \cdot dW_t = \sum_{i=1}^{d} \sigma_Q^i \, dW_t^i.$$

Also, we write $|\cdot|$ to denote the Euclidean norm in $\mathbb{R}^d$. Using this notation, we can make a clear distinction between models based on a one-dimensional Brownian motion, and models in which the multidimensional character of the underlying noise process is essential (for an alternative convention, see Sect. 7.1.9, in which one-dimensional correlated Brownian motions are examined).

4.1.1 Domestic Martingale Measure

In view of (4.1), the exchange rate at time t equals

$$Q_t = Q_0 \exp \left(\sigma_Q \cdot W_t + (\mu_Q - \tfrac{1}{2} |\sigma_Q|^2)t \right).$$

Let us introduce an auxiliary process $Q_t^* = B_t^f Q_t / B_t^d = e^{(r_f - r_d)t} Q_t$ representing the value at time t of the foreign savings account, when converted into the domestic currency, and discounted by the current value of the domestic savings account. It is useful to observe that Q^* satisfies

$$Q_t^* = Q_0 \exp \left(\sigma_Q \cdot W_t + (\mu_Q + r_f - r_d - \tfrac{1}{2} |\sigma_Q|^2)t \right)$$

or equivalently, that the dynamics of Q^* are

$$dQ_t^* = Q_t^* \left((\mu_Q + r_f - r_d) \, dt + \sigma_Q \cdot dW_t \right). \tag{4.2}$$

We thus see that the process Q^* is a martingale under the original probability measure $\mathbb{P}$ if and only if the drift coefficient μ_Q satisfies $\mu_Q = r_d - r_f$.

In order to rule out arbitrage between investments in domestic and foreign bonds, we have to postulate the existence of a martingale measure, which is defined as follows.

Definition 4.1.1 A probability measure $\mathbb{P}^*$, equivalent to $\mathbb{P}$ on $(\Omega, \mathcal{F}_{T^*})$, is a *martingale measure of the domestic market* (or briefly, a *domestic martingale measure*) if the process Q^* is a $\mathbb{P}^*$-martingale.

Intuitively, the domestic martingale measure $\mathbb{P}^*$ is a risk-neutral probability as seen from the perspective of a domestic investor – i.e., an investor who constantly denominates the prices of all assets in units of domestic currency. In view of Girsanov's theorem, any martingale measure $\mathbb{P}^*$ is associated with some vector $\zeta \in \mathbb{R}^d$ that satisfies

$$\mu_Q + r_f - r_d + \sigma_Q \cdot \zeta = 0. \tag{4.3}$$

It worth observing that a martingale measure $\mathbb{P}^*$ is not unique, in general. Indeed, the uniqueness of a solution to (4.3) need not hold, in general. Hence for any $\hat{\eta}$ satisfying (4.3), the probability measure $\mathbb{P}^*$, given by the usual exponential formula

$$\frac{d\mathbb{P}^*}{d\mathbb{P}} = \exp\left(\zeta \cdot W_{T^*} - \tfrac{1}{2}|\zeta|^2 T^*\right), \quad \mathbb{P}\text{-a.s.},$$

can play the role of a domestic martingale measure. In addition, the process W^*, given by $W_t^* = W_t - \zeta t$ for $t \in [0, T^*]$, is a d-dimensional Brownian motion under $\mathbb{P}^*$. Since $Q_t = Q_t^* B_t^d / B_t^f$, we see that the dynamics of the exchange rate Q under the domestic martingale measure $\mathbb{P}^*$ are

$$dQ_t = Q_t\left((r_d - r_f)\,dt + \sigma_Q \cdot dW_t^*\right), \quad Q_0 > 0. \tag{4.4}$$

Completeness. The uniqueness of the martingale measure can be gained by introducing the possibility of trading in additional foreign or domestic assets, foreign or domestic *stocks,* say. In other words, the uniqueness of a martingale measure holds if the number of (non-redundant) traded assets, including the domestic savings account, equals $d + 1$, where d stands for the dimensionality of the underlying Brownian motion. For instance, if no domestic stocks are traded and only one foreign stock is considered, to guarantee the uniqueness of a martingale measure $\mathbb{P}^*$, and thus the completeness of the market, it is enough to assume that W, and thus also W^*, is a two-dimensional Brownian motion. In such a case, the market model involves three primary securities – the domestic and foreign savings accounts (or equivalently, domestic and foreign bonds) and a foreign stock.

Risk-neutral valuation. Assume that the model is complete. Then the arbitrage price $\pi_t(X)$, in units of domestic currency, of any contingent claim X that settles at time T and is also denominated in the domestic currency, equals

$$\pi_t(X) = e^{-r_d(T-t)}\, \mathbb{E}_{\mathbb{P}^*}(X \mid \mathcal{F}_t). \tag{4.5}$$

If Y is denominated in units of foreign currency, its arbitrage price at time t, expressed in units of domestic currency, is

$$\pi_t(Y) = e^{-r_d(T-t)}\, \mathbb{E}_{\mathbb{P}^*}(Q_T Y \mid \mathcal{F}_t).$$

4.1.2 Foreign Martingale Measure

The arbitrage price of such a claim can be alternatively evaluated using the martingale measure associated with the foreign market, and ultimately converted into domestic currency using the current exchange rate Q_t. To this end, we need to introduce an arbitrage-free probability measure associated with the foreign market, termed a *foreign martingale measure*.

We now take the perspective of a foreign-based investor – that is, an investor who consistently denominates her profits and losses in units of foreign currency. Since Q_t is the price at time t of one unit of foreign currency in the domestic currency, it is evident that the price at time t of one unit of the domestic currency, expressed in units of foreign currency, equals $R_t = 1/Q_t$. From Itô's formula, we have

$$dQ_t^{-1} = -Q_t^{-2} dQ_t + Q_t^{-3} d\langle Q, Q \rangle_t,$$

or more explicitly

$$dR_t = -R_t \big((r_d - r_f) dt + \sigma_Q \cdot dW_t^* \big) + R_t |\sigma_Q|^2 dt.$$

The dynamics of R under $\mathbb{P}^*$ are thus given by the expression

$$dR_t = R_t \big((r_f - r_d) dt - \sigma_Q \cdot (dW_t^* - \sigma_Q dt) \big).$$

Equivalently,

$$dR_t^* = -R_t^* \sigma_Q \cdot (dW_t^* - \sigma_Q dt), \tag{4.6}$$

where R^* stands for the following process

$$R_t^* = B_t^d R_t / B_t^f = e^{(r_d - r_f)t} R_t, \quad \forall t \in [0, T^*].$$

Observe that R^* represents the price process of the domestic savings account, expressed in units of foreign currency, and discounted using the foreign risk-free interest rate. By virtue of (4.6), it is easily seen that R^* is a martingale under a probability measure $\tilde{\mathbb{P}}$, equivalent to $\mathbb{P}^*$ on $(\Omega, \mathcal{F}_{T^*})$, and given by the formula

$$\frac{d\tilde{\mathbb{P}}}{d\mathbb{P}^*} = \eta_{T^*}, \quad \mathbb{P}^*\text{-a.s.} \tag{4.7}$$

where the process η equals

$$\eta_t = \exp \big(\sigma_Q \cdot W_t^* - \tfrac{1}{2} |\sigma_Q|^2 t \big), \quad \forall t \in [0, T^*]. \tag{4.8}$$

Any probability measure $\tilde{\mathbb{P}}$ defined in this way is referred to as a *foreign martingale measure*. If the uniqueness of a domestic martingale measure $\mathbb{P}^*$ is not valid, the uniqueness of a foreign market martingale measure $\tilde{\mathbb{P}}$ does not hold either.

However, under any foreign martingale measure $\tilde{\mathbb{P}}$, we have

$$dR_t^* = -R_t^* \sigma_Q \cdot d\tilde{W}_t, \tag{4.9}$$

where $\tilde{W}_t = W_t^* - \sigma_Q t$, $t \in [0, T^*]$, follows a d-dimensional Brownian motion under $\tilde{\mathbb{P}}$. It is useful to observe that the dynamics of R under the martingale measure $\tilde{\mathbb{P}}$ are given by the following counterpart of (4.4)

$$dR_t = R_t \left((r_f - r_d)\, dt - \sigma_Q \cdot d\tilde{W}_t\right). \tag{4.10}$$

For any attainable contingent claim X, that settles at time T and is denominated in units of domestic currency, the arbitrage price at time t in units of foreign currency is given by the equality

$$\tilde{\pi}_t(X) = e^{-r_f(T-t)}\, \mathbb{E}_{\tilde{\mathbb{P}}}(R_T X \mid \mathcal{F}_t), \quad \forall\, t \in [0, T].$$

We are now in a position to establish a relationship linking a conditional expectation evaluated under the foreign market martingale measure $\tilde{\mathbb{P}}$ to its counterpart evaluated under the domestic martingale measure $\mathbb{P}^*$. We assume here that $\tilde{\mathbb{P}}$ is associated with $\mathbb{P}^*$ through (4.7).

Proposition 4.1.1 *The following formula is valid for any $\mathcal{F}_T$-measurable random variable X (provided that the conditional expectation is well-defined)*

$$\mathbb{E}_{\tilde{\mathbb{P}}}(X \mid \mathcal{F}_t) = \mathbb{E}_{\mathbb{P}^*}\left(X \exp\left(\sigma_Q \cdot (W_T^* - W_t^*) - \tfrac{1}{2}|\sigma_Q|^2(T-t)\right)\Big| \mathcal{F}_t\right). \tag{4.11}$$

Proof. Applying the abstract Bayes formula, we obtain

$$\mathbb{E}_{\tilde{\mathbb{P}}}(X \mid \mathcal{F}_t) = \frac{\mathbb{E}_{\mathbb{P}^*}(\eta_T X \mid \mathcal{F}_t)}{\mathbb{E}_{\mathbb{P}^*}(\eta_T \mid \mathcal{F}_t)},$$

and thus (note that η follows a martingale under $\mathbb{P}^*$)

$$\mathbb{E}_{\tilde{\mathbb{P}}}(X \mid \mathcal{F}_t) = \eta_t^{-1}\, \mathbb{E}_{\mathbb{P}^*}(\eta_T X \mid \mathcal{F}_t) = \mathbb{E}_{\mathbb{P}^*}(\eta_T \eta_t^{-1} X \mid \mathcal{F}_t).$$

It is now enough to make use of (4.8). $\qquad\square$

4.1.3 Foreign Stock Price Dynamics

Let S_t^f be the foreign currency price at time t of a foreign traded stock which pays no dividends. In order to exclude arbitrage, we assume that the dynamics of the price process S^f under the foreign martingale measure $\tilde{\mathbb{P}}$ are

$$dS_t^f = S_t^f \left(r_f\, dt + \sigma_{Sf} \cdot d\tilde{W}_t\right), \quad S_0^f > 0, \tag{4.12}$$

with a constant volatility coefficient $\sigma_{Sf} \in \mathbb{R}^d$. This means that, under the domestic martingale measure $\mathbb{P}^*$ associated with $\tilde{\mathbb{P}}$, the stock price process S^f is governed by the equation

$$dS_t^f = S_t^f \left((r_f - \sigma_Q \cdot \sigma_{Sf})\, dt + \sigma_{Sf} \cdot dW_t^*\right) \tag{4.13}$$

For the purpose of pricing foreign equity options, we will sometimes find it useful to convert the price of the underlying foreign stock into the domestic currency. We write $\tilde{S}_t^f = Q_t S_t^f$ to denote the price of a foreign stock S^f expressed in units of domestic currency. Using Itô's formula, and the dynamics under $\mathbb{P}^*$ of the exchange rate Q, namely,

$$dQ_t = Q_t \left((r_d - r_f) \, dt + \sigma_Q \cdot dW_t^* \right), \tag{4.14}$$

one finds that under the domestic martingale measure $\mathbb{P}^*$, the process $\tilde{S}^f$ satisfies

$$d\tilde{S}_t^f = \tilde{S}_t^f \left(r_d \, dt + (\sigma_{S^f} + \sigma_Q) \cdot dW_t^* \right). \tag{4.15}$$

The last equality shows that the price process $\tilde{S}^f$ behaves as the price process of a domestic stock in the classical Black-Scholes framework; however, the corresponding volatility coefficient is equal to the superposition $\sigma_{S^f} + \sigma_Q$ of two volatilities – the foreign stock price volatility and the exchange rate volatility. By defining in the usual way the class of admissible trading strategies, one may now easily construct a market model in which there is no arbitrage between investments in foreign and domestic bonds and stocks. Since this can be easily done, we leave the details to the reader.

4.2 Currency Forward Contracts and Options

In this section, we consider derivative securities whose value depends exclusively on the fluctuations of exchange rate Q, as opposed to those securities that depend also on some foreign equities. Currency options, forward contracts and futures contracts provide an important financial instrument through which to control the risk exposure induced by the uncertain future exchange rate. The deliverable instrument in a classical foreign exchange option is a fixed amount of underlying foreign currency.

The valuation formula that provides the arbitrage price of foreign exchange European-style options was established independently by Biger and Hull (1983) and Garman and Kohlhagen (1983). They have shown that if the domestic and foreign risk-free rates are constant, and the evolution of the exchange rate is described by (4.1), then a foreign currency option may be valued by means of a suitable variant of the Black-Scholes option valuation formula. More precisely, one may apply formula (3.62) of Sect. 3.2.1, which gives the arbitrage price of a European option written on a stock paying a constant dividend yield. The stock price S_t should be replaced by the exchange rate Q_t, and the dividend yield κ with the foreign interest rate r_f. In view of the arguments presented in Sect. 4.1, we may and do assume throughout this section that the Brownian motion W is one-dimensional.

4.2.1 Forward Exchange Rate

Let us first consider a foreign exchange forward contract written at time t that settles at the future date T. The asset to be delivered by the party assuming a short position in the contract is a predetermined amount of foreign currency, say 1 unit. The party who assumes a long position in a currency forward contract is obliged to pay a certain number of units of a domestic currency, the *delivery price*. As usual, the delivery price that makes the forward contract worthless at time $t \leq T$ is called the *forward price* at time t of one unit of the foreign currency to be delivered at the settlement date T. In the present context, it is natural to refer to this forward price as the *forward exchange rate*. We write $F_Q(t, T)$ to denote the forward exchange rate.

Proposition 4.2.1 *The forward exchange rate $F_Q(t, T)$ at time t for the settlement date T is given by the following formula*

$$F_Q(t, T) = e^{(r_d - r_f)(T-t)} Q_t, \quad \forall t \in [0, T]. \tag{4.16}$$

Proof. It is easily seen that if (4.16) does not hold, risk-free profitable opportunities arise between the domestic and the foreign market.   □

Relationship (4.16), commonly known as the *interest rate parity,* asserts that the forward exchange premium must equal, in the market equilibrium, the interest rate differential $r_d - r_f$. A relatively simple version of the interest rate parity still holds even when the domestic and foreign interest rates are no longer deterministic constants, but follow stochastic processes. Under uncertain interest rates, we need to to introduce the price processes $B_d(t, T)$ and $B_f(t, T)$ of the domestic and foreign zero-coupon bonds with maturity T. A zero-coupon bond with a given maturity T is a financial security paying one unit of the corresponding currency at the future date T. Suppose that zero-coupon bonds with maturity T are traded in both domestic and foreign markets. Then equality (4.16) may be extended to cover the case of stochastic interest rates. Indeed, it is not hard to show, by means of no-arbitrage arguments, that

$$F_Q(t, T) = \frac{B_f(t, T)}{B_d(t, T)} Q_t, \quad \forall t \in [0, T], \tag{4.17}$$

where $B_d(t, T)$ and $B_f(t, T)$ stand for the respective time t prices of the domestic and foreign zero-coupon bonds with maturity T. Note that in (4.17), both $B_d(t, T)$ and $B_f(t, T)$ are to be seen as the domestic and foreign discount factors rather than the prices. Indeed, it is natural to express prices in units of the corresponding currencies, while discount factors are simply the corresponding real numbers. Finally, it follows immediately from (4.14) that for any fixed settlement date T, the forward price dynamics under the martingale measure (of the domestic economy) $\mathbb{P}^*$ are

$$dF_Q(t, T) = F_Q(t, T) \sigma_Q \cdot dW_t^*, \tag{4.18}$$

and $F_Q(T, T) = Q_T$. In what follows, we will use the last formula as a convenient argument to show that the forward and futures exchange rates agree (at least as long as the interest rates are deterministic).

4.2.2 Currency Option Valuation Formula

As a first example of a currency option, we consider a standard European call option, whose payoff at the expiry date T equals

$$C_T^Q \overset{\text{def}}{=} N(Q_T - K)^+,$$

where Q_T is the spot price of the deliverable currency (i.e., the spot exchange rate at the option's expiry date), K is the strike price in units of domestic currency per foreign unit, and $N > 0$ is the nominal value of the option, expressed in units of the underlying foreign currency. It is clear that payoff from the option is expressed in the domestic currency; also, there is no loss of generality if we assume that $N = 1$. Summarizing, we consider an option to buy one unit of a foreign currency at a pre-specified price K, which may be exercised at the date T only.

We find it convenient to introduce the process $\tilde{B}^f$, defined as follows

$$\tilde{B}_t^f = B_t^f Q_t = e^{r_f t} Q_t, \quad \forall t \in [0, T^*].$$

Note that $\tilde{B}_t^f$ represents the value at time t of a unit investment in a foreign savings account, expressed in units of the domestic currency.

Proposition 4.2.2 *The arbitrage price, in units of domestic currency, of a currency European call option is given by the risk-neutral valuation formula*

$$C_t^Q = e^{-r_d(T-t)} \, \mathbb{E}_{\mathbb{P}^*}\big((Q_T - K)^+ \,\big|\, \mathcal{F}_t\big), \quad \forall t \in [0, T]. \tag{4.19}$$

Moreover, the price C_t^Q is given by the following expression

$$C_t^Q = Q_t e^{-r_f(T-t)} \, N\big(h_1(Q_t, T-t)\big) - K e^{-r_d(T-t)} N\big(h_2(Q_t, T-t)\big),$$

where N is the standard Gaussian cumulative distribution function, and

$$h_{1,2}(q, t) = \frac{\ln(q/K) + (r_d - r_f \pm \tfrac{1}{2}\sigma_Q^2)t}{\sigma_Q \sqrt{t}}.$$

Proof. Let us first examine a trading strategy in risk-free domestic and foreign bonds, called a *currency trading strategy* in what follows. Formally, by a currency trading strategy we mean an adapted stochastic process $\phi = (\phi^1, \phi^2)$. In financial interpretation, $\phi^1 \tilde{B}_t^f$ and $\phi^2 B_t^d$ represent the amounts of money invested at time t in foreign and domestic bonds respectively. It is important to note that both amounts are expressed in units of domestic currency (see, in particular, (4.2)).

A currency trading strategy ϕ is said to be self-financing if its wealth process $V(\phi)$, given by the formula

$$V_t(\phi) = \phi_t^1 \tilde{B}_t^f + \phi_t^2 B_t^d, \quad \forall t \in [0, T],$$

where $\tilde{B}_t^f = B_t^f Q_t$, $B_t^d = e^{r_d t}$, satisfies the following relationship

$$dV_t(\phi) = \phi_t^1 \, d\tilde{B}_t^f + \phi_t^2 \, dB_t^d.$$

For the discounted wealth process $V_t^*(\phi) = e^{-r_d t} V_t(\phi)$ of a self-financing currency trading strategy, we easily get

$$dV_t^*(\phi) = \phi_t^1 \, d(e^{-r_d t} \tilde{B}_t^f) = \phi_t^1 \, dQ_t^*.$$

On the other hand, by virtue of (4.14), the dynamics of the process Q^*, under the domestic martingale measure $\mathbb{P}^*$, are given by the expression

$$dQ_t^* = \sigma_Q\, Q_t^*\, dW_t^*.$$

Therefore, the discounted wealth $V^*(\phi)$ of any self-financing currency trading strategy ϕ follows a martingale under $\mathbb{P}^*$. This justifies the risk-neutral valuation formula (4.19). Taking into account the equality $Q_T = \tilde{B}_T^f e^{-r_f T}$, one gets also

$$
\begin{aligned}
C_t^Q &= e^{-r_d(T-t)}\, \mathbb{E}_{\mathbb{P}^*}\big((Q_T - K)^+ \mid \mathcal{F}_t\big) \\
&= e^{-r_f T} e^{-r_d(T-t)}\, \mathbb{E}_{\mathbb{P}^*}\big((\tilde{B}_T^f - K e^{r_f T})^+ \mid \mathcal{F}_t\big) \\
&= e^{-r_f T} c(\tilde{B}_t^f, T - t, K e^{r_f T}, r_d, \sigma_Q),
\end{aligned}
$$

where $c(s, t, K, r, \sigma)$ stands for the standard Black-Scholes call option pricing function (see (3.56)). More explicitly, we have

$$
\begin{aligned}
C_t^Q &= e^{-r_f T}\left(\tilde{B}_t^f\, N\big(d_1(\tilde{B}_t^f, T - t)\big) - K e^{r_f T} e^{-r_d(T-t)}\, N\big(d_2(\tilde{B}_t^f, T - t)\big)\right) \\
&= Q_t e^{-r_f(T-t)} N\big(d_1(\tilde{B}_t^f, T - t)\big) - K e^{-r_d(T-t)}\, N\big(d_2(\tilde{B}_t^f, T - t)\big).
\end{aligned}
$$

This proves the formula we wish to show, since

$$d_i(\tilde{B}_t^f, T - t, K e^{r_f T}, r_d, \sigma_Q) = h_i(Q_t, T - t)$$

for $i = 1, 2$. Finally, one finds immediately that the first component of the self-financing currency trading strategy that replicates the option equals

$$\phi_t^1 = e^{-r_f T} N\big(d_1(\tilde{B}_t^f, T - t)\big) = e^{-r_f T} N\big(h_1(Q_t, T - t)\big).$$

This means that to hedge a short position, the writer of the currency call should invest at time $t \le T$ the amount (expressed in units of foreign currency)

$$\phi_t^1 B_t^f = e^{-r_f(T-t)} N\big(h_1(Q_t, T - t)\big)$$

in foreign market risk-free bonds (or equivalently, in the foreign savings account). In addition, she should also invest the amount (denominated in domestic currency)

$$C_t^Q - Q_t e^{-r_f(T-t)} N\big(h_1(Q_t, T - t)\big)$$

in the domestic savings account. $\hfill\square$

Remarks. (a) As was mentioned earlier, a comparison of the currency option valuation formula established in Proposition 4.2.2 with expression (3.62) of Sect. 3.2 shows that the exchange rate Q can be formally seen as the price of a fictitious domestic "stock". Under such a convention, the foreign interest rate r_f can be interpreted as a dividend yield that is continuously paid by this fictitious stock.

(b) It is easy to derive the put-call relationship for currency options. Indeed, the pay-off in domestic currency of a portfolio composed of one long call option and one short put option is

$$C_T^Q - P_T^Q = (Q_T - K)^+ - (K - Q_T)^+ = Q_T - K,$$

where we assume, as before, that the options are written on one unit of foreign currency. Consequently, for any $t \in [0, T]$ we have

$$C_t^Q - P_t^Q = e^{-r_f(T-t)} Q_t - e^{-r_d(T-t)} K. \tag{4.20}$$

Suppose that the strike level K is equal to the current value of the forward exchange rate $F_Q(t, T)$. By substituting (4.16) in (4.20), we get

$$C_t^Q - P_t^Q = 0, \quad \forall t \in [0, T],$$

so that the arbitrage price of the call option with exercise price K equal to the forward exchange rate $F_Q(t, T)$ coincides with the value of the corresponding put option.
(c) We may also rewrite the currency option valuation formula of Proposition 4.2.2 in the following way

$$C_t^Q = e^{-r_d(T-t)} \Big(F_t N\big(\tilde{d}_1(F_t, T - t)\big) - K N\big(\tilde{d}_2(F_t, T - t)\big) \Big), \tag{4.21}$$

where $F_t = F_Q(t, T)$ and

$$\tilde{d}_{1,2}(F, t) = \frac{\ln(F/K) \pm \frac{1}{2}\sigma_Q^2 t}{\sigma_Q \sqrt{t}}, \quad \forall (F, t) \in \mathbb{R}_+ \times (0, T].$$

This shows that the currency option valuation formula can be seen as a variant of the Black futures formula (3.86) of Sect. 3.4. Furthermore, it is possible to re-express the replicating strategy of the option in terms of domestic bonds and currency forward contracts. Let us mention that under the present assumptions of deterministic domestic and foreign interest rates, the distinction between the currency futures price and forward exchange rate is not essential. In market practice, currency options are frequently hedged by taking positions in forward and futures contracts, rather than by investing in foreign risk-free bonds. Similar strategies are used to hedge the risk associated with options written on a foreign stock. Hence, there is a need to analyze investments in foreign market futures contracts in some detail.

The reader should be advised that, unfortunately, the empirical studies devoted to the constant interest rates model introduced in this chapter have been disappointing. First, Goodman et al. (1985) (see also Shastri and Tandon (1986)) found that the lognormal model of exchange rate gives biased predictions of the actual prices of currency options. Various subsequent modifications of the model, taking into account the possibility of early exercise in the case of American options (as in Bodurtha and Courtadon (1987, 1995)), as well as allowing for alternative exchange rate processes,[1] provided only slightly better empirical results. More accurate option prices were obtained by Shastri and Wethyavivorn (1987), who examined the jump-diffusion model of the exchange rate. Note that since both domestic and foreign interest rates are highly correlated with inflation and other dominant factors in each economy, it seems reasonable to expect that a model allowing for stochastic interest rates correlated with the exchange rate process would provide a more adequate description of real-world currency markets. Such a model of the cross-economy is examined in Chap. 14. Interested readers are referred also to the papers of Grabbe (1983), Adams and Wyatt (1987), and Amin and Jarrow (1991).

4.3 Foreign Equity Forward Contracts

In a global equity market, an investor may link his foreign stock and currency exposures in a large variety of ways. More specifically, he may choose to combine his investments in foreign equities with differing degrees of protection against adverse moves in exchange rates and stock prices, using forward and futures contracts as well as a variety of options.

4.3.1 Forward Price of a Foreign Stock

We will first consider an ordinary forward contract with a foreign stock being the underlying asset to be delivered – that is, an agreement to buy a stock on a certain date at a certain delivery price in a specified currency. It is natural to distinguish between the two following cases: (a) when the delivery price K^f is denominated in the foreign currency, and (b) when it is expressed in the domestic currency; in the latter case the delivery price will be denoted by K^d. Let us clarify that in both situations, the value of the forward contract at the settlement date T is equal to the spread between the stock price at this date and the delivery price expressed in foreign currency. The terminal payoff is then converted into units of domestic currency at the exchange rate that prevails at the settlement date T. Summarizing, in units of domestic currency, the terminal payoffs from the long positions are

$$V_T^d(K^f) = Q_T(S_T^f - K^f)$$

in the first case, and

$$V_T^d(K^d) = Q_T(S_T^f - Q_T^{-1}K^d) = (Q_T S_T^f - K^d) = (\tilde{S}_T^f - K^d)$$

if the second case is considered.

Case (a). Observe that the foreign-currency payoff at settlement of the forward contract equals $X_T = S_T^f - K^f$. Therefore its value at time t, denominated in the foreign currency, is

---

[1] Let us mention here the CEV diffusion of Cox and Ross (1976b), which was analyzed in the context of currency option valuation by Tucker et al. (1988).

$$V_t^f(K^f) = e^{-r_f(T-t)} \mathbb{E}_{\hat{\mathbb{P}}}(S_T^f - K^f \mid \mathcal{F}_t) = S_t^f - e^{-r_f(T-t)} K^f.$$

Consequently, when expressed in the domestic currency, the value of the contract at time t equals

$$V_t^d(K^f) = Q_t(S_t^f - e^{-r_f(T-t)} K^f).$$

We conclude that the forward price of the stock S^f, expressed in units of foreign currency, equals

$$F_{S^f}^f(t, T) = e^{r_f(T-t)} S_t^f, \quad \forall t \in [0, T].$$

Case (b). In this case, equality (4.5) yields immediately

$$V_t^d(K^d) = e^{-r_d(T-t)} \mathbb{E}_{\mathbb{P}^*}(\tilde{S}_T^f - K^d \mid \mathcal{F}_t);$$

hence, by virtue of (4.15), the domestic-currency value of the forward contract with the delivery price K^d denominated in domestic currency equals

$$V_t^d(K^d) = Q_t S_t^f - e^{-r_d(T-t)} K^d.$$

This implies that the forward price of a foreign stock in domestic currency equals

$$F_{S^f}^d(t, T) = e^{r_d(T-t)} \tilde{S}_t^f, \quad \forall t \in [0, T],$$

so that, somewhat surprisingly, it is independent of the foreign risk-free interest rate r_f. To explain this apparent paradox, observe that in order to determine the forward price $F_{S^f}^d(t, T)$, one can find first the delivery price from the perspective of the foreign-based investor, $F_{S^f}^f(t, T)$, and then convert its value into domestic currency using the appropriate forward exchange rate. Indeed, such calculations yield

$$F_{S^f}^f(t, T) F_Q(t, T) = e^{r_f(T-t)} S_t^f e^{(r_d - r_f)(T-t)} Q_t = e^{r_d(T-t)} Q_t S_t = F_{S^f}^d(t, T),$$

so that the interest rate r_f drops out from the final result.

4.3.2 Quanto Forward Contracts

The aim of this section is to examine a *quanto forward contract* on a foreign stock.[2] Such a contract is also known as a *guaranteed exchange rate forward contract* (a *GER forward contract* for short). To describe the intuition that underpins the concept of a quanto forward contract, let us consider an investor who expects a certain foreign stock to appreciate significantly over the next period, and who wishes to capture this appreciation in his portfolio. Buying the stock, or taking a long position in it through a forward contract or call option, leaves the investor exposed to exchange rate risk. To avoid having his return depend on the performance of the foreign currency, he needs a guarantee that he can close his foreign stock position at an exchange rate close to

---

[2] Generally speaking, a financial asset is termed to be a *quanto* product if it is denominated in a currency other than that in which it is usually traded.

the one that prevails at present. This can be done by entering a quanto forward or option contract in a foreign stock. We will first concentrate on the study of quanto forward contracts, leaving the analysis of quanto options to the next section. We start by defining precisely what is meant by a quanto forward contract in a foreign stock S^f. As before, the payoff of a guaranteed exchange rate forward contract on a foreign stock at settlement date T is the difference between the stock price at time T and the delivery price denominated in the foreign currency, say K^f. However, this payoff is converted into domestic currency at a predetermined exchange rate, denoted by $\bar{Q}$ in what follows. More formally, denoting by $V_t^d(K^f, \bar{Q})$ the time t value in domestic currency of the quanto forward contract, we have

$$V_T^d(K^f, \bar{Q}) = \bar{Q}(S_T^f - K^f).$$

We wish to determine the right value of such a contract at time t before the settlement. Note that the terminal payoff of a quanto forward contract does not accounts for the future exchange rate fluctuations during the life of a contract. Nevertheless, as we shall see in what follows, its value $V_t^d(K^f, \bar{Q})$ depends on the volatility coefficient σ_Q of the exchange rate process Q – more precisely, on the scalar product $\sigma_Q \cdot \sigma_{Sf}$ that determines the instantaneous covariance between the logarithmic returns of the stock price and the exchange rate. By virtue of the risk-neutral valuation formula, the value at time t of the quanto forward contract equals (in domestic currency)

$$V_t^d(K^f, \bar{Q}) = \bar{Q}e^{-r_d(T-t)}\left(\mathbb{E}_{\mathbb{P}^*}(S_T^f \mid \mathcal{F}_t) - K^f\right).$$

To find the conditional expectation $\mathbb{E}_{\mathbb{P}^*}(S_T^f \mid \mathcal{F}_t)$, observe that by virtue of (4.13), the process $\hat{S}_t = e^{-\delta t} S_t^f$ follows a martingale under $\mathbb{P}^*$, provided that we take $\delta = r_f - \sigma_Q \cdot \sigma_{Sf}$. Consequently, we find easily that

$$\mathbb{E}_{\mathbb{P}^*}(S_T^f \mid \mathcal{F}_t) = e^{\delta T}\mathbb{E}_{\mathbb{P}^*}(\hat{S}_T \mid \mathcal{F}_t) = e^{\delta T}\hat{S}_t = e^{\delta(T-t)}S_t^f,$$

and thus

$$V_t^d(K^f, \bar{Q}) = \bar{Q}e^{-r_d(T-t)}\left(e^{(r_f - \sigma_Q \cdot \sigma_{Sf})(T-t)}S_t^f - K^f\right). \tag{4.22}$$

This in turn implies that the forward price at time t associated with the quanto forward contract that settles at time T equals (in units of foreign currency)

$$\hat{F}_{Sf}^f(t, T) = e^{(r_f - \sigma_Q \cdot \sigma_{Sf})(T-t)}S_t^f = \mathbb{E}_{\mathbb{P}^*}(S_T^f \mid \mathcal{F}_t).$$

It is interesting to note that $\hat{F}_{Sf}^f(t, T)$ is simply the conditional expectation of the stock price at the settlement date T, as seen at time t from the perspective of a domestic-based investor. Furthermore, at least when $\kappa = \sigma_Q \cdot \sigma_{Sf} \geq 0$, it can also be interpreted as the forward price of a fictitious dividend-paying stock, with $\kappa = \sigma_Q \cdot \sigma_{Sf}$ playing the role of the dividend yield (cf. formula (3.66) in Sect. 3.2).

4.4 Foreign Market Futures Contracts

Let us now consider an investor who assumes positions in foreign market futures contracts. We need to translate, in an appropriate way, the marking to market feature of futures contracts. To this end we assume that the profits or losses from futures positions are immediately (i.e., continuously) converted into domestic currency. Let us start by examining investments in foreign market futures contracts in a discrete-time framework. Suppose that $f_t = f_{S^f}(t, T)$ represents the foreign market futures price of some asset S^f. We consider a finite collection of dates, $t_0 = 0 < t_1 < \cdots < t_N = T$, that is assumed to represent the set of dates when the futures contracts are marked to market. If at any date t_i an investor assumes α_{t_i} positions in foreign market futures contracts, and then holds the portfolio unchanged up to the date $t_{i+1} > t_i$, the cumulative profits or losses incurred by the investor up to the terminal date T are given by the expression

$$\sum_{i=0}^{N-1} \alpha_{t_i} Q_{t_{i+1}} (f_{t_{i+1}} - f_{t_i}),$$

where we assume that all cash flows resulting from the marking to market procedure are immediately converted into domestic currency. The last sum may be given the following equivalent form

$$\sum_{i=0}^{N-1} \alpha_{t_i} Q_{t_i} (f_{t_{i+1}} - f_{t_i}) + \sum_{i=0}^{N-1} \alpha_{t_i} (Q_{t_{i+1}} - Q_{t_i})(f_{t_{i+1}} - f_{t_i}).$$

Assume now that the number of resettlement dates tends to infinity, the time interval $[0, T]$ being fixed. In view of properties of the cross-variation of continuous semi-martingales, the equality above leads to the following self-financing condition for continuously rebalanced portfolios.

Definition 4.4.1 An adapted process $\tilde{\phi} = (\tilde{\phi}^1, \tilde{\phi}^2)$ is a *self-financing* trading strategy in foreign market futures contracts with the price process $f_t = f_{S^f}(t, T)$, and in a domestic savings account with the price process B_t^d, if

$$dV_t(\tilde{\phi}) = \tilde{\phi}_t^1 \left(Q_t \, df_t + d\langle Q, f \rangle_t \right) + \tilde{\phi}_t^2 \, dB_t^d, \tag{4.23}$$

where the wealth process equals $V_t(\tilde{\phi}) = \tilde{\phi}_t^2 B_t^d$ for $t \in [0, T]$.

In the sequel, we will assume that the futures price $f = f_{S^f}$ follows a martingale under the foreign market martingale measure $\tilde{\mathbb{P}}$. More specifically, the dynamics of the foreign market futures price f_{S^f} are given by the following expression

$$df_{S^f}(t, T) = f_{S^f}(t, T) \sigma_f \cdot d\tilde{W}_t, \tag{4.24}$$

where $\sigma_f = \sigma_{S^f}$ is a constant volatility vector. Combining (4.14) with (4.24), we get the following equivalent form of (4.23)

$$dV_t(\tilde{\phi}) = \tilde{\phi}_t^1 Q_t f_t \left(\sigma_f \cdot d\tilde{W}_t + \sigma_f \cdot \sigma_Q \, dt \right) + \tilde{\phi}_t^2 r_d B_t^d \, dt,$$

and finally
$$dV_t(\tilde{\phi}) = \tilde{\phi}_t^1 Q_t f_t \sigma_f \cdot dW_t^* + \tilde{\phi}_t^2 r_d B_t^d \, dt.$$

Let $f_Q(t, T)$ stand for the futures exchange rate at time t for the settlement date T – that is, the domestic price at time t of the T-maturity futures contract with terminal price Q_T. It is natural to postulate that $f_Q(t, T) = \mathbb{E}_{\mathbb{P}^*}(Q_T \mid \mathcal{F}_t)$. It follows easily that $f_Q(t, T) = F_Q(t, T)$, i.e., the forward and futures exchange rates agree. Equivalently, the dynamics of the futures exchange rate under the domestic martingale measure $\mathbb{P}^*$ are (cf. (4.18))

$$df_Q(t, T) = f_Q(t, T) \sigma_Q \cdot dW_t^*, \tag{4.25}$$

with the terminal condition $f_Q(T, T) = Q_T$.

Definition 4.4.2 By the *domestic futures price* of the foreign asset S^f we mean the process $f_{S^f}^d(t, T)$, denominated in units of domestic currency, that satisfies the terminal condition

$$f_{S^f}^d(T, T) = Q_T f_{S^f}(T, T) = Q_T S_T^f, \tag{4.26}$$

and is such that for an arbitrary self-financing futures trading strategy $\tilde{\phi}$ in foreign market futures contracts, there exists a self-financing (in the usual sense) futures strategy ϕ such that $dV_t(\phi) = dV_t(\tilde{\phi})$.

More explicitly, we postulate that for any strategy $\tilde{\phi}$ satisfying the conditions of Definition 4.4.1, there exists a trading strategy $\phi = (\phi^1, \phi^2, \phi^3)$ such that the corresponding wealth process $V_t(\phi) = \phi_t^3 B_t^d$, $t \in [0, T]$, satisfies the standard self-financing condition for the futures market, namely

$$dV_t(\phi) = \phi_t^1 \, df_{S^f}^d(t, T) + \phi_t^2 \, df_Q(t, T) + \phi_t^3 \, dB_t^d,$$

and the instantaneous gains or losses from both strategies are identical. Intuitively, using the domestic futures price $f_{S^f}^d(t, T)$ and futures exchange rate $f_Q(t, T)$, we are able to mimic positions in foreign market futures contracts on foreign assets by entering contracts on the corresponding domestic futures market. This in turn implies that the valuation and hedging of foreign market futures options can be, in principle, reduced to pricing of domestic futures options.

Example 4.4.1 As a real-life example of the domestic futures price $f_{S^f}^d(t, T)$ and the corresponding foreign market futures price $f_{S^f}(t, T)$, one can consider the dollar-denominated gold futures contracts traded on the COMEX and the yen-denominated gold futures contracts traded on the Tokyo Commodity Exchange. This shows that the underlying asset of both futures contracts is not necessarily a foreign equity, but virtually any tradable good.

Our aim is now to show that it is possible to express $f_{S^f}^d(t, T)$ in terms of the futures exchange rate $f_Q(t, T)$ and the foreign market futures price $f_{S^f}(t, T)$. Indeed, we have the following simple result.

Lemma 4.4.1 *The domestic futures price* $f_{S^f}^d(t, T)$ *of the foreign market asset* S^f *for the settlement date* T *satisfies*

$$f_{S^f}^d(t, T) = f_Q(t, T) f_{S^f}(t, T), \quad \forall t \in [0, T]. \tag{4.27}$$

Proof. It is clear that (4.27) implies the terminal equality (4.26). Furthermore, using (4.24)–(4.25) and Itô's formula, we get

$$dZ_t = Z_t(\sigma_f + \sigma_Q) \cdot dW_t^*, \tag{4.28}$$

where we write $Z_t = f_Q(t, T) f_{S^f}(t, T)$. We conclude that Z follows a $\mathbb{P}^*$-martingale, and satisfies the terminal condition $Z_T = Q_T S_T^f$. It is now not difficult to check that Z is the domestic futures price of the foreign asset S^f. The details are left to the reader. □

In order to justify equality (4.27) in a more intuitive way, let us consider a specific trading strategy. We combine here a self-financing trading strategy in foreign market futures contracts with a dynamic portfolio of exchange rate futures contracts. Let $f_{S^f}^d(t, T)$ be given by (4.27). Suppose that at any time $t \leq T$ the strategy $\eta = (\eta^1, \eta^2, \eta^3)$ is long in η_t^1 foreign market futures contracts and long in η_t^2 exchange rate futures contracts, where[3]

$$\eta_t^1 = \frac{\psi_t f_{S^f}^d(t, T)}{Q_t f_{S^f}(t, T)}, \quad \eta_t^2 = \frac{\psi_t f_{S^f}^d(t, T)}{f_Q(t, T)}, \quad \forall t \in [0, T],$$

for some process ψ, and the instantaneous marked-to-market profits and losses from the foreign positions are continually converted into domestic currency.

Furthermore, let η_t^3 be the bonds component of the portfolio, so that the portfolio's wealth in domestic currency is $V_t(\eta) = \eta_t^3 B_t^d$, and we may assume that the portfolio is self-financing. This means that the incremental profits and losses from η satisfy

$$dV_t(\eta) = \eta_t^1 \left(Q_t \, df_t + d\langle Q, f \rangle_t \right) + \eta_t^2 \, df_Q(t, T) + \eta_t^3 \, dB_t^d,$$

where, as before, we write $f_t = f_{S^f}(t, T)$. By applying (4.24) and (4.25), we get

$$dV_t(\eta) = \psi_t f_{S^f}^d(t, T)(\sigma_f \cdot d\tilde{W}_t + \sigma_f \cdot \sigma_Q \, dt + \sigma_Q \cdot dW_t^*) + \eta_t^3 \, dB_t^d.$$

In view of equality $\tilde{W}_t = W_t^* - \sigma_Q t$, the last equality can be simplified as follows

$$dV_t(\eta) = \psi_t f_{S^f}^d(t, T)(\sigma_f + \sigma_Q) \cdot dW_t^* + \eta_t^3 \, dB_t^d.$$

Finally, using (4.28), we arrive at the following equality

$$dV_t(\eta) = \psi_t \, df_{S^f}^d(t, T) + \eta_t^3 \, dB_t^d.$$

---

[3] This choice of trading strategy was discussed in Jamshidian (1994a).

This shows that the strategy η_t is essentially equivalent to ψ_t positions in the domestic futures contract, whose price process is $f_{Sf}^d(t, T)$.

Jamshidian (1994b) exploits similar arguments to evaluate and hedge various foreign market futures contracts using their domestic (quanto) counterparts. Let us set $\tilde{\phi}_t^1 = g_t Q_t^{-1} f_t^{-1}$, where

$$g_t = f_t\, e^{-\sigma_Q \cdot \sigma_f (T-t)}, \quad \forall t \in [0, T], \tag{4.29}$$

and f_t represents the foreign market futures price of a certain contract. The specific nature of the futures contract in question has no relevance; we need only assume that the dynamics of f under the foreign market martingale measure are

$$df_t = f_t\, \sigma_f \cdot d\tilde{W}_t$$

with $f_T = X$, where σ_f is a constant (or at least deterministic) volatility vector, and X is an arbitrary $\mathcal{F}_T$-measurable random variable. If the bond component $\tilde{\phi}^2$ of this strategy is chosen in such a way that the strategy $\tilde{\phi} = (\tilde{\phi}^1, \tilde{\phi}^2)$ is self-financing in the sense of Definition 4.4.1, then we have

$$\begin{aligned}
dV_t(\tilde{\phi}) &= \frac{g_t}{Q_t f_t}\left(Q_t\, df_t + d\langle Q, f\rangle_t\right) + \tilde{\phi}_t^2\, dB_t^d \\
&= g_t\left(\sigma_f \cdot d\tilde{W}_t + \sigma_f \cdot \sigma_Q\, dt\right) + \tilde{\phi}_t^2\, dB_t^d \\
&= dg_t + \tilde{\phi}_t^2\, dB_t^d,
\end{aligned}$$

since an application of Itô's rule to (4.29) yields

$$dg_t = g_t\left(\sigma_f \cdot d\tilde{W}_t + \sigma_f \cdot \sigma_Q\, dt\right) = g_t\sigma_f \cdot dW_t^*.$$

Since, in addition, we have $g_T = f_T = X$ (note the distinction with condition (4.26)), it is clear that the process g represents the quanto counterpart of the foreign market futures price f. Put differently, in order to hedge a domestic-denominated futures contract whose terminal value in units of the domestic currency equals X, we may take positions either in foreign market futures contracts with price process f, or in their domestic counterparts with price process g.

Example 4.4.2 The method presented above may be used, for example, to evaluate and hedge the dollar-denominated futures contract of the Japanese equity Nikkei index traded on the CME, whose terminal dollar price is the yen level of Nikkei 225 at time T, using the yen-denominated Nikkei 225 futures contract traded on the Osaka Stock Exchange.

4.5 Foreign Equity Options

In this section, we shall study examples of *foreign equity options* – that is, options whose terminal payoff (in units of domestic currency) depends not only on the future behavior of the exchange rate, but also on the price fluctuations of a certain foreign stock. Recall that we have examined the dynamics of the price process S^f under the domestic and the foreign martingale measure in Sect. 4.1.3. To value options related to foreign market equities, we shall use either the domestic martingale measure $\mathbb{P}^*$ or the foreign martingale measure $\tilde{\mathbb{P}}$, whichever will be more convenient.

4.5.1 Options Struck in a Foreign Currency

Assume first that an investor wants to participate in gains in foreign equity, desires protection against losses in that equity, but is unconcerned about the translation risk arising from the potential drop in the exchange rate. We denote by T the expiry date and by K^f the exercise price of an option. It is essential to note that K^f is expressed in units of foreign currency. The terminal payoff from a *foreign equity call struck in foreign currency* equals

$$C_T^1 \stackrel{\text{def}}{=} Q_T(S_T^f - K^f)^+.$$

This means, in particular, that the terminal payoff is assumed to be converted into domestic currency at the spot exchange rate that prevails at the expiry date. By reasoning in much the same way as in the previous section, one can check that the arbitrage price of a European call option at time t equals

$$C_t^1 = e^{-r_d(T-t)} \mathbb{E}_{\mathbb{P}^*}\big(Q_T(S_T^f - K^f)^+ \big| \mathcal{F}_t\big).$$

Using (4.14), we find that

$$C_t^1 = e^{-r_d(T-t)} Q_t \mathbb{E}_{\mathbb{P}^*}\Big\{ (S_T^f - K^f)^+ \exp\big(\sigma_Q \cdot (W_T^* - W_t^*) + \lambda(T-t)\big) \Big| \mathcal{F}_t \Big\},$$

where $\lambda = r_d - r_f - \frac{1}{2}|\sigma_Q|^2$. Equivalently, using (4.11), we get

$$C_t^1 = e^{-r_f(T-t)} Q_t \mathbb{E}_{\tilde{\mathbb{P}}}\big((S_T^f - K^f)^+ \big| \mathcal{F}_t\big).$$

Since $\tilde{\mathbb{P}}$ is the arbitrage-free measure of the foreign economy, it is not hard to establish the following expression

$$C_t^1 = Q_t\Big(S_t^f N\big(g_1(S_t^f, T-t)\big) - K^f e^{-r_f(T-t)} N\big(g_2(S_t^f, T-t)\big)\Big),$$

where

$$g_{1,2}(s, t) = \frac{\ln(s/K^f) + (r_f \pm \frac{1}{2}|\sigma_{S^f}|^2)t}{|\sigma_{S^f}|\sqrt{t}}.$$

An inspection of the valuation formula above makes clear that a hedging portfolio involves at any instant t the number $N(g_1(S_t^f, T-t))$ shares of the underlying stock; this stock investment demands the additional borrowing of

$$\beta_t^d = Q_t K^f e^{-r_f(T-t)} N(g_2(S_t^f, T-t))$$

units of the domestic currency, or equivalently, the borrowing of

$$\beta_t^f = K^f e^{-r_f(T-t)} N(g_2(S_t^f, T - t))$$

units of the foreign currency.

Remarks. The valuation result established above is in fact quite natural. Indeed, seen from the foreign market perspective, the foreign equity option struck in foreign currency can be priced directly by means of the standard Black-Scholes formula. Put another way, if the domestic price of such an option were different from its foreign market price, converted into units of domestic currency at the current exchange rate, it would be possible to make arbitrage profits by buying (or selling) the option in the domestic market, and simultaneously selling (or buying) it in the foreign market.

As was explained in Sect. 4.1, when dealing with foreign equity options, one can either do the computations with reference to the domestic economy, or equivalently, one may work within the framework of the foreign economy and then convert the final result into units of domestic currency. The choice of the method depends on the particular form of an option's payoff; to some extent it is also a matter of taste.

In the case considered above, in order to complete the calculations in the domestic economy, one needs to compute directly the following expectation (for the sake of notational simplicity we put $t = 0$)

$$C_0^1 = e^{-r_d T} Q_0 \, \mathbb{E}_{\mathbb{P}^*} \left(S_0^f e^{(\sigma_{sf}+\sigma_Q) \cdot W_T^* + (r_d - \frac{1}{2}|\sigma_{sf}+\sigma_Q|^2)T} - K^f e^{\sigma_Q \cdot W_T^* + \lambda T} \right)^+,$$

where, as before, we write $\lambda = r_d - r_f - \frac{1}{2}|\sigma_Q|^2$. This can be done using the following elementary lemma, whose proof is left to the reader.

Lemma 4.5.1 *Let (ξ, η) be a zero-mean, jointly Gaussian (non-degenerate), two-dimensional random variable on a probability space $(\Omega, \mathcal{F}, \mathbb{P})$. Then for arbitrary positive real numbers a and b we have*

$$\mathbb{E}_{\mathbb{P}} \left(a e^{\xi - \frac{1}{2} \operatorname{Var} \xi} - b e^{\eta - \frac{1}{2} \operatorname{Var} \eta} \right)^+ = a N(h) - b N(h - k), \qquad (4.30)$$

where

$$h = \tfrac{1}{k} \ln(a/b) + \tfrac{1}{2}k, \quad k = \sqrt{\operatorname{Var}(\xi - \eta)}.$$

4.5.2 Options Struck in Domestic Currency

Assume now that an investor wishes to receive any positive returns from the foreign market, but wants to be certain that those returns are meaningful when translated back into his own currency. In this case he might be interested in a *foreign equity call struck in domestic currency,* with payoff at expiry

$$C_T^2 \overset{\text{def}}{=} (S_T^f Q_T - K^d)^+ = (\tilde{S}_T^f - K^d)^+,$$

where the strike price K^d is expressed in domestic currency. Due to the particular form of the option's payoff, it is clear that it is now convenient to study the option from the domestic perspective. To find the arbitrage price of the option at time t, it is sufficient to evaluate the following conditional expectation

$$C_t^2 = e^{-r_d(T-t)} \mathbb{E}_{\mathbb{P}^*}\big((\tilde{S}_T^f - K^d)^+ \big| \mathcal{F}_t\big),$$

and by virtue of (4.14)–(4.13), the stock price expressed in units of domestic currency $\tilde{S}_t^f$ has the following dynamics under $\mathbb{P}^*$

$$d\tilde{S}_t^f = \tilde{S}_t^f \big(r_d\, dt + (\sigma_{S^f} + \sigma_Q) \cdot dW_t^*\big).$$

Hence, arguing as in the proof of the classical Black-Scholes formula, one finds easily that the option's price, expressed in units of domestic currency, is given by the formula

$$C_t^2 = \tilde{S}_t^f\, N\big(l_1(\tilde{S}_t^f, T-t)\big) - e^{-r_d(T-t)} K^d\, N\big(l_2(\tilde{S}_t^f, T-t)\big),$$

where

$$l_{1,2}(s, t) = \frac{\ln(s/K^d) + (r_d \pm \tfrac{1}{2}|\sigma_{S^f} + \sigma_Q|^2)t}{|\sigma_{S^f} + \sigma_Q|\sqrt{t}}.$$

Once again, the value $N(l_1(\tilde{S}_t^f, T-t))$ represents the number of shares of the underlying stock held in the replicating portfolio at time t. To establish the replicating portfolio, an additional borrowing of

$$\beta_t^d = e^{-r_d(T-t)} K^d\, N\big(l_2(\tilde{S}_t^f, T-t)\big)$$

units of the domestic currency is required.

4.5.3 Quanto Options

Assume, as before, that an investor wishes to capture positive returns on his foreign equity investment, but now desires to eliminate all exchange risk by fixing an advance rate at which the option's payoff will be converted into domestic currency. At the intuitive level, such a contract can be seen as a combination of a foreign equity option with a currency forward contract. By definition, the payoff of a *quanto call* (i.e., a *guaranteed exchange rate foreign equity call option*) at expiry is set to be

$$C_T^3 \stackrel{\text{def}}{=} \bar{Q}(S_T^f - K^f)^+,$$

where $\bar{Q}$ is the prespecified exchange rate at which the conversion of the option's payoff is made. Note that the quantity $\bar{Q}$ is denominated in the domestic currency per unit of foreign currency, and the strike price K^f is expressed in units of foreign currency. Since the payoff from the quanto option is expressed in units of domestic currency, its arbitrage price equals

$$C_t^3 = \bar{Q}e^{-r_d(T-t)} \mathbb{E}_{\mathbb{P}^*}\big((S_T^f - K^f)^+ \big| \mathcal{F}_t\big). \tag{4.31}$$

The next proposition shows that a closed-form solution for the price of a quanto option is easily available.

Proposition 4.5.1 *The arbitrage price at time t of a European quanto call option with expiry date T and strike price K^f equals (in units of domestic currency)*

$$C_t^3 = \bar{Q} e^{-r_d(T-t)} \left(S_t^f e^{\delta(T-t)} N\left(c_1(S_t^f, T-t)\right) - K^f N\left(c_2(S_t^f, T-t)\right)\right),$$

where $\delta = r_f - \sigma_Q \cdot \sigma_{Sf}$ and

$$c_{1,2}(s, t) = \frac{\ln(s/K^f) + (\delta \pm \frac{1}{2}|\sigma_{Sf}|^2)t}{|\sigma_{Sf}|\sqrt{t}}.$$

Proof. Using (4.31), we obtain

$$C_t^3 = \bar{Q} e^{(\delta - r_d)(T-t)} \mathbb{E}_{\mathbb{P}*}\left(e^{-\delta(T-t)}(S_T^f - K^f)^+ \big| \mathcal{F}_t\right).$$

Since the dynamics of S^f under $\mathbb{P}^*$ are (cf. (4.13))

$$dS_t^f = S_t^f \left(\delta \, dt + \sigma_{Sf} \cdot dW_t^*\right),$$

in order to evaluate the conditional expectation, we can make use of the classical form of the Black-Scholes formula. Namely, we have

$$\mathbb{E}_{\mathbb{P}*}\left(e^{-\delta(T-t)}(S_T^f - K^f)^+ \big| \mathcal{F}_t\right) = c(S_t^f, T - t, K^f, \delta, \sigma_{Sf}),$$

where the function $c(s, t, K, r, \sigma)$ is given by (3.56). Upon rearranging, this yields the desired equality.

    The quanto option can also be examined under the foreign market martingale probability; the domestic market method is slightly more convenient, however. To this end, note that, expressed in units of foreign currency, the price at time t of the quanto option is given by the expression

$$e^{-r_f(T-t)} \bar{Q} \mathbb{E}_{\tilde{\mathbb{P}}}\left(R_T (S_T^f - K^f)^+ \big| \mathcal{F}_t\right).$$

By virtue of (4.10), we have

$$R_T = R_0 \exp\left(-\sigma_Q \cdot \tilde{W}_T + (r_f - r_d - \tfrac{1}{2}|\sigma_Q|^2)T\right).$$

On the other hand, (4.12) yields

$$S_T^f = S_0^f \exp\left(\sigma_{Sf} \cdot \tilde{W}_T + (r_f - \tfrac{1}{2}|\sigma_{Sf}|^2)T\right).$$

By combining the last two equalities, we arrive at the following expression

$$R_T S_T^f = R_0 S_0^f \exp\left((r_f - \gamma)T\right) \exp\left(\sigma \cdot \tilde{W}_T - \tfrac{1}{2}|\sigma|^2 T\right),$$

where
$$\sigma = \sigma_{Sf} - \sigma_Q, \quad \gamma = r_d - r_f + \sigma_Q \cdot \sigma_{Sf}.$$

Let us consider the case $t = 0$. To find the value of the option in units of domestic currency, it is enough to evaluate
$$C_0^3 = \bar{Q}\, Q_0\, \mathbb{E}_{\tilde{\mathbb{P}}}\big(e^{-r_f T} R_T S_T^f - e^{-r_f T} K^f R_T\big)^+.$$

For this purpose, one can make use of Lemma 4.5.1 with constants
$$a = S_0^f e^{-\gamma T}, \quad b = K^f e^{-r_d T}$$

and with the random variables
$$\xi = (\sigma_{Sf} - \sigma_Q) \cdot \tilde{W}_T, \quad \eta = -\sigma_Q \cdot \tilde{W}_T.$$

It is easy to check that $k = |\sigma_{Sf}|\sqrt{T}$, and
$$h = \frac{\ln(S_0^f / K^f) + \big(\delta + \frac{1}{2}|\sigma_{Sf}|^2\big)T}{|\sigma_{Sf}|\sqrt{T}},$$

where $\delta = r_f - \sigma_Q \cdot \sigma_{Sf}$. Consequently, by virtue of (4.30), we get
$$C_0^3 = \bar{Q}\big(aN(h) - bN(h-k)\big) = \bar{Q}e^{-r_d T}\big(S_0^f e^{\delta T} N(h) - K^f N(h-k)\big).$$

The last equality is easily seen to agree with the option valuation formula established in Proposition 4.5.1. By proceeding along the same lines, one can also find the price P_t^3 of a quanto put option with the strike price K^f and the guaranteed exchange rate $\bar{Q}$. Furthermore, we have the following version of the put-call parity relationship $C_t^3 - P_t^3 = V_t^d(K^f, \bar{Q})$, where $V_t^d(K^f, \bar{Q})$ represents the value at time t of the corresponding guaranteed exchange rate forward contract in domestic currency (see (4.22)).

4.5.4 Equity-linked Foreign Exchange Options

Finally, assume that an investor desires to hold foreign equity regardless of whether the stock price rises or falls (that is, he is indifferent to the foreign equity exposure), however, wishes to place a floor on the exchange rate risk of his foreign investment. An *equity-linked foreign exchange call* (an *Elf-X call,* for short) with payoff at expiry (in units of domestic currency)
$$C_T^4 \overset{\text{def}}{=} (Q_T - K)^+ S_T^f,$$

where K is a *strike exchange rate* expressed in domestic currency per unit of foreign currency, is thus a combination of a currency option with an equity forward. The arbitrage price (in units of domestic currency) of a European Elf-X call with expiry date T equals
$$C_t^4 = e^{-r_d(T-t)}\, \mathbb{E}_{\mathbb{P}^*}\big((Q_T - K)^+ S_T^f \,\big|\, \mathcal{F}_t\big).$$

We shall first value an equity-linked foreign exchange call option using the domestic martingale measure.

Proposition 4.5.2 *The arbitrage price, expressed in domestic currency, of a European equity-linked foreign exchange call option, with strike exchange rate K and expiry date T, is given by the following formula*

$$C_t^4 = S_t^f \left(Q_t N\big(w_1(Q_t, T - t)\big) - K\, e^{-\gamma(T-t)} N\big(w_2(Q_t, T - t)\big) \right), \qquad (4.32)$$

where $\gamma = r_d - r_f + \sigma_Q \cdot \sigma_{Sf}$ *and*

$$w_{1,2}(q, t) = \frac{\ln(q/K) + (\gamma \pm \frac{1}{2} |\sigma_Q|^2) t}{|\sigma_Q| \sqrt{t}}.$$

Proof. As usual, it is sufficient to consider the case of $t = 0$. In view of (4.13), we have

$$S_T^f = S_0^f \exp\left(\sigma_{Sf} \cdot W_T^* - \tfrac{1}{2} |\sigma_{Sf}|^2 T + (r_f - \sigma_Q \cdot \sigma_{Sf}) T \right).$$

We define a probability measure $\mathbb{Q}$ on $(\Omega, \mathcal{F}_T)$ by setting

$$\frac{d\mathbb{Q}}{d\mathbb{P}^*} = \exp\left(\sigma_{Sf} \cdot W_T^* - \tfrac{1}{2} |\sigma_{Sf}|^2 T \right), \quad \mathbb{P}^*\text{-a.s.}$$

Notice that the process $U_t = W_t^* - \sigma_{Sf} t$ follows a Brownian motion under $\mathbb{Q}$. Furthermore,

$$C_0^4 = S_0^f\, e^{(r_f - r_d - \sigma_Q \cdot \sigma_{Sf})T} \, \mathbb{E}_{\mathbb{Q}}\big((Q_T - K)^+\big),$$

and the dynamics of the process Q under the probability measure $\mathbb{Q}$ are

$$dQ_t = Q_t \left((r_d - r_f + \sigma_Q \cdot \sigma_{Sf})\, dt + \sigma_Q \cdot dU_t \right).$$

The expectation $\mathbb{E}_{\mathbb{Q}}(Q_T - K)^+$ can be evaluated along the same lines as in the standard Black-Scholes model, yielding the following equality

$$C_0^4 = S_0^f c(Q_0, T, K, \gamma, \sigma_Q),$$

where, as usual, the function $c(s, t, K, r, \sigma)$ is given by (3.56). □

The Elf-X option can also be valued by taking the perspective of the foreign market-based investor. The option price at time 0, when expressed in units of foreign currency, is given by the risk-neutral valuation formula

$$C_0^4 = e^{-r_f T} \, \mathbb{E}_{\tilde{\mathbb{P}}}(S_T^f - K R_T S_T^f)^+.$$

To derive equality (4.32) from the last formula, one can directly apply Lemma 4.5.1 with the constants

$$a = S_0^f, \quad b = K R_0 S_0^f\, e^{-\gamma T},$$

and the random variables

$$\xi = \sigma_{Sf} \cdot \tilde{W}_T, \quad \eta = (\sigma_{Sf} - \sigma_Q) \cdot \tilde{W}_T.$$

Note that the standard deviation of the Gaussian random variable $\xi - \eta$ is $k = |\sigma_Q|\sqrt{T}$, and thus h equals

$$h = \tfrac{1}{k} \ln(a/b) + \tfrac{1}{2}k = \frac{-\ln(R_0 K) + \left(\gamma + \tfrac{1}{2}|\sigma_Q|^2\right)T}{|\sigma_Q|\sqrt{T}}.$$

This implies that the option's price, expressed in units of domestic currency, equals

$$C_0^4 = S_0^f \left(Q_0 N(h) - K e^{-\gamma T} N(h - k)\right).$$

Formula (4.32) now follows by standard arguments.

Remarks. Note that as S^f we may take the price process of any foreign security, not necessarily a stock. For instance, to find the currency option valuation formula, it is enough to take as S^f the price process of a foreign bond paying one unit of the foreign currency at maturity date T. Under the present assumptions the bond price equals $B^f(t, T) = \exp(-r_f(T - t))$, so that the bond price volatility vanishes – that is, $\sigma_{S^f} = 0$. To derive from (4.32) formula (4.19), which gives the arbitrage price C_t^Q of a European currency call option, it is enough to observe that the relationship $C_t^4 = \exp(-r_f(T - t))C_t^Q$ is satisfied for every $t \in [0, T]$.

Let us conclude this chapter by mentioning that extensions of several results of this chapter to the case of a financial market model that makes account for uncertainty of interest rate can be found in Chap. 14. For more information on exotic foreign exchange products and their valuation in other models than the lognormal model examined in this chapter, the reader is referred to the monograph by Lipton (2001).

5

American Options

In contrast to the holder of a European option, the holder of an *American option* is allowed to exercise his right to buy (or sell) the underlying asset at any time before or at the expiry date. This special feature of American-style options – and more generally of American claims – makes the arbitrage pricing of American options much more involved than the valuation of standard European claims. We know already that arbitrage valuation of American claims is closely related to specific optimal stopping problems. Intuitively, one might expect that the holder of an American option will choose her exercise policy in such a way that the expected payoff from the option will be maximized. Maximization of the expected discounted payoff under subjective probability would lead, of course, to non-uniqueness of the price. It appears, however, that for the purpose of arbitrage valuation, the maximization of the expected discounted payoff should be done under the martingale measure (that is, under risk-neutral probability). Thus, the uniqueness of the arbitrage price of an American claim holds.

One of the earliest works to examine the relationship between the early exercise feature of American options and optimal stopping problems was the paper by McKean (1965). As the arbitrage valuation of derivative securities was not yet discovered at this time, the optimal stopping problem associated with the optimal exercise of American put was studied by McKean (1965) under an actual probability $\mathbb{P}$, rather than under the martingale measure $\mathbb{P}^*$, as is done nowadays. For further properties of the optimal stopping boundary, we refer the reader to Van Moerbeke (1976). Basic features of American options, within the framework of arbitrage valuation theory, were already examined in Merton (1973). However, mathematically rigorous valuation results for American claims were first established by means of arbitrage arguments in Bensoussan (1984) and Karatzas (1988, 1989). An exhaustive survey of results and techniques related to the arbitrage pricing of American options was given by Myneni (1992). For an innovative approach to American options and related issues, see Bank and Föllmer (2003).

The purpose of this chapter is to provide the most fundamental results concerning the arbitrage valuation of American claims within the continuous-time framework of the Black-Scholes financial model.

Firstly, we discuss the concept of the arbitrage price of American contingent claims and its basic properties. As a consequence, we present the well-known result that an American call option with a constant strike price, written on a non-dividend-paying stock, is equivalent to the corresponding European call option. Subsequently, we focus on the features of the optimal exercise policy associated with the American put option.

Next, the analytical approach to the pricing of American options is presented. The free boundary problem associated with the optimal exercise of American put options was studied by, among others, McKean (1965) and Van Moerbeke (1976). More recently, Jaillet et al. (1990) applied the general theory of variational inequalities to study the optimal stopping problem associated with American claims.

Finally, the most widely used numerical procedures related to the approximate valuation of American contingent claims are reviewed. An analytic approximation of the American put price on a non-dividend-paying stock was examined by Brennan and Schwartz (1977a), Johnson (1983) and MacMillan (1986). We close this chapter with an analysis of an American call written on a dividend-paying stock (this was examined in Roll (1977)).

5.1 Valuation of American Claims

We place ourselves within the classical Black-Scholes set-up. Hence the prices of primary securities – that is, the stock price, S, and the savings account, B – are modelled by means of the following differential equations

$$dS_t = \mu S_t \, dt + \sigma S_t \, dW_t, \quad S_0 > 0,$$

where $\mu \in \mathbb{R}$ and $\sigma > 0$ are real numbers, and

$$dB_t = r B_t \, dt, \quad B_0 = 1,$$

with $r \in \mathbb{R}$. As usual, we denote by W the standard Brownian motion defined on a filtered probability space $(\Omega, \mathbb{F}, \mathbb{P})$, where $\mathbb{F} = \mathbb{F}^W$. For ease of notation, we assume here that the underlying Brownian motion W is one-dimensional.

In the context of arbitrage valuation of American contingent claims, it is convenient (although not necessary) to assume that an individual may withdraw funds to finance his consumption needs. For any fixed t, we denote by A_t the cumulative amount of funds that are withdrawn and consumed[1] by an investor up to time t. The process A is assumed to be progressively measurable with non-decreasing sample paths; also, by convention, $A_0 = A_{0-} = 0$.

We assume also that A is an RCLL process – that is, almost all sample paths of A are right-continuous functions, with finite left-hand limits. We say that A represents the *consumption strategy,* as opposed to the *trading strategy* ϕ. It is thus natural

---

[1] The term "consumed" refers to the fact that the wealth is dynamically diminished according to the process A.

to call a pair (ϕ, A) a *trading and consumption strategy* in (S, B). The formal definition of a self-financing strategy now reads as follows.

Definition 5.1.1 A trading and consumption strategy (ϕ, A) in (S, B) is *self-financing* on $[0, T]$ if its wealth process $V(\phi, A)$, given by the formula $V_t(\phi, A) = \phi_t^1 S_t + \phi_t^2 B_t$ for $t \in [0, T]$, satisfies for every $t \in [0, T]$

$$V_t(\phi, A) = V_0(\phi, A) + \int_0^t \phi_u^1 \, dS_u + \int_0^t \phi_u^2 \, dB_u - A_t. \tag{5.1}$$

In view of (5.1), it is clear that A models the flow of funds that are not reinvested in primary securities, but rather are put aside forever.[2] By convention, we say that the amount of funds represented by A_t is *consumed* by the holder of the dynamic portfolio (ϕ, A) up to time t. Using the equality $V_t(\phi, A) = \phi_t^1 S_t + \phi_t^2 B_t$, we can eliminate the component ϕ^2. This yields the following equivalent form of (5.1)

$$dV_t = r V_t \, dt + \phi_t^1 S_t \big((\mu - r) \, dt + \sigma \, dW_t\big) - dA_t,$$

where we write V to denote the wealth process $V(\phi, A)$. Equivalently,

$$dV_t = r V_t \, dt + \zeta_t (\mu - r) \, dt + \sigma \zeta_t \, dW_t - dA_t, \tag{5.2}$$

where $\zeta_t = \phi_t^1 S_t$ represents the amount of cash invested in shares at time t. It is known that the unique solution of the linear SDE (5.2) is given by the formula

$$V_t = B_t \left(V_0 + \int_0^t (\mu - r) \zeta_u B_u^{-1} \, du - \int_0^t B_u^{-1} \, dA_u + \int_0^t \sigma \zeta_u B_u^{-1} \, dW_u \right),$$

which holds for every $t \in [0, T]$. We conclude that the wealth process of any self-financing trading and consumption strategy is uniquely determined by the following quantities: the initial endowment V_0, the consumption process A, and the process ζ representing the amount of cash invested in shares. In other words, given an initial endowment V_0, there is one-to-one correspondence between self-financing trading and consumption strategies (ϕ, A) and two-dimensional processes (ζ, A). We will sometimes find it convenient to identify a self-financing trading and consumption strategy (ϕ, A) with the corresponding pair (ζ, A), where $\zeta = \phi_t^1 S_t$. Recall that the unique martingale measure $\mathbb{P}^*$ for the Black-Scholes spot market satisfies

$$\frac{d\mathbb{P}^*}{d\mathbb{P}} = \exp \left(\frac{r - \mu}{\sigma} W_{T^*} - \frac{1}{2} \frac{(r - \mu)^2}{\sigma^2} T^* \right), \quad \mathbb{P}\text{-a.s.}$$

It is easily seen that the dynamics of the wealth process V under the martingale measure $\mathbb{P}^*$ are given by the following expression

$$dV_t = r V_t \, dt + \sigma \zeta_t \, dW_t^* - dA_t,$$

---

[2] Let us stress that, in contrast to the trading-consumption optimization problems, the role of the consumption process is not essential in the present context. Indeed, one can alternatively assume that these funds are invested in risk-free bonds.

where W^* follows the standard Brownian motion under $\mathbb{P}^*$, and thus

$$V_t = B_t \left(V_0 - \int_0^t B_u^{-1} \, dA_u + \int_0^t \sigma \zeta_u B_u^{-1} \, dW_u^* \right).$$

Hence an auxiliary process Z, given by the formula

$$Z_t \overset{\text{def}}{=} V_t^* + \int_0^t B_u^{-1} \, dA_u = V_0 + \int_0^t \sigma \zeta_u B_u^{-1} \, dW_u^*,$$

where $V_t^* = V_t / B_t$, follows a local martingale under $\mathbb{P}^*$. We say that a self-financing trading and consumption strategy (ϕ, A) is *admissible* if the condition

$$\mathbb{E}_{\mathbb{P}^*} \left(\int_0^T \zeta_u^2 \, du \right) = \mathbb{E}_{\mathbb{P}^*} \left(\int_0^T (\phi_u^1 S_u)^2 \, du \right) < \infty$$

is satisfied so that Z is a $\mathbb{P}^*$-martingale. Similarly to Sect. 3.1, this assumption is imposed in order to exclude pathological examples of arbitrage opportunities from the market model. We are now in a position to formally introduce the concept of a American style contingent claim. To this end, we take an arbitrary continuous *reward function* $g : \mathbb{R}_+ \times [0, T] \to \mathbb{R}$ satisfying the linear growth condition $|g(s, t)| \leq k_1 + k_2 s$ for some constants k_1, k_2. An *American claim* with the reward function g and expiry date T is a financial security that pays to its holder the amount $g(S_t, t)$ when exercised at time t.

Note that we deliberately restrict our attention to path-independent American contingent claims – that is, to American claims whose payoff at exercise depends on the value of the underlying asset at the exercise date only.

The writer of an American claim with the reward function g accepts the obligation to pay the amount $g(S_t, t)$ at any time t. It should be emphasized that the choice of the exercise time is at discretion of the holder of an American claim (that is, of a party assuming a long position). In order to formalize the concept of an American claim, we need to introduce first a suitable class of admissible exercise times. Since we exclude clairvoyance, the admissible exercise time τ is assumed to be a stopping time[3] of filtration $\mathbb{F}$. Since in the Black-Scholes model we have $\mathbb{F} = \mathbb{F}^W = \mathbb{F}^{W^*} = \mathbb{F}^S$, any stopping time of the filtration $\mathbb{F}$ is also a stopping time of the filtration $\mathbb{F}^S$ generated by the stock price process S. In intuitive terms, it is assumed throughout that the decision to exercise an American claim at time t is based on the observations of stock price fluctuations up to time t, but not after this date. This interpretation is consistent with our general assumption that $\mathcal{F}_t$ represents the information available to all investors at time t. Let us denote by $\mathcal{T}_{[t,T]}$ the set of all stopping times of the filtration $\mathbb{F}$ that satisfy $t \leq \tau \leq T$ (with probability 1).

Definition 5.1.2 An American contingent claim X^a with the reward function $g :$ $\mathbb{R}_+ \times [0, T] \to \mathbb{R}$ is a financial instrument consisting of: (a) an expiry date T; (b) the selection of a stopping time $\tau \in \mathcal{T}_{[0,T]}$; and (c) a payoff $X_\tau^a = g(S_\tau, \tau)$ on exercise.

---

[3] A random variable $\tau : (\Omega, \mathcal{F}_T, \mathbb{P}) \to [0, T]$ is a *stopping time of filtration* $\mathbb{F}$ if, for every $t \in [0, T]$ the event $\{\tau \leq t\}$ belongs to the σ-field $\mathcal{F}_t$.

The most typical examples of American claims are American options with constant strike price K and expiry date T. The payoffs of American call and put options, when exercised at the random time τ, are equal to $X_\tau = (S_\tau - K)^+$ and $Y_\tau = (K - S_\tau)^+$ respectively. In a slightly more general case, when the strike price is allowed to vary in time, the corresponding reward functions are $g^c(s, t) = (s - K_t)^+$ and $g^p(s, t) = (K_t - s)^+$ for American call and put options respectively. Here, $K : [0, T] \to \mathbb{R}_+$ is a deterministic function representing the variable level of the strike price. Our aim is to derive the "rational" price and to determine the "rational" exercise time of an American contingent claim by means of purely arbitrage arguments. To this end, we shall first introduce a specific class of trading strategies. For expositional simplicity, we concentrate on the price of an American claim X^a at time 0; the general case can be treated similarly, but is more cumbersome from the notational viewpoint. It will be sufficient to consider a very special class of trading strategies associated with the American contingent claim X^a, namely the *buy-and-hold* strategies.

By a *buy-and-hold* strategy associated with an American claim X^a we mean a pair (c, τ), where $\tau \in T_{[0,T]}$ and c is a real number. In financial interpretation, a buy-and-hold strategy (c, τ) assumes that $c > 0$ units of the American security X^a are acquired (or shorted, if $c < 0$) at time 0, and then held in the portfolio[4] up to the exercise time τ.

Let us assume that there exists a "market" price, say U_0, at which the American claim X^a trades in the market at time 0. Our first task is to find the right value of U_0 by means of no-arbitrage arguments (as was mentioned above, the arguments leading to the arbitrage valuation of the claim X^a at time $t > 0$ are much the same as in the case of $t = 0$, therefore the general case is left to the reader).

Definition 5.1.3 By a *self-financing trading strategy in* (S, B, X^a) we mean a collection (ϕ, A, c, τ), where (ϕ, A) is a trading and consumption strategy in (S, B) and (c, τ) is a buy-and-hold strategy associated with X^a. In addition, we assume that on the random interval $(\tau, T]$ we have

$$\phi_t^1 = 0, \quad \phi_t^2 = \phi_\tau^1 S_\tau B_\tau^{-1} + \phi_\tau^2 + cg(S_\tau, \tau)B_\tau^{-1}. \tag{5.3}$$

It will soon become apparent that it is enough to consider the cases of $c = 1$ and $c = -1$. In other words, we shall consider the long and short positions in the American claim X^a.

An analysis of condition (5.3) shows that the definition of a self-financing strategy (ϕ, A, c, τ) implicitly assumes that the American claim is exercised at a random time τ, existing positions in shares are closed at time τ, and all the proceeds are invested in risk-free bonds. For brevity, we shall sometimes write $\tilde{\psi}$ to denote the dynamic portfolio (ϕ, A, c, τ) in what follows. Note that the wealth process $V(\tilde{\psi})$ of any self-financing strategy in (S, B, X^a) satisfies the following initial and terminal conditions

---

[4] Observe that such a strategy excludes trading in the American claim after the initial date. In other words, dynamic trading in the American claim is not considered at this stage.

$$V_0(\tilde{\psi}) = \phi_0^1 S_0 + \phi_0^2 + cU_0$$

and

$$V_T(\tilde{\psi}) = e^{r(T-\tau)}(\phi_\tau^1 S_\tau + cg(S_\tau, \tau)) + e^{rT}\phi_\tau^2.$$

In what follows, we shall restrict our attention to the class of *admissible* trading strategies $\tilde{\psi} = (\phi, A, c, \tau)$ in (S, B, X^a) that are defined in the following way.

Definition 5.1.4 A self-financing trading strategy (ϕ, A, c, τ) in (S, B, X^a) is said to be *admissible* if a trading and consumption strategy (ϕ, A) is admissible and $A_T = A_\tau$. The class of all admissible strategies (ϕ, A, c, τ) is denoted by $\tilde{\Psi}$.

Let us introduce the class $\tilde{\Psi}^0$ of those admissible trading strategies $\tilde{\psi}$ for which the initial wealth satisfies $V_0(\tilde{\psi}) < 0$ and the terminal wealth has the nonnegative value; that is[5] $V_T(\tilde{\psi}) = \phi_T^2 B_t \geq 0$. In order to precisely define an arbitrage opportunity, we have to take into account the early exercise feature of American claims. It is intuitively clear that it is enough to consider two cases – a long and a short position in one unit of an American claim. This is due to the fact that we need to exclude the existence of arbitrage opportunities for both the seller and the buyer of an American claim. Indeed, the position of both parties involved in a contract of American style is no longer symmetric, as it was in the case of European claims. The holder of an American claim can actively choose his exercise policy. The seller of an American claim, on the contrary, should be ready to meet his obligations at any (random) time. We therefore set down the following definition of arbitrage and an arbitrage-free market model.

Definition 5.1.5 There is *arbitrage* in the market model with trading in the American claim X^a with initial price U_0 if either (a) there is *long arbitrage,* i.e., there exists a stopping time τ such that for some trading and consumption strategy (ϕ, A) the strategy $(\phi, A, 1, \tau)$ belongs to the class $\tilde{\Psi}^0$, or (b) there is *short arbitrage,* i.e., there exists a trading and consumption strategy (ϕ, A) such that for any stopping time τ the strategy $(\phi, A, -1, \tau)$ belongs to the class $\tilde{\Psi}^0$. In the absence of arbitrage in the market model, we say that the model is *arbitrage-free.*

Definition 5.1.5 can be reformulated as follows: there is *absence of arbitrage* in the market if the following conditions are satisfied: (a) for any stopping time τ and any trading and consumption strategy (ϕ, A), the strategy $(\phi, A, 1, \tau)$ is not in $\tilde{\Psi}^0$; and (b) for any trading and consumption strategy (ϕ, A), there exists a stopping time τ such that the strategy $(\phi, A, -1, \tau)$ is not in $\tilde{\Psi}^0$.

Intuitively, under the absence of arbitrage in the market, the holder of an American claim is unable to find an exercise policy τ and a trading and consumption strategy (ϕ, A) that would yield a risk-free profit. Also, under the absence of arbitrage, it is not possible to make risk-free profit by selling the American claim at time 0, provided that the buyer makes a clever choice of the exercise date. More precisely,

---

[5] Since the existence of a strictly positive savings account is assumed, one can alternatively define the class Ψ^0 as the set of those strategies $\tilde{\psi}$ from $\tilde{\Psi}$ for which $V_0(\tilde{\psi}) = 0$, $V_T(\tilde{\psi}) = \phi_T^2 B_T \geq 0$, and the latter inequality is strict with positive probability.

there exists an exercise policy for the long party that prevents the short party from locking in a risk-free profit.

By definition, the *arbitrage price* at time 0 of the American claim X^a, denoted by $\pi_0(X^a)$, is that level of the price U_0 that makes the model arbitrage-free. Our aim is now to show that the assumed absence of arbitrage in the sense of Definition 5.1.5 leads to a unique value for the arbitrage price $\pi_0(X^a)$ of X^a (as was mentioned already, it is not hard to extend this reasoning in order to determine the arbitrage price $\pi_t(X^a)$ of the American claim X^a at any date $t \in [0, T]$). Also, we shall find the *rational* exercise policy of the holder – that is, the stopping time that excludes the possibility of short arbitrage.

The following auxiliary result relates the value process associated with the specific optimal stopping problem to the wealth process of a certain admissible trading strategy. For any reward function g, we define an adapted process V by setting

$$V_t = \operatorname{ess\,sup}_{\tau \in \mathcal{T}_{[t,T]}} \mathbb{E}_{\mathbb{P}^*}\left(e^{-r(\tau-t)} g(S_\tau, \tau) \mid \mathcal{F}_t\right) \tag{5.4}$$

for every $t \in [0, T]$, provided that the right-hand side of (5.4) is well-defined.

Proposition 5.1.1 *Let V be an adapted process defined by formula (5.4) for some reward function g. Then there exists an admissible trading and consumption strategy (ϕ, A) such that $V_t = V_t(\phi, A)$ for every $t \in [0, T]$.*

Proof. We shall give the outline of the proof (for technical details, we refer to Karatzas (1988) and Myneni (1992)). Let us introduce the Snell envelope J of the discounted reward process $Z_t^* = e^{-rt} g(S_t, t)$. By definition, the process J is the smallest supermartingale majorant to the process Z^*. From the general theory of optimal stopping, we know that

$$J_t = \operatorname{ess\,sup}_{\tau \in \mathcal{T}_{[t,T]}} \mathbb{E}_{\mathbb{P}^*}\left(e^{-r\tau} g(S_\tau, \tau) \mid \mathcal{F}_t\right) = \operatorname{ess\,sup}_{\tau \in \mathcal{T}_{[t,T]}} \mathbb{E}_{\mathbb{P}^*}\left(Z_\tau^* \mid \mathcal{F}_t\right)$$

for every $t \in [0, T]$, and thus $V_t = e^{rt} J_t$. Since J is a RCLL *regular* supermartingale of class DL,[6] it follows from general results that J admits the unique Doob-Meyer decomposition $J = M - H$, where M is a (square-integrable) martingale and H is a continuous, non-decreasing process with $H_0 = 0$. Consequently,

$$d\left(e^{rt} J_t\right) = r e^{rt} J_t \, dt + e^{rt} \, dM_t - e^{rt} \, dH_t.$$

The predictable representation property (see Theorem A.11.1) yields

$$M_t = M_0 + \int_0^t \xi_u \, dW_u^*, \quad \forall t \in [0, T],$$

---

[6] Basically, one needs to check that the family $\{J_\tau \mid \tau \in \mathcal{T}_{[0,T]}\}$ of random variables is *uniformly integrable* under $\mathbb{P}^*$. We refer the reader to Sect. 1.4 in Karatzas and Shreve (1998a) for the definition of a *regular* process and for the concept of the *Doob-Meyer decomposition* of a semimartingale. Regularity of J, which follows from the results of Bismut and Skalli (1977), implies the continuity of H and thus it is not essential (otherwise, H is known to be *predictable*).

for some progressively measurable process ξ with $\mathbb{E}_{\mathbb{P}^*}\big(\int_0^T \xi_u^2\, du\big) < \infty$. Hence, upon setting

$$\phi_t^1 = e^{rt}\xi_t\sigma^{-1}S_t^{-1}, \quad \phi_t^2 = J_t - \xi_t\sigma^{-1}, \quad A_t = \int_0^t e^{ru}\, dH_u, \tag{5.5}$$

we conclude that the process V represents the wealth process of some (admissible) trading and consumption strategy.   □

By the general theory of optimal stopping, we know also that the random time τ_t that maximizes the expected discounted reward after the date t is the first instant at which the process J drops to the level of the discounted reward, that is (by convention $\inf \emptyset = T$)

$$\tau_t = \inf\{u \in [t, T] \mid J_u = Z_u^*\}.$$

In other words, the optimal (under $\mathbb{P}^*$) exercise policy of the American claim with reward function g is given by the equality

$$\tau_0 = \inf\{u \in [0, T] \mid J_u = e^{-ru}g(S_u, u)\}.$$

Observe that the stopping time τ_0 is well-defined (i.e., the set on the right-hand side is non-empty with probability 1), and necessarily

$$V_{\tau_0} = g(S_{\tau_0}, \tau_0). \tag{5.6}$$

In addition, the stopped process $J_{t \wedge \tau_0}$ is a martingale, so that the process H is constant on the interval $[0, \tau_0]$. This means also that $A_t = 0$ on the random interval $[0, \tau_0]$, so that no consumption is present before time τ_0. We find it convenient to introduce the following definition.

Definition 5.1.6 An admissible trading and consumption strategy (ϕ, A) is said to be a *perfect hedging* against the American contingent claim X^a with reward function g if, with probability 1,

$$V_t(\phi) \geq g(S_t, t), \quad \forall t \in [0, T].$$

We write $\Phi(X^a)$ to denote the class of all perfect hedging strategies against the American contingent claim X^a.

From the majorizing property of the Snell envelope, we infer that the trading and consumption strategy (ϕ, A) introduced in the proof of Proposition 5.1.1 is a perfect hedging against the American claim with reward function g. Moreover, this strategy has the special property of minimal initial endowment amongst all admissible perfect hedging strategies against the American claim. Our goal is now to determine $\pi_0(X^a)$ explicitly, by assuming that trading in the American claim X^a would not destroy the arbitrage-free features of the Black-Scholes model.

Theorem 5.1.1 *There is absence of arbitrage (in the sense of Definition 5.1.5) in the market model with trading in an American claim if and only if the price $\pi_0(X^a)$ is given by the formula*

$$\pi_0(X^a) = \sup_{\tau \in \mathcal{T}_{[0,T]}} \mathbb{E}_{\mathbb{P}^*}\big(e^{-r\tau}g(S_\tau, \tau)\big).$$

More generally, the arbitrage price at time t of an American claim with reward function g equals

$$\pi_t(X^a) = \operatorname{ess\,sup}_{\tau \in \mathcal{T}_{[t,T]}} \mathbb{E}_{\mathbb{P}^*}\big(e^{-r(\tau-t)} g(S_\tau, \tau) \,|\, \mathcal{F}_t\big).$$

Proof. We shall follow Myneni (1992). Let us assume that the "market" price U_0 of the option satisfies $U_0 > V_0$. We shall show that in this case, the American claim is overpriced – that is, a short arbitrage is possible. Let (ϕ, A) be the trading and consumption strategy considered in the proof of Proposition 5.1.1 (see formula (5.5)). Suppose that the option's buyer selects an arbitrary stopping time $\tau \in \mathcal{T}_{[0,T]}$ as his exercise policy. Let us consider the following strategy $(\hat{\phi}, \hat{A}, -1, \tau)$ (observe that in implementing this strategy, we do not need to assume that the exercise time τ is known in advance)

$$\hat{\phi}_t^1 = \phi_t^1 \mathbb{1}_{[0,\tau]}(t),$$
$$\hat{\phi}_t^2 = \phi_t^2 \mathbb{1}_{[0,\tau]}(t) + \big(\phi_\tau^2 + \phi_\tau^1 S_\tau B_\tau^{-1} - g(S_\tau, \tau) B_\tau^{-1}\big) \mathbb{1}_{(\tau,T]}(t),$$

and $\hat{A}_t = A_{t \wedge \tau}$. Since (ϕ, A) is assumed to be a perfect hedging, we have $\hat{\phi}_\tau^1 S_\tau + \hat{\phi}_\tau^2 B_\tau \geq g(S_\tau, \tau)$, so that $\hat{\phi}_\tau^2 B_\tau \geq 0$, $\mathbb{P}^*$-a.s. Moreover, by construction, the initial wealth of $(\hat{\phi}, \hat{A}, -1, \tau)$ satisfies $\hat{\phi}_0^1 S_0 + \hat{\phi}_0^2 - U_0 = V_0 - U_0 < 0$. We conclude that the strategy $(\hat{\phi}, \hat{A}, -1, \tau)$ is a (short) arbitrage opportunity – that is, a risk-free profitable strategy for the seller of the American claim X^a. Suppose now that $U_0 < V_0$, so that the American claim is underpriced. We shall now construct an arbitrage opportunity for the buyer of this claim. In this case, we may and do assume that he chooses the stopping time τ_0 as an exercise time. In addition, we assume that he holds a dynamic portfolio $(-\hat{\phi}, -\hat{A})$. Note that the process $\hat{A}$ vanishes identically, since $\tau = \tau_0$. This means that no consumption is involved in the strategy chosen by the buyer. Furthermore, the initial wealth of his portfolio satisfies

$$-\hat{\phi}_0^1 S_0 - \hat{\phi}_0^2 + U_0 = U_0 - V_0 < 0$$

and the terminal wealth is zero, since in view of (5.6) the wealth of the portfolio at the exercise time τ_0 vanishes. This shows that by making a clever choice of exercise policy, the buyer of the American claim is able to lock in a risk-free profit. We conclude that the arbitrage price $\pi_0(X^a)$ necessarily coincides with V_0, since otherwise arbitrage opportunities would exist in the model.   □

5.2 American Call and Put Options

From now on we restrict our attention to the case of American call and put options. We allow the strike price to vary in time; the strike price is represented by a deterministic function $K : [0, T] \to \mathbb{R}_+$ that satisfies

$$K_t = K_0 + \int_0^t k_u \, du, \quad \forall t \in [0, T],$$

for a bounded function $k : [0, T] \to \mathbb{R}$. The reward functions we shall study in what follows are $g^c(s, t) = (s - K_t)^+$ and $g^p(s, t) = (K_t - s)^+$, where the rewards g^c and g^p correspond to the call and the put options respectively. It will be convenient to introduce the discounted rewards $X_t^* = (S_t - K_t)^+/B_t$ and $Y_t^* = (K_t - S_t)^+/B_t$. For a continuous semimartingale Z and a fixed $a \in \mathbb{R}$, we denote by $L_t^a(Z)$ the (right) semimartingale *local time* of Z, given explicitly by the formula

$$L_t^a(Z) \stackrel{\text{def}}{=} |Z_t - a| - |Z_0 - a| - \int_0^t \text{sgn}\,(Z_u - a)\,dZ_u$$

for every $t \in [0, T]$ (by convention we set $\text{sgn}(0) = -1$). It is well known that the local time $L^a(Z)$ of a continuous semimartingale Z is an adapted process whose sample paths are almost all continuous, non-decreasing functions. Moreover, for an arbitrary convex function $f : \mathbb{R} \to \mathbb{R}$ the following decomposition, referred to as the Itô-Tanaka-Meyer formula, is valid

$$f(Z_t) = f(Z_0) + \int_0^t f_l'(Z_u)\,dZ_u + \frac{1}{2} \int_{\mathbb{R}} L_t^a(Z)\,\mu(da),$$

where f_l' is the left-hand side derivative of f, and the measure $\mu = f''$ represents the second derivative of f, in the sense of distributions. An application of the Itô-Tanaka-Meyer formula yields

$$X_t^* = X_0^* + \int_0^t \mathbb{1}_{\{S_u > K_u\}}\,B_u^{-1}\left(\zeta_u\,du + \sigma\,S_u\,dW_u^*\right) + \frac{1}{2} \int_0^t B_u^{-1}\,dL_u^0(S - K),$$

where $\zeta_u = rK_u - k_u$. Similarly, for the process Y^* we get

$$Y_t^* = Y_0^* - \int_0^t \mathbb{1}_{\{S_u < K_u\}}\,B_u^{-1}\left(\zeta_u\,du - \sigma\,S_u\,dW_u^*\right) + \frac{1}{2} \int_0^t B_u^{-1}\,dL_u^0(K - S).$$

In particular, if the strike price $K_t = K > 0$ is a constant, then the discounted rewards X^* and Y^* satisfy[7]

$$X_t^* = X_0^* + \int_0^t \mathbb{1}_{\{S_u > K\}}\,B_u^{-1}\left(\sigma\,S_u\,dW_u^* + rK\,du\right) + \frac{1}{2} \int_0^t B_u^{-1}\,dL_u^K(S)$$

and

$$Y_t^* = Y_0^* - \int_0^t \mathbb{1}_{\{S_u < K\}}\,B_u^{-1}\left(\sigma\,S_u\,dW_u^* + rK\,du\right) + \frac{1}{2} \int_0^t B_u^{-1}\,dL_u^K(S).$$

Since the local time is known to be an increasing process, it is evident that if the strike price is a positive constant and the interest rate is nonnegative, then the discounted

---

[7] Obviously we have $L_t^0(S - K) = L_t^K(S)$ and $L_t^0(S - K) = L_t^{-K}(-S)$. It is also possible to show that $L_t^{-K}(-S) = L_t^K(S)$.

reward process X^* follows a submartingale under the martingale measure[8] $\mathbb{P}^*$, that is,

$$\mathbb{E}_{\mathbb{P}^*}(X_t^* \mid \mathcal{F}_u) \geq X_u^*, \quad \forall u \leq t \leq T.$$

On the other hand, the discounted reward Y^* of the American put option with constant exercise price is a submartingale under $\mathbb{P}^*$, provided that $r \leq 0$. Summarizing, we have the following useful result.

Corollary 5.2.1 *The discounted reward X^* (respectively, Y^*) of the American call option (respectively, put option) with constant strike price follows a submartingale under $\mathbb{P}^*$ if $r \geq 0$ (respectively, if $r \leq 0$).*

Let us now examine the rational exercise policy of the holder of an option of American style. We shall use throughout the superscripts c and p to denote the quantities associated with the reward functions g^c and g^p respectively. In particular,

$$V_t^c = \operatorname{ess\,sup}_{\tau \in \mathcal{T}_{[t,T]}} \mathbb{E}_{\mathbb{P}^*}\big(e^{-r(\tau-t)}(S_\tau - K_\tau)^+ \mid \mathcal{F}_t\big)$$
$$= e^{rt} \operatorname{ess\,sup}_{\tau \in \mathcal{T}_{[t,T]}} \mathbb{E}_{\mathbb{P}^*}\big(X_\tau^* \mid \mathcal{F}_t\big),$$

and

$$V_t^p = \operatorname{ess\,sup}_{\tau \in \mathcal{T}_{[t,T]}} \mathbb{E}_{\mathbb{P}^*}\big(e^{-r(\tau-t)}(K_\tau - S_\tau)^+ \mid \mathcal{F}_t\big)$$
$$= e^{rt} \operatorname{ess\,sup}_{\tau \in \mathcal{T}_{[t,T]}} \mathbb{E}_{\mathbb{P}^*}\big(Y_\tau^* \mid \mathcal{F}_t\big).$$

We know already that in the case of a constant strike price and under a nonnegative interest rate r, the discounted reward X^* of a call option follows a $\mathbb{P}^*$-submartingale. Consequently, the Snell envelope J^c of X^* equals

$$J_t^c = \operatorname{ess\,sup}_{\tau \in \mathcal{T}_{[t,T]}} \mathbb{E}_{\mathbb{P}^*}\big(X_\tau^* \mid \mathcal{F}_t\big) = \mathbb{E}_{\mathbb{P}^*}(X_T^* \mid \mathcal{F}_t). \qquad (5.7)$$

This in turn implies that for every date t the rational exercise time after time t of the American call option with a constant strike price K is the option's expiry date T (the same property holds for the American put option, provided that $r \leq 0$). In other words, under a nonnegative interest rate, an American call option with constant strike price should never be exercised before its expiry date, and thus its arbitrage price coincides with the Black-Scholes price of a European call. Put differently, in usual circumstances, the American call option written on a non-dividend-paying stock is always worth more alive than dead, hence it is equivalent to the European call option with the same contractual features. (Let us qualify here that this remark does not apply, for instance, to American currency call options.) This shows that in the case of constant strike price and nonnegative interest rate r, only an American put option written on a non-dividend-paying stock requires further examination.

---

[8] This property follows directly from Jensen's conditional inequality.

5.3 Early Exercise Representation of an American Put

The following result is a consequence of a much more general theorem due to El Karoui and Karatzas (1991). For the sake of concreteness, we shall focus on the case of the American put option. The proof of Proposition 5.3.1 can be found in Myneni (1992).

Proposition 5.3.1 *The Snell envelope J^p admits the following decomposition*

$$ J_t^p = \mathbb{E}_{\mathbb{P}^*}\left(e^{-rT}(K_T - S_T)^+ \,\Big|\, \mathcal{F}_t \right) + \mathbb{E}_{\mathbb{P}^*}\left(\int_t^T e^{-ru} \mathbb{1}_{\{\tau_u = u\}} (r K_u - k_u)\, du \,\Big|\, \mathcal{F}_t \right). $$

Recall that $V_t^p = e^{rt} J_t^p$ for every t. Consequently, the price P_t^a of an American put option satisfies

$$ P_t^a = V_t^p = \mathbb{E}_{\mathbb{P}^*}\left(e^{-r(T-t)}(K_T - S_T)^+ \,\Big|\, \mathcal{F}_t \right) $$
$$ + \mathbb{E}_{\mathbb{P}^*}\left(\int_t^T e^{-r(u-t)} \mathbb{1}_{\{\tau_u = u\}} (r K_u - k_u)\, du \,\Big|\, \mathcal{F}_t \right). $$

Furthermore, in view of the Markov property of the stock price process S, it is clear that for any stopping time $\tau \in \mathcal{T}_{[t,T]}$, we have

$$ \mathbb{E}_{\mathbb{P}^*}\left(e^{-r(\tau-t)}(K_\tau - S_\tau)^+ \mid \mathcal{F}_t \right) = \mathbb{E}_{\mathbb{P}^*}\left(e^{-r(\tau-t)}(K_\tau - S_\tau)^+ \mid S_t \right). $$

We conclude that the price of an American put equals $P_t^a = P^a(S_t, T - t)$ for a certain function $P^a : \mathbb{R}_+ \times [0, T] \to \mathbb{R}$. To be a bit more explicit, we define the function $P^a(s, t)$ by setting

$$ P^a(s, T - t) = \sup\nolimits_{\tau \in \mathcal{T}_{[t,T]}} \mathbb{E}_{\mathbb{P}^*}\left(e^{-r(\tau-t)}(K_\tau - S_\tau)^+ \,\Big|\, S_t = s \right). $$

We assume from now on that the strike price K is constant; that is, $K_t = K > 0$ for every $t \in [0, T]$. It is possible to show that the function $P^a(s, u)$ is decreasing and convex in s, and increasing in u. Let us denote by $\mathcal{C}$ and $\mathcal{D}$ the *continuation region* and *stopping region* respectively. The stopping region $\mathcal{D}$ is defined as that subset of $\mathbb{R}_+ \times [0, T]$ for which the stopping time τ_t satisfies

$$ \tau_t = \inf\{u \in [t, T] \mid (S_u, u) \in \mathcal{D}\} $$

for every $t \in [0, T]$. The continuation region $\mathcal{C}$ is, of course, the complement of $\mathcal{D}$ in $\mathbb{R}_+ \times [0, T]$. Note that in terms of the function P^a, we have

$$ \mathcal{D} = \{(s, t) \in \mathbb{R}_+ \times [0, T] \mid P^a(s, T - t) = (K - s)^+\} $$

and

$$ \mathcal{C} = \{(s, t) \in \mathbb{R}_+ \times [0, T] \mid P^a(s, T - t) > (K - s)^+\}. $$

Let us define the function $b^* : [0, T] \to \mathbb{R}_+$ by setting

$$b^*(T - t) = \sup \{s \in \mathbb{R}_+ \mid P^a(s, T - t) = (K - s)^+\}.$$

It can be shown that the graph of b^* is contained in the stopping region $\mathcal{D}$. This means that it is only rational to exercise the put option at time t if the current stock price S_t is at or below the level $b^*(T - t)$. For this reason, the value $b^*(T - t)$ is commonly referred to as the *critical stock price* at time t. It is sometimes convenient to consider the function $c^* : [0, T] \to \mathbb{R}_+$ which is given by the equality $c^*(t) = b^*(T - t)$. For any $t \in [0, T]$, the optimal exercise time τ_t after time t satisfies

$$\tau_t = \inf \{u \in [t, T] \mid (K - S_u)^+ = P^a(S_u, T - u)\},$$

or equivalently,

$$\tau_t = \inf \{u \in [t, T] \mid S_u \le b^*(T - u)\} = \inf \{u \in [t, T] \mid S_u \le c^*(u)\}.$$

We quote the following lemma from Van Moerbeke (1976).

Lemma 5.3.1 *The function c^* is non-decreasing, infinitely smooth over $(0, T)$ and* $\lim_{t \uparrow T} c^*(t) = \lim_{T - t \downarrow 0} b^*(T - t) = K$.

By virtue of the next result, the price of an American put option may be represented as the sum of the arbitrage price of the corresponding European call option and the so-called early exercise premium.

Corollary 5.3.1 *The following decomposition of the price of an American put option is valid*

$$P_t^a = P_t + \mathbb{E}_{\mathbb{P}^*}\left(\int_t^T e^{-r(u-t)} \mathbb{1}_{\{S_u < b^*(T-u)\}} r K \, du \, \Big| \, \mathcal{F}_t \right),$$

where $P_t^a = P^a(S_t, T - t)$ is the price of the American put option, and $P_t = P(S_t, T - t)$ is the Black-Scholes price of a European put with strike K.

Decomposition of the price provided by Corollary 5.3.1 is commonly referred to as the *early exercise premium representation* of an American put. It was derived independently, by different means and at various levels of strictness, by Kim (1990), Jacka (1991) and Jamshidian (1992) (see also Carr et al. (1992) for other representations of the price of an American put). Using Corollary 5.3.1, we get for $t = 0$

$$P_0^a = P_0 + \mathbb{E}_{\mathbb{P}^*}\left(\int_0^T e^{-ru} \mathbb{1}_{\{S_u < b^*(T-u)\}} r K \, du \right), \tag{5.8}$$

where P_0 is the price at time 0 of the European put. Observe that if $r = 0$, then the early exercise premium vanishes, which means that the American put is equivalent to the European put and thus should not be exercised before the expiry date. Taking into account the dynamics of the stock price under $\mathbb{P}^*$, we can make representation (5.8) more explicit, namely

$$P^a(S_0, T) = P(S_0, T) + rK \int_0^T e^{-ru} N \left(\frac{\ln(b^*(T-u)/S_0) - \rho u}{\sigma \sqrt{u}} \right) du,$$

where $\rho = r - \sigma^2/2$ and N denotes the standard Gaussian cumulative distribution function. A similar decomposition is valid for any instant $t \in [0, T]$, provided that the current stock price S_t belongs to the continuation region $\mathcal{C}$; that is, the option should not be exercised immediately. For any date $t \in [0, T]$, we have

$$P^a(S_t, T - t) = P(S_t, T - t)$$
$$+ rK \int_t^T e^{-r(u-t)} N \left(\frac{\ln(b^*(T-u)/S_t) - \rho(u-t)}{\sigma \sqrt{u-t}} \right) du,$$

where $P(S_t, T - t)$ stands for the price of a European put option of maturity $T - t$, and S_t is the current level of the stock price. A change of variables leads to the following equivalent expression

$$P^a(s, t) = P(s, t) + rK \int_0^t e^{-ru} N \left(\frac{\ln(b^*(t-u)/s) - \rho u}{\sigma \sqrt{u}} \right) du,$$

which is valid for every $(s, t) \in \mathcal{C}$. If $S_t = b^*(T - t)$, then we have necessarily $P^a(S_t, T - t) = K - b^*(T - t)$, so that clearly $P^a(b^*(t), t) = K - b^*(t)$ for every $t \in [0, T]$. This simple observation leads to the following integral equation satisfied by the optimal boundary function b^*

$$K - b^*(t) = P(b^*(t), t) + rK \int_0^t e^{-ru} N \left(\frac{\ln(b^*(t-u)/b^*(t)) - \rho u}{\sigma \sqrt{u}} \right) du.$$

Unfortunately, a solution to this integral equation is not known explicitly, and thus it needs to be solved numerically. On the other hand, the following bounds for the price $P^a(s, t)$ are easy to derive

$$P^a(s, t) - P(s, t) \le rK \int_0^t e^{-ru} N \left(\frac{\ln(K/s) - \rho u}{\sigma \sqrt{u}} \right) du$$

and

$$P^a(s, t) - P(s, t) \ge rK \int_0^t e^{-ru} N \left(\frac{\ln(b_\infty^*/s) - \rho u}{\sigma \sqrt{u}} \right) du,$$

where b_∞^* stands for the optimal exercise boundary of a *perpetual put* – that is, an American put option with expiry date $T = \infty$. To this end, it is enough to show that for any maturity T, the values of the optimal stopping boundary b^* lie between the strike price K and the level b_∞^*, i.e., $K \le b^*(t) \le b_\infty^*$ for every $t \in [0, T]$. The value of b_∞^* is known to be (see McKean (1965), Merton (1973), or the next section)

$$b_\infty^* = \frac{2rK}{2r + \sigma^2}. \tag{5.9}$$

The properties of the critical stock price near the option expiry were studied by, among others, Barles et al. (1995) and Lamberton (1995). Their main result states that the critical price satisfies

$$K - b^*(T - t) \sim K\sigma\sqrt{(t - T)\ln(T - t)}$$

as t approaches the expiry date T, i.e., when $T - t \downarrow 0$.

5.4 Analytical Approach

The aim of this section is to give a short overview of analytical methods related to the optimal stopping problem for diffusion processes.

The *free boundary problem* related to the optimal stopping problem for an American put option was first examined by McKean (1965) and Van Moerbeke (1976). As usual, we shall focus on the valuation within the Black-Scholes framework of an American put written on a stock paying no dividends during the option's lifetime. For a fixed expiry date T and constant strike price K, we denote by $P^a(S_t, T - t)$ the price of an American put at time $t \in [0, T]$; in particular, $P^a(S_T, 0) = (K - S_T)^+$. It will be convenient to denote by $\mathcal{L}$ the following differential operator

$$\mathcal{L}v = \frac{1}{2}\sigma^2 s^2 \frac{\partial^2 v}{\partial s^2} + rs \frac{\partial v}{\partial s} - rv = \mathcal{A}v - rv,$$

where $\mathcal{A}$ stands for the infinitesimal generator of the one-dimensional diffusion process S (considered under the martingale measure $\mathbb{P}^*$). Also, let $\mathcal{L}_t$ stand for the following differential operator

$$\mathcal{L}_t v = \frac{\partial v}{\partial t} + \mathcal{L}v = v_t + \mathcal{A}v - rv.$$

The proof of the following proposition, focusing on the properties of the price function $P^a(s, t)$ of an American put, can be found in van Moerbeke (1976) (see also McKean (1965), Jacka (1991), and Salopek (1997)).

Recall that we write $c^*(t) = b^*(T - t)$.

Proposition 5.4.1 *The American put value function $P^a(s, t)$ is smooth on the continuation egion $\mathcal{C}$ with $-1 \leq P_s^a(s, t) \leq 0$ for all $(s, t) \in \mathcal{C}$. The optimal stopping boundary b^* is a continuous and non-increasing function on $(0, T]$, hence c^* is a non-decreasing function on $[0, T)$, and*

$$\lim_{T-t \downarrow 0} b^*(T - t) = \lim_{t \uparrow T} c^*(t) = K.$$

On $\mathcal{C}$, the function P^a satisfies $P_t^a(s, t) = \mathcal{L}P^a(s, t)$; that is,

$$P_t^a(s, t) = \tfrac{1}{2}\sigma^2 s^2 P_{ss}^a(s, t) + rs P_s^a(s, t) - r P^a(s, t), \quad \forall (s, t) \in \mathcal{C}.$$

Furthermore, we have

$$\lim_{s \downarrow b^*(t)} P^a(s, t) = K - b^*(t), \quad \forall t \in (0, T],$$

$$\lim_{t \to 0} P^a(s, t) = (K - s)^+, \quad \forall s \in \mathbb{R}_+,$$

$$\lim_{s \to \infty} P^a(s, t) = 0, \quad \forall t \in (0, T],$$

$$P^a(s, t) \geq (K - s)^+, \quad \forall (s, t) \in \mathbb{R}_+ \times (0, T].$$

In order to determine the optimal stopping boundary, one needs to impose an additional condition, known as the *smooth fit principle*, that reads as follows (for the proof of the next result, we refer the reader to Van Moerbeke (1976)).

Proposition 5.4.2 *The partial derivative $P_s^a(s, t)$ is continuous a.e. across the stopping boundary c^*; that is*

$$\lim_{s \downarrow c^*(t)} P_s^a(s, t) = -1$$

for almost every $t \in [0, T]$.

We are now in a position to state (without proof) the result characterizing the price of an American put option as the solution to the free boundary problem (see Van Moerbeke (1976), Jacka (1991), and Salopek (1997)).

Theorem 5.4.1 *Let $\mathcal{G}$ be an open domain in $\mathbb{R}_+ \times [0, T)$ with continuously differentiable boundary c. Assume that $v : \mathbb{R}_+ \times [0, T]$ is a continuous function such that $u \in C^{3,1}(\mathcal{G})$, the function $g(s, t) = v(e^s, t)$ has Tychonov growth,[9] and v satisfies $\mathcal{L}_t v = 0$ on $\mathcal{G}$; that is*

$$v_t(s, t) + \tfrac{1}{2}\sigma^2 s^2 \, v_{ss}(s, t) + rs \, v_s(s, t) - rv(s, t) = 0 \tag{5.10}$$

for every $(s, t) \in \mathcal{G}$, and

$$v(s, t) > (K - s)^+, \quad \forall (s, t) \in \mathcal{G},$$

$$v(s, t) = (K - s)^+, \quad \forall (s, t) \in \mathcal{G}^c,$$

$$v(s, T) = (K - s)^+, \quad \forall s \in \mathbb{R}_+,$$

$$\lim_{s \downarrow c(t)} v_s(s, t) = -1, \quad \forall t \in [0, T).$$

Then the function $P^a(s, t) = v(s, T - t)$ for every $(s, t) \in \mathbb{R}_+ \times [0, T]$ is the value function of the American put option with strike price K and maturity T. Moreover, the set $\mathcal{C} = \mathcal{G}$ is the option's continuation region, and the function $b^*(t) = c(T - t)$, $t \in [0, T]$, represents the critical stock price.

We shall now apply the above theorem to a *perpetual put* – that is, an American put option that has no expiry date (i.e., with maturity $T = \infty$). Since the time to

---

[9] A function $g \in \mathbb{R}_+ \times (0, T]$ has *Tychonov growth* if g and partial derivatives g_s, g_{ss}, g_{st}, and g_{sss} have growth at most $\exp(o(s^2))$, uniformly on compact sets, as s tends to infinity.

expiry of a perpetual put is always infinite, the critical stock price becomes a real number $b_\infty^* \le K$, and the PDE (5.10) becomes the following ordinary differential equation (ODE) for the function $v_\infty(s) = v(s, \infty)$

$$\frac{1}{2}\sigma^2 s^2 \frac{d^2 v_\infty}{ds^2}(s) + rs \frac{dv_\infty}{ds}(s) - r v_\infty(s) = 0, \quad \forall s \in (b_\infty^*, \infty), \quad (5.11)$$

with $v_\infty(s) = K - s$ for every $s \in [0, b_\infty^*]$. Our aim is to show that equality (5.9) is valid. For this purpose, observe that ODE (5.11) admits a general solution of the form

$$v_\infty(s) = c_1 s^{d_1} + c_2 s^{d_2}, \quad \forall s \in (b_\infty^*, \infty),$$

where c_1, c_2, d_1 and d_2 are constants. Using the boundary condition

$$v_\infty(b_\infty) = K - b_\infty,$$

and the smooth fit condition

$$\lim_{s \downarrow b_\infty} \frac{dv_\infty}{ds}(s) = -1,$$

we find that

$$v_\infty(s) = (K - b_\infty)\left(\frac{b_\infty}{s}\right)^{2r\sigma^{-2}}, \quad \forall s \in (b_\infty^*, \infty),$$

and thus, in particular, equality (5.9) is valid.

In contrast to the free boundary approach, the formulation of the optimal stopping problem in terms of *variational inequalities* allows us to treat the domain of the option as an entire region. In other words, we do not need to introduce explicitly here the stopping boundary b^*.

Jaillet et al. (1990) have exploited the general theory of variational inequalities – as exposed in a monograph by Bensoussan and Lions (1978) – to study the optimal stopping problem associated with American claims. They show, in particular, that the techniques of variational inequalities provide an adequate framework for the study of numerical methods related to American options.

Theorem 5.4.2 *Assume that $v : \mathbb{R}_+ \times [0, T]$ is a continuous function such that the function $g(s, t) = v(e^s, t)$ satisfies certain growth conditions. Suppose that for every $(s, t) \in \mathbb{R}_+ \times [0, T]$ we have*

$$\mathcal{L}_t v(s, t) \le 0,$$
$$v(s, t) \ge (K - s)^+,$$
$$v(s, T) = (K - s)^+,$$
$$\left((K - s)^+ - v(s, t)\right)\mathcal{L}_t v(s, t) = 0.$$

Equivalently, we have $v(s, T) = (K - s)^+$ and

$$\max\left\{(K - s)^+ - v(s, t), \mathcal{L}_t v(s, t)\right\} = 0.$$

Then v is unique and $v(s, t) = P^a(s, T - t)$ for every $(s, t) \in \mathbb{R}_+ \times [0, T]$.

As one might expect, a solution to the problem above is not known explicitly (for the proof of existence and uniqueness we refer to Bensoussan and Lions (1978)). Numerical methods of solving variational inequalities associated with American options were developed by Jaillet et al. (1990).

5.5 Approximations of the American Put Price

Since no closed-form expression for the value of an American put is available, in order to value American options one needs to use a numerical procedure. It appears that the use of the CRR binomial tree, although remarkably simple, is far from being the most efficient way of pricing American options. Various approximations of the American put price on a non-dividend-paying stock were examined by Brennan and Schwartz (1977a), Johnson (1983), MacMillan (1986), Leisen (1996), and Broadie and Detemple (1996, 1997b). Dividend-paying stocks and other types of American options were studied using analytical methods in Geske (1979b), Blomeyer (1986), Barone-Adesi and Waley (1986, 1987), Omberg (1987a), Broadie and Detemple (1995), and Huang et al. (1996). Numerical integration was applied to American option valuation by Parkinson (1977). A compound option approach to American put options was examined in Geske and Johnson (1984). Geske and Roll (1984) analyzed the valuation of American call options with the Black-Scholes formula.

Let us comment briefly on the approximate valuation method proposed by Geske and Johnson (1984). Basically, the *Geske-Johnson approximation* relies on the discretization of the time parameter and the application of backward induction, as in any other standard discrete-time approach. However, in contrast to the space-time discretization used in the multinomial trees approach or in the finite difference methods, the approach of Geske and Johnson makes use of the exact distribution of the vector of stock prices $(S_{t_1}, \ldots, S_{t_n})$, where $t_1 < \cdots < t_n = T$ are the only admissible (deterministic) exercise times. [10] In other words, the decision to exercise an option can be made at any of the dates $t_1, \ldots, t_n$ only. Let us start by considering the special case when $n = 2$ and $t_1 = T/2$, $t_2 = T$. Note that if $n = 1$ then the option can be exercised at $t_1 = T$ only, so that it is equivalent to a European put. To find an approximate value for an American put, we argue by backward induction. Suppose that the option was not exercised at time t_1. Then the value of the option at time t_1 is equal to the value of a European put option with maturity $t_2 - t_1 = T/2$, given the initial stock price $S_{t_1} = S_{T/2}$. The price of a European put is given, of course, by the standard Black-Scholes formula, denoted by $P(S_{T/2}, T/2)$. The critical stock price b_1^* at time $t_1 = T/2$ solves the equation $K - S_{T/2} = P(S_{T/2}, T/2)$, hence it can be found by numerical methods. Moreover, it is clear that the value $V_{T/2}$ of the option at time $T/2$ satisfies

$$V_{T/2} = \begin{cases} P(S_{T/2}, T/2) & \text{if } S_{T/2} > b_1^*, \\ K - S_{T/2} & \text{if } S_{T/2} \leq b_1^*. \end{cases}$$

---

[10] An option that may be exercised early, but only on predetermined dates, is commonly referred to as *Bermudan option*.

Note that it is optimal to exercise the option at time $t_1 = T/2$ if and only if $S_{T/2} \leq b_1^*$. To find the value of the option at time 0, we need first to evaluate the expectation $V_0(S_0) = \mathbb{E}_{\mathbb{P}^*}(e^{-rT/2} V_{T/2})$, or equivalently,

$$V_0(S_0) = e^{-rT/2} \mathbb{E}_{\mathbb{P}^*}\left(e^{-rT/2}(K - S_T)^+ \mathbb{1}_{\{S_{T/2} > b_1^*\}} + (K - S_{T/2})\mathbb{1}_{\{S_{T/2} \leq b_1^*\}}\right).$$

Notice that the latter expectation can be expressed in terms of the probability distribution of the two-dimensional random variable $(S_{T/2}, S_T)$; equivalently, one exploit the joint distribution of $W_{T/2}$ and W_T. A specific, quasi-explicit representation of $V_0(S_0)$ in terms of two-dimensional Gaussian cumulative distribution function is in fact a matter of convenience. The approximate value of an American put with two admissible exercise times, $T/2$ and T, equals $P_2^a(S_0, T) = V_0(S_0)$. The same iterative procedure may be applied to an arbitrary finite sequence of times $t_1 < \cdots < t_n = T$. In this case, the Geske-Johnson approximation formula involves integration with respect to a n-dimensional Gaussian probability density function. It appears that for three admissible exercise times, $T/3$, $2T/3$ and T, the approximate quasi-analytical valuation formula provided by the Geske and Johnson method is roughly as accurate as the binomial tree with 150 time steps.

For any natural n, let us denote by $P_n^a(S_0, T)$ the Geske-Johnson option's approximate value associated with admissible dates $t_i = Ti/n$, $i = 1, \ldots, n$. It is possible to show that the sequence $P_n^a(S_0, T)$ converges to the option's exact price $P^a(S_0, T)$ when the number of steps tends to infinity, so that the step length tends to zero. To estimate the limit $P^a(S_0, T)$, one can make use of any extrapolation technique, for instance Richardson's approximation scheme. Let us briefly describe the latter technique. Suppose that the function F satisfies

$$F(h) = F(0) + c_1 h + c_2 h^2 + o(h^2)$$

in the neighborhood of zero, so that

$$F(kh) = F(0) + c_1 kh + c_2 k^2 h^2 + o(h^2)$$

and

$$F(lh) = F(0) + c_1 lh + c_2 l^2 h^2 + o(h^2)$$

for arbitrary $l > k > 1$. Ignoring the term $o(h^2)$, and solving the above system of equations for $F(0)$, we obtain ("$\approx$" denotes approximate equality)

$$F(0) \approx F(h) + \frac{a}{c}\left(F(h) - F(kh)\right) + \frac{b}{c}\left(F(kh) - F(lh)\right),$$

where $a = l(l-1) - k(k-1)$, $b = k(k-1)$ and $c = l^2(k-1) - l(k^2-1) + k(k-1)$. Let us write P_n^a to denote $P_n^a(S_0, T)$ for $n = 1, 2, 3$ (in particular, $P_1^a(S_0, T)$ is the European put price $P(S_0, T)$). For $n = 3$ upon setting $k = 3/2$, $l = 3$ and $P_1^a = F(lh)$, $P_2^a = F(kh)$, $P_3^a = F(h)$, we get the following approximate formula

$$P^a(S_0, T) \approx P_3^a + \tfrac{7}{2}(P_3^a - P_2^a) - \tfrac{1}{2}(P_2^a - P_1^a).$$

Bunch and Johnson (1992) argue that the Geske-Johnson method can be further improved if the exercise times are chosen iteratively in such a way that the option's approximate value is maximized.

The *analytic method of lines*, developed in Carr and Faguet (1994), relies on the approximate solution of the free-boundary problem of Sect. 5.4 by means of a suitable time discretization (the quadratic approximation developed by MacMillan (1986) and Barone-Adesi and Whaley (1987) can be seen as a special case of this method). Let us write $v(s, t) = P^a(s, T - t)$ and $c^*(t) = b^*(T - t)$ so that $\mathcal{L}_t v = 0$ (cf. (5.10)) on the open domain $\{(s, t) \in \mathbb{R}_+ \times (0, T) \mid s > c(t)\}$. Moreover, v is known to satisfy the following set of conditions

$$v(s, T) = (K - s)^+, \quad \forall s \in \mathbb{R}_+,$$
$$\lim_{s \downarrow c(t)} v(s, t) = K - c(t), \quad \forall t \in [0, T],$$
$$\lim_{s \downarrow c(t)} v_s(s, t) = -1, \quad \forall t \in [0, T],$$
$$\lim_{s \to \infty} v(s, t) = 0, \quad \forall t \in [0, T].$$

For arbitrary natural n and fixed horizon date T, we write $\Delta_n = T/n$ and $t_i = i\Delta_n$ for $i = 0, \dots, n$. Denote by $v^i(s)$ the approximate value of $v(s, i\Delta_n)$. Then the PDE (5.10) becomes the following ODE

$$\Delta_n^{-1}(v^{i+1}(s) - v^i(s)) + \frac{1}{2}\sigma^2 s^2 \frac{d^2 v^i}{ds^2}(s) + rs \frac{dv^i}{ds}(s) - rv^i(s) = 0$$

for $s \in (c(i\Delta_n), \infty)$. The last equation can be solved recursively for both the function $v(s, i\Delta_n)$ and the optimal stopping boundary $c(i\Delta_n)$ $i = 0, \dots, n$. Since the explicit formulas are rather involved, we refer the reader to the original paper, Carr and Faguet (1994), for details.

5.6 Option on a Dividend-paying Stock

Since most traded options on stocks are *unprotected* American call options written on dividend-paying stocks, it is worthwhile to comment briefly on the valuation of these contracts. A call option is said to be *unprotected* if it has no contracted "protection" against the stock price decline that occurs when a dividend is paid. It is intuitively clear that an unprotected American call written on a dividend-paying stock is not equivalent to the corresponding option of European style, in general. Suppose that a known dividend, D, will be paid to each shareholder with certainty at a predetermined date T_D during the option's lifetime. Furthermore, assume that the ex-dividend stock price decline equals δD for a given constant $\delta \in [0, 1]$. Let us denote by S_{T_D} and $P_{T_D} = S_{T_D} - \delta D$ respectively the cum-dividend and ex-dividend stock prices at time T_D. It is clear that the option should eventually be exercised just before the dividend is paid – that is, an instant before T_D. Consequently, as first noted by Black (1975), the lower bound for the price of such an option is the price of the European

call option with expiry date T_D and strike price K. This lower bound is a good estimate of the exact value of the price of the American option whenever the probability
of early exercise is large – that is, when the probability $\mathbb{P}\{C_{T_D} < S_{T_D} - K\}$ is large,
where $C_{T_D} = C(P_{T_D}, T - T_D, K)$ is the Black-Scholes price of the European call
option with maturity $T - T_D$ and exercise price K. Hence early exercise of the American call is more likely the larger the dividend, the higher the stock price S_{T_D} relative
to the strike price K, and the shorter the time-period $T - T_D$ between expiry and dividend payment dates. An analytic valuation formula for unprotected American call
options on stocks with known dividends was established by Roll (1977). However,
it seems to us that Roll's original reasoning, which refers to options that expire an
instant before the ex-dividend date, assumes implicitly that the holder of an option
may exercise it before the ex-dividend date, but apparently is not allowed to sell it
before the ex-dividend date. To avoid this discrepancy, we prefer instead to consider
European options which expire on the ex-dividend date – i.e., after the ex-dividend
stock price decline.

Before formulating the next result, we need to introduce some notation. Let us
denote by b^* the cum-dividend stock price level above which the original American
option will be exercised at time T_D, so that

$$C(b^* - \delta D, T - T_D, K) = b^* - K. \tag{5.12}$$

It is worthwhile to observe that $C(s - \delta D, T - T_D, K) < s - K$ when $s \in (b^*, \infty)$,
and $C(s - \delta D, T - T_D, K) > s - K$ for every $s \in (0, b^*)$. Note that the first
two terms on the right-hand side of equality (5.13) below represent the values of
European options, written on a stock S, that expire at time T and on the ex-dividend
date T_D respectively. The last term, $\mathbf{CO}_t(T_D, b^* - K)$, represents the price of a
so-called *compound option* (see Sect. 6.4). To be more specific, we deal here with a
European call option with strike price $b^* - K$ that expires on the ex-dividend date T_D,
and whose underlying asset is the European call option, written on S, with maturity
T and strike price K. The compound option will be exercised by its holder at the
ex-dividend date T_D if and only if he is prepared to pay $b^* - K$ for the underlying
European option. Since the value of the underlying option after the ex-dividend stock
price decline equals $C(P_{T_D}, T - T_D, K)$, the compound option is exercised whenever

$$C(P_{T_D}, T - T_D, K) = C(S_{T_D} - \delta D, T - T_D, K) > b^* - K,$$

that is, when the cum-dividend stock price exceeds b^* (this follows from the fact that
the price of a standard European call option is an increasing function of the stock
price, combined with equality (5.12)).

Proposition 5.6.1 *The arbitrage price $\tilde{C}_t^a(T, K)$ of an unprotected American call
option with expiry date $T > T_D$ and strike price K, written on a stock paying a
known dividend D at time T_D, equals*

$$\tilde{C}_t^a(T, K) = \tilde{C}_t(T, K) + C_t(T_D, b^*) - \mathbf{CO}_t(T_D, b^* - K) \tag{5.13}$$

for $t \in [0, T_D]$, where b^* is the solution to (5.12).

Proof. Note that the first term in (5.13) represents the price of an option written on a dividend-paying stock, hence it is not given by the standard Black-Scholes formula. On the other hand, on the ex-dividend date T_D we have

$$\tilde{C}_{T_D}(T, K) = C(S_{T_D} - \delta D, T - T_D, K) = C(P_{T_D}, T - T_D, K).$$

From the reasoning above, it is clear that in order to check the validity of (5.13), it is enough to consider the value of the portfolio of options on the ex-dividend date T_D. Let us assume first that the cum-dividend stock price S_{T_D} is above the early exercise level b^*. The value of the portfolio is

$$C(P_{T_D}, T - T_D, K) + (S_{T_D} - b^*) - \big(C(P_{T_D}, T - T_D, K) - (b^* - K)\big) = S_{T_D} - K,$$

as expected. Assume now that the stock is below the level b^*. In this case, the right-hand side of (5.13) equals simply $C(P_{T_D}, T - T_D, K)$, as the remaining options are worthless. This completes the derivation of (5.13).   □

5.7 Game Contingent Claims

In this short section, we will present an extension to a continuous-time framework some pricing results of Sect. 2.9.2, in which we dealt with the game contingent claims in the setup of an arbitrage-free and complete finite market model. Since the arguments used to establish pricing results in a continuous-time set-up are rather technical, we will restrict ourselves to a brief survey of fundamental ideas and results. For the proofs, the interested reader is referred to original papers (see, for instance, Kiefer (1971) and Lepeltier and Maingueneau (1984)), as well as to the monograph by Friedman (1976).

We continue to consider the Black and Scholes market model consisting of two primary traded assets: the stock S and the savings account B. This means that the short-term rate is constant and the stock price S is the unique solution to the stochastic differential equation

$$dS_t = S_t\big(r\,dt + \sigma\,dW_t^*\big),$$

where W^* is the standard Brownian motion under the unique spot martingale measure $\mathbb{P}^*$. We assume that the filtration $\mathbb{F} = \mathbb{F}^W = \mathbb{F}^{W^*}$, so that also $\mathbb{F} = \mathbb{F}^S$. As usual, we write $\mathcal{T}_{[t,T]}$ to denote the class of all $\mathbb{F}$-stopping times τ taking values in $[t, T]$.

Definition 5.7.1 A *game contingent claim* $X^g = (L, H, \mathcal{T}^e, \mathcal{T}^c)$ expiring at time T consists of $\mathbb{F}$-adapted payoff processes L and H, a set $\mathcal{T}^e$ admissible exercise times, and a set $\mathcal{T}^c$ of admissible cancelation times.

We postulate that $L \le H$, meaning that the inequality $L_t \le H_t$ holds for every $t \in [0, T]$. Also, the sets of admissible exercise and cancelation times are restricted to the class $\mathcal{T}_{[0,T]}$ of all stopping times of the filtration $\mathbb{F}^S$ with values in $[0, T]$, so that $\mathcal{T}^e = \mathcal{T}^c = \mathcal{T}_{[0,T]}$.

We interpret L_t as the payoff received by the holder upon exercising a game contingent claim at time t. The random variable H_t represents the payoff received by the holder if the claim is canceled by its issuer at time t. More formally, if a game contingent claim is exercised by either of the two parties at some date $t \leq T$, that is, on the event $\{\tau = t\} \cup \{\sigma = t\}$, the random payoff equals

$$X_t = \mathbb{1}_{\{\tau=t\leq\sigma\}} L_t + \mathbb{1}_{\{\sigma=t<\tau\}} H_t, \tag{5.14}$$

where τ and σ are the exercise and cancelation times, respectively.

In the present set-up, it is natural to postulate that $L_t = L(S_t, t)$ and $H_t = H(S_t, t)$ for some functions $L, H : \mathbb{R}_+ \times [0, T] \to \mathbb{R}$ satisfying suitable regularity conditions and such that $L(s, t) \leq H(s, t)$.

As in Section 2.9.2, we can associate with a game contingent claim the Dynkin game with the payoff

$$Z^*(\sigma, \tau) = \mathbb{1}_{\{\tau\leq\sigma\}} L^*(S_\tau, \tau) + \mathbb{1}_{\{\sigma<\tau\}} H^*(S_\sigma, \sigma),$$

where $L^*(S_t, t) = B_t^{-1} L(S_t, t)$ and $H^*(S_t, t) = B_t^{-1} H(S_t, t)$. We assume that

$$\inf_{\sigma\in\mathcal{T}_{[t,T]}} \sup_{\tau\in\mathcal{T}_{[t,T]}} \mathbb{E}_{\mathbb{P}^*}\big(Z^*(\sigma, \tau) \mid S_t = s\big) = \sup_{\tau\in\mathcal{T}_{[t,T]}} \inf_{\sigma\in\mathcal{T}_{[t,T]}} \mathbb{E}_{\mathbb{P}^*}\big(Z^*(\sigma, \tau) \mid S_t = s\big).$$

This ensures that the Dynkin game associated with a game contingent claim has a well-defined value process $\bar{Y}$, which is given by the formula

$$\bar{Y}_t = v(S_t, t) = \inf_{\sigma\in\mathcal{T}_{[t,T]}} \sup_{\tau\in\mathcal{T}_{[t,T]}} \mathbb{E}_{\mathbb{P}^*}\big(Z^*(\sigma, \tau) \mid S_t = s\big).$$

Let us denote

$$\mathcal{D}_L = \big\{(s, t) \in \mathbb{R}_+ \times [0, T] \mid v(s, t) = L(s, t)\big\}$$

and

$$\mathcal{D}_H = \big\{(s, t) \in \mathbb{R}_+ \times [0, T] \mid v(s, t) = H(s, t)\big\}.$$

For any fixed $t \in [0, T]$, we define

$$\sigma_t^* = \inf\big\{u \in [t, T] \mid (S_u, u) \in \mathcal{D}_H\big\} \wedge T$$

and

$$\tau_t^* = \inf\big\{u \in [t, T] \mid (S_u, u) \in \mathcal{D}_L\big\} \wedge T.$$

It is clear that σ_t^* and τ_t^* are stopping times from the class $\mathcal{T}_{[t,T]}$.

Suppose that the Nash equilibrium holds for the Dynkin game, meaning that

$$\mathbb{E}_{\mathbb{P}^*}\big(Z^*(\sigma_t^*, \tau) \mid S_t = s\big) \leq \mathbb{E}_{\mathbb{P}^*}\big(Z^*(\sigma_t^*, \tau_t^*) \mid S_t = s\big) \leq \mathbb{E}_{\mathbb{P}^*}\big(Z^*(\sigma, \tau_t^*) \mid S_t = s\big)$$

for all $\sigma, \tau \in \mathcal{T}_{[t,T]}$ and for all $(s, t) \in \mathbb{R}_+ \times [0, T]$. Then we say that (σ_t^*, τ_t^*) is a *saddle point* for the Dynkin game.

Recall that $\mathcal{L}_t$ stands for the following differential operator associated with the dynamics of S

$$\mathcal{L}_t v = \frac{\partial v}{\partial t} + \frac{1}{2}\sigma^2 s^2 \frac{\partial^2 v}{\partial s^2} + rs \frac{\partial v}{\partial s} - rv.$$

The following result, which is a synthesis of Theorems 9.1 and 12.1 in Friedman (1976), extends Theorem 5.4.2 to the case of a game contingent claim. Let us assume, for simplicity, that L and H are bounded functions (for sufficient integrability conditions, see Theorem 12.1 in Friedman (1976)).

Theorem 5.7.1 *The value process $\bar{Y}$ the Dynkin game associated with a game contingent claim satisfies $\bar{Y}_t = v(S_t, t)$, where the function $v(s, t)$ is a solution to the linear complementarity problem*

$$\begin{pmatrix} \mathcal{L}_t v(s, t) = 0 \\ L(s, t) \le v(s, t) \le H(s, t) \end{pmatrix} \vee \begin{pmatrix} \mathcal{L}_t v(s, t) \ge 0 \\ v(s, t) = L(s, t) \end{pmatrix} \vee \begin{pmatrix} \mathcal{L}_t v(s, t) \le 0 \\ v(s, t) = H(s, t) \end{pmatrix}$$

with $v(s, T) = L(s, T)$, where the notation $(a) \vee (b) \vee (c)$ means that v satisfies at least one of the conditions (a), (b) or (c). The pair (σ_t^*, τ_t^*) forms a saddle point for the Dynkin game.

In view of Theorem 5.7.1, the stopping times σ_t^* and τ_t^* are rational concelation and exercise times as of time t for a game contingent claim and its arbitrage price equals $\pi_t(X^g) = v(S_t, t)$ for every $t \in [0, T]$. The arbitrage price processes is an $\mathbb{F}$-martingale under $\mathbb{P}^*$, when discounted by the savings account and stopped at time $\sigma_0^* \wedge \tau_0^*$. In general, the arbitrage price of a game contingent claim can be shown to be a solution to the so-called *backward stochastic differential equation*. For a detailed analysis of a relationship between Dynkin games and backward stochastic differential equations, the reader may consult the paper by Cvitanić and Karatzas (1996).

For more information on the valuation and hedging of game contingent claims (in particular, convertible bonds), the interested reader is referred to Ayache et al. (2003), Kallsen and Kühn (2004, 2005), Sîrbu et al. (2004), and Sîrbu and Shreve (2005).

6

Exotic Options

In the preceding chapters, we have focused on the two standard classes of options – that is, call and put options of European and American style. The aim of this chapter is to study examples of more sophisticated option contracts. Although the payoffs of *exotic options* are given by similar expressions for both spot and futures options, the corresponding valuation formulas would not agree. We restrict here our attention to the case of exotic spot options. We find it convenient to classify the large family of exotic options as follows:

(a) *packages* – options that are equivalent to a portfolio of standard European options, cash and the underlying asset (stock, say);

(b) *forward-start options* – options that are paid for in the present but received by holders at a prespecified future date;

(c) *chooser options* – option contracts that are chosen by their holders to be call or put at a prescribed future date;

(d) *compound options* – option contracts with other options playing the role of the underlying assets;

(e) *binary options* – contracts whose payoff is defined by means of some binary function;

(f) *barrier options* – options whose payoff depends on whether the underlying asset price reaches some barrier during the option's lifetime;

(g) *lookback options* – options whose payoff depends, in particular, on the minimum or maximum price of the underlying asset during options' lifetimes;

(h) *Asian options* – options whose payoff depends on the average price of the underlying asset during a prespecified period;

(i) *basket options* – options with a payoff depending on the average of prices of several assets;

(j) α-*quantile options* – options whose payoff depends on the percentage of time that the price of the underlying asset remains below some level;

(k) *combined options on several assets* – these include, for instance, options on the minimum or maximum price of two risky assets;

(l) *Russian option* – a "user friendly" variant of a standard American option;

(i) *passport option* – an option related to the performance of a traded account.

6.1 Packages

An arbitrary financial contract whose terminal payoff is a piecewise linear function of the terminal price of the underlying asset may be seen as a *package option* – that is, a combination of standard options, cash and the underlying asset. As usual, we denote by S the price process of the underlying asset; we shall refer to S as a stock price. Unless explicitly stated otherwise, we shall place ourselves within the classical Black-Scholes framework (or its direct multidimensional generalization).

Collars. Let $K_2 > K_1 > 0$ be fixed real numbers. The payoff at expiry date T from the long position in a *collar option* equals

$$\mathbf{CL}_T \stackrel{\text{def}}{=} \min\big\{\max\{S_T, K_1\}, K_2\big\}.$$

It is easily seen that the payoff $\mathbf{CL}_T$ can be represented as follows

$$\mathbf{CL}_T = K_1 + (S_T - K_1)^+ - (S_T - K_2)^+,$$

so that a collar option can be seen as a portfolio of cash and two standard call options.

This implies that the arbitrage price of a collar option at any date t before expiry equals

$$\mathbf{CL}_t = K_1 e^{-r(T-t)} + C(S_t, T - t, K_1) - C(S_t, T - t, K_2),$$

where $C(s, T - t, K) = c(s, T - t, K, r, \sigma)$ stands for the Black-Scholes call option price at time t, where the current level of the stock price is s, and the exercise price of the option equals K (see formula (3.56)).

Break forwards. By a *break forward* we mean a modification of a typical forward contract, in which the potential loss from the long position is limited by some pre-specified number. More explicitly, the payoff from the long *break forward* is defined by the equality

$$\mathbf{BF}_T \stackrel{\text{def}}{=} \max\{S_T, F\} - K,$$

where $F = F_S(0, T) = S_0 e^{rT}$ is the forward price of a stock for settlement at time T, and $K > F$ is some constant. The delivery price K is set in such a manner that the break forward contract is worthless when it is entered into. Since

$$\mathbf{BF}_T = (S_T - F)^+ + F - K,$$

it is clear that for every $t \in [0, T]$,

$$\mathbf{BF}_t = C(S_t, T - t, F) + (F - K)e^{-r(T-t)}.$$

In particular, the right level of K, K_0 say, is given by the expression

$$K_0 = e^{rT}\big(S_0 + C(S_0, T, S_0 e^{rT})\big).$$

Using the Black-Scholes valuation formula, we end up with the following equality

$$K_0 = e^{rT} S_0 \Big(1 + N\big(d_1(S_0, T)\big) - N\big(d_2(S_0, T)\big) \Big),$$

where d_1 and d_2 are given by (3.18)–(3.19) with $K_0 = S_0 e^{rT}$.

Range forwards. A *range forward* may be seen as a special case of a collar – one with zero initial cost. Its payoff at expiry is

$$\mathbf{RF}_T \overset{\text{def}}{=} \max\big\{ \min\{S_T, K_2\}, K_1 \big\} - F = \max\big\{ \min\{S_T - F, K_2 - F\}, K_1 - F \big\},$$

where $K_1 < F < K_2$, and as before $F = F_S(0, T) = S_0 e^{rT}$. It appears convenient to decompose the payoff of a range forward in the following way

$$\mathbf{RF}_T = S_T - F + (K_1 - S_T)^+ - (S_T - K_2)^+.$$

Indeed, the above representation of the payoff implies directly that a range forward may be seen as a portfolio composed of a long forward contract, a long put option with strike price K_1, and finally a short call option with strike price K_2. Furthermore, its price at t equals

$$\mathbf{RF}_t = S_t - S_0 e^{rt} + P(S_t, T - t, K_1) - C(S_t, T - t, K_2).$$

As was mentioned earlier, the levels K_1 and K_2 are chosen in such a way that the initial value of a range forward equals 0.

6.2 Forward-start Options

Let us consider two dates, say T_0 and T, with $T_0 < T$. A *forward-start option* is a contract in which the holder receives, at time T_0 (at no additional cost), an option with expiry date T and exercise price equal to $K S_{T_0}$ for some $K > 0$. On the other hand, the holder must pay at time 0 an up-front fee, the price of a forward-start option. Let us consider the case of a forward-start call option, with terminal payoff

$$\mathbf{FS}_T \overset{\text{def}}{=} (S_T - K S_{T_0})^+.$$

To find the price at time $t \in [0, T_0]$ of such an option, it suffices to consider its value at the delivery date T_0, that is, the price at time T_0 of the at-the-money option with expiry date T. Thus, we have

$$\mathbf{FS}_{T_0} = C(S_{T_0}, T - T_0, K S_{T_0}).$$

Since we restrict our attention to the classical Black-Scholes model, it is easily seen that

$$C(S_{T_0}, T - T_0, K S_{T_0}) = S_{T_0} C(1, T - T_0, K).$$

Since $C(1, T - T_0, K)$ is non-random, the option's value at time 0 equals

$$\mathbf{FS}_0 = S_0 C(1, T - T_0, K) = C(S_0, T - T_0, S_0 K).$$

If a stock continuously pays dividends at a constant rate κ, the above equality should be modified as follows

$$\mathbf{FS}_0^\kappa = e^{-\kappa T_0} C^\kappa(S_0, T - T_0, K S_0),$$

where C^κ stands for the call option price derived in Proposition 3.2.1. Similar formulas can be produced for the case of a forward-start put option. We shall return to the analysis of forward-start options in Sect. 7.1.10, in which the assumption of a constant volatility will be relaxed.

6.3 Chooser Options

As suggested by the name of the contract, a *chooser option* is an agreement in which one party has the right to choose at some future date T_0 whether the option is to be a call or put option with a common exercise price K and remaining time to expiry $T - T_0$. The payoff at T_0 of a standard chooser option is

$$\mathbf{CH}_{T_0} \stackrel{\text{def}}{=} \max \left\{ C(S_{T_0}, T - T_0, K), P(S_{T_0}, T - T_0, K) \right\},$$

while its terminal payoff is given by the expression

$$\mathbf{CH}_{T_0} = (S_T - K)^+ \mathbb{1}_A + (K - S_T)^+ \mathbb{1}_{A^c},$$

where A stands for the following event belonging to the σ-field $\mathcal{F}_{T_0}$

$$A = \{ \omega \in \Omega \mid C(S_{T_0}, T - T_0, K) > P(S_{T_0}, T - T_0, K) \}$$

and A^c is the complement of A in Ω. Recall that the call-put parity implies that

$$P(S_{T_0}, T - T_0, K) = C(S_{T_0}, T - T_0, K) - S_{T_0} + K e^{-r(T-T_0)},$$

and thus

$$\mathbf{CH}_{T_0} = \max \left\{ C(S_{T_0}, T - T_0, K), C(S_{T_0}, T - T_0, K) - S_{T_0} + K e^{-r(T-T_0)} \right\},$$

or finally

$$\mathbf{CH}_{T_0} = C(S_{T_0}, T - T_0, K) + (K e^{-r(T-T_0)} - S_{T_0})^+.$$

The last equality implies immediately that the standard chooser option is equivalent to the portfolio composed of a long call option and a long put option (with different exercise prices and different expiry dates), so that its arbitrage price equals, for every $t \in [0, T_0]$,

$$\mathbf{CH}_t = C(S_t, T - t, K) + P(S_t, T_0 - t, K e^{-r(T-T_0)}).$$

In particular, using the Black-Scholes formula, we get for $t = 0$

$$\mathbf{CH}_0 = S_0 \left(N(d_1) - N(-\bar{d}_1) \right) + K e^{-rT} \left(N(-\bar{d}_2) - N(d_2) \right),$$

where

$$d_{1,2} = \frac{\ln(S_0/K) + (r \pm \frac{1}{2}\sigma^2)T}{\sigma\sqrt{T}}$$

and

$$\bar{d}_{1,2} = \frac{\ln(S_0/K) + rT \pm \frac{1}{2}\sigma^2 T_0}{\sigma\sqrt{T_0}}.$$

6.4 Compound Options

A *compound option* (see Geske (1977, 1979a, 1979b), Selby and Hodges (1987)) is a standard option with another standard option being the underlying asset. One can distinguish four basic types of compound options: call on a call, put on a call, call on a put, and, finally, put on a put. Let us consider, for instance, the case of a *call on a call* compound option. For two future dates T_0 and T, with $T_0 < T$, and two exercise prices K_0 and K, consider a call option with exercise price K_0 and expiry date T_0 on a call option with strike price K and maturity T. It is clear that the payoff of the compound option at time T_0 is

$$\mathbf{CO}_{T_0} \overset{\text{def}}{=} \left(C(S_{T_0}, \tau, K) - K_0 \right)^+,$$

where $C(S_{T_0}, \tau, K)$ stands for the value at time T_0 of a standard call option with strike price K and expiry date $T = T_0 + \tau$. In the Black-Scholes framework, we obtain the following equality

$$C(s, \tau, K) = s N \left(d_1(s, \tau, K) \right) - K e^{-r\tau} N \left(d_2(s, \tau, K) \right).$$

Moreover, since under $\mathbb{P}^*$ we have

$$S_{T_0} = S_0 \exp \left(\sigma\sqrt{T_0}\, \xi + (r - \frac{1}{2}\sigma^2)\, T_0 \right),$$

where ξ has a standard Gaussian distribution under $\mathbb{P}^*$, the price of the compound option at time 0 equals

$$\mathbf{CO}_0 = e^{-rT_0} \int_{x_0}^{\infty} \left(g(x) N(\hat{d}_1) - K e^{-r\tau} N(\hat{d}_2) - K_0 \right) n(x)\, dx,$$

where $\hat{d}_i = d_i(g(x), \tau, K)$ for $i = 1, 2$, the function $g : \mathbb{R} \to \mathbb{R}$ is given by the formula

$$g(x) = S_0 \exp \left(\sigma\sqrt{T_0}\, x + (r - \frac{1}{2}\sigma^2)\, T_0 \right)$$

and, finally, the constant x_0 is defined implicitly by the equation

$$x_0 = \inf\{x \in \mathbb{R} \mid C(g(x), \tau, K) \geq K_0\}.$$

Straightforward calculations yield

$$d_1(g(x), \tau, K) = \frac{\ln(S_0/K) + \sigma\sqrt{T_0}\,x + rT - \sigma^2 T_0 + \frac{1}{2}\sigma^2 T}{\sigma\sqrt{T - T_0}}$$

and

$$d_2(g(x), \tau, K) = \frac{\ln(S_0/K) + \sigma\sqrt{T_0}\,x + rT - \frac{1}{2}\sigma^2 T}{\sigma\sqrt{T - T_0}}.$$

6.5 Digital Options

By a *digital* (or *binary*) option we mean a contract whose payoff depends in a discontinuous way on the terminal price of the underlying asset. The simplest examples of binary options are *cash-or-nothing* options and *asset-or-nothing* options. The payoffs at expiry of a cash-or-nothing call and put options are

$$\mathbf{BCC}_T \stackrel{\text{def}}{=} X\mathbb{1}_{\{S_T > K\}}, \quad \mathbf{BCP}_T \stackrel{\text{def}}{=} X\mathbb{1}_{\{S_T < K\}},$$

where in both cases X stands for a predetermined amount of cash. Similarly, for the asset-or-nothing option we have

$$\mathbf{BAC}_T \stackrel{\text{def}}{=} S_T\mathbb{1}_{\{S_T > K\}}, \quad \mathbf{BAP}_T \stackrel{\text{def}}{=} S_T\mathbb{1}_{\{S_T < K\}}$$

for a call and put respectively. All options introduced above may be easily priced by means of the risk-neutral valuation formula. Somewhat more complex binary options are the so-called *gap options,* whose payoff at expiry equals

$$\mathbf{GC}_T \stackrel{\text{def}}{=} (S_T - X)\mathbb{1}_{\{S_T > K\}} = \mathbf{BAC}_T - \mathbf{BCC}_T$$

for the call option, and

$$\mathbf{GP}_T \stackrel{\text{def}}{=} (X - S_T)\mathbb{1}_{\{S_T < K\}} = \mathbf{BCP}_T - \mathbf{BAP}_T$$

for the corresponding put option. Once again, pricing these options involves no difficulties. As a last example of a binary option, let us mention a *supershare,* whose payoff is

$$\mathbf{SS}_T \stackrel{\text{def}}{=} \frac{S_T}{K_1}\mathbb{1}_{\{K_1 < S_T < K_2\}}$$

for some positive constants $K_1 < K_2$. The price of such an option at time 0 is easily seen to equal

$$\mathbf{SS}_0 = \frac{S_0}{K_1}\Big(N\big(h_1(S_0, T)\big) - N\big(h_2(S_0, T)\big)\Big),$$

where

$$h_i(s, t) = \frac{\ln(S_0/K_i) + (r + \frac{1}{2}\sigma^2)t}{\sigma\sqrt{t}}.$$

6.6 Barrier Options

The generic term *barrier options* refers to the class of options whose payoff depends on whether or not the underlying prices hit a prespecified barrier during the options' lifetimes. For closed-form expressions for prices of various barrier options and numerical methods, we refer to Rubinstein and Reiner (1991), Kunitomo and Ikeda (1992), Douady (1994), Rich (1994), Carr (1995), Cheuk and Vorst (1996), Ritchken (1995), Broadie et al. (1997), and Roberts and Shortland (1997). Heynen and Kat (1994) examine *partial barrier options* – that is, barrier options in which the underlying price is monitored for barrier hits only during a prespecified period during an option's lifetime. So-called *double barrier options* are treated in Kunitomo and Ikeda (1992), Geman and Yor (1996), Hui (1996), Jamshidian (1997b), and Pelsser (2000b).

To give the flavor of the mathematical techniques used when dealing with barrier options, we will examine a specific kind of currency barrier option, namely the *down-and-out call option*. The payoff at expiry of a down-and-out call option equals (in units of domestic currency)

$$C_T^1 \stackrel{\text{def}}{=} (Q_T - K)^+ \mathbb{1}_{\{\min_{0 \leq t \leq T} Q_t \geq H\}},$$

where K and H are constants. It follows from the formula above that the down-and-out option becomes worthless (or is knocked out) if, at any time t prior to the expiry date T, the current exchange rate Q_t falls below a predetermined level H. It is thus evident that a down-and-out option is less valuable than a standard currency option. Our aim is to find an explicit formula for the so-called *knock-out discount*.

Out-of-the-money knock-out option. Suppose first that the inequalities $H < K$ and $H < Q_0$ are satisfied. From the general features of a down-and-out call, it is clear that the option is knocked out when it is out-of-the-money. Recall that under the domestic martingale measure $\mathbb{P}^*$ we have (cf. (4.4))

$$Q_t = Q_0 \, e^{\sigma_Q W_t^* + \lambda t} = Q_0 \, e^{X_t},$$

where $X_t = \sigma_Q W_t^* + \lambda t$ for $t \in [0, T]$, and $\lambda = r_d - r_f - \frac{1}{2}\sigma_Q^2$. Therefore,

$$\{\omega \in \Omega \mid \min_{0 \leq t \leq T} Q_t \geq H\} = \{\omega \in \Omega \mid m_T \geq \ln(H/Q_0)\},$$

where $m_T = \min_{0 \leq t \leq T} X_t$, and thus

$$C_T^1 = (Q_T - K)\mathbb{1}_{\{Q_T \geq K, \, \min_{0 \leq t \leq T} Q_t \geq H\}} = Q_0 \, e^{X_T} \mathbb{1}_D - K\mathbb{1}_D,$$

where D stands for the set

$$D = \{\omega \in \Omega \mid X_T \geq \ln(K/Q_0), \, m_T \geq \ln(H/Q_0)\}.$$

We conclude that the price at time 0 of a down-and-out call option admits the following representation

$$C_0^1 = e^{-r_d T} Q_0 \, \mathbb{E}_{\mathbb{P}^*}\big(e^{X_T} \mathbb{1}_D\big) - e^{-r_d T} K \mathbb{P}^*\{D\},$$

where $\mathbb{P}^*$ is the martingale measure of the domestic market. In order to directly evaluate C_0^1 by means of integration, we need to find first the joint probability distribution of random variables X_T and m_T. One can show (see Harrison (1985) or Sect. A.18) that for all x, y such that $y \le 0$ and $y \le x$, we have

$$\mathbb{P}^*\{X_T \ge x, \, m_T \ge y\} = N\left(\frac{-x + \lambda T}{\sigma\sqrt{T}}\right) - e^{2\lambda y \sigma^{-2}} N\left(\frac{-x + 2y + \lambda T}{\sigma\sqrt{T}}\right),$$

where, for ease of notation, we write σ in place of σ_Q. Consequently, the probability density function of (X_T, m_T) equals

$$f(x, y) = \frac{-2(2y - x)}{\sigma^3 T^{3/2}} e^{2\lambda y \sigma^{-2}} n\left(\frac{-x + 2y + \lambda T}{\sigma\sqrt{T}}\right)$$

for $y \le 0$, $y \le x$, where n stands for the standard Gaussian density function. From the above it follows, in particular, that

$$\mathbb{P}^*\{D\} = N\left(\frac{\ln(Q_0/K) + \lambda T}{\sigma\sqrt{T}}\right) - (H/Q_0)^{2\lambda\sigma^{-2}} N\left(\frac{\ln(H^2/Q_0 K) + \lambda T}{\sigma\sqrt{T}}\right).$$

To find the expectation

$$I_1 \stackrel{\text{def}}{=} \mathbb{E}_{\mathbb{P}^*}\big(e^{X_T} \mathbb{1}_D\big) = \mathbb{E}_{\mathbb{P}^*}\Big(e^{X_T} \mathbb{1}_{\{X_T \ge \ln(K/Q_0), \, m_T \ge \ln(H/Q_0)\}}\Big),$$

we need to evaluate the double integral

$$\iint_A e^x f(x, y) \, dx \, dy,$$

where $A = \{(x, y); \; x \ge \ln(K/Q_0), \; y \ge \ln(H/Q_0), \; y \le 0, \; y \le x\}$. Straightforward (but rather cumbersome) integration leads to the following result

$$I_1 = e^{(r_d - r_f)T} \Big(N\big(h_1(Q_0, T)\big) - (H/Q_0)^{2\lambda\sigma^{-2}+2} N\big(c_1(Q_0, T)\big)\Big),$$

where

$$h_{1,2}(q, t) = \frac{\ln(q/K) + (r_d - r_f \pm \frac{1}{2}\sigma^2)t}{\sigma\sqrt{t}}$$

and

$$c_{1,2}(q, t) = \frac{\ln(H^2/qK) + (r_d - r_f \pm \frac{1}{2}\sigma^2)t}{\sigma\sqrt{t}}.$$

By collecting and rearranging the formulas above, we conclude that the price at initiation of the knock-out option admits the following representation (recall that we write $\sigma = \sigma_Q$)

$$C_0^1 = C_0^Q - J_0 = \quad \textit{Standard Call Price Knockout Discount,} \qquad (6.1)$$

where (cf. Proposition 4.2.2)

$$C_0^Q = Q_0 e^{-r_f T} N(h_1) - K e^{-r_d T} N(h_2)$$

and

$$J_0 = Q_0 e^{-r_f T} (H/Q_0)^{2\lambda\sigma^{-2}+2} N(c_1) - K e^{-r_d T} (H/Q_0)^{2\lambda\sigma^{-2}} N(c_2),$$

where $h_{1,2} = h_{1,2}(Q_0, T)$ and $c_{1,2} = c_{1,2}(Q_0, T)$. Notice that the proof of this formula can be substantially simplified by an application of Girsanov's theorem. We define an auxiliary probability measure $\bar{\mathbb{P}}$ by setting

$$\frac{d\bar{\mathbb{P}}}{d\mathbb{P}^*} = \exp\left(\sigma W_T^* - \tfrac{1}{2}\sigma^2 T\right) = \eta_T, \quad \mathbb{P}^*\text{-a.s.}$$

It follows from the Girsanov theorem that the process $\bar{W}_t = W_t^* - \sigma t$ is a standard Brownian motion under the probability measure $\bar{\mathbb{P}}$. Moreover, taking into account the definition of X, we find that

$$\mathbb{E}_{\mathbb{P}^*}\left(e^{X_T} \mathbb{1}_D\right) = e^{(r_d - r_f)T} \mathbb{E}_{\mathbb{P}^*}\left(\eta_T \mathbb{1}_D\right)$$

and thus

$$I_1 = e^{(r_d - r_f)T} \bar{\mathbb{P}}\{D\} = e^{(r_d - r_f)T} \bar{\mathbb{P}}\left\{X_T \geq \ln(K/Q_0), \ m_T \geq \ln(H/Q_0)\right\}.$$

Finally, the semimartingale decomposition of the process X under $\bar{\mathbb{P}}$ is

$$X_t = \sigma \bar{W}_t + (r_d - r_f + \tfrac{1}{2}\sigma^2)t, \quad \forall t \in [0, T],$$

hence, for every $y \leq 0$, $y \leq x$, we have

$$\bar{\mathbb{P}}\{D\} = N\left(h_1(Q_0, T)\right) - (H/Q_0)^{2\lambda\sigma^{-2}+2} N\left(c_1(Q_0, T)\right).$$

Representation (6.1) of the option's price now follows easily.

In-the-money knock-out option. When $K \leq H$ and $H < Q_0$, the option is knocked out when it is in-the-money. In this case we have

$$D = \{\omega \in \Omega \mid m_T \geq \ln(H/Q_0)\},$$

since $\{m_T \geq \ln(H/Q_0)\} \subset \{X_T \geq \ln(K/Q_0)\}$. It is well known (see Harrison (1985) or Sect. A.18) that for every $y \leq 0$

$$\mathbb{P}^*\{m_T \geq y\} = N\left(\frac{-y + \lambda T}{\sigma\sqrt{T}}\right) - e^{2\lambda y \sigma^{-2}} N\left(\frac{y + \lambda T}{\sigma\sqrt{T}}\right)$$

and thus

$$\mathbb{P}^*\{D\} = N\left(\frac{\ln(Q_0/H) + \lambda T}{\sigma\sqrt{T}}\right) - (H/Q_0)^{2\lambda\sigma^{-2}} N\left(\frac{-\ln(Q_0/H) + \lambda T}{\sigma\sqrt{T}}\right).$$

On the other hand,

$$I_1 = \mathbb{E}_{\mathbb{P}^*}\left(e^{X_T} \mathbb{1}_D\right) = e^{(r_d - r_f)T} \,\bar{\mathbb{P}}\big\{m_T \geq \ln(H/Q_0)\big\},$$

so that

$$I_1 = e^{(r_d - r_f)T}\left(N\big(\hat{h}_1(Q_0, T)\big) - (H/Q_0)^{2\lambda\sigma^{-2}+2}\, N\big(\hat{c}_1(Q_0, T)\big)\right),$$

where

$$\hat{h}_{1,2}(q, t) = \frac{\ln(q/H) + (r_d - r_f \pm \frac{1}{2}\sigma^2)t}{\sigma\sqrt{t}}$$

and

$$\hat{c}_{1,2}(q, t) = \frac{\ln(H/q) + (r_d - r_f \pm \frac{1}{2}\sigma^2)t}{\sigma\sqrt{t}}.$$

Consequently, the option price at time 0 equals $C_0^1 = \hat{C}_0 - \hat{J}_0$, where

$$\hat{C}_0 = Q_0 e^{-r_f T}\, N\big(\hat{h}_1(Q_0, T)\big) - K e^{-r_d T}\, N\big(\hat{h}_2(Q_0, T)\big)$$

is the price of the standard currency call with strike H. Denoting $\hat{c}_{1,2} = \hat{c}_{1,2}(Q_0, T)$, we get

$$\hat{J}_0 = Q_0 e^{-r_f T}\,(H/Q_0)^{2\lambda\sigma^{-2}+2}\, N(\hat{c}_1) - K e^{-r_d T}\,(H/Q_0)^{2\lambda\sigma^{-2}}\, N(\hat{c}_2).$$

6.7 Lookback Options

Lookback options are another example of path-dependent options – i.e., option contracts whose payoff at expiry depends not only on the terminal prices of the underlying assets, but also on asset price fluctuations during the options' lifetimes. We will examine the two following cases: that of a *standard lookback call option*, with payoff at expiry

$$\mathbf{LC}_T \overset{\text{def}}{=} (S_T - m_T^S)^+ = S_T - m_T^S, \tag{6.2}$$

where $m_T^S = \min_{t \in [0,T]} S_t$; and that of a *standard lookback put option*, whose terminal payoff equals

$$\mathbf{LP}_T \overset{\text{def}}{=} (M_T^S - S_T)^+ = M_T^S - S_T, \tag{6.3}$$

where $M_T^S = \max_{t \in [0,T]} S_t$. Note that a *lookback option* is not a genuine option contract since the (European) lookback option is always exercised by its holder at its expiry date. It is clear that the arbitrage prices of a lookback option are

$$\mathbf{LC}_t = e^{-r(T-t)}\, \mathbb{E}_{\mathbb{P}^*}(S_T \mid \mathcal{F}_t) - e^{-r(T-t)}\, \mathbb{E}_{\mathbb{P}^*}(m_T^S \mid \mathcal{F}_t) = I_1 - I_2$$

and

$$\mathbf{LP}_t = e^{-r(T-t)}\, \mathbb{E}_{\mathbb{P}^*}(M_T^S \mid \mathcal{F}_t) - e^{-r(T-t)}\, \mathbb{E}_{\mathbb{P}^*}(S_T \mid \mathcal{F}_t) = J_1 - J_2$$

for the lookback call and put respectively. Standard lookback options were first studied by Goldman et al. (1979a, 1979b). So-called *limited risk* and *partial* lookback options were examined by Conze and Viswanathan (1991).

Proposition 6.7.1 *Assume that $r > 0$. Then the price at time $t < T$ of a European lookback call option equals*

$$\mathbf{LC}_t = sN\left(\frac{\ln(s/m) + r_1\tau}{\sigma\sqrt{\tau}}\right) - me^{-r\tau}N\left(\frac{\ln(s/m) + r_2\tau}{\sigma\sqrt{\tau}}\right)$$
$$- \frac{s\sigma^2}{2r}N\left(\frac{\ln(m/s) - r_1\tau}{\sigma\sqrt{\tau}}\right) + e^{-r\tau}\frac{s\sigma^2}{2r}\left(\frac{m}{s}\right)^{2r\sigma^{-2}}N\left(\frac{\ln(m/s) + r_2\tau}{\sigma\sqrt{\tau}}\right),$$

where $s = S_t$, $m = m_t^S$, $\tau = T - t$ and $r_{1,2} = r \pm \frac{1}{2}\sigma^2$. Equivalently,

$$\mathbf{LC}_t = sN(\tilde{d}) - me^{-r\tau}N(\tilde{d} - \sigma\sqrt{\tau}) - \frac{s\sigma^2}{2r}N(-\tilde{d})$$
$$+ e^{-r\tau}\frac{s\sigma^2}{2r}\left(\frac{m}{s}\right)^{2r\sigma^{-2}}N(-\tilde{d} + 2r\sigma^{-1}\sqrt{\tau}),$$

where

$$\tilde{d} = \frac{\ln(s/m) + r_1\tau}{\sigma\sqrt{\tau}}.$$

In particular, if $s = S_t = m = m_t^S$ then by setting $d = r_1\sqrt{\tau}/\sigma$, we get

$$\mathbf{LC}_t = s\left(N(d) - e^{-r\tau}N(d - \sigma\sqrt{\tau}) - \frac{\sigma^2}{2r}N(-d) + e^{-r\tau}\frac{\sigma^2}{2r}N(d - \sigma\sqrt{\tau})\right).$$

Proof. Since the discounted stock price S^* is a martingale under $\mathbb{P}^*$, we have

$$I_1 = e^{-r(T-t)}\,\mathbb{E}_{\mathbb{P}^*}(S_T \mid \mathcal{F}_t) = e^{rt}\,\mathbb{E}_{\mathbb{P}^*}(S_T^* \mid \mathcal{F}_t) = S_t.$$

To evaluate I_2, observe first that for every $u \in [t, T]$, we have

$$S_u = S_t \exp\left(\sigma(W_u^* - W_t^*) + r_2(u - t)\right) = S_t e^{-(X_u - X_t)},$$

where $X_t = -\sigma W_t^* + vt$ with $v = \frac{1}{2}\sigma^2 - r$, and W^* follows the standard Brownian motion under $\mathbb{P}^*$. Hence,

$$m_{t,T}^S \overset{\text{def}}{=} \min_{u \in [t,T]} S_u = S_t e^{-M_{t,T}^X},$$

where $M_{t,T}^X = \max_{u \in [t,T]}(X_u - X_t)$. From the properties of the Brownian motion it is clear that the random variable $M_{t,T}^X$ is independent of $\mathcal{F}_t$. Moreover, the probability distribution of $M_{t,T}^X$ under $\mathbb{P}^*$ coincides with the distribution under $\mathbb{P}^*$ of M_τ^X, where $M_\tau^X = \max_{u \in [0,\tau]} X_u$ and $\tau = T - t$. Finally, we have

$$m_T^S = \min\{m_t^S, m_{t,T}^S\} = \min\{m_t^S, S_t e^{-M_{t,T}^X}\},$$

where both m_t^S and S_t are $\mathcal{F}_t$-measurable random variables, and $M_{t,T}^X$ is independent of the σ-field $\mathcal{F}_t$. It thus suffices to compute the expectation

$$L(s, m) \overset{\text{def}}{=} \mathbb{E}_{\mathbb{P}^*}\left(\min\{m, se^{-M_{t,T}^X}\}\right) = \mathbb{E}_{\mathbb{P}^*}\left(\min\{m, se^{-M_\tau^X}\}\right)$$

for fixed real numbers $s \geq m > 0$.

Indeed, we have

$$I_2 = e^{-r(T-t)} \mathbb{E}_{\mathbb{P}^*}(m_T^S \mid \mathcal{F}_t) = e^{-r\tau} L(S_t, m_t^S).$$

To find $L(s, m)$ explicitly, note first that

$$L(s, m) - m = \mathbb{E}_{\mathbb{P}^*}\left(\min\{m, se^{-M_\tau^X}\} \right) - m = \mathbb{E}_{\mathbb{P}^*}\left((se^{-M_\tau^X} - m)\mathbb{1}_{\{M_\tau^X \geq z\}}\right),$$

where $z = -\ln(m/s)$. Consequently,

$$L(s, m) - m = s\mathbb{E}_{\mathbb{P}^*}\left((e^{-M_\tau^X} - e^{-z})\mathbb{1}_{\{M_\tau^X \geq z\}}\right) = -s \int_z^\infty e^{-y} \mathbb{P}^*\{M_\tau^X \geq y\}dy.$$

By virtue of equality (A.92) in Sect. A.18, we have

$$\mathbb{P}^*\{M_\tau^X \geq y\} = \mathbb{P}^*\{X_\tau \geq y\} + e^{2v\sigma^{-2}y} \mathbb{P}^*\{X_\tau \geq y + 2v\tau\}.$$

Using the trivial equality $2v\sigma^{-2} - 1 = -2r\sigma^{-2}$, we obtain

$$L(s, m) - m = -s \int_z^\infty e^{-y} \mathbb{P}^*\{X_\tau \geq y\}\,dy$$

$$- s \int_z^\infty e^{-2r\sigma^{-2}y} \mathbb{P}^*\{X_\tau \geq y + 2v\tau\}\,dy = L_1(s, m) + L_2(s, m).$$

The first integral can be represented in the following way

$$L_1(s, m) = -s \int_z^\infty e^{-y} \mathbb{P}^*\{X_\tau \geq y\}\,dy = s\mathbb{E}_{\mathbb{P}^*}\left((e^{-X_\tau} - e^{-z})\mathbb{1}_{\{X_\tau \geq z\}}\right)$$

and thus

$$L_1(s, m) = se^{r\tau} \mathbb{E}_{\mathbb{P}^*}\left(e^{\sigma W_\tau^* - \sigma^2\tau/2}\mathbb{1}_{\{X_\tau \geq z\}}\right) - m\mathbb{P}^*\{X_\tau \geq z\}.$$

Equivalently, we have

$$L_1(s, m) = se^{r\tau}\mathbb{Q}\{X_\tau \geq z\} - m\mathbb{P}^*\{X_\tau \geq z\},$$

where the probability measure $\mathbb{Q}$ satisfies on $(\Omega, \mathcal{F}_T)$

$$\frac{d\mathbb{Q}}{d\mathbb{P}^*} = \exp\left(\sigma W_\tau^* - \tfrac{1}{2}\sigma^2\tau\right), \quad \mathbb{P}^*\text{-a.s.}$$

It is now easily seen that

$$L_1(s, m) = se^{r\tau} N\left(\frac{\ln(m/s) - r_1\tau}{\sigma\sqrt{\tau}}\right) - mN\left(\frac{\ln(m/s) - r_2\tau}{\sigma\sqrt{\tau}}\right).$$

For $L_2(s, m)$, by setting $\tilde{X}_\tau = X_\tau - 2v\tau$, we get

$$L_2(s, m) = -s \int_z^\infty e^{-2r\sigma^{-2}y} \, \mathbb{P}^*\{\tilde{X}_\tau \geq y\} \, dy.$$

Consequently, we have

$$L_2(s, m) = \frac{s\sigma^2}{2r} \, \mathbb{E}_{\mathbb{P}^*}\left((e^{-2r\sigma^{-2}\tilde{X}_\tau} - e^{-2r\sigma^{-2}z})\mathbb{1}_{\{\tilde{X}_\tau \geq z\}}\right)$$

$$= \frac{s\sigma^2}{2r}\left[\mathbb{E}_{\mathbb{P}^*}\left(e^{-2r\sigma^{-2}\tilde{X}_\tau}\mathbb{1}_{\{\tilde{X}_\tau \geq z\}}\right) - e^{-2r\sigma^{-2}z}\mathbb{P}^*\{\tilde{X}_\tau \geq z\}\right].$$

Since $X_\tau = -\sigma W_\tau^* + \nu\tau$, by elementary algebra we get

$$e^{-2r\sigma^{-2}\tilde{X}_\tau} = e^{r\tau} \, e^{2r\sigma^{-1}W_\tau^* - 2r^2\sigma^{-2}\tau}.$$

Let us define the probability measure $\tilde{\mathbb{Q}}$ by setting

$$\frac{d\tilde{\mathbb{Q}}}{d\mathbb{P}^*} = \exp\left(2r\sigma^{-1}W_\tau^* - 2r^2\sigma^{-2}\tau\right), \quad \mathbb{P}^*\text{-a.s.}$$

Note that the process $\tilde{W}_t = W_t^* - 2r\sigma^{-1}t$ follows a Brownian motion under $\tilde{\mathbb{Q}}$. Furthermore,

$$\frac{s\sigma^2}{2r} \, \mathbb{E}_{\mathbb{P}^*}\left(e^{-2r\sigma^{-2}\tilde{X}_\tau}\mathbb{1}_{\{\tilde{X}_\tau \geq z+\}}\right) = e^{r\tau}\frac{s\sigma^2}{2r} \, \tilde{\mathbb{Q}}\{\tilde{X}_\tau \geq z\},$$

and finally

$$L_2(s, m) = e^{r\tau}\frac{s\sigma^2}{2r}N\left(\frac{-z - r_1\tau}{\sigma\sqrt{\tau}}\right) - e^{-2r\sigma^{-2}z}\frac{s\sigma^2}{2r}N\left(\frac{-z + r_2\tau}{\sigma\sqrt{\tau}}\right),$$

where, as before, $z = -\ln(m/s)$. The asserted formula is now an immediate consequence of the relationship

$$\mathbf{LC}_t = s - e^{-r\tau}(L_1(s, m) + L_2(s, m) + m),$$

with $s = S_t$ and $m = m_t^S$. $\qquad\qquad\qquad\qquad\qquad\qquad\qquad\square$

The next result, which is stated without proof, deals with the lookback put option (see Goldman et al. (1979a, 1979b)).

Proposition 6.7.2 *Assume that $r > 0$. The price of a European lookback put option at time $t < T$ equals*

$$\mathbf{LP}_t = -sN\left(-\frac{\ln(s/M) + r_1\tau}{\sigma\sqrt{\tau}}\right) + Me^{-r\tau}N\left(-\frac{\ln(s/M) + r_2\tau}{\sigma\sqrt{\tau}}\right)$$

$$+ \frac{s\sigma^2}{2r}N\left(\frac{\ln(s/M) + r_1\tau}{\sigma\sqrt{\tau}}\right) - e^{-r\tau}\frac{s\sigma^2}{2r}\left(\frac{M}{s}\right)^{2r\sigma^{-2}}N\left(\frac{\ln(s/M) - r_2\tau}{\sigma\sqrt{\tau}}\right),$$

where we write $s = S_t$, $M = M_t^S$, $\tau = T - t$, and $r_{1,2} = r \pm \frac{1}{2}\sigma^2$. Equivalently,

$$\mathbf{LP}_t = -sN(-\hat{d}) + Me^{-r\tau}N(-\hat{d} + \sigma\sqrt{\tau}) + s\,\frac{\sigma^2}{2r}\,N(\hat{d})$$
$$- e^{-r\tau}\frac{s\sigma^2}{2r}\Big(\frac{M}{s}\Big)^{2r\sigma^{-2}} N(\hat{d} - 2r\sigma^{-1}\sqrt{\tau}),$$

where

$$\hat{d} = \frac{\ln(s/M) + (r + \frac{1}{2}\sigma^2)\tau}{\sigma\sqrt{\tau}}.$$

In particular, if $s = S_t = M = M_t^S$ then, denoting $d = r_1\sqrt{\tau}/\sigma$, we obtain

$$\mathbf{LP}_t = s\Big(-N(-d) + e^{-r\tau}N(-d + \sigma\sqrt{\tau}) + \frac{\sigma^2}{2r}\,N(d) - e^{-r\tau}\frac{\sigma^2}{2r}\,N(-d + \sigma\sqrt{\tau})\Big).$$

Notice that an American lookback call option is equivalent to its European counterpart. Indeed, the process $Z_t = e^{-rt}(S_t - m_t^S) = S_t^* - A_t$ is a submartingale, since the process A has non-increasing sample paths with probability 1. American and European lookback put options are not equivalent, however. The following bounds for the price $\mathbf{LP}_t^a$ of an American lookback put option can be established

$$\mathbf{LP}_t \le \mathbf{LP}_t^a \le e^{r\tau}\mathbf{LP}_t + S_t(e^{r\tau} - 1).$$

Numerical approximations of the price of an American lookback put option are examined in Barraquand and Pudet (1996). For a robust hedging of a lookback option, the reader may consult Hobson (1998a). A new method of pricing lookback options is proposed in Buchen and Konstandatos (2005).

6.8 Asian Options

An *Asian option* (or an *average option*) is a generic name for the class of options (of European or American style) whose terminal payoff is based on average asset values during some period within the options' lifetimes. Due to their averaging feature, Asian options are particularly suitable for thinly traded assets (or commodities). Actually, in contrast to standard options, Asian options are more robust with respect to manipulations near their expiry dates. Typically, they are also less expensive than standard options. Let T be the exercise date, and let $0 \le T_0 < T$ stand for the beginning date of the averaging period. Then the payoff at expiry of an Asian call option equals

$$C_T^A \stackrel{\text{def}}{=} \big(A_S(T_0, T) - K\big)^+, \tag{6.4}$$

where

$$A_S(T_0, T) = \frac{1}{T - T_0}\int_{T_0}^{T} S_u\, du \tag{6.5}$$

is the arithmetic average of the asset price over the time interval $[T_0, T]$, K is the fixed strike price, and the price S of the stock is assumed to follow a geometric

Brownian motion. The main difficulty in pricing and hedging Asian options is due to the fact that the random variable $A(T_0, T)$ does not have a lognormal distribution. This feature makes the task of finding an explicit formula for the price of an Asian option surprisingly involved.

For this reason, early studies of Asian options were based either on approximations or on the direct application of the Monte Carlo method. The numerical approach to the valuation of Asian options, proposed independently by Ruttiens (1990) and Vorst (1992), is based on the approximation of the arithmetic average using the geometric average. Note first that it is natural to substitute for the continuous-time average $A(T_0, T)$ its discrete-time counterpart

$$A_S^n(T_0, T) = \frac{1}{n} \sum_{i=0}^{n-1} S_{T_i}, \qquad (6.6)$$

where $T_i = T_0 + i(T - T_0)/n$. Furthermore, the arithmetic average $A_S^n(T_0, T)$ can be replaced with the geometric average, denoted by $G_S^n(T_0, T)$ in what follows. Recall that the random variables S_{T_i}, $i = 1, \ldots, n$ are given explicitly by the expression

$$S_{T_i} = S_{T_0} \exp \left(\sigma (W_{T_i}^* - W_{T_0}^*) + (r - \tfrac{1}{2}\sigma^2)(T_i - T_0) \right),$$

where W^* is a standard Brownian motion under the martingale measure $\mathbb{P}^*$. Hence the geometric average admits the following representation

$$G_S^n(T_0, T) = \left(\prod_{i=0}^{n-1} S_{T_i} \right)^{1/n} = c S_{T_0} \exp \left(\frac{\sigma}{n} \sum_{i=0}^{n-1} (n - i - 1)(W_{T_{i+1}}^* - W_{T_i}^*) \right)$$

for a strictly positive constant c. In view of the independence of increments of the Brownian motion, the last formula makes clear that the geometric average $G_S^n(T_0, T)$ has a lognormal distribution under $\mathbb{P}^*$. The approximate Black-Scholes-like formula for the price of an Asian call option can thus be easily found by the direct evaluation of the conditional expectation

$$\tilde{C}_{T_0}^n = e^{-r(T - T_0)} \, \mathbb{E}_{\mathbb{P}^*} \left((G_S^n(T_0, T) - K)^+ \,\middle|\, \mathcal{F}_{T_0} \right).$$

It appears, however, that such an approach significantly underprices Asian call options. To overcome this deficiency, one may directly approximate the true distribution of the arithmetic average using an approximate distribution, typically a lognormal distribution with the appropriate parameters[1] (see Levy (1992), Turnbull and Wakeman (1991), and Bouaziz et al. (1994)). Another approach, initiated by Carverhill and Clewlow (1990), relies on the use of the fast Fourier transform to evaluate the density of the sum of random variables as the convolution of individual densities. The second step in this method involves numerical integration of the option's payoff function with respect to this density function. Kemna and Vorst (1990) apply the

---

[1] It should be noted that explicit formulas for all moments of the arithmetic average are available (see Geman and Yor (1992, 1993) and Yor (1992a, 1992b)).

Monte Carlo simulation with variance reduction to price Asian options. They replace $A(T_0, T)$ with the arithmetic average (6.6), so that the approximate value of an Asian call option is given by the formula

$$\bar{C}_{T_0}^n = e^{-r(T-T_0)} \, \mathbb{E}_{\mathbb{P}*}\left\{ \left(\frac{1}{n} \sum_{i=0}^{n-1} S_{T_i} - K \right)^+ \Big| \mathcal{F}_{T_0} \right\}.$$

Since the random variables S_{T_i}, $i = 1, \ldots, n$ are given by an explicit formula, they can easily be generated using any standard procedure. A lower bound for the price of an Asian option was found in Rogers and Shi (1995). Nielsen and Sandmann (1996) examined numerical pricing of Asian options under stochastic interest rates.

The most efficient analytical tools leading to quasi-explicit pricing formulas were developed in recent years by Geman and Yor (1992, 1993) (see also Hoffman (1993)). We shall now give a short account of their results. Consider first the special case of an Asian option that is already known to be in-the-money, i.e., assume that $t < T$ belongs to the averaging period, and the past values of stock price are such that

$$A_S(T_0, T) = \frac{1}{T - T_0} \int_{T_0}^{T} S_u \, du > \frac{1}{T - T_0} \int_{T_0}^{t} S_u \, du \geq K. \tag{6.7}$$

In this case, the value at time t of the Asian option equals

$$C_t^A = \frac{S_t(1 - e^{-r(T-t)})}{r(T - T_0)} - e^{-r(T-t)} \left(K - \frac{1}{T - T_0} \int_{T_0}^{t} S_u \, du \right). \tag{6.8}$$

Indeed, under (6.7), the price of the option satisfies

$$C_t^A = e^{-r(T-t)} \, \mathbb{E}_{\mathbb{P}*}\left(\frac{1}{T - T_0} \int_{T_0}^{T} S_u \, du - K \Big| \mathcal{F}_t \right),$$

or equivalently,

$$C_t^A = \frac{e^{-r(T-t)}}{T - T_0} \, \mathbb{E}_{\mathbb{P}*}\left(\int_{t}^{T} S_u \, du \Big| \mathcal{F}_t \right) + e^{-r(T-t)} \left(\frac{1}{T - T_0} \int_{T_0}^{t} S_u \, du - K \right).$$

Furthermore, we have

$$\mathbb{E}_{\mathbb{P}*}\left(\int_{t}^{T} S_u \, du \Big| \mathcal{F}_t \right) = r^{-1} S_t (e^{r(T-t)} - 1),$$

since (recall that $S_t^* = e^{-rt} S_t$ and $\mathbb{E}_{\mathbb{P}*}(S_T^* \mid \mathcal{F}_t) = S_t^*$)

$$e^{rT} S_T^* - e^{rt} S_t^* = \int_{t}^{T} d(e^{ru} S_u^*) = \int_{t}^{T} r S_u \, du + \int_{t}^{T} e^{ru} \, dS_u^*.$$

For an Asian option that is not known at time t to be in-the-money at time T, no explicit valuation formula is available. We close this section with a result which gives a partial answer to the valuation problem (for a proof we refer to original papers by Geman and Yor). Let us only mention that the proof of the quasi-explicit formula (6.9) is based on a connection between an exponential Brownian motion and time-changed Bessel processes.

Proposition 6.8.1 *The price of an Asian call option admits the representation*

$$C_t^A = \frac{4e^{-r(T-t)} S_t}{\sigma^2(T - T_0)} \, C^\nu(h, q), \tag{6.9}$$

where

$$\nu = \frac{2r}{\sigma^2} - 1, \quad h = \frac{\sigma^2}{4}(T - t), \quad q = \frac{\sigma^2}{4S_t}\left(K(T - T_0) - \int_{T_0}^{t} S_u \, du\right).$$

Moreover, $C^\nu(h, q)$ has the Laplace transform with respect to h given by the formula

$$\int_0^\infty e^{-\lambda h} C^\nu(h, q) \, dh = \int_0^{1/2q} d e^{-x} x^{\gamma - 2} (1 - 2qx)^{\gamma + 1} \, dx \stackrel{\text{def}}{=} g(\lambda),$$

where $\mu = \sqrt{2\lambda + \nu^2}$, $\gamma = \frac{1}{2}(\mu - \nu)$, and $d = \left(\lambda(\lambda - 2 - 2\nu)\Gamma(\gamma - 1)\right)^{-1}$.

In order to apply the last result to price the option, one needs to find (at least numerically) the inverse Laplace transform of the function $g(\lambda)$. For this purpose, we write $f(h) = C^\nu(h, q)$ and we introduce an auxiliary function $\tilde{f}(h) = e^{-\alpha h} f(h)$, with $\alpha > 2\nu + 2$. Then the Laplace transform of $\tilde{f}$ equals $g(\lambda + \alpha) = \tilde{g}(\lambda)$. Moreover, since the function $\tilde{g}$ is regular for $\lambda \geq 0$, we are in a position to make use of the inverse Fourier transform – that is

$$\tilde{f}(h) = \frac{1}{2\pi i} \int_{0-i\infty}^{0+i\infty} e^{-i\lambda h} \, \tilde{g}(\lambda) \, d\lambda.$$

This can be done numerically, using the fast Fourier transform – a detailed description of this approach (and numerical examples) is provided in Geman and Eydeland (1995). For other approaches to the valuation of an Asian option, we refer to Alziary et al. (1997), Chesney et al. (1997a), and Lipton (2001).

6.9 Basket Options

A *basket option*, as suggested by its name, is a kind of option contract that serves to hedge against the risk exposure of a basket of assets – that is, a predetermined portfolio of assets. Generally speaking, a basket option is more cost-effective than a portfolio of single options, as the latter over-hedges the exposure, and costs more than a basket option. An intuitive explanation for this feature is that the basket option takes into account the correlation between different risk factors. For instance, in the case of a strong negative correlation between two or more underlying assets, the total risk exposure may almost vanish, and this nice feature is not reflected in payoffs and prices of single options.

From the analytical viewpoint, there is a close analogy between basket options and Asian options. Let us denote by S^i, $i = 1, \ldots, k$ the price processes of k underlying assets; they will be referred to as stocks in what follows. In this case, it seems natural to refer to such a basket option as the *stock index option* (in market practice, options on a basket of currencies are also quite common). The payoff at expiry of a basket call option is defined in the following way

$$C_T^B \stackrel{\text{def}}{=} \left(\sum_{i=1}^k w_i S_T^i - K \right)^+ = (A_T - K)^+, \tag{6.10}$$

where $w_i \geq 0$ is the weight of the ith asset, so that $\sum_{i=1}^k w_i = 1$. Note that A_T stands for the weighted arithmetic average $A_T = \sum_{i=1}^k w_i S_T^i$. We assume that each stock price S^i follows a geometric Brownian motion. More explicitly, under the martingale measure $\mathbb{P}^*$ we have

$$dS_t^i = S_t^i \left(r \, dt + \hat{\sigma}_i \cdot dW_t^* \right) \tag{6.11}$$

for some non-zero vectors $\hat{\sigma}_i \in \mathbb{R}^k$, where $W^* = (W^{1*}, \ldots, W^{k*})$ stands for a k-dimensional Brownian motion under $\mathbb{P}^*$, and r is the risk-free interest rate. Observe that for any fixed i, we can find a standard one-dimensional Brownian motion $\tilde{W}^i$ such that

$$dS_t^i = S_t^i \left(r \, dt + \sigma_i \, d\tilde{W}_t^i \right) \tag{6.12}$$

and $\sigma_i = |\hat{\sigma}_i|$, where $|\hat{\sigma}_i|$ is the Euclidean norm of $\hat{\sigma}_i$. Let us denote by $\rho_{i,j}$ the instantaneous correlation coefficient

$$\rho_{i,j} = \frac{\hat{\sigma}_i \cdot \tilde{\sigma}_j}{\sigma_i \sigma_j} = \frac{\hat{\sigma}_i \cdot \hat{\sigma}_j}{|\hat{\sigma}_i| |\hat{\sigma}_j|}.$$

We may thus alternatively assume that the dynamics of price processes S^i are given by (6.12), where $\tilde{W}^i$, $i = 1, \ldots, k$ are one-dimensional Brownian motions, whose cross-variations satisfy $\langle \tilde{W}^i, \tilde{W}^j \rangle_t = \rho_{i,j} t$ for every $i, j = 1, \ldots, k$. Let us return to the problem of valuation of basket options. For similar reasons as those applying to Asian options, basket options are rather intractable analytically. Rubinstein (1991a) developed a simple technique of pricing basket options on a bivariate binomial lattice, thus generalizing the standard Cox-Ross-Rubinstein methodology. Unfortunately, this numerical method is very time-consuming, especially where there are several underlying assets. To overcome this, Gentle (1993) proposed valuation of a basket using an approximation of the weighted arithmetic mean in the form of its geometric counterpart (this follows the approach of Ruttiens (1990) and Vorst (1992) to Asian options). For a fixed $t \leq T$, let us denote by $\hat{w}_i$ the modified weights

$$\hat{w}_i = \frac{w_i S_t^i}{\sum_{j=1}^k w_j S_t^j} = \frac{w_i F_{S^i}(t, T)}{\sum_{j=1}^k w_j F_{S^j}(t, T)}, \tag{6.13}$$

where $F_{S^i}(t, T)$ is the forward price at time t of the ith asset for the settlement date T.

We may rewrite (6.10) as follows[2]

$$C_T^B = \left(\sum_{j=1}^k w_j F_{S^j}(t,T)\right)\left(\sum_{i=1}^k \hat{w}_i \tilde{S}_T^i - \tilde{K}\right)^+ = \left(\sum_{j=1}^k w_j F_{S^j}(t,T)\right)(\tilde{A}_T - \tilde{K})^+,$$

where $\tilde{S}_T^i = S_T^i/F_{S^i}(t,T)$, $\tilde{A}_T = \sum_{i=1}^k \hat{w}_i \tilde{S}_T^i$, and

$$\tilde{K} = \frac{K}{\sum_{j=1}^k w_j F_{S^j}(t,T)} = \frac{e^{-r(T-t)} K}{\sum_{j=1}^k w_j S_t^j}. \tag{6.14}$$

The arbitrage price at time t of a basket call option thus equals

$$C_t^B = e^{-r(T-t)} \left(\sum_{j=1}^k w_j F_{S^j}(t,T)\right) \mathbb{E}_{\mathbb{P}^*}\left((\tilde{A}_T - \tilde{K})^+ \,\middle|\, \mathcal{F}_t\right),$$

or equivalently,

$$C_t^B = \left(\sum_{j=1}^k w_j S_t^j\right) \mathbb{E}_{\mathbb{P}^*}\left((\tilde{A}_T - \tilde{K})^+ \,\middle|\, \mathcal{F}_t\right). \tag{6.15}$$

The next step relies on an approximation of the weighted arithmetic mean $\sum_{i=1}^k \hat{w}_i \tilde{S}_T^i$ using a similarly weighted geometric mean. More specifically, we approximate the price C_0^B of the basket option using the number $\hat{C}_0^B$ given by the formula (for ease of notation, we set $t = 0$ in what follows)

$$\hat{C}_0^B = \left(\sum_{j=1}^k w_j S_0^j\right) \mathbb{E}_{\mathbb{P}^*}(\tilde{G}_T - \hat{K})^+, \tag{6.16}$$

where $\tilde{G}_T = \prod_{i=1}^k (\tilde{S}_T^i)^{\hat{w}_i}$ and

$$\hat{K} = \tilde{K} + \mathbb{E}_{\mathbb{P}^*}(\tilde{G}_T - \tilde{A}_T). \tag{6.17}$$

In view of (6.11), we have (recall that $F_{S^i}(0,T) = e^{rT} S_0^i$)

$$\tilde{S}_T^i = S_T^i/F_{S^i}(0,T) = e^{\hat{\sigma}_i \cdot W_T^* - \sigma_i^2 T/2},$$

and thus the weighted geometric average $\tilde{G}_T$ equals

$$\tilde{G}_T = e^{c_1 \cdot W_T^* - c_2 T/2} = e^{\eta_T - c_2 T/2}, \tag{6.18}$$

with $\eta_T = c_1 \cdot W_T^*$, where we write $c_1 = \sum_{i=1}^k \hat{w}_i \hat{\sigma}_i$ and $c_2 = \sum_{i=1}^k \hat{w}_i \sigma_i^2$. We conclude that the random variable $\tilde{G}_T$ is lognormally distributed under $\mathbb{P}^*$.

---

[2] This representation is introduced because it appears to give a better approximation of the price of a basket option than formula (6.10).

More precisely, the random variable η_T in (6.18) has Gaussian distribution with expected value 0 and variance

$$\text{Var}_{\mathbb{P}^*}(\eta_T) = \mathbb{E}_{\mathbb{P}^*}\left\{ \left(\sum_{i,l=1}^{k} \hat{w}_i \hat{\sigma}_{il} W_T^{l*} \right)\left(\sum_{j,m=1}^{k} \hat{w}_j \hat{\sigma}_{jm} W_T^{m*} \right) \right\}$$

$$= \sum_{i,j,l,m=1}^{k} \hat{w}_i \hat{w}_j \hat{\sigma}_{il} \hat{\sigma}_{jm} \mathbb{E}_{\mathbb{P}^*}\left(W_T^{l*} W_T^{m*} \right) = \sum_{i,j=1}^{k} \hat{w}_i \hat{w}_j \hat{\sigma}_i \cdot \hat{\sigma}_j \, T = v^2 T,$$

where we write $v^2 = \sum_{i,j=1}^{k} \rho_{i,j} \hat{w}_i \hat{w}_j \sigma_i \sigma_j$. We have used here the equality $\mathbb{E}_{\mathbb{P}^*}\left(W_T^{l*} W_T^{m*} \right) = \delta_{lm} T$, where δ_{lm} stands for Kronecker's delta – that is, δ_{lm} equals 1 if $l = m$ and zero otherwise. Note that

$$\mathbb{E}_{\mathbb{P}^*}(\tilde{A}_T) = \sum_{j=1}^{k} \hat{w}_j S_0^j \, \mathbb{E}_{\mathbb{P}^*}\left(e^{-rT} S_T^i \right) = \sum_{j=1}^{k} \hat{w}_j = 1,$$

and

$$\mathbb{E}_{\mathbb{P}^*}(\tilde{G}_T) = e^{(v^2-c_2)T/2} \, \mathbb{E}_{\mathbb{P}^*}\left(e^{\eta_T - \frac{1}{2}\text{Var}(\eta_T)} \right) = e^{(v^2-c_2)T/2} \stackrel{\text{def}}{=} c.$$

We conclude that $\hat{K} = \tilde{K} + c - 1$. The expectation in (6.16) can now be evaluated explicitly, using the following simple lemma (cf. Lemma 4.5.1).

Lemma 6.9.1 *Let ξ be a Gaussian random variable on $(\Omega, \mathcal{F}, \mathbb{P})$ with expected value 0 and variance $\sigma^2 > 0$. For any real numbers $a, b > 0$, we have*

$$\mathbb{E}_{\mathbb{P}}\left(a e^{\xi - \frac{1}{2}\sigma^2} - b \right)^+ = a N(h) - b N(h - \sigma), \tag{6.19}$$

where $h = \sigma^{-1} \ln(a/b) + \frac{1}{2}\sigma$.

We have

$$\mathbb{E}_{\mathbb{P}^*}\left(\tilde{G}_T - \hat{K} \right)^+ = \mathbb{E}_{\mathbb{P}^*}\left(c e^{\eta_T - \frac{1}{2}\text{Var}(\eta_T)} - (\tilde{K} + c - 1) \right)^+,$$

so that we may set $\xi = \eta_T$, $a = c$ and $b = \tilde{K} + c - 1$. In view of Lemma 6.9.1, the following result is straightforward.

Proposition 6.9.1 *The approximate value $\hat{C}_t^B$ of the price C_t^B of a basket call option with strike price K and expiry date T equals*

$$\hat{C}_t^B = \left(\sum_{j=1}^{k} w_j S_t^j \right)\left(c N\left(l_1(T-t) \right) - (\tilde{K} + c - 1) N\left(l_2(T-t) \right) \right), \tag{6.20}$$

where

$$c = \exp\left\{ \left(\frac{1}{2} \sum_{i,j=1}^{k} \rho_{i,j} \hat{w}_i \hat{w}_j \sigma_i \sigma_j - \sum_{j=1}^{k} \hat{w}_j \sigma_j^2 \right)(T-t) \right\},$$

and where the modified weights $\hat{w}_i$ are given by (6.13), $\tilde{K}$ is given by (6.14), and

$$l_{1,2}(t) = \frac{\ln c - \ln(\tilde{K} + c - 1) \pm \frac{1}{2}v^2 t}{v\sqrt{t}}.$$

Suppose that $k = 1$. In this case, $w_1 = 1$, and the arithmetic average agrees with the geometric one. Consequently, $c = 1$ and (6.20) reduces to the standard Black-Scholes formula. For a slightly different approach to approximate valuation of basket options, the reader is referred to Huynh (1994). In this approach, the weighted average $\tilde{A}_T$ in (6.15) is directly replaced by a lognormally distributed random variable. Then the series expansion of the true distribution of $\tilde{A}_T$ is derived in terms of an approximating lognormal distribution. An estimate of the basket option price can thus be obtained by direct integration. The idea of using the series expansion, known as the generalized Edgeworth series expansion, in the context of option valuation goes back at least to Jarrow and Rudd (1982). Turnbull and Wakeman (1991) applied this method to approximate valuation of Asian options.

6.10 Quantile Options

A new type of path-dependent option, the so-called α-*quantile* option, was proposed recently by Miura (1992). Let us fix a strictly positive number $0 < \alpha < 1$. For a fixed horizon date T, the α-quantile of a continuous, strictly positive semimartingale S over the time interval $[0, T]$ is the random variable $Q_T^\alpha(S)$, which is given by the formula

$$Q_T^\alpha(S) = \inf \left\{ x \in \mathbb{R}_+ \,\middle|\, l\,\{t \in [0, T] \mid S_t < x\} \geq \alpha T \right\},$$

where l denotes the Lebesgue measure, or equivalently,

$$Q_T^\alpha(S) = \inf \left\{ x \in \mathbb{R}_+ \,\middle|\, \frac{1}{T} \int_0^T \mathbb{1}_{\{S_t < x\}} \, dt \geq \alpha \right\}.$$

It is apparent that the random variable $Q_T^\alpha(S)$ is defined pathwise; hence, the term α-*percentile* seems to be a more adequate name for $Q_T^\alpha(S)$ than the term α-*quantile* that usually refers to probabilistic properties. In financial interpretation, if the process S represents the stock price, then the α-quantile is that price level at which the percentage of time the stock price remains below this level before time T equals α. By definition, the payoff at expiry of the α-quantile call option equals $C_T^\alpha = (Q_T^\alpha(S) - K)^+$. Assume, more specifically, that under the martingale measure $\mathbb{P}^*$, the price S equals

$$S_t = S_0 \exp\left(\sigma W_t^* + (r - \tfrac{1}{2}\sigma^2)t\right), \quad \forall t \in [0, T],$$

for a certain $\sigma > 0$. Note that by virtue of the scaling property of the Brownian motion, it is sufficient to consider the case $T = 1$ and $\sigma = 1$.

To value the α-quantile option explicitly within the Black-Scholes framework, we need to determine the probability distribution of the following random variable (W stands for a one-dimensional Brownian motion)

$$Q(\alpha, v, t) = \inf \left\{ x \in \mathbb{R} \; \Big| \; \frac{1}{t} \int_0^t \mathbb{1}_{\{W_u + vu \, < \, x\}} \, du \geq \alpha \right\}$$

for any fixed $0 < \alpha < 1$, $t > 0$ and drift coefficient $v \in \mathbb{R}$. The relevant computations were done using different methods by Yor (1995), Akahori (1995), Dassios (1995), and Embrechts et al. (1995). Let us set $\beta = \sqrt{(1 - \alpha)/\alpha}$. Then for $v = 0$ and $t = 1$ the probability density function f of the distribution of $Q(\alpha, 0, 1)$ is given by the following formula, due to Yor (1995),

$$f(x) = \sqrt{\frac{2}{\pi}} \, \exp \left(-\frac{x^2}{2} \right) \hat{N}(\beta x), \quad \forall x > 0,$$

and

$$f(x) = \sqrt{\frac{2}{\pi}} \, \exp \left(-\frac{x^2}{2} \right) \hat{N}(\beta^{-1}|x|), \quad \forall x < 0,$$

where

$$\hat{N}(x) = 2N(-x) = \sqrt{\frac{2}{\pi}} \int_x^\infty e^{-z^2/2} \, dz, \quad \forall x > 0.$$

As observed in Yor (1995), the generalization of the formula above to the case of a non-zero drift term is not straightforward. On the other hand, Dassios (1995) showed that for any $0 < \alpha < 1$, $t > 0$ and $v \in \mathbb{R}$ the random variable $Q(\alpha, v, t)$ admits the probability density function f^v, given by the formula

$$f^v(x) = \int_{-\infty}^\infty f_1(x - y, \alpha t) f_2(y, (1 - \alpha)t) \, dy, \tag{6.21}$$

where

$$f_1(x, t) = \sqrt{\frac{2}{\pi t}} \, \exp \left(-\frac{(x - vt)^2}{2t} \right) - 2v \exp(2vx) \tilde{N} \left(\frac{x + vt}{\sqrt{t}} \right), \quad \text{if } x > 0,$$

and $f_1(x, t) = 0$ otherwise;

$$f_2(x, t) = \sqrt{\frac{2}{\pi t}} \, \exp \left(-\frac{(x - vt)^2}{2t} \right) + 2v \exp(2vx) N \left(\frac{x + vt}{\sqrt{t}} \right), \quad \text{if } x > 0,$$

and $f_2(x, t) = 0$ otherwise. Here N is the cumulative distribution function of the standard Gaussian distribution, and $\tilde{N}$ denotes the tail of N, that is

$$\tilde{N}(x) = 1 - N(x) = \frac{1}{\sqrt{2\pi}} \int_x^\infty e^{-z^2/2} \, dz, \quad \forall x \in \mathbb{R}.$$

In the special case of $v = 0$, the convolution (6.21) can be evaluated explicitly, yielding Yor's formula. Relationship (6.21) leads to the following identity, which was originally derived by Dassios (1995 (an alternative proof of Dassios' result is also given in Embrechts et al. (1995))

$$Q(\alpha, v) \overset{\text{(d)}}{=} \sup_{s \leq \alpha} (W_s + vs) + \inf_{s \leq 1-\alpha} (\tilde{W}_s + vs), \qquad (6.22)$$

where $\overset{\text{(d)}}{=}$ denotes equality of probability distributions, and it is assumed that the one-dimensional standard Brownian motions W and $\tilde{W}$ are mutually independent. On the other hand, (6.21) can also be used to evaluate the price of the α-quantile option, at least for $t = 0$. Let us denote by g the function $g(t, x, v) = \mathbb{P}\{\eta(t, v) < x\}$, where the random variable $\eta(t, v)$ is given by the formula

$$\eta(t, v) = \frac{1}{t} \int_0^t \mathbb{1}_{\{W_u + vu < 0\}} \, du,$$

and W follows a one-dimensional standard Brownian motion under $\mathbb{P}$. The function g is known to be given by the quasi-explicit formula

$$g(t, x, v) = \frac{1}{2} \int_0^{tx} \left(\sqrt{\frac{2}{\pi s}} \exp\left(-\frac{v^2}{2}s\right) - 2v \tilde{N}(v\sqrt{s}) \right)$$
$$\times \left[\left(2v + \sqrt{\frac{2}{\pi(t-s)}} \exp\left(-\frac{v^2}{2}(t-s)\right) \right) - 2v \tilde{N}(v\sqrt{t-s}) \right] ds,$$

where $\tilde{N}$ denotes the tail of the standard Gaussian distribution. Let us state without proof a valuation result due to Akahori (1995). Basically, the calculations that lead to the pricing formula of Proposition 6.10.1 are equivalent to those based on Dassios's results (we refer to the original papers for details).

Proposition 6.10.1 *The price at time 0 of the α-quantile call option in the Black-Scholes framework equals*

$$C_0^\alpha = e^{-rT} \int_K^\infty G\left(T, \sigma^{-1} \ln \frac{y}{S_0}, \alpha, v\right) dy + e^{-rT} K G\left(T, \sigma^{-1} \ln \frac{K}{S_0}, \alpha, v\right),$$

where $v = \sigma^{-1}(r - \frac{1}{2}\sigma^2)$ and

$$G(t, x, \alpha, v) = \int_0^{\alpha t} h(s, x, v) g(t - s, \alpha - t^{-1}s, v) \, ds$$

for every $x \neq 0$, where $h(s, x, v)$ is the density of the first hitting time of x by the process $W_t + vt$, that is

$$h(s, x, v) = \frac{|x|}{\sqrt{2\pi s}} \exp\left(\frac{(|x| - vs)^2}{2s}\right), \quad \forall s > 0.$$

6.11 Other Exotic Options

Multi-asset options. By a *multi-asset option* (or a *combined option*) we mean an option contract in which the terminal payoff depends on the price processes of several risky assets. Let us mention that options on a maximum of two risky assets enter the payoff functions of some traded securities in a straightforward way. As a typical example of such a security, we may cite the so-called *currency option-bond* that offers to its bearer the option to choose the currency in which payment is to be made. For instance, a currency option-bond could let the bearer choose at maturity between U.S. dollars and British pounds at a predetermined exchange rate written in the indenture of the bond. Many other types of commonly used financial contracts (such as *risk-sharing contracts, secured debts, compensation plans*) are contingent claims whose payoff functions include calculation of the maximum or minimum of two (or more) risky assets.

Combined options were studied by Feiger and Jacquillat (1979), who consider the *currency option-bonds*; Stulz (1982), who works out explicit formulas for prices of European options on the minimum or maximum of two risky assets using the partial differential equation method in the two-dimensional Black-Scholes framework; and Margrabe (1978), who examines options to exchange one asset for another. A general approach to the valuation of European and American multi-asset contingent claims can be found in Stapleton and Subrahmanyam (1984), Boyle et al. (1989), Cheyette (1990), Broadie and Detemple (1997a), and Veiga (2004).

Russian options. As in the case of a standard American option, the owner of a *Russian option* has the right to choose the exercise time τ. However, the Russian option pays the owner either S_τ or the maximum stock price achieved up to the exercise date, whichever is larger, discounted by $e^{-r\tau}$.

An analysis of an optimal stopping problem associated with the valuation of the Russian option was done by Shepp and Shiryaev (1993, 1994) and Kramkov and Shiryaev (1994). The formal justification of the option's arbitrage price can be found in Duffie and Harrison (1993).

Passport options. A *passport option* is a specific option contract that has the put-like payoff directly linked to the performance of a portfolio of assets held by an investor (the option's buyer). The rationale for this kind of option is to provide a protection of the holder of a traded account against the downside risk without sacrificing upside potential. The valuation of a passport option (and other options on a traded account) is related to a solution of an optimization problem. For more information on the valuation of passport options, we refer to Hyer et al. (1997), Andersen et al. (1998), Lipton (1999), Delbaen and Yor (2002), Henderson and Hobson (2000, 2001), Shreve and Vecer (2000), and Henderson et al. (2002).

7

Volatility Risk

Loosely speaking, a model of financial security can be often identified with the following three components: an underlying variable, a mechanism used to reflect the uncertainty with regard to the future value of this variable and, last but not least, arbitrage-free considerations.

Let us consider the simple case of a stock price model. The underlying variable is today's observed stock price. The simplest way to describe uncertainty with regard to its future value is to think about the (infinitesimal) standard deviation around the current level. This corresponds to the concept of the *absolute* (or *normal*) volatility. Assume that the absolute volatility of the stock price is constant. Then the usual assumption of the sample path continuity, combined with the no arbitrage requirement, lead to the Bachelier (1900) model for the stock price. Alternatively, one could define uncertainty with regard to the future stock value by concentrating on the standard deviation of logarithmic returns; that is, the *relative* (or *lognormal*) volatility. The assumption of constancy of relative volatility results in the Black and Scholes (1973) model. The choice of a particular quantity to measure the uncertainty is thus almost equivalent to specification of basic probabilistic features of a model. Within this convention, it is quite natural to link informally the risk of *model's specification* with the *volatility risk*. We shall follow this idea here, although we would like to make it clear that such a terminological link can be justified only for certain relatively restrictive classes of models, such as diffusion-type models or binomial trees, and it does not hold in general.

The chapter is organized as follows. First, in Sect. 7.1, we introduce the concepts of a *historical volatility* and of an *implied volatility*, and we present a preliminary analysis of their importance in a real-life implementation of arbitrage pricing theory. We present briefly a commonly used method of dealing with volatility risk, based on the sensitivities of the price with respect to volatility. We examine the issue of valuation of exotic options (specifically, forward-start options) under stochastic volatility. Our goal is to illustrate the well-known fact that the knowledge of the implied volatility surface is not sufficient for the unique valuation of exotic options.

In Sect. 7.2, we present a model, commonly known as the *CEV model*, in which both normal and lognormal volatilities are random. The time-homogeneous version of this model, first proposed by Cox (1975), assumes that the level of the volatility depends on the current price of the underlying asset. The model can thus accommodate for the property that the volatility may increase (or decrease) when the stock price falls (or rises). The functional dependence of the volatility on the stock price means also that the risk of the underlying and the volatility risk are closely tied to each other. This is an undesirable feature if we wish to use this model to hedge volatility risk.

Subsequently, in Sect. 7.3, we present an innovative approach, originated by Dupire (1993a, 1993b, 1994) and Derman and Kani (1994), that is based on the observation that the risk-neutral probability distribution of the underlying asset at any future date T is uniquely specified by market prices of European call (or put) options with expiration date T and strikes ranging over all possible states of the process. In the discrete-time set-up, one may build an *implied tree* for an underlying asset, as in Derman et al. (1996a). In the continuous-time framework, the method advocated by the afore-mentioned authors results in the concept of a *local volatility* (LV). Sect. 7.3.3 is devoted to a particular class of LV models, known as the *mixture models*.

In Sect. 7.4, we give an overview of diffusion-type models of *stochastic volatility* (SV). Since no specific restrictions are imposed on the exogenously given dynamics of the stochastic process representing the volatility, a number of alternate SV models were proposed by various researchers; to mention a few, Hull and White (1987a), Scott (1987, 1991), Wiggins (1987), Stein and Stein (1991), Heston (1993). Typically, the stock price dynamics and the volatility dynamics are governed by two (possibly correlated) Brownian motions, and thus the price risk and the volatility risk are partially separated (they may even be "orthogonal" if the two driving Brownian motions are independent as, for instance, in the case of Hull and White (1987a) model).

The presence of two driving Brownian motions results in a model incompleteness, and the associated phenomenon of non-uniqueness of a martingale measure. Hence the issue of the correct specification of a risk-neutral probability used for pricing of derivatives arises. In the context of classical examples of SV models, such as that of Hull and White model or Heston's model, the choice of a martingale measure is usually done by a judicious specification of the *market price of volatility risk*. We also present in this section the SABR model, which is a stochastic volatility extension of the CEV model. It is known to provide a relatively good fit to observed prices of some interest rate options. More importantly, it also seems to generate the dynamics of the implied volatility smile more consistently with its historical behavior.

Sect. 7.6 concludes the chapter with a brief overview of alternative approaches to volatility risk, such as: random time change, modelling of an underlying asset by means of a Lévy process, as well as direct modelling of the implied volatility surface.

7.1 Implied Volatilities of Traded Options

All potential practical applications of the Black-Scholes formula hinge on knowledge of the volatility parameter of the stock price returns. Indeed, of the five variables necessary to specify the model, all are directly observable except for the stock price volatility. The most natural, albeit somewhat naive, approach to the volatility specification uses an estimate of the standard deviation based upon an ex-post series of returns from the underlying stock. Needless to say that this approach is based on the implicit belief that the real-world dynamics of the stock price are described by means of a continuous semimartingale, with a specific form of the quadratic variation, namely, $d\langle S \rangle_t = \sigma^2 S_t^2 \, dt$ for some constant $\sigma > 0$.

7.1.1 Historical Volatility

In the first empirical tests of the Black-Scholes option pricing formula, performed by Black and Scholes (1972), the authors used over-the-counter data covering the 1966-1969 period. The stock volatilities were estimated from daily data over the year preceding each option price observation. They concluded that the model overpriced (respectively, underpriced) options on stocks with high (respectively, low) historical volatilities. More generally, they suggested that the usefulness of the model depends to a great extent upon investors' abilities to make good forecasts of the volatility. In subsequent years, the model was tested by several authors on exchange-traded options (see, for instance, Galai (1977)) confirming the bias in theoretical option prices observed originally by Black and Scholes. Although the estimation of stock price volatility from historical data is a fairly straightforward procedure, some important points should be mentioned. Firstly, to reduce the estimation risk arising from the sampling error, it seems natural to increase the sample size, e.g., by using a longer series of historical observations or by increasing the frequency of observations. It is thus important to mention that, unfortunately, there is strong empirical evidence to suggest that the variance is non-stationary, so that extending the observation period may make matters even worse. Furthermore, in many cases only daily data are available, so that there is a limit on the number of observations available within a given period. Finally, since the option pricing formula is non-linear in the standard deviation, an unbiased estimate of the standard deviation does not produce an unbiased estimate of the option price. To summarize, since the volatility is usually unstable through time, historical precedent is a poor guide for estimating future volatility. Moreover, estimates of option prices based on historical volatilities are systematically biased. For detailed studies of the historical volatility method, we refer to Boyle and Ananthanarayanan (1977), Parkinson (1980), Ball and Torous (1984), Butler and Schachter (1986), French and Roll (1986), Marsh and Rosenfeld (1986), Skinner (1989), Rogers and Satchell (1991), Chesney et al. (1993b), and Levy and Yoder (1996).

7.1.2 Implied Volatility

Alternatively, one can infer the market's consensus outlook as to the volatility of a given asset by examining the prices at which options on that asset trade. Since the Black-Scholes call (respectively, put) option price appears to be a strictly increasing (respectively, strictly decreasing) function of the underlying stock volatility, and all other factors determining the option price are known with certainty, one can infer the volatility that is implicit in the observed market price of an option.

More specifically, let us consider at time t a European call option with exercise price K and the time $\tau = T - t$ to the expiration date T. Suppose that the current market price C_t^m of the option is observed, and thus it can be taken as an input. Then the *implied volatility* at time t, denoted as $\hat{\sigma}_t$, is derived from the non-linear equation

$$C_t^m = S_t N\big(d_1(S_t, \tau, \hat{\sigma}_t, K, r)\big) - K e^{-r\tau} N\big(d_2(S_t, \tau, \hat{\sigma}_t, K, r)\big), \qquad (7.1)$$

where the only unknown quantity is $\hat{\sigma}_t$. This means that the implied volatility $\hat{\sigma}$ is the value that, when put in the Black-Scholes formula, results in a model price equal to the current market price of a call option. Of course, the same reasoning may by applied to a put option; in this case, equation (7.1) becomes

$$P_t^m = K e^{-r\tau} N\big(-d_2(S_t, \tau, \hat{\sigma}_t, K, r)\big) - S_t N\big(-d_1(S_t, \tau, \hat{\sigma}_t, K, r)\big), \qquad (7.2)$$

where P_t^m is the market price of a put. Due to the put-call parity relationship (which, as well known, does not depend on the choice of model, but hinges on the no-arbitrage arguments only) it is to be expected that the implied volatility of a call option, and that of a put option with the same strike and expiration date, should co-incide (or at least should be very close to each other). Let us finally, mention that the quantity $\hat{\sigma}_t$ obtained from (7.1) should be called the *market-based Black-Scholes implied volatility*, rather than simply the *implied volatility*. In the special case of $r = 0$ (this corresponds to forward price dynamics), it is more convenient to term $\hat{\sigma}_t$ the *Black implied volatility*. The distinction between the Black-Scholes implied volatility and the Black implied volatility is, mathematically speaking, absolutely unnecessary in any model in which interest rates are deterministic, however.

Other conventions related to the right-hand side of (7.1) are also possible, and used in the literature. For instance, if instead of the Black-Scholes result as a conventional options pricing formula, we use the Bachelier formula (3.77), so that (7.1) becomes

$$C_t^m = (S_t - K)N\big(d(S_t, \tau, \hat{\sigma}_t, K, r)\big) + \hat{\sigma}_t \sqrt{\tau} n\big(d(S_t, \tau, \hat{\sigma}_t, K, r)\big), \qquad (7.3)$$

then the unique value $\hat{\sigma}_t$ that satisfies (7.3) will, of course, be termed the *Bachelier implied volatility*. Let us mention in this regard that some authors prefer to use the terms: relative (or lognormal) volatility and absolute (normal) volatility for the Black-Scholes and the Bachelier volatility respectively.

7.1.3 Implied Volatility Versus Historical Volatility

Before we proceed to modelling issues, let us say few words about econometric studies of the Black-Scholes implied volatility. Early research (Latané and Rendleman (1976), Schmalensee and Trippi (1978), Beckers (1981)) found that the estimates of the actual volatility based on market implied volatilities outperform, at least in terms of their predictive power, more straightforward estimates based on historical data. Contrary to these findings, subsequent studies of stock index options (reported in Canina and Figlewski (1992), Day and Lewis (1992), and Lamoureux and Lastrapes (1993)) suggest that the implied volatility has virtually no correlation with future volatility. These results may be explained by the presence of relatively high transaction costs in the case of stock options; the market prices may thus deviate substantially from the theoretical prices based on a model assuming the absence of transaction costs.

In a recent study by Szakmary et al. (2003), the authors focus on futures prices and options. They conclude that the implied volatilities outperform historical volatility as a predictor of realized volatility in the underlying futures prices, thus confirming Jorion's (1995) conjecture. This can be related to the fact that futures options markets, due to minimal trading frictions, are more efficient than other options markets.

Interesting conclusions were also obtained by researchers who studied the impact of transaction costs on implied volatilities and Black-Scholes option prices. More specifically, Jarrow and O'Hara (1989) considered the relation between the implied volatility and transaction costs. They conjectured that the difference between an option's implied volatility and the historical volatility of the underlying stock should reflect the transaction costs of a dynamic hedge. Swidler and Diltz (1992) research, which takes into account the bid-ask spread, suggests that using a nonconstant volatility model – such as the CEV model (see Sect. 7.4) – would be more appropriate to price long-term options. The implied volatility, considered as a function of the option's strike price, sometimes exhibits a specific U-shape (cf. Rubinstein (1985), Shastri and Wethyavivorn (1987) or Taylor and Xu (1993)). One of the long-standing problems has been how to reconcile this peculiar feature of empirical option prices, referred to as the *smile effect,* with the Black-Scholes model. A typical solution to this problem relies on a judicious choice of a discrete- or continuous-time model for stock price returns. Let us mention in this context the growing interest in financial modelling based on *stable processes*, first proposed by Mandelbrot (1963) to address the departure of returns from normality. In the first place, these alternative models are focused on fitting the observed asset returns. Thus, models produced within this stream of research are econometric models based on statistical data. For more information on applications of stable distribution in financial modelling, we refer to Blattberg and Gonedes (1974), Hsu et al. (1974), Mittnik and Rachev (1993), Cheng and Rachev (1995), Popova et al. (1995), and Hurst et al. (1999).

7.1.4 Approximate Formulas

We continue the study of the notion of Black-Scholes implied volatility. We start by considering a rather elementary, but practically important, question of solving equation (7.1) for an unknown value of the volatility parameter σ, all other variables being fixed. The following properties of the Black-Scholes price of a call option are easy to check

$$\lim_{\sigma \to 0} C_0 = (S_0 - Ke^{-rT})^+, \quad \lim_{\sigma \to \infty} C_0 = \infty.$$

Moreover, as shown in Proposition 1.8.1, the value $(S_0 - Ke^{-rT})^+$ is actually the no-arbitrage lower bound for the price of a call option (notice this bound is universal, that is, it is a model-free quantity), it is natural to expect that the market price will never fall below this level. Finally, the partial derivative of the option price with respect to the volatility parameter σ is strictly positive, since it equals (see Sect. 3.1.11)

$$c_\sigma(S_0, T, K, \sigma, r) = S_0\sqrt{T}n(d_1(S_0, T, K, \sigma, r)) > 0,$$

where n is the standard Gaussian probability density function. We conclude that for any observed market price $C_0^m \geq (S_0 - Ke^{-rT})^+$ of the option with strike K and maturity T, the non-linear equation

$$C_0^m = S_0 N(d_1(S_0, T, \sigma, K, r)) - Ke^{-rT} N(d_2(S_0, T, \sigma, K, r))$$

possesses a unique solution $\hat{\sigma} = \hat{\sigma}_0(T, K)$ (recall that the interest rate r is assumed here to be fixed and known).

Unfortunately, there is no explicit solution to this equation, and thus one needs to make use of some numerical method, such as, for instance, the Newton-Raphson method (let us mention that Manaster and Koehler (1982) discuss the choice of a starting value for the first iteration). It is thus important to notice that some authors, including Brenner and Subrahmanyam (1988) and Corrado and Miller (1996), provide explicit approximate formulas for the Black-Scholes implied volatility. To derive the Corrado-Miller formula, we start with the expansion of the standard Gaussian cumulative distribution function

$$N(x) = \frac{1}{2} + \frac{1}{\sqrt{2\pi}}\left(x - \frac{x^3}{6} + \frac{x^5}{40} + \dots\right).$$

By substituting this expansion into the Black-Scholes pricing formula, we obtain the following approximation for the call price (note that cubic and higher order terms have been ignored)

$$C_0 \approx S_0\left(\frac{1}{2} + \frac{d_1(S_0, T)}{\sqrt{2\pi}}\right) - Ke^{-rT}\left(\frac{1}{2} + \frac{d_2(S_0, T)}{\sqrt{2\pi}}\right).$$

Let us set $\tilde{K} = Ke^{-rT}$. After standard manipulations, we arrive at the following quadratic equation in the quantity $\tilde{\sigma} = \sigma\sqrt{T}$

$$\tilde{\sigma}^2(S_0 + \tilde{K}) - \tilde{\sigma}\sqrt{8\pi}(C_0 - \tfrac{1}{2}(S_0 - \tilde{K})) + 2(S_0 - \tilde{K})\ln(S_0/\tilde{K}) = 0.$$

The larger root is given by

$$\tilde{\sigma} = \sqrt{2\pi}\left(\frac{C_0 - \frac{1}{2}\tilde{S}_0}{S_0 + \tilde{K}}\right) + \sqrt{2\pi\left(\frac{C_0 - \frac{1}{2}\tilde{S}_0}{S_0 + \tilde{K}}\right)^2 - \frac{2\tilde{S}_0\ln(S_0/\tilde{K})}{S_0 + \tilde{K}}},$$

where $\tilde{S}_0 = S_0 - \tilde{K}$. In particular, when the current stock price S_0 equals the discounted exercise price (i.e., we deal with an at-the-money forward call), the formula above reduces to the original Brenner-Subrahmanyam formula, which reads

$\tilde{\sigma} = \sqrt{2\pi}\, C_0/S_0$. It appears that a further gain in accuracy can be obtained by using the following approximation

$$\ln(S_0/\tilde{K}) \approx \frac{2(S_0 - \tilde{K})}{S_0 + K} = \frac{2\tilde{S}_0}{S_0 + K}$$

and by substituting the number 4 with the parameter α. In this way, we arrive at the following improved approximate formula (which still reduces to the Brenner-Subrahmanyam formula if $S_0 = Ke^{-rT}$)

$$\tilde{\sigma} \approx \sqrt{2\pi} \left(\frac{C_0 - \frac{1}{2}\tilde{S}_0}{S_0 + \tilde{K}} \right) + \sqrt{2\pi \left(\frac{C_0 - \frac{1}{2}\tilde{S}_0}{S_0 + \tilde{K}} \right)^2 - \alpha \left(\frac{\tilde{S}_0}{S_0 + \tilde{K}} \right)^2}.$$

A judicious choice of an auxiliary parameter α allows us to improve the overall accuracy of the last approximation, without affecting at-the-money accuracy. Corrado and Miller (1996) argue that $\alpha = 2$ is a reasonable choice. By inserting this value into last formula, we obtain yet another approximation for the implied volatility of a call option

$$\tilde{\sigma} \approx \frac{\sqrt{2\pi}}{S_0 + \tilde{K}} \left(C_0 - \tfrac{1}{2}(S_0 - \tilde{K}) + \sqrt{\left(C_0 - \tfrac{1}{2}(S_0 - \tilde{K}) \right)^2 - \tfrac{1}{\pi}(S_0 - \tilde{K})^2} \right).$$

For a discussion of the accuracy of the last formula, we refer to the original paper by Corrado and Miller (1996).

7.1.5 Implied Volatility Surface

So far, it has been implicitly assumed that the strike and maturity of an option are fixed, and thus equality (7.1) (or formula (7.3), depending on our choice of a market convention) yielded a single number, called the implied volatility. Note, however, that the actual value of the implied volatility $\hat{\sigma}$ of a call (or a put) option determined in this way may depend, in general, not only on a choice of a conventional model, but also on the option's contractual features – that is, on the strike K and the time to maturity T. Given a cross-section of current market prices at time $t = 0$ of call options with different strikes and maturities, we therefore find a parametrized family of implied volatilities $\hat{\sigma}_0(T, K)$, where the parameter K ranges over all available strikes and the parameter T runs over all available maturities of traded options. It is convenient to assume that the strike K is any positive number, and the maturity T is any date prior to some horizon date T^*. In this way, we arrive at the concept of the *implied volatility surface*.

Definition 7.1.1 Let $C_0^m(T, K)$ stand for the family of market prices of European call options with all strikes $K > 0$ and all maturities $0 < T \le T^*$ for some $T^* > 0$. Then the *market-based Black-Scholes implied volatility surface* $\hat{\sigma}_0(T, K)$ is implicitly defined through the equation

$$C_0^m(T, K) = c(S_0, T, K, r, \hat{\sigma}_0(T, K)), \qquad (7.4)$$

where $c(S_0, T, K, r, \sigma)$ is the Black-Scholes price of a call option. If $C_0^m(T, K)$ is replaced by prices $C_0(T, K)$ given by some stochastic model, we say that the implied volatility surface is *model-based*.

The parameter r is assumed to be constant in this chapter, and thus it plays no essential role. Definition 7.1.1 can be easily adapted to cover the case of random interest rates. Rather than focus on the spot price, it suffices to consider the forward price of the underlying asset and the associated implied volatility. In such a case, one deals with a variant of Black's formula, and thus r disappears from the right-hand side of (7.4).

According to Definition 7.1.1, the implied volatility surface $\hat{\sigma}_0$ is a mapping $\hat{\sigma}_0$: $(0, T^*] \times \mathbb{R}_+ \rightarrow \mathbb{R}_+$. Let us note, however, that the implied volatility formally depends also on the current level of the price of the underlying asset (we still maintain our standing assumption that the interest rate r is fixed). Hence the more adequate notation in such a case would be $\hat{\sigma}_0(S_0, T, K)$, rather than $\hat{\sigma}_0(T, K)$, especially when one is interested in random fluctuations of this surface when time elapses.

In this subsection, we shall examine only briefly the issues related to the static shape of a model-based implied volatility surface $\hat{\sigma}_0(T, K)$. The issue of the arbitrage-free dynamics of a model-based implied volatility surface is postponed to Sect. 7.6, in which some preliminary results related to the modelling of dynamics of implied volatility surface are mentioned.

Observe that Definition 7.1.1 can be easily reformulated to be conform with the Bachelier model. To this end, it suffices to replace the Black-Scholes pricing function c by the Bachelier pricing function c_B. Since in the Bachelier model the stock price ranges over all real values, the implied volatility surface $\hat{\sigma}_0$ should now be interpreted as a mapping from $(0, T^*] \times \mathbb{R}$ to $\mathbb{R}_+$.

More generally, if a considered model is based on a parameter other than the volatility, denoted by α say, then we may equally well define the corresponding implied surface $\hat{\alpha}_0(T, K)$. Nevertheless, since we adopt the convention of treating the Black-Scholes model (equivalently, the Black model) as a benchmark continuous-time model, unless explicitly stated otherwise we shall mean by the *implied volatility surface* the Black-Scholes implied volatility surface, as given by Definition 7.1.1.

Assume first that the maturity date T is fixed. Then the mapping $K \mapsto \hat{\sigma}_0(T, K)$ is termed the *implied volatility curve* for the maturity date T. If implied volatilities for low strikes are higher than implied volatilities for higher strikes, the shape is termed a *skew*. The term *smile* is used when the function $K \mapsto \hat{\sigma}_0(T, K)$ has a minimum, usually for the value of the strike K lying not far away from the forward value of the underlying asset. Consequently, the dependence of the shape of the implied volatility curve on the maturity date T is referred to as either as the *term structure of volatility smiles* or the *term structure of volatility skews*.

Consider the rather unlikely situation when the implied volatility $\hat{\sigma}_0(T, K)$ inferred from call option prices is flat in K; that is, for each maturity date T the implied volatility $\hat{\sigma}_0(T, K)$ does not depend on strike K. We thus deal with the term structure of volatilities given by some function $\hat{\sigma} : (0, T^*] \rightarrow \mathbb{R}_+$. Note that in this case, in order to ensure an exact match to market data it suffices to consider a simple extension of the Black-Scholes model with time-dependent (but state-independent) volatility function $\hat{\sigma} : \mathbb{R}_+ \to \mathbb{R}_+$

$$dS_t = S_t \big(r \, dt + \hat{\sigma}(t) \, dW_t^* \big),$$

where the volatility function $\hat{\sigma}(t)$ satisfies, for every $T \in [0, T^*]$,

$$\hat{\sigma}_0^2(T, K) = \frac{1}{T} \int_0^T \hat{\sigma}^2(u) \, du. \tag{7.5}$$

Of course, the equality above is valid for any choice of $K > 0$, since under the present assumptions the right-hand side does not depend on K. The existence of the function $\hat{\sigma}$ satisfying (7.5) is obvious, provided that the function $\hat{\sigma}_0^2(T, K)T$ is increasing in T and sufficiently smooth (recall that $\hat{\sigma}_0(T, K)$ is now assumed to not depend on K). It is a simple exercise to check that the Black-Scholes price of each call option on a stock is equal to its market price, as specified by the implied volatility. Formally, we have that

$$\mathbb{E}_{\mathbb{P}^*}\big(e^{-rT}(S_T - K)^+\big) = c\big(S_0, T, K, r, \hat{\sigma}_0(T, K)\big) = C_0^m(T, K),$$

where c is the Black-Scholes pricing function for a call option. We conclude that the model-based implied volatilities match exactly the market-based volatilities for all call options. Let us stress that the derivation of the implied volatility surface from the market data is a highly non-trivial problem. Since only a finite collection of prices of call and put options are observed in practice, equality (7.4) yields the implied volatility parameter for a finite family of dates and maturities only. Consequently, it is not possible to determine uniquely the whole surface of implied volatilities on the basis of available market data. Alternative methodologies of inferring the shape of the implied volatility surface were examined by, among others, Andersen and Brotherton-Ratcliffe (1997), Avellaneda et al. (1997), Lagnado and Osher (1997), Dumas et al. (1998), Bodurtha and Jermakyan (1999), Tompkins (2001), Berestycki et al. (2002, 2004), Cont (2002), Cont and Tankov (2002, 2003), Samperi (2002), and Crépey (2003).

7.1.6 Asymptotic Behavior of the Implied Volatility

It is a challenging theoretical and practical issue how to describe an arbitrage-free shape of the implied volatility surface. Let us fix T and let us assume that $S_0 > 0$ and that we are given a nonnegative random variable S_T under a probability measure $\mathbb{P}^*$ that is interpreted as a risk-neutral probability. Thus, we have $S_0 = B_0 \, \mathbb{E}_{\mathbb{P}^*}(B_T^{-1} S_T)$ and for every strike $K > 0$, the call price $C_0(T, K)$ at time 0 is given by the formula

$$C_0(T, K) = B_0 \, \mathbb{E}_{\mathbb{P}^*}\big(B_T^{-1}(S_T - K)^+\big).$$

Our first goal is to derive no-arbitrage bounds on the Black-Scholes implied volatility under fairly general assumptions. Let us mention that a similar set-up will be used in Sect. 7.3 to derive the so-called local volatility function.

We may and do assume without loss of generality that $r = 0$, or equivalently, that $B_0 = B_T = 1$. In the case of random interest rates, it can be shown (see Sect. 9.6) that there exists a probability measure $\mathbb{P}_T$ equivalent to $\mathbb{P}^*$ on $(\Omega, \mathcal{F}_T)$, termed the *forward measure*, such that the forward price of S for the settlement date T is a martingale under $\mathbb{P}_T$ and

$$C_0(T, K) = B(0, T)\,\mathbb{E}_{\mathbb{P}_T}(S_T - K)^+.$$

The shape of the implied volatility surface was recently examined by Hodges (1996), Gatheral (1999), Gatheral et al. (2000), and Lee (2004a). Let us briefly describe the findings of the last author. Lee (2004a) focuses on the asymptotic behavior of the implied volatility in terms of the moments of the random variable S_T. He first establishes the following auxiliary result (similar upper bounds were previously derived by Broadie et al. (1998)).

Proposition 7.1.1 *For any $p > 0$ and $K > 0$, for a call option we have*

$$C_0(T, K) \leq \frac{\mathbb{E}_{\mathbb{P}^*}\left(S_T^{p+1}\right)}{(p+1)K^p}\left(\frac{p}{p+1}\right)^p \tag{7.6}$$

and the price of a put option satisfies

$$P_0(T, K) \leq \frac{\mathbb{E}_{\mathbb{P}^*}\left(S_T^{-p}\right)}{(p+1)K^{p+1}}\left(\frac{p}{p+1}\right)^p, \tag{7.7}$$

where, by convention, $S_T^{-p} = \infty$ on the event $\{S_T = 0\}$.

Proof. It is not difficult to check that

$$(s - K)^+ \leq \frac{s^{p+1}}{(p+1)K^p}\left(\frac{p}{p+1}\right)^p.$$

Indeed, both sides have the same value for $s = (p+1)K/p$, and the right-hand side has positive second order derivative. Similarly,

$$(K - s)^+ \leq \frac{s^{-p}}{(p+1)K^{p+1}}\left(\frac{p}{p+1}\right)^p.$$

By taking expected values, we achieve the proof.   □

Remarks. Note that if $\mathbb{E}_{\mathbb{P}^*}\left(S_T^{p+1}\right) < \infty$ then $C_0(T, K) = O(K^{-p})$ as $K \to \infty$. Similarly, $\mathbb{E}_{\mathbb{P}^*}\left(S_T^{-p}\right) < \infty$ then $P_0(T, K) = O(K^{p+1})$ as $K \to 0$.

To analyze the asymptotic behavior of the implied volatility, it is convenient to introduce an auxiliary variable $y = \ln(S_0/K)$, commonly referred to as the *log-moneyness*, and to set $x = -y = \ln(K/S_0)$ (recall that we assume that $r = 0$; in general, the spot price S_0 should be replaced by the forward price of the stock for the settlement date T).

For any fixed maturity date $T > 0$, the Black-Scholes implied volatility at log-moneyness x is defined as the unique solution $\hat{\sigma}_0(x)$ of the equation $C_0(T, S_0 e^x) = c(S_0, T, S_0 e^x, \hat{\sigma}_0(x))$. The first result furnishes the bound on the slope of the right-hand tail of the square of the implied volatility.

Lemma 7.1.1 *There exists a real number $x^* > 0$ such that for every $x > x^*$ we have $\hat{\sigma}_0(x) < \sqrt{2x/T}$.*

Proof. Since the Black-Scholes price $c(S_0, T, K, \sigma)$ is increasing in σ, it suffices to check that $c(S_0, T, S_0 e^x, \hat{\sigma}_0(x)) < c(S_0, T, S_0 e^x, \sqrt{2x/T})$ for x large enough. This inequality holds since, on the one hand, by the dominated convergence theorem, we get

$$\lim_{x \to \infty} c(S_0, T, S_0 e^x, \hat{\sigma}_0(x)) = \lim_{x \to \infty} C_0(T, S_0 e^x) = \lim_{x \to \infty} \mathbb{E}_{\mathbb{P}^*}(S_T - S_0 e^x)^+ = 0,$$

and, on the other hand, using the Black-Scholes formula, we obtain

$$\lim_{x \to \infty} c(S_0, T, S_0 e^x, \sqrt{2x/T}) = S_0\big(N(0) - \lim_{x \to \infty} e^x N(-\sqrt{2x})\big) = S_0/2 > 0,$$

where the second equality follows by l'Hôpital rule. □

The next result yields the first *moment formula* for implied volatility. Let us denote

$$\tilde{p} = \sup\{p > 0 \mid \mathbb{E}_{\mathbb{P}^*}\big(S_T^{p+1}\big) < \infty\}, \quad \tilde{\beta} = \limsup_{x \to \infty} T x^{-1} \hat{\sigma}_0^2(x).$$

Since the proof of the next result is elementary, but rather tedious, it is omitted (see Lee (2004a, 2004b)).

Proposition 7.1.2 *For any $\tilde{p} < \infty$, we have that*

$$\tilde{\beta} = 2 - 4\big(\sqrt{\tilde{p}^2 + \tilde{p}} - \tilde{p}\big).$$

Moreover, $\tilde{\beta} = 0$ if $\tilde{p} = \infty$.

Remarks. Hodges (1996) and Gatheral (1999) focus on the slope of the implied volatility, and they derive the bound by analyzing an ordinary differential equation satisfied by the implied volatility. The approach developed by Lee (2004a) is more straightforward, and his results are sharper.

Let us now consider the left-hand (i.e., small strikes) tail of the square of implied volatility. Again, the interested reader is referred to Lee (2004a) for the proofs of foregoing two results.

Lemma 7.1.2 *For any $\beta > 2$, there exists a real number x^* such that for every $x < x^*$ we have $\hat{\sigma}_0(x) < \sqrt{\beta|x|/T}$. For $\beta = 2$, the same conclusion is valid if and only if $\mathbb{P}^*\{S_T = 0\} < 1/2$.*

Let

$$\hat{p} = \sup\{p > 0 \mid \mathbb{E}_{\mathbb{P}^*}\big(S_T^{-p}\big) < \infty\}, \quad \hat{\beta} = \limsup_{x \to -\infty} T|x|^{-1}\hat{\sigma}_0^2(x).$$

The second *moment formula* for implied volatility is provided by the following result.

Proposition 7.1.3 *For any $\hat{p} < \infty$, we have that*

$$\hat{\beta} = 2 - 4\big(\sqrt{\hat{p}^2 + \hat{p}} - \hat{p}\big).$$

In addition, $\hat{\beta} = 0$ if $\hat{p} = \infty$.

Let us first observe, that the existence of moments of the underlying random variable can be deduced from the properties of a postulated model. The maximal moment exponents $\tilde{p}$ and $\hat{p}$ are therefore readily available for many popular models.

The asymptotic formulas given above can be applied to extrapolation of the implied volatility skew (smile), as well as to model calibration.

First, when dealing with an extrapolation of the volatility smile beyond the actively traded strikes, it is not recommended, in view of Propositions 7.1.2 and 7.1.3, to use functional forms that allow either tail to grow faster than $|x|^{1/2}$. In addition, unless the underlying variable has all moments, it is advisable not to use functional forms which allow either tail to grow slower than $|x|^{1/2}$.

Second, the moment formulas facilitate the calibration of model parameters to observed volatility skews. Suppose that we can observe the tail slopes of the skew, so that the maximal moment exponents $\tilde{p}$ and $\hat{p}$ are available. For a chosen model for the underlying asset, this produces two identifying restrictions on the model's parameters. To illustrate this important feature, Lee (2004a) examines the following two particular examples: the double-exponential jump diffusion model (see Kou (2002)) and the normal inverse Gaussian model (see Barndorff-Nielsen (1998)).

In the former case, the asset price follows a geometric Brownian motion with jumps at event times of a Poisson process. The sizes of up-jumps and down-jumps in returns are exponentially distributed with the parameters η_1 and η_2 respectively. It appears that the equalities $\tilde{p} = \eta_1 - 1$ and $\hat{p} = \eta_2$ are valid. We are therefore able to infer the values of η_1 and η_2 from the values of maximal moment exponents $\tilde{p}$ and $\hat{p}$, and thus also from the tail slopes of the implied volatility skew that are, in principle, observable.

In the latter example, the asset returns have the normal inverse Gaussian distribution with four parameters, denoted as $NIG(a, b, c, d)$. It appears that two of them, namely b and c, can be inferred from $\tilde{p}$ and $\hat{p}$. The other two parameters have no influence on the tail slopes.

7.1.7 Marked-to-Market Models

By an *actively traded option* (or a *liquid option*) we will invariably mean an option contract for which there exists a liquid market, so that an option can be sold or purchased at any time and the bid-ask spread can be neglected. It is thus clear that the valuation of liquid options does not require a mathematical model. However, hedging of positions in liquid options appears to be an issue that is not much simpler than hedging of these derivative assets for which a liquid market does not exists. Hence one of major goals of financial modelling is to develop an efficient method of hedging positions in liquid options. Another important goal is the development of a reliable pricing procedure for exotic products, based on market prices of liquid derivatives (in most cases, plain-vanilla options). This imposes a natural requirement: any mathematical model used for pricing exotic products should be *marked-to-market*; that is, at any given date it should reproduce with the desired precision the current market prices of liquid assets.

Marking-to-market is typically associated with some procedure of specifying the levels of a model's variables and parameters; for this reason it is commonly referred to as a model's *calibration*. A marked-to-market model is considered to have stability properties, from the point of view of calibration, if after fitting it to the current market prices of liquid assets the remaining parameters can be kept constant (or almost constant) over a reasonably long period of time. It should be stressed, however, that no matter how sophisticated a model is used, we will need to update its parameters as soon as market conditions change dramatically. We summarize the considerations above by making an attempt to identify the most important features of an acceptable model for pricing and hedging derivatives. Modern practical implementations of arbitrage pricing theory seem to rely on the following paradigms:

1. Market inputs
Liquid derivative contracts play almost identical role as the underlying asset in the classical Black-Scholes approach. In particular, the prices of these contracts are considered to be given by the market, together with the price of the underlying.

2. Marking-to-market
A model used for pricing of exotic derivatives should at any given date be fitted to liquid market prices of related contracts.

3. Relative pricing
Non-liquid (exotic) products are priced and hedged relatively to the liquid price information with regard to the underlying and relevant derivatives.

4. Dynamical features
A model used for pricing and hedging is expected to have a potential to reproduce the observed typical behavior of all pertinent risk factors. Ideally, it should take as an underlying variable all liquid assets and aim at stability over time of parameters used to describe the uncertainty about the future value of this variable.

7.1.8 Vega Hedging

Suppose that a trader has opted for the use of the classical Black-Scholes model. Then the following pertinent questions arise:

– how to reconcile hedging of volatility risk with the assumption that the volatility is constant in the Black-Scholes world?

– which implied volatility should one use for pricing and hedging exotic derivatives such as, for instance, barrier options?

We shall first address these questions from the practical, rather than the theoretical, perspective. We thus acknowledge that the Black-Scholes model is not correct, but we insist at the same time on the use of the Black-Scholes formula.

Vega. As was mentioned in Sect. 3.1.11, the derivative of the option price with respect to (implied or actual) volatility is known as the *vega* of an option. In the classical Black-Scholes model, it measures the sensitivity of the option value to small changes in the volatility parameter σ. We also know

$$c_\sigma(s, \tau, K, r, \sigma) = S_t \sqrt{\tau}\, n\big(d_1(s, \tau, K, r, \sigma)\big) = p_\sigma(s, \tau, K, r, \sigma),$$

where $c(s, \tau, K, r, \sigma)$ and $p(s, \tau, K, r, \sigma)$ stand for the Black-Scholes pricing functions for a call and a put options, and n is the standard Gaussian probability density function. Hence the vega of both a call and a put is strictly positive at any time over the lifetime of the option. This means simply that the prices of both options always rise with the increase of the volatility σ.

An intuitive meaning of the vega parameter is thus rather clear: a positive vega position should result in profits from increases in volatility; similarly, a negative vega means a strategy should profit from falling volatility. It is definitely less obvious how to reconcile formally the use of Black-Scholes vega as a tool to study the impact of random fluctuations of volatility on the value of a derivative asset, with the assumption of constancy of the volatility within the Black-Scholes set-up.

Volga and vanna. Following the above logic, it is also natural to make an attempt to quantify the second order and cross effects on the value if the derivative security due to the fluctuations of σ. For this purpose, it is quite common to use the following partial derivatives

$$c_{\sigma\sigma}(s, \tau, K, r, \sigma) = \frac{s \sqrt{\tau}\, n(d_1) d_1 d_2}{\sigma} = p_{\sigma\sigma}(s, \tau, K, r, \sigma)$$

and

$$c_{s\sigma}(s, \tau, K, r, \sigma) = -\frac{n(d_1) d_2}{\sigma} = -p_{s\sigma}(s, \tau, K, r, \sigma),$$

where $d_1 = d_1(s, \tau, K, r, \sigma)$ and $d_2 = d_2(s, \tau, K, r, \sigma)$ (see (3.58)). Recall also that we have $sn(d_1) = Ke^{-r\tau} n(d_2)$. Partial derivatives $c_{\sigma\sigma}$ and $c_{s\sigma}$ are known among practitioners as *volga* and *vanna*. They are perceived as intuitive measures of fluctuations of vega as market conditions change.

Hedging of vega risk. It is widely acknowledged that in a securities market with liquid plain-vanilla options, the volatility risk can be hedged by buying or selling these options in much the same way as delta hedging is used to neutralize the risks to changes in the price of the underlying asset.

The claim that the incompleteness is a major deficiency of a mathematical model should thus be taken with caution. The observation that some options do not have unique prices in an incomplete model is not relevant, if they are among liquid assets, so that their valuation is not our goal at all. Moreover, by postulating that plain-vanilla options are traded, we may formally complete an incomplete model, so that exotic options will have unique prices and can be replicated. Theoretical research in this vein was done, among others, by Romano and Touzi (1997), who focused on the PDE approach. Derman et al. (1995), Carr et al. (1998), and Brown et al. (2001a, 2001b) examine a related issue of static hedging of exotic options.

From the practical perspective, it it common to immunize the vega, the vanna and the volga of an exotic option (a barrier currency option, say) with a static combination of plain-vanilla options, and at the same time to derive the hedge ratio using a model with constant volatility. This practical approach can be seen as an ad hoc substitute of a strict procedure of a market's completion by an enlargement of a class of traded assets.

Modified delta. If the random character of σ is acknowledged, it seems intuitively justified to use the so-called *modified hedge ratio*. It aims to represent, as the standard Black-Scholes delta, the partial derivative of the option price with respect to the underlying asset. However, it is now calculated in the following way (for an example, see (7.19))

$$\Delta = \frac{\partial c}{\partial s} + \frac{\partial c}{\partial \hat{\sigma}} \frac{\partial \hat{\sigma}}{\partial s}. \tag{7.8}$$

The appearance of an additional term is due to the observation that the implied volatility depends on the level of the underlying asset. The modified delta is expected to yield a better hedging strategy than the classical Black-Scholes delta, which ignores the volatility risk completely.

From the discussion above, it is rather clear that the classical Black-Scholes model with constant volatility cannot be seen as a realistic option pricing model. Thus, its unquestionable popularity among traders requires a deeper analysis. It can be justified, for instance, by the following arguments.

First, it provides a trader with a rule that links the move of the underlying (and other risk factors) to a well specified rule of action. The concept of sensitivities (delta, gamma, vega, etc.) yields an effective tool for hedging options through dynamic revisions of a portfolio.

Second, the concept of an implied volatility provides traders with an efficient common language that synthetizes option prices in a single number. It can be used as a reference point to an ideal model in which all liquid relevant market information is taken as an underlying variable.

7.1.9 Correlated Brownian Motions

We shall now describe two alternative well-known constructions of correlated one-dimensional Brownian motions that will be used in what follows.

First method. Let $W = (W^1, W^2)$ be a two-dimensional standard Brownian motion with respect to some filtration $\mathbb{F}$. Let ρ be an arbitrary $\mathbb{F}$-progressively measurable stochastic process taking values in $[-1, 1]$. We set $W_t^* = W_t^1$ and

$$\tilde{W}_t = \int_0^t \rho_u \, dW_u^1 + \int_0^t \sqrt{1 - \rho_u^2} \, dW_u^2$$

for every $t \in \mathbb{R}_+$. One can check, using Lévy's characterization theorem (see Theorem A.10.1), that the processes W^* and $\tilde{W}$ are standard one-dimensional Brownian motions under $\mathbb{P}^*$ with respect to $\mathbb{F}$. Moreover, their cross-variation satisfies $d\langle W^*, \tilde{W} \rangle_t = \rho_t \, dt$ for $t \in \mathbb{R}_+$. Indeed, we have

$$\langle W^*, \tilde{W} \rangle_t = \int_0^t \rho_u \, d\langle W^1, W^1 \rangle_u + \int_0^t \sqrt{1 - \rho_u} \, d\langle W^1, W^2 \rangle_u = \int_0^t \rho_u \, du$$

since by the assumed independence of W^1 and W^2 their cross-variation $\langle W^1, W^2 \rangle$ vanishes, i.e., $\langle W^1, W^2 \rangle_t = 0$ for every $t \in \mathbb{R}_+$.

Second method. An alternative, slightly more general, construction runs as follows. Let $W = (W^1, \ldots, W^d)$ be a d-dimensional standard Brownian motion with respect to $\mathbb{F}$, and let b_t^i, $i = 1, \ldots, n$ be a family of $\mathbb{R}^d$-valued, $\mathbb{F}$-progressively measurable processes. Assume that $|b_t^i| \neq 0$ for every $t \in \mathbb{R}_+$ and $i = 1, \ldots, n$, where $| \cdot |$ stands for the Euclidean norm in $\mathbb{R}^d$. For each i, we define a bounded, $\mathbb{R}^d$-valued, $\mathbb{F}$-progressively measurable process $\tilde{b}^i$ by setting $\tilde{b}_t^i = b_t^i |b_t^i|^{-1}$ for every $t \in \mathbb{R}_+$. Then for each $i = 1, \ldots, n$ the process $\tilde{W}^i$ given as

$$\tilde{W}_t^i = \int_0^t \tilde{b}_u^i \cdot dW_u, \quad \forall t \in \mathbb{R}_+,$$

is a one-dimensional standard Brownian motion with respect to $\mathbb{F}$ (again, by Lévy's characterization theorem). Moreover, for every $i, j = 1, \ldots, n$ we have

$$\langle \tilde{W}^i, \tilde{W}^j \rangle_t = \int_0^t \tilde{b}_u^i \cdot \tilde{b}_u^j \, du, \quad \forall t \in \mathbb{R}_+,$$

where the dot $\cdot$ stands for the inner product in $\mathbb{R}^d$. We conclude that the infinitesimal correlation between $\tilde{W}^i$ and $\tilde{W}^j$ equals, for every $t \in \mathbb{R}_+$,

$$\rho_t^{ij} = \tilde{b}_t^i \cdot \tilde{b}_t^j = \frac{b_t^i \cdot b_t^j}{|b_t^i||b_t^j|} \in [-1, 1].$$

In the financial literature, it is common to refer to the matrix $[\rho_t^{ij}]_{1 \leq i, j \leq n}$ as the *correlation matrix* of the n-dimensional process $\tilde{W}$.

Change of a probability measure. We shall now analyze the behavior of correlated Brownian motions under an equivalent change of a probability measure. To this end, shall use a two-dimensional version of Girsanov's theorem (see Theorem A.15.1). Let $W = (W^1, W^2)$ be a standard two-dimensional Brownian motion on a filtered probability space $(\Omega, \mathbb{F}, \mathbb{P}^*)$. Let us fix $T > 0$ and let us assume that $\lambda = (\lambda^1, \lambda^2)$ is an adapted process such that

$$\mathbb{E}_{\mathbb{P}^*} \left\{ \exp \left(\int_0^T \lambda_u \cdot dW_u - \int_0^T |\lambda_u|^2 \, du \right) \right\} = 1. \tag{7.9}$$

where $|\lambda_u|^2 = (\lambda_u^1)^2 + (\lambda_u^2)^2$ and the dot "$\cdot$" stands for the usual inner product in $\mathbb{R}^2$. We define the probability measure $\hat{\mathbb{P}}$, equivalent to $\mathbb{P}$ on $(\Omega, \mathcal{F}_T)$, by postulating that the Radon-Nikodým derivative of $\hat{\mathbb{P}}$ with respect to $\mathbb{P}^*$ equals

$$\frac{d\hat{\mathbb{P}}}{d\mathbb{P}^*} = \exp\left(\int_0^T \lambda_u \cdot dW_u - \int_0^T |\lambda_u|^2 \, du\right), \quad \mathbb{P}^*\text{-a.s.} \tag{7.10}$$

Then, by virtue of Girsanov's theorem, the two-dimensional process

$$\left(W_t^1 - \int_0^t \lambda_u^1 \, du, \; W_t^2 - \int_0^t \lambda_u^2 \, du\right), \quad \forall t \in [0, T],$$

is a standard two-dimensional Brownian motion on the space $(\Omega, \mathbb{F}, \hat{\mathbb{P}})$.

Now, let W^* and $\tilde{W}$ be the two one-dimensional correlated Brownian motions, as defined at the beginning of this section. Let us assume that an adapted process $\lambda = (\lambda^1, 0)$ satisfies condition (7.9). We define the probability $\hat{\mathbb{P}}$ on $(\Omega, \mathcal{F}_T)$ through formula (7.10). Then the process

$$\hat{W}_t = W_t^* - \int_0^t \lambda_u^1 \, du = W_t^1 - \int_0^t \lambda_u^1 \, du, \quad \forall t \in [0, T],$$

as well as the process W^2, are one-dimensional Brownian motions with respect to $\hat{\mathbb{P}}$. Moreover, the process $\tilde{W}_t$ admits the following representation

$$\tilde{W}_t = \int_0^t \rho_u \, d(\hat{W}_u + \lambda_u^1 \, du) + \int_0^t \sqrt{1 - \rho_u^2} \, dW_u^2.$$

From the last formula, it is easy to conclude that the process $\bar{W}$ given by the equality

$$\bar{W}_t = \tilde{W}_t - \int_0^t \rho_u \lambda_u^1 \, du, \quad \forall t \in [0, T],$$

is also a one-dimensional Brownian motion under $\hat{\mathbb{P}}$. Finally, the cross-variation of $\hat{W}$ and $\bar{W}$ equals

$$\langle \hat{W}, \bar{W} \rangle_t = \langle W^*, \tilde{W} \rangle_t = \int_0^t \rho_u \, du, \quad \forall t \in [0, T].$$

As expected, it is invariant with respect to an equivalent change of the underlying probability measure.

7.1.10 Forward-start Options

We shall now address the second question formulated at the beginning of Sect. 7.1.8, that is, the impact of the choice of an (implied) volatility on the value of an exotic option. To illustrate the utmost importance of a volatility specification on the valuation of exotic options, we shall re-examine a *forward-start option* (see Sect. 6.2). Recall that, for two dates $0 < T_0 < T$, a *forward-start call option* is a a call option with expiry date T and strike equal to S_{T_0} set at time T_0. The contract can be formally identified with the payoff

$$\mathbf{FS}_T = (S_T - K S_{T_0})^+$$

at time T. Equivalently, it may be seen as the payoff

$$\mathbf{FS}_{T_0} = C_{T_0}(S_{T_0}, T - T_0, K S_{T_0})$$

at time T_0. Note that in the right-hand side of the last formula, we have the price of a plain-vanilla call option evaluated at time T_0.

Case of a constant volatility. In Sect. 6.2, we have shown that, within the framework of the classical Black-Scholes model, the price at time T_0 of a forward-start option satisfies

$$C_{T_0}(S_{T_0}, T - T_0, K S_{T_0}) = S_{T_0} c(1, T - T_0, K, r, \sigma),$$

and thus its value at time 0 equals

$$\mathbf{FS}_0 = S_0 c(1, T - T_0, K, r, \sigma) = c(S_0, T - T_0, K S_0, r, \sigma).$$

Case of a deterministic volatility. Suppose now that we deal with a generalized Black-Scholes model with deterministic volatility $\sigma(t)$. We have argued in Sect. 7.1.5 that such a model is capable of explaining an implied volatility surface with flat smile. We now obtain

$$C_{T_0}(S_{T_0}, T - T_0, K S_{T_0}) = S_{T_0} c(1, T - T_0, K, r, \sigma(T_0, T)),$$

where the *average future volatility* $\sigma(T_0, T)$ is such that

$$\sigma^2(T_0, T) = \frac{1}{T - T_0} \int_{T_0}^{T} \sigma^2(t) \, dt.$$

Consequently,

$$\mathbf{FS}_0 = S_0 c(1, T - T_0, K, r, \sigma(T_0, T)) = c(S_0, T - T_0, K S_0, r, \sigma(T_0, T)).$$

Assuming flat implied volatility smiles for T_0 and T, we may infer the *forward implied volatility* $\hat{\sigma}(T_0, T)$ from the implied volatility surface $\hat{\sigma}_0(T, K)$, since (see (7.5))

$$\hat{\sigma}^2(T_0, T) = \frac{T \hat{\sigma}_0^2(T, K) - T_0 \hat{\sigma}_0^2(T_0, K)}{T - T_0},$$

where, by assumption, the implied volatilities $\hat{\sigma}_0^2(T, K)$ and $\hat{\sigma}_0^2(T_0, K)$ are independent of the strike K.

Case of a random volatility. In the presence of the volatility smile, it is no longer possible to derive uniquely the forward volatility from the implied volatility surface. To deal with this case, we make the generic assumption that S satisfies

$$dS_t = S_t \left(r \, dt + \sigma_t \, dW_t^* \right)$$

for some stochastic volatility process σ that is not necessarily adapted to the filtration generated by a one-dimensional standard Brownian motion W^* under a martingale measure $\mathbb{P}^*$ (for explicit examples of such models, see Sect. 7.4 below). We observe that the discounted stock price $S_t^* = S_t/B_t$ is a $\mathbb{P}^*$-martingale.

Suppose that the volatility process σ is given, and we are in a position to find a closed-form solution for the price $C(S_{T_0}, T - T_0, K S_{T_0})$ in this model. In order to find the price of a forward-start option at time $t \in [0, T_0]$, we need to compute the following conditional expectation

$$\mathbf{FS}_t = e^{-r(T_0-t)} \, \mathbb{E}_{\mathbb{P}^*}\big(C(S_{T_0}, T - T_0, K S_{T_0}) \,\big|\, \mathcal{F}_t\big)$$

and this appears to be a non-trivial problem, in general. Note, however, that the terminal payoff can be represented as follows

$$\mathbf{FS}_T = (S_T - K S_{T_0})^+ = S_{T_0}(Y - K)^+ = \hat{S}_T (Y - K)^+,$$

where the auxiliary process $\hat{S}$ is given as $\hat{S}_t = S_{t \wedge T_0}$ for every $t \in [0, T]$, and where we set $Y = S_T/S_{T_0} = S_T/\hat{S}_T$. We define a probability measure $\hat{\mathbb{P}}$, equivalent to $\mathbb{P}^*$ on $(\Omega, \mathcal{F}_T)$, by setting

$$\eta_T \stackrel{\text{def}}{=} \frac{d\hat{\mathbb{P}}}{d\mathbb{P}^*} = a \frac{\hat{S}_T B_0}{\hat{S}_0 B_T} = \frac{S_{T_0} B_0}{S_0 B_{T_0}}, \quad \mathbb{P}^*\text{-a.s.} \tag{7.11}$$

where we set $a = e^{-r(T-T_0)}$. Moreover, for every $t \in [0, T_0]$ we have

$$\eta_t \stackrel{\text{def}}{=} \frac{d\hat{\mathbb{P}}}{d\mathbb{P}^*}\Big|_{\mathcal{F}_t} = \mathbb{E}_{\mathbb{P}^*}\Big(\frac{S_{T_0} B_0}{S_0 B_{T_0}} \,\Big|\, \mathcal{F}_t\Big) = \frac{S_t B_0}{S_0 B_t}, \quad \mathbb{P}^*\text{-a.s.}$$

Let us write $b_t = a S_t$. The next lemma shows that by changing a probability measure we may simplify the calculations.

Lemma 7.1.3 *We have*

$$\mathbf{FS}_t = b_t \, \mathbb{E}_{\hat{\mathbb{P}}}\big((Y - K)^+ \,\big|\, \mathcal{F}_t\big). \tag{7.12}$$

Proof. We have, for every $t \in [0, T_0]$

$$\mathbf{FS}_t = B_t \, \mathbb{E}_{\mathbb{P}^*}\big(B_T^{-1}(S_T - K S_{T_0})^+ \,\big|\, \mathcal{F}_t\big) = a S_0 B_t \, \mathbb{E}_{\mathbb{P}^*}\big(\eta_T (Y - K)^+ \,\big|\, \mathcal{F}_t\big)$$
$$= b_t (\eta_t)^{-1} \mathbb{E}_{\mathbb{P}^*}\big(\eta_T (Y - K)^+ \,\big|\, \mathcal{F}_t\big) = b_t \, \mathbb{E}_{\hat{\mathbb{P}}}\big((Y - K)^+ \,\big|\, \mathcal{F}_t\big),$$

where the last equality is an immediate consequence of the abstract Bayes formula (see Lemma A.1.4). $\qquad\square$

Since the process $S_t/\hat{S}_t$ is obviously constant for $t \in [0, T_0]$, the random variable $Y = S_T/S_{T_0} = S_T/\hat{S}_T$ corresponds to the value at time T of a process with the volatility vanishing for every $t \in [0, T_0]$. This simple observation underpins the proof of the following result.

Proposition 7.1.4 *The arbitrage price at time* $t \in [0, T_0]$ *of a forward-start option equals*

$$\mathbf{FS}_t = S_t \, \mathbb{E}_{\hat{\mathbb{P}}} \Big(e^{\int_{T_0}^{T} \sigma_t \, d\hat{W}_t - \frac{1}{2} \int_{T_0}^{T} \sigma_t^2 \, dt} - K e^{-r(T-T_0)} \, \Big| \, \mathcal{F}_t \Big)^+,$$

where the process

$$\hat{W}_t = W_t^* - \int_0^t \sigma_u \mathbb{1}_{[0,T_0]}(u) \, du$$

is a standard Brownian motion under $\hat{\mathbb{P}}$ *with respect to the filtration* $\mathbb{F}$.

Proof. The fact that $\hat{W}$ is a standard Brownian motion is a consequence of equality (7.11) and Girsanov's theorem. Indeed, (7.11) yields

$$\eta_T = \exp\Big(\int_0^T \sigma_t \mathbb{1}_{[0,T_0]}(t) \, dW_t^* - \frac{1}{2} \int_0^T \sigma_t^2 \mathbb{1}_{[0,T_0]}(t) \, dt \Big).$$

In view of Lemma 7.1.3, to establish the valuation formula, it suffices to observe that the random variable $Y = S_t/S_{T_0}$ can be represented as follows

$$Y = \exp\Big(r(T - T_0) + \int_{T_0}^{T} \sigma_t \, d\hat{W}_t - \frac{1}{2} \int_{T_0}^{T} \sigma_t^2 \, dt \Big)$$

since obviously $dW_t^* = d\hat{W}_t$ for every $t \in [T_0, T]$. $\qquad\square$

In order to proceed with explicit calculation of the option price, we need to find the dynamics of the volatility process σ under $\hat{\mathbb{P}}$. This is not a difficult task provided, for instance, that the SDE governing σ under $\mathbb{P}^*$ is given. Suppose that σ satisfies under $\mathbb{P}^*$

$$d\sigma_t = \tilde{a}(\sigma_t, t) \, dt + b(\sigma_t, t) \, d\tilde{W}_t,$$

where $\tilde{W}$ is a one-dimensional standard Brownian motion, possibly correlated with W^*, so that $d\langle W^*, \tilde{W} \rangle_t = \rho_t \, dt$ for some $\mathbb{F}$-progressively measurable process ρ taking values in $[-1, 1]$. Then under $\hat{\mathbb{P}}$ the volatility σ_t is governed by

$$d\sigma_t = \bar{a}(\sigma_t, t) \, dt + b(\sigma_t, t) \, d\bar{W}_t,$$

where $\bar{W}_t = \tilde{W}_t - \int_0^t \rho_u \sigma_u \mathbb{1}_{[0,T_0]}(u) \, du$ and the adjusted drift coefficient $\bar{a}(t, \sigma_t)$ is given by

$$\bar{a}(\sigma_t, t) = \tilde{a}(\sigma_t, t) + \sigma_t \rho_t b(\sigma_t, t) \mathbb{1}_{[0,T_0]}(t).$$

Lucic (2004) and Kruse and Nögel (2005) have independently applied the method described above to arbitrage valuation of a forward-start option in the framework of a stochastic volatility model proposed by Heston (1993) (for a description of this model, see Sect. 7.4). Since the pricing formulas produced in these papers are rather heavy, they are not reported here.

7.2 Extensions of the Black-Scholes Model

We start the study of volatility modelling by examining a few examples of models with random Black-Scholes (i.e., relative) volatility, in which the same one-dimensional Brownian motion governs the stock price and the stochastic volatility of this price. In other words, we assume here that the relative volatility of the underlying asset is state-dependent, so that $\sigma_t = \sigma(S_t)$ for some function σ. Particular models within this approach can be easily produced by postulating a priori that the stock price is described by the SDE

$$dS_t = \tilde{\mu}(S_t, t)\, dt + g(S_t)\, dW_t \qquad (7.13)$$

for some strictly increasing (or strictly decreasing) and smooth function $g : \mathbb{R}_+ \to \mathbb{R}_+$. By applying Itô's formula to the process $v_t = g(S_t)$, we obtain

$$dv_t = g'(S_t)\, dS_t + \tfrac{1}{2} g''(S_t)\, d\langle S \rangle_t.$$

If we denote by h the inverse function of g, then we get

$$dv_t = \left(g'\big(h(v_t)\big)\tilde{\mu}\big(h(v_t), t\big) + \tfrac{1}{2} g''\big(h(v_t)\big) v_t^2 \right) dt + g'\big(h(v_t)\big) v_t\, dW_t.$$

7.2.1 CEV Model

Cox (1975)[1] proposed the following particular choice of the function g in (7.13)

$$g(s) = \alpha s^{\beta}, \quad \forall s \in \mathbb{R}_+,$$

where $\alpha > 0$ and $0 \le \beta \le 1$ are constants. According to this choice, the stock price satisfies under $\mathbb{P}^*$

$$dS_t = S_t \left(r\, dt + \alpha S_t^{\beta-1}\, dW_t^* \right) \qquad (7.14)$$

so that the relative volatility of S equals $\sigma_t = f(S_t) = \alpha S_t^{\beta-1}$. For $\beta = 0$, (7.14) yields the Bachelier model, while for $\beta = 1$ it coincides with the Black-Scholes model. It is thus convenient to exclude the extreme cases, and to assume in what follows that $0 < \beta < 1$. Notice that for any $0 < \beta < 1$ the infinitesimal conditional variance of the logarithmic rate of return of the stock equals $\sigma_t^2 = \alpha^2 S_t^{2(\beta-1)}$, and thus it changes inversely with the price. This feature was found to be characteristic of actual stock price movements by several econometric studies. Furthermore, we have formally

$$\frac{dv_t}{dS_t} \frac{S_t}{v_t} = \frac{g'(S_t) S_t}{g(S_t)} = \frac{\alpha \beta S_t^{\beta-1} S_t}{\sigma S_t^{\beta}} = \beta$$

and

---

[1] For related results, see also Cox and Ross (1976b), Beckers (1980), Ang and Peterson (1984), Schroder (1989), Goldenberg (1991), Bibby and Sørensen (1997), and Lo et al. (2000).

$$\frac{d\sigma_t}{dS_t}\frac{S_t}{\sigma_t} = \frac{f'(S_t)S_t}{f(S_t)} = \frac{\alpha(\beta-1)S_t^{\beta-2}S_t}{\alpha S_t^{\beta-1}} = \beta - 1.$$

The last two equalities serve as a justification for the following commonly standard terminology.

Definition 7.2.1 A stock price model given by (7.14) is called the *Cox model* or the *constant elasticity of variance model* (the *CEV model*, for short).

Before addressing the issue of option pricing, let us examine the properties of solutions to (7.14). It is worth noting that this equation corresponds to the classical Girsanov example in the theory of stochastic differential equations. Assume first that $r = 0$. Then it is well known that the SDE (7.14) has a unique[2] solution for any $\beta \geq 1/2$; uniqueness of solutions fails to hold, however, for any $0 < b < 1/2$ (although it still holds in the class of positive solutions). We shall therefore assume in what follows that the inequalities $1/2 \leq \beta < 1$ and $r \geq 0$ are valid. Under these conditions, the solution to the SDE (7.14) is known to be a strictly positive process, provided that $S_0 > 0$.

Remarks. The choice of the parameter β can be motivated not only by mathematical concerns, such as the existence and uniqueness of a solution to (7.14), strict positivity of a solution, etc., but also by some financial considerations. Some authors argue along the following lines: they postulate that the process that represents the total value of the firm has constant volatility; the value of the stock (equity) is then derived through a variant of the Black-Scholes option's pricing formula (for instance, by assuming, as in Merton (1974), that the firm's debt is in the form of a single zero-coupon bond of maturity T issued by the firm). In this case, the volatility of the stock is not constant, but it depends on the current level of the value of the firm (and thus on the current level of the stock price as well). It should be acknowledged, however, that it is rather difficult to justify the dynamics (7.14) using purely financial arguments.

Let us stress once again that in this case (and indeed in any model of Sect. 7.2 and 7.3) the random character of the volatility is somewhat conventional, since it is derived from the randomness of the underlying asset. To clarify this point, let us consider the Bachelier model

$$dS_t = rS_t\,dt + \sigma\,dW_t^*$$

and the Black-Scholes model

$$dS_t = rS_t\,dt + \sigma S_t\,dW_t^*.$$

---

[2] By uniqueness we mean here *pathwise uniqueness* (i.e., *strong uniqueness*) of solutions, which is known to imply *uniqueness in distribution* (i.e., *weak uniqueness*). Pathwise uniqueness of solutions implies also that any solution is *strong*, i.e., it is adapted to the filtration generated by the driving noise W^* (also when W^* is a Brownian motion with respect to some larger filtration).

In the first case, the absolute volatility is constant, but the relative volatility is random, since it can be represented as σ/S_t. In the Black-Scholes set-up, the relative volatility is constant, and the absolute volatility σS_t is random.

Completeness of the CEV model. Let us make a standard assumption that the reference filtration $\mathbb{F}$ is generated by the driving Brownian motion W^*. Then the CEV model is complete, so that any European contingent claim that is $\mathcal{F}_T$-measurable and $\mathbb{P}^*$-integrable possesses a unique arbitrage price given by the risk-neutral valuation formula

$$\pi_t(X) = B_t \, \mathbb{E}_{\mathbb{P}^*}\big(B_T^{-1}h(S_T) \,|\, \mathcal{F}_t\big).$$

By applying the Feynman-Kac formula (see Sect. A.17), we deduce that $\pi_t(X) = v(S_t, t)$, where the function v solves the following PDE (cf. (3.100)

$$v_t(s, t) + \tfrac{1}{2}\alpha^2 s^{2\beta} v_{ss}(s, t) + rs v_s(s, t) - rv(s, t) = 0 \qquad (7.15)$$

with terminal condition $v(s, T) = h(s)$.

Option prices. Option prices for the CEV model were examined, among others, by Beckers (1980), Schroder (1989), Goldenberg (1991), and Lo et al. (2000). Let us mention that the transition probability density function for the stock price governed by the CEV model can be explicitly expressed in term of the modified Bessel functions (see, e.g., Sect. 10.7.3 in Lipton (2001)). Consequently, the arbitrage price of any European contingent claim can be found, at least in principle, by integration of the payoff function with respect to the transition density. The derivation of the option pricing formula is beyond the scope of this text. Let us only mention that Schroder (1989) established the following computationally convenient representation for the call price in the CEV model

$$C_t(S_t, T - t) = S_t\left(1 - \sum_{n=1}^{\infty} g(n + 1 + \gamma, \tilde{K}_t) \sum_{m=1}^{n} g(m, \tilde{F}_t)\right)$$

$$- Ke^{-r(T-t)} \sum_{n=1}^{\infty} g(n + \gamma, \tilde{F}_t) \sum_{m=1}^{n} g(m, \tilde{K}_t),$$

where $g(p, x)$ stands for the density function of the Gamma distribution $g(p, x) = x^{p-1}e^{-x}/\Gamma(p)$ and where we set $\gamma = 1/2(1 - \beta)$. Moreover, we denote

$$\tilde{F}_t = \frac{F_t^{2(1-\beta)}}{2\chi(t)(1 - \beta)^2}, \qquad \tilde{K}_t = \frac{K^{2(1-\beta)}}{2\chi(t)(1 - \beta)^2},$$

where in turn $F_t = S_t/B(t, T)$ is the forward price of a stock, and where

$$\chi(t) = \frac{\sigma^2\big(e^{2r(1-\beta)(T-t)} - 1\big)}{2r(1 - \beta)} = \sigma^2 \int_t^T e^{2r(1-\beta)u} \, du$$

is the scaled time to maturity of an option.

Implied volatility smile in the CEV model. A rather obvious advantage of the CEV model over the classical Black-Scholes model stems from the presence of an additional parameter β that makes the former model more flexible than the latter one. By an appropriate choice of α and β, it is possible to get a better fit to observed market prices of options, than by using a single parameter σ (see, e.g., Beckers (1980) or Lauterbach and Schultz (1990)). Let us stress that as soon as we have decided to use the CEV model to value options, the Black-Scholes formula is reduced to a conventional tool that allows us to quote market prices in terms of Black-Scholes implied volatilities. Hence the Black-Scholes formula is not formally needed for the calibration of the CEV model to market data. It is, of course, useful in an equivalent problem of fitting the observed surface of (Black-Scholes) implied volatilities. It is rather clear that for any choice of parameters $\beta \neq 1$ and $\alpha \neq 0$, the CEV model yields prices of European options that correspond to a certain smile in the Black-Scholes implied volatility surface. To be more specific, it appears that for a fixed maturity T, the implied volatility of a call option is a decreasing function of the strike K.

Dynamic behavior of the implied volatility surface predicted by the CEV model was recently examined by Hagan and Woodward (1999a). They have considered the case when the stock price S is governed by equation (7.14) with time-dependent coefficient σ. Thus, the forward price of a stock

$$F_t = F_S(t, T) = S_t / B(t, T) = e^{r(T-t)} S_t$$

satisfies, under the martingale measure $\mathbb{P}^*$,

$$dF_t = \alpha(t) F_t^\beta \, dW_t^*. \tag{7.16}$$

Since the SDE above is a straightforward generalization of the Black model, it is natural to focus here on the associated Black implied volatility.

Hagan and Woodward (1999a) (see also Sect. 10.7.5 in Lipton (2001)) have shown that the implied volatility $\hat{\sigma}_0(T, K)$ predicted by the model (7.16) is, when expressed in terms of the current level F_0 of the forward price, fairly accurately described by the following approximate formula

$$\hat{\sigma}_0(T, K) = \frac{\alpha_a}{F_a^{1-\beta}} \left(1 + \frac{(1 - \beta)(2 + \beta)(F_0 - K)^2}{24 F_a^2} + \frac{(1 - \beta)^2 \alpha_a^2 T}{24 F_a^{2(1-\beta)}} + \cdots \right)$$

where $F_a = (F_0 + K)/2$ and α_a is given by the formula (in the classical CEV model in which α is constant, we have $\alpha_a = \alpha$)

$$\alpha_a = \left(\frac{1}{T} \int_0^T \alpha^2(u) \, du \right)^{1/2}.$$

To derive their formula, Hagan and Woodward (1999a) employed singular perturbation techniques[3] that are beyond the scope of the present text. We thus refer to the

---

[3] The same techniques were subsequently used by Hagan et al. (2002) to analyze a stochastic volatility extension of the CEV model (see Sect. 7.4.5).

original paper for the derivation of the approximation given above, and for more accurate formulas with higher-order terms. Let us only mention, that Hagan and Woodward (1999a) examined also a more general set-up, namely, they consider a model in which the forward price F solves the SDE

$$dF_t = \alpha(t)A(F_t)\,dW_t^*, \tag{7.17}$$

where the function A is assumed to be twice differentiable.

Notice that the Black implied volatility $\hat{\sigma}_0(T, K)$ depends also on the level F_0 of the forward price, so that we have $\hat{\sigma}_0(T, K) = \hat{\sigma}(F_0, T, K)$. This trivial observation allows us to write the option's price in the following way (cf. (7.8))

$$c_{\alpha,\beta}(F, T, K) = c\big(F, T, K, \hat{\sigma}(F, T, K)\big), \tag{7.18}$$

where $c_{\alpha,\beta}(F, T, K)$ is the pricing function for a call option within the CEV model with parameters α and β, and $c(F, T, K, \sigma)$ is the Black pricing function. Consequently, the delta of a call option can be represented as follows

$$\frac{\partial c_{\alpha,\beta}}{\partial F} = \frac{\partial c}{\partial F} + \frac{\partial c}{\partial \hat{\sigma}}\frac{\partial \hat{\sigma}}{\partial F}. \tag{7.19}$$

This shows that the value of delta predicted by the CEV model has two components: the Black delta and an additional term representing the impact of the change of the implied volatility caused by changes in the forward price.

Using their approximate formula, Hagan and Woodward (1999a) draw an important conclusion that the CEV model has an inherent flaw of incorrectly predicting the future movements of the Black (or Black-Scholes) implied volatility. This undesirable feature is shared by any model based on the local volatility function technique that is presented in Sect. 7.3.

7.2.2 Shifted Lognormal Models

A relatively simple way to fit the implied volatility skew is to use the so-called *shifted lognormal model* (see Rubinstein (1983) and Marris (1999)). According to this approach, the dynamics of S under the martingale measure $\mathbb{P}^*$ are

$$dS_t = rS_t\,dt + (\sigma_1 S_t + \sigma_2)\,dW_t^*.$$

The model offers the advantage of simple closed-form expressions for plain-vanilla call and put options and hence it is easier to work with than many other models such as, for instance, the CEV model. It should be noticed, however, that a model constructed in this way has a potential to fit the observed volatility surface, but is unlikely to adequately reproduce dynamical features of this surface (for reasons mentioned above). Let us note that deterministic-shift extensions of *short-term rate models* (such models are examined in Chap. 10) were successfully applied to provide an exact fit of a term structure model to the observed yield curve (see Sect. 3.8 in Brigo and Mercurio (2001a) and the references therein). The same idea was recently applied to other term structure models, such as the model of LIBORs (see Rebonato and Joshi (2001), Joshi and Rebonato (2003), and Sect. 12.7 below).

7.3 Local Volatility Models

Suppose that the implied volatility surface that is currently observed in the market is not flat with respect to K. Then we are forced to conclude that market prices of plain-vanilla options cannot be explained by a simple generalization of the Black-Scholes model with time-dependent deterministic volatility. It appears, however, that it is possible to construct a complete diffusion-type model that yields an exact fit to the observed term structure of volatility smiles, provided that no static arbitrage opportunities are present in the market. Specifically, we may postulate that the martingale dynamics of S are

$$dS_t = S_t\big(r\,dt + \sigma(S_t, t)\,dW_t^*\big)$$

for some diffusion coefficient $\sigma(s, t)$, referred to as the *local volatility function* in what follows.

7.3.1 Implied Risk-Neutral Probability Law

Breeden and Litzenberger (1978) observed that, for any given future date T, the one-dimensional risk-neutral probability distribution of the underlying asset is uniquely determined by prices of European call options with all possible strikes and maturity date T. Before we proceed to the proof of this result, let us stress that when prices of all European call (or put) options with different strikes and maturities are known, we can only establish the uniqueness of one-dimensional marginal distributions of the stock price under the risk-neutral probability. No higher-dimensional marginal probability distributions of the underlying asset can be uniquely inferred if the only available data are prices of standard European calls and/or puts.

In this paragraph, we fix T and we assume that we observe prices of all call options with different strikes. Our goal is to show that, under mild technical conditions, we can recover the probability distribution of S_T uniquely under a probability measure $\mathbb{P}^*$ that is interpreted as a risk-neutral probability.

Suppose, in addition, that we know a priori that the random variable S_T is non-negative and admits a probability density function $f(\,\cdot\,, T)$ under $\mathbb{P}^*$. We fix S_0 and we postulate that, for a fixed exercise date $T > 0$ and every strike $K \in \mathbb{R}$, the call price $C_0(T, K)$ at time 0 is given by the formula

$$C_0(T, K) = B_0\,\mathbb{E}_{\mathbb{P}^*}\big(B_T^{-1}(S_T - K)^+\big) = e^{-rT}\int_{\mathbb{R}}(s - K)^+ f(s, T)\,ds = c(K, T)$$

for some pricing function c. We shall show (see Corollary 7.3.1) that the density function $f(\,\cdot\,, T)$ satisfies

$$f(s, T) = e^{rT}c_{KK}(s, T). \tag{7.20}$$

This implies, in particular, that the function $c(\,\cdot\,, T) : \mathbb{R} \to \mathbb{R}$ is twice differentiable with respect to K. To be more specific, the function $c(\,\cdot\,, T)$ is continuously differentiable with respect to K, and the partial derivative $c_K(\cdot, T)$ is an absolutely continuous function with respect to the Lebesgue measure on $\mathbb{R}$. In other words, we have

$$c_K(K, T) = c_K(0, T) + \int_0^K c_{KK}(s, T)\, ds, \quad \forall K \in \mathbb{R}_+,$$

for some (locally integrable) function $c_{KK}(s, T)$.

It appears that formula (7.20) can be essentially generalized. Suppose that, for a fixed maturity $T > 0$ and arbitrary strike K, we have

$$C_0(T, K) = e^{-rT} \int_{\mathbb{R}} (s - K)^+ \, dF(s, T) = c(K, T), \tag{7.21}$$

where $F(\cdot, T)$ denotes the cumulative distribution function of the random variable S_T under $\mathbb{P}^*$. Note that $c(\cdot, T) : \mathbb{R} \to \mathbb{R}$ is a nonnegative, convex function (in particular, it is continuous). Hence, as a locally integrable function, it defines a distribution μ^c, which is a nonnegative measure on the measurable space $(\mathbb{R}, \mathcal{B}(\mathbb{R}))$. Specifically, for any continuous function h with compact support (that is, vanishing outside some bounded interval) we have

$$\mu^c(h) = \int_{\mathbb{R}} h(s) c(s, T)\, ds,$$

where $\mu^c(h)$ stands for the value of the distribution μ^c represented by the function $c(\cdot, T)$ on the element h.

Let us now analyze the first and second order distributional derivatives of μ^c, denoted by μ^c_K and μ^c_{KK} in what follows. By definition, the first order distributional derivative of μ^c in the distributional sense is a distribution, denoted as μ^c_K, such that for any continuously differentiable function h with compact support we have

$$\mu^c_K(h) = -\mu^c(h').$$

In order to make the last formula more explicit, let us observe that since the function $c(\cdot, T)$ is convex, it admits non-decreasing (and thus locally bounded) right-hand-side and left-hand-side derivatives $c'_r(\cdot, T)$ and $c'_l(\cdot, T)$. Thus, the first order distributional derivative μ^c_K can be identified with the distribution determined by either $c'_r(\cdot, T)$ or $c'_l(\cdot, T)$. For ease of notation, we shall frequently refer to the symbol $c_K(\cdot, T)$ rather than μ^c_K.

In particular, for any continuous function h with compact support the following chain of equalities holds

$$\mu^c_K(h) = \int_{\mathbb{R}} h(s) c_K(s, T)\, ds = \int_{\mathbb{R}} h(s) c'_r(s, T)\, ds = \int_{\mathbb{R}} h(s) c'_l(s, T)\, ds.$$

Moreover, for any continuously differentiable function h with compact support we have, as expected,

$$\mu^c_K(h) = -\int_{\mathbb{R}} h'(s) c(s, T)\, ds = -\mu^c(h').$$

Let us now analyze the second order distributional derivative of μ^c, denoted as μ^c_{KK}. It is represented by a nonnegative measure on $(\mathbb{R}, \mathcal{B}(\mathbb{R}))$ determined by the first order distributional derivative of the non-decreasing function $c^r_K(\,\cdot\,, T)$ (or $c^l_K(\,\cdot\,, T)$).

For this reason, we shall frequently write $\mu^c_{KK} = c_{KK}(\,\cdot\,, T)$, so that for any continuous function h with compact support we have

$$\mu^c_{KK}(h) = \int_{\mathbb{R}} h(s) c_{KK}(ds, T),$$

and for any twice continuously differentiable function h with compact support we have

$$\mu^c_{KK}(h) = -\mu^c_K(h') = -\int_{\mathbb{R}} h'(s) c_K(s, T)\, ds = \int_{\mathbb{R}} h''(s) c(s, T)\, ds.$$

Lemma 7.3.1 *For any bounded Borel measurable function $g : \mathbb{R} \to \mathbb{R}$, we have*

$$\int_{\mathbb{R}} g(s)\, dF(s, T) = e^{rT} \int_{\mathbb{R}} g(s) c_{KK}(ds, T), \qquad (7.22)$$

and thus for every Borel set A in $\mathbb{R}$

$$\mathbb{P}^*\{S_T \in A\} = e^{rT} \mu^c_{KK}(A) = e^{rT} c_{KK}(A, T). \qquad (7.23)$$

Before proving the lemma, let us emphasize that the derivative c_{KK} in the last two formulas should be understood in the sense of distributions. In view of equality (7.23), this derivative is represented by a probability measure (up to a normalizing constant).

Proof of Lemma 7.3.1. Let us take an arbitrary twice continuously differentiable function h with compact support. As explained above we have, by virtue of the definition of the distributional derivative,

$$\mu^c(h'') = -\mu^c_K(h') = \mu^c_{KK}(h),$$

or equivalently,

$$\int_{\mathbb{R}} h''(s) c(s, T)\, ds = -\int_{\mathbb{R}} h'(s) c_K(s, T)\, ds = \int_{\mathbb{R}} h(s)\, c_{KK}(ds, T).$$

Using equality (7.21) and Fubini's theorem, we obtain

$$\int_{\mathbb{R}} h''(s) c(s, T)\, ds = \int_{\mathbb{R}} h''(s) e^{-rT} \int_{\mathbb{R}} (u - s)^+\, dF(u, T)\, ds$$

$$= e^{-rT} \int_{\mathbb{R}} \int_{\mathbb{R}} h''(s)(u - s)^+\, ds\, dF(u, T) = e^{-rT} \int_{\mathbb{R}} h(u)\, dF(u, T),$$

since the integration by parts formula yields

$$\int_{\mathbb{R}} h''(s)(u - s)^+\, ds = -\int_u^\infty h'(s)\, ds = h(u).$$

We conclude that for any twice continuously differentiable function h with compact support we have

$$e^{-rT} \int_{\mathbb{R}} h(s) \, dF(s, T) = \int_{\mathbb{R}} h(s) c_{KK}(ds, T).$$

The assertion now follows by the standard approximation arguments.   □

Corollary 7.3.1 *Suppose that the random variable S_T admits a continuous probability density function $f(\cdot, T)$. Then the function $c(\cdot, T)$ is twice continuously differentiable with respect to K and*

$$c_{KK}(s, T) = e^{-rT} f(s, T), \quad \forall s \in \mathbb{R}.$$

By virtue of Lemma 7.3.1, we conclude that, under mild technical conditions, all one-dimensional marginal distributions of the price process S are uniquely determined by the prices $C_0(T, K)$ of European call options (or equivalently, by the knowledge of the Black-Scholes implied volatility surface $\hat{\sigma}_0(T, K)$). For more information on estimation of implied probability density functions, see Aït-Sahalia (1996a), Aït-Sahalia and Lo (1998), Bliss and Panigirtzoglou (2002) and the references therein.

Note that the knowledge of all one-dimensional marginal distributions is not sufficient for the unique specification of a stochastic process. It is known that, even when the price process S is a Markov martingale, this is not sufficient for the uniqueness of the distribution of this process (see Madan and Yor (2002)). Put another way, given all prices of plain-vanilla call (or put) options, it is possible to construct various alternative models for the underlying asset that exactly reproduce prices of all plain-vanilla call (or put) options, but in which the prices of exotic options and the corresponding hedging strategies are essentially different.

7.3.2 Local Volatility

From the previous section, we know that a complete collection of prices of all plain-vanilla call (or put) options uniquely determine one-dimensional marginal distributions of the price process S under a probability measure $\mathbb{P}^*$, interpreted as a risk-neutral probability. Dupire (1993a, 1993b, 1994) and, independently, Derman and Kani (1994) proposed to move one step further, and to find a fully specified stochastic process S that is capable of reproducing the observed prices of call (put) options. To ensure the uniqueness of such an *implied process*, Dupire proposed to restrict attention to the class of diffusion processes. Specifically, he postulated that under $\mathbb{P}^*$ we have

$$dS_t = S_t \left(r \, dt + \sigma(S_t, t) \, dW_t^* \right), \tag{7.24}$$

where $S_0 > 0$ and $\sigma : \mathbb{R}_+ \times \mathbb{R}_+ \to \mathbb{R}$ is a yet unspecified *local volatility* (LV, for short). Of course, a yet unknown function σ should be sufficiently regular, in order to guarantee the existence and uniqueness of a strictly positive solution to (7.24), such that $S_t^* = e^{-rt} S_t$ is a $\mathbb{P}^*$-martingale.

Remarks. In a discrete-time set-up, a similar approach was initiated by Rubinstein (1994) (see also Jackwerth and Rubinstein (1996)). Their method is based on the concept of an implied tree, that is, a binomial (or trinomial) model for the asset price that, owing to some specific smoothness assumptions, gives the best fit to market data. Andersen and Brotherton-Ratcliffe (1997) and Britten-Jones and Neuberger (2000) (see also Rossi (2002)) examine an alternative discrete-time approach, in which a finite-difference grid is constructed. In the present text, we shall not discuss these methods, and thus the interested reader is referred to original papers for details (see, for instance, Derman and Kani (1998), Davis and Hobson (2004), and Bühler (2006)).

As before, the prices of call options given by the market are denoted as $C_0^m(T, K)$. Dupire's goal was to show that knowledge of the family $C_0^m(T, K)$ is indeed sufficient for the unique specification of the process S within the considered class of diffusion processes.

Let us stress that it is by no means obvious whether the observed family of market prices $C_0^m(T, K)$ corresponds to an arbitrage-free model; for some preliminary results in this direction, we refer to Carr and Madan (2004) and Davis (2004b). Instead of dealing directly with market prices, we shall therefore postulate from the outset that we are given the family $C_0(T, K)$ of prices of plain-vanilla call options that were obtained within an (unknown) arbitrage-free model of the stock price. Our goal is to show that if this is the case, then we can find a function $\sigma(s, t)$ such that the model (7.24) will exactly reproduce these prices. As one might guess, this statement is not necessarily true, in general, unless some regularity conditions on the pricing function $c(K, T) = C_0(T, K)$ are imposed.

Formally, we shall assume from now on that we are given the family $C_0(T, K)$ such that there exists a probability space $(\Omega, \mathcal{F}, \mathbb{P}^*)$ and a martingale M on this space, for which

$$C_0(T, K) = \mathbb{E}_{\mathbb{P}^*}\left(M_T - Ke^{-rT}\right)^+ \stackrel{\text{def}}{=} c(K, T).vs2 \tag{7.25}$$

We adopt the following definition of the local volatility.

Definition 7.3.1 A function $\sigma : \mathbb{R}_+ \times \mathbb{R}_+ \to \mathbb{R}$ such that the prices of all plain-vanilla call options given by the model (7.24) coincide with a given family of prices $C_0(T, K)$, that is,

$$C_0(T, K) = B_0\, \mathbb{E}_{\mathbb{P}^*}\left(B_T^{-1}(S_T - K)^+\right),$$

is called the *local volatility.*

Before stating the main result of this subsection, let us consider once again the standard Black-Scholes model. Let $c(S_0, T, K)$ be the family of Black-Scholes call prices, for a fixed initial value S_0 of the stock price and arbitrary strikes and maturities. It is not difficult to show by straightforward calculations that, for any fixed

$S_0 > 0$, the function $c(K, T) = c(S_0, K, T)$ satisfies the following Black-Scholes-like PDE

$$-c_T(K, T) + \tfrac{1}{2}\sigma^2 K^2 c_{KK}(K, T) - rKc_K(K, T) = 0 \qquad (7.26)$$

with the initial condition $c(K, 0) = (S_0 - K)^+$. It is interesting to notice that Dupire's PDE (7.27) is a direct generalization of the equation above.

Proposition 7.3.1 *Assume that the local volatility σ is such that the SDE (7.24) has a unique solution S and for any $t \in \mathbb{R}_+$ the random variable S_t admits a continuous density function $f(s, t)$. Let $c(K, T)$ be the pricing function for European call options given by the model (7.24). If c is of class $C^{1,2}(\mathbb{R} \times [0, T^*], \mathbb{R})$, then it satisfies the following PDE*

$$-c_T(K, T) + \tfrac{1}{2}\sigma^2(K, T)K^2 c_{KK}(K, T) - rKc_K(K, T) = 0 \qquad (7.27)$$

with the initial condition $c(K, 0) = (S_0 - K)^+$.

Proof. We assume that for every T and K

$$c(K, T) = B_0\, \mathbb{E}_{\mathbb{P}^*}\big(B_T^{-1}(S_T - K)^+\big) = \mathbb{E}_{\mathbb{P}^*}\big(S_T^* - e^{-rT}K\big)^+,$$

where the process S satisfies (7.24) and where we write $S_T^* = e^{-rT}S_T$. Note that the process $S_t^* = e^{-rt}S_t$ is a continuous (local) martingale under $\mathbb{P}^*$, since it satisfies

$$dS_t^* = S_t^*\sigma(S_t, t)\, dW_t^* = S_t^*\sigma(e^{rt}S_t^*, t)\, dW_t^*.$$

For any $t > 0$, let $f^*(\cdot, t)$ stand for the probability density of the random variable S_t^*. The obvious equality

$$c(Ke^{rT}, T) = \mathbb{E}_{\mathbb{P}^*}(S_T^* - K)^+$$

yields

$$f^*(K, T) = e^{2rT}c_{KK}(Ke^{rT}, T). \qquad (7.28)$$

so that the function $c(\cdot, T)$ is twice differentiable with respect to K.

By applying the Itô-Tanaka-Meyer formula (see Sect. A.9) to the convex function $(x - K)^+$ and a continuous (local) martingale S^*, we obtain

$$(S_t^* - K)^+ = (S_0^* - K)^+ + \int_0^t \mathbb{1}_{\{S_u^* > K\}}\, dS_u^* + \tfrac{1}{2}L_t^K(S^*).$$

Consequently, we have

$$c(Ke^{rT}, T) = \mathbb{E}_{\mathbb{P}^*}(S_T^* - K)^+ = (S_0^* - K)^+ + \tfrac{1}{2}\mathbb{E}_{\mathbb{P}^*}(L_T^K(S^*)). \qquad (7.29)$$

Recall that for any bounded (or nonnegative) Borel measurable function h the following density of occupation time formula is valid, for every $t \in \mathbb{R}_+$,

$$\int_{\mathbb{R}} h(a)L_t^a(S^*)\, da = \int_0^t h(S_u^*)\, d\langle S^*\rangle_u.$$

In our case, the last equality yields

$$\mathbb{E}_{\mathbb{P}*}\left(\int_{\mathbb{R}} h(a) L_t^a(S^*) \, da\right) = \mathbb{E}_{\mathbb{P}*}\left(\int_0^t h(S_u^*)(S_u^*)^2 \sigma^2 (e^{ru} S_u^*, u) \, du\right).$$

Let h be a continuous function with compact support. By applying the last equality, formula (7.29), and Fubini's theorem, we obtain, for every T,

$$\int_{\mathbb{R}} h(K) c(Ke^{rT}, T) \, dK = \alpha + \frac{1}{2} \int_{\mathbb{R}} h(K) \mathbb{E}_{\mathbb{P}*}(L_T^K(S^*)) \, dK$$

$$= \alpha + \frac{1}{2} \mathbb{E}_{\mathbb{P}*}\left(\int_0^T h(S_t^*)(S_t^*)^2 \sigma^2 (e^{rt} S_t^*, t) \, dt\right)$$

$$= \alpha + \frac{1}{2} \int_0^T \int_{\mathbb{R}} h(K) f^*(K, t) K^2 \sigma^2 (e^{rt} K, t) \, dK \, dt$$

$$= \alpha + \frac{1}{2} \int_{\mathbb{R}} h(K) \int_0^T f^*(K, t) K^2 \sigma^2 (e^{rt} K, t) \, dt \, dK,$$

where

$$\alpha = \int_{\mathbb{R}} h(K)(S_0^* - K)^+ \, dK.$$

Since the last equality is valid for any continuous function h with compact support, we obtain

$$c(Ke^{rT}, T) = (S_0^* - K)^+ + \frac{1}{2} \int_0^T f^*(K, t) K^2 \sigma^2 (e^{rt} K, t) \, dt.$$

Differentiation of both sides with respect to T yields

$$-c_T(Ke^{rT}, T) + \tfrac{1}{2} \sigma^2 (Ke^{rT}, T) K^2 f^*(K, T) - rKe^{rT} c_K(Ke^{rT}, T) = 0.$$

In view of (7.28) and upon substituting Ke^{rT} with K, we obtain

$$-c_T(K, T) + \tfrac{1}{2} \sigma^2 (K, T) K^2 c_{KK}(K, T) - rK c_K(K, T) = 0.$$

Since obviously $c(K, 0) = (S_0 - K)^+$, this completes the proof.   □

The following corollary to Proposition 7.3.1 shows that, under suitable technical assumptions, there exists a unique implied local volatility $\sigma(s, t)$ compatible with a full collection $C_0(T, K)$ of prices of European calls.

Corollary 7.3.2 *Let the call prices $C_0(T, K) = c(K, T)$ be given by (7.25) for some martingale M. We assume that the function $c(K, T)$ is once differentiable with respect to T and twice differentiable with respect to K. Suppose, in addition, that $c_{KK}(s, t) \neq 0$ for every $t > 0$ and $s > 0$. Then the unique implied local volatility is given by the expression*

$$\sigma^2(s, t) = \frac{2(c_T(s, t) + rsc_K(s, t))}{s^2 c_{KK}(s, t)} \tag{7.30}$$

provided that the function $\sigma(s, t)$, *implicitly defined by the last equation, is sufficiently regular to guarantee the existence and uniqueness of solutions to* (7.24) *and the existence of a continuous probability density function* $f(s, t)$.

Before proceeding, let us mention that the original proof of Proposition 7.3.1, given by Dupire (1993a), was based on the PDE approach. Let us sketch Dupire's arguments briefly. We denote by $f(s, t)$ the density of S_t under $\mathbb{P}^*$. It is known that, under suitable regularity conditions, it satisfies the following PDE (commonly known as the Kolmogorov forward equation)

$$-\frac{\partial f}{\partial t} + \frac{1}{2}\frac{\partial^2}{\partial s^2}\left(s^2\sigma^2(s, t)f(s, t)\right) - \frac{\partial}{\partial s}\left(rsf(s, t)\right) = 0.$$

On the other hand, we already know from (7.20) that for every $t > 0$ we have $f(s, t) = e^{rt}c_{ss}(s, t)$. Consequently,

$$f_t(s, t) = re^{rt}c_{ss}(s, t) + e^{rt}c_{sst}(s, t).$$

Substituting and dividing by e^{rt}, we obtain

$$-2rc_{ss}(s, t) - c_{sst}(s, t) + \frac{1}{2}\frac{\partial^2}{\partial s^2}\left(s^2\sigma^2(s, t)c_{ss}(s, t)\right) - rsc_{sss}(s, t) = 0.$$

We note that

$$2rc_{ss}(s, t) + rsc_{sss}(s, t) = \frac{\partial^2}{\partial s^2}\left(rsc_s(s, t)\right).$$

We therefore obtain

$$\frac{1}{2}\frac{\partial^2}{\partial s^2}\left(s^2\sigma^2(s, t)c_{ss}(s, t)\right) = \frac{\partial^2}{\partial s^2}\left(rsc_s(s, t) + c_t(s, t)\right).$$

Integrating the last equation twice with respect to s, and using again the equality $f(s, t) = e^{rt}c_{ss}(s, t)$, we get

$$-c_t(s, t) + \frac{1}{2}s^2\sigma^2(s, t)c_{ss}(s, t) - rsc_s(s, t) = g(t)s + h(t)$$

for some functions g and h of the time parameter. Taking into account the condition $\lim_{s \to \infty} g(t)s + h(t) = 0$ for every t, we find that $g = h = 0$. This concludes the original derivation of Dupire's PDE.

Extension to a dividend-paying stock. Local volatility approach can be easily extended to the case of a dividend-paying stock, which was considered in Sect. 3.2.1. Specifically, if a stock pays dividends continuously at a fixed rate κ then the Black-Scholes pricing function for a call option can be shown to satisfy

$$-c_T(K, T) + \frac{1}{2}\sigma^2K^2c_{KK}(K, T) - (r - \kappa)Kc_K(K, T) - \kappa C(K, T) = 0.$$

It is thus easy to guess that equation (7.30) should now be replaced by the following generalized formula

$$\sigma^2(s, t) = \frac{2(c_T(s, t) + (r - \kappa)sc_K(s, t) + \kappa c(s, t))}{s^2c_{KK}(s, t)}. \tag{7.31}$$

Interpretation of the local volatility. Let us mention another important feature of the local volatility, first noted by Dupire (1996) and Derman and Kani (1998), and formally established by Klebaner (2002) and Savine (2002). We shall focus now on the 'true' dynamics of the underlying asset. Namely, we no longer postulate that S is governed by (7.24), but we assume instead that the process S^* is a continuous martingale of class $\mathcal{H}^1$ on a finite interval $[0, T^*]$ (that is, the random variable $\max_{t \leq T^*} |S_t^*|$ is assumed to be $\mathbb{P}^*$-integrable). Assume, in addition, that the random variable S_t (or equivalently, S_t^*) admits a probability density function for every $0 < t \leq T^*$ and the quadratic variation $\langle S \rangle$ of the process S is pathwise differentiable. Then the function

$$c(K, T) = B_0 \, \mathbb{E}_{\mathbb{P}^*}\big(B_T^{-1}(S_T - K)^+\big) = \mathbb{E}_{\mathbb{P}^*}\big(S_T^* - Ke^{-rT}\big)^+$$

satisfies the following version of Dupire's PDE

$$c_T(K, T) + \tfrac{1}{2} V(K, T)c_{KK}(K, T) - rKc_K(K, T) = 0 \qquad (7.32)$$

with the initial condition $c(K, 0) = (S_0 - K)^+$, where the auxiliary function $V(K, T)$ is defined as follows

$$V(K, T) = \mathbb{E}_{\mathbb{P}^*}\left(\frac{d\langle S \rangle_T}{dT} \,\Big|\, S_T\right)_{S_T = K} = \mathbb{E}_{\mathbb{P}^*}\left(\frac{d\langle S^* \rangle_T}{dT} \,\Big|\, S_T^*\right)_{S_T^* = e^{-rT}K}. \qquad (7.33)$$

Let us observe that if the underlying filtration $\mathbb{F}$ is generated by a Brownian motion then, by virtue of the predictable representation theorem, the process $\langle S \rangle$ is necessarily absolutely continuous with respect to the Lebesgue measure. Let us write $d\langle S \rangle_t = \sigma_t^2 \, dt$ for some (sufficiently regular) stochastic process σ and let $V(K, T) = \sigma^2(K, T)K^2$. Then we obtain

$$\sigma^2(K, T) = \mathbb{E}_{\mathbb{P}^*}(\sigma_T^2 \mid S_T = K).$$

We thus conclude that the *squared local volatility* $\sigma^2(K, T)$ is the risk-neutral expectation of the future squared stochastic volatility of the stock price S, conditioned on the event $\{S_T = K\}$.

To establish (7.32)-(7.33), it suffices to mimic the proof of Proposition 7.3.1 and to observe that under the present assumptions the following equalities are valid:

$$\mathbb{E}_{\mathbb{P}^*}\left(\int_0^T h(S_t^*)d\langle S^* \rangle_t\right) = \int_0^T \mathbb{E}_{\mathbb{P}^*}\left(h(S_t^*)\mathbb{E}_{\mathbb{P}^*}\left(\frac{d\langle S^* \rangle_t}{dt} \,\Big|\, S_t^*\right)\right) dt$$

$$= \int_0^T \int_{\mathbb{R}} h(K)f^*(K, t)V(e^{rt}K, t) \, dK \, dt.$$

Klebaner (2002) gives a simple example of a stochastic volatility model of Bachelier type in which the function $V(K, T)$ can be computed explicitly (such examples are rather rare, however). The interesting problem of finding the local volatility model associated with a given stochastic volatility model was recently addressed by several researchers (see, e.g., Atlan (2006)).

7.3.3 Mixture Models

In this section, we present the so-called *mixture models* that can formally be seen as particular cases of a local volatility model.

Let us consider a finite family F_i, $i = 1, \ldots, m$ of cumulative distribution functions (c.d.f.) on $\mathbb{R}$ or, more generally, on $\mathbb{R}^d$. By a mixture of probability distributions on $\mathbb{R}$ (on $\mathbb{R}^d$) we mean the probability distribution associated with the c.d.f. F of the form $F = \sum_{i=1}^m p_i F_i$, where the weights p_i are such that $p_i > 0$ and $\sum_{i=1}^m p_i = 1$. It is clear that we may interpret F as a result of picking at random a c.d.f. F_i with the corresponding probability p_i. A generic mixture model is an attempt to extend this simple procedure to the case of stochastic processes, rather then random variables.

It is well known that a mixture of two or more normal densities on $\mathbb{R}$ yields a probability distribution with heavy tails. This observation underpins an attempt to reproduce the shape of the volatility smile through a judiciously chosen mixture of either normal or lognormal distributions. Before proceeding, let us make a trivial observation that a mixture of normal (respectively, lognormal) laws is not a normal (respectively, lognormal) distribution.

Ritchey (1990), Melick and Thomas (1997), Guo (1998) were among the first to examine modelling of stock returns[4] based on mixtures of lognormal distributions. Formally, they constructed in this way a model with 'random' volatility σ, in which the level of the volatility is chosen at random at time 0 from the given set of volatility functions $\{\sigma_1, \ldots, \sigma_m\}$ in accordance with probabilities $p_1, \ldots, p_m$. The probabilities $p_1, \ldots, p_m$ can thus be seen as additional parameters appearing in the mixture model for the stock price. The main goal of the papers mentioned above was the static calibration of an option pricing model to observed market data, rather than dynamical hedging or pricing of exotic options.

Let us stress that in its classical version, a mixture model is simply a family of reference models used to specify the one-dimensional marginal distributions of a mixture model. Dynamical features of each reference model are used within each of these models to value exotic options, but they are not fully employed in order to uniquely specify the dynamics (equivalently, multidimensional marginal distributions) of a mixture model.

Before we present some results in this vein, let us comment on the advantages and drawbacks of this method. On the positive side, it provides a simple and efficient method of fitting the market data. It is not at all clear, however, whether a mixture model is capable of correctly addressing the issue of volatility risk. Piterbarg (2003b) points out that since the method does not specify a dynamical model of the underlying asset, it should be used with care for the valuation and hedging of these exotic products that exhibit a strong dependence on multidimensional distributions of the underlying asset.

To be more specific, if we consider a forward-start option or a compound option then, in principle, we can value such a contract in each of the reference models

---

[4] The method is by no means limited to modelling of the stock price. Andersen and Brotherton-Ratcliffe (2001) and Gątarek (2003) applied this method to produce extended Libor market models with stochastic volatility.

(for instance, Black-Scholes models with different, but constant, volatilities), and subsequently specify its price in a mixture model as the weighted value of these prices. Such a procedure clearly neglects the random character of the volatility, and its impact on the future value of a contract. Indeed, the future value of many exotic options depends on the future value of the volatility, and thus such options should be priced and hedged in a model that is flexible enough to account for this feature.

Within the framework of mixture models, this issue was addressed by Brigo and Mercurio (2001c, 2002a, 2002b) and Brigo et al. (2003), who made an important next step, by deriving a dynamical diffusion-type model, associated with a fairly general mixture model. In this way, they obtained a diffusion-type model, that can be easily calibrated to market prices of European options. In addition, it can be used for the valuation and hedging of exotic (path-dependent) products.

Mixture of lognormal models. Let us consider a family S^i, $i = 1, \ldots, m$ of one-dimensional diffusion processes

$$dS_t^i = r S_t^i \, dt + g_i(S_t^i, t) \, dW_t^* \tag{7.34}$$

with $S_0^i = S_0$ for every i, and let us denote by $f^i(\cdot, t)$ the probability density function of the random variable S_t^i under the martingale measure $\mathbb{P}^*$. We assume that the diffusion coefficients $g_i(s, t)$ are appropriately chosen, so that the SDE above admits a unique, strong solution for each $i = 1, \ldots, m$.

Although the method developed by Brigo and Mercurio is fairly general, and thus covers various probability distributions, for concreteness, we shall focus on the special case of a mixture of lognormal probability distributions. This case corresponds to the choice $g_i(s, t) = \sigma_i(t)s$ for a deterministic function σ_i defined on the time interval $[0, T]$.

In this case, for any $t > 0$ the probability density function $f_t^i(s)$, $s > 0$, of the random variable S_t^i is given by the following well-known expression

$$f^i(s, t) = \frac{1}{\sqrt{2\pi} v_i(t)s} \exp\left\{-\frac{1}{2v_i^2(t)}\left(\ln\frac{s}{S_0^i} - rt + \frac{1}{2}v_i^2(t)\right)^2\right\},$$

where we set

$$v_i^2(t) = \int_0^t \sigma_i^2(u) \, du.$$

The goal is to find a diffusion coefficient $\tilde{\sigma}(s, t)$ such that the process S, which is a (unique, strong) solution to the SDE

$$dS_t = S_t\big(r \, dt + \tilde{\sigma}(S_t, t) \, dW_t\big), \tag{7.35}$$

has the following property: there exists a family p_i, $i = 1, \ldots, m$ of strictly positive constants, such that $\sum_{i=1}^m p_i = 1$ and for any $t \in [0, T]$ we have

$$f(s, t) = \sum_{i=1}^m p_i f^i(s, t), \quad \forall s \in \mathbb{R}_+, \tag{7.36}$$

where $f(\cdot, t)$ is the probability density function of the random variable S_t under the martingale measure $\mathbb{P}^*$. The following result is due to Brigo and Mercurio (2001c). As becomes apparent from the proof, their method can be seen as an extension of Dupire's approach.

Proposition 7.3.2 *Assume that the weights* $p_1, \ldots, p_m$ *are given and* $r = 0$. *Let the function* $\tilde{\sigma}(s, t)$ *satisfy*

$$\tilde{\sigma}^2(s, t) = \frac{\sum_{i=1}^m p_i g_i^2(s, t) f^i(s, t)}{\sum_{i=1}^m p_i s^2 f^i(s, t)}. \tag{7.37}$$

If the diffusion coefficient $\tilde{\sigma}(s, t)$ *defined in this way is sufficiently regular to guarantee the existence and uniqueness of a strong solution* S *to* (7.35), *then the probability density function of* S_t *satisfies* (7.36) *for every* $t \in [0, T]$.

Proof. Let us sketch the proof. For any maturity $T > 0$ and strike $K > 0$, we define

$$C_0^i(T, K) = B_0 \, \mathbb{E}_{\mathbb{P}^*} \big(B_T^{-1} (S_T^i - K)^+ \big),$$

where the process S^i satisfies (7.34) with $S_0^i = S_0$. Proceeding along similar lines to the proof of Proposition 7.3.1, we obtain

$$C_0^i(T, K e^{rT}) = (S_0^* - K)^+ + \frac{1}{2} \int_0^T f_i^*(K, t) g_i^2(e^{rt} K, t) \, dt,$$

where $f_i^*(\cdot, t)$ is the probability density function of $S_t^{i*} = e^{-rt} S_t^i$. Let for every T and K

$$C_0(T, K) = B_0 \, \mathbb{E}_{\mathbb{P}^*} \big(B_T^{-1} (S_T - K)^+ \big),$$

where S is given by (7.35). We have

$$C_0(T, K e^{rT}) = (S_0^* - K)^+ + \frac{1}{2} \int_0^T f^*(K, t) K^2 \tilde{\sigma}^2(e^{rt} K, t) \, du,$$

where $f^*(\cdot, t)$ is the density of $S_t^* = e^{-rt} S_t$. Equality (7.36) holds if and only if we have

$$C_0(K, T) = \sum_{i=1}^m p_i C_0^i(T, K)$$

for every $T > 0$ and $K > 0$, or more precisely, if we have

$$\sum_{i=1}^m \int_0^T p_i f_i^*(K, t) g_i^2(e^{rt} K, t) \, dt = \sum_{i=1}^m \int_0^T p_i f_i^*(K, t) K^2 \tilde{\sigma}^2(e^{rt} K, t) \, dt.$$

The last equality is indeed satisfied, since we have

$$\sum_{i=1}^m p_i f_i^*(s, t) g_i^2(e^{rt} s, t) = \sum_{i=1}^m p_i f_i^*(s, t) s^2 \tilde{\sigma}^2(e^{rt} s, t).$$

For $r = 0$, the formula above is an immediate consequence of (7.37). $\qquad\square$

7.3.4 Advantages and Drawbacks of LV Models

From the mathematical perspective, a generic local volatility (LV) model is self-consistent, arbitrage-free, and it can be calibrated to perfectly fit the observed market smiles and skews provided, that the market is arbitrage-free. Indeed, under the assumptions of Proposition 7.3.1, there exist a unique LV model reproducing market prices of standard options. In addition, any model obtained through the LV method is complete, provided that $\sigma(s, t) > 0$. Thus, any contingent claim can be exactly replicated by trading in bond and stock and thus all derivative assets (including plain-vanilla options) are formally redundant. Despite these nice features, the concept of a local volatility is by no means an ideal tool for hedging of volatility risk.

First, it is clear that the LV methodology hinges on an unrealistic assumption that we observe option prices at a continuum of strikes and maturities. Since this assumption is never satisfied in practice, the computation of the implied local volatility is a non-trivial mathematical issue, and the final expression for the local volatility heavily relies on a numerical procedure at hand (for examples, see, e.g., Avellaneda et al. (1997), Andersen and Brotherton-Ratcliffe (1997), or Carr and Madan (1998a)). Note that searching for an implied local volatility function is essentially equivalent to searching for an implied volatility surface. Indeed, although the two implied volatilities are conceptually different, they are linked via a one-to-one correspondence. Practical consequences of this property for a robust specification of an implied volatility surface were examined by Berestycki et al. (2000, 2002, 2004).

Second, according to the LV paradigm, given the market data for Europeans calls, we pretend to be able not only to uniquely specify all one-dimensional marginal distributions, but also to recover the risk-neutral dynamics of the underlying asset. As a consequence, the LV methodology is based on an implicit belief that the values of all contingent claim, including exotic options, are uniquely determined by market prices of standard European calls (or puts). This unrealistic feature may serve as an argument that the LV methodology is unlikely to adequately address the issue of volatility risk. Dumas et al. (1998) performed empirical tests of implied volatility functions; they concluded that the LV approach yields rather poor hedging results. Related numerical studies of performance of hedging strategies that are based on *sticky-strike* and *sticky implied tree* models are reported in Lee (2001). To improve pricing and hedging properties of LV method, Dupire (1996) and Derman and Kani (1998) proposed to introduce dynamical behavior of the local volatility. In this version of the LV method, the local volatility surface is no longer 'static', but it is assumed to obey a stochastic process, driven by some additional Brownian motion. The pertinent mathematical issue is the derivation of the drift coefficient in the risk-neutral dynamics of the local volatility, by making use of the martingale property of discounted option prices. It appears that the dynamics of the *stochastic local volatility* is rather complicated, and thus its implementation is rather difficult (see Sect. 7.5).

7.4 Stochastic Volatility Models

In a continuous-time framework, the random volatility σ_t is usually assumed to obey a diffusion-type process. Let the stock price S be given as

$$dS_t = \mu(S_t, t)\,dt + \sigma_t S_t\,dW_t$$

with the *stochastic volatility* σ (also known as the *instantaneous volatility* or the *spot volatility*) satisfying

$$d\sigma_t = a(\sigma_t, t)\,dt + b(\sigma_t, t)\,d\hat{W}_t$$

where W and $\hat{W}$ are standard one-dimensional Brownian motions defined on some filtered probability space $(\Omega, \mathbb{F}, \mathbb{P})$, with the cross-variation satisfying $d\langle W, \hat{W}\rangle_t = \rho\,dt$ for some constant $\rho \in [-1, 1]$. Recall that the Brownian motions W and $\hat{W}$ are mutually independent if and only if they are uncorrelated – that is, when $\rho = 0$. More generally, we may assume that ρ_t is a stochastic process adapted to the filtration $\mathbb{F}$ generated by W and $\hat{W}$. For a fixed horizon date T, a martingale measure $\mathbb{P}^*$ for the process $S_t^* = S_t/B_t$ is defined as a probability measure equivalent to $\mathbb{P}$ on on $(\Omega, \mathcal{F}_T)$ such that S^* is a (local) martingale under $\mathbb{P}^*$. Under any martingale measure $\mathbb{P}^*$, we have

$$dS_t = r S_t\,dt + \sigma_t S_t\,dW_t^*, \tag{7.38}$$

with the spot volatility σ satisfying

$$d\sigma_t = \tilde{a}_t\,dt + b(\sigma_t, t)\,d\tilde{W}_t, \tag{7.39}$$

for some drift coefficient $\tilde{a}_t$. We shall adopt a commonly standard convention that

$$\tilde{a}_t = \tilde{a}(\sigma_t, t) = a(\sigma_t, t) + \lambda(\sigma_t, t)b(\sigma_t, t) \tag{7.40}$$

for some (sufficiently regular) function $\lambda(\sigma, t)$. The presence of the additional term in the drift of the stochastic spot volatility σ under an equivalent martingale measure is an immediate consequence of Girsanov's theorem. A specific form of this term, as given above, is a matter of convenience and its choice is motivated by practical considerations.

Definition 7.4.1 A generic model of the form (7.38)–(7.39) is referred to as a *stochastic volatility model* (a SV model, for short).

Under suitable regularity conditions, a unique solution (S, σ) to (non-linear) stochastic differential equations (7.38)–(7.39) is known to follow a two-dimensional diffusion process; results concerning the existence and uniqueness of the SDEs can be found, e.g., in Ikeda and Watanabe (1981) or Karatzas and Shreve (1998a). The existence of an equivalent probability measure under which the process $S_t^* = S_t/B_t$ is a martingale (as opposed to a local martingale) is a non-trivial issue, however, and thus it needs to be examined on a case-by-case basis for each particular SV model (see, e.g., Sin (1998)).

Stochastic volatility models of the stock price are also supported by empirical studies of stock returns. Early studies of market stock prices (reported in Mandelbrot

(1963), Fama (1965), Praetz (1972), and Blattberg and Gonedes (1974)) concluded that the lognormal law is an inadequate descriptor of stock returns. More recent studies (see, for instance, Hsu et al. (1974) and Kon (1984)) have found that the mixture of Gaussian distributions better describes the leptokurtic empirical distributions. Ball and Torous (1985) have empirically estimated models of returns as mixtures of continuous and jump processes. Empirical studies of Black (1976a), Schmalensee and Trippi (1978), and Christie (1982) uncovered an inverse correlation between stock returns and changes in volatility. This peculiar feature of stock returns supports the conjecture that the stock price volatility should be modelled by means of an autonomous stochastic process, rather then as a function of the underlying asset price.

7.4.1 PDE Approach

Generally speaking, stochastic volatility models are not complete, and thus a typical contingent claim (such as a European option) cannot be priced by arbitrage. In other words, the standard replication arguments can no longer be applied to most contingent claims. For this reason, the issue of valuation of derivative securities under market incompleteness has attracted considerable attention in recent years, and various alternative approaches to this problem were subsequently developed. Seen from a different perspective, the incompleteness of a generic SV model is reflected by the fact that the class of all martingale measures for the process $S_t^* = S_t/B_t$ comprises more than one probability measure, and thus the necessity of specifying a single pricing probability arises.

Since under (7.38)–(7.40) we deal with a two-dimensional diffusion process, it is possible to derive, under mild additional assumptions, the partial differential equation satisfied by the value function of a European contingent claim. For this purpose, one needs first to specify the *market price of volatility risk* $\lambda(\sigma, t)$. Mathematically speaking, the market price for risk is associated with the Girsanov transformation of the underlying probability measure leading to a particular martingale measure. Let us observe that pricing of contingent claims using the market price of volatility risk is not preference-free, in general (typically, one assumes that the representative investor is risk-averse and has a constant relative risk-aversion utility function).

To illustrate the PDE approach mentioned above, assume that the dynamics of two-dimensional diffusion process (S, σ) under a martingale measure are given by (7.38)–(7.40), with Brownian motions W^* and $\tilde{W}$ such that $d\langle W^*, \tilde{W}\rangle_t = \rho\, dt$ for some constant $\rho \in [-1, 1]$. Suppose also that both processes, S and σ, are nonnegative. Then the price function $v = v(s, \sigma, t)$ of a European contingent claim is well known to satisfy a specific PDE (see, for instance, Garman (1976a, 1976b) or Hull and White (1987a)).

Since the proof of Proposition 7.4.1 is based on a straightforward application of the Feynman-Kac formula (see Sect. A.17), it is omitted.

Proposition 7.4.1 *Consider a European contingent claim* $X = g(S_T)$ *that settles at time* T. *Assume that the price of* X *is given by the risk-neutral valuation formula under* $\mathbb{P}^*$

$$\pi_t(X) = B_t \, \mathbb{E}_{\mathbb{P}^*}(B_T^{-1} g(S_T) \,|\, \mathcal{F}_t) = v(S_t, \sigma_t, t), \quad \forall\, t \in [0, T],$$

for some choice of a martingale probability $\mathbb{P}^*$ for the process $S_t^* = S_t/B_t$. Then the pricing function $v : \mathbb{R}_+ \times \mathbb{R}_+ \times [0, T] \to \mathbb{R}$ solves the PDE

$$v_t + \tfrac{1}{2}\sigma^2 s^2 v_{ss} + r s v_s - r v + \tfrac{1}{2}b^2(\sigma, t) v_{\sigma\sigma} + \rho\sigma s b(\sigma, t) v_{s\sigma}$$
$$+ \big(a(\sigma, t) + \lambda(\sigma, t) b(\sigma, t)\big) v_\sigma = 0,$$

with the terminal condition $v(s, \sigma, T) = g(s)$ for every $s \in \mathbb{R}_+$ and $\sigma \in \mathbb{R}_+$.

Let us stress once again that we do not claim here that $\mathbb{P}^*$ is a unique martingale measure for a given model. Hence, unless volatility-based derivatives are assumed to be among primary assets, the market price of volatility risk needs to be exogenously specified. For some specifications of stochastic volatility dynamics and the market price of volatility risk, a closed-form expression for the option's price is available. In other cases, suitable numerical procedures need to be employed. Since we deal here with a multidimensional diffusion process, deterministic methods based on the discretization of the partial differential equation satisfied by the pricing function appear excessively time-consuming. An alternative Monte Carlo approach for stochastic volatility models was examined by Fournié et al. (1997b).

7.4.2 Examples of SV Models

Various SV model are obtained by making different choices of dynamics for the stochastic volatility process σ_t (or its square σ_t^2). For instance, Wiggins (1987) postulated the following set-up

$$dS_t = \mu S_t \, dt + \sigma_t S_t \, dW_t,$$

where the (spot) volatility process σ_t satisfies

$$d\sigma_t = a(\sigma_t) \, dt + \theta \sigma_t \, d\hat{W}_t, \tag{7.41}$$

while W and $\hat{W}$ are correlated Brownian motions. As a special case of dynamics (7.41), he considered the mean-reverting Ornstein-Uhlenbeck process; more precisely, he assumed that $\ln \sigma_t$ satisfies

$$d \ln \sigma_t = \kappa(\nu - \ln \sigma_t) \, dt + \theta \, d\hat{W}_t$$

for some constants κ, ν and θ. The same specification of the dynamics of $\ln \sigma_t$ was assumed also by Scott (1987, 1991) and Chesney and Scott (1989).

Specifications of stochastic volatility proposed by other authors include the following SDEs:

$$d\sigma_t = \kappa(\nu - \sigma_t) \, dt + \theta \, d\hat{W}_t,$$
$$d\sigma_t = \kappa\sigma_t(\nu - \sigma_t) \, dt + \theta\sigma_t \, d\hat{W}_t,$$
$$d\sigma_t = \kappa\sigma_t \, dt + \theta\sigma_t \, d\hat{W}_t,$$
$$d\sigma_t = \sigma_t^{-1}(\nu - \kappa\sigma_t^2) \, dt + \theta \, d\hat{W}_t,$$
$$d\sigma_t^2 = \kappa(\nu - \sigma_t^2) \, dt + \theta\sigma_t \, d\hat{W}_t.$$

A non-exhaustive list of papers devoted to stochastic volatility models includes: Johnson and Shanno (1987), Stein and Stein (1991), Hofmann et al. (1992), Heston (1993), Ball and Roma (1994), Heynen et al. (1994), Renault and Touzi (1996), Frey (1996), Bakshi et al. (1997, 2000), Frey and Stremme (1997), Romano and Touzi (1997), Scott (1997), Zhu and Avellaneda (1998), Sircar and Papanicolaou (1999), Fouque et al. (2000b), Lee (2001), Hagan et al. (2002), Sabanis (2002), Berestycki et al. (2004), and Andersen and Piterbarg (2004). A comprehensive survey of closed-form results for diffusion-type stochastic volatility models was provided by Leblanc (1996) (see also Pitman and Yor (1996) for related theoretical results). Let us finally mention that stochastic volatility models are also used in the framework of term structure modelling (see Sect. 11.7 and 12.7 in this regard).

7.4.3 Hull and White Model

It is not possible to deal in detail with each of stochastic volatility models mentioned above. We shall only analyze rather briefly a few most widely popular, and proven to be successful in practical implementations, models.

We start by a concise presentation of the general framework which covers, in particular, the classical Hull and White (1987a) model. We postulate that the stock price S obeys the SDE

$$dS_t = S_t(\mu_t \, dt + \sigma_t \, dW_t),$$

with the volatility σ satisfying the diffusion equation

$$d\sigma_t = a(\sigma_t, t) \, dt + b(\sigma_t, t) \, d\hat{W}_t,$$

where W and $\hat{W}$ are independent, one-dimensional standard Brownian motions under $\mathbb{P}$. In other words, the process $(W, \hat{W})$ follows a standard two-dimensional Brownian motion under the real-world probability $\mathbb{P}$. We require that the market price of volatility risk is chosen in such a way that under a (non-unique) martingale measure $\mathbb{P}^*$ we have

$$dS_t = S_t(r \, dt + \sigma_t \, dW_t^*), \tag{7.42}$$

and

$$d\sigma_t = \tilde{a}(\sigma_t, t) \, dt + b(\sigma_t, t) \, d\tilde{W}_t, \tag{7.43}$$

for some function $\tilde{a}(\sigma, t)$, where $(W^*, \tilde{W})$ is a standard two-dimensional Brownian motion under $\mathbb{P}^*$.

To achieve this goal, we postulate that the market price of volatility risk is specified by a process $\lambda(\sigma_t, t)$ for some function $\lambda(\sigma, t)$. Of course, an equivalent change of a probability measure should also correctly modify the drift term in the dynamics of S. This drift correction be done in a similar way as in the classical Black-Scholes model.

Despite the fact that W^* and $\tilde{W}$ are assumed to be independent standard Brownian motions under $\mathbb{P}^*$, it is obvious that the processes S and σ are not independent under a martingale measure $\mathbb{P}^*$. It is rather clear, however, that dynamics of the stock price S, when conditioned on a particular sample path of the volatility process σ, is lognormal under $\mathbb{P}^*$. Consequently, a stochastic volatility model driven by two independent Brownian motions is relatively easy to handle.

Before we proceed to valuation of contingent claims, let us explain how to derive formally the drift coefficient in (7.43). To this end, we shall make use of the multidimensional version of the Girsanov theorem (see Theorem A.15.1).

Let us fix a finite horizon date T^*. It is straightforward to check that the Radon-Nikodým derivative of $\mathbb{P}^*$ with respect to $\mathbb{P}$ on $(\Omega, \mathcal{F}_{T^*})$ is given by the following expression

$$\frac{d\mathbb{P}^*}{d\mathbb{P}} = \mathcal{E}_{T^*}\Big(\int_0^{\cdot} \frac{r - \mu_u}{\sigma_u} \, dW_u\Big)\mathcal{E}_{T^*}\Big(\int_0^{\cdot} \lambda(\sigma_u, u) \, d\hat{W}_u\Big), \quad \mathbb{P}\text{-a.s.}, \qquad (7.44)$$

where

$$\mathcal{E}_{T^*}\Big(\int_0^{\cdot} \frac{r - \mu_u}{\sigma_u} \, dW_u\Big) = \exp\Big(\int_0^{T^*} \frac{r - \mu_u}{\sigma_u} \, dW_u - \frac{1}{2}\int_0^{T^*} \frac{(r - \mu_u)^2}{\sigma_u^2} \, du\Big)$$

and

$$\mathcal{E}_{T^*}\Big(\int_0^{\cdot} \lambda(\sigma_u, u) \, d\hat{W}_u\Big) = \exp\Big(\int_0^{T^*} \lambda(\sigma_u, u) \, d\hat{W}_u - \frac{1}{2}\int_0^{T^*} \lambda^2(\sigma_u, u) \, du\Big)$$

where the processes μ, σ and $\lambda(\sigma_t, t)$ satisfy suitable technical assumptions ensuring that $\mathbb{P}^*(\Omega) = 1$. We conclude that the drift term in (7.43) takes the following form (cf. (7.40))

$$\tilde{a}(\sigma_t, t) = a(\sigma_t, t) + \lambda(\sigma_t, t)b(\sigma_t, t).$$

We assume from now on that the coefficients $\tilde{a}(\sigma, t)$ and $b(\sigma, t)$ are sufficiently regular to ensure the existence and uniqueness of a strong solution to stochastic differential equation (7.43). It is now possible to show that the discounted stock price is a martingale (as opposed to a local martingale) under any probability measure $\mathbb{P}^*$ formally given by expression (7.44).

Proposition 7.4.2 *The process S^* given by, for $t \in [0, T^*]$,*

$$S_t^* = B_t^{-1} S_t = S_0 \exp\Big(\int_0^t \sigma_u \, dW_u^* - \frac{1}{2}\int_0^t \sigma_u^2 \, du\Big)$$

is a martingale with respect to $\mathbb{F}$ under $\mathbb{P}^*$.

Proof. First, by virtue of continuity of sample paths of σ, it is clear that

$$\mathbb{P}^*\Big\{\int_0^{T^*} \sigma_t^2 \, dt < \infty\Big\} = 1.$$

Thus, the process S^* is well defined and it is a local martingale under $\mathbb{P}^*$. Since, in addition, the processes W^* and σ are mutually independent under $\mathbb{P}^*$, we may compute the conditional expectation $\mathbb{E}_{\mathbb{P}^*}(S_t^* \,|\, \mathcal{F}_u)$ by conditioning first on the enlarged σ-field $\mathcal{G} = \mathcal{F}_u \vee \sigma(\sigma_s \,|\, u \leq s \leq t)$. Since $\mathbb{E}_{\mathbb{P}^*}(S_t^* \,|\, \mathcal{G}) = S_u^*$ and S_u^* is $\mathcal{F}_u$-measurable, it is clear that $\mathbb{E}_{\mathbb{P}^*}(S_t^* \,|\, \mathcal{F}_u) = S_u^*$, as desired. $\qquad\square$

To examine another interesting feature of the Hull and White model, let us denote by A the increasing stochastic process representing the integrated squared volatility:

$$A_t = \int_0^t \sigma_u^2 \, du, \quad \forall t \in [0, T^*].$$

Let us assume that $\mathbb{P}^*\{A_t > 0\} = 1$ for every $t > 0$. Then for any $t \in [0, T^*]$ we have

$$S_t^* = B_t^{-1} S_t = S_0 \exp\left(\xi_t \sqrt{A_t} - \tfrac{1}{2} A_t\right),$$

where we set

$$\xi_t = (A_t)^{-1/2} \int_0^t \sigma_u \, dW_u^*.$$

One can show that the random variable ξ_t is independent of the filtration $\mathbb{F}^\sigma$ generated by the volatility process σ and has for any $t \in [0, T^*]$ the standard Gaussian law $N(0, 1)$.

The model given by (7.42)–(7.43) is not complete (we shall return to the issue of completeness of stochastic volatility models later in this section), and thus the standard replication argument is no longer valid. Alternative ad hoc solution to the option pricing problem is to choose any probability measure $\mathbb{P}^*$ of the form (7.44) to play the role of the risk-neutral probability that will serve to the valuation of contingent claims. In particular, one may apply Proposition 7.4.1, if the PDE approach appears to be more computationally efficient than direct probabilistic calculations (or the Monte Carlo method).

We assume from now on that the martingale measure $\mathbb{P}^*$ has been selected, and we define the price of a European contingent claim X that settles at T using the risk-neutral valuation formula

$$\pi_t(X) = B_0 \, \mathbb{E}_{\mathbb{P}^*}(B_T^{-1} X \mid \mathcal{F}_t), \quad \forall t \in [0, T].$$

Let F_S stand for the cumulative distribution function of S_T under $\mathbb{P}^*$, so that

$$F_S(s) = \int_0^\infty F_{S|A}(s|w) \, dF_A(w),$$

where $F_{S|A}(s|w)$ denotes the conditional cumulative distribution function of S_T given that $A_T = w$, and F_A is the cumulative distribution function of A_T under $\mathbb{P}^*$.

The conditional cumulative distribution function $F_{S|A}(s|w)$ has a lognormal density, specifically, $dF_{S|A}(s|w) = f_{S|A}(s|w) \, ds$, where for any fixed $w > 0$ and any $s > 0$

$$f_{S|A}(s|w) = \frac{1}{\sqrt{2\pi w}\, s} \exp\left\{ -\frac{\left(\ln(s/S_0) - rT + \tfrac{1}{2}w \right)^2}{2w} \right\}.$$

We are in a position to state a result that furnishes a simple representation for the price of a path-independent European claim.

Proposition 7.4.3 *The price at time 0 of any contingent claim* $X = h(S_T)$ *settling at time* T, *satisfies*

$$\pi_0(X) = \int_0^\infty h(s) \, dF_S(s) = \int_0^\infty h(s) f_{S|A}(s|w) \, ds \, dF_A(w),$$

or more explicitly

$$\pi_0(X) = \int_0^\infty \int_0^\infty h(s) \frac{1}{\sqrt{2\pi w s}} \exp\left\{ -\frac{\left(\ln(s/S_0) - rT + \frac{1}{2}w\right)^2}{2w} \right\} ds \, dF_A(w).$$

From the proposition above, it is clear that valuation of any contingent claim $X = h(S_T)$ is rather straightforward, provided that we know how to value this claim analytically in the Black-Scholes set-up, and the cumulative distribution function F_A is given explicitly. Indeed, we can then use the following simple recipe

$$\pi_0(X) = \int_0^\infty \pi_0^{BS}(X|w) \, dF_A(w),$$

where $\pi_0^{BS}(X|w)$ stands for the price of X in the standard Black-Scholes model with the volatility parameter $\sigma = \sqrt{w/T}$. Put equivalently, the price of a path-independent contingent claim in a stochastic volatility framework with independent driving Brownian motions is the expected Black-Scholes price, where the expected value is taken over the probability distribution of the integrated squared volatility. For example, the price at time 0 of a European call option can be represented as follows

$$C_0(T, K) = \int_0^\infty \hat{c}(S_0, T, K, w) \, dF_A(w), \tag{7.45}$$

where $\hat{c}(S_0, T, K, w)$ is the Black-Scholes pricing function of a call option, expressed in terms of the stock price S_0, strike K and the integrated squared volatility in the Black-Scholes model, that is, $w = \sigma^2 T$. It is clear that

$$\hat{c}(S_0, T, K, w) = S_0 N\left(\hat{d}_1(S_0, T, K, w)\right) - K e^{-r(T-t)} N\left(\hat{d}_2(S_0, T, K, w)\right),$$

where

$$\hat{d}_{1,2}(S_0, T, K, w) = \frac{\ln(S_0/K) + rT \pm \frac{1}{2}w}{\sqrt{w}}.$$

The last step in a model specification relies on a choice of $\tilde{a}$ and b in the dynamics (7.43) of the volatility process σ_t. Hull and White (1987a) concentrated on the *instantaneous variance* $v_t = \sigma_t^2$ and they postulated that

$$dv_t = \bar{\mu}(v_t, t) v_t \, dt + \eta(v_t, t) v_t \, d\bar{W}_t.$$

Unfortunately, even in the case of constant $\bar{\mu}$ and η, a closed-form expression for the cumulative distribution function F_A is not available. For this reason, Hull and White (1987a) computed the moments of the random variable A_T, and used them to derive an approximate formula for the option price.

As observed by Romano and Touzi (1997), the pricing formula of Proposition 7.4.3 can be extended to the case of correlated Brownian motions W^* and $\tilde{W}$ in (7.42)–(7.43). Specifically, let us assume that $d\langle W^*, \tilde{W}\rangle_t = \rho_t\, dt$ for some process $\rho_t = \rho(\sigma_t, t)$ taking values in $[-1, 1]$. In this case the price at time t of a European call option with strike K and exercise date T can be computed from the following representation

$$C_t = \mathbb{E}_{\mathbb{Q}^*}\Big(\hat{c}\big(S_t e^{Z_t}, t, T, K, w(t, T)\big)\,\Big|\, \mathcal{F}_t\Big), \tag{7.46}$$

where $\hat{c}(s, t, T, K, w)$ is the Black-Scholes call pricing function expressed in terms of the integrated squared volatility $w = \sigma^2(T - t)$, and where we set

$$Z_t = \int_t^T \rho_u \sigma_u\, d\tilde{W}_u - \frac{1}{2}\int_t^T \rho_u^2 \sigma_u^2\, du$$

and

$$w(t, T) = \int_t^T (1 - \rho_u^2)\sigma_u^2\, du.$$

The derivation of representation (7.46) is based on similar arguments as the proof of Proposition 7.4.3, and thus it is left to the reader.

Using the pricing formula (7.46) it is possible to show (see Romano and Touzi (1997)) that the arbitrage price of a call (or put) option is a strictly convex function of the current level of the stock price. This property proves to be important if we wish to complete a stochastic volatility model by formally postulating that a particular option is among primary traded assets.

Renault and Touzi (1996) observed that a generic stochastic volatility model yields a smile effect in the shape of the implied volatility. Specifically, they proved that when the Brownian motions driving a stock and its volatility are independent then, under mild technical conditions, the implied volatility is locally convex around the minimum that corresponds to the forward price of the stock. The original proof of this result was later simplified by Sircar and Papanicolaou (1999), who also dealt with the case of correlated Brownian motions. Another result in this vein was recently obtained by Lee (2001). He showed that if the price and volatility are driven by independent Brownian motions then local and implied volatilities are symmetric functions of log-moneyness, and thus any SV model of this kind yields a symmetric smile.

Sircar and Papanicolaou (1999) and Fouque et al. (2000a, 2000b) examine the asymptotic expansion of implied volatility under *slowly-varying* stochastic volatility. Lee (2001) applies various asymptotic expansions to assess the bias in option pricing and hedging strategies under stochastic volatility.

Let us return to the issue of the model incompleteness. It is generally agreed that the real-world markets are incomplete. Clearly, in the case of the complete securities market, options (and other derivative securities) are redundant. Hence, strictly speaking, trading in derivatives has no economic justification in such a model. As was pointed out above, to value contingent claims in an incomplete stochastic volatility model, it is necessary and sufficient to specify the market price for volatility risk.

Alternatively, as shown by Bajeux-Besnainou and Rochet (1996) and Romano and Touzi (1997), it is possible to complete a stochastic volatility model by enlarging the class of primary traded assets. Usually, this is done by postulating that some derivatives (for instance, a finite family of plain-vanilla options) are liquidly traded. Davis (2004a) develops a complete stochastic volatility model based on a finite family of traded options. Hobson and Rogers (1998) take a different approach: they produce an example of a complete stochastic volatility model, driven by a single Brownian motion, in which the stochastic volatility is defined in terms of weighted moments of historic log-price.

7.4.4 Heston's Model

A widely popular stochastic volatility model, proposed by Heston (1993), assumes that the asset price S satisfies

$$dS_t = S_t\left(\mu_t\,dt + \sqrt{v_t}\right)dW_t \tag{7.47}$$

with the instantaneous variance v governed by the SDE

$$dv_t = \hat{\kappa}(\hat{v} - v_t)\,dt + \eta\sqrt{v_t}\,d\hat{W}_t, \tag{7.48}$$

where W and $\hat{W}$ are standard one-dimensional Brownian motions defined on a filtered probability space $(\Omega, \mathbb{F}, \mathbb{P})$, with the cross-variation $\langle W, \hat{W}\rangle_t = \rho t$ for some constant $\rho \in [-1, 1]$. In this case, it is more convenient to express the pricing function v and the market price of volatility risk λ in terms of variables (s, v, t), rather than (s, σ, t). We now make a judicious choice of the market price of volatility risk; specifically, we set $\lambda(v_t, t) = \alpha\sqrt{v_t}$ for some constant α such that $\alpha\eta \neq \hat{\kappa}$. Hence, under a martingale measure $\mathbb{P}^*$, equations (7.47)-(7.48) become

$$dS_t = S_t\left(r\,dt + \sqrt{v_t}\right)dW_t^*, \tag{7.49}$$

and

$$dv_t = \kappa(\tilde{v} - v_t)\,dt + \eta\sqrt{v_t}\,d\tilde{W}_t, \tag{7.50}$$

where we set $\kappa = \hat{\kappa}^2(\hat{\kappa} - \alpha\eta)^{-1}$, $\tilde{v} = \hat{v}\hat{\kappa}(\hat{\kappa} - \alpha\eta)^{-1}$, and where W^* and $\tilde{W}$ are standard one-dimensional Brownian motions such that $d\langle W^*, \tilde{W}\rangle_t = \rho\,dt$.

It is now easily seen that the pricing PDE in Heston's model has the following form

$$v_t + \tfrac{1}{2}vs^2 v_{ss} + rsv_s - rv + \tfrac{1}{2}\eta^2 vv_{vv} + \rho\eta vsv_{sv} + \kappa(\tilde{v} - v)v_v = 0$$

with the terminal condition $v(s, v, T) = g(s)$ for every $s \in \mathbb{R}_+$ and $v \in \mathbb{R}_+$. We take here for granted the existence and uniqueness of (nonnegative) solutions S and v to Heston's SDE. It is common to assume that $2\kappa\tilde{v} > \sigma^2$ so that the solution v is strictly positive if $v_0 > 0$. Let us note that (7.50) has exactly the same form as the Cox-Ingersoll-Ross equation for the short-term rate (see Sect. 10.1 and 10.3).

In order to value a European call option, we shall first simplify the PDE at hand. To this end, let us introduce a new variable $x = \ln(F/K)$, where $F = s/B(t, T)$. In other words, we consider the log-moneyness of an option in terms of the forward price of the underlying asset. Also, we shall focus on the forward price of an option, so that the pricing function is $f = v/B(t, T)$. Straightforward calculations show that the function $f = f(x, v, t)$ satisfies the following PDE

$$f_t + \tfrac{1}{2}v v_{xx} + \tfrac{1}{2}v v_x + \tfrac{1}{2}\eta^2 v v_{vv} + \rho\eta v v_{xv} + \kappa(\tilde{v} - v)v_v = 0. \tag{7.51}$$

As in Duffie et al. (2000), we search for a solution in the form

$$f(x, v, t) = e^x f^1(x, v, t) - f^0(x, v, t).$$

After substituting the proposed equation to (7.51), we obtain, for $j = 0, 1$,

$$f_t^j + \tfrac{1}{2}v v_{xx}^j + \left(\tfrac{1}{2} - j\right)v v_x^j + \tfrac{1}{2}\eta^2 v v_{vv}^j + \rho\eta v v_{xv}^j + \left(a - b_j v\right)v_v^j = 0, \tag{7.52}$$

where

$$a = \kappa\tilde{v}, \quad b_j = \kappa - j\rho\eta$$

and the terminal condition becomes $f^j(x, v, T) = 1$ if $x > 0$ and 0 otherwise. To complete a solution of the option valuation problem, it is convenient to use the Fourier transform technique. Before stating the final result, we need to introduce an auxiliary notation

$$\alpha_j(k) = -\tfrac{1}{2}k^2 - \tfrac{1}{2}ik - ijk,$$
$$\beta_j(k) = \kappa - \rho\eta j - \rho\eta ik,$$

and

$$r_{1,2}^j(k) = \frac{\beta_j(k) \pm \sqrt{\beta_j^2(k) - 2\alpha_j(k)\eta^2}}{\eta^2} = \frac{\beta_j(k) \pm d_j(k)}{\eta^2}.$$

As a final result, we obtain the following formula (see Heston (1993))

$$f^j(x, v, \tau) = \frac{1}{2} + \frac{1}{\pi}\int_0^\infty \mathrm{Re}\left\{\frac{\exp\left(C_j(k, \tau)\tilde{v} + D_j(k, \tau)v + ikx\right)}{ik}\right\}dk,$$

where we set $\tau = T - t$.

Moreover, we set

$$C_j(k, \tau) = \kappa\left\{r_2^j(k)\tau - \frac{2}{\eta^2}\ln\left(\frac{1 - g_j(k)e^{-d_j(k)\tau}}{1 - g_j(k)}\right)\right\},$$

$$D_j(k, \tau) = r_2^j(k)\frac{1 - e^{-d_j(k)}}{1 - g_j(k)e^{-d_j(k)}},$$

where in turn $g_j(k) = r_2^j(k)/r_1^j(k)$. The integration of the real-valued function appearing in the formula for f^j can be done by standard numerical methods. As was mentioned in Sect. 7.1.10, Lucic (2004) and Kruse and Nögel (2005) examined the valuation of forward-start options within the framework of Heston's model.

From Lewis (2001), it is possible to derive the level, slope and curvature of at-the-money forward (ATMF) implied volatility from the parameters of Heston's model. It appears that they are, respectively,

$$\frac{4\kappa'\bar{v}'\gamma}{\eta^2(1-\rho)^2}, \quad \frac{4\rho\gamma}{\eta^2(1-\rho)^2 T}, \quad \frac{2\gamma}{\kappa'\bar{v}'T^2},$$

where $\kappa' = \kappa - \rho\eta/2$, $\kappa'\bar{v}' = \kappa\bar{v}$ and $\gamma = \sqrt{4\kappa'^2 + \eta^2(1-\rho)^2} - 2\kappa'$.

Furthermore, Gatheral et al. (2000) found the asymptotics of the far-from-the-money implied volatility for Heston's model. Finally, using the results of Lee (2004a) (see Sect. 7.1.6), it is possible to find the asymptotes of the square of the implied volatility, specifically,

$$v(\infty)T \to \frac{2\gamma}{\eta(1-\rho)}, \quad v(-\infty)T \to \frac{2\gamma}{\eta(1+\rho)},$$

as T tends to infinity, where

$$v(\infty) = \lim_{x\to\infty} (\sigma_0^2(x))', \quad v(-\infty) = \lim_{x\to-\infty} (\sigma_0^2(x))'.$$

7.4.5 SABR Model

In a recent paper by Hagan et al. (2002) (see also Hagan and Woodward (1999a) for related research), the authors examine the issue of dynamics of the implied volatility smile. They argue that any model based on the local volatility function incorrectly predicts the future behavior of the smile: when the price of the underlying decreases, LV models predict that the smile shifts to higher prices. Similarly, an increase of the price results in a shift of the smile to lower prices. It was observed that the market behavior of the smile is precisely the opposite. Thus, the local volatility model has an inherent flaw of predicting the wrong dynamics of the Black-Scholes implied volatility. Consequently, hedging strategies based on such a model may be worse than the hedging strategies evaluated for the 'naive' model with constant volatility – that is, the Black-Scholes or Black's models.

A challenging issue is to identify a class of models that has the following essential features: a model should be easily and effectively calibrated and it should correctly capture the dynamics of the implied volatility smile.

A particular model proposed and analyzed by Hagan et al. (2002) is specified as follows: under the martingale measure $\mathbb{P}^*$ the forward price $F_t = F_S(t, T)$ of some asset S is assumed to obey the SDE

$$dF_t = \hat{\alpha}_t F_t^\beta \, dW_t^*, \tag{7.53}$$

where

$$d\hat{\alpha}_t = v\hat{\alpha}_t \, d\tilde{W}_t, \quad \hat{\alpha}_0 = \alpha, \tag{7.54}$$

where W^* and $\tilde{W}$ are Brownian motions with respect to a common filtration $\mathbb{F}$, with a constant correlation coefficient ρ. The model given by (7.53)-(7.54) is termed the SABR model.[5] Of course, it can be seen as a natural extension of the classical CEV model, proposed by Cox (1975).

The model of Hagan et al. (2002) can be accurately fitted to the observed implied volatility curve for a single maturity T. A more complicated version of this model is needed, however, if we wish to fit volatility smiles at several different maturities.

More importantly, the model seems to predict the correct dynamics of the implied volatility skews (as opposed to the CEV model or any model based on the concept of a local volatility function). To support this claim, Hagan et al. (2002) derive and study the approximate formulas for the implied Black and Bachelier volatilities in the SABR model. It appears that the Black implied volatility in this model can be represented as follows

$$
\hat{\sigma}_0(K, T) = \frac{\alpha}{(F_0 K)^{\hat{\beta}/2} \left\{ 1 + \frac{\hat{\beta}^2}{24} \ln^2(F_0/K) + \frac{\hat{\beta}^4}{1920} \ln^4(F_0/K) + \ldots \right\}} \left(\frac{z}{x(z)} \right)
$$
$$
\times \left\{ 1 + \left[\frac{\hat{\beta}^2 \alpha^2}{24(F_0 K)^{\hat{\beta}}} + \frac{\rho \beta \alpha \nu}{4(F_0 K)^{\hat{\beta}/2}} + \frac{(2 - 3\rho^2)\nu^2}{24} \right] T + \ldots \right\},
$$

where we write $\hat{\beta} = 1 - \beta$,

$$
z = \frac{\nu}{\alpha}(F_0 K)^{\hat{\beta}/2} \ln(F_0/K),
$$

and

$$
x(z) = \ln \left\{ \frac{\sqrt{1 - 2\rho z + z^2} + z - \rho}{1 - \rho} \right\}.
$$

In the case of at-the-money option, the formula above reduces to

$$
\hat{\sigma}_0(F_0, T) = \frac{\alpha}{F_0^{\hat{\beta}}} \left\{ 1 + \left[\frac{\hat{\beta}^2 \alpha^2}{24 F_0^{2\hat{\beta}}} + \frac{\rho \beta \alpha \nu}{4 F_0^{\hat{\beta}}} + \frac{(2 - 3\rho^2)\nu^2}{24} \right] T + \ldots \right\}.
$$

7.5 Dynamical Models of Volatility Surfaces

Let us recall that a local volatility (LV) model by construction yields exact prices of European options, but the prices of most other options, including many options with path- and volatility-dependent payoffs, are only approximate. Indeed, exact prices for most contingent claim depend on the full stochastic behavior of (unobserved and possibly autonomous) stochastic volatility process, while an LV model implicitly

---

[5] The acronym SABR refers, presumably, to the commonly used notation for the model's parameters: stochastic-$\alpha\beta\rho$ model.

postulates that the spot volatility satisfies $\sigma_t = \sigma(S_t, t)$, so that it is fully specified by the static local volatility surface and the dynamics of the stock price.

To resolve the inherent shortcomings of the LV approach that were mentioned above, it is natural to allow for stochastic evolution of the local volatility surface, thus defining a dynamical model of the local volatility surface. As was mentioned already, Dupire (1996) and Derman and Kani (1998) proposed an extension of the LV methodology from a static local volatility surface to a fully dynamical framework. To emphasize the restrictive character of financial models based on the concept of a static local volatility, Derman and Kani (1998) attribute to the LV approach the name of *effective theory of volatility*. They argue that the (static) local volatility can be seen as a counterpart of the concept of a static yield curve in term structure modelling.

7.5.1 Dynamics of the Local Volatility Surface

Neuberger (1990b) observed that if we apply a standard delta-hedging strategy to a contract that pays the logarithm of the price of the underlying asset (such a contract is termed the *log-price*), then the hedging error accumulates to the difference between the realized variance and the fixed variance used for hedging. Starting from this observation, Dupire (1996) introduced the concept of a *forward variance*, which is a quantity associated with a calendar spread of two log-price contracts. He then directly modelled the evolution of the term structure of forward variances, by proceeding along similar lines to the Heath et al. (1992a) approach to modelling of the term structure of interest rates (for the latter approach, see Chap. 11). In this way, he constructed a stochastic volatility model in which an exogenous specification of the market price of volatility risk is not required.

In an independent study by Derman and Kani (1998), the authors developed a dynamical model of a local volatility surface. Using the martingale property of discounted option prices, they derived the no-arbitrage drift restriction on the forward local volatility surface. Since this condition is rather involved, it is not reported here. Let us only mention that practical applications of their ideas are based on simulated stochastic implied trees in which the level of the underlying asset and the local volatility, as reflected by transition probabilities, change in a simulation step (see Derman and Kani (1998)). In a related paper, Carr and Madan (1998b) analyze the consequences of stochastic local volatility on volatility trading.

7.5.2 Dynamics of the Implied Volatility Surface

A closely related, but slightly more straightforward, approach to stochastic volatility modelling was initiated by Schönbucher (1999). His idea was to directly prescribe the stochastic development of the implied Black-Scholes volatility for a given family of plain-vanilla options with different strikes and maturities. In the first step, Schönbucher (1999) focused on arbitrage-free dynamics of the (model-based) *stochastic implied volatility* (SIV) $\hat{\sigma}_t(T, K)$ for a finite collection of maturity dates and strikes. Subsequently, he has also examined the case of a continuum of strikes and maturities.

Formally, the *Black-Scholes stochastic implied volatility surface* $\hat{\sigma}_t(T, K)$ is implicitly defined through the equation

$$C_t(T, K) = c(S_t, T - t, K, r, \hat{\sigma}_t(T, K)), \tag{7.55}$$

where $C_t(T, K)$ is the price at time t of a call option predicted by some stochastic model for the price of the underlying asset S, and c stands for the Black-Scholes pricing function of a call option. Let us stress that T represents the actual maturity date of an option, as opposed to the time to maturity that will be denoted as $\tau = T - t$ in what follows.

Schönbucher (1999) approach. To illustrate the main ideas underpinning Schönbucher's approach, let us consider the special case of a fixed maturity date T and a given strike level K. Suppose that the stock price S satisfies

$$dS_t = S_t(r\,dt + \sigma_t\,dW_t^*),$$

where W^* is a Brownian motion under the martingale measure $\mathbb{P}^*$ and σ_t represents the spot volatility. We assume that under $\mathbb{P}^*$

$$d\hat{\sigma}_t(T, K) = \alpha_t(T, K)\,dt + \beta_t(T, K)\,d\tilde{W}_t + \gamma_t(T, K)\,dW_t^*, \tag{7.56}$$

where W^* and $\tilde{W}$ are independent Brownian motions, and coefficients α, β and γ are adapted stochastic processes.

In order to correctly specify the risk-neutral drift in the dynamics (7.56) of the implied volatility surface, expression (7.56) is substituted into Black-Scholes pricing formula. Since the discounted option price necessarily follows a $\mathbb{P}^*$-martingale, using straightforward calculations based on Itô's formula and the Black-Scholes PDE, we obtain the following relationship

$$\alpha_t(T, K) = \frac{1}{2c_\sigma}\Big(\big(\hat{\sigma}_t(T, K) - \sigma_t\big)^2 S_t^2 c_{ss} - c_{\sigma\sigma}\beta_t^2(T, K) - 2\gamma_t(T, K)\sigma_t S_t c_{s\sigma}\Big),$$

where c_σ, c_{ss}, etc. are sensitivities of the Black-Scholes pricing function of a call option, and d_1, d_2 are the corresponding coefficients; note that the usual variables $(S_t, T - t, K, r, \hat{\sigma}_t(T, K))$ were suppressed. The last formula can be further simplified, to give

$$\hat{\sigma}_t(T, K)\alpha_t(T, K) = \frac{1}{\tau}\big(\hat{\sigma}_t(T, K) - \sigma_t\big)^2 - \frac{d_1 d_2}{2}\beta_t^2(T, K) + \frac{d_2}{\tau}\sigma_t\gamma_t(T, K),$$

where we denote $T - t$ by τ.

Starting from the last equation, one can try to derive the dynamics of the stochastic volatility σ_t under $\mathbb{P}^*$, and to study the pertinent relationships between the stochastic implied volatility surface $\hat{\sigma}_t(T, K)$ and the instantaneous volatility σ_t. For an informal discussion of these properties, we refer to the original paper by Schönbucher (1999).

Furthermore, by considering options with different maturities, it is possible to derive drift constraints for the forward implied volatility $\hat{\sigma}_t(K, T, U)$, which is given as, for any dates $t \leq T < U$,

$$\hat{\sigma}_t^2(K, T, U) = \frac{(U - t)\hat{\sigma}_t^2(K, U) - (T - t)\hat{\sigma}_t^2(K, T)}{U - T}.$$

Since we do not present these results here, the interested reader may consult the original paper for details.

Stochastic implied volatility. Following Schönbucher's ideas, Brace et al. (2001b, 2002) focused on mathematical features of stochastic model of implied volatility surface. Their research emphasized the fact that not only the risk-neutral drift, but also the volatility of the stochastic implied volatility cannot be chosen arbitrarily, but should satisfy rather stringent condition.

Brace et al. (2002) conjecture that the initial implied volatility surface and the volatility of the stochastic implied volatility fully specify the underlying stochastic volatility model, via the relationship $\sigma_t = \hat{\sigma}_t(t, S_t)$. To formalize this conjecture, they represent the financial problem at hand as a purely mathematical problem, and impose suitable restrictions on the underlying processes.

For fixed T and K, let us write $\xi_t(T, K) = \hat{\sigma}_t^2(T, K)(T - t)$ for $t \in [0, T]$. Without loss of generality, we may and do assume that $r = 0$ in what follows. The Bachelier version of the *stochastic volatility model* (SIV) developed by Brace et al. (2001b, 2002) relies on the following standing assumptions (SIV.1)–(SIV.4). The first assumption is in fact the choice of convention for the instantaneous volatility.

(SIV.1) Under $\mathbb{P}^*$, the price process S is governed by the Bachelier equation

$$dS_t = \sigma_t \, dW_t^* \tag{7.57}$$

for some process σ modelling the instantaneous absolute volatility.

The next postulate imposes a particular form of the generic dynamics of the family of processes $\xi(T, K)$.

(SIV.2) For every maturity T and strike K, the process $\xi = \xi(T, K)$ obeys

$$d\xi_t(T, K) = m_t \, dt + 2\xi_t(T, K)u_t^1(T, K) \, d\tilde{W}_t + 2\xi_t(T, K)u_t^2(T, K) \, dW_t^*,$$

where the processes m, u^1 and u^2 are given as: $m_t = m(T, K, S_t, \sigma_t, \xi_t)$ and $u_t^i = u^i(T, K, S_t, \sigma_t, \xi_t)$ for some regular functions m and u^i, $i = 1, 2$. At this stage, regularity is understood in a fairly general sense, and thus more specific conditions should be given in a further study of a particular SIV model.

The next two assumptions, (SIV.3) and (SIV.4), deal with the initial and terminal conditions respectively.

(SIV.3) The initial condition is $\xi_0(T, K) = \hat{\sigma}_0^2(T, K)T$, where $\hat{\sigma}_0(T, K)$ is the initial implied volatility surface.

The initial condition $\xi_0(T, K)$ is said to be *admissible* if the associated volatility surface $\hat{\sigma}_0(T, K)$ is *feasible* in the sense that there exists a $\mathbb{P}^*$-martingale S such that

$\hat{\sigma}_0(T, K)$ represents the implied volatility surface associated with option prices corresponding to S. In other words, we impose the natural requirement that the implied volatility surface $\hat{\sigma}_0(T, K)$ can be realized by some arbitrage-free model S. Although this property is essential, rather little is known about explicit necessary and sufficient conditions for this property to hold (see Sect. 7.1.6).

Remarks. Following an earlier work by Figlewski (2002), Henderson et al. (2003) consider a related problem of admissible pricing functions for European call options. They propose specific explicit expressions for valuation formulas, and check that they satisfy the necessary conditions for the no-arbitrage property.

(SIV.4)   The following terminal condition is satisfied $\lim_{t \to T} \xi_t(T, K) = \xi_T(T, K) = 0$.

Terminal condition of assumption (SIV.4) is called the *feedback condition* in Brace et al. (2001b, 2002). We shall see below that this terminology is indeed justified. Let us remark that exactly the feedback condition makes the problem of explicit specification of a model satisfying (SIV.1)–(SIV.4) more difficult that this might appear at the first glance.

Arguing along similar lines to Schönbucher (1999), Brace et al. (2001b) derive the following generic expression for risk-neutral dynamics of $\xi = \xi(T, K)$

$$d\xi_t = \xi_t |u_t|^2 \, dt - |\bar{\sigma}_t - u_t(K - S_t)|^2 \, dt + 2\xi_t u_t \cdot dW_t, \qquad (7.58)$$

where $u_t = (u_t^1, u_t^2)$, $\bar{\sigma}_t = (\sigma_t, 0)$ and $W_t = (W_t^*, \tilde{W}_t)$ is a two-dimensional standard Brownian motion under $\mathbb{P}^*$. Note that $|\cdot|$ stands here for the Euclidean norm in $\mathbb{R}^2$. As was mentioned already, the correct specification of the drift is equivalent to the martingale property of the discounted option price under $\mathbb{P}^*$.

Remarks. Suppose that we adopt the Black-Scholes convention, rather than the Bachelier convention, so that (7.57) is replaced by

$$dS_t = \sigma_t S_t \, dW_t^*$$

for some instantaneous relative volatility process σ. Then the associated process $\xi_t = \xi_t(T, K)$ can be shown to satisfy under $\mathbb{P}^*$ (see Brace et al. (2001b))

$$d\xi_t = \xi_t \left(|u_t|^2 \left(1 + \tfrac{1}{4}\xi_t\right) - \sigma_t u_t^1 \right) dt - |\bar{\sigma}_t - u_t \ln(K/S_t)|^2 \, dt + 2\xi_t u_t \cdot dW_t.$$

Let us return to the Bachelier formulation, as given by (SIV.1)–(SIV.4). In this case, any solution to (7.58) (if exists) necessarily satisfies

$$\xi_t = e^{2M_t} \xi_0(T, K) - \int_0^t e^{2(M_t - M_s)} |\bar{\sigma}_s - u_s(K - S_s)|^2 \, ds$$

where we denote

$$M_t = \int_0^t u_s(T, K) \, dW_s - \frac{1}{2} \int_0^t |u_s(T, K)|^2 \, ds.$$

Consequently, in view of the terminal condition (SIV.4), by setting $t = T$ we obtain

$$\xi_0(T, K) = \int_0^T e^{-2M_s} |\bar{\sigma}_s - u_s(K - S_s)|^2 \, ds$$

This in turn implies that for any maturity T and strike K we have that

$$\partial_T \xi_t(T, K)|_{t=T} = \hat{s}_T^2(T, K) = |\bar{\sigma}_T + (K - S_T)u_T(T, K)|^2$$

Finally, by setting $T = t$ and $K = S_t$, we get $\sigma_t^2 = \hat{\sigma}_t^2(t, S_t)$. The last equality makes it clear why condition (SIV.4) was referred to as the feedback condition. Essentially, it implies that the instantaneous volatility σ is specified endogenously as being equal to the implied volatility of the immediately maturing at-the-money option.

It is not clear, however, whether for an arbitrary specification of the volatility process u in the dynamics (7.58) the price process S will not only be a martingale under $\mathbb{P}^*$, but will also return the initial option prices. If this is the case, the process u is called *admissible*. The choice of an admissible volatility process u is the most difficult issue in the present framework.

Let us conclude these general considerations by stating the following three challenging (and rather difficult to solve in all generality) theoretical problems that need to be examined in the context of modelling stochastic implied volatility surface:
(a) the determination of the class of all admissible initial conditions for the implied volatility surface,
(b) the determination of the class of admissible volatilities of the implied volatility surface, in the sense described above,
(c) the conjecture that the initial implied volatility surface and the volatility of stochastic implied volatility surface fully specify the dynamics of the underlying asset.

Apart from theoretical studies, Brace et al. (2002) report also result of Monte Carlo experiments based on a particular choice of a stochastic volatility model. As a result of these experiments, they conclude that the marginal distributions of the underlying asset are determined (as they should) on the implied volatilities only, but the joint distributions depend also on the choice of the volatility of the stochastic implied volatility surface. In this way, they partially substantiate their conjecture mentioned above. Some related theoretical issues were recently addressed by Durrleman (2004).

7.6 Alternative Approaches

In this section, we provide a brief overview of alternative approaches to the issue of stochastic volatility and the related problem of fitting a model to market data. These alternative approaches include various mathematical techniques permitting to construct stochastic models that are focused on at least one of the following modelling goals a better fit to empirical return data, a more realistic shapes of the implied volatility surface, or a better forecast of the future volatility (for the latter issue, see, e.g., Figlewski (1997)).

7.6.1 Modelling of Asset Returns

Let us first focus on various methodologies that are primarily directed towards forecasting the volatility of various macroeconomic variables. Models of this kind are tailored to reflect at least some of several well documented features of time series of logarithmic daily asset returns, such as: skewness of distributions, heavy tails (leptokurtosis), volatility clustering, or the mean reversion in the price and volatility and their negative correlation.

Autoregressive models. A discretization of a diffusion-type stochastic volatility model may lead to an *autoregressive random variance* models, an ARV model for short. For instance, the evolution of a discrete-time, two-dimensional process (S_t, σ_t) may be described by the following recurrence relation

$$\ln \left(\frac{S_{t+1}}{S_t} \right) = \mu + \sigma_t \xi_{t+1}$$

and

$$\ln \sigma_{t+1} = \nu - \kappa (\nu - \ln \sigma_t) + \theta \eta_{t+1},$$

where (ξ_t, η_t), $t \in \mathbb{N}$, are independent, identically distributed random variables with Gaussian distribution. The ARV models of asset returns were examined by Wiggins (1987), Chesney and Scott (1989), Melino and Turnbull (1990), Duffie and Singleton (1993), and others. A related approach to the modelling of stochastic volatility in a discrete-time framework, is based on so-called ARCH or GARCH models – that is, models with the property of (*general*) *autoregressive conditional heteroskedasticity*. These models were first introduced by Engle (1982) and Bollerslev (1986) respectively.

Engle and Ng (1993) and Duan (1995) analyze the following extended version of a GARCH model. The stock price S is assumed to satisfy, under the real-life probability $\mathbb{P}$,

$$\ln \left(\frac{S_{t+1}}{S_t} \right) = r + \lambda \sqrt{h_t} - \frac{1}{2} h_t + \sqrt{h_t}\, v_{t+1}, \qquad (7.59)$$

where

$$h_{t+1} = \beta_0 + \beta_1 h_t + \beta_2 h_t (v_{t+1} - c)^2 \qquad (7.60)$$

where β_0, β_1 and β_2 are nonnegative parameters and the constant λ represents the unit risk premium.

Furthermore, the random variable v_{t+1} has, conditionally on the σ-field $\mathcal{F}_t$ representing the information available at time t, the standard Gaussian distribution. Finally, the nonnegative parameter c is aimed to capture the negative correlation between return and volatility innovations. In the special case of $c = 0$, the model reduces to the standard GARCH model introduced by Bollerslev (1986).

It follows from (7.59)–(7.60) that, when conditioned on the σ-field $\mathcal{F}_t$, the stock price has a lognormal distribution with conditional expected value

$$\mathbb{E}_{\mathbb{P}}(S_{t+1} \mid \mathcal{F}_t) = S_t e^{r + \lambda \sqrt{h_t}}$$

and conditional variance

$$\text{Var}_{\mathbb{P}}(S_{t+1} \mid \mathcal{F}_t) = \big(\mathbb{E}_{\mathbb{P}}(S_{t+1} \mid \mathcal{F}_t)\big)^2 \big(e^{h_t} - 1\big).$$

Duan (1995) develops an equilibrium approach to the valuation of contingent claims in this set-up, by postulating specific risk preferences of investors. According to his method (see also Kallsen and Taqqu (1998) for an alternative justification), the risk-neutral dynamics of the stock price are

$$\ln \left(\frac{S_{t+1}}{S_t} \right) = r - \frac{1}{2} h_t + \sqrt{h_t} \varepsilon_{t+1}$$

with the process h satisfying

$$h_{t+1} = \beta_0 + \beta_1 h_t + \beta_2 h_t (\varepsilon_{t+1} - c^*)^2,$$

where $c^* = c + \lambda$ and the random variable ε_{t+1} has, conditionally on the σ-field $\mathcal{F}_t$, the standard Gaussian distribution under $\mathbb{P}^*$. In principle, we are now in a position to use Monte Carlo simulations to value contingent claims. A simple lattice algorithm for pricing claims under the model described above was developed by Ritchken and Trevor (1997).

It is important to observe that GARCH models can be formally linked to stochastic volatility models through a limiting procedure. Specifically, most existing bivariate diffusion models of the stock price and its volatility can be obtained as limits of sequences of GARCH models (see Nelson (1990) and Duan (1996)). As a consequence, the lattice-based method of Ritchken and Trevor (1997) can also be used to value options in many existing continuous-time diffusion-type models of stochastic volatility.

Empirical studies reported in Taylor (1994) support the conclusion that both discrete-time aforementioned methodologies lead to reasonably close values for option prices. Lehar et al. (2002) show, however, that although the GARCH model (7.59)–(7.60) dominates the Hull and White (1987a, 1987b) stochastic volatility in option pricing, there are no notable differences between the models, when they are used for risk management. For more details on GARCH models, we refer to Engle and Bollerslev (1986), Nelson (1990, 1991), Bollerslev et al. (1992), Engle and Mustafa (1992), Engle and Ng (1993), Duan (1995), Shephard (1996), Ritchken and Trevor (1997), Kallsen and Taqqu (1998) and to the monograph by Gouriéroux (1997).

Jump-diffusion processes. Merton (1976) proposed an extension of the Black-Scholes model in which the stock returns no longer obey a stochastic process with continuous sample paths, but are subject to unpredictable jumps. It appears that in the case of a stock price following a jump-diffusion process with random jump sizes we deal with an incomplete model, in general, so that the exact replication of contingent claims is no longer possible. To solve the valuation problem, Merton (1976) therefore postulated that the jump risk is diversifiable, and thus the jump risk should not be rewarded. In other words, the market price for the jump risk should vanish. Formally, the dynamics of S under the martingale measure $\mathbb{P}^*$ in Merton's model are

$$dS_t = S_{t-}\big((r - \lambda\mu)\,dt + \sigma\,dW_t^* + dN_t^U\big),$$

where N^U is a pure jump process. The moments of jumps of N coincide with the jumps the Poisson process N with intensity λ, and the size of the ith jump is determined by a random variable U_i. Specifically, $(U_i)_{i \geq 1}$ is a sequence of independent, identically distributed random variables with the finite expectation $\mu = \mathbb{E}_{\mathbb{P}^*} U_i$. Hence the jump process N^U is a *marked Poisson process*, given as

$$N_t^U = \sum_{i=1}^{N_t} U_i, \quad \forall\, t \in \mathbb{R}_+.$$

It is not difficult to check that the relative price $S_t^* = S_t / B_t$ is indeed a $\mathbb{P}^*$-martingale. It is common to assume that the jump sizes U_i are lognormally distributed random variables independent of the Brownian motion W^* and the Poisson process N. Under this set of assumptions, one may derive a closed-form expression for the price of a European call option.

Unfortunately, an analysis of theoretical and practical issues related to modelling returns through discontinuous semimartingales would require a separate chapter. The interested reader is thus referred to original papers by, among others, Ball and Torous (1983a, 1985), Colwell and Elliott (1993), Amin (1993), Bardhan and Chao (1993, 1995), Mercurio and Runggaldier (1993), Lando (1995), Scott (1997), Zhang (1997), Jiang (1999), Kou (2002), and Gatheral (2003). Let us only mention that empirical studies of performance of alternative option pricing models done by Bakshi et al. (1997), Andersen and Andreasen (2000a), Duffie et al. (2000), and Cont and Tankov (2002) seem to indicate that a jump-diffusion (possibly with stochastic volatility) model fits better the observed volatility smile than a continuous stochastic volatility model, especially for shorter expirations. In a recent paper by Matytsin (1999), the author argues that a good fit to the observed volatility smile and, more importantly, reasonable values of parameters can be obtained if a jump-diffusion model is combined with a stochastic volatility model in such a way that jumps of the stock price and jumps of its volatility occur simultaneously (more precisely, each jump of a stock price is accompanied by a positive jump in volatility).

Lévy processes. A *Lévy process* is a process with right-continuous sample paths, which has stationary (that is, time-homogeneous), independent increments and, by convention, starts from 0 at time 0. An almost immediate important consequence of the definition of a Lévy process is the property of *infinite divisibility* of its marginal distributions. The characteristic function of an infinitely divisible distribution is well-known to be given by the classical Lévy-Khintchine formula that can be found in any advanced textbook on probability theory. The marginal distributions of a particular Lévy process are therefore relatively easy to identify. Since we do not go into details here, for an introduction to the theory of Lévy processes and an analysis of financial models driven by Lévy processes, we refer to Shiryaev (1999), Schoutens (2003), Cont and Tankov (2003), and Miyahara (2006). A nice survey of Lévy processes in the context of financial modelling is given by Geman (2002).

A Brownian motion is a classical example of a Lévy process. It should be stressed, however, that a Brownian motion with drift is the only Lévy process with continuous sample paths, and thus stock price given by a generic Lévy process has typically discontinuities. A Lévy process that does not possess a non-trivial Gaussian component is referred to as *pure jump* Lévy process. Such processes may have sample paths of finite variation, as opposed to Lévy processes with non-zero diffusion component. Some recent studies provide arguments in favor of pure jump Lévy processes as good models for the prices of all European options across all strikes and maturities at a given moment of time (see, e.g., Leblanc and Yor (1998) or Carr et al. (2003)).

Let us also observe that stable processes, briefly mentioned in Sect. 7.1.3, form a subclass of Lévy processes. However, except for the special case of a Gaussian process, stable processes are known to have infinite variance (and sometimes infinite expectation), which seems to be an undesirable feature for financial modelling. Hence we shall focus on two other particular classes of Lévy processes that attracted attention of researchers in recent years.

First, in order to account for skewness and kurtosis of market returns, several authors proposed to use alternative probability distributions leading to specific classes of Lévy models, such as the *hyperbolic* model or the *normal inverse Gaussian* model. Since both hyperbolic and normal inverse Gaussian probability distributions are known to be infinitely divisible, it is possible to use the corresponding Lévy processes (instead of a standard Brownian motion) as the driving noise in the dynamics of asset returns.

Studies of these models were done by, among others, Eberlein and Keller (1995), Barndorff-Nielsen et al. (1996, 2002), Barndorff-Nielsen (1997, 1998), Bibby and Sørensen (1997), Eberlein et al. (1998), Rydberg (1999), and Bingham and Kiesel (2002). Eberlein and Keller (1995) advocate the use of the class of *generalized hyperbolic Lévy motions* that includes both the hyperbolic motion and the normal inverse Gaussian motion as special cases. We refer to original papers for explicit representations of characteristic functions and the so-called Lévy densities of the aforementioned probability distributions.

Another pure jump Lévy model is the *CGMY* model proposed by Geman et al. (2001a) and Carr et al. (2002a). They show that a pure jump version of this model yields an excellent fit to market data for the SPX index and reasonably well reproduces the implied volatility surface for stock index options. Let us mention that Carr et al. (2003) conduct a detailed theoretical and empirical study of stochastic volatility models driven by a Lévy process.

From standard results of stochastic analysis, it follows that the only Lévy processes that have market completeness are the Brownian motion and the Poisson process, but not the combination of both these processes (see Jeanblanc-Picqué and Pontier (1990)). It is thus clear that most security market models driven by a Lévy process are incomplete, so that the valuation of derivative securities becomes a non-trivial issue in this set-up (see, for instance, Bühlmann et al. (1996) or Chan (1999)). If we take for granted that the price of a claim is given as the expected value under a (judiciously chosen) martingale measure, then the option price can be found by the general method of inverting the characteristic function of a given stochastic process, provided that it is explicitly known (see Carr and Madan (1999), Duffie et al. (2000), Lewis (2001), Gatheral (2003) or Lee (2004b)).

In regard to incompleteness of financial models driven by a Lévy process, let us mention that Dritschel and Protter (2000) introduced a family of discontinuous stochastic processes, indexed by a parameter $-2 \leq \beta \leq 0$, that are not Lévy processes since they do not have independent increments (except for the extreme case $\beta = 0$ corresponding to a Brownian motion), but are strong Markov processes under the unique martingale measure. For values of β close to zero, the sample paths of a process have similar features as sample paths of a Brownian motion, except for the fact that occasional jumps occur. We thus deal here with a parametrized family of complete models of the discontinuous stock price.

Random time change. A still another way of addressing the issue of skewness and fat tails of daily asset returns is based on the notion of a *subordinated* stochastic process. A subordinated process is obtained from a primitive stochastic process by using an independent random time change process, referred to as a *subordinator* (formally, a *subordinator* is an increasing Lévy process). It is thus rather clear that probability distributions corresponding to marginal distributions of subordinated stochastic processes are closely related to mixtures of probability distributions (cf. Sect. 7.3.3)

Clark (1973) was the first to propose a subordinated lognormal process in order to produce a model of stock returns with heavy tails, essentially by mixing the underlying Gaussian probability distributions with different variances. His ideas subsequently developed by, among others, Madan and Seneta (1990), Madan and Milne (1991), Eberlein and Jacod (1997), Leblanc and Yor (1996), and Bibby and Sørensen (1997). In particular, Madan and Seneta (1990) (see also Madan and Milne (1991)) propose to use a *gamma process* as a subordinator, thus producing the class of so-called *variance gamma* models.

Lesne and Prigent (2001) justify a continuous-time subordinated model by considering a sequence of Cox-Ross-Rubinstein models with random number of time steps, given by a sequence of stochastic processes $(Y^n)_{n \geq 1}$. Formally, the random variable Y_t^n specifies the number of price changes during the period $[0, t]$ in the nth discrete-time model. As expected, the random time change Y^n is assumed to be independent of the process describing the jump sizes of the stock price. They show that, under suitable assumptions, the associated sequence $(S^n)_{n \geq 1}$ converges in distribution to the price process S that satisfies $S_t = S_0 e_t^Z$, where the log-return process Z obeys the following SDE

$$dZ_t = \mu \, dY_t + \sigma \, dW_{Y_t}$$

where in turn W is a standard Brownian motion and Y is an increasing process independent of W. Hurst et al. (1997) and Lesne and Prigent (2001) analyze also the impact of the choice of a subordinator on option prices.

Following the conclusions of econometric studies reported in Jones et al. (1994) and Ghysels et al. (1995), in recent papers by Geman and Ané (1996), Geman et al. (1998), and Ané and Geman (2000) the authors focus on the concept of *business time* (or *transaction clock*), as opposed to the calendar time. Essentially, the business time is here understood as a stochastic process determined by the cumulative number of trades realized in a given time period. They argue that a Brownian motion subjected to a general random time change (not necessarily time-homogeneous), as opposed to a subordinated process, is a better tool to model asset returns. Their approach leads to a model with stochastic volatility that is directly related to the business time process.

7.6.2 Modelling of Volatility and Realized Variance

In the context of stochastic volatility models, it is worth mentioning that the (implied or instantaneous) volatility process may also be considered as the "underlying asset" in derivative contracts. The valuation of volatility futures and volatility futures options was examined by Grünbichler and Longstaff (1996), who postulated that the volatility process σ is governed by the following stochastic differential equation, previously encountered in Heston's model,

$$d\sigma_t = \kappa(v - \sigma_t)\, dt + \theta \sqrt{\sigma_t}\, dW_t.$$

In this regard, it is interesting to note that some exchanges have introduced daily quotations of a volatility index.[6] As a generic rule, the current value of such an index is evaluated on the basis of some kind of implied volatilities of a given portfolio of exchange traded options. Thus, it is based on the concept of the implied volatility, rather than the instantaneous (realized) volatility.

A discussion of the importance of hedging the volatility risk through trading (local) volatility derivatives (such as *volatility swaps*[7] and *volatility gadgets*) and, more generally, volatility-sensitive contracts (e.g., *compound forward-start straddle options*) can be found in Brenner and Galai (1989), Whaley (1993), Kani et al. (1996), Carr and Madan (1998b), and Brenner et al. (2000). In a recent work by Carr and Lee (2003) (see also Friz and Gatheral (2005)), the authors examine a robust valuation and hedging of derivatives on quadratic variation.

Swap contracts on *realized variation*, defined as the sum of squared daily logarithmic returns, have been trading for some years, so that there is a good reason to develop specific models for this path-dependent random quantity. Heston and Nandi (2000) proposed a version of Heston's model for pricing options on realized variance. In a more recent paper by Carr et al. (2005), the authors focus on a direct modelling of quadratic variation of logarithmic asset returns. Their goal is to develop a model for the valuation of options on the future realized quadratic variation of asset returns (see also Dufresne (2001) in this regard). Let us stress that for a pure jump Lévy process the quadratic variation is random, as opposed to the case of a Brownian motion, so that processes from this class are suitable for modelling realized variance. Carr et al. (2003) consider also additive processes of the type introduced by Sato (1991) (see

---

[6] In 1993, the Chicago Board Options Exchange has introduced the CBOE Volatility Index (VIX). It measures market expectations of near term volatility conveyed by S&P 500 option prices. CBOE offers also futures and options on VIX.

[7] Essentially, a *volatility swap* pays the difference between the realized volatility and the fixed swap rate determined at the outset of the contract.

also Sato (1999)) that are associated with the so-called *self-decomposable* distributions. Recall that a random variable X is *self-decomposable* if for every $0 < c < 1$ there exists an independent random variable Y^c such that the distribution of X and the distribution of $cX + Y^c$ coincide. A self-decomposable probability distribution is known to be infinitely divisible. Let us also recall that a process X is called γ *self-similar* if the distribution of the process X_{ct}, $t \in \mathbb{R}_+$, is identical as the distribution of the process $c^\gamma X_t$, $t \in \mathbb{R}_+$.[8]

Sato (1991) showed that for every self-decomposable probability distribution μ and any constant $\gamma > 0$ there exists an additive self-similar process X such that the distribution of the random variable X_1 coincides with μ. Carr et al. (2005) examine in detail relevant properties of Lévy and Sato processes (such as: infinite activity, infinite variation, complete monotonicity) and they empirically test the CGMY model. They conclude that this model outperforms the corresponding subordinated process (a time-changed Brownian motion with drift) in pricing options on realized variance.

For more detailed information on derivative pricing under uncertain and/or stochastic volatility, see Chriss (1996), Fouque et al. (2000a), Lewis (2000), Buff (2002), or Cont and Tankov (2003). Aït-Sahalia (2002, 2004) and Aït-Sahalia and Mykland (2003, 2004) examine a related statistical issue of identification of a continuous-time process from discretely observed data.

---

[8] A Brownian motion is manifestly self-similar, since for any $c > 0$ the distribution of the process W_{ct} coincides with the distribution of the process $\sqrt{c}\,W_t$.

8

Continuous-time Security Markets

This chapter furnishes a summary of basic results associated with continuous-time financial modelling. The first section deals with a continuous-time model, which is based on the Itô stochastic integral with respect to a semimartingale. Such a model of financial market, in which the arbitrage-free property hinges on the chosen class of admissible trading strategies, is termed the *standard market model* hereafter. We discuss the relevance of a judicious choice of a numeraire asset. On a more theoretical side, we briefly comment on the class of results – informally referred to as a *fundamental theorem of asset pricing* – which say, roughly, that the absence of arbitrage opportunities is equivalent to the existence of a martingale measure. The theory developed in this chapter applies both to stock markets and bond markets. It can thus be seen as a theoretical background to the second part of this text.

For simplicity, we restrict ourselves to the case of processes with continuous sample paths. Putting aside a somewhat higher level of technical complexity, jump-diffusion or Lévy-type models of discontinuous prices can be dealt with along the same lines. As was mentioned already, in a typical jump-diffusion model, price discontinuities are introduced through a Poisson component. In this regard, we refer to Cox and Ross (1975), Aase (1988), Madan et al. (1989), Elliott and Kopp (1990), Naik and Lee (1990), Shirakawa (1991), Ahn (1992), Mercurio and Runggaldier (1993), Cutland et al. (1993a), Björk (1995), Mulinacci (1996) or Scott (1997).

The second section deals with a particular example of a market model – the *multi-dimensional Black-Scholes market*. In contrast to Chap. 3 and 6, we focus on general questions such as market completeness, rather than on explicit valuation of contingent claims. Since the pricing of particular claims such as options is not examined in detail, let us mention here that in a complete multidimensional Black-Scholes model with constant interest rate and stock price volatility matrix, it is straightforward to derive a PDE – analogous to the Black-Scholes PDE – which is satisfied by the price of any path-independent European claim.

8.1 Standard Market Models

Consider a continuous-time economy with a trading interval $[0, T^*]$ for a fixed horizon date $T^* > 0$. Uncertainty in the economy is modelled by means of a family of complete filtered probability spaces $(\Omega, \mathbb{F}, \mathbb{P})$, $\mathbb{P} \in \mathcal{P}$, where $\mathcal{P}$ is a collection of mutually equivalent probability measures on $(\Omega, \mathcal{F}_{T^*})$. Each individual in the economy is characterized by a subjective probability measure $\mathbb{P}$ from $\mathcal{P}$. Events in our economy are revealed over time – simultaneously to all individuals – according to the filtration $\mathbb{F} = (\mathcal{F}_t)_{t \in [0, T^*]}$, which is assumed to satisfy the "usual conditions", meaning that (a) the underlying filtration $\mathbb{F}$ is right-continuous, i.e., $\mathcal{F}_t = \cap_{u>t} \mathcal{F}_u$ for every $t < T^*$; (b) $\mathcal{F}_0$ contains all null sets, i.e., if $B \subset A \in \mathcal{F}_0$ and $\mathbb{P}\{A\} = 0$, then $B \in \mathcal{F}_0$. We find it convenient to assume that the σ-field $\mathcal{F}_0$ is $\mathbb{P}$-trivial (for some, and thus for all, $\mathbb{P} \in \mathcal{P}$); that is, for every $A \in \mathcal{F}_0$ either $\mathbb{P}\{A\} = 0$ or $\mathbb{P}\{A\} = 1$.

There are k primary traded securities whose *price processes* are given by stochastic processes $Z^1, \ldots, Z^k$. We assume that $Z = (Z^1, \ldots, Z^k)$ follows a continuous, $\mathbb{R}^k$-valued semimartingale on $(\Omega, \mathbb{F}, \mathbb{P})$ for some – and thus for all – $\mathbb{P} \in \mathcal{P}$. This means that each process Z^i admits a unique decomposition $Z^i = Z_0^i + M^i + A^i$, where M^i is a continuous local martingale, and A^i is a continuous, adapted process of finite variation, with $M_0^i = A_0^i = 0$. For the definition and properties of the vector- and component-wise stochastic integrals with respect to a multidimensional semimartingale, we refer to Jacod (1979) and Protter (2003).

8.1.1 Standard Spot Market

We assume first that processes $Z^1, \ldots, Z^k$ represent the spot prices of some traded assets. It is convenient to assume that Z^k (and thus also $1/Z^k$) follows a continuous, strictly positive semimartingale. We take Z^k as a benchmark security; in other words, we choose Z^k as the numeraire asset. Following the seminal paper of Harrison and Pliska (1981) (see also Harrison and Kreps (1979)), we say that an $\mathbb{R}^k$-valued *predictable*[1] stochastic process $\phi_t = (\phi_t^1, \ldots, \phi_t^k)$, $t \in [0, T]$, is a *self-financing (spot) trading strategy* over time interval $[0, T]$ if the *wealth process* $V(\phi)$, which equals

$$V_t(\phi) \stackrel{\text{def}}{=} \phi_t \cdot Z_t = \sum_{i=1}^{k} \phi_t^i Z_t^i, \quad \forall t \in [0, T], \tag{8.1}$$

satisfies $V_t(\phi) = V_0(\phi) + G_t(\phi)$ for every $t \leq T$, where $G_t(\phi)$ stands for the *gains process*

$$G_t(\phi) \stackrel{\text{def}}{=} \int_0^t \phi_u \cdot dZ_u, \quad \forall t \in [0, T]. \tag{8.2}$$

---

[1] For the definition of a continuous-time *predictable* process, see Protter (2003) or Revuz and Yor (1999). Basically, *predictability* is a slight extension of the left-continuity of the sample paths of the process. In the case of the Itô integral with respect to continuous local martingales, it is actually enough to assume that the integrand is progressively measurable.

Observe that since $G_t(\phi)$ models the gains or losses realized up to and including time t, it is clear that we implicitly assume that the securities do not generate any revenue such as dividends. If the dividends of the risky security are paid continuously at the rate $\kappa_t = (\kappa_t^1, \ldots, \kappa_t^k)$, then the gains process includes also the accumulated dividend gains and thus

$$G_t(\phi) = \int_0^t \phi_u \cdot dZ_u + \int_0^t \phi_u \cdot Z_u^\kappa \, du, \quad \forall \, t \in [0, T],$$

where $Z_t^\kappa = (\kappa_t^1 Z_t^1, \ldots, \kappa_t^k Z_t^k)$. For convenience of exposition, we assume that $\kappa = 0$.

The financial interpretation of equalities (8.1)–(8.2) is that all changes in the wealth of the portfolio are due to capital gains, as opposed to withdrawals of cash or infusions of new funds. They reflect also the fact that the market that is the object of our studies is implicitly assumed to be frictionless, meaning that there are no transaction costs and no restrictions on short-selling. Equality (8.2) assumes that the process ϕ is sufficiently regular so that the stochastic integral in (8.2) is well-defined. The last property of ϕ is also invariant with respect to an equivalent change of the underlying probability measure. More exactly, if for some predictable process ϕ the Itô stochastic integral in (8.2) exists under some probability measure $\mathbb{P}$, then it exists also under any probability measure $\mathbb{Q}$ equivalent to $\mathbb{P}$; furthermore, the integrals evaluated under $\mathbb{P}$ and $\mathbb{Q}$ coincide (see Theorem IV.25 in Protter (2003)); that is

$$\mathbb{P} - \int_0^t \phi_u \cdot dZ_u = \mathbb{Q} - \int_0^t \phi_u \cdot dZ_u, \quad \forall \, t \in [0, T].$$

To avoid technicalities, we assume throughout, unless otherwise specified,[2] that the trading process ϕ is locally bounded. Let us comment briefly on the implications of this assumption. First, if Z is a $\mathbb{P}$-local martingale, then the integral

$$N_t = \int_0^t \phi_u \cdot dZ_u, \quad \forall \, t \in [0, T],$$

is known (see Theorem IV.29 in Protter (2003)) to follow a $\mathbb{P}$-local martingale (in general, the Itô integral of a predictable process with respect to a local martingale is not necessarily a local martingale). Second, if ϕ is a locally bounded process, then the Itô stochastic integral in (8.2) is component-wise; that is

$$\int_0^t \phi_u \cdot dZ_u = \sum_{i=1}^k \phi_u^i \, dZ_u^i, \quad \forall \, t \in [0, T].$$

We denote by Φ_T the class of all self-financing spot trading strategies over the time interval $[0, T]$. Similarly, $\Phi = \cup_{T \le T^*} \Phi_T$ stands for the class of all self-financing trading strategies.

---

[2] This restriction becomes inconvenient when the concept of *completeness* of a market model is examined.

Definition 8.1.1 A strategy $\phi \in \Phi_T$ is called an *arbitrage opportunity* if the wealth process $V(\phi)$ satisfies, for some (hence for all) $\mathbb{P} \in \mathcal{P}$, the following set of conditions

$$V_0(\phi) = 0, \quad \mathbb{P}\{V_T(\phi) \geq 0\} = 1, \quad \text{and} \quad \mathbb{P}\{V_T(\phi) > 0\} > 0.$$

Arbitrage opportunities represent the limitless creation of wealth through risk-free profit and thus they should not exist in a well-functioning market (in practice, they should disappear rapidly). In the case of a continuous-time model, the class Φ of all self-financing strategies is usually too large; that is, arbitrage opportunities are not ruled out a priori from Φ. Put another way, the "natural" market model $\tilde{\mathcal{M}} = (Z, \Phi)$ is not well suited to the purpose of arbitrage pricing. To circumvent this drawback, for any $T \leq T^*$, it is necessary to restrict attention to a certain subclass $\Psi_T \subset \Phi_T$, referred to as admissible trading strategies (the class Ψ_T will be defined later). Given a collection $\Psi = \cup_{T \leq T^*} \Psi_T$ of admissible trading strategies, we say that a security market $\mathcal{M} = (Z, \Psi)$ is *arbitrage-free* if there are no arbitrage opportunities in Ψ. As usual, a *European contingent claim* that settles at time T is modelled by means of an $\mathcal{F}_T$-measurable random variable X.

Definition 8.1.2 A strategy $\phi \in \Psi_T$ *replicates* a European contingent claim X settling at time T if $V_T(\phi) = X$. If a claim X admits at least one replicating strategy ϕ from Ψ_T, it is said to be *attainable* in $\mathcal{M}$ and the wealth process $V(\phi)$ is referred to as the *replicating process* of X.

Two following issues should be addressed. First, it is essential to ensure the uniqueness of a replicating process[3] for any claim attainable in $\mathcal{M}$. By definition, a contingent claim X is *uniquely replicated* in $\mathcal{M}$ if the replicating process of X in $\mathcal{M}$ is unique, up to indistinguishability[4] of stochastic processes. Second, we have to exclude those replicating processes that constitute arbitrage opportunities. As was mentioned already, to develop a theory of arbitrage pricing in a continuous-time setting, we need first to make a judicious choice of the class Ψ of admissible trading strategies. The definition of admissibility hinges on the notion of a martingale measure for relative prices.

Let Z^* stand for the relative price process of primary securities when the price Z^k is chosen as a numeraire, i.e., $Z_t^* = Z_t/Z_t^k = (Z_t^1/Z_t^k, \ldots, 1)$. For any trading strategy ϕ, we denote by $V^*(\phi)$ its relative wealth

$$V_t^*(\phi) = V_t(\phi)/Z_t^k = \phi_t \cdot Z_t^*, \quad \forall\, t \in [0, T^*].$$

For brevity, we say that a probability measure $\mathbb{Q}$ is *equivalent* to $\mathcal{P}$ if it is equivalent to some probability measure $\mathbb{P}$ from $\mathcal{P}$, and thus, of course, to any probability measure $\mathbb{P}$ from $\mathcal{P}$.

---

[3] If the replicating process for X were not uniquely determined, it would be tempting to search for the *minimal* replicating process, in a suitable sense.

[4] Two processes Y^1 and Y^2, defined on a common probability space $(\Omega, \mathbb{F}, \mathbb{P})$, are said to be *indistinguishable* if $\mathbb{P}\{\omega \in \Omega \mid Y_t^1(\omega) = Y_t^2(\omega), \ \forall\, t \in [0, T]\} = 1$.

Definition 8.1.3 A probability measure $\mathbb{P}^*$ on $(\Omega, \mathcal{F}_{T^*})$, equivalent to $\mathcal{P}$, is called (a) a *martingale measure for* Z^* if Z^* follows a local martingale under $\mathbb{P}^*$; (b) a *martingale measure for* $\mathcal{M}^k$ if the relative wealth process $V^*(\phi)$ of any strategy $\phi \in \Phi$ follows a local martingale under $\mathbb{P}^*$.

Note that the superscript k in $\mathcal{M}^k$ corresponds to the choice of the kth security as a numeraire asset. Let us denote by $\mathcal{P}(Z^*)$ and by $\mathcal{P}(\mathcal{M}^k)$ the class, possibly empty, of all martingale measures for Z^* and for $\mathcal{M}^k$ respectively. Then we have the following simple result, which shows that $\mathcal{P}(Z^*)$ and $\mathcal{P}(\mathcal{M}^k)$ coincide.

Lemma 8.1.1 *A probability measure* $\mathbb{P}^*$ *is a martingale measure for* Z^* *if and only if it is a martingale measure for* $\mathcal{M}^k$.

Proof. For the inclusion $\mathcal{P}(Z^*) \subseteq \mathcal{P}(\mathcal{M}^k)$, note that using Itô's integration by parts formula, we obtain

$$
\begin{aligned}
dV_t^*(\phi) &= (1/Z_t^k)\, dV_t(\phi) + V_t(\phi)\, d(1/Z_t^k) + d\langle V(\phi), 1/Z^k \rangle_t \\
&= \phi_t \cdot \big((1/Z_t^k)\, dZ_t + Z_t\, d(1/Z_t^k) + d\langle Z, 1/Z^k \rangle_t \big) \\
&= \phi_t \cdot dZ_t^*,
\end{aligned}
$$

since $dV_t(\phi) = \phi_t \cdot dZ_t$. Put differently, we have

$$
V_t^*(\phi) = V_0^*(\phi) + \int_0^t \phi_u \cdot dZ_u^* = V_0^*(\phi) + G_t^*(\phi), \tag{8.3}
$$

where $G^*(\phi)$ follows a local martingale under $\mathbb{P}^*$ (recall that ϕ is a locally bounded process). For the converse inclusion, it is enough to observe that buy-and-hold strategies of the form $(0, \dots, 1, \dots, 0)$ belong to Φ. □

We assume from now on that the class $\mathcal{P}(\mathcal{M}^k)$ is non-empty, i.e., that there exists at least one martingale measure for the market model $\mathcal{M}^k$. For a fixed, but otherwise arbitrary, martingale measure $\mathbb{P}^* \in \mathcal{P}(\mathcal{M}^k)$ we find it convenient to introduce the class of $\mathbb{P}^*$-admissible trading strategies in the following way.

Definition 8.1.4 A trading strategy $\phi \in \Phi_T$ is said to be $\mathbb{P}^*$-*admissible* if the relative gains process

$$
G_t^*(\phi) = \int_0^t \phi_u \cdot dZ_u^*, \quad \forall\, t \in [0, T], \tag{8.4}
$$

follows a (true) martingale under $\mathbb{P}^*$. The class of all $\mathbb{P}^*$-admissible trading strategies from Φ_T is denoted by $\Phi_T^k(\mathbb{P}^*)$. We write $\Phi^k(\mathbb{P}^*) = \cup_{T \leq T^*} \Phi_T^k(\mathbb{P}^*)$.

We denote by $\mathcal{M}^k(\mathbb{P}^*)$ the spot market model $(Z, \Phi^k(\mathbb{P}^*))$. We say that a contingent claim X is *attainable* in $\mathcal{M}^k(\mathbb{P}^*)$ if X is $\Phi^k(\mathbb{P}^*)$-attainable. Under our hypotheses, if a strategy ϕ is $\mathbb{P}^*$-admissible, then the process G^* given by (8.4) follows a local martingale under any martingale measure for $\mathcal{M}^k$. If we relax the assumption that the trading strategy is locally bounded, then G^* is not necessarily a local martingale under a probability measure $\mathbb{Q}$ equivalent to $\mathbb{P}^*$.

Proposition 8.1.1 *For any martingale measure* $\mathbb{P}^* \in \mathcal{P}(\mathcal{M}^k)$, *the spot market model* $\mathcal{M}^k(\mathbb{P}^*)$ *is arbitrage-free. Any contingent claim* X *attainable in* $\mathcal{M}^k(\mathbb{P}^*)$ *admits a unique replicating process in* $\Phi^k(\mathbb{P}^*)$.

Proof. To prove the first statement, it is enough to verify that the class $\Phi^k(\mathbb{P}^*)$ of trading strategies does not contain arbitrage opportunities. For any $\phi \in \Phi^k(\mathbb{P}^*)$, with $V_0(\phi) = 0$, we have $V^*(\phi) = G^*(\phi)$, where $G^*(\phi)$ follows a martingale under $\mathbb{P}^*$, with $G_0^*(\phi) = 0$. This immediately yields equality $\mathbb{E}_{\mathbb{P}^*}(V_{T*}^*) = 0$. Since $V_T(\phi) = Z_T^k V_T^*(\phi)$, and Z_T^k is a strictly positive random variable $\mathbb{P}^*$-a.s., it is easily seen that if $V_0(\phi) = 0$ and $\mathbb{P}^*\{V_T \geq 0\} = 1$ then $\mathbb{P}^*\{V_T(\phi) = 0\} = 1$. The absence of arbitrage opportunities in $\Phi^k(\mathbb{P}^*)$ thus follows immediately from the equivalence of $\mathbb{P}^*$ and $\mathcal{P}$. For the second assertion, note that if the uniqueness of a replicating process were violated, it would be possible to construct an arbitrage opportunity along similar lines to the proof of Proposition 2.6.1, by investing in the kth asset (note that the relative price of this asset is manifestly a $\mathbb{P}^*$-martingale). $\quad\square$

Let X be a European contingent claim attainable in $\mathcal{M}^k(\mathbb{P}^*)$. The wealth process $V(\phi)$ of any $\mathbb{P}^*$-admissible trading strategy which replicates X is called the *arbitrage price of* X *in* $\mathcal{M}^k(\mathbb{P}^*)$. We denote it by $\pi^k(X \mid \mathbb{P}^*)$. For any claim X (not necessarily attainable in $\mathcal{M}^k(\mathbb{P}^*)$), we introduce the following definition.

Definition 8.1.5 Let X be a European contingent claim settling at time T such that the random variable X/Z_T^k is $\mathbb{P}^*$-integrable. The *expected value process of* X *relative to* Z^k *under* $\mathbb{P}^*$ is defined by setting

$$v_t^k(X \mid \mathbb{P}^*) = Z_t^k \, \mathbb{E}_{\mathbb{P}^*}(X/Z_T^k \mid \mathcal{F}_t), \quad \forall \, t \in [0, T].$$

The next result shows that the arbitrage price of any contingent claim agrees with the associated expected value process.

Proposition 8.1.2 *For any European contingent claim* X *that settles at time* T *and is attainable in* $\mathcal{M}(\mathbb{P}^*)$, *we have*

$$\pi_t^k(X \mid \mathbb{P}^*) = v_t^k(X \mid \mathbb{P}^*), \quad \forall \, t \in [0, T]. \tag{8.5}$$

Proof. Let us define the relative price of X by setting

$$\pi_t^{k,*}(X \mid \mathbb{P}^*) \stackrel{\text{def}}{=} \pi_t^k(X \mid \mathbb{P}^*)/Z_t^k. \tag{8.6}$$

For any replicating strategy $\phi \in \Phi^k(\mathbb{P}^*)$, we have

$$\pi_t^{k,*}(X \mid \mathbb{P}^*) = V_t^*(\phi) = \mathbb{E}_{\mathbb{P}^*}(X/Z_T^k \mid \mathcal{F}_t), \tag{8.7}$$

since $V_T^*(\phi) = X/Z_T^k$. In particular, it is clear that the random variable X/Z_T^k is $\mathbb{P}^*$-integrable. Combining (8.6) with (8.7), we get (8.5). $\quad\square$

An apparent drawback of Definition 8.1.4 is the dependence of the class of admissible trading strategies on the choice of a martingale measure.[5] To circumvent this deficiency, we might postulate that a self-financing strategy is admissible if it belongs to the class $\Phi^k(\mathbb{P}^*)$ for some $\mathbb{P}^* \in \mathcal{P}(\mathcal{M}^k)$. For this choice of the class of admissible trading strategies, the spot market model is manifestly arbitrage-free. However, it would be unsatisfactory to have two distinct arbitrage prices for an attainable claim. We therefore need to show that for any two martingale measures, $\mathbb{P}_1^*$ and $\mathbb{P}_2^*$ say, the arbitrage prices $\pi^k(X \mid \mathbb{P}_1^*)$ and $\pi^k(X \mid \mathbb{P}_2^*)$ agree. Since this problem is rather difficult to handle without additional assumptions, we shall restrict ourselves from now on to strategies whose wealth, expressed in units of the kth asset, is bounded from below by a constant (this idea goes back at least to Dybvig and Huang (1988)). Intuitively, this means that the maximal leverage – whose level is arbitrarily large but finite – is known in advance.

Definition 8.1.6 A strategy $\phi \in \Phi_T$ is said to be *tame relative to* Z^k if there exists $m \in \mathbb{R}$ such that the relative wealth $V^*(\phi) = V(\phi)/Z^k$ satisfies $V_t^*(\phi) \geq m$ for every $t \in [0, T]$.

Note that the class of tame strategies is manifestly invariant with respect to an equivalent change of a probability measure, however it is not invariant with respect to the choice of the numeraire asset.

Definition 8.1.7 A trading strategy $\phi \in \Phi_T$ is said to be *admissible relative to* Z^k if it is a tame strategy and is $\mathbb{P}^*$-admissible for some martingale measure $\mathbb{P}^* \in \mathcal{P}(\mathcal{M}^k)$. We denote by Φ_0^k the class of all strategies that are admissible relative to Z^k. The pair $\mathcal{M}^k = (Z, \Phi_0^k)$ is referred to as the *standard spot market model*.

Proposition 8.1.3 *The standard spot market model $\mathcal{M}^k$ is arbitrage-free. The arbitrage price of any contingent claim attainable in $\mathcal{M}^k$ is well-defined. If a European contingent claim X settling at time T is $\Phi_0^k(\mathbb{P}_i^*)$-attainable for $i = 1, 2$, then for every $t \in [0, T]$*

$$\pi_t^k(X \mid \mathbb{P}_1^*) = \pi_t^k(X \mid \mathbb{P}_2^*),$$

or equivalently,

$$\mathbb{E}_{\mathbb{P}_1^*}\big(X/Z_T^k \mid \mathcal{F}_t\big) = \mathbb{E}_{\mathbb{P}_2^*}\big(X/Z_T^k \mid \mathcal{F}_t\big).$$

Proof. Under any probability measure $\mathbb{P}^*$ from $\mathcal{P}(\mathcal{M}^k)$, the relative wealth $V^*(\phi)$ of a strategy ϕ, admissible relative to Z^k, follows a local martingale bounded from below. Hence it is a $\mathbb{P}^*$-supermartingale (this property is an immediate consequence of Fatou's lemma and the definition of a local martingale; see Proposition A.7.1); that is

$$\mathbb{E}_{\mathbb{P}^*}(V_u^*(\phi) \mid \mathcal{F}_t) \leq V_t^*(\phi), \quad \forall\, t \leq u \leq T.$$

---

[5] As was mentioned already, the martingale property of the process of relative wealth is not invariant with respect to an equivalent change of a martingale measure, in general.

In particular, if $V_0(\phi) = 0$ then

$$\mathbb{E}_{\mathbb{P}^*}\left(V_T^*(\phi)\right) \leq V_0^*(\phi) = 0,$$

and thus, for any $\mathbb{P} \in \mathcal{P}$, if $\mathbb{P}\{V_T(\phi) \geq 0\} = 1$ then $\mathbb{P}\{V_T(\phi) > 0\} = 0$. This shows that there are no arbitrage opportunities in Φ_0^k. To prove the second statement, consider a Φ_0^k-attainable contingent claim X that settles at time T. Let $\phi, \psi \in \Phi_0^k$ be two strategies such that $V_T(\phi) = V_T(\psi) = X$. Let $\phi \in \Phi(\mathbb{P}_1^*)$ and $\psi \in \Phi(\mathbb{P}_2^*)$. Then

$$V_t(\phi) = \pi_t^k(X \mid \mathbb{P}_1^*) = Z_t^k \, \mathbb{E}_{\mathbb{P}_1^*}\left(X/Z_T^k \mid \mathcal{F}_t\right)$$

and

$$V_t(\psi) = \pi_t^k(X \mid \mathbb{P}_2^*) = Z_t^k \, \mathbb{E}_{\mathbb{P}_2^*}\left(X/Z_T^k \mid \mathcal{F}_t\right).$$

On the other hand, from (8.3) we get

$$V_T^*(\phi) = V_t^*(\phi) + \int_t^T \phi_u \cdot dZ_u^* = V_t^*(\phi) + G_T^*(\phi) - G_t^*(\phi),$$

where $G^*(\phi)$ is a $\mathbb{P}_1^*$-martingale. Since $G^*(\phi)$ follows a supermartingale under $\mathbb{P}_2^*$, we have

$$V_t^*(\psi) = \mathbb{E}_{\mathbb{P}_2^*}\left(X/Z_T^k \mid \mathcal{F}_t\right) = V_t^*(\phi) + \mathbb{E}_{\mathbb{P}_2^*}\left(G_T^* - G_t^* \mid \mathcal{F}_t\right) \leq V_t^*(\phi),$$

and thus $V_t(\psi) \leq V_t(\phi)$. Interchanging the roles, we find that $V_t(\phi) \leq V_t(\psi)$ and thus the equality $V_t(\phi) = V_t(\psi)$ is satisfied for every $t < T$. $\qquad\square$

The common value of an arbitrage price $\pi^k(X \mid \mathbb{P}^*)$ is denoted $\pi^k(X)$ and is referred to as the *arbitrage price* of X in $\mathcal{M}^k$. The following result shows that the arbitrage price in $\mathcal{M}^k$ equals the maximal expected value over all martingale measures. Furthermore, it corresponds also to the minimal cost of a tame replicating strategy.

Proposition 8.1.4 *Let X be a contingent claim that settles at time T and is attainable in the market model $\mathcal{M}^k$. The arbitrage price of X satisfies*

$$\pi_0^k(X) = \sup_{\mathbb{P}^* \in \mathcal{P}(\mathcal{M}^k)} Z_0^k \, \mathbb{E}_{\mathbb{P}^*}(X/Z_T^k) = \inf_{\phi \in \Theta(X)} V_0(\phi), \qquad (8.8)$$

where we denote by $\Theta(X)$ the class of all tame trading strategies that replicate a claim X.

Proof. Since a claim X is $\Phi_0^k(\mathbb{P}_1^*)$-attainable for some martingale measure $\mathbb{P}_1^* \in \mathcal{P}(\mathcal{M}^k)$, there exists a strategy $\phi \in \Phi_0^k(\mathbb{P}_1^*)$ such that

$$X/Z_T^k = V_0^*(\phi) + \int_0^T \phi_u \cdot dZ_u^* = \pi_0^k(X)/Z_0^k + G_T^*(\phi).$$

Under our assumptions, the process $G^*(\phi)$ follows a supermartingale under any martingale measure $\mathbb{P}^* \in \mathcal{P}(\mathcal{M}^k)$, and thus

$$\mathbb{E}_{\mathbb{P}^*}(X/Z_T^k) \leq \pi_0^k(X)/Z_0^k = \mathbb{E}_{\mathbb{P}_1^*}(X/Z_T^k). \qquad (8.9)$$

This ends the proof of the first equality. For the second, assume that ψ is an arbitrary tame strategy belonging to $\Theta(X)$, so that

$$X/Z_T^k = V_0^*(\psi) + \int_0^T \psi_u \cdot dZ_u^* = V_0^*(\psi) + G_T^*(\psi).$$

Once again, $G^*(\psi)$ is a supermartingale under any martingale measure $\mathbb{P}^* \in \mathcal{P}(\mathcal{M}^k)$, so that $\mathbb{E}_{\mathbb{P}^*} G_T^*(\psi) \le 0$. Using (8.9), we obtain

$$V_0^*(\psi) \ge \sup_{\mathbb{P}^* \in \mathcal{P}(\mathcal{M}^k)} \mathbb{E}_{\mathbb{P}^*}(X/Z_T^k). \tag{8.10}$$

Since X is attainable in $\mathcal{M}^k$, there exists a tame strategy ψ and a martingale measure $\mathbb{P}^*$ such that the equality holds in (8.10). This proves the second equality in (8.8). □

In view of the last result, it is tempting to conjecture that the arbitrage price of a bounded attainable claim satisfies, for any $t \in [0, T]$,

$$\pi_t^k(X) = \text{ess sup}_{\mathbb{P}^* \in \mathcal{P}(\mathcal{M}^k)} Z_t^k \, \mathbb{E}_{\mathbb{P}^*}\big(X/Z_T^k \mid \mathcal{F}_t\big) \tag{8.11}$$

and

$$\pi_t^k(X) = \text{ess inf}_{\phi \in \Theta(X)} V_t(\phi). \tag{8.12}$$

It is not at all clear, however, whether the right-hand sides of (8.11)–(8.12) represent stochastic processes (they are defined almost surely, for any fixed t). Nevertheless, it is possible to show, in some circumstances, that the right-hand side of (8.11) is a well-defined stochastic process, with almost all sample paths right-continuous and with finite left-hand limits (that is, an RCLL process) and equality (8.11) is valid (cf. El Karoui and Quenez (1995)). For the second equality, we can show that the (relative) arbitrage price of an attainable claim can be characterized as the smallest supermartingale, under any martingale measure $\mathbb{P}^* \in \mathcal{P}(\mathcal{M}^k)$, that equals X/Z_T^k at time T. More precisely, if $\mathbb{P}^*$ is a fixed, but arbitrary, martingale measure and Y is an arbitrary $\mathbb{P}^*$-supermartingale that satisfies $Y_T = X/Z_T^k$ then for any $t \in [0, T]$ we obtain (for the definition of $\pi_t^{k,*}(X)$ see the proof of Proposition 8.1.2)

$$Y_t \ge \mathbb{E}_{\mathbb{P}^*}(X/Z_T^k \mid \mathcal{F}_t) = \pi_t^{k,*}(X).$$

If a contingent claim X is $\mathbb{P}^*$-attainable for some martingale measure $\mathbb{P}^*$, and the random variable X/Z_T^k is bounded from below by a constant, then it is attainable by means of a tame strategy; that is, X is attainable in $\mathcal{M}^k$. Clearly, any claim attainable in $\mathcal{M}^k$ is bounded from below in this sense.

Let X be a $\mathbb{P}^*$-attainable contingent claim such that X/Z_T^k is bounded from below. Then, in view of (8.8), for any martingale measure $\mathbb{Q}^* \in \mathcal{P}(\mathcal{M}^k)$, we have

$$\mathbb{E}_{\mathbb{Q}^*}(X/Z_T^k) \le \mathbb{E}_{\mathbb{P}^*}(X/Z_T^k) < \infty. \tag{8.13}$$

Suppose that a claim X that is attainable in $\mathcal{M}^k$, is also bounded from above, meaning that $X/Z_T^k \le m$ for some $m \in \mathbb{R}$. Let ϕ be a tame replicating strategy for X, so that $V^*(\phi)$ is a $\mathbb{P}^*$-martingale under some martingale measure $\mathbb{P}^*$. In this case, the relative wealth $V^*(\phi)$ follows a martingale under any martingale measure $\mathbb{Q}^* \in \mathcal{P}(\mathcal{M}^k)$. Indeed, using standard arguments, the process $V^*(\phi)$ can be shown to be simultaneously a supermartingale and a submartingale under any probability measure $\mathbb{Q}^*$ from $\mathcal{P}(\mathcal{M}^k)$. Consequently, we have equality in (8.13). This leads to the following corollary.

Corollary 8.1.1 *Let X be a contingent claim such that the random variable $|X|/Z_T^k$ is bounded by a constant. Then X is $\mathbb{P}^*$-attainable for some $\mathbb{P}^* \in \mathcal{P}(\mathcal{M}^k)$ if and only if X is $\mathbb{P}^*$-attainable for every $\mathbb{P}^* \in \mathcal{P}(\mathcal{M}^k)$. Furthermore, X is attainable in $\mathcal{M}^k$ if and only if the mapping $\mathbb{P}^* \mapsto \mathbb{E}_{\mathbb{P}^*}(X/Z_T^k)$ from $\mathcal{P}(\mathcal{M}^k)$ to $\mathbb{R}$ is constant.*

Proof. The first assertion is clear. The "only if" clause of the second assertion is also evident (cf. Proposition 8.1.3). It remains to show that the "if" clause is valid. To this end, we employ a result due to Jacka (1992) (for related results, see also Stricker (1984), Ansel and Stricker (1994) and Delbaen (1992)). It states[6] that if X is a claim bounded from below,[7] and $\mathbb{P}^*$ is a martingale measure such that

$$\mathbb{E}_{\mathbb{Q}^*}(X/Z_T^k) \le \mathbb{E}_{\mathbb{P}^*}(X/Z_T^k) < \infty, \quad \forall \mathbb{Q}^* \in \mathcal{P}(\mathcal{M}),$$

then X is $\mathbb{P}^*$-attainable. In particular, if H is a constant mapping (with finite value), then X is $\mathbb{P}^*$-attainable for any martingale measure $\mathbb{P}^*$. $\square$

Remarks. (a) Consider a contingent claim X_A that settles at time T^* and has the form $X_A = Z_{T^*}^k \mathbb{1}_A$, where A is an event from $\mathcal{F}_{T^*}$. In view of Corollary 8.1.1, the claim X_A is attainable in the model $\mathcal{M}^k$ if and only if the mapping $\mathbb{P}^* \mapsto \mathbb{E}_{\mathbb{P}^*}(X_A/Z_{T^*}^k) = \mathbb{P}^*(A)$, from $\mathcal{P}(\mathcal{M}^k)$ to $\mathbb{R}$, is constant. This shows immediately that if the uniqueness of a martingale measure does not hold, then there exists an event $A \in \mathcal{F}_{T^*}$ such that the claim X_A is not attainable in $\mathcal{M}^k$.

(b) At first glance, Corollary 8.1.1 might also suggest that the choice of a martingale measure is not relevant if we wish to price a claim that is attainable in $\mathcal{M}^k$. Unfortunately, this is not the case, in general; that is, the inequality in (8.13) can be strict. This means that there exists a contingent claim X, attainable in $\mathcal{M}^k$, and such that the mapping $\mathbb{P}^* \mapsto \mathbb{E}_{\mathbb{P}^*}(X/Z_T^k)$ from $\mathcal{P}(\mathcal{M}^k)$ to $\mathbb{R}$ is not constant (for an example, see Schachermayer (1994)). Put another way, there exist (unbounded) contingent claims X that are $\mathbb{P}^*$-attainable under some martingale measure $\mathbb{P}^*$, and are not $\mathbb{Q}^*$-attainable under a martingale measure $\mathbb{Q}^*$. On the other hand, the characterization of attainable claims given in Corollary 8.1.1 is also valid for certain classes of unbounded claims (cf. El Karoui and Quenez (1995)).

(c) Suppose that we have chosen the class of all tame strategies as the class of all admissible trading strategies. From the first part of the proof of Proposition 8.1.3,

---

[6] Since the proof relies on the duality between certain linear topological spaces of stochastic processes (which will not be introduced here), it is omitted.

[7] Actually, Jacka focuses on the case of nonnegative claims. The case of contingent claims that are bounded from below can be derived easily from his results.

it follows that such a market model is arbitrage-free. The uniqueness of the replicating process fails to hold, in general; that is, there exists an attainable contingent claim X and two tame replicating strategies ϕ and ψ such that $V_0(\psi) \neq V_0(\phi)$.[8] To circumvent this deficiency, we may define the arbitrage price of X as the minimal cost of a tame replicating strategy, provided that there exists a tame trading strategy which realizes the minimum (in this case, a claim is said to be attainable). Such an approach to admissibility of strategies leads to the same class of attainable claims as in the market model (Z, Φ_0^k), and to the same prices for all attainable claims.

(d) Attainability of contingent claims and characterization of the arbitrage price for the case of the multidimensional version of the Black-Scholes model are examined in a number of papers; to mention a few: Pagès (1987), Karatzas et al. (1991) and El Karoui and Quenez (1995). For more general results in this vein - proved in an abstract semimartingale setting – the interested reader is referred to Delbaen and Schachermayer (1994a, 1994b, 1997a, 1997b), Kramkov (1996), and Föllmer and Kramkov (1997).

8.1.2 Futures Market

Generally speaking, given an arbitrage-free model of the spot market, the futures price of any traded security can be, in principle, derived using no-arbitrage arguments. In some circumstances, however, on may find it preferable to impose conditions directly on the futures price dynamics of certain assets. For this reason, we shall now comment on a direct construction of an arbitrage-free model involving the futures prices of $k - 1$ assets and, in addition, the spot price of one traded security. Let us write $(Z^1, \ldots, Z^{k-1}) = (f^1, \ldots, f^{k-1}) = f$. In financial interpretation, each process f^i is assumed to represent the futures price of a certain asset (corresponding to the delivery date $T^i \geq T^*$). As before, the process Z^k stands for the spot price of some traded security. For convenience, we assume that Z^k and $1/Z^k$ follow continuous, strictly positive semimartingales. In view of specific features of futures contracts, we need modify the definition of a self-financing spot trading strategy as follows.

Definition 8.1.8 An $\mathbb{R}^k$-valued predictable process $\phi = (\phi^1, \ldots, \phi^k)$ is a *self-financing* futures trading strategy if the *wealth process* $V^f(\phi)$, defined as $V_t^f(\phi) = \phi_t^k Z_t^k$, satisfies $V_t^f(\phi) = V_0^f(\phi) + G_t^f(\phi)$, where *gains process* $G^f(\phi)$ is given by the formula

$$G_t^f(\phi) \stackrel{\text{def}}{=} \int_0^t \phi_u^f \cdot df_u + \int_0^t \phi_u^k \, dZ_u^k, \quad \forall\, t \in [0, T],$$

where $\phi^f = (\phi^1, \ldots, \phi^{k-1})$. We write Φ^f to denote the class of all self-financing futures trading strategies.

---

[8] For instance, in the Black-Scholes framework, it is enough to take $X = 0$, the strategy ϕ as in Example 3.1.1, and ψ equal to 0 identically. Note that the strategy $\psi - \phi = -\phi$ is not an arbitrage opportunity, since it is not a tame strategy.

The arbitrage-free futures market model $\mathcal{M}^f = (Z, \Psi^f)$ relies on the specification of the class Ψ^f of admissible trading strategies. Of course, the concepts of a contingent claim, replicating process, arbitrage opportunity and an arbitrage-free market remain valid. For the reader's convenience, we shall formulate the definitions of martingale measures explicitly. First, the *martingale measure for f* is any probability measure $\tilde{\mathbb{P}}$ equivalent to $\mathcal{P}$ such that f follows a local martingale under $\tilde{\mathbb{P}}$. Second, a probability measure $\tilde{\mathbb{P}}$, equivalent to $\mathcal{P}$, is called a *martingale measure for the futures market model* $\mathcal{M}^f$ if, for arbitrary futures trading strategy $\phi \in \Phi^f$, the relative wealth process $\tilde{V}_t^f(\phi) = V_t^f(\phi)/Z_t^k$ follows a local martingale under $\tilde{\mathbb{P}}$. We denote by $\mathcal{P}(f)$ and $\mathcal{P}(\mathcal{M}^f)$ the class of all martingale measures for the process f and for the market model $\mathcal{M}^f$ respectively. The following result is a direct counterpart of Proposition 8.1.1.

Proposition 8.1.5 *Assume that for every* $i = 1, \dots, k - 1$ *we have*

$$\langle f^i, 1/Z^k \rangle_t = 0, \quad \forall t \in [0, T^*]. \tag{8.14}$$

Then any probability measure $\tilde{\mathbb{P}}$ *from* $\mathcal{P}(f)$ *is a martingale measure for the futures market model* $\mathcal{M}^f$*; that is,* $\mathcal{P}(f) \subseteq \mathcal{P}(\mathcal{M}^f)$.

Proof. From Itô's formula, we have

$$d(1/Z_t^k) = -(1/Z_t^k)^2 \, dZ_t^k + (1/Z_t^k)^3 \, d\langle Z^k, Z^k \rangle_t.$$

This implies that $d\langle Z^k, 1/Z^k \rangle_t = -(Z_t^k)^{-2} \, d\langle Z^k, Z^k \rangle_t$. An application of Itô's integration by parts formula yields

$$\begin{aligned}
d\tilde{V}_t^f &= (1/Z_t^k) \, dV_t^f + V_t^f \, d(1/Z_t^k) + d\langle V^f, 1/Z^k \rangle_t \\
&= (Z_t^k)^{-1}(\phi_t^f \cdot df_t + \phi_t^k \, dZ_t^k) + \phi_t^k Z_t^k \, d(1/Z_t^k) \\
&\quad + \phi_t^f \cdot d\langle f, 1/Z^k \rangle_t + \phi_t^k \, d\langle Z^k, 1/Z^k \rangle_t \\
&= (1/Z_t^k)\phi_t^f \cdot df_t + \phi_t^f \cdot d\langle f, 1/Z^k \rangle_t,
\end{aligned}$$

where

$$\phi_t^f \cdot d\langle f, 1/Z^k \rangle_t = \sum_{i=1}^{k-1} \phi_t^i \, d\langle f^i, 1/Z^k \rangle_t.$$

Since by assumption the process $\langle f^i, 1/Z^k \rangle$ vanishes for every i, we have

$$\tilde{V}_t^f = \tilde{V}_0 + \int_0^t \phi_u^f \cdot df_u = \tilde{V}_0^f + N_t^f, \quad \forall t \in [0, T^*],$$

where the process N^f follows a $\tilde{\mathbb{P}}$-local martingale. □

The Black model of the futures market assumes deterministic interest, so that (8.14) is trivially satisfied, and indeed we know that $\mathcal{P}(f) = \mathcal{P}(\mathcal{M}^f)$. Condition (8.14) need not hold, in general, when the stochastic character of the interest rates is acknowledged. Thus, under uncertainty of interest rates, the property $\mathcal{P}(f) \subseteq \mathcal{P}(\mathcal{M}^f)$ is not necessarily valid.

8.1.3 Choice of a Numeraire

For simplicity of exposition, we assume throughout this section that a contingent claim is represented by a nonnegative $\mathcal{F}_T$-measurable random variable. The price processes $Z^1, \ldots, Z^k$ are assumed to be continuous, strictly positive semimartingales, so that processes $1/Z^1, \ldots, 1/Z^k$ also follow continuous semimartingales. It is clear that any security Z^i can be chosen as the numeraire asset;[9] in this way, we formally obtain not a single market model, but rather a finite family of standard spot market models $\mathcal{M}^i$, $i = 1, \ldots, k$. A natural question that arises in this context is the dependence (or, more to the point, independence) of the arbitrage price on the choice of the numeraire asset, for a contingent claim attainable in two market models $\mathcal{M}^i$ and $\mathcal{M}^j$, where $i \neq j$.

By a *positive* trading strategy we mean a strategy $\phi \in \Phi$ such that the wealth $V(\phi)$ follows a nonnegative process. We denote by Φ_+ the class of all positive strategies. The class Φ_+ is invariant not only with respect to an equivalent change of probability measure, but also with respect to the choice of a numeraire asset. In particular, if ϕ belongs to Φ_+ then, for any fixed i, the relative wealth $V(\phi)/Z^i$ follows a nonnegative local martingale under any martingale measure for the process Z/Z^i, since (cf. the proof of Lemma 8.1.1)

$$V_t(\phi)/Z_t^i = V_0(\phi)/Z_0^i + \int_0^t \phi_u \cdot d(Z_u/Z_u^i) \tag{8.15}$$

for every $t \in [0, T]$. In financial interpretation, equality (8.15) means that a self-financing trading strategy remains self-financing after the change of the benchmark asset in which we express the prices of all other securities (such a property is intuitively clear). Furthermore, if a strategy ϕ replicates the claim X that settles at time T then it replicates the claim X/Z_T^i. Therefore, attainability of a given claim in a particular market model $\mathcal{M}^i$ hinges on the martingale property of the relative wealth $V(\phi)/Z^i$ of a replicating strategy under a martingale measure from the class $\mathcal{P}(\mathcal{M}^i)$. Of course, the class $\mathcal{P}(\mathcal{M}^i)$ coincides with the collection of all martingale measures for the relative price process Z/Z^i. Let us observe that, if the class $\mathcal{P}(\mathcal{M}^i)$ is nonempty, arbitrage opportunities are manifestly excluded from the market model $\mathcal{M}^i$.

Example 8.1.1 In the context of stochastic interest rate models, it is frequently convenient to take the price of the T-maturity zero-coupon bond as the numeraire asset. The market model obtained in this way is referred to as the T-*forward market,* and the corresponding martingale measure is called the *forward measure* for the date T. Since the analysis based on the concept of a forward measure is presented at some length in the second part of this text (cf. Sect. 9.6), we prove here only a simple

---

[9] Provided that a suitable class of admissible trading strategies is introduced, so that the change of numeraire does not destroy the no-arbitrage property. For an extensive discussion of this point, the interested reader may consult Delbaen and Schachermayer (1995a, 1995b, 1995c).

auxiliary result.[10] Let $F_{Z^i}(t, T) = Z_t^i / B(t, T)$ stand for the forward price of the ith asset for settlement at time T. For brevity, we sometimes write $F_t^i = F_{Z^i}(t, T)$ and $F_Z(t, T) = (F^i(t, T), \ldots, F^k(t, T))$. We write $\tilde{V}(\psi)$ to denote the *forward wealth* of ψ; that is $\tilde{V}_t(\psi) = \psi_t \cdot F_Z(t, T) = V_t(\phi)/B(t, T)$, where $\psi = \phi$ is the corresponding spot strategy and, as usual, $V_t(\phi) = \phi_t \cdot Z_t$.

Lemma 8.1.2 *A trading strategy ψ is a self-financing forward strategy; that is, the forward wealth $\tilde{V}(\psi)$ satisfies*

$$d\tilde{V}_t(\psi) = \sum_{i=1}^k \psi_t^i \, dF^i(t, T) = \psi_t \cdot dF_Z(t, T), \tag{8.16}$$

if and only if the wealth process $V_t(\phi) = \phi_t \cdot Z_t$ of the spot trading strategy $\phi = \psi$ satisfies

$$dV_t(\phi) = \sum_{i=1}^k \phi_t^i \, dZ_t^i = \phi_t \cdot dZ_t.$$

Proof. We shall prove the "only if" clause. We need to show that

$$dV_t(\psi) = \psi_t \cdot d(B(t, T) F_Z(t, T)). \tag{8.17}$$

Since $V_t(\psi) = B(t, T) \tilde{V}_t(\psi)$, the Itô formula yields

$$dV_t(\psi) = B(t, T) d\tilde{V}_t(\psi) + \tilde{V}_t(\psi) \, dB(t, T) + d\langle B(\cdot, T), \tilde{V}(\psi) \rangle_t = I_1 + I_2 + I_3.$$

From (8.16), it follows that $I_1 = B(t, T) \psi_t \cdot dF_Z(t, T)$.
    Furthermore,

$$I_2 = \tilde{V}_t(\psi) \, dB(t, T) = \psi_t \cdot F_Z(t, T) \, dB(t, T).$$

Finally, once again by (8.16), we obtain

$$I_3 = d\langle B(\cdot, T), \tilde{V}(\psi) \rangle_t = \psi_t \cdot d\langle B(\cdot, T), F_Z(\cdot, T) \rangle_t,$$

where $\langle B(\cdot, T), F_Z(\cdot, T) \rangle = (\langle B(\cdot, T), F^1(\cdot, T) \rangle, \ldots, \langle B(\cdot, T), F^k(\cdot, T) \rangle)$. Since for any $i = 1, \ldots, k$

$$d(B(t, T) F_t^i) = B(t, T) \, dF_t^i + F_t^i \, dB(t, T) + d\langle B(\cdot, T), F^i \rangle_t,$$

it is clear that equality (8.17) is indeed satisfied.   □

    Let us stress that we may take the value process of any dynamic portfolio of primary securities as the numeraire asset, provided that it follows a strictly positive process (or a strictly negative process). We thus have the following definition.

---

[10] It deals with the special case of invariance of the class of self-financing strategies under the change of the numeraire asset, as discussed above. Nevertheless, for the reader's convenience, we provide its proof.

Definition 8.1.9 A *numeraire* is a strictly positive (or strictly negative) wealth process $V(\psi)$ of an arbitrary self-financing trading strategy ψ.

Given such a strategy ψ, it is not difficult to verify that

$$V_t(\phi)/V_t(\psi) = V_0(\phi)/V_0(\psi) + \int_0^t \phi_u \cdot d\big(Z_u/V_u(\psi)\big),$$

where ϕ is an arbitrary self-financing strategy. Summarizing, we have a considerable degree of freedom in the choice of the numeraire asset.

Proposition 8.1.6 *Let the class $\mathcal{P}(\mathcal{M}^i)$ be non-empty for some $i \leq k$. More precisely, we assume that relative prices Z^j/Z^i are $\mathbb{P}^*$-martingales for some $\mathbb{P}^* \in \mathcal{P}(\mathcal{M}^i)$. Then: (i) for any $j \neq i$, if the process Z^j/Z^i follows a $\mathbb{P}^*$-martingale, the class $\mathcal{P}(\mathcal{M}^j)$ of martingale measures is non-empty; (ii) for any trading strategy $\phi \in \Phi$ such that the relative wealth process $V(\phi)/Z^i$ follows a strictly positive martingale under $\mathbb{P}^*$, there exists a martingale measure for the relative price process $Z/V(\phi)$.*

Proof. Since buy-and-hold strategies are manifestly self-financing, the first statement follows from the second. To prove (ii), we may in fact take an arbitrary strictly positive process N such that N/Z^i is a $\mathbb{P}^*$-martingale. We define a measure $\mathbb{Q}^*$ on $(\Omega, \mathcal{F}_{T^*})$ by setting

$$\frac{d\mathbb{Q}^*}{d\mathbb{P}^*} = \frac{Z_0^i N_{T^*}}{N_0 Z_{T^*}^i} \stackrel{\text{def}}{=} \eta_{T^*}, \quad \mathbb{P}^*\text{-a.s.} \tag{8.18}$$

The random variable η_{T^*} is strictly positive and $\mathbb{E}_{\mathbb{P}^*}(\eta_{T^*}) = 1$ (this follows from the martingale property of the process N/Z^i under $\mathbb{P}^*$). Consequently, $\mathbb{Q}^*$ is indeed a probability measure equivalent to $\mathbb{P}^*$. We wish to show that all relative price processes Z^j/N follow (local) martingales under $\mathbb{Q}^*$.

Let us fix j, and assume, for simplicity, that the random variable $Z_{T^*}^j/N_{T^*}$ is $\mathbb{Q}^*$-integrable. An application of the Bayes rule yields

$$\mathbb{E}_{\mathbb{Q}^*}(Z_{T^*}^j/N_{T^*} \mid \mathcal{F}_t) = \frac{\mathbb{E}_{\mathbb{P}^*}(\eta_{T^*} Z_{T^*}^j/N_{T^*} \mid \mathcal{F}_t)}{\mathbb{E}_{\mathbb{P}^*}(\eta_{T^*} \mid \mathcal{F}_t)} = \frac{J_1}{J_2}.$$

Using (8.18), we get

$$J_1 = \mathbb{E}_{\mathbb{P}^*}\left(\frac{Z_0^i Z_{T^*}^j}{N_0 Z_{T^*}^i} \,\bigg|\, \mathcal{F}_t\right) = \frac{Z_0^i Z_t^j}{N_0 Z_t^i},$$

where we have used the martingale property of Z^j/Z^i under $\mathbb{P}^*$. For J_2, we have $J_2 = (N_t Z_0^i)/(N_0 Z_t^i)$, and thus $J_1/J_2 = Z_t^j/N_t$, as expected.   □

Proposition 8.1.7 *Let X be a contingent claim that can be priced by arbitrage in standard market models $\mathcal{M}^i$ and $\mathcal{M}^j$. More specifically, we assume that X is $\mathbb{P}^*_i$-attainable and $\mathbb{P}^*_j$-attainable, where $\mathbb{P}^*_i \in \mathcal{P}(\mathcal{M}^i)$ and $\mathbb{P}^*_j \in \mathcal{P}(\mathcal{M}^j)$. Then for every $t \in [0, T]$*

$$\pi^i_t(X) = Z^i_t \, \mathbb{E}_{\mathbb{P}^*_i}\big(X/Z^i_T \mid \mathcal{F}_t\big) = Z^j_t \, \mathbb{E}_{\mathbb{P}^*_j}\big(X/Z^j_T \mid \mathcal{F}_t\big) = \pi^j_t(X).$$

Proof. The proof is quite standard, and thus it is left to the reader.   □

Example 8.1.2 To illustrate the way in which Proposition 8.1.7 can be applied to facilitate the valuation of derivative securities, we place ourselves once again within the standard Black-Scholes framework. Suppose that we wish to price a European call option maturing at time T. For convenience, as the second security we take a unit zero-coupon bond maturing at time T (instead of the savings account B with the price process $B_t = e^{rt}$). Since the bond pays one unit of cash at time T, its price process is easily seen to be $B(t, T) = e^{-r(T-t)} = e^{-rT} B_t$. Formally, we have $Z^1_t = S_t$ and $Z^2_t = B(t, T)$ for every $t \in [0, T]$. The option payoff admits the representation

$$C_T = (S_T - K)^+ = S_T \mathbb{1}_A - K B(T, T) \mathbb{1}_A,$$

where $A = \{S_T > K\}$. Consequently,

$$\pi_t(C_T) = \pi_t(S_T \mathbb{1}_A) - \pi_t(K \mathbb{1}_A) = \pi^1_t(S_T \mathbb{1}_A) - K \pi^2_t(B(T, T) \mathbb{1}_A)$$

so that

$$\pi_t(C_T) = S_t \, \bar{\mathbb{P}} \{S_T > K \mid \mathcal{F}_t\} - K B(t, T) \, \mathbb{P}^* \{S_T > K \mid \mathcal{F}_t\},$$

where $\bar{\mathbb{P}}$ and $\mathbb{P}^*$ are the (unique) martingale measures that correspond to the choice of stock and bond as the numeraire respectively. Since such an approach was already implicitly used in the second proof of Theorem 3.1.1, we do not go into details.

For further examples of applications of the change of numeraire technique, we refer to Geman et al. (1995). Let us only note that virtually all European options (including currency options, exotic options and interest rate options) can be priced using this method. In fact, we use it – either explicitly or implicitly – in many places throughout this text.

8.1.4 Existence of a Martingale Measure

Let us examine the problem of the existence of a martingale measure for a given process Z that follows a continuous semimartingale under $\mathbb{P}$. For more general results, the interested reader may consult Stricker (1990), Ansel and Stricker (1992), Christopeit and Musiela (1994), Schachermayer (1994), Delbaen and Schachermayer (1994a, 1994b, 1995b), and Rydberg (1997).

We start by studying the form of the Radon-Nikodým derivative process. Since the underlying filtration is not necessarily Brownian, the density process may be discontinuous, in general. Consider a real-valued RCLL semimartingale U, defined on $(\Omega, \mathbb{F}, \mathbb{P})$, with $U_0 = U_{0-} = 0$. We denote by $\mathcal{E}(U)$ the *Doléans exponential* of U; that is, the unique solution of the SDE

$$d\mathcal{E}_t(U) = \mathcal{E}_{t-}(U)\, dU_t, \tag{8.19}$$

with $\mathcal{E}_0(U) = 1$. Let us define the *quadratic variation* of U by setting

$$[U]_t = U_t^2 - 2\int_0^t U_{u-}\, dU_u, \quad \forall\, t \in [0, T^*].$$

The solution to equation (8.19) is known explicitly, namely

$$\mathcal{E}_t(U) = \exp\left(U_t - \tfrac{1}{2}[U]_t^c\right) \prod_{u \leq t}(1 + \Delta_u U)\exp(-\Delta_u U),$$

where $\Delta_u U = U_u - U_{u-}$ and $[U]^c$ is the path-by-path continuous part of $[U]$. [11]
Suppose that

$$\Delta_t U > -1 \quad \text{and} \quad \mathbb{E}_{\mathbb{P}}\big(\mathcal{E}_{T^*}(U)\big) = 1. \tag{8.20}$$

Then $\mathcal{E}(U)$ follows a strictly positive, uniformly integrable martingale under $\mathbb{P}$. In such a case, we may introduce a probability measure $\mathbb{Q}^U$ on $(\Omega, \mathcal{F}_{T^*})$, equivalent to $\mathbb{P}$, by postulating that the Radon-Nikodým derivative equals

$$\frac{d\mathbb{Q}^U}{d\mathbb{P}} = \mathcal{E}_{T^*}(U), \quad \mathbb{P}\text{-a.s.}$$

Conversely, if $\mathbb{Q}$ is a probability measure on $(\Omega, \mathcal{F}_{T^*})$ equivalent to $\mathbb{P}$ then we denote by η the RCLL version of the conditional expectation

$$\eta_t = \mathbb{E}_{\mathbb{P}}\left(\frac{d\mathbb{Q}}{d\mathbb{P}}\,\Big|\,\mathcal{F}_t\right), \quad \forall\, t \in [0, T^*]. \tag{8.21}$$

The process η follows a strictly positive, uniformly integrable martingale. Also, it coincides with the Doléans exponential $\mathcal{E}(U)$ of a local martingale U, which is given by the formula

$$U_t = \int_0^t \eta_{u-}^{-1}\, d\eta_u, \quad \forall\, t \in [0, T^*].$$

To summarize, a probability measure $\mathbb{Q}$ on $(\Omega, \mathcal{F}_{T^*})$ is equivalent to the underlying probability measure $\mathbb{P}$ if and only if $\mathbb{Q} = \mathbb{Q}^U$ for some local martingale U, such that condition (8.20) is satisfied and $U_0 = U_{0-} = 0$. Let $\mathbb{Q}$ be a probability measure equivalent to $\mathbb{P}$ and let η be given by (8.21). Consider a continuous, real-valued $\mathbb{P}$-semimartingale Z, with the canonical decomposition $Z = Z_0 + M + A$. Since M follows a continuous martingale, it is well known that the cross-variation $\langle \eta, M \rangle$ exists. By virtue of Girsanov's theorem (cf. Theorem III.20 in Protter (2003), Z is a continuous semimartingale under $\mathbb{Q}$, and its canonical decomposition under $\mathbb{Q}$ is $Z = Z_0 + N + B$, where $N = M - \langle M, U \rangle$ follows a continuous local martingale under $\mathbb{Q}$, and $B = A + \langle M, U \rangle$ is a continuous process of finite variation. The following result provides the basic criteria for the existence of a martingale measure.

---

[11] It is well known that $[U]^c = \langle U^c \rangle$, where U^c is the *continuous local martingale part* of U (we refer to Protter (2003) for more details).

Proposition 8.1.8 *Let Z be a real-valued special[12] semimartingale under $\mathbb{P}$, with the canonical decomposition $Z = Z_0 + M + A$. Then a local martingale U, satisfying (8.20), defines a martingale measure for Z if and only if*

$$A_t + \langle M, U \rangle_t = 0, \quad \forall t \in [0, T^*]. \tag{8.22}$$

If (8.22) holds, there exists a real-valued predictable process ζ satisfying

$$\mathbb{P}\left(\int_0^{T^*} |\zeta_u| \, d \langle M \rangle_u < +\infty \right) = 1,$$

and such that A admits the representation $A_t = \int_0^t \zeta_u \, d \langle M \rangle_u$.

Proof. The first assertion follows from Girsanov's theorem. The second can be proved using the Kunita-Watanabe inequality (for the latter, see Protter (2003)).   □

8.1.5 Fundamental Theorem of Asset Pricing

Following the fundamental ideas of Arrow (1964), many authors have contributed to the development of the general equilibrium approach to asset pricing under uncertainty; to mention just a few: Harrison and Kreps (1979), Kreps (1981), and Duffie and Huang (1985, 1986). By a *first fundamental theorem of asset pricing,* we mean a result that establishes the equivalence of the absence of an arbitrage opportunity in the stochastic model of financial market, and the existence of a martingale measure. In a discrete-time framework, such an equivalence result is established, by different methods and under various assumptions, in Harrison and Pliska (1981), Taqqu and Willinger (1987), Dalang et al. (1990), Back and Pliska (1990), Schachermayer (1992), Rogers (1994), and Kabanov and Kramkov (1994a).

In a continuous-time set-up, the first results in this direction were obtained by Stricker (1990), and Ansel and Stricker (1992). Subsequently, they were improved and extended in various directions by Delbaen (1992), Schweizer (1992b), Lakner (1993), Delbaen and Schachermayer (1994a, 1994b, 1995b, 1995c, 1996a, 1997b), Frittelli and Lakner (1994), Jouini and Kallal (1995), Levental and Skorohod (1995), Klein and Schachermayer (1996a), Kabanov and Kramkov (1994a, 1998), and Cherny (2003).

The definition of an arbitrage opportunity can be formulated in many different ways; in particular, it depends essentially on the choice of topology on the space of random variables. We shall quote here only a result that deals with continuous – but possibly unbounded – processes. Let us say that a process Z admits a *strict martingale measure* if it admits a martingale measure $\mathbb{P}^*$ such that Z is a martingale under $\mathbb{P}^*$. Note that we do not assume a priori that (relative) security prices follow semimartingales, therefore the Itô integration theory is not at hand. To circumvent

---

[12] A semimartingale Z is *special* if it admits a (unique, i.e., *canonical*) decomposition $Z = Z_0 + M + A$, where M is a local martingale and A is a predictable process of finite variation. A continuous semimartingale is special.

this difficulty, by a trading strategy we mean a *simple predictable* process (i.e., piece-wise constant between predictable stopping times), so that the integral is trivially well-defined. Formally, a simple predictable trading strategy is a predictable process, which can be represented as a (finite) linear combination of stochastic processes of the form $\psi \mathbb{1}_{]\tau_1, \tau_2]}$, where τ_1, τ_2 are stopping times, and ψ is an $\mathcal{F}_{\tau_1}$-measurable random variable. A simple predictable trading strategy is δ-*admissible* if the relative wealth satisfies $V_t(\phi) \geq -\delta$ for every t.

Definition 8.1.10 A price process Z satisfies NFLBR (*no free lunch with bounded risk*) if there does not exist a sequence of simple strategies $\{\phi_n\}_{n \geq 1}$ and an $\mathcal{F}_{T^*}$-measurable random variable X with values in $\mathbb{R}_+ \cup \{+\infty\}$ such that $V_{T^*}(\phi_n) \geq -1$, $\lim_{n \to \infty} V_{T^*}(\phi_n) = X$ and $\mathbb{P}\{X > 0\} > 0$.

Delbaen (1992) showed that for a bounded, continuous process, condition NFLBR is equivalent to the existence of a strict martingale measure (in a discrete-time setting, this equivalence was established by Schachermayer (1993), without the boundedness assumption). If Z is an unbounded, continuous process, condition NFLBR implies the existence of a martingale measure (not necessarily strict). A slightly stronger no-arbitrage condition, referred to as NFLVR (*no free lunch with vanishing risk*), is obtained if the inequality $V_{T^*}(\phi_n) \geq -1$ in Definition 8.1.10 is replaced by the following condition

$$V_{T^*}(\phi_n) \geq -\delta_n, \quad \text{for some sequence } \delta_n \downarrow 0.$$

Put another way, for any sequence $\{\phi_n\}_{n \geq 1}$ of simple predictable strategies such that ϕ_n is δ_n-admissible and the sequence δ_n tends to zero, we have $V_{T^*}(\phi_n) \to 0$ (in probability) as $n \to \infty$. Delbaen and Schachermayer (1994b) showed that NFLVR is equivalent to the existence of a martingale measure, even when the continuity assumption is relaxed (we still need to assume that Z is a locally bounded semimartingale). The case of unbounded processes was examined in Delbaen and Schachermayer (1997b).

8.2 Multidimensional Black-Scholes Model

In this section, the classical Black-Scholes model is extended in several directions. First, the number of risky primary securities (stocks) can be greater than 1. Second, the underlying Brownian motion is assumed to be multidimensional, rather than one-dimensional. Finally, the volatility coefficient is no longer assumed to be constant (or deterministic). For any $i = 1, \ldots, k$, the price process $Z^i = S^i$ of the ith stock is modelled as an Itô process

$$dS_t^i = S_t^i \left(\mu_t^i \, dt + \sigma_t^i \cdot dW_t \right),$$

with $S_0^i > 0$, or equivalently,

$$dS_t^i = S_t^i \left(\mu_t^i \, dt + \sum_{j=1}^{d} \sigma_t^{ij} \, dW_t^j \right),$$

where $W = (W^1, \ldots, W^d)$ is a standard d-dimensional Brownian motion, defined on a filtered probability space $(\Omega, \mathbb{F}, \mathbb{P})$. For simplicity, we assume that the underlying filtration $\mathbb{F}$ coincides with $\mathbb{F}^W$, i.e., the filtration $\mathbb{F}$ is generated by the Brownian motion W. The coefficients σ^i and μ^i follow bounded progressively measurable processes on the space $(\Omega, \mathbb{F}, \mathbb{P})$, with values in $\mathbb{R}^d$ and $\mathbb{R}$ respectively. A special, but important, case is obtained by postulating that for every i the volatility coefficient σ^i is represented by a fixed vector in $\mathbb{R}^d$ and the appreciation rate μ^i is a real number. For brevity, we write $\sigma = \sigma_t$ to denote the volatility matrix – that is, the time-dependent random matrix $[\sigma_t^{ij}]$, whose i^{th} row represents the volatility of the ith stock. The last primary security, the savings account, has the price process $Z^{k+1} = B$ satisfying

$$dB_t = r_t B_t \, dt, \quad B_0 = 1,$$

for a bounded, nonnegative, progressively measurable interest rate process r. To ensure the absence of arbitrage opportunities, we postulate the existence a d-dimensional, progressively measurable process γ such that the equality

$$r_t - \mu_t^i = \sum_{j=1}^{d} \sigma_t^{ij} \gamma_t^j = \sigma_t^i \cdot \gamma_t \tag{8.23}$$

is satisfied simultaneously for every $i = 1, \ldots, k$ (for Lebesgue a.e. $t \in [0, T^*]$, with probability one). Note that the *market price for risk* γ is not uniquely determined, in general. Indeed, the uniqueness of a solution γ to (8.23) holds only if $d \leq k$ and the volatility matrix σ has full rank for every $t \in [0, T^*]$. For instance, if $d = k$ and the volatility matrix σ is non-singular (for Lebesgue a.e. $t \in [0, T^*]$, with probability one), then[13]

$$\gamma_t = \sigma_t^{-1}(r_t \mathbf{1} - \mu_t), \quad \forall t \in [0, T^*],$$

where $\mathbf{1}$ denotes the d-dimensional vector with every component equal to one, and μ_t is the vector with components μ_t^i. Given any process γ satisfying (8.23), we introduce a measure $\mathbb{P}^*$ on $(\Omega, \mathcal{F}_{T^*})$ by setting

$$\frac{d\mathbb{P}^*}{d\mathbb{P}} = \exp \left(\int_0^{T^*} \gamma_u \cdot dW_u - \frac{1}{2} \int_0^{T^*} |\gamma_u|^2 \, du \right), \quad \mathbb{P}\text{-a.s.}, \tag{8.24}$$

provided that the right-hand side of (8.24) is well-defined. The Doléans exponential

$$\eta_t = \mathcal{E}_t \left(\int_0^{\cdot} \gamma_u \cdot dW_u \right) = \exp \left(\int_0^t \gamma_u \cdot dW_u - \frac{1}{2} \int_0^t |\gamma_u|^2 \, du \right)$$

is known to follow a strictly positive supermartingale (but not necessarily a martingale) under $\mathbb{P}$, so that the case $\mathbb{E}_{\mathbb{P}^*}(\eta_{T^*}) < 1$ is not excluded a priori. We conclude

---

[13] Note that no specific symbol is used for the matrix product.

that formula (8.24) defines a probability measure $\mathbb{P}^*$ equivalent to $\mathbb{P}$ if and only if η follows a $\mathbb{P}$-martingale. It is thus essential to check whether a given process γ gives rise to an associated equivalent martingale measure.[14] We assume from now on that the class of martingale measures is non-empty. By virtue of Girsanov's theorem, the process W^*, which equals

$$W_t^* = W_t - \int_0^t \gamma_u \, du, \quad \forall \, t \in [0, T^*],$$

is a d-dimensional standard Brownian motion on $(\Omega, \mathbb{F}, \mathbb{P}^*)$. Using Itô's formula, it is easy to verify that the discounted stock price S_t^i / B_t satisfies, under $\mathbb{P}^*$,

$$d(S_t^i / B_t) = (S_t^i / B_t) \, \sigma_t^i \cdot dW_t^*$$

for any $i = 1, \ldots, k$. This mean that the discounted prices of all stocks follow local martingales under $\mathbb{P}^*$, so that any probability measure defined by means of (8.23)–(8.24) is a martingale measure for our model (corresponding to the choice of the savings account as the numeraire). For any fixed $\mathbb{P}^*$, we define the class $\Phi(\mathbb{P}^*)$ of admissible trading strategies, and the class of tame strategies relative to B, as in Sect. 8.1. Let us emphasize, however, that we no longer assume that a process ϕ representing a trading strategy is necessarily locally bounded.[15] The standard market model obtained in this way is referred to as the *multidimensional Black-Scholes market*.

Remarks. The classical multidimensional Black-Scholes model assumes that $d = k$, the constant volatility matrix σ is non-singular and the appreciation rates μ_i and the continuously compounded interest rate r are constant. It is easily seen that under these assumptions, the martingale measure exists and is unique.

8.2.1 Market Completeness

Let us now focus on the concept of market *completeness*. Basically, it is defined in much the same way as for a finite market model, except that certain technical restrictions are imposed on contingent claims.

Definition 8.2.1 The multidimensional Black-Scholes model is *complete* if any $\mathbb{P}^*$-integrable contingent claim X that is bounded from below is attainable; that is, if for any such claim X there exists an admissible trading strategy $\phi \in \Phi_0^{k+1}$ such that $X = V_T(\phi)$. In the opposite case, the market model is said to be *incomplete*.

---

[14] For the last property to hold, it is enough (but not necessary) that γ follows a bounded process.

[15] In the present set-up, if the wealth process follows a (local) martingale under some martingale measure $\mathbb{P}^*$ then it necessarily follows a local martingale under any equivalent martingale measure.

The superscript $k+1$ refers to the fact that we have chosen the savings account as the numeraire. Recall also that, by assumption, the interest rate process r is nonnegative and bounded. The integrability and boundedness of X is therefore equivalent to the integrability and boundedness of the discounted claim X/B_T. On the other hand, we do not postulate a priori that uniqueness of a martingale measure holds, hence the $\mathbb{P}^*$-integrability of X refers to an arbitrary martingale measure for the model. The next result establishes necessary and sufficient conditions for the completeness of the Black-Scholes market. In the general semimartingale framework, the equivalence of the uniqueness of a martingale measure and the completeness of a market model was conjectured and analyzed by Harrison and Pliska (1981, 1983) (see also Müller (1989)). The case of the Brownian filtration was examined in Jarrow and Madan (1991). Finally, Chatelain and Stricker (1994, 1995) provided rigorous and definitive results for the case of continuous local martingales (see also Artzner and Heath (1995) and Pratelli (1996) for related results) They focus on the important distinction between the vector- and component-wise stochastic integrals. In the next result, we consider the Itô vector-wise stochastic integral.

Proposition 8.2.1 *The following are equivalent:* (i) *the multidimensional Black-Scholes model is complete;* (ii) *inequality $d \leq k$ holds and the volatility matrix σ has full rank for Lebesgue a.e. $t \in [0, T^*]$, with probability 1;* (iii) *there exists a unique martingale measure $\mathbb{P}^*$ for discounted stock prices S^i/B, $i = 1, \ldots, k$.*

Proof. We shall merely outline the proof. Let us first examine the implication (ii) $\Rightarrow$ (i). Essentially, it is a consequence of the representation theorem for the filtration of a multidimensional Brownian motion,[16] combined with the possibility of expressing the underlying Brownian motions $W^1, \ldots, W^d$ in terms of stock price processes $S^1, \ldots, S^k$. The last property is a consequence of postulated features of the volatility process σ.

For the implication (i) $\Rightarrow$ (iii), it is enough to show that if non-uniqueness of the martingale measure occurs, then the market model is incomplete. Indeed, if (iii) is not satisfied, then manifestly there exist two martingale measures, $\mathbb{P}^*$ and $\mathbb{Q}^*$, and an event $A \in \mathcal{F}_{T^*}$ such that $\mathbb{E}_{\mathbb{P}^*}(\mathbb{1}_A/B_{T^*}) \neq \mathbb{E}_{\mathbb{Q}^*}(\mathbb{1}_A/B_{T^*})$. This in turn implies that the claim $X = \mathbb{1}_A$ that settles at time T^* is not attainable (cf. Corollary 8.1.1), and thus the market model is incomplete. It remains to check the implication (iii) $\Rightarrow$ (ii). Assume, on the contrary, that property (ii) does not hold, and let us show that the non-uniqueness of the martingale measure is valid. For this purpose, it is enough to show that there exists a bounded progressively measurable process $\tilde{\gamma}$ that does not vanish identically, and such that (cf. (8.23)) $\sigma_t^i \cdot \tilde{\gamma}_t = 0$, for every $i = 1, \ldots, k$ and for Lebesgue a.e. $t \in [0, T^*]$, with probability 1. In other words, the values of the bounded progressively measurable process $\tilde{\gamma}$ belong to the kernel of the (random) linear mapping associated with σ; that is, $\tilde{\gamma}_t \in \text{Ker}\,\sigma_t$ for Lebesgue a.e. $t \in [0, T^*]$, with probability 1. To exclude a trivial solution, we may postulate, in addition, that

---

[16] See Chap. V in Revuz and Yor (1999). It is important to observe that the process W^* has predictable representation property with respect to the filtration $\mathbb{F}^W$, since W has (this can be shown using Bayes's formula).

$\tilde{\gamma}_t \neq 0$ if $\text{Ker}\, \sigma_t \neq \{0\}$. The existence of a process $\tilde{\gamma}$ with desired properties can be achieved in an explicit way – for instance, by projecting the vector of the orthonormal basis in $\mathbb{R}^d$ on the orthogonal complement of $\text{Ker}\, \sigma_t$; details are left to the reader (special attention should be paid to the measurability of $\tilde{\gamma}$). □

Suppose that $k = d$, that is, the number of stocks equals the dimension of the Brownian motion. If we restrict our attention to the component-wise stochastic integral (i.e., we consider a smaller class of trading strategies), then (as shown by Chatelain and Stricker (1994)) to ensure the completeness of the Black-Scholes market, it is sufficient to assume, in addition, that the process $g^{li} \sigma^{ij}$ is bounded (for every $i, j, l = 1, \ldots, d$), where the matrix $G_t = [g_t^{ij}] = \sigma_t^{-1}$ is the inverse of the (non-singular) matrix σ. Chatelain and Stricker (1994) show that if the discounted prices follow continuous local martingales, then vector-wise completeness implies component-wise completeness, provided that a suitable condition is satisfied by the coefficients of the decomposition of discounted prices on an orthonormal martingale basis.

The one-dimensional Black-Scholes market model, examined in Chap. 3, is, of course, a special case of multidimensional market. Hence Proposition 8.2.1 applies also to the classical Black-Scholes market model, in which the martingale measure $\mathbb{P}^*$ is well known to be unique. We conclude that the one-dimensional Black-Scholes market model is complete; that is, any $\mathbb{P}^*$-integrable contingent claim is $\mathbb{P}^*$-attainable, and thus is priced by arbitrage.

Note that we have examined the completeness of the market model in which trading was restricted to primary securities. Completeness of a model of financial market with traded call and put options (and related topics, such as *static hedging* of derivatives) is examined by several authors; to mention a few: Ross (1976b), Breeden and Litzenberger (1978), Greene and Jarrow (1987), Nachman (1989), Madan and Milne (1993), and Bajeux-Besnainou and Rochet (1996).

8.2.2 Variance-minimizing Hedging

We shall now examine rather succinctly the mean-variance approach to the hedging of non-attainable claims in an incomplete market. We associate the generic term of *mean-variance hedging* with an arbitrary method for the hedging of non-attainable contingent claims that is based on the expected value and variance. In a continuous-time framework, typical optimization problems are the following:

(MV.1) For a fixed $c \in \mathbb{R}$, minimize

$$J_1(\phi) = \mathbb{E}_{\mathbb{P}} \left\{ \left(X^* - c - \int_0^T \phi_u \cdot dS_u^* \right)^2 \right\} \tag{8.25}$$

over all self-financing trading strategies ϕ.

(MV.2) Minimize

$$J_2(\phi) = \mathbb{E}_{\mathbb{P}} \left\{ \left(X^* - c - \int_0^T \phi_u \cdot dS_u^* \right)^2 \right\} \tag{8.26}$$

over all $c \in \mathbb{R}$ and all self-financing trading strategies ϕ.

(MV.3) Minimize

$$J_3(\phi) = \mathrm{Var}_{\mathbb{P}}\left(X^* - \int_0^T \phi_u \cdot dS_u^*\right) \tag{8.27}$$

over all self-financing trading strategies ϕ.

The problems above have been extensively treated, at various levels of generality, by Duffie and Richardson (1991), Schweizer (1992a, 1994a, 1995b), Monat and Stricker (1993, 1995), Delbaen and Schachermayer (1996a, 1996b), Gouriéroux et al. (1998), Pham et al. (1998), and Laurent and Pham (1999). For an excellent survey of ideas and results related to the mean-variance hedging, we refer to Schweizer (2001).

Note that problem (MV.2) is related to the closedness, in the space $L^2(\mathbb{P})$ of $\mathcal{F}_T$-measurable random variables, of the following set

$$\mathcal{G}_T = \left\{ \int_0^T \phi_u \cdot dS_u^* \;\middle|\; \int_0^{\cdot} \phi_u \cdot dS_u^* \in \mathcal{H}_{\mathbb{P}}^2 \right\}.$$

In the last formula, we write $\mathcal{H}_{\mathbb{P}}^2$ to denote the class of all real-valued special semi-martingales with finite $\mathcal{H}_{\mathbb{P}}^2$ norm, where

$$\|Z\|_{\mathcal{H}_{\mathbb{P}}^2} \overset{\text{def}}{=} \||Z_0| + \langle M, M \rangle_T^{1/2}\|_{L^2(\mathbb{P})} + \left\| \int_0^T |dA_u| \right\|_{L^2(\mathbb{P})}$$

and $Z = Z_0 + M + A$ is the canonical decomposition of Z. If the discounted price S^* follows a square-integrable martingale under $\mathbb{P}$, then the closedness of the space $\mathcal{G}_T$ of stochastic integrals is straightforward, since Itô's integral defines an isometry. In the general case, the closedness of $\mathcal{G}_T$ is examined by Schweizer (1994a) and Monat and Stricker (1993).

8.2.3 Risk-minimizing Hedging

A different approach to the hedging of non-attainable claims, referred to as *risk-minimizing hedging*, starts by enlarging the class of trading strategies in order to allow for additional transfers of funds (referred to as *costs*). This method of hedging in incomplete markets, originated by Föllmer and Sondermann (1986), was subsequently developed by Föllmer and Schweizer (1989, 1991) and Schweizer (1990, 1991).

We fix a probability measure $\mathbb{P}$ and we assume that the canonical decomposition of S^* under $\mathbb{P}$ is $S^* = S_0^* + M + A$. For a given claim X, we consider the class of all replicating strategies, not necessarily self-financing, such that the discounted wealth process follows a real-valued semimartingale of class $\mathcal{H}_{\mathbb{P}}^2$. By definition, the *(discounted) cost process* $C(\phi)$ of a strategy ϕ equals[17]

$$C_t(\phi) = V_t^*(\phi) - \int_0^t \phi_u \cdot dS_u^*, \quad \forall\, t \in [0, T^*], \tag{8.28}$$

---

[17] Note that the cost process does not involve the initial cost, but only the additional transfers of funds (i.e., withdrawals and infusions).

where $V_t^*(\phi) = \phi_t \cdot S_t^*$. As in a discrete-time setting, a *local risk minimization* is based on the suitably defined minimality of the cost process. Since in a continuous-time setting the formal definition is rather involved, we refer the reader to Schweizer (1991) for details. Schweizer (1991) shows also that, under mild technical assumptions, a replicating strategy ϕ^* is *locally risk-minimizing* if it is mean self-financing and the cost process $C(\phi^*)$ follows a square-integrable martingale strongly orthogonal[18] to the martingale part of S^*. A strategy ϕ is said to be *mean self-financing* under $\mathbb{P}$ if the cost process $C(\phi)$ follows a martingale under $\mathbb{P}$. The following generalization of the Kunita-Watanabe decomposition, introduced by Föllmer and Schweizer (1989, 1991), appears to be useful.

Definition 8.2.2 We say that a square-integrable random variable X^* admits the *Föllmer-Schweizer decomposition* with respect to S^* under $\mathbb{P}$ if

$$X^* = c_0 + \int_0^T \xi_u \cdot dS_u^* + L_T, \tag{8.29}$$

where c_0 is a real number, ξ is a predictable process such that the stochastic integral $\int_0^t \xi_u \cdot dS_u^*$ belongs to the class $\mathcal{H}_{\mathbb{P}}^2$ and L is a square-integrable martingale strongly orthogonal to M under $\mathbb{P}$, with $L_0 = 0$.

The following result, stated without proof, emphasizes the relevance of the Föllmer-Schweizer decomposition in the present context. Basically, it says that a locally risk-minimizing replicating strategy is determined by the process ξ in representation (8.29).

Proposition 8.2.2 *Let X be a European claim, settling at time T, such that X^* is square-integrable under $\mathbb{P}$. The following are equivalent: (i) X admits a locally risk-minimizing replicating strategy; and (ii) X^* admits a Föllmer-Schweizer decomposition with respect to the stock price S^* under $\mathbb{P}$. Moreover, if (ii) holds, then there exists a locally risk-minimizing replicating strategy ϕ^* such that $(\phi^*)^i = \xi^i$ for $i = 1, \ldots, k$, where ξ is given by (8.29). Finally, for any locally risk-minimizing strategy ϕ, we have*

$$\int_0^t \phi_u \cdot dS_u^* = \int_0^t \xi_u \cdot dS_u^*, \quad \forall t \in [0, T^*].$$

In view of Proposition 8.2.2, it is clear that the Föllmer-Schweizer decomposition provides a neat method of searching for the locally risk-minimizing strategy (as the last component of ϕ^* can be found, in principle, using the mean self-financing condition). This in turn raises the issue of effectively finding the Föllmer-Schweizer decomposition of a given payoff. It appears that the Föllmer-Schweizer decomposition under $\mathbb{P}$ corresponds to the Kunita-Watanabe decomposition under the so-called *minimal martingale measure* $\hat{\mathbb{P}}$ associated with $\mathbb{P}$. Furthermore, if X

---

[18] Basically, two martingales are said to be *strongly orthogonal* if their product follows a martingale.

admits the locally risk-minimizing replicating strategy ϕ^* then the initial cost equals $V_0(\phi^*) = C_0(\phi^*) = \mathbb{E}_{\hat{\mathbb{P}}}(X^*)$ (recall that $B_0 = 1$). More generally, the following version of the risk-neutral valuation formula holds

$$V_t(\phi^*) = B_t \, \mathbb{E}_{\hat{\mathbb{P}}}(X^* \mid \mathcal{F}_t), \quad \forall t \in [0, T^*]. \tag{8.30}$$

Despite its resemblance to the standard risk-neutral valuation formula, equality (8.30) is essentially weaker, since the right-hand side manifestly depends on the choice of the minimal martingale measure (through the choice of the actual probability $\mathbb{P}$). On the other hand, when applied to an attainable contingent claim, it gives the right result – that is, the arbitrage price of X in the Black-Scholes market. We introduce the notion of a minimal martingale measure for a continuous, $\mathbb{R}^k$-valued semimartingale S^*, with the canonical decomposition $S^* = S_0^* + M + A$.

Definition 8.2.3 A martingale measure $\hat{\mathbb{P}}$ for S^* is called a *minimal martingale measure* associated with $\mathbb{P}$ if any local $\mathbb{P}$-martingale strongly orthogonal (under $\mathbb{P}$) to each local martingale M^i for $i = 1, \ldots, k$ remains a local martingale under $\hat{\mathbb{P}}$.

General results regarding the existence and uniqueness of a minimal martingale measure can be found in Föllmer and Schweizer (1991) and Ansel and Stricker (1992) (see also Schachermayer (1993) for an important counter-example, and Hofmann et al. (1992) for applications to option pricing in a stochastic volatility model). The last result of this section provides necessary and sufficient conditions for the existence of a minimal martingale measure in the multidimensional Black-Scholes framework.

Proposition 8.2.3 *A minimal martingale measure $\hat{\mathbb{P}}$ associated with $\mathbb{P}$ exists if and only if there exists a progressively measurable $\mathbb{R}^d$-valued process $\hat{\gamma}$ such that:* (i) *for every $i = 1, \ldots, k$*

$$r_t - \mu_t^i = \sigma_t^i \cdot \hat{\gamma}_t, \quad l \otimes \mathbb{P}\text{-a.e. on } [0, T^*] \times \Omega;$$

(ii) *the Doléans exponential $\mathcal{E}(U^{\hat{\gamma}})$ of the process $U_t^{\hat{\gamma}} = \int_0^t \hat{\gamma}_u \cdot dW_u$ is a martingale under $\mathbb{P}$; and* (iii) *with probability 1, for almost every t we have $\hat{\gamma}_t \in \operatorname{Im} \sigma_t^* = \left(\operatorname{Ker} \sigma_t \right)^{\perp}$, where σ^* is the transpose of σ.*

The proof of Proposition 8.2.3 is omitted (see Ansel and Stricker (1992) and El Karoui and Quenez (1995)). Note that conditions (i)–(ii) are sufficient for the existence of a martingale measure, as we may define the martingale measure $\hat{\mathbb{P}}$ using formula (8.24). Condition (iii) that corresponds to the concept of minimality of $\hat{\mathbb{P}}$, says, essentially, that there exists a process ζ such that $\tilde{\gamma}_t = \sigma_t^* \zeta_t$.

In a typical example, when one starts with the complete Black-Scholes model and assumes that only some stocks are accessible for trading, under the minimal martingale measure the returns on traded stocks equal the risk-free rate of return, and the returns on non-traded stocks remain unchanged – that is, they are the same under the original probability measure $\mathbb{P}$ and under the associated minimal martingale measure $\hat{\mathbb{P}}$ (cf. Lamberton and Lapeyre (1993) and Rutkowski (1996b)). We shall now present such an example in some detail.

Example 8.2.1 We place ourselves in the multidimensional Black-Scholes setting with constant coefficients, so that the price B of a risk-free bond satisfies $dB_t = rB_t\,dt$, and prices S^i, $i = 1, \ldots, k$ of risky stocks satisfy

$$dS_t^i = S_t^i\left(\mu^i dt + \sigma^i \cdot dW_t\right), \quad S_0^i > 0,$$

where W follows a k-dimensional standard Brownian motion on a probability space $(\Omega, \mathbb{F}, \mathbb{P})$, the appreciation rates μ^i are constants, and the volatility coefficients σ^i, $i = 1, \ldots, k$ are linearly independent vectors in $\mathbb{R}^k$. We define the *stock index process* I by setting

$$I_t = \sum_{i=1}^k w_i S_t^i, \quad t \in [0, T],$$

where $w_i > 0$ are constants such that $\sum_{i=1}^k w_i = 1$. The stock index is thus the weighted arithmetic average of prices of all traded stocks. We consider a European call option written on the stock index, formally defined as the contingent claim $C_T = (I_T - K)^+$. Assume first that $r = 0$ (this assumption will be subsequently relaxed). It is clear that under the present hypotheses, there exists the unique martingale measure $\mathbb{P}^*$ for the multidimensional Black-Scholes model. $\mathbb{P}^*$ is determined by the unique solution $\gamma^* \in \mathbb{R}^k$ to the system of equations $\mu^i + \sigma^i \cdot \gamma^* = 0$, $i = 1, \ldots, k$, through the Doléans exponential

$$\frac{d\mathbb{P}^*}{d\mathbb{P}} = \exp\left(\gamma^* \cdot W_T - \tfrac{1}{2}|\gamma^*|^2 T\right), \quad \mathbb{P}\text{-a.s.}$$

Assume for the moment that all stocks can be used for hedging; we thus deal with a complete model of a security market. It is clear that the arbitrage price of the stock index call option in the complete Black-Scholes market equals

$$\pi_t(C_T) = \mathbb{E}_{\mathbb{P}^*}(C_T | \mathcal{F}_t) = h(S_t^1, \ldots, S_t^k, T - t), \quad \forall t \in [0, T],$$

where $h : \mathbb{R}^k \times [0, T] \to \mathbb{R}$ is a certain smooth function. Moreover, the strategy replicating the option satisfies

$$\phi_t^i = h_{s_i}(S_t^1, \ldots, S_t^k, T - t),$$

for $i = 1, \ldots, k$ and $\phi_t^0 = h(S_t^1, \ldots, S_t^k, T - t) - \sum_{i=1}^k \phi_t^i S_t^i$. The option's price and the replicating strategy ϕ are, of course, independent of the drift coefficients μ^i. This follows immediately from the fact that the dynamics of stock prices under the martingale measure $\mathbb{P}^*$ are

$$dS_t^i = S_t^i \sigma^i \cdot dW_t^*,$$

where the process W^* satisfying $W_t^* = W_t - \gamma^* t$ follows a k-dimensional standard Brownian motion under $\mathbb{P}^*$.

From now on, we assume that only some stocks, say $S^1, \ldots, S^m$ (with $1 \le m < k$) are accessible for a particular trader, whose assessment of the future market behavior is reflected by the subjective probability measure $\mathbb{P}$. Such a specification is now essential, since a solution to the locally risk-minimization problem will depend on appreciation rates μ^i, which in turn depend on the choice of $\mathbb{P}$. Let us observe that the martingale measure for the m-dimensional process $(S^1, \ldots, S^m)$ is not unique. Indeed, any solution $\gamma \in \mathbb{R}^k$ to the system of equations $\mu^i + \sigma^i \cdot \gamma = 0$, $i = 1, \ldots, m$ defines a martingale measure $\mathbb{P}_\gamma$ for the process $(S^1, \ldots, S^m)$, namely,

$$\frac{d\mathbb{P}_\gamma}{d\mathbb{P}} = \exp\left(\gamma \cdot W_T - \tfrac{1}{2}|\gamma|^2 T\right), \quad \mathbb{P}\text{-a.s.}$$

The minimal martingale measure $\hat{\mathbb{P}}$ for the process $\hat{S} = (S^1, \ldots, S^m)$, associated with $\mathbb{P}$, corresponds to that vector $\hat{\gamma} \in \mathbb{R}^k$ that satisfies

$$\begin{cases} \mu^i + \sigma^i \cdot \hat{\gamma} = 0, & i = 1, \ldots, m, \\ \hat{\gamma} \in \operatorname{Im} \tilde{\sigma}^*, \end{cases}$$

where $\tilde{\sigma}$ stands for the matrix with rows $\sigma^1, \ldots, \sigma^m$.

Let us now examine the Föllmer-Schweizer decomposition of the stock index process[19] I with respect to $\hat{S}$ under the probability measure $\mathbb{P}$.

First, observe that under $\mathbb{P}^*$ we have

$$dI_t = \sum_{i=1}^k w_i \, dS_t^i = \sigma_t^I \cdot dW_t^*,$$

where $\sigma^I = \sum_{i=1}^k w_i S_t^i \sigma^i$. Let $v^{m+1}, \ldots, v^k \in \mathbb{R}^k$ be any orthonormal basis in $\operatorname{Ker} \tilde{\sigma}$. Then, we have (note that $\tilde{\sigma}$ represents a linear mapping $\tilde{\sigma} : \mathbb{R}^k \to \mathbb{R}^m$, and thus $\tilde{\sigma}^* : \mathbb{R}^m \to \mathbb{R}^k$)

$$\sigma_t^I = v_t^\perp + v_t = v_t^\perp + \hat{\sigma}_t^* \psi_t,$$

where

$$v_t^\perp = \sum_{j=m+1}^k (\sigma_t^I \cdot v^j) v^j = \sum_{i,j=m+1}^k w_i S_t^i (\sigma^i \cdot v^j) v^j,$$

and ψ is some $\mathbb{R}^m$-valued adapted process. Consequently, under the martingale measure $\mathbb{P}^*$ we have

$$dI_t = v_t^\perp \cdot dW_t^* + \sum_{i=1}^m \psi_t^i \, dS_t^i.$$

On the other hand, under the original probability measure $\mathbb{P}$ the dynamics of I are

$$dI_t = v_t^\perp \cdot dW_t - v_t^\perp \cdot (\gamma^*)^\perp \, dt + \sum_{i=1}^m \psi_t^i \, dS_t^i,$$

where $(\gamma^*)^\perp \in \mathbb{R}^k$ stands for the orthogonal projection of γ^* on $\operatorname{Ker} \sigma$, that is, $(\gamma^*)^\perp = \sum_{j=m+1}^k (\gamma^* \cdot v^j) v^j$. Concluding, under the minimal martingale measure $\hat{\mathbb{P}}$ associated with $\mathbb{P}$ we have

---

[19] Essentially, a process X_t, $t \in [0, T]$, admits the Föllmer-Schweizer decomposition if representation of form (8.29) is valid for any $t \in [0, T]$.

$$dI_t = v_t^\perp \cdot d\hat{W}_t - v_t^\perp \cdot (\gamma^*)^\perp \, dt + \sum_{i=1}^{m} \psi_t^i \, dS_t^i,$$

since $\hat{\gamma} \perp v_t^\perp$. Similar decompositions can be established for the process $\pi_t(C_T)$ representing the option's price. Note, however, that the knowledge of these representations is not sufficient to determine the locally risk-minimizing hedging of an option. Note that the locally risk-minimizing value of an option can be found from the formula

$$\hat{\pi}_{t|\mathbb{P}}(X) = \mathbb{E}_{\hat{\mathbb{P}}}(C_T | \mathcal{F}_t)$$

since the dynamics of $(S^1, \dots, S^k)$ under the minimal martingale measure $\hat{\mathbb{P}}$ are easily seen to be

$$\begin{cases} dS_t^i = S_t^i \sigma^i \cdot d\hat{W}_t, & i = 1, \dots, m, \\ dS_t^i = S_t^i \left((\mu^i + \sigma^i \cdot \hat{\gamma}) \, dt + \sigma^i \cdot d\hat{W}_t \right), & i = m+1, \dots, k. \end{cases}$$

To find a locally risk-minimizing replicating strategy under $\mathbb{P}$, we proceed as follows. First, we introduce auxiliary stochastic processes G^j, $j = m+1, \dots, k$ by setting

$$dG_t^j = G_t^j v^j \cdot d\hat{W}_t, \; G_0^j > 0.$$

It is essential to observe that the unique martingale measure for the k-dimensional semimartingale $(S^1, \dots, S^m, G^{m+1}, \dots, G^k)$ is easily seen to coincide with the minimal martingale measure $\hat{\mathbb{P}}$. The self-financing replicating strategy for C_T, with S^i, $i = 1, \dots, m$ and G^j, $j = m+1, \dots, k$ playing the role of hedging assets, can be found by proceeding along the same lines as in the case of a complete market. Indeed, the arbitrage price of the option in this fictitious market equals

$$\hat{\pi}_{t|\mathbb{P}}(X) = \mathbb{E}_{\hat{\mathbb{P}}}(C_T | \mathcal{F}_t) = \hat{h}(S_t^1, \dots, S_t^m, G_t^{m+1}, \dots, G_t^k, T - t)$$

for some function $\hat{h}(x_1, \dots, x_k, T - t)$. Furthermore, the self-financing replicating strategy $\psi = (\psi^0, \psi^1, \dots, \psi^k)$ equals

$$\psi_t^i = \frac{\partial \hat{h}}{\partial x_i}(S_t^1, \dots, S_t^m, G_t^{m+1}, \dots, G_t^k, T - t)$$

for $i = 1, \dots, k$, and

$$\psi_t^0 = \hat{\pi}_{t|\mathbb{P}}(X) - \sum_{i=1}^{m} \psi_t^i S_t^i - \sum_{j=m+1}^{k} \psi_t^j G_t^j.$$

In this way, we arrive at the following formula

$$C_T = \hat{\pi}_{0|\mathbb{P}}(X) + \sum_{i=1}^{m} \int_0^t \psi_u^i \, dS_u^i + \sum_{j=m+1}^{k} \int_0^t \psi_u^j \, dG_u^j,$$

which is the Kunita-Watanabe decomposition of C_T under $\hat{\mathbb{P}}$ (notice that for every i and j the processes S^i and G^j are square-integrable martingales mutually orthogonal under $\hat{\mathbb{P}}$, since $\sigma^i \perp v^j$ if $i \leq m$ and $j > m$). We conclude that the process $(\psi^1, \ldots, \psi^m)$ represents the locally risk-minimizing replicating strategy of the option, associated with the original probability measure $\mathbb{P}$. Furthermore, the corresponding cost process $C(\psi)$ satisfies

$$C_t(\psi) = \hat{\pi}_{0|\mathbb{P}}(X) + \sum_{j=m+1}^{k} \int_0^t \psi_u^j \, dG_u^j, \quad \forall t \in [0, T].$$

From now on, we relax the assumption that $r = 0$. In the general case, we find easily that

$$dS_t^i = S_t^i(\hat{\mu}^i \, dt + \sigma^i \cdot d\hat{W}_t),$$

under the minimal martingale measure $\hat{\mathbb{P}}$, where $\hat{\mu}^i = r$ for $i = 1, \ldots, m$ and

$$\hat{\mu}^i = \mu^i + \sigma^i \cdot \hat{\gamma}$$

for $i = m + 1, \ldots, k$. Furthermore, we now set

$$dG_t^j = G_t^j (r \, dt + v^j \cdot d\hat{W}_t), \quad \forall i = m + 1, \ldots, k.$$

Our goal is to find a quasi-explicit expression for the valuation function $\hat{h}$. The first step is to evaluate the following conditional expectation

$$\mathbb{E}_{\hat{\mathbb{P}}} \left\{ e^{-r(T-t)} \left(\sum_{i=1}^{k} w_i S_T^i - K \right)^+ \Big| \mathcal{F}_t \right\} = g(S_t^1, \ldots, S_t^k, T - t).$$

As soon as the function g is known, to find the function $\hat{h}$, it is sufficient to express processes $S^{m+1}, \ldots, S^k$ in terms of $G^{m+1}, \ldots, G^k$. This can be done using explicit formulas that are available for all these processes.

First step. Note that processes S^i are given by the following explicit formula

$$S_t^i = S_0^i \exp \left(\left(\mu^i - \tfrac{1}{2} |\sigma^i|^2 \right)t + \sigma^i \cdot \hat{W}_t \right).$$

Let us denote by ρ the following function

$$\rho(w_1, \ldots, w_k, \sigma^1, \ldots, \sigma^k, K) = \mathbb{Q} \left\{ \sum_{i=1}^{k} w_i \, e^{\sigma^i \cdot \xi} \geq K \right\},$$

where the random variable $\xi = (\xi_1, \ldots, \xi_k)$ has the standard k-dimensional Gaussian probability distribution under $\mathbb{Q}$. Using Proposition 2.2 in Lamberton and Lapeyre (1993) (or by direct computations), we obtain

$$g(s_1, \ldots, s_k, T - t) = e^{-r(T-t)} f(\tilde{s}_1, \ldots, \tilde{s}_k, \sqrt{T - t}\, \sigma^1, \ldots, \sqrt{T - t}\, \sigma^k, K),$$

where

$$\tilde{s}_i = s_i \exp\left((\mu^i - \tfrac{1}{2}|\sigma^i|^2)(T - t)\right)$$

and

$$f(x_1, \ldots, x_k, b^1, \ldots, b^k, K)$$

$$= \sum_{i=1}^{k} w_i x_i \, e^{\,|b^i|^2/2} \, \rho(w_1 x_1 \, b^1 \cdot b^i, \ldots, w_k x_k \, b^k \cdot b^i, b^1, \ldots, b^k, K)$$

$$K\rho(w_1 x_1, \ldots, w_k x_k, b^1, \ldots, b^k, K)$$

for every $x_1, \ldots, x_k \in \mathbb{R}$ and $b^1, \ldots, b^k \in \mathbb{R}^k$.

Second step. As was mentioned already, to find the function $\hat{h}$ it is sufficient to express processes $S^{m+1}, \ldots, S^k$ in terms of auxiliary processes $G^{m+1}, \ldots, G^k$. Since this involves no difficulties, the second step is left to the reader.

Remarks. It is worth mentioning that a utility-based approach to the valuation and hedging of contingent claims under market incompleteness is not presented in this text. Hence for the so-called *indifference pricing* of non-attainlable claims we refer to, for instance, Davis (1997) or Musiela and Zariphopoulou (2004b) and the references therein.

8.2.4 Market Imperfections

Under market imperfections – such as the presence of transaction costs, different lending/borrowing rates or short sales constraints – the problem of arbitrage pricing becomes much more involved. We shall comment briefly on the two most relevant techniques used in the context of the Black-Scholes model with imperfections, namely *backward stochastic differential equations* (BSDEs) and *stochastic optimal control*.

Backward SDEs. Assume first that the market is perfect, but possibly incomplete. Let ϕ be a self-financing trading strategy. Then the wealth process $V = V(\phi)$ satisfies (cf. (5.2))

$$dV_t = r_t V_t \, dt + \zeta_t \cdot \left((\mu_t - r_t \mathbf{1}) \, dt + \sigma_t \, dW_t\right), \tag{8.31}$$

where the ith component of ζ_t denotes the amount of cash invested in the ith stock at time t. Given a process ζ, we may consider (8.31) as a linear SDE, with one unknown process, V. It is well known that such an equation can be solved explicitly for any initial condition V_0 (cf. Sect. 5.1). However, in replication of contingent claims, we are given instead the terminal condition $V_T = X$. Also, the stock portfolio ζ is not known a priori. It is therefore more appropriate to treat (8.31) as a *backward* SDE, with two unknown processes, V and ζ. Observe that, at the intuitive level, the concept of a BSDE combines the predictable representation property with the linear (or, more generally, non-linear) SDE. One may thus argue that no essential gain can be achieved by introducing this notion within the framework of a perfect (complete or incomplete) market. On the other hand, there is no doubt that the notion of a BSDE appears to be a useful tool when dealing with market imperfections. To this end, one needs first to develop a theoretical background, including the existence and uniqueness results as well as the so-called *comparison* theorem. A comparison theorem that basically states that solutions of BSDEs are ordered if the drift coefficients are, allows one to deal with a situation where a perfect hedging strategy is not available as a solution to a particular BSDE, but can be described as a limit of a monotone sequence of solutions of simpler BSDEs. Let us write down an example of a BSDE that arises in the study of imperfect markets. If the lending and borrowing rates are different, say $R_t \geq r_t$ for every t, then (8.31) becomes

$$dV_t = r_t V_t \, dt + \left(R_t - r_t \right) \left(\sum_{i=1}^{k} \zeta_t^i - V_t \right)^+ dt + \zeta_t \cdot \left((\mu_t - r_t \mathbf{1}) \, dt + \sigma_t \, dW_t \right).$$

Note that the non-linearity in the last equation appears in the drift term only, but depends on both the wealth process V and the stock portfolio ζ. Similar equations arise when other kinds of market imperfections are examined. For further information, in particular for references to original papers, the reader may consult El Karoui and Quenez (1997) and El Karoui et al. (1997b).

Stochastic optimal control. The optimal control (or *dynamic programming*) technique provides all the necessary tools to deal with arbitrage pricing, and indeed has already been used in some places in this text. It should be acknowledged, however, that in the present text much emphasis is given to the modelling of perfect markets and finding explicit solutions to the valuation problems. Hence the stochastic control methodology – which is particularly well suited for the theoretical study of market imperfections – is not fully exploited. For comprehensive studies of pricing contingent claims via optimization techniques in the multidimensional Black-Scholes market, we refer the interested reader to the monograph[20] by Karatzas and Shreve (1998b) (recent literature in this vein includes also Karatzas (1989, 1996), Shreve (1991), and Sethi (1997)).

Let us only indicate some papers in the area.[21] Karatzas et al. (1991) deal with market incompleteness by introducing a fictitious complete market – adding some stocks, and then penalizing, in an appropriate way, investments in these stocks (see also Karatzas and Xue (1991) for a similar approach). Cvitanić and Karatzas (1992, 1993) and Karatzas and Kou (1996, 1998) extend these studies, by considering the problem of super-replication of European and American contingent claims with constrained portfolios. A relatively concise exposition of their results can be found in

---

[20] It provides also an exhaustive treatment of optimal consumption/investment problems under constraints; such problems are not covered by the present text.

[21] All papers that are mentioned in what follows deal with the continuous-time Black-Scholes model. The dynamic programming approach to the valuation of European and American claims in a discrete-time multinomial model can be found in Tessitore and Zabczyk (1996).

Cvitanić (1997). A stochastic control approach to the perfect hedging of contingent claims in an incomplete market is also developed in El Karoui and Quenez (1995).

Let us finally mention these works, in which the dynamic programming technique is applied to analyze continuous-time financial model under the presence transaction costs. Super-replication of contingent claims and the portfolio optimization problem for the Black-Scholes model with proportional transaction costs are treated in Cvitanić and Karatzas (1996a), Broadie et al. (1998), and Cvitanić et al. (1999b). They provide, in particular, a quasi-explicit martingale characterization of the seller's price of an option. Using slightly different techniques, Soner et al. (1995) solve explicitly a particular problem, namely, they show that in the Black-Scholes market with proportional transaction costs, no non-trivial super-replicating strategy for a European call option exists.[22] Put another way, the trivial buy-and-hold strategy is "optimal" and thus the seller's price of the option equals, at any time before the option's expiry date, the price of the underlying stock. It is interesting to note that this specific feature of a European call option cannot be easily deduced from general results established in Cvitanić and Karatzas (1996a).

---

[22] This result confirms the conjecture formulated in Davis and Clark (1994). For related results, see Barles and Soner (1998) and Levental and Skorohod (1997).

Part II

Fixed-income Markets

9

Interest Rates and Related Contracts

By a *fixed-income market* we mean that sector of the global financial market on which various interest rate-sensitive instruments, such as *bonds, swaps, swaptions, caps,* etc. are traded. In real-world practice, several fixed-income markets operate; as a result, many concepts of interest rates have been developed. There is no doubt that management of interest rate risk, by which we mean the control of changes in value of a stream of future cash flows resulting from changes in interest rates, or more specifically the pricing and hedging of interest rate products, is an important and complex issue. It creates a demand for mathematical models capable of covering all sorts of interest rate risks.

Due to the somewhat peculiar way in which fixed-income securities and their derivatives are quoted in existing markets, theoretical term structure models are often easier to formulate and analyze in terms of interest rates that are different from the conventional market rates. In this chapter, we first give an overview of various concepts of interest rates. We also describe the most important financial contracts related to interest rates. Subsequently, in Sect. 9.5 we examine the most basic features of a generic stochastic term structure model driven by a Brownian motion, and we present the so-called forward measure technique of valuation of interest-rate derivatives.

9.1 Zero-coupon Bonds

Let $T^* > 0$ be a fixed horizon date for all market activities. By a *zero-coupon bond* (a *discount bond*) of maturity T we mean a financial security paying to its holder one unit of cash at a prespecified date T in the future. This means that, by convention, the bond's *principal* (known also as *face value* or *nominal value*) is one dollar. We assume throughout that bonds are *default-free*; that is, the possibility of default by the bond's issuer is excluded. The price of a zero-coupon bond of maturity T at any instant $t \leq T$ will be denoted by $B(t, T)$; it is thus obvious that $B(T, T) = 1$ for any maturity date $T \leq T^*$. Since there are no other payments to the holder, in practice a discount bond sells for less than the principal before maturity – that is, at a discount (hence its name). This is because one could carry cash at virtually no cost, and thus would have no incentive to invest in a discount bond costing more than its face value.

9.1.1 Term Structure of Interest Rates

Let us consider a zero-coupon bond with maturity date $T \leq T^*$. We assume that, for any fixed maturity $T \leq T^*$, the bond price $B(t, T)$, $t \in [0, T]$, is a strictly positive and adapted process on a filtered probability space $(\Omega, \mathbb{F}, \mathbb{P})$. The simple rate of return from holding the bond over the time interval $[t, T]$ equals

$$\frac{1 - B(t, T)}{B(t, T)} = \frac{1}{B(t, T)} - 1.$$

The equivalent rate of return, with continuous compounding, is commonly referred to as a (continuously compounded) *yield-to-maturity* on a bond.

Definition 9.1.1 An adapted process $Y(t, T)$ defined by the formula

$$Y(t, T) = -\frac{1}{T - t} \ln B(t, T), \quad \forall t \in [0, T), \tag{9.1}$$

is called the *yield-to-maturity* on a zero-coupon bond maturing at time T.

The *term structure of interest rates,* known also as the *yield curve,* is the function that relates the yield $Y(t, T)$ to maturity T. It is obvious that, for arbitrary fixed maturity date T, there is a one-to-one correspondence between the bond price process $B(t, T)$ and its yield-to-maturity process $Y(t, T)$. Given the yield-to-maturity process $Y(t, T)$, the corresponding bond price process $B(t, T)$ is uniquely determined by the formula

$$B(t, T) = e^{-Y(t,T)(T-t)}, \quad \forall t \in [0, T]. \tag{9.2}$$

The *discount function* relates the discount bond price $B(t, T)$ to maturity T. At the theoretical level, the initial term structure of interest rates may be represented either by the family of current bond prices $B(0, T)$, or by the initial *yield curve* $Y(0, T)$, as

$$B(0, T) = e^{-Y(0,T)T}, \quad \forall T \in [0, T^*]. \tag{9.3}$$

In practice, the term structure of interest rates is derived from the prices of several actively traded interest rate instruments, such as Treasury bills, Treasury bonds, swaps and futures. Note that the yield curve at any given day is determined exclusively by market prices quoted on that day. The shape of an historically observed yield curve varies over time; the observed yield curve may be upward sloping, flat, descending, or humped.

There is also strong empirical evidence that the movements of yields of different maturities are not perfectly correlated. Also, the short-term interest rates fluctuate more than long-term rates; this may be partially explained by the typical shape of the term structure of yield volatilities that is downward sloping. These features mean that the construction of a reliable model for stochastic behavior of the term structure of interest rates is a task of considerable complexity.

9.1.2 Forward Interest Rates

Let $f(t, T)$ be the forward interest rate at date $t \leq T$ for instantaneous risk-free borrowing or lending at date T. Intuitively, $f(t, T)$ should be interpreted as the interest rate over the infinitesimal time interval $[T, T + dT]$ as seen from time t. As such, $f(t, T)$ will be referred to as the *instantaneous, continuously compounded forward rate,* or shortly, *instantaneous forward rate.* In contrast to bond prices, the concept of an instantaneous forward rate is a mathematical idealization rather than a quantity observable in practice. Still, a widely accepted approach to the bond price modelling, due to Heath, Jarrow and Morton, is actually based on the exogenous specification of a family $f(t, T)$, $t \leq T \leq T^*$, of forward rates. Given such a family $f(t, T)$, the bond prices are then defined by setting

$$B(t, T) = \exp\left(-\int_t^T f(t, u)\, du\right), \quad \forall t \in [0, T]. \tag{9.4}$$

On the other hand, if the family of bond prices $B(t, T)$ is sufficiently smooth with respect to maturity T, the implied instantaneous forward interest rate $f(t, T)$ is given by the formula

$$f(t, T) = -\frac{\partial \ln B(t, T)}{\partial T} \tag{9.5}$$

which, indeed, can be seen as the formal definition of the instantaneous forward rate $f(t, T)$.

Alternatively, the instantaneous forward rate can be seen as a limit case of a *forward rate* $f(t, T, U)$ that prevails at time t for risk-free borrowing or lending over the future time interval $[T, U]$. The rate $f(t, T, U)$ is in turn tied to the zero-coupon bond price by means of the formula

$$\frac{B(t, U)}{B(t, T)} = e^{-f(t,T,U)(U-T)}, \quad \forall t \leq T \leq U,$$

or equivalently,

$$f(t, T, U) = \frac{\ln B(t, T) - \ln B(t, U)}{U - T}. \tag{9.6}$$

Observe that $Y(t, T) = f(t, t, T)$, as expected – indeed, investing at time t in T-maturity bonds is equivalent to lending money over the time interval $[t, T]$.

On the other hand, under suitable technical assumptions, the convergence $f(t, T) = \lim_{U \downarrow T} f(t, T, U)$ holds for every $t \leq T$. For convenience, we focus on interest rates that are subject to continuous compounding. In practice, interest rates are quoted on an *actuarial basis,* rather than as continuously compounded rates. For instance, the *actuarial rate* (or *effective rate*) $a(t, T)$ at time t for maturity T (i.e., over the time interval $[t, T]$) is given by the following relationship

$$(1 + a(t, T))^{T-t} = e^{f(t,t,T)(T-t)} = e^{Y(t,T)(T-t)}, \quad \forall t \in [0, T].$$

This means, of course, that the bond price $B(t, T)$ equals

$$B(t, T) = \frac{1}{(1 + a(t, T))^{T-t}}, \quad \forall t \in [0, T].$$

Similarly, the *forward actuarial rate* $a(t, T, U)$ prevailing at time t over the future time period $[T, U]$ is set to satisfy

$$(1 + a(t, T, U))^{U-T} = \exp\big(f(t, T, U)(U - T)\big) = \frac{B(t, T)}{B(t, U)}.$$

9.1.3 Short-term Interest Rate

Most traditional stochastic interest rate models are based on the exogenous specification of a short-term rate of interest. We write r_t to denote the *instantaneous interest rate* (also referred to as a *short-term interest rate,* or *spot interest rate*) for riskfree borrowing or lending prevailing at time t over the infinitesimal time interval $[t, t + dt]$.

In a stochastic set-up, the short-term interest rate is modelled as an adapted process, say r, defined on a filtered probability space $(\Omega, \mathbb{F}, \mathbb{P})$ for some $T^* > 0$. We assume throughout that r is a stochastic process with almost all sample paths integrable on $[0, T^*]$ with respect to the Lebesgue measure. We may then introduce an adapted process B of finite variation and with continuous sample paths, given by the formula

$$B_t = \exp\left(\int_0^t r_u \, du\right), \quad \forall t \in [0, T^*]. \tag{9.7}$$

Equivalently, for almost all $\omega \in \Omega$, the function $B_t = B_t(\omega)$ solves the differential equation $dB_t = r_t B_t dt$, with the conventional initial condition $B_0 = 1$.

In financial interpretation, B represents the price process of a risk-free security that continuously compounds in value at the rate r. The process B is referred to as an *accumulation factor* or a *savings account* in what follows. Intuitively, B_t represents the amount of cash accumulated up to time t by starting with one unit of cash at time 0, and continually rolling over a bond with infinitesimal time to maturity.

9.2 Coupon-bearing Bonds

A *coupon-bearing bond* is a financial security that pays to its holder the amounts $c_1, \ldots, c_m$ at the dates $T_1, \ldots, T_m$. Unless explicitly stated otherwise, we assume that the time variable is expressed in years. Obviously the bond price, say $B_c(t)$, at time t can be expressed as a sum of the cash flows $c_1, \ldots, c_m$ discounted to time t, namely

$$B_c(t) = \sum_{j=1}^{m} c_j B(t, T_j). \tag{9.8}$$

A real bond typically pays a fixed coupon c and repays the principal N. We therefore have $c_j = c$ for $j = 1, \ldots, m-1$ and $c_m = c+N$. The main difficulty in dealing with bond portfolios is due to the fact that most bonds involved are coupon-bearing bonds, rather than zero-coupon bonds. Although the coupon payments and the relevant dates are preassigned in a bond contract, the future cash flows from holding a bond are reinvested at rates that are not known in advance. The total return on a coupon-bearing bond that is held to maturity (or for a lesser period of time) thus appears to be uncertain. As a result, bonds with different coupons and cash flow dates may not be easy to compare.

The standard way to circumvent this difficulty is to extend the notion of a yield-to-maturity to coupon-bearing bonds. We give below two versions of the definition of a yield-to-maturity. The first corresponds to the case of discrete compounding (such an assumption reflects more accurately the market practice); the second assumes continuous compounding.

9.2.1 Yield-to-Maturity

Consider a bond that pays m identical yearly coupons c at the dates $1, \ldots, m$ and the principal N after m years. Its yield-to-maturity at time 0, denoted by $\tilde{Y}_c(0)$, may be found from the following relationship

$$B_c(0) = \sum_{j=1}^{m} \frac{c}{(1 + \tilde{Y}_c(0))^j} + \frac{N}{(1 + \tilde{Y}_c(0))^m}.$$

Since the coupon payments are usually determined by a preassigned interest rate $r_c > 0$ (known as a *coupon rate*), this may be rewritten as follows

$$B_c(0) = \sum_{j=1}^{m} \frac{r_c N}{(1 + \tilde{Y}_c(0))^j} + \frac{N}{(1 + \tilde{Y}_c(0))^m}.$$

It is clear that in this case the yield does not depend on the face value of the bond. Notice that the price $B_c(0)$ equals the bond's face value N whenever the coupon rate equals the bond's yield-to-maturity – that is, $r_c = \tilde{Y}_c(0)$; in this case a bond is said to be priced *at par*. Similarly, we say that a bond is priced below par (i.e., *at a discount*) when its current price is lower than its face value, $B_c(0) < N$, or equivalently, when its yield-to-maturity exceeds the coupon rate, $\tilde{Y}_c(0) > r_c$. Finally, a bond is priced above par (i.e., *at a premium*) when $B_c(0) > N$, that is, when $\tilde{Y}_c(0) < r_c$. In the case of continuous compounding, the corresponding yield-to-maturity $Y_c(0)$ satisfies

$$B_c(0) = \sum_{j=1}^{m} c e^{-jY_c(0)} + N e^{-mY_c(0)},$$

where $B_c(0)$ stands for the current market price of the bond.

Let us now focus on zero-coupon bonds (i.e., $c = 0$ and $N = 1$). The initial price $B(0, m)$ of a zero-coupon bond can easily be found provided its yield-to-maturity $\tilde{Y}(0, m)$ is known. Indeed, we have

$$B(0, m) = \frac{1}{(1 + \tilde{Y}(0, m))^m}.$$

Similarly, in a continuous-time framework, we have $B(0, T) = e^{-Y(0,T)T}$, where $B(0, T)$ is the initial price of a unit zero-coupon bond of maturity T, and $Y(0, T)$ stands for its yield-to-maturity. We adopt the following definitions of the yield-to-maturity of a coupon-bearing bond in a discrete-time and in a continuous-time setting.

Definition 9.2.1 The discretely compounded *yield-to-maturity* $\tilde{Y}_c(i)$ at time i on a coupon-bearing bond that pays the positive deterministic cash flows $c_1, \ldots, c_m$ at the dates $1 < \cdots < m$ is given implicitly by means of the formula

$$B_c(i) = \sum_{j=i+1}^{m} \frac{c_j}{(1 + \tilde{Y}_c(i))^{j-i}}, \qquad (9.9)$$

where $B_c(i)$ stands for the price of a bond at the date $i < m$.

Definition 9.2.2 In a continuous-time framework, if a bond pays the positive cash flows $c_1, \ldots, c_m$ at the dates $T_1 < \cdots < T_m$, then its continuously compounded *yield-to-maturity* $Y_c(t) = Y_c(t; c_1, \ldots, c_m, T_1, \ldots, T_m)$ is uniquely determined by the following relationship

$$B_c(t) = \sum_{T_j > t} c_j e^{-Y_c(t)(T_j - t)}, \qquad (9.10)$$

where $B_c(t)$ denotes the bond price at time $t < T_m$.

Note that on the right-hand side of (9.9) (respectively, (9.10)), the coupon payment at time i (respectively, at time t) is not taken into account. Consequently, the price $B_c(i)$ (respectively, $B_c(t)$) is the price of a bond after the coupon at time i (respectively, at time t) has been paid. We focus mainly on the continuously compounded yield-to-maturity $Y_c(t)$. It is common to interpret the yield-to-maturity $Y_c(t)$ as a proxy for the uncertain return on a bond; this means that it is implicitly assumed that all coupon payments occurring after the date t are reinvested at the rate $Y_c(t)$. Since this cannot, of course, be guaranteed, the yield-to-maturity should be seen as a very rough approximation of the uncertain return on a coupon-bearing bond. On the other hand, the return on a discount bond is certain, therefore the yield-to-maturity determines exactly the return on a discount bond. It is worth noting that for every t an $\mathcal{F}_t$-adapted random variable $Y_c(t)$ is uniquely determined for any given collection of positive cash flows $c_1, \ldots, c_m$ and dates $T_1, \ldots, T_m$, provided that the bond price at time t is known.

Let us conclude this section by observing that the bond price moves inversely to the bond's yield-to-maturity. Moreover, it can also be checked that the moves are asymmetric, so that a decrease in yields raises bond prices more than the same increase lowers bond prices. This specific feature of the bond price is referred to as *convexity*. Finally, it should be stressed that the uncertain return on a bond comes from both the interest paid and from the potential capital gains (or losses) caused by the future fluctuations of the bond price. Hence the term *fixed-income security* should not be taken literally, unless we consider a bond that is held to its maturity.

9.2.2 Market Conventions

The market conventions related to U.S. government debt securities differ slightly from our generic definitions adopted in the preceding section. Debt securities issued by the U.S. Treasury are divided into three classes: *bonds, notes* and *bills*. The Treasury bill (T-bill, for short) is a discount bond – it pays no coupons, and the investor receives the face value at maturity. Maturity of a T-bill is no longer than one year. Treasury notes and bonds, T-notes and T-bonds for short, are coupon securities. T-bonds have more than 10 years to maturity when issued, T-notes have shorter times to maturity; bonds and notes are otherwise identical. The U.S. Treasury pays bond-holders total annual interest equal to the coupon rate, however a m-year government bond pays coupons semi-annually in equal instalments, say at times $T_j = j\delta$ where $\delta = 1/2$ and $j = 1, 2, \ldots, 2m$. The quoted "yield-to-maturity" $\hat{Y}_c(0)$ on a government bond, more correctly called a *bond equivalent yield,* is based on the following relationship

$$B_c(0) = \sum_{j=1}^{2m-1} \frac{r_c N/2}{(1 + \hat{Y}_c(0)/2)^j} + \frac{(1 + r_c/2)N}{(1 + \hat{Y}_c(0)/2)^{2m}}, \qquad (9.11)$$

where r_c is the coupon rate of a bond and N stands for its face value (note that, for simplicity, we consider only government bonds with a round number of coupon periods to maturity). By simple algebra, one finds that formula (9.11) may be rewritten as follows

$$B_c(0) = \frac{r_c N}{\hat{Y}_c(0)} + \frac{N(1 - r_c/\hat{Y}_c(0))}{(1 + \hat{Y}_c(0)/2)^{2m}}.$$

The yield at time i is implicitly defined by means of the relationship

$$B_c(i) = \sum_{j=i+1}^{2m-1} \frac{r_c N/2}{(1 + \hat{Y}_c(i)/2)^j} + \frac{(1 + r_c/2)N}{(1 + \hat{Y}_c(i)/2)^{2m}},$$

where $\hat{Y}_c(i)$ is the yield-to-maturity on a bond at time i, after the ith interest payment has been made. Note that the interest rate $\hat{Y}_c(i)$ is annualized with no compounding.

The compounded annualized yield $\hat{Y}_c^e(i)$, defined as

$$\hat{Y}_c^e(i) = \left(1 + \hat{Y}_c(i)/2\right)^2 - 1,$$

is termed the *effective annual yield.* A government bond is traded in terms of its price, which is quoted as a percentage of face value – unless it is trading as WI (i.e., *when issued*) before an auction; a WI bond is quoted in yield terms. The quoted price of a bond does not coincide with the price a customer has to pay for it; the *invoice price* of a bond is the quoted price plus accrued interest from the last interest payment back to the purchase date.

9.3 Interest Rate Futures

The most heavily traded interest rate futures contracts are those related either to fixed-income securities (such as, Treasury bonds, notes and bills), or to specific interest rates (such as, LIBORs or swap rates). Typical contracts from the first category are: *Treasury bond futures, Treasury note futures*, and *Treasury bill futures*. The *Eurodollar futures* and *LIBOR futures*, which have as the underlying instrument the 3-month LIBOR and the 1-month LIBOR respectively, are examples of futures contracts from the second category. The LIBOR – that is, the London Interbank Offer Rate – is the rate of interest offered by banks on deposits from other banks in Eurocurrency markets. LIBOR represents the interest rate at which banks lend money to each other; it is also the floating rate widely used in interest rate swap agreements in international financial markets. LIBOR is determined by trading between banks, and changes continuously as economic conditions change.

The most important interest rate futures options are: *T-bond futures options, T-note futures options, Eurodollar futures options*, and *LIBOR futures options*. The nominal size of the option contract usually agrees with the size of the underlying futures contract; for instance, it amounts to $100,000 for both T-bond futures and T-bond futures options, and to $1 million for Eurodollar futures and the corresponding options. Let us now describe these contracts in some detail.

9.3.1 Treasury Bond Futures

Until the introduction of financial futures, the futures market consisted only of contracts for delivery of commodities. In 1975, the Chicago Board of Trade (CBOT) created the first financial futures contract, a futures contract for so-called *mortgage-backed securities*. Mortgage-backed securities are bonds collaterized with a pool of government-guaranteed home mortgages. Since these securities are issued by the Government National Mortgage Association (GNMA), the corresponding futures contract is commonly referred to as *Ginnie Mae futures*.

Treasury bond futures contracts were introduced on the CBOT two years later, that is, in 1977. Nominally, the underlying instrument of a T-bond futures contract is a 15-year T-bond with an 6% coupon. T-bond futures contracts and T-bond futures options trade with up to one year to maturity. As usual, the futures contract specifies precisely the time and other relevant conditions of delivery. Delivery is made on any business day of the delivery month, two days after the *delivery notice* (i.e., the declaration of intention to make delivery) is passed to the exchange. The *invoice price* received by the party with a short position equals the bond futures settlement price multiplied by the *delivery factor* for the bond to be delivered, plus the accrued interest. The delivery factor, determined for each deliverable bond issue, is based on the coupon rate and the time to the bond's expiry date. Basically, it equals the price of a unit bond with the same coupon rate and maturity, assuming that the yield-to-maturity of the bond equals 8%. For instance, for a bond with m years to maturity and coupon rate r_c, the *conversion factor* δ equals

$$\delta = \sum_{j=1}^{2m} \frac{r_c/2}{(1+0.04)^j} + \frac{1}{(1+0.04)^{2m}},$$

so that $\delta > 1$ (respectively, $\delta < 1$) whenever $r_c > 0.08$ (respectively, $r_c < 0.08$). Note that the adjustment factor δ makes the yields of each deliverable bond roughly equal for a party paying the invoice price. In particular, if the settlement futures price[1] is close to 100, this yield is approximately 8%. At any given time, there are about 30 bonds that can be delivered in the T-bond futures contract (basically, any bond with at least 15 years to maturity). The *cheapest-to-deliver* bond is that deliverable issue for which the difference

Quoted bond price − Settlement futures price × Conversion factor

is least. Put differently, the cheapest-to-deliver bond is the one for which the *basis* $b_t^i = B_c^i(t) - f_t \delta^i$ is minimal, where $B_c^i(t)$ is the current price of the ith deliverable bond, f_t is the bond futures settlement price, and δ^i is the conversion factor of the ith bond. Usually, the market is able to forecast the cheapest-to-deliver bond for a given delivery month. A change in the shape of the yield curve or a change in the level of yields often means a switch in cheapest-to-deliver bond, however. This is because, as yields change, a security with a slightly different coupon or maturity may become cheaper for market makers to deliver. Before the delivery month, the determination of the cheapest-to-deliver issue also involves the *cost of carry* (net financing cost) of a given bond; the top delivery choice is the issue with the lowest after-carry basis. Due to the change of yield level (or new bond issue) as time passes, the top delivery choice also changes. The possibility of such an event, which may be seen as an additional source of risk, makes the valuation of futures contracts and their use for hedging purposes more involved.

9.3.2 Bond Options

Currently traded bond-related options split into two categories: OTC bond options and T-bond futures options. The market for the first class of bond options is made by primary dealers and some active trading firms. The long (i.e., 30-year) bond is the most popular underlying instrument of OTC bond options; however, options on shorter-term issues are also available to customers. Since a large number of different types of OTC bond options exists in the market, the market is rather illiquid. Most options are written with one month or less to expiry. They usually trade at-the-money. This convention simplifies quotation of bond option prices. Options with exercise

---

[1] It is customary to quote both the bond price and the bond futures price for a $100 face value bond.

prices that are up to two points out-of-the-money are also common. Bond options are used by traders to immunize their positions from the direction of future price changes. For instance, if a dealer buys call options from a client, he usually sells cash bonds in the open market at the same time.

Like all typical exchange-traded options, T-bond futures options have fixed strike prices and expiry dates. Strike prices come in two-point increments. The options are written on the first four delivery months of a futures contract (note that the delivery of the T-bond futures contract occurs only every three months). In addition, a 1-month option is traded (unless the next month is the delivery month of the futures contract). The options stop trading a few days before the corresponding delivery month of the underlying futures contract. The T-bond futures option market is highly liquid. An open interest in one option contract may amount to $5 billion in face value (this corresponds to 50,000 option contracts).

9.3.3 Treasury Bill Futures

The *Treasury bill* (*T-bill*, for short) is a bill of exchange issued by the U.S. Treasury to raise money for temporary needs. It pays no coupons, and the investor receives the face value at maturity. T-bills are issued on a regular schedule in 3-month, 6-month and 1-year maturities. In the T-bill futures contract, the underlying asset is a 90-day T-bill. The common market practice is to quote a discount bond, such as the T-bill, not in terms of the yield-to-maturity, but rather in terms of so-called *discount rates*. The discount rate represents the size of the price reduction for a 360-day period (for instance, a bill of face value 100 that matures in 90 days and is sold at a discount rate 10% is priced at 97.5).

Formally, a *discount rate* $R_b(t, T)$ (known also as *bankers' discount yield*) of a security that pays a deterministic cash flow X_T at the future date T, and has the price X_t at time $t < T$, equals

$$R_b(t, T) = \frac{X_T - X_t}{X_T} \frac{360}{T - t},$$

where $T - t$ is now expressed in days. In particular, for a discount bond this gives

$$R_b(t, T) = \frac{100 - P(t, T)}{100} \frac{360}{T - t} = (1 - B(t, T)) \frac{360}{T - t},$$

where $P(t, T)$ (respectively, $B(t, T)$) stands for the cash price of a bill with face value 100 (respectively, with the unit face value) and $T - t$ days to maturity. Conversely, given a discount rate $R_b(t, T)$ of a bill, we find its cash price from the following formula

$$P(t, T) = 100\left(1 - R_b(t, T)\frac{T - t}{360}\right).$$

For a just-issued 90-day T-bill, the above formulas can be further simplified. Indeed, we have

$$R_b(0, 90) = 4(1 - P(0, 90)/100)$$

and

$$P(0, 90) = 100\left(1 - \tfrac{1}{4}R_b(0, 90)\right).$$

The *bill yield* on a T-bill equals

$$Y_b(t, T) = \frac{100 - P(t, T)}{P(t, T)} \frac{360}{T - t} = \frac{1 - B(t, T)}{B(t, T)} \frac{360}{T - t} = \frac{R_b(t, T)}{B(t, T)},$$

so that it represents the annualized (with no compounding) interest rate earned by the bill owner. In terms of a bill yield $Y_b(t, T)$, its price $B(t, T)$ equals

$$B(t, T) = \frac{1}{1 + Y_b(t, T)(T - t)/360}.$$

Note that the above conventions concerning bill yields assume a 360-day year (definitions assuming a 365-day year are not uncommon, however). A widely used practical formula for the bond equivalent yield on a T-bill with 182 or fewer days to maturity is

$$\tilde{Y}_b(t, T) = Y_b(t, T) \frac{365}{360} = \frac{R_b(t, T)}{B(t, T)} \frac{365}{360} = \frac{1 - B(t, T)}{B(t, T)} \frac{365}{T - t}.$$

Taking the yield on a government bond as a benchmark, we define the *bond equivalent yield* $\hat{Y}_b(t, T)$ on a T-bill by setting

$$\hat{Y}_b(t, T) = 2\left((1/B(t, T))^{365/2(T-t)} - 1\right),$$

where the time period $T - t$ is expressed in days. The above equality is a consequence of the following implicit definition of the bond equivalent yield $\hat{Y}_b(t, T)$ on a T-bill

$$B(t, T) = \frac{1}{\left(1 + \tfrac{1}{2}\hat{Y}_b(t, T)\right)^{2(T-t)/365}}.$$

Finally, the *effective annual yield* $\hat{Y}_b^e(t, T)$ on a T-bill, that is directly comparable with an effective annual yield $\hat{Y}_c^e$ on a T-bond, equals

$$\hat{Y}_b^e(t, T) = \left(1 + \tfrac{1}{2}\hat{Y}_b(t, T)\right)^2 - 1 = (1/B(t, T))^{365/(T-t)} - 1.$$

In contrast to T-bills, which are quoted in terms of the discount rate, T-bill futures are quoted in terms of the price. In particular, the T-bill futures price at maturity equals 100 minus the T-bill quote. The marking to market procedure is based, however, on the corresponding cash price of a given futures contract. For instance, if the quotation for T-bill futures is $f_t^b = 95.02$, the implied discount rate equals $R^f(t, T) = 4.98$, and thus the corresponding cash futures price, which is used in marking to the market, equals

$$P^f(t, T) = 100 - \tfrac{1}{4}(100 - f_t^b) = 100 - \tfrac{1}{4}R^f(t, T) = 98.755$$

per \$100 face value bill, or equivalently, \$987,550 per futures contract (the nominal size of one T-bill futures contract that trades on the Chicago Mercantile Exchange (CME) is \$1 million).

9.3.4 Eurodollar Futures

Since conventions associated with market quotations of the LIBOR and *Eurodollar time deposit futures* (*Eurodollar futures*, for short) are close to those examined above, we shall describe them in a rather succinct way. Eurodollar futures contracts and related Eurodollar futures options have traded on the Chicago Mercantile Exchange since 1981 and 1985 respectively. Eurodollar futures and Eurodollar futures options trade with up to five years to maturity. A Eurodollar futures option is of American style; one option covers one futures contract and it expires at the settlement date of the underlying Eurodollar futures contract.

Formally, the underlying instrument of a Eurodollar time deposit futures contract is the 3-month LIBOR. At the settlement date of a Eurodollar futures contract, the CME surveys 12 randomly selected London banks, which are asked to give their perception of the rate at which 3-month Eurodollar deposit funds are currently offered by the market to prime banks. A suitably rounded average of these quotes serves to determine the Eurodollar futures price at the settlement date of a Eurodollar futures contract (cf. Amin and Morton (1994)).

Let us emphasize that the LIBOR is defined as the *add-on yield*; that is, the actual interest payment on a 3-month Eurodollar time deposit equals "LIBOR × numbers of days for investment/360" per unit investment. In our framework, the (spot) LIBOR at time t on a Eurodollar deposit with a maturity of τ days is formally defined as

$$l(t, t + \tau) = \frac{360}{\tau} \left(\frac{1}{B(t, t + \tau)} - 1 \right).$$

Equivalently, the bond price is given by the following expression

$$B(t, t + \tau) = \frac{1}{1 + l(t, t + \tau)\tau/360}.$$

In particular, for a 3-month LIBOR the relationship simplifies to

$$l(t, t + 90) = 4 \left(B^{-1}(t, t + 90) - 1 \right).$$

A Eurodollar futures price $f_l(T, T)$ on the settlement day T is

$$f_l(T, T) = 100\left(1 - l(T, T + 90)\right) = 100 \left(1 - 4 \left(B^{-1}(T, T + 90) - 1 \right) \right),$$

hence $100 - f_l(t, T)$ converges to a 3-month LIBOR (in percentage terms) as the time argument t tends to the delivery date T. A market quotation of Eurodollar futures contracts is based on the same rule as the quotation of T-bill futures. Explicitly, if q stands for the market quotation of Eurodollar futures then the value of a contract is

$$100 - \tfrac{1}{4}\left(100 - q\right) = 100\left(1 - \tfrac{1}{4} f_l(t, t + 90)\right)$$

per $100 of nominal value. The nominal value of one Eurodollar futures contract is $1 million; one basis point is thus worth $25 when the contract is marked to the market daily.

For instance, the quoted Eurodollar futures price 94.47 corresponds to a 3-month LIBOR futures rate of 5.53 %, and to the price $986,175 of one Eurodollar futures contract. If the next day the quoted price rises to 94.48 (i.e., the LIBOR futures rate declines to 5.52%), the value of one contract appreciates by $25 to $986,200.

Let us briefly describe market conventions related to Eurodollar futures options. The owner of a Eurodollar futures call option obtains a long position in the futures contract with a futures price equal to the option's exercise price; the call writer obtains a short futures position. On marking to market, the call owner receives the cash difference between the marked-to-market futures price and the strike price.

9.4 Interest Rate Swaps

Generally speaking, a *swap contract* (or a *swap*) is an agreement between two parties to exchange cash flows at some future dates according to a prearranged formula. In a classical swap contract, the value of the swap at the time it is entered into, as well as at the end of its life, is zero. The two most popular kinds of swap agreements are standard *interest rate swaps* and *cross-currency swaps* (known also as *differential swaps*).

In a *plain vanilla interest rate swap,* one party, say A, agrees to pay to the other party, say B, amounts determined by a fixed interest rate on a *notional principal* at each of the payment dates. At the same time, the party B agrees to pay to the party A interest at a floating reference rate on the same notional principal for the same period of time. Thus an interest rate swap can be used to transform a floating-rate loan into a fixed-rate loan or vice versa. In essence, a swap is a long position in a fixed-rate coupon bond combined with short positions in floating-rate notes (alternatively, it can be seen as a portfolio of specific forward contracts). In a *payer swap,* the fixed rate is paid at the end (or, depending on contractual features of the swap, at the beginning) of each period, and the floating rate is received (therefore, it may also be termed a *fixed-for-floating swap*). Similarly, a *receiver swap* (or a *floating-for-fixed swap*) is one in which an investor pays a floating rate and receives a fixed rate on the same notional principal.

In a payer swap settled *in arrears,* the floating rate paid at the end of a period is set at the beginning of this period. We say that a swap is settled *in advance* if payments are made at the beginning of each period. Notice that payments of a swap settled in advance are the payments, discounted to the beginning of each period, of the corresponding swap settled in arrears. However, the discounting conventions vary from country to country. In some cases, both sides of a swap are discounted using the same floating rate; in others, the floating is discounted using the floating and the fixed using the fixed.

Let us consider a collection $T_0 = T < T_1 < \cdots < T_n$ of future dates. Formally, a *forward start swap* (or briefly, a *forward swap*) is a swap agreement entered at the trade date $t \leq T_0$ with payment dates $T_1 < \ldots < T_n$ (if a swap is settled in arrears) or $T_0 < \ldots < T_{n-1}$ (if a swap is settled in advance). For most swaps, a fee (the up-front cost) is negotiated between the counter-parties at the trade date t. The *forward swap rate* is that value of the fixed rate that makes the value of the forward swap zero.

The market gives quotes on *swap rates,* i.e., the fixed rates at which financial institutions offer to their clients interest rate swap contracts of differing maturities, with fixed quarterly, semi-annual or annual payment schedules. Some benchmark swap rates are the underlying instruments in swap futures contracts. The most typical option contract associated with swaps is a *swaption* – that is, an option on the value of the underlying swap or, equivalently, on the *swap rate.*

9.4.1 Forward Rate Agreements

Let us comment briefly on a more conventional class of contracts, widely used by companies to hedge the interest rate risk. Consider a company that forecasts that it will need to borrow cash at a future date, say T, for the period $[T, U]$. The company will be, of course, unhappy if the interest rate rises by the time the loan is required. Interest rates for floating-rate loans are usually set by reference to some floating benchmark interest rate. For instance, if a company raises a loan at "LIBOR+p%", the company will pay a rate of interest on its loan equal to whatever LIBOR is, plus an extra p%.

A commonly used contract that serves to reduce interest rate risk exposure by locking into a rate of interest, is a *forward rate agreement.* A forward rate agreement (an FRA) is a contract in which two parties (a seller of a contract and a buyer) agree to exchange, at some future date, interest payments on the notional principal of a contract. It will be convenient to assume that this payment is made at the end of the period, say at time U. The cash flow is determined by the length of the time-period, say $[T, U]$, and by two relevant interest rates: the predetermined rate of interest and the risk-free rate of interest prevailing at time T. The buyer of an FRA thus pays interest at a preassigned rate and receives interest at a floating reference rate that prevails at time T. A typical use of a forward rate agreement is a long position in an FRA combined with a loan taken at time T over the period $[T, U]$. A synthetic version of such a strategy is a *forward-forward loan* – that is, a combination of a longer-term loan and a shorter-term deposit (a company just takes a loan over $[0, U]$ and makes a deposit over $[0, T]$). Assuming a frictionless market, the rate of interest a company manages to lock into on its loan using the above strategy will coincide with the preassigned rate of interest in forward rate agreements proposed to customers by financial institutions at no additional charge. Indeed, instead of manufacturing a forward-forward loan, a company may alternatively buy (at no charge) an FRA and take at time T a loan on the spot market.

We shall examine first a forward rate agreement written at time 0 with the reference period $[T, U]$. We may and do assume, without loss of generality, that the notional principal of the contract is 1. Let $r(0, T)$ stand for the continuously compounded interest rate for risk-free borrowing and lending over the time-period $[0, T]$. It is clear that, barring arbitrage opportunities between bank deposits and the zero-coupon bond market, the T-maturity spot rate $r(0, T)$ should satisfy $e^{r(0,T)T} = B^{-1}(0, T)$. In other words, the interest rate $r(0, T)$ coincides with the continuously compounded yield on a T-maturity discount bond – that is, $r(0, T) = Y(0, T)$ for every T. As was mentioned earlier, the buyer of an FRA receives at time U a cash flow corresponding to an interest rate set at time T, and pays interest according to a rate preassigned at time 0. The level of the predetermined rate, loosely termed a *forward interest rate,* is chosen in such a way that the contract is worthless at the date it is entered into. Let us denote by $f(0, T, U)$ this level of interest rate, corresponding to an FRA written at time 0 and referring to the future time period $[T, U]$. The *forward rate* $f(0, T, U)$ may alternatively be seen as a continuously compounded interest rate, prevailing at time 0, for risk-free borrowing or lending over the time period $[T, U]$. The "right" level of the forward rate $f(0, T, U)$ can be determined by standard no-arbitrage arguments. By considering two alternative trading strategies, it is easy to establish the following relationship

$$e^{Ur(0,U)} = e^{Tr(0,T)} e^{f(0,T,U)(U-T)}.$$

More generally, the forward rate $f(t, T, U)$ satisfies, for every $t \le T \le U$,

$$f(t, T, U) = \frac{(U - t)r(t, U) - (T - t)r(t, T)}{U - T}$$

where $r(t, T)$ is the future spot rate, as from time t, for risk-free borrowing or lending over the time period $[t, T]$. Note that $f(0, T, T) = f(0, T)$, i.e., the rate $f(0, T, T)$ (if well-defined) equals the instantaneous forward interest rate $f(0, T)$. For similar reasons, the equality $f(t, T, T) = f(t, T)$ is valid. If $r(t, T) = Y(t, T)$, then in terms of bond prices we have (cf. (9.6))

$$f(t, T, U) = \frac{\ln B(t, T) - \ln B(t, U)}{U - T}.$$

Bid-offer spread. In order to derive formulas that more closely reflect the market practice, one needs to take into account the bid-offer spread – that is, the spread between the borrowing and lending rates. Banks quote forward loan and deposit rates to their customers if requested. Denote by $r_b(0, T)$ (respectively, $r_o(0, T)$) the bid (respectively, offered) interest rate prevailing on the money market at time 0. The bid rate $r_b(0, T)$ is the interest rate that financial institutions, say banks, are ready to pay on deposits, while the offered rate $r_o(0, T)$ is the rate charged by banks on loans. The *offered forward rate* $f_o(0, T, U)$ (i.e., the forward loan rate) corresponds to a short position in a forward rate agreement – that is, $f_o(0, T, U)$ is the rate charged by banks for loans over $[T, U]$. It is not difficult to check that

$$f_o(0, T, U) = \frac{Ur_o(0, U) - Tr_b(0, T)}{U - T}.$$

In fact, the last equality means that the rate $f_o(0, T, U)$ is the actual rate over the period $[T, U]$ that a company may lock into by applying a forward-forward loan (the company has to pay an offered rate of $r_o(0, U)$ for its longer-term loan; it receives interest determined by the bid rate $r_b(0, T)$ on the shorter-term deposit). Summarizing, the offered forward rate $f_o(0, T, U)$ is equal to the rate of interest on loans implied by a synthetic forward-forward loan over the same period of time. By similar arguments, one can check that the bid forward interest rate equals

$$f_b(0, T, U) = \frac{U r_b(0, U) - T r_o(0, T)}{U - T}.$$

9.5 Stochastic Models of Bond Prices

In this section, we examine the basic technical issues related to generic continuous-time term structure models. We start by introducing the concept of an arbitrage-free family of bond prices associated with a given process modelling the short-term interest rate. This is subsequently specified to the most widely popular and practically important case of Itô processes.

9.5.1 Arbitrage-free Family of Bond Prices

Recall that, by convention, a zero-coupon bond pays one unit of cash at a prescribed date T in the future. The price at any instant $t \leq T$ of a zero-coupon bond of maturity T is denoted by $B(t, T)$; it is thus clear that, necessarily, $B(T, T) = 1$ for any maturity date $T \leq T^*$. Furthermore, since there are no intervening interest payments, in market practice the bond sells for less than the principal before the maturity date.

Essentially, this follows from the fact that it is possible to invest money in a risk-free savings account yielding a nonnegative interest rate (or at least to carry cash at virtually no cost). However, the assumption that $B(t, T) \leq 1$ is not covenient, since it would exclude some Gaussian term structure models from our considerations. Thus, we make the following standing assumption.

Assumption (B). We postulate that for any fixed maturity $T \leq T^*$, the price process $B(t, T)$, $t \in [0, T]$, is a strictly positive and $\mathbb{F}$-adapted process, defined on a filtered probability space $(\Omega, \mathbb{F}, \mathbb{P})$, where the filtration $\mathbb{F}$ is the $\mathbb{P}$-completed version of the filtration generated by the underlying d-dimensional standard Brownian motion W.

The first question that we will address is the absence of arbitrage opportunities between all bonds with different maturities and a savings account. Suppose that an adapted process r, given on a filtered probability space $(\Omega, \mathbb{F}, \mathbb{P})$, models the short-term interest rate, meaning that the savings account process B satisfies

$$B_t = \exp \left(\int_0^t r_u \, du \right), \quad \forall \, t \in [0, T^*].$$

Then we have the following definition of an arbitrage-free property.

Definition 9.5.1 A family $B(t, T)$, $t \leq T \leq T^*$, of adapted processes is called an *arbitrage-free family of bond prices relative to a short-term interest rate process r* if the following conditions are satisfied:

(a) $B(T, T) = 1$ for every $T \in [0, T^*]$,

(b) there exists a probability measure $\mathbb{P}^*$ on $(\Omega, \mathcal{F}_{T^*})$ equivalent to $\mathbb{P}$, and such that for any maturity $T \in [0, T^*]$ the relative bond price

$$Z^*(t, T) = B(t, T)/B_t, \quad \forall t \in [0, T], \tag{9.12}$$

is a martingale under $\mathbb{P}^*$.

Any probability measure $\mathbb{P}^*$ of Definition 9.5.1 is called a *martingale measure for the family* $B(t, T)$ *relative to* r, or briefly, a *martingale measure for the family* $B(t, T)$ if no confusion may arise. In what follows, we shall distinguish between *spot* and *forward* martingale measures. In this context, the martingale measure of Definition 9.5.1 should be seen as a *spot martingale measure* for the family $B(t, T)$.

The reader might wonder why it is assumed that the relative price Z^* follows a martingale, and not merely a local martingale, under $\mathbb{P}^*$. The main reason is that under the former assumption we have trivially $Z^*(t, T) = \mathbb{E}_{\mathbb{P}^*}(Z^*(T, T) \mid \mathcal{F}_t)$ for $t \leq T$, so that the bond price satisfies

$$B(t, T) = B_t \, \mathbb{E}_{\mathbb{P}^*}(B_T^{-1} \mid \mathcal{F}_t), \quad \forall t \in [0, T]. \tag{9.13}$$

In other words, for any martingale measure $\mathbb{P}^*$ of an arbitrage-free family of bond prices, we have

$$B(t, T) = \mathbb{E}_{\mathbb{P}^*}\left(e^{-\int_t^T r_u \, du} \,\Big|\, \mathcal{F}_t\right), \quad \forall t \in [0, T]. \tag{9.14}$$

Conversely, given any nonnegative short-term interest rate r defined on a probability space $(\Omega, \mathbb{F}, \mathbb{P})$, and any probability measure $\mathbb{P}^*$ on $(\Omega, \mathcal{F}_{T^*})$ equivalent to $\mathbb{P}$, the family $B(t, T)$ given by (9.14) is easily seen to be an arbitrage-free family of bond prices relative to r. Let us observe that if a family $B(t, T)$ satisfies Definition 9.5.1 then necessarily the bond price process $B(\cdot, T)$ is a $\mathbb{P}^*$-semimartingale, as a product of a martingale and a process of finite variation (that is, a product of two $\mathbb{P}^*$-semimartingales). Thus, it is also a $\mathbb{P}$-semimartingale, since the probability measures $\mathbb{P}$ and $\mathbb{P}^*$ are assumed to be mutually equivalent (see Theorem III.20 in Protter (2003)).

9.5.2 Expectations Hypotheses

Suppose that equality (9.14) is satisfied under the actual probability measure $\mathbb{P}$, that is

$$B(t, T) = \mathbb{E}_{\mathbb{P}}\left(e^{-\int_t^T r_u \, du} \,\Big|\, \mathcal{F}_t\right), \quad \forall t \in [0, T]. \tag{9.15}$$

Equality (9.15) is traditionally referred to as the *local expectations hypothesis* ((L-EH) for short), or a *risk-neutral expectations hypothesis*. The term "local expectations" refers to the fact that under (9.15), the current bond price equals the expected value, under the actual probability, of the bond price in the next (infinitesimal) period, discounted at the current short-term rate. This property can be made more explicit in a discrete-time setting (see Ingersoll (1987) or Jarrow (1995)). In our framework, given an arbitrage-free family of bond prices relative to a short-term rate r, it is evident that (9.14) holds, by definition, under any martingale measure $\mathbb{P}^*$. This does not mean, however, that the local expectations hypothesis, or any other traditional form of expectations hypothesis, is satisfied under the actual probability $\mathbb{P}$.

An alternative hypothesis, known as the *return-to-maturity expectations hypothesis* (RTM-EH) assumes that the return from holding any discount bond to maturity is equal to the return expected from rolling over a series of single-period bonds. Its continuous-time counterpart reads as follows

$$\frac{1}{B(t, T)} = \mathbb{E}_\mathbb{P}\left(e^{\int_t^T r_u \, du} \,\Big|\, \mathcal{F}_t\right), \quad \forall\, t \in [0, T],$$

for any maturity date $T \leq T^*$.

Finally, the *yield-to-maturity expectations hypothesis* (YTM-EH) asserts that the yield from holding any bond is equal to the yield expected from rolling over a series of single-period bonds. In a continuous-time framework, this means that for any maturity date $T \leq T^*$, we have

$$B(t, T) = \exp\left\{-\mathbb{E}_\mathbb{P}\left(\int_t^T r_u \, du \,\Big|\, \mathcal{F}_t\right)\right\}, \quad \forall\, t \in [0, T].$$

The last formula may also be given the following equivalent form

$$Y(t, T) = \frac{1}{T - t} \mathbb{E}_\mathbb{P}\left(\int_t^T r_u \, du \,\Big|\, \mathcal{F}_t\right),$$

or finally

$$f(t, T) = \mathbb{E}_\mathbb{P}(r_T \,|\, \mathcal{F}_t), \quad \forall\, t \in [0, T]. \tag{9.16}$$

In view of (9.16), under the yield-to-maturity expectations hypothesis, the forward interest rate $f(t, T)$ is an unbiased estimate, under the actual probability $\mathbb{P}$, of the future short-term interest rate r_T. For this reason, the YTM-EH is also frequently referred to as the *unbiased expectations hypothesis*. We will see in what follows that condition (9.16) is always satisfied – not under the actual probability, however, but under the so-called *forward martingale measure* for the given date T.

Remarks. If the short-term rate r is a deterministic function, then all expectations hypotheses coincide, and follow easily from the absence of arbitrage. Several authors have examined the validity of various forms of expectations hypotheses under the actual probability, usually within the framework of a general equilibrium approach (see, e.g., Cox et al. (1981a), Fama (1984b), Campbell (1986), Longstaff (1990b), Stigler (1990), and McCulloch (1993)).

9.5.3 Case of Itô Processes

In a continuous-time framework, it is customary to model the short-term rate of interest by means of an Itô process, or more specifically, as a one-dimensional diffusion process.[2] We will first examine the general case of a short-term interest rate follow-ing an Itô process. We thus assume that the dynamics of r are given in a differential form

$$dr_t = \mu_t \, dt + \sigma_t \cdot dW_t, \quad r_0 > 0, \tag{9.17}$$

where W is a d-dimensional Brownian motion, and μ and σ are adapted stochastic processes with values in $\mathbb{R}$ and $\mathbb{R}^d$ respectively.

Recall that expression (9.17) is a shorthand form of the following integral repre-sentation

$$r_t = r_0 + \int_0^t \mu_u \, du + \int_0^t \sigma_u \cdot dW_u, \quad \forall \, t \in [0, T].$$

It is thus implicitly assumed that μ and σ satisfy the suitable integrability condi-tions, so that the process r is well-defined. In financial interpretation, the underlying probability measure $\mathbb{P}$ is believed to reflect a subjective assessment of the "market" upon the future behavior of the short-term interest rate. In other words, the under-lying probability $\mathbb{P}$ is the actual probability, as opposed to a martingale probability measure for the bond market, which we are now going to construct. Let us recall, for the reader's convenience, a few basic facts concerning the notion of equivalence of probability measures on a filtration of a Brownian motion. Firstly, it is well known that any probability measure $\mathbb{Q}$ equivalent to $\mathbb{P}$ on $(\Omega, \mathcal{F}_{T^*})$ is given by the Radon-Nikodým derivative

$$\frac{d\mathbb{Q}}{d\mathbb{P}} = \mathcal{E}_{T^*}\left(\int_0^{\cdot} \lambda_u \cdot dW_u\right) \stackrel{\text{def}}{=} \eta_{T^*}, \quad \mathbb{P}\text{-a.s.} \tag{9.18}$$

for some predictable $\mathbb{R}^d$-valued process λ. The member on the right-hand side of (9.18) is the Doléans exponential, which is given by the following expression (see Sect. A.13)

$$\eta_t \stackrel{\text{def}}{=} \mathcal{E}_t\left(\int_0^{\cdot} \lambda_u \cdot dW_u\right) = \exp\left(\int_0^t \lambda_u \cdot dW_u - \frac{1}{2}\int_0^t |\lambda_u|^2 \, du\right).$$

Given an adapted process λ, we write $\mathbb{P}^\lambda$ to denote the probability measure whose Radon-Nikodým derivative with respect to $\mathbb{P}$ is given by the right-hand side of (9.18). In view of Girsanov's theorem, the process

$$W_t^\lambda = W_t - \int_0^t \lambda_u \, du, \quad \forall \, t \in [0, T^*], \tag{9.19}$$

follows a d-dimensional standard Brownian motion under $\mathbb{P}^\lambda$. Let us mention that the natural filtrations of the Brownian motions W and W^λ do not coincide, in gen-eral. The following result deals with the behavior of the short-term interest rate r and the bond price $B(t, T)$ under a probability measure $\mathbb{P}^\lambda$ equivalent to $\mathbb{P}$ – more specifically, under a probability measure $\mathbb{P}^\lambda$ that is a martingale measure in the sense of Definition 9.5.1 (see Artzner and Delbaen (1989) for related results).

---

[2] Generally speaking, a *diffusion process* is an arbitrary strong Markov process with contin-uous sample paths. In our framework, a diffusion process is given as a strong solution of a stochastic differential equation (SDE) driven by the underlying Brownian motion W.

Proposition 9.5.1 *Assume the short-term interest rate r follows an Itô process under the actual probability $\mathbb{P}$, as specified by (9.17). Let $B(t, T)$ be an arbitrage-free family of bond prices relative to r. For any martingale measure $\mathbb{P}^* = \mathbb{P}^\lambda$ of Definition 9.5.1, the following statements are true.*
(i) The process r satisfies under $\mathbb{P}^\lambda$

$$dr_t = (\mu_t + \sigma_t \cdot \lambda_t)\, dt + \sigma_t \cdot dW_t^\lambda.$$

(ii) There exists an adapted $\mathbb{R}^d$-valued process $b^\lambda(t, T)$ such that

$$dB(t, T) = B(t, T)\big(r_t\, dt + b^\lambda(t, T) \cdot dW_t^\lambda\big). \tag{9.20}$$

Consequently, for any fixed maturity $T \in (0, T^*]$ we have

$$B(t, T) = B(0, T)\, B_t\, \mathcal{E}_t\Big(\int_0^\cdot b^\lambda(u, T) \cdot dW_u^\lambda \Big).$$

Proof. To show (i), it is enough to combine (9.17) with (9.19). For (ii), it suffices to observe that the process $M = Z^*\eta$ follows a (local) martingale under $\mathbb{P}$ (recall that Z^* is given by (9.12)). In view of Theorem A.11.1, we have

$$M_t = Z^*(t, T)\eta_t = Z^*(0, T) + \int_0^t \gamma_u \cdot dW_u, \quad \forall\, t \in [0, T],$$

for some $\mathbb{F}$-adapted process γ. Applying Itô's formula, we obtain

$$dZ^*(t, T) = d(M_t \eta_t^{-1}) = \eta_t^{-1}\big(\gamma_t - M_t \lambda_t\big) \cdot dW_u^\lambda, \tag{9.21}$$

where we have used

$$d\eta_t^{-1} = -\eta_t^{-1}\lambda_t \cdot dW_t^\lambda = -\eta_t^{-1}\lambda_t \cdot (dW_t - \lambda_t\, dt).$$

Equality (9.20) now follows easily from (9.21), once again by Itô's formula. The last asserted formula is also evident. □

It is important to note that the process $b^\lambda(t, T)$ does not depend in fact on the choice of λ (this is an immediate consequence of Girsanov's theorem). This observation allows us to formulate the following definition.

Definition 9.5.2 The process $b(t, T) = b^\lambda(t, T)$ is called the *volatility* of a zero-coupon bond of maturity T.

The next result provides a link between representations of the price $B(t, T)$ associated with different, but mutually equivalent, probability measures.

Corollary 9.5.1 *Let $\mathbb{P}^\lambda$ and $\mathbb{P}^{\tilde\lambda}$ be two probability measures equivalent to the underlying probability measure $\mathbb{P}$. Assume that the bond price $B(t, T)$ is given by formula (9.14), with $\mathbb{P}^* = \mathbb{P}^\lambda$. Then for every $t \in [0, T]$*

$$B(t, T) = \mathbb{E}_{\mathbb{P}^{\tilde\lambda}}\Big(e^{-\int_t^T r_u\, du}\, e^{\int_t^{T^*}(\lambda_u - \tilde\lambda_u)\cdot dW_u^{\tilde\lambda} - \frac{1}{2}\int_t^{T^*} |\lambda_u - \tilde\lambda_u|^2\, du}\, \Big|\, \mathcal{F}_t\Big).$$

Proof. For any process λ, let U^λ stand for the integral

$$U_t^\lambda = \int_0^t \lambda_u \cdot dW_u, \quad \forall t \in [0, T^*].$$

Straightforward calculations show that

$$\frac{d\mathbb{P}^\lambda}{d\mathbb{P}^{\tilde\lambda}} = \frac{\mathcal{E}_{T^*}(U^\lambda)}{\mathcal{E}_{T^*}(U^{\tilde\lambda})} = \mathcal{E}_{T^*}\left(\int_0^\cdot \gamma_u \cdot dW_u^{\tilde\lambda}\right) = \mathcal{E}_{T^*}(U^\gamma),$$

where $\gamma_t = \lambda_t - \tilde\lambda_t$. Applying the abstract Bayes rule, we get

$$B(t, T) = \mathbb{E}_{\mathbb{P}^\lambda}\left(e^{-\int_t^T r_u\,du} \mid \mathcal{F}_t\right) = \mathbb{E}_{\mathbb{P}^{\tilde\lambda}}\left(\frac{\mathcal{E}_{T^*}(U^\gamma)}{\mathcal{E}_t(U^\gamma)} e^{-\int_t^T r_u\,du} \mid \mathcal{F}_t\right).$$

This yields the asserted equality upon simplification. □

9.5.4 Market Price for Interest Rate Risk

Let us comment on the consequences of the results above. Suppose that the short-term interest rate r satisfies (9.17) under a probability measure $\mathbb{P}$. Let $\mathbb{P}^* = \mathbb{P}^\lambda$ be an arbitrary probability measure equivalent to $\mathbb{P}$. Then we may define a bond price $B(t, T)$ by setting

$$B(t, T) = \mathbb{E}_{\mathbb{P}^*}\left(e^{-\int_t^T r_u\,du} \mid \mathcal{F}_t^{W^\lambda}\right), \quad \forall t \in [0, T]. \tag{9.22}$$

It follows from (9.20) that the bond price $B(t, T)$ satisfies, under the actual probability $\mathbb{P}$,

$$dB(t, T) = B(t, T)\Big(\big(r_t - \lambda_t \cdot b^\lambda(t, T)\big)\,dt + b^\lambda(t, T) \cdot dW_t\Big).$$

This means that the instantaneous returns from holding the bond differ, in general, from the short-term interest rate r. In financial literature, the additional term is commonly referred to as the *risk premium* or the *market price for interest rate risk*. It is usually argued that due to the riskiness of a zero-coupon bond, it is reasonable to expect that the instantaneous return from holding the bond will exceed that of a risk-free security (i.e, of a savings account) in a market equilibrium.[3] Unfortunately, since our arguments refer only to the absence of arbitrage between primary securities and derivatives (that is, we place ourselves in a partial equilibrium framework), we are unable to identify the risk premium.

Summarizing, we have a certain degree of freedom; if the short-term rate r is given by (9.17) then any probability measure $\mathbb{P}^*$ equivalent to $\mathbb{P}$ can formally be used to construct an arbitrage-free family of bond prices through formula (9.22).

---

[3] The general equilibrium approach to the modelling of the term structure of interest rates is beyond the scope of this text. The interested reader is referred to the fundamental papers by Cox et al. (1985a, 1985b).

Notice, however, that if the actual probability measure $\mathbb{P}$ is used to define the bond price through formula (9.22), the market prices for interest rate risk vanish.

We end this section by mentioning an important problem of matching the initial yield curve. Given a short-term interest rate process r and a probability measure $\mathbb{P}^*$, the initial term structure $B(0, T)$ is uniquely determined by the formula

$$B(0, T) = \mathbb{E}_{\mathbb{P}^*}\left(e^{-\int_0^T r_u\,du}\right), \quad \forall\, T \in [0, T^*]. \tag{9.23}$$

This feature of bond price models based on a specified short-term interest rate process makes the problem of matching the current yield curve much more cumbersome than in the case of models that incorporate the initial term structure as an input of the model. The latter methodology, commonly known as the Heath-Jarrow-Morton approach (see Heath et al. (1992a)), will be presented in detail in Chap. 11.

9.6 Forward Measure Approach

The aim of this section is to describe the specific features that distinguish the arbitrage valuation of contingent claims within the classical Black-Scholes framework from the pricing of options on stocks and bonds under stochastic interest rates. We assume throughout that the price $B(t, T)$ of a zero-coupon bond of maturity $T \le T^*$ ($T^* > 0$ is a fixed horizon date) follows an Itô process under the martingale measure $\mathbb{P}^*$[4]

$$dB(t, T) = B(t, T)\left(r_t\,dt + b(t, T) \cdot dW_t^*\right), \tag{9.24}$$

with $B(T, T) = 1$, where W_t^*, $t \in [0, T^*]$, denotes a d-dimensional standard Brownian motion defined on a filtered probability space $(\Omega, \mathbb{F}, \mathbb{P}^*)$, and r_t, $t \in [0, T^*]$, stands for the instantaneous, continuously compounded rate of interest. In other words, we take for granted the existence of an arbitrage-free family $B(t, T)$ of bond prices associated with a certain process r modeling the short-term interest rate.

Moreover, it is implicitly assumed that we have already constructed an arbitrage-free model of a security market in which all bonds of different maturities, as well as a certain number of other assets (called *stocks* in what follows), are primary traded securities. It should be stressed that the way in which such a construction is achieved is not relevant for the results presented in what follows. In particular, the concept of the instantaneous forward interest rate, which is known to play an essential role in the HJM methodology, is not employed.

As was mentioned already, in addition to zero-coupon bonds, we will also consider other primary assets, referred to as *stocks* in what follows. The dynamics of a stock price S^i, $i = 1, \ldots, M$, under the martingale measure $\mathbb{P}^*$ are given by the following expression

$$dS_t^i = S_t^i\left(r_t\,dt + \sigma_t^i \cdot dW_t^*\right), \quad S_0^i > 0, \tag{9.25}$$

---

[4] The reader may find it convenient to assume that the probability measure $\mathbb{P}^*$ is the unique martingale measure for the family $B(t, T)$, $T \le T^*$; this is not essential, however.

where σ^i represents the volatility of the stock price S^i. Unless explicitly stated otherwise, for every T and i, the bond price volatility $b(t, T)$ and the stock price volatility σ^i_t are assumed to be $\mathbb{R}^d$-valued, bounded, adapted processes. Generally speaking, we assume that the prices of all primary securities follow strictly positive processes with continuous sample paths. Note, however, that certain results presented in this section are independent of the particular form of bond and stock prices introduced above. We denote by $\pi_t(X)$ the arbitrage price at time t of an attainable contingent claim X at time T. By virtue of the standard risk-neutral valuation formula

$$\pi_t(X) = B_t \, \mathbb{E}_{\mathbb{P}^*}(X B_T^{-1} \mid \mathcal{F}_t), \quad \forall t \in [0, T]. \tag{9.26}$$

In (9.26), B represents the savings account given by (9.7). Recall that the price $B(t, T)$ of a zero-coupon bond maturing at time T admits the following representation (cf. (9.13))

$$B(t, T) = B_t \, \mathbb{E}_{\mathbb{P}^*}(B_T^{-1} \mid \mathcal{F}_t), \quad \forall t \in [0, T], \tag{9.27}$$

for any maturity $0 \le T \le T^*$. Suppose now that we wish to price a European call option, with expiry date T, that is written on a zero-coupon bond of maturity $U > T$. The option's payoff at expiry equals

$$C_T = (B(T, U) - K)^+,$$

so that the option price C_t at any date $t \le T$ is

$$C_t = B_t \, \mathbb{E}_{\mathbb{P}^*}\big(B_T^{-1}(B(T, U) - K)^+ \mid \mathcal{F}_t\big).$$

To find the option's price using the last equality, we need to know the joint (conditional) probability distribution of $\mathcal{F}_T$-measurable random variables B_T and $B(T, U)$. The technique that was developed to circumvent this step is based on an equivalent change of probability measure. It appears that it is possible to find a probability measure $\mathbb{P}_T$ such that the following holds

$$C_t = B(t, T)\mathbb{E}_{\mathbb{P}_T}\big((B(T, U) - K)^+ \mid \mathcal{F}_t\big).$$

Consequently, as will be shown in what follows, we also have

$$C_t = B(t, T)\mathbb{E}_{\mathbb{P}_T}\big((F_{B(T,U)}(T, T) - K)^+ \mid \mathcal{F}_t\big),$$

where $F_{B(T,U)}(t, T)$ is the forward price at time t, for settlement at the date T, of the U-maturity zero-coupon bond (see formula (9.29)). If $b(t, U) - b(t, T)$ is a deterministic function, then the forward price $F_{B(T,U)}(t, T)$ can be shown to follow a lognormal martingale under $\mathbb{P}_T$; a Black-Scholes-like expression for the option's price is thus available.

9.6.1 Forward Price

Recall that a *forward contract* is an agreement, established at the date $t < T$, to pay or receive on settlement date T a preassigned payoff, say X, at an agreed forward price. Let us emphasize that there is no cash flow at the contract's initiation and the contract is not marked to market. We may and do assume, without loss of generality, that a forward contract is settled by cash on date T. Hence a forward contract written at time t with the underlying contingent claim X and prescribed settlement date $T > t$ may be summarized by the following two basic rules: (a) a cash amount X will be received at time T, and a preassigned amount $F_X(t, T)$ of cash will be paid at time T; (b) the amount $F_X(t, T)$ should be predetermined at time t (according to the information available at this time) in such a way that the arbitrage price of the forward contract at time t is zero.

In fact, since nothing is paid up front, it is natural to admit that a forward contract is worthless at its initiation. We thus adopt the following formal definition of a forward contract.

Definition 9.6.1 Let us fix $0 \le t \le T \le T^*$. A *forward contract* written at time t on a time T contingent claim X is represented by the time T contingent claim $G_T = X - F_X(t, T)$ that satisfies the following conditions: (a) $F_X(t, T)$ is an $\mathcal{F}_t$-measurable random variable; (b) the arbitrage price at time t of a contingent claim G_T equals zero, i.e., $\pi_t(G_T) = 0$.

The random variable $F_X(t, T)$ is referred to as the *forward price* of a contingent claim X at time t for the settlement date T. The contingent claim X may be defined in particular as a preassigned amount of the underlying financial asset to be delivered at the settlement date. For instance, if the underlying asset of a forward contract is one share of a stock S then clearly $X = S_T$. Similarly, if the asset to be delivered at time T is a zero-coupon bond of maturity $U \ge T$, we have $X = B(T, U)$. Note that both S_T and $B(T, U)$ are attainable contingent claims in our market model. The following well-known result expresses the forward price of a claim X in terms of its arbitrage price $\pi_t(X)$ and the price $B(t, T)$ of a zero-coupon bond maturing at time T.

Lemma 9.6.1 *The forward price $F_X(t, T)$ at time $t \le T$, for the settlement date T, of an attainable contingent claim X equals*

$$F_X(t, T) = \frac{\mathbb{E}_{\mathbb{P}^*}(X B_T^{-1} \mid \mathcal{F}_t)}{\mathbb{E}_{\mathbb{P}^*}(B_T^{-1} \mid \mathcal{F}_t)} = \frac{\pi_t(X)}{B(t, T)}. \qquad (9.28)$$

Proof. It is sufficient to observe that

$$\pi_t(G_T) = B_t \, \mathbb{E}_{\mathbb{P}^*}(G_T B_T^{-1} \mid \mathcal{F}_t)$$

$$= B_t \left(\mathbb{E}_{\mathbb{P}^*}(X B_T^{-1} \mid \mathcal{F}_t) - F_X(t, T) \mathbb{E}_{\mathbb{P}^*}(B_T^{-1} \mid \mathcal{F}_t) \right) = 0,$$

where the last equality is a consequence of condition (b) of Definition 9.6.1. This proves the first equality in (9.28); the second equality in (9.28) follows immediately from (9.25)–(9.27). $\qquad \square$

Let us examine the two typical cases of forward contracts mentioned above. If the underlying asset for delivery at time T is a zero-coupon bond of maturity $U \geq T$, then (9.28) becomes

$$F_{B(T,U)}(t, T) = \frac{B(t, U)}{B(t, T)}, \quad \forall t \in [0, T]. \tag{9.29}$$

On the other hand, the forward price of a stock S (S stands hereafter for S^i for some i) equals

$$F_{S_T}(t, T) = \frac{S_t}{B(t, T)}, \quad \forall t \in [0, T]. \tag{9.30}$$

For brevity, we will write $F_B(t, U, T) = F_{B(T,U)}(t, T)$ and $F_S(t, T) = F_{S_T}(t, T)$. More generally, for any tradable asset Z we denote by $F_Z(t, T)$ the forward price of the asset – that is, $F_Z(t, T) = Z_t / B(t, T)$ for $t \in [0, T]$.

9.6.2 Forward Martingale Measure

To the best of our knowledge, within the framework of arbitrage valuation of interest rate derivatives, the method of a forward risk adjustment was pioneered under the name of a *forward risk-adjusted process* by Jamshidian (1987) (the corresponding equivalent change of probability measure was then used by Jamshidian (1989a) in the Gaussian framework). The formal definition of a forward probability measure was introduced in Geman (1989) under the name of *forward neutral probability.*

In particular, Geman (1989) observed that the forward price of any financial asset follows a (local) martingale under the forward neutral probability associated with the settlement date of a forward contract. For further developments of the forward measure approach, we refer the reader, in particular, to El Karoui and Rochet (1989) and Geman et al. (1995). Most results in this section do not rely on specific assumptions imposed on the dynamics of bond and stock prices. We assume that we are given an arbitrage-free family $B(t, T)$ of bond prices and the related savings account B. Note that by assumption, $0 < B(0, T) = \mathbb{E}_{\mathbb{P}^*}(B_T^{-1}) < \infty$.

Definition 9.6.2 A probability measure $\mathbb{P}_T$ on $(\Omega, \mathcal{F}_T)$ equivalent to $\mathbb{P}^*$, with the Radon-Nikodým derivative given by the formula

$$\frac{d\mathbb{P}_T}{d\mathbb{P}^*} = \frac{B_T^{-1}}{\mathbb{E}_{\mathbb{P}^*}(B_T^{-1})} = \frac{1}{B(0, T)B_T}, \quad \mathbb{P}^*\text{-a.s.,} \tag{9.31}$$

is called the *forward martingale measure* (or briefly, the *forward measure*) for the settlement date T.

Notice that the above Radon-Nikodým derivative, when restricted to the σ-field $\mathcal{F}_t$, satisfies for every $t \in [0, T]$

$$\eta_t \stackrel{\text{def}}{=} \frac{d\mathbb{P}_T}{d\mathbb{P}^*}\bigg|_{\mathcal{F}_t} = \frac{B_0}{B(0, T)} \mathbb{E}_{\mathbb{P}^*}\left(\frac{B(T, T)}{B_T}\bigg|\mathcal{F}_t\right) = \frac{B_0}{B(0, T)}\frac{B(t, T)}{B_t},$$

where we have used the martingale property of $B(t, T)/B_t$ under $\mathbb{P}^*$ (or equivalently, formula (9.27)). The following result shows that forward measure is in fact the martingale measure associated with the choice of the T-maturity zero-coupon bond as a numeraire.

Proposition 9.6.1 *The relative prices $B_t/B(t, T)$ and $B(t, U)/B(t, T)$ are martingales under $\mathbb{P}_T$.*

Proof. The result is an immediate consequence of the martingale property of $B(t, T)/B_t$ and $B(t, U)/B_t$ under $\mathbb{P}^*$ and Lemma A.14.1.   □

In view of Proposition 9.6.1, it is natural to expect that the arbitrage price of any (non-dividend-paying) asset should follow a martingale under $\mathbb{P}_T$. This is indeed the case, as will be shown in Proposition 9.6.2.

When the bond price is governed by (9.24), a more explicit representation for the density process η_t is available, namely,

$$\eta_t = \exp \left(\int_0^t b(u, T) \cdot dW_u^* - \frac{1}{2} \int_0^t |b(u, T)|^2 \, du \right). \tag{9.32}$$

In other words, $\eta_t = \mathcal{E}_t(U^T)$, where we denote

$$U_t^T = \int_0^t b(u, T) \cdot dW_u^*.$$

Furthermore, the process W^T given by the formula

$$W_t^T = W_t^* - \int_0^t b(u, T) \, du, \quad \forall t \in [0, T], \tag{9.33}$$

is a standard Brownian motion under the forward measure $\mathbb{P}_T$ (this is, of course, a consequence of Girsanov's theorem). We will sometimes refer to W^T as the *forward Brownian motion* for the date T.

The next result shows that the forward price of a European contingent claim X at time T can be expressed easily in terms of the conditional expectation under the forward measure $\mathbb{P}_T$.

Lemma 9.6.2 *The forward price at time t for the delivery date T of an attainable contingent claim X, settling at time T, equals*

$$F_X(t, T) = \mathbb{E}_{\mathbb{P}_T}(X \mid \mathcal{F}_t), \quad \forall t \in [0, T], \tag{9.34}$$

provided that X is $\mathbb{P}_T$-integrable. In particular, the forward price process $F_X(t, T)$, $t \in [0, T]$, is a martingale under the forward measure $\mathbb{P}_T$.

Proof. The abstract Bayes rule (see Lemma A.1.4) yields

$$\mathbb{E}_{\mathbb{P}_T}(X \mid \mathcal{F}_t) = \frac{\mathbb{E}_{\mathbb{P}^*}(\eta_T X \mid \mathcal{F}_t)}{\mathbb{E}_{\mathbb{P}^*}(\eta_T \mid \mathcal{F}_t)} = \mathbb{E}_{\mathbb{P}^*}(\eta_T \eta_t^{-1} X \mid \mathcal{F}_t), \tag{9.35}$$

where

$$\eta_T = \frac{d\mathbb{P}_T}{d\mathbb{P}^*} = \frac{1}{B_T B(0, T)}$$

and $\eta_t = \mathbb{E}_{\mathbb{P}^*}(\eta_T \mid \mathcal{F}_t)$. Combining (9.28) with (9.35), we obtain the desired result. □

Under (9.24), the last term in (9.35) can be given a more explicit form, namely, we have

$$\mathbb{E}_{\mathbb{P}_T}(X \mid \mathcal{F}_t) = \mathbb{E}_{\mathbb{P}^*}\left\{ X \exp\left(\int_t^T b(u, T) \cdot dW_u^* - \frac{1}{2} \int_t^T |b(u, T)|^2 \, du \right) \Big| \mathcal{F}_t \right\}$$

for every $t \in [0, T]$.

The equalities

$$F_B(t, U, T) = \mathbb{E}_{\mathbb{P}_T}(B(T, U) \mid \mathcal{F}_t), \quad \forall \, 0 < T \le U \le T^*,$$

and $F_S(t, T) = \mathbb{E}_{\mathbb{P}_T}(S_T \mid \mathcal{F}_t)$ for every $t \in [0, T^*]$, are immediate consequences of Lemma 9.6.2. More generally, the relative price of any traded security (which pays no coupons or dividends) follows a local martingale under the forward probability measure $\mathbb{P}_T$, provided that the price of a bond maturing at time T is taken as a numeraire. The next next establishes a version of the risk-neutral valuation formula that is tailored to the stochastic interest rate framework.

Proposition 9.6.2 *The arbitrage price at time t of an attainable contingent claim X settling at time T is given by the formula*

$$\pi_t(X) = B(t, T) \, \mathbb{E}_{\mathbb{P}_T}(X \mid \mathcal{F}_t), \quad \forall \, t \in [0, T]. \tag{9.36}$$

Proof. Equality (9.36) is an immediate consequence of (9.28) combined with (9.34). For a more direct proof, note that the price $\pi_t(X)$ can be re-expressed as follows

$$\pi_t(X) = B_t \, \mathbb{E}_{\mathbb{P}^*}(X B_T^{-1} \mid \mathcal{F}_t) = B_t B(0, T) \, \mathbb{E}_{\mathbb{P}^*}(\eta_T X \mid \mathcal{F}_t).$$

An application of Bayes's rule yields

$$\pi_t(X) = B_t B(0, T) \, \mathbb{E}_{\mathbb{P}_T}(X \mid \mathcal{F}_t) \, \mathbb{E}_{\mathbb{P}^*}(\eta_T \mid \mathcal{F}_t)$$
$$= B_t B(0, T) \, \mathbb{E}_{\mathbb{P}_T}(X \mid \mathcal{F}_t) \, \mathbb{E}_{\mathbb{P}^*}\left(\frac{1}{B_T B(0, T)} \Big| \mathcal{F}_t \right)$$
$$= B(t, T) \, \mathbb{E}_{\mathbb{P}_T}(X \mid \mathcal{F}_t),$$

as expected. □

The following corollary deals with a contingent claim that settles at time $U \ne T$. We wish to express the value of this claim in terms of the forward measure for the date T.

Corollary 9.6.1 *Consider an attainable contingent claim X settling at time U. (i) If $U \leq T$, then the price of X at time $t \leq U$ equals*

$$\pi_t(X) = B(t, T)\, \mathbb{E}_{\mathbb{P}_T}(X B^{-1}(U, T) \mid \mathcal{F}_t). \qquad (9.37)$$

(ii) If $U \geq T$ and X is $\mathcal{F}_T$-measurable, then for any $t \leq U$ we have

$$\pi_t(X) = B(t, T)\, \mathbb{E}_{\mathbb{P}_T}(X B(T, U) \mid \mathcal{F}_t). \qquad (9.38)$$

Proof. Both equalities are intuitively clear. In case (i), we invest at time U an $\mathcal{F}_U$-measurable payoff X in zero-coupon bonds which mature at time T. For the second case, observe that in order to replicate an $\mathcal{F}_T$-measurable claim X at time U, it is enough to purchase, at time T, X units of a zero-coupon bond maturing at time U. Both strategies are manifestly self-financing, and thus the result follows.

An alternative way of deriving (9.37) is to observe that since X is $\mathcal{F}_U$-measurable, we have for every $t \in [0, U]$

$$B_t\, \mathbb{E}_{\mathbb{P}^*}\!\left(\frac{X}{B_T B(U, T)} \,\Big|\, \mathcal{F}_t \right) = B_t\, \mathbb{E}_{\mathbb{P}^*}\!\left\{ \frac{X}{B_U B(U, T)}\, \mathbb{E}_{\mathbb{P}^*}\!\left(\frac{B_U}{B_T} \,\Big|\, \mathcal{F}_U \right) \Big|\, \mathcal{F}_t \right\}$$
$$= B_t\, \mathbb{E}_{\mathbb{P}^*}\!\left(\frac{X}{B_U} \,\Big|\, \mathcal{F}_t \right).$$

This means that the claim X settling at time U has, at any date $t \in [0, U]$, an identical arbitrage price to the claim $Y = X B^{-1}(U, T)$ that settles at time T. Formula (9.37) now follows from relation (9.36) applied to the claim Y. Similarly, to prove the second formula, we observe that since X is $\mathcal{F}_T$-measurable, we have for $t \in [0, T]$

$$B_t\, \mathbb{E}_{\mathbb{P}^*}\!\left(\frac{X}{B_U} \,\Big|\, \mathcal{F}_t \right) = B_t\, \mathbb{E}_{\mathbb{P}^*}\!\left\{ \frac{X}{B_T}\, \mathbb{E}_{\mathbb{P}^*}\!\left(\frac{B_T}{B_U} \,\Big|\, \mathcal{F}_T \right) \Big|\, \mathcal{F}_t \right\}$$
$$= B_t\, \mathbb{E}_{\mathbb{P}^*}\!\left(\frac{X B(T, U)}{B_T} \,\Big|\, \mathcal{F}_t \right).$$

We conclude once again that an $\mathcal{F}_T$-measurable claim X at time $U \geq T$ is essentially equivalent to a claim $Y = X B(T, U)$ at time T. $\qquad \square$

9.6.3 Forward Processes

Let us now consider any two maturities $T, U \in [0, T^*]$. We define the forward process $F_B(t, T, U)$ by setting (note that we do not assume that $U \leq T$)

$$F_B(t, T, U) \stackrel{\text{def}}{=} \frac{B(t, T)}{B(t, U)}, \quad \forall\, t \in [0, T \wedge U]. \qquad (9.39)$$

Note that we have

$$F_B(t, T, U) = \frac{F_B(t, T, T^*)}{F_B(t, U, T^*)}, \quad \forall\, t \in [0, T \wedge U].$$

Suppose first that $U > T$; then the amount

$$f_s(t, T, U) = (U - T)^{-1}(F_B(t, T, U) - 1) \qquad (9.40)$$

is the *add-on (annualized) forward rate* over the future time interval $[T, U]$ prevailing at time t, and

$$f(t, T, U) = \frac{\ln F_B(t, T, U)}{U - T}$$

is the (continuously compounded) forward rate at time t over this interval (cf. (9.6)). On the other hand, if $U < T$ then $F_B(t, T, U)$ represents the value at time t of the *forward price* of a T-maturity bond in a forward contract with settlement date U.

Let us assume that $\mathbb{P} = \mathbb{P}_{T^*}$, so that $W = W^{T^*}$. We thus recognize the underlying probability measure $\mathbb{P}$ as a forward martingale measure associated with the horizon date T^*. Assume that, for any maturity $T \leq T^*$

$$dF_B(t, T, T^*) = F_B(t, T, T^*) \gamma(t, T, T^*) \cdot dW_t.$$

The following lemma can be proved by a straightforward application of Itô's formula.

Lemma 9.6.3 *For any maturities* $T, U \in [0, T^*]$, *the dynamics under* $\mathbb{P}$ *of the forward process are given by the following expression*

$$dF_B(t, T, U) = F_B(t, T, U) \gamma(t, T, U) \cdot \left(dW_t - \gamma(t, U, T^*) \, dt\right),$$

where $\gamma(t, T, U) = \gamma(t, T, T^*) - \gamma(t, U, T^*)$ *for every* $t \in [0, U \wedge T]$.

Combining Lemma 9.6.3 with Girsanov's theorem, we obtain

$$dF_B(t, T, U) = F_B(t, T, U) \gamma(t, T, U) \cdot dW_t^U, \qquad (9.41)$$

where for every $t \in [0, U]$

$$W_t^U = W_t - \int_0^t \gamma(u, U, T^*) \, du. \qquad (9.42)$$

The process W^U is a standard Brownian motion on the filtered probability space $(\Omega, (\mathcal{F}_t)_{t \in [0, U]}, \mathbb{P}_U)$, where the probability measure $\mathbb{P}_U$ equivalent to $\mathbb{P}$ is defined on $(\Omega, \mathcal{F}_U)$ by means of its Radon-Nikodým derivative with respect to the underlying probability measure $\mathbb{P}$, i.e.,

$$\frac{d\mathbb{P}_U}{d\mathbb{P}} = \mathcal{E}_U \left(\int_0^{\cdot} \gamma(u, U, T^*) \cdot dW_u \right) \quad \mathbb{P}\text{-a.s.} \qquad (9.43)$$

9.6.4 Choice of a Numeraire

Before we end this introductory chapter, let us examine briefly the basic theoretical ideas that underpin recently developed approaches to term structure modelling (in particular, the so-called *market models* of term structure of interest rates). For the reader's convenience, we shall restrict here our attention to the case of two bond portfolios serving as numeraire assets. It should be stressed, however, that the choice of a numeraire asset is by no means limited to this specific situation. As seen from Definition 8.1.9, any wealth process can be chosen as a numeraire asset (see also Geman et al. (1995), Delbaen and Schachermayer (1995b), Artzner (1997), and Davis (1998)).

By focusing directly on numeraire portfolios and the associated martingale measures, the forward measure approach that proved to be successful in the valuation of bond options under deterministic bonds volatilities, was extended to cover virtually all practically important situations.

Let us comment, in a rather informal manner, on the rationale behind a specific choice of a numeraire. To this end, we consider two particular portfolios of zero-coupon bonds with respective strictly positive wealth processes V^1 and V^2. It appears that a typical standard interest rate derivative (such as, a bond option, a cap or a swaption) can be formally seen as an option to exchange one portfolio of zero-coupon bonds for another, with an expiry date T, say. Hence we deal with a generic payoff of the following form

$$C_T = (V_T^1 - K V_T^2)^+ = V_T^1 \mathbb{1}_D - K V_T^2 \mathbb{1}_D,$$

where $K > 0$ is a constant strike level, and $D = \{V_T^1 > K V_T^2\}$ stands for the exercise event. It is not difficult to check, using the abstract Bayes rule (see Lemma A.1.4), that the martingale measures $\mathbb{P}^1$ and $\mathbb{P}^2$ associated with the choice of wealth processes V^1 and V^2 as numeraire assets are linked to each other through the following equality:

$$\frac{d\mathbb{P}^1}{d\mathbb{P}^2} = \frac{V_0^2}{V_0^1} \frac{V_T^1}{V_T^2}, \quad \mathbb{P}^2\text{-a.s.} \tag{9.44}$$

Note that we consider here both probability measures on the measurable space $(\Omega, \mathcal{F}_T)$. Furthermore, the arbitrage price of the option admits the following representation (provided, of course, that the option is attainable, so that it can be replicated)

$$C_t = V_t^1 \mathbb{P}^1(D \mid \mathcal{F}_t) - K V_t^2 \mathbb{P}^2(D \mid \mathcal{F}_t), \quad \forall t \in [0, T].$$

To obtain the Black-Scholes-like formula for the price C_t, it is enough to assume that the relative value V^1/V^2 is a lognormal martingale under $\mathbb{P}^2$, so that

$$d (V_t^1/V_t^2) = (V_t^1/V_t^2) \gamma_t^{1,2} \cdot dW_t^{1,2}$$

for some function $\gamma^{1,2} : [0, T] \rightarrow \mathbb{R}^d$ (for simplicity, we also assume that the function $\gamma^{1,2}$ is bounded), where $W^{1,2}$ is a Brownian motion under $\mathbb{P}^2$.

In view of (9.44), the Radon-Nikodým derivative of $\mathbb{P}^1$ with respect to $\mathbb{P}^2$ equals

$$\frac{d\mathbb{P}^1}{d\mathbb{P}^2} = \mathcal{E}_T \left(\int_0^\cdot \gamma_u^{1,2} \cdot dW_u^{1,2} \right), \quad \mathbb{P}^2\text{-a.s.,}$$

and thus the process

$$W_t^{2,1} = W_t^{1,2} - \int_0^t \gamma_u^{1,2} \, du, \quad \forall t \in [0, T],$$

is a standard Brownian motion under $\mathbb{P}^1$.

Proceeding along similar lines to the derivation of the Black-Scholes formula (see the proof of Theorem 3.1.1), we obtain the following extension of the classical Black-Scholes result

$$C_t = V_t^1 N\big(d_1(t, T)\big) - K V_t^2 N\big(d_2(t, T)\big), \tag{9.45}$$

where

$$d_{1,2}(t, T) = \frac{\ln(V_t^1 / V_t^2) - \ln K \pm \frac{1}{2} v_{1,2}^2(t, T)}{v_{1,2}(t, T)}$$

and

$$v_{1,2}^2(t, T) = \int_t^T |\gamma_u^{1,2}|^2 \, du, \quad \forall t \in [0, T].$$

Using formula (9.45), it is not difficult to show the considered option can be replicated, provided that portfolios V^1 and V^2 are tradable primary securities. For instance, it suffices to assume that the reference zero-coupon bonds are among liquid assets. It is also possible to justify formula (9.45) by showing the existence of a replicating strategy based on forward contracts.

Let us emphasize that the valuation formulas for interest rate caps and swaptions in respective *lognormal market models* of forward LIBOR and swap rates that will be presented in Chap. 12 and 13 respectively, can also be seen as special cases of the generic valuation formula (9.45). Assume that we are given a collection $T_0 < T_1 < \cdots < T_n$ of reset/settlement dates. Then for the jth caplet we will take

$$V_t^1 = B(t, T_j) - B(t, T_{j+1}), \quad V_t^2 = \delta_{j+1} B(t, T_{j+1}),$$

and in the case of the jth swaption we will choose

$$V_t^1 = B(t, T_j) - B(t, T_n), \quad V_t^2 = \sum_{k=j+1}^{n} \delta_k B(t, T_k),$$

where we denote $\delta_j = T_j - T_{j-1}$ for $j = 1, \ldots, n$.

Remarks. It is interesting to notice also that in order to get the valuation result (9.45) for $t = 0$, it is enough to assume that the random variable V_T^1 / V_T^2 has a lognormal probability distribution under the martingale measure $\mathbb{P}^2$. This simple observation underpins the construction of the so-called *Markov-functional interest rate model*. This alternative approach to term structure modelling, developed by Hunt et al. (1996, 2000), is presented in Sect. 13.7.

10

Short-Term Rate Models

The aim of this chapter is to survey the most popular models of the short-term interest rate. For convenience, we will work throughout within a continuous-time framework; a detailed presentation of a discrete-time approach to term structure modelling is done in Jarrow (1995). The continuous-time short-term interest rate is usually modelled as a one-dimensional diffusion process. In this text, we provide only a brief survey of the most widely accepted examples of diffusion processes used to model the short-term rate. The short-term rate approach to bond price modelling is not developed in subsequent chapters. This is partially explained by the abundance of literature taking this approach, and partially by the difficulty of fitting the observed term structure of interest rates and volatilities within a simple diffusion model (see Pelsser (2000a) or Brigo and Mercurio (2001a)). Instead, we develop the term structure theory for a much larger class of models that includes diffusion-type models as special cases. Nevertheless, it should be made clear that diffusion-type modelling of the short-term interest rate is still the most popular method for the valuing and hedging of interest rate-sensitive derivatives.

Generally speaking, existing stochastic models of the term structure can be classified either according to the number of factors or with respect to the number of state variables. The first classification refers to the number of sources of uncertainty in a model; it will usually correspond to the dimensionality of the underlying Brownian motion. Single-factor models assume perfect correlation among different points on the yield curve; multi-factor models have the potential to explain the lack of perfect correlation. The second classification refers to the dimensionality of a certain Markov process embedded in a model – deterministic functions of this Markov process define the yield curve. Under such a classification, the number of factors is always not greater than the number of state variables. Let us stress that this classification is not generally accepted in existing literature. In effect, factors are frequently identified with state variables, especially when dealing with a model that directly postulates the dynamics of the short-term interest rate. In that case, the dimensionality of the Brownian motion usually coincides with the number of state variables.

10.1 Single-factor Models

In this section, we provide a survey of most widely accepted single-factor models of the short-term interest rate. It is assumed throughout that dynamics of r are specified under the martingale probability measure $\mathbb{P}^*$. As a consequence, the risk premium does not appear explicitly in our formulas. The underlying Brownian motion W^* is assumed to be one-dimensional. In this sense, the models considered in this section are based on a single source of uncertainty, so that they belong to the class of single-factor models.

10.1.1 Time-homogeneous Models

Merton's model. Merton (1973) proposed to model the short-term rate process r through the formula:

$$r_t = r_0 + at + \sigma W_t^*, \quad \forall t \in [0, T^*], \tag{10.1}$$

where W^* is a one-dimensional standard Brownian motion under the spot martingale measure $\mathbb{P}^*$, and r_0, a, σ are positive constants. The following elementary lemma will prove useful.

Lemma 10.1.1 *Let the random variable ξ have the Gaussian distribution $N(\mu, \sigma^2)$. Then the random variable $\zeta = e^{\xi}$ has the expected value $e^{\mu + \sigma^2/2}$ and the variance $e^{2\mu + \sigma^2}(e^{\sigma^2} - 1)$.*

The next result furnishes the valuation formula for the zero-coupon bond in Merton's model.

Proposition 10.1.1 *The arbitrage price $B(t, T)$ of a T-maturity zero-coupon bond in Merton's model equals*

$$B(t, T) = e^{-r_t(T-t) - \frac{1}{2}a(T-t)^2 + \frac{1}{6}\sigma^2(T-t)^3}. \tag{10.2}$$

The dynamics of the bond price under $\mathbb{P}^*$ are

$$dB(t, T) = B(t, T)\big(r_t\, dt - \sigma(T - t)\, dW_t^*\big). \tag{10.3}$$

Proof. Let us evaluate the price $B(0, T)$ using the formula

$$B(0, T) = \mathbb{E}_{\mathbb{P}^*}\Big(e^{-\int_0^T r_u\, du}\Big), \quad \forall\, T \in [0, T^*]. \tag{10.4}$$

To this end, we need to find the probability distribution of the integral $\xi_T = \int_0^T r_u\, du$. Let us set $Y_t = r_t(T - t)$ so that $Y_T = 0$. Using Itô's product rule, we obtain

$$dY_t = (T - t)\, dr_t - r_t\, dt,$$

that is,

$$Y_T - Y_0 = -r_0 T = \int_0^T (T - t)\, dr_t - \int_0^T r_t\, dt.$$

In view of (10.1), we find that

$$\xi_T = r_0 T + \int_0^T (T - t)\, dr_t = r_0 T + \int_0^T (T - t)(a\, dt + \sigma\, dW_t^*),$$

or equivalently,

$$\xi_T = r_0 T + \frac{1}{2} a T^2 + \int_0^T \sigma (T - t)\, dW_t^*.$$

We shall now use the fact that for any continuous function $f : [0, T] \to \mathbb{R}$ the Itô integral $\int_0^T f(t)\, dW_t^*$ has the Gaussian distribution with mean zero and the variance $\int_0^T f^2(t)\, dt$. In our case $f(t) = \sigma (T - t)$, so that

$$\int_0^T f^2(t)\, dt = \int_0^T \sigma^2 (T - t)^2\, dt = \frac{1}{3} T^3 \sigma^2.$$

We conclude that ξ_T has under $\mathbb{P}^*$ the Gaussian distribution with the expected value $r_0 T + \frac{1}{2} a T^2$ and the variance $\frac{1}{3}\sigma^2 T^3$.

In view of (10.4), we have $B(0, T) = \mathbb{E}_{\mathbb{P}^*}(e^{-\xi_T})$. By virtue of Lemma 10.1.1, we obtain

$$B(0, T) = e^{-r_0 T - \frac{1}{2} a T^2 + \frac{1}{6}\sigma^2 T^3}.$$

The general result (10.2) is a consequence of the Markov property of r_t and the formula

$$B(t, T) = \mathbb{E}_{\mathbb{P}^*}\left(e^{-\int_t^T r_u du} \,\Big|\, \mathcal{F}_t\right), \quad \forall\, T \in [0, T^*].$$

Using (10.2), it is not difficult to derive the dynamics of $B(t, T)$. We have $B(t, T) = f(Z_t)$, where $f(z) = e^z$ and

$$Z_t = -r_t (T - t) - \frac{1}{2} a (T - t)^2 + \frac{1}{6}\sigma^2 (T - t)^3.$$

Using Itô's formula, we get

$$dB(t, T) = B(t, T)(r_t\, dt - \sigma (T - t)\, dW_t^*).$$

This completes the proof of the proposition.   □

Note that

$$B(t, T) = e^{m(t,T) - n(t,T) r_t},$$

where we denote

$$m(t, T) = -\frac{1}{2} a (T - t)^2 + \frac{1}{6}\sigma^2 (T - t)^3, \quad n(t, T) = T - t.$$

Let us also mention that since the volatility of the bond's price appears to be a deterministic function in Merton's model, the valuation formula for a bond option is a special case of the general result of Proposition 11.3.1.

Vasicek's model. The model analyzed by Vasicek (1977) is one of the earliest models of term structure (see also Brennan and Schwartz (1977c), Richard (1978), and Dothan (1978)). The diffusion process proposed by Vasicek is a mean-reverting version of the Ornstein-Uhlenbeck process. The short-term interest rate r is defined as the unique strong solution of the SDE

$$dr_t = \left(a - br_t\right)dt + \sigma\,dW_t^*, \tag{10.5}$$

where a, b and σ are strictly positive constants. It is well-known that the solution to (10.5) is a Markov process with continuous sample paths and Gaussian increments. In fact, equation (10.5) can be solved explicitly, as the following lemma shows.

Lemma 10.1.2 *The unique solution to the SDE* (10.5) *is given by the formula*

$$r_t = r_s e^{-b(t-s)} + \frac{a}{b}\left(1 - e^{-b(t-s)}\right) + \sigma\int_s^t e^{-b(t-u)}\,dW_u^*.$$

For any $s < t$ *the conditional distribution of* r_t *with respect to the* σ-*field* $\mathcal{F}_s$ *is Gaussian, with the conditional expected value*

$$\mathbb{E}_{\mathbb{P}^*}(r_t\mid\mathcal{F}_s) = r_s e^{-b(t-s)} + \frac{a}{b}\left(1 - e^{-b(t-s)}\right),$$

and the conditional variance

$$\mathrm{Var}_{\mathbb{P}^*}(r_t\mid\mathcal{F}_s) = \frac{\sigma^2}{2b}\left(1 - e^{-2b(t-s)}\right).$$

The limits of $\mathbb{E}_{\mathbb{P}^*}(r_t\mid\mathcal{F}_s)$ *and* $\mathrm{Var}_{\mathbb{P}^*}(r_t\mid\mathcal{F}_s)$ *when* t *tends to infinity are*

$$\lim_{t\to\infty}\mathbb{E}_{\mathbb{P}^*}(r_t\mid\mathcal{F}_s) = \frac{a}{b}$$

and

$$\lim_{t\to\infty}\mathrm{Var}_{\mathbb{P}^*}(r_t\mid\mathcal{F}_s) = \frac{\sigma^2}{2b}.$$

Proof. Let us fix $s > 0$ and let us consider the process $Y_t = r_t e^{b(t-s)}$ where $t \geq s$. Using Itô's formula and (10.5), we obtain

$$\begin{aligned}
dY_t &= e^{b(t-s)}\,dr_t + be^{b(t-s)}r_t\,dt \\
&= e^{b(t-s)}\left(a\,dt - br_t\,dt + \sigma\,dW_t^* + br_t\,dt\right) \\
&= e^{b(t-s)}\left(a\,dt + \sigma\,dW_t^*\right).
\end{aligned}$$

Hence we have

$$r_t e^{b(t-s)} - r_s = \int_s^t dY_u = \int_s^t ae^{b(u-s)}\,du + \sigma\int_s^t e^{b(u-s)}\,dW_u^*,$$

and consequently

$$r_t = r_s e^{-b(t-s)} + e^{-b(t-s)}\int_s^t ae^{b(u-s)}\,du + \sigma e^{-b(t-s)}\int_s^t e^{b(u-s)}\,dW_u^*.$$

Finally, we obtain

$$r_t = r_s e^{-b(t-s)} + \frac{a}{b}\left(1 - e^{-b(t-s)}\right) + \sigma \int_s^t e^{-b(t-u)} \, dW_u^*$$

since

$$\int_s^t e^{b(u-s)} \, du = \frac{1}{b}\left(e^{b(t-s)} - 1\right).$$

It is well-known that for any square-integrable function g the Itô integral $\int_s^t g(u) \, dW_u^*$ is a random variable independent of the σ-field $\mathcal{F}_s$ and has the Gaussian distribution $N(0, \int_s^t g^2(u) \, du)$. In our case, we have that

$$\int_s^t g^2(u) \, du = \int_s^t \sigma^2 e^{-2b(t-u)} \, du = \frac{\sigma^2}{2b}\left(1 - e^{-2b(t-s)}\right).$$

We conclude that for any $s < t$ the conditional distribution of r_t with respect to the σ-field $\mathcal{F}_s$ is Gaussian, with the conditional expected value

$$\mathbb{E}_{\mathbb{P}^*}(r_t \mid \mathcal{F}_s) = r_s e^{-b(t-s)} + \frac{a}{b}\left(1 - e^{-b(t-s)}\right)$$

and the conditional variance

$$\mathrm{Var}_{\mathbb{P}^*}(r_t \mid \mathcal{F}_s) = \frac{\sigma^2}{2b}\left(1 - e^{-2b(t-s)}\right).$$

This ends the proof of the lemma.   □

Note that the solution to (10.5) admits a stationary distribution, namely, the Gaussian distribution with expected value a/b and variance $\sigma^2/2b$.

Our next goal is to find the price of a bond in Vasicek's model. To this end, we shall use the risk-neutral valuation formula directly. An alternative derivation of bond pricing formula, based on a PDE approach, is given in Proposition 10.1.3.

Proposition 10.1.2 *The price at time t of a zero-coupon bond in Vasicek's model equals:*

$$B(t, T) = e^{m(t,T) - n(t,T)r_t}, \tag{10.6}$$

where

$$n(t, T) = \frac{1}{b}\left(1 - e^{-b(T-t)}\right) \tag{10.7}$$

and

$$m(t, T) = \frac{\sigma^2}{2} \int_t^T n^2(u, T) \, du - a \int_t^T n(u, T) \, du. \tag{10.8}$$

The bond price volatility is a function $b(\cdot, T) : [0, T] \to \mathbb{R}$, specifically, $b(t, T) = -\sigma n(t, T)$, and thus the dynamics of the bond price under $\mathbb{P}^*$ are

$$dB(t, T) = B(t, T)\left(r_t \, dt - \sigma n(t, T) \, dW_t^*\right). \tag{10.9}$$

Proof. We shall first evaluate $B(0, T)$ using the formula:

$$B(0, T) = \mathbb{E}_{\mathbb{P}^*}(B_T^{-1}) = \mathbb{E}_{\mathbb{P}^*}\left(e^{-\int_0^T r_t \, dt}\right).$$

We know already that

$$r_t = r_0 e^{-bt} + \frac{a}{b}(1 - e^{-bt}) + \sigma \int_0^t e^{-b(t-u)} \, dW_u^*,$$

so that

$$\int_0^T r_t \, dt = r_0 \int_0^T e^{-bt} \, dt + \frac{a}{b} T - \frac{a}{b} \int_0^T e^{-bt} \, dt + \sigma \int_0^T \int_0^t e^{-b(t-u)} \, dW_u^* \, dt.$$

Observe that

$$\int_0^T e^{-bt} \, dt = \frac{1}{b}(1 - e^{-bT}) = n(0, T).$$

Using the stochastic Fubini theorem (see Theorem IV.45 in Protter (2003)), we obtain

$$\int_0^T \int_0^t e^{-b(t-u)} \, dW_u^* \, dt = \int_0^T \int_0^t e^{-bt} e^{bu} \, dW_u^* \, dt = \int_0^T \int_u^T e^{-bt} e^{bu} \, dt \, dW_u^*$$

$$= \int_0^T e^{bu} \int_u^T e^{-bt} \, dt \, dW_u^* = \frac{1}{b} \int_0^T \left(1 - e^{-b(T-u)}\right) dW_u^*.$$

This means that

$$\int_0^T r_t \, dt = r_0 n(0, T) + \frac{a}{b} T - an(0, T) + \sigma \int_0^T n(u, T) \, dW_u^*.$$

It is easy to check that

$$\int_0^T n(u, T) \, du = \frac{T}{b} - n(0, T),$$

and thus

$$\xi_T \overset{\text{def}}{=} \int_0^T r_t \, dt = n(0, T) r_0 + a \int_0^T n(u, T) \, du + \sigma \int_0^T n(u, T) \, dW_u^*.$$

The random variable ξ_T has under $\mathbb{P}^*$ the Gaussian distribution with expected value $n(0, T) r_0 + a \int_0^T n(u, T) \, du$ and variance $\sigma^2 \int_0^T n^2(u, T) \, du$. In view of Lemma 10.1.1, we obtain

$$B(0, T) = \mathbb{E}_{\mathbb{P}^*}\left(e^{-\xi_T}\right) = e^{-n(0,T)r_0 - a \int_0^T n(u,T) \, du + \frac{\sigma^2}{2} \int_0^T n^2(u,T) \, du}.$$

Since $m(0, T)$ is given by

$$m(0, T) = \frac{\sigma^2}{2} \int_0^T n^2(u, T) \, du - a \int_0^T n(u, T) \, du,$$

we conclude that

$$B(0, T) = e^{m(0,T)-n(0,T)r_0}.$$

The general valuation formula (10.6) is an easy consequence of the Markov property of r. Finally, to establish (10.9), it suffices to apply Itô's rule.   □

By combining (10.7) with (10.8), we obtain a more explicit representation for $m(t, T)$

$$m(t, T) = \left(\frac{a}{b} - \frac{\sigma^2}{2b^2}\right)\left(n(t, T) - T + t\right) - \frac{\sigma^2}{4b}n^2(t, T)$$

$$= \left(\frac{a}{b} - \frac{\sigma^2}{2b^2}\right)\left(\frac{1}{b}\left(1 - e^{-b(T-t)}\right) - T + t\right) - \frac{\sigma^2}{4b^3}\left(1 - e^{-b(T-t)}\right)^2.$$

Let us consider any security whose payoff depends on the short-term rate r as the only state variable. More specifically, we assume that this security is of European style, pays dividends continuously at a rate $h(r_t, t)$, and yields a terminal payoff $G_T = g(r_T)$ at time T. Its arbitrage price $\pi_t(X)$ is given by the following version of the risk-neutral valuation formula:

$$\pi_t(X) = \mathbb{E}_{\mathbb{P}^*}\left(\int_t^T h(r_u, u)e^{-\int_t^u r_v\,dv}\,du + g(r_T)\,e^{-\int_t^T r_u\,du}\,\Big|\,\mathcal{F}_t\right) = v(r_t, t),$$

where v is some function $v : \mathbb{R} \times [0, T^*] \to \mathbb{R}$. From the well-known connection[1] between diffusion processes and partial differential equations, it follows that the function $v : \mathbb{R} \times [0, T^*] \to \mathbb{R}$ solves the following valuation PDE

$$\frac{\partial v}{\partial t}(r, t) + \frac{1}{2}\sigma^2\frac{\partial^2 v}{\partial r^2}(r, t) + (a - br)\frac{\partial v}{\partial r}(r, t) - rv(r, t) + h(r, t) = 0$$

subject to the terminal condition $v(r, T) = g(r)$. Solving this equation with $h = 0$ and $g(r) = 1$, Vasicek (1977) proved the following result (recall that we have already derived formulae (10.10)–(10.12) through the probabilistic approach; see Proposition 10.1.2).

Proposition 10.1.3 *The price at time t of a zero-coupon bond in Vasicek's model equals*

$$B(t, T) = v(r_t, t, T) = e^{m(t,T)-n(t,T)r_t}, \tag{10.10}$$

where

$$n(t, T) = \frac{1}{b}\left(1 - e^{-b(T-t)}\right) \tag{10.11}$$

and

$$m(t, T) = \frac{\sigma^2}{2}\int_t^T n^2(u, T)\,du - a\int_t^T n(u, T)\,du. \tag{10.12}$$

---

[1] Recall that this result is known as the *Feynman-Kac formula* (see Sect. A.17 or Theorem 5.7.6 in Karatzas and Shreve (1998a)).

Proof. To establish the bond pricing formula through the PDE approach, it is enough to assume that the bond price is given by (10.10), with the functions m and n satisfying $m(T, T) = n(T, T) = 0$, and to make use of the valuation PDE. By separating terms that do not depend on r, and those that are linear in r, we arrive at the following system of differential equations:

$$m_t(t, T) = an(t, T) - \tfrac{1}{2}\sigma^2 n^2(t, T) \tag{10.13}$$

and

$$n_t(t, T) = bn(t, T) - 1 \tag{10.14}$$

with $m(T, T) = n(T, T) = 0$, that in turn yields easily the expressions (10.11)–(10.12). Using Itô's formula, one can check that

$$dB(t, T) = B(t, T)\big(r_t \, dt + \sigma n(t, T) \, dW_t^*\big), \tag{10.15}$$

so that the bond price volatility equals $b(t, T) = -\sigma n(t, T)$, where $n(t, T)$ is given by (10.11). ☐

If the bond price admits representation (10.10), then obviously

$$Y(t, T) = \frac{n(t, T)r_t - m(t, T)}{T - t},$$

and thus the bond's yield, $Y(t, T)$, is an affine function of the short-term rate r_t. For this reason, Markovian models of the short-term rate in which the bond price satisfies (10.10) for some functions m and n are termed *affine models of the term structure*.

Jamshidian (1989a) obtained closed-form solutions for the prices of a European option written on a zero-coupon and on a coupon-bearing bond for Vasicek's model. He showed that the arbitrage price at time t of a call option on a U-maturity zero-coupon bond, with strike price K and expiry $T \leq U$, equals (let us mention that Jamshidian implicitly used the *forward measure* technique, presented in Sect. 9.6)

$$C_t = B(t, T)\mathbb{E}_{\mathbb{Q}}\big((\xi\eta - K)^+\big|\mathcal{F}_t\big),$$

where $\eta = B(t, U)/B(t, T)$ and $\mathbb{Q}$ stands for some probability measure equivalent to the spot martingale measure $\mathbb{P}^*$.

The random variable ξ is independent of the σ-field $\mathcal{F}_t$ under $\mathbb{Q}$, and has under $\mathbb{Q}$ a lognormal distribution such that the variance $\mathrm{Var}_{\mathbb{Q}}(\ln\xi)$ equals $v_U(t, T)$, where

$$v_U^2(t, T) = \int_t^T |b(t, T) - b(t, U)|^2 \, du,$$

or explicitly

$$v_U^2(t, T) = \frac{\sigma^2}{2b^3}\big(1 - e^{-2b(T-t)}\big)\big(1 - e^{-b(U-t)}\big)^2.$$

The bond option valuation formula established in Jamshidian (1989a) reads as follows

$$C_t = B(t, U)N\big(h_1(t, T)\big) - KB(t, T)N\big(h_2(t, T)\big), \tag{10.16}$$

where for every $t \leq T \leq U$

$$h_{1,2}(t, T) = \frac{\ln\left(B(t, U)/B(t, T)\right) - \ln K \pm \frac{1}{2} v_U^2(t, T)}{v_U(t, T)}. \tag{10.17}$$

As in the case of Merton's model, the valuation formula for a bond option in Vasicek's model is a special case of the general result of Proposition 11.3.1. This is due to the fact that the bond volatility $b(t, T)$ is deterministic.

It is important to observe that the coefficient a that is present in the dynamics (10.5) of r under $\mathbb{P}^*$, does not enter the bond option valuation formula. This suggests that the actual value of the risk premium has no impact whatsoever on the bond option price (at least if it is deterministic); the only relevant quantities are in fact the bond price volatilities $b(t, T)$ and $b(t, U)$. To account for the risk premium, it is enough to make an equivalent change of the probability measure in (10.15). Since the volatility of the bond price is invariant with respect to such a transformation of the underlying probability measure, the bond option price is independent of the risk premium, provided that the bond price volatility is deterministic.

The determination of the risk premium may thus appear irrelevant, if we concentrate on the valuation of derivatives. This is not the case, however, if our aim is to model the actual behavior of bond prices. Let us stress that stochastic term structure models presented in this text have derivative pricing as a primary goal (as opposed to, for instance, asset management).

Remarks. Let us analyze the instantaneous forward rate $f(t, T)$. Note that

$$f(t, T) = -\frac{\partial \ln B(t, T)}{\partial T} = -\frac{\partial m(t, T)}{\partial T} + r_t \frac{\partial n(t, T)}{\partial T},$$

where $n(t, T)$ and $m(t, T)$ are given by (10.7) and (10.8) respectively. Let us denote $g(t) = 1 - e^{-bt}$. Elementary calculations show that

$$\frac{\partial m(t, T)}{\partial T} = g(T - t)\left(\tfrac{1}{2}\sigma^2 b^{-2} g(T - t) - ab^{-1}\right)$$

and, of course,

$$\frac{\partial n(t, T)}{\partial T} = e^{-b(T-t)}.$$

Consequently, we find that

$$f(t, T) = g(T - t)\left(ab^{-1} - \tfrac{1}{2}\sigma^2 b^{-2} g(T - t)\right) + r_t e^{-b(T-t)}.$$

It is interesting to note that the instantaneous forward rate $f(t, T)$ tends to 0 when maturity date T tends to infinity (the current date t being fixed).

Dothan's model. It is evident that Vasicek's model, as indeed any Gaussian model, allows for negative values of (nominal) interest rates. This property is manifestly incompatible with no-arbitrage in the presence of cash in the economy. This important drawback of Gaussian models was examined, among others, by Rogers (1996). To overcome this shortcoming of Gaussian models, Dothan (1978) proposed to specify the short-term rate process r through the expression:

$$dr_t = \sigma r_t \, dW_t^*,$$

so that in his model

$$r_t = r_0 \exp \left(\sigma W_t^* - \tfrac{1}{2}\sigma^2 t \right), \quad \forall t \in [0, T^*],$$

where W^* is a one-dimensional standard Brownian motion under the spot martingale measure $\mathbb{P}^*$, and σ is a positive constant. It is clear that r_t is lognormally distributed under $\mathbb{P}^*$, and thus the short-term interest rate cannot become negative. On the other hand, however, the probability distribution of the integral $\int_t^T r_u \, du$ under $\mathbb{P}^*$ is not known, and explicit formula for the bond price is not available in Dothan's model.

Brennan and Schwartz model. Brennan and Schwartz (1980b) proposed a mean-reverting version of Dothan's model, namely,

$$dr_t = \left(a - br_t \right) dt + \sigma r_t \, dW_t^*.$$

As in the case of Dothan's model, the probability distribution of the integral $\int_t^T r_u \, du$ under $\mathbb{P}^*$ is not known, so that numerical procedures are required to value zero-coupon bonds and bond options.

Cox-Ingersoll-Ross model. The general equilibrium approach to term structure modelling developed by Cox et al. (1985b) (CIR, for short) leads to the following modification of the mean-reverting diffusion of Vasicek, known as the *square-root process*

$$dr_t = \left(a - br_t \right) dt + \sigma \sqrt{r_t} \, dW_t^*, \tag{10.18}$$

where a, b and σ are strictly positive constants. Due to the presence of the square-root in the diffusion coefficient, the CIR diffusion takes only positive values. More precisely, it can reach zero, but it never becomes negative. In a way similar to the previous case, the price process $G_t = v(r_t, t)$ of any standard European interest rate derivative can be found, in principle, by solving the valuation PDE (for the origin of this PDE, see Sect. 10.1.3)

$$\frac{\partial v}{\partial t}(r, t) + \frac{1}{2}\sigma^2 r \frac{\partial^2 v}{\partial r^2}(r, t) + (a - br)\frac{\partial v}{\partial r}(r, t) - rv(r, t) + h(r, t) = 0,$$

subject to the terminal condition $v(r, T) = g(r, T)$. Cox et al. (1985b) found closed-form solutions for the price of a zero-coupon bond by a PDE approach.

We conjecture that the bond price $B(t, T)$ in the CIR model satisfies (10.10) for some functions m and n. Then, using the valuation PDE above, we find that the function n solves, for each fixed maturity date T, the Riccati equation

$$n_t(t, T) - \tfrac{1}{2}\sigma^2 n^2(t, T) - bn(t, T) + 1 = 0 \tag{10.19}$$

with the terminal condition $m(T, T) = 0$, and the function m satisfies

$$m_t(t, T) = an(t, T) \qquad (10.20)$$

with the terminal condition $m(T, T) = 0$. Solving this system of ordinary differential equations, we obtain

$$m(t, T) = \frac{2a}{\sigma^2} \ln \left\{ \frac{\gamma e^{b\tau/2}}{\gamma \cosh \gamma \tau + \frac{1}{2} b \sinh \gamma \tau} \right\} \qquad (10.21)$$

and

$$n(t, T) = \frac{\sinh \gamma \tau}{\gamma \cosh \gamma \tau + \frac{1}{2} b \sinh \gamma \tau}, \qquad (10.22)$$

where we denote $\tau = T - t$ and $2\gamma = (b^2 + 2\sigma^2)^{1/2}$. We conclude that the CIR model belongs to the class of affine term structure models.

Closed-form expressions for the price of an option on a zero-coupon bond and an option on a coupon-bearing bond in the CIR framework were derived in Cox et al. (1985b) and in Longstaff (1993), respectively.

In Sect. 10.3, we shall analyze in some detail the time-dependent version of the CIR model. Let us only mention here that the CIR bond option pricing formulas involves the cumulative non-central chi-square distribution, and it depend on the deterministic risk premium (it is easily seen that the bond price volatility in the CIR framework is stochastic, rather than deterministic).

Remarks. As pointed out by Rogers (1995), the dynamics of the form (10.18) can be obtained by a simple transformation of a d-dimensional Gaussian process. Let X be a diffusion process solving the SDE

$$dX_t = \tfrac{1}{2}\alpha(t)X_t \, dt + \tfrac{1}{2}\sigma(t) \, dW_t, \qquad (10.23)$$

where W is a d-dimensional Brownian motion, and $\beta, \sigma : [0, T^*] \to \mathbb{R}$ are bounded functions. Then the process $r_t = |X_t|^2$ satisfies

$$dr_t = \left(\tfrac{1}{4}\sigma^2(t)d - a(t)r_t \right) dt + \sigma(t)\sqrt{r_t} \, dW_t^*, \qquad (10.24)$$

where W^* is a one-dimensional Brownian motion. This shows the connection between the properties of the (generalized) CIR model and the theory of Bessel processes. We shall analyze this issue in more detail in Sect. 10.3 in which we shall examine the extended CIR model, i.e., the time-inhomogeneous version of the CIR model in which r satisfies

$$dr_t = \left(a(t) - b(t)r_t \right) dt + \sigma(t)\sqrt{r_t} \, dW_t^*.$$

Longstaff (1990a) has shown how to value European call and put *options on yields* in the CIR model. For a fixed time to maturity, the yield on a zero-coupon bond in the CIR framework is, of course, a linear function of the short-term rate, since

$$Y(t, t + \tau) = \tilde{Y}(r_t, \tau) = \tilde{m}(\tau) + \tilde{n}(\tau)r_t,$$

where $\tau = T - t$ is fixed. The number $\tilde{Y}(t, \tau)$ represents the yield at time t for zero-coupon bonds with a constant maturity τ, provided that the current level of the short-term rate is $r_t = r$. In accordance with the contractual features, for a fixed τ, a European *yield call option* entitles its owner to receive the payoff C_T^Y, which is expressed in monetary units and equals $C_T^Y = (\tilde{Y}(r_T, \tau) - K)^+$, where K is the fixed level of the yield.

Longstaff's model. Longstaff (1989) modified the CIR model by postulating the following dynamics for the short-term rate

$$dr_t = a(b - c\sqrt{r_t}) dt + \sigma \sqrt{r_t} dW_t^*, \qquad (10.25)$$

referred to as the *double square-root* (DSR) process. Longstaff derived a closed-form expression for the price of a zero-coupon bond

$$B(t, T) = v(r_t, t, T) = e^{m(t,T) - n(t,T)r_t - p(t,T)\sqrt{r_t}}$$

for some functions m, n and p that are known explicitly (we do not reproduce them here). The bond's yield is thus a non-linear function of the short-term rate. Also, the bond price is not a monotone function of the current level of the short-term rate. This feature makes the valuation of a bond option less straightforward than usual. Indeed, it is, typically, possible to represent the exercise set of a bond option in terms of r as the interval $[r^*, \infty)$ or $(-\infty, r^*]$ for some constant r^*, depending on whether an option is a put or a call (see Sect. 10.1.4).

An empirical comparison of the CIR model and the DSR model, which was done by Longstaff (1989), suggests that the DSR model outperforms the CIR model in most circumstances. Empirical studies of the CIR model are reported also in Brown and Dybvig (1986), Gibbons and Ramaswamy (1993), and Pearson and Sun (1994). Note that, albeit in a continuous-time setting, the short-term interest rate is defined theoretically as the instantaneous rate with continuous compounding; in empirical studies it is common to use the 1-month Treasury bill yield as a proxy for the short-term interest rate. A general conclusion from these studies is that the actual behavior of the bond price cannot be fully explained within the framework of a single-factor model. As one might easily guess, in order to increase the explanatory power of short-term models it is tempting to increase the number of the underlying state variables.[2] Another possible way of improving the explanatory power of a model is to allow the coefficients to depend explicitly on the time parameter.

10.1.2 Time-inhomogeneous Models

All term structure models should preferably match liquid market information concerning historically observed interest rates and their volatilities. In this regard, let us observe that fitting of the initial yield curve is done by construction in some models;

---

[2] See, however, Chen and Scott (1992), Maghsoodi (1996), and Jamshidian (1995) and Sect. 10.3 below for further single-factor extensions of the CIR model.

however, it may prove to be a rather cumbersome task in others. Fitting a model to the term structure of market volatilities – that is, to the historically observed volatilities of forward rates – is typically even more difficult. Advanced optimization techniques are necessary to search for the parameters that return market prices for liquid interest rate derivatives such as bond options, futures options, caps and swaptions. For this reason, it is easier to work with models for which closed-form valuation formulas are available, at least for zero-coupon bonds and the most typical European options.

Hull and White model. Note that both Vasicek's and the CIR models are special cases of the following mean-reverting diffusion process

$$dr_t = a(b - cr_t)dt + \sigma r_t^\beta \, dW_t^*,$$

where $0 \le \beta \le 1$ is a constant. These models of the short-term rate are thus built upon a certain diffusion process with constant (i.e., time-independent) coefficients. In practical applications, it is more reasonable to expect that in some situations, the market's expectations with regard to future interest rates involve time-dependent coefficients. Also, it would be a plausible feature if a model fitted not merely the initial value of the short-term rate, but rather the whole initial yield curve. This desirable property of a bond price model motivated Hull and White (1990a) to propose an essential modification of the models above. The fairly general interest rate model they proposed extends both Vasicek's model and the CIR model in such a way that the model is able, in principle, to fit exactly any initial yield curve. In some circumstances, it leads also to a closed-form solution for the price of a European bond option. Let us describe the main steps of this approach. In its most general form, the Hull and White methodology assumes that

$$dr_t = \big(a(t) - b(t)r_t\big) dt + \sigma(t)r_t^\beta \, dW_t^* \tag{10.26}$$

for some constant $\beta \ge 0$, where W^* is a one-dimensional Brownian motion, and $a, b, \sigma : \mathbb{R}_+ \to \mathbb{R}$ are locally bounded functions. By setting $\beta = 0$ in (10.26), we obtain the *generalized Vasicek model,* in which the dynamics of r are[3]

$$dr_t = \big(a(t) - b(t)r_t\big) dt + \sigma(t) \, dW_t^*. \tag{10.27}$$

To solve this equation explicitly, let us denote $l(t) = \int_0^t b(u) \, du$. Then we have

$$d\big(e^{l(t)}r_t\big) = e^{l(t)}\big(a(t) \, dt + \sigma(t) \, dW_t^*\big).$$

Consequently

$$r_t = e^{-l(t)}\Big(r_0 + \int_0^t e^{l(u)}a(u) \, du + \int_0^t e^{l(u)}\sigma(u) \, dW_u^*\Big).$$

---

[3] A special case of such a model, with $b = 0$, was considered by Merton (1973).

It is thus not surprising that closed-form solutions for bond and bond option prices are not hard to derive in this setting. On the other hand, if we put $\beta = 1/2$ then we obtain the generalized CIR model

$$dr_t = \big(a(t) - b(t)r_t\big)\,dt + \sigma(t)\sqrt{r_t}\,dW_t^*.$$

In this case, however, the closed-form expressions for the bond and option prices are not easily available, in general (this would require solving (10.19)–(10.20) with time-dependent coefficients $a(t)$, $b(t)$ and $\sigma(t)$). For special cases, we refer the reader to Jamshidian (1995), where a generalized CIR model with closed-form solutions is examined, and to Schlögl and Schlögl (1997), who consider the case of piecewise constant coefficients. As was mentioned already, the extended CIR model (more specifically, the special case of the *constant dimension model*) is examined in Sect. 10.3.

The most important feature of the Hull and White approach is the possibility of the exact fit of the initial term structure and, in some circumstances, also of the term structure of forward rate volatilities. This can be done, for instance, by means of a judicious choice of the functions a, b and σ. Since the details of the fitting procedure depend on the particular model (i.e., on the choice of the exponent β), let us illustrate this point by restricting our attention to the generalized Vasicek model. We start by assuming that the bond price $B(t, T)$ can be represented in the following way

$$B(t, T) = B(r_t, t, T) = e^{m(t,T) - n(t,T)r_t} \tag{10.28}$$

for some functions m and n, with $m(T, T) = 0$ and $n(T, T) = 0$. Plugging (10.28) into the fundamental PDE for the zero-coupon bond, which is

$$\frac{\partial v}{\partial t}(r, t) + \frac{1}{2}\sigma^2(t)\frac{\partial^2 v}{\partial r^2}(r, t) + \big(a(t) - b(t)r\big)\frac{\partial v}{\partial r}(r, t) - rv(r, t) = 0,$$

we obtain

$$m_t(t, T) - a(t)n(t, T) + \tfrac{1}{2}\sigma^2(t)n^2(t, T) - \big(1 + n_t(t, T) - b(t)n(t, T)\big)r = 0.$$

Since the last equation holds for every t, T and r, we deduce that m and n satisfy the following system of differential equations (cf. (10.13)–(10.14))

$$n_t(t, T) = b(t)n(t, T) - 1, \quad n(T, T) = 0,$$

and

$$m_t(t, T) = a(t)n(t, T) - \tfrac{1}{2}\sigma^2(t)n^2(t, T), \quad m(T, T) = 0. \tag{10.29}$$

Fitting the yield curve. Suppose that an initial term structure $P(0, T)$ is given exogenously. We adopt the convention to denote by $P(0, T)$ the initial term structure, which is given (it can, for instance, be inferred from the market data), as opposed to the initial term structure $B(0, T)$ implied by a particular stochastic model of the term structure. Assume also that the forward rate volatility is not predetermined. In this case, we may and do assume that $b(t) = b$ and $\sigma(t) = \sigma$ are given constants; only the function a is thus unknown. Since b and σ are constants, n is given by (10.11). Furthermore, in view of (10.29), m equals

$$m(t, T) = \frac{1}{2} \int_t^T \sigma^2 n^2(u, T) \, du - \int_t^T a(u) n(u, T) \, du. \tag{10.30}$$

Since the forward rates implied by the model equal (cf. (9.5))

$$f(0, T) = -\frac{\partial \ln B(0, T)}{\partial T} = n_T(0, T) r_0 - m_T(0, T),$$

easy calculations involving (10.11) and (10.30) show that the unknown function $a(t)$ is implicitly given by the following equation:

$$\hat{f}(0, T) \overset{\text{def}}{=} \frac{\partial \ln P(0, T)}{\partial T} = e^{-bT} r_0 + \int_0^T e^{-b(T-u)} a(u) \, du - \frac{\sigma^2}{2b^2} \left(1 - e^{-bT}\right)^2.$$

Put another way, $\hat{f}(0, T) = g(T) - h(T)$, where $g'(T) = -bg(T) + a(T)$ with $g(0) = r_0$, and

$$h(T) = (1/2) b^{-2} \sigma^2 (1 - e^{-bT})^2.$$

Consequently, we obtain

$$a(T) = g'(T) + bg(T) = \hat{f}_T(0, T) + h'(T) + b(\hat{f}(0, T) + h(T)),$$

and thus the function a is indeed uniquely determined. This terminates the fitting procedure. Although, at least theoretically, this procedure can be extended to fit the volatility structure, is should be stressed that the possibility of an exact match with the historical data is only one of several desirable properties of a model of the term structure. If the forward rate volatilities are also fitted, the Hull and White approach becomes close to the Heath, Jarrow and Morton methodology, which is treated in the next chapter.

In the matching procedure, the exact knowledge of the first derivative of the initial term structure $P(0, T)$ with respect to T is assumed. In practice, the yield curve is known only at a finite number of points, corresponding to maturities of traded bonds, and the accuracy of data is also largely limited. The actual shape of the yield curve is therefore known only approximately (see, for instance, McCulloch (1971, 1975), Vasicek and Fong (1982), Shea (1984, 1985), Nelson and Siegel (1987), Delbaen and Lorimier (1992), Adams and van Deventer (1994), or Marangio et al. (2002)). Furthermore, in fitting additional initial data, we would typically need to use also the higher derivatives of the initial yield curve.

Lognormal model. The *lognormal model* of the short-term rate derives from a single-factor discrete-time model of term structure, put forward independently by Black et al. (1990) and Sandmann and Sondermann (1989) (in financial literature, it is usually referred to as the Black-Derman-Toy model). Generally speaking, the discrete-time model is based on judicious construction of a binary tree for a one-period interest rate. In principle, the value of the short-term interest rate at each node of the tree, and the corresponding transition probabilities, can be chosen in such a way that the model matches not only the initial forward rate curve, but also the initial volatility structure of forward rates (see Sandmann (1993)). In its general form, the lognormal model of the short-term rate is described by the following dynamics of the process r:

$$d \ln r_t = \left(a(t) - b(t) \ln r_t\right) dt + \sigma(t) \, dW_t^* \tag{10.31}$$

for some deterministic functions a, b and σ. Put differently, $r_t = e^{Y_t}$, where the process Y follows a time-inhomogeneous Vasicek's model:

$$dY_t = \left(a(t) - b(t)Y_t\right) dt + \sigma(t) \, dW_t^*.$$

For this reason, it is also known as the *exponential Vasicek model*.

In particular, the continuous-time limit of the Black-Derman-Toy model corresponds to the choice $b(t) = -\sigma'(t)/\sigma(t)$. Black and Karasinski (1991) examine practical aspects of the model given by (10.31). They postulate that the model fits the yield curve, the volatility curve and the cap curve.

Example 10.1.1 Consider the special case of the Black and Karasinski (1991) model, specifically,

$$d \ln r_t = \left(a(t) - b \ln r_t\right) dt + \sigma \, dW_t^*, \tag{10.32}$$

where W^* follows a one-dimensional Brownian motion under $\mathbb{P}^*$, the coefficient $a : \mathbb{R}_+ \to \mathbb{R}_+$ is a deterministic function, and $b \in \mathbb{R}, \sigma > 0$ are constants. As was mentioned already, the process $Y_t = \ln r_t$ satisfies Vasicek's (1977) dynamics. Consequently, the unique solution to (10.32) is given by the following expression

$$r_t = \exp\left(e^{-bt} \ln r_0 + \int_0^t e^{-b(t-u)} a(u) \, du + \sigma \int_0^t e^{-b(t-u)} \, dW_u^*\right).$$

It is easy to check that the probability distribution of r_t under $\mathbb{P}^*$ is lognormal with the following parameters

$$\mathbb{E}_{\mathbb{P}^*}(r_t) = \exp\left(e^{-bt} \ln r_0 + \int_0^t e^{-b(t-u)} a(u) \, du + \frac{\sigma^2}{4b}\left(1 - e^{-2bt}\right)\right)$$

and

$$\mathbb{E}_{\mathbb{P}^*}(r_t^2) = \exp\left(2e^{-bt} \ln r_0 + 2\int_0^t e^{-b(t-u)} a(u) \, du + \frac{\sigma^2}{b}\left(1 - e^{-2bt}\right)\right).$$

If $a(t) = a \in \mathbb{R}$ and $b > 0$, then for any $t > s$ the conditional first and second moments of r_t given the σ-field $\mathcal{F}_s$ are

$$\lim_{s \to \infty} \mathbb{E}_{\mathbb{P}^*}(r_t \mid \mathcal{F}_s) = \exp\left(\frac{a}{b} + \frac{\sigma^2}{4b}\right)$$

and

$$\lim_{s \to \infty} \mathrm{Var}_{\mathbb{P}^*}(r_t \mid \mathcal{F}_s) = \exp\left(\frac{2a}{b} + \frac{\sigma^2}{2b}\right)\left(\exp\left(\frac{\sigma^2}{2b}\right) - 1\right).$$

Remarks. Hogan and Weintraub (1993) showed that the dynamics of (10.31) lead to infinite prices for Eurodollar futures, and this is, of course, an undesirable property. To overcome this drawback, Sandmann and Sondermann (1993b, 1997) proposed a focus on effective annual rates, rather than on simple rates over shorter periods. In the continuous-time limit, they found the following dynamics for the annual effective rate $\hat{r}_t$

$$d\hat{r}_t = \mu(t)\hat{r}_t \, dt + \sigma(t)\hat{r}_t \, dW_t^*.$$

Consequently, the annual, continuously compounded rate, which is given by the equality $\tilde{r}_t = \ln(1 + \hat{r}_t)$, satisfies

$$d\tilde{r}_t = (1 - e^{-\tilde{r}_t})\Big(\big(\mu(t) - \tfrac{1}{2}(1 - e^{-\tilde{r}_t})\sigma^2(t)\big) \, dt + \sigma(t) \, dW_t^*\Big).$$

Remarks. In terms of the continuously compounded and actuarial forward rates of Sect. 9.1, we have $\tilde{r}_t = f(t, t, t-1)$ and $\hat{r}_t = a(t, t+1)$, where time is expressed in years.

10.1.3 Model Choice

Let us first summarize the most important features of term structure models that assume the diffusion-type dynamics of the short-term rate. Suppose that the dynamics of r under the actual probability $\mathbb{P}$ satisfy

$$dr_t = \mu(r_t, t) \, dt + \sigma(r_t, t) \, dW_t \tag{10.33}$$

for some sufficiently regular functions μ and σ. Assume, in addition, that the risk premium process equals $\lambda_t = \lambda(r_t, t)$ for some function $\lambda = \lambda(r, t)$. In financial interpretation, the last condition means that the excess rate of return of a given zero-coupon bond depends only on the current short-term rate and the price volatility of this bond. Using (10.33) and Girsanov's theorem, we conclude that under the martingale measure $\mathbb{P}^* = \mathbb{P}^\lambda$, the short-term rate process r satisfies

$$dr_t = \mu^\lambda(r_t, t)dt + \sigma(r_t, t) \, dW_t^*, \tag{10.34}$$

where the risk-neutral drift coefficient equals

$$\mu^\lambda(r, t) = \mu(r, t) + \lambda(r, t)\sigma(r, t).$$

Let us stress once again that it is necessary to assume that the functions μ, σ and λ are sufficiently regular (for instance, locally Lipschitz with respect to the first variable, and satisfying the linear growth condition), so that the SDE (10.34), taken with initial condition $r_0 > 0$, admits a unique global strong solution. Under such assumptions, the process r is known to follow, under the martingale measure $\mathbb{P}^*$, a strong Markov process with continuous sample paths. The latter property allows for the use of techniques developed in the theory of Markov processes.

Consider an arbitrary attainable contingent claim X of European style, with the settlement date T. Let the payoff at time T associated with X has the form $g(r_T)$ for some function $g : \mathbb{R} \to \mathbb{R}$. In addition, we assume that this claim pays dividends continuously at a state- and time-dependent rate $h(r_t, t)$. Under this set of assumptions, the ex-dividend price[4] of X is given by the following version of the risk-neutral valuation formula:

$$\pi_t(X) = \mathbb{E}_{\mathbb{P}^*}\left(\int_t^T h(r_u, u)e^{-\int_t^u r_v \, dv} \, du \,\Big|\, \mathcal{F}_t\right)$$

$$+ \mathbb{E}_{\mathbb{P}^*}\left(g(r_T)\, e^{-\int_t^T r_u \, du} \,\Big|\, \mathcal{F}_t\right) = v(r_t, t),$$

where v is some function $v : \mathbb{R} \times [0, T^*] \to \mathbb{R}$. It follows from the general theory of diffusion processes, more precisely from the result known as the Feynman-Kac formula (see Theorem 5.7.6 in Karatzas and Shreve (1998a)), that under mild technical assumptions, if a security pays continuously at a rate $h(r_t, t)$, and yields a terminal payoff $G_T = g(r_T)$ at time T, then the valuation function v solves the following *fundamental PDE*

$$\frac{\partial v}{\partial t}(r, t) + \frac{1}{2}\sigma^2(r, t)\frac{\partial^2 v}{\partial r^2}(r, t) + \mu^\lambda(r, t)\frac{\partial v}{\partial r}(r, t) - rv(r, t) + h(r, t) = 0,$$

subject to the terminal condition $v(r, T) = g(r)$. Existence of a closed-form solution of this equation for the most typical derivative securities (in particular, for a zero-coupon bond and an option on such a bond) is, of course, a desirable property of a term structure model of diffusion type. Otherwise, the efficiency of numerical procedures used to solve the fundamental PDE becomes an important practical issue.

A time-discretization of a diffusion-type model of the short-term rate becomes a substantially simpler task if a model exhibits the nice feature of path-independence. Formally, a diffusion-type term structure model is said to be *path-independent* if a solution r of (10.34) admits the representation $r_t = g(W_t, t)$ for some (deterministic) function g. This particular form of path-independence is a rather strong assumption, however, and thus it is unlikely to be satisfied by a short-term rate model.

As shown by Schmidt (1997) (see also Jamshidian (1991b)), most of the short-term models used in practice (these include, among others, Vasicek's model, the CIR model, the lognormal model) possess a weaker property; namely, r can be represented in the following way

$$r_t = H\big(g(t) + h(t)W^*_{u(t)}\big), \tag{10.35}$$

where $g, h : \mathbb{R}_+ \to \mathbb{R}$ are continuous functions, and the functions $H : \mathbb{R} \to \mathbb{R}$ and $u : \mathbb{R}_+ \to \mathbb{R}_+$ (with $u(0) = 0$) are strictly increasing and continuous. Since, typically, all of these functions are known explicitly, representation (10.35), which is indeed a weaker form of path-independence, leads to a considerable simplification of numerical procedures used in implementing single-factor models.

---

[4] The so-called *ex-dividend price* of an asset reflects the expected future dividends only. In other words, we do not assume that the dividends received prior to time t were reinvested in some way, but we simply ignore them.

In view of the abundance of existing single-factor models (see Brennan and Schwartz (1977c), Rich (1994), Courtadon (1982a), Ball and Torous (1983b), and Evans et al. (1994) for further examples) the choice of model is an issue of primary importance. Leaving analytical tractability aside, it is desirable for a model to be able to explain the actual behavior of the term structure, market prices of derivatives, or simply the observed short-term interest. For hedging purposes, it is essential that a model is able to generate a sufficiently rich family of yield curves (for instance, increasing, decreasing, or humped) and the variations of the yield curve predicted by the model are consistent with typical real-life fluctuations of this curve. More details on this issue can be found in papers by Schlögl and Sommer (1994, 1998) and, especially, Rogers (1995).

10.1.4 American Bond Options

Let us denote by $P^a(r_t, t, T)$ the price at time t of an American put option, with strike price K and expiry date T, written on a zero-coupon bond of maturity $U \geq T$. Let us denote by $B(r_t, t, U)$ the price at time $t \leq U$ of this bond. Arguing along similar lines as in Chap. 5, it is possible to show that

$$P^a(r_t, t, T) = \operatorname{ess\,sup}_{\tau \in \mathcal{T}_{[t,T]}} \mathbb{E}_{\mathbb{P}^*}\left(e^{-\int_t^\tau r_v \, dv}\left(K - B(r_\tau, \tau, U)\right)^+ \,\Big|\, r_t\right),$$

where $\mathcal{T}_{[t,T]}$ is the class of all stopping times with values in the time interval $[t, T]$. For a detailed justification of the application of standard valuation procedures to American contingent claims under uncertain interest rates, we refer the reader to Amin and Jarrow (1992). For any $t \in [0, T]$ the optimal exercise time τ_t equals

$$\tau_t = \inf\{u \in [t, T] \mid (K - B(r_u, u, U))^+ = P^a(r_u, u, T)\}.$$

Assume that the bond price $B(r, u, U)$ is a decreasing function of the rate r, all other variables being fixed (this holds in most, but not all, single-factor models).
Then the rational exercise time satisfies

$$\tau_t = \inf\{u \in [t, T] \mid r_u \geq r_u^*\}$$

for a certain adapted stochastic process r^* representing the critical level of the short-term interest rate. Using this, one can derive the following early exercise premium representation of the price of an American put option on a zero-coupon bond

$$P^a(r_t, t, T) = P(r_t, t, T) + \mathbb{E}_{\mathbb{P}^*}\left(\int_t^T e^{-\int_t^u r_v \, dv} \mathbb{1}_{\{r_u \geq r_u^*\}} r_u K \, du \,\Big|\, \mathcal{F}_t\right),$$

where $P(r_t, t, T)$ stands for the price at time $t \leq T$ of the corresponding European-style put option. The quasi-analytical forms of this representation for Vasicek's model and the CIR model were found by Jamshidian (1992). Chesney et al. (1993a) have studied bond and yield options of American style for the CIR model, using the properties of the so-called *Bessel bridges*.

10.1.5 Options on Coupon-bearing Bonds

A coupon-bearing bond is formally equivalent to a portfolio of discount bonds with different maturities. To value European options on coupon-bearing bonds, we take into account the fact that the zero-coupon bond price is typically a decreasing function of the short-term rate r. This implies that an option on a portfolio of zero-coupon bonds is equivalent to a portfolio of options on zero-coupon bonds with appropriate strike prices. Let us consider, for instance, a European call option with exercise price K and expiry date T on a coupon-bearing bond that pays coupons $c_1, \ldots, c_m$ at dates $T_1 < \cdots < T_m \le T^*$. The payoff of the option at expiry equals

$$C_T = \Big(\sum_{j=1}^{m} c_j \, B(r_T, T, T_j) - K \Big)^+.$$

The option will be therefore exercised if and only if $r_t < r^*$, where the critical interest rate r^* solves the equation

$$\sum_{j=1}^{m} c_j \, B(r^*, T, T_j) = K.$$

The option's payoff can be represented in the following way

$$C_T = \sum_{j=1}^{m} c_j \big(B(r_T, T, T_j) - K_j \big)^+,$$

where $K_j = B(r^*, T, T_j)$. The valuation of a call option on a coupon-bearing bond thus reduces, in the present setting, to the pricing of options on zero-coupon bonds.

10.2 Multi-factor Models

All models considered in the previous section belong to the class of single-factor models. Since the short-term rate was assumed to follow a (one-dimensional) Markov process, it could also be identified with a unique state variable of the model. A more general approach to term structure modelling incorporates multi-factor models; that is, those term structure models that involve several sources of uncertainty (typically represented by a multidimensional Brownian motion).

In most two-factor models, the term structure is inferred from the evolution of the short-term interest rate and some other economic variable (for instance, the long-term interest rate, the spread between the short-term and long-term rates, the yields on a fixed number of bonds, etc.). Since the analysis of two-factor models of the term structure of interest rates is rather lengthy, and is frequently based on the general equilibrium arguments, we refer the reader to the original papers, such as Brennan and Schwartz (1979, 1980a, 1980b, 1982), Schaefer and Schwartz (1984), Duffie and

Kan (1994, 1996), and Longstaff and Schwartz (1992a, 1992b). Richard (1978) and Cox et al. (1985b) also consider models with two state variables; the underlying state variables (such as the rate of inflation) are not easily observed, however.

Empirical tests of two-factor and multi-factor models, reported in Stambaugh (1988), Longstaff and Schwartz (1992a), and Pearson and Sun (1994), show that introducing additional state variables (and thus also additional factors) significantly improves the fit. This positive feature is counterbalanced, however, by the need to use much more cumbersome numerical procedures and by the difficult problem of identifying additional factors.

10.2.1 State Variables

From the theoretical point of view, a general multi-factor model is based on the specification of a multidimensional Markov process X. Assume, for instance, that $X = (X^1, \ldots, X^d)$ is a multidimensional diffusion process, defined as a unique strong solution of the following SDE

$$dX_t = \mu(X_t, t)\, dt + \sigma(X_t, t)\, dW_t, \qquad (10.36)$$

where W is a d-dimensional Brownian motion, and the coefficients μ and σ take values in $\mathbb{R}^d$ and $\mathbb{R}^{d \times d}$ respectively. Equation (10.36) may be extended by inserting also a jump term; in this case the state variable process X is termed a *jump-diffusion* (cf. Sect. 7.6.1).

Components $X^1, \ldots, X^d$ of the process X are referred to as the *state variables*; their economic interpretation, if any, is not always apparent, however. A special class of multi-factor models is obtained by postulating that the state variables X^i are directly related to yields of some bonds, with a preassigned finite set of maturities, as in Duffie and Kan (1996).

The short-term rate is now defined by setting $r_t = g(X_t)$ for some function $g : \mathbb{R}^d \to \mathbb{R}$. A special class of such models is obtained by postulating that X solves a linear SDE with time-dependent coefficients

$$dX_t = (a(t) + b(t)X_t)\, dt + \sigma(t)\, dW_t,$$

where $a : [0, T^*] \to \mathbb{R}^d$ and $b, \sigma : [0, T^*] \to \mathbb{R}^{d \times d}$ are bounded functions. Furthermore, it is common to set either $r_t = \frac{1}{2}|X_t|^2$ or $r_t = \gamma \cdot X_t$ for some vector $\gamma \in \mathbb{R}^d$. In the former case, we deal with the so-called *squared-Gauss-Markov* process; in the latter, the short-term rate is manifestly a Gaussian process. In the Gauss-Markov setting, multi-factor models were studied by Langetieg (1980), Jamshidian (1991a), El Karoui and Lacoste (1992)), and Duffie and Kan (1996) (see also El Karoui et al. (1992, 1995) and Jamshidian (1996) for related results).

Note that the generalized CIR model, which is essentially a single-factor model, can also be seen, at least for some parameters, as a multi-factor squared-Gauss-Markov (cf. (10.23)–(10.24)). This shows that the distinction between single-factor and multi-factor models is a bit ambiguous. Indeed, it refers to the postulated information structure (that is, to the underlying filtration), rather than to purely distributional properties of a model.

10.2.2 Affine Models

Recall that a Markovian model of the short term rate is called *affine* if the bond price admits the following representation

$$B(t, T) = e^{m(t,T) - n(t,T)r_t}, \quad \forall t \in [0, T],$$

for some functions m and n. Duffie and Kan (1996) show that a multi-factor model of the term structure is affine if and only if the coefficients of the dynamics of X, and the function g, are affine, in a suitable sense. To be more specific, let us assume that X satisfies

$$dX_t = \mu(X_t)\, dt + \sigma(X_t)\, dW_t, \tag{10.37}$$

where $\mu : \mathbb{R}^d \to \mathbb{R}^d$ and $\sigma : \mathbb{R}^d \to \mathbb{R}^{d \times d}$ are sufficiently regular, so that the SDE (10.37) has a unique, global solution. Then X is a time-homogeneous Markov process with some state space $D \subseteq \mathbb{R}^d$. We postulate that $r_t = g(X_t)$ for some affine function $g : D \to \mathbb{R}$. In other words, $r_t = \gamma_0 + \gamma_1 \cdot X_t$, where $\gamma_0 \in \mathbb{R}$ and $\gamma_1 \in \mathbb{R}^d$ are fixed. It appears that such a model is affine if and only if $\mu(x) = \mu_0 + \mu_1 x$ for some $\mu_0 \in \mathbb{R}^d$ and $\mu_1 \in \mathbb{R}^{d \times d}$. Finally, we postulate that $\sigma(x)\sigma^t(x) = \sigma_{ij}^0 + \sigma_{ij}^1 x$, where $\sigma^t(x)$ is the transpose of $\sigma(x)$. Furthermore, σ^0 belongs to $\mathbb{R}^{d \times d}$ and σ^1 is in $\mathbb{R}^{d \times d \times d}$.

Duffie et al. (2000) (see also Duffie and Kan (1996), Dai and Singleton (2000), Singleton and Umantsev (2002), Duffie et al. (2003)) examine the case of an affine jump-diffusion X. They show that the valuation of bonds and bond options can be achieved via the inversion of the Fourier transform of X_t (or a related random variable). Filipović (2001b) provides a complete characterization of all (not necessarily continuous) time-homogeneous Markovian short rate models that are affine.

10.2.3 Yield Models

As was indicated already, a detailed analysis of multi-factor models is beyond the scope of this text. We would like, however, to focus the reader's attention on specific mathematical problems which arise in the context of two-factor models based on the short-term rate and the *consol rate*.

At the intuitive level, the *consol yield* can be defined as the yield on a bond that has a constant continuous coupon and infinite maturity. To make this concept precise, we need to consider an economy with an infinite horizon date – that is, we set $T^* = +\infty$. A *consol* (or *consolidated bond*) is a special kind of coupon-bearing bond with no final maturity date. An investor purchasing a consol is entitled to receive coupons from the issuer forever.

In a continuous-time framework, it is convenient to view a consol as a risk-free security that continuously pays dividends at a constant rate, κ say. In the framework of term structure models based on a (nonnegative) short-term interest rate r, the price of a consol at time 0 equals

$$B^{\kappa}(0, \infty) = \mathbb{E}_{\mathbb{P}^*}\left(\int_0^{\infty} \kappa e^{-\int_0^s r_u\, du}\, ds \right) = \int_0^{\infty} \kappa B(0, s)\, ds.$$

The last result can be carried over to the case of arbitrary $t \in \mathbb{R}_+$, yielding the following equality for the price of a consol at time t

$$B^\kappa(t, \infty) = \mathbb{E}_{\mathbb{P}^*}\left(\int_t^\infty \kappa e^{-\int_t^s r_u\, du}\, ds \,\Big|\, \mathcal{F}_t \right) = \int_t^\infty \kappa B(t, s)\, ds. \qquad (10.38)$$

It is clear that we may and do assume, without loss of generality, that the rate κ equals 1. For brevity, we denote $B^1(t, \infty)$ by D_t in what follows. A substitution of r_u by an $\mathcal{F}_t$-measurable random variable η in the last formula gives

$$\int_t^\infty e^{-\int_t^s \eta\, du}\, ds = \frac{1}{\eta}.$$

For this reason, the yield on a consol, called also the *consol rate,* is simply defined as the reciprocal of its price, $Y_t = D_t^{-1}$. The consol rate can be seen as a proxy of a long-term rate of interest. To extend the short-term rate model, we may take into account the natural interdependence between both rates. In this way, we arrive at a two-factor diffusion-type model, in which the short-term rate r and the consol yield Y are intertwined. Since $D_t = Y_t^{-1}$, we may work directly with the price D of the consol equally well. Assume that the two-dimensional process (r_t, D_t) solves the following SDEs (cf. Brennan and Schwartz (1979))

$$dr_t = \mu(r_t, D_t)\, dt + \sigma(r_t, D_t) \cdot dW_t^*,$$

and

$$dD_t = \nu(r_t, D_t)\, dt + \gamma(r_t, D_t) \cdot dW_t^*,$$

where W^* is a two-dimensional standard Brownian motion under the martingale measure $\mathbb{P}^*$. Note that the drift coefficient ν in the dynamics of D cannot be chosen arbitrarily. Indeed, an application of Itô's formula to (10.38) gives the following SDE governing a consol's price dynamics

$$dD_t = (r_t D_t - \kappa)\, dt + \gamma(r_t, D_t) \cdot dW_t^* \qquad (10.39)$$

under the martingale measure $\mathbb{P}^*$. SDEs for r and D should still be considered simultaneously with equality (10.38) linking the price D to the short-term rate r, since otherwise the term structure model would be inconsistent. A reasonable approach is therefore not to specify all coefficients in these SDEs, but rather to leave a certain degree of freedom, for instance in the choice of the diffusion coefficient γ. It was conjectured by Black that, under mild technical assumptions, for any choice of short-term rate coefficients μ and σ, a consistent choice of γ is indeed possible.

Duffie et al. (1995) have confirmed Black's conjecture, by applying the technique of forward-backward SDEs over an infinite horizon. Furthermore, it appears that the consol price, and thus also the consol rate, is necessarily of the form $D_t = g(r_t)$ for some function g. The function $g = g(x)$ in question is shown to be a unique solution of the following ordinary differential equation

$$g'(x)\mu(x, g(x)) - xg(x) + \tfrac{1}{2}g''(x)\sigma^2(x, g(x)) + 1 = 0.$$

Concluding, the model can be reduced to a model with a single state variable r, in which the dynamics of the short-term rate are given by the SDE above, with $D_t = g(r_t)$. Let us finally observe that if we start by postulating that the consol rate follows

$$dY_t = \alpha(r_t, Y_t)\, dt + \beta(r_t, Y_t) \cdot d\tilde{W}_t$$

under some probability measure $\mathbb{P}$, where $\tilde{W}$ is a Brownian motion and r is an as yet unspecified short-term rate process, then we necessarily have

$$dD_t = (r_t D_t - \kappa)\, dt - D_t^2 \beta(r_t, D_t^{-1}) \cdot dW_t^*$$

under the martingale measure $\mathbb{P}^*$. As pointed out by Hogan (1993), for some traditional choices of the drift coefficient in the dynamics of the short-term rate r, a solution of the last SDE explodes in a finite time. Y reaches infinity in a finite time with positive probability).

Remarks. An interesting issue which is not treated here is the asymptotic behavior of long-term yields. More exactly, we are interested in the dynamics of the process $\tilde{Y}_t$ obtained from the zero-coupon yield $Y(t, T)$ by passing to the limit when the maturity date T tends to infinity. It appears that in typical term structure models the *long-term yield* process $\tilde{Y}$ is either constant or has non-decreasing sample paths. For alternative proofs of this property, see Dybvig et al. (1996), El Karoui et al. (1996), and Hubalek et al. (2002).

10.3 Extended CIR Model

In this section, in which we follow Szatzschneider (1998), we shall analyze the non-homogenous CIR model of the short-term rate, referred to as the *extended CIR model*. By assumption, the dynamics of the short-term rate r under $\mathbb{P}^*$ are

$$dr_t = (a(t) - b(t)r_t)\, dt + \sigma(t)\sqrt{r_t}\, dW_t^*, \tag{10.40}$$

where W^* is a standard Brownian motion under $\mathbb{P}^*$ and a, b, σ are positive and locally bounded functions. This model was considered by Hull and White (1990a, 1990b), who generalized the model put forward by Cox et al. (1985b). We shall examine this model under an additional condition:

$$\frac{a(t)}{\sigma^2(t)} = \text{constant}, \tag{10.41}$$

referred to as the *constant dimension condition* – this condition was considered in the case of the extended CIR by Jamshidian (1995) (see also Rogers (1995) and Maghsoodi (1996)). In particular, we shall check that the model is affine in the short rate, in the sense that $B(t, T) = e^{m(t,T) - n(t,T)r_t}$, where $B(t, T)$ is the price at time t of the zero-coupon bond maturing at time T.

10.3.1 Squared Bessel Process

Let $W = (W^1, \ldots, W^n)$ be a standard n-dimensional standard Brownian motion on $(\Omega, \mathcal{F}, \mathbb{P})$ started at $x = (x^1, \ldots, x^n) \in \mathbb{R}^n$. Then the process $X_t = |W_t|^2 = (W_t^1)^2 + (W_t^2)^2 + \ldots + (W_t^n)^2$ can be written as the solution of the stochastic differential equation

$$X_t = |x|^2 + nt + 2 \int_0^t \sqrt{X_u}\, d\tilde{W}_u,$$

where

$$\tilde{W}_t \stackrel{\text{def}}{=} \sum_{i=1}^n \int_0^t \frac{W_u^i}{\sqrt{X_u}}\, dW_u^i.$$

It follows easily from Lévy's characterization theorem (see Theorem A.10.1) that the process $\tilde{W}$ is a one-dimensional standard Brownian motion under $\mathbb{P}$. For a fixed $\delta \geq 0$, let $X^{\delta,x}$ be a δ-dimensional squared Bessel process BESQ_x^δ started at $x \geq 0$, that is, the unique strong solution of the stochastic differential equation:

$$X_t = x + \delta t + 2 \int_0^t \sqrt{X_u}\, dW_u, \quad \forall t \in \mathbb{R}_+, \tag{10.42}$$

where W follows a one-dimensional standard Brownian motion under $\mathbb{P}$. By applying a suitable version of a comparison theorem for SDEs, we can deduce that the unique solution to the equation (10.42) is nonnegative, and thus we may omit the absolute value under the square root sign.

The Shiga and Watanabe theorem (see Shiga and Watanabe (1973) or Theorem 1.2 in Chap. XI in Revuz and Yor (1999)) states that we have: $X^{\delta_1,x_1} \oplus X^{\delta_2,x_2} \stackrel{d}{=} X^{\delta_1+\delta_2,x_1+x_2}$, where $\oplus$ stands for the sum of two independent processes and $\stackrel{d}{=}$ denotes the equality in the sense of probability distributions. Let us mention that, for any $d \geq 0$ and $y \geq 0$, the square root of the process $\mathrm{BESQ}_{y^2}^\delta$ is called the δ-dimensional Bessel process started at y and it is denoted as BES_y^δ.

It is worth noting that in case of fixed coefficients a, b and σ, the process r can be represented as $r_t = e^{-bt} X_{f(t)}$, where X is the δ-dimensional squared Bessel process with $\delta = 4a\sigma^{-2}$ and $f(t) = (4b)^{-1}\sigma^2(e^{bt} - 1)$.

10.3.2 Model Construction

Assume that X is a δ-dimensional squared Bessel process under $\mathbb{P}$, so that

$$dX_t = \delta\, dt + 2\sqrt{X_t}\, dW_t, \tag{10.43}$$

where W is a standard Brownian motion under $\mathbb{P}$. We will show in Sect. 10.3.3 that, by applying a suitable change of an underlying probability measure, we obtain under $\mathbb{P}^*$

$$dX_t = (\delta + 2\beta(t)X_t)\, dt + 2\sqrt{X_t}\, dW_t^*, \tag{10.44}$$

where $\beta(t) < 0$ is a locally bounded and continuously differentiable function (the last condition will be dropped later on).

In the next step, we multiply the process X by a strictly positive, locally bounded function $s : \mathbb{R}_+ \to \mathbb{R}_+$. If the function s is continuously differentiable, then the process r, given as $r_t = s(t)X_t$, satisfies

$$dr_t = \left\{ \delta s(t) + \left(2\beta(t) + \frac{s'(t)}{s(t)} \right) r_t \right\} dt + 2\sqrt{s(t)r_t}\, dW_t^*. \tag{10.45}$$

Note that condition (10.41) is satisfied in the present framework.

10.3.3 Change of a Probability Measure

Since X is a δ-dimensional squared Bessel process under $\mathbb{P}$, we have

$$dM_t = d(X_t - \delta t) = 2\sqrt{X_t}\, dW_t,$$

where we set $M_t = X_t - \delta t$ for $t \in \mathbb{R}_+$. Consider the exponential continuous local martingale η given by

$$\eta_t = \mathcal{E}_t\left(\int_0^{\cdot} \beta(u)\sqrt{X_u}\, dW_u \right) = \exp\left\{ \frac{1}{2}\int_0^t \beta(u)\, dM_u - \frac{1}{2}\int_0^t \beta^2(u)X_u\, du \right\}.$$

It is easily seen that

$$\eta_t = \exp\left\{ \frac{1}{2}\left(\beta(t)X_t - \beta(0)X_0 - \delta\int_0^t \beta(u)\, du - \int_0^t (\beta^2(u) + \beta'(u))X_u\, du \right) \right\}.$$

If the condition $\beta^2(u) + \beta'(u) \geq 0$ is satisfied, then the local martingale η is bounded (since $X_t \geq 0$ and $\beta(t) < 0$), and thus it is a martingale. In the general situation, we can use a slightly more elaborated argument, known as Novikov's condition, for small time intervals. We conclude that η is always a martingale and thus the change of drift via Girsanov's theorem is justified. We introduce a probability measure $\mathbb{P}^*$ by setting $d\mathbb{P}^* = \eta_T\, d\mathbb{P}$ on $(\Omega, \mathcal{F}_T)$. The proof of the next lemma is based on a straightforward application of the martingale property of the Doléans exponential process η.

Lemma 10.3.1 *Let X be a δ-dimensional squared Bessel process under $\mathbb{P}$. Then*

$$\mathbb{E}_{\mathbb{P}}\left\{ \exp\left[\frac{1}{2}\left(\beta(t)X_t - \int_0^t (\beta^2(u) + \beta'(u))X_u\, du \right) \right] \right\}$$

$$= \exp\left[\frac{1}{2}\left(\beta(0)X_0 + \delta\int_0^t \beta(u)\, du \right) \right].$$

More generally, for any $t < T$ we have

$$\mathbb{E}_{\mathbb{P}}\left\{ \exp\left[\frac{1}{2}\left(\beta(T)X_T - \int_t^T (\beta^2(u) + \beta'(u))X_u\, du \right) \right] \,\Big|\, \mathcal{F}_t \right\}$$

$$= \exp\left[\frac{1}{2}\left(\beta(t)X_t + \delta\int_t^T \beta(u)\, du \right) \right].$$

10.3.4 Zero-coupon Bond

We shall now derive the valuation formula for a zero-coupon bond in the extended CIR model. For ease of notation, we consider the case of $t = 0$. Since $r_t = s(t)X_t$, we obtain

$$B(0, T) = \mathbb{E}_{\mathbb{P}*}\left(e^{-\int_0^T s(u)X_u du}\right) = \mathbb{E}_{\mathbb{P}}\left(\eta_T e^{-\int_0^T s(u)X_u du}\right).$$

After simple manipulations, we obtain

$$B(0, T) = e^{\frac{1}{2}\left(-\beta(0)X_0 - \delta\int_0^T \beta(u)\, du\right)} \mathbb{E}_{\mathbb{P}}\left\{e^{\frac{1}{2}\left(\beta(T)X_T - \int_0^T g(u)X_u\, du\right)}\right\},$$

where we denote

$$g(u) = \beta^2(u) + \beta'(u) + 2s(u).$$

For a fixed, but otherwise arbitrary, $T > 0$, we search for an auxiliary function $F_T(u) = F(u)$ that solves the following ordinary differential equation (known as the *Riccati equation*)

$$F^2(u) + F'(u) = \beta^2(u) + \beta'(u) + 2s(u), \tag{10.46}$$

for $u \in [0, T)$, with the terminal condition $F(T) = \beta(T)$.

If the function F is as described above, then we obtain

$$B(0, T) = e^{\frac{1}{2}\left(-\beta(0)X_0 - \delta\int_0^T \beta(u)\, du\right)} \mathbb{E}_{\mathbb{P}}\left\{e^{\frac{1}{2}\left(F(T)X_T - \int_0^T (F^2(u)+F'(u))X_u\, du\right)}\right\}.$$

We are in the position to apply Lemma 10.3.1 with the function β replaced by F. Note that $X_0 = r_0/s(0)$.

Proposition 10.3.1 *Assume that $\beta < 0$. Then for any fixed $T > 0$ we have*

$$B(0, T) = \exp\left\{\frac{1}{2}\left((F_T(0) - \beta(0))X_0 + \delta\int_0^T (F_T(u) - \beta(u))\, du\right)\right\},$$

where the function $F_T : [0, T] \to \mathbb{R}$ solves the ODE (10.46) with the terminal condition $F_T(T) = \beta(T)$.

The most convenient approach to the Riccati equation (10.46) is through the corresponding Sturm-Liouville equation. Let us write $F(u) = \frac{\varphi'(u)}{\varphi(u)}$ for $u \in [0, T)$, where $\varphi(0) = 1$. Then for $\varphi(\cdot) = \varphi_T(\cdot)$ we obtain

$$\frac{\varphi''(u)}{\varphi(u)} = \beta^2(u) + \beta'(u) + 2s(u), \quad \forall u \in [0, T),$$

with the terminal condition $F(T) = \frac{\varphi'_l(T)}{\varphi(T)} = \beta(T)$, where φ'_l stands for the left-hand side derivative. Since the function s is strictly positive, it is not difficult to check that

$$F(u) < \beta(u) < 0, \quad \forall u \in [0, T). \tag{10.47}$$

Observe also that

$$\int_0^T F_T(u)\, du = \int_0^T (\ln \varphi_T(u))'\, du = \ln \varphi_T(T).$$

We thus obtain the following result.

Corollary 10.3.1 *The price at time 0 of a zero-coupon bond with maturity date T satisfies*

$$B(0, T) = \exp\left\{ \frac{1}{2}\big((\varphi_T'(0) - \beta(0))X_0 + \delta \ln \varphi_T(T) - \delta \int_0^T \beta(u)\, du\big) \right\},$$

where $X_0 = r_0/s(0)$.

We write here φ_T rather than φ since we wish to stress the dependence on the maturity date T of the solution φ in $[0, T]$. From (10.47), we obtain the intuitively obvious fact that the bond price is a decreasing function of the short-term rate, as well as the explicit form of its affine representation

$$B(0, T) = e^{m(0,T) - n(0,T)r_0}.$$

10.3.5 Case of Constant Coefficients

Note that the ordinary differential equation $\varphi'' = c^2 \varphi$, where c is a positive constant, can be solved easily. Hence, if

$$\beta^2(u) + \beta'(u) + 2s(u) = c^2$$

is a constant function, the bond price can be found explicitly. The solution of the second-order ODE

$$\varphi''(u) = c^2 \varphi(u)$$

in $[0, T)$ is clearly $\varphi(u) = Ae^{cu} + Be^{-cu}$, where the constants A and B satisfy the following conditions: $A + B = 1$ (since $\varphi(0) = 1$) and

$$c\left(\frac{Ae^{cT} - Be^{-cT}}{Ae^{cT} + Be^{-cT}} \right) = \beta(T).$$

Remarks. Assume that the dynamics of r under $\mathbb{P}^*$ are

$$dr_t = (a - br_t)\, dt + \sigma \sqrt{r_t}\, dW_t^*,$$

where W^* is a standard Brownian motion under $\mathbb{P}^*$ and a, b, σ are strictly positive constants. It is easy to check that we deal here with the following case

$$\beta(t) = -\tfrac{1}{2}b, \quad s(t) = \tfrac{1}{4}\sigma^2, \quad \delta = 4a\sigma^{-2}, \quad c^2 = \tfrac{1}{4}(b^2 + 2\sigma^2).$$

We conclude that (cf. (10.21)–(10.22))

$$m(0, T) = \frac{2a}{\sigma^2} \ln \left\{ \frac{\gamma e^{bT/2}}{\gamma \cosh \gamma T + \frac{1}{2}b \sinh \gamma T} \right\}$$

and

$$n(0, T) = \frac{\sinh \gamma T}{\gamma \cosh \gamma T + \frac{1}{2}b \sinh \gamma T},$$

where in turn $2\gamma = 2\sqrt{c} = (b^2 + 2\sigma^2)^{1/2}$.

10.3.6 Case of Piecewise Constant Coefficients

Our next goal is to show how to deal with the case of piecewise-constant functions β and s. This case was also studied by Schlögl and Schlögl (1997), who used a different approach involving several χ^2 distributions. We will show how to find the bond price in the case of $\beta(u) = \beta_i$ and $s(u) = s_i$ for $u \in [t_i, t_{i+1})$, $i = 0, \ldots, n-1$, where $t_0 = 0$, $t_n = T$ and $\beta(T) = \beta_n$. As before $\beta_i < 0$ and $s_i > 0$. Setting $\Delta\beta_i = \beta_i - \beta_{i-1}$, we obtain

$$\int_0^T X_u \, d\beta(u) = \sum_{i=1}^n (\beta_i - \beta_{i-1}) X_{t_i} = \sum_{i=1}^n \Delta\beta_i X_{t_i}.$$

The bond pricing formula

$$B(0, T) = e^{\frac{1}{2}\left(-\beta(0)X_0 - \delta \int_0^T \beta(u)\,du\right)} \mathbb{E}_\mathbb{P}\left\{ e^{\frac{1}{2}\left(\beta(T)X_T - \int_0^T g(u)X_u\,du\right)} \right\},$$

where, as before,

$$g(u) = \beta^2(u) + \beta'(u) + 2s(u),$$

becomes

$$B(0, T) = e^{\frac{1}{2}\left(-\beta(0)X_0 - \delta \int_0^T \beta(u)\,du\right)}$$
$$\mathbb{E}_\mathbb{P}\left\{ e^{\frac{1}{2}\left(\beta(T)X_T - \sum_{i=1}^n \left(\gamma_{i-1}\int_{t_{i-1}}^{t_i} X_u\,du\right) - \sum_{i=1}^n \Delta\beta_i X_{t_i}\right)} \right\},$$

where we denote $\gamma_{i-1} = \beta_{i-1}^2 + 2s_{i-1}$. Notice that, as expected, $\beta_n = \beta(T)$ has disappeared. Observe also that we have chosen the right-continuous version of the function β.

To completely solve the problem of bond pricing, it thus suffices to to find the function $\varphi(u)$, $u \in [0, T]$, satisfying the following conditions:
(a) φ is continuous, with $\varphi(0) = 1$,
(b) for every $i = 1, \ldots, n$ we have

$$\frac{\varphi''(u)}{\varphi(u)} = \beta_{i-1}^2 + 2s_{i-1}, \quad \forall u \in (t_{i-1}, t_i),$$

(c) for every $i = 1, \ldots, n-1$ we have

$$\frac{\varphi'_r(t_i) - \varphi'_l(t_i)}{\varphi(t_i)} = \beta_i - \beta_{i-1},$$

(d) for $i = n$ we have

$$\frac{\varphi'_l(t_n)}{\varphi(t_n)} = \beta_{n-1}.$$

We write here φ'_r and φ'_l to denote the right-hand-side and left-hand-side derivatives respectively. The general solution of the second-order ODE

$$\varphi''(u) = c_{i-1}^2 \varphi(u)$$

is given by the formula

$$A_{i-1} e^{c_{i-1} u} + B_{i-1} e^{-c_{i-1} u}.$$

Thus, the bond valuation problem is now reduced to the solution of a system of $2n$ linear equations for the coefficients A_{i-1}, B_{i-1}, $i = 1, \ldots, n$. Recalling our general formula for $B(0, T)$, we need only find A_0 and A_{n-1}, since $B_0 = 1 - A_0$, and the constant B_{n-1} is recovered from condition (d).

10.3.7 Dynamics of Zero-coupon Bond

To deal with derivative contracts, such as bond options, we need to examine the dynamics of the bond price. Recall that for every $0 \le t \le T$ we set

$$B(t, T) = \mathbb{E}_{\mathbb{P}*}\left(e^{-\int_t^T r_u du} \,\Big|\, \mathcal{F}_t \right) = \mathbb{E}_{\mathbb{P}*}\left(e^{-\int_t^T s(u) X_u du} \,\Big|\, \mathcal{F}_t \right),$$

or equivalently,

$$B(t, T) = \mathbb{E}_{\mathbb{P}}\left(\eta_T \eta_t^{-1} e^{-\int_t^T s(u) X_u du} \,\Big|\, \mathcal{F}_t \right).$$

More explicitly

$$B(t, T) = e^{\frac{1}{2}\left(-\beta(t) X_t - \delta \int_t^T \beta(u)\, du\right)} \mathbb{E}_{\mathbb{P}}\left\{ e^{\frac{1}{2}\left(\beta(T) X_T - \int_t^T g(u) X_u\, du\right)} \right\},$$

where $g(u) = \beta^2(u) + \beta'(u) + 2s(u)$. The proof of the next result is analogous to the proof of Lemma 10.3.1 and thus it is omitted.

Proposition 10.3.2 *For any dates $0 \le t \le T$, we have*

$$B(t, T) = \exp\left\{ \frac{1}{2}\left((F_T(t) - \beta(t)) X_t + \delta \int_t^T (F_T(u) - \beta(u))\, du \right) \right\},$$

where the function $F_T : [0, T] \to \mathbb{R}$ solves (10.46) with the terminal condition $F_T(T) = \beta(T)$. Put another way, the bond price equals

$$B(t, T) = e^{m(t,T) - n(t,T) r_t},$$

where

$$m(t, T) = \frac{\delta}{2} \int_t^T (F_T(u) - \beta(u))\, du$$

and

$$n(t, T) = \frac{1}{2s(t)}\left(\beta(t) - F_T(t) \right).$$

The following corollary is a straightforward consequence of Itô's formula.

Corollary 10.3.2 *The dynamics of the bond price under* $\mathbb{P}^*$ *are*

$$dB(t, T) = B(t, T)\big(r_t\, dt + b(t, T)\, dW_t^*\big),$$

where the process

$$W_t^* = W_t - \int_0^t \beta(u)\sqrt{X_u}\, du$$

is a standard Brownian motion under $\mathbb{P}^*$ *and the bond price volatility* $b(t, T)$ *is given by the formula*

$$b(t, T) = -2n(t, T)s(t)\sqrt{X_t} = \big(F_T(t) - \beta(t)\big)\sqrt{X_t}.$$

Let us introduce the forward martingale measure $\mathbb{P}_T$ on $(\Omega, \mathcal{F}_T)$ by setting

$$\frac{d\mathbb{P}_T}{d\mathbb{P}^*} = \frac{1}{B_T B(0, T)} = \exp\left(\int_0^T b(u, T)\, dW_u^* - \frac{1}{2}\int_0^T b^2(u, T)\, du\right).$$

Then the process

$$W_t^T = W_t^* - \int_0^t b(u, T)\, du, \quad \forall t \in [0, T],$$

is a standard Brownian motion under $\mathbb{P}_T$. Moreover, by combining the last equality with formula (10.44), we conclude that the following result holds.

Lemma 10.3.2 *The dynamics of the process* X *under the forward martingale measure* $\mathbb{P}_T$ *are*

$$dX_t = (\delta + 2F_T(t)X_t)\, dt + 2\sqrt{X_t}\, dW_t^T.$$

For a fixed $t < T$, let $\tilde{\varphi} : [t, T] \to \mathbb{R}$ be the unique solution to the following ordinary differential equation:

$$\frac{\tilde{\varphi}''(u)}{\tilde{\varphi}(u)} = \beta^2(u) + \beta'(u) + 2s(u)$$

subject to the boundary conditions: $\tilde{\varphi}(t) = 1$, $\frac{\tilde{\varphi}'(T)}{\tilde{\varphi}(T)} = \beta(T)$. It is clear that for every $u \in [t, T]$ we have $\tilde{\varphi}(u) = \frac{\varphi_T(u)}{\varphi_T(t)}$. Applying the same method as in the case $t = 0$, we obtain the following result.

Proposition 10.3.3 *The arbitrage price at time* $t \in [0, T]$ *of a zero-coupon bond with maturity date* T *equals*

$$B(t, T) = \exp\left\{\frac{1}{2}\big((\tilde{\varphi}'(t) - \beta(t))X_t + \delta \ln \tilde{\varphi}(T) - \delta \int_t^T \beta(u)\, du\big)\right\}.$$

10.3.8 Transition Densities

Let us fix maturity date $T > 0$. We shall now find the transition density of an auxiliary process U, which is defined through the following stochastic differential equation:

$$U_t = x + \int_0^t 2\sqrt{U_s}\, dW_s + \int_0^t \left(2F_T(s)U_s + \delta\right) ds,$$

where W is a standard Brownian motion under $\mathbb{P}$. Notice that the probability distribution of the process U_t, $t \le T$ under $\mathbb{P}$ coincides with the probability distribution of the process X under the forward martingale measure $\mathbb{P}_T$. Hence to determine the probability distribution of the bond price under $\mathbb{P}_T$, it suffices to find the distribution of U. To this end, we shall use a deterministic time change. For brevity, we shall write F and ϕ, rather than F_T and ϕ_T.

Case $\delta = 1$. Let $\bar{W}$ be a standard Brownian motion under $\mathbb{P}$. Assume that V satisfies

$$V_t = v_0 + \bar{W}_t + \int_0^t F(s)V_s\, ds,$$

or more explicitly

$$V_t = \varphi(t)\left(v_0 + \int_0^t \varphi^{-1}(s)\, d\bar{W}_s\right), \quad \frac{\varphi'(s)}{\varphi(s)} = F(s), \quad \varphi(0) = 1.$$

Because the stochastic integral of a deterministic function is a Gaussian random variable, it is clear that V_t is a Gaussian random variable with expected value $v_0\varphi(t)$ and variance $g(t)$, where the increasing function $g : \mathbb{R}_+ \to \mathbb{R}_+$ is given by the formula $g(t) = \varphi^2(t) \int_0^t \varphi^{-2}(s)\, ds$. Let us set $U_t = V_t^2$ for $t \in [0, T]$. Then the process U satisfies

$$U_t = v_0^2 + \int_0^t 2\sqrt{U_s}\, \mathrm{sgn} V_s\, d\bar{W}_s + \int_0^t 2F(s)U_s\, ds + t,$$

or equivalently

$$U_t = v_0^2 + \int_0^t 2\sqrt{U_s}\, dW_s + \int_0^t (2F(s)U_s + 1)\, ds,$$

where $W_t = \int_0^t \mathrm{sgn} V_s\, d\bar{W}_s$ is a standard Brownian motion under $\mathbb{P}$. We conclude that, given the initial condition $U_0 = x \ge 0$, the probability density function of U_t under $\mathbb{P}$ can be written as $p^1(g(t), \varphi^2(t)x, y)$, where $p^1(t, x, y)$ is the transition density of the square of a standard Brownian motion – that is, the squared Bessel process of dimension 1.

Remarks. Recall that here $F = F_T$ and thus also $\varphi = \varphi_T$. For $F = F_U$, the corresponding function $\varphi = \varphi_U$ will be denoted briefly as ψ in what follows.

General case. In general, the transition density of the process U under $\mathbb{P}$ is $p^\delta(g(t), \varphi^2(t)x, y)$, where $p^\delta(t, x, y)$ is the transition density of the squared δ-dimensional

Bessel process. This fact results easily from the Shiga and Watanabe theorem applied to the probability distribution of the process U. The proof is exactly as for BESQ processes. Let us recall the well-known formula for the transition probability density function $p^\delta(t, x, y)$ of the squared δ-dimensional Bessel process (see, for instance, Göing-Jaeschke and Yor (1999) or Revuz and Yor (1999)). For every $t > 0$ and $x > 0$, we have

$$p^\delta(t, x, y) = \frac{1}{2t}\left(\frac{y}{x}\right)^{\frac{\nu}{2}} \exp\left(-\frac{(x+y)}{2t}\right) I_\nu\left(\frac{\sqrt{xy}}{t}\right),$$

where I_ν is the Bessel function of index $\nu = \frac{\delta}{2} - 1$, that is,

$$I_\nu(x) = \left(\frac{x}{2}\right)^\nu \sum_{n=0}^{\infty} \frac{\left(\frac{x}{2}\right)^{2n}}{n!\,\Gamma(\nu + n + 1)}.$$

10.3.9 Bond Option

Consider a European call option with the expiration date T and exercise price K, which is written on a U-maturity zero-coupon bond. According to Proposition 10.3.3, for arbitrary two dates $T < U$, we have

$$B(T, U) = \exp\left\{\frac{1}{2}\left((F_U(T) - \beta(T))X_T + \delta \int_T^U (F_U(s) - \beta(s))\,ds\right)\right\},$$

where the function $F_U : [0, U] \to \mathbb{R}$ solves (10.46) with the terminal condition $F_U(U) = \beta(U)$. The option's payoff at expiry can be represented as follows

$$C_T = \left(B(T, U) - K\right)^+ = \left(B(T, U) - K\right)\mathbb{1}_{\{B(T,U)>K\}},$$

or equivalently,

$$C_T = B(T, U)\mathbb{1}_{\{X_T < \bar{K}\}} - K B(T, T)\mathbb{1}_{\{X_T < \bar{K}\}},$$

where $X_T = r_T/s(T)$ and the constant $\bar{K} > 0$ is given by the formula

$$\bar{K} = \frac{2\left(\ln K + \frac{1}{2}\delta \int_T^U (\beta(s) - F_U(s))\,ds\right)}{\beta(T) - F_U(T)}.$$

Forward martingale measures $\mathbb{P}_T$ and $\mathbb{P}_U$ have the following Radon-Nikodým derivatives on $(\Omega, \mathcal{F}_T)$ with respect to the spot martingale measure $\mathbb{P}^*$

$$\frac{d\mathbb{P}_T}{d\mathbb{P}^*} = \frac{1}{B_T B(0, T)}, \quad \frac{d\mathbb{P}_U}{d\mathbb{P}^*} = \frac{B(T, U)}{B_T B(0, U)}.$$

Proposition 10.3.4 *The arbitrage price in the extended CIR model of a call option written on the U-maturity zero-coupon bond equals*

$$C_0 = B(0, U)\, \mathbb{P}_U\{X_T < \bar{K}\} - K B(0, T)\, \mathbb{P}_T\{X_T < \bar{K}\},$$

where under $\mathbb{P}_T$ the random variable X_T has the density $p^\delta(g(T), \varphi^2(T)x_0, y)$ with $x_0 = r(0)/s(0)$ and $g(T) = \varphi^2(T) \int_0^T \varphi^{-2}(s)\,ds$. In an analogous formula for the density of X_T under $\mathbb{P}_U$ the function φ is replaced by ψ.

Proof. The statement follows from results of Sect. 10.3.8.   □

A similar valuation result holds for any European claim, as the next result shows.

Proposition 10.3.5 *Consider a contingent claim $Y = h(r_T)$ at time T. Its arbitrage price at time $t \in [0, T]$ equals*

$$\pi_0(X) = B_0\, \mathbb{E}_{\mathbb{P}^*}\big(B_T^{-1} h(r_T)\big) = B(0, T)\, \mathbb{E}_{\mathbb{P}_T}\big(h(s(T)X_T)\big),$$

where the random variable X_T has under the forward measure $\mathbb{P}_T$ the probability density function $p^\delta(g(T), \varphi^2(T)x_0, y)$.

11

Models of Instantaneous Forward Rates

The Heath, Jarrow and Morton approach to term structure modelling is based on an exogenous specification of the dynamics of instantaneous, continuously compounded forward rates $f(t, T)$. For any fixed maturity $T \leq T^*$, the dynamics of the forward rate $f(t, T)$ are (cf. Heath et al. (1990a, 1992a))

$$df(t, T) = \alpha(t, T) \, dt + \sigma(t, T) \cdot dW_t, \tag{11.1}$$

where α and σ are adapted stochastic processes with values in $\mathbb{R}$ and $\mathbb{R}^d$ respectively, and W is a d-dimensional standard Brownian motion with respect to the underlying probability measure $\mathbb{P}$ (to be interpreted as the actual probability). For any fixed maturity date T, the initial condition $f(0, T)$ is determined by the current value of the continuously compounded forward rate for the future date T that prevails at time 0. The price $B(t, T)$ of a zero-coupon bond maturing at the date $T \leq T^*$ can be recovered from the formula (cf. (9.4))

$$B(t, T) = \exp\left(- \int_t^T f(t, u) \, du \right), \quad \forall t \in [0, T], \tag{11.2}$$

provided that the integral on the right-hand side of (11.2) exists (with probability 1). Leaving the technical assumptions aside, the first question that should be addressed is the absence of arbitrage in a financial market model involving all bonds with differing maturities as primary traded securities. As expected, the answer to this question can be formulated in terms of the existence of a suitably defined martingale measure. It appears that in an arbitrage-free setting – that is, under the martingale probability – the drift coefficient α in the dynamics (11.1) of the forward rate is uniquely determined by the volatility coefficient σ, and a stochastic process that can be interpreted as the *risk premium*. More importantly, if σ follows a deterministic function then the valuation results for interest rate-sensitive derivatives appear to be independent of the choice of the risk premium. In this sense, the choice of a particular model from the broad class of Heath-Jarrow-Morton (HJM) models hinges uniquely on the specification of the volatility coefficient σ. For this specific feature of continuous-time forward rate modelling to hold, we need to restrict our attention to the class of HJM models with deterministic coefficient σ; that is, to *Gaussian HJM models*.

In this chapter, the forward measure methodology is employed in arbitrage pricing of interest rate derivative securities in a Gaussian HJM framework. By a Gaussian HJM framework we mean in fact any model of the term structure, either based on the short-term rate or on forward rates, in which all bond price volatilities (as well as the volatility of any other underlying asset) follow deterministic functions. This assumption is made for expositional simplicity; it is not a necessary condition in order to obtain a closed-form solution for the price of a particular option, however. For instance, when a European option on a specific asset is examined in order to obtain an explicit expression for its arbitrage price, it is in fact enough to assume that the volatility of the forward price of the underlying asset for the settlement date coinciding with the option's maturity date is deterministic.

This chapter is organized as follows. In the first two sections, we examine the general HJM set-up and the Gaussian HJM set-up respectively. Sect. 11.3 deals with issues related to the valuation of European options on stocks, zero-coupon bonds and coupon-bearing bonds. As was indicated already, we postulate that the bond price volatilities, as well as the volatility of the option's underlying asset, follow deterministic functions. The next section is devoted to the study of futures prices and to arbitrage valuation of futures options. We focus on a straightforward derivation of partial differential equations associated with the arbitrage price of spot and futures contingent claims in a framework of stochastic interest rates. The fundamental valuation formulas for European options, established previously by means of the forward measure approach, are re-derived by solving the corresponding terminal value problems. It is worth observing that the standard Black-Scholes PDE can be seen as a special case of the PDEs obtained in Sect. 11.6.

Let us note that an efficient valuation of American-style options under uncertainty of interest rates is a rather difficult problem,[1] and relatively little is known on this topic. It appears that the rational exercise policy of an American bond option or an American swaption cannot, in general, be described in terms of the short-term interest rate. For more details on this issue, we refer to Tanudjaja (1996), who examined various approximation techniques associated with the pricing of American-style options in a Gaussian HJM framework.

11.1 Heath-Jarrow-Morton Methodology

Let us first present briefly a discrete-time predecessor of the HJM model, introduced by Ho and Lee (1986). Basically, the idea is to model the uncertain behavior of the yield curve as a whole, as opposed to modelling the short-term rate representing a single point on this curve (specifically, its short side end). In view of its discrete-time feature, the Ho and Lee model of the forward rate dynamics can also be seen as a distant relative of the CRR binomial model. In contrast to the real-valued CRR process, Ho and Lee take the class of all continuous functions on $\mathbb{R}_+$ as the state-space for their model; any such function represents a particular shape of the yield

---

[1] Putting aside some particular cases when the short-term interest rate follows a specific diffusion process (cf. Sect. 10.1.4).

curve. Yield curve deformations over time are modelled by means of a binomial lattice.

11.1.1 Ho and Lee Model

To obtain relatively simple formulas for the values of bond options, Ho and Lee ensure that their model is path-independent. However, path-dependent versions of their model are considered by several authors. The original Ho and Lee approach is based on the assumption that, given an initial term structure $B(0, T)$, the family $B(n, T)$, $n \geq 1$, is defined by means of the recurrence relation

$$B(n, T) = \frac{B(n - 1, T)}{B(n - 1, n)} h(n, \xi_{n-1}, T), \quad \forall T \in [n, \infty), \tag{11.3}$$

where $\{\xi_n\}_{n \geq 0}$ is a sequence of independent Bernoulli random variables with the probability distributions $\mathbb{P}^*\{\xi_n = u\} = p_n = 1 - \mathbb{P}^*\{\xi_n = d\}$ for some $u \neq d$. Furthermore, $h(n, d, T)$ and $h(n, u, T)$ are given perturbation functions that cannot be chosen freely. Actually, in order to make the family $B(n, T)$ arbitrage-free, the following condition must be imposed on the perturbation functions

$$p_n h(n, u, T) + (1 - p_n)h(n, d, T) = 1, \quad \forall T \in [n, \infty),$$

provided that the sequence (p_n) represents the martingale measure for the model. For example, if we take as perturbation functions

$$h(n, d, T) = \frac{\delta^n}{p_* + (1 - p_*)\delta^n} < 1, \quad h(n, u, T) = \frac{1}{p_* + (1 - p_*)\delta^n} > 1,$$

for some constants $0 < \delta$, $p_* < 1$, then the martingale measure is determined by the binomial probabilities $p_n = p_*$ for $n \geq 0$. The major drawback of the Ho and Lee approach is the perfect correlation between the prices of zero-coupon bonds of different maturities. Indeed, this property manifestly contradicts the actual behavior of the yield curve. For a detailed study of the Ho and Lee model, we refer the reader to Turnbull and Milne (1991), Jensen and Nielsen (1991), and El Karoui and Saada (1992). Let us mention only that Ji and Yin (1993) examine the convergence in distribution of the sequence of Ho and Lee models; they provide necessary and sufficient conditions for weak convergence of the appropriate sequence of the Ho and Lee models, and show that the limit corresponds to the following dynamics of the short-term rate r

$$dr_t = b(t) dt + \sigma dW_t^*,$$

where b is some function and σ a constant (see also Dybvig (1988) and Jamshidian (1988) in this regard).

11.1.2 Heath-Jarrow-Morton Model

Let W be a d-dimensional standard Brownian motion given on a filtered probability space $(\Omega, \mathbb{F}, \mathbb{P})$. As usual, the filtration $\mathbb{F} = \mathbb{F}^W$ is assumed to be the right-continuous and $\mathbb{P}$-completed version of the natural filtration of W. We consider a continuous-time trading economy with a trading interval $[0, T^*]$ for the fixed horizon date T^*. We are in a position to formulate the basic postulates of the HJM approach.

(HJM.1) For every fixed $T \leq T^*$, the dynamics of the instantaneous forward rate $f(t, T)$ are given as

$$f(t, T) = f(0, T) + \int_0^t \alpha(u, T) \, du + \int_0^t \sigma(u, T) \cdot dW_u, \quad \forall t \in [0, T], \quad (11.4)$$

for a Borel-measurable function $f(0, \cdot) : [0, T^*] \to \mathbb{R}$, and some applications $\alpha : C \times \Omega \to \mathbb{R}$ and $\sigma : C \times \Omega \to \mathbb{R}^d$, where $C = \{(u, t) \mid 0 \leq u \leq t \leq T^*\}$. It is common to use the differential form of (11.4), namely,

$$df(t, T) = \alpha(t, T) \, dt + \sigma(t, T) \cdot dW_t.$$

(HJM.2) For any maturity T, $\alpha(\cdot, T)$ and $\sigma(\cdot, T)$ follow adapted processes, such that

$$\int_0^T |\alpha(u, T)| \, du + \int_0^T |\sigma(u, T)|^2 \, du < \infty, \quad \mathbb{P}\text{-a.s.}$$

The coefficients α and σ can be introduced as the functions of the forward rate, so that

$$\alpha(u, T) = \alpha(u, T, f(t, T)), \quad \sigma(u, T) = \sigma(u, T, f(t, T)).$$

In that case, the problem of the existence and uniqueness of solutions of SDE (11.4) appears in a natural way. Since in what follows we do not pay much attention to such an approach, we refer to Morton (1989) and Miltersen (1994) for details.

Though it is unnecessary, one may find it useful to introduce also a savings account as an additional primary security. For this purpose, assume that there exists a measurable version of the process $f(t, t)$, $t \in [0, T^*]$. It is then natural to postulate that the short-term interest rate satisfies $r_t = f(t, t)$ for every t. Consequently, the savings account equals (cf. (9.7))

$$B_t = \exp \left(\int_0^t f(u, u) \, du \right), \quad \forall t \in [0, T^*]. \quad (11.5)$$

The following auxiliary lemma deals with the dynamics of the bond price process $B(t, T)$ under the actual probability $\mathbb{P}$. It is interesting to observe that the drift and volatility coefficients in the dynamics of $B(t, T)$ can be expressed in terms of the coefficients α and σ of forward rate dynamics, and the short-term interest rate $f(t, t)$.

Lemma 11.1.1 *The dynamics of the bond price $B(t, T)$ are determined by the expression*

$$dB(t, T) = B(t, T)\big(a(t, T) \, dt + b(t, T) \cdot dW_t\big), \quad (11.6)$$

where a and b are given by the following formulas

$$a(t, T) = f(t, t) - \alpha^*(t, T) + \tfrac{1}{2} |\sigma^*(t, T)|^2, \quad b(t, T) = -\sigma^*(t, T), \quad (11.7)$$

and for any $t \in [0, T]$ *we have*

$$\alpha^*(t, T) = \int_t^T \alpha(t, u)\, du, \quad \sigma^*(t, T) = \int_t^T \sigma(t, u)\, du. \quad (11.8)$$

Proof. Let us denote $I_t = \ln B(t, T)$. According to formulas (11.2) and (11.4), we have that

$$I_t = -\int_t^T f(0, u)\, du - \int_t^T \int_0^t \alpha(v, u)\, dv\, du - \int_t^T \int_0^t \sigma(v, u) \cdot dW_v\, du.$$

Applying Fubini's standard and stochastic theorems (for the latter, see Theorem IV.45 in Protter (2003)), we find that

$$I_t = -\int_t^T f(0, u)\, du - \int_0^t \int_t^T \alpha(v, u)\, du\, dv - \int_0^t \int_t^T \sigma(v, u)\, du \cdot dW_v,$$

or equivalently,

$$I_t = -\int_0^T f(0, u)\, du - \int_0^t \int_v^T \alpha(v, u)\, du\, dv - \int_0^t \int_v^T \sigma(v, u)\, du \cdot dW_v$$
$$+ \int_0^t f(0, u)\, du + \int_0^t \int_v^t \alpha(v, u)\, du\, dv + \int_0^t \int_v^t \sigma(v, u)\, du \cdot dW_v.$$

Consequently,

$$I_t = I_0 + \int_0^t r_u\, du - \int_0^t \int_u^T \alpha(u, v)\, dv\, du - \int_0^t \int_u^T \sigma(u, v)\, dv \cdot dW_u,$$

where we have used the representation

$$r_u = f(u, u) = f(0, u) + \int_0^u \alpha(v, u)\, dv + \int_0^u \sigma(v, u) \cdot dW_v. \quad (11.9)$$

Taking into account equations (11.8), we obtain

$$I_t = I_0 + \int_0^t r_u\, du - \int_0^t \alpha^*(u, T)\, du - \int_0^t \sigma^*(u, T) \cdot dW_u.$$

To check that (11.6) holds, it suffices to apply Itô's formula. $\qquad \square$

11.1.3 Absence of Arbitrage

In the present setting, a continuum of bonds with different maturities is available for trade. We shall assume, however, that any particular portfolio involves investments in an arbitrary, but finite, number of bonds. An alternative approach, in which infinite portfolios are also allowed, can be found in Björk et al. (1997a, 1997b). For any collection of maturities $0 < T_1 < T_2 < \cdots < T_k = T^*$, we write $\mathcal{T}$ to denote the vector $(T_1, \ldots, T_k)$; similarly, $B(\cdot, \mathcal{T})$ stands for the $\mathbb{R}^k$-valued process $(B(t, T_1), \ldots, B(t, T_k))$. We find it convenient to extend the $\mathbb{R}^k$-valued process $B(\cdot, \mathcal{T})$ over the time interval $[0, T^*]$ by setting $B(t, T) = 0$ for any $t \in (T, T^*]$ and any maturity $0 < T < T^*$.

By a *bond trading strategy* we mean a pair $(\phi, \mathcal{T})$, where ϕ is a predictable $\mathbb{R}^k$-valued stochastic process that satisfies $\phi_t^i = 0$ for every $t \in (T_i, T^*]$ and any $i = 1, \ldots, k$. A bond trading strategy $(\phi, \mathcal{T})$ is said to be *self-financing* if the wealth process $V(\phi)$, which equals

$$V_t(\phi) \stackrel{\text{def}}{=} \phi_t \cdot B(t, \mathcal{T}) = \sum_{i=1}^{k} \phi_t^i B(t, T_i),$$

satisfies

$$V_t(\phi) = V_0(\phi) + \int_0^t \phi_u \cdot dB(u, \mathcal{T}) = V_0(\phi) + \sum_{i=1}^{k} \int_0^t \phi_u^i dB(u, T_i)$$

for every $t \in [0, T^*]$. To ensure the arbitrage-free properties of the bond market model, we need to examine the existence of a martingale measure for a suitable choice of a numeraire; in the present set-up, we can take either the bond price $B(t, T^*)$ or the savings account B.

Assume, for simplicity, that the coefficient σ in (11.4) is bounded. We are looking for a condition ensuring the absence of arbitrage opportunities across all bonds of different maturities. Let us introduce an auxiliary process F_B by setting

$$F_B(t, T, T^*) \stackrel{\text{def}}{=} \frac{B(t, T)}{B(t, T^*)}, \quad \forall t \in [0, T].$$

In view of (11.6), the dynamics of the process $F_B(\cdot, T, T^*)$ are given by

$$dF_B(t, T, T^*) = F_B(t, T, T^*)\Big(\tilde{a}(t, T) \, dt + \big(b(t, T) - b(t, T^*)\big) \cdot dW_t\Big),$$

where for every $t \in [0, T]$

$$\tilde{a}(t, T) = a(t, T) - a(t, T^*) - b(t, T^*) \cdot \big(b(t, T) - b(t, T^*)\big).$$

We know that any probability measure equivalent to $\mathbb{P}$ on $(\Omega, \mathcal{F}_{T^*})$ satisfies

$$\frac{d\hat{\mathbb{P}}}{d\mathbb{P}} = \mathcal{E}_{T^*}\Big(\int_0^\cdot h_u \cdot dW_u \Big), \quad \mathbb{P}\text{-a.s.} \tag{11.10}$$

for some predictable $\mathbb{R}^d$-valued process h.

Let us fix a maturity date T. It is easily seen from Girsanov's theorem and the dynamics of $F_B(t, T, T^*)$ that $F_B(t, T, T^*)$ follows a martingale[2] under $\hat{\mathbb{P}}$, provided that for every $t \in [0, T]$

$$a(t, T) - a(t, T^*) = \big(b(t, T^*) - h_t\big) \cdot \big(b(t, T) - b(t, T^*)\big). \tag{11.11}$$

---

[2] The martingale property of $F_B(t, T, T^*)$, as opposed to the local martingale property, follows from the assumed boundedness of σ.

In order to rule out arbitrage opportunities between all bonds with different maturities, it suffices to assume that a martingale measure $\hat{\mathbb{P}}$ can be chosen simultaneously for all maturities. The following condition is thus sufficient for the absence of arbitrage between all bonds. As usual, we restrict our attention to the class of *admissible* trading strategies (cf. Sect. 8.1).

Condition (M.1) There exists an adapted $\mathbb{R}^d$-valued process h such that

$$\mathbb{E}_\mathbb{P}\left\{\mathcal{E}_{T^*}\left(\int_0^\cdot h_u \cdot dW_u\right)\right\} = 1$$

and, for every $T \leq T^*$, equality (11.11) is satisfied, or equivalently,

$$\int_T^{T^*} \alpha(t, u) \, du + \frac{1}{2}\left|\int_T^{T^*} \sigma(t, u) \, du\right|^2 + h_t \cdot \int_T^{T^*} \sigma(t, u) \, du = 0.$$

By taking the partial derivative with respect to T, we obtain

$$\alpha(t, T) + \sigma(t, T) \cdot \left(h_t + \int_T^{T^*} \sigma(t, u) \, du\right) = 0, \tag{11.12}$$

for every $0 \leq t \leq T \leq T^*$. For any process h of condition (M.1), the probability measure $\hat{\mathbb{P}}$ given by (11.10) will later be interpreted as the *forward martingale measure* for the date T^* (see Sect. 9.6). Assume now, in addition, that one may invest also in the savings account given by (11.5). In view of (11.6), the relative bond price $Z^*(t, T) = B(t, T)/B_t$ satisfies under $\mathbb{P}$

$$dZ^*(t, T) = -Z^*(t, T)\left(\left(\alpha^*(t, T) - \tfrac{1}{2}|\sigma^*(t, T)|^2\right) dt + \sigma^*(t, T) \cdot dW_t\right).$$

The following no-arbitrage condition excludes arbitrage not only across all bonds, but also between bonds and the savings account.

Condition (M.2) There exists an adapted $\mathbb{R}^d$-valued process λ such that

$$\mathbb{E}_\mathbb{P}\left\{\mathcal{E}_{T^*}\left(\int_0^\cdot \lambda_u \cdot dW_u\right)\right\} = 1$$

and, for any maturity $T \leq T^*$, we have

$$\alpha^*(t, T) = \tfrac{1}{2}|\sigma^*(t, T)|^2 - \sigma^*(t, T) \cdot \lambda_t.$$

Differentiation of the last equality with respect to T yields the equality

$$\alpha(t, T) = \sigma(t, T) \cdot (\sigma^*(t, T) - \lambda_t), \quad \forall t \in [0, T], \tag{11.13}$$

which holds for any $T \leq T^*$. A probability measure $\mathbb{P}^*$ satisfying

$$\frac{d\mathbb{P}^*}{d\mathbb{P}} = \mathcal{E}_{T^*}\left(\int_0^\cdot \lambda_u \cdot dW_u\right), \quad \mathbb{P}\text{-a.s.}$$

for some process λ satisfying (M.2), can be seen as a *spot martingale measure* for the HJM model; in this context, the process λ is associated with the *risk premium*. Define a $\mathbb{P}^*$-Brownian motion W^* by setting

$$W_t^* = W_t - \int_0^t \lambda_u \, du, \quad \forall t \in [0, T].$$

The next result, whose proof is straightforward, deals with the dynamics of bond prices and interest rates under the spot martingale measure $\mathbb{P}^*$.

Corollary 11.1.1 *For any fixed maturity $T \leq T^*$, the dynamics of the bond price $B(t, T)$ under the spot martingale measure $\mathbb{P}^*$ are*

$$dB(t, T) = B(t, T)\big(r_t \, dt - \sigma^*(t, T) \cdot dW_t^*\big), \tag{11.14}$$

and the forward rate $f(t, T)$ satisfies

$$df(t, T) = \sigma(t, T) \cdot \sigma^*(t, T) \, dt + \sigma(t, T) \cdot dW_t^*. \tag{11.15}$$

Finally, the short-term interest rate $r_t = f(t, t)$ is given by the expression

$$r_t = f(0, t) + \int_0^t \sigma(u, t) \cdot \sigma^*(u, t) \, du + \int_0^t \sigma(u, t) \cdot dW_u^*. \tag{11.16}$$

It follows from (11.16) that the expectation of the future short-term rate under the spot martingale measure $\mathbb{P}^*$ does not equal the current value $f(0, T)$ of the instantaneous forward rate; that is, $f(0, T) \neq \mathbb{E}_{\mathbb{P}^*}(r_T)$, in general. We shall see soon that $f(0, T)$ equals the expectation of r_T under the *forward martingale measure* for the date T (see Corollary 11.3.1). In view of (11.14), the relative bond price $Z^*(t, T) = B(t, T)/B_t$ satisfies

$$dZ^*(t, T) = -Z^*(t, T)\sigma^*(t, T) \cdot dW_t^*, \tag{11.17}$$

and thus

$$Z^*(t, T) = B(0, T) \exp\Big(-\int_0^t \sigma^*(u, T) \cdot dW_u^* - \frac{1}{2}\int_0^t |\sigma^*(u, T)|^2 \, du\Big),$$

or equivalently

$$\ln B(t, T) = \ln B(0, T) + \int_0^t \big(r_u - \tfrac{1}{2}|\sigma^*(u, T)|^2\big) \, du - \int_0^t \sigma^*(u, T) \cdot dW_u^*.$$

Remarks. It is possible to go the other way around; that is, to assume a priori that the dynamics of $B(t, T)$ under the martingale measure $\mathbb{P}^*$ are

$$dB(t, T) = B(t, T)\big(r_t \, dt + b(t, T) \cdot dW_t^*\big),$$

where r is the short-term rate process, and the volatility coefficient b is differentiable with respect to the maturity T. Then clearly

$$\ln B(t, T) = \ln B(0, T) + \int_0^t \left(r_u - \tfrac{1}{2} |b(u, T)|^2 \right) du + \int_0^t b(u, T) \cdot dW_u^*.$$

By differentiating this equality with respect to T, we find the dynamics of the forward rate under $\mathbb{P}^*$, namely

$$f(t, T) = f(0, T) + \frac{1}{2} \int_0^t \frac{\partial k(u, T)}{\partial T} du + \int_0^t \frac{\partial b(u, T)}{\partial T} \cdot dW_u^*, \qquad (11.18)$$

where we write $k(t, T) = |b(t, T)|^2$. Note that we have interchanged the differentiation with the stochastic integration. This is essentially equivalent to an application of the stochastic Fubini theorem.

It is not hard to check that, under mild technical assumptions, the no-arbitrage conditions (M.1) and (M.2) are equivalent.

Lemma 11.1.2 *Conditions* (M.1) *and* (M.2) *are equivalent.*

Proof. Let us sketch the proof. Suppose first that condition (M.1) holds. Let us define λ by setting

$$\lambda_t \stackrel{\text{def}}{=} h_t - b(t, T^*) = h_t + \int_t^{T^*} \sigma(t, u) \, du, \quad \forall t \in [0, T^*],$$

and check that for such a process λ condition (11.13) is satisfied. Indeed, for the right-hand side of (11.13), we obtain

$$\sigma(t, T) \cdot (\sigma^*(t, T) - \lambda_t) = -\sigma(t, T) \cdot \left(h_t + \int_T^{T^*} \sigma(t, u) \, du \right) = \alpha(t, T),$$

where the last equality follows from (11.12). Conversely, given a process λ that satisfies (11.13), we define the process h by setting

$$h_t \stackrel{\text{def}}{=} \lambda_t + b(t, T^*), \quad \forall t \in [0, T^*].$$

Then h satisfies (11.12), since

$$\sigma(t, T) \cdot \left(h_t + \int_T^{T^*} \sigma(t, u) \, du \right) = \sigma(t, T) \cdot (\lambda_t - \sigma^*(t, T)) = -\alpha(t, T),$$

where the last equality follows from (11.13). $\qquad \square$

We assume from now on that the following assumption is satisfied.

(HJM.3) No-arbitrage condition (M.1) (or equivalently, (M.2)) is satisfied.

It is not essential to assume that the martingale measure for the bond market is unique, so long as we are not concerned with the completeness of the model. Recall that if a market model is arbitrage-free, any attainable claim admits a unique arbitrage price anyway (it is uniquely determined by the replicating strategy), whether a market model is complete or not. For instance, in a Gaussian HJM setting, several claims – such as options on a bond or a stock – are attainable, with replicating strategies given by closed-form expressions. One may argue that from the practical viewpoint, the possibility of explicit replication of most typical claims is a more important feature of a model than its theoretical completeness.

Remarks. As was mentioned before, it seems plausible that σ might depend explicitly on the level of the forward rate. This corresponds to the following forward interest rate dynamics

$$df(t, T) = \alpha\big(t, T, f(t, \cdot)\big)\, dt + \sigma\big(t, T, f(t, T)\big) \cdot dW_t, \tag{11.19}$$

where the deterministic function $\sigma : C \times \mathbb{R} \to \mathbb{R}$ is sufficiently regular, so that SDE (11.19) (with the drift coefficient α given by expression (11.20)) admits a unique strong solution for any fixed maturity T. Recall that the drift coefficient α satisfies, under the martingale measure $\mathbb{P}^*$,

$$\alpha\big(t, T, f(t, \cdot)\big) = \sigma\big(t, T, f(t, T)\big) \cdot \int_t^T \sigma\big(t, u, f(t, u)\big)\, du. \tag{11.20}$$

Consequently, under the martingale measure $\mathbb{P}^*$, SDE (11.19) may be rewritten in the following way

$$df(t, T) = \sigma\big(t, T, f(t, T)\big) \cdot \left(\int_t^T \sigma\big(t, u, f(t, u)\big)\, du\, dt + dW_t^* \right). \tag{11.21}$$

Let us assume, for instance, that the coefficient σ satisfies

$$\sigma\big(t, T, f(t, T)\big) = \gamma f(t, T), \quad \forall t \in [0, T],$$

where γ is a fixed vector in $\mathbb{R}^d$. Under such an assumption, (11.21) becomes

$$f(t, T) = f(0, T) + \int_0^t |\gamma|^2 f(u, T) \int_u^T f(u, v)\, dv\, du + \int_0^t f(u, T)\gamma \cdot dW_u^*$$

for any maturity date $T \in [0, T^*]$. Since the drift coefficient grows rapidly as a function of the forward rate, the last stochastic differential equation does not admit a global solution; indeed, it can be shown that its local solution explodes in a finite time. For more details on this important issue, we refer to Morton (1989) and Miltersen (1994).

11.1.4 Short-term Interest Rate

It is interesting to note that under some regularity assumptions, the short-term interest rate process specified by the HJM model of instantaneous forward rates follows a continuous semimartingale, or even a diffusion process.[3] the following proposition can be seen as the first step in this direction. Further results and typical examples are given in Sect. 11.2.1.

Proposition 11.1.1 *Suppose that the coefficients $\alpha(t, T)$ and $\sigma(t, T)$ and the initial forward curve $f(0, T)$ are differentiable with respect to maturity T, with bounded partial derivatives $\alpha_T(t, T)$, $\sigma_T(t, T)$ and $f_T(0, T)$. Then the short-term interest rate r follows a continuous semimartingale under $\mathbb{P}$. More specifically, for any $t \in [0, T^*]$ we have*

$$r_t = r_0 + \int_0^t \zeta_u \, du + \int_0^t \sigma(u, u) \cdot dW_u, \tag{11.22}$$

where ζ stands for the following process

$$\zeta_t = \alpha(t, t) + f_T(0, t) + \int_0^t \alpha_T(u, t) \, du + \int_0^t \sigma_T(u, t) \cdot dW_u.$$

Proof. Observe first that r satisfies

$$r_t = f(t, t) = f(0, t) + \int_0^t \alpha(u, t) \, du + \int_0^t \sigma(u, t) \cdot dW_u.$$

Applying the stochastic Fubini theorem to the Itô integral, we obtain

$$
\begin{aligned}
\int_0^t \sigma(u, t) \cdot dW_u &= \int_0^t \sigma(u, u) \cdot dW_u + \int_0^t \big(\sigma(u, t) - \sigma(u, u)\big) \cdot dW_u \\
&= \int_0^t \sigma(u, u) \cdot dW_u + \int_0^t \int_u^t \sigma_T(u, s) \, ds \cdot dW_u \\
&= \int_0^t \sigma(u, u) \cdot dW_u + \int_0^t \int_0^s \sigma_T(u, s) \cdot dW_u \, ds.
\end{aligned}
$$

Furthermore,

$$\int_0^t \alpha(u, t) \, du = \int_0^t \alpha(u, u) \, du + \int_0^t \int_0^u \alpha_T(s, u) \, ds \, du,$$

and finally

$$f(0, t) = r_0 + \int_0^t f_T(0, u) \, du.$$

Combining these formulas, we obtain (11.22).   □

---

[3] If the regularity of α and σ with respect to maturity T is not assumed, the behavior of the short-term rate is rather difficult to study.

11.2 Gaussian HJM Model

In this section, we assume that the volatility σ of the forward rate is deterministic; such a case will be referred to as the Gaussian HJM model. This terminology refers to the fact that the forward rate $f(t, T)$ and the spot rate r_t have Gaussian probability distributions under the martingale measure $\mathbb{P}^*$ (cf. (11.15)–(11.16)). Our aim is to show that the arbitrage price of any attainable interest rate-sensitive claim can be evaluated by each of the following procedures.

(I) We start with arbitrary dynamics of the forward rate such that condition (M.1) (or (M.2)) is satisfied. We then find a martingale measure $\mathbb{P}^*$, and apply the risk-neutral valuation formula.

(II) We assume instead that the underlying probability measure $\mathbb{P}$ is actually the spot (respectively, forward) martingale measure. In other words, we assume that condition (M.2) (respectively, condition (M.1)) is satisfied, with the process λ (respectively, h) equal to zero.

Since both procedures give the same valuation results, we conclude that the specification of the risk premium is not relevant in the context of arbitrage valuation of interest rate-sensitive derivatives in the Gaussian HJM framework. Put differently, when the coefficient σ is deterministic, we can assume, without loss of generality, that $\alpha(t, T) = \sigma(t, T)\sigma^*(t, T)$. Observe that by combining the last equality with (11.7), we find immediately that $a(t, T) = f(t, t) = r_t$. To state a result that formally justifies the considerations above, we need to introduce some additional notation. Let a function α_0 be given by (11.13), with $\lambda = 0$, i.e.,

$$\alpha_0(t, T) = \sigma(t, T) \cdot \sigma^*(t, T), \quad \forall t \in [0, T], \tag{11.23}$$

so that

$$\alpha_0^*(t, T) = \int_t^T \alpha_0(u, T) \, du = \frac{1}{2} \, | \sigma^*(t, T) |^2.$$

Finally, we denote by $B_0(t, T)$ the bond price specified by the equality

$$B_0(t, T) = \exp\left(- \int_t^T f_0(t, u) \, du \right),$$

where the dynamics under $\mathbb{P}$ of the instantaneous forward rate $f_0(t, T)$ are

$$f_0(t, T) = f(0, T) + \int_0^t \alpha_0(u, T) \, du + \int_0^t \sigma(u, T) \cdot dW_u.$$

Let us put

$$Z^*(t, T) = B(t, T)/B_t, \quad Z_0^*(t, T) = B_0(t, T)/B_t,$$

where $\mathcal{T} = (T_1, \ldots, T_k)$ is any finite collection of maturity dates.

Proposition 11.2.1 *Suppose that the coefficient σ is deterministic. Then for any choice $\mathcal{T}$ of maturity dates and of a spot martingale measure $\mathbb{P}^*$, the probability* *distribution of the process $Z^*(t, \mathcal{T})$, $t \in [0, T^*]$, under the martingale measure $\mathbb{P}^*$ coincides with the probability distribution of the process $Z_0^*(t, \mathcal{T})$, $t \in [0, T^*]$, under $\mathbb{P}$.*

Proof. The assertion follows easily by Girsanov's theorem. Indeed, for any fixed $0 < T \leq T^*$ the dynamics of $Z^*(t, T)$ under a spot martingale measure $\mathbb{P}^* = \mathbb{P}^\lambda$ are

$$dZ^*(t, T) = -Z^*(t, T)\, \sigma^*(t, T) \cdot dW_t^\lambda, \tag{11.24}$$

where W^λ follows a standard Brownian motion under $\mathbb{P}^\lambda$. On the other hand, under $\mathbb{P}$ we have

$$dZ_0^*(t, T) = -Z_0^*(t, T)\, \sigma^*(t, T) \cdot dW_t. \tag{11.25}$$

Moreover, for every $0 < T \leq T^*$

$$Z^*(0, T) = B(0, T) = e^{-\int_0^T f(0, u)\, du} = B_0(0, T) = Z_0^*(0, T).$$

Since σ is deterministic, the assertion follows easily from (11.24)–(11.25).   □

To show the independence of the arbitrage pricing of the market prices for risk in a slightly more general setting, it is convenient to make use of the savings account. Under any martingale measure $\mathbb{P}^*$ we have

$$df(t, T) = \sigma\big(t, T, f(t, T)\big) \cdot \left(\int_t^T \sigma\big(t, u, f(t, u)\big)\, du\, dt + dW_t^* \right). \tag{11.26}$$

We assume that the coefficient σ satisfies regularity assumptions, so that the SDE (11.26) admits a unique strong global solution. Since

$$B(t, T) = \exp\left(-\int_t^T f(t, u)\, du \right), \quad \forall\, t \in [0, T],$$

and

$$B_t = \exp\left(\int_0^t f(u, u)\, du \right), \quad \forall\, t \in [0, T^*],$$

it is clear that the joint probability distribution of processes $B(\cdot, T)$ and B is uniquely determined under $\mathbb{P}^*$, and thus the arbitrage price $\pi_t(X)$ of any attainable European claim X depending on short-term rate and bond prices (or on the forward rates), which equals

$$\pi_t(X) = B_t\, \mathbb{E}_{\mathbb{P}^*}(X B_T^{-1} \mid \mathcal{F}_t),$$

is obviously independent of the market prices for risk. This does not mean that the market prices for risk are neglected altogether in the HJM approach. They are still present in the specification of the actual bond price fluctuations.

However, in contrast to traditional models of the short-term rate, in which the dynamics of the short-term rate and bond processes are not jointly specified, the HJM methodology assumes a simultaneous influence of market prices for risk on the dynamics of the short-term rate and all bond prices. Consequently in this case, the market prices for risk drop out altogether from arbitrage values of interest rate-sensitive derivatives.

11.2.1 Markovian Case

The HJM approach typically produces models in which the short-term rate has path-dependent features. A discrete-time approximation within the HJM framework is therefore usually less efficient than in the path-independent case, since the number of operations rises exponentially with the number of steps. However, in some circumstances (see Jamshidian (1991a), Carverhill (1994), Jeffrey (1995), Ritchken and Sankarasubramanian (1995), and Hagan and Woodward (1999b)) the implied short-term interest rate r has the Markov property. Assume that the bond price volatility $b(t, T)$ is a deterministic function that is twice continuously differentiable with respect to maturity date T (equivalently, that $\sigma(t, T)$ is continuously differentiable with respect to T). Combining (11.9) with (11.13), we find that r satisfies

$$r_t = f(0, t) + \int_0^t \sigma(u, t) \cdot \sigma^*(u, t) \, du + \int_0^t \sigma(u, t) \cdot dW_u^*. \tag{11.27}$$

To show that the short-term rate r has the Markov property with respect to the filtration $\mathbb{F}^{W^*}$ under $\mathbb{P}^*$, it is enough to show that for any bounded Borel-measurable function $h : \mathbb{R} \to \mathbb{R}$, and any $t \leq s \leq T^*$, we have

$$\mathbb{E}_{\mathbb{P}^*}\big(h(r_s) \mid \mathcal{F}_t^{W^*}\big) = \mathbb{E}_{\mathbb{P}^*}\big(h(r_s) \mid r_t\big),$$

where on the right-hand side we have the conditional expectation with respect to the σ-field $\sigma(r_t)$; that is, the σ-field generated by the random variable r_t. For the sake of brevity, we say that r is *Markovian* if it has the Markov property with respect to the filtration $\mathbb{F}^{W^*}$ under $\mathbb{P}^*$.

Proposition 11.2.2 *Suppose that the short-term rate r is Markovian. Assume, in addition, that for any maturity $T \leq T^*$ we have $\sigma(t, T) \neq 0$ for every $t \in [0, T]$. Then there exist functions $g : [0, T^*] \to \mathbb{R}$ and $h : [0, T^*] \to \mathbb{R}^d$ such that*

$$\sigma(t, T) = h(t)g(T), \quad \forall \, t \leq T \leq T^*. \tag{11.28}$$

Moreover, we have

$$b^i(t, T) = \frac{\sigma^i(t, T^*)}{\sigma^i(s, T^*)} \cdot (b^i(s, T) - b^i(s, t)), \quad \forall \, t \leq s \leq T,$$

for every $T \in [0, T^*]$ and every $i = 1, \ldots, d$. If the volatility $\sigma(t, T)$ is also time-homogeneous, that is, $\sigma(t + s, T + s) = \sigma(t, T)$ for all $s \geq 0$, then there exist constants α^i and γ such that for every $i = 1, \ldots, d$ we have

$$b^i(t, T) = \alpha^i \, (1 - e^{-\gamma(T-t)}), \quad \forall \, t \leq T \leq T^*. \tag{11.29}$$

Proof. In view of (11.27), it is clear that the short-term rate r is Markovian if and only if the process $D_t = \int_0^t \sigma(u, t) \cdot dW_u^*$ is Markovian – that is, if

$$\mathbb{E}_{\mathbb{P}^*}(h(D_s) \mid \mathcal{F}_t^D) = \mathbb{E}_{\mathbb{P}^*}(h(D_s) \mid D_t), \quad \forall \, t \leq s \leq T^*, \tag{11.30}$$

for any bounded Borel-measurable function $h : \mathbb{R} \to \mathbb{R}$. Observe that

$$D_s = D_t + \int_0^s \sigma(u, s) \cdot dW_u^* - \int_0^t \sigma(u, t) \cdot dW_u^*,$$

and thus also

$$D_s = D_t + \int_t^s \sigma(u, s) \cdot dW_u^* + \int_0^t \left(\sigma(u, s) - \sigma(u, t)\right) \cdot dW_u^*.$$

In view of the last formula, (11.30) is valid if and only if, given the random variable D_t, the integral

$$I(t, s) = \int_0^t \left(\sigma(u, s) - \sigma(u, t)\right) \cdot dW_u^* = \int_0^t \sigma(u, s) \cdot dW_u^* - D_t$$

depends only on the increments of the Brownian motion W^* on $[t, s]$. But $I(t, s)$ is, for obvious reasons, independent of these increments, therefore we conclude that the integral $J(t, s) = \int_0^t \sigma(u, s) \cdot dW_u^*$ is completely determined when D_t is given. Since the joint distribution of $(J(t, s), D_t)$ is Gaussian, this means that the correlation coefficient satisfies $|\mathrm{Corr}\,(J(t, s), D_t)| = 1$, or more explicitly

$$\left(\mathbb{E}_{\mathbb{P}^*}\left(J(t, s)D_t\right)\right)^2 = \mathbb{E}_{\mathbb{P}^*}\left(J^2(t, s)\right) \mathbb{E}_{\mathbb{P}^*}(D_t^2).$$

Using the standard properties of the Itô integral, we deduce from this that

$$\left(\int_0^t \sigma(u, t) \cdot \sigma(u, s)\, du\right)^2 = \int_0^t |\sigma(u, t)|^2\, du \int_0^t |\sigma(u, s)|^2\, du.$$

Hence the $\mathbb{R}^d$-valued functions $\sigma(\cdot, t)$ and $\sigma(\cdot, s)$ are collinear for any fixed $0 < t \le s \le T^*$. This implies that we have, for arbitrary $T \in [0, T^*]$,

$$\sigma(t, T) = g(T)\sigma(t, T^*), \quad \forall t \in [0, T],$$

for some function $g : [0, T^*] \to \mathbb{R}$. Upon setting $h(t) = \sigma(t, T^*)$, $t \in [0, T^*]$, we obtain (11.28). For the second statement, it is sufficient to note that, for any $T \le T^*$ and $t \in [0, T]$,

$$b(t, T) = -\int_t^T \sigma(t, u)\, du = -h(t) \int_t^T g(u)\, du.$$

If, in addition, the coefficient $\sigma^i(t, T)$ is time-homogeneous then

$$z^i(T - t) \stackrel{\text{def}}{=} \ln \sigma^i(t, T) = \ln g(T) + \ln h^i(t) = \tilde{g}(T) + \tilde{h}^i(t).$$

By differentiating separately with respect to t and T, we find that

$$(\tilde{h}^i)'(t) = -(z^i)'(T - t), \quad \tilde{g}'(T) = (z^i)'(T - t),$$

and thus

$$(z^i)'(T - t) = -(\tilde{h}^i)'(t) = \tilde{g}'(T) = \text{const.} \tag{11.31}$$

It is now easy to check that (11.31) implies (11.29).   □

Example 11.2.1 Let us assume that the volatility of each forward rate is constant, i.e., independent of the maturity date and the level of the forward interest rate. Taking $d = 1$, we thus have $\sigma(t, T) = \sigma$ for a strictly positive constant $\sigma > 0$. By virtue of (11.15), the dynamics of the forward rate process $f(t, T)$ under the martingale measure are given by the expression

$$df(t, T) = \sigma^2(T - t)\,dt + \sigma\,dW_t^*, \tag{11.32}$$

so that the dynamics of the bond price $B(t, T)$ are

$$dB(t, T) = B(t, T)\big(r_t\,dt - \sigma(T - t)\,dW_t^*\big),$$

where the short-term rate of interest r satisfies

$$r_t = f(0, t) + \tfrac{1}{2}\sigma^2 t^2 + \sigma\,W_t^*.$$

It follows from the last formula that

$$dr_t = \big(f_T(0, t) + \sigma^2 t\big)\,dt + \sigma\,dW_t^*.$$

Since this agrees with the general form of the continuous-time Ho and Lee model, we conclude that in the HJM framework, the Ho and Lee model corresponds to the constant volatility of forward rates. Dynamics (11.32) make apparent that the only possible movements of the yield curve in the Ho and Lee model are parallel shifts; that is, all rates along the yield curve fluctuate in the same way. For the price $B(t, T)$ of a bond maturing at T, we have that

$$B(t, T) = \frac{B(0, T)}{B(0, t)}\exp\Big(-\tfrac{1}{2}(T - t)Tt\sigma^2 - (T - t)\sigma W_t^*\Big), \tag{11.33}$$

and it follows from (11.33) that yields on bonds of differing maturities are perfectly correlated. It can also be expressed in terms of r, namely

$$B(t, T) = \frac{B(0, T)}{B(0, t)}\exp\Big((T - t)f(0, t) - \tfrac{1}{2}t(T - t)^2\sigma^2 - (T - t)r_t\Big).$$

Example 11.2.2 It is a conventional wisdom that forward rates of longer maturity fluctuate less than rates of shorter maturity. To account for this feature in the HJM framework, we assume now that the volatility of a forward rate is a decreasing function of the time to its effective date. For instance, we may assume that the volatility structure is exponentially dampened, specifically, $\sigma(t, T) = \sigma e^{-b(T-t)}$ for strictly positive real numbers $\sigma, b > 0$.

Then $\sigma^*(t, T)$ equals

$$\sigma^*(t, T) = \int_t^T \sigma e^{-b(u-t)}\,du = -\sigma b^{-1}\big(e^{-b(T-t)} - 1\big), \tag{11.34}$$

and consequently

$$df(t, T) = -\sigma^2 b^{-1} e^{-b(T-t)}\big(e^{-b(T-t)} - 1\big)\,dt + \sigma e^{-b(T-t)}\,dW_t^*. \tag{11.35}$$

It is thus clear that for any maturity T the bond price $B(t, T)$ satisfies

$$dB(t, T) = B(t, T)\Big(r_t dt + \sigma b^{-1}\big(e^{-b(T-t)} - 1\big) dW_t^*\Big). \tag{11.36}$$

Substituting (11.34) into (11.16), we obtain

$$r_t = f(0, t) - \int_0^t \sigma^2 b^{-1} e^{-b(t-u)}\big(e^{-b(t-u)} - 1\big) du + \int_0^t \sigma e^{-b(t-u)} dW_u^*,$$

so that, as in the previous example, the negative values of the short-term interest rate are not excluded. Denoting

$$m(t) = f(0, t) + \frac{\sigma^2}{2b^2}\big(e^{-2bt} - 1\big) - \frac{\sigma^2}{b^2}\big(e^{-bt} - 1\big),$$

we arrive at the following formula

$$r_t = m(t) + \int_0^t \sigma e^{-b(t-u)} dW_u^*,$$

so that finally

$$dr_t = (a(t) - br_t)dt + \sigma dW_t^*, \tag{11.37}$$

where $a(t) = bm(t) + m'(t)$. This means that (11.34) leads to a generalized version of Vasicek's model (cf. (10.5)). Note that in the present framework, the perfect fit of the initial term structure is a trivial consequence.

Example 11.2.3 By combining the models examined in preceding examples, we arrive at the Ho and Lee/Vasicek two-factor version of the Gaussian HJM model, originally put forward by Heath et al. (1992a). We assume that the underlying Brownian motion is two-dimensional, and the volatility coefficient σ has two deterministic components σ_1 and σ_2 that correspond to Ho and Lee and Vasicek's models respectively. More explicitly, for any maturity $T \leq T^*$ we put

$$\sigma(t, T) = (\sigma_1, \sigma_2 e^{-b(T-t)}), \quad \forall t \in [0, T],$$

where σ_1, σ_2 and γ are strictly positive constants. Similarly, if σ equals

$$\sigma(t, T) = (\sigma_1 e^{-b_1(T-t)}, \sigma_2 e^{-b_2(T-t)}), \quad \forall t \in [0, T],$$

for some strictly positive constants σ_1, σ_2, b_1 and b_2, the model is referred to as the two-factor Vasicek model. Closed-form expressions for the prices of bonds and bond options are also easily available in both cases. Since general valuation formulas for the Gaussian HJM set-up (see next section) are directly applicable to the present models, we do not go into details here.

11.3 European Spot Options

The first step towards explicit valuation of European options is to observe that Proposition 9.6.2 provides a simple formula expressing the price of a European call option written on a tradable asset, Z say, in terms of the forward price process $F_Z(t, T)$ and the forward probability measure $\mathbb{P}_T$. Indeed, we have for every $t \in [0, T]$

$$\pi_t\big((Z_T - K)^+\big) = B(t, T)\, \mathbb{E}_{\mathbb{P}_T}\big((F_Z(T, T) - K)^+ \mid \mathcal{F}_t\big), \tag{11.38}$$

since manifestly $Z_T = F_Z(T, T)$. To evaluate the conditional expectation on the right-hand side of (11.38), we need first to find the dynamics, under the forward probability measure $\mathbb{P}_T$, of the forward price $F_Z(t, T)$. The following auxiliary result is an easy consequence of (9.29)–(9.30) and (9.32).

Lemma 11.3.1 *For any fixed $T > 0$, the process W^T given by the formula*

$$W_t^T = W_t^* - \int_0^t b(u, T)\, du, \quad \forall\, t \in [0, T], \tag{11.39}$$

is a standard d-dimensional Brownian motion under the forward measure $\mathbb{P}_T$. The forward price process for the settlement date T of a zero-coupon bond that matures at time U satisfies

$$dF_B(t, U, T) = F_B(t, U, T)\big(b(t, U) - b(t, T)\big) \cdot dW_t^T, \tag{11.40}$$

subject to the terminal condition $F_B(T, U, T) = B(T, U)$. The forward price of a stock S satisfies $F_S(T, T) = S_T$, and

$$dF_S(t, T) = F_S(t, T)\big(\sigma_t - b(t, T)\big) \cdot dW_t^T, \tag{11.41}$$

where the $\mathbb{R}^d$-valued process σ is the instantaneous volatility of the stock price.

The next result, which uses the HJM framework, shows that the yield-to-maturity expectations hypothesis (cf. Sect. 9.5.1) is satisfied for any fixed maturity T under the corresponding forward probability measure $\mathbb{P}_T$. This feature is merely a distant reminder of the classical hypothesis that for every maturity T, the instantaneous forward rate $f(0, T)$ is an unbiased estimate, under the actual probability $\mathbb{P}$, of the future short-term rate r_T.

Corollary 11.3.1 *For any fixed $T \in [0, T^*]$, the instantaneous forward rate $f(0, T)$ is equal to the expected value of the spot rate r_T under the forward probability measure $\mathbb{P}_T$.*

Proof. Observe that in view of (11.16), we have

$$r_T = f(0, T) + \int_0^T \sigma(t, T) \cdot \big(\sigma^*(t, T)\, dt + dW_t^*\big) = f(0, T) + \int_0^T \sigma(t, T) \cdot dW_t^T,$$

since $\sigma^*(t, T) = -b(t, T)$. Consequently, $\mathbb{E}_{\mathbb{P}_T}(r_T) = f(0, T)$, as expected. $\square$

11.3.1 Bond Options

For the reader's convenience, we shall examine separately options written on zero-coupon bonds and on stocks (a general valuation formula is given in Proposition 11.3.3). At expiry date T, the payoff of a European call option written on a zero-coupon bond maturing at time $U \geq T$ equals

$$C_T = \big(B(T, U) - K\big)^+. \tag{11.42}$$

Since $B(T, U) = F_B(T, U, T)$, the payoff C_T can alternatively be re-expressed in the following way

$$C_T = (F_B(T, U, T) - K)^+ = F_B(T, U, T)\mathbb{1}_D - K\mathbb{1}_D,$$

where

$$D = \{B(T, U) > K\} = \{F_B(T, U, T) > K\}$$

is the exercise set of the option.

The next proposition provides a closed-from expression for the arbitrage price of a European bond option. Valuation results of this form were derived previously by several authors, including El Karoui and Rochet (1989), Amin and Jarrow (1992), Brace and Musiela (1994), and Madsen (1994a, 1994b), who all worked within the Gaussian HJM framework. It should be pointed out, however, that the bond option valuation formula for Vasicek's model (see formula (10.16)), established by Jamshidian (1989a), can also be seen as a special case of equality (11.43).

For the sake of expositional simplicity, we assume that the volatilities are bounded; however, this assumption can be weakened.

Proposition 11.3.1 *Assume that the bond price volatilities $b(\cdot, T)$ and $b(\cdot, U)$ are $\mathbb{R}^d$-valued, bounded, deterministic functions. The arbitrage price at time $t \in [0, T]$ of a European call option with expiry date T and exercise price K, written on a zero-coupon bond maturing at time $U \geq T$, equals*

$$C_t = B(t, U)N\big(h_1(B(t, U), t, T)\big) - KB(t, T)N\big(h_2(B(t, U), t, T)\big), \tag{11.43}$$

where, for every $(b, t) \in \mathbb{R}_+ \times [0, T]$,

$$h_{1,2}(b, t, T) = \frac{\ln(b/K) - \ln B(t, T) \pm \frac{1}{2} v_U^2(t, T)}{v_U(t, T)} \tag{11.44}$$

and

$$v_U^2(t, T) = \int_t^T |b(u, U) - b(u, T)|^2 \, du, \quad \forall t \in [0, T]. \tag{11.45}$$

The arbitrage price of the corresponding European put option written on a zero-coupon bond equals

$$P_t = KB(t, T)N\big(-h_2(B(t, U), t, T)\big) - B(t, U)N\big(-h_1(B(t, U), t, T)\big).$$

Proof. In view of the general valuation formula (11.38), it is clear that we have to evaluate the conditional expectations

$$C_t = B(t, T)\, \mathbb{E}_{\mathbb{P}_T}\big(F_B(T, U, T)\, \mathbb{1}_D \,|\, \mathcal{F}_t\big) - K B(t, T)\, \mathbb{P}_T(D \,|\, \mathcal{F}_t) = I_1 - I_2.$$

We know from Lemma 11.3.1 that the dynamics of $F_B(t, U, T)$ under $\mathbb{P}_T$ are given by formula (11.40), so that

$$F_B(T, U, T) = F_B(t, U, T)\exp\left(\int_t^T \gamma(u, U, T)\cdot dW_u^T - \frac{1}{2}\int_t^T |\gamma(u, U, T)|^2\, du \right)$$

with $\gamma(u, U, T) = b(u, U) - b(u, T)$. This can be rewritten as follows

$$F_B(T, U, T) = F_B(t, U, T)\exp\big(\zeta(t, T) - \tfrac{1}{2}v_U^2(t, T) \big),$$

where $F_B(t, U, T)$ is $\mathcal{F}_t$-measurable and $\zeta(t, T) = \int_t^T \gamma(u, U, T)\cdot dW_u^T$ is under $\mathbb{P}_T$ a real-valued Gaussian random variable, independent of the σ-field $\mathcal{F}_t$, with expected value 0 and variance $\mathrm{Var}_{\mathbb{P}_T}(\zeta(t, T)) = v_U^2(t, T)$. Using the properties of conditional expectation, we obtain

$$\mathbb{P}_T\{D \,|\, \mathcal{F}_t\} = \mathbb{P}_T\Big\{ \zeta(t, T) < \ln\big(F/K\big) - \tfrac{1}{2}v_U^2(t, T) \Big\},$$

where $F = F_B(t, U, T)$, so that

$$I_2 = K B(t, T)\, N\left(\frac{\ln\big(F_B(t, U, T)/K\big) - \tfrac{1}{2}v_U^2(t, T)}{v_U(t, T)} \right).$$

To evaluate I_1, we introduce an auxiliary probability measure $\tilde{\mathbb{P}}_T$ equivalent to $\mathbb{P}_T$ on $(\Omega, \mathcal{F}_T)$ by setting

$$\frac{d\tilde{\mathbb{P}}_T}{d\mathbb{P}_T} = \exp\left(\int_0^T \gamma(u, U, T) \cdot dW_u^T - \frac{1}{2}\int_0^T |\gamma(u, U, T)|^2\, du \right) \stackrel{\text{def}}{=} \tilde{\eta}_T.$$

By Girsanov's theorem, it is clear that the process $\tilde{W}^T$, which equals

$$\tilde{W}_t^T = W_t^T - \int_0^t \gamma(u, U, T)\, du, \quad \forall\, t \in [0, T],$$

is a standard Brownian motion under $\tilde{\mathbb{P}}_T$. Note also that the forward price $F_B(T, U, T)$ admits the following representation under $\tilde{\mathbb{P}}_T$

$$F_B(T, U, T) = F_B(t, U, T)\exp\left(\int_t^T \gamma(u, U, T)\cdot d\tilde{W}_u^T + \frac{1}{2}\int_t^T |\gamma(u, U, T)|^2\, du \right)$$

so that

$$F_B(T, U, T) = F_B(t, U, T)\, \exp\big(\tilde{\zeta}(t, T) + \tfrac{1}{2}v_U^2(t, T) \big), \tag{11.46}$$

where we denote $\tilde{\zeta}(t, T) = \int_t^T \gamma(u, U, T) \cdot d\tilde{W}_u^T$. The random variable $\tilde{\zeta}(t, T)$ has under $\tilde{\mathbb{P}}_T$ a Gaussian distribution with expected value 0 and variance $v_U^2(t, T)$; it is also independent of the σ-field $\mathcal{F}_t$. Furthermore, once again using Lemma 11.3.1, we obtain

$$I_1 = B(t, U)\mathbb{E}_{\mathbb{P}_T}\left\{\mathbb{1}_D \exp\left(\int_t^T \gamma(u, U, T) \cdot dW_u^T - \frac{1}{2}\int_t^T |\gamma(u, U, T)|^2 \, du\right) \middle| \mathcal{F}_t\right\}$$

and thus

$$I_1 = B(t, T)\,\mathbb{E}_{\mathbb{P}_T}\left(\tilde{\eta}_T\, \tilde{\eta}_t^{-1}\, \mathbb{1}_D \mid \mathcal{F}_t\right).$$

By virtue of Bayes's rule (see (9.35)), we find that $I_1 = B(t, U)\,\tilde{\mathbb{P}}_T\{D \mid \mathcal{F}_t\}$. Taking into account (11.46), we conclude that

$$\tilde{\mathbb{P}}_T(D \mid \mathcal{F}_t) = \tilde{\mathbb{P}}_T\left(\tilde{\zeta}(t, T) \le \ln\left(F_B(t, U, T)/K\right) + \tfrac{1}{2}v_U^2(t, T)\right),$$

and thus

$$I_1 = B(t, U)\,N\left(\frac{\ln\left(F_B(t, U, T)/K\right) + \tfrac{1}{2}v_U^2(t, T)}{v_U(t, T)}\right).$$

This completes the proof of the valuation formula (11.43). The formula that gives the price of the put option can be established along the same lines. Alternatively, to find the price of a European put option written on a zero-coupon bond, one may combine formula (11.43) with the put-call parity relationship (11.57). □

Formula (11.43) can be re-expressed as follows

$$C_t = B(t, T)\left(F_t N(\tilde{d}_1(F_t, t, T)) - K N(\tilde{d}_2(F_t, t, T))\right), \tag{11.47}$$

where we write briefly F_t to denote the forward price $F_B(t, U, T)$, and

$$\tilde{d}_{1,2}(F, t, T) = \frac{\ln(F/K) \pm \tfrac{1}{2}v_U^2(t, T)}{v_U(t, T)} \tag{11.48}$$

for $(F, t) \in \mathbb{R}_+ \times [0, T]$, where $v_U(t, T)$ is given by (11.45). Note also that we have

$$\frac{d\tilde{\mathbb{P}}_T}{d\mathbb{P}^*} = \frac{d\tilde{\mathbb{P}}_T}{d\mathbb{P}_T}\frac{d\mathbb{P}_T}{d\mathbb{P}^*} = \exp\left(\int_0^T b(u, U) \cdot dW_u - \frac{1}{2}\int_0^T |b(u, U)|^2 \, du\right).$$

It is thus apparent that the auxiliary probability measure $\tilde{\mathbb{P}}_T$ is in fact the restriction of the forward measure $\mathbb{P}_U$ to the σ-field $\mathcal{F}_T$. Since the exercise set D belongs to the σ-field $\mathcal{F}_T$, we have $\tilde{\mathbb{P}}_T(D \mid \mathcal{F}_t) = \mathbb{P}_U(D \mid \mathcal{F}_t)$. Formula (11.43) thus admits the following alternative representation

$$C_t = B(t, U)\mathbb{P}_U(D \mid \mathcal{F}_t) - K B(t, T)\mathbb{P}_T(D \mid \mathcal{F}_t), \tag{11.49}$$

which is the special case of the general formula established in Chap. 8.

11.3.2 Stock Options

The payoff at expiry of a European call option written on a stock S equals $C_T = (S_T - K)^+$, where T is the expiry date and K denotes the exercise price. The next result, which is a straightforward generalization of the Black-Scholes formula, provides an explicit formula for the arbitrage price of a stock call option (for related results, see Merton (1973) and Jarrow (1987)). We assume that the dynamics of S under the martingale measure $\mathbb{P}^*$ are

$$dS_t = S_t \left(r_t \, dt + \sigma_t \cdot dW_t^* \right),$$

where $\sigma : [0, T^*] \to \mathbb{R}^d$ is a deterministic function.

Proposition 11.3.2 *Assume that the bond price volatility $b(\cdot, T)$ and the stock price volatility σ are $\mathbb{R}^d$-valued, bounded, deterministic functions. Then the arbitrage price of a European call option with expiry date T and exercise price K, written on a stock S, equals*

$$C_t = S_t N\big(h_1(S_t, t, T)\big) - K B(t, T) N\big(h_2(S_t, t, T)\big), \tag{11.50}$$

where

$$h_{1,2}(s, t, T) = \frac{\ln(s/K) - \ln B(t, T) \pm \frac{1}{2} v^2(t, T)}{v(t, T)} \tag{11.51}$$

for $(s, t) \in \mathbb{R}_+ \times [0, T]$ and

$$v^2(t, T) = \int_t^T |\sigma_u - b(u, T)|^2 \, du, \quad \forall t \in [0, T]. \tag{11.52}$$

Proof. The proof goes along similar lines to the proof of Proposition 11.3.1, we shall therefore merely sketch its main steps. It is clear that

$$C_T = (F_S(T, T) - K)^+ = F_S(T, T) \mathbb{1}_D - K \mathbb{1}_D,$$

where $D = \{S_T > K\} = \{F_S(T, T) > K\}$. It is therefore enough to evaluate the conditional expectations

$$C_t = B(t, T) \mathbb{E}_{\mathbb{P}_T}\big(F_S(T, T) \mathbb{1}_D \,\big|\, \mathcal{F}_t\big) - K B(t, T) \mathbb{P}_T(D \,|\, \mathcal{F}_t) = I_1 - I_2,$$

where $F_S(T, T)$ is given by the formula

$$F_S(T, T) = F_S(t, T) \exp \left(\int_t^T \gamma_S(u, T) \cdot dW_u^T - \frac{1}{2} \int_t^T |\gamma_S(u, T)|^2 \, du \right)$$

and $\gamma_S(u, T) = \sigma_u - b(u, T)$. Proceeding as for the proof of Proposition 11.3.1, one finds that

$$I_2 = K B(t, T) \, N \left(\frac{\ln(S_t/K) - \ln B(t, T) - \frac{1}{2} v^2(t, T)}{v(t, T)} \right),$$

where $v(t, T)$ is given by (11.52).

We now define an auxiliary probability measure $\hat{\mathbb{P}}_T$ by setting

$$\frac{d\hat{\mathbb{P}}_T}{d\mathbb{P}_T} = \exp\left(\int_0^T \gamma_S(u, T) \cdot dW_u^T - \frac{1}{2}\int_0^T |\gamma_S(u, T)|^2 \, du\right).$$

Then the process $\tilde{W}^T$ given by the formula

$$\tilde{W}_t^T = W_t^T - \int_0^t \gamma_S(u, T) \, du, \quad \forall t \in [0, T],$$

follows the standard Brownian motion under $\hat{\mathbb{P}}_T$. Furthermore, the forward price $F_S(T, T)$ satisfies

$$F_S(T, T) = \frac{S_t}{B(t, T)} \exp\left(\int_t^T \gamma_S(u, T) \cdot d\tilde{W}_u^T + \frac{1}{2}\int_t^T |\gamma_S(u, T)|^2 \, du\right)$$

for every $t \in [0, T]$. Since we have

$$I_1 = S_t \, \mathbb{E}_{\mathbb{P}_T}\left\{ \mathbb{1}_D \exp\left(\int_t^T \gamma_S(u, T) \cdot dW_u^T - \frac{1}{2}\int_t^T |\gamma_S(u, T)|^2 \, du\right) \Big| \mathcal{F}_t\right\},$$

from Bayes's rule we get $I_1 = S_t \, \hat{\mathbb{P}}_T(D \mid \mathcal{F}_t)$. Consequently, we obtain

$$I_1 = S_t \, N\left(\frac{\ln(S_t/K) - \ln B(t, T) + \frac{1}{2}v^2(t, T)}{v(t, T)}\right).$$

This completes the proof of the proposition. □

Example 11.3.1 Let us examine a very special case of the pricing formula established in Proposition 11.3.2. Let $W^* = (W^{1*}, W^{2*})$ be a two-dimensional standard Brownian motion given on a probability space $(\Omega, \mathbb{F}, \mathbb{P}^*)$. We assume that the bond price $B(t, T)$ satisfies, under $\mathbb{P}^*$,

$$dB(t, T) = B(t, T)\left(r_t \, dt + \hat{b}(t, T)(\rho, \sqrt{1 - \rho^2}) \cdot dW_t^*\right),$$

where $\hat{b}(\cdot, T) : [0, T] \to \mathbb{R}$ is a real-valued, bounded deterministic function, and the dynamics of the stock price S are

$$dS_t = S_t\left(r_t \, dt + (\hat{\sigma}(t), 0) \cdot dW_t^*\right)$$

for some function $\hat{\sigma} : [0, T^*] \to \mathbb{R}$. Let us introduce the real-valued stochastic processes $\hat{W}^1$ and $\hat{W}^2$ by setting $\hat{W}_t^1 = W_t^{1*}$ and $\hat{W}_t^2 = \rho W_t^{1*} + \sqrt{1 - \rho^2}\, W_t^{2*}$. It is not hard to check that $\hat{W}^1$ and $\hat{W}^2$ follow standard one-dimensional Brownian motions under the martingale measure $\mathbb{P}^*$, and their cross-variation equals $\langle \hat{W}^1, \hat{W}^2 \rangle_t = \rho t$ for $t \in [0, T^*]$. It is evident that

$$dB(t, T) = B(t, T)\left(r_t \, dt + \hat{b}(t, T) \, d\hat{W}_t^2\right) \tag{11.53}$$

and

$$dS_t = S_t \left(r_t \, dt + \hat{\sigma}(t) \, d\hat{W}_t^1 \right). \tag{11.54}$$

An application of Proposition 11.3.2 yields the following result, first established in Merton (1973).

Corollary 11.3.2 *Assume that the dynamics of a bond and a stock price are given by (11.53) and (11.54) respectively. If the volatility coefficients $\hat{b}$ and $\hat{\sigma}$ are deterministic functions, then the arbitrage price of a European call option written on a stock S is given by (11.50)–(11.51), with*

$$v^2(t, T) = \int_t^T \left(\hat{\sigma}^2(u) - 2\rho\hat{\sigma}(u)\hat{b}(u, T) + \hat{b}^2(u, T) \right) du.$$

We are in a position to formulate a result that encompasses both cases studied above. The dynamics of the spot price Z of a tradable asset are assumed to be given by the expression

$$dZ_t = Z_t \left(r_t \, dt + \xi_t \cdot dW_t^* \right).$$

It is essential to assume that the volatility $\xi_t - b(t, T)$ of the forward price of Z for the settlement date T is deterministic.

Proposition 11.3.3 *The arbitrage price of a European call option with expiry date T and exercise price K, written on an asset Z, is given by the formula*

$$C_t = B(t, T)\left(F_Z(t, T)N\big(\tilde{d}_1(F_Z(t, T), t, T)\big) - KN\big(\tilde{d}_2(F_Z(t, T), t, T)\big) \right),$$

where, for $(F, t) \in \mathbb{R}_+ \times [0, T]$,

$$\tilde{d}_{1,2}(F, t, T) = \frac{\ln(F/K) \pm \frac{1}{2} v^2(t, T)}{v(t, T)} \tag{11.55}$$

and

$$v^2(t, T) = \int_t^T |\xi_u - b(u, T)|^2 \, du, \quad \forall t \in [0, T]. \tag{11.56}$$

Let P_t stand for the price at time $t \leq T$ of a European put option written on an asset Z, with expiry date T and strike price K. Then the following useful result is valid. The reader may find it instructive to derive (11.57) by constructing particular trading portfolios.

Corollary 11.3.3 *The following put-call parity relationship is valid*

$$C_t - P_t = Z_t - B(t, T)K, \quad \forall t \in [0, T]. \tag{11.57}$$

Proof. We make use of the forward measure method. We have

$$C_t - P_t = B(t, T)\,\mathbb{E}_{\mathbb{P}_T}\big(F_Z(T, U, T) - K \,|\, \mathcal{F}_t\big),$$

and thus we obtain

$$C_t - P_t = B(t, T)F_Z(t, U, T) - B(t, T)K = Z_t - B(t, T)K$$

for every $t \in [0, T]$. $\qquad\qquad\qquad\qquad\qquad\qquad\qquad\qquad\square$

11.3.3 Option on a Coupon-bearing Bond

For a given selection of dates $T_1 < \ldots < T_m \leq T^*$, we consider a coupon-bearing bond whose value Z_t at time $t \leq T_1$ is

$$Z_t = \sum_{j=1}^{m} c_j B(t, T_j), \quad \forall t \in [0, T_1],$$

where c_j are real numbers. We shall study a European call option with expiry date $T \leq T_1$, whose payoff at expiry has the following form

$$C_T = (Z_T - K)^+ = \Big(\sum_{j=1}^{m} c_j B(T, T_j) - K \Big)^+.$$

Proposition 11.3.4 *Let $(\zeta_1, \ldots, \zeta_m)$ be a random variable whose distribution under $\mathbb{Q}$ is Gaussian, with zero expected value and the following variance-covariance matrix, for $k, l = 1, \ldots, m$ (recall that $\gamma(t, T_k, T) = b(t, T_k) - b(t, T)$)*

$$\mathrm{Cov}_{\mathbb{Q}}(\zeta_k, \zeta_l) = v_{kl} = \int_{t}^{T} \gamma(u, T_k, T) \cdot \gamma(u, T_l, T) \, du.$$

Let $J_1^j = J_1^j\big(B(t, T_1), \ldots, B(t, T_m), B(t, T)\big)$ where

$$J_1^j(x_1, \ldots, x_m, y) = \mathbb{Q} \Big\{ \sum_{l=1}^{m} c_l x_l \, e^{\zeta_l + v_{lj} - v_{ll}/2} > K y \Big\}$$

for $j = 1, \ldots, m$, and let $J_2 = J_2\big(B(t, T_1), \ldots, B(t, T_m), B(t, T)\big)$ where

$$J_2(x_1, \ldots, x_m, y) = \mathbb{Q} \Big\{ \sum_{l=1}^{m} c_l x_l \, e^{\zeta_l - v_{ll}/2} > K y \Big\}.$$

The arbitrage price of a European call option on a coupon-bearing bond is given by the formula

$$C_t = \sum_{j=1}^{m} c_j B(t, T_j) J_1^j - K B(t, T) J_2. \tag{11.58}$$

Proof. We need to evaluate the conditional expectation

$$C_t = B(t, T) \sum_{j=1}^{m} c_j \mathbb{E}_{\mathbb{P}_T}(F_B(T, T_j, T) \mathbb{1}_D | \mathcal{F}_t) - K B(t, T) \mathbb{P}_T\{D \,|\, \mathcal{F}_t\} = I_1 - I_2,$$

where D stands for the exercise set

$$D = \Big\{ \sum_{j=1}^{m} c_j B(T, T_j) > K \Big\} = \Big\{ \sum_{j=1}^{m} c_j F_B(T, T_j, T) > K \Big\}.$$

Let us first examine the conditional probability $\mathbb{P}_T\{D\,|\,\mathcal{F}_t\}$. By virtue of Lemma 11.3.1, the process $F_B^l(t) = F_B(t, T_l, T)$ satisfies

$$F_B^l(T) = F_B^l(t)\,\exp\left(\int_t^T \gamma(u, T_l, T)\cdot dW_u^T - \frac{1}{2}\int_t^T |\gamma(u, T_l, T)|^2\, du\right),$$

where $\gamma(u, T_l, T) = b(u, T_l) - b(u, T)$. In other words,

$$F_B(T, T_l, T) = F_B(t, T_l, T)\, e^{\xi_l^T - v_{ll}/2},$$

where ξ_l^T is a random variable independent of the σ-field $\mathcal{F}_t$, and such that the probability distribution of ξ_l^T under $\mathbb{P}_T$ is the Gaussian distribution $N(0, v_{ll})$. Hence,

$$\mathbb{P}_T\{D\,|\,\mathcal{F}_t\} = \mathbb{P}_T\left\{\sum_{l=1}^m c_l x_l\, e^{\xi_l^T - v_{ll}^2/2} > Ky\right\}$$

with $x_l = B(t, T_l)$ and $y = B(t, T)$. This proves that $I_2 = K B(t, T) J_2$. Let us show that $I_1 = \sum_{j=1}^m c_j\, B(t, T_j) J_1^j$. To this end, it is sufficient to check that for any fixed j we have

$$B(t, T)\mathbb{E}_{\mathbb{P}_T}\big(F_B(T, T_j, T)\mathbb{1}_D\,|\,\mathcal{F}_t\big) = B(t, T_j) J_1^j.$$

This can be done by proceeding in much the same way as in the proof of Proposition 11.3.1. Let us fix j and introduce an auxiliary probability measure $\tilde{\mathbb{P}}_{T_j}$ on $(\Omega, \mathcal{F}_T)$ by setting

$$\frac{d\tilde{\mathbb{P}}_{T_j}}{d\mathbb{P}_T} = \mathcal{E}_T\left(\int_0^{\cdot}\gamma(u, T_j, T)\cdot dW_u^T\right).$$

Then the process

$$\tilde{W}_t^j = W_t^T - \int_0^t \gamma(u, T_j, T)\, du$$

is a standard Brownian motion under $\tilde{\mathbb{P}}_{T_j}$. Recall that $\tilde{\mathbb{P}}_{T_j} = \mathbb{P}_{T_j}$ on $\mathcal{F}_T$, hence we shall write simply $\mathbb{P}_{T_j}$ in place of $\tilde{\mathbb{P}}_{T_j}$ in what follows. For any l, the forward price $F_B^l(t)$ has the following representation under $\mathbb{P}_{T_j}$

$$dF_B^l(t) = F_B^l(t)\gamma(t, T_l, T)\cdot\big(\tilde{W}_t^j + \gamma(t, T_j, T)\, dt\big). \tag{11.59}$$

For a fixed j, we define the random variable $(\xi_1, \dots, \xi_m)$ by the formula

$$\xi_l = \int_t^T \gamma(u, T_l, T)\cdot d\tilde{W}_u^j.$$

The random variable $(\xi_1, \dots, \xi_m)$ is independent of $\mathcal{F}_t$, with Gaussian distribution under $\mathbb{P}_{T_j}$. In addition, the expected value of each random variable ξ_i is zero, and for every $k, l = 1, \dots, m$ we have

$$v_{kl} = \mathrm{Cov}_{\mathbb{P}_{T_j}}(\xi_k, \xi_l) = \int_t^T \gamma(u, T_k, T)\cdot\gamma(u, T_l, T)\, du.$$

On the other hand, using (11.59) we find that

$$F_B^l(T) = F_B^l(t) \exp\left(\xi_l - \tfrac{1}{2}v_{ll} + v_{lj}\right)$$

for every $l = 1, \ldots, m$. The abstract Bayes formula yields

$$B(t, T)\mathbb{E}_{\mathbb{P}_T}\left(F_B(T, T_j, T)\mathbb{1}_D \,\big|\, \mathcal{F}_t\right) = B(t, T_j)\mathbb{P}_{T_j}\{D \,|\, \mathcal{F}_t\}.$$

Furthermore,

$$\mathbb{P}_{T_j}\{D \,|\, \mathcal{F}_t\} = \mathbb{P}_{T_j}\left\{ \sum_{l=1}^m c_l x_l \exp\left(\xi_l - \tfrac{1}{2}v_{ll} + v_{lj}\right) > K y \right\}$$

with $x_l = B(t, T_l)$ and $y = B(t, T)$. This completes the proof. □

The next result suggests an alternative way to prove Proposition 11.3.4.

Lemma 11.3.2 *Let us denote $D = \{Z_T > K\}$. Then the arbitrage price of a European call option written on a coupon bond satisfies*

$$C_t = \sum_{j=1}^m c_j B(t, T_j)\mathbb{P}_{T_j}(D \,|\, \mathcal{F}_t) - K B(t, T)\mathbb{P}_T(D \,|\, \mathcal{F}_t).$$

Proof. We have $Z_T = \sum_{j=1}^m c_j B_T \, \mathbb{E}_{\mathbb{P}^*}\left(B_{T_j}^{-1} \,|\, \mathcal{F}_T\right)$, and thus

$$C_t = B_t \, \mathbb{E}_{\mathbb{P}^*}\left\{ \mathbb{1}_D B_T^{-1} \left(\sum_{j=1}^m c_j B_T \, \mathbb{E}_{\mathbb{P}^*}(B_{T_j}^{-1} \,|\, \mathcal{F}_T) - K \right) \,\Big|\, \mathcal{F}_t \right\}$$

$$= \sum_{j=1}^m c_j B_t \, \mathbb{E}_{\mathbb{P}^*}\left(B_{T_j}^{-1}\mathbb{1}_D \,|\, \mathcal{F}_t\right) - K B_t \mathbb{E}_{\mathbb{P}^*}\left(B_T^{-1}\mathbb{1}_D \,|\, \mathcal{F}_t\right),$$

since $D \in \mathcal{F}_T$. Using (9.35), we get for every j

$$B_t \, \mathbb{E}_{\mathbb{P}^*}\left(B_{T_j}^{-1}\mathbb{1}_D \,|\, \mathcal{F}_t\right) = B_t \, B(0, T_j) \, \mathbb{E}_{\mathbb{P}^*}\left(\eta_{T_j}\mathbb{1}_D \,|\, \mathcal{F}_t\right)$$

$$= B_t \, \mathbb{P}_{T_j}\{D \,|\, \mathcal{F}_t\} \mathbb{E}_{\mathbb{P}^*}(B_{T_j}^{-1} \,|\, \mathcal{F}_t)$$

$$= B(t, T_j) \, \mathbb{P}_{T_j}\{D \,|\, \mathcal{F}_t\}.$$

Since a similar relation holds for the last term, this ends the proof. □

In practice, call options are also embedded in some bond issues – the so-called *callable bonds*. An issuer of a callable bond has the right, but not the obligation, to buy back the bond after the call date. The call date gives the issuer the right to refinance some of its debt once interest rates fall. The date at which a callable bond will be redeemed is uncertain. Call provisions are not usually operative during the first few years of a bond's life, however. This feature is referred to as a *deferred call*, and the bond is said to be *call protected* during this period. The exercise price is variable (usually decreasing) in time.

It appears that a call provision generally reduces the sensitivity of a bond's value to changes in the level of interest rates. From the theoretical viewpoint, a long position in a callable bond can be seen as a combination of a long position in a (straight) coupon-bearing bond with a short position in a deferred American call option written on the underlying bond (cf. Jarrow (1995)).

Other typical classes of bond issues that involve option provisions include *savings* (or *parity*) bonds, *retractable* bonds, *extendible* bonds and *convertible* bonds. A *savings* bond can be cashed at par at the discretion of the holder at any time before maturity. A *retractable* bond is a long-term bond that may be redeemed at par at a specified date before maturity. It is thus clear that the holder should retract the bond if, on the retraction date, the value of the unretracted bond is below par value. The option embedded in a savings bond is therefore of American type, while a retractable bond involves a European-style option. An *extendible* bond is a nominally shorter-term instrument that may be extended over a longer period at the holder's discretion. Finally, a *convertible* bond is one that may be converted into another form of security, typically common stock, at the discretion of the holder at a specified price for a fixed period of time.

We refer the interested reader to Brennan and Schwarz (1977b, 1977c), Ingersoll (1977), Bodie and Taggart (1978), Athanassakos (1996), Büttler and Waldvogel (1996), Nyborg (1996), Kifer (2000), Andersen and Buffum (2003), Lau and Kwok (2004), Sîrbu et al. (2004), Kallsen and Kühn (2004, 2005), and Sîrbu and Shreve (2005) for further exploration of these classes of bonds.

11.3.4 Pricing of General Contingent Claims

Consider a European contingent claim X, at time T, of the form $X = g(Z_T^1, \ldots, Z_T^n)$, where $g : \mathbb{R}^n \to \mathbb{R}$ is a bounded Borel-measurable function. Assume that the price process Z^i of the ith asset satisfies, under $\mathbb{P}^*$,

$$dZ_t^i = Z_t^i \left(r_t \, dt + \xi_t^i \cdot dW_t^* \right). \tag{11.60}$$

Then

$$F_{Z^i}(T, T) = F_{Z^i}(t, T) \exp \left(\int_t^T \gamma_i(u, T) \cdot dW_u^T - \frac{1}{2} \int_t^T |\gamma_i(u, T)|^2 \, du \right),$$

where $\gamma_i(u, T) = \xi_u^i - b(u, T)$, or in short

$$F_{Z^i}(T, T) = F_{Z^i}(t, T) \exp \left(\zeta_i(t, T) - \tfrac{1}{2} \gamma_{ii} \right),$$

where $\zeta_i(t, T) = \int_t^T \gamma_i(u, T) \cdot dW_u^T$ and $\gamma_{ii} = \int_t^T |\gamma_i(u, T)|^2 \, du$. The forward price $F_{Z^i}(t, T)$ is a random variable measurable with respect to the σ-field $\mathcal{F}_t$, while the random variable $\zeta_i(t, T)$ is independent of this σ-field. Moreover, it is clear that the probability distribution under the forward measure $\mathbb{P}_T$ of the vector-valued random variable

$$(\zeta_1(t, T), \ldots, \zeta_n(t, T)) = \left(\int_t^T \gamma_1(u, T) \cdot dW_u^T, \ldots, \int_t^T \gamma_n(u, T) \cdot dW_u^T \right)$$

is Gaussian $N(0, \Gamma)$, where the entries of the $n \times n$ matrix Γ are

$$\gamma_{ij} = \int_t^T \gamma_i(u, T) \cdot \gamma_j(u, T) \, du.$$

Introducing a $k \times n$ matrix $\Theta = [\theta_1, \ldots, \theta_n]$ such that $\Gamma = \Theta^t \Theta$ leads to the following valuation result (cf. Brace and Musiela (1994)).

Proposition 11.3.5 *Assume that γ_i is a deterministic function for $i = 1, \ldots, n$. Then the arbitrage price at time $t \in [0, T]$ of a European contingent claim $X = g(Z_T^1, \ldots, Z_T^n)$ at time T equals*

$$\pi_t(X) = B(t, T) \int_{\mathbb{R}^k} g \left(\frac{Z_t^1 n_k(x + \theta_1)}{B(t, T) n_k(x)}, \ldots, \frac{Z_t^n n_k(x + \theta_n)}{B(t, T) n_k(x)} \right) n_k(x) \, dx,$$

where n_k is the standard k-dimensional Gaussian density

$$n_k(x) = (2\pi)^{-k/2} e^{-|x|^2/2}, \quad \forall x \in \mathbb{R}^k,$$

and the vectors $\theta_1, \ldots, \theta_n \in \mathbb{R}^k$ are such that for every $i, j = 1, \ldots, n$, we have

$$\theta_i \cdot \theta_j = \int_t^T \gamma_i(u, T) \cdot \gamma_j(u, T) \, du.$$

Proof. We have

$$\pi_t(X) = B_t \, \mathbb{E}_{\mathbb{P}^*} \left(B_T^{-1} g(Z_T^1, \ldots, Z_T^n) \mid \mathcal{F}_t \right)$$

$$= B(t, T) \, \mathbb{E}_{\mathbb{P}_T} \left(g(F_{Z^1}(T, T), \ldots, F_{Z^n}(T, T)) \mid \mathcal{F}_t \right) = B(t, T) J.$$

In view of the definition of the matrix Θ, it is clear that

$$J = \int_{\mathbb{R}^k} g \left(F_{Z^1}(t, T) \, e^{\theta_1 \cdot x - |\theta_1|^2/2}, \ldots, F_{Z^n}(t, T) \, e^{\theta_n \cdot x - |\theta_n|^2/2} \right) n_k(x) \, dx$$

$$= \int_{\mathbb{R}^k} g \left(\frac{Z_t^1 n_k(x + \theta_1)}{B(t, T) n_k(x)}, \ldots, \frac{Z_t^n n_k(x + \theta_n)}{B(t, T) n_k(x)} \right) n_k(x) \, dx.$$

This ends the proof of the proposition.   □

Representation $\Gamma = \Theta^t \Theta$, where $\Theta = [\theta_1, \ldots, \theta_n]$ is a $k \times n$ matrix and Θ^t is the transpose of Θ, can be obtained easily from the eigenvalues and the eigenvectors of the matrix Γ. Let $\delta_1, \ldots, \delta_n$ be the eigenvalues of Γ and $w_1, \ldots, w_n$

the corresponding orthonormal eigenvectors. Then with $D = \mathrm{diag}\,(\delta_1, \ldots, \delta_n)$ and $V = [w_1, \ldots, w_n]$, we have $\Gamma = VDV^t = VD^{1/2}(VD^{1/2})^t$, where V^t is the transpose of V. Let $k \le n$ be such that $\delta_1 \ge \delta_2 \ge \ldots \ge \delta_k$ are strictly positive numbers, and $\delta_{k+1} = \ldots = \delta_n = 0$. Then

$$VD^{1/2} = \left[\sqrt{\delta_1}\, w_1, \ldots, \sqrt{\delta_k}\, w_k, 0, \ldots, 0\right]$$

and with $\Theta^t = \left[\sqrt{\delta_1}\, w_1, \ldots, \sqrt{\delta_k}\, w_k\right]$ we have $\Gamma = \Theta^t \Theta$, as desired.

11.3.5 Replication of Options

In preceding sections, we have valued options using a risk-neutral valuation approach, assuming implicitly that options correspond to attainable claims. In this section, we focus on the construction of a replicating portfolio. Consider a contingent claim X that settles at time T and is represented by a $\mathbb{P}_T$-integrable, strictly positive random variable X. The forward price of X for the settlement date T satisfies

$$F_X(t, T) = \mathbb{E}_{\mathbb{P}_T}(X \mid \mathcal{F}_t) = F_X(0, T) + \int_0^t F_X(u, T)\gamma_u \cdot dW_u^T \qquad (11.61)$$

for some predictable process γ. Assume, in addition, that γ is a deterministic function. Let us denote $F_X(t, T)$ simply by F_t. Our aim is to show, by means of a replicating strategy, that the arbitrage price of a European call option written on a claim X, with expiry date T and strike price K, equals

$$C_t = B(t, T)\Big(F_t N\big(\tilde{d}_1(F_t, t, T)\big) - K N\big(\tilde{d}_2(F_t, t, T)\big)\Big), \qquad (11.62)$$

where $\tilde{d}_1$ and $\tilde{d}_2$ are given by (11.55) with $v^2(t, T) = v_X^2(t, T) = \int_t^T |\gamma_u|^2\, du$. Equality (11.62) yields the following formula for the forward price of the option

$$F_C(t, T) = F_t N\big(\tilde{d}_1(F_t, t, T)\big) - K N\big(\tilde{d}_2(F_t, t, T)\big). \qquad (11.63)$$

Note that by applying Itô's formula to (11.63), we obtain

$$dF_C(t, T) = N(\tilde{d}_1(F_t, t, T))\, dF_t. \qquad (11.64)$$

Forward asset/bond market. Let us consider a T-forward market, i.e., a financial market in which the forward contracts for settlement at time T play the role of primary securities. Consider a forward strategy $\psi = (\psi^1, \psi^2)$, where ψ^1 and ψ^2 stand for the number of forward contracts on the underlying claim X and on the zero-coupon bond with maturity T respectively. Observe that the T-forward market differs in an essential way from a futures market. The forward wealth process $\tilde{V}$ of a T-forward market portfolio ψ equals

$$\tilde{V}_t(\psi) = \psi_t^1 F_X(t, T) + \psi_t^2 F_B(t, T, T). \qquad (11.65)$$

Since clearly $F_B(t, T, T) = 1$ for any $t \in [0, T]$, a portfolio ψ is self-financing in the T-forward market if its forward wealth satisfies

$$d\tilde{V}_t(\psi) = \psi_t^1 dF_X(t, T) = \psi_t^1 F_X(t, T)\gamma_t \cdot dW_t^T,$$

where the last equality follows from (11.61). Our aim is to find the forward portfolio ψ that replicates the forward contract written on the option, and subsequently to re-derive pricing formulas (11.62)–(11.63).

To replicate the forward contract written on the option, it is enough to take positions in forward contracts on a claim X and in forward contracts on T-maturity bonds. Suppose that the option's forward price equals $F_C(t, T) = u(F_X(t, T), t)$ for some function u. Arguing along similar lines to the first proof of Theorem 3.1.1, with constant interest rate $r = 0$ and time-variable deterministic volatility γ_t, one may derive the following PDE

$$u_t(x, t) + \tfrac{1}{2} |\gamma_t|^2 x^2 u_{xx}(x, t) = 0$$

with $u(x, T) = (x - K)^+$ for $x \in \mathbb{R}_+$. The solution u to this problem is given by the formula

$$u(x, t) = x N\big(\tilde{d}_1(x, t, T)\big) - K N\big(\tilde{d}_2(x, t, T)\big).$$

The corresponding strategy $\psi = (\psi^1, \psi^2)$ in the T-forward market is

$$\psi_t^1 = u_x\big(F_X(t, T), t\big) = N\big(\tilde{d}_1(F_X(t, T), t, T)\big) \tag{11.66}$$

and $\psi_t^2 = u(F_X(t, T), t) - \psi_t^1 F_X(t, T)$. It can be checked, using Itô's formula, that the strategy ψ is self-financing in the T-forward market; moreover, $\tilde{V}_T(\psi) = V_T(\psi) = (X - K)^+$. The forward price of the option is thus given by (11.63), and consequently its spot price at time t equals

$$C_t = B(t, T)\tilde{V}_t(\psi) = B(t, T)u(F_X(t, T), t). \tag{11.67}$$

The last formula coincides with (11.62).

Forward/spot asset/bond market. It may be convenient to replicate the terminal payoff of an option by means of a combined spot/forward trading strategy (cf. Jamshidian (1994a)). Let the date t be fixed, but arbitrary. Consider an investor who purchases at time t the number $F_C(t, T)$ of T-maturity bonds and holds them to maturity. In addition, at any date $s \geq t$ she takes ψ_s^1 positions in T-maturity forward contracts on the underlying claim, where ψ_s^1 is given by (11.66). The terminal wealth of such a strategy at the date T equals

$$F_C(t, T) + \int_t^T \psi_s^1 dF_X(s, T) = F_C(t, T) + \tilde{V}_T(\psi) - \tilde{V}_t(\psi) = (X - K)^+,$$

since $\tilde{V}_t(\psi) = F_C(t, T)$ and $\tilde{V}_T(\psi) = (X - K)^+$.

Spot asset/bond market. To replicate an option in a spot market, we need to assume that it is written on an asset that is tradable in the spot market. As the second asset, we use a T-maturity bond, with the spot price $B(t, T)$. Assume that a claim X corresponds to the value Z_T of a tradable asset, whose spot price at time t equals Z_t. To replicate an option in the spot market, we consider the spot trading strategy $\phi = \psi$, where Z and a T-maturity bond are primary securities. We deduce easily from (11.67) that the wealth $V(\phi)$ equals

$$V_t(\phi) = \phi_t^1 Z_t + \phi_t^2 B(t, T) = B(t, T)V_t(\psi) = C_t.$$

The last equality shows that the strategy ϕ replicates the option value at any date $t \leq T$. It remains to check that ϕ is self-financing. The following property is a generic feature of self-financing strategies in the T-forward market: a T-forward trading strategy ψ is self-financing if and only if the spot market strategy $\phi = \psi$ is self-financing (see Lemma 8.1.2). Replication of a European call option with terminal payoff $(Z_T - K)^+$ can thus be done using the spot trading strategy $\phi = (\phi^1, \phi^2)$, where $\phi_t^1 = N(\tilde{d}_1(F_Z(t, T), t, T))$ and

$$\phi_t^2 = (C_t - \phi_t^1 Z_t)/B(t, T) = -KN(\tilde{d}_2(F_t, t, T)).$$

Here, ϕ_t^1 and ϕ_t^2 represent the number of units of the underlying asset and of T-maturity bonds held at time t respectively.

Spot asset/cash market. Let us show that since a savings account follows a process of finite variation, replication of an option written on Z in the spot asset/cash market is not always possible. Suppose that $\hat{\phi} = (\hat{\phi}^1, \hat{\phi}^2)$ is an asset/cash self-financing trading strategy that replicates an option. In particular, we have

$$\hat{\phi}_t^1 dZ_t + \hat{\phi}_t^2 dB_t = dC_t. \tag{11.68}$$

On the other hand, from the preceding paragraph, we know that

$$\phi_t^1 dZ_t + \phi_t^2 dB(t, T) = dC_t = C_t(r_t\, dt + \xi_t^C \cdot dW_t^*), \tag{11.69}$$

where

$$\xi_t^C = (\phi_t^1 Z_t \xi_t + \phi_t^2 Z_t b(t, T))/C_t.$$

A comparison of martingale parts in (11.68) and (11.69) yields

$$\phi_t^1 Z_t \xi_t \cdot dW_t^* + \phi_t^2 B(t, T)b(t, T) \cdot dW_t^* = \hat{\phi}_t^1 Z_t \xi_t \cdot dW_t^*.$$

When the underlying Brownian motion is multidimensional, we cannot solve the last equality for $\hat{\phi}_t^1$, in general. If, however, W^* is one-dimensional and processes Z and ξ are strictly positive then we have

$$\hat{\phi}_t^1 = \phi_t^1 + \phi_t^2 b(t, T)B(t, T)/(\xi_t Z_t).$$

We put, in addition, $\hat{\phi}_t^2 = B_t^{-1}(C_t - \hat{\phi}_t^1 Z_t)$. It is clear that such a strategy replicates the option. Moreover, it is self-financing, since simple calculations show that

$$\hat{\phi}_t^1 dZ_t + \hat{\phi}_t^2 dB_t = r_t C_t\, dt + \xi_t^C C_t \cdot dW_t^* = dC_t = dV_t(\hat{\phi}).$$

For instance, the stock/cash trading strategy that involves at time t

$$\hat{\phi}_t^1 = N\big(\tilde{d}_1(F_t, t, T)\big) - K \,\frac{b(t, T) B(t, T)}{\xi_t Z_t}\, N\big(\tilde{d}_2(F_t, t, T)\big)$$

shares of stock, and the amount $C_t - \hat{\phi}_t^1 Z_t$ held in a savings account, is a self-financing strategy replicating a European call option written on Z. The possibility of replication using the cash market is also examined in Sect. 11.6.

11.4 Volatilities and Correlations

Let is comment briefly on empirical studies related to the validity of the HJM methodology of term structure modelling. Needless to say, the volatility coefficient $\sigma(t, T)$ is not directly observed in the market. On the contrary, the market usually quotes interest rate derivatives in terms of implied Black volatilities. Practitioners typically work with a matrix of Black swaption volatilities; one axis is the length of the underlying swap, the other is the expiry of the swaption. In real-world markets only few swaptions are liquid, so that the market provides only a few entries of this matrix – the remaining entries are calculated by some method of interpolation.

11.4.1 Volatilities

Since $\sigma(t, T)$ represents the volatility of the whole family $f(t, T)$ of forward interest rates, one should ideally use all traded instruments whose prices are sensitive with respect to the volatility of forward interest rates. In practical studies, it is essential to focus on some of them only, and thus a judicious choice of a family of instruments under study is of great importance.

To the best of our knowledge, the first empirical studies of the HJM model were undertaken by Flesaker (1993a), who examined the single-factor Ho and Lee model. Subsequently, various HJM models were tested by, among others, Cohen (1991), who focused on Treasury bond futures and futures options; Amin and Morton (1994), who examined the prices of Eurodollar futures and futures options; Brace and Musiela (1994), who used exchange-traded options on bill futures and caps in the Australian market; and Heitmann and Trautmann (1995), who used data from the German bond market. At first glance, a major shortcoming of the Gaussian HJM model is the fact that it allows interest rates to become negative with positive probabilities. However, in a judiciously constructed Gaussian HJM model, the probabilities of negative rates appear to be almost negligible (or, at least, are under control).

In all of these papers, the authors postulate a priori a volatility $\sigma(t, T)$ that depends on $T - t$ only; also, the implied volatility structure is typically assumed to be a smooth function. In Amin and Morton (1994) specific functional forms of σ depending on a finite number of parameters are assumed. They examine six particular cases of the function σ that can all be expressed in the following form

$$\sigma\big(t, T, f(t, T)\big) = \big(\sigma_0 + \sigma_1(T - t)\big) \exp(-\lambda(T - t))(f(t, T))^\gamma,$$

where σ_0, σ_1, λ and γ are unknown parameters, to be estimated on the basis of market data. The smoothness of σ is here automatically guaranteed, and the whole problem is thus reduced to the estimation of the unknown parameters. It is important to observe that the underlying Brownian motion is one-dimensional; in this sense, all term structure models examined by Amin and Morton belong to the class of single-factor HJM models.

The first, essential and rather cumbersome step is the estimation of the parameters of the model, using the historical market prices of Eurodollar futures options. First, one needs to estimate the forward interest rates using Eurodollar futures prices. The implied volatility is then derived by fitting the option prices predicted by the model to the market prices of Eurodollar futures options. The implied volatility parameters appear to vary over time, which of course is not surprising. The predictive power of each model is then tested by means of a day-by-day comparison of the option prices, which are calculated on the basis of the previous day's implied forward rate volatility and the estimated term structure, using the current market prices of Eurodollar futures options. Since Eurodollar futures options are American, the option pricing procedure used by Amin and Morton was based on a discrete-time approximation of the HJM model by means of a non-recombining tree (cf. Heath et al. (1992a) and Amin and Bodurtha (1995)).

A slightly different approach was adopted by Heitmann and Trautmann (1995), who examined the implementation of the HJM model using data from the German bond market. They focused on single-factor Ho and Lee and Vasicek's models, as well as on two-factor combinations of these models. Firstly, they concluded that the single-factor Vasicek model describes the behavior of the observed forward rates better than the single-factor Ho-Lee model. More importantly, they found that the two-factor Vasicek model has greater explanatory power than the combined Ho and Lee/Vasicek model.

At the intuitive level, from the models examined by Heitmann and Trautmann, only the two-factor Vasicek model was able to produce three patterns of yield curve movements: parallel shift, reversion and a twist. The presence of the latter effect is related to the specific feature of forward rate volatility in the two-factor Vasicek model – namely, to the fact that the short-term and long-term volatilities can simultaneously dominate the medium-term volatility of forward rates. In this sense, the Ho and Lee/Vasicek model also has a greater ability than other HJM models to explain the historical volatility structure of the German bond market.

Brace and Musiela (1994) assume that the volatility is piecewise constant over some period of time. Such an approach gives more flexibility, but at the obvious cost of losing the smoothness of $\sigma(t, T)$. They use the prices of caps and options on bank bill futures in the Australian market to determine the parameters of the Gaussian HJM model. For some market data, a commonly postulated dependence of $\sigma(t, T)$ on the difference $T - t$ (that is, on the time to maturity) is found to be too restrictive. A second factor and an explicit dependence of the coefficient $\sigma(t, T)$ on both variables, t and T, are introduced to deal with such cases.

Amin and Ng (1997) examine the possibility of inferring future volatility from the information in implied volatility in Eurodollar options. They consider various specifications of the volatility function in the HJM framework, as was done in the aforementioned paper by Amin and Morton (1994).

A detailed comparison of the most popular forward and spot interest rate models is reported in Moraleda and Pelsser (1996) (see also Bühler et al. (1999)). They found – somewhat surprisingly – that short-term interest rate models usually outperform their direct counterparts that are based on the modelling of the forward rate dynamics.

Let us finally mention the paper by Brace et al. (1997), who analyze a two-factor version of the lognormal model of the forward LIBOR, with piecewise constant volatility functions. The model is tested using the U.K. market prices of caps and swaptions and the historically estimated correlation between forward rates. It appears that the implied volatilities are not uniquely determined; that is, a given set of market data can be explained by different piecewise constant volatility functions.

11.4.2 Correlations

So far, we have assumed that the volatilities are $\mathbb{R}^d$-valued deterministic functions and a term structure model is driven be a d-dimensional Brownian motion $W = (W^1, \ldots, W^d)$, so that the driving processes $W^1, \ldots, W^d$ are mutually independent one-dimensional standard Brownian motions. Correlations between various interest rate processes were thus implicitly encoded in the corresponding d-dimensional volatilities, and they were not even mentioned explicitly.

When calibrating a term structure model driven by a multidimensional Brownian motion, it is more convenient to take a slightly different approach, and to examine one-dimensional (implied) volatilities and (historical) correlations separately. Let us present briefly the method of introducing correlations.

Let $W = (W^1, \ldots, W^d)$ be a d-dimensional Brownian motion. Suppose that we deal with a finite family of processes $X^1, \ldots, X^n$ governed by

$$dX_t^i = X_t^i (\mu_t^i \, dt + \zeta_t^i \cdot dW_t)$$

with one-dimensional drift coefficients $\mu^1, \ldots, \mu^n$ and d-dimensional volatilities $\zeta^1, \ldots, \zeta^n$. In our set-up, we take bond prices $B(t, T^i)$ as processes X^i.

Our goal is to find real-valued volatilities $\sigma^1, \ldots, \sigma^n$ and a family of correlated one-dimensional standard Brownian motions $\tilde{W}^1, \ldots, \tilde{W}^n$ such that the processes $X^1, \ldots, X^n$ can be represented as follows:

$$dX_t^i = X_t^i (\mu_t^i \, dt + \sigma_t^i \, d\tilde{W}_t^i). \tag{11.70}$$

To this end, we define $\sigma_t^i = |\zeta_t^i|$ and $\tilde{\zeta}_t^i = \zeta_t^i |\zeta_t^i|^{-1}$ (by convention $0/0 = v$ where v is any vector in $\mathbb{R}^d$ such that $|v| = 1$) and we set

$$\tilde{W}_t^i = \int_0^t \tilde{\zeta}_u^i \cdot dW_u.$$

The processes $\tilde{W}^1, \ldots, \tilde{W}^n$ are one-dimensional Brownian motions and they satisfy $d\langle \tilde{W}^i, \tilde{W}^j \rangle_t = \rho_t^{i,j} \, dt$, where the instantaneous (or *local*) correlation $\rho_t^{i,j}$ equals $\tilde{\zeta}_t^i \cdot \tilde{\zeta}_t^j$ (see Sect. 7.1.9). Finally, equality (11.70) is manifestly valid.

11.5 Futures Price

Let us consider an arbitrary attainable contingent claim X that settles at time T. Since the martingale measure $\mathbb{P}^*$ for the spot market is assumed to be unique, it is natural to introduce the futures price by means of the following definition. In a discrete-time framework, this definition is formally justified by means of no-arbitrage arguments in Jarrow (1995) (see pp. 134-135 therein). For related studies of forward and futures prices, see Cornell and French (1983a, 1983b), Cakici and Chatterjee (1991), Amin and Jarrow (1992), Flesaker (1993b), and Grinblatt and Jegadeesh (1996).

Definition 11.5.1 The *futures price* $f_X(t, T)$ of a claim X, in the futures contract that expires at time T, is given by the formula

$$f_X(t, T) = \mathbb{E}_{\mathbb{P}^*}(X \mid \mathcal{F}_t), \quad \forall t \in [0, T]. \tag{11.71}$$

We are in a position to find the spread between the forward and futures prices of an arbitrary asset. Recall that the forward price of a claim X equals $F_X(t, T) = \pi_t(X)/B(t, T)$. Let B represent the savings account. The next result is fairly general, but not always convenient for explicit calculations.

Lemma 11.5.1 *The forward and futures prices of a claim X satisfy*

$$F_X(t, T) - f_X(t, T) = \frac{\mathrm{Cov}_{\mathbb{P}^*}(X, B_t B_T^{-1} \mid \mathcal{F}_t)}{B(t, T)}, \quad \forall t \in [0, T]. \tag{11.72}$$

Proof. Indeed, the conditional covariance is defined as follows

$$\mathrm{Cov}_{\mathbb{P}^*}(X, B_t B_T^{-1} \mid \mathcal{F}_t) = \mathbb{E}_{\mathbb{P}^*}(X B_t B_T^{-1} \mid \mathcal{F}_t) - \mathbb{E}_{\mathbb{P}^*}(X \mid \mathcal{F}_t)\mathbb{E}_{\mathbb{P}^*}(B_t B_T^{-1} \mid \mathcal{F}_t).$$

Thus, in our case, we obtain

$$\mathrm{Cov}_{\mathbb{P}^*}(X, B_t B_T^{-1} \mid \mathcal{F}_t) = \pi_t(X)\mathbb{E}_{\mathbb{P}^*}(X B_t B_T^{-1} \mid \mathcal{F}_t) - \mathbb{E}_{\mathbb{P}^*}(X \mid \mathcal{F}_t)B(t, T).$$

Formula (11.72) now follows easily. □

To get a more explicit representation of the spread $F_X(t, T) - f_X(t, T)$, let us now an arbitrary tradable asset, whose spot price Z has the dynamics given by the expression (11.60) (with index i omitted). The forward price of Z for settlement at the date T is already known to satisfy

$$F_Z(T, T) = F_Z(t, T) \exp\left(\int_t^T \gamma_Z(u, T) \cdot dW_u^T - \frac{1}{2} \int_t^T |\gamma_Z(u, T)|^2 \, du \right),$$

where $\gamma_Z(u, T) = \xi_u - b(u, T)$, and $W_t^T = W_t^* - \int_0^t b(u, T) \, du$ is a Brownian motion under the forward measure $\mathbb{P}_T$. The futures price is given by formula (11.71), that is, $f_Z(t, T) = \mathbb{E}_{\mathbb{P}^*}(Z_T \mid \mathcal{F}_t)$ for $t \in [0, T]$.

Our next goal is to establish a volatility-based relationship between the forward and the futures price of Z, due to Jamshidian (1993a).

Proposition 11.5.1 *Assume that the volatility* $\gamma_Z(\cdot, T) = \xi - b(\cdot, T)$ *of the forward price process* $F_Z(t, T)$ *and the bond volatility* $b(\cdot, T)$ *are deterministic. Then the futures price* $f_Z(t, T)$ *equals*

$$f_Z(t, T) = F_Z(t, T) \exp\left(\int_t^T \left(b(u, T) - \xi_u\right) \cdot b(u, T) \, du\right). \qquad (11.73)$$

Proof. It is clear that

$$F_Z(T, T) = F_Z(t, T) \zeta_t \exp\left(\int_t^T \left(b(u, T) - \xi_u\right) \cdot b(u, T) \, du\right),$$

where ζ_t stands for the following random variable

$$\zeta_t = \exp\left(\int_t^T \left(\xi_u - b(u, T)\right) \cdot dW_u^* - \frac{1}{2} \int_t^T |\xi_u - b(u, T)|^2 \, du\right).$$

The random variable ζ_t is independent of the σ-field $\mathcal{F}_t$, and its expectation under $\mathbb{P}^*$ is equal to 1 – that is, $\mathbb{E}_{\mathbb{P}^*}(\zeta_t) = 1$. Since by definition

$$f_Z(t, T) = \mathbb{E}_{\mathbb{P}^*}(Z_T \mid \mathcal{F}_t) = \mathbb{E}_{\mathbb{P}^*}(F_Z(T, T) \mid \mathcal{F}_t),$$

using the well-known properties of conditional expectation, we obtain

$$f_Z(t, T) = F_Z(t, T) \exp\left(\int_t^T \left(b(u, T) - \xi_u\right) \cdot b(u, T) \, du\right) \mathbb{E}_{\mathbb{P}^*}(\zeta_t),$$

which is the desired result. $\qquad\qquad\square$

Observe that the dynamics of the futures price process $f_Z(t, T)$, $t \in [0, T]$, under the martingale measure $\mathbb{P}^*$ are

$$df_Z(t, T) = f_Z(t, T)\left(\xi_t - b(t, T)\right) \cdot dW_t^*. \qquad (11.74)$$

It is interesting to note that the dynamics of the forward price $F_Z(t, T)$ under the forward measure $\mathbb{P}_T$ are given by the analogous expression

$$dF_Z(t, T) = F_Z(t, T)\left(\xi_t - b(t, T)\right) \cdot dW_t^T. \qquad (11.75)$$

11.5.1 Futures Options

We shall now focus on an explicit formula for the arbitrage price of a European call option written on a futures contract on a zero-coupon bond. Let us denote by $f_B(t, U, T)$ the futures price for settlement at the date T of a U-maturity zero-coupon bond. From (11.74), we have

$$df_B(t, U, T) = f_B(t, U, T)\left(b(t, U) - b(t, T)\right) \cdot dW_t^*$$

subject to the terminal condition $f_B(T, U, T) = B(T, U)$.

The wealth process $V^f(\psi)$ of any futures trading strategy $\psi = (\psi^1, \psi^2)$ equals $V_t^f(\psi) = \psi_t^2 B(t, T)$. A futures trading strategy $\psi = (\psi^1, \psi^2)$ is said to be *self-financing* if its wealth process $V^f = V^f(\psi)$ satisfies the standard relationship

$$V_t^f(\psi) = V_0^f(\psi) + \int_0^t \psi_u^1 \, df_u + \int_0^t \psi_u^2 \, dB(u, T).$$

We fix U and T, and we write briefly f_t instead of $f_B(t, U, T)$ in what follows. Let us consider the relative wealth process $\tilde{V}_t^f = V_t^f(\psi) B^{-1}(t, T)$. As one might expect, the relative wealth of a self-financing futures trading strategy follows a local martingale under the forward measure $\mathbb{P}_T$. Indeed, using Itô's formula we get

$$
\begin{aligned}
d\tilde{V}_t^f &= B^{-1}(t, T) \, dV_t^f + V_t^f \, dB^{-1}(t, T) + d \langle V^f, B^{-1}(\cdot, T) \rangle_t, \\
&= B^{-1}(t, T) \psi_t^1 \, df_t + B^{-1}(t, T) \psi_t^2 \, dB(t, T) + \psi_t^2 B(t, T) \, dB^{-1}(t, T) \\
&\quad + \psi_t^1 \, d \langle f, B^{-1}(\cdot, T) \rangle_t + \psi_t^2 \, d \langle B(\cdot, T), B^{-1}(\cdot, T) \rangle_t \\
&= B^{-1}(t, T) \psi_t^1 \, df_t + \psi_t^1 \, d \langle f, B^{-1}(\cdot, T) \rangle_t.
\end{aligned}
$$

On the other hand, we have

$$d \langle f, B^{-1}(, T) \rangle_t = -f_t B^{-1}(t, T) \big(b(t, U) - b(t, T)\big) \cdot b(t, T) \, dt.$$

Combining these formulas, we arrive at the expression

$$d\tilde{V}_t^f(\psi) = \psi_t^1 f_t B^{-1}(t, T) \big(b(t, U) - b(t, T)\big) \cdot (dW_t^* - b(t, T) \, dt),$$

which is valid under $\mathbb{P}^*$, or equivalently, at the formula

$$d\tilde{V}_t^f(\psi) = \psi_t^1 f_t B^{-1}(t, T) \big(b(t, U) - b(t, T)\big) \cdot dW_t^T,$$

which in turn is satisfied under $\mathbb{P}_T$. We conclude that the relative wealth of any self-financing futures strategy follows a local martingale under the forward measure for the date T. Hence, to find the arbitrage price $\pi_t^f(X)$ at time $t \in [0, T]$ of any $\mathbb{P}_T$-integrable (attainable) contingent claim X of the form $X = g(f_T, T)$, we can make use of the formula

$$\pi_t^f(X) = B(t, T) \, \mathbb{E}_{\mathbb{P}_T} \big(X \,|\, \mathcal{F}_t\big), \quad \forall \, t \in [0, T]. \tag{11.76}$$

To check this, note that if ψ is a futures trading strategy replicating X, then the process $\tilde{V}^f(\psi)$ is a $\mathbb{P}_T$-martingale, and thus

$$\mathbb{E}_{\mathbb{P}_T} \big(X \,|\, \mathcal{F}_t\big) = \mathbb{E}_{\mathbb{P}_T} \big(\tilde{V}_T^f(\psi) \,|\, \mathcal{F}_t\big) = B^{-1}(t, T) \, V_t^f(\psi) = B^{-1}(t, T) \pi_t^f(X),$$

as expected. One may argue that equality (11.76) is trivial, since the equality

$$\pi_t(X) = B(t, T) \, \mathbb{E}_{\mathbb{P}_T}(X \,|\, \mathcal{F}_t)$$

was already established in Proposition 9.6.2, and the arbitrage price of any attainable contingent claim is independent of the choice of financial instruments used in replication. For this argument to be valid, it is necessary to construct first a consistent financial model that encompasses both spot and futures contracts. In this section, for the sake of expositional simplicity, we assume that the expiry date of an option coincides with the settlement date of the underlying futures contract (this restriction is slightly modified in the next section, where the PDE approach is presented).

Proposition 11.5.2 *Assume that $U \geq T$. The arbitrage price at time $t \in [0, T]$ of a European call option with expiry date T and exercise price K, written on the futures contract for a U-maturity zero-coupon bond with delivery date T, equals*

$$C_t^f = B(t, T)\Big(f_t h_U(t, T)N\big(g_1(f_t, t, T)\big) - KN\big(g_2(f_t, t, T)\big)\Big),$$

where

$$g_1(f, t, T) = \frac{\ln(f/K) + \frac{1}{2}\int_t^T (|b(u, U)|^2 - |b(u, T)|^2)du}{v_U(t, T)},$$

$g_2(f, t, T) = g_1(f, t, T) - v_U(t, T)$, *the function $v_U(t, T)$ is given by* (11.45), *and*

$$h_U(t, T) = \exp\Big(\int_t^T \big(b(u, U) - b(u, T)\big) \cdot b(u, T)\, du\Big).$$

Proof. We need to evaluate

$$\begin{aligned}
C_t^f &= B(t, T)\, \mathbb{E}_{\mathbb{P}_T}\big((f_B(T, U, T) - K)^+ \,\big|\, \mathcal{F}_t\big) \\
&= B(t, T)\, \mathbb{E}_{\mathbb{P}_T}\big((F_B(T, U, T) - K)^+ \,\big|\, \mathcal{F}_t\big).
\end{aligned}$$

Proceeding as in the proof of Proposition 11.3.1, we find that (cf. (11.47)–(11.48))

$$C_t^f = B(t, T)\Big(F_t N\big(\tilde{d}_1(F_t, t, T)\big) - KN\big(\tilde{d}_2(F_t, t, T)\big)\Big), \tag{11.77}$$

where $F_t = F_B(t, U, T)$,

$$\tilde{d}_1(F, t, T) = \frac{\ln(F/K) + \frac{1}{2}v_U^2(t, T)}{v_U(t, T)},$$

$\tilde{d}_2(F, t, T) = \tilde{d}_1(F, t, T) - v_U(t, T)$, and $v_U(t, T)$ is given by (11.45). On the other hand, (11.73) yields

$$\begin{aligned}
F_B(t, U, T) &= f_B(t, U, T)\exp\Big(-\int_t^T \big(b(u, T) - b(u, U)\big) \cdot b(u, T)\, du\Big) \\
&= f_B(t, U, T)h_U(t, T).
\end{aligned}$$

Substituting the last formula into (11.77), we find the desired formula. □

For ease of notation, we write f_t to denote the futures price $f_Z(t, T)$ of a tradable security Z. We assume that the volatilities $b(\cdot, T)$ and $\gamma_Z(\cdot, T)$ are deterministic functions. The proof of the next result is similar to that of Proposition 11.5.2, and is thus omitted.

Proposition 11.5.3 *The arbitrage price of a European call option with expiry date T and exercise price K, written on a futures contract settling at time T for delivery of one unit of security Z, is given by the formula*

$$C_t^f = B(t, T)\Big(f_t h(t, T) N\big(g_1(f_t, t, T)\big) - K N\big(g_2(f_t, t, T)\big)\Big),$$

where, for $(f, t) \in \mathbb{R}_+ \times [0, T]$,

$$g_{1,2}(f, t, T) = \frac{\ln(f/K) + \ln h(t, T) \pm \frac{1}{2} v^2(t, T)}{v(t, T)},$$

the function $v(t, T)$ is given by (11.56), and

$$h(t, T) = \exp\Big(\int_t^T \gamma_Z(u, T) \cdot b(u, T)\, du \Big).$$

Assume that the bond volatility $b(u, T)$ vanishes identically for $u \in [0, T]$ and the asset volatility is constant; that is, $\xi_u = \sigma$ for some real number σ. In this case, the valuation formula provided by Proposition 11.5.3 reduces to the classical Black futures option formula. Let us examine the put-call parity for futures options. We write P_t^f to denote the price of a European futures put option with expiry date T and strike price K. Arguing as in the proof of Proposition 11.3.3, one may establish the following result.

Proposition 11.5.4 *Under the assumptions of Proposition 11.5.3, the following put-call parity relationship is valid*

$$C_t^f - P_t^f = B(t, T)\Big(f_Z(t, T) \exp\Big(\int_t^T \gamma_Z(u, T) \cdot b(u, T)\, du \Big) - K \Big).$$

The put-call parity relationship (11.57) for European spot options can be easily established by a direct construction of two portfolios. Let us consider the following two portfolios: long call and short put option; and one unit of the underlying asset and K units of T-maturity zero-coupon bonds. The portfolios have manifestly the same value at time T, namely $X_T - K$. Consequently, if no arbitrage opportunities exist, they have the same value at any time $t \leq T$; that is, (11.57) holds.

Due to the specific daily marking to market feature of the futures market, this simple reasoning is no longer valid for futures options under stochastic interest rates. It appears that the put-call parity for futures options has the form given in Proposition 11.5.4. Note that this relationship depends on the assumed dynamics of the underlying price processes – that is, on the choice of a market model.

11.6 PDE Approach to Interest Rate Derivatives

This section presents the PDE approach to the hedging and valuation of contingent claims in the Gaussian HJM setting. As discussed in Chap. 9.5, PDEs play an important role in the pricing of term-structure derivatives in the framework of diffusion models of short-term interest rates. In such a case, one works with the PDE satisfied by the price process of an interest rate-sensitive security, considered as a function of the time parameter t and the current value of a short-term rate r_t. In the present setting, however, it is not assumed that the short-term rate follows a diffusion process. The PDEs examined in this section are directly related to the price dynamics of bonds and underlying assets. To be more specific, the arbitrage price of a derivative security is expressed in terms of the time parameter t, the current price of an underlying asset and the price of a certain zero-coupon bond. For ease of exposition, we focus on the case of spot and futures European call options.

11.6.1 PDEs for Spot Derivatives

We start by examining the case of a European call option with expiry date T written on a tradable asset Z. We assume throughout that the dynamics of the (spot) price process of Z are governed under a probability measure $\mathbb{P}$ by the expression[4]

$$dZ_t = Z_t \left(\mu_t \, dt + \xi_t \cdot dW_t \right), \tag{11.78}$$

where μ is a stochastic process. For a fixed date $D \geq T$, the price of a bond maturing at time D is assumed to follow, under $\mathbb{P}$

$$dB(t, D) = B(t, D) \left(\kappa_t \, dt + b(t, D) \cdot dW_t \right), \tag{11.79}$$

where κ is a stochastic process. Volatilities ξ and $b(\cdot, D)$ can also follow stochastic processes; we shall assume, however, that the volatility $\xi_t - b(t, D)$ of the forward price of Z is deterministic.

We consider a European option, with expiry date T, written on the forward price of Z for the date D, where $D \geq T$. More precisely, by definition the option's payoff at expiry equals

$$C_T = B(T, D)(F_Z(T, D) - K)^+ = (Z_T - K B(T, D))^+.$$

When $D = T$, we deal with a standard option written on Z. For $D > T$, the option can be interpreted either as an option written on the forward price of Z, with deferred payoff at time D, or simply as an option to exchange one unit of an asset Z for K units of D-maturity bonds. Our goal is to express the option price in terms of the spot prices of the underlying asset and the D-maturity bond.

---

[4] We assume implicitly that Z follows a strictly positive process. It should be stressed that $\mathbb{P}$ is not necessarily a martingale measure.

Suppose that the price C_t admits the representation $C_t = v(Z_t, B(t, D), t)$, where $v : \mathbb{R}_+ \times [0, 1] \times [0, T] \to \mathbb{R}$ is an unknown function satisfying the terminal condition (it is implicitly assumed that $B(T, D) \le 1$; this restriction is not essential, however)

$$v(x, y, T) = y(x/y - K)^+, \quad \forall (x, y) \in \mathbb{R}_+ \times (0, 1].$$

Let $\phi = (\phi^1, \phi^2)$ be a self-financing trading strategy that assumes continuous trading in the option's underlying asset and in D-maturity bonds. As expected, ϕ_t^1 and ϕ_t^2 stand for the number of shares of the underlying asset and the number of units of D-maturity bonds respectively. Assume that the terminal wealth of the portfolio ϕ replicates the option payoff. Then the following chain of equalities is valid for any $t \in [0, T]$

$$V_t(\phi) = \phi_t^1 Z_t + \phi_t^2 B(t, D) = v(Z_t, B(t, D), t) = C_t.$$

Since ϕ is self-financing, its wealth process also satisfies

$$dV_t(\phi) = \phi_t^1 dZ_t + \phi_t^2 dB(t, D).$$

Substituting (11.78)–(11.79) into the last formula, we obtain

$$dV_t(\phi) = \left(\phi_t^1 \mu_t Z_t + \phi_t^2 \kappa_t B(t, D)\right) dt + \left(\phi_t^1 Z_t \xi_t + \phi_t^2 B(t, D)b(t, D)\right) \cdot dW_t.$$

It is useful to note that the second component of the portfolio ϕ is uniquely determined as soon as that the first component of ϕ and the function v are known. Indeed, ϕ^2 can be then found from the equality

$$\phi_t^2 = B^{-1}(t, D)\left(v(Z_t, B(t, D), t) - \phi_t^1 Z_t\right). \tag{11.80}$$

In order to proceed further, we need to assume that the function $v = v(x, y, t)$ is sufficiently smooth on the open domain $(0, \infty) \times (0, 1) \times (0, T)$. Using Itô's formula, we obtain[5]

$$dC_t = v_t \, dt + v_x \, dZ_t + v_y \, dB(t, D) + \tfrac{1}{2}\left(v_{xx} \, d\langle Z \rangle_t + v_{yy} \, d\langle B(\cdot, D) \rangle_t\right)$$
$$+ v_{xy} \, d\langle Z, B(\cdot, D) \rangle_t,$$

where the argument $(Z_t, B(t, T), t)$ is suppressed. Consequently, substitution of the dynamics of Z and B (cf. (11.78)–(11.79)) yields

$$dC_t = \left(v_t + Z_t \mu_t v_x + B(t, D)\kappa_t v_y + Z_t B(t, D) \xi_t \cdot b(t, D)v_{xy}\right) dt$$
$$+ \tfrac{1}{2}\left(Z_t^2 |\xi_t|^2 v_{xx} + B^2(t, D)|b(t, D)|^2 v_{yy}\right) dt$$
$$+ (Z_t \xi_t v_x + B(t, D)b(t, D)v_y) \cdot dW_t.$$

By equating the differential dC_t given by the last formula and the previously found differential $dV_t(\phi)$, we obtain

---

[5] Subscripts on v denote partial derivatives with respect to the corresponding variables.

$$\int_0^t \left(\phi_u^1 Z_u \xi_u + \phi_u^2 B(u, D) b(u, D) - Z_u \xi_u v_x - B(u, D) b(u, D) v_y \right) \cdot dW_u = 0$$

for every $t \in [0, T]$. Eliminating ϕ^2 and rearranging, we get

$$\int_0^t \left(Z_u \xi_u (\phi_u^1 - v_x) + b(u, D)(v - \phi_u^1 Z_u - B(u, D) v_y) \right) \cdot dW_u = 0 \quad (11.81)$$

for every $t \in [0, T]$. Suppose now that

$$\phi_t^1 = v_x(Z_t, B(t, D), t), \quad \forall t \in [0, T]. \quad (11.82)$$

For (11.81) to hold, it suffices to assume, in addition, that the equality

$$v(Z_t, B(t, D), t) = Z_t v_x(Z_t, B(t, D), t) + B(t, D) v_y(Z_t, B(t, D), t)$$

is satisfied for every $t \in [0, T]$. In terms of the function v, we thus have

$$v(x, y, t) = x v_x(x, y, t) + y v_y(x, y, t), \quad (11.83)$$

which should hold for every $(x, y, t) \in (0, \infty) \times (0, 1) \times (0, T)$. Combining (11.83) with (11.80), we find immediately that

$$\phi_t^2 = v_y(Z_t, B(t, D), t), \quad \forall t \in [0, T]. \quad (11.84)$$

Furthermore, by taking partial derivatives of relationship (11.83) with respect to x and y, we obtain

$$\begin{cases} x v_{xx}(x, y, t) + y v_{xy}(x, y, t) = 0, \\ x v_{xy}(x, y, t) + y v_{yy}(x, y, t) = 0. \end{cases} \quad (11.85)$$

Using (11.82) and (11.84), we find that

$$C_t - V_t(\phi) = \int_0^t \left(v_t + \tfrac{1}{2} Z_u^2 |\xi_u|^2 v_{xx} + \tfrac{1}{2} B^2(u, D) |b(u, D)|^2 v_{yy} \right) du$$
$$+ \int_0^t Z_u B(u, D) \xi_u \cdot b(u, D) v_{xy} \, du.$$

Since we have assumed that $C_t = V_t(\phi)$, we arrive at the following PDE that is satisfied by the function $v = v(x, y, t)$

$$v_t + \tfrac{1}{2} |\xi_t|^2 x^2 v_{xx} + \tfrac{1}{2} |b(t, D)|^2 y^2 v_{yy} + \xi_t \cdot b(t, D) xy \, v_{xy} = 0.$$

It should be stressed that we need to consider the last PDE together with the PDE (11.83). Making use of (11.85), we find that the function v solves

$$v_t + \tfrac{1}{2} |\xi_t - b(t, D)|^2 x^2 v_{xx} = 0, \quad (11.86)$$

subject to the terminal condition $v(x, y, T) = y(x/y - K)^+$ for every $(x, y) \in \mathbb{R}_+ \times (0, 1)$.

Note that (11.83) implies that the function v admits the representation $v(x, y, t) = yH(x/y, t)$, where $H : \mathbb{R}_+ \times [0, T] \to \mathbb{R}$ (it is enough to set $x = zy$, and to check that the function $h(z, y, t) = y^{-1}v(zy, y, t)$ does not depend on the second argument, since $h_y(z, y, t) = 0$). Putting $v(x, y, t) = yH(x/y, t)$ into (11.86), we find the PDE satisfied by H

$$H_t(z, t) + \tfrac{1}{2} |\xi_t - b(t, D)|^2 z^2 H_{zz}(z, t) = 0,$$

with the terminal condition $H(z, T) = (z - K)^+$ for every $z \in \mathbb{R}_+$. The solution to the last problem is well known to be given by the formula

$$H(z, t) = zN\big(d_1(z, t, T)\big) - KN\big(d_2(z, t, T)\big),$$

where

$$d_{1,2}(z, t, T) = \frac{\ln(z/K) \pm \tfrac{1}{2}v_Z^2(t, T)}{v_Z(t, T)}$$

and

$$v_Z^2(t, T) = \int_t^T |\xi_u - b(u, D)|^2 \, du.$$

It is now straightforward to derive the following formula[6] for the function $v(x, y, t)$

$$v(x, y, t) = xN\big(k_1(x, y, t, T)\big) - KyN\big(k_2(x, y, t, T)\big),$$

where

$$k_{1,2}(x, y, t, T) = \frac{\ln(x/K) - \ln y \pm \tfrac{1}{2} v_Z^2(t, T)}{v_Z(t, T)}.$$

The replicating strategy ϕ of a call option equals: $\phi_t^1 = v_x\big(Z_t, B(t, D), t\big)$ and $\phi_t^2 = v_y\big(Z_t, B(t, D), t\big)$. Put more explicitly, at any time $t \le T$, the replicating portfolio involves $\phi_t^1 = N\big(k_1(Z_t, B(t, D), t, T)\big)$ units of the underlying asset, combined with $\phi_t^2 = -KN\big(k_2(Z_t, B(t, D), t, T)\big)$ units of D-maturity zero-coupon bonds.

The PDE approach can be carried over to cover the case of more general European contingent claims which may depend on several assets – for instance, to European options on coupon-bearing bonds. We end this section by formulating a result that can be seen as a first step in this direction. For simplicity, we still consider only one underlying asset however.

Proposition 11.6.1 *Assume that the price processes Z and $B(t, D)$ follow (11.78) and (11.79) respectively, and the volatility $\xi_t - b(t, D)$ of the forward price is deterministic. Consider a European contingent claim X of the form $X = B(T, D)g(Z_T / B(T, D))$ settling at time T. The arbitrage price of X equals, for every $t \in [0, T]$,*

$$\pi_t(X) = v\big(Z_t, B(t, D), t\big) = B(t, D)H\big(Z_t B^{-1}(t, D), t\big)$$

---

[6] For $U = T$, this result was already established by means of the forward measure approach (cf. Proposition 11.3.2).

where the function $H : \mathbb{R}_+ \times [0, T] \to \mathbb{R}$ *solves the following PDE*

$$\frac{\partial H}{\partial t}(z, t) + \frac{1}{2} |\xi_t - b(t, D)|^2 z^2 \frac{\partial^2 H}{\partial z^2}(z, t) = 0,$$

with the terminal condition $H(z, T) = g(z)$ *for every* $z \in \mathbb{R}_+$.

Remarks. A savings account can be used in the replication of European claims that settle at time T and have the form $X = B_T g(Z_T / B_T)$ for some function g. Let us consider, for instance, a European option with expiry date T and terminal payoff $(Z_T - K B_T)^+$. If the volatility of the underlying asset is deterministic, Proposition 11.6.1 is in force, and thus replication of such an option involves $N(k_1(Z_t, B_t, t, T))$ units of the underlying asset, combined with the amount $-K B_t N(k_2(Z_t, B_t, t, T))$ held in a savings account. In general, a standard European option cannot be replicated using a savings account.

11.6.2 PDEs for Futures Derivatives

Let us fix three dates T, D and R, such that $T \le \min\{D, R\}$. The futures price of an asset Z in a contract at time R satisfies

$$df_Z(t, R) = f_Z(t, R)(\xi_t - b(t, R)) \cdot dW_t = f_Z(t, R)\zeta_t \cdot dW_t, \qquad (11.87)$$

where $\zeta_t = \xi_t - b(t, R)$. We have assumed that the drift coefficient in the dynamics of $f_Z(t, R)$ vanishes. This is not essential however. Indeed, suppose that a non-zero drift in the dynamics of the futures price is present. Then we may either modify all foregoing considerations in a suitable way, or, more conveniently, we may first make, using Girsanov's theorem, an equivalent change of an underlying probability measure in such a way that the drift of the futures price will disappear. The drift of bond price will thus change – however, it is arbitrary (cf. (11.89)), and it does not enter the final result anyway. It is convenient to assume that the volatility ζ of the futures price and the bond price volatility $b(\cdot, D)$ are deterministic functions.

For convenience, we shall write f_t instead of $f_Z(t, R)$. Consider a European option with the terminal payoff $C_T^f = B(T, D)(f_t - K)^+$ at time T, or equivalently, the payoff $(f_T - K)^+$ at time D (hence it is a standard futures option with deferred payoff). Assume that the price at time t of such an option admits the representation $C_t^f = v(f_t, B(t, D), t)$ for some function $v : \mathbb{R}_+ \times [0, 1] \times [0, T] \to \mathbb{R}$ that satisfies the terminal condition $v(x, y, T) = y(x - K)^+$ for $(x, y) \in \mathbb{R}_+ \times (0, 1]$.

We now consider a self-financing trading strategy $\psi = (\psi^1, \psi^2)$, where ψ^1 and ψ^2 represent positions in futures contracts and D-maturity zero-coupon bonds respectively. It is apparent that the wealth process $V^f(\psi)$ satisfies

$$V_t^f(\psi) = \psi_t^2 B(t, D) = v(f_t, B(t, D), t), \qquad (11.88)$$

where the second equality is a consequence of the assumption that the trading strategy ψ replicates the value of the option.

Furthermore, since ψ is self-financing, its wealth process $V^f(\psi)$ also satisfies

$$dV_t^f(\psi) = \psi_t^1 \, df_t + \psi_t^2 \, dB(t, D),$$

so that

$$dV_t^f(\psi) = \kappa_t \psi_t^2 B(t, D) \, dt + \left(\psi_t^1 f_t \zeta_t + \psi_t^2 B(t, D) b(t, D) \right) \cdot dW_t. \qquad (11.89)$$

On the other hand, assuming that the function $v = v(x, y, t)$ is sufficiently smooth, and applying the Itô formula, we get (the argument $(f_t, B(t, D), t)$ is suppressed)

$$dC_t^f = v_t \, dt + v_x \, df_t + v_y \, dB(t, D) + \tfrac{1}{2}\left(v_{xx} \, d\langle f \rangle_t + v_{yy} \, d\langle B(\cdot, D) \rangle_t \right)$$
$$+ v_{xy} \, d\langle f, B(\cdot, D) \rangle_t.$$

Using the dynamics of f and $B(t, D)$ (cf. (11.79) and (11.87)), we get

$$dC_t^f = \left(v_t + B(t, D)\kappa_t v_y + f_t B(t, D)\zeta_t \cdot b(t, D)v_{xy} \right) dt$$
$$+ \tfrac{1}{2}\left(f_t^2 |\zeta_t|^2 v_{xx} + B^2(t, D)|b(t, D)|^2 v_{yy} \right) dt$$
$$+ \left(f_t \zeta_t v_x + B(t, D)b(t, D)v_y \right) \cdot dW_t.$$

Comparing this formula with (11.89), we obtain the following equality

$$\int_0^t \left(\psi_u^1 f_u \zeta_u + \psi_u^2 B(u, D)b(u, D) - f_u \zeta_u v_x - B(u, D)b(u, D)v_y \right) \cdot dW_u = 0,$$

which is valid for every $t \in [0, T]$. Using (11.88), we find that

$$\int_0^t \left(\psi_u^1 f_u \zeta_u + b(u, D)v - f_u \zeta_u v_x - B(u, D)b(u, D)v_y \right) \cdot dW_u = 0. \qquad (11.90)$$

Suppose now that

$$\psi_t^1 = v_x\left(f_t, B(t, D), t \right), \quad \forall t \in [0, T]. \qquad (11.91)$$

Then for (11.90) to hold, it is enough to assume in addition that

$$v(f_t, B(t, D), t) - B(t, D)v_y\left(f_t, B(t, D), t \right) = 0, \qquad (11.92)$$

or equivalently, that the equality

$$v(x, y, t) = y v_y(x, y, t) \qquad (11.93)$$

is satisfied. Combining (11.88) and (11.92), we find that

$$\phi_t^2 = v_y(f_t, B(t, D), t), \quad \forall t \in [0, T].$$

By taking partial derivatives of (11.93) with respect to x and y, we get

$$\begin{cases} y v_{xy}(x, y, t) = v_x(x, y, t), \\ y v_{yy}(x, y, t) = 0. \end{cases} \qquad (11.94)$$

Using (11.91) and (11.92), we find that

$$C_t^f - V_t^f(\psi) = \int_0^t \left(v_t + \tfrac{1}{2} f_u^2 |\zeta_u|^2 v_{xx} + \tfrac{1}{2} B^2(u, D)|b(u, D)|^2 v_{yy} \right) du$$

$$+ \int_0^t f_u B(u, D)\zeta_u \cdot b(u, D)v_{xy}\, du$$

for every $t \in [0, T]$. Since processes C^f and $V^f(\psi)$ coincide, this leads to the following PDE

$$v_t + \tfrac{1}{2} |\zeta_t|^2 x^2 v_{xx} + \tfrac{1}{2} |b(t, D)|^2 y^2 v_{yy} + \zeta_t \cdot b(t, D)xy v_{xy} = 0.$$

Finally, taking into account (11.94), we find that v satisfies

$$v_t + \tfrac{1}{2} |\zeta_t|^2 x^2 v_{xx} + \zeta_t \cdot b(t, D)x v_x = 0, \tag{11.95}$$

subject to the terminal condition $v(x, y, T) = y(x - K)^+$ for $(x, y) \in \mathbb{R}_+ \times (0, 1]$. In view of (11.93), we restrict our attention to those solutions of (11.95) admitting the representation $v(x, y, t) = yH(x, t)$ for a certain function $H : \mathbb{R}_+ \times [0, T] \to \mathbb{R}$ (indeed, if a function v satisfies (11.93) then the function $h(x, y, t) = y^{-1}v(x, y, t)$ does not depend on the second argument, since the partial derivative $h_y(x, y, t)$ vanishes). Using (11.95), we deduce that H satisfies

$$H_t(x, t) + \tfrac{1}{2} |\zeta_t|^2 x^2 H_{xx}(x, t) + \zeta_t \cdot b(t, D)x H_x(x, t) = 0,$$

with the terminal condition $H(x, T) = (x - K)^+$. In order to simplify the last equation, we define the function $L : \mathbb{R}_+ \times [0, T] \to \mathbb{R}$ by setting $L(z, t) = H(ze^{-\eta(t,T)}, t)$, where

$$\eta(t, T) = \int_t^T \zeta_u \cdot b(u, D)\, du, \quad \forall t \in [0, T].$$

It is straightforward to verify that L solves the PDE

$$L_t(z, t) + \tfrac{1}{2} |\zeta_t|^2 z^2 L_{zz}(z, t) = 0,$$

with the terminal condition $L(z, T) = (z - K)^+$. Arguing along the same lines as in the previous section, we find that

$$L(z, t) = zN\big(d_1(z, t, T)\big) - KN\big(d_2(z, t, T)\big),$$

where

$$d_{1,2}(z, t, T) = \frac{\ln(z/K) \pm \tfrac{1}{2} v_f^2(t, T)}{v_f(t, T)},$$

with

$$v_f^2(t, T) = \int_t^T |\zeta_u|^2\, du = \int_t^T |\xi_u - b(u, R)|^2\, du.$$

Recalling that $v(x, y, t) = yH(x, t) = yL\big(xe^{\eta(t,T)}, t\big)$, we obtain

$$v(x, y, t) = y\Big(xe^{\eta(t,T)}N\big(l_1(x, t, T)\big) - KN\big(l_2(x, t, T)\big)\Big)$$

for $(x, y, t) \in \mathbb{R}_+ \times [0, 1] \times [0, T]$, where

$$l_{1,2}(x, t, T) = \frac{\ln(x/K) + \eta(t, T) \pm \frac{1}{2}v_f^2(t, T)}{v_f(t, T)}.$$

At any date t, the replicating strategy ψ involves ψ_t^1 positions in the underlying futures contract, where $\psi_t^1 = v_x(f_t, B(t, D), t)$, combined with holding ψ_t^2 D-maturity zero-coupon bonds, where

$$\psi_t^2 = B^{-1}(t, D)v\big(f_t, B(t, D), t\big) = v_y\big(f_t, B(t, D), t\big).$$

More explicitly, we have $\psi_t^1 = B(t, D)\, e^{\eta(t,T)}\, N\big(l_1(f_t, t, T)\big)$ and

$$\psi_t^2 = f_t\, e^{\eta(t,T)}\, N\big(l_1(f_t, t, T)\big) - K\, N\big(l_2(f_t, t, T)\big).$$

Example 11.6.1 Assume that a futures contract has a zero-coupon bond maturing at time $U \geq R$ as the underlying asset. Then $\xi_u = b(u, U)$ and thus

$$v_f^2(t, T) = \int_t^T |b(u, U) - b(u, R)|^2\, du.$$

Moreover, in this case we have (writing f in place of x)

$$l_1(f, t, T) = \frac{\ln(f/K) + \int_t^T \gamma(u, U, R) \cdot b(u, D)\, du + \frac{1}{2}\int_t^T |\gamma(u, U, R)|^2\, du}{v(t, T)},$$

where $\gamma(u, U, R) = b(u, U) - b(u, R)$. In particular, if $D = R = T$ we obtain $l_1(f, t, T) = g_1(f, t, T)$ and $\eta(t, T) = h(t, T)$, where g_1 and h are given by (11.5.2) and (11.5.2) respectively.

Proposition 11.6.2 *Suppose that the futures price $f_t = F_Z(t, R)$ of an asset Z satisfies (11.87), where the volatility $\zeta_t = \xi_t - b(t, R)$ is such that $\zeta_t \cdot b(t, D)$ is a deterministic function. Let X be a European contingent claim that settles at time T and has the form $X = B(T, D)g(f_T)$ for some function $g : \mathbb{R}_+ \to \mathbb{R}$. The arbitrage price of X equals*

$$\pi_t^f(X) = v\big(f_t, B(t, D), t\big) = B(t, D)L\big(f_t e^{\eta(t,T)}, t\big)$$

for every $t \in [0, T]$, where

$$\eta(t, T) = \int_t^T (\xi_u - b(u, R)) \cdot b(u, D)\, du, \quad \forall t \in [0, T],$$

and the function $L = L(z, t)$ solves the PDE

$$\frac{\partial L}{\partial t}(z, t) + \frac{1}{2} |\zeta_t|^2 z^2 \frac{\partial^2 L}{\partial z^2}(z, t) = 0,$$

with the terminal condition $L(z, T) = g(z)$ for every $z \in \mathbb{R}_+$.

11.7 Recent Developments

Since we are not in a position to cover in detail recent developments of the HJM methodology, we shall only present a brief overview of existing literature. An important stream of research in existing financial literature is associated with the so-called *geometric properties* of the HJM model. In order to deal with these properties, it is convenient to introduce a different parametrization of the model, specifically, the maturity date T is replaced by the time to maturity, denoted by x in what follows; this is frequently referred to as the *Musiela parametrization* (see Musiela (1993)).

Let us set $r(t, x) = f(t, t + x)$ for $t \geq 0$ and $x \geq 0$, so that $r(t, x)$ is the instantaneous forward rate as seen at time t for the future date $t + x$. Recall that for every $t \geq 0$ and $x \geq 0$ we have

$$B(t, t + x) = \exp \left(\int_0^x r(t, u) \, du \right) = \exp \left(- Y(t, t + x)x \right).$$

Let us introduce an infinite-dimensional stochastic process $r(t, \cdot)$ describing the fluctuations of the whole forward rate curve. The process $r(t, \cdot)$ solves the following stochastic partial differential equation (SPDE)[7]

$$dr(t, x) = \left(\frac{\partial}{\partial x} r(t, x) + \sigma(t, x) \int_0^x \sigma(t, u) \, du \right) dt + \sigma(t, x) \cdot dW_t^*. \quad (11.96)$$

Fundamental theoretical properties of this equation, such as, the existence and uniqueness of solutions in an appropriate functional space H and the existence of invariant measures, were examined by Goldys and Musiela (1996, 2001), Vargiolu (1999), Goldys et al. (2000), and Tehranchi (2005). Carmona and Tehranchi (2004) examined hedging strategies for *infinite-factor HJM models* (see also Kennedy (1994, 1997), Santa Clara and Sornette (2001), and Galluccio et al. (2003b) for related studies).

Consistency. Practical considerations have attracted attention to such features of solutions to (11.96) as the *consistency* and the *finite-dimensional realization* property. Suppose that we are given a finite-dimensional family $\mathcal{K}$ of forward rate curves. We say that a term structure model is *consistent* with $\mathcal{K}$ if, under the assumption that it starts from a curve belonging to $\mathcal{K}$, it produces only yield curves belonging to this family. Practitioners are interested in consistency of a particular HJM model with a

---

[7] For the general theory of SPDEs, see Da Prato and Zabczyk (1992).

given family of yield curves that is used to estimate in a parsimonious way the term structure of interest rates.For instance, the problem of consistency of a HJM model with the Nelson and Siegel (1987) family, which can be represented as follows:

$$f(t, T) = \alpha_0 + \alpha_1 e^{-\lambda(T-t)} + \alpha_2 \lambda(T - t) e^{-\lambda(T-t)},$$

was completely solved by Filipović (1999). For more details on consistency of a HJM model, see Björk and Christensen (1999), Filipović (2000, 2001a), as well as the survey by Björk (2004).

Finite-dimensional realizations. A HJM term structure model is said to have the *finite-dimensional realization* property if there exists an n-dimensional diffusion process Z (interpreted as the state vector process), which satisfies

$$dZ_t = a(Z_t) dt + b(Z_t) \cdot dW_t,$$

and a map G from $\mathbb{R}^n$ to the Hilbert (or, more generally, Fréchet) space H such that $r(t, x) = G(Z_t)$ for every $t \in \mathbb{R}_+$. Let us note that the finite-dimensional realization property is important both for calibration and hedging purposes, especially in the case where Z has a financial interpretation (for instance, Z may represent a vector of benchmark instantaneous forward rates). For more information on the finite-dimensional realization property of term structure models, we refer to Björk and Gombani (1999), Björk and Svensson (2001), Filipović and Teichmann (2002, 2003), Björk and Landén (2002), Björk (2004), and Björk et al. (2004).

Stochastic volatility. Though formally the coefficient $\sigma(t, T)$ in the dynamics (11.4) may follow a stochastic process, it is rather difficult to make a reasonable choice of stochastic volatility in a general HJM framework.

A particular stochastic volatility extension of the HJM model was examined by Andreasen et al. (1998), who formally introduced futures contracts on the *instantaneous variance* $|b(T, U)|^2$ of zero-coupon bonds. Let us present the main lines of their approach. For any $t \le T \le U$, let $V(t, T, U)$ stand for the futures prices at time t for delivery of the claim $|b(T, U)|^2$ at time T. Andreasen et al. (1998) postulate that, under the real-life probability $\mathbb{P}$, we have

$$dV(t, T, U) = V(t, T, U)\big(\mu_V(t, T, U) dt + \sigma_V(t, T, U) dW_t\big).$$

They argue that although futures contracts on the instantaneous variance are not traded, they can be formally replicated by trading in futures on continuously compounded yields, and the latter contracts are indeed traded. They deduce that, under the spot martingale measure $\mathbb{P}^*$, we necessarily have

$$dV(t, T, U) = V(t, T, U)\sigma_V(t, T, U) dW_t^*.$$

Finally, they investigate the effects of the stochastic volatility $b^2(t, T) = V(t, t, T)$ on prices of bond options for various specifications of volatilities and correlations.

In a recent paper by Björk et al. (2004), the authors take a completely different approach. They assume that the coefficient $\sigma(t, T)$ has the form $\sigma(t, f(t, T), Y_t)$;

that is, they allow for explicit dependence of the bond volatility on some (hidden) Markov process Y that obeys

$$dY_t = \alpha(Y_t)\, dt + \beta(Y_t) \cdot dW_t.$$

Filipović and Teichmann (2004) extend this model by postulating that

$$dY_t = \alpha(r_t, Y_t)\, dt + \beta(r_t, Y_t) \cdot dW_t,$$

where r_t stands for the infinite-dimensional process $r(t, \cdot)$.

12

Market LIBOR Models

As was mentioned already, the acronym LIBOR stands for the *London Interbank Offered Rate*. It is the rate of interest offered by banks on deposits from other banks in eurocurrency markets. Also, it is the floating rate commonly used in interest rate swap agreements in international financial markets (in domestic financial markets as the reference interest rate for a floating rate loans it is customary to take a *prime rate* or a *base rate*). LIBOR is determined by trading between banks and changes continuously as economic conditions change. For more information on market conventions related to the LIBOR and Eurodollar futures, we refer to Sect. 9.3.4.

In this chapter, we present an overview of recently developed methodologies related to the arbitrage-free modelling of *market rates,* such as LIBORs. In contrast to more traditional aproaches, term structure models developed recently by, among others, (Miltersen et al. 1997), Brace et al. (1997), Musiela and Rutkowski (1997), Jamshidian (1997a), Hunt et al. (1996, 2000), Hunt and Kennedy (1996, 1997, 1998), and Andersen and Andreasen (2000b), are tailored to handle the most actively traded interest-rate options, such as caps and swaptions. For this reason, they typically enjoy a higher degree of tractability than the classical term structure models based on the diffusion-type behavior of instantaneous (spot or forward) rates.

Recall that the Heath-Jarrow-Morton methodology of term structure modelling is based on the arbitrage-free dynamics of instantaneous, continuously compounded forward rates. The assumption that instantaneous rates exist is not always convenient, since it requires a certain degree of smoothness with respect to the *tenor* (i.e., maturity) of bond prices and their volatilities. An alternative construction of an arbitrage-free family of bond prices, making no reference to the instantaneous, continuously compounded rates, is in some circumstances more suitable.

The first step in this direction was taken by Sandmann and Sondermann (1993b), who focused on the effective annual interest rate (cf. Sect. 10.1.2). This idea was further developed by Goldys et al. (2000), Musiela (1994), Sandmann et al. (1995), Miltersen et al. (1997) and Brace et al. (1997).

It is worth pointing out that in all these papers, the HJM framework is adopted (at least implicitly). For instance, Goldys et al. (2000) introduce a HJM-type model based on the rate $j(t, T)$, which is related to the instantaneous forward rate through the formula $1 + j(t, T) = e^{f(t,T)}$. The model put forward in this paper assumes a deterministic volatility function for the process $j(t, T)$. A slightly more general case of nominal annual rates $q(t, T)$, which satisfy (δ representing the duration of each compounding period)

$$(1 + \delta q(t, T))^{1/\delta} = e^{f(t,T)},$$

was studied by Musiela (1994), who assumes the deterministic volatility $\gamma(t, T)$ of each nominal annual rate $q(t, T)$. This implies the following form of the coefficient σ in the dynamics of the instantaneous forward rate

$$\sigma(t, T) = \delta^{-1}(1 - e^{-\delta f(t,T)})\gamma(t, T),$$

so that the model is indeed well-defined (that is, instantaneous forward rates, and thus also the nominal annual rates, do not explode). Unfortunately, these models do not give closed-form solutions for zero-coupon bond options, and thus a numerical approach to option pricing is required. Miltersen et al. (1997) focus on the actuarial (or effective) forward rates $a(t, T, U)$ satisfying

$$\left(1 + a(t, T, T + \delta)\right)^{\delta} = \exp\left(\int_{T}^{T+\delta} f(t, u)\, du\right).$$

They show that a closed-form solution for the bond option price is available when $\delta = 1$. More specifically, an *interest rate cap* is priced according to the market standard. However, the model is not explicitly identified and its arbitrage-free features are not examined, thus leaving open the question of pricing other interest rate derivatives. These problems were addressed in part in a paper by Sandmann et al. (1995), where a lognormal-type model based on an add-on forward rate (*add-on yield*) $f_s(t, T, T + \delta)$, defined by

$$1 + \delta f_s(t, T, T + \delta) = \exp\left(\int_{T}^{T+\delta} f(t, u)\, du\right),$$

was analyzed. Finally, using a different approach, Brace et al. (1997) explicitly identify the dynamics of all rates $f_s(t, T, T + \delta)$ under the martingale measure $\mathbb{P}^*$ and analyze the properties of the model.

Let us summarize the content of this chapter. We start by describing in Sect. 12.1 forward and futures LIBORs. The properties of the LIBOR in the Gaussian HJM model are also dealt with in this section. Subsequently, in Sect. 12.4 we present various approaches to LIBOR market models. Further properties of these models are examined in Sect. 12.5. In Sect. 12.2 we describe interest rate cap and floor agreements. Next, we provide in Sect. 12.3 the valuation results for these contracts within the framework of the Gaussian HJM model. In Sect. 12.6, we deal with the valuation of contingent claims within the framework of the lognormal LIBOR market model. In the last section, we present briefly some extensions of this model.

12.1 Forward and Futures LIBORs

We shall frequently assume that we are given a prespecified collection of reset/ settlement dates $0 \leq T_0 < T_1 < \cdots < T_n$, referred to as the *tenor structure*. Also, we shall write $\delta_j = T_j - T_{j-1}$ for $j = 1, \ldots, n$. As usual, $B(t, T)$ stands for the price at time t of a T-maturity zero-coupon bond, $\mathbb{P}^*$ is the spot martingale measure, while $\mathbb{P}_{T_j}$ (respectively, $\mathbb{P}_{T_{j+1}}$) is the forward martingale measure associated with the date T_j (respectively, T_{j+1}). The corresponding Brownian motions are denoted by W^* and W^{T_j} (respectively, $W^{T_{j+1}}$). Also, we write $F_B(t, T, U) = B(t, T)/B(t, U)$. Finally, $\pi_t(X)$ is the value (that is, the arbitrage price) at time t of a European claim X.

Our first task is to examine those properties of interest rate forward and futures contracts that are universal, in the sense that do not rely on specific assumptions imposed on a particular model of the term structure of interest rates. To this end, we fix an index j, and we consider various interest rates related to the period $[T_j, T_{j+1}]$.

12.1.1 One-period Swap Settled in Arrears

Let us first consider a one-period swap agreement settled in arrears; i.e., with the *reset date* T_j and the *settlement date* T_{j+1} (more realistic multi-period swap agreements are examined in Chap. 13). By the contractual features, the long party pays $\delta_{j+1}\kappa$ and receives $B^{-1}(T_j, T_{j+1}) - 1$ at time T_{j+1}. Equivalently, he pays an amount $Y_1 = 1 + \delta_{j+1}\kappa$ and receives $Y_2 = B^{-1}(T_j, T_{j+1})$ at this date. The values of these payoffs at time $t \leq T_j$ are

$$\pi_t(Y_1) = B(t, T_{j+1})\left(1 + \delta_{j+1}\kappa\right), \quad \pi_t(Y_2) = B(t, T_j).$$

The second equality above is trivial, since the payoff Y_2 is equivalent to the unit payoff at time T_j. Consequently, for any fixed $t \leq T_j$, the value of the *forward swap rate* that makes the contract worthless at time t can be found by solving for $\kappa = \kappa_t$ the following equation

$$\pi_t(Y_1) = B(t, T_{j+1})\left(1 + \delta_{j+1}\kappa_t\right) = B(t, T_j) = \pi_t(Y_2).$$

It is thus apparent that

$$\kappa_t = \frac{B(t, T_j) - B(t, T_{j+1})}{\delta_{j+1} B(t, T_{j+1})}, \quad \forall\, t \in [0, T_j].$$

Note that κ_t coincides with the *forward LIBOR* $L(t, T_j)$ which, by convention, is set to satisfy

$$1 + \delta_{j+1} L(t, T_j) \overset{\text{def}}{=} \frac{B(t, T_j)}{B(t, T_{j+1})}.$$

It is also useful to observe that

$$1 + \delta_{j+1} L(t, T_j) = F_B(t, T_j, T_{j+1}) = \mathbb{E}_{\mathbb{P}_{T_{j+1}}}(B^{-1}(T_j, T_{j+1}) \,|\, \mathcal{F}_t), \quad (12.1)$$

where the last equality is a consequence of the definition of the forward measure $\mathbb{P}_{T_{j+1}}$. We conclude that in order to determine the forward LIBOR $L(\cdot, T_j)$, it is enough to find the forward price of the claim $B^{-1}(T_j, T_{j+1})$ for the settlement date T_{j+1}. Furthermore, it is evident that the process $L(\cdot, T_j)$ necessarily follows a martingale under the forward probability measure $\mathbb{P}_{T_{j+1}}$.

Recall that in the HJM framework, we have

$$dF_B(t, T_j, T_{j+1}) = F_B(t, T_j, T_{j+1})\big(b(t, T_j) - b(t, T_{j+1})\big) \cdot dW_t^{T_{j+1}} \qquad (12.2)$$

under $\mathbb{P}_{T_{j+1}}$, where $b(\cdot, T)$ is the price volatility of the T-maturity zero-coupon bond. On the other hand, $L(\cdot, T_j)$ can be shown to admit the following representation

$$dL(t, T_j) = L(t, T_j)\lambda(t, T_j) \cdot dW_t^{T_{j+1}}$$

for a certain adapted process $\lambda(\cdot, T_j)$. Combining the last two formulas with (12.1), we arrive at the following fundamental relationship

$$\frac{\delta_{j+1}L(t, T_j)}{1 + \delta_{j+1}L(t, T_j)}\, \lambda(t, T_j) = b(t, T_j) - b(t, T_{j+1}), \quad \forall\, t \in [0, T_j]. \qquad (12.3)$$

It is worth stressing that equality (12.3) will play an essential role in the construction of the so-called *lognormal LIBOR market model*. For instance, in the construction based on the backward induction, relationship (12.3) will allow us to specify uniquely the forward measure for the date T_j, provided that $\mathbb{P}_{T_{j+1}}$, $W^{T_{j+1}}$ and the volatility $\lambda(t, T_j)$ are known (we may postulate, for instance, that $\lambda(\cdot, T_j)$ is a given deterministic function).

Recall also that in the HJM framework the Radon-Nikodým derivative of $\mathbb{P}_{T_j}$ with respect to $\mathbb{P}_{T_{j+1}}$ is known to satisfy

$$\frac{d\mathbb{P}_{T_j}}{d\mathbb{P}_{T_{j+1}}} = \mathcal{E}_{T_j}\bigg(\int_0^\cdot \big(b(t, T_j) - b(t, T_{j+1})\big) \cdot dW_t^{T_{j+1}}\bigg). \qquad (12.4)$$

In view of (12.3), we thus have

$$\frac{d\mathbb{P}_{T_j}}{d\mathbb{P}_{T_{j+1}}} = \mathcal{E}_{T_j}\bigg(\int_0^\cdot \frac{\delta_{j+1}L(t, T_j)}{1 + \delta_{j+1}L(t, T_j)}\, \lambda(t, T_j) \cdot dW_t^{T_{j+1}}\bigg).$$

For our further purposes, it is also useful to observe that this density admits the following representation

$$\frac{d\mathbb{P}_{T_j}}{d\mathbb{P}_{T_{j+1}}} = cF_B(T_j, T_j, T_{j+1}) = c\big(1 + \delta_{j+1}L(T_j, T_j)\big), \quad \mathbb{P}_{T_{j+1}}\text{-a.s.,} \qquad (12.5)$$

where $c > 0$ is a normalizing constant, and thus we have that

$$\frac{d\mathbb{P}_{T_j}}{d\mathbb{P}_{T_{j+1}}}\bigg|_{\mathcal{F}_t} = cF_B(t, T_j, T_{j+1}) = c\big(1 + \delta_{j+1}L(t, T_j)\big), \quad \mathbb{P}_{T_{j+1}}\text{-a.s.,}$$

for any date $t \leq T_j$.

Finally, the dynamics of the process $L(\cdot, T_j)$ under the probability measure $\mathbb{P}_{T_j}$ are given by a somewhat involved stochastic differential equation

$$dL(t, T_j) = L(t, T_j) \left(\frac{\delta_{j+1} L(t, T_j) |\lambda(t, T_j)|^2}{1 + \delta_{j+1} L(t, T_j)} \, dt + \lambda(t, T_j) \cdot dW_t^{T_j} \right).$$

As we shall see in what follows, it is nevertheless not hard to determine the probability distribution of $L(\cdot, T_j)$ under the forward measure $\mathbb{P}_{T_j}$ – at least in the case of the deterministic volatility function $\lambda(\cdot, T_j)$.

12.1.2 One-period Swap Settled in Advance

Consider now a similar swap that is, however, settled in advance – that is, at time T_j. Our first goal is to determine the forward swap rate implied by such a contract. Note that under the present assumptions, the long party (formally) pays an amount $Y_1 = 1 + \delta_{j+1}\kappa$ and receives $Y_2 = B^{-1}(T_j, T_{j+1})$ at the settlement date T_j (which coincides here with the reset date).

It is easy to see that the values of these payoffs at time $t \le T_j$ admit the following representations

$$\pi_t(Y_1) = B(t, T_j)\big(1 + \delta_{j+1}\kappa\big),$$

and

$$\pi_t(Y_2) = B(t, T_j)\mathbb{E}_{\mathbb{P}_{T_j}}\big(B^{-1}(T_j, T_{j+1}) \,|\, \mathcal{F}_t\big).$$

The value $\kappa = \tilde{\kappa}_t$ of the *modified forward swap rate* that makes the swap agreement settled in advance worthless at time t can be found from the equality $\pi_t(Y_1) = \pi_t(Y_2)$, where

$$\pi_t(Y_1) = B(t, T_j)\big(1 + \delta_{j+1}\kappa\big)$$

and

$$\pi_t(Y_2) = B(t, T_j)\mathbb{E}_{\mathbb{P}_{T_j}}\big(B^{-1}(T_j, T_{j+1}) \,|\, \mathcal{F}_t\big).$$

It is clear that

$$\tilde{\kappa}_t = \delta_{j+1}^{-1}\big(\mathbb{E}_{\mathbb{P}_{T_j}}\big(B^{-1}(T_j, T_{j+1}) \,|\, \mathcal{F}_t\big) - 1\big).$$

We are in a position to introduce the *modified forward LIBOR* $\tilde{L}(t, T_j)$ by setting, for $t \in [0, T_j]$,

$$\tilde{L}(t, T_j) \overset{\text{def}}{=} \delta_{j+1}^{-1}\big(\mathbb{E}_{\mathbb{P}_{T_j}}\big(B^{-1}(T_j, T_{j+1}) \,|\, \mathcal{F}_t\big) - 1\big).$$

Let us make two remarks. First, it is clear that finding the modified LIBOR $\tilde{L}(\cdot, T_j)$ is essentially equivalent to pricing the claim $B^{-1}(T_j, T_{j+1})$ at T_j (more precisely, we need to know the forward price of this claim for the date T_j).

Second, it is useful to observe that

$$\tilde{L}(t, T_j) = \mathbb{E}_{\mathbb{P}_{T_j}}\left(\frac{1 - B(T_j, T_{j+1})}{\delta_{j+1} B(T_j, T_{j+1})} \,\bigg|\, \mathcal{F}_t \right) = \mathbb{E}_{\mathbb{P}_{T_j}}\big(L(T_j, T_j) \,|\, \mathcal{F}_t\big).$$

In particular, it is evident that at the reset date T_j the two forward LIBORs introduced above coincide, since manifestly

$$\tilde{L}(T_j, T_j) = \frac{1 - B(T_j, T_{j+1})}{\delta_{j+1} B(T_j, T_{j+1})} = L(T_j, T_j).$$

To summarize, the standard forward LIBOR $L(\cdot, T_j)$ satisfies

$$L(t, T_j) = \mathbb{E}_{\mathbb{P}_{T_{j+1}}} (L(T_j, T_j) \,|\, \mathcal{F}_t), \quad \forall t \in [0, T_j],$$

with the initial condition

$$L(0, T_j) = \frac{B(0, T_j) - B(0, T_{j+1})}{\delta_{j+1} B(0, T_{j+1})},$$

while for the modified LIBOR $\tilde{L}(\cdot, T_j)$ we have

$$\tilde{L}(t, T_j) = \mathbb{E}_{\mathbb{P}_{T_j}} (\tilde{L}(T_j, T_j) \,|\, \mathcal{F}_t), \quad \forall t \in [0, T_j],$$

with the initial condition

$$\tilde{L}(0, T_j) = \delta_{j+1}^{-1} \big(\mathbb{E}_{\mathbb{P}_{T_j}} (B^{-1}(T_j, T_{j+1})) - 1 \big).$$

Note that the last condition depends not only on the initial term structure, but also on the volatilities of bond prices (see, e.g., formula (12.11) below).

12.1.3 Eurodollar Futures

As was mentioned in Sect. 9.3, *Eurodollar futures contract* is a futures contract in which the LIBOR plays the role of an underlying asset. We assume here that, by convention, at the contract's maturity date T_j, the quoted Eurodollar futures price $E(T_j, T_j)$ is set to satisfy

$$E(T_j, T_j) \stackrel{\text{def}}{=} 1 - \delta_{j+1} L(T_j).$$

Equivalently, in terms of the price of a zero-coupon bond we have $E(T_j, T_j) = 2 - B^{-1}(T_j, T_{j+1})$. From the general properties of futures contracts, it follows that the Eurodollar futures price at time $t \leq T_j$ equals

$$E(t, T_j) \stackrel{\text{def}}{=} \mathbb{E}_{\mathbb{P}^*}(E(T_j, T_j)) = 1 - \delta_{j+1} \mathbb{E}_{\mathbb{P}^*} (L(T_j, T_j) \,|\, \mathcal{F}_t)$$

and thus

$$E(t, T_j) = 2 - \mathbb{E}_{\mathbb{P}^*} \big(B^{-1}(T_j, T_{j+1}) \,|\, \mathcal{F}_t \big). \tag{12.6}$$

Recall that the probability measure $\mathbb{P}^*$ represents the spot martingale measure in a given model of the term structure. It seems natural to introduce the concept of the *futures LIBOR*, associated with the Eurodollar futures contract, through the following definition.

Definition 12.1.1 Let $E(t, T_j)$ be the Eurodollar futures price at time t for the settlement date T_j. The implied *futures LIBOR* $L^f(t, T_j)$ satisfies

$$E(t, T_j) = 1 - \delta_{j+1} L^f(t, T_j), \quad \forall t \in [0, T_j]. \tag{12.7}$$

It follows immediately from (12.6)-(12.7) that the following equality is valid

$$1 + \delta_{j+1} L^f(t, T_j) = \mathbb{E}_{\mathbb{P}^*}\big(B^{-1}(T_j, T_{j+1}) \,\big|\, \mathcal{F}_t\big).$$

Equivalently, we have

$$L^f(t, T_j) = \delta_{j+1}^{-1}\left(\mathbb{E}_{\mathbb{P}^*}\big(B^{-1}(T_j, T_{j+1}) \,\big|\, \mathcal{F}_t\big) - 1\right) = \mathbb{E}_{\mathbb{P}^*}(L(T_j, T_j) \,|\, \mathcal{F}_t).$$

It is thus clear that the futures LIBOR follows a martingale under the spot martingale measure $\mathbb{P}^*$.

12.1.4 LIBOR in the Gaussian HJM Model

In this section, we make a standing assumption that the bond price volatilities $b(t, T_j)$ are deterministic functions, that is, we place ourselves within the Gaussian HJM framework. In this case, it is not hard to express forward and futures LIBORs in terms of bond prices and bond price volatilities. Furthermore, as soon as the dynamics of various rates under forward probability measures are known explicitly, it is straightforward to value interest-rate sensitive derivatives. Recall that in the HJM framework we have

$$dF_B(t, T_j, T_{j+1}) = F_B(t, T_j, T_{j+1})\big(b(s, T_j) - b(s, T_{j+1})\big) \cdot dW_t^{T_{j+1}}$$

with the terminal condition $F_B(T_j, T_j, T_{j+1}) = B^{-1}(T_j, T_{j+1})$. Also, the spot and forward Brownian motions are known to satisfy

$$dW_t^{T_j} = dW_t^{T_{j+1}} - \big(b(s, T_j) - b(s, T_{j+1})\big) dt, \tag{12.8}$$

and

$$dW_t^* = dW_t^{T_j} + b(s, T_j) \, dt. \tag{12.9}$$

In view of the relationships above, it is quite standard to establish the following proposition (see Flesaker (1993b) for related results). It is worth pointing out that in the present framework there are no ambiguities in the definition of the spot probability measure (this should be contrasted with the case of the discrete-tenor lognormal model of forward LIBORs, in which the spot measure is not uniquely defined).

For conciseness, we shall frequently write $F_B(t) = F_B(t, T_j, T_{j+1})$. Also, we write, as usual,

$$\gamma(t, T_j, T_{j+1}) = b(t, T_j) - b(t, T_{j+1})$$

to denote the volatility of the process $F_B(t, T_j, T_{j+1})$.

Proposition 12.1.1 *Assume the Gaussian HJM model of the term structure of interest rates. Then the following relationships are valid*

$$1 + \delta_{j+1} L(t, T_j) = F_B(t, T_j, T_{j+1}), \tag{12.10}$$

$$1 + \delta_{j+1} \tilde{L}(t, T_j) = F_B(t, T_j, T_{j+1}) \, e^{\int_t^{T_j} |\gamma(u, T_j, T_{j+1})|^2 \, du}, \tag{12.11}$$

$$1 + \delta_{j+1} L^f(t, T_j) = F_B(t, T_j, T_{j+1}) \, e^{-\int_t^{T_j} b(u, T_{j+1}) \cdot \gamma(u, T_j, T_{j+1}) \, du}. \tag{12.12}$$

Proof. For brevity, we shall write $F_B(t) = F_B(t, T_j, T_{j+1})$. The first formula is in fact universal (see (12.1)). For the second, note that (cf. (12.2))

$$dF_B(t) = F_B(t)\gamma(t, T_j, T_{j+1}) \cdot \left(dW_t^{T_j} + \gamma(t, T_j, T_{j+1}) \, dt \right).$$

Consequently,

$$F_B(T_j) = F_B(t) \exp \left(\int_t^{T_j} \gamma_u \cdot \left(dW_u^{T_j} + \gamma_u \, du \right) - \frac{1}{2} \int_t^{T_j} |\gamma_u|^2 du \right),$$

where we write $\gamma_u = \gamma(u, T_j, T_{j+1})$. Since $B^{-1}(T_j, T_{j+1}) = F_B(T_j)$, upon taking conditional expectation with respect to the σ-field $\mathcal{F}_t$, we obtain (12.11). Furthermore, we have

$$dF_B(t) = F_B(t)\gamma_t \cdot \left(dW_t^* - b(t, T_{j+1}) \, dt \right)$$

and thus

$$F_B(T_j) = F_B(t) \exp \left(\int_t^{T_j} \gamma_u \cdot \left(dW_u^* - b(u, T_{j+1}) \, du \right) - \frac{1}{2} \int_t^{T_j} |\gamma_u|^2 du \right).$$

This leads to equality (12.12).   □

Dynamics of the forward LIBORs are also easy to find, as the following corollary shows.

Corollary 12.1.1 *We have*

$$dL(t, T_j) = \delta_{j+1}^{-1} \left(1 + \delta_{j+1} L(t, T_j) \right) \gamma(t, T_j, T_{j+1}) \cdot dW_t^{T_{j+1}}, \tag{12.13}$$

$$d\tilde{L}(t, T_j) = \delta_{j+1}^{-1} \left(1 + \delta_{j+1} \tilde{L}(t, T_j) \right) \gamma(t, T_j, T_{j+1}) \cdot dW_t^{T_j}, \tag{12.14}$$

$$dL^f(t, T_j) = \delta_{j+1}^{-1} \left(1 + \delta_{j+1} L^f(t, T_j) \right) \gamma(t, T_j, T_{j+1}) \cdot dW_t^*. \tag{12.15}$$

Proof. Formula (12.13) is an immediate consequence of (12.1) combined with (12.2). Expressions (12.14) and (12.15) can be derived by applying Itô's rule to equalities (12.11) and (12.12) respectively.   □

From Corollary 12.1.1, it is rather clear that closed-form expressions for values of options written on forward or futures LIBORs are not available in the Gaussian HJM framework.

12.2 Interest Rate Caps and Floors

An *interest rate cap* (known also as a *ceiling rate agreement,* or briefly *CRA*) is a contractual arrangement where the grantor (seller) has an obligation to pay cash to the holder (buyer) if a particular interest rate exceeds a mutually agreed level at some future date or dates. Similarly, in an *interest rate floor,* the grantor has an obligation to pay cash to the holder if the interest rate is below a preassigned level. When cash is paid to the holder, the holder's net position is equivalent to borrowing (or depositing) at a rate fixed at that agreed level. This assumes that the holder of a cap (or floor) agreement also holds an underlying asset (such as a deposit) or an underlying liability (such as a loan). Finally, the holder is not affected by the agreement if the interest rate is ultimately more favorable to him than the agreed level. This feature of a cap (or floor) agreement makes it similar to an option.

Specifically, a *forward start cap* (or a *forward start floor*) is a strip of caplets (floorlets), each of which is a call (put) option on a forward rate respectively. Let us denote by κ and by δ_j the cap *strike rate* and the length of the *accrual period* of a caplet, respectively. We shall check that an interest rate caplet (i.e., one leg of a cap) may also be seen as a put option with strike price 1 (per dollar of principal) that expires at the caplet start day on a discount bond with face value $1 + \kappa\delta_j$ maturing at the caplet end date. This property makes the valuation of a cap relatively simple; essentially, it can be reduced to the problem of option pricing on zero-coupon bonds.

Similarly to the swap agreements examined in the next chapter, interest rate caps and floors may be settled either *in arrears* or *in advance.* In a forward cap or floor with the notional principal N settled in arrears at dates T_j, $j = 1, \ldots, n$, where $T_j - T_{j-1} = \delta_j$ the cash flows at times T_j are

$$N\big(L(T_{j-1}, T_{j-1}) - \kappa\big)^+\delta_j$$

and

$$N\big(\kappa - L(T_{j-1}, T_{j-1})\big)^+\delta_j$$

respectively, where the (spot) LIBOR $L(T_{j-1}, T_{j-1})$ is determined at the reset date T_{j-1}, and it formally satisfies

$$B(T_{j-1}, T_j)^{-1} = 1 + \delta_j L(T_{j-1}, T_{j-1}). \tag{12.16}$$

The arbitrage price at time $t \leq T_0$ of a *forward cap,* denoted by $\mathbf{FC}_t$, is

$$\mathbf{FC}_t = \sum_{j=1}^{n} \mathbb{E}_{\mathbb{P}^*}\Big(\frac{B_t}{B_{T_j}}\big(L(T_{j-1}, T_{j-1}) - \kappa\big)^+\delta_j \,\Big|\, \mathcal{F}_t\Big). \tag{12.17}$$

We have assumed here, without loss of generality, that the notional principal $N = 1$. This convention will be in force throughout the rest of this chapter. Let us consider a *caplet* (i.e., one leg of a cap) with reset date T_{j-1} and settlement date $T_j = T_{j-1}+\delta_j$.

The value at time t of a caplet equals (for simplicity, we write $\tilde{\delta}_j = 1 + \kappa\delta_j$)

$$\textbf{Cpl}_t = \mathbb{E}_{\mathbb{P}^*}\left\{ \frac{B_t}{B_{T_j}}\left(\delta_j^{-1}\big(B(T_{j-1}, T_j)^{-1} - 1 \big) - \kappa \right)^+ \delta_j \,\Big|\, \mathcal{F}_t \right\}$$

$$= \mathbb{E}_{\mathbb{P}^*}\left\{ \frac{B_t}{B_{T_j}}\left(\frac{1}{B(T_{j-1}, T_j)} - \tilde{\delta}_j \right)^+ \Big|\, \mathcal{F}_t \right\}$$

$$= \mathbb{E}_{\mathbb{P}^*}\left\{ \frac{B_t}{B_{T_{j-1}}}\left(\frac{1}{B(T_{j-1}, T_j)} - \tilde{\delta}_j \right)^+ \mathbb{E}_{\mathbb{P}^*}\!\left(\frac{B_{T_{j-1}}}{B_{T_j}} \,\Big|\, \mathcal{F}_{T_{j-1}} \right) \Big|\, \mathcal{F}_t \right\}$$

$$= \mathbb{E}_{\mathbb{P}^*}\left\{ \frac{B_t}{B_{T_{j-1}}}\left(1 - \tilde{\delta}_j B(T_{j-1}, T_j) \right)^+ \Big|\, \mathcal{F}_t \right\}$$

$$= B(t, T_{j-1})\, \mathbb{E}_{\mathbb{P}_{T_{j-1}}}\left\{ \left(1 - \tilde{\delta}_j B(T_{j-1}, T_j) \right)^+ \Big|\, \mathcal{F}_t \right\},$$

where the last equality was deduced from Proposition 9.6.2. It is apparent that a caplet is essentially equivalent to a put option on a zero-coupon bond; it may also be seen as an option on a one-period forward swap.

Since the cash flow of the jth caplet at time T_j is an $\mathcal{F}_{T_{j-1}}$-measurable random variable, we may also use Corollary 9.6.1 to express the value of the caplet in terms of expectation under a forward measure. Indeed, from (9.38) we have

$$\textbf{Cpl}_t = B(t, T_{j-1})\, \mathbb{E}_{\mathbb{P}_{T_{j-1}}}\left(B(T_{j-1}, T_j)\big(L(T_{j-1}, T_{j-1}) - \kappa \big)^+ \delta_j \,\Big|\, \mathcal{F}_t \right).$$

Consequently, using (12.16) we get once again the equality

$$\textbf{Cpl}_t = B(t, T_{j-1})\, \mathbb{E}_{\mathbb{P}_{T_{j-1}}}\left(\big(1 - \tilde{\delta}_j B(T_{j-1}, T_j) \big)^+ \,\Big|\, \mathcal{F}_t \right),$$

which is valid for every $t \in [0, T_{j-1}]$.

Finally, the equivalence of a cap and a put option on a zero-coupon bond can be explained in an intuitive way. For this purpose, it is enough to examine two basic features of both contracts: the exercise set and the payoff value. Let us consider the jth caplet. A caplet is exercised at time T_{j-1} if and only if $L(T_{j-1}) - \kappa > 0$, or equivalently, if

$$B(T_{j-1}, T_j)^{-1} = 1 + L(T_{j-1}, T_{j-1})(T_j - T_{j-1}) > 1 + \kappa\delta_j = \tilde{\delta}_j.$$

The last inequality holds whenever $B(T_{j-1}, T_j) < \tilde{\delta}_j^{-1}$. This shows that both of the considered options are exercised in the same circumstances. If exercised, the caplet pays $\delta_j(L(T_{j-1}, T_{j-1}) - \kappa)$ at time T_j, or equivalently,

$$\delta_j B(T_{j-1}, T_j)\big(L(T_{j-1}, T_{j-1}) - \kappa \big) = \tilde{\delta}_j \big(\tilde{\delta}_j^{-1} - B(T_{j-1}, T_j) \big)$$

at time T_{j-1}. This shows once again that the jth caplet, with strike level κ and nominal value 1, is essentially equivalent to a put option with strike price $(1 + \kappa\delta_j)^{-1}$ and nominal value $(1 + \kappa\delta_j)$ written on the corresponding zero-coupon bond with maturity T_j.

The price of a *forward floor* at time $t \in [0, T]$ equals

$$\mathbf{FF}_t = \sum_{j=1}^{n} \mathbb{E}_{\mathbb{P}^*}\left(\frac{B_t}{B_{T_j}}\left(\kappa - L(T_{j-1}, T_{j-1})\right)^+ \delta_j \,\Big|\, \mathcal{F}_t\right). \qquad (12.18)$$

Using a trivial equality

$$\left(\kappa - L(T_{j-1}, T_{j-1})\right)^+ \delta_j = \left(L(T_{j-1}, T_{j-1}) - \kappa\right)^+ \delta_j - \left(L(T_{j-1}, T_{j-1}) - \kappa\right)\delta_j,$$

we find that the following cap-floor parity relationship is satisfied at any time $t \in [0, T]$ (the three contracts are assumed to have the same payment dates)

$$Forward\ Cap\,(t) - Forward\ Floor\,(t) = Forward\ Swap\,(t),$$

For a description of a (multi-period) forward swap, we refer to the next chapter. This relationship can also be verified by a straightforward comparison of the corresponding cash flows of both portfolios. Let us finally mention that by a *cap* (respectively, *floor*), we mean a forward cap (respectively, forward floor) with $t = T$.

12.3 Valuation in the Gaussian HJM Model

We assume that the bond price volatility is a deterministic function, that is, we place ourselves within the Gaussian HJM framework. Recall that for any two maturity dates U, T we write $F_B(t, T, U) = B(t, T)/B(t, U)$, so that the function $\gamma(t, T, U) = b(t, T) - b(t, U)$ represents the volatility of $F_B(t, T, U)$.

12.3.1 Plain-vanilla Caps and Floors

The following lemma is an immediate consequence of Proposition 11.3.1 and the equivalence of a caplet and a specific put option on a zero-coupon bond.

Lemma 12.3.1 *Consider a caplet with settlement date T, accrual period δ, and strike level κ, that pays at time $T + \delta$ the amount $(L(T, T) - \kappa)^+ \delta$. Its arbitrage price at time $t \in [0, T]$ in the Gaussian HJM set-up equals*

$$\mathbf{Cpl}_t = B(t, T)\Big(N\big(e_1(t, T)\big) - \tilde{\delta}F_B(t, T + \delta, T)N\big(e_2(t, T)\big)\Big),$$

where $\tilde{\delta} = 1 + \kappa\delta$ and

$$e_{1,2}(t, T) = \frac{\ln F_B(t, T, T + \delta) - \ln \tilde{\delta} \pm \frac{1}{2}v^2(t, T)}{v(t, T)}$$

with

$$v^2(t, T) = \int_t^T |\gamma(u, T, T + \delta)|^2 \, du.$$

The next result, which is an almost immediate consequence of Lemma 12.3.1, provides a generic pricing formula for a forward cap in the Gaussian HJM set-up.

Proposition 12.3.1 *Assume the Gaussian HJM framework, so that the volatilities* $\gamma(t, T_{j-1}, T_j)$ *are deterministic. Then the arbitrage price at time* $t \le T_0$ *of an interest rate cap with strike level* κ, *settled in arrears at times* T_j, $j = 1, \dots, n$, *equals*

$$\mathbf{FC}_t = \sum_{j=1}^{n} B(t, T_{j-1}) \left(N(e_1(t, T_{j-1})) - \tilde{\delta}_j F_B(t, T_j, T_{j-1}) N(e_2^j(t, T_{j-1})) \right)$$

where $\tilde{\delta}_j = 1 + \kappa \delta_j$ *and*

$$e_{1,2}(t, T_{j-1}) = \frac{\ln F_B(t, T_{j-1}, T_j) - \ln \tilde{\delta}_j \pm \frac{1}{2} v^2(t, T_{j-1})}{v(t, T_{j-1})}$$

with

$$v^2(t, T_{j-1}) = \int_t^{T_{j-1}} |\gamma(u, T_{j-1}, T_j)|^2 \, du.$$

Proof. We represent the price of a forward cap in the following way

$$\mathbf{FC}_t = \sum_{j=1}^{n} \mathbb{E}_{\mathbb{P}^*} \left\{ \frac{B_t}{B_{T_j}} \left(L(T_{j-1}, T_{j-1}) - \kappa \right)^+ \delta_j \,\Big|\, \mathcal{F}_t \right\}$$

$$= \sum_{j=1}^{n} \mathbb{E}_{\mathbb{P}^*} \left\{ \frac{B_t}{B_{T_j}} \left((B(T_{j-1}, T_j)^{-1} - 1) \delta_j^{-1} - \kappa \right)^+ \delta_j \,\Big|\, \mathcal{F}_t \right\}$$

$$= \sum_{j=1}^{n} \mathbb{E}_{\mathbb{P}^*} \left\{ \frac{B_t}{B_T} \left(1 - \tilde{\delta}_j B(T, T_j) \right)^+ \,\Big|\, \mathcal{F}_t \right\} = \sum_{j=1}^{n} \mathbf{Cpl}_t^j,$$

where $\mathbf{Cpl}_t^j$ stands for the price at time t of the jth caplet. The assertion now follows from Lemma 12.3.1. $\qquad\square$

To derive the valuation formula for a floor, it is enough to make use of the capfloor parity, that is, the universal relationship $\mathbf{FC}_t - \mathbf{FF}_t = \mathbf{FS}_t$. By combining the valuation formulas for caps and swaps, we find easily that, under the assumptions of Proposition 12.3.1, the arbitrage price of a floor is given by the expression

$$\mathbf{FF}_t = \sum_{j=1}^{n} \left(\tilde{\delta}_j B(t, T_j) N\left(-e_2^j(t) \right) - B(t, T_{j-1}) N\left(-e_1^j(t) \right) \right).$$

In the derivation of the last formula we have used, in particular, the universal (i.e., model independent) valuation formula (13.2) for swaps, which will be established in Sect. 13.1.1 below.

12.3.2 Exotic Caps

A large variety of *exotic caps* is offered to institutional clients of financial institutions. In this section, we develop pricing formulas for some of them within the Gaussian HJM set-up.

Dual-strike caps. The *dual-strike cap* (known also as a *N-cap*) is an interest rate cap that has a lower strike κ_1, an upper strike κ_2 (with $\kappa_1 \le \kappa_2$), and a trigger, say l. So long as the floating rate L is below the level l, the N-cap owner enjoys protection at the lower strike κ_1. For periods when L is at or above the level l, the N-cap owner has protection at the upper strike level κ_2. Let us consider an N-cap on principal 1 settled in arrears at times T_j, $j = 1, \ldots, n$, where $T_j - T_{j-1} = \delta_j$ and $T_0 = T$. It is clear that the cash flow of the N-cap at time T_j equals

$$c_j = \left(L(T_{j-1}) - \kappa_1\right)^+ \delta_j \, \mathbb{1}_{\{L(T_{j-1}) < l\}} + \left(L(T_{j-1}) - \kappa_2\right)^+ \delta_j \mathbb{1}_{\{L(T_{j-1}) \ge l\}},$$

where $L(T_{j-1}) = L(T_{j-1}, T_{j-1})$. It is not hard to check that the N-cap price at time $t \in [0, T]$ is

$$\mathrm{NC}_t = \sum_{j=0}^{n-1} B(t, T_j) \left(N\!\left(h_2^j(t, i)\right) - N\!\left(h_2^j(t, \kappa_1 \wedge l)\right) + N\!\left(- h_2^j(t, \kappa_2 \vee l)\right)\right)$$

$$- \sum_{j=0}^{n-1} (1 + \kappa_1 \delta_{j+1}) B(t, T_{j+1}) \left(N\!\left(h_1^j(t, l)\right) - N\!\left(h_1^j(t, \kappa_1 \wedge l)\right)\right)$$

$$- \sum_{j=0}^{n-1} (1 + \kappa_2 \delta_{j+1}) B(t, T_{j+1}) N\!\left(- h_1^j(t, \kappa_2 \vee l)\right),$$

where

$$h_{1,2}^j(t, \kappa) = \frac{\ln(1 + \kappa \delta_{j+1}) - \ln F_B(t, T_j, T_{j+1}) \pm \frac{1}{2} v^2(t, T_j)}{v(t, T_j)}$$

and $v^2(t, T_j)$ is given in Proposition 12.3.1.

Bounded caps. A *bounded cap* (or a *B-cap*) consists of a sequence of caplets in which the difference between the fixed and floating levels is paid only if the total payments to date are less than some prescribed level b (let us stress that other kinds of B-caps exist). Let us first consider a particular *B-caplet* maturing at a reset date T_{j-1}. The corresponding cash flow will be paid in arrears at time T_j only if the accumulated cash flows at time T_{j-1}, due to resets at times T_k and cash flows of the B-cap paid at times T_{k+1}, $k = 0, \ldots, j - 2$, are still less than b. More formally, the cash flow of a B-caplet maturing at T_{j-1} equals

$$c_j(\kappa, b) = (L(T_{j-1}) - \kappa)^+ \delta_j \, \mathbb{1}_D,$$

where D stands for the following set

$$D = \left\{ \sum_{k=1}^{j} \left(L(T_{k-1}) - \kappa\right)^+ \delta_k \le b \right\} = \left\{ \sum_{k=1}^{j} \left(B(T_{k-1}, T_k)^{-1} - \tilde{\delta}_k\right)^+ \le b \right\},$$

where, as usual, $\tilde{\delta}_k = 1 + \kappa \delta_k$. The amount $c_j(\kappa, b)$ is paid at time T_j. The arbitrage price of a B-caplet at time $t \le T$ therefore equals

$$\mathbf{BCpl}_t = \mathbb{E}_{\mathbb{P}^*} \left\{ \frac{B_t}{B_{T_j}} \left(L(T_{j-1}) - \kappa \right)^+ \delta_j \, \mathbb{1}_D \, \Big| \, \mathcal{F}_t \right\},$$

or equivalently,

$$\mathbf{BCpl}_t = \mathbb{E}_{\mathbb{P}^*} \left\{ \frac{B_t}{B_{T_{j-1}}} \left(1 - \tilde{\delta}_j B(T_{j-1}, T_j) \right)^+ \mathbb{1}_D \, \Big| \, \mathcal{F}_t \right\}.$$

Using the standard forward measure method, the last equality can be given the following form

$$\mathbf{BCpl}_t = B(t, T_{j-1}) \mathbb{E}_{\mathbb{P}_{T_{j-1}}} \left\{ \left(1 - \tilde{\delta}_j B(T_{j-1}, T_j) \right)^+ \mathbb{1}_D \, \Big| \, \mathcal{F}_t \right\},$$

where $\mathbb{E}_{\mathbb{P}_{T_{j-1}}}$ stands for the expectation under the forward measure $\mathbb{P}_{T_{j-1}}$. Furthermore,

$$B(T_{k-1}, T_k) = \frac{B(t, T_k)}{B(t, T_{k-1})} \exp\left(- \int_t^{T_{k-1}} \gamma_k(u) \cdot dW_u^{T_{j-1}} \right.$$
$$\left. - \frac{1}{2} \int_t^{T_{k-1}} |\gamma_k(u)|^2 \, du - \int_t^{T_{k-1}} \gamma_k(u) \cdot \gamma(u, T_{k-1}, T_{j-1}) \, du \right),$$

where $\gamma_k(u) = \gamma(u, T_{k-1}, T_k)$. The random variable $(\xi_1(t), \ldots, \xi_j(t))$, where

$$\xi_k(t) = \int_t^{T_{k-1}} \gamma_k(u) \cdot dW_u^{T_{j-1}}$$

for $k = 1, \ldots, j$, is independent of the σ-field $\mathcal{F}_t$ under the forward measure $\mathbb{P}_{T_{j-1}}$. Furthermore, its probability distribution under $\mathbb{P}_{T_{j-1}}$ is Gaussian $N(0, \Gamma)$, where the entries of the matrix Γ are (notice that $\gamma_{kk}(t) = v_k^2(t)$)

$$\gamma_{kl}(t) = \int_t^{T_{k-1} \wedge T_{l-1}} \gamma_k(u) \cdot \gamma_l(u) \, du.$$

It is apparent that

$$\mathbf{BCpl}_t = \mathbb{E}_{\mathbb{P}_{T_{j-1}}} \left\{ \left(B(t, T_{j-1}) - \tilde{\delta}_j B(t, T_j) e^{-\xi_{j-1}(t) - v_{j-1}^2(t)/2} \right)^+ \mathbb{1}_{D_{j-1}} \, \Big| \, \mathcal{F}_t \right\},$$

where D_{j-1} stands for the set

$$D_{j-1} = \left\{ \sum_{k=1}^j \left(\frac{B(t, T_{k-1})}{B(t, T_k)} e^{\xi_k(t) + \alpha_{kj}(t) + v_k^2(t)/2} - \tilde{\delta}_j \right)^+ \le b \right\}.$$

and

$$\alpha_{kj}(t) = \int_t^{T_{k-1}} \gamma_k(u) \cdot \gamma(u, T_{k-1}, T_{j-1}) \, ds.$$

Denoting

$$A_{j-1} = \left\{ \xi_{j-1} \geq \ln \frac{\tilde{\delta}_j B(t, T_j)}{B(t, T_{j-1})} + \frac{1}{2} v_{j-1}^2(t) \right\},$$

we arrive at the following expression

$$\mathbf{BCpl}_t = B(t, T_{j-1}) \mathbb{P}_{T_{j-1}} \{ A_{j-1} \cap D_{j-1} \}$$
$$- \tilde{\delta}_j B(t, T_j) \, \mathbb{E}_{\mathbb{P}_{T_{j-1}}} \left(\exp \left(- \xi_{j-1}(t) - v_{j-1}^2(t)/2 \right) \mathbb{1}_{A_{j-1} \cap D_{j-1}} \right).$$

12.3.3 Captions

Since a caplet is essentially a put option on a zero-coupon bond, a European call option on a caplet is an example of a compound option. More exactly, it is a call option on a put option with a zero-coupon bond as the underlying asset of the put option. Hence the valuation of a call option on a caplet can be done similarly as in Chap. 6 (provided, of course, that the model of a zero-coupon bond price has sufficiently nice properties). A call option on a cap, or a *caption,* is thus a call on a portfolio of put options. To price a caption observe that its payoff at expiry is

$$\mathbf{CC}_T = \left(\sum_{j=1}^n \mathbf{Cpl}_T^j - K \right)^+,$$

where as usual $\mathbf{Cpl}_T^j$ stands for the price at time T of the jth caplet of the cap, T is the call option's expiry date and K is its strike price. Suppose that we place ourselves within the framework of the spot rate models of Chap. 9.5 – for instance, the Hull and White model. Typically, the caplet price is an increasing function of the current value of the spot rate r_t. Let r^* be the critical level of interest rate, which is implicitly determined by the equality $\sum_{j=1}^n \mathbf{Cpl}_T^j(r^*) = K$. It is clear that the option is exercised when the rate r_T is greater than r^*. Let us introduce numbers K_j by setting $K_j = \mathbf{Cpl}_T^j(r^*)$ for $j = 1, \ldots, n$. It is easily seen that the caption's payoff is equal to the sum of the payoffs of n call options on particular caplets, with K_j being the corresponding strike prices. Consequently, the caption's price $\mathbf{CC}_t$ at time $t \leq T_1$ is given by the formula $\mathbf{CC}_t = \sum_{j=1}^n C_t(\mathbf{Cpl}^j, T, K_j)$, where $C_t(\mathbf{Cpl}^j, T, K_j)$ is the price at time t of a call option with expiry date T and strike level K_j written on the jth caplet (see Hull and White (1994)). An option on a cap (or floor) can also be studied within the Gaussian HJM framework (see Brace and Musiela (1997)). However, results concerning caption valuation within this framework are less explicit than in the case of the Hull and White model.

12.4 LIBOR Market Models

The goal of this section is to present various approaches to the direct modelling of forward LIBORs. We focus here on the model's construction, its basic properties, and the valuation of typical derivatives. For further details, the reader is referred to the papers by Musiela and Sondermann (1993), Sandmann and Sondermann (1993a, 1993b), Goldys et al. (2000), Sandmann et al. (1995), Brace et al. (1997), Jamshidian (1997a), Miltersen et al. (1997), Musiela and Rutkowski (1997), Rady (1997), Sandmann and Sondermann (1997), Rutkowski (1998b, 1999a), Yasuoka (2001, 2005), Glasserman and Kou (2003), Galluccio and Hunter (2004a, 2004b), and Eberlein and Özkan (2005).

Issues related to the model's implementation, including model calibration and the valuation of exotic LIBOR and swap derivatives, are treated in Brace (1996), Brace et al. (1998, 2001a) Hull and White (2000), Schlögl (2002), Uratani and Utsunomiya (1999), Lotz and Schlögl (2000), Schoenmakers and Coffey (1999), Andersen (2000), Andersen and Andreasen (2000b), Brace and Womersley (2000), Dun et al. (2000), Hull and White (2000), Glasserman and Zhao (2000), Sidenius (2000), Andersen and Brotherton-Ratcliffe (2001), De Jong et al. (2001a, 2001b), Longstaff et al. (2001a), Pelsser et al. (2002), Wu (2002), Wu and Zhang (2002), d'Aspremont (2003), Glasserman and Merener (2003), Galluccio et al. (2003a), Jäckel and Rebonato (2003), Kawai (2003), Pelsser and Pietersz (2003), Piterbarg (2003a, 2003c), and Pietersz (2005).

The main motivation for the introduction of the lognormal LIBOR model was the market practice of pricing caps (and swaptions) by means of Black-Scholes-like formulas. For this reason, we shall first describe how market practitioners value caps. The formulas commonly used by practitioners assume that the underlying instrument follows a geometric Brownian motion under some probability measure, $\mathbb{Q}$ say. Since the strict definition of this probability measure is not available in the present context, we shall informally refer to $\mathbb{Q}$ as the *market probability*.

12.4.1 Black's Formula for Caps

Let us consider an interest rate cap with expiry date T and fixed strike level κ. Market practice is to price the option assuming that the underlying forward interest rate process is lognormally distributed with zero drift. Let us first consider a caplet – that is, one leg of a cap. Assume that the forward LIBOR $L(t, T)$, $t \in [0, T]$, for the accrual period of length δ follows a geometric Brownian motion under the 'market probability', $\mathbb{Q}$ say. More specifically,

$$dL(t, T) = L(t, T)\sigma \, dW_t, \tag{12.19}$$

where W follows a one-dimensional standard Brownian motion under $\mathbb{Q}$, and σ is a strictly positive constant. The unique solution of (12.19) is

$$L(t, T) = L(0, T) \exp\left(\sigma W_t - \tfrac{1}{2}\sigma^2 t^2\right), \quad \forall t \in [0, T].$$

The 'market price' at time t of a caplet with expiry date T and strike level κ is now found from the formula

$$\mathbf{Cpl}_t = \delta B(t, T + \delta) \, \mathbb{E}_{\mathbb{Q}}\big((L(T, T) - \kappa)^+ \,|\, \mathcal{F}_t\big).$$

More explicitly, for any $t \in [0, T]$ we have

$$\mathbf{Cpl}_t = \delta B(t, T + \delta)\Big(L(t, T)N\big(\hat{e}_1(t, T)\big) - \kappa N\big(\hat{e}_1(t, T)\big)\Big),$$

where

$$\hat{e}_{1,2}(t, T) = \frac{\ln(L(t, T)/\kappa) \pm \frac{1}{2} \hat{v}_0^2(t, T)}{\hat{v}_0(t, T)}$$

and $\hat{v}_0^2(t, T) = \sigma^2(T - t)$. This means that market practitioners price caplets using Black's formula, with discount from the settlement date $T + \delta$. A cap settled in arrears at times T_j, $j = 1, \ldots, n$, where $T_j - T_{j-1} = \delta_j$, $T_0 = T$, is priced by the formula

$$\mathbf{FC}_t = \sum_{j=1}^{n} \delta_j B(t, T_j)\Big(L(t, T_{j-1})N\big(\hat{e}_1(t, T_{j-1})\big) - \kappa N\big(\hat{e}_2(t, T_{j-1})\big)\Big),$$

where for every $j = 0, \ldots, n - 1$

$$\hat{e}_{1,2}(t, T_{j-1}) = \frac{\ln(L(t, T_{j-1})/\kappa) \pm \frac{1}{2} \hat{v}^2(t, T_{j-1})}{\hat{v}(t, T_{j-1})}$$

and $\hat{v}^2(t, T_{j-1}) = \sigma_j^2(T_{j-1} - t)$ for some constants σ_j, $j = 1, \ldots, n$. The constant σ_j is referred to as the *implied volatility* of the jth caplet. Hence for a fixed strike κ we obtain in this way the term structure of caplet volatilities. Since the implied caplet volatilities usually depend on the strike level, we also observe the volatility smile in the caplets market. In practice, caps are quoted in terms of implied volatilities, assuming a flat term structure for underlying caplets. The term structure of caplets volatilities can be stripped from market prices of caps.

The market convention described above implicitly assumes (at least in the case of flat caplet volatilities) that for any maturity T_j the corresponding forward LIBOR has a lognormal probability distribution under the 'market probability'. As we shall see in what follows, the valuation formula obtained for caps (and floors) in the lognormal LIBOR market model agrees with market practice.

Recall that in the general framework of stochastic interest rates, the price of a *forward cap* equals (see formula (12.17))

$$\mathbf{FC}_t = \sum_{j=1}^{n} \mathbb{E}_{\mathbb{P}^*}\Big(\frac{B_t}{B_{T_j}}(L(T_{j-1}) - \kappa)^+ \delta_j \,\Big|\, \mathcal{F}_t\Big) = \sum_{j=1}^{n} \mathbf{Cpl}_t^j, \qquad (12.20)$$

where

$$\mathbf{Cpl}_t^j = B(t, T_j) \, \mathbb{E}_{\mathbb{P}_{T_j}}\Big((L(T_{j-1}) - \kappa)^+ \delta_j \,\Big|\, \mathcal{F}_t\Big) \qquad (12.21)$$

for every $j = 1, \ldots, n$.

12.4.2 Miltersen, Sandmann and Sondermann Approach

The first attempt to provide a rigorous construction a lognormal model of forward LI-BORs was done by Miltersen, Sandmann and Sondermann in their paper published in 1997 (see also Musiela and Sondermann (1993), Goldys et al. (2000), and Sandmann et al. (1995) for related studies). As a starting point of their analysis, Miltersen et al. (1997) postulate that the forward LIBOR process $L(t, T)$ satisfies

$$dL(t, T) = \mu(t, T) \, dt + L(t, T)\lambda(t, T) \cdot dW_t^*$$

with a deterministic volatility function $\lambda(t, T + \delta)$. It is not difficult to deduce from the last formula that the forward price of a zero-coupon bond satisfies

$$dF(t, T + \delta, T) = -F(t, T + \delta, T)\big(1 - F(t, T + \delta, T)\big)\lambda(t, T) \cdot dW_t^T.$$

Subsequently, they focus on the partial differential equation satisfied by the function $v = v(t, x)$ that expresses the forward price of the bond option in terms of the forward bond price. The PDE for the option's price is

$$\frac{\partial v}{\partial t} + \frac{1}{2}|\lambda(t, T)|^2 x^2 (1 - x)^2 \frac{\partial^2 v}{\partial x^2} = 0 \qquad (12.22)$$

with the terminal condition, $v(T, x) = (K - x)^+$.

It is interesting to note that the PDE (12.22) was previously solved by Rady and Sandmann (1994) who worked within a different framework, however. In fact, they were concerned with the valuation of a bond option for the Bühler and Käsler (1989) model.

By solving the PDE (12.22), Miltersen et al. (1997) derived not only the closed-form solution for the price of a bond option (this goal was already achieved in Rady and Sandmann (1994)), but also the "market formula" for the caplet's price. It should be stressed, however, that the existence of a lognormal family of LIBORs $L(t, T)$ with different maturities T was not formally established in a definitive manner by Miltersen et al. (1997) (although some partial results were provided). The positive answer to the problem of existence of such a model was given by Brace et al. (1997), who also start from the continuous-time HJM framework.

12.4.3 Brace, Gątarek and Musiela Approach

To introduce formally the notion of a *forward LIBOR,* we assume that we are given a family $B(t, T)$ of bond prices, and thus also the collection $F_B(t, T, U)$ of forward processes. Let us fix a horizon date T^*. In contrast to the previous section, we shall now assume that a strictly positive real number $\delta < T^*$ representing the length of the accrual period, is fixed throughout. By definition, the forward δ-LIBOR rate $L(t, T)$ for the future date $T \leq T^* - \delta$ prevailing at time t is given by the conventional market formula

$$1 + \delta L(t, T) = F_B(t, T, T + \delta), \quad \forall t \in [0, T].$$

Comparing this formula with (9.40), we find that $L(t, T) = f_s(t, T, T + \delta)$, so that the forward LIBOR $L(t, T)$ represents in fact the add-on rate prevailing at time t over the future time period $[T, T + \delta]$. We can also re-express $L(t, T)$ directly in terms of bond prices, as for any $T \in [0, T^* - \delta]$ we have

$$1 + \delta L(t, T) = \frac{B(t, T)}{B(t, T + \delta)}, \quad \forall t \in [0, T]. \tag{12.23}$$

In particular, the initial term structure of forward LIBORs satisfies

$$L(0, T) = \frac{B(0, T) - B(0, T + \delta)}{\delta B(0, T + \delta)}. \tag{12.24}$$

Assume that we are given a family $F_B(t, T, T^*)$ of forward processes satisfying

$$dF_B(t, T, T^*) = F_B(t, T, T^*)\gamma(t, T, T^*) \cdot dW_t^{T^*}$$

on $(\Omega, (\mathcal{F}_t)_{t \in [0, T^*]}, \mathbb{P}_{T^*})$, where W^{T^*} is a d-dimensional Brownian motion under $\mathbb{P}_{T^*}$. Then it is not hard to derive the dynamics of the associated family of forward LIBORs. For instance, one finds that under the forward measure $\mathbb{P}_{T+\delta}$ we have

$$dL(t, T) = \delta^{-1} F_B(t, T, T + \delta) \, \gamma(t, T, T + \delta) \cdot dW_t^{T+\delta},$$

where $W_t^{T+\delta}$ and $\mathbb{P}_{T+\delta}$ are defined by (see (9.42))

$$W_t^{T+\delta} = W_t^{T^*} - \int_0^t \gamma(u, T + \delta, T^*) \, du.$$

The process $W^{T+\delta}$ is a d-dimensional Brownian motion with respect to the probability measure $\mathbb{P}_{T+\delta}$ equivalent to $\mathbb{P}_{T^*}$, defined on $(\Omega, \mathcal{F}_T)$ by means of the Radon-Nikodým derivative (see (9.43)

$$\frac{d\mathbb{P}_{T+\delta}}{d\mathbb{P}_{T^*}} = \mathcal{E}_{T+\delta}\left(\int_0^{\cdot} \gamma(u, T + \delta, T^*) \cdot dW_u^{T^*}\right), \quad \mathbb{P}\text{-a.s.}$$

This means that $L(t, T)$ solves the equation

$$dL(t, T) = \delta^{-1}(1 + \delta L(t, T)) \gamma(t, T, T + \delta) \cdot dW_t^{T+\delta} \tag{12.25}$$

with the initial condition (12.24). Suppose that forward LIBORs $L(t, T)$ are strictly positive. Then formula (12.25) can be rewritten as follows

$$dL(t, T) = L(t, T)\lambda(t, T) \cdot dW_t^{T+\delta}, \tag{12.26}$$

where for every $t \in [0, T]$ we have

$$\lambda(t, T) = \frac{1 + \delta L(t, T)}{\delta L(t, T)} \gamma(t, T, T + \delta). \tag{12.27}$$

We thus see that the collection of forward processes uniquely specifies the family of forward LIBORs.

The construction of a model of forward LIBORs relies on the following set of assumptions.

(LR.1) For any maturity $T \leq T^* - \delta$, we are given a $\mathbb{R}^d$-valued, bounded $\mathbb{F}$-adapted process $\lambda(t, T)$ representing the volatility of the forward LIBOR process $L(t, T)$.

(LR.2) We assume a strictly decreasing and strictly positive initial term structure $B(0, T)$, $T \in [0, T^*]$, and thus an initial term structure $L(0, T)$ of forward LIBORs

$$L(0, T) = \frac{B(0, T) - B(0, T + \delta)}{\delta B(0, T + \delta)}, \quad \forall\, T \in [0, T^* - \delta].$$

Note that the volatility λ is a stochastic process, in general. In the special case when $\lambda(t, T)$ is a (bounded) deterministic function, a model we are going to construct is termed *lognormal LIBOR market model* for a fixed accrual period.

Remarks. Needless to say that the boundedness of λ can be weakened substantially. In fact, we shall frequently postulate, to simplify the exposition, that a volatility process (or function) is bounded but it is clear that a suitable integrability conditions are sufficient.

To construct a model satisfying (LR.1)-(LR.2), Brace et al. (1997) place themselves in the HJM set-up and they assume that for every $T \in [0, T^*]$, the volatility $b(t, T)$ vanishes for every $t \in [(T - \delta) \vee 0, T]$.

The construction presented in Brace et al. (1997) relies on forward induction, as opposed to the backward induction, which will be used in what follows. They start by postulating that the dynamics of $L(t, T)$ under the martingale measure $\mathbb{P}^*$ are governed by the following SDE

$$dL(t, T) = \mu(t, T)\, dt + L(t, T)\lambda(t, T) \cdot dW_t^*,$$

where λ is known, but the drift coefficient μ is unspecified. Recall that the arbitrage-free dynamics of the instantaneous forward rate $f(t, T)$ are

$$df(t, T) = \sigma(t, T) \cdot \sigma^*(t, T)\, dt + \sigma(t, T) \cdot dW_t^*.$$

In addition, we have the following relationship (cf. (12.23))

$$1 + \delta L(t, T) = \exp\left(\int_T^{T+\delta} f(t, u)\, du \right). \tag{12.28}$$

Applying Itô's formula to both sides of (12.28), and comparing the diffusion terms, we find that

$$\sigma^*(t, T + \delta) - \sigma^*(t, T) = \int_T^{T+\delta} \sigma(t, u)\, du = \frac{\delta L(t, T)}{1 + \delta L(t, T)}\, \lambda(t, T).$$

To solve the last equation for σ^* in terms of L, it is necessary to impose some kind of 'initial condition' on the process σ^*, or, equivalently, on the coefficient σ in the dynamics of $f(t, T)$.

For instance, by setting $\sigma(t, T) = 0$ for $0 \leq t \leq T \leq t + \delta$ (this choice was postulated in Brace et al. (1997)), we obtain the following relationship

$$b(t, T) = -\sigma^*(t, T) = -\sum_{k=1}^{[\delta^{-1}T]} \frac{\delta L(t, T - k\delta)}{1 + \delta L(t, T - k\delta)}\, \lambda(t, T - k\delta). \tag{12.29}$$

The existence and uniqueness of solutions to the SDEs that govern the instantaneous forward rate $f(t, T)$ and the forward LIBOR $L(t, T)$ for σ^* given by (12.29) can be shown rather easily, using the forward induction. Taking this result for granted, we conclude that the process $L(t, T)$ satisfies, under the spot martingale measure $\mathbb{P}^*$,

$$dL(t, T) = L(t, T)\sigma^*(t, T) \cdot \lambda(t, T)\, dt + L(t, T)\lambda(t, T) \cdot dW_t^*,$$

or equivalently,

$$dL(t, T) = L(t, T)\lambda(t, T) \cdot dW_t^{T+\delta}$$

under the forward measure $\mathbb{P}_{T+\delta}$. In this way, Brace et al. (1997) were able to specify completely their generic model of forward LIBORs. In particular, in the case of deterministic volatilities $\lambda(t, T)$, we obtain the lognormal model of forward LIBORs, that is, the model in which the process $L(t, T)$ is lognormal under $\mathbb{P}_{T+\delta}$ for any maturity $T > 0$ and for a fixed δ.

Let us note that this model is sometimes referred to as the LLM model (that is, Lognormal LIBOR Market model) model or as the BGM model (that is, Brace-Gątarek-Musiela model).

12.4.4 Musiela and Rutkowski Approach

As an alternative to forward induction, we describe the backward induction approach to the modelling of forward LIBORs. We shall now focus on the modelling of a finite family of forward LIBORs that are associated with a pre-specified collection $T_0 < \cdots < T_n$ of reset/settlement dates. The construction presented below is based on the one given by Musiela and Rutkowski (1997).

Let us start by recalling the notation. We assume that we are given a predetermined collection of reset/settlement dates $0 \leq T_0 < T_1 < \cdots < T_n$ referred to as the *tenor structure*. Let us write $\delta_j = T_j - T_{j-1}$ for $j = 1, \ldots, n$, so that $T_j = T_0 + \sum_{i=1}^{j} \delta_i$ for every $j = 0, \ldots, n$.

Since δ_j is not necessarily constant, the assumption of a fixed accrual period δ is now relaxed, and thus a model will be more suitable for practical purposes. Indeed, in most LIBOR derivatives the accrual period (day-count fraction) varies over time, and thus it is essential to have a model of LIBORs that is capable of mimicking this important real-life feature.

We find it convenient to set $T^* = T_n$ and

$$T_m^* = T^* - \sum_{j=n-m+1}^{n} \delta_j = T_{n-m}, \quad \forall\, m = 0, \ldots, n.$$

For any $j = 0, \ldots, n - 1$, we define the forward LIBOR $L(t, T_j)$ by setting

$$L(t, T_j) = \frac{B(t, T_j) - B(t, T_{j+1})}{\delta_{j+1} B(t, T_{j+1})}, \quad \forall\, t \in [0, T_j].$$

Let us introduce the notion of a martingale probability associated with the forward LIBOR $L(t, T_{j-1})$.

Definition 12.4.1 Let us fix $j = 1, \ldots, n$. A probability measure $\mathbb{P}_{T_j}$ on $(\Omega, \mathcal{F}_{T_j})$, equivalent to $\mathbb{P}$, is said to be the *forward LIBOR measure* for the date T_j if, for every $k = 1, \ldots, n$, the relative bond price

$$U_{n-j+1}(t, T_k) \stackrel{\text{def}}{=} \frac{B(t, T_k)}{B(t, T_j)}, \quad \forall t \in [0, T_k \wedge T_j],$$

is a local martingale under $\mathbb{P}_{T_j}$.

It is clear that the notion of forward LIBOR measure is formally identical with that of a forward martingale measure for a given date. The slight modification of our previous terminology emphasizes our intention to make a clear distinction between various kinds of forward probabilities, which we are going to study in the sequel. Also, it is trivial to observe that the forward LIBOR $L(t, T_j)$ is necessarily a local martingale under the forward LIBOR measure for the date T_{j+1}. If, in addition, it is a strictly positive process and the underlying filtration is generated by a Brownian motion, the existence of the associated volatility process can be justified easily.

In our further development, we shall go the other way around; that is, we will assume that for any date T_j, the volatility $\lambda(t, T_j)$ of the forward LIBOR $L(t, T_j)$ is exogenously given. Basically, it can be a deterministic $\mathbb{R}^d$-valued function of time, an $\mathbb{R}^d$-valued function of the underlying forward LIBORs, or a d-dimensional stochastic process adapted to a Brownian filtration. For simplicity, we assume that the volatilities of forward LIBORs are bounded, but, of course, this assumption can be relaxed.

Our aim is to construct a family $L(t, T_j)$, $j = 0, \ldots, n-1$ of forward LIBORs, a collection of mutually equivalent probability measures $\mathbb{P}_{T_j}$, $j = 1, \ldots, n$, and a family W^{T_j}, $j = 0, \ldots, n-1$ of processes that satisfy:
(i) for any $j = 1, \ldots, n$ the process W^{T_j} is a d-dimensional standard Brownian motion under the probability measure $\mathbb{P}_{T_j}$,
(ii) for any $j = 1, \ldots, n-1$, the forward LIBOR $L(t, T_j)$ satisfies the SDE

$$dL(t, T_j) = L(t, T_j) \lambda(t, T_j) \cdot dW_t^{T_{j+1}}, \quad \forall t \in [0, T_j],$$

with the initial condition

$$L(0, T_j) = \frac{B(0, T_j) - B(0, T_{j+1})}{\delta_{j+1} B(0, T_{j+1})}.$$

As was mentioned already, the construction of the model is based on backward induction. We start by defining the forward LIBOR with the longest maturity, T_{n-1}. We postulate that $L(t, T_{n-1}) = L(t, T_1^*)$ is governed under the underlying probability measure $\mathbb{Q}$ by the following SDE (note that, for simplicity, we have chosen the underlying probability measure $\mathbb{Q}$ to play the role of the forward LIBOR measure for the date T^*)

$$dL(t, T_1^*) = L(t, T_1^*) \lambda(t, T_1^*) \cdot dW_t,$$

with the initial condition

$$L(0, T_1^*) = \frac{B(0, T_1^*) - B(0, T^*)}{\delta_n B(0, T^*)}.$$

Put another way, we have, for every $t \in [0, T_1^*]$,

$$L(t, T_1^*) = \frac{B(0, T_1^*) - B(0, T^*)}{\delta_n B(0, T^*)} \, \mathcal{E}_t \left(\int_0^{\cdot} \lambda(u, T_1^*) \cdot dW_u \right).$$

Since $B(0, T_1^*) > B(0, T^*)$, it is clear that the $L(t, T_1^*)$ follows a strictly positive martingale under $\mathbb{P}_{T^*} = \mathbb{Q}$.

The next step is to define the forward LIBOR for the date T_2^*. For this purpose, we need to introduce first the forward martingale measure for the date T_1^*. By definition, it is a probability measure $\mathbb{P}_{T_1^*}$ equivalent to $\mathbb{P}_{T^*}$ and such that processes

$$U_2(t, T_k^*) = \frac{B(t, T_k^*)}{B(t, T_1^*)}$$

are $\mathbb{P}_{T_1^*}$-local martingales. It is important to observe that the process $U_2(t, T_k^*)$ admits the following representation

$$U_2(t, T_k^*) = \frac{U_1(t, T_k^*)}{\delta_n L(t, T_1^*) + 1}.$$

The following auxiliary result is a straightforward consequence of Itô's rule.

Lemma 12.4.1 *Let G and H be real-valued adapted processes, such that*

$$dG_t = \alpha_t \cdot dW_t, \quad dH_t = \beta_t \cdot dW_t.$$

Assume, in addition, that $H_t > -1$ for every t and write $Y_t = (1 + H_t)^{-1}$. Then

$$d(Y_t G_t) = Y_t (\alpha_t - Y_t G_t \beta_t) \cdot (dW_t - Y_t \beta_t \, dt).$$

It follows immediately from Lemma 12.4.1 that

$$dU_2(t, T_k^*) = \eta_t^k \cdot \left(dW_t - \frac{\delta_n L(t, T_1^*)}{1 + \delta_n L(t, T_1^*)} \lambda(t, T_1^*) \, dt \right)$$

for a certain process η^k.

It is therefore enough to find a probability measure under which the process

$$W_t^{T_1^*} \stackrel{\text{def}}{=} W_t - \int_0^t \frac{\delta_n L(u, T_1^*)}{1 + \delta_n L(u, T_1^*)} \lambda(u, T_1^*) \, du = W_t - \int_0^t \gamma(u, T_1^*) \, du,$$

where $t \in [0, T_1^*]$, follows a standard Brownian motion (the definition of $\gamma(t, T_1^*)$ is clear from the context). This can easily be achieved using Girsanov's theorem, as we may put

$$\frac{d\mathbb{P}_{T_1^*}}{d\mathbb{P}_{T^*}} = \mathcal{E}_{T_1^*} \left(\int_0^{\cdot} \gamma(u, T_1^*) \cdot dW_u \right), \quad \mathbb{P}_{T^*}\text{-a.s.}$$

We are in a position to specify the dynamics of the forward LIBOR for the date T_2^* under $\mathbb{P}_{T_1^*}$, namely we postulate that $L(t, T_2^*)$ solves the SDE

$$dL(t, T_2^*) = L(t, T_2^*) \lambda(t, T_2^*) \cdot dW_t^{T_1^*}$$

with the initial condition

$$L(0, T_2^*) = \frac{B(0, T_2^*) - B(0, T_1^*)}{\delta_{n-1} B(0, T_1^*)}.$$

Let us assume that we have already found processes $L(t, T_1^*), \ldots, L(t, T_m^*)$. In particular, the forward LIBOR measure $\mathbb{P}_{T_{m-1}^*}$ and the associated Brownian motion $W^{T_{m-1}^*}$ are already specified. Our aim is to determine the forward LIBOR measure $\mathbb{P}_{T_m^*}$. It is easy to check that

$$U_{m+1}(t, T_k^*) = \frac{U_m(t, T_k^*)}{\delta_{n-m} L(t, T_m^*) + 1}.$$

Using Lemma 12.4.1, we obtain the following relationship, for $t \in [0, T_m^*]$,

$$W_t^{T_m^*} = W_t^{T_{m-1}^*} - \int_0^t \frac{\delta_{n-m} L(u, T_m^*)}{1 + \delta_{n-m} L(u, T_m^*)} \lambda(u, T_m^*) \, du.$$

The forward LIBOR measure $\mathbb{P}_{T_m^*}$ can thus be found easily using Girsanov's theorem. Finally, we define the process $L(t, T_{m+1}^*)$ as the solution to the SDE

$$dL(t, T_{m+1}^*) = L(t, T_{m+1}^*) \lambda(t, T_{m+1}^*) \cdot dW_t^{T_m^*}$$

with the initial condition

$$L(0, T_{m+1}^*) = \frac{B(0, T_{m+1}^*) - B(0, T_m^*)}{\delta_{n-m} B(0, T_m^*)}.$$

Remarks. If the volatility coefficient $\lambda(t, T_m) : [0, T_m] \to \mathbb{R}^d$ is deterministic, then, for each date $t \in [0, T_m]$, the random variable $L(t, T_m)$ has a lognormal probability distribution under the forward martingale measure $\mathbb{P}_{T_{m+1}}$. In this case, the model is referred to as the lognormal LIBOR model.

12.4.5 SDEs for LIBORs under the Forward Measure

Let a collection of reset/settlement dates $0 < T_0 < T_1 < \cdots < T_n$ be given. We assume that each LIBOR $L_i(t) = L(t, T_i)$, $i = 0, \ldots, n-1$, solves under the actual probability $\mathbb{P}$ the following stochastic differential equation

$$dL_i(t) = L_i(t)\left(\mu_t^i \, dt + \lambda_i(L_i(t), t) \, dW_t^i\right) \tag{12.30}$$

for some real-valued drift processes μ^i and some volatility functions $\lambda_i : \mathbb{R}_+ \times [0, T_j] \to \mathbb{R}$. In general, we may have

$$\mu_t^i = \mu_i(L_0(t), L_1(t), \ldots, L_{n-1}(t), t),$$

but we postulate, of course, that the drift coefficient μ^i only depends on the forward LIBORs L_j, $j = 0, \ldots, n - 1$ existing at time t, and that it is sufficiently regular in order to ensure the existence and uniqueness of a solution to the SDE (12.30). By assumption, the one-dimensional Brownian motions $W^0, W^1, \ldots, W^{n-1}$ in (12.30) have the instantaneous correlations given by

$$d\langle W^i, W^j \rangle_t = \rho_t^{i,j} \, dt \tag{12.31}$$

for every $i, j = 0, \ldots, n - 1$. Recall that the forward measure $\mathbb{P}_{T_n}$ corresponds to the choice of the zero-coupon bond maturing at T_n as a numeraire. Using this new numeraire, the relative bond prices are

$$U_i(t) \stackrel{\text{def}}{=} U_1(t, T_i) = B(t, T_i)/B(t, T_n),$$

and thus they can also be expressed in terms of forward LIBORs, since

$$U_i(t) = \prod_{j=i}^{n-1} \left(1 + \delta_{j+1} L_j(t)\right).$$

The next proposition follows easily from the construction presented in the preceding section. However, we sketch also an alternative proof of this result.

Proposition 12.4.1 *The drift term* $\hat{\mu}_i(t)$ *in the dynamics of* $L_i(t) = L(t, T_i)$ *under the forward measure* $\mathbb{P}_{T_n}$ *equals*

$$\hat{\mu}_i(t) = -\sum_{j=i+1}^{n-1} \frac{\delta_{j+1} L_j(t)}{1 + \delta_{j+1} L_j(t)} \lambda_j(L_j(t), t) \lambda_i(L_i(t), t) \rho_t^{i,j}. \tag{12.32}$$

The forward LIBORs $L_0, L_1, \ldots, L_{n-1}$ *satisfy under the forward measure* $\mathbb{P}_{T_n}$ *the following stochastic differential equations*

$$dL_i(t) = L_i(t)\left(\hat{\mu}_i(t) + \lambda_i(L_i(t), t) \, d\hat{W}_t^i\right),$$

where $\hat{W}^0, \hat{W}^1, \ldots, \hat{W}^{n-1}$ *are one-dimensional Brownian motions under* $\mathbb{P}_{T_n}$ *with the instantaneous correlations given by*

$$d\langle \hat{W}^i, \hat{W}^j \rangle_t = d\langle W^i, W^j \rangle_t = \rho_t^{i,j} \, dt \tag{12.33}$$

for $i, j = 0, \ldots, n - 1$.

Proof. Using Girsanov's theorem, we get from (12.30), for $i = 0, \ldots, n - 1$,

$$dL_i(t) = L_i(t)\big(\hat{\mu}_i(t)\,dt + \lambda_i(L_i(t), t)\,d\hat{W}_t^i\big), \tag{12.34}$$

where $\hat{W}^0, \hat{W}^1, \ldots, \hat{W}^{n-1}$ are Brownian motions under $\mathbb{P}_{T_n}$ with the instantaneous correlations given by (12.33) (recall that an equivalent change of a probability measure preserves the correlation between Brownian motions; see Sect. 7.1.9). The derivation of drift coefficients $\hat{\mu}_i(t)$ hinges on the requirement that relative bond prices U_i enjoy the martingale property under the forward measure $\mathbb{P}_{T_n}$. By applying the Itô formula to the equality

$$U_i(t) = U_{i+1}(t)\,(1 + \delta_{i+1}L_i(t)), \tag{12.35}$$

we obtain

$$dU_i(t) = (1 + \delta_{i+1}L_i(t))\,dU_{i+1}(t) + \delta_{i+1}U_{i+1}(t)\,dL_i(t) + \delta_{i+1}d\langle U_{i+1}, L_i\rangle_t.$$

The relative price U_i follows a (local) martingale under $\mathbb{P}_{T_n}$, and thus the finite variation term in the Itô differential vanishes. Hence the drift $\hat{\mu}_i(t)$ necessarily satisfies

$$U_{i+1}(t)\hat{\mu}_i(t)L_i(t)\,dt = -d\langle U_{i+1}, L_i\rangle_t. \tag{12.36}$$

To establish (12.32), it thus suffices to compute the cross-variation $\langle U_{i+1}, L_i\rangle$. To this end, we need first to find the martingale component in the canonical decomposition of U_{i+1}. Since clearly (by convention, $\prod_{j=n}^{n-1} \ldots = 1$)

$$U_{i+1}(t) = \prod_{j=i+1}^{n-1} \big(1 + \delta_{j+1}L_j(t)\big),$$

the Itô formula yields

$$\begin{aligned}
dU_{i+1}(t) &= \sum_{j=i+1}^{n-1} \prod_{k=i+1,\,k\neq j}^{n-1} (1 + \delta_{k+1}L_k(t))\,d\big(1 + \delta_{j+1}L_j(t)\big) + A_t \\
&= \sum_{j=i+1}^{n-1} \prod_{k=i+1,\,k\neq j}^{n-1} (1 + \delta_{k+1}L_k(t))\,\delta_{j+1}\,dL_j(t) + A_t \\
&= U_{i+1}(t) \sum_{j=i+1}^{n-1} \frac{\delta_{j+1}L_j(t)}{1 + \delta_{j+1}L_j(t)}\,\lambda_j(L_j(t), t)\,d\hat{W}_t^j + \tilde{A}_t,
\end{aligned}$$

where A and $\tilde{A}$ stand for some continuous processes of finite variation. Consequently,

$$d\langle U_{i+1}, L_i\rangle_t = U_{i+1}(t)L_i(t) \sum_{j=i+1}^{n-1} \frac{\delta_{j+1}L_j(t)}{1 + \delta_{j+1}L_j(t)}\,\lambda_j(L_j(t), t)\lambda_i(L_i(t), t)\rho_t^{i,j}\,dt.$$

Inserting the last equality into (12.36), we conclude that (12.32) holds.   □

12.4.6 Jamshidian's Approach

The backward induction approach to modelling of forward LIBORs presented in the preceding section was re-examined and modified by Jamshidian (1997a). In this section, we present briefly his alternative approach to the modelling of forward LIBORs.

As was made apparent in the previous section, in the direct modelling of LIBORs, no explicit reference is made to the bond price processes, which are used to define formally a forward LIBOR through equality (12.23). Nevertheless, to explain the idea that underpins Jamshidian's approach, we shall temporarily assume that we are given a family of bond prices $B(t, T_j)$ for the dates T_j, $j = 0, \ldots, n$. By definition, the *spot LIBOR measure* is any probability measure equivalent to $\mathbb{P}$, under which all relative bond prices are local martingales when the wealth process obtained by rolling over one-period bonds is taken as a numeraire. The existence of such a measure can be either postulated, or derived from other conditions. Let us define the process G by

$$G_t = B(t, T_{m(t)}) \prod_{j=1}^{m(t)} B^{-1}(T_{j-1}, T_j), \qquad (12.37)$$

for every $t \in [0, T^*]$, where we set

$$m(t) = \inf\{k = 0, 1, \ldots \mid T_0 + \sum_{i=1}^{k} \delta_i \geq t\} = \inf\{k = 0, 1, \ldots \mid T_k \geq t\}.$$

It is easily seen that G_t represents the wealth at time t of a portfolio that starts at time 0 with one unit of cash invested in a zero-coupon bond of maturity T_0, and whose wealth is then reinvested at each date T_j, $j = 0, \ldots, n - 1$, in zero-coupon bonds maturing at the next date; that is, at time T_{j+1}.

Definition 12.4.2 A *spot LIBOR measure* $\mathbb{P}^L$ is any probability measure on $(\Omega, \mathcal{F}_{T^*})$ equivalent to a reference probability $\mathbb{P}$, and such that the relative prices $B(t, T_j)/G_t$, $j = 0, \ldots, n$, are local martingales under $\mathbb{P}^L$.

Note that

$$\frac{B(t, T_{k+1})}{G_t} = \prod_{j=1}^{m(t)} \left(1 + \delta_j L(T_{j-1}, T_{j-1})\right)^{-1} \prod_{j=m(t)+1}^{k} \left(1 + \delta_j L(t, T_{j-1})\right)^{-1},$$

so that all relative bond prices $B(t, T_j)/G_t$, $j = 0, \ldots, n$ are uniquely determined by a collection of forward LIBORs. In this sense, the *rolling bond* (or the *LIBOR numeraire*) G is the correct choice of the numeraire asset in the present set-up. Let us note that when we use the probability measure $\mathbb{P}^L$ for risk-neutral valuation of contingent claims, their arbitrage prices are expressed in units of the rolling bond G.

We shall now concentrate on the derivation of the dynamics under $\mathbb{P}^L$ of forward LIBOR processes $L(t, T_j)$, $j = 0, \ldots, n - 1$. We shall show that it is possible to define the whole family of forward LIBORs simultaneously under a single probability measure (of course, this feature can also be deduced from the previously examined recursive construction). To facilitate the derivation of the dynamics of $L(t, T_j)$, we postulate temporarily that bond prices $B(t, T_j)$ follow Itô processes under the underlying probability measure $\mathbb{P}$, more explicitly

$$dB(t, T_j) = B(t, T_j)\big(a(t, T_j)\, dt + b(t, T_j) \cdot dW_t\big) \qquad (12.38)$$

for every $j = 0, \ldots, n$, where, as before, W is a d-dimensional standard Brownian motion under an underlying probability measure $\mathbb{P}$ (it should be stressed, however, that we do not assume here that $\mathbb{P}$ is a forward (or spot) martingale measure). Combining (12.37) with (12.38), we obtain

$$dG_t = G_t\big(a(t, T_{m(t)})\, dt + b(t, T_{m(t)}) \cdot dW_t\big).$$

Furthermore, by applying Itô's rule to equality

$$1 + \delta_{j+1} L(t, T_j) = \frac{B(t, T_j)}{B(t, T_{j+1})}, \qquad (12.39)$$

we find that

$$dL(t, T_j) = \mu(t, T_j)\, dt + \zeta(t, T_j) \cdot dW_t,$$

where

$$\mu(t, T_j) = \frac{B(t, T_j)}{\delta_{j+1} B(t, T_{j+1})}\big(a(t, T_j) - a(t, T_{j+1})\big) - \zeta(t, T_j) b(t, T_{j+1})$$

and

$$\zeta(t, T_j) = \frac{B(t, T_j)}{\delta_{j+1} B(t, T_{j+1})}\big(b(t, T_j) - b(t, T_{j+1})\big). \qquad (12.40)$$

Using (12.39) and (12.40), we arrive at the following relationship

$$b(t, T_{m(t)}) - b(t, T_{j+1}) = \sum_{k=m(t)}^{j} \frac{\delta_{k+1} \zeta(t, T_k)}{1 + \delta_{k+1} L(t, T_k)}. \qquad (12.41)$$

By the definition of a spot LIBOR measure $\mathbb{P}^L$, each relative price process $B(t, T_j)/G_t$ follows a local martingale under $\mathbb{P}^L$. Since, in addition, $\mathbb{P}^L$ is assumed to be equivalent to $\mathbb{P}$, it is clear (because of Girsanov's theorem) that it is given by the Doléans exponential, that is,

$$\frac{d\mathbb{P}^L}{d\mathbb{P}} = \mathcal{E}_{T*}\bigg(\int_0^{\cdot} h_u \cdot dW_u\bigg), \quad \mathbb{P}\text{-a.s.}$$

for some adapted process h. It it not hard to check, using Itô's rule, that h needs to satisfy, for every $t \in [0, T_j]$ and every $j = 0, \ldots, n$,

$$a(t, T_j) - a(t, T_{m(t)}) = \big(b(t, T_{m(t)}) - h_t\big) \cdot \big(b(t, T_j) - b(t, T_{m(t)})\big).$$

Combining (12.40) with the formula above, we obtain

$$\frac{B(t, T_j)}{\delta_{j+1} B(t, T_{j+1})} \left(a(t, T_j) - a(t, T_{j+1})\right) = \zeta(t, T_j) \cdot \left(b(t, T_{m(t)}) - h_t\right),$$

and this in turn yields

$$dL(t, T_j) = \zeta(t, T_j) \cdot \left(\left(b(t, T_{m(t)}) - b(t, T_{j+1}) - h_t\right) dt + dW_t\right).$$

Using the last formula, (12.41), and Girsanov's theorem, we arrive at the following result due to Jamshidian (1997a).

Proposition 12.4.2 *For any $j = 0, \ldots, n - 1$, the process $L(t, T_j)$ satisfies*

$$dL(t, T_j) = \sum_{k=m(t)}^{j} \frac{\delta_{k+1}}{1 + \delta_{k+1} L(t, T_k)} \zeta(t, T_k) \cdot \zeta(t, T_j) \, dt + \zeta(t, T_j) \cdot dW_t^L,$$

where the process $W_t^L = W_t - \int_0^t h_u \, du$ is a d-dimensional Brownian motion under the spot LIBOR measure $\mathbb{P}^L$.

To further specify the model, we postulate that the processes $\zeta(t, T_j)$, $j = 0, \ldots, n - 1$ are exogenously given. Specifically, let

$$\zeta(t, T_j) = \lambda_j\left(t, L(t, T_j), L(t, T_{j+1}), \ldots, L(t, T_{n-1})\right), \quad \forall \, t \in [0, T_j],$$

where $\lambda_j : [0, T_j] \times \mathbb{R}^{n-j} \to \mathbb{R}^d$ are given functions. This leads to the following system of SDEs

$$dL(t, T_j) = \sum_{k=m(t)}^{j} \frac{\delta_{k+1} \lambda_k(t, \bar{L}_k(t)) \cdot \lambda_j(t, \bar{L}_j(t))}{1 + \delta_{k+1} L(t, T_k)} \, dt + \lambda_j(t, \bar{L}_j(t)) \cdot dW_t^L,$$

where, for brevity, we write

$$\bar{L}_j(t) = (L(t, T_j), L(t, T_{j+1}), \ldots, L(t, T_{n-1})).$$

Under standard regularity assumptions imposed on the set of coefficients λ_j, this system of SDEs can be solved recursively, starting from the SDE for the process $L(t, T_{n-1})$. In this way, one can produce a large variety of alternative versions of a forward LIBOR model, including the CEV LIBOR model and a simple version of displaced-diffusion model (these stochastic volatility versions of a LIBOR model are presented briefly in Sect. 12.7).

Let us finally observe that the lognormal LIBOR market model corresponds to the choice of $\zeta(t, T_j) = \lambda(t, T_j) L(t, T_j)$, where $\lambda(t, T_j) : [0, T_j] \to \mathbb{R}^d$ is a deterministic function for every $j = 0, \ldots, n - 1$. In that case, we deal with the following family of SDEs

$$\frac{dL(t, T_j)}{L(t, T_j)} = \sum_{k=m(t)}^{j} \frac{\delta_{k+1} L(t, T_k)}{1 + \delta_{k+1} L(t, T_k)} \lambda(t, T_k) \cdot \lambda(t, T_j) \, dt + \lambda(t, T_j) \cdot dW_t^L.$$

12.4.7 Alternative Derivation of Jamshidian's SDE

We place ourselves in the set-up of Sect. 12.4.5. We present below an alternative derivation of dynamics of forward LIBORs under the spot LIBOR measure. Let us first observe that, for $i = 0, \ldots, n-1$ and $t \in [0, T_i]$,

$$\tilde{U}_i(t) \stackrel{\text{def}}{=} \frac{B(t, T_i)}{G_t} = \prod_{j=0}^{m(t)-1} \frac{1}{1 + \delta_j L_j(T_j)} \prod_{j=m(t)}^{i-1} \frac{1}{1 + \delta_j L_j(t)}. \tag{12.42}$$

Note that the processes $\tilde{U}_i$, $i = 0, \ldots, n-1$ are (local) martingales under $\mathbb{P}^L$.

Proposition 12.4.3 *Assume that the forward LIBORs $L_0, L_1, \ldots, L_{n-1}$ satisfy (12.30) under the actual probability $\mathbb{P}$. Then they satisfy the following stochastic differential equations under the spot LIBOR measure $\mathbb{P}^L$*

$$dL_i(t) = L_i(t)\big(\tilde{\mu}_i(t)\, dt + \lambda_i(L_i(t), t)\, d\tilde{W}_t^i\big), \tag{12.43}$$

where

$$\tilde{\mu}_i(t) = \sum_{j=m(t)}^{i} \frac{\delta_{j+1} L_j(t)}{1 + \delta_{j+1} L_j(t)} \lambda_j(L_j(t), t)\lambda_i(L_i(t), t)\rho_t^{i,j}$$

and $\tilde{W}^0, \tilde{W}^1, \ldots, \tilde{W}^{n-1}$ are one-dimensional Brownian motions under $\mathbb{P}^L$ with the instantaneous correlations given by

$$d\langle \tilde{W}^i, \tilde{W}^j \rangle_t = d\langle W^i, W^j \rangle_t = \rho_t^{i,j}\, dt \tag{12.44}$$

for $i, j = 0, \ldots, n-1$.

Proof. Since the relative bond price $\tilde{U}_i$ should follow a (local) martingale under the spot LIBOR measure $\mathbb{P}^L$, we require that the drift term in the dynamics of $\tilde{U}_i$ under $\mathbb{P}^L$ equals zero. As expected, the proof is based on an application of the Girsanov theorem. We do not present the detailed proof here, but we focus on the derivation of the drift coefficient $\tilde{\mu}_i$ in (12.43).

We start by combining (12.42) with (12.30). Itô's formula yields

$$d\tilde{U}_{i+1}(t) = d\bigg(\prod_{j=0}^{m(t)-1} \frac{1}{1 + \delta_{j+1} L_j(T_j)} \prod_{j=m(t)}^{i} \frac{1}{1 + \delta_{j+1} L_j(t)} \bigg) + A_t$$

$$= -\tilde{U}_{i+1}(t) \sum_{j=m(t)}^{i} \bigg(\frac{\delta_{j+1}}{1 + \delta_{j+1} L_j(t)}\, dL_j(t) - \frac{\delta_j^2}{(1 + \delta_{j+1} L_j(t))^2}\, d\langle L_j \rangle_t \bigg) + A_t$$

$$= -\tilde{U}_{i+1}(t) \sum_{j=m(t)}^{i} \frac{\delta_{j+1} L_j(t)}{1 + \delta_{j+1} L_j(t)} \Big(\mu_t^j\, dt + \lambda_j(L_j(t), t)\, dW_t^j \Big)$$

$$+ \tilde{U}_{i+1}(t) \sum_{j=m(t)}^{i} \frac{\delta_{j+1}^2 L_j^2(t)}{(1 + \delta_{j+1} L_j(t))^2} \lambda_j^2(L_j(t), t)\, dt + A_t$$

for some continuous process A of finite variation. Hence we have

$$\tilde{U}_{i+1}(t) = -\tilde{U}_{i+1}(t) \sum_{j=m(t)}^{i} \frac{\delta_{j+1}L_j(t)}{1 + \delta_{j+1}L_j(t)} \lambda_j(L_j(t), t)\, dW_t^j + \tilde{A}_t$$

for some continuous process $\tilde{A}$ of finite variation. Assuming that we can apply Girsanov's theorem, we see that the requirement that $\tilde{U}_{i+1}$ is a (local) martingale under $\mathbb{P}^L$ leads to the following stochastic differential equation satisfied by $\tilde{U}_{i+1}$ under $\mathbb{P}^L$

$$d\tilde{U}_{i+1}(t) = -\tilde{U}_{i+1}(t) \sum_{j=m(t)}^{i} \frac{\delta_{j+1}L_j(t)}{1 + \delta_{j+1}L_j(t)} \lambda_j(L_j(t), t)\, d\tilde{W}_t^j, \qquad (12.45)$$

where $\tilde{W}^1, \tilde{W}^2, \ldots, \tilde{W}^n$ are one-dimensional Brownian motions under $\mathbb{P}^L$ with the instantaneous correlations given by formula (12.44).

To derive the drift term in the dynamics of the forward LIBOR L_i under $\mathbb{P}^L$, we shall make use of the relationship

$$\tilde{U}_i(t) = \tilde{U}_{i+1}(t)(1 + \delta_{i+1}L_i(t)),$$

which holds for $t \in [0, T_i]$. From this identity, we find that $\tilde{U}_i$ is a (local) martingale under $\mathbb{P}^L$ if and only if the process $L_i\tilde{U}_{i+1}$ is a (local) martingale under $\mathbb{P}^L$.

In view of (12.43) and (12.45), using Itô's formula, we obtain

$$d(L_i(t)\tilde{U}_{i+1}(t)) = \tilde{U}_{i+1}(t)\, dL_i(t) + L_i(t)\, d\tilde{U}_{i+1}(t)$$

$$- L_i(t)\tilde{U}_{i+1}(t) \sum_{j=m(t)}^{i} \frac{\delta_{j+1}}{1 + \delta_{j+1}L_j(t)} \lambda_j(L_j(t), t)\lambda_i(L_i(t), t)\rho_t^{i,j}\, dt$$

$$= L_i(t)\tilde{U}_{i+1}(t)\Big(\tilde{\mu}_i(t) - \sum_{j=m(t)}^{i} \frac{\delta_{j+1}L_j(t)}{1 + \delta_{j+1}L_j(t)} \lambda_j(L_j(t), t)\lambda_i(L_i(t), t)\rho_t^{i,j}\Big) dt$$

$$+ L_i(t)\tilde{U}_{i+1}(t)\lambda_i(L_i(t), t)\, d\tilde{W}_t^i + L_i(t)\, d\tilde{U}_{i+1}(t).$$

Recall that we assumed that the process $\tilde{U}_{i+1}$ is a (local) martingale under $\mathbb{P}^L$ and $\tilde{W}^i$ is a Brownian motion under $\mathbb{P}^L$. Hence the process $L_i\tilde{U}_{i+1}$ is a (local) martingale if and only the continuous process of bounded variation in the above equation vanishes, i.e., when

$$\tilde{\mu}_i(t) = \sum_{j=m(t)}^{i} \frac{\delta_{j+1}L_j(t)}{1 + \delta_{j+1}L_j(t)} \lambda_j(L_j(t), t)\lambda_i(L_i(t), t)\rho_t^{i,j}.$$

We have thus identified the drift term in the dynamics of L_i under $\mathbb{P}^L$, which was our goal. $\qquad\square$

12.5 Properties of the Lognormal LIBOR Model

We make the standing assumptions the volatilities of forward LIBORs $L(t, T_j)$ for $j = 0, \ldots, n - 1$ are deterministic. In other words, we place ourselves within the framework of the lognormal LIBOR model. It is interesting to note that in all approaches, there is a uniquely determined correspondence between forward measures (and forward Brownian motions) associated with different dates (it is based on relationships (12.3) and (12.8)). On the other hand, however, there is a considerable degree of ambiguity in the way in which the spot martingale measure is specified (in some instances, it is not introduced at all). Consequently, the futures LIBOR $L^f(t, T_j)$, which equals (cf. Sect. 12.1.3)

$$L^f(t, T_j) = \mathbb{E}_{\mathbb{P}^*}(L(T_j, T_j) \mid \mathcal{F}_t) = \mathbb{E}_{\mathbb{P}^*}(\tilde{L}(T_j, T_j) \mid \mathcal{F}_t), \tag{12.46}$$

is not necessarily specified in the same way in various approaches to the LIBOR market model.

For a given function $g : \mathbb{R} \to \mathbb{R}$ and a fixed date $u \le T_j$, we are interested in the payoff of the form $X = g(L(u, T_j))$ that settles at time T_j. Particular cases of such payoffs are

$$X_1 = g(B^{-1}(T_j, T_{j+1})), \quad X_2 = g(B(T_j, T_{j+1})), \quad X_3 = g(F_B(u, T_{j+1}, T_j)).$$

Recall that

$$B^{-1}(T_j, T_{j+1}) = 1 + \delta_{j+1} L(T_j) = 1 + \delta_{j+1} \tilde{L}(T_j) = 1 + \delta_{j+1} L^f(T_j).$$

The choice of the "pricing measure" is thus largely matter of convenience. Similarly, we have

$$B(T_j, T_{j+1}) = \frac{1}{1 + \delta_{j+1} L(T_j, T_j)} = F_B(T_j, T_{j+1}, T_j). \tag{12.47}$$

More generally, the forward price of a T_{j+1}-maturity bond for the settlement date T_j equals

$$F_B(u, T_{j+1}, T_j) = \frac{B(u, T_{j+1})}{B(u, T_j)} = \frac{1}{1 + \delta_{j+1} L(u, T_j)}. \tag{12.48}$$

To value a European contigent claim $X = g(L(u, T_j)) = \tilde{g}(F_B(u, T_{j+1}, T_j))$ settling at time T_j we may use the risk-neutral valuation formula

$$\pi_t(X) = B(t, T_j) \mathbb{E}_{\mathbb{P}_{T_j}}(X \mid \mathcal{F}_t), \quad \forall t \in [0, T_j].$$

It is thus clear that to value a claim in the case $u \le T_j$, it is enough to know the dynamics of either $L(t, T_j)$ or $F_B(t, T_{j+1}, T_j)$ under the forward martingale measure $\mathbb{P}_{T_j}$. When $u = T_j$, we may equally well use the dynamics under $\mathbb{P}_{T_j}$ of either the process $\tilde{L}(t, T_j)$, or the $L^f(t, T_j)$.

For instance,

$$\pi_t(X_1) = B(t, T_j)\mathbb{E}_{\mathbb{P}_{T_j}}(B^{-1}(T_j, T_{j+1}) \,|\, \mathcal{F}_t),$$

or equivalently,

$$\pi_t(X_1) = B(t, T_j)\mathbb{E}_{\mathbb{P}_{T_j}}(F_B^{-1}(T_j, T_{j+1}, T_j) \,|\, \mathcal{F}_t),$$

but also

$$\pi_t(X_1) = B(t, T_j)\big(1 + \delta_{j+1}\mathbb{E}_{\mathbb{P}_{T_j}}(Z(T_j) \,|\, \mathcal{F}_t)\big),$$

where $Z(T_j) = L(T_j) = \tilde{L}(T_j) = L^f(T_j)$.

12.5.1 Transition Density of the LIBOR

We shall now derive the transition probability density function of the process $L(t, T_j)$ under the forward martingale measure $\mathbb{P}_{T_j}$. Let us first prove the following related result that is of independent interest (it was first established by Jamshidian (1993a, 1993b)).

Proposition 12.5.1 *Let $t \leq u \leq T_j$. Then we have*

$$\mathbb{E}_{\mathbb{P}_{T_j}}\big(L(u, T_j) \,|\, \mathcal{F}_t\big) = L(t, T_j) + \frac{\delta_{j+1}\mathrm{Var}_{\mathbb{P}_{T_{j+1}}}\big(L(u, T_j) \,|\, \mathcal{F}_t\big)}{1 + \delta_{j+1}L(t, T_j)}.$$

In the case of the lognormal model of forward LIBORs, we have

$$\mathbb{E}_{\mathbb{P}_{T_j}}\big(L(u, T_j) \,|\, \mathcal{F}_t\big) = L(t, T_j)\left(1 + \frac{\delta_{j+1}L(t, T_j)\big(e^{v_j^2(t,u)} - 1\big)}{1 + \delta_{j+1}L(t, T_j)}\right),$$

where

$$v_j^2(t, u) = \mathrm{Var}_{\mathbb{P}_{T_{j+1}}}\left(\int_t^u \lambda(s, T_j) \cdot dW_s^{T_{j+1}}\right) = \int_t^u |\lambda(s, T_j)|^2\, ds.$$

Proof. Combining (12.5) with the martingale property of $L(t, T_j)$ under $\mathbb{P}_{T_{j+1}}$, we obtain

$$\mathbb{E}_{\mathbb{P}_{T_j}}\big(L(u, T_j) \,|\, \mathcal{F}_t\big) = \frac{\mathbb{E}_{\mathbb{P}_{T_{j+1}}}\big((1 + \delta_{j+1}L(u, T_j))L(u, T_j) \,|\, \mathcal{F}_t\big)}{1 + \delta_{j+1}L(t, T_j)},$$

so that

$$\mathbb{E}_{\mathbb{P}_{T_j}}\big(L(u, T_j) \,|\, \mathcal{F}_t\big) = L(t, T_j) + \frac{\delta_{j+1}\mathbb{E}_{\mathbb{P}_{T_{j+1}}}\big((L(u, T_j) - L(t, T_j))^2 \,|\, \mathcal{F}_t\big)}{1 + \delta_{j+1}L(t, T_j)}.$$

In the case of the lognormal model, we have

$$L(u, T_j) = L(t, T_j)\, e^{\eta(t,u) - \frac{1}{2}v_j^2(t,u)},$$

where

$$\eta(t, u) = \int_t^u \lambda(s, T_j) \, dW_s^{T_{j+1}}. \tag{12.49}$$

Consequently,

$$\mathbb{E}_{\mathbb{P}_{T_{j+1}}} \left((L(u, T_j) - L(t, T_j))^2 \mid \mathcal{F}_t \right) = L^2(t, T_j) \left(e^{v_j^2(t,u)} - 1 \right).$$

This gives the desired equality. □

To derive the transition probability density function of the process $L(t, T_j)$, note that for any $t \le u \le T_j$ and any bounded Borel measurable function $g : \mathbb{R} \to \mathbb{R}$ we have

$$\mathbb{E}_{\mathbb{P}_{T_j}} \left(g(L(u, T_j)) \mid \mathcal{F}_t \right) = \frac{\mathbb{E}_{\mathbb{P}_{T_{j+1}}} \left(g(L(u, T_j)) \left(1 + \delta_{j+1} L(u, T_j) \right) \mid \mathcal{F}_t \right)}{1 + \delta_{j+1} L(t, T_j)}.$$

The following simple lemma appears to be useful in what follows.

Lemma 12.5.1 *Let ζ be a nonnegative random variable on a probability space $(\Omega, \mathcal{F}, \mathbb{P})$ with the probability density function $f_{\mathbb{P}}$. Let $\mathbb{Q}$ be a probability measure equivalent to $\mathbb{P}$. Suppose that for any bounded Borel measurable function $g : \mathbb{R} \to \mathbb{R}$ we have*

$$\mathbb{E}_{\mathbb{P}}(g(\zeta)) = \mathbb{E}_{\mathbb{Q}}((1 + \zeta)g(\zeta)).$$

Then the probability density function $f_{\mathbb{Q}}$ of ζ under $\mathbb{Q}$ satisfies $f_{\mathbb{P}}(y) = (1 + y) f_{\mathbb{Q}}(y)$.

Proof. The assertion is in fact trivial since, by assumption,

$$\int_{-\infty}^{\infty} g(y) f_{\mathbb{P}}(y) \, dy = \int_{-\infty}^{\infty} g(y)(1 + y) f_{\mathbb{Q}}(y) \, dy$$

for any bounded Borel measurable function $g : \mathbb{R} \to \mathbb{R}$. □

Assume the lognormal LIBOR model, and fix $x \in \mathbb{R}$. Recall that for any $t \ge u$ we have

$$L(u, T_j) = L(t, T_j) \, e^{\eta(t,u) - \frac{1}{2} \mathrm{Var}_{\mathbb{P}_{T_{j+1}}} (\eta(t,u))},$$

where $\eta(t, u)$ is given by (12.49) (so that it is independent of the σ-field $\mathcal{F}_t$). The Markovian property of $L(t, T_j)$ under the forward measure $\mathbb{P}_{T_{j+1}}$ is thus apparent. Denote by $p_L(t, x, u, y)$ the transition probability density function under $\mathbb{P}_{T_{j+1}}$ of the process $L(t, T_j)$. Elementary calculations involving Gaussian densities yield

$$p_L(t, x, u, y) = \frac{1}{\sqrt{2\pi} v_j(t, u) y} \exp \left\{ -\frac{\left(\ln(y/x) + \frac{1}{2} v_j^2(t, u) \right)^2}{2 v_j^2(t, u)} \right\}$$

for any $x, y > 0$ and arbitrary $t < u$, where

$$p_L(t, x, u, y) = \mathbb{P}_{T_{j+1}} \{ L(u, T_j) = y \mid L(t, T_j) = x \}.$$

Taking into account Lemma 12.5.1, we conclude that the transition probability density function of the process[1] $L(t, T_j)$, under the forward martingale measure $\mathbb{P}_{T_j}$, satisfies

$$\tilde{p}_L(t, x, u, y) = \mathbb{P}_{T_j}\{L(u, T_j) = y \mid L(t, T_j) = x\} = \frac{1 + \delta_{j+1} y}{1 + \delta_{j+1} x}\, p_L(t, x, u, y).$$

We are now in a position to state the following result.

Corollary 12.5.1 *The transition probability density function under $\mathbb{P}_{T_j}$ of the forward LIBOR $L(t, T_j)$ equals*

$$\tilde{p}_L(t, x, u, y) = \frac{1 + \delta_{j+1} y}{\sqrt{2\pi}\, v_j(t, u)\, y(1 + \delta_{j+1} x)} \exp\left\{ -\frac{\left(\ln(y/x) + \frac{1}{2} v_j^2(t, u)\right)^2}{2 v_j^2(t, u)} \right\}$$

for any $t < u$ and arbitrary $x, y > 0$.

12.5.2 Transition Density of the Forward Bond Price

Observe that the forward bond price $F_B(t, T_{j+1}, T_j)$ satisfies

$$F_B(t, T_{j+1}, T_j) = \frac{B(t, T_{j+1})}{B(t, T_j)} = \frac{1}{1 + \delta_{j+1} L(t, T_j)}. \tag{12.50}$$

First, this implies that in the lognormal LIBOR model, the dynamics of the forward bond price $F_B(t, T_{j+1}, T_j)$ are governed by the following stochastic differential equation, under $\mathbb{P}_{T_j}$,

$$d F_B(t) = -F_B(t)\big(1 - F_B(t)\big)\lambda(t, T_j) \cdot dW_t^{T_j}, \tag{12.51}$$

where we write $F_B(t) = F_B(t, T_{j+1}, T_j)$. If the initial condition in (12.51) satisfies $0 < F_B(0) < 1$, then this equation can be shown to admit a unique strong solution (it satisfies $0 < F_B(t) < 1$ for every $t > 0$). This makes it clear that the forward bond price $F_B(t, T_{j+1}, T_j)$, and thus also the LIBOR $L(t, T_j)$, are Markovian under $\mathbb{P}_{T_j}$.

Using the formula established in Corollary 12.5.1 and relationship (12.50), one can find the transition probability density function of the Markov process $F_B(t, T_{j+1}, T_j)$ under $\mathbb{P}_{T_j}$; that is,

$$p_B(t, x, u, y) = \mathbb{P}_{T_j}\{F_B(u, T_{j+1}, T_j) = x \mid F_B(t, T_{j+1}, T_j) = y\}.$$

We have the following result (see Rady and Sandmann (1994), Miltersen et al. (1997) and Jamshidian (1997a)).

---

[1] The Markov property of $L(t, T_j)$ under $\mathbb{P}_{T_j}$ follows from the properties of the forward price, which will be established in Sect. 12.5.2.

Corollary 12.5.2 *The transition probability density function under* $\mathbb{P}_{T_j}$ *of the forward bond price* $F_B(t, T_{j+1}, T_j)$ *equals*

$$
p_B(t, x, u, y) = \frac{x}{\sqrt{2\pi} v_j(t, u) y^2 (1 - y)} \exp\left\{ -\frac{\left(\ln \frac{x(1-y)}{y(1-x)} + \frac{1}{2} v_j^2(t, u) \right)^2}{2 v_j^2(t, u)} \right\}
$$

for any $t < u$ *and arbitrary* $0 < x, y < 1$.

Proof. Let us fix $x \in (0, 1)$. Using (12.50), it is easy to show that

$$
p_B(t, x, u, y) = y^{-2} \tilde{p}_L\left(t, \frac{1 - x}{\delta x}, u, \frac{1 - y}{\delta y} \right),
$$

where $\delta = \delta_{j+1}$. The formula now follows from Corollary 12.5.1.   □

Goldys (1997) established the following result (we find it convenient to defer the proof of equality (12.53) to Sect. 12.6.3).

Proposition 12.5.2 *Let X be a solution of the following stochastic differential equation*

$$
dX_t = -X_t(1 - X_t)\lambda(t) \cdot dW_t, \quad X_0 = x, \tag{12.52}
$$

where W follows a standard Brownian motion under $\mathbb{P}$, *and* $\lambda : \mathbb{R}_+ \to \mathbb{R}$ *is a locally square integrable function. Then for any nonnegative Borel measurable function* $g : \mathbb{R} \to \mathbb{R}$ *and any* $u > 0$ *we have*

$$
\mathbb{E}_{\mathbb{P}}(g(X_u)) = (1 - x)\mathbb{E}_{\mathbb{Q}}\big(g(h_1(\zeta))\big) + x\mathbb{E}_{\mathbb{Q}}\big(g(h_2(\zeta))\big), \tag{12.53}
$$

where ζ *has, under* $\mathbb{Q}$, *the Gaussian probability distribution with zero mean value and variance*

$$
\mathrm{Var}_{\mathbb{Q}}(\zeta) = v^2(0, u) = \int_0^u |\lambda(s)|^2 \, ds,
$$

and the auxiliary functions h_1 *and* h_2 *are given by*

$$
\frac{1}{h_{1,2}(y)} = 1 + \exp\left(y + \ln \frac{1 - x}{x} \pm \frac{1}{2} v^2(0, u) \right), \quad \forall y \in \mathbb{R}.
$$

Before we end this chapter, we shall check that formula (12.50), which gives the transition probability density function of the forward bond price, can be re-derived using formula (12.53). According to (12.53), we have $\mathbb{E}_{\mathbb{P}_{T_j}}(g(X_u)) = I_1 + I_2$, where

$$
I_1 = \frac{1 - x}{\sqrt{2\pi} v(0, u)} \int_{-\infty}^{\infty} g(h_1(z)) \, e^{-z^2/2v^2(0, u)} \, dz
$$

and

$$
I_2 = \frac{x}{\sqrt{2\pi} v(0, u)} \int_{-\infty}^{\infty} g(h_2(z)) \, e^{-z^2/2v^2(0, u)} \, dz.
$$

First, let us set $y = h_1(z)$ in I_1, so that $dy = y(1 - y)dz$. Then we obtain

$$I_1 = \frac{1 - x}{\sqrt{2\pi} v(0, u)} \int_0^1 \frac{g(y)}{y(1 - y)} k(x, y)\, dy,$$

where we set

$$k(x, y) = \exp\left\{ -\frac{\left(\ln \frac{x(1-y)}{y(1-x)} - \frac{1}{2}v^2(0, u) \right)^2}{2v^2(0, u)} \right\}.$$

Equivalently, we have

$$I_1 = \frac{1 - x}{\sqrt{2\pi} v(0, u)} \int_0^1 \frac{xg(y)}{(1 - x)y^2} \tilde{k}(x, y)\, dy,$$

where

$$\tilde{k}(x, y) = \exp\left\{ -\frac{\left(\ln \frac{x(1-y)}{y(1-x)} + \frac{1}{2}v^2(0, u) \right)^2}{2v^2(0, u)} \right\}.$$

Similarly, the change of variable $y = h_2(z)$ in I_2 (so that once again $dy = y(1 - y)\, dz$) leads to the following equality

$$I_2 = \frac{x}{\sqrt{2\pi} v(0, u)} \int_0^1 \frac{g(y)}{y(1 - y)} \tilde{k}(x, y)\, dy.$$

Simple algebra now yields (recall that $\mathbb{E}_{\mathbb{P}_{T_j}}(g(X_u)) = I_1 + I_2$)

$$I_1 + I_2 = \frac{1}{\sqrt{2\pi} v(0, u)} \int_0^1 \frac{xg(y)}{y^2(1 - y)} \exp\left\{ -\frac{\left(\ln \frac{x(1-y)}{y(1-x)} + \frac{1}{2}v^2(0, u) \right)^2}{2v^2(0, u)} \right\} dy.$$

It is interesting to note that the formula above generalizes easily to the case $0 \le t < u$. Indeed, it is enough to consider the stochastic differential equation (12.52) with the initial condition $X_t = x$ at time t. As a result, we obtain the following formula for the conditional expectation

$$\mathbb{E}_{\mathbb{P}_{T_j}}(g(X_u) \mid X_t = x)$$

$$= \frac{1}{\sqrt{2\pi} v(t, u)} \int_0^1 \frac{xg(y)}{y^2(1 - y)} \exp\left\{ -\frac{\left(\ln \frac{x(1-y)}{y(1-x)} + \frac{1}{2}v^2(t, u) \right)^2}{2v^2(t, u)} \right\} dy.$$

It is worth noting that an application of the last formula to the process $X_t = F_B(t, T_{j+1}, T_j)$ leads to an alternative derivation of the formula established in Corollary 12.5.2.

12.6 Valuation in the Lognormal LIBOR Model

We start by considering the valuation of plain-vanilla caps and floors. Subsequently, we shall study the case of a generic path-independent claim.

12.6.1 Pricing of Caps and Floors

We shall now examine the valuation of caps within the lognormal LIBOR model of Sect. 12.4.4. To this end, we formally assume that $k \leq n$. Dynamics of the forward LIBOR process $L(t, T_{j-1})$ under the forward martingale measure $\mathbb{P}_{T_j}$ are known to be

$$dL(t, T_{j-1}) = L(t, T_{j-1}) \lambda(t, T_{j-1}) \cdot dW_t^{T_j}, \qquad (12.54)$$

where W^{T_j} is a d-dimensional Brownian motion under the forward measure $\mathbb{P}_{T_j}$, and $\lambda(t, T_{j-1}) : [0, T_{j-1}] \to \mathbb{R}^d$ is a deterministic function. Consequently, for every $t \in [0, T_{j-1}]$ we have

$$L(t, T_{j-1}) = L(0, T_{j-1}) \mathcal{E}_t \left(\int_0^{\cdot} \lambda(u, T_{j-1}) \cdot dW_u^{T_j} \right).$$

To the best of our knowledge, the cap valuation formula (12.55) was first established in a rigorous way by Miltersen et al. (1997), who focused on the dynamics of the forward LIBOR for a given date. Equality (12.55) was subsequently re-derived independently in Goldys (1997) and Rady (1997), who used probabilistic methods (though they dealt with a European bond option, their results are essentially equivalent to equality (12.55)). Finally, the same was established by means of the forward measure approach in Brace et al. (1997), where an arbitrage-free continuous-time model of all forward LIBORs was presented. It is instructive to compare the valuation formula (12.55) with the formula of Proposition 12.3.1, which holds for a Gaussian HJM case. The following proposition is an immediate consequence of formulas (12.20)-(12.21), combined with dynamics (12.54). Since the proof of the next result is rather standard, it is provided for the sake of completeness.

Proposition 12.6.1 *Consider an interest rate cap with strike level κ, settled in arrears at times T_j, $j = 1, \ldots, k$. Assuming the lognormal LIBOR model, the price of a cap at time $t \in [0, T]$ equals*

$$\mathbf{FC}_t = \sum_{j=1}^{k} \delta_j B(t, T_j) \Big(L(t, T_{j-1}) N\big(\tilde{e}_1^j(t)\big) - \kappa N\big(\tilde{e}_2^j(t)\big) \Big), \qquad (12.55)$$

where

$$\tilde{e}_{1,2}^j(t) = \frac{\ln(L(t, T_{j-1})/\kappa) \pm \frac{1}{2} \tilde{v}_j^2(t)}{\tilde{v}_j(t)}$$

and

$$\tilde{v}_j^2(t) = \int_t^{T_{j-1}} |\lambda(u, T_{j-1})|^2 \, du.$$

Proof. We fix j and we consider the jth caplet, with the payoff at time T_j

$$\mathbf{Cpl}^{j}_{T_j} = \delta_j (L(T_{j-1}) - \kappa)^+ = \delta_j L(T_{j-1}) \mathbb{1}_D - \delta_j \kappa \mathbb{1}_D, \qquad (12.56)$$

where $D = \{L(T_{j-1}) > \kappa\}$ is the exercise set. Since the caplet settles at time T_j, it is convenient to use the forward measure $\mathbb{P}_{T_j}$ to find its arbitrage price. We have

$$\mathbf{Cpl}^{j}_t = B(t, T_j) \, \mathbb{E}_{\mathbb{P}_{T_j}} \big(\mathbf{Cpl}^{j}_{T_j} \, | \, \mathcal{F}_t \big), \quad \forall t \in [0, T_j].$$

Obviously, it is enough to find the value of a caplet for $t \in [0, T_{j-1}]$. In view of (12.56), it suffices to compute the following conditional expectations

$$\begin{aligned}\mathbf{Cpl}^{j}_t &= \delta_j B(t, T_j) \, \mathbb{E}_{\mathbb{P}_{T_j}} \big(L(T_{j-1}) \mathbb{1}_D \, | \, \mathcal{F}_t \big) - \kappa \delta_j B(t, T_j) \, \mathbb{P}_{T_j} (D \, | \, \mathcal{F}_t) \\ &= \delta_j B(t, T_j)(I_1 - I_2),\end{aligned}$$

where the meaning of I_1 and I_2 is clear from the context. Recall that the spot LIBOR $L(T_{j-1}) = L(T_{j-1}, T_{j-1})$ is given by the formula

$$L(T_{j-1}) = L(t, T_{j-1}) \exp \bigg(\int_t^{T_{j-1}} \lambda_u^{j-1} \cdot dW_u^{T_j} - \frac{1}{2} \int_t^{T_{j-1}} |\lambda_u^{j-1}|^2 \, du \bigg),$$

where we set $\lambda_u^{j-1} = \lambda(u, T_{j-1})$. Since λ^{j-1} is a deterministic function, the probability distribution under $\mathbb{P}_{T_j}$ of the Itô integral

$$\zeta(t, T_{j-1}) = \int_t^{T_{j-1}} \lambda_u^{j-1} \cdot dW_u^{T_j}$$

is Gaussian, with zero mean and the variance

$$\mathrm{Var}_{\mathbb{P}_{T_j}} (\zeta(t, T_{j-1})) = \int_t^{T_{j-1}} |\lambda_u^{j-1}|^2 \, du.$$

It is thus straightforward to check that

$$I_2 = \kappa \, N \left(\frac{\ln(L(t, T_{j-1}) - \ln \kappa - \frac{1}{2} v_j^2(t)}{v_j(t)} \right).$$

To evaluate I_1, we introduce an auxiliary probability measure $\hat{\mathbb{P}}_{T_j}$, equivalent to $\mathbb{P}_{T_j}$ on $(\Omega, \mathcal{F}_{T_{j-1}})$, by setting

$$\frac{d\hat{\mathbb{P}}_{T_j}}{d\mathbb{P}_{T_j}} = \mathcal{E}_{T_{j-1}} \bigg(\int_0^\cdot \lambda_u^{j-1} \cdot dW_u^{T_j} \bigg).$$

Then the process $\hat{W}^{T_j}$, given by the formula

$$\hat{W}_t^{T_j} = W_t^{T_j} - \int_0^t \lambda_u^{j-1} \, du, \quad \forall t \in [0, T_{j-1}],$$

is the d-dimensional standard Brownian motion under $\hat{\mathbb{P}}_{T_j}$.

Furthermore, the forward price $L(T_{j-1})$ admits the following representation under $\hat{\mathbb{P}}_{T_j}$, for $t \in [0, T_{j-1}]$,

$$L(T_{j-1}) = L(t, T_{j-1}) \exp \left(\int_t^{T_{j-1}} \lambda_u^{j-1} \cdot d\hat{W}_u^{T_j} + \frac{1}{2} \int_t^{T_{j-1}} |\lambda_u^{j-1}|^2 du \right).$$

Since

$$I_1 = L(t, T_{j-1}) \mathbb{E}_{\mathbb{P}_{T_j}} \left(\mathbb{1}_D \exp \left(\int_t^{T_{j-1}} \lambda_u^{j-1} \cdot dW_u^{T_j} - \frac{1}{2} \int_t^{T_{j-1}} |\lambda_u^{j-1}|^2 du \right) \Big| \mathcal{F}_t \right)$$

from the abstract Bayes rule we get $I_1 = L(t, T_{j-1}) \hat{\mathbb{P}}_{T_j}(D \mid \mathcal{F}_t)$. Arguing in much the same way as for I_2, we thus obtain

$$I_1 = L(t, T_{j-1}) N \left(\frac{\ln L(t, T_{j-1}) - \ln \kappa + \frac{1}{2} v_j^2(t)}{v_j(t)} \right).$$

This completes the proof of the proposition.   □

As before, to derive the valuation formula for a floor, it is enough to make use of the cap-floor parity.

12.6.2 Hedging of Caps and Floors

It is clear the replicating strategy for a cap is a simple sum of replicating strategies for caplets. It is therefore enough to focus on a particular caplet. Let us denote by $F_C(t, T_j)$ the forward price of the jth caplet for the settlement date T_j. From (12.55), it is clear that

$$F_C(t, T_j) = \delta_j \left(L(t, T_{j-1}) N(\tilde{e}_1^j(t)) - \kappa N(\tilde{e}_2^j(t)) \right),$$

so that an application of Itô's formula yields (the calculations here are essentially the same as in the classical Black-Scholes model)

$$dF_C(t, T_j) = \delta_j N(\tilde{e}_1^j(t)) dL(t, T_{j-1}).$$

Let us consider the following self-financing trading strategy in the T_j-forward market, that is, with values of all securities expressed in units of T_j-maturity zero-coupon bonds. We start our trade at time 0 with $F_C(0, T_j)$ units of T_j-maturity zero-coupon bonds; we need thus to invest at time 0 the amount $\mathbf{Cpl}_0^j = F_C(0, T_j) B(0, T_j)$ of cash.

Subsequently, at any date $t \leq T_{j-1}$, we take $\psi_t^j = N(\tilde{e}_1^j(t))$ positions in one-period forward swaps over the period $[T_{j-1}, T_j]$. The associated gains process, $\tilde{G}$, in the T_j-forward market, satisfies $\tilde{G}_0 = 0$ and

$$d\tilde{G}_t = \delta_j \psi_t^j dL(t, T_{j-1}) = \delta_j N(\tilde{e}_1^j(t)) dL(t, T_{j-1}) = dF_C(t, T).$$

Consequently,

$$F_C(T_{j-1}, T_j) = F_C(0, T_j) + \int_0^{T_{j-1}} \delta_j \psi_t^j \, dL(t, T_{j-1}) = F_C(0, T_j) + \tilde{G}_{T_{j-1}}.$$

It should be stressed that dynamic trading is restricted to the interval $[0, T_{j-1}]$ only. The gains/losses (involving the initial investment) are incurred at time T_j, however. All quantities in the last formula are expressed in units of T_j-maturity zero-coupon bonds. Also, the caplet's payoff is known already at time T_{j-1}, so that it is completely specified by its forward price $F_C(T_{j-1}, T_j) = \mathbf{Cpl}_{T_{j-1}}^j / B(T_{j-1}, T_j)$. It is thus clear that the strategy ψ^j introduced above replicates the jth caplet.

Formally, the replicating strategy has also the second component, η_t^j say, representing the number of forward contracts, with the settlement date T_j, on T_j-maturity bond. Note that $F_B(t, T_j, T_j) = 1$ for every $t \le T_j$, and thus $dF_B(t, T_j, T_j) = 0$. Hence for the T_j-forward value of our strategy, we get

$$\tilde{V}_t(\psi^j, \eta^j) = \delta_j \psi_t^j L(t, T_{j-1}) + \eta_t^j = F_C(t, T_j)$$

and

$$d\tilde{V}_t(\psi^j, \eta^j) = \delta_j \psi_t^j \, dL(t, T_{j-1}) + \eta_t^j \, dF_B(t, T_j, T_j) = \delta_j N(\tilde{e}_1^j(t)) \, dL(t, T_{j-1}).$$

It should be stressed that, with the exception for the initial investment at time 0 in T_j-maturity bonds, no trading in bonds is required for replication of a caplet. In practical terms, the hedging of a cap within the framework of the lognormal LIBOR model in done exclusively through dynamic trading in the underlying one-period forward swaps. In this interpretation, the component η^j represents simply the future effects of continuous trading in forward contracts. The same remarks (and similar calculations) apply also to floors.

Alternatively, replication of a caplet can be done in the spot (that is, cash) market, using two simple portfolios of bonds. Indeed, it is easily seen that for the process

$$V_t(\psi^j, \eta^j) = B(t, T_j) \tilde{V}_t(\psi^j, \eta^j) = \mathbf{Cpl}_t^j$$

we have

$$V_t(\psi^j, \eta^j) = \psi_t^j \big(B(t, T_{j-1}) - B(t, T_j) \big) + \eta_t^j B(t, T_j)$$

and

$$\begin{aligned} dV_t(\psi^j, \eta^j) &= \psi_t^j \, d\big(B(t, T_{j-1}) - B(t, T_j) \big) + \eta_t^j \, dB(t, T_j) \\ &= N(\tilde{e}_1^j(t)) \, d\big(B(t, T_{j-1}) - B(t, T_j) \big) + \eta_t^j \, dB(t, T_j). \end{aligned}$$

In this interpretation, the components ψ^j and η^j represent the number of units of portfolios $B(t, T_{j-1}) - B(t, T_j)$ and $B(t, T_j)$ that are held at time t.

12.6.3 Valuation of European Claims

We follow here the approach due to Goldys (1997). Let X be a solution to the stochastic differential equation

$$dX_t = -X_t(1 - X_t)\lambda(t) \cdot dW_t, \quad X_0 = x, \quad (12.57)$$

where we assume that the function $\lambda : [0, T^*] \to \mathbb{R}^d$ is bounded and measurable and, as usual, W is a standard Brownian motion defined on a filtered probability space $(\Omega, \mathbb{F}, \mathbb{P})$. For any $x \in (0, 1)$, the existence of a unique global solution to (12.57) can be deduced easily from the general theory of stochastic differential equations. However, in Lemma 12.6.1 below we provide a direct proof by means of a simple transformation which is also crucial for the further calculations. Consider the following stochastic differential equation

$$dZ_t = \frac{1}{2} \frac{1 - e^{-Z_t}}{1 + e^{-Z_t}} |\lambda(t)|^2 dt - \lambda(t) \cdot dW_t \quad (12.58)$$

with the initial condition $Z_0 = z$. Since the drift term in this equation is represented by a bounded and globally Lipschitz continuous function, equation (12.58) is known to have a unique strong non-exploding solution for any initial condition $z \in \mathbb{R}$ (see, e.g., Theorem 5.2.9 in Karatzas and Shreve (1998a)).

Lemma 12.6.1 *For any $x \in (0, 1)$, the process*

$$X_t = \left(1 + e^{-Z_t}\right)^{-1}, \quad \forall t \in [0, T^*], \quad (12.59)$$

where z satisfies

$$Z_0 = z = \ln \frac{x}{1 - x},$$

is the unique strong and non-exploding solution to equation (12.57). Moreover, the inequalities $0 < X_t < 1$ hold for every $t \in [0, T^*]$.

Proof. It is easy to see that equation (12.58) can be rewritten in the form

$$dZ_t = \left(X_t - \tfrac{1}{2}\right)|\lambda(t)|^2 dt - \lambda(t) \cdot dW_t.$$

Hence, applying the Itô formula to the process X given by formula (12.59), we find that

$$dX_t = -X_t(1 - X_t)\lambda(t) \cdot dW_t.$$

Hence the process X is indeed a solution to (12.57). Conversely, if X is any local (weak) solution to (12.57) then it is in fact a strong solution because the diffusion coefficient in (12.57) is a locally Lipschitz continuous function. Moreover, using the Itô formula again, one can check that the process

$$Z_t = \ln \frac{X_t}{1 - X_t}, \quad \forall t \in [0, T^*],$$

is a strong solution of (12.58).

Hence it can be continued to a global solution, which is then unique. The last part of the lemma follows from the definition of the process X and the uniqueness of solutions to (12.57).   □

The next result provides a representation of the expected value $\mathbb{E}_{\mathbb{P}}(g(X_t))$ in terms of the integral with respect to the Gaussian distribution.

Proposition 12.6.2 *Let $g : [0, 1] \to \mathbb{R}$ be a nonnegative Borel function such that the random variable $g(X_t)$ is $\mathbb{P}$-integrable for some $t > 0$, where X is the unique solution (12.57) with the initial condition $x \in (0, 1)$. Then the expected value $\mathbb{E}_{\mathbb{P}}(g(X_t))$ is given by the formula*

$$\mathbb{E}_{\mathbb{P}}(g(X_t)) = \sqrt{x(1-x)}e^{-\frac{v^2(0,t)}{8}}\,\mathbb{E}_{\mathbb{Q}}\left\{\left(\eta^{-1} + \eta\right)g\left((1 + \eta^2)^{-1}\right)\right\},$$

where

$$\eta = \exp\left(-\frac{z + \zeta}{2}\right), \quad z = \ln\frac{x}{1 - x},$$

and the random variable ζ has under $\mathbb{Q}$ the Gaussian probability distribution with zero mean value and variance

$$v^2(0, t) = \int_0^t |\lambda(u)|^2\,du.$$

Proof. The proof of the proposition, due to Goldys (1997), is based on a simple idea that for any $\mathbb{P}$-integrable random variable, U say, we have

$$\mathbb{E}_{\mathbb{P}}U = \mathbb{E}_{\mathbb{P}}(X_t U) + \mathbb{E}_{\mathbb{P}}(Y_t U), \tag{12.60}$$

where $Y_t = 1 - X_t$. Next, it is essential to observe that X_t and Y_t can be represented as follows

$$X_t = x\,\mathcal{E}_t\left(-\int_0^{\cdot} Y_u\lambda(u) \cdot dW_u\right) \tag{12.61}$$

and

$$Y_t = (1 - x)\,\mathcal{E}_t\left(-\int_0^{\cdot} X_u\lambda(u) \cdot dW_u\right). \tag{12.62}$$

Furthermore, from (12.58) it follows that

$$dZ_t = -\lambda(t) \cdot \left(dW_t - (X_t - \tfrac{1}{2})\lambda(t)\,dt\right),$$

or equivalently,

$$dZ_t = -\lambda(t) \cdot \left(dW_t - (\tfrac{1}{2} - Y_t)\lambda(t)\,dt\right).$$

Let us introduce the auxiliary probability measures $\tilde{\mathbb{P}}$ and $\hat{\mathbb{P}}$ on $(\Omega, \mathcal{F}_t)$ by setting

$$\frac{d\tilde{\mathbb{P}}}{d\mathbb{P}} = \mathcal{E}_t\left(\int_0^{\cdot} (X_u - \tfrac{1}{2})\lambda(u) \cdot dW_u\right) \stackrel{\text{def}}{=} \tilde{\eta}_t$$

and

$$\frac{d\hat{\mathbb{P}}}{d\mathbb{P}} = \mathcal{E}_t \left(\int_0^\cdot (\tfrac{1}{2} - Y_u)\lambda(u) \cdot dW_u \right) \overset{\text{def}}{=} \hat{\eta}_t.$$

Note that

$$Z_t = z - \int_0^t \lambda(u) \cdot d\tilde{W}_u = z - \int_0^t \lambda(u) \cdot d\hat{W}_u,$$

where the processes

$$\tilde{W}_u = W_u - \int_0^u \left(X_v - \tfrac{1}{2} \right)\lambda(v)\, dv$$

$$\hat{W}_u = W_u - \int_0^u \left(\tfrac{1}{2} - Y_v \right)\lambda(v)\, dv$$

are known to be standard Brownian motions under $\tilde{\mathbb{P}}$ and $\hat{\mathbb{P}}$ respectively. Simple manipulations show that

$$\tilde{\eta}_t = \mathcal{E}_t \left(\int_0^\cdot X_u\lambda(u) \cdot dW_u \right) \exp\left(-\frac{1}{2} \int_0^t \lambda(u) \cdot d\tilde{W}_u + \frac{1}{8} \int_0^t |\lambda(u)|^2 du \right),$$

and thus, using (12.62), we obtain

$$Y_t = (1 - x)\, e^{-\frac{v^2(0,t)}{8}} \tilde{\eta}_t \exp\left(\frac{1}{2} \int_0^t \lambda(u) \cdot d\tilde{W}_u \right). \tag{12.63}$$

Similarly, one can check that

$$\hat{\eta}_t = \mathcal{E}_t \left(\int_0^\cdot Y_u\lambda(u) \cdot dW_u \right) \exp\left(\frac{1}{2} \int_0^t \lambda(u) \cdot d\hat{W}_u + \frac{1}{8} \int_0^t |\lambda(u)|^2 du \right).$$

Hence, in view of (12.61), we get

$$X_t = x\, e^{-\frac{v^2(0,t)}{8}} \hat{\eta}_t \exp\left(-\frac{1}{2} \int_0^t \lambda(u) \cdot d\hat{W}_u \right).$$

The last equality yields

$$\mathbb{E}_{\mathbb{P}}(X_t g(X_t)) = x\, e^{-\frac{v^2(0,t)}{8}} \mathbb{E}_{\hat{\mathbb{P}}} \left\{ g(X_t) \exp\left(-\frac{1}{2} \int_0^t \lambda(u) \cdot d\hat{W}_u \right) \right\},$$

whereas (12.63) gives

$$\mathbb{E}_{\mathbb{P}}(Y_t g(X_t)) = (1 - x)\, e^{-\frac{v^2(0,t)}{8}} \mathbb{E}_{\tilde{\mathbb{P}}} \left\{ g(X_t) \exp\left(\frac{1}{2} \int_0^t \lambda(u) \cdot d\tilde{W}_u \right) \right\}.$$

Using (12.59) and (12.60), we conclude that

$$\mathbb{E}_{\mathbb{P}}(g(X_t)) = e^{-\frac{v^2(0,t)}{8}} \mathbb{E}_{\mathbb{Q}} \left\{ g\left(\frac{1}{1 + e^{-(z+\zeta)}} \right) \left(x e^{\frac{1}{2}\zeta} + (1 - x)e^{-\frac{1}{2}\zeta} \right) \right\},$$

where ζ is, under $\mathbb{Q}$, a Gaussian random variable with zero mean value and variance $v^2(0, t)$. The last formula is equivalent to the formula in the statement of the proposition.   □

To establish directly formula (12.53), a slightly different change of the underlying probability measure $\mathbb{P}$ is convenient. Namely, we put

$$\frac{d\tilde{\mathbb{P}}}{d\mathbb{P}} = \mathcal{E}_t\left(\int_0^{\cdot} X_u\lambda(u) \cdot dW_u\right)$$

$$\frac{d\hat{\mathbb{P}}}{d\mathbb{P}} = \mathcal{E}_t\left(\int_0^{\cdot} Y_u\lambda(u) \cdot dW_u\right).$$

Now, under $\tilde{\mathbb{P}}$ and $\hat{\mathbb{P}}$ we have

$$dZ_t = -\tfrac{1}{2}\lambda(t)\,dt - \lambda(t) \cdot d\tilde{W}_t,$$

$$dZ_t = \tfrac{1}{2}\lambda(t)\,dt - \lambda(t) \cdot d\hat{W}_t,$$

respectively, where the processes $\tilde{W}$ and $\hat{W}$, given by, for $u \in [0, t]$,

$$\tilde{W}_u = W_u - \int_0^u X_v\lambda(v)\,dv, \quad \hat{W}_u = W_u - \int_0^t Y_v\lambda(v)\,dv,$$

are standard Brownian motions under $\tilde{\mathbb{P}}$ and $\hat{\mathbb{P}}$ respectively. Formula (12.53) can thus be derived easily from representation (12.60) combined with equalities (12.61)-(12.62).

Corollary 12.6.1 *Let* $g : \mathbb{R} \to \mathbb{R}$ *be a nonnegative Borel function. Then for every* $x \in (0, 1)$ *and any* $0 < t < T$ *the conditional expectation* $\mathbb{E}_{\mathbb{P}}(g(X_T) \mid \mathcal{F}_t)$ *equals* $\mathbb{E}_{\mathbb{P}}(g(X_T) \mid \mathcal{F}_t) = k(X_t)$, *where the function* $k : (0, 1) \to \mathbb{R}$ *is given by the formula*

$$k(x) = \sqrt{x(1-x)}e^{-\frac{v^2(t,T)}{8}} \mathbb{E}_{\mathbb{Q}}\left\{g\left(\frac{1}{1+e^{-(z+\zeta)}}\right)\left(e^{(z+\zeta)/2} + e^{-(z+\zeta)/2}\right)\right\}$$

with $z = \ln x/(1-x)$, *and the random variable* ζ *has under* $\mathbb{Q}$ *a Gaussian distribution with zero mean value and variance* $v^2(t, T) = \int_t^T |\lambda(u)|^2 du$.

12.6.4 Bond Options

Our next goal is to establish the bond option valuation formula within the framework of the lognormal LIBOR model. It is interesting to notice that an identical formula was previously established by Rady and Sandmann (1994), who adopted the PDE approach, and who worked with a different model, however. In this section, in which we follow Goldys (1997), we present a probabilistic approach to the bond option valuation. His derivation of the bond option price is based on Corollary 12.6.1. Note that this result can be used to the valuation of an arbitrary European claim.

Proposition 12.6.3 *The price C_t at time $t \leq T_{j-1}$ of a European call option, with expiration date T_{j-1} and strike price $0 < K < 1$, written on a zero-coupon bond maturing at $T_j = T_{j-1} + \delta_j$, equals*

$$C_t = (1 - K)B(t, T_j)N\big(l_1^j(t)\big) - K(B(t, T_{j-1}) - B(t, T_j))N\big(l_2^j(t)\big),$$

where

$$l_{1,2}^j(t) = \frac{\ln((1 - K)B(t, T_j)) - \ln\big(K\big(B(t, T_{j-1}) - B(t, T_j)\big)\big) \pm \frac{1}{2}\tilde{v}_j(t)}{\tilde{v}_j(t)}$$

and

$$\tilde{v}_j^2(t) = \int_t^{T_{j-1}} |\lambda(u, T_{j-1})|^2 du.$$

Proof. Let us write $T = T_{j-1}$ and $T + \delta = T_j$. In view of Corollary 12.6.1, it is clear that

$$C_t = B(t, T)\,\mathbb{E}_{\mathbb{P}_T}\big((F(T, T + \delta, T) - K)^+ \mid \mathcal{F}_t\big) = B(t, T)k(x),$$

where $x = F(t, T + \delta, T)$ and $g(y) = (y - K)^+$. Using the notation $\tilde{v} = \tilde{v}_j(t)$ and

$$l = \tilde{v}^{-1}\left(\ln \frac{x(1 - K)}{K(1 - x)}\right),$$

we obtain

$$k(x) = \sqrt{x(1 - x)}\,e^{-\frac{1}{8}\tilde{v}^2} \int_{-l}^{\infty} \left(\frac{1}{1 + e^{-\tilde{y}}} - K\right)\left(e^{-\frac{1}{2}\tilde{y}} + e^{\frac{1}{2}\tilde{y}}\right) n(y)\,dy,$$

where $\tilde{y} = z + \tilde{v}y$, and n stands for the standard normal density. Let us set

$$h(y) = e^{-\frac{1}{2}(z+\tilde{v}y)} + e^{\frac{1}{2}(z+\tilde{v}y)}.$$

Then

$$k(x) = \sqrt{x(1 - x)}\,e^{-\frac{1}{8}\tilde{v}^2} \left(\int_{-l}^{\infty} \frac{h(y)n(y)}{1 + e^{-(z+\tilde{v}y)}}\,dy - K \int_{-l}^{\infty} h(y)n(y)\,dy\right).$$

Equivalently,

$$k(x) = \sqrt{x(1 - x)}\,e^{-\frac{1}{8}\tilde{v}^2}\,(I_1 - KI_2),$$

where

$$I_1 = \int_{-l}^{\infty} e^{\frac{1}{2}(z+\tilde{v}y)} n(y)\,dy = e^{\frac{1}{8}\tilde{v}^2 + \frac{1}{2}z}\big(1 - N(-l - \tfrac{1}{2}\tilde{v})\big)$$

and

$$I_2 = I_1 + e^{\frac{1}{8}\tilde{v}^2 - \frac{1}{2}z}\big(1 - N(-l + \tfrac{1}{2}\tilde{v})\big).$$

Consequently, we find that

$$k(x) = \sqrt{x(1-x)}(1-K)e^{\frac{1}{2}z}\left(1 - N(-l - \tfrac{1}{2}\tilde{v})\right)$$
$$- K\sqrt{x(1-x)}e^{-\frac{1}{2}z}\left(1 - N(-l + \tfrac{1}{2}\tilde{v})\right),$$

or, after simplification,

$$k(x) = x(1-K)N(l + \tfrac{1}{2}\tilde{v}) - K(1-x)N(l - \tfrac{1}{2}\tilde{v}).$$

Since

$$x = F(t, T + \delta, T) = B(t, T + \delta)/B(t, T),$$

the proof of the proposition is completed.   □

Using the put-call parity relationship for options written on a zero-coupon bond,

$$C_t - P_t = B(t, T_j) - K B(t, T_{j-1}),$$

it is easy to check that the price of the corresponding put option equals

$$P_t = (K - 1)B(t, T_j)N\left(-l_1(t)\right) - K\left(B(t, T_j) - B(t, T_{j-1})\right)N\left(-l_2(t)\right).$$

Recall that the jth caplet is equivalent to the put option written on a zero-coupon bond, with expiry date T_{j-1} and strike price $K = \tilde{\delta}_j^{-1} = (1 + \kappa d_j)^{-1}$. More precisely, the option's payoff should be multiplied by the nominal value $\tilde{\delta}_j = 1 + \kappa d_j$. Hence, using the last formula, we obtain

$$\mathbf{Cpl}_t^j = \left(B(t, T_{j-1}) - B(t, T_j)\right)N\left(-l_2^j(t)\right) - \kappa\delta_j B(t, T_j)N\left(-l_1^j(t)\right),$$

since clearly $K - 1 = -\kappa\delta_j\tilde{\delta}_j^{-1}$. To show that the last formula coincides with (12.55), it is enough to check that if $K = \tilde{d}_j^{-1}$, then the terms $-l_2^j(t)$ and $-l_1^j(t)$ coincide with the terms $\tilde{e}_1^j(t)$ and $\tilde{e}_2^j(t)$ of Proposition 12.6.1. In this way, we obtain an alternative probabilistic derivation of the cap valuation formula within the lognormal LIBOR market model.

Remarks. Proposition 12.6.3 shows that the replication of the bond option using the underlying bonds of maturity T_{j-1} and T_j is not straightforward. This should be contrasted with the case of the Gaussian HJM set-up in which hedging of bond options with the use of the underlying bonds is done in a standard way. This illustrates our general observation that each particular model of the term structure should be tailored to a specific class of derivatives and hedging instruments.

12.7 Extensions of the LLM Model

Let us emphasize that alternative constructions of the LIBOR market model presented in Sect. 12.4.4-12.4.6 do not require that the volatilities of LIBORs be deterministic functions. We have seen that they may be adapted stochastic processes, or some (deterministic or random) functions of the underlying forward LIBORs. The most commonly used choice of deterministic volatilities for forward LIBORs leads inevitably to the so-called lognormal LIBOR market model (also known as the LLM model or the BGM model).

Empirical studies have shown that the implied volatilities of market prices of caplets (and swaptions) tend to be decreasing functions of the strike level. Hence it is of practical interest to develop stochastic volatility versions of the LIBOR market model capable of matching the observed volatility smiles of caplets.

CEV LIBOR model. A straightforward generalization of the lognormal LIBOR market model was examined by Andersen and Andreasen (2000b). In their approach, the assumption that the volatilities are deterministic functions was replaced by a suitable functional form of the volatility coefficient. The main emphasis in Andersen and Andreasen (2000b) is put on the use of the CEV process[2] as a model of a forward LIBOR. To be more specific, they postulate that, for every $t \in [0, T_j]$,

$$dL(t, T_j) = L^\beta(t, T_j)\lambda(t, T_j) \cdot dW_t^{T_{j+1}},$$

where $\beta > 0$ is a strictly positive constant. Under this specification of the dynamics of forward LIBORs with the exponent $\beta \neq 1$, they derive closed-form solutions for caplet prices in terms of the cumulative distribution function of a non-central χ^2 probability distribution. They show also that, depending on the choice of the parameter β, the implied Black volatilities for caplets, when considered as a function of the strike level $\kappa > 0$, exhibit downward- or upward-sloping skew.

Stochastic volatility LIBOR model. In a recent paper by Joshi and Rebonato (2003), the authors examine a stochastic volatility displaced-diffusion extension of a LIBOR market model. Recall that the stochastic volatility displaced-diffusion approach to the modelling of stochastic volatility was presented in Sect. 7.2.2. Basically, Joshi and Rebonato (2003) postulate that

$$d(f(t, T_i) + \alpha) = (f(t, T_i) + \alpha)(\mu^\alpha(t, T_i)\, dt + \sigma^\alpha(t, T_i)\, dW_t),$$

where the drift term can be found explicitly, and where by assumption the volatility $\sigma^\alpha(t, T_i)$ is given by the expression:

$$\sigma^\alpha(t, T_i) = \big(a_t + b_t(T_i - t)\big)e^{-c_t(T_i - t)} + d_t$$

for some mean-reverting diffusion processes a_t, b_t, $\ln c_t$ and $\ln d_t$. They argue that such a model is sufficiently flexible to be capable of describing in a realistic way not only the today's implied volatility surface, but also the observed changes in the market term structure of volatilities.

For other examples of stochastic volatility extensions of a LIBOR market model, the interested reader is referred to Rebonato and Joshi (2001), Gątarek (2003), and Piterbarg (2003a).

---

[2] In the context of equity options, the CEV (*constant elasticity of variance*) process was introduced by Cox and Ross (1976a, 1976b). For more details, see Sect. 7.2.

13

Alternative Market Models

The BGM market model, examined in some detail in the previous chapter, is clearly oriented toward a particular interest rate, LIBOR, and thus it serves well for the valuation and hedging of the LIBOR related derivatives, such as plain-vanilla and exotic caps and floors. It thus may be see as a good candidate for the role of the Black-Scholes-like benchmark model for this particular sector of the fixed-income market. Interest rate swaps are another class of interest-rate-sensitive contracts of great practical importance. In a generic *fixed-for-floating swap*, a fixed rate of interest is exchanged periodically for some preassigned variable (floating) rate. Since swaps and derivative securities on the swap rate, termed *swap derivatives*, are in many markets more liquidly traded than LIBOR derivatives, there is an obvious demand for specific models capable of efficient handling this class of interest rate products. Our goal in this chapter is to present some recent research focused on market models that are alternatives to the market model for LIBORs.

In Sect. 13.1, we give a general description of interest rate swaps and we derive the universal valuation formula for a standard fixed-for-floating swap. We also examine the most typical examples of over-the-counter option contracts related to swap rates; that is, plain-vanilla and exotic swaptions. In the next section, the valuation formulas for some of these instruments are established within the framework of the Gaussian HJM model. In Sect. 13.3-13.5, three alternative versions of Jamshidian's model of (co-terminal, co-initial and co-sliding) forward swap rates are presented. In all three models, the plain-vanilla swaptions are priced and hedged in a similar way as caps in the lognormal model of LIBORs.

In the next section, we compare the lognormal model of forward LIBORs with the lognormal model of co-terminal forward swap rates, and we study the issue of swaption (respectively, caplet) valuation in the lognormal model of forward LIBORs (respectively, co-terminal forward swap rates).

The final two sections are devoted to the *Markov-functional approach*, which was put forward by Hunt et al. (1996), and to the *rational lognormal model*, which was proposed by Flesaker and Hughston (1996a, 1996b).

13.1 Swaps and Swaptions

We first consider a *fixed-for-floating forward start swap* settled *in arrears,* with notional principal N. For conciseness, we shall frequently refer to such a contract as the *payer swap.* A long position in a payer swap corresponds to the situation when an investor makes periodic payments determined by a fixed interest rate, and receives in exchange payments specified by some floating rate. A short position in a payer swap defines a closely related contract, known as the *receiver swap.*

13.1.1 Forward Swap Rates

Let a finite collection of future dates T_j, $j = 0, \ldots, n$, where $T_0 \geq 0$ and $T_j - T_{j-1} = \delta_j > 0$ for every $j = 1, \ldots, n$ be given. The floating rate $L(T_j)$ received at time T_{j+1} is set at time T_j by reference to the price of a zero-coupon bond over that period. Specifically, $L(T_j)$ satisfies

$$B(T_j, T_{j+1})^{-1} = 1 + (T_{j+1} - T_j)L(T_j) = 1 + \delta_{j+1}L(T_j). \tag{13.1}$$

Formula (13.1) thus agrees with market quotations of LIBOR; we recognize that $L(T_j)$ is the spot LIBOR that prevails at time T_j and refers to the period of length $\delta_{j+1} = T_{j+1} - T_j$.

In a payer swap, at any settlement date T_j, $j = 1, \ldots, n$ the cash flows are $L(T_{j-1})\delta_{j+1}N$ and $-\kappa\delta_{j+1}N$, where κ is a pre-agreed fixed rate of interest, and N is the *notional principal.* It is obvious that the cash flows of a receiver swap are of the same size as that of a payer swap, but have the opposite signs.

The number n that coincides with the number of payments, is referred to as the *length* of a swap; for instance, the length of a 3-year swap with quarterly settlement equals $n = 12$. The dates $T_0, \ldots, T_{n-1}$ are known as *reset dates,* and the dates $T_1, \ldots, T_n$ are termed *settlement dates.* The first reset date, T_0, is called the *start date* of a swap. Finally, the time interval $[T_{j-1}, T_j]$ is referred to as the jth *accrual period* (it is also known as *day-count fraction* or *coverage*).

Before we proceed to the swap valuation, let us make a simple observation that the cash flows in a swap are linear functions of the notional principal N. Hence we may and do assume, without loss of generality, that $N = 1$.

Let us place ourselves within the framework of some arbitrage-free term structure model. Then the value at time $t \leq T_0$ of a forward start payer swap, denoted by $\mathbf{FS}_t$ or $\mathbf{FS}_t(\kappa)$, equals

$$\mathbf{FS}_t(\kappa) = \mathbb{E}_{\mathbb{P}^*}\left\{ \sum_{j=1}^{n} \frac{B_t}{B_{T_j}} (L(T_{j-1}) - \kappa)\delta_j \,\Big|\, \mathcal{F}_t \right\}$$

$$= \sum_{j=1}^{n} \mathbb{E}_{\mathbb{P}^*}\left\{ \frac{B_t}{B_{T_j}} \left(B(T_{j-1}, T_j)^{-1} - (1 + \kappa\delta_j) \right) \Big|\, \mathcal{F}_t \right\}.$$

Consequently, writing $\tilde{\delta}_j = 1 + \kappa\delta_j$ and noting that $t \leq T_{j-1}$, we get

$$\mathbf{FS}_t(\kappa) = \sum_{j=1}^{n} \mathbb{E}_{\mathbb{P}^*} \left\{ B(T_{j-1}, T_j)^{-1} \frac{B_t}{B_{T_{j-1}}} \mathbb{E}_{\mathbb{P}^*} \left(\frac{B_{T_{j-1}}}{B_{T_j}} \Big| \mathcal{F}_{T_{j-1}} \right) \Big| \mathcal{F}_t \right\}$$

$$- \sum_{j=1}^{n} \tilde{\delta}_j \, \mathbb{E}_{\mathbb{P}^*} \left\{ \frac{B_t}{B_{T_j}} \Big| \mathcal{F}_t \right\} = \sum_{j=1}^{n} \Big(B(t, T_{j-1}) - \tilde{\delta}_j B(t, T_j) \Big).$$

After rearranging, we obtain the following simple, but important, result.

Lemma 13.1.1 *We have, for every* $t \in [0, T_0]$,

$$\mathbf{FS}_t(\kappa) = B(t, T_0) - \sum_{j=1}^{n} c_j B(t, T_j), \qquad (13.2)$$

where $c_j = \kappa \delta_j$ *for* $j = 1, \ldots, n-1$ *and* $c_n = \tilde{\delta}_n = 1 + \kappa \delta_n$.

Alternatively, formula (13.2) can be established using the forward measure approach. Indeed, since we have

$$L(t, T_{j-1}) = \frac{B(t, T_{j-1}) - B(t, T_j)}{\delta_j B(t, T_j)},$$

the process $L(t, T_{j-1})$ is a martingale under the forward measure $\mathbb{P}_{T_j}$, and thus

$$\mathbf{FS}_t(\kappa) = \sum_{j=1}^{n} B(t, T_j) \, \mathbb{E}_{\mathbb{P}_{T_j}} \big((L(T_{j-1}) - \kappa) \delta_j \, \big| \, \mathcal{F}_t \big)$$

$$= \sum_{j=1}^{n} B(t, T_j) \delta_j \big(L(t, T_{j-1}) - \kappa \big)$$

$$= \sum_{j=1}^{n} \big(B(t, T_{j-1}) - B(t, T_j) - \kappa \delta_j B(t, T_j) \big)$$

$$= B(t, T_0) - \sum_{j=1}^{n} c_j B(t, T_j).$$

Equality (13.2) makes it clear that a forward swap settled in arrears is, essentially, a contract to deliver a specific coupon-bearing bond and to receive at the same time a zero-coupon bond. Hence relationship (13.2) may also be inferred by a straightforward comparison of the future cash flows from these bonds. Note that (13.2) provides a simple method for the replication of a swap contract. Let us emphasize that this method of replication of a forward swap (and consequently, the method of arbitrage valuation of a forward swap) is thus independent of the term structure model. For this reason, it is referred to as a *universal* swap pricing formula.

In the forward start payer swap settled *in advance* – that is, in which each reset date is also a settlement date – the discounting method varies from country to country.

Under one market convention, the cash flows of a swap settled in advance at reset dates T_j, $j = 0, \ldots, n - 1$ are

$$L(T_j)\delta_{j+1}(1 + L(T_j)\delta_{j+1})^{-1}$$

and

$$-\kappa\delta_{j+1}(1 + L(T_j)\delta_{j+1})^{-1}.$$

The value $\mathbf{FS}_t^*(\kappa)$ at time t of this swap therefore equals

$$\mathbf{FS}_t^*(\kappa) = \mathbb{E}_{\mathbb{P}^*}\left\{ \sum_{j=0}^{n-1} \frac{B_t}{B_{T_j}} \frac{\delta_{j+1}(L(T_j) - \kappa)}{1 + \delta_{j+1}L(T_j)} \,\Big|\, \mathcal{F}_t \right\}$$

$$= \mathbb{E}_{\mathbb{P}^*}\left\{ \sum_{j=0}^{n-1} \frac{B_t}{B_{T_j}}(L(T_j) - \kappa)\delta_{j+1}B(T_j, T_{j+1}) \,\Big|\, \mathcal{F}_t \right\}$$

$$= \mathbb{E}_{\mathbb{P}^*}\left\{ \sum_{j=0}^{n-1} \frac{B_t}{B_{T_{j+1}}}(L(T_j) - \kappa)\delta_{j+1} \,\Big|\, \mathcal{F}_t \right\},$$

which coincides with the value of the swap settled in arrears. Once again, this is by no means surprising, since the payoffs $L(T_j)\delta_{j+1}(1 + L(T_j)\delta_{j+1})^{-1}$ and $-\kappa\delta_{j+1}(1 + L(T_j)\delta_{j+1})^{-1}$ at time T_j are easily seen to be equivalent to payoffs $L(T_j)\delta_{j+1}$ and $-\kappa\delta_{j+1}$ respectively at time T_{j+1} (recall that $1 + L(T_j)\delta_{j+1} = B^{-1}(T_j, T_{j+1})$). Under an alternative convention, the cash flows at each reset date T_j are

$$L(T_j)\delta_{j+1}(1 + L(T_j)\delta_{j+1})^{-1}$$

and

$$-\kappa\delta_{j+1}(1 + \kappa\delta_{j+1})^{-1}.$$

The value of such a forward swap at time t, denoted by $\mathbf{FS}_t^{**}(\kappa)$, equals

$$\mathbf{FS}_t^{**}(\kappa) = \mathbb{E}_{\mathbb{P}^*}\left\{ \sum_{j=0}^{n-1} \frac{B_t}{B_{T_j}}\left(\frac{\delta_{j+1}L(T_j)}{1 + \delta_{j+1}L(T_j)} - \frac{\kappa\delta_{j+1}}{1 + \kappa\delta_{j+1}} \right) \,\Big|\, \mathcal{F}_t \right\}$$

$$= \sum_{j=0}^{n-1} \tilde{\delta}_{j+1}^{-1}\left\{ B(t, T_j) - \tilde{\delta}_{j+1}\,\mathbb{E}_{\mathbb{P}^*}\left(\frac{B_t}{B_{T_j}} B(T_j, T_{j+1}) \,\Big|\, \mathcal{F}_t \right) \right\}$$

$$= \sum_{j=1}^{n} \tilde{\delta}_{j+1}^{-1}\left(B(t, T_{j-1}) - \tilde{\delta}_{j+1}B(t, T_j) \right),$$

which is, of course, the value of the forward start payer swap settled in arrears, discounted at the fixed rate κ.

In what follows, we shall restrict our attention to interest rate swaps settled in arrears. As was mentioned already, a swap agreement is worthless at initiation. This important feature of a swap leads to the following definition that refers in fact to the more general concept of a forward swap.

Definition 13.1.1 The *forward swap rate* $\kappa(t, T_0, n)$ at time t for the date T_0 is that value of the fixed rate κ that makes the value of the n-period forward swap zero, i.e., that value of κ for which $\mathbf{FS}_t(\kappa) = 0$.

Using (13.2), we obtain an explicit formula for the n-period forward swap rate in terms of zero-coupon bonds

$$\kappa(t, T_0, n) = (B(t, T_0) - B(t, T_n)) \left(\sum_{j=1}^{n} \delta_j B(t, T_j) \right)^{-1}. \tag{13.3}$$

A *swap* (respectively, *swap rate*) is the forward swap (respectively, forward swap rate) in which we set $t = T_0$. It is thus clear that the swap rate, $\kappa(T_0, n) = \kappa(T_0, T_0, n)$, is given by the formula

$$\kappa(T_0, n) = (1 - B(T_0, T_n)) \left(\sum_{j=1}^{n} \delta_j B(T_0, T_j) \right)^{-1}. \tag{13.4}$$

Note that the definition of a forward swap rate implicitly refers to a swap contract of length n starting at time T. It would thus be more correct to refer to $\kappa(t, T, n)$ as the *n-period forward swap rate* prevailing at time t for the future date T. A forward swap rate is a rather theoretical concept, as opposed to swap rates, which are quoted daily (subject to an appropriate bid-ask spread) by financial institutions who offer interest rate swap contracts to their institutional clients. In practice, swap agreements of various lengths are offered. Also, typically, the length of the reference period varies over time; for instance, a 5-year swap may be settled quarterly during the first three years, and semi-annually during the last two.

Remarks. Let us examine one leg of a swap – that is, an interest rate swap agreement with only one payment date. For $n = 1$, from (13.3) we get

$$\kappa(t, T, U) = \frac{B(t, T) - B(t, U)}{(U - T)B(t, U)},$$

where we write T, U and $\kappa(t, T, U)$ instead of T_0, T_1 and $\kappa(t, T_0, 1)$ respectively. Using (9.3), we find that for $t = 0$

$$\kappa(0, T, U) = \frac{\exp(-TY(0, T)) - \exp(-UY(0, U))}{(U - T)\exp(-UY(0, U))},$$

and thus

$$\kappa(0, T, U) \approx \frac{UY(0, U) - TY(0, T)}{U - T} = f(0, T, U),$$

where "$\approx$" denotes approximate equality. This shows that the swap rate does not coincide with the forward interest rate $f(t, T, U)$ determined by a forward rate agreement. If instead of forward rates, one makes use of futures rates (implied by the Eurodollar futures contracts) to determine the swap rate, a systematic bias leading to arbitrage opportunities arises (cf. Burghardt and Hoskins (1995)).

13.1.2 Swaptions

The owner of a *payer* (respectively, *receiver*) *swaption* with strike rate κ, maturing at time $T = T_0$, has the right to enter at time T the underlying forward payer (respectively, receiver) swap settled in arrears. Because $\mathbf{FS}_T(\kappa)$ is the value at time T of the payer swap with the fixed interest rate κ, it is clear that the price of the payer swaption at time t equals

$$\mathbf{PS}_t = \mathbb{E}_{\mathbb{P}^*}\left\{ \frac{B_t}{B_T} \left(\mathbf{FS}_T(\kappa) \right)^+ \Big| \mathcal{F}_t \right\}.$$

More explicitly, we have

$$\mathbf{PS}_t = \mathbb{E}_{\mathbb{P}^*}\left\{ \frac{B_t}{B_T} \left(\mathbb{E}_{\mathbb{P}^*}\left(\sum_{j=1}^{n} \frac{B_T}{B_{T_j}} (L(T_{j-1}) - \kappa)\delta_j \Big| \mathcal{F}_T \right) \right)^+ \Big| \mathcal{F}_t \right\}. \tag{13.5}$$

For the receiver swaption, we have

$$\mathbf{RS}_t = \mathbb{E}_{\mathbb{P}^*}\left\{ \frac{B_t}{B_T} \left(-\mathbf{FS}_T(\kappa) \right)^+ \Big| \mathcal{F}_t \right\},$$

that is

$$\mathbf{RS}_t = \mathbb{E}_{\mathbb{P}^*}\left\{ \frac{B_t}{B_T} \left(\mathbb{E}_{\mathbb{P}^*}\left(\sum_{j=1}^{n} \frac{B_T}{B_{T_j}} (\kappa - L(T_{j-1}))\delta_j \Big| \mathcal{F}_T \right) \right)^+ \Big| \mathcal{F}_t \right\}, \tag{13.6}$$

where we denote by $\mathbf{RS}_t$ the price at time t of a receiver swaption. We will now focus on a payer swaption. In view of (13.5), it is apparent that a payer swaption is exercised at time T if and only if the value of the underlying swap is positive at this date. A swaption may be exercised by its owner only at its maturity date T. If exercised, a swaption gives rise to a sequence of cash flows at prescribed future dates. By considering the future cash flows from a swaption and from the corresponding *market swap*[1] available at time T, it is easily seen that the owner of a swaption is protected against the adverse movements of the swap rate that may occur before time T.

Suppose, for instance, that the swap rate at time T is greater than κ. Then by combining the swaption with a market swap, the owner of a swaption with exercise rate κ is entitled to enter at time T, at no additional cost, a swap contract in which the fixed rate is κ. If, on the contrary, the swap rate at time T is less than κ then the swaption is worthless, but its owner is, of course, able to enter a market swap contract based on the current swap rate $\kappa(T, T, n) \leq \kappa$. We thus conclude that the fixed rate paid by the owner of a swaption who intends to initiate a swap contract at time T will never be above the preassigned level κ.

---

[1] At any time t, a *market swap* is that swap whose current market value equals zero. In other words, it is the swap in which the fixed rate κ equals the current swap rate.

Swaption as a put option on a coupon-bond. Since we may rewrite (13.5) as follows

$$PS_t = \mathbb{E}_{\mathbb{P}^*}\left\{ \frac{B_t}{B_T}\left(1 - \sum_{j=1}^{n} c_j B(T, T_j)\right)^+ \Big| \mathcal{F}_t \right\}, \tag{13.7}$$

the payer swaption may also be seen as a put option on a coupon-bearing bond. Similar remarks are valid for the receiver swaption. We conclude that formally a payer (respectively, receiver) swaption may be seen as a put (respectively, call) option on a coupon bond with strike price 1 and coupon rate equal to the strike rate κ of the underlying forward swap. Therefore, the arbitrage price of payer and receiver swaptions can be evaluated by applying the general valuation formula of Proposition 11.3.5 to the functions

$$g^P(x_1, \ldots, x_n) = \left(1 - \sum_{j=1}^{n} c_j x_j\right)^+, \quad g^r(x_1, \ldots, x_n) = \left(\sum_{j=1}^{n} c_j x_j - 1\right)^+$$

for a payer and a receiver swaption respectively.

It follows easily from (13.5)–(13.6) that $PS_t - RS_t = FS_t$, i.e.,

$$\textit{Payer Swaption}\,(t) \;-\; \textit{Receiver Swaption}\,(t) \;=\; \textit{Forward Swap}\,(t)$$

provided that both swaptions expire at the same date T (and have the same contractual features).

Swaption as a call option on a swap rate. We shall now show that a payer (respectively, receiver) swaption can also be viewed as a sequence of call (respectively, put) options on a swap rate that are not allowed to be exercised separately. At time T the long party receives the value of a sequence of cash flows, discounted from time T_j, $j = 1, \ldots, n$ to the date T, defined by $d_j^P = \delta_j\,(\kappa(T, T, n) - \kappa)^+$ and $d_j^r = \delta_j\,(\kappa - \kappa(T, T, n))^+$ for the payer option and the receiver option respectively, where

$$\kappa(T, T, n) = (1 - B(T, T_n))\left(\sum_{j=1}^{n} \delta_j B(T, T_j)\right)^{-1}$$

is the corresponding swap rate at the option's expiry. Indeed, the price at time t of the call (payer) option on a swap rate is

$$C_t = \mathbb{E}_{\mathbb{P}^*}\left\{ \frac{B_t}{B_T} \mathbb{E}_{\mathbb{P}^*}\left(\sum_{j=1}^{n} \frac{B_T}{B_{T_j}}(\kappa(T, T, n) - \kappa)^+\delta_j \Big| \mathcal{F}_T \right) \Big| \mathcal{F}_t \right\}$$

$$= \mathbb{E}_{\mathbb{P}^*}\left\{ \frac{B_t}{B_T}\left(1 - \sum_{j=1}^{n} c_j B(T, T_j)\right)^+ \Big| \mathcal{F}_t \right\}$$

$$= \mathbb{E}_{\mathbb{P}^*}\left\{ \frac{B_t}{B_T}\left(\mathbb{E}_{\mathbb{P}^*}\left(\sum_{j=1}^{n} \frac{B_T}{B_{T_j}}(L(T_{j-1}) - \kappa)\delta_j \Big| \mathcal{F}_T \right)\right)^+ \Big| \mathcal{F}_t \right\},$$

which is the payer swaption price PS_t.

Equality $C_t = \mathbf{PS}_t$ may also be derived by directly verifying that the future cash flows from the following portfolios established at time T are identical: portfolio A – a swaption and a market swap; and portfolio B – an option on a swap rate and a market swap. Indeed, both portfolios correspond to a payer swap with the fixed rate equal to κ. Similarly, for every $t \leq T$, the price of the put (receiver) option on a swap rate is (as before, $c_j = \kappa \delta_j$, $j = 1, \ldots, n-1$ and $c_n = 1 + \kappa \delta_n$)

$$
P_t = \mathbb{E}_{\mathbb{P}*} \left\{ \frac{B_t}{B_T} \mathbb{E}_{\mathbb{P}*} \left(\sum_{j=1}^{n} \frac{B_T}{B_{T_j}} (\kappa - \kappa(T, T, n))^+ \delta_j \,\Big|\, \mathcal{F}_T \right) \Big| \mathcal{F}_t \right\}
$$

$$
= \mathbb{E}_{\mathbb{P}*} \left\{ \frac{B_t}{B_T} \left(\sum_{j=1}^{n} c_j B(T, T_j) - 1 \right)^+ \Big| \mathcal{F}_t \right\}
$$

$$
= \mathbb{E}_{\mathbb{P}*} \left\{ \frac{B_t}{B_T} \left(\mathbb{E}_{\mathbb{P}*} \left(\sum_{j=1}^{n} \frac{B_T}{B_{T_j}} (\kappa - L(T_{j-1})) \delta_j \,\Big|\, \mathcal{F}_T \right) \right)^+ \Big| \mathcal{F}_t \right\},
$$

which equals the price $\mathbf{RS}_t$ of the receiver swaption.

Let us re-derive the valuation formula for the payer swaption in a more intuitive way. Recall that a payer swaption is essentially a sequence of fixed payments $d_j = \delta_j (\kappa(T, T, n) - \kappa)^+$ that are received at settlement dates $T_1, \ldots, T_n$, but whose value is known already at the expiry date T. The random variable d_j is therefore $\mathcal{F}_T$-measurable, and thus we may apply Corollary 9.6.1, directly obtaining

$$
\mathbf{PS}_t = B(t, T) \sum_{j=1}^{n} \mathbb{E}_{\mathbb{P}_T} \left(B(T, T_j)(\kappa(T, T, n) - \kappa)^+ \delta_j \,\Big|\, \mathcal{F}_t \right)
$$

for every $t \in [0, T]$. After simple manipulations this yields, as expected,

$$
\mathbf{PS}_t = B(t, T) \mathbb{E}_{\mathbb{P}_T} \left\{ \left(1 - \sum_{j=1}^{n} c_j B(T, T_j) \right)^+ \Big| \mathcal{F}_t \right\}.
$$

13.1.3 Exotic Swap Derivatives

Within the framework of swap derivatives, the standard swaptions described in the preceding section play the role of plain-vanilla options. Needless to say, a plethora of more sophisticated contracts on swap rates are currently traded on financial markets. We shall now describe some of them, and we shall later show that the risks involved in each of these contracts are of a different nature, and thus they require product-specific approaches to the modelling of forward swap rates. In other words, owing to the complexity of a swap derivative market, it is unlikely that one could construct a universal model of forward swap rates that were sufficiently flexible to serve all purposes.

Forward start swaptions. In a *forward start swaption* (also known as the *mid-curve option*) the expiry date T of the swaption precedes the initiation date T_0 of the underlying payer swap – that is, $T \leq T_0$. Note that if κ is a fixed strike level then we have always

$$\mathbf{FS}_t(\kappa) = \mathbf{FS}_t(\kappa) - \mathbf{FS}_t(\kappa(t, T_0, n)),$$

as by the definition of the forward swap rate we have $\mathbf{FS}_t(\kappa(t, T_0, n)) = 0$. A direct application of valuation formula (13.2) to both members on the right-hand side of the last equality yields

$$\mathbf{FS}_t(\kappa) = \sum_{j=1}^{n} \delta_j B(t, T_j)\big(\kappa(t, T_0, n) - \kappa\big)$$

for $t \in [0, T_0]$. It is thus clear that the payoff $\mathbf{PS}_T$ at expiry T of the forward swaption (with strike 0) is either 0, if $\kappa \geq \kappa(T, T_0, n)$, or

$$\mathbf{PS}_T = \sum_{j=1}^{n} \delta_j B(T, T_j)\big(\kappa(T, T_0, n) - \kappa\big)$$

if, on the contrary, inequality $\kappa(T, T_0, n) > \kappa$ holds. We conclude that the following result is valid.

Lemma 13.1.2 *The payoff* $\mathbf{PS}_T$ *at expiration date* $T \leq T_0$ *of a forward swaption equals*

$$\mathbf{PS}_T = \sum_{j=1}^{n} \delta_j B(T, T_j)\big(\kappa(T, T_0, n) - \kappa\big)^+. \tag{13.8}$$

Formula (13.8) shows that, if exercised, the forward swaption gives rise to a sequence of payments $\delta_j(\kappa(T, T_0, n) - \kappa)$ at each settlement date T_j, $j = 1, \ldots, n$. By setting $T = T_0$, we recover, in a more general framework, the previously observed dual nature of the swaption: it may be seen either as an option on the value of a particular (forward) swap or, equivalently, as an option on the corresponding (forward) swap rate. It is also clear that the owner of a forward swaption is able to enter at time T (at no additional cost) into a forward payer swap with preassigned fixed interest rate κ.

Constant maturity swaps. Similarly as in the case of a plain-vanilla fixed-for-floating swap, in a constant maturity swap the fixed and floating payments occur at regularly spaced dates. The floating payments are not based on the LIBOR, but on some other swap rate. Formally, at any of the settlement dates T_j, $j = 1, \ldots, n$ the fixed payment κ is exchanged for the variable payment $\kappa(T_{j-1}, T_{j-1}, m)$ for a preassigned length $m \geq 1$. A constant maturity swap gives rise to an interest rate, which is termed the constant maturity swap rate, or briefly, the CMS rate. Let us first consider the case where $n = 1$ and let us fix m. Then the following definition can be justified easily.

Definition 13.1.2 For any date $t \leq T_0$, the *one-period forward constant maturity swap rate* $\mathbf{CMS}(t, T_0, m)$ is formally defined as follows

$$\mathbf{CMS}(t, T_0, m) = \mathbb{E}_{\mathbb{P}_{T_1}}\big(\kappa(T_0, T_0, m) \mid \mathcal{F}_t\big).$$

Note that $\mathbf{CMS}(T_0, T_0, m) = \kappa(T_0, T_0, m)$ and that $\mathbf{CMS}(t, T_0, n)$ represents the forward price at time t, for the settlement date T_1, of the payoff $\kappa(T_0, T_0, n)$ occurring at time T_1. In general, for any natural number n, the *n-period forward constant maturity swap rate* $\mathbf{CMS}(t, T_0, m, n)$ equals

$$\mathbf{CMS}(t, T_0, m, n) = \frac{\sum_{j=1}^{n} \delta_j B(t, T_j)\, \mathbb{E}_{\mathbb{P}_{T_j}}\big(\kappa(T_{j-1}, T_{j-1}, m) \mid \mathcal{F}_t\big)}{\sum_{j=1}^{n} \delta_j B(t, T_j)}$$

It is rather obvious (and easy to check) that $\mathbf{CMS}(t, T_0, 1, n) = \kappa(t, T_0, n)$.

CMS options. For ease of notation, we assume that $n = 1$. A holder of a *CMS spread option* has the right to exchange for one period of time the difference between two CMS rates minus a spread κ. The payoff at time $T = T_0$ thus equals (we set $N = \delta = 1$)

$$\big(\kappa(T, T, m_1) - \kappa(T, T, m_2) - \kappa\big)^+$$

for some $m_1 \neq m_2$. A CMS spread option can be seen as a special case of a *CMS basket option*. A generic CMS basket option, written on k CMS rates that reset at the option's expiry date $T = T_0$, has the payoff

$$\Big(\sum_{i=1}^{k} w_i \kappa(T, T, m_i) - \kappa\Big)^+,$$

where w_i and κ are constants, and $m_1, \ldots, m_k$ are preassigned lengths of reference swaps. For instance, a CMS basket option on three underlying CMS rates may have the payoff given as

$$\big(2\kappa(T, T, m_1) - \kappa(T, T, m_2) - \kappa(T, T, m_3) - \kappa\big)^+.$$

Bermudan swaptions. Traded swaptions are of American rather than European style. More exactly, they typically have semi-American features, since exercising is allowed on a finite number of dates (for instance, on reset dates). As a simple example of such a contract, let us consider a *Bermudan swaption*. Consider a fixed collection of reset dates $T_0, \ldots, T_{n-1}$ and an associated family of exercise dates $t_1, \ldots, t_l$ with $t_i \leq T_{j_i}$. It should be stressed that the exercise dates are known in advance; that is, they cannot be chosen freely by the long party. A Bermudan swaption gives its holder the right to enter at time t_i a forward swap that starts at T_{j_i} and ends at time T_n, provided that this right has not already been exercised at a previous time t_i for some $i < m$. Bermudan swaptions are the most actively traded exotic swaptions; they also arise as embedded options in cancellable swaps. We shall analyze the valuation of Bermudan swaptions in Sect. 13.3.4.

13.2 Valuation in the Gaussian HJM Model

In this section, plain-vanilla swaptions and some exotic swap derivatives are examined within the framework of the Gaussian HJM model.

13.2.1 Swaptions

Our next goal is to provide a quasi-explicit formula for the arbitrage price of a payer swaption in the Gaussian HJM framework (note that the price of a receiver swaption is given by an analogous formula). Recall that $T = T_0$ stands for the start date of an underlying swap, as well as for the expiration date of a swaption. Using Proposition 11.3.5, we obtain without difficulties the following result.

Proposition 13.2.1 *Assume the Gaussian HJM model of the term structure of interest rates. For $t \in [0, T]$, the price of a payer swaption equals*

$$\mathbf{PS}_t = \int_{\mathbb{R}^k} \left(B(t, T) n_k(x) - \sum_{i=1}^{n} c_i B(t, T_i) n_k(x + \theta_i) \right)^+ dx, \qquad (13.9)$$

where n_k is the standard k-dimensional Gaussian probability density function, and vectors $\theta_1, \ldots, \theta_n \in \mathbb{R}^k$ satisfy for every $i, j = 1, \ldots, n$

$$\theta_i \cdot \theta_j = \int_{t}^{T} \gamma(u, T_i, T) \cdot \gamma(u, T_j, T) \, du. \qquad (13.10)$$

Formula (13.9) can be generalized easily to the case of a forward swaption. To this end, it is enough to consider the following claim that settles at time $T < T_0$ (cf. (13.8))

$$\mathbf{PS}_T = \sum_{j=1}^{n} \delta_j B(T, T_j) \left(\frac{B(T, T_0) - B(T, T_n)}{\sum_{i=1}^{n} \delta_i B(T, T_i)} - \kappa \right)^+.$$

13.2.2 CMS Spread Options

As was mentioned already, in contrast to a standard swaption, the payoff from a CMS spread option, maturing at $T = T_0$, is related to swap rates associated with interest rate swaps that have different number of payment dates. Let us consider the owner of a call option on a spread between the swap rates $\kappa(T, T, m_1)$ and $\kappa(T, T, m_2)$,

$$\kappa(T, T, m_1) = (1 - B(T, T_{m_1})) \left(\sum_{j=1}^{m_1} \delta_j B(T, T_j) \right)^{-1}$$

and the swap rate

$$\kappa(T, T, m_2) = (1 - B(T, T_{m_2})) \left(\sum_{j=1}^{m_2} \delta_j B(T, T_j) \right)^{-1},$$

where $m_1 \neq m_2$. If the strike level is κ, then the holder of a CMS spread option is entitled to receive at time T the following amount (as usual, the principal is set to be equal to 1)

$$C_T(\kappa, m_1, m_2) = \Big(\kappa(T, T, m_1) - \kappa(T, T, m_2) - \kappa\Big)^+.$$

It is thus evident that the arbitrage price of this contract at time t equals

$$\mathbf{CMSC}_t = \mathbb{E}_{\mathbb{P}^*}\Big\{ \frac{B_t}{B_T}\Big(\kappa(T, T, m_1) - \kappa(T, T, m_2) - \kappa\Big)^+ \Big| \mathcal{F}_t \Big\}.$$

By applying the general valuation formula of Proposition 11.3.5 to the function

$$g(x_1, \ldots, x_{m_1 \vee m_2}) = \Big(\frac{1 - x_{m_1}}{\sum_{j=1}^{m_1} \delta_j x_j} - \frac{1 - x_{m_2}}{\sum_{j=1}^{m_2} \delta_j x_j} - \kappa \Big)^+,$$

where $m_1 \vee m_2 = \max\{m_1, m_2\}$, we get the following result.

Proposition 13.2.2 *Assume the Gaussian HJM model of the term structure of interest rates. Then the arbitrage price at time t of a CMS spread call option equals*

$$\mathbf{CMSC}_t = B(t, T) \int_{\mathbb{R}^k} (f_1(x) - f_2(x) - \kappa)^+ n_k(x)\, dx,$$

where for $l = 1, 2$

$$f_l(x) = \frac{B(t, T)n_k(x) - B(t, T_{m_l})n_k(x + \theta_{m_l})}{\sum_{j=1}^{m_l} \delta_j B(t, T_j)n_k(x + \theta_j)}$$

and the vectors $\theta_1, \ldots, \theta_{m_1 \vee m_2}$ are implicitly specified by the following relationships

$$\theta_i \cdot \theta_j = \int_t^T \gamma(u, T_i, T) \cdot \gamma(u, T_j, T)\, du.$$

The corresponding put option pays at time T the amount

$$P_T(\kappa, m_1, m_2) = \Big(\kappa - \kappa(T, T, m_1) + \kappa(T, T, m_2)\Big)^+.$$

Consequently, the arbitrage price at time $t \in [0, T]$ of this contract is given by the following formula

$$\mathbf{CMSP}_t = \mathbb{E}_{\mathbb{P}^*}\Big\{ \frac{B_t}{B_T}\Big(\kappa - \kappa(T, T, m_1) + \kappa(T, T, m_2)\Big)^+ \Big| \mathcal{F}_t \Big\},$$

or more explicitly

$$\mathbf{CMSP}_t = B(t, T) \int_{\mathbb{R}^k} (\kappa - f_1(x) + f_2(x))^+ n_k(x)\, dx.$$

13.2.3 Yield Curve Swaps

A *one-period yield curve swap* is a contract in which, at time T, one party pays $\kappa(T, T, m_2) + \kappa$ and receives $\kappa(T, T, m_1)$ on some notional principal. The payoff of the one-period yield-curve swap therefore satisfies (for the principal equal to 1)

$$\kappa_1 - \kappa_2 - \kappa = (\kappa_1 - \kappa_2 - \kappa)^+ - (\kappa - \kappa_1 + \kappa_2)^+,$$

where we write $\kappa_l = \kappa(T, T, m_l)$ for $l = 1, 2$. It is thus easily seen that its value at time $t \in [0, T]$ can be found from the following relationship

$$\mathbf{YCS}_t = \mathbf{CMSC}_t - \mathbf{CMSP}_t.$$

That is, the value $\mathbf{YCS}_t$ of a yield curve swap at time t is equal to the price difference between a call and a put CMS spread option. The *margin rate* is that value of a constant κ that makes the value of the one-period yield curve swap zero. Using the results of Sect. 13.2.2, one can check directly that in the Gaussian HJM framework, the margin rate equals

$$\hat{\kappa} = \int_{\mathbb{R}^k} (g_1(x) - g_2(x)) n_k(x)\, dx,$$

where

$$g_l(x) = \frac{B(t, T) n_k(x) - B(t, T_{m_l}) n_k(x + \theta_{m_l})}{\sum_{j=1}^{m_l} \delta_j B(t, T_j) n_k(x + \theta_l)}, \quad \forall x \in \mathbb{R}^k,$$

for $l = 1, 2$. More generally, a *multi-period yield curve swap* consists of n payments at times T_i, $i = 1, \dots, n$. At each settlement date T_i, the swap rate

$$\kappa(T_{i-1}, T_{i-1}, m_1) = \left(1 - B(T_{i-1}, T_{i-1+m_1})\right) \left(\sum_{j=1}^{m_1} \delta_{i-1+j} B(T_{i-1}, T_{i-1+j}) \right)^{-1}$$

is received, and the swap rate $\kappa(T_{i-1}, T_{i-1}, m_2)$ plus the margin rate κ is paid. Under the assumptions of Proposition 13.2.2, the following result is valid. The vectors $\theta_1, \dots, \theta_{n-1+m}$ are determined, as usual, through (13.10).

Proposition 13.2.3 *The value of a yield curve swap at time t is given by the risk-neutral valuation formula*

$$\mathbf{YCS}_t = \mathbb{E}_{\mathbb{P}^*}\left(\sum_{i=0}^{n-1} \frac{B_t}{B_{T_i}} \left(\kappa(T_i, T_i, m_1) - \kappa(T_i, T_i, m_2) - \kappa \right) \,\Big|\, \mathcal{F}_t \right).$$

Consequently, within the framework of the Gaussian HJM model we have

$$\mathbf{YCS}_t = \sum_{i=0}^{n-1} B(t, T_{i+1}) \int_{\mathbb{R}^k} \left(g_{1,i}(x) - g_{2,i}(x) - \kappa \right) n_k(x)\, dx,$$

where for $l = 1, 2$ and every $x \in \mathbb{R}^k$ we set

$$g_{l,i}(x) = \frac{B(t, T_i) n_k(x) - B(t, T_{i+m_l}) n_k(x + \theta_{i+m_l})}{\sum_{j=1}^{m_l} \delta_{i+j} B(t, T_{i+j}) n_k(x + \theta_{i+j})}.$$

Proof. The proof of the proposition relies on a straightforward application of the valuation formula established in Proposition 11.3.5.   □

13.3 Co-terminal Forward Swap Rates

The commonly used formula for pricing swaptions, based on the assumption that the underlying swap rate follows a geometric Brownian motion under the "market probability" $\mathbb{Q}$, is given by the *Black swaptions formula*

$$\mathbf{PS}_t = \sum_{j=1}^{n} B(t, T_j)\delta_j\Big(\kappa(t, T, n)N\big(h_1(t, T)\big) - \kappa N\big(h_2(t, T)\big)\Big), \qquad (13.11)$$

where

$$h_{1,2}(t, T) = \frac{\ln(\kappa(t, T, n)/\kappa) \pm \frac{1}{2}\sigma^2(T - t)}{\sigma\sqrt{T - t}}$$

for some constant volatility $\sigma > 0$. For a long time, the Black swaptions formula was merely a popular market convention that used by practitioners as a convenient tool to quote swaption prices in terms of implied volatilities. Indeed, the use of this formula was not supported by any specific term structure model.

On the theoretical side, Neuberger (1990a) analyzed the valuation and hedging of swaptions based on Black's formula. However, the formal derivation of these heuristic results within the set-up of a well-specified term structure model was first achieved in Jamshidian (1997a) (see also, Rutkowski (1999a), Galluccio et al. (2003a), Galluccio and Hunter (2004a, 2004b), Pietersz and van Regenmortel (2005), and Davis and Mataix-Pastor (2005)).

The goal of this section is to present Jamshidian's model of forward swap rates and to show that Black's formula for swaptions can be easily obtained within this model. We assume, as before, that the tenor structure $0 \le T_0 < T_1 < \cdots < T_n = T^*$ is given. Recall that we denote $\delta_j = T_j - T_{j-1}$ for $j = 1, \ldots, n$, so that $T_j = T_0 + \sum_{i=0}^{j} \delta_i$ for every $j = 0, \ldots, n$. For ease of notation, we shall sometimes write $T_l^* = T_{n-l}$ for $l = 0, \ldots, n$.

Co-terminal forward swaps. For any fixed $j = 0, \ldots, n - 1$, we consider a fixed-for-floating forward (payer) swap that starts at time T_j and has $n - j$ accrual periods, whose consecutive lengths are $\delta_{j+1}, \ldots, \delta_n$. The last settlement date of each swap is thus invariably T_n. This feature justifies the name of the *co-terminal* (or the *fixed-maturity*) forward swaps.

In the jth co-terminal forward swap, the fixed interest rate κ is paid at each of reset dates T_l for $l = j + 1, \ldots, n$. The corresponding floating payments are based on LIBORs $L(T_l)$, $l = j + 1, \ldots, n$.

It is easy to see that the value of the jth co-terminal forward swap equals, for $t \in [0, T_j]$ (by convention, the notional principal equals 1)

$$\mathbf{FS}_t(\kappa) = B(t, T_j) - \sum_{l=j+1}^{n} c_l B(t, T_l),$$

where $c_l = \kappa \delta_l$ for $l = j+1, \ldots, n-1$ and $c_n = 1 + \kappa \delta_n$. The associated co-terminal forward swap rate $\kappa(t, T_j, n - j)$, which makes the jth forward swap worthless at time t, is given by the formula

$$\kappa(t, T_j, n - j) = \frac{B(t, T_j) - B(t, T_n)}{\delta_{j+1} B(t, T_{j+1}) + \cdots + \delta_n B(t, T_n)} \tag{13.12}$$

for every $t \in [0, T_j]$, $j = 0, \ldots, n - 1$. In this section, we shall deal with the family of co-terminal forward swap rates $\kappa(t, T_j, n - j)$, $j = 0, \ldots, n - 1$. Let us stress once again that the reference swaps have different lengths, but they all have a common expiration date $T_n = T^*$.

Suppose momentarily that we are given a family of bond prices $B(t, T_l)$, $l = 1, \ldots, n$, defined on a filtered probability space $(\Omega, \mathbb{F}, \mathbb{Q})$ equipped with a d-dimensional Brownian motion W. We find it convenient to postulate that $\mathbb{Q} = \mathbb{P}_{T^*}$ is the forward measure for the date T^* and the process $W = W^{T^*}$ is the corresponding Brownian motion.

The next step is to introduce the *level process*, which is also termed the *swap annuity* or the *present value of basis point* (PVBP or PV01). Formally, for any $m = 1, \ldots, n - 1$, the level process $G(m)$ is defined by setting

$$G_t(m) = \sum_{l=n-m+1}^{n} \delta_l B(t, T_l) = \sum_{k=0}^{m-1} \delta_{n-k} B(t, T_k^*) \tag{13.13}$$

for $t \in [0, T_{n-m+1}]$. A forward swap measure (also known as a *level measure*) is a probability measure equivalent to $\mathbb{Q}$ corresponding to the choice of the level process as a numeraire.

Definition 13.3.1 For every $j = 0, \ldots, n$, a probability measure $\tilde{\mathbb{P}}_{T_j}$ on $(\Omega, \mathcal{F}_{T_j})$, equivalent to $\mathbb{Q}$, is said to be the *co-terminal forward swap measure* for the date T_j if, for every $k = 0, \ldots, n$, the relative bond price

$$Z_{n-j+1}(t, T_k) \stackrel{\text{def}}{=} \frac{B(t, T_k)}{G_t(n - j + 1)} = \frac{B(t, T_k)}{\delta_j B(t, T_j) + \cdots + \delta_n B(t, T_n)},$$

where $t \in [0, T_k \wedge T_j]$, is a local martingale under $\tilde{\mathbb{P}}_{T_j}$.

Put differently, for any fixed $m = 1, \ldots, n + 1$, the relative bond prices

$$Z_m(t, T_k^*) = \frac{B(t, T_k^*)}{G_t(m)} = \frac{B(t, T_k^*)}{\delta_{n-m+1} B(t, T_{m-1}^*) + \cdots + \delta_n B(t, T^*)},$$

where $t \in [0, T_k^* \wedge T_{m-1}^*]$, are local martingales under the co-terminal forward swap measure $\tilde{\mathbb{P}}_{T_{m-1}^*}$.

Let us write $\tilde{\kappa}(t, T_j) = \kappa(t, T_j, n - j)$. It follows immediately from (13.12) that the forward swap rate for the date T_m^* equals, for $t \in [0, T_m^*]$,

$$\tilde{\kappa}(t, T_m^*) = \frac{B(t, T_m^*) - B(t, T^*)}{\delta_{n-m+1} B(t, T_{m-1}^*) + \cdots + \delta_n B(t, T^*)},$$

or equivalently,

$$\tilde{\kappa}(t, T_m^*) = Z_m(t, T_m^*) - Z_m(t, T^*).$$

Therefore $\tilde{\kappa}(t, T_m^*)$ is a local martingale under the forward swap measure $\tilde{\mathbb{P}}_{T_{m-1}^*}$ as well. Moreover, since obviously $G_t(1) = \delta_n B(t, T^*)$, it is evident that $Z_1(t, T_k^*) = \delta_n^{-1} F_B(t, T_k^*, T^*)$, and thus the probability measure $\tilde{\mathbb{P}}_{T^*}$ can be interpreted as the forward martingale measure $\mathbb{P}_{T^*}$. Our aim is to construct a model of forward swap rates through backward induction. As one might expect, the underlying bond price processes will not be specified explicitly. We make the following standing assumption.

Assumption (SR). We assume that we are given a family of $\mathbb{R}^d$-valued (bounded) adapted processes $v(t, T_j)$, $j = 0, \ldots, n - 1$ representing the volatilities of forward swap rates $\tilde{\kappa}(t, T_j)$. We are also given an initial term structure of forward swap rates $\tilde{\kappa}(0, T_j) > 0$ for $j = 0, \ldots, n - 1$ (it can be specified by a family $B(0, T_j)$, $j = 0, \ldots, n$ of prices of zero-coupon bonds, such that $B(0, T_j) > B(0, T_n)$ for $j = 0, \ldots, n - 1$).

The lognormal model of forward swap rates was first constructed by Jamshidian (1997a) by means of suitable multidimensional SDEs. In this section, we start by presenting an equivalent construction given by Rutkowski (1999a). We wish to construct a family of forward swap rates in such a way that

$$d\tilde{\kappa}(t, T_j) = \tilde{\kappa}(t, T_j) v(t, T_j) \cdot d\tilde{W}_t^{T_{j+1}} \tag{13.14}$$

for any $j = 0, \ldots, n - 1$, where each process $\tilde{W}^{T_{j+1}}$ is a d-dimensional Brownian motion under the corresponding forward swap measure $\tilde{\mathbb{P}}_{T_{j+1}}$. The model should also be consistent with the initial term structure of interest rates, meaning that

$$\tilde{\kappa}(0, T_j) = \frac{B(0, T_j) - B(0, T^*)}{\delta_{j+1} B(0, T_{j+1}) + \cdots + \delta_n B(0, T_n)}. \tag{13.15}$$

First step. We proceed by backward induction. The first step is to introduce the forward swap rate for the date T_1^* by postulating that the forward swap rate $\tilde{\kappa}(t, T_1^*)$ solves the SDE

$$d\tilde{\kappa}(t, T_1^*) = \tilde{\kappa}(t, T_1^*) v(t, T_1^*) \cdot d\tilde{W}_t^{T^*}, \quad \forall t \in [0, T_1^*], \tag{13.16}$$

where $\tilde{W}^{T^*} = W^{T^*} = W$, with the initial condition

$$\tilde{\kappa}(0, T_1^*) = \frac{B(0, T_1^*) - B(0, T^*)}{\delta_n B(0, T^*)}.$$

To specify the process $\tilde{\kappa}(t, T_2^*)$, we need first to introduce a forward swap measure $\tilde{\mathbb{P}}_{T_1^*}$ and an associated d-dimensional Brownian motion $\tilde{W}^{T_1^*}$.

To this end, notice that each process $Z_1(t, T_k^*) = B(t, T_k^*)/\delta_n B(t, T^*)$ is a strictly positive local martingale under $\tilde{\mathbb{P}}_{T^*} = \mathbb{P}_{T^*}$. More specifically, we have

$$dZ_1(t, T_k^*) = Z_1(t, T_k^*)\gamma_1(t, T_k^*) \cdot d\tilde{W}_t^{T^*} \tag{13.17}$$

for some adapted process $\gamma_1(t, T_k^*)$. According to the definition of a co-terminal forward swap measure, we postulate that for every k the process

$$Z_2(t, T_k^*) = \frac{B(t, T_k^*)}{\delta_{n-1}B(t, T_1^*) + \delta_n B(t, T^*)} = \frac{Z_1(t, T_k^*)}{1 + \delta_{n-1}Z_1(t, T_1^*)}$$

follows a local martingale under $\tilde{\mathbb{P}}_{T_1^*}$. By applying Lemma 12.4.1 to processes

$$G_t = Z_1(t, T_k^*), \quad H_t = \delta_{n-1}Z_1(t, T_1^*)$$

it is easy to see that for this property to hold, it suffices to assume that the process $\tilde{W}^{T_1^*}$, which is given by the formula

$$\tilde{W}_t^{T_1^*} = \tilde{W}_t^{T^*} - \int_0^t \frac{\delta_{n-1}Z_1(u, T_1^*)}{1 + \delta_{n-1}Z_1(u, T_1^*)} \gamma_1(u, T_1^*)\,du,$$

$t \in [0, T_1^*]$, follows a Brownian motion under $\tilde{\mathbb{P}}_{T_1^*}$ (the probability measure $\tilde{\mathbb{P}}_{T_1^*}$ is as yet unspecified, but will be soon found through Girsanov's theorem). Note that

$$Z_1(t, T_1^*) = \frac{B(t, T_1^*)}{\delta_n B(t, T^*)} = \tilde{\kappa}(t, T_1^*) + Z_1(t, T^*) = \tilde{\kappa}(t, T_1^*) + \delta_n^{-1}. \tag{13.18}$$

Differentiating both sides of the last equality, we get (cf. (13.16) and (13.17))

$$Z_1(t, T_1^*)\gamma_1(t, T_1^*) = \tilde{\kappa}(t, T_1^*)v(t, T_1^*).$$

Consequently, $\tilde{W}^{T_1^*}$ is given explicitly by the formula

$$\tilde{W}_t^{T_1^*} = \tilde{W}_t^{T^*} - \int_0^t \frac{\delta_{n-1}\tilde{\kappa}(u, T_1^*)}{1 + \delta_{n-1}\delta_n^{-1} + \delta_{n-1}\tilde{\kappa}(u, T_1^*)} v(u, T_1^*)\,du$$

for $t \in [0, T_1^*]$. We are now in a position to define, using Girsanov's theorem, the associated forward swap measure $\tilde{\mathbb{P}}_{T_1^*}$. Subsequently, we introduce the process $\tilde{\kappa}(t, T_2^*)$ by postulating that it solves the SDE

$$d\tilde{\kappa}(t, T_2^*) = \tilde{\kappa}(t, T_2^*)v(t, T_2^*) \cdot d\tilde{W}_t^{T_1^*}$$

with the initial condition

$$\tilde{\kappa}(0, T_2^*) = \frac{B(0, T_2^*) - B(0, T^*)}{\delta_{n-1}B(0, T_1^*) + \delta_n B(0, T^*)}.$$

Second step. For the reader's convenience, let us consider one more inductive step, in which we are looking for $\tilde{\kappa}(t, T_3^*)$. We now consider the following processes:

$$Z_3(t, T_k^*) = \frac{B(t, T_k^*)}{\delta_{n-2}B(t, T_2^*) + \delta_{n-1}B(t, T_1^*) + \delta_n B(t, T^*)} = \frac{Z_2(t, T_k^*)}{1 + \delta_{n-2}Z_2(t, T_2^*)}.$$

Consequently, the process $\tilde{W}^{T_2^*}$ satisfies

$$\tilde{W}_t^{T_2^*} = \tilde{W}_t^{T_1^*} - \int_0^t \frac{\delta_{n-2}Z_2(u, T_2^*)}{1 + \delta_{n-2}Z_2(u, T_2^*)}\, \gamma_2(u, T_2^*)\, du$$

for $t \in [0, T_2^*]$. It is useful to note that

$$Z_2(t, T_2^*) = \frac{B(t, T_2^*)}{\delta_{n-1}B(t, T_1^*) + \delta_n B(t, T^*)} = \tilde{\kappa}(t, T_2^*) + Z_2(t, T^*), \qquad (13.19)$$

where in turn

$$Z_2(t, T^*) = \frac{Z_1(t, T^*)}{1 + \delta_{n-1}Z_1(t, T^*) + \delta_{n-1}\tilde{\kappa}(t, T_1^*)}$$

and the process $Z_1(t, T^*)$ is already known from the previous step (clearly, $Z_1(t, T^*) = 1/\delta_n$). Differentiating the last equality, we may thus find the volatility of the process $Z_2(t, T^*)$, and consequently, to define $\tilde{\mathbb{P}}_{T_2^*}$.

General step. We proceed by induction with respect to m. Suppose that we have found forward swap rates $\tilde{\kappa}(t, T_1^*), \ldots, \tilde{\kappa}(t, T_m^*)$, the forward swap measure $\tilde{\mathbb{P}}_{T_{m-1}^*}$ and the associated Brownian motion $\tilde{W}^{T_{m-1}^*}$. Our aim is to determine the forward swap measure $\tilde{\mathbb{P}}_{T_m^*}$, the associated Brownian motion $\tilde{W}^{T_m^*}$, and the forward swap rate $\tilde{\kappa}(t, T_{m+1}^*)$. To this end, we postulate that processes

$$Z_{m+1}(t, T_k^*) = \frac{B(t, T_k^*)}{G_t(m+1)} = \frac{B(t, T_k^*)}{\delta_{n-m}B(t, T_m^*) + \cdots + \delta_n B(t, T^*)}$$
$$= \frac{Z_m(t, T_k^*)}{1 + \delta_{n-m}Z_m(t, T_m^*)}$$

follow local martingales under $\tilde{\mathbb{P}}_{T_m^*}$. In view of Lemma 12.4.1, applied to processes $G = Z_m(t, T_k^*)$ and $H = Z_m(t, T_m^*)$, it is clear that we may set

$$\tilde{W}_t^{T_{m\delta}^*} = \tilde{W}_t^{T^*} - \int_0^t \frac{\delta_{n-m}Z_m(u, T_m^*)}{1 + \delta_{n-m}Z_m(u, T_m^*)}\, \gamma_m(u, T_m^*)\, du, \qquad (13.20)$$

for $t \in [0, T_m^*]$. Therefore it is sufficient to analyze the process

$$Z_m(t, T_m^*) = \frac{B(t, T_m^*)}{\delta_{n-m+1}B(t, T_{m-1}^*) + \cdots + \delta_n B(t, T^*)} = \tilde{\kappa}(t, T_m^*) + Z_m(t, T^*).$$

To conclude, it is enough to notice that

$$Z_m(t, T^*) = \frac{Z_{m-1}(t, T^*)}{1 + \delta_{n-m+1} Z_{m-1}(t, T^*) + \delta_{n-m+1} \tilde{\kappa}(t, T^*_{m-1})}.$$

Indeed, from the preceding step, we know that the process $Z_{m-1}(t, T^*)$ is a (rational) function of forward swap rates $\tilde{\kappa}(t, T^*_1), \ldots, \tilde{\kappa}(t, T^*_{m-1})$. Consequently, the process under the integral sign on the right-hand side of (13.20) can be expressed using the terms $\tilde{\kappa}(t, T^*_1), \ldots, \tilde{\kappa}(t, T^*_{m-1})$ and their volatilities (for the explicit formula, see Jamshidian's approach below).

Having found the process $\tilde{W}^{T^*_m}$ and probability measure $\tilde{\mathbb{P}}_{T^*_m}$, we introduce the forward swap rate $\tilde{\kappa}(t, T^*_{m+1})$ through (13.14)–(13.15), and so forth. If all volatility processes $v(t, T_j)$, $j = 0, \ldots, n-1$ are taken to be deterministic functions, the model is termed the *lognormal model of co-terminal forward swap rates*.

13.3.1 Jamshidian's Model

Let us now present the original construction of an arbitrage-free model of forward swap rates proposed by Jamshidian (1997a). Recall that we write $\tilde{\kappa}(t, T_j) = \kappa(t, T_j, n-j)$ for every $j = 0, \ldots, n-1$. Let us set, for arbitrary $1 \le j \le k \le n-1$,

$$g_t^{jk} = \sum_{i=k}^{n-1} \delta_{i+1} \prod_{l=j+1}^{i} \left(1 + \delta_l \tilde{\kappa}(t, T_l)\right), \tag{13.21}$$

where, by convention, the product $\prod_{l=j}^{i} \ldots = 1$ if $j > i$. For brevity, we write $g_t^j = g_t^{jj}$. Let us first prove the following auxiliary lemma.

Lemma 13.3.1 *Equality* $G_t(n - j) = B(t, T_n) g_t^j$ *holds for* $j = 1, \ldots, n-1$.

Proof. We shall proceed by induction with respect to j. Let us first check that the equality holds for $j = n - 1$. We have

$$G_t(n - (n-1)) = G_t(1) = \delta_n B(t, T_n) = B(t, T_n) g_t^{n-1},$$

since (13.21) yields

$$g_t^{n-1} = \delta_n \prod_{l=n}^{n-1} \left(1 + \delta_l \tilde{\kappa}(t, T_l)\right) = \delta_n.$$

Let us now assume that the equality is valid for some $j = 1, \ldots, n-1$. We wish to show that it also holds for $j - 1$. On the one hand, we have that

$$G_t(n - (j-1)) = G_t(n - j + 1) = \sum_{l=j}^{n} \delta_l B(t, T_l).$$

On the other hand, we obtain

$$
\begin{aligned}
g_t^{j-1} &= \sum_{i=j-1}^{n-1} \delta_{i+1} \prod_{l=j}^{i} \left(1 + \delta_l \tilde{\kappa}(t, T_l)\right) \\
&= \delta_j \prod_{l=j}^{j-1} \left(1 + \delta_l \tilde{\kappa}(t, T_l)\right) + \sum_{i=j}^{n-1} \delta_{i+1} \prod_{l=j}^{i} \left(1 + \delta_l \tilde{\kappa}(t, T_l)\right) \\
&= \delta_j + \left(1 + \delta_j \tilde{\kappa}(t, T_j)\right) \sum_{i=j}^{n-1} \delta_{i+1} \prod_{l=j+1}^{i} \left(1 + \delta_l \tilde{\kappa}(t, T_l)\right) \\
&= \delta_j + \left(1 + \delta_j \frac{B(t, T_j) - B(t, T_n)}{\delta_{j+1} B(t, T_{j+1}) + \ldots + \delta_n B(t, T_n)}\right) g_t^j \\
&= \delta_j + \left(\frac{\delta_j B(t, T_j) + \ldots + \delta_n B(t, T_n) - \delta_j B(t, T_n)}{\delta_{j+1} B(t, T_{j+1}) + \ldots + \delta_n B(t, T_n)}\right) \frac{G_t(n-j)}{B(t, T_n)} \\
&= \delta_j + \frac{\delta_j B(t, T_j) + \ldots + \delta_n B(t, T_n) - \delta_j B(t, T_n)}{B(t, T_n)} = \frac{G_t(n-j+1)}{B(t, T_n)},
\end{aligned}
$$

where the induction assumption was used in penultimate equality. □

Assume that forward swap rates are continuous semimartingales. Then we have the following result.

Lemma 13.3.2 *Let us denote* $\tilde{\kappa}(t, T_i) = \tilde{\kappa}_t^i$ *and*

$$
z_t^j = \ln \frac{B(t, T_n)}{G_t(n-j)} = -\ln g_t^j.
$$

Then we have for $j = 0, \ldots, n-1$ *and* $t \in [0, T_j]$

$$
\langle \tilde{\kappa}^j, z^j \rangle_t = -\int_0^t \sum_{k=j+1}^{n-1} \frac{\delta_k g_u^{jk}}{(1 + \delta_k \tilde{\kappa}_u^k) g_u^j} \, d\langle \tilde{\kappa}^j, \tilde{\kappa}^k \rangle_u. \tag{13.22}
$$

Proof. Observe that

$$
\ln g_t^j = \ln \left(\sum_{i=j}^{n-1} \delta_{i+1} \prod_{k=j+1}^{i} (1 + \delta_k \tilde{\kappa}_t^k) \right).
$$

The result thus follows by an application of Itô's formula to the right-hand side of the formula above. □

The last lemma provides a useful link between the forward swap measures $\tilde{\mathbb{P}}_{T_{j+1}}$ and $\tilde{\mathbb{P}}_{T_n}$. Indeed, it can be easily deduced from general considerations (see Sect. 9.6.4) that, for every $t \in [0, T_{j+1}]$,

$$
\frac{d\tilde{\mathbb{P}}_{T_n}}{d\tilde{\mathbb{P}}_{T_{j+1}}}\bigg|_{\mathcal{F}_t} = \frac{B(t, T_n)}{G_t(n-j)}, \quad \tilde{\mathbb{P}}_{T_{j+1}}\text{-a.s.} \tag{13.23}
$$

In order, the construct an arbitrage-free model of forward swap rates, it thus suffices to compute the drifts of forward swap rates $\kappa(t, T_0), \ldots, \kappa(t, T_{n-1})$ under a single probability measure, for instance, the forward swap measure $\tilde{\mathbb{P}}_{T_n}$. We postulate that

$$d\tilde{\kappa}(t, T_j) = \mu_t^j \, dt + \phi_t^j \cdot d\hat{W}_t,$$

under the actual probability measure, or equivalently, that under $\tilde{\mathbb{P}}_{T_n}$

$$d\tilde{\kappa}(t, T_j) = \tilde{\mu}_t^j \, dt + \phi_t^j \cdot d\tilde{W}_t^{T_n},$$

where $\tilde{\mu}_t^{n-1} = 0$ and the drifts $\tilde{\mu}_t^j$, $j = 0, \ldots, n-2$ are yet unspecified. Note that $\tilde{\mathbb{P}}_{T_n} = \mathbb{P}_{T_n}$ is simply the forward measure for the date T_n. It is thus natural to set $\tilde{W}^{T_n} = W^{T_n} = W$ for some underlying d-dimensional Brownian motion W. The following result furnishing the no-arbitrage drifts was first established by Jamshidian (1997a).

Proposition 13.3.1 *The following recursive relationship is valid for every $j = 0, \ldots, n-1$ and $t \in [0, T_j]$*

$$d\tilde{\kappa}(t, T_j) = -\sum_{k=j+1}^{n-1} \frac{\delta_k g_t^{jk} \phi_t^j \cdot \phi_t^k}{\left(1 + \delta_k \tilde{\kappa}(t, T_k)\right) g_t^j} \, dt + \phi_t^j \cdot d\tilde{W}_t^{T_n}. \tag{13.24}$$

Proof. The process $\tilde{\kappa}(t, T_j)$ is a martingale under the forward swap measure, and thus

$$d\tilde{\kappa}(t, T_j) = \phi_t^j \cdot d\tilde{W}_t^{T_{j+1}}, \tag{13.25}$$

But from equality

$$\langle \tilde{\kappa}^j, z^j \rangle_t = -\int_0^t \sum_{k=j+1}^{n-1} \frac{\delta_k g_u^{jk} \phi_u^j \cdot \phi_u^k}{\left(1 + \delta_k \tilde{\kappa}(u, T_k)\right) g_u^j} \, du$$

combined with (13.25) and, of course, Girsanov's theorem, we can deduce that

$$\tilde{W}_t^{T_{j+1}} = \tilde{W}_t^{T_n} - \int_0^t \sum_{k=j+1}^{n-1} \frac{\delta_k g_u^{jk} \phi_u^k}{\left(1 + \delta_k \tilde{\kappa}(u, T_k)\right) g_u^j} \, du.$$

Upon substitution to (13.25) we obtain (13.24).   □

Lognormal model of co-terminal swap rates. In the special case of the lognormal model of co-terminal swap rates, Jamshidian's construction runs as follows. We postulate that

$$d\tilde{\kappa}(t, T_j) = \tilde{\mu}_t^j \, dt + v(t, T_j)\tilde{\kappa}(t, T_j) \cdot dW_t,$$

so that $\phi_t^j = v(t, T_j)\tilde{\kappa}(t, T_j)$. Using the recursive relationship of Proposition 13.3.1, we obtain explicit expressions for forward swap rates under the forward measure $\mathbb{P}_{T_n}$, specifically, for $j = n - 1$ we have

$$\tilde{\kappa}(t, T_{n-1}) = \tilde{\kappa}(0, T_{n-1}) \, \mathcal{E}_t \left(\int_0^{\cdot} v(u, T_{n-1}) \cdot dW_u^{T_n} \right)$$

and for every $j = 0, \ldots, n - 2$

$$\tilde{\kappa}(t, T_j) = \tilde{\kappa}(0, T_j) h_t^j \, \mathcal{E}_t \left(\int_0^{\cdot} v(u, T_j) \cdot dW_u^{T_n} \right),$$

where the auxiliary process h_t^j is given by the formula

$$h_t^j = \exp \left(- \int_0^t \sum_{k=j+1}^{n-1} \frac{\delta_k g_u^{jk} v(u, T_j) v(u, T_k) \tilde{\kappa}(u, T_k)}{\left(1 + \delta_k \tilde{\kappa}(u, T_k) \right) g_u^j} \, du \right).$$

13.3.2 Valuation of Co-terminal Swaptions

For a fixed, but otherwise arbitrary, date T_j, $j = 0, \ldots, n - 1$, we consider a swaption with expiry date T_j, written on a forward payer swap settled in arrears. The underlying forward payer swap starts at date T_j, has the fixed rate κ, and $n - j$ accrual periods. Such a swaption is referred to as the jth co-terminal swaption in what follows. The jth co-terminal swaption can be seen as a contract that pays to its holder the amount $\delta_k (\kappa(T_j, T_j, n - j) - \kappa)^+$ at each settlement date T_k, where $k = j + 1, \ldots, n$ (recall that we assume that the notional principal $N = 1$). Equivalently, the jth swaption pays the amount

$$\tilde{Y} = \sum_{k=j+1}^{n} \delta_k B(T_j, T_k) \left(\tilde{\kappa}(T_j, T_j) - \kappa \right)^+$$

at maturity date T_j. It is useful to observe that $\tilde{Y}$ admits the following representation in terms of the level process $G(n - j)$ introduced in Sect. 13.3 (cf. formula (13.13))

$$\tilde{Y} = G_{T_j}(n - j) \left(\tilde{\kappa}(T_j, T_j) - \kappa \right)^+.$$

Recall that the model of co-terminal forward swap rates specifies the dynamics of the process $\tilde{\kappa}(t, T_j)$ through the following SDE:

$$d\tilde{\kappa}(t, T_j) = \tilde{\kappa}(t, T_j) v(t, T_j) \cdot d\tilde{W}_t^{T_{j+1}},$$

where $\tilde{W}^{T_{j+1}}$ is a d-dimensional Brownian motion under the corresponding forward swap measure $\tilde{\mathbb{P}}_{T_{j+1}}$. Recall that by the definition of $\tilde{\mathbb{P}}_{T_{j+1}}$ any process of the form $B(t, T_k)/G_t(n - j)$, $k = 0, \ldots, n$, is a local martingale under this probability. Furthermore, from the general considerations concerning the choice of a numeraire it is easy to see that the arbitrage price $\pi_t(X)$ of an attainable contingent claim $X = g(B(T_j, T_{j+1}), \ldots, B(T_j, T_n))$ equals, for $t \in [0, T_j]$,

$$\pi_t(X) = G_t(n - j)\, \mathbb{E}_{\tilde{\mathbb{P}}_{T_{j+1}}}\left(G_{T_j}^{-1}(n - j)X \,\big|\, \mathcal{F}_t\right),$$

provided that X settles at time T_j. Applying the last formula to the swaption's payoff $\tilde{Y}$, we obtain the following representation for the arbitrage price $\mathbf{PS}_t^j$ at time $t \in [0, T_j]$ of the jth swaption

$$\mathbf{PS}_t^j = \pi_t(\tilde{Y}) = G_t(n - j)\, \mathbb{E}_{\tilde{\mathbb{P}}_{T_{j+1}}}\left((\tilde{\kappa}(T_j, T_j) - \kappa)^+ \,\big|\, \mathcal{F}_t\right).$$

Assume that the swap rate volatility $v(t, T_j) : [0, T_j] \to \mathbb{R}^d$ is deterministic for any $j = 0, \ldots, n - 1$. The following result is due to Jamshidian (1997a).

Proposition 13.3.2 *Assume the lognormal market model of co-terminal forward swap rates. For any $j = 0, \ldots, n - 1$, the arbitrage price $\mathbf{PS}_t^j$ of the jth co-terminal swaption equals, for every $t \in [0, T_j]$,*

$$\mathbf{PS}_t^j = \sum_{k=j+1}^{n} \delta_k B(t, T_k)\left(\tilde{\kappa}(t, T_j)N\big(\tilde{h}_1(t, T_j)\big) - \kappa N\big(\tilde{h}_2(t, T_j)\big)\right),$$

where N stands for the standard Gaussian cumulative distribution function and

$$\tilde{h}_{1,2}(t, T_j) = \frac{\ln(\tilde{\kappa}(t, T_j)/\kappa) \pm \frac{1}{2}v^2(t, T_j)}{v(t, T_j)},$$

with $v^2(t, T_j) = \int_t^{T_j} |v(u, T_j)|^2\, du$.

Proof. The proof of the proposition is similar to that of Proposition 12.6.1, and thus it is omitted.   □

13.3.3 Hedging of Swaptions

The replicating strategy for a swaption within the present framework has features similar to those of the replicating strategy for a cap in the lognormal LIBOR model. For this reason, we shall focus mainly on differences between these two cases. Let us fix j, and let us denote by $F_{S^j}(t, T)$ the relative price at time $t \leq T_j$ of the jth swaption, when the level process

$$G_t(n - j) = \sum_{k=j+1}^{n} \delta_k B(t, T_k)$$

is chosen as a numeraire asset. From Proposition 13.3.2, we find easily that for every $t \leq T_j$

$$F_{S^j}(t, T_j) = \tilde{\kappa}(t, T_j)N\big(\tilde{h}_1(t, T_j)\big) - \kappa N\big(\tilde{h}_2(t, T_j)\big).$$

Applying Itô's formula to the last expression, we obtain

$$dF_{S^j}(t, T_j) = N\big(\tilde{h}_1(t, T_j)\big)\, d\tilde{\kappa}(t, T_j). \tag{13.26}$$

Let us consider the following self-financing trading strategy. We start our trade at time 0 with the amount $\mathbf{PS}_0^j$ of cash, which is then immediately invested in the portfolio $G(n - j)$. Recall that one unit of portfolio $G(n - j)$ costs $\sum_{k=j+1}^{n} \delta_k B(0, T_k)$ at time 0. At any time $t \le T_j$ we assume $\psi_t^j = N\big(h_1(t, T_j)\big)$ positions in market forward swaps.

Of course, the market swaps should have the same starting date and tenor structure as the underlying forward swap. The associated gains process $\tilde{G}(\psi^j)$, expressed in units of the swap annuity $G(n - j)$, satisfies $\tilde{G}_0(\psi^j) = 0$ and

$$d\tilde{G}_t(\psi^j) = \psi_t^j\, d\tilde{\kappa}(t, T_j) = N\big(\tilde{h}_1(t, T_j)\big)\, d\tilde{\kappa}(t, T_j) = dF_{S^j}(t, T_j).$$

Consequently,

$$F_{S^j}(T_j, T_j) = F_{S^j}(0, T_j) + \int_0^{T_j} \psi_t^j\, d\tilde{\kappa}(t, T_j) = F_{S^j}(0, T_j) + \tilde{G}_{T_j}.$$

Here the dynamic trading in market forward swaps is done at any date $t \in [0, T_j]$, and all gains/losses from trading (including the initial investment) are expressed in units of the swap annuity $G(n - j)$. The last equality makes it clear that the strategy ψ^j described above does indeed replicate the jth co-terminal swaption.

13.3.4 Bermudan Swaptions

One of the major motivations for the development of Jamshidian's model of swap rates was the desire to have a model for valuation and hedging of Bermudan swaptions (see Jamshidian (1997a) or Joshi and Theis (2002)). A *Bermudan receiver swaption* is the option, which at each given date $t_j \le T_j$ gives the holder the right to enter the jth swap, provided this right has not yet been exercised at a previous date t_l, $l = 0, \ldots, j - 1$.

As was mentioned already, Bermudan swaptions frequently arise as embedded options in cancellable (callable) swaps. Since a Bermudan swaption can be seen as an example of a pseudo-American option, it is obvious that its valuation and hedging are closely related to an optimal stopping problem associated with the rational exercise policy.

To start the analysis of a Bermudan swaption, let us notice that the jth swap is, at time t_j, worth the amount $G_{t_j}(n - j)(\kappa - \tilde{\kappa}(t_j, T_j))$. An equivalent payoff V^j at time T_n is thus given by the expression

$$V^j = G_{t_j}(n - j)(\kappa - \tilde{\kappa}(t_j, T_j)) / B(t_j, T_n).$$

Assume that a holder decides to exercise a Bermudan swaption at time t_j. We may formally assume that he will receive the equivalent payoff V^j at time T_n. Let us define inductively a sequence C^j, $j = 1, \ldots, n$ of random variables by setting $C^n = \max(V^n, 0) = (V^n)^+$ and for $j = 1, \ldots, n$

$$C^j = \mathbb{1}_{\{V^j \geq \mathbb{E}_{\tilde{\mathbb{P}}_{T_n}}(C^{j+1} \,|\, \mathcal{F}_{t_j})\}} V^j + \mathbb{1}_{\{V^j < \mathbb{E}_{\tilde{\mathbb{P}}_{T_n}}(C^{j+1} \,|\, \mathcal{F}_{t_j})\}} C^{j+1}.$$

It is not difficult to check, in view of the usual optimality criterion for pseudo-American claims, that a Bermudan swaption can be formally represented as a single payoff C^1 that settles at time T_n. Consequently, the arbitrage price at any date $t \leq t_0$ of a Bermudan swaption can be found by evaluating the conditional expectation $B(t, T_n) \mathbb{E}_{\tilde{\mathbb{P}}_{T_n}}(C^1 \,|\, \mathcal{F}_t)$.

13.4 Co-initial Forward Swap Rates

In this section, we no longer postulate that the reference swaps have the same maturity date. We focus instead on a finite family of swaps with a common start date, but with different lengths. It is thus natural to use the term *co-initial swaps* for this family of swap contracts. The presentation given below is largely based on Galluccio and Hunter (2004a).

Formally, given the family of reset/settlement dates $T = T_0 < \ldots < T_n$ we define the *co-initial forward swap rate* $\kappa(t, T_0, j)$ by setting

$$\kappa(t, T_0, j) = \frac{B(t, T_0) - B(t, T_j)}{\delta_1 B(t, T_1) + \cdots + \delta_j B(t, T_j)} \tag{13.27}$$

for every $t \in [0, T_0]$ and $j = 1, \ldots, n$. We wish to examine the family of co-initial forward swap rates $\bar{\kappa}(t, j) = \kappa(t, T_0, j)$ for $j = 1, \ldots, n$. The demand for a corresponding model is motivated by the desire to value and hedge such exotic contracts as CMS spread options (see Sect. 13.2.2) or forward start swaptions (see Sect. 13.1.3) in an efficient and practically appealing way.

In order to deal with such swap derivatives, it is convenient to derive the joint arbitrage-free dynamics of processes $\bar{\kappa}(t, 1), \ldots, \bar{\kappa}(t, n)$. Let us define the family of level processes

$$\bar{G}_t(j) = \delta_1 B(t, T_1) + \cdots + \delta_j B(t, T_j).$$

As expected, the jth level process $\bar{G}(j)$ will play the role of the numeraire for the jth rate $\bar{\kappa}(t, j)$ in the sense that $\bar{\kappa}(t, j)$ is a (local) martingale under the corresponding probability $\bar{\mathbb{P}}^j$. In order to determine the joint dynamics of co-initial forward swap rates, we shall choose the T-maturity bond as a numeraire. This simply means that we will work under the forward measure $\mathbb{P}_T$. As usual, W^T stands for a d-dimensional standard Brownian motion under $\mathbb{P}_T$. Our goal is to determine the drift coefficients in the expression

$$d\bar{\kappa}(t, j) = \bar{\mu}_t^j \, dt + \phi_t^j \cdot dW_t^T, \tag{13.28}$$

specifying the dynamics of $\tilde{\kappa}(t, T_j)$ under the forward measure $\mathbb{P}_T$. To this end, let us first observe that manifestly

$$\langle \bar{\kappa}(\cdot, i), \bar{\kappa}(\cdot, j) \rangle_t = \int_0^t \phi_u^i \cdot \phi_u^j \, du = \int_0^t v_u^{i,j} \, du,$$

where we set $v_t^{i,j} = \phi_t^i \cdot \phi_t^j$.

Before proceeding to the general result, let us notice that none of the forward rates $\bar{\kappa}(t, j)$ is a martingale under $\mathbb{P}_T$. For instance, the rate $\bar{\kappa}(t, 1)$ equals the LIBOR $L(t, T_0)$ and thus it is a martingale under $\mathbb{P}_{T_1}$. We know already that the dynamics of this process under $\mathbb{P}_T$ are

$$d\bar{\kappa}(t, 1) = \frac{\delta_1 v_t^{1,1}}{1 + \delta_1 \bar{\kappa}(t, 1)} \, dt + \phi_t^1 \cdot dW_t^T. \tag{13.29}$$

The following result, borrowed from Galluccio and Hunter (2004a), furnishes the general formula for the drift $\bar{\mu}^j$. The reader is invited to check that for $j = 1$ the expression of Proposition 13.4.1 yields the drift term in (13.29).

Proposition 13.4.1 *Assume that for each* $j = 1, \ldots n$ *the dynamics of the co-initial forward swap rate* $\bar{\kappa}(t, j)$ *under the forward martingale measure* $\mathbb{P}_T$ *are governed by* (13.28). *Then the drift coefficient* $\bar{\mu}^j$ *is given by the expression*

$$\bar{\mu}_t^j = \frac{B(t, T)}{\bar{G}_t(j)} \sum_{i=1}^j \delta_i \left[\frac{\delta_i v_t^{i,j}}{(1 + \delta_i \bar{\kappa}(t, i))^2} \right.$$
$$+ \left. \sum_{l=1}^{i-1} \frac{\delta_l}{\prod_{k=l}^i (1 + \delta_k \bar{\kappa}(t, k))} \left(v_t^{i,j} - \bar{\kappa}(t, i) \sum_{m=l}^i \frac{\delta_m v_t^{m,j}}{1 + \delta_m \bar{\kappa}(t, m)} \right) \right].$$

Proof. We shall sketch the main steps of the derivation of drift formula. Let us fix the index j and let us focus on the process $\bar{\kappa}(t, j)$. Since the product $\bar{\kappa}(t, j)\bar{G}(j)$ represents the value of a portfolio of zero-coupon bonds, it is clear that the process X^j, given as

$$X_t^j = \frac{\bar{\kappa}(t, j)\bar{G}_t(j)}{B(t, T)} = \bar{\kappa}(t, j)Y_t^j, \quad \forall t \in [0, T],$$

is necessarily a (local) martingale under $\mathbb{P}_T$. Itô's integration by parts formula yields

$$dX_t^j = \bar{\kappa}(t, j) \, dY_t^j + Y_t^j \, d\bar{\kappa}(t, j) + d\langle \bar{\kappa}(\cdot, j), Y^j \rangle_t,$$

where the process $Y_t^j = \bar{G}_t(j)/B(t, T)$ is a strictly positive (local) martingale under $\mathbb{P}_T$. It is therefore easily seen that the finite variation part of X disappears if and only if

$$Y_t^j \bar{\mu}_t^j \, dt + d\langle \bar{\kappa}(\cdot, j), Y^j \rangle_t = 0, \quad \forall t \in [0, T],$$

or more explicitly, if

$$\bar{\mu}_t^j \, dt = -(Y_t^j)^{-1} \sum_{i=1}^j \delta_i \, d\langle \bar{\kappa}(\cdot, j), F_B(\cdot, T_i, T) \rangle_t, \tag{13.30}$$

where $F_B(t, T_i, T) = B(t, T_i)/B(t, T)$ is the forward bond price. To continue the calculations, we need first to find the representation for the forward bond price $F_B(t, T_i, T)$ in terms of co-initial forward swap rates. To this end, we observe that

$$\bar{\kappa}(t, i)\bar{G}_t(i) = \bar{\kappa}(t, i) \sum_{l=1}^{i} \delta_l B(t, T_l) = B(t, T) - B(t, T_i)$$

and thus the equality

$$B(t, T_i)(1 + \delta_i \bar{\kappa}(t, i)) + \sum_{l=1}^{i-1} \delta_l \bar{\kappa}(t, i) B(t, T_l) = B(t, T)$$

is valid for every $i = 1, \ldots, n$.

Finally, we obtain, for $i = 1, \ldots, n$,

$$F_B(t, T_i, T)(1 + \delta_i \bar{\kappa}(t, T_i)) + \sum_{l=1}^{i-1} \delta_l \bar{\kappa}(t, i) F_B(t, T_l, T) = 1.$$

By solving these equations iteratively for $F_B(t, T_i, T)$, we obtain

$$F_B(t, T_i, T) = \frac{1}{1 + \delta_i \bar{\kappa}(t, i)} - \sum_{l=1}^{i-1} \frac{\delta_l \bar{\kappa}(t, i)}{\prod_{k=l}^{i}(1 + \delta_k \bar{\kappa}(t, k))}. \tag{13.31}$$

Let us write, for $l = 1, \ldots, i$,

$$Z_t^{i,l} = \left(\prod_{k=l}^{i}(1 + \delta_k \bar{\kappa}(t, k)) \right)^{-1}. \tag{13.32}$$

Then clearly

$$F_B(t, T_i, T) = Z^{i,i} - \sum_{l=1}^{i-1} \delta_l \bar{\kappa}(t, T_i) Z_t^{i,l},$$

and thus (13.30) becomes

$$\bar{\mu}_t^j \, dt = -(Y_t^j)^{-1} \sum_{i=1}^{j} \delta_i \left(d\langle \bar{\kappa}(\cdot, j), Z^{i,i} \rangle_t - \sum_{l=1}^{i-1} \delta_l d\langle \bar{\kappa}(\cdot, j), \bar{\kappa}(\cdot, T_i) Z^{i,l} \rangle_t \right).$$

An application of Itô's formula yields

$$d\langle \bar{\kappa}(\cdot, j), Z^{i,i} \rangle_t = -\delta_i (Z_t^{i,i})^2 \, d\langle \bar{\kappa}(\cdot, j), \bar{\kappa}(\cdot, T_i) \rangle_t = -\frac{\delta_i v_t^{i,j} \, dt}{(1 + \delta_i \bar{\kappa}(t, i))^2}.$$

Furthermore, again by Itô's formula, we obtain

$$d\langle \bar{\kappa}(\cdot, j), \bar{\kappa}(\cdot, i) Z^{i,l} \rangle_t = Z_t^{i,l} \, d\langle \bar{\kappa}(\cdot, j), \bar{\kappa}(\cdot, i) \rangle_t + \bar{\kappa}(t, i) \, d\langle \bar{\kappa}(\cdot, j), Z^{i,l} \rangle_t,$$

or more explicitly

$$d\langle \bar{\kappa}(\cdot, j), \bar{\kappa}(\cdot, i) Z^{i,l} \rangle_t = Z_t^{i,l} v_t^{i,j} \, dt - \bar{\kappa}(t, i) Z_t^{i,l} \sum_{m=l}^{i} \frac{\delta_m v_t^{m,j}}{1 + \delta_m \bar{\kappa}(t, m)} \, dt.$$

To conclude it suffices to combine the formulas above (in particular, we make use of equality (13.32)). $\qquad\square$

Suppose that $\phi_t^i = \bar{\kappa}(t, j) \bar{v}(t, j)$ for some volatility processes (or functions) $\bar{v}(\cdot, j) : [0, T] \to \mathbb{R}^d$. Then each swap rate process $\bar{\kappa}(t, j)$, $j = 1, \ldots, n$ is easily seen to satisfy

$$d\bar{\kappa}(t, j) = \bar{\kappa}(t, j) \bar{v}(t, j) \cdot d\bar{W}_t^j,$$

where $\bar{W}^j$ is a d-dimensional standard Brownian motion under the martingale measure $\bar{\mathbb{P}}^j$ corresponding to the choice of the level process $\bar{G}(j)$ as a numeraire. Notice that the probability measures $\bar{\mathbb{P}}^j$ are all defined on $(\Omega, \mathcal{F}_T)$ and they are equivalent to the forward measure $\mathbb{P}_{T_{j+1}} = \bar{\mathbb{P}}^1$.

13.4.1 Valuation of Co-initial Swaptions

For any date $j = 1, \ldots, n$, we consider a payer swaption with expiration date T, which is written on a payer swap, with a fixed rate κ, that starts at time $T_0 = T$ and has j accrual periods. As usual, we assume that the notional principal equals $N = 1$. Hence the jth swaption can be seen as a European claim paying at time T the following amount:

$$\bar{Y} = \sum_{l=1}^{j} \delta_l B(T, T_l) \big(\bar{\kappa}(T, j) - \kappa \big)^+.$$

It is clear that the payoff $\bar{Y}$ has the following equivalent representation in terms of the level process $\bar{G}(j)$

$$\hat{Y} = \bar{G}_T(j) \big(\bar{\kappa}(T, j) - \kappa \big)^+.$$

From the general theory, we know that the price of any attainable claim X that settles at time T equals, for every $t \in [0, T]$,

$$\pi_t(X) = \bar{G}_t(j) \, \mathbb{E}_{\bar{\mathbb{P}}^j} \big(\bar{G}_T^{-1}(j) X \mid \mathcal{F}_t \big).$$

In particular, for the jth co-initial swaption we obtain

$$\bar{\mathbf{PS}}_t^j = \pi_t(\bar{Y}) = \bar{G}_t(j) \, \mathbb{E}_{\bar{\mathbb{P}}^j} \big((\bar{\kappa}(T, j) - \kappa)^+ \mid \mathcal{F}_t \big).$$

Within the framework of the co-initial model of forward swap rates, the dynamics of the process $\bar{\kappa}(t, j)$ are

$$d\bar{\kappa}(t, j) = \bar{\kappa}(t, j) \bar{v}(t, j) \cdot d\bar{W}_t^j,$$

where $\bar{W}^j$ is a d-dimensional standard Brownian motion under the probability measure $\bar{\mathbb{P}}^j$ corresponding to the choice of the level process $\bar{G}(j)$ as a numeraire asset. It is clear that if the volatilities $\bar{v}(t, j)$, $j = 1, \ldots, n$ are deterministic, then the probability distribution of $\bar{\kappa}(T, j)$ under $\bar{\mathbb{P}}^j$ is lognormal. The proof of the following valuation result is therefore straightforward.

Proposition 13.4.2 *Assume the lognormal market model of co-initial forward swap rates. For any $j = 1, \ldots, n$, the arbitrage price at time $t \in [0, T]$ of the jth co-initial swaption equals*

$$\bar{PS}_t^j = \sum_{l=1}^{l=j} \delta_l B(t, T_l) \Big(\bar{\kappa}(t, j) N\big(\bar{h}_1(t, j)\big) - \kappa N\big(\bar{h}_2(t, j)\big) \Big),$$

where N is the standard Gaussian probability distribution function and

$$\bar{h}_{1,2}(t, j) = \frac{\ln(\bar{\kappa}(t, j)/\kappa) \pm \frac{1}{2} \bar{v}^2(t, j)}{\bar{v}(t, j)},$$

with $\bar{v}^2(t, j) = \int_t^T |\bar{v}(u, j)|^2 \, du$.

13.4.2 Valuation of Exotic Options

Recall that an option on a swap rate spread (see Sect. 13.2.2) with expiry date $T = T_0$ is a contract that pays at time T the (positive) difference between the rates $\kappa(T, T, j)$ and $\kappa(T, T, k)$ for some $j \neq k$. Using the notation introduced in this section, it can be represented as follows

$$\text{CMSC}_T(\kappa, j, k) = \big(\bar{\kappa}(T, j) - \bar{\kappa}(T, k) - \kappa\big)^+,$$

where j and k are lengths of the underlying swaps, and κ is a constant. It is clear that the valuation of this claim relies on the computation of the (conditional) expected value

$$\text{CMSC}_T(\kappa, j, k) = \mathbb{E}_{\mathbb{P}_T}\Big(\big(\bar{\kappa}(T, j) - \bar{\kappa}(T, k) - \kappa\big)^+ \,\Big|\, \mathcal{F}_t\Big).$$

Since drift terms of processes $\bar{\kappa}(T, j)$ and $\bar{\kappa}(T, k)$, derived in Proposition 13.4.1, are rather complicated, no closed-form solution to the latter problem is available and thus an application of some numerical procedure (such as Monte Carlo simulation) is required.

From Sect. 13.1.3, we know that the payoff of a forward-start swaption (known also as a mid-curve option), with expiration date $T = T_0$ written on a swap with start date T_1, can be represented in the following way (see formula (13.8))

$$\text{PS}_T = \sum_{j=2}^{n} \delta_j B(T, T_j) \big(\kappa(T, T_1, n - 1) - \kappa\big)^+. \tag{13.33}$$

To deal with this payoff within the framework of a (Gaussian or lognormal, say) model of co-initial forward swap rates, it is convenient to rewrite this payoff as follows

$$\mathbf{PS}_T = \Big(\bar{G}_T(n)\big(\bar{\kappa}(T,n) - \kappa\big) - \bar{G}_T(1)\big(\bar{\kappa}(T,1) - \kappa\big)\Big)^+.$$

In addition, we shall use the following representation of the level process

$$\bar{G}_T(k) = \sum_{i=1}^{k} \delta_i \left(\frac{1}{1 + \delta_i \bar{\kappa}(T,i)} - \sum_{l=1}^{i-1} \frac{\delta_l \bar{\kappa}(T,i)}{\prod_{k=l}^{i}(1 + \delta_k \bar{\kappa}(T,k))} \right),$$

which is an almost immediate consequence of (13.31). By combining the preceding two formulas, we conclude that the forward start swaption was reexpressed as a relatively simple European claim that settles at time T and has the form $X = g(\bar{\kappa}(T,1), \ldots, \bar{\kappa}(T,i))$. Again, the availability of a closed-form solution to the valuation problem is not evident. Nevertheless, the model is likely to be the most convenient tool to value and hedge (through numerical procedures) these swap derivatives that depend on the value at a given date of several swap rates. It is not suitable, of course, to deal with claims with path-dependence and/or American features.

13.5 Co-sliding Forward Swap Rates

In this section, in which we follow Rutkowski (1999a) and Galluccio et al. (2003a), we no longer assume that the underlying swap agreements have different lengths but the same start or maturity date. On the contrary, the length K of a swap will now be fixed, but the start (and thus also maturity) date will vary. As before, we assume that we are given a family of reset/settlement dates $T_0, \ldots, T_n$. For a forward swap which starts at time T_j, $j = 1, \ldots, n - K$, the first settlement date is T_{j+1}, and its maturity date equals T_{K+j}. It is clear that the value at time $t \in [0, T_j]$ of a swap with these features equals

$$\mathbf{FS}_t^j(\kappa) = B(t, T_j) - \sum_{l=j+1}^{j+K} c_l B(t, T_l),$$

where $c_l = \kappa \delta_l$ for $l = j + 1, \ldots, j + K - 1$ and $c_{j+K} = 1 + \kappa \delta_{j+K}$. The forward swap rate of length K, for the date T_j, is that value of the fixed rate that makes the underlying forward swap starting at T_j worthless at time t. Using the last formula, we find easily that

$$\kappa(t, T_j, K) = \frac{B(t, T_j) - B(t, T_{j+K})}{\delta_{j+1} B(t, T_{j+1}) + \cdots + \delta_{j+K} B(t, T_{j+K})}.$$

Recall that the forward LIBOR $L(t, T_j)$ coincides with the one-period forward swap rate over $[T_j, T_{j+1}]$. Hence the model of forward LIBORs presented in Sect. 12.4.4 can be seen as a special case of the model of co-sliding forward swap rates (since in that case $K = 1$, its construction is simpler).

13.5.1 Modelling of Co-sliding Swap Rates

In this section, we examine a particular model of co-sliding swap rates. We shall write $\hat{\kappa}(t, T_j)$ instead of $\kappa(t, T_j, K)$ and we set $\hat{n} = n - K + 1$. For any $j = 1, \ldots, \hat{n}$, we define the *sliding level process* $\hat{G}(j)$ by setting

$$\hat{G}_t(j) = \sum_{l=j}^{j+K-1} \delta_l B(t, T_l), \quad \forall\, t \in [0, T_j].$$

The forward swap measure is a martingale measure corresponding to the choice of a sliding coupon process as a numeraire asset.

Definition 13.5.1 For a fixed $j = 1, \ldots, \hat{n}$, a probability measure $\hat{\mathbb{P}}_{T_j}$ on $(\Omega, \mathcal{F}_{T_j})$, equivalent to the underlying probability measure $\mathbb{P}$, is called the *co-sliding forward swap measure* for the date T_j if, for every $k = 1, \ldots, n$, the relative bond price

$$U_j(t, T_k) \stackrel{\text{def}}{=} \frac{B(t, T_k)}{\hat{G}_t(j)} = \frac{B(t, T_k)}{\sum_{l=j}^{j+K-1} \delta_l B(t, T_l)}, \quad \forall\, t \in [0, T_k \wedge T_j],$$

is a local martingale under $\hat{\mathbb{P}}_{T_j}$.

Since $\hat{\kappa}(t, T_j) = U_j(t, T_j) - U_j(t, T_{j+K})$, it is clear that each co-sliding forward swap rate $\hat{\kappa}(t, T_j)$, $j = 0, \ldots, n - K$, follows a local martingale under the co-sliding forward swap measure for the date T_{j+1}. If we assume, in addition, that $\hat{\kappa}(t, T_j)$ is a strictly positive process, it is natural to postulate that it satisfies

$$d\hat{\kappa}(t, T_j) = \hat{\kappa}(t, T_j)\hat{v}(t, T_j) \cdot d\hat{W}_t^{T_{j+1}},$$

where $\hat{W}_t^{T_{j+1}}$ is a standard Brownian motion under $\hat{\mathbb{P}}_{T_{j+1}}$. The initial condition is

$$\hat{\kappa}(0, T_j) = \frac{B(0, T_j) - B(0, T_{j+K})}{\delta_{j+1} B(0, T_{j+1}) + \cdots + \delta_{j+K} B(0, T_{j+K})}.$$

The construction of co-sliding forward swap rates model presented in this section relies on backward induction.

We start by postulating that the underlying probability measure $\mathbb{P}$ represents the forward swap measure for the date $T_{\hat{n}}$. Our goal is to define recursively the forward swap rates and probability measures corresponding to the dates $T_{\hat{n}-1}, T_{\hat{n}-2}, \ldots, T_0$. To proceed further, we need to assume that we are given a family of processes $\hat{v}(t, T_j)$, $j = 1, \ldots, n - K$ representing the volatilities of forward swap rates $\hat{\kappa}(t, T_j)$.

In addition, the initial term structure of swap rates $\hat{\kappa}(0, T_j)$, $j = 0, \ldots, n - K$ is known. In the first step, we postulate that $\hat{\kappa}(t, T_{\hat{n}-1}) = \hat{\kappa}(t, T_{n-K})$, which equals

$$\hat{\kappa}(t, T_{\hat{n}-1}) = \frac{B(t, T_{n-K}) - B(t, T_n)}{\sum_{l=n-K+1}^{n} \delta_l B(t, T_l)} = \frac{B(t, T_{\hat{n}-1}) - B(t, T_n)}{\hat{G}_t(\hat{n})},$$

satisfies

$$d\hat{\kappa}(t, T_{\hat{n}-1}) = \hat{\kappa}(t, T_{\hat{n}-1})\hat{v}(t, T_{\hat{n}-1}) \cdot dW_t.$$

This means, in particular, that we have $\hat{\mathbb{P}}_{T_{\hat{n}}} = \mathbb{P}$ and $\hat{W}^{T_{\hat{n}}} = W$. Our next goal is to introduce the forward swap measure for the date $T_{\hat{n}-2}$.

Case of equal accrual periods. Assume first, for simplicity, that $\delta_j = \delta$ is constant. In this case the backward induction goes through without difficulty. Recall that

$$\hat{\kappa}(t, T_{\hat{n}-2}) = \frac{B(t, T_{n-K-1}) - B(t, T_{n-1})}{\delta \sum_{l=n-K}^{n-1} B(t, T_l)} = \frac{B(t, T_{\hat{n}-2}) - B(t, T_{n-1})}{\hat{G}_t(\hat{n} - 1)}.$$

We note that the relative bond price

$$U_{\hat{n}-1}(t, T_k) = \frac{B(t, T_k)}{\hat{G}_t(\hat{n} - 1)} = \frac{B(t, T_k)}{\delta B(t, T_{\hat{n}-1}) + \cdots + \delta B(t, T_{n-1})}$$

admits the following representation

$$U_{\hat{n}-1}(t, T_k) = \frac{U_{\hat{n}}(t, T_k)}{1 + \delta\hat{\kappa}(t, T_{\hat{n}-1})}.$$

In view of Lemma 12.4.1, it is obvious that the process $U_{\hat{n}-1}(t, T_k)$ is a local martingale under an equivalent probability measure $\hat{\mathbb{P}}_{T_{\hat{n}-1}}$ if and only if the process

$$\hat{W}_t^{T_{\hat{n}-1}} = \hat{W}_t^{T_{\hat{n}}} - \int_0^t \frac{\delta\hat{\kappa}(u, T_{\hat{n}-1})}{1 + \delta\hat{\kappa}(u, T_{\hat{n}-1})} \hat{v}(u, T_{\hat{n}-1}) \, du$$

is a standard Brownian motion under this probability. To find $\hat{\mathbb{P}}_{T_{\hat{n}-1}}$, it is thus enough to apply Girsanov's theorem.

Similarly, in a general induction step we use the following relationship

$$U_m(t, T_k) = \frac{B(t, T_k)}{\hat{G}_t(m)} = \frac{U_{m+1}(t, T_k)}{1 + \delta\hat{\kappa}(t, T_m)}$$

that allows us to define the forward swap measure $\hat{\mathbb{P}}_{T_m}$ and the Brownian motion process $\hat{W}^{T_m}$, provided that the forward swap measure $\hat{\mathbb{P}}_{T_{m+1}}$, the forward swap rate $\hat{\kappa}(t, T_m)$, and the Brownian motion $W^{T_{m+1}}$ are already known. We set

$$\hat{W}_t^{T_m} = \hat{W}_t^{T_{m+1}} - \int_0^t \frac{\delta\hat{\kappa}(u, T_m)}{1 + \delta\hat{\kappa}(u, T_m)} \hat{v}(u, T_m) \, du,$$

for $t \in [0, T_m]$, and we specify $\hat{\mathbb{P}}_{T_m}$ through Girsanov's theorem. Given $\hat{W}^{T_m}$ and $\hat{\mathbb{P}}_{T_m}$, we postulate that the forward swap rate $\hat{\kappa}(t, T_{m-1})$ is governed by the SDE

$$d\hat{\kappa}(t, T_{m-1}) = \hat{\kappa}(t, T_{m-1})\hat{v}(t, T_{m-1}) \cdot d\hat{W}_t^{T_m}.$$

This shows that the knowledge of all swap rate volatilities (and, of course, the initial term structure) is sufficient if the purpose is to determine uniquely an arbitrage-free family of forward swap rates. Somewhat surprisingly, this is no longer the case if the length of the accrual period varies over time.

General case. We no longer assume that the accrual periods are equal. Assume that $K \geq 2$ (the case $K = 1$ was examined in Sect. 12.4.4). For the date $T_{\hat{n}-2}$, we now have

$$\hat{\kappa}(t, T_{\hat{n}-2}) = \frac{B(t, T_{\hat{n}-2}) - B(t, T_{n-1})}{\sum_{l=n-K}^{n-1} \delta_l B(t, T_l)} = \frac{B(t, T_{\hat{n}-2}) - B(t, T_{n-1})}{\hat{G}_t(\hat{n} - 1)}.$$

Furthermore, for any k we have

$$U_{\hat{n}-1}(t, T_k) = \frac{B(t, T_k)}{\hat{G}_t(\hat{n} - 1)} = \frac{B(t, T_k)}{\delta_{\hat{n}-1} B(t, T_{\hat{n}-1}) + \cdots + \delta_{n-1} B(t, T_{n-1})},$$

so that

$$U_{\hat{n}-1}(t, T_k) = \frac{U_{\hat{n}}(t, T_k)}{1 + \Delta_{\hat{n}} U_{\hat{n}}(t, T_n) + \delta_{\hat{n}-1} \hat{\kappa}(t, T_{\hat{n}-1})} = \frac{U_{\hat{n}}(t, T_k)}{1 + V_{\hat{n}}(t)},$$

where we write $\Delta_{\hat{n}} = \delta_{\hat{n}-1} - \delta_n$, and the definition of the process $V_{\hat{n}}$ is clear from the context. To continue, we need to assume that the dynamics of the process $U_{\hat{n}}(t, T_n) = B(t, T_n)/\hat{G}_t(\hat{n})$ under $\hat{\mathbb{P}}_{T_{\hat{n}}}$ are known.

More precisely, it is enough to specify volatility of this process, since it is clear that $U_{\hat{n}}(t, T_n)$ satisfies

$$dU_{\hat{n}}(t, T_n) = U_{\hat{n}}(t, T_n) \zeta_{\hat{n}}(t, T_n) \cdot d\hat{W}_t^{T_{\hat{n}}}$$

for some process $\zeta_{\hat{n}}(t, T_n)$. Suppose that the volatility process $\zeta_{\hat{n}}(t, T_n)$, and thus also the process $U_{\hat{n}}(t, T_n)$, are known. Let us write

$$v_{\hat{n}}(t) = \Delta_{\hat{n}} U_{\hat{n}}(t, T_n) \zeta_{\hat{n}}(t, T_n) + \delta_{\hat{n}-1} \hat{\kappa}(t, T_{\hat{n}-1}) \hat{v}(t, T_{\hat{n}-1}).$$

Then, using again Lemma 12.4.1, we conclude that

$$\hat{W}_t^{T_{\hat{n}-1}} = \hat{W}_t^{T_{\hat{n}}} - \int_0^t \frac{v_{\hat{n}}(u)}{1 + V_{\hat{n}}(u)} \, du, \quad \forall t \in [0, T_{\hat{n}-1}],$$

and thus the forward swap measure $\hat{\mathbb{P}}_{T_{\hat{n}-1}}$ can be found explicitly. Next, the swap rate $\hat{\kappa}(t, T_{\hat{n}-2})$ is defined by setting

$$d\hat{\kappa}(t, T_{\hat{n}-2}) = \hat{\kappa}(t, T_{\hat{n}-2}) \hat{v}(t, T_{\hat{n}-2}) \cdot d\hat{W}_t^{T_{\hat{n}-1}}.$$

Let us now consider a general induction step. We assume that we have already found $\hat{\mathbb{P}}_{T_{m+1}}$ and $\hat{W}^{T_{m+1}}$, and thus also the forward swap rate $\hat{\kappa}(t, T_m)$. In order to define $\hat{\kappa}(t, T_{m-1})$, we need first to specify $\hat{\mathbb{P}}_{T_m}$ and $\hat{W}^{T_m}$. We have the following relationship

$$U_m(t, T_k) = \frac{B(t, T_k)}{\hat{G}_t(m)} = \frac{U_{m+1}(t, T_k)}{1 + V_{m+1}(t)},$$

where

$$V_{m+1}(t) = \Delta_{m+1} U_{m+1}(t, T_{m+K}) + \delta_m \hat{\kappa}(t, T_m)$$

and $\Delta_{m+1} = \delta_m - \delta_{m+K}$. We have, as expected

$$\hat{W}_t^{T_m} = \hat{W}_t^{T_{m+1}} - \int_0^t \frac{v_{m+1}(u)}{1 + V_{m+1}(u)}\, du, \quad \forall t \in [0, T_m],$$

where

$$v_{m+1}(t) = \Delta_{m+1} U_{m+1}(t, T_{m+K}) \zeta_{m+1}(t, T_{m+K}) + \delta_{m+1} \hat{\kappa}(t, T_{m+1}) \hat{v}(t, T_{m+1})$$

and the process $\zeta_{m+1}(t, T_{m+K})$ is the volatility of the $\hat{\mathbb{P}}_{T_{m+1}}$-martingale

$$U_{m+1}(t, T_{m+K}) = \hat{G}^{-1}(m+1) B(t, T_{m+K}).$$

Concluding, in order to construct a model of co-sliding forward swap rates through backward induction, we need to assume that not only the volatilities of co-sliding forward rates, but also the volatilities of processes

$$U_m(t, T_{m+K}) = \hat{G}^{-1}(m) B(t, T_{m+K}), \quad m = 1, \ldots, n - K - 1,$$

are exogenously given.

As seen from above, in the case of variable accrual periods, the exogenous specification of volatilities (for instance, deterministic functions) of co-sliding forward swap rates $\hat{\kappa}(t, T_0), \ldots, \kappa(t, T_{n-K})$ does not uniquely determine the joint probability distribution of these processes, except for the special case $K = 1$. This should be contrasted with the case of equal accrual periods, where the latter property was valid for any choice of the swap length K.

Galluccio et al. (2003a) consider the following related question: under which assumptions does a given family of forward swap rates allow us to recover all relative bond prices for a given tenor structure $T_0 < T_1 < \cdots < T_n$? It appears that the following definition of *admissibility* of a family of forward swap rates with respect to a given tenor structure is useful in this context.

Definition 13.5.2 Given a preassigned collection of reset/settlement dates $T_0 < T_1 < \cdots < T_n$, a family $\mathcal{B}$ of forward swap rates is said to be *admissible* if it satisfies the following properties:
(a) the cardinality of the set of rates is equal to n,
(b) any date T_j, $j = 0, \ldots n$ coincides with a reset/settlement date of at least one forward swap rate from $\mathcal{B}$.

The following result, due to Galluccio et al. (2003a), gives the answer to the aforementioned question. The proof of Proposition 13.5.1 is based on algebraic considerations, and thus is left to the reader.

Proposition 13.5.1 *The following conditions are equivalent:*
(i) *a family of swap rates $\mathcal{B}$ is admissible with respect to reset/settlement dates $T_0 < T_1 < \cdots < T_n$,*

(ii) *we can recover the relative bond prices* $F_B(t, T_k, T_j) = B(t, T_k)/B(t, T_j)$, $j, k = 0, \ldots, n$, *uniquely in terms of forward swap rates belonging to the family* $\mathcal{B}$.

Note that in a model of co-sliding forward swap rates, we deal with $n - K$ rates and n reset-settlement dates, so that manifestly the definition of admissibility is not satisfied (unless, of course, $K = 1$).

13.5.2 Valuation of Co-sliding Swaptions

For any date T_j, $j = 0, \ldots, n - K$, we consider a payer swaption with expiration date T_j, written on a payer swap with fixed rate κ, that starts at time T_j and has the length K. As usual, we set $N = 1$. Thus the jth co-sliding swaption may be seen as a contract paying at time T_j the amount

$$\hat{Y} = \sum_{l=j+1}^{j+K} \delta_l B(T_j, T_l)\big(\hat{\kappa}(T_j, T_j) - \kappa\big)^+,$$

or equivalently,

$$\hat{Y} = \hat{G}_{T_j}(j + 1)\big(\hat{\kappa}(T_j, T_j) - \kappa\big)^+.$$

In the framework of a model of co-sliding swap rates, the forward swap rate $\hat{\kappa}(t, T_j)$ satisfies

$$d\hat{\kappa}(t, T_j) = \hat{\kappa}(t, T_j)\hat{v}(t, T_j) \cdot d\hat{W}_t^{T_{j+1}},$$

where $\hat{W}^{T_{j+1}}$ is a standard d-dimensional Brownian motion under $\hat{\mathbb{P}}_{T_{j+1}}$. In view of the definition of the forward swap measure $\hat{\mathbb{P}}_{T_{j+1}}$, any process of the form $B(t, T_k)/\hat{G}_t(j+1)$ is a local martingale under $\hat{\mathbb{P}}_{T_{j+1}}$. Hence the price of an attainable claim X that settles at time T_j and depends on bond prices equals

$$\pi_t(X) = \hat{G}_t(j + 1)\,\mathbb{E}_{\hat{\mathbb{P}}_{T_{j+1}}}\big(\hat{G}_{T_j}^{-1}(j + 1)X \mid \mathcal{F}_t\big)$$

for $t \in [0, T_j]$. In particular, for the jth co-sliding swaption we get, for every $t \in [0, T_j]$,

$$\widehat{\mathbf{PS}}_t^j = \pi_t(\hat{Y}) = \hat{G}_t(j + 1)\,\mathbb{E}_{\hat{\mathbb{P}}_{T_{j+1}}}\big((\hat{\kappa}(T_j, T_j) - \kappa)^+ \mid \mathcal{F}_t\big).$$

Let us place ourselves within the framework of the lognormal market model of co-sliding forward swap rates, in which the volatilities $\hat{v}(t, T_j)$ of forward swap rates are assumed to be bounded deterministic functions. Then we have the following result, which is a counterpart of Proposition 13.3.2. It is useful to note that if $K = 1$ then the jth swaption may be identified with the jth caplet. Thus, Proposition 13.5.2 covers also the valuation of caplets (and caps) in the lognormal LIBOR market model presented in Sect. 12.4.4.

Proposition 13.5.2 *Assume the lognormal market model of co-sliding forward swap rates. For any $j = 1, \ldots, n - K$, the arbitrage price at time $t \in [0, T - j]$ of the jth co-sliding swaption equals*

$$\widehat{PS}^j_t = \sum_{l=j+1}^{l=j+K} \delta_l B(t, T_l)\Big(\hat{\kappa}(t, T_j)N\big(\hat{h}_1(t, T_j)\big) - \kappa N\big(\hat{h}_2(t, T)\big)\Big),$$

where N is the standard Gaussian probability distribution function and

$$\hat{h}_{1,2}(t, T_j) = \frac{\ln(\hat{\kappa}(t, T_j)/\kappa) \pm \frac{1}{2}\hat{v}^2(t, T_j)}{\hat{v}(t, T_j)},$$

with $\hat{v}^2(t, T_j) = \int_t^{T_j} |\hat{v}(u, T_j)|^2 \, du$.

Note that the swaption price at time 0 depends only on the initial term structure and the volatility of the underlying forward swap rate. In particular, the volatilities $\zeta_{m+1}(t, T_{m+K})$, $m = 1, \ldots, n - K$ of relative bond prices that were used as auxiliary inputs in the model's construction, do not enter the swaption valuation formula. As a consequence, in the valuation of plain-vanilla swaptions, we may disregard these volatilities altogether. The choice of volatilities $\zeta_{m+1}(t, T_{m+K})$ becomes relevant in the valuation of claims that depend simultaneously on several forward swap rates.

13.6 Swap Rate Model Versus LIBOR Model

It should be emphasized that the lognormal version of a model of co-terminal forward swap rates and the lognormal LIBOR model are incompatible with each other. Indeed, it is not difficult to check that forward LIBORs and co-terminal swap rates are linked to each other through the following relationship

$$\tilde{\kappa}(t, T_j) = \frac{\prod_{i=j}^{n-1}(1 + \delta_{i+1}L(t, T_i)) - 1}{\sum_{i=j}^{n} \delta_i \prod_{k=i+1}^{n-1}(1 + \delta_{k+1}L(t, T_k))}. \tag{13.34}$$

Conversely

$$1 + \delta_{j+1}L(t, T_j) = \frac{1 + g_t^j \tilde{\kappa}(t, T_j)}{1 + g_t^{j+1}\tilde{\kappa}(t, T_{j+1})} \tag{13.35}$$

or equivalently

$$L(t, T_j) = w_t^{j,j}\tilde{\kappa}(t, T_j) + w_t^{j,j+1}\tilde{\kappa}(t, T_{j+1}), \tag{13.36}$$

where $w_t^{j,j} = g_t^j\big(g_t^j - g_t^{j+1}\big)^{-1}$, $w_t^{j,j+1} = -g_t^{j+1}\big(g_t^j - g_t^{j+1}\big)^{-1}$ with (cf. (13.21))

$$g_t^j = \sum_{i=j}^{n-1} \delta_{i+1} \prod_{l=j+1}^{i} (1 + \delta_l\tilde{\kappa}(t, T_l)).$$

Since, obviously, $w_t^{j,j} + w_t^{j,j+1} = 1$ for every $t \in [0, T_j]$, we see that the LIBOR rate $L(t, T_j)$ can be represented as the weighted sum of two consecutive swap rates (note, however, that the weights vary randomly and the second one is negative). Formulas (13.34) and (13.35) make it clear that forward LIBORs and forward swap rates cannot have simultaneously deterministic volatilities. We conclude that lognormal market models of forward LIBORs and forward swap rates are inherently inconsistent with each other.

A challenging practical question of the choice of a benchmark model for simultaneous pricing and hedging of LIBOR and swap derivatives thus arises. Existing research in this area was predominantly focused either on the calibration of the LLM model (and its extensions) to caplets and swaptions, or the derivation of approximate valuation formulas for swaptions within the LLM model. In other words, the LLM model has apparently been accepted as a natural candidate for a benchmark model for LIBOR and swap derivatives. Since the analysis of calibration techniques is beyond the scope of this text, we refer the interested reader to original papers by Brace et al. (1997), Pedersen (1998), Rebonato (1999a, 1999b), Schoenmakers and Coffey (1999), Andersen and Andreasen (2000b, 2001), Hull and White (2000), Wu (2002), Andersen and Brotherton-Ratcliffe (2001), Longstaff et al. (2001a, 2001b), De Malherbe (2002), Gątarek (2002), Wu and Zhang (2002), d'Aspremont (2003), Pelsser et al. (2002), Brigo and Mercurio (2003), Glasserman and Merener (2003), Jäckel and Rebonato (2003), Joshi and Rebonato (2003), and Piterbarg (2003c).

In a recent paper by Galluccio et al. (2003a), the authors advocate a different approach, by arguing that the choice of a (lognormal) market model of forward swap rates as a benchmark has some advantages over the choice of the LLM model. They motivate this choice not only by an easier and more stable calibration procedure, but also by pertinent arguments related to the hedging of exotic products. The procedure put forward by Galluccio et al. (2003a) relies on the calibration of the lognormal model of co-terminal swap rates to at-the-money swaptions and caplets. Let us stress that in the calibration we need to deal with the following issues that can be dealt with separately: the specification of the volatility and the choice of the correlation matrix. At the practical level, the first issue can be addressed by taking prices of swaptions as inputs, and the second problem can be resolved by adjusting the historical correlation matrix in a simultaneous calibration to swaptions and caplets. In this way, the joint distribution of forward swap rates and thus also, by virtue of relationship (13.34), of LIBORs can be inferred with a greater degree of confidence. The correct specification of the correlation structure of a model has a great importance for the valuation of exotic LIBOR and swap derivatives.

13.6.1 Swaptions in the LLM Model

In this section, we deal with approximate valuation of a swaption in the lognormal LIBOR model. To this end, let us recall that, within the general framework, the price at time $t \in [0, T_0]$ of a payer swaption[2] with expiry date $T = T_0$ and strike level κ

---

[2] Since the relationship $\mathbf{PS}_t - \mathbf{RS}_t = \mathbf{FS}_t$ is always valid, and the value of a forward swap is given by (13.2), it is enough to examine the case of a payer swaption.

equals

$$\mathbf{PS}_t = \mathbb{E}_{\mathbb{P}^*}\left\{\frac{B_t}{B_T}\left(\mathbb{E}_{\mathbb{P}^*}\left(\sum_{j=1}^{n}\frac{B_T}{B_{T_j}}(L(T_{j-1})-\kappa)\delta_j \,\Big|\, \mathcal{F}_T\right)\right)^+ \Big|\, \mathcal{F}_t\right\}.$$

Let $D \in \mathcal{F}_T$ be the exercise set of a swaption; that is

$$D = \{\omega \in \Omega \mid (\kappa(T,T,n) - \kappa)^+ > 0\} = \left\{\omega \in \Omega \mid \sum_{j=1}^{n} c_j B(T, T_j) < 1\right\}.$$

Lemma 13.6.1 *The following equality holds for every* $t \in [0, T]$

$$\mathbf{PS}_t = \sum_{j=1}^{n} \delta_j B(t, T_j)\, \mathbb{E}_{\mathbb{P}_{T_j}}\Big((L(T, T_{j-1}) - \kappa)\mathbb{1}_D \,\Big|\, \mathcal{F}_t\Big). \tag{13.37}$$

Proof. Since

$$\mathbf{PS}_t = \mathbb{E}_{\mathbb{P}^*}\left\{\frac{B_t}{B_T}\mathbb{1}_D\, \mathbb{E}_{\mathbb{P}^*}\left(\sum_{j=1}^{n}\frac{B_T}{B_{T_j}}(L(T_{j-1})-\kappa)\delta_j \,\Big|\, \mathcal{F}_T\right)\Big|\, \mathcal{F}_t\right\},$$

we have

$$\mathbf{PS}_t = \mathbb{E}_{\mathbb{P}^*}\left\{\mathbb{E}_{\mathbb{P}^*}\left(\sum_{j=1}^{n}\frac{B_t}{B_{T_j}}(L(T_{j-1})-\kappa)\delta_j\mathbb{1}_D \,\Big|\, \mathcal{F}_T\right)\Big|\, \mathcal{F}_t\right\}$$

$$= \sum_{j=1}^{n} B(t, T_j)\, \mathbb{E}_{\mathbb{P}_{T_j}}\Big((L(T_{j-1}) - \kappa)\delta_j\mathbb{1}_D \,\Big|\, \mathcal{F}_t\Big),$$

where $L(T_{j-1}) = L(T_{j-1}, T_{j-1})$. For any $j = 1, \ldots, n$, we have

$$\mathbb{E}_{\mathbb{P}_{T_j}}\Big((L(T_{j-1}) - \kappa)\mathbb{1}_D \,\Big|\, \mathcal{F}_t\Big) = \mathbb{E}_{\mathbb{P}_{T_j}}\Big(\mathbb{E}_{\mathbb{P}_{T_j}}(L(T_{j-1}) - \kappa \mid \mathcal{F}_T)\mathbb{1}_D \,\Big|\, \mathcal{F}_t\Big)$$

$$= \mathbb{E}_{\mathbb{P}_{T_j}}\Big((L(T, T_{j-1}) - \kappa)\mathbb{1}_D \,\Big|\, \mathcal{F}_t\Big),$$

since $\mathcal{F}_t \subset \mathcal{F}_T$ and the process $L(t, T_{j-1})$ is a $\mathbb{P}_{T_j}$-martingale. $\qquad\square$

For any $k = 1, \ldots, n$, we define the random variable $\zeta_k(t)$ by setting

$$\zeta_k(t) = \int_t^T \lambda(u, T_{k-1}) \cdot dW_u^{T_k}, \quad \forall\, t \in [0, T], \tag{13.38}$$

and we write

$$\lambda_k^2(t) = \int_t^T |\lambda(u, T_{k-1})|^2\, du, \quad \forall\, t \in [0, T]. \tag{13.39}$$

Note that for every $k = 1, \ldots, n$ and $t \in [0, T]$ we have

$$L(T, T_{k-1}) = L(t, T_{k-1})\, e^{\zeta_k(t) - \lambda_k^2(t)/2}.$$

Recall also that the processes W^{T_k} satisfy the following relationship

$$W_t^{T_{k+1}} = W_t^{T_k} + \int_0^t \frac{\delta_{k+1} L(u, T_k)}{1 + \delta_{k+1} L(u, T_k)}\, \lambda(u, T_k)\, du \tag{13.40}$$

for $t \in [0, T_k]$ and $k = 0, \ldots, n-1$. For ease of notation, we formulate the next result for $t = 0$ only; the general case can be treated along the same lines. For any fixed j, we denote by G_j the joint probability distribution function of the n-dimensional random variable $(\zeta_1(0), \ldots, \zeta_n(0))$ under the forward measure $\mathbb{P}_{T_j}$.

Proposition 13.6.1 *Assume the lognormal LIBOR model. The price at time 0 of a payer swaption with expiry date $T = T_0$ and strike level κ equals*

$$\mathbf{PS}_0 = \sum_{j=1}^n \delta_j B(0, T_j) \int_{\mathbb{R}^n} \left(L(0, T_{j-1}) e^{y_j - \lambda_j^2(0)/2} - \kappa \right) \mathbb{1}_{\tilde{D}}\, dG_j(y_1, \ldots, y_n),$$

where $\mathbb{1}_{\tilde{D}} = \mathbb{1}_{\tilde{D}}(y_1, \ldots, y_n)$, and $\tilde{D}$ stands for the set

$$\tilde{D} = \left\{ (y_1, \ldots, y_n) \in \mathbb{R}^n \,\Big|\, \sum_{j=1}^n c_j \prod_{k=1}^j \left(1 + \delta_k L(0, T_{k-1})\, e^{y_k - \lambda_k^2(0)/2} \right)^{-1} < 1 \right\}.$$

Proof. Let us start by considering arbitrary $t \in [0, T]$. Notice that

$$\frac{B(t, T_j)}{B(t, T)} = \prod_{k=1}^j \frac{B(t, T_k)}{B(t, T_{k-1})} = \prod_{k=1}^j (F_B(t, T_{k-1}, T_k))^{-1}$$

and thus, in view of (12.23), we have

$$B(T, T_j) = \prod_{k=1}^j \left(1 + \delta_k L(T, T_{k-1}) \right)^{-1}.$$

Consequently, the exercise set D can be re-expressed in terms of forward LIBORs. Indeed, we have

$$D = \left\{ \omega \in \Omega \,\Big|\, \sum_{j=1}^n c_j \prod_{k=1}^j \left(1 + \delta_k L(T, T_{k-1}) \right)^{-1} < 1 \right\},$$

or more explicitly

$$D = \left\{ \omega \in \Omega \,\Big|\, \sum_{j=1}^n c_j \prod_{k=1}^j \left(1 + \delta_k L(t, T_{k-1})\, e^{\zeta_k(t) - \lambda_k^2(t)/2} \right)^{-1} < 1 \right\}.$$

Let us put $t = 0$. In view of Lemma 13.6.1, to find the arbitrage price of a swaption at time 0, it is sufficient to determine the joint distribution under the forward measure $\mathbb{P}_{T_j}$ of the random variable $(\zeta_1(0), \ldots, \zeta_n(0))$, where $\zeta_1(0), \ldots, \zeta_n(0)$ are given by (13.38). Note also that

$$D = \left\{ \omega \in \Omega \;\middle|\; \sum_{j=1}^{n} c_j \prod_{k=1}^{j} \left(1 + \delta_k L(0, T_{k-1}) \, e^{\zeta_k(0) - \lambda_k^2(0)/2} \right)^{-1} < 1 \right\}.$$

This shows the validity of the valuation formula for $t = 0$. It is clear that it admits a rather straightforward generalization to arbitrary $0 < t \le T$. When $t > 0$, one needs to examine the conditional probability distribution of $(\zeta_1(t), \ldots, \zeta_n(t))$ with respect to the σ-field $\mathcal{F}_t$.   □

We shall now examine a closed-form approximation of the swaption price proposed in Brace et al. (1997). Since

$$\mathbf{PS}_t = \sum_{j=1}^{n} \delta_j B(t, T_j) \, \mathbb{E}_{\mathbb{P}_{T_j}} \left((L(T, T_{j-1}) - \kappa) \mathbb{1}_D \;\middle|\; \mathcal{F}_t \right), \tag{13.41}$$

we shall focus on a suitable approximation of the conditional expectation

$$E_j(t) = \mathbb{E}_{\mathbb{P}_{T_j}} \left((L(T, T_{j-1}) - \kappa) \mathbb{1}_D \;\middle|\; \mathcal{F}_t \right)$$

for a fixed, but otherwise arbitrary, j.

Note first that for any k we have

$$\zeta_k(t) = \int_t^T \lambda(u, T_{k-1}) \cdot dW_u^{T_j} + \int_t^T \lambda(u, T_{k-1}) \cdot d(W_u^{T_k} - W_u^{T_j}),$$

and by virtue of (13.40)

$$W_t^{T_k} - W_t^{T_j} = \int_0^t \sum_{i=k \wedge j}^{k \vee j - 1} \frac{\delta_{i+1} L(u, T_i)}{1 + \delta_{i+1} L(u, T_i)} \lambda(u, T_i) \, du, \quad \forall t \in [0, T_j \wedge T_k].$$

By combining these formulas, we get the following equality that holds for any $k = 1, \ldots, n$ and every $t \in [0, T]$

$$\zeta_k(t) = \int_t^T \lambda(u, T_{k-1}) \cdot dW_u^{T_j} + \int_t^T h_{k,j}(u) \lambda(u, T_{i-1}) \cdot \lambda(u, T_{k-1}) \, du,$$

where

$$h_{k,j}(u) = \sum_{i=k \wedge j + 1}^{k \vee j} \frac{\delta_i L(u, T_{i-1})}{1 + \delta_i L(u, T_{i-1})}.$$

For any fixed j and arbitrary $k = 1, \ldots, n$, we approximate the random variable $\zeta_k(t)$ by

$$\zeta_k^j(t) = \int_t^T \lambda(u, T_{k-1}) \cdot dW_u^{T_j} + \sum_{i=k\wedge j+1}^{k\vee j} \frac{\delta_i L(t, T_{i-1})}{1 + \delta_i L(t, T_{i-1})} \lambda_{i,k}(t),$$

where

$$\lambda_{i,k}(t) = \int_t^T \lambda(u, T_{i-1}) \cdot \lambda(u, T_{k-1}) \, du, \quad \forall\, t \in [0, T].$$

It is clear that for any fixed j the conditional probability distribution of the n-dimensional random variable $(\zeta_1^j(t), \dots, \zeta_n^j(t))$ under $\mathbb{P}_{T_j}$, given the σ-field $\mathcal{F}_t$, is Gaussian $N(\mu^j(t), \Lambda(t))$ with the matrix $\Lambda(t) = [\lambda_{i,k}(t)]_{1 \le i,k \le n}$ and the expected value $\mu^j(t) \in \mathbb{R}^n$ having the following components

$$\mu_k^j(t) = \sum_{i=k\wedge j+1}^{k\vee j} \frac{\delta_i L(t, T_{i-1})}{1 + \delta_i L(t, T_{i-1})} \lambda_{i,k}(t) \tag{13.42}$$

for $k = 1, \dots, n$.

It is now straightforward to find a quasi-explicit formula that gives an approximate value of a swaption. Unfortunately, this formula involves integration with respect to n-dimensional Gaussian density.

An alternative approach to the valuation of swaptions in the LLM model, first proposed by Rebonato (1999b) and later refined by Hull and White (2000), relies on the derivation of the approximate lognormal dynamics of forward swap rates in the LLM model, by the method of "freezing coefficients." For the sake of simplicity, we present briefly these methods in the context of valuation of caplets in a model of swap rates.

13.6.2 Caplets in the Co-terminal Swap Market Model

We place ourselves in the lognormal model of co-terminal forward swap rates. Recall that we write $\tilde{\kappa}(t, T_j) = \kappa(t, T_j, n - j)$. In order to derive the approximate dynamics of LIBOR, we start from the representation (see (13.36))

$$L(t, T_j) = w_t^{j,j} \tilde{\kappa}(t, T_j) + w_t^{j,j+1} \tilde{\kappa}(t, T_{j+1}). \tag{13.43}$$

Following the idea of Rebonato (1999b), we freeze the weights $w_t^{j,j}$, $w_t^{j,j}$ at time 0 and we differentiate this equation with respect to t. We obtain

$$dL(t, T_j) = w_0^{j,j} \, d\tilde{\kappa}(t, T_j) + w_0^{j,j+1} \, d\tilde{\kappa}(t, T_{j+1}).$$

Hence

$$dL(t, T_j) = w_0^{j,j} \tilde{\kappa}(t, T_j) v(t, T_j) \, d\tilde{W}^{T_{j+1}} + w_0^{j,j+1} \tilde{\kappa}(t, T_{j+1}) v(t, T_{j+1}) \, d\tilde{W}^{T_{j+2}}.$$

The last equality shows that under the forward measure $\mathbb{P}_{T_{j+1}}$ we have

$$dL(t, T_j) = L(t, T_j) \Big(\hat{w}_0^{j,j} v(t, T_j) + \hat{w}_0^{j,j+1} v(t, T_{j+1}) \Big) dW^{T_{j+1}},$$

where we set (see (13.35))

$$\hat{w}_0^{j,j} = \frac{g_0^j \tilde{\kappa}(0, T_j)}{L(0, T_j)(g_0^j - g_0^{j+1})}, \quad \hat{w}_0^{j,j+1} = -\frac{g_0^{j+1} \tilde{\kappa}(0, T_{j+1})}{L(0, T_j)(g_0^j - g_0^{j+1})}.$$

In this way, we obtain a rather crude lognormal approximation of forward LIBOR $L(t, T_j)$ in the lognormal model of co-terminal forward swap rates. It was implicitly assumed here that the variability of stochastic weights $w_t^{j,j}$ and $w_t^{j,j+1}$ is small with respect to the volatilities of forward swap rates.

A more precise approximation can be obtained by first differentiating (13.43) and subsequently freezing all stochastic terms in the volatility of $L(t, T_j)$. This method was proposed by Hull and White (2000). It ensues that

$$dL(t, T_j) = L(t, T_j) \sum_{l=j}^{n-1} \bar{w}_0^{j,l} v(t, T_l) dW^{T_{j+1}},$$

where $\bar{w}_0^{j,l} = \tilde{w}_0^{j,l} \tilde{\kappa}(0, T_l)/L(0, T_j)$ and (for the definition of $g_0^{j+1,l}$ see formula (13.21))

$$\tilde{w}_0^{j,l} = \begin{cases} \dfrac{g_0^j}{g_0^j - g_0^{j+1}}, & \text{if } l = j, \\[2ex] -\dfrac{\delta_{j+1} g_0^{j+1}\left(1 + g_0^{j+1}\tilde{\kappa}(0,T_j)\right)}{\left(g_0^j - g_0^{j+1}\right)^2}, & \text{if } l = j+1, \\[2ex] \dfrac{\delta_l \delta_j g_0^{j+1,l}\left(\tilde{\kappa}(0,T_j) - \tilde{\kappa}(0,T_{j+1})\right)}{\left(1 + \delta_l \tilde{\kappa}(0,T_l)\right)\left(g_0^j - g_0^{j+1}\right)^2}, & \text{if } j+2 \le j \le n-1, \\[2ex] 0, & \text{otherwise.} \end{cases}$$

From the representation above, it is possible to find the volatility parameter to be plugged into Black's caplet formula.

13.7 Markov-functional Models

As was shown in Sect. 12.4.6, in a lognormal market model the forward LIBORs (or forward swap rates) follow multidimensional Markov process under any of the associated forward (or swap) measures. In principle, lognormal market models can be calibrated in a natural way to market prices of caps (or swaptions) and this is a nice feature of this class of term structure models, as opposed to classical models based on the specification of the dynamics of (spot or forward) instantaneous rates. However, due to the high dimensionality of the underlying Markov process, the implementation of market models is not as straightforward as is might appear at the first glance.

An alternative approach was recently developed in a series of papers by Hunt and Kennedy (1997, 1998) and Hunt et al. (1996, 2000) (for an independent research, see Balland and Hughston (2000)). It hinges on a specification of a low-dimensional Markov process that by assumption governs, through a simple functional dependence, the dynamics of all other relevant processes. For this reason, term structure models of this type are referred to as *Markov-functional interest rate models*. In the economic interpretation, the underlying Markov process is assumed to represent the state of the economy. This can be seen as a drawback of this approach with respect to market models, since the natural interpretation of risk factors (i.e., volatilities and correlations) is lost. In fact, Markov-functional constructions are based on marginal distributions of market models, rather than on their volatilities.

Formally, one starts by introducing a one- or multidimensional process M possessing the Markov property under the *terminal measure,* where the generic term *terminal measure* is intended to cover not only cases considered in previous sections, but also other suitable choices of the numeraire portfolio. As was mentioned already, the relevant processes such as, in particular, the value process of the numeraire portfolio and zero-coupon bond prices, are assumed to be functions of M. For instance, if $T^* > 0$ is the horizon date then for any $t \leq T$ we have

$$\frac{B(t, T, M_t)}{V_t(M_t)} = \mathbb{E}_{\hat{\mathbb{P}}} \left(\frac{B(s, T, M_s)}{V_s(M_s)} \,\middle|\, \mathcal{F}_t \right)$$

where $V_t(M_t)$, $t \leq T^*$, is the value process of the numeraire portfolio, and $\hat{\mathbb{P}}$ is the associated martingale measure. The notation $B(t, T, M_t)$ emphasizes the dependence of the bond price on time variables, t and T, and on the state variable represented by the random variable M_t. Note that the functional form $B(t, T, M_t)$ is not known explicitly, except for some very special choices of dates t and T. For instance, if may appear convenient to postulate that

$$\frac{B(t, T, M_t)}{V_t(M_t)} = A + C_T M_t$$

for some constants A and C_T, and to derive further properties of a model from the martingale feature of relative prices.

We shall now present a particular example of such an approach, in which we focus on the derivation of a simple expression for the so-called *convexity correction*. Subsequently, we shall discuss the problem of calibration of the Markov-functional model.

13.7.1 Terminal Swap Rate Model

The terminal swap rate model – put forward by Hunt et al. (1996) – was primarily designed for the purpose of the comparative pricing of exotic swaps with respect to plain vanilla swaps. Let us consider a given collection of reset/settlements dates $T_0, \ldots, T_n$. As usual, we shall write $T = T_0$. We shall examine the family of bonds $B(T, S_i)$, where the maturity date $S_i \geq T$, $i = 1, \ldots, m$ belongs to some set $\mathcal{S} = \{S_1, \ldots, S_m\}$ of dates.

In what follows, we shall focus on the so-called linear model. Thus, we postulate that there exist constants A and C_{S_i} such that for any $S_i \in \mathcal{S}$

$$D(T, S_i) \overset{\text{def}}{=} B(T, S_i)G_T^{-1}(n) = A + C_{S_i}\kappa(T, T, n). \tag{13.44}$$

Recall that for every $t \leq T$ we have $G_t(n) = \sum_{j=1}^{n} \delta_j B(t, T_j)$ and (cf. (13.12))

$$\kappa(t, T, n) = \frac{B(t, T) - B(t, T_n)}{\delta_1 B(t, T_1) + \cdots + \delta_n B(t, T_n)} = \frac{B(t, T) - B(t, T_n)}{G_t(n)}.$$

Using the martingale property of the discounted bond price $D(t, S_i)$ and the forward swap rate $\kappa(t, T, n)$ under the corresponding forward swap measure associated with the choice of the process $G_t(n)$ as a numeraire, we get

$$D(t, S_i) = A + C_{S_i} \kappa(t, T, n),$$

or equivalently,

$$B(t, S_i) = A G_t(n) + C_{S_i}(1 - B(t, T_n))$$

for every $t \in [0, T]$. We thus see that condition (13.44) is rather stringent; it implies that the price of any bond of maturity S_i from S can by represented as a linear combination of values of two particular portfolios of bonds, with one coefficient independent of maturity date S_i. The problem whether such an assumption can be supported by an arbitrage-free model of the term structure is not addressed in Hunt et al. (1996).

Let us now focus on the derivation of values of constants A and C_{S_i}. To this end, we assume that equality (13.44) is satisfied for any maturity T_j, $j = 1, \ldots, n$. In other words, we now take $S = \{T_1, \ldots, T_n\}$. Then

$$A \sum_{j=1}^{n} \delta_j + \sum_{j=1}^{n} \delta_j C_{T_j} \kappa(T, T, n) = A(T_n - T_0) + \sum_{j=1}^{n} \delta_j C_{T_j} \kappa(T, T, n) = 1,$$

and thus

$$A = (T_n - T_0)^{-1}, \quad \sum_{j=1}^{n} \delta_j C_{T_j} = 0. \tag{13.45}$$

Consequently, using the first equality above and the martingale property of processes $D(T, T_j)$ and $\kappa(t, T, n)$, we obtain

$$B(0, T_j) G_0^{-1}(n) = (T_n - T_0)^{-1} + C_{T_j} \kappa(0, T, n), \tag{13.46}$$

so that for each maturity in question the constant C_{T_j} is also uniquely determined. Note that the second equality in (13.45) is also satisfied for this choice of C_{T_j} for $j = 1, \ldots, n$.

Hunt and Kennedy (1997) argue that under (13.44) the problem of pricing irregular cash flows becomes relatively easy to handle. To illustrate this point, let us assume that we wish to value a contingent claim X that settles at time T and has the following representation

$$X = \sum_{i=1}^{m} c_i B(T, S_i) F,$$

where c_is are constants, and $S_i \in \mathcal{S}$ for $i = 1, \ldots, m$. We assume that an $\mathcal{F}_T$-measurable random variable F has the form $F = \tilde{F}\big(B(T, S_1), \ldots, B(T, S_m)\big)$ for some function $\tilde{F} : \mathbb{R}_+^m \to \mathbb{R}$. To be conform with the notation introduced in Sect. 9.6.4, we denote

$$V_t^1 = B(t, T) - B(t, T_n), \quad V_t^2 = \sum_{j=1}^n \delta_j B(t, T_j) = G_t(n).$$

Using (13.44) and (13.45)–(13.46), we obtain

$$X = \sum_{i=1}^m c_i \big(A(1 - B(T, T_n)) + C_{S_i} G_T(n)\big) F = w_1 V_T^1 F + w_2 V_T^2 F,$$

where

$$w_1 = \sum_{i=1}^m c_i A, \quad w_2 = \sum_{i=1}^m c_i C_{S_i}.$$

In view of the discussion in Sect. 9.6.4, it is clear that

$$\pi_t(X) = w_1 V_t^1 \, \mathbb{E}_{\mathbb{P}^1}(F \mid \mathcal{F}_t) + w_2 V_t^2 \, \mathbb{E}_{\mathbb{P}^2}(F \mid \mathcal{F}_t). \tag{13.47}$$

Under the assumption that the forward rate $\kappa(t, T, n)$ follows a geometric Brownian motion under the forward swap measure $\mathbb{P}^2$, it follows also a lognormally distributed process under $\mathbb{P}^1$ (see the discussion in Sect. 9.6.4). Consequently, under (13.44), the joint (conditional) probability distribution of random variables $B(T, S_1), \ldots, B(T, S_m)$ under probability measures $\mathbb{P}^1$ and $\mathbb{P}^2$ are known explicitly. We conclude that the conditional expectations in (13.47) can, in principle, be evaluated.

Constant maturity swap. We assume that the market price at time 0 of the (plain-vanilla) fixed-for-floating swaption is known. As an example, we shall consider a fixed-for-floating constant maturity swap.

To value one leg of the floating side of a constant maturity swap, consider a cash flow proportional to $\kappa(T, T, n)$ that occurs at some date $U > T$ (notice that $U = T_j$ for some j). Ignoring the constant, such a payoff is equivalent to the claim $X = B(T, U)\kappa(T, T, n)$ that settles at time T. Using representation (13.47) of the price, we obtain

$$\pi_t(X) = C_U V_t^1 \, \mathbb{E}_{\mathbb{P}^1}(\kappa(T, T, n) \mid \mathcal{F}_t) + A V_t^2 \, \mathbb{E}_{\mathbb{P}^2}(\kappa(T, T, n) \mid \mathcal{F}_t).$$

Consequently, at time 0 we have

$$\pi_0(X) = C_U(B(0, T) - B(0, T_n))\kappa(0, T, n)e^{\sigma^2 T} + A G_0(n)\kappa(0, T, n),$$

where the constant σ represents the implied volatility of the traded swaption, with maturity date T. Using the formula for the constant C_U, we obtain

$$\pi_0(X) = \big(B(0, U) - A G_0(n)\big)\kappa(0, T, n)e^{\sigma^2 T} + A G_0(n)\kappa(0, T, n),$$

so finally

$$\pi_0(X) = B(0, U)\kappa(0, T, n)\big(1 + (1 - w)e^{\sigma^2 T}\big), \tag{13.48}$$

where we write $w = AG_0(n)B^{-1}(0, U)$. It should be stressed that the simple valuation result (13.48) hinges on the strong assumption (13.44).

13.7.2 Calibration of Markov-functional Models

The most important feature of a generic Markov-functional model is the fact that its calibration to market prices of plain-vanilla derivatives is relatively easy to achieve. For convenience, we shall focus here on the calibration of the Markov-functional model of fixed-maturity forward swap rates. The case of forward LIBORs can be dealt with analogously. A more extensive discussion of issues examined in this section can be found in Hunt et al. (2000).

To start the calibration procedure, we postulate that the forward swap rate for the date T_{n-1} follows a lognormal martingale under the corresponding forward measure $\mathbb{P}_{T_n}$. More specifically, we assume that the process $\tilde{\kappa}(t, T_{n-1}) = \kappa(t, T_{n-1}, 1)$ obeys the equation

$$d\tilde{\kappa}(t, T_{n-1}) = \tilde{\kappa}(t, T_{n-1})v(t, T_{n-1}) \, dW_t, \tag{13.49}$$

where W is a Brownian motion under $\mathbb{P}_{T_n}$ and $v(t, T_{n-1})$ is a strictly positive deterministic function. Let us take the process

$$M_t = \int_0^t v(u, T_{n-1}) \, dW_u$$

as the driving Markov process for our model, and let us denote

$$v_M^2(0, T_{n-1}) = \int_0^{T_{n-1}} v^2(u, T_{n-1}) \, du.$$

Then clearly

$$\tilde{\kappa}(T_{n-1}, T_{n-1}) = \tilde{\kappa}(0, T_{n-1}) \, e^{M_{T_{n-1}} - \frac{1}{2}v_M^2(0, T_{n-1})} = h_{n-1}(M_{T_{n-1}}), \tag{13.50}$$

where the function $h_{n-1} : \mathbb{R} \to \mathbb{R}_+$ is known explicitly.

Consequently, the bond price can be represented as follows

$$B(T_{n-1}, T_n, M_{T_{n-1}}) = \Big(1 + \delta_n \tilde{\kappa}(0, T_{n-1}) \, e^{M_{T_{n-1}} - \frac{1}{2}v_M^2(0, T_{n-1})}\Big)^{-1}. \tag{13.51}$$

This also means that

$$B(T_{n-1}, T_n, M_{T_{n-1}}) = \Big(1 + g_{n-1}(M_{T_{n-1}})h_{n-1}(M_{T_{n-1}})\Big)^{-1}, \tag{13.52}$$

where in turn $g_{n-1}(M_{T_{n-1}}) = \delta_n$, so that the function $g_{n-1} : \mathbb{R} \to \mathbb{R}_+$ is in fact constant.

Suppose that we observe market prices of digital swaptions for all strikes $\kappa > 0$ and expiration dates $T_0, \ldots, T_{n-1}$. Our goal is to find the joint probability distribution under $\mathbb{P}_{T_n}$ of the $(n-1)$-dimensional random variable $(\tilde{\kappa}(T_0, T_0), \ldots,$ $\tilde{\kappa}(T_{n-1}, T_{n-1}))$. This goal can be achieved by deriving the functional dependence of each rate $\tilde{\kappa}(T_j, T_j)$ on the underlying Markov process. To be more specific, we shall search for the function $h_j : \mathbb{R}_+ \to \mathbb{R}_+$ such that $\tilde{\kappa}(T_j, T_j) = h_j(M_{T_j})$. To this end, we assume that for any $j = 0, \ldots, n-1$ there exists a strictly increasing function h_j such that this property is valid (notice that in view of (13.50), this statement is obviously true for $j = n-1$).

By the definition of the probability measure $\mathbb{P}_{T_n}$, we have for $i = j+1, \ldots, n$

$$\frac{B(T_j, T_i)}{B(T_j, T_n)} = \mathbb{E}_{\mathbb{P}_{T_n}}\left(\frac{B(T_i, T_i)}{B(T_i, T_n)}\,\Big|\,\mathcal{F}_{T_j}\right) = \mathbb{E}_{\mathbb{P}_{T_n}}\left(\frac{B(T_i, T_i)}{B(T_i, T_n)}\,\Big|\,M_{T_j}\right)$$

since $\mathcal{F}_{T_j} = \mathcal{F}_{T_j}^W = \mathcal{F}_{T_j}^M$. Hence, if $B(T_i, T_n) = B(T_i, T_n, M_{T_i})$, we obtain

$$\frac{B(T_j, T_i)}{B(T_j, T_n)} = \mathbb{E}_{\mathbb{P}_{T_n}}\left(\frac{1}{B(T_i, T_n, M_{T_i})}\,\Big|\,M_{T_j}\right).$$

This means, of course, that the right-hand side of the last formula is a function of M_{T_j}. Consequently, for

$$G_{T_j}(n-j) = \sum_{i=j+1}^{n} \delta_i B(T_j, T_i)$$

we get for $j = 1, \ldots, n-1$

$$\frac{G_{T_j}(n-j)}{B(T_j, T_n)} = \sum_{i=j+1}^{n} \mathbb{E}_{\mathbb{P}_{T_n}}\left(\frac{\delta_i}{B(T_i, T_n, M_{T_i})}\,\Big|\,M_{T_j}\right) = g_j(M_{T_j}), \qquad (13.53)$$

where $g_j : \mathbb{R} \to \mathbb{R}$ is a measurable function with strictly positive values.

In particular, for $j = n-1$

$$\frac{G_{T_{n-1}}(1)}{B(T_{n-1}, T_n)} = \delta_n = g_{n-1}(M_{T_{n-1}}),$$

and for $j = n-2$

$$\frac{G_{T_{n-2}}(2)}{B(T_{n-2}, T_n)} = \mathbb{E}_{\mathbb{P}_{T_n}}\left(\frac{\delta_{n-1}}{B(T_{n-1}, T_n, M_{T_{n-1}})}\,\Big|\,M_{T_{n-2}}\right) + \delta_n$$

where $B(T_{n-1}, T_n, M_{T_{n-1}})$ is given by (13.52). In general, the right-hand side of (13.53) can be evaluated using the transition probability density function $p_M(t, m; u, x)$ of the Markov process M, provided that the functional form of

$B(T_i, T_n, M_{T_i})$ is known for every $i = j + 1, \ldots, n$. More explicitly

$$g_j(m) = \sum_{i=j+1}^{n} \int_{\mathbb{R}} \frac{\delta_i \, p_M(T_j, m; T_i, x)}{B(T_i, T_n, x)} \, dx. \qquad (13.54)$$

We work back iteratively from the last relevant date T_{n-1}. Recall that in the first step (that is, when $j = n - 2$) the functional form of $B(T_{n-1}, T_n, M_{T_{n-1}})$ is given by (13.52).

Assume now that the functional form of $B(T_i, T_n, M_{T_i})$ has already been found for $i = j + 1, \ldots, n - 1$. To determine $B(T_j, T_n, M_{T_j})$, it is enough to find the functional form of the swap rate $\tilde{\kappa}(T_j, T_j)$. Indeed, we have

$$\tilde{\kappa}(T_j, T_j) = \frac{1 - B(T_j, T_n)}{G_{T_j}(n - j)}$$

and thus

$$B^{-1}(T_j, T_n) = 1 + \tilde{\kappa}(T_j, T_j) \frac{G_{T_j}(n - j)}{B(T_j, T_n)} = 1 + h_j(M_{T_j}) g_j(M_{T_j}). \qquad (13.55)$$

Our next goal is to show how to find the function h_j, under the assumption that the functional forms of bond prices $B(T_i, T_n, M_{T_i})$ are known for every $i = j+1, \ldots, n$. To this end, we assume that we are given all market prices of *digital swaptions* with expiration date T_j and all strictly positive strikes κ. By definition, the jth digital swaption, with unit notional principal, pays the amount δ_i at times T_i, $i = j + 1, \ldots, n$ provided that $\tilde{\kappa}(T_j, T_j) > \kappa$. It is convenient to represent the price at time 0 of the jth digital swaption, with expiration date T_j and strike κ, in the following way

$$\mathbf{DS}_0^j(\kappa) = B(0, T_n) \, \mathbb{E}_{\mathbb{P}_{T_n}} \left(\frac{G_{T_j}(n - j)}{B(T_j, T_n)} \mathbb{1}_{\{\tilde{\kappa}(T_j, T_j) > \kappa\}} \right)$$

for $j = 0, \ldots, n - 2$. Under the present assumptions, we obtain

$$\mathbf{DS}_0^j(\kappa) = B(0, T_n) \, \mathbb{E}_{\mathbb{P}_{T_n}} \left(g_j(M_{T_j}) \mathbb{1}_{\{h_j(M_{T_j}) > \kappa\}} \right),$$

or equivalently,

$$\mathbf{DS}_0^j(\kappa) = B(0, T_n) \, \mathbb{E}_{\mathbb{P}_{T_n}} \left(g_j(M_{T_j}) \mathbb{1}_{\{M_{T_j} > h_j^{-1}(\kappa)\}} \right).$$

Finally, if we denote by $f_M(x) = p_M(0, 0; T_j, x)$ the (Gaussian) p.d.f. of M_{T_j} under $\mathbb{P}_{T_n}$ then

$$\mathbf{DS}_0^j(\kappa) = B(0, T_n) \int_{\mathbb{R}} g_j(x) \mathbb{1}_{\{x > \hat{h}_j(\kappa)\}} f_M(x) \, dx, \qquad (13.56)$$

where we write $\hat{h}_j = h_j^{-1}$. It is natural to assume that the function[3] $\mathbf{DS}_0^j : \mathbb{R}_+ \to \mathbb{R}_+$ is strictly decreasing as a function of the strike level κ, with

$$\mathbf{DS}_0^j(0) = \sum_{i=j+1}^{n} \delta_i B(0, T_i) = G_0(n - j)$$

and $\mathbf{DS}_0^j(+\infty) = 0$. Since

$$\mathbb{E}_{\mathbb{P}_{T_n}} \left(g_j(M_{T_j}) \right) = G_0(n - j) B^{-1}(0, T_n),$$

it can be deduced from (13.56) that $\hat{h}_j(0) = -\infty$. Furthermore, condition $\mathbf{DS}_0^j(+\infty) = 0$ implies that $\hat{h}_j(+\infty) = +\infty$. Finally, the function $\hat{h}_j$ implicitly defined through equality (13.56) is strictly increasing, so that it admits the inverse function h_j with the desired properties. To wit, for $h_j = \hat{h}_j^{-1}$ we have: $h_j : \mathbb{R} \to \mathbb{R}_+$ is strictly increasing, with $h_j(-\infty) = 0$ and $h_j(+\infty) = +\infty$. This shows that the procedure above leads to a reasonable specification of the functional form $\tilde{\kappa}(T_j, T_j) = h_j(M_{T_j})$.

Let us recapitulate the main steps of the calibration procedure. In the first step, we find the function h_{n-2} expressing $\tilde{\kappa}(T_{n-2}, T_{n-2})$ in terms of $M_{T_{n-2}}$ numerically. To this end, we need first to evaluate the function g_{n-2} using formula (13.54) with $B(T_n, T_n, x) = 1$ and $B(T_{n-1}, T_n, x)$ given by (13.52)

In the second step, we first determine $B(T_{n-2}, T_n, x)$ using relationship (13.55), that is,

$$B^{-1}(T_{n-2}, T_n, x) = 1 + h_{n-2}(x) g_{n-2}(x).$$

Next, we find g_{n-3} using (13.54), and subsequently we determine the rate $\tilde{\kappa}(T_{n-3}, T_{n-3})$, or rather the corresponding function h_{n-3}.

Continuing this procedure, we end up with the following representation of the finite family of swap rates

$$\left(\tilde{\kappa}(T_0, T_0), \dots, \tilde{\kappa}(T_{n-1}, T_{n-1}) \right) = \left(h_0(M_{T_0}), \dots, h_{n-1}(M_{T_{n-1}}) \right).$$

This representation specifies the probability distribution of the considered family of swap rates uniquely under the terminal measure $\mathbb{P}_{T_n}$.

13.8 Flesaker and Hughston Approach

We shall now analyze an approach proposed by Flesaker and Hughston (1996a, 1996b) (see also Goldberg (1998), Rogers (1997b), Rutkowski (1997), Brody and Hughston (2003, 2004), and Hughston and Rafailidis (2005)).

---

[3] The function $\mathbf{DS}_0^j$ represents the observed market prices of digital swaptions, and thus the foregoing assumptions concerning the behavior of this function are natural.

From the theoretical viewpoint, their approach is rather close to the classical methodology that hinges on the specification of the short-term interest rate. An interesting feature of this method is the fact that it leads to closed-form solutions for the prices of caps and swaptions of all maturities.

Let A_t, $t \in [0, T^*]$, be a strictly positive supermartingale, defined on a filtered probability space $(\Omega, \mathbb{F}, \mathbb{P})$. For any maturity $T \in [0, T^*]$, the bond price $B(t, T)$ is specified by means of the following *pricing formula*

$$B(t, T) \stackrel{\text{def}}{=} A_t^{-1} \mathbb{E}_{\mathbb{P}}(A_T \mid \mathcal{F}_t), \quad \forall t \in [0, T]. \tag{13.57}$$

For simplicity, the σ-field $\mathcal{F}_0$ is assumed to be trivial, so that A_0 is a strictly positive constant; there is no loss of generality in setting $A_0 = 1$. The following statements are trivial: (a) $B(T, T) = 1$ for any maturity T; (b) $B(t, U) \leq B(t, T)$ for every $t \in [0, T]$, and all maturities U, T such that $U \geq T$. Note that in order to match the initial term structure, we need to specify A in such a way that $B(0, T) = \mathbb{E}_{\mathbb{P}}(A_T)$ for every $t \in [0, T^*]$. Putting $D_t = A_t^{-1}$, $t \in [0, T^*]$, we obtain

$$B(t, T) = D_t \mathbb{E}_{\mathbb{P}}(D_T^{-1} \mid \mathcal{F}_t), \quad \forall t \in [0, T]. \tag{13.58}$$

Note that, under mild technical assumptions, D is a strictly positive submartingale under $\mathbb{P}$. In particular, if D follows an increasing process then it can be interpreted as a savings account.

Let us show that we may use an equivalent change of probability measure in order to get the standard risk-neutral valuation formula

$$B(t, T) = B_t \mathbb{E}_{\mathbb{P}^*}(B_T^{-1} \mid \mathcal{F}_t), \quad \forall t \in [0, T], \tag{13.59}$$

where B is an increasing process, to be specified later (of course, B admits a natural interpretation as a savings account implied by the family of bond prices). More generally, we will show that

$$B_t \mathbb{E}_{\mathbb{P}^*}(X B_T^{-1} \mid \mathcal{F}_t) = A_t^{-1} \mathbb{E}_{\mathbb{P}}(X A_T \mid \mathcal{F}_t), \quad \forall t \in [0, T],$$

for any integrable contingent claim X that settles at time T. For any probability measure $\mathbb{P}^*$ on $(\Omega, \mathcal{F}_{T^*})$, equivalent to $\mathbb{P}$, we write

$$\frac{d\mathbb{P}^*}{d\mathbb{P}} = \eta_{T^*}, \quad \mathbb{P}\text{-a.s.}, \tag{13.60}$$

so that

$$\frac{d\mathbb{P}^*}{d\mathbb{P}}\bigg|_{\mathcal{F}_t} = \eta_t = \mathbb{E}_{\mathbb{P}}(\eta_{T^*} \mid \mathcal{F}_t), \quad \forall t \in [0, T^*].$$

We have the following auxiliary lemma. It should be stressed that the strictly positive supermartingale A, defined on a filtered probability space $(\Omega, \mathbb{F}, \mathbb{P})$, is fixed throughout.

Lemma 13.8.1 *Let $\mathbb{P}^*$ be an arbitrary probability measure equivalent to $\mathbb{P}$. Define the process B by setting $B_t = \eta_t A_t^{-1}$ for every $t \in [0, T^*]$, where η is the Radon-Nikodým derivative of $\mathbb{P}^*$ with respect to $\mathbb{P}$. Then*

$$B(t, T) = A_t^{-1} \mathbb{E}_{\mathbb{P}}(A_T \mid \mathcal{F}_t) = B_t \mathbb{E}_{\mathbb{P}^*}(B_T^{-1} \mid \mathcal{F}_t), \quad \forall\, t \in [0, T^*],$$

for any maturity $T \in [0, T^*]$.

Proof. We have

$$B_t \mathbb{E}_{\mathbb{P}^*}(B_T^{-1} \mid \mathcal{F}_t) = \frac{\mathbb{E}_{\mathbb{P}}(\eta_T B_T^{-1} \mid \mathcal{F}_t)}{\eta_t B_t^{-1}} = \frac{\mathbb{E}_{\mathbb{P}}(A_T \mid \mathcal{F}_t)}{A_t} = B(t, T),$$

where the first equality follows from Bayes's rule, the second is a consequence of the definition of B, and the last follows from (13.57). $\qquad\square$

We shall now focus on the absence of arbitrage between bonds with different maturities and cash. In the present context, we find it convenient to say that a model is *arbitrage-free* if there exists a probability measure $\mathbb{P}^*$ equivalent to $\mathbb{P}$ such that the corresponding process $B_t = \eta_t A_t^{-1}$ is of finite variation (and thus B is an increasing process). In this case, B can be identified as a savings account – that is, an additional tradable asset – and we are back in a standard set-up. It follows from (13.57) that all "discounted" processes

$$\tilde{B}(t, T) = A_t^{-1} B(t, T) = \mathbb{E}_{\mathbb{P}}(A_T \mid \mathcal{F}_t), \quad \forall\, t \in [0, T],$$

are martingales under $\mathbb{P}$. This does not imply immediately that the model is arbitrage-free, however, since the process A cannot in general be identified with the price process of a tradable asset.

Proposition 13.8.1 *Let A be a strictly positive supermartingale. Then there exists a unique strictly positive martingale η, with $\eta_0 = 1$, such that the process $B_t = \eta_t A_t^{-1}$ is an increasing process. The bond price model $B(t, T)$ is arbitrage-free, and the arbitrage price $\pi_t(X)$ of any attainable contingent claim X settling at time T equals*

$$\pi_t(X) = B_t \mathbb{E}_{\mathbb{P}^*}(X B_T^{-1} \mid \mathcal{F}_t) = A_t^{-1} \mathbb{E}_{\mathbb{P}}(X A_T \mid \mathcal{F}_t) \tag{13.61}$$

for every $t \in [0, T]$, where the probability measure $\mathbb{P}^*$ is given by formula (13.60).

Proof. The first statement follows directly from the multiplicative decomposition of a strictly positive supermartingale A. The first equality in (13.61) is standard; the second is an immediate consequence of Bayes's rule. $\qquad\square$

From Proposition 13.8.1, we see that the bond pricing model based on (13.57) is equivalent to the construction that assumes the existence of an increasing savings account B.

13.8.1 Rational Lognormal Model

Assume that the supermartingale A is given by the formula $A_t = f(t) + g(t)M_t$ for $t \in [0, T^*]$, where $f, g : [0, T^*] \to \mathbb{R}_+$ are strictly positive decreasing functions, and M is a strictly positive martingale, with $M_0 = 1$. This model is commonly known as the *rational lognormal model*. It follows from (13.61) that

$$B(t, T) = \frac{f(T) + g(T)M_t}{f(t) + g(t)M_t}, \quad \forall t \in [0, T], \tag{13.62}$$

for any maturity $T \in [0, T^*]$. To match the initial yield curve, it is sufficient to choose strictly positive decreasing functions f and g in such a way that

$$B(0, T) = \frac{f(T) + g(T)}{f(0) + g(0)}, \quad \forall T \in [0, T^*].$$

To get explicit valuation formulas, we need to further specify the model. We postulate, in addition, that M solves the SDE $dM_t = \sigma_t M_t \, dW_t$, with $M_0 = 1$, for some deterministic function $\sigma : [0, T^*] \to \mathbb{R}^d$, where W is a d-dimensional standard Brownian motion on $(\Omega, \mathbb{F}, \mathbb{P})$. Equivalently, we assume that

$$M_t = \mathcal{E}_t \left(\int_0^\cdot \sigma_u \cdot dW_u \right).$$

As we shall check now, such a specification of M leads to closed-form expressions for the prices of all caplets and swaptions. To this end, assume that $\mathbb{F} = \mathbb{F}^W$. Then for any probability measure $\mathbb{P}^*$ on $(\Omega, \mathcal{F}_{T^*})$ equivalent to the underlying probability measure $\mathbb{P}$, there exists an adapted process γ such that

$$\frac{d\mathbb{P}^*}{d\mathbb{P}} = \eta_{T^*} = \mathcal{E}_{T^*} \left(\int_0^\cdot \gamma_u \cdot dW_u \right), \quad \mathbb{P}\text{-a.s.}$$

Suppose that the functions f and g are differentiable. It is not hard to check, using Itô's formula and Girsanov's theorem, that the process γ, which equals

$$\gamma_t = \frac{\sigma_t g(t)M_t}{f(t) + g(t)M_t}, \quad \forall t \in [0, T^*],$$

gives the unique right choice of $\mathbb{P}^*$ – that is, the probability measure $\mathbb{P}^*$ for which the associated process $B_t = \eta_t A_t^{-1}$ is increasing. Under $\mathbb{P}^*$, we have $dB_t = r_t B_t \, dt$, where r equals

$$r_t = -\frac{f'(t) + g'(t)M_t}{f(t) + g(t)M_t} > 0, \quad \forall t \in [0, T^*]. \tag{13.63}$$

Also, it is possible to check by direct differentiation that $r_t = f(t, t)$, where

$$f(t, T) = -\frac{\partial \ln B(t, T)}{\partial T}$$

and $B(t, T)$ is given by (13.62).

13.8.2 Valuation of Caps and Swaptions

From now on, we work within the set-up of the lognormal rational model. Let us fix the expiry date T and the settlement date $T + \delta$, where $\delta > 0$ is a fixed number. A caplet pays $X = \delta(L(T, T) - \kappa)^+$ at the settlement date $T + \delta$, or equivalently,

$$Y = B(T, T + \delta)X = B(T, T + \delta)\delta(L(T, T) - \kappa)^+$$

at the expiry date T. Straightforward calculations show that

$$Y = \left(1 - (1 + \kappa\delta)B(T, T + \delta)\right)^+ = \left(1 - (1 + \kappa\delta)A_T^{-1} \mathbb{E}_{\mathbb{P}}(A_{T+\delta} \mid \mathcal{F}_T)\right)^+,$$

and thus the arbitrage price $\mathbf{Cpl}_t$ of such a caplet equals

$$\mathbf{Cpl}_t = A_t^{-1} \mathbb{E}_{\mathbb{P}}\left\{\left(A_T - (1 + \kappa\delta)\mathbb{E}_{\mathbb{P}}(A_{T+\delta} \mid \mathcal{F}_T)\right)^+ \mid \mathcal{F}_t\right\}. \qquad (13.64)$$

Consequently, the arbitrage price $\mathbf{FC}_t$ of a cap, at any date t before expiry $T = T_0$, is

$$\mathbf{FC}_t = A_t^{-1} \sum_{j=1}^{n} \mathbb{E}_{\mathbb{P}}\left\{\left(A_{T_{j-1}} - (1 + \kappa\delta_j)\mathbb{E}_{\mathbb{P}}(A_{T_j} \mid \mathcal{F}_{T_{j-1}})\right)^+ \mid \mathcal{F}_t\right\},$$

where, as usual, $T_j = T_{j-1} + \delta_j$ for $j = 1, \ldots, n - 1$. Since M is a martingale, for every $u \leq T^* - \delta$ we have

$$\mathbb{E}_{\mathbb{P}}(A_{u+\delta} \mid \mathcal{F}_u) = f(u + \delta) + g(u + \delta)M_u, \qquad (13.65)$$

for any $\delta \geq 0$, and thus

$$\mathbf{Cpl}_t = \frac{1}{f(t) + g(t)M_t} \mathbb{E}_{\mathbb{P}}\left\{\left(f(T) - \tilde{\delta}f(T + \delta) - (\tilde{\delta}g(T + \delta) - g(T))M_T\right)^+ \mid \mathcal{F}_t\right\},$$

where $\tilde{\delta} = 1 + \kappa\delta$, or equivalently,

$$\mathbf{Cpl}_t = (f(t) + g(t)M_t)^{-1} \mathbb{E}_{\mathbb{P}}\left((a_0 - b_0 M_T)^+ \mid \mathcal{F}_t\right),$$

where

$$a_0 = f(T) - \tilde{\delta}f(T + \delta), \quad b_0 = \tilde{\delta}g(T + \delta) - g(T).$$

Consequently, the price of a cap admits the following representation

$$\mathbf{FC}_t = (f(t) + g(t)M_t)^{-1} \sum_{j=0}^{n-1} \mathbb{E}_{\mathbb{P}}\left((a_j - b_j M_{T_j})^+ \mid \mathcal{F}_t\right),$$

where the constants a_j and b_j satisfy, for $j = 0, \ldots, n - 1$,

$$a_j = f(T_j) - \tilde{\delta}_{j+1}f(T_{j+1}), \quad b_j = \tilde{\delta}_{j+1}g(T_{j+1}) - g(T_j).$$

Recall that a (payer) swaption is essentially equivalent to a contract that pays $X_j = \delta_j(\kappa(T, T, n) - \kappa)^+$ at each future date $T_1, \ldots, T_n$. By discounting those payments to the date T, we observe that a swaption corresponds to the European claim X settling at time $T = T_0$ and given as

$$X = \left(\kappa(T, T, n) - \kappa\right)^+ \sum_{j=1}^{n} \delta_j B(T, T_j) = \left(1 - B(T, T_n) - \kappa \sum_{j=1}^{n} \delta_j B(T, T_j)\right)^+.$$

In terms of the supermartingale A, we have

$$X = \left(1 - A_T^{-1} \mathbb{E}_{\mathbb{P}}(A_{T_n} \mid \mathcal{F}_T) - \kappa \sum_{j=1}^{n} \delta_j A_T^{-1} \mathbb{E}_{\mathbb{P}}(A_{T_j} \mid \mathcal{F}_T)\right)^+,$$

and thus the arbitrage price of a swaption equals

$$\mathbf{PS}_t = A_t^{-1} \mathbb{E}_{\mathbb{P}} \left\{ \left(A_T - \mathbb{E}_{\mathbb{P}}(A_{T_n} \mid \mathcal{F}_T) - \kappa \sum_{j=1}^{n} \delta_j \mathbb{E}_{\mathbb{P}}(A_{T_j} \mid \mathcal{F}_T) \right)^+ \Big| \mathcal{F}_t \right\}.$$

Using (13.65), we find that

$$\mathbf{PS}_t = (f(t) + g(t)M_t)^{-1} \mathbb{E}_{\mathbb{P}} \left\{ \left(h(T) - h(T_n) - \kappa \sum_{j=1}^{n} \delta_j h(T_j) \right)^+ \Big| \mathcal{F}_t \right\},$$

where we write $h(t) = f(t) + g(t)M_T$ for every $t \in [0, T^*]$. After rearranging, this yields

$$\mathbf{PS}_t = (f(t) + g(t)M_t)^{-1} \mathbb{E}_{\mathbb{P}}\left((a - bM_T)^+ \mid \mathcal{F}_t\right),$$

where the constants a and b satisfy

$$a = f(T) - f(T_n) - \kappa \sum_{j=1}^{n} \delta_j f(T_j), \quad b = g(T) - g(T_n) - \kappa \sum_{j=1}^{n} \delta_j g(T_j).$$

We shall evaluate the cap price $\mathbf{FC}_t$ and the swaption price $\mathbf{PS}_t$ in terms of the underlying martingale M. Note that from (13.62) we obtain

$$M_t = \frac{f(t)B(t, T) - f(T)}{g(T) - g(t)B(t, T)}, \quad \forall\, t \in [0, T].$$

Similarly, simple algebra reveals that

$$M_t = \frac{(1 + \delta_1 L(t, T))f(T_1) - f(T)}{g(T) - (1 + \delta_1 L(t, T))g(T_1)}, \quad \forall\, t \in [0, T_1].$$

We thus conclude that in order to express cap and swaption prices in terms of the bond price $B(t, T)$ (or in terms of the forward LIBOR $L(t, T)$), it is enough to find closed-form solutions for these prices in terms of the driving martingale M.

It is worth noting that for small values of κ (all other variables being fixed), we have $a_0 > 0$ and $b_0 < 0$; that is, a caplet is always exercised. We assume from now on that a_0 and b_0 are strictly positive (the case of negative values is left to the reader). We set

$$v^2(t, T) = \int_t^T |\sigma_u|^2 \, du.$$

Recall that we have assumed here the set-up of the rational lognormal model.

Proposition 13.8.2 *Assume that the coefficients a_j and b_j are strictly positive. Then the price of a caplet maturing at time T_j equals, for $t \in [0, T_j]$,*

$$\mathbf{Cpl}_t^j = (f(t) + g(t)M_t)^{-1} \left(a_j N(d_1(t, T_j)) - b_j M_t N(d_2(t, T_j)) \right),$$

where

$$d_{1,2}(t, T_j) = \frac{\ln c_t \pm \frac{1}{2} v^2(t, T_j)}{v(t, T_j)}$$

and

$$c_t = M_t^{-1} \frac{a_j}{b_j} = M_t^{-1} \frac{f(T_j) - (1 + \kappa \delta_{j+1}) f(T_{j+1})}{(1 + \kappa \delta_{j+1}) g(T_{j+1}) - g(T_j)}.$$

Proof. We have

$$\mathbb{E}_\mathbb{P} \left((a_j - b_j M_{T_j})^+ \mid \mathcal{F}_t \right) = \mathbb{E}_\mathbb{P} \left((a_j - b_j M_t \zeta)^+ \mid \mathcal{F}_t \right),$$

where the random variable ζ, which is given by the formula

$$\zeta = \exp \left(\int_t^{T_j} \sigma_u \cdot dW_u - \frac{1}{2} \int_t^{T_j} |\sigma_u|^2 du \right),$$

is independent of the σ-field $\mathcal{F}_t$. The asserted formula follows by standard calculations. □

In order to price a cap, it is sufficient to add the prices of the underlying caplets (it is essential to determine first the signs of a_js and b_js).

The standard proof of the next result is left to the reader.

Proposition 13.8.3 *Suppose that the coefficients a and b are strictly positive. Then the price $\mathbf{PS}_t$ of a payer swaption, with expiry date T and strike level κ, equals*

$$\mathbf{PS}_t = (f(t) + g(t)M_t)^{-1} \left(aN(\tilde{d}_1(t, T)) - bM_t N(\tilde{d}_2(t, T)) \right),$$

where

$$\tilde{d}_{1,2}(t, T) = \frac{\ln \tilde{c}_t \pm \frac{1}{2} v^2(t, T)}{v(t, T)}$$

and

$$\tilde{c}_t = M_t^{-1} \frac{a}{b} = M_t^{-1} \frac{f(T) - f(T_n) - \kappa \sum_{j=1}^n \delta_j f(T_j)}{g(T) - g(T_n) - \kappa \sum_{j=1}^n \delta_j g(T_j)}.$$

14

Cross-currency Derivatives

In this chapter, we deal with derivative securities related to at least two economies (a domestic market and a foreign market, say). Any such security will be referred to as a *cross-currency derivative*. In contrast to the model examined in Chap. 4, all interest rates and exchange rates are assumed to be random. It seems natural to expect that the fluctuations of interest rates and exchange rates will be highly correlated. This feature should be reflected in the valuation and hedging of foreign and cross-currency derivative securities in the domestic market. Feiger and Jacquillat (1979) (see also Grabbe (1983)) were probably the first to study, in a systematic way, the valuation of currency options within the framework of stochastic interest rates (they do not provide a closed-form solution for the price, however). More recently, Amin and Jarrow (1991) extended the HJM approach by incorporating foreign economies. Frachot (1995) examined a special case of the HJM model with stochastic volatilities, in which the bond price and the exchange rate are assumed to be deterministic functions of a single state variable.

The first section introduces the basic assumptions of the model along the same lines as in Amin and Jarrow (1991). In the next section, the model is further specified by postulating deterministic volatilities for all bond prices and exchange rates. We examine the arbitrage valuation of foreign market derivatives such as currency options, foreign equity options, cross-currency swaps and swaptions, and basket options (see Jamshidian (1988, 1994a), Turnbull (1994), Frey and Sommer (1996), Brace and Musiela (1997), Dempster and Hutton (1997), Mikkelsen (2002), and Schlögl (2002)).

Let us explain briefly the last three contracts. A *cross-currency swap* is an interest rate swap agreement in which at least one of the reference interest rates is taken from a foreign market; the payments of a cross-currency swap can be denominated in units of any foreign currency, or in domestic currency. As one might guess, a *cross-currency swaption* is an option contract written on the value of a cross-currency swap. Finally, by a *basket option* we mean here an option written on a basket (i.e., weighted average) of foreign interest rates. Typical examples of such contracts are *basket caps* and *basket floors*.

The final section is devoted to the valuation of foreign market interest rate derivatives in the framework of the lognormal model of forward LIBOR rates. It appears that closed-form expressions for the prices of such interest rate derivatives as quanto caps and cross-currency swaps are not easily available in this case, since the bond price volatilities follow stochastic processes.

14.1 Arbitrage-free Cross-currency Markets

To analyze cross-currency derivatives within the HJM framework, or in a general stochastic interest rate model, we need to expand our model so that it includes foreign assets and indices. Generally speaking, the superscript i indicates that a given process represents a quantity related to the ith *foreign market*. The *exchange rate* Q_t^i of currency i, which is denominated in domestic currency per unit of the currency i, establishes the direct link between the spot domestic market and the ith spot foreign market. As usual, we write $\mathbb{P}^*$ to denote the domestic martingale measure, and W^* stands for the d-dimensional standard Brownian motion under $\mathbb{P}^*$. Our aim is to construct an arbitrage-free model of foreign markets in a similar way to that of Chap. 4. In order to avoid rather standard Girsanov-type transformations, we prefer to start by postulating the "right" (that is, arbitrage-free) dynamics of all relevant processes. For instance, in order to prevent arbitrage between investments in domestic and foreign bonds, we assume that the dynamics of the ith exchange rate Q^i under the measure $\mathbb{P}^*$ are

$$dQ_t^i = Q_t^i\big((r_t - r_t^i)\,dt + v_t^i \cdot dW_t^*\big), \quad Q_0^i > 0, \tag{14.1}$$

where r_t and r_t^i stand for the spot interest rate in the domestic and the ith foreign market, respectively. The rationale behind expression (14.1) is similar to that which leads to formula (4.14) of Chap. 4 (see also formulas (14.8)–(14.9) below). In the cross-currency HJM approach, the interest rate risk is modelled by the domestic and foreign market instantaneous forward rates, denoted by $f(t, T)$ and $f^i(t, T)$ respectively. We postulate that for any maturity $T \le T^*$, the dynamics under $\mathbb{P}^*$ of the *foreign forward rate* $f^i(t, T)$ are

$$df^i(t, T) = \sigma_i(t, T) \cdot (\sigma_i^*(t, T) - v_t^i)\,dt + \sigma_i(t, T) \cdot dW_t^*, \tag{14.2}$$

where

$$\sigma_i^*(t, T) = \int_t^T \sigma_i(t, u)\,du, \quad \forall\, t \in [0, T].$$

We assume also that for every i we are given an *initial foreign term structure* $f^i(0, T)$, $T \in [0, T^*]$, and that the foreign spot rates r^i satisfy $r_t^i = f^i(t, t)$ for every $t \in [0, T^*]$. The price $B^i(t, T)$ of a T-maturity *foreign zero-coupon bond*, denominated in foreign currency, is

$$B^i(t, T) = \exp\Big(-\int_t^T f^i(t, u)\,du\Big), \quad \forall\, t \in [0, T].$$

Consequently, the dynamics of $B^i(t, T)$ under the domestic martingale measure $\mathbb{P}^*$ are

$$dB^i(t, T) = B^i(t, T)\left(\left(r_t^i + v_t^i \cdot \sigma_i^*(t, T)\right) dt - \sigma_i^*(t, T) \cdot dW_t^*\right), \tag{14.3}$$

with $B^i(T, T) = 1$. Denoting $b^i(t, T) = -\sigma_i^*(t, T)$, we thus have

$$dB^i(t, T) = B^i(t, T)\left(\left(r_t^i - v_t^i \cdot b^i(t, T)\right) dt + b^i(t, T) \cdot dW_t^*\right). \tag{14.4}$$

Similarly, we assume that the price of an arbitrary *foreign asset* Z^i that pays no dividend satisfies[1]

$$dZ_t^i = Z_t^i\left(\left(r_t^i - v_t^i \cdot \xi_t^i\right) dt + \xi_t^i \cdot dW_t^*\right), \quad Z_0^i > 0, \tag{14.5}$$

for some process ξ^i. For simplicity, the adapted volatility processes $\sigma_i(t, T)$, v^i and ξ^i, taking values in $\mathbb{R}^d$, are assumed to be bounded.

Remarks. Let us denote $dW_t^i = dW_t^* - v_t^i\, dt$. Then (14.2) and (14.3) become

$$df^i(t, T) = \sigma_i(t, T) \cdot \sigma_i^*(t, T)\, dt + \sigma_i(t, T) \cdot dW_t^i \tag{14.6}$$

and

$$dB^i(t, T) = B^i(t, T)\left(r_t^i\, dt - \sigma_i^*(t, T) \cdot dW_t^i\right) \tag{14.7}$$

respectively, where W^i follows a Brownian motion under the spot probability measure $\mathbb{P}^i$ of the ith market, and where the probability measure $\mathbb{P}^i$ is obtained from Girsanov's theorem (cf. (14.14)). It is instructive to compare (14.6)–(14.7) with formulas (11.14)–(11.15) of Corollary 11.1.1.

Let us verify that under (14.4)–(14.5), the combined market is arbitrage-free for both domestic and foreign-based investors. In particular, the processes $B^i(t, T)Q_t^i$ and $Z_t^i Q_t^i$ representing prices of foreign assets expressed in domestic currency satisfy

$$d\left(B^i(t, T)Q_t^i\right) = B^i(t, T)Q_t^i\left(r_t\, dt + \left(v_t^i + b^i(t, T)\right) \cdot dW_t^*\right) \tag{14.8}$$

and

$$d\left(Z_t^i Q_t^i\right) = Z_t^i Q_t^i\left(r_t\, dt + \left(v_t^i + \xi_t^i\right) \cdot dW_t^*\right). \tag{14.9}$$

Let B_t represent a domestic savings account. It follows immediately from (14.8)–(14.9) that the relative prices $B^i(t, T)Q_t^i/B_t$ and $Z_t^i Q_t^i/B_t$ of foreign assets, expressed in units of domestic currency, are local martingales under the domestic martingale measure $\mathbb{P}^*$. Because of this property, it is clear that by proceeding along the same lines as in Chap. 8, it is possible to construct an arbitrage-free model of

---

[1] Recall that the superscript i refers to the fact that Z_t^i is the price of a given asset at time t, expressed in units of the ith foreign currency.

the cross-currency market after making a judicious choice of the class of admissible trading strategies.

Remarks. The existence of short-term rates in all markets is not an essential condition if one wishes to construct an arbitrage-free model of a cross-currency market under uncertain interest rates. It is enough to postulate suitable dynamics for all zero-coupon bonds in all markets, as well as for the corresponding exchange rates. In that case, it is natural to make use of forward measures, rather than spot martingale measures. Assume, for instance, that $B(t, T^*)$ models the price of a domestic bond for the horizon date T^*, and $\mathbb{P}_{T^*}$ is the domestic forward measure for this date. For any fixed i, we need to specify the dynamics of the foreign bond price $B^i(t, T)$, expressed in units of the ith currency, and the exchange rate process Q^i. In such an approach, it is sufficient to assume that for every T, the process

$$\tilde{F}_B^i(t, T, T^*) = \frac{B^i(t, T)Q_t^i}{B(t, T^*)}, \quad \forall t \in [0, T],$$

is a local martingale under $\mathbb{P}_{T^*}$. One needs to impose the standard conditions that rule out arbitrage between foreign bonds, as seen from the perspective of a foreign-based investor (a similar remark applies to any foreign market asset).

14.1.1 Forward Price of a Foreign Asset

Let us start by analyzing the forward price of a foreign bond in the domestic market. It is not hard to check that for any maturities $T \leq U$, the dynamics of the forward price $F_{B^i}(t, U, T) = B^i(t, U)/B^i(t, T)$, under the domestic martingale measure $\mathbb{P}^*$, expressed in the ith foreign currency, satisfy

$$dF_{B^i}(t, U, T) = F_{B^i}(t, U, T)\gamma^i(t, U, T) \cdot \left(dW_t^* - \left(v_t^i + b^i(t, T)\right)dt\right), \quad (14.10)$$

where $\gamma^i(t, U, T) = b^i(t, U) - b^i(t, T)$. On the other hand, when expressed in units of the domestic currency, the forward price at time t for settlement at date T of the U-maturity zero-coupon bond of the ith foreign market equals[2]

$$\tilde{F}_B^i(t, U, T) = B^i(t, U)Q_t^i/B(t, T), \quad \forall t \in [0, T]. \quad (14.11)$$

Relationship (14.11) is in fact a universal property, meaning that it can be derived by simple no-arbitrage arguments, independently of the model of term structure. Notice that

$$\tilde{F}_B^i(t, U, T) = \frac{B^i(t, U)Q_t^i}{B(t, T)} \neq \frac{B^i(t, U)Q_t^i}{B^i(t, T)} = Q_t^i F_{B^i}(t, U, T),$$

in general. This means that the domestic forward price of a foreign bond does not necessarily coincide with the foreign market forward price of the bond, when the

---

[2] It should be made clear that we consider here a forward contract in which a U-maturity foreign bond is delivered at time T, in exchange for $\tilde{F}_B^i(t, U, T)$ units of the domestic currency.

latter is converted into domestic currency at the current exchange rate. It is useful to observe that in the special case when $T = U$, the forward price $\tilde{F}_B^i(t, T, T)$ satisfies

$$\tilde{F}_B^i(t, T, T) = B^i(t, T)Q_t^i/B(t, T) = F_{Q^i}(t, T), \quad \forall t \in [0, T], \quad (14.12)$$

i.e., it agrees with the *forward exchange rate* for the settlement date T (cf. formula (4.16) in Chap. 4). More generally, we have the following result, which is valid for any foreign market asset Z^i (recall that the price Z_t^i is expressed in units of the ith foreign currency).

Lemma 14.1.1 *The domestic forward price $\tilde{F}_{Z^i}(t, T)$ for the settlement at time T of the foreign market security Z^i (which pays no dividends) satisfies*

$$\tilde{F}_{Z^i}(t, T) = \frac{Z_t^i Q_t^i}{B(t, T)} = F_{Z^i}(t, T)F_{Q^i}(t, T). \quad (14.13)$$

Proof. The first equality follows by standard no-arbitrage arguments. For the second, notice that

$$\frac{Z_t^i Q_t^i}{B(t, T)} = \frac{Z_t^i}{B^i(t, T)} \frac{B^i(t, T)}{B(t, T)} Q_t^i = F_{Z^i}(t, T)F_{Q^i}(t, T),$$

where $F_{Z^i}(t, T)$ is the foreign forward price (in units of the ith currency). $\square$

For our further purposes, it is useful to examine the dynamics of the forward price of a foreign market asset. Let us start by analyzing the case of a foreign zero-coupon bond. It is easily seen that for any choice of maturities $T \leq U \leq T^*$, the dynamics of the forward price process $\tilde{F}_B^i(t, U, T)$ under the domestic martingale measure $\mathbb{P}^*$ are given by the expression

$$d\tilde{F}_B^i(t, U, T) = \tilde{F}_B^i(t, U, T)\Big(v_t^i + b^i(t, U) - b(t, T)\Big) \cdot \Big(dW_t^* - b(t, T)dt\Big),$$

or, in the standard HJM framework

$$d\tilde{F}_B^i(t, U, T) = \tilde{F}_B^i(t, U, T)\Big(v_t^i + \sigma^*(t, T) - \sigma_i^*(t, U)\Big) \cdot \Big(dW_t^* + \sigma^*(t, T)\,dt\Big),$$

since $b(t, T) = -\sigma^*(t, T)$. Similarly, the dynamics of the forward price $\tilde{F}_{Z^i}(t, T)$ under the domestic martingale measure $\mathbb{P}^*$ are

$$d\tilde{F}_{Z^i}(t, T) = \tilde{F}_{Z^i}(t, T)\Big(v_t^i + \xi_t^i - b(t, T)\Big) \cdot \Big(dW_t^* - b(t, T)dt\Big),$$

that is

$$d\tilde{F}_{Z^i}(t, T) = \tilde{F}_{Z^i}(t, T)\Big(v_t^i + \xi_t^i + \sigma^*(t, T)\Big) \cdot \Big(dW_t^* + \sigma^*(t, T)\,dt\Big).$$

Let $\mathbb{P}^i$ be the probability measure on $(\Omega, \mathcal{F}_{T^*})$ defined by the Doléans exponential

$$\frac{d\mathbb{P}^i}{d\mathbb{P}^*} = \mathcal{E}_{T^*}(U^i) = \mathcal{E}_{T^*}\Big(\int_0^\cdot v_u^i \cdot dW_u^*\Big), \quad \mathbb{P}^*\text{-a.s.} \quad (14.14)$$

By virtue of Girsanov's theorem, the process W^i, which is given by the formula

$$W_t^i = W_t^* - \int_0^t v_u^i \, du, \quad \forall t \in [0, T^*],$$

follows a Brownian motion under the probability measure $\mathbb{P}^i$. Since

$$df^i(t, T) = \sigma_i(t, T) \cdot \sigma_i^*(t, T) \, dt + \sigma_i(t, T) \cdot dW_t^i,$$

$$dB^i(t, T) = B^i(t, T)\left(r_t^i \, dt - \sigma_i^*(t, T) \cdot dW_t^i\right),$$

and

$$dZ_t^i = Z_t^i\left(r_t^i \, dt + \xi_t^i \cdot dW_t^i\right),$$

we conclude that the probability measure $\mathbb{P}^i$ is the (spot) martingale measure of the ith foreign market (cf. formulas (4.7)–(4.8) of Chap. 4). Let us now examine the corresponding forward probability measures. Recall that the forward measure $\mathbb{P}_T$ in the domestic market is given on $(\Omega, \mathcal{F}_T)$ by means of the following expression (cf. formula (9.32) in Sect. 9.6)

$$\frac{d\mathbb{P}_T}{d\mathbb{P}^*} = \mathcal{E}_T(U^T) = \mathcal{E}_T\left(\int_0^\cdot b(u, T) \cdot dW_u^*\right), \quad \mathbb{P}^*\text{-a.s.}$$

Moreover, under the domestic forward measure $\mathbb{P}_T$, the process W^T, which equals

$$W_t^T = W_t^* - \int_0^t b(u, T) \, du = W_t^* + \int_0^t \sigma^*(u, T) \, du,$$

is a d-dimensional standard Brownian motion. Analogously, the forward measure for the ith foreign market, denoted by $\mathbb{P}_T^i$, is defined on $(\Omega, \mathcal{F}_T)$ by the formula

$$\frac{d\mathbb{P}_T^i}{d\mathbb{P}^i} = \mathcal{E}_T(U^{T,i}) = \mathcal{E}_T\left(\int_0^\cdot b^i(u, T) \, dW_u^i\right), \quad \mathbb{P}^*\text{-a.s.} \tag{14.15}$$

The process $W^{T,i}$, which satisfies

$$W_t^{T,i} = W_t^i - \int_0^t b^i(u, T) \, du = W_t^* - \int_0^t \left(v_u^i + b^i(u, T)\right) du, \tag{14.16}$$

follows a d-dimensional standard Brownian motion under $\mathbb{P}_T^i$. Furthermore, the foreign market forward rate $f^i(\cdot, T)$ follows a local martingale under $\mathbb{P}_T^i$, more explicitly

$$df^i(t, T) = \sigma_i(t, T) \cdot dW_t^{T,i}.$$

The next result links the forward measure of a foreign market to the domestic spot martingale measure.

Lemma 14.1.2 *The Radon-Nikodým derivative on $(\Omega, \mathcal{F}_T)$ of the forward measure $\mathbb{P}_T^i$ of the ith foreign market with respect to the domestic spot martingale measure $\mathbb{P}^*$ equals*

$$\frac{d\mathbb{P}_T^i}{d\mathbb{P}^*} = \mathcal{E}_T(V^{T,i}), \quad \mathbb{P}^*\text{-a.s.}, \tag{14.17}$$

where

$$V_t^{T,i} = \int_0^t \left(v_u^i + b^i(u,T)\right) \cdot dW_u^*, \quad \forall t \in [0,T].$$

Proof. For any two continuous semimartingales X, Y defined on a probability space $(\Omega, \mathbb{F}, \mathbb{Q})$, with $X_0 = Y_0 = 0$, we have (see Theorem II.37 in Protter (2003))

$$\mathcal{E}_t(X)\,\mathcal{E}_t(Y) = \mathcal{E}_t(X + Y + \langle X, Y \rangle), \quad \forall t \in [0,T].$$

Applying this equality to the density

$$\frac{d\mathbb{P}_T^i}{d\mathbb{P}^*} = \frac{d\mathbb{P}_T^i}{d\mathbb{P}^i}\frac{d\mathbb{P}^i}{d\mathbb{P}^*} = \mathcal{E}_T(U^{T,i})\,\mathcal{E}_T(U^i),$$

we obtain

$$\frac{d\mathbb{P}_T^i}{d\mathbb{P}^*} = \mathcal{E}_T(U^{T,i} + U^i + \langle U^{T,i}, U^i \rangle). \tag{14.18}$$

Furthermore, by virtue of (14.14) and (14.15), we find that

$$\langle U^{T,i}, U^i \rangle_t = \int_0^t b^i(u,T) \cdot v_u^i \, du,$$

and thus for every $t \in [0,T]$

$$U_t^{T,i} + U_t^i + \langle U^{T,i}, U^i \rangle_t = \int_0^t b^i(u,T) \cdot (dW_u^* - v_u^i \, du)$$

$$+ \int_0^t v_u^i \cdot dW_u^* + \int_0^t b^i(u,T) \cdot v_u^i \, du = \int_0^t (v_u^i + b^i(u,T)) \cdot dW_u^* = V_t^{T,i}.$$

Combining the last equality with (14.18), we obtain (14.17). $\qquad\square$

The next auxiliary result, which gives the density of the foreign forward measure with respect to the domestic forward measure, can be proved along similar lines.

Lemma 14.1.3 *The following formula holds*

$$\frac{d\mathbb{P}_T^i}{d\mathbb{P}_T} = \mathcal{E}_T(Z^{T,i}), \quad \mathbb{P}_T\text{-a.s.},$$

where

$$Z_t^{T,i} = \int_0^t \left(v_u^i + b^i(u,T) - b(u,T)\right) \cdot dW_u^T, \quad \forall t \in [0,T].$$

14.1.2 Valuation of Foreign Contingent Claims

In this section, we deal with the valuation of general contingent claims denominated in foreign currency. Consider a time T *contingent claim* Y^i in the ith foreign market – that is, a contingent claim denominated in the currency of market i. We assume as usual that Y^i is a random variable, measurable with respect to the σ-field $\mathcal{F}_T$. Under appropriate integrability conditions, its arbitrage price at time t, expressed in domestic currency, is

$$\pi_t(Y^i) = B_t \, \mathbb{E}_{\mathbb{P}^*}\big(Y^i Q_T^i / B_T \mid \mathcal{F}_t\big) = B(t, T) \, \mathbb{E}_{\mathbb{P}_T}(Y^i Q_T^i \mid \mathcal{F}_t),$$

where the second equality is a consequence of Proposition 9.6.2. Indeed, a claim $X_T = Y^i Q_T^i$, which is denominated in units of domestic currency, can be priced as any "usual" domestic contingent claim. An alternative way of valuing Y^i is to first determine the price $\pi_t^i(Y^i)$ in units of foreign currency, which is

$$\pi_t^i(Y^i) = B_t^i \, \mathbb{E}_{\mathbb{P}^i}\big(Y^i / B_T^i \mid \mathcal{F}_t\big) = B^i(t, T) \, \mathbb{E}_{\mathbb{P}_T^i}(Y^i \mid \mathcal{F}_t), \tag{14.19}$$

and then to convert it into domestic currency, using the current exchange rate. This means that we have

$$\pi_t(Y^i) = Q_t^i \pi_t^i(Y^i) = Q_t^i B^i(t, T) \, \mathbb{E}_{\mathbb{P}_T^i}(Y^i \mid \mathcal{F}_t). \tag{14.20}$$

The former method for the valuation of foreign market contingent claims is frequently referred to as the *domestic market method,* while the latter is known as the *foreign market method.* Since the arbitrage price is uniquely defined, both methods must necessarily give the same price for any given foreign claim. A comparison of (14.19) and (14.20) yields immediately an interesting equality

$$B_t \, \mathbb{E}_{\mathbb{P}^*}\big(Y^i Q_T^i B_T^{-1} \mid \mathcal{F}_t\big) = Q_t^i B_t^i \, \mathbb{E}_{\mathbb{P}^i}\big(Y^i (B_T^i)^{-1} \mid \mathcal{F}_t\big), \tag{14.21}$$

which can alternatively be proved by standard arguments. To show more directly that (14.21) holds, observe that

$$\frac{d\mathbb{P}^i}{d\mathbb{P}^*} = \mathcal{E}_T(U^i) \overset{\text{def}}{=} \xi_T^i, \quad \mathbb{P}^*\text{-a.s.}$$

On the other hand, the exchange rate Q^i is easily seen to satisfy

$$Q_t^i = Q_0^i B_t (B_t^i)^{-1} \mathcal{E}_t(U^i), \quad \forall\, t \in [0, T^*],$$

so that Q_T^i and Q_t^i satisfy the following relationship

$$Q_T^i = Q_t^i B_T B_t^i \xi_T^i \big(B_t B_T^i \xi_t\big)^{-1}. \tag{14.22}$$

Consequently, from Bayes rule we get

$$\mathbb{E}_{\mathbb{P}^i}\big(Y^i Q_t^i B_t^i (B_T^i)^{-1} \mid \mathcal{F}_t\big) = \frac{\mathbb{E}_{\mathbb{P}^*}(Y^i Q_t^i B_t^i (B_T^i)^{-1} \xi_T^i \mid \mathcal{F}_t)}{\mathbb{E}_{\mathbb{P}^*}(\xi_T^i \mid \mathcal{F}_t)}.$$

Finally, taking (14.22) into account, we obtain

$$Q_t^i B_t^i \, \mathbb{E}_{\mathbb{P}^i} \left(Y^i / B_T^i \mid \mathcal{F}_t \right) = \mathbb{E}_{\mathbb{P}^*} \left(\frac{Y^i \, Q_T^i B_T^i \xi_T^i}{\xi_t^i \, B_T^i} \,\bigg|\, \mathcal{F}_t \right) = B_t \, \mathbb{E}_{\mathbb{P}^*} (Y^i \, Q_T^i / B_T \mid \mathcal{F}_t),$$

as expected.

14.1.3 Cross-currency Rates

In some instances it will be convenient to consider a *cross-currency rate,* which is simply the exchange rate between two foreign currencies. Consider two foreign markets, say l and m, and denote by $Q^{m/l}$ the corresponding cross-currency rate. More specifically, we assume that the exchange rate $Q^{m/l}$ is the price of one unit of currency l denominated in currency m. In terms of our previous notation, we have

$$Q_t^{m/l} \stackrel{\text{def}}{=} Q_t^l / Q_t^m, \quad \forall t \in [0, T^*],$$

hence by the Itô formula

$$dQ_t^{m/l} = Q_t^{m/l} \left(\left(r_t^m - r_t^l - v_t^m \cdot (v_t^l - v_t^m) \right) dt + (v_t^l - v_t^m) \cdot dW_t^* \right),$$

which, after rearranging, gives

$$dQ_t^{m/l} = Q_t^{m/l} \left(\left(r_t^m - r_t^l \right) dt + \left(v_t^l - v_t^m \right) \cdot dW_t^m \right),$$

where W^m is a Brownian motion of the mth foreign market under the spot martingale measure $\mathbb{P}^m$. Concluding, we can identify the volatility $v^{m/l}$ of the exchange rate $Q^{m/l}$ in terms of the volatilities v^l and v^m of the exchange rates Q^l and Q^m, respectively, as

$$v_t^{m/l} = v_t^l - v_t^m \tag{14.23}$$

for every $t \in [0, T^*]$.

14.2 Gaussian Model

We do not present here a systematic study of various option contracts based on foreign currencies, bonds and equities. We consider instead just a few typical examples of foreign market options (cf. Chap. 4). For simplicity, we assume throughout that the volatilities of all prices and exchange rates involved in a given contract follow deterministic functions. This assumption, which can be substantially weakened in some circumstances, leads to closed-form solutions for the prices of typical cross-currency options. Results obtained in this section are straightforward generalizations of option valuation formulas established in Chap. 4.

14.2.1 Currency Options

The first task is to examine the arbitrage valuation of European currency options in a stochastic interest rate framework. Recall that the forward exchange rate $F_{Q^i}(t, T)$ may be interpreted as the forward price for the settlement date T of one unit of foreign currency (i.e., of a foreign zero-coupon bond that matures at T). This implies the martingale property of F_Q^i under the domestic forward probability measure. More precisely, for any fixed T, we have under the domestic forward measure $\mathbb{P}_T$

$$dF_{Q^i}(t, T) = F_{Q^i}(t, T)\sigma_{Q^i}(t, T) \cdot dW_t^T \qquad (14.24)$$

for a deterministic function $\sigma_{Q^i}(\cdot, T) : [0, T] \to \mathbb{R}$. In view of (14.12), the volatility σ_{Q^i} can be expressed in terms of bond price volatilities and the volatility of the exchange rate. For any maturity $T \in [0, T^*]$, we have

$$\sigma_{Q^i}(t, T) = v_t^i + b^i(t, T) - b(t, T), \quad \forall t \in [0, T]. \qquad (14.25)$$

Our goal is to value a European currency call option with the payoff at expiry date T

$$C_T^{Q^i} \stackrel{\text{def}}{=} N(Q_T^i - K)^+ = N(F_{Q^i}(T, T) - K)^+,$$

where N is a preassigned number of units of foreign currency (we set $N = 1$ in what follows), K is the strike exchange rate, and T is the option expiry date. The arbitrage price of such an option under deterministic interest rates was found in Chap. 4 (see Proposition 4.2.2). Under the present assumption – that is, when $\sigma_{Q^i}(t, T)$ is deterministic – the closed-form expression for the price of a currency option can be established using the forward measure approach. Since $C_T^{Q^i}$ is expressed in domestic currency, it is enough to find the expected value of the option's payoff under the domestic forward probability measure $\mathbb{P}_T$ for the date T. Since this involves no difficulties, we prefer instead to apply a simple approach to the replication of currency options, based on the idea employed in Sect. 11.3.5. We claim that for every $t \in [0, T]$, we have

$$C_t^{Q^i} = B^d(t, T)\Big(F_t^i N\big(\tilde{d}_1(F_t^i, t, T)\big) - KN\big(\tilde{d}_2(F_t^i, t, T)\big)\Big), \qquad (14.26)$$

where $F_t^i = F_{Q^i}(t, T)$ is the forward exchange rate,

$$\tilde{d}_{1,2}(F, t, T) = \frac{\ln(F/K) \pm \frac{1}{2} v_{Q^i}^2(t, T)}{v_{Q^i}(t, T)},$$

and $v_{Q^i}(t, T)$ represents the volatility of the forward exchange rate integrated over the time interval $[t, T]$ – that is

$$v_{Q^i}^2(t, T) = \int_t^T |\sigma_{Q^i}(u, T)|^2 \, du = \int_t^T |v_u^i + b^i(u, T) - b(u, T)|^2 \, du.$$

Formula (14.26) can be rewritten as follows

$$C_t^{Q^i} = B^i(t, T)Q_t^i N\big(h_1(Q_t^i, t, T)\big) - KB(t, T) N\big(h_2(Q_t^i, t, T)\big), \qquad (14.27)$$

where

$$h_{1,2}(Q_t^i, t, T) = \frac{\ln(Q_t^i/K) + \ln(B^i(t, T)/B(t, T)) \pm \frac{1}{2} v_{Q^i}^2(t, T)}{v_{Q^i}(t, T)}.$$

To check, in an intuitive way, the validity of (14.26) for $t = 0$, let us consider the following combined spot-forward trading strategy: at time 0 we purchase $F_C(0, T) = C_0^{Q^i}/B^d(0, T)$ zero-coupon domestic bonds maturing at T; in addition, at any time $t \in [0, T]$, we are long $\psi_t^1 = N(\tilde{d}_1(F_t^i, t, T))$ forward currency contracts. The wealth of this portfolio at expiry equals

$$F_C(0, T) + \int_0^T \psi_t^1 \, dF_{Q^i}(t, T) = (Q_T^i - K)^+,$$

since direct calculations yield

$$N(\tilde{d}_1(F_t^i, t, T)) \, dF_{Q^i}(t, T) = dF_C(t, T),$$

where $F_C(t, T) = C_t^{Q^i}/B^d(t, T)$ is the forward price of the option. We conclude that formula (14.26) is valid for $t = 0$. A general formula can be established by similar arguments. For the valuation formula (14.27) to hold, it is sufficient to assume that the volatility of the forward exchange rate follows a deterministic function.

14.2.2 Foreign Equity Options

The following examples deal with various kinds of European options written on a foreign market asset.

Option on a foreign asset struck in foreign currency. Let Z^i stand for the price of a foreign asset (for instance, a bond or a stock). We consider a European call option with the payoff at expiry

$$C_T^1 \stackrel{\text{def}}{=} Q_T^i(Z_T^i - K^i)^+,$$

where K^i is the strike price, denominated in the ith foreign currency. To price this option, it is convenient to apply the foreign market method. It appears that it is sufficient to convert the foreign price of the option into domestic currency at the current exchange rate. Therefore, we get an intuitively obvious result (cf. Sect. 4.5 and Corollary 11.3.3 of Sect. 11.3)

$$C_t^1 = Q_t^i \Big(Z_t^i \, N\big(g_1(Z_t^i, t, T)\big) - K^i \, B^i(t, T) \, N\big(g_2(Z_t^i, t, T)\big) \Big), \tag{14.28}$$

where

$$g_{1,2}(Z_t^i, t, T) = \frac{\ln(Z_t^i/K^i) - \ln B^i(t, T) \pm \frac{1}{2} v_{Z^i}^2(t, T)}{v_{Z^i}(t, T)}.$$

and

$$v_{Z^i}^2(t, T) = \int_t^T |\xi_u^i - b^i(u, T)|^2 \, du.$$

Note that this result remains valid even if the volatility of the exchange rate is random, provided that the volatility function of the asset's foreign market forward price is deterministic.

Option on a foreign asset struck in domestic currency. Suppose now that the option on a foreign asset has its strike price expressed in domestic currency, so that the payoff from the option at expiry equals

$$C_T^2 \overset{\text{def}}{=} (Q_T^i Z_T^i - K)^+,$$

where K is expressed in units of the domestic currency. By applying the domestic market method to the synthetic domestic asset $\tilde{Z}_t^i = Q_t^i Z_t^i$, it is not hard to check that the arbitrage price of this option at time $t \in [0, T]$ is

$$C_t^2 = \tilde{Z}_t^i N(l_1(\tilde{Z}_t^i, t, T)) - K B(t, T) N(l_2(\tilde{Z}_t^i, t, T)), \qquad (14.29)$$

where

$$l_{1,2}(\tilde{Z}_t^i, t, T) = \frac{\ln(\tilde{Z}_t^i / K) - \ln B(t, T) \pm \frac{1}{2} \tilde{v}_{Z^i}^2(t, T)}{\tilde{v}_{Z^i}(t, T)}$$

and

$$\tilde{v}_{Z^i}^2(t, T) = \int_t^T |v_u^i + \xi_u^i - b(u, T)|^2 \, du.$$

For instance, if the underlying asset of the option is a foreign zero-coupon bond with maturity $U \geq T$, we obtain

$$C_t^2 = Q_t^i B^i(t, U) N(\hat{l}_1(B^i(t, U), t, T)) - K B(t, T) N(\hat{l}_2(B^i(t, U), t, T)),$$

where

$$\hat{l}_{1,2}(B^i(t, U), t, T) = \frac{\ln(Q_t^i / K) + \ln(B^i(t, U) / B(t, T)) \pm \frac{1}{2} \tilde{v}_U^2(t, T)}{\tilde{v}_U(t, T)}$$

and

$$\tilde{v}_U^2(t, T) = \int_t^T |v_u^i + b^i(u, U) - b(u, T)|^2 \, du.$$

It is not difficult to check that if we choose the maturity date U equal to the expiry date T, then the formula above agrees, as expected, with the currency option valuation formula (14.27). Also, it is clear that to establish equality (14.29), it is sufficient to assume that the volatility $\tilde{v}_{Z^i}(t, T)$ of the domestic forward price of the foreign asset Z^i follows a deterministic function.

Quanto option. As usual, let Z^i denote the price process of a certain foreign asset. The payoff at expiry of a quanto call equals (in domestic currency)

$$C_T^3 \stackrel{\text{def}}{=} \bar{Q}^i (Z_T^i - K^i)^+, \tag{14.30}$$

where $\bar{Q}^i$ is the prescribed exchange rate that is used eventually to convert the terminal payoff into domestic currency. Therefore, $\bar{Q}^i$ is specified in domestic currency per unit of the ith foreign currency. Moreover, the exercise price K^i is expressed in units of the ith foreign currency. Let $F_{Z^i}(t, T)$ be the forward price of the asset Z^i in the foreign market. Recall that we write $F_{Q^i}(t, T)$ to denote the forward exchange rate for the ith currency. Observe that the cross-variation of these processes satisfies

$$d\langle F_{Q^i}(\cdot, T), F_{Z^i}(\cdot, T)\rangle_t = F_{Q^i}(t, T) F_{Z^i}(t, T) \sigma_{Q^i, Z^i}(t, T)\, dt,$$

where $\sigma_{Q^i, Z^i}(\cdot, T)$ is a deterministic function. We find it convenient to denote

$$v_{Q^i, Z^i}(t, T) = \int_t^T \sigma_{Q^i, Z^i}(u, T)\, du, \quad \forall t \in [0, T].$$

Assume, in addition, that the volatility ξ^i of an underlying asset Z^i is also deterministic, and put

$$v_{Z^i}^2(t, T) = \int_t^T (\xi_u^i)^2\, du.$$

Then the arbitrage price of a quanto call option at time $t \in [0, T]$ equals

$$C_t^3 = \bar{Q}^i B(t, T)\Big(F_{Z^i}(t, T)\, e^{-v_{Q^i, Z^i}(t, T)}\, N\big(c_1(Z_t^i, t, T)\big) - K^i N\big(c_2(Z_t^i, t, T)\big)\Big),$$

where

$$c_{1,2}(z, t, T) = \frac{\ln(z/K^i) - \ln B^i(t, T) - v_{Q^i, Z^i}(t, T) \pm \frac{1}{2} v_{Z^i}^2(t, T)}{v_{Z^i}(t, T)}.$$

The reader may find it instructive to compare this result with the formula established in Proposition 4.5.1.

Equity-linked foreign exchange option. The payoff at expiry of an Elf-X option equals (see Sect. 4.5)

$$C_T^4 \stackrel{\text{def}}{=} (Q_T^i - K)^+ Z_T^i = (Q_T^i - K)^+ F_{Z^i}(T, T), \tag{14.31}$$

where K is a fixed level of the ith exchange rate, and $F_{Z^i}(t, T)$ is the foreign market forward price of a foreign asset Z^i. The dynamics of the price of the foreign asset Z^i and foreign bond $B^i(t, T)$ under the domestic martingale measure $\mathbb{P}^*$ are (see (14.4)–(14.5))

$$dZ_t^i = Z_t^i \Big((r_t^i - v_t^i \cdot \xi_t^i)\, dt + \xi_t^i \cdot dW_t^*\Big)$$

and

$$dB^i(t, T) = B^i(t, T)\Big((r_t^i - v_t^i \cdot b^i(t, T))\, dt + b^i(t, T) \cdot dW_t^*\Big)$$

respectively.

Using Itô's formula, we find the dynamics of the foreign market forward price $F_{Z^i}(t, T)$ under the domestic martingale measure $\mathbb{P}^*$, namely

$$dF_{Z^i}(t, T) = F_{Z^i}(t, T)\big(b^i(t, T) - \xi_t^i\big) \cdot \big(v_t^i + b^i(t, T)\big)\, dt$$
$$+ F_{Z^i}(t, T)\big(\xi_t^i - b^i(t, T)\big) \cdot dW_t^*.$$

Consequently, under the domestic forward measure $\mathbb{P}_T$, we have

$$dF_{Z^i}(t, T) = F_{Z^i}(t, T)\Big(\big(b^i(t, T) - \xi_t^i\big) \cdot \sigma_{Q^i}(t, T)\, dt + \big(\xi_t^i - b^i(t, T)\big) \cdot dW_t^T\Big),$$

where (cf. (14.25))

$$\sigma_{Q^i}(t, T) = v_t^i + b^i(t, T) - b(t, T), \quad \forall\, t \in [0, T].$$

For the sake of notational simplicity, we consider the case $t = 0$. Let us define an auxiliary probability measure $\mathbb{Q}_T$ by setting

$$\frac{d\mathbb{Q}_T}{d\mathbb{P}_T} = \mathcal{E}_T\Big(\int_0^{\cdot} \zeta_u \cdot dW_u^T\Big) = \eta_T, \quad \mathbb{P}_T\text{-a.s.},$$

where $\zeta_t = \xi_t^i - b^i(t, T)$ for $t \in [0, T]$. It is easily seen that Z_T^i equals

$$Z_T^i = F_{Z^i}(T, T) = F_{Z^i}(0, T)\,\eta_T\, e^{\theta(0, T)},$$

where we write

$$\theta(0, T) = \int_0^T (b^i(u, T) - \xi_u^i) \cdot \sigma_{Q^i}(t, T)\, du.$$

The price of the option at time 0 equals

$$C_0^4 = B(0, T)\mathbb{E}_{\mathbb{P}_T}\big((Q_T^i - K)^+ Z_T^i\big),$$

or equivalently,

$$C_0^4 = B(0, T)F_{Z^i}(0, T)e^{\theta(0, T)}\, \mathbb{E}_{\mathbb{Q}_T}\big(F_Q^i(T, T) - K\big)^+. \tag{14.32}$$

To evaluate the expectation in (14.32), we need to analyze the dynamics of the forward exchange rate $F_{Q^i}(t, T)$ under the auxiliary probability measure $\mathbb{Q}_T$. We know already that $F_{Q^i}(t, T)$ satisfies, under $\mathbb{P}_T$, the following SDE (cf. (14.24))

$$dF_{Q^i}(t, T) = F_{Q^i}(t, T)\sigma_{Q^i}(t, T) \cdot dW_t^T.$$

Therefore, under $\mathbb{Q}_T$ we have

$$dF_{Q^i}(t, T) = F_{Q^i}(t, T)\sigma_{Q^i}(t, T) \cdot \Big(\big(\xi_t^i - b^i(t, T)\big)\, dt + d\hat{W}_t\Big),$$

where the process $\hat{W}$, given by the formula

$$\hat{W}_t = W_t^T - \int_0^t \zeta_u \, du, \quad \forall t \in [0, T],$$

follows a Brownian motion under $\mathbb{Q}_T$. Consequently, the forward exchange rate $F_{Q^i}(t, T)$ can be represented as follows

$$F_{Q^i}(T, T) = F_{Q^i}(0, T) e^{-\theta(0,T)} \mathcal{E}_T \left(\int_0^{\cdot} \sigma_{Q^i}(u, T) \, d\hat{W}_u \right).$$

Putting the last equality into (14.32), we obtain

$$C_0^4 = B(0, T) F_{Z^i}(0, T) \, \mathbb{E}_{\mathbb{Q}_T} \left((F_{Q^i}(0, T) e^{\xi} - K e^{-\theta(0,T)})^+ \right), \tag{14.33}$$

where ξ is a Gaussian random variable, with zero mean and the variance under $\mathbb{Q}_T$

$$\mathrm{Var}_{\mathbb{Q}_T}(\xi) = v_{Q^i}^2(0, T) = \int_0^T |\sigma_{Q^i}(u, T)|^2 \, du.$$

Calculation of the expected value in (14.33) is standard. In general, we find that the price at time $t \in [0, T]$ of the Elf-X call option equals

$$C_t^4 = B(t, T) F_{Z^i}(t, T) \Big(F_t^i N\big(w_1(F_t^i, t, T)\big) - K e^{\theta(t,T)} N\big(w_2(F_t^i, t, T)\big) \Big),$$

where $F_t^i = F_{Q^i}(t, T)$,

$$w_{1,2}(F, t, T) = \frac{\ln(F/K) - \theta(t, T) \pm \frac{1}{2} v_{Q^i}^2(t, T)}{v_{Q^i}(t, T)},$$

and

$$\theta(t, T) = \int_t^T (b^i(u, T) - \xi_u^i) \cdot \sigma_{Q^i}(u, T) \, du, \quad \forall t \in [0, T].$$

After simple manipulations, we find that

$$C_t^4 = Z_t^i \Big(Q_t^i N\big(\tilde{w}_1(Q_t^i, t, T)\big) - K e^{\theta(t,T)} \frac{B(t, T)}{B^i(t, T)} N\big(\tilde{w}_1(Q_t^i, t, T)\big) \Big),$$

where

$$\tilde{w}_{1,2}(q, t, T) = \frac{\ln(q/K) + \ln(B^i(t, T)/B(t, T)) - \theta(t, T) \pm \frac{1}{2} v_{Q^i}^2(t, T)}{v_{Q^i}(t, T)}.$$

This ends the derivation of the option's pricing formula.

Remarks. Assume that the domestic and foreign interest rates r_t and r_t^i are deterministic for every $t \in [0, T]$; that is, $b(t, T) = b^i(t, T) = 0$. In this case, the value of C_t^4 given by the formula above agrees with the formula established in Proposition 4.5.2. Furthermore, if we take the foreign bond that pays one unit of the foreign currency at time T as the underlying foreign asset of the option, then θ vanishes identically, and we recover the currency option valuation formula (14.27).

14.2.3 Cross-currency Swaps

Cross-currency swaps are financial instruments that allow financial managers to capture existing and expected floating or money market rate spreads between alternative currencies without incurring foreign exchange exposure. Let us briefly describe a typical cross-currency swap. The party entering into such a swap will typically agree to receive payment in a particular currency on a specific principal amount, for a specific term, at the prevailing floating money market rate in that currency (such as, e.g., the LIBOR). In exchange, this party will make payments on the same principal amount, in the same currency, for the same term, based on the prevailing floating money market rate in another currency. Therefore, the major features of a typical cross-currency swap are that: (a) both payments and receipts (which are based on the same notional principal) are on a floating-rate basis, with the rate reset at specified intervals (usually quarterly or semi-annually); (b) all payments under the transaction are made in the preassigned currency, thereby eliminating foreign exchange exposure; and (c) consistent with the transaction's single-currency nature, no exchange of principal amounts is required. Our aim is to find valuation formulas for cross-currency swaps as well as for their derivatives, such as *cross-currency swaptions* – that is, options written on cross-currency swaps.

Formally, by a *cross-currency* (or *differential*) *swap* we mean an interest rate swap agreement in which at least one of the interest rates involved is related to a foreign market. In contrast to a classic fixed-for-floating (single-currency) swap agreement, in a typical cross-currency swap, both underlying interest rates are preassigned floating rates from two markets. To be more specific, a *floating-for-floating cross-currency* $(k, l; m)$ *swap* per unit of mth currency consists of swapping the floating rates of another two currencies. At each of the payment dates T_j, $j = 1, \ldots, n$, the floating rate $L^k(T_{j-1})$ of currency k is received and the corresponding floating rate $L^l(T_{j-1})$ of currency l is paid. Let us emphasize that in the most general form of a swap, the payments are made in units of still another foreign currency, say m. Similarly, by a *fixed-for-floating cross-currency* $(k; m)$ *swap* we mean a cross-currency swap with payments in the mth foreign currency, in which one of the underlying rates of interest is a prespecified fixed rate, while the other is a reference floating rate from currency k.

Floating-for-floating $(k, 0; 0)$ **swaps.** Let us first consider a floating-for-floating cross-currency $(k, 0; 0)$ swap between two parties in which, at each of the payment dates, the *buyer* pays the *seller* a U.S. dollar[3] amount equal to a fixed notional principal times the then level of a prespecified U.S. floating interest rate. The seller pays the buyer a U.S. dollar amount equal to the same principal times the then level of a prespecified foreign (e.g., Japanese, German, Australian) floating interest rate. If foreign interest rates are higher than U.S. interest rates, one may expect that the buyer should pay the seller a positive up-front fee, negotiated between the counterparts at the time the contract is entered into. Our goal is to determine – following, in partic-

---

[3] For ease of exposition, we assume hereafter that U.S. dollars play the role of the domestic currency.

ular, Jamshidian (1993b, 1994a) and Brace and Musiela (1997) – this up-front cost, called the value of the cross-currency swap. We will also examine a corresponding hedging portfolio. It is clear that at each of the payment dates T_j, $j = 1, \ldots, n$, the interest determined by the floating rate $L^k(T_{j-1})$ of the foreign currency k is received and the interest corresponding to the floating rate $L(T_{j-1})$ of the domestic currency is paid. In our framework, the rate levels $L^i(T_{j-1})$, $j = 1, \ldots, n$, are set by reference to the zero-coupon bond prices; namely, we have

$$B^i(T_{j-1}, T_j)^{-1} = 1 + L^i(T_{j-1})(T_j - T_{j-1}) = 1 + \delta_j L^i(T_{j-1}) \qquad (14.34)$$

for $i = 0, k$. The time t value, in units of the domestic currency, of a floating-for-floating $(k, 0; 0)$ cross-currency forward swap is

$$\mathbf{CCFS}_t(k, 0; 0) = \mathbb{E}_{\mathbb{P}^*}\Big\{ \sum_{j=1}^n \frac{B_t}{B_{T_j}} \Big(L^k(T_{j-1}) - L(T_{j-1}) \Big) \delta_j \,\Big|\, \mathcal{F}_t \Big\},$$

or equivalently

$$\mathbf{CCFS}_t(k, 0; 0) = \mathbb{E}_{\mathbb{P}^*}\Big\{ \sum_{j=1}^n \frac{B_t}{B_{T_j}} \Big(\frac{1}{B^k(T_{j-1}, T_j)} - \frac{1}{B(T_{j-1}, T_j)} \Big) \,\Big|\, \mathcal{F}_t \Big\}.$$

We define a (T, U) *roll bond* to be a dollar cash security that pays $1/B(T, U)$ dollars at its maturity U. Similarly, by a (T, U) *quanto roll bond* we mean a security that pays $1/B^k(T, U)$ dollars at time U. In view of the last equality, it is evident that a long position in a cross-currency swap is equivalent to being long a portfolio of (T_j, T_{j+1}) quanto roll bonds, and short a portfolio of (T_j, T_{j+1}) roll bonds. Therefore, we need to examine the following conditional expectation

$$\mathbb{E}_{\mathbb{P}^*}\Big\{ \frac{B_t}{B_{T_j}} \Big(\frac{1}{B^k(T_{j-1}, T_j)} - \frac{1}{B(T_{j-1}, T_j)} \Big) \,\Big|\, \mathcal{F}_t \Big\}$$

for any $t \le T_{j-1}$. One can easily check that

$$\mathbb{E}_{\mathbb{P}^*}\Big(\frac{B_t}{B_{T_j} B(T_{j-1}, T_j)} \,\Big|\, \mathcal{F}_t \Big) = B(t, T_{j-1}).$$

Indeed, to replicate the payoff of a (T_{j-1}, T_j) roll bond it is sufficient to buy at time $t \le T_{j-1}$ one bond with maturity T_{j-1}, and then reinvest the principal received at time T_{j-1} by purchasing $1/B(T_{j-1}, T_j)$ units of bonds with maturity T_j. The problem of the replication of a cross-currency swap thus reduces to replication of a quanto roll bond for $t \le T_{j-1}$, supplemented by a simple netting of positions at payment dates. Observe that for $t \ge T_{j-1}$, we have simply

$$\mathbb{E}_{\mathbb{P}^*}\Big(\frac{B_t}{B_{T_j} B^k(T_{j-1}, T_j)} \,\Big|\, \mathcal{F}_t \Big) = \frac{B(t, T_j)}{B^k(T_{j-1}, T_j)}.$$

In particular, for $t = T_{j-1}$ this yields

$$\mathbb{E}_{\mathbb{P}^*}\Big(\frac{B_{T_{j-1}}}{B_{T_j} B^k(T_{j-1}, T_j)} \,\Big|\, \mathcal{F}_{T_{j-1}} \Big) = \frac{B(T_{j-1}, T_j)}{B^k(T_{j-1}, T_j)}.$$

Therefore, our goal is now to find a replicating strategy for the contingent claim X that settles at time T_{j-1} and whose value is

$$X = \frac{B(T_{j-1}, T_j)}{B^k(T_{j-1}, T_j)}. \tag{14.35}$$

To simplify the notation, we denote $T = T_{j-1}$ and $U = T_j$. Let us consider a dynamic portfolio composed at any time $t \leq T$ of ϕ_t^1 units of U-maturity domestic bonds, ϕ_t^2 units of U-maturity foreign bonds, and finally ϕ_t^3 units of T-maturity foreign bonds. The wealth of such a portfolio at time $t \leq T$, expressed in domestic currency, equals

$$V_t(\phi) = \phi_t^1 B(t, U) + Q_t^k(\phi_t^2 B^k(t, U) + \phi_t^3 B^k(t, T)),$$

or in short

$$V_t(\phi) = \phi_t^1 B(t, U) + \phi_t^2 \tilde{B}^k(t, U) + \phi_t^3 \tilde{B}^k(t, T),$$

where for any maturity date T, we write

$$\tilde{B}^k(t, T) = Q_t^k B^k(t, T).$$

Note that $\tilde{B}^k(t, T)$ and $\tilde{B}^k(t, U)$ stand for the price at time t of the foreign market zero-coupon bond, expressed in units of domestic currency, with maturities T and U respectively. As usual, we say that a portfolio ϕ is self-financing when the relationship

$$dV_t(\phi) = \phi_t^1 \, dB(t, U) + \phi_t^2 \, d\tilde{B}^k(t, U) + \phi_t^3 \, d\tilde{B}^k(t, T) \tag{14.36}$$

is valid. To provide an intuitive argument supporting (14.36), observe that in the case of a discretely adjusted portfolio we have

$$\phi_{t_1}^1 B(t_2, U) + \phi_{t_1}^2 \tilde{B}^k(t_2, U) + \phi_{t_1}^3 \tilde{B}^k(t_2, T)$$
$$= \phi_{t_2}^1 B(t_2, U) + \phi_{t_2}^2 \tilde{B}^k(t_2, U) + \phi_{t_2}^3 \tilde{B}^k(t_2, T)$$

if a portfolio is held fixed over the interval $[t_1, t_2)$, and revised at time t_2. This shows that processes $\tilde{B}^k(t, T)$ and $\tilde{B}^k(t, U)$ can be formally seen as prices of domestic securities. Recall that

$$dB(t, T) = B(t, T) \left(r_t \, dt + b(t, T) \cdot dW_t^* \right) \tag{14.37}$$

and

$$dB^k(t, T) = B^k(t, T) \left(r_t^k \, dt + b^k(t, T) \cdot dW_t^* \right),$$

where W^* follows a Brownian motion under $\mathbb{P}^*$, and the exchange rate Q^k satisfies (see formula (14.1))

$$dQ_t^k = Q_t^k \left((r_t - r_t^k) \, dt + v_t^k \cdot dW_t^* \right).$$

Finally, recall that the forward exchange rate for the settlement date U is

$$F_{Q^k}(t, U) = \frac{B^k(t, U)}{B(t, U)} Q_t^k, \quad \forall t \in [0, U],$$

and the forward price of a T-maturity foreign market bond for settlement at time U equals

$$F_{B^k}(t, T, U) = \frac{B^k(t, T)}{B^k(t, U)}, \quad \forall t \in [0, U],$$

where $U \leq T$. Observe that we have the following expression for the forward price $F_{B^k}(t, T, U)$, under the domestic martingale measure $\mathbb{P}^*$,

$$dF_{B^k}(t, T, U) = F_{B^k}(t, T, U)\gamma^k(t, T, U) \cdot d\big(W_t^* - (v_t^k + b^k(t, U)) \, dt\big),$$

where $\gamma^k(t, T, U) = b^k(t, T) - b^k(t, U)$. We will show that to replicate a short position in a cross-currency swap, it is enough to hold a continuously rebalanced portfolio involving domestic and foreign zero-coupon bonds with maturities corresponding to the payment dates of the underlying swap. The net value of positions in foreign bonds is assumed to be zero – that is, the instantaneous profits or losses from foreign market positions are immediately converted into domestic currency and invested in domestic bonds. We start with an auxiliary lemma.

Lemma 14.2.1 *Let* $V^k(T, U)$ *stand for the following process*

$$V_t^k(T, U) = \frac{B(t, U)B^k(t, T)G_t^k(T, U)}{B^k(t, U)}, \quad \forall t \in [0, T], \tag{14.38}$$

where

$$G_t^k(T, U) = \exp\left(\int_t^T g_u^k(T, U) \, du\right) \tag{14.39}$$

and

$$g_t^k(T, U) = \gamma^k(t, U, T) \cdot (v_t^k + b^k(t, U) - b(t, U)), \tag{14.40}$$

where $\gamma^k(t, U, T) = b^k(t, U) - b^k(t, T)$. *Suppose that the process* $G^k(T, U)$ *is adapted. Then the Itô differential of* $V^k(T, U)$ *is given by the following expression*

$$\frac{dV_t^k(T, U)}{V_t^k(T, U)} = \frac{dB(t, U)}{B(t, U)} - \frac{dB^k(t, U)}{B^k(t, U)} + \frac{dB^k(t, T)}{B^k(t, T)} + \gamma^k(t, T, U) \cdot v_t^k \, dt.$$

Proof. Since $G^k(T, U)$ is an adapted process of finite variation, Itô's formula yields

$$dV_t^k(T, U) = G_t^k(T, U)\Big(F_t \, dB(t, U) + B(t, U) \, dF_t + d\langle B(\cdot, U), F \rangle_t\Big)$$
$$+ \gamma^k(t, T, U) \cdot (v_t^k + b^k(t, U) - b(t, U))V_t^k(T, U) \, dt,$$

where we write $F_t = F_{B^k}(t, T, U)$. The asserted formula now follows easily from equality (14.37), combined with the dynamics of the forward price $F_{B^k}(t, T, U)$ under the domestic martingale measure. □

Note that the adapted process $g^k(T, U)$ is linked to the instantaneous covariance between the U-delivery forward exchange rate and the T-delivery forward price of a U-maturity foreign market bond. More explicitly, the cross-variation equals

$$\langle F_{Q^k}(\cdot, U), F_{B^k}(\cdot, T, U)\rangle_t = \int_0^t g_u^k(T, U) F_{Q^k}(u, U) F_{B^k}(u, T, U)\, du.$$

Assumption. We assume from now on that $G^k(T, U)$ follows an adapted process of finite variation (in particular, it can be a deterministic function).

The above assumption is motivated by the following two arguments, each of a different nature. First, we can make use of Lemma 14.2.1. Second, it is evident that if $G^k(T, U)$, and consequently the process $V^k(T, U)$, were not adapted, then the process ϕ that is given by formula (14.41) below, would fail to satisfy the definition of a trading strategy.

Proposition 14.2.1 *Let us consider the portfolio* $\phi = (\phi^1, \phi^2, \phi^3)$ *that equals*

$$\phi_t^1 = \frac{V_t^k(T, U)}{B(t, U)}, \quad \phi_t^2 = -\frac{V_t^k(T, U)}{\tilde{B}^k(t, U)}, \quad \phi_t^3 = \frac{V_t^k(T, U)}{\tilde{B}^k(t, T)}. \tag{14.41}$$

Then the strategy ϕ *is self-financing and the wealth process* $V(\phi)$ *equals* $V^k(T, U)$.

Proof. For the last claim, it is enough to check that

$$V_t(\phi) = \phi_t^1 B(t, U) + Q_t^k(\phi_t^2 B^k(t, U) + \phi_t^3 B^k(t, T)) = V_t^k(T, U)$$

for every $t \in [0, T]$. It remains to verify that the trading strategy ϕ is self-financing. By virtue of (14.36) and (14.41), it is clear that we need to show the following equality

$$dV_t^k(T, U) = V_t^k(T, U)\left(\frac{dB(t, U)}{B(t, U)} - \frac{d\tilde{B}(t, U)}{\tilde{B}^k(t, U)} + \frac{d\tilde{B}^k(t, T)}{\tilde{B}^k(t, T)}\right). \tag{14.42}$$

For any maturity T, we have

$$d\tilde{B}^k(t, T) = Q_t^k\, dB^k(t, T) + B^k(t, T)\, dQ_t^k + \langle Q^k, B(\cdot, T)\rangle_t,$$

so that

$$d\tilde{B}^k(t, T) = \tilde{B}^k(t, T)\left(\left(r_t + b^k(t, T) \cdot v_t^k\right) dt + \left(b^k(t, T) + v_t^k\right) \cdot dW_t^*\right).$$

A substitution of this relationship into the right-hand side of (14.42) gives

$$\frac{dV_t^k(T, U)}{V_t^k(T, U)} = \frac{dB(t, U)}{B(t, U)} - b^k(t, U) \cdot dW_t^* + b^k(t, T) \cdot dW_t^* + \gamma^k(t) \cdot v_t^k\, dt$$

$$= \frac{dB(t, U)}{B(t, U)} - \frac{dB^k(t, U)}{B^k(t, U)} + \frac{dB^k(t, T)}{B^k(t, T)} + \gamma^k(t) \cdot v_t^k\, dt,$$

where $\gamma^k(t) = \gamma^k(t, T, U)$. Comparing this with the formula established in Lemma 14.2.1, we conclude that ϕ is self-financing. $\qquad\square$

Corollary 14.2.1 *The arbitrage price at time $t \in [0, T_{j-1}]$ of a contingent claim X that is given by formula (14.35), equals*

$$\pi_t(X) = V_t^k(T_{j-1}, T_j) = \frac{B(t, T_j)B^k(t, T_{j-1})G_t^k(T_{j-1}, T_j)}{B^k(t, T_j)}, \qquad (14.43)$$

where

$$G_t^k(T_{j-1}, T_j) = \exp\left(\int_t^{T_{j-1}} \gamma^k(u, T_j, T_{j-1}) \cdot \left(v_u^k + b^k(u, T_j) - b(u, T_j)\right) du\right)$$

and $\gamma^k(u, T_j, T_{j-1}) = b^k(u, T_j) - b^k(u, T_{j-1})$.

Proposition 14.2.2 *The arbitrage price of the floating-for-floating cross-currency $(k, 0; 0)$ swap at time $t \in [0, T_0]$ equals*

$$\mathbf{CCFS}_t(k, 0; 0) = \sum_{j=1}^{n} \left(\frac{B(t, T_j)B^k(t, T_{j-1})G_t^k(T_{j-1}, T_j)}{B^k(t, T_j)} - B(t, T_{j-1}) \right).$$

Proof. It is enough to observe that

$$\mathbf{CCFS}_t(k, 0; 0) = \mathbb{E}_{\mathbb{P}^*}\left\{ \sum_{j=1}^{n} \frac{B_t}{B_{T_j}} \left(\frac{1}{B^k(T_{j-1}, T_j)} - \frac{1}{B(T_{j-1}, T_j)} \right) \Big| \mathcal{F}_t \right\}$$

$$= \sum_{j=1}^{n} \mathbb{E}_{\mathbb{P}^*}\left\{ \frac{B_t B(T_{j-1}, T_j)}{B_{T_{j-1}} B^k(T_{j-1}, T_j)} \Big| \mathcal{F}_t \right\} - \sum_{j=1}^{n} B(t, T_{j-1})$$

$$= \sum_{j=1}^{n} \left(V_t^k(T_{j-1}, T_j) - B(t, T_{j-1}) \right)$$

and to apply Corollary 14.2.1. □

Floating-for-floating $(k, l; 0)$ swaps. The next step is to examine the slightly more general case of a floating-for-floating cross-currency $(k, l; 0)$ swap. The contractual conditions of a $(k, l; 0)$ swap agreement imply immediately that its arbitrage price $\mathbf{CCFS}_t(k, l; 0)$ at time t satisfies

$$\mathbf{CCFS}_t(k, l; 0) = \mathbb{E}_{\mathbb{P}^*}\left\{ \sum_{j=1}^{n} \frac{B_t}{B_{T_j}} \left(L^k(T_{j-1}) - L^l(T_{j-1}) \right) \delta_j \Big| \mathcal{F}_t \right\}, \qquad (14.44)$$

or equivalently

$$\mathbf{CCFS}_t(k, l; 0) = \mathbb{E}_{\mathbb{P}^*}\left\{ \sum_{j=1}^{n} \frac{B_t}{B_{T_j}} \left(\frac{1}{B^k(T_{j-1}, T_j)} - \frac{1}{B^l(T_{j-1}, T_j)} \right) \Big| \mathcal{F}_t \right\}.$$

It is easily seen that the arbitrage price $\mathbf{CCFS}(k, l; 0)$ also admits the following representation

$$\mathbf{CCFS}_t(k, l; 0) = \mathbf{CCFS}_t(k, 0; 0) - \mathbf{CCFS}_t(l, 0; 0).$$

Since we wish to apply the results of the previous section, we assume that for every $j = 1, \ldots, n$, both $G^k(T_{j-1}, T_j)$ and $G^l(T_{j-1}, T_j)$ follow adapted processes of finite variation. Arguing along the same lines as in the proof of Proposition 14.2.2, we find the following equality, which holds for $t \in [0, T_0]$

$$\mathbf{CCFS}_t(k, l; 0) = \sum_{j=1}^{n} \left(V_t^k(T_{j-1}, T_j) - V_t^l(T_{j-1}, T_j) \right), \qquad (14.45)$$

where $V_t^k(T_{j-1}, T_j)$ and $V_t^l(T_{j-1}, T_j)$ are given by the expressions

$$V_t^k(T_{j-1}, T_j) = \frac{B(t, T_j) B^k(t, T_{j-1}) G_t^k(T_{j-1}, T_j)}{B^k(t, T_j)}$$

and

$$V^l(T_{j-1}, T_j) = \frac{B(t, T_j) B^l(t, T_{j-1}) G_t^l(T_{j-1}, T_j)}{B^l(t, T_j)}$$

respectively. To visualize the replicating portfolio of the $(k, l; 0)$ swap agreement, let us consider an arbitrary payment date T_j. Then the portfolio $\phi = (\phi^1, \phi^2, \phi^3, \phi^4, \phi^5)$ that replicates a particular payoff of a swap that occurs at time T_j involves, at time t, ϕ_t^1 units of T_j-maturity domestic bonds, where

$$\phi_t^1 = \frac{V_t^k(T_{j-1}, T_j) - V_t^l(T_{j-1}, T_j)}{B(t, T_j)},$$

and the following positions in foreign bonds $B^k(t, T_j)$, $B^k(t, T_{j-1})$, $B^l(t, T_j)$ and $B^l(t, T_{j-1})$ respectively

$$\phi_t^2 = -\frac{V_t^k}{\tilde{B}^k(t, T_j)}, \quad \phi_t^3 = \frac{V_t^k}{\tilde{B}^k(t, T_{j-1})}, \quad \phi_t^4 = \frac{V_t^l}{\tilde{B}^l(t, T_j)}, \quad \phi_t^5 = -\frac{V_t^l}{\tilde{B}^l(t, T_{j-1})},$$

where $V_t^k = V_t^k(T_{j-1}, T_j)$ and $V_t^l = V_t^l(T_{j-1}, T_j)$. It is not hard to verify that the trading strategy ϕ given by the last formula is self-financing. Moreover, for every $t \in [0, T_{j-1}]$, the wealth of such a portfolio, expressed in units of domestic currency, equals

$$V_t(\phi) = \phi_t^1 B(t, T_{j-1}) + Q_t^l \big(\phi_t^1 B^l(t, T_j) + \phi_t^2 B^l(t, T_{j-1})\big)$$

$$+ Q_t^k \big(\phi_t^3 B^k(t, T_j) + \phi_t^4 B^k(t, T_{j-1})\big) = V_t^k(T_{j-1}, T_j) - V_t^l(T_{j-1}, T_j),$$

as expected.

Fixed-for-floating $(k; m)$ swaps. Before studying the case of floating-for-floating cross-currency $(k, l; m)$ swaps, we find it convenient to examine the case of a *fixed-for-floating cross-currency $(k; m)$ swap*. In such a contract, a floating rate $L^k(T_{j-1})$

is received at each payment date, and a prescribed fixed interest rate κ is paid. Let us stress that the payments are made in units of the mth foreign currency. Consequently the price of the fixed-for-floating swap, expressed in units of domestic currency, equals (cf. (14.34))

$$\mathbf{CCFS}_t^{\kappa}(k; m) = \mathbb{E}_{\mathbb{P}*}\left\{ \sum_{j=1}^{n} \frac{B_t}{B_{T_j}} \left(L^k(T_{j-1}) - \kappa \right) Q_{T_j}^m \, \delta_j \, \Big| \, \mathcal{F}_t \right\}$$

for every $t \in [0, T_0]$. Equivalently,

$$\mathbf{CCFS}_t^{\kappa}(k; m) = \mathbb{E}_{\mathbb{P}*}\left\{ \sum_{j=1}^{n} \frac{B_t}{B_{T_j}} \left(B^k(T_{j-1}, T_j)^{-1} - \tilde{\delta}_j \right) Q_{T_j}^m \, \Big| \, \mathcal{F}_t \right\}, \qquad (14.46)$$

where $\tilde{\delta}_j = 1 + \kappa \delta_j$. Let us write the last representation as $\mathbf{CCFS}_t^{\kappa}(k; m) = I_t - J_t$, where the meaning of I_t and J_t is apparent from the context. We define

$$G_t^{km}(T, U) = \exp\left(\int_t^T g_u^{km}(T, U) \, du \right), \qquad (14.47)$$

where

$$g_t^{km}(T, U) = \gamma^k(t, U, T) \cdot (v_t^{m/k} + b^k(t, U) - b^m(t, U)).$$

Observe that in the special case when $m = 0$, the process $G^{k0}(T, U)$ given by (14.47) coincides with the process $G^k(T, U)$ introduced in the preceding section (see formula (14.39)). For notational simplicity, we write

$$g_t^{kmj} = g_t^{km}(T_{j-1}, T_j), \quad \forall t \in [0, T_0],$$

and

$$\delta_t^*(k, m, j) = b^k(t, T_j) - b^m(t, T_j), \quad \forall t \in [0, T_j], \qquad (14.48)$$

in what follows. Notice that

$$g_t^{kmj} = \gamma^k(u, T_j, T_{j-1}) \cdot (v_t^{m/k} + \delta_t^*(k, m, j)).$$

Lemma 14.2.2 *The following equalities hold for every $t \in [0, T_0]$*

$$J_t = \sum_{j=1}^{n} \tilde{\delta}_j Q_t^m B^m(t, T_j) \qquad (14.49)$$

and

$$I_t = \sum_{j=1}^{n} \frac{Q_t^m B^m(t, T_j) B^k(t, T_{j-1}) G_t^{km}(T_{j-1}, T_j)}{B^k(t, T_j)}. \qquad (14.50)$$

Proof. For the first formula, it is enough to observe that the equality

$$\mathbb{E}_{\mathbb{P}^*}\Big\{ \frac{B_t}{B_{T_j}}\, Q^m_{T_j} \,\Big|\, \mathcal{F}_t \Big\} = Q^m_t\, B^m(t, T_j)$$

is valid for every $t \in [0, T_j]$. To establish (14.50), observe first that

$$I_t = \mathbb{E}_{\mathbb{P}^*}\Big\{ \sum_{j=1}^{n} \frac{B_t}{B_{T_j}}\, B^k(T_{j-1}, T_j)^{-1}\, Q^m_{T_j} \,\Big|\, \mathcal{F}_t \Big\}$$

$$= \mathbb{E}_{\mathbb{P}^*}\Big\{ \sum_{j=1}^{n} \frac{B_t}{B_{T_{j-1}}}\, B^k(T_{j-1}, T_j)^{-1}\, \mathbb{E}_{\mathbb{P}^*}\Big(\frac{B_{T_{j-1}}}{B_{T_j}}\, Q^m_{T_j} \,\Big|\, \mathcal{F}_{T_{j-1}} \Big) \,\Big|\, \mathcal{F}_t \Big\}$$

$$= \mathbb{E}_{\mathbb{P}^*}\Big\{ \sum_{j=1}^{n} \frac{B_t}{B_{T_{j-1}}}\, B^k(T_{j-1}, T_j)^{-1}\, B^m(T_{j-1}, T_j) Q^m_{T_{j-1}} \,\Big|\, \mathcal{F}_t \Big\}$$

$$= \sum_{j=1}^{n} Q^m_t\, \mathbb{E}_{\mathbb{P}^m}\Big\{ \frac{B^m_t}{B^m_{T_{j-1}}}\, B^k(T_{j-1}, T_j)^{-1}\, B^m(T_{j-1}, T_j) \,\Big|\, \mathcal{F}_t \Big\}.$$

Consequently, we obtain the following equality

$$I_t = \sum_{j=1}^{n} Q^m_t\, B^m(t, T_{j-1})\, \mathbb{E}_{\mathbb{P}^m_{T_{j-1}}}\Big\{ \frac{B^m(T_{j-1}, T_j)}{B^k(T_{j-1}, T_j)} \,\Big|\, \mathcal{F}_t \Big\}, \tag{14.51}$$

which expresses I_t in terms of the forward measure $\mathbb{P}^m_{T_{j-1}}$ for the market m. In order to evaluate this conditional expectation, observe that the dynamics of $B^m(t, T_j)$ and $B^k(t, T_j)$, $k \neq m$, as seen in the market m, are given by the following expressions

$$dB^m(t, T_j) = B^m(t, T_j)\big(r^m_t\, dt + b^m(t, T_j) \cdot dW^m_t\big)$$

and

$$dB^k(t, T_j) = B^k(t, T_j) \Big(\big(r^k_t - b^k(t, T_j) \cdot v^{m/k}_t\big) dt + b^k(t, T_j) \cdot dW^m_t \Big),$$

since $dW^k_t = dW^m_t - v^{m/k}_t\, dt$. Let us denote

$$H^1_t = \frac{B^m(t, T_j)}{B^k(t, T_j)}, \qquad H^2_t = \frac{B^k(t, T_{j-1})}{B^m(t, T_{j-1})}, \qquad \forall t \in [0, T_0].$$

Using Itô's formula, we get

$$dH^1_t = H^1_t \Big(r^m_t - r^k_t + b^k(t, T_j) \cdot \big(v^{m/k}_t + \delta^*_t(k, m, j) \big) \Big) dt - H^1_t \delta^*_t(k, m, j) \cdot dW^m_t$$

and

$$dH^2_t = H^2_t \big(r^k_t - r^m_t - b^k(t, T_{j-1}) \cdot v^{m/k}_t - b^m(t, T_{j-1}) \cdot \delta^*_t(k, m, j-1) \big) dt$$
$$+ H^2_t \delta^*_t(k, m, j-1) \cdot dW^m_t.$$

Consequently, for the process H, which equals

$$H_t = H_t^1 H_t^2 = \frac{B^m(t, T_j) B^k(t, T_{j-1})}{B^k(t, T_j) B^m(t, T_{j-1})},$$

we obtain

$$dH_t = H_t \left(g_t^{kmj} \, dt + \left(\gamma^k(t, T_{j-1}, T_j) + \gamma^m(t, T_j, T_{j-1}) \right) \cdot dW_t^{T_{j-1}, m} \right)$$

for every $j = 1, \dots, n$. Since the quantities g^{kmj} are assumed to follow deterministic functions, taking into account the relationship

$$\mathbb{E}_{\mathbb{P}_{T_{j-1}}^m} \left\{ \frac{B^m(T_{j-1}, T_j)}{B^k(T_{j-1}, T_j)} \,\Big|\, \mathcal{F}_t \right\} = \mathbb{E}_{\mathbb{P}_{T_{j-1}}^m} \left\{ \frac{B^m(T_{j-1}, T_j) B^k(T_{j-1}, T_{j-1})}{B^k(T_{j-1}, T_j) B^m(T_{j-1}, T_{j-1})} \,\Big|\, \mathcal{F}_t \right\}$$

and using the just-established expression for the differential of H, we obtain

$$\mathbb{E}_{\mathbb{P}_{T_{j-1}}^m} \left\{ \frac{B^m(T_{j-1}, T_j)}{B^k(T_{j-1}, T_j)} \,\Big|\, \mathcal{F}_t \right\} = \frac{B^m(t, T_j) B^k(t, T_{j-1})}{B^k(t, T_j) B^m(t, T_{j-1})} \, e^{\int_t^{T_{j-1}} g_u^{kmj} \, du}.$$

Note that the last equality is valid for every $j = 1, \dots, n$. Combining it with (14.51), we arrive at equality (14.50). □

The next result follows from Lemma 14.2.2 and formula (14.46).

Proposition 14.2.3 *The arbitrage price at time $t \in [0, T_0]$ of a fixed-for-floating cross-currency $(k; m)$ swap with the underlying fixed interest rate equal to κ is given by the expression (recall that $\tilde{\delta}_j = 1 + \kappa \delta_j$)*

$$\mathbf{CCFS}_t^\kappa(k; m) = \sum_{j=1}^n Q_t^m B^m(t, T_j) \left(\frac{B^k(t, T_{j-1}) G_t^{km}(T_{j-1}, T_j)}{B^k(t, T_j)} - \tilde{\delta}_j \right).$$

Consider a fixed-for-floating cross-currency $(k; k)$ swap – that is, a swap in which the floating rate and the currency used for payments are those of the market k. Proposition 14.2.3 yields the following price of such a swap, expressed in units of domestic currency (cf. Sect. 13.1)

$$\mathbf{CCFS}_t^\kappa(k; k) = \sum_{j=1}^n Q_t^k \left(B^k(t, T_{j-1}) - \tilde{\delta}_j B^k(t, T_j) \right). \tag{14.52}$$

Similarly, for the $(k; 0)$ fixed-for-floating swap (that is, an agreement in which the underlying floating rate is that of the market k, but where payments are made in domestic currency), we get

$$\mathbf{CCFS}_t^\kappa(k; 0) = \sum_{j=1}^n B(t, T_j) \left(\frac{B^k(t, T_{j-1}) G_t^k(T_{j-1}, T_j)}{B^k(t, T_j)} - \tilde{\delta}_j \right). \tag{14.53}$$

Floating-for-floating $(k, l; m)$ **swaps.** Let us now examine the case of a general floating-for-floating cross-currency $(k, l; m)$ swap per unit of the currency m, which consists of swapping the floating rates of another two currencies, say k and l. At each of the payment dates T_j, $j = 1, \ldots, n$, the floating rate $L^k(T_{j-1})$ of currency k is received and the corresponding floating rate $L^l(T_{j-1})$ of the currency l is paid. As usual, the underlying floating rates $L^i(T_{j-1})$, $i = k, l$; $j = 1, \ldots, n$ are set by reference to the price of a zero-coupon bond, so that

$$B^i(T_{j-1}, T_j)^{-1} = 1 + L^i(T_{j-1})(T_j - T_{j-1}) = 1 + \delta_j L^i(T_{j-1})$$

for $j = 1, \ldots, n$. The value at time t, in the domestic currency, of the floating-for-floating forward $(k, l; m)$ swap equals

$$\mathbf{CCFS}_t(k, l; m) = \mathbb{E}_{\mathbb{P}^*}\left\{ \sum_{j=1}^{n} \frac{B_t}{B_{T_j}} \left(L^k(T_{j-1}) - L^l(T_{j-1}) \right) Q^m_{T_j} \delta_j \,\Big|\, \mathcal{F}_t \right\}. \quad (14.54)$$

Therefore, the following useful relationship

$$\mathbf{CCFS}_t(k, l; m) = \mathbf{CCFS}^\kappa_t(k; m) - \mathbf{CCFS}^\kappa_t(l; m) \quad (14.55)$$

holds for any fixed level κ. The next result follows immediately from formula (14.55) and Proposition 14.2.3.

Proposition 14.2.4 *The price* $\mathbf{CCFS}_t = \mathbf{CCFS}_t(k, l; m)$ *at time* $t \in [0, T_0]$ *of the floating-for-floating cross-currency forward swap of type* $(k, l; m)$ *in a Gaussian HJM model equals*

$$\mathbf{CCFS}_t = \sum_{j=1}^{n} Q^m_t B^m(t, T_j) \left(\frac{B^k(t, T_{j-1})}{B^k(t, T_j)} e^{kmj}_t - \frac{B^l(t, T_{j-1})}{B^l(t, T_j)} e^{lmj}_t \right),$$

where

$$g^{kmj}_t = \gamma^k(t, T_j, T_{j-1}) \cdot \left(v^k_t - v^m_t + \delta^*_t(k, m, j) \right)$$

for every $k, m = 0, \ldots, N$, $j = 1, \ldots, n$, *and*

$$e^{pmj}_t = \exp \left(\int_t^{T_{j-1}} g^{pmj}_u \, du \right)$$

for $p = k, l$.

For the special case of a floating-for-floating cross-currency $(k, l; 0)$ swap (i.e., a swap agreement with payments in domestic currency), the above proposition yields the pricing result (14.45), which was previously derived by means of a replicating portfolio.

14.2.4 Cross-currency Swaptions

By a *cross-currency swaption* (or *differential swaption*) we mean an option contract with a cross-currency forward swap being the option's underlying asset. Note that the option expiry date, T, precedes the initial date (i.e., the first reset date) of the underlying swap agreement – that is, $T \leq T_0$. We first examine the case of a fixed-for-floating swaption, subsequently we turn our attention to a floating-for-floating swaption.

A *fixed-for-floating* $(k; 0)$ *cross-currency swaption* is an option, with expiry date $T \leq T_0$, whose holder has the right to decide whether he wishes to pay at some future dates $T_1, \ldots, T_n$ a fixed interest rate, say κ, on some notional principal, and receive simultaneously a floating rate L^k of currency k. Note that we assume here that all payments are made in units of the domestic currency. Using our terminological conventions, a cross-currency fixed-for-floating $(k; 0)$ swaption may be seen as a call option (whose exercise price equals zero) with a fixed-for-floating $(k; 0)$ cross-currency swap being the underlying financial instrument. Therefore, it is clear that the value at time $t \in [0, T]$ of a fixed-for-floating $(k; 0)$ cross-currency swaption, denoted by $\mathbf{CCS}_t^{\kappa,\lambda}(k; 0)$ or shortly $\mathbf{CCS}_t(k; 0)$, equals

$$\mathbf{CCS}_t^{\kappa,\lambda}(k; 0) = \mathbb{E}_{\mathbb{P}^*}\left(\frac{B_t}{B_T}\left(\mathbf{CCFS}_T^{\kappa}(k; 0) - 0\right)^+ \Big| \mathcal{F}_t\right). \tag{14.56}$$

More explicitly, we have

$$\mathbf{CCS}_t(k; 0) = \mathbb{E}_{\mathbb{P}^*}\left\{ \frac{B_t}{B_T}\left[\mathbb{E}_{\mathbb{P}^*}\left(\sum_{j=1}^{n} \frac{B_t}{B_{T_j}}(L^k(T_{j-1}) - \kappa)\delta_j \Big| \mathcal{F}_T\right)\right]^+ \Big| \mathcal{F}_t\right\}.$$

In view of (14.53), the formula above can also be rewritten in the following way

$$\mathbf{CCS}_t(k; 0) = \mathbb{E}_{\mathbb{P}^*}\left\{ \frac{B_t}{B_T}\left[\sum_{j=1}^{n} B(T, T_j)\left(\frac{B^k(T, T_{j-1})G_T^{kj}}{B^k(T, T_j)} - \tilde{\delta}_j\right)\right]^+ \Big| \mathcal{F}_t\right\},$$

where $\tilde{\delta}_j = 1 + \kappa\delta_j$ and (cf. (14.39)–(14.40))

$$G_t^{kj} \stackrel{\text{def}}{=} G_t^k(T_{j-1}, T_j) = \exp\left(\int_t^{T_{j-1}} g_u^k(T_{j-1}, T_j)\, du\right) \tag{14.57}$$

for $t \in [0, T]$. Under the forward measure $\mathbb{P}_T$, we have

$$\mathbf{CCS}_t(k; 0) = B(t, T) \mathbb{E}_{\mathbb{P}_T}\left\{ \left[\sum_{j=1}^{n} \frac{B(T, T_j)}{B(T, T)}\left(\frac{B^k(T, T_{j-1})G_T^{kj}}{B^k(T, T_j)} - \tilde{\delta}_j\right)\right]^+ \Big| \mathcal{F}_t\right\},$$

or equivalently $\mathbf{CCS}_t(k; 0) = B(t, T) \mathbb{E}_{\mathbb{P}_T}(X \mid \mathcal{F}_t)$, where

$$X = \left(\sum_{j=1}^{n} F_B(T, T_j, T)\left(F_{B^k}(T, T_{j-1}, T_j) G_T^{kj} - \tilde{\delta}_j\right)\right)^+.$$

Recall that the dynamics of the process $F_t = F_B(t, T_j, T)$ under the forward measure $\mathbb{P}_T$ of the domestic market are (cf. (11.40))

$$dF_t = F_t \, \gamma(t, T_j, T) \cdot dW_t^T,$$

so that for any $t \in [0, T]$, we have

$$F_T = F_t \, \exp\left(\int_t^T \gamma(u, T_j, T) \cdot dW_u^T - \frac{1}{2} \int_t^T |\gamma(u, T_j, T)|^2 \, du \right).$$

Furthermore, by virtue of (14.10), we get the following expression for the dynamics under $\mathbb{P}_T$ of the process $F_t^{kj} = F_{B^k}(t, T_{j-1}, T_j)$

$$dF_t^{kj} = F_t^{kj} \gamma^k(t, T_{j-1}, T_j) \cdot \left(dW_t^T + (b(t, T) - v_t^k - b^k(t, T_j)) \, dt \right).$$

Consequently,

$$F_T^{kj} = L_t F_t^{kj} \, \exp\left(\int_t^T \gamma_k(u, T_{j-1}, T_j) \cdot dW_u^T - \frac{1}{2} \int_t^T |\gamma^k(u, T_{j-1}, T_j)|^2 \, du \right),$$

where L is given by the following expression

$$L_t = \exp\left(\int_t^T \gamma^k(u, T_{j-1}, T_j) \cdot (b(u, T) - b^k(u, T_j) - v_u^k) \, du \right).$$

By straightforward calculations, we obtain

$$G_T^{kj} L_t = G_t^{kj} \, \exp\left(\int_t^T \gamma^k(u, T_{j-1}, T_j) \cdot \gamma(u, T, T_j) \, du \right).$$

We conclude that

$$\mathbf{CCS}_t(k; 0) = \mathbb{E}_{\mathbb{P}_T}\left\{ \left(\sum_{j=1}^n B(t, T_j)\left(G_t^{kj} F_t^{kj} N_t^T - \tilde{\delta}_j \, M_t^T \right) \right)^+ \Big| \mathcal{F}_t \right\},$$

where

$$N_t^T = \exp\left(- \int_t^T \Gamma_u^{kj} \cdot dW_u^T - \frac{1}{2} \int_t^T |\Gamma_u^{kj}|^2 du \right)$$

with

$$\Gamma_u^{kj} = \gamma(u, T_j, T) + \gamma^k(u, T_{j-1}, T_j), \qquad (14.58)$$

and M_t^T equals

$$M_t^T = \exp\left(\int_t^T \gamma(u, T_j, T) \cdot dW_u^T - \frac{1}{2} \int_t^T |\gamma(u, T_j, T)|^2 \, du \right).$$

We are in a position to state the following result, whose proof presents no difficulties.

Proposition 14.2.5 *The arbitrage price* $\mathbf{CCS}_t^{\kappa,\lambda}(k;0) = \mathbf{CCS}_t(k;0)$ *of a fixed-for-floating cross-currency swaption equals*

$$\mathbf{CCS}_t(k;0) = \int_{\mathbb{R}^d} \Big(\sum_{j=1}^n B(t,T_j)\big(G_t^{kj} F_t^{kj} n_d(x+y_j^k) - \tilde{\delta}_j n_d(x+z_j)\big) \Big)^+ dx,$$

where the vectors $y_1^k, \ldots, y_n^k, z_1, \ldots, z_n \in \mathbb{R}^d$ *satisfy for* $i, j = 1, \ldots, n$

$$y_i^k \cdot y_j^k = \int_t^T \Gamma_u^{ki} \cdot \Gamma_u^{kj}\, du, \quad y_i^k \cdot z_j = \int_t^T \Gamma_u^{ki} \cdot \gamma(u,T_j,T)\, du,$$

with Γ^{ki} *given by* (14.58), *and* $z_i \cdot z_j = \int_t^T \gamma(u,T_i,T) \cdot \gamma(u,T_j,T)\, du.$

By a *floating-for-floating cross-currency swaption* we mean a call option, with strike price equal to zero, to receive the floating rate L^k of currency k plus margin μ, and to pay simultaneously the floating rate L^l of currency l. At any time t before the swaption's expiry date T, the value of a floating-for-floating cross-currency swaption is

$$\mathbf{CCS}_t^\mu(k,l;0) = \mathbb{E}_{\mathbb{P}^*}\Big\{ \frac{B_t}{B_T}\Big[\mathbb{E}_{\mathbb{P}^*}\Big(\sum_{j=1}^n \frac{B_T}{B_{T_j}} \Delta_{j-1} \Big| \mathcal{F}_T \Big)\Big]^+ \Big| \mathcal{F}_t \Big\},$$

where $\Delta_{j-1} = \delta_j\big(L^k(T_{j-1}) + \mu - L^l(T_{j-1})\big)$. From the definition of a floating-for-floating cross-currency swap, it is easily seen that the value $\mathbf{CCS}_t^\mu(k,l;0)$ also equals

$$\mathbf{CCS}_t^\mu(k,l;0) = \mathbb{E}_{\mathbb{P}^*}\Big\{ \frac{B_t}{B_T}\Big(\mathbf{CCFS}_T(k,l;0) + \mu\sum_{j=1}^n \delta_j B(T,T_j)\Big)^+ \Big| \mathcal{F}_t \Big\}.$$

Therefore, using representation (14.45), we get

$$\mathbf{CCS}_t^\mu(k,l;0) = \mathbb{E}_{\mathbb{P}^*}\Big\{ \frac{B_t}{B_T}\Big(\sum_{j=1}^n B(T,T_j)\big(\tilde{\Delta}_T^{j-1} + \mu\delta_j\big)\Big)^+ \Big| \mathcal{F}_t \Big\},$$

where

$$\tilde{\Delta}_T^{j-1} = \frac{B^k(T,T_{j-1})}{B^k(T,T_j)} G_T^{kj} - \frac{B^l(T,T_{j-1})}{B^l(T,T_j)} G_T^{lj} = F_T^{kj} G_T^{kj} - F_T^{lj} G_T^{lj}.$$

Proposition 14.2.6 *The arbitrage price at time* $t \in [0,T]$ *of a floating-for-floating cross-currency swaption equals*

$$\mathbf{CCS}_t^\mu(k,l;0) = \int_{\mathbb{R}^d} \Big(\sum_{j=1}^n B(t,T_j) f_j(x) \Big)^+ dx,$$

where for every $x \in \mathbb{R}^d$

$$f_j(x) = F_t^{kj} G_t^{kj} n_d(x+y_j^k) - F_t^{lj} G_t^{lj} n_d(x+y_j^l) + \mu\delta_j n_d(x+z_j)$$

and n_d is the standard d-dimensional Gaussian probability density function. More-over, the vectors $y_1^k, \ldots, y_n^k, y_1^l, \ldots, y_n^l, z_1, \ldots, z_n \in \mathbb{R}^d$ satisfy for $h = k, l$ and $i, j = 1, \ldots, n$

$$y_i^h \cdot y_j^h = \int_t^T \Gamma_u^{hi} \cdot \Gamma_u^{hj} \, du, \qquad y_i^h \cdot z_j = \int_t^T \Gamma_u^{hi} \cdot \gamma(u, T_j, T) \, du,$$

with Γ^{hi} given by (14.58), and $z_i \cdot z_j = \int_t^T \gamma(u, T_i, T) \cdot \gamma(u, T_j, T) \, du$.

Proof. The proof goes along the same lines as in the case of a cross-currency fixed-for-floating swaption. $\square$

14.2.5 Basket Caps

As the next example of a foreign market interest rate derivative, we shall now con-sider a cap (settled in arrears) on a basket of floating rates L^k, $k = 1, \ldots, N$ of foreign markets; such an option is usually referred to as a *basket cap*. In this agree-ment, the cash flows received at times T_j, $j = 1, \ldots, n$, are

$$c_j = \Big(\sum_{k=1}^N w_k L^k(T_{j-1}) - \kappa \Big)^+ \delta_j,$$

where the weights w_k, $k = 1, \ldots, N$ are assumed to be strictly positive constants, and κ is a preassigned rate of interest. The value of each particular basket caplet at time t equals

$$\mathbf{BCpl}_t^\kappa = \mathbb{E}_{\mathbb{P}^*} \Big\{ \frac{B_t}{B_{T_j}} \Big(\sum_{k=1}^N w_k L^k(T_{j-1}) - \kappa \Big)^+ \delta_j \,\Big|\, \mathcal{F}_t \Big\}$$

$$= \mathbb{E}_{\mathbb{P}^*} \Big\{ \frac{B_t}{B_{T_{j-1}}} \Big(\sum_{k=1}^N w_k \frac{B(T_{j-1}, T_j)}{B^k(T_{j-1}, T_j)} - \tilde{\delta}_j B(T_{j-1}, T_j) \Big)^+ \,\Big|\, \mathcal{F}_t \Big\},$$

where $\tilde{\delta}_j = \kappa \delta_j + \sum_{k=1}^N w_k$. Therefore,

$$\mathbf{BCpl}_t^\kappa = B(t, T_{j-1}) \, \mathbb{E}_{\mathbb{P}_{T_{j-1}}} \Big\{ \Big(\sum_{k=1}^N w_k \frac{B(T_{j-1}, T_j)}{B^k(T_{j-1}, T_j)} - \tilde{\delta}_j B(T_{j-1}, T_j) \Big)^+ \,\Big|\, \mathcal{F}_t \Big\}.$$

It is convenient to denote $F_t^{0j} = F_B(t, T_{j-1}, T_j)$ and $F_t^{kj} = F_{B^k}(t, T_{j-1}, T_j)$. Rea-soning along the same lines as in Sect. 14.2.4, we find that (recall that G_t^{kj} is defined by (14.39)–(14.40), see also (14.57))

$$\mathbf{BCpl}_t^\kappa = B(t, T_{j-1}) \, \mathbb{E}_{\mathbb{P}_{T_{j-1}}} \Big\{ \Big(\sum_{k=1}^N w_k \frac{F_t^{kj} G_t^{kj} M_t^k}{F_t^{0j}} - \tilde{\delta}_j \frac{M_t^0}{F_t^{kj}} \Big)^+ \,\Big|\, \mathcal{F}_t \Big\},$$

where

$$M_t^k = \exp\left(-\int_t^{T_{j-1}} \zeta_u^{kj} \cdot dW_u^{T_{j-1}} - \frac{1}{2} \int_t^{T_{j-1}} |\zeta_u^{kj}|^2 \, du\right),$$

$\zeta_u^{kj} = \gamma^k(u, T_j, T_{j-1}) - \gamma(u, T_j, T_{j-1})$, and

$$M_t^0 = \exp\left(\int_t^{T_{j-1}} \gamma(u, T_{j-1}, T_j) \cdot dW_u^{T_{j-1}} - \frac{1}{2} \int_t^{T_{j-1}} |\gamma(u, T_{j-1}, T_j)|^2 \, du\right).$$

The following result can thus be proved by standard arguments.

Proposition 14.2.7 *The arbitrage price of a basket cap at time t equals*

$$\mathbf{BC}_t^\kappa = \sum_{j=1}^n B(t, T_j) \int_{\mathbb{R}^d} \left(\sum_{k=1}^N w_k F_t^{kj} G_t^{kj} n_d(x + y_k) - \tilde{\delta}_j n_d(x + y_0)\right)^+ dx,$$

where $\tilde{\delta}_j = \kappa \delta_j + \sum_{k=1}^N w_k$, the vectors $y_j^0, \ldots, y_j^N \in \mathbb{R}^d$ are such that for every $k, l = 1, \ldots, N$ and $j = 1, \ldots, n$

$$y_j^k \cdot y_j^l = \int_t^{T_{j-1}} \zeta_u^{k,j} \cdot \zeta_u^{l,j} \, du, \quad y_j^0 \cdot y_j^k = \int_t^{T_{j-1}} \zeta_u^{kj} \cdot \gamma(u, T_{j-1}, T_j) \, du,$$

and

$$|y_j^0|^2 = \int_t^{T_{j-1}} |\gamma(u, T_{j-1}, T_j)|^2 \, du.$$

14.3 Model of Forward LIBOR Rates

A cross-currency extension of a market model of forward LIBOR rates can be constructed by proceeding along the same lines as in Sect. 14.1 – the details are omitted (see Mikkelsen (2002) and Schlögl (2002)). Let us only mention that the main notions we shall employ in what follows are the forward measures and the corresponding Brownian motions. For each date $T_0, T_1, \ldots, T_n$, the dynamics of the forward LIBOR rate $L^i(t, T_j)$ of the ith market are governed by the SDE

$$dL^i(t, T_{j-1}) = L^i(t, T_{j-1}) \lambda^i(t, T_{j-1}) \cdot dW_t^{T_j, i} \tag{14.59}$$

under the corresponding forward probability measure $\mathbb{P}_{T_j}^i$, where the volatilities $\lambda^i(t, T_{j-1})$ are assumed to follow deterministic functions. Furthermore, the process $W^{T_j, i}$ is related to the domestic forward Brownian motion W^{T_j} through the formula (cf. (14.16))

$$W_t^{T_j, i} = W_t^{T_j} - \int_0^t \left(v_u^i + b^i(u, T) - b(u, T)\right) du. \tag{14.60}$$

14.3.1 Quanto Cap

For simplicity, we consider only two dates, $T_j = T$ and $T_{j+1} = T + \delta$; that is, we deal with a single caplet with reset date T and settlement date $T + \delta$ (note that we consider here the case of a fixed-length accrual period). A *quanto caplet* with expiry date T pays to its holder at time $T + \delta$ the amount (expressed in domestic currency)

$$\mathbf{Cplq}_{T+\delta} = \delta \bar{Q}(L^i(T) - \kappa)^+,$$

where $\bar{Q}$ is the preassigned level of the exchange rate and $L^i(T)$ is the (spot) LIBOR rate at time T in the ith market. Consequently, the domestic arbitrage price of a quanto caplet at time t equals

$$\mathbf{Cplq}_t = \delta \bar{Q} B(t, T + \delta) \, \mathbb{E}_{\mathbb{P}_{T+\delta}} \big((L^i(T) - \kappa)^+ \big| \mathcal{F}_t\big)$$

per one unit of nominal value (the nominal value of a quanto cap is expressed in foreign currency). It is thus clear that to value such a contract, we need to examine the dynamics of the foreign LIBOR rate under the domestic forward measure. The payoff of a quanto caplet, expressed in foreign currency, equals $\delta \bar{Q}(L^i(T) - \kappa)^+ / Q^i_{T+\delta}$. Consequently, its price at time t, in domestic currency, admits the following representation

$$\mathbf{Cplq}_t = \delta \bar{Q} B(t, T + \delta) Q^i_t \, \mathbb{E}_{\mathbb{P}^i_{T+\delta}} \left(\frac{(L^i(T) - \kappa)^+}{F_{Q^i}(T + \delta, T + \delta)} \bigg| \mathcal{F}_t \right),$$

where $\mathbb{P}^i_{T+\delta}$ is the foreign market forward measure. The last representation makes clear that the volatility of the forward exchange rate $F_{Q^i}(t, T)$, and thus also bond price volatilities $b^i(t, T + \delta)$ and $b(t, T + \delta)$ (cf. (14.24)–(14.25)), will enter the valuation formula. To uniquely specify these volatilities, we may adopt, for instance, the approach of Brace et al. (1997). In their approach, bond price volatilities are linked to forward LIBOR rates by means of the formula (cf. (12.29))

$$b^k(t, T + \delta) = -\sum_{m=0}^{n(t)} \frac{\delta L^k(t, T - m\delta)}{1 + \delta L^k(t, T - m\delta)} \lambda^k(t, T - m\delta), \tag{14.61}$$

where $n(t) = [\delta^{-1}(T - t)]$. It is thus clear that bond price volatilities $b^i(t, T + \delta)$ follow necessarily stochastic processes, and thus the Gaussian methodology examined in preceding sections is no longer applicable, even under deterministic volatilities of (spot) exchange rates. Therefore, we will examine an approximation of the caplet's price. Combining the dynamics (14.59), which read

$$dL^i(t, T) = L^i(t, T) \lambda^i(t, T) \cdot dW_t^{T+\delta, i},$$

with (14.60), we obtain (recall that $L^i(T) = L^i(T, T)$)

$$L^i(T) = L^i(t, T) e^{-\eta^i(t, T)} \exp \left(\int_t^T \lambda^i(u, T) \cdot dW_u^{T+\delta} - \frac{1}{2} \int_t^T |\lambda^i(u, T)|^2 du \right),$$

where

$$\eta^i(t, T) = \int_t^T \lambda^i(u, T) \cdot \left(v_u^i + b^i(u, T + \delta) - b(u, T + \delta)\right) du.$$

To derive a simple approximate formula for the price of a quanto caplet at time t, it is convenient to substitute into (14.62) the term $\eta^i(t, T)$ with $\tilde{\eta}^i(t, T)$, where

$$\tilde{\eta}^i(t, T) = \int_t^T \lambda^i(u, T) \cdot \left(v_u^i + \tilde{b}^i(u, T + \delta) - \tilde{b}(u, T + \delta)\right) du \qquad (14.62)$$

and $\tilde{b}^i(u, T+\delta)$ and $\tilde{b}(u, T+\delta)$ are $\mathcal{F}_t$-measurable random variables, namely (recall that $\lambda^k(t, T)$ is a deterministic function)

$$\tilde{b}^k(u, T + \delta) = \sum_{m=0}^{n(u)} \frac{\delta L^k(t, T - m\delta)}{1 + \delta L^k(t, T - m\delta)} \lambda^k(u, T - m\delta) \qquad (14.63)$$

for $k = 0, i$ and $u \in [t, T]$. The following result provides an approximate valuation formula for a quanto caplet in a lognormal model of forward LIBOR rates.

Proposition 14.3.1 *The arbitrage price of a quanto caplet written on a LIBOR rate of the ith market satisfies*

$$\mathbf{Cplq}_t \approx \bar{Q}\delta B(t, T + \delta)\left(L^i(t, T)e^{-\tilde{\eta}^i(t,T)} N\big(\tilde{e}_1(t, T)\big) - \kappa N\big(\tilde{e}_2(t, T)\big)\right),$$

where

$$\tilde{e}_{1,2}^i(t, T) = \frac{\ln(L^i(t, T)/\tilde{\kappa}) \pm \frac{1}{2}\tilde{v}_i^2(t, T)}{\tilde{v}_i(t, T)}$$

and

$$\tilde{v}_i^2(t, T) = \int_t^T |\lambda^i(u, T)|^2 du.$$

Proof. It is enough to observe that

$$\mathbf{Cplq}_t \approx \bar{Q}\delta B(t, T + \delta)e^{-\tilde{\eta}^i(t,T)} \mathbb{E}_{\mathbb{P}_{T+\delta}}\left\{\left(L^i(t, T)\Lambda(t, T) - \tilde{\kappa}\right)^+ \Big| \mathcal{F}_t\right\},$$

where

$$\Lambda(t, T) = \exp\left(\int_t^T \lambda^i(u, T) \cdot dW_u^{T+\delta} - \frac{1}{2}\int_t^T |\lambda^i(u, T)|^2 du\right)$$

and $\tilde{\kappa} = \kappa e^{\tilde{\eta}^i(0,T)}$. The expected value can be evaluated in exactly the same way as in the case of a domestic caplet (see Proposition 12.6.1). $\qquad \square$

14.3.2 Cross-currency Swap

As the next example, let us consider a floating-for-floating $(k, l; 0)$ swap. Recall that the payoffs of such a swap are made in domestic currency. Therefore, its price at time t, denoted by $\mathbf{CCFS}_t(k, l; 0)$, equals (cf. (14.44))

$$\mathbf{CCFS}_t(k, l; 0) = \sum_{j=1}^{n} \delta\, B(t, T_j)\, \mathbb{E}_{\mathbb{P}_{T_j}} \left(L^k(T_{j-1}) - L^l(T_{j-1}) \,\middle|\, \mathcal{F}_t \right)$$

per one unit of nominal value (in domestic currency), where $\mathbb{P}_{T_j}$ is the forward measure of the domestic market for the date T_j. Notice that

$$L^i(T_j) = L^i(t, T_j)\, e^{-\eta^i(t, T_j)}\, M^i(t, T_j)$$

for $i = k, l$ and $j = 0, \ldots, n-1$, where

$$M^i(t, T_j) = \exp\left(\int_t^{T_j} \lambda^i(u, T_j) \cdot dW_u^{T_{j+1}} - \frac{1}{2} \int_t^{T_j} |\lambda^i(u, T_j)|^2\, du \right)$$

and

$$\eta^i(t, T_j) = \int_t^{T_j} \lambda^i(u, T_j) \cdot \left(v_u^i + b^i(u, T_j) - b(u, T_j) \right) du.$$

Applying an approximation similar to that of the previous section, we find that

$$\mathbf{CCFS}_t(k, l; 0) \approx \sum_{j=0}^{n-1} \delta\, B(t, T_{j+1}) \left(L^k(t, T_j) e^{-\tilde{\eta}^k(t, T_j)} - L^l(t, T_j) e^{-\tilde{\eta}^l(t, T_j)} \right),$$

where $\tilde{\eta}^k(t, T_j)$ and $\tilde{\eta}^l(t, T_j)$ are given by (14.62)–(14.63). Let us end this section by commenting on the valuation of a floating-for-floating $(k, l; m)$ swap, which is denominated in units of the mth currency. The price of such a swap at time t, in domestic currency, equals (cf. (14.54))

$$\mathbf{CCFS}_t(k, l; m) = \sum_{j=0}^{n-1} \delta\, B(t, T_{j+1})\, \mathbb{E}_{\mathbb{P}_{T_j}} \left\{ Q_{T_{j+1}}^m \left(L^k(T_j) - L^l(T_j) \right) \,\middle|\, \mathcal{F}_t \right\}.$$

From the results of Sect. 14.2.1, we know that the forward exchange rate satisfies $F_{Q^m}(T_{j+1}, T_{j+1}) = Q_{T_{j+1}}^m$ and (see (14.24)–(14.25))

$$dF_{Q^m}(t, T_{j+1}) = F_{Q^m}(t, T_{j+1}) \sigma_{Q^m}(t, T_{j+1}) \cdot dW_t^{T_{j+1}},$$

where $\sigma_{Q^m}(t, T_{j+1}) = v_t^m + b^m(t, T_{j+1}) - b(t, T_{j+1})$. Replacing the term $\sigma_{Q^m}(t, T_{j+1})$ with

$$\tilde{\sigma}_{Q^m}(t, T_{j+1}) = v_t^m + \tilde{b}^m(t, T_{j+1}) - \tilde{b}(t, T_{j+1}),$$

it is possible to derive an approximate formula for the price of a floating-for-floating $(k, l; m)$ swap in a lognormal model of forward LIBOR rates.

14.4 Concluding Remarks

Let us acknowledge that term structure models examined in this text do not cover all kinds of risks occurring in actual fixed-income markets. Indeed, we assumed throughout that all primary fixed-income securities (and derivative financial contracts) are default-free; that is, we concentrated on the *market risk* due to the uncertain future behavior of asset prices, as opposed to the *credit risk* (or *default risk*) that is known to play a non-negligible role in some sectors of financial markets. The latter kind of risk relates to the possibility of default by one party to a contract. If default risk is accounted for, one has to deal with such financial contracts as, for instance, defaultable (or credit-risky) bonds, vulnerable options, or options on credit spreads.

In this regard, let us mention that in recent years, term structure models that make explicit account for default risk has attracted growing interest. Generally speaking, this involves the study of the impact of credit ratings on the yield spread – that is, modelling the term structure of defaultable debt – as well as the valuation of other contingent claims that are subject to default risk. Mathematical tools that are used in this context have attracted attention of several researchers and, as a result, a considerable progress in the theory of default risk was achieved in recent years. From the practical perspective, the demand for more sophisticated mathematical models was further enhanced by a rapid growth of trading in *credit derivatives*, that is, financial contracts that are capable of transferring the credit (default) risk of some reference entity between the two counterparties.

Since neither the modelling of defaultable term structure nor the valuation and hedging of credit derivatives were covered in this text, we refer to the recent monographs by Ammann (1999), Cossin and Pirotte (2000), Bielecki and Rutkowski (2002), Duffie and Singleton (2003), Schönbucher (2003), and Lando (2004) for an introduction to this field.

The interested reader may also consult original papers by Merton (1974), Geske (1977), Jonkhart (1979), Ho and Singer (1982, 1984), Titman and Torous (1989), Chance (1990), Artzner and Delbaen (1995, 1992), Kim et al. (1993), Lando (1994), Hull and White (1995), Jarrow and Turnbull (1995), Cooper and Martin (1996), Duffee (1996, 1998), Longstaff and Schwartz (1995), Duffie (1998a, 1998b), Duffie and Huang (1996), Duffie and Singleton (1997, 1999), Jarrow et al. (1997), Duffie et al. (1997), Huge and Lando (1999), Kusuoka (1999), Elliott et al. (2000), Jeanblanc and Rutkowski (2002), Bélanger et al. (2004), and Bielecki et al. (2004a, 2004b, 2005a, 2005b).

We have also implicitly assumed that the inflation plays no role in the valuation of interest rate derivatives. In fact, the distinction between *nominal* and *real* interest rates is not important for most interest rate derivatives, but, obviously, this is not true for the so-called *inflation-based derivatives*, such as *inflation-linked* bonds or options. Let us only mention that a few models aiming the valuation and hedging of inflation-based derivatives were recently developed (see, for instance, Jarrow and Yildirim (2003) or Mercurio (2005)).

Part III

APPENDIX

A

An Overview of Itô Stochastic Calculus

The appendix provides a brief account of Itô stochastic integration theory. It is not our intention to compete here with any full-scale textbook on stochastic analysis. We merely gathered here the most relevant results that are frequently referred to in previous chapters. Hence the choice of results is definitely subjective and their exposition as concise as possible.

We start by recalling some classic concepts from the theory of stochastic processes. Next, we define the standard Brownian motion and we introduce the notion of the Itô stochastic integral. Subsequently, we present the fundamental formula of stochastic calculus, the *Itô formula*, and its most important applications. Finally, the probability distributions of certain functionals of a Brownian motion are examined. For more details on Itô stochastic integration with respect to the Brownian motion (and, more generally, continuous semimartingales) we refer the reader, for instance, to monographs by Krylov (1995), Durrett (1996), Karatzas and Shreve (1998a), Mikosch (1999), Revuz and Yor (1999), Steele (2000), Øksendal (2003), and Jeanblanc et al. (2006).

A.1 Conditional Expectation

Let us start by considering a finite decomposition of the underlying probability space. We say that a finite collection $\mathcal{D} = \{D_1, \ldots, D_k\}$ of non-empty subsets of Ω is a *decomposition* of Ω if the sets $D_1, \ldots, D_k$ are pairwise disjoint; that is, if $D_i \cap D_j = \emptyset$ for every $i \neq j$, and the equality $D_1 \cup D_2 \cup \ldots \cup D_k = \Omega$ holds. A random variable ψ on Ω is called *simple* if it admits a representation

$$\psi(\omega) = \sum_{i=1}^{m} x_i \mathbb{1}_{D_i(\psi)}(\omega), \tag{A.1}$$

where $D_i(\psi) = \{\omega \in \Omega \mid \psi(\omega) = x_i\}$ and x_i, $i = 1, \ldots, m$ are real numbers satisfying $x_i \neq x_j$ for $i \neq j$. For a simple random variable ψ, we denote by $\mathcal{D}(\psi)$

the decomposition $\{D_1(\psi), \ldots, D_m(\psi)\}$ generated by ψ. It is clear that if Ω is a finite set then any random variable $\psi : \Omega \to \mathbb{R}$ is simple.

Definition A.1.1 For any decomposition $\mathcal{D}$ of Ω and any event $A \in \mathcal{F}$, the *conditional probability* of A with respect to $\mathcal{D}$ is defined by the formula

$$\mathbb{P}(A \mid \mathcal{D}) = \sum_{j=1}^{k} \mathbb{P}(A \mid D_j) \mathbb{1}_{D_j}. \tag{A.2}$$

Moreover, if ψ is a simple random variable with the representation (A.1), its *conditional expectation* given $\mathcal{D}$ equals

$$\mathbb{E}_{\mathbb{P}}(\psi \mid \mathcal{D}) = \sum_{i=1}^{m} x_i \, \mathbb{E}_{\mathbb{P}}(\mathbb{1}_{D_i(\psi)} \mid \mathcal{D}) = \sum_{i=1}^{m} \sum_{j=1}^{k} x_i \, \mathbb{P}(D_i(\psi) \mid D_j) \mathbb{1}_{D_j}. \tag{A.3}$$

Observe that the conditional expectation $\mathbb{E}_{\mathbb{P}}(\psi \mid \mathcal{D})$ is constant on each set D_j from $\mathcal{D}$. Let η be another simple random variable on Ω, i.e.,

$$\eta(\omega) = \sum_{l=1}^{r} y_l \mathbb{1}_{D_l(\eta)}(\omega), \tag{A.4}$$

where $D_l(\eta) = \{\omega \in \Omega \mid \eta(\omega) = y_l\}$ and $y_i \neq y_j$ for $i \neq j$.

Suppose that η and ψ are simple random variables given from (A.1) and (A.4) respectively. Then, by the definition of conditional expectation, $\mathbb{E}_{\mathbb{P}}(\psi \mid \eta)$ coincides with $\mathbb{E}_{\mathbb{P}}(\psi \mid \mathcal{D}(\eta))$, and thus

$$\mathbb{E}_{\mathbb{P}}(\psi \mid \eta) = \mathbb{E}_{\mathbb{P}}(\psi \mid \mathcal{D}(\eta)) = \sum_{i=1}^{m} x_i \, \mathbb{E}_{\mathbb{P}}(\mathbb{1}_{D_k(\psi)} \mid \mathcal{D}(\eta)), \tag{A.5}$$

where the second equality follows by (A.3). Consequently,

$$\mathbb{E}_{\mathbb{P}}(\psi \mid \eta) = \sum_{i=1}^{m} \sum_{l=1}^{r} x_i \, \mathbb{P}(D_i(\psi) \mid D_l(\eta)) \mathbb{1}_{D_l(\eta)} = \sum_{l=1}^{r} c_l \mathbb{1}_{D_l(\eta)}, \tag{A.6}$$

where $c_l = \sum_{i=1}^{m} x_i \, \mathbb{P}(D_i(\psi) \mid D_l(\eta))$ for $l = 1, \ldots, r$. Note that $\mathbb{E}_{\mathbb{P}}(\psi \mid \eta)$ does not depend on the particular values of η. More precisely, if for two random variables η_1 and η_2 we have $\mathcal{D}(\eta_1) = \mathcal{D}(\eta_2)$ then $\mathbb{E}_{\mathbb{P}}(\psi \mid \eta_1) = \mathbb{E}_{\mathbb{P}}(\psi \mid \eta_2)$.

Note, however, that in view of (A.6) we also have that $\mathbb{E}_{\mathbb{P}}(\psi \mid \eta) = g(\eta)$, where the function $g : \{y_1, \ldots, y_r\} \to \mathbb{R}$ is given by the formula $g(y_l) = c_l$ for $l = 1, 2, \ldots, r$.

It is not difficult to check that the conditional expectation of a simple random variable with respect to a decomposition $\mathcal{D}$ has the following properties (ψ and η stand here for arbitrary simple random variables):

(i) if $\mathcal{D}(\psi) \subseteq \mathcal{D}$ then $\mathbb{E}_{\mathbb{P}}(\psi \mid \mathcal{D}) = \psi$;

(ii) for arbitrary real numbers c, d we have

$$\mathbb{E}_{\mathbb{P}}(c\psi + d\eta \mid \mathcal{D}) = c \mathbb{E}_{\mathbb{P}}(\psi \mid \mathcal{D}) + d \mathbb{E}_{\mathbb{P}}(\eta \mid \mathcal{D});$$

(iii) if $\mathcal{D}_1 \subseteq \mathcal{D}_2$ then

$$\mathbb{E}_{\mathbb{P}}(\mathbb{E}_{\mathbb{P}}(\psi \mid \mathcal{D}_1) \mid \mathcal{D}_2) = \mathbb{E}_{\mathbb{P}}(\mathbb{E}_{\mathbb{P}}(\psi \mid \mathcal{D}_2) \mid \mathcal{D}_1) = \mathbb{E}_{\mathbb{P}}(\psi \mid \mathcal{D}_1);$$

(iv) if ψ is independent of $\mathcal{D}$ – that is, if for every $A \in \mathcal{D}(\psi)$ and every $B \in \mathcal{D}$ we have $\mathbb{P}(A \cap B) = \mathbb{P}(A)\mathbb{P}(B)$ – then $\mathbb{E}_{\mathbb{P}}(\psi \mid \mathcal{D}) = \mathbb{E}_{\mathbb{P}}\psi$;

(v) if ψ is $\mathcal{D}$-measurable and η is independent of $\mathcal{D}$ then for any function $h : \mathbb{R}^2 \to \mathbb{R}$ we have

$$\mathbb{E}_{\mathbb{P}}(h(\psi, \eta) \mid \mathcal{D}) = H(\psi),$$

where the function $H : \mathbb{R} \to \mathbb{R}$ is given by the formula $H(x) = \mathbb{E}_{\mathbb{P}}h(x, \eta)$.

We shall now define the conditional expectation of a random variable with respect to an arbitrary σ-field $\mathcal{G}$.

Definition A.1.2 Let ψ be an integrable random variable on $(\Omega, \mathcal{F}, \mathbb{P})$ (i.e., $\mathbb{E}_{\mathbb{P}}(|\psi|) < \infty$). For an arbitrary σ-field $\mathcal{G}$ of subsets of Ω satisfying $\mathcal{G} \subseteq \mathcal{F}$ (i.e., a *sub-σ-field* of $\mathcal{F}$), the *conditional expectation* $\mathbb{E}_{\mathbb{P}}(\psi \mid \mathcal{G})$ of ψ with respect to $\mathcal{G}$ is defined by the following conditions: (i) $\mathbb{E}_{\mathbb{P}}(\psi \mid \mathcal{G})$ is $\mathcal{G}$-measurable, i.e., $\{\omega \in \Omega \mid \mathbb{E}_{\mathbb{P}}(\psi \mid \mathcal{G})(\omega) \leq a\} \in \mathcal{G}$ for any $a \in \mathbb{R}$; (ii) for an arbitrary event $A \in \mathcal{G}$, we have

$$\int_A \mathbb{E}_{\mathbb{P}}(\psi \mid \mathcal{G}) \, d\mathbb{P} = \int_A \psi \, d\mathbb{P}. \tag{A.7}$$

It is well known that the conditional expectation exists and is unique (up to the $\mathbb{P}$-a.s. equivalence of random variables).

Remarks. If $\mathcal{D}$ is a (finite) decomposition of Ω, then the family of all unions of sets from $\mathcal{D}$, together with an empty set, forms a σ-field of subsets of Ω. We denote it by $\sigma(\mathcal{D})$ and we call it the σ-field *generated* by the decomposition[1] $\mathcal{D}$. It can be easily checked that if $\mathcal{G} = \sigma(\mathcal{D})$ then $\mathbb{P}(A \mid \mathcal{G}) = \mathbb{P}(A \mid \mathcal{D})$, so that the conditional expectation with respect to the σ-field $\sigma(\mathcal{D})$ coincides with the conditional expectation with respect to the decomposition $\mathcal{D}$.

Definition A.1.3 For an arbitrary simple random variable η, the σ-field $\mathcal{F}(\eta) = \sigma(\mathcal{D}(\eta))$ is called the σ-*field generated by* η, or briefly, the *natural σ-field of* η. More generally, for any random variable η, the σ-*field generated by* η is the least σ-field of subsets of Ω with respect to which η is measurable. It is denoted by either $\sigma(\eta)$ or $\mathcal{F}(\eta)$.

For a real-valued random variable η, it can be shown that the σ-field $\mathcal{F}(\eta)$ is the smallest σ-field that contains all events of the form $\{\omega \in \Omega \mid \eta(\omega) \leq x\}$, where x

---

[1] In the case of a finite Ω, any σ-field $\mathcal{G}$ of subsets of Ω is of this form; that is, $\mathcal{G} = \sigma(\mathcal{D})$ for some decomposition $\mathcal{D}$ of Ω.

is an arbitrary real number. More generally, if $\eta = (\eta^1, \ldots, \eta^d)$ is a d-dimensional random variable then

$$\mathcal{F}(\eta) = \sigma\big(\{\omega \in \Omega \mid \eta^1(\omega) \leq x_1, \ldots, \eta^d(\omega) \leq x_d\} \mid x_1, \ldots, x_d \in \mathbb{R}\big).$$

For any $\mathbb{P}$-integrable random variable ψ and any random variable η, we define the conditional expectation of ψ with respect to η by setting $\mathbb{E}_{\mathbb{P}}(\psi \mid \eta) = \mathbb{E}_{\mathbb{P}}(\psi \mid \mathcal{F}(\eta))$. It is possible to show that for arbitrary real-valued random variables ψ and η, there exists a Borel measurable function $g : \mathbb{R} \to \mathbb{R}$ such that $\mathbb{E}_{\mathbb{P}}(\psi \mid \eta) = g(\eta)$.

Definition A.1.4 We say that the σ-fields $\mathcal{G}$ and $\mathcal{H}$ are *independent* under $\mathbb{P}$ whenever $\mathbb{P}(A \cap B) = \mathbb{P}(A)\mathbb{P}(B)$ for any $A \in \mathcal{G}$ and $B \in \mathcal{H}$.

Let us summarize the basic properties of a conditional expectation.

Lemma A.1.1 *Let ψ and η be integrable random variables on $(\Omega, \mathcal{F}, \mathbb{P})$. Also let $\mathcal{G}$ and $\mathcal{H}$ be some sub-σ-fields of $\mathcal{F}$. Then*
(i) if ψ is $\mathcal{G}$-measurable (or equivalently if $\mathcal{F}(\psi) \subseteq \mathcal{G}$) then $\mathbb{E}_{\mathbb{P}}(\psi \mid \mathcal{G}) = \psi$;
(ii) for arbitrary real numbers c, d we have

$$\mathbb{E}_{\mathbb{P}}(c\psi + d\eta \mid \mathcal{G}) = c\mathbb{E}_{\mathbb{P}}(\psi \mid \mathcal{G}) + d\mathbb{E}_{\mathbb{P}}(\eta \mid \mathcal{G});$$

(iii) if $\mathcal{H} \subseteq \mathcal{G}$ then

$$\mathbb{E}_{\mathbb{P}}\big(\mathbb{E}_{\mathbb{P}}(\psi \mid \mathcal{G}) \mid \mathcal{H}\big) = \mathbb{E}_{\mathbb{P}}\big(\mathbb{E}_{\mathbb{P}}(\psi \mid \mathcal{H}) \mid \mathcal{G}\big) = \mathbb{E}_{\mathbb{P}}(\psi \mid \mathcal{H});$$

in particular, if $\mathbb{E}_{\mathbb{P}}(\psi \mid \mathcal{G})$ is $\mathcal{H}$-measurable then $\mathbb{E}_{\mathbb{P}}(\psi \mid \mathcal{G}) = \mathbb{E}_{\mathbb{P}}(\psi \mid \mathcal{H})$;
(iv) if ψ is independent of $\mathcal{G}$, i.e., the σ-fields $\mathcal{F}(\psi)$ and $\mathcal{G}$ are independent under $\mathbb{P}$, then $\mathbb{E}_{\mathbb{P}}(\psi \mid \mathcal{G}) = \mathbb{E}_{\mathbb{P}}(\psi)$;
(v) if ψ is $\mathcal{G}$-measurable and η is independent of $\mathcal{G}$ then for any Borel measurable function $h : \mathbb{R}^2 \to \mathbb{R}$ we have $\mathbb{E}_{\mathbb{P}}(h(\psi, \eta) \mid \mathcal{G}) = H(\psi)$, where $H(x) = \mathbb{E}_{\mathbb{P}}h(x, \eta)$, provided that the inequality $\mathbb{E}_{\mathbb{P}}|h(\psi, \eta)| < \infty$ holds.

The next result is the conditional form of Fatou's lemma.

Lemma A.1.2 *Let $\xi_1, \xi_2 \ldots$ be integrable random variables on a probability space $(\Omega, \mathcal{F}, \mathbb{P})$. Assume that for every n we have $\xi_n \geq \eta$, where η is some random variable with $\mathbb{E}_{\mathbb{P}}\eta > -\infty$. Then for any sub-$\sigma$-field $\mathcal{G}$ of $\mathcal{F}$ we have*

$$\mathbb{E}_{\mathbb{P}}(\liminf_{n\to\infty} \xi_n \mid \mathcal{G}) \leq \liminf_{n\to\infty} \mathbb{E}_{\mathbb{P}}(\xi_n \mid \mathcal{G}).$$

The following result, referred to as (conditional) *Jensen's inequality*, is also well known.

Lemma A.1.3 *Let $h : \mathbb{R} \to \mathbb{R}$ be a convex function, and ξ a random variable on $(\Omega, \mathcal{F}, \mathbb{P})$ such that $\mathbb{E}_{\mathbb{P}}|g(\xi)| < \infty$. Then for any sub-$\sigma$-field $\mathcal{G}$ of $\mathcal{F}$, Jensen's inequality holds – that is, $g\big(\mathbb{E}_{\mathbb{P}}(\xi \mid \mathcal{G})\big) \leq \mathbb{E}_{\mathbb{P}}(g(\xi) \mid \mathcal{G})$.*

The last result of this section refers to the situation where two mutually equivalent probability measures, $\mathbb{P}$ and $\mathbb{Q}$ say, are defined on a common measurable space $(\Omega, \mathcal{F})$. Suppose that the Radon-Nikodým density of $\mathbb{Q}$ with respect to $\mathbb{P}$ equals (cf. Sect. A.14)

$$\frac{d\mathbb{Q}}{d\mathbb{P}} = \eta, \quad \mathbb{P}\text{-a.s.} \tag{A.8}$$

Note that the random variable η is strictly positive $\mathbb{P}$-a.s., moreover η is $\mathbb{P}$-integrable with $\mathbb{E}_{\mathbb{P}}\eta = \mathbb{Q}(\Omega) = 1$. Finally, in view of (A.8), it is clear that equality $\mathbb{E}_{\mathbb{Q}}\psi = \mathbb{E}_{\mathbb{P}}(\psi\eta)$ holds for any $\mathbb{Q}$-integrable random variable ψ.

We are in the position prove the so-called *abstract Bayes formula*.

Lemma A.1.4 *Let $\mathcal{G}$ be a sub-σ-field of the σ-field $\mathcal{F}$, and let ψ be a random variable integrable with respect to $\mathbb{Q}$. Then the following version of Bayes's formula holds*

$$\mathbb{E}_{\mathbb{Q}}(\psi \mid \mathcal{G}) = \frac{\mathbb{E}_{\mathbb{P}}(\psi\eta \mid \mathcal{G})}{\mathbb{E}_{\mathbb{P}}(\eta \mid \mathcal{G})}. \tag{A.9}$$

Proof. It can be easily checked that $\mathbb{E}_{\mathbb{P}}(\eta \mid \mathcal{G})$ is strictly positive $\mathbb{P}$-a.s. so that the right-hand side of (A.9) is well-defined. By our assumption, the random variable $\xi = \psi\eta$ is $\mathbb{P}$-integrable, it is therefore enough to show that

$$\mathbb{E}_{\mathbb{P}}(\xi \mid \mathcal{G}) = \mathbb{E}_{\mathbb{Q}}(\psi \mid \mathcal{G})\mathbb{E}_{\mathbb{P}}(\eta \mid \mathcal{G}).$$

Since the right-hand side of the last formula defines a $\mathcal{G}$-measurable random variable, we need to verify that for any set $A \in \mathcal{G}$, we have

$$\int_A \psi\eta \, d\mathbb{P} = \int_A \mathbb{E}_{\mathbb{Q}}(\psi \mid \mathcal{G})\mathbb{E}_{\mathbb{P}}(\eta \mid \mathcal{G}) \, d\mathbb{P}.$$

But for every $A \in \mathcal{G}$, we get

$$\int_A \psi\eta \, d\mathbb{P} = \int_A \psi \, d\mathbb{Q} = \int_A \mathbb{E}_{\mathbb{Q}}(\psi \mid \mathcal{G}) \, d\mathbb{Q} = \int_A \mathbb{E}_{\mathbb{Q}}(\psi \mid \mathcal{G})\eta \, d\mathbb{P}$$

$$= \int_A \mathbb{E}_{\mathbb{P}}(\mathbb{E}_{\mathbb{Q}}(\psi \mid \mathcal{G})\eta \mid \mathcal{G}) \, d\mathbb{P} = \int_A \mathbb{E}_{\mathbb{Q}}(\psi \mid \mathcal{G})\,\mathbb{E}_{\mathbb{P}}(\eta \mid \mathcal{G}) \, d\mathbb{P}. \quad \square$$

A.2 Filtrations and Adapted Processes

Let us consider a probability space $(\Omega, \mathcal{F}, \mathbb{F}, \mathbb{P})$ equipped with a filtration $\mathbb{F} = (\mathcal{F}_t)_{t \in [0,T]}$. A *filtration* is an increasing family of σ-fields, that is, we have that $\mathcal{F}_u \subseteq \mathcal{F}_t$ for any $0 \le u \le t \le T$. For concreteness, we assume that T is a finite strictly positive real number and we set $\mathcal{F} = \mathcal{F}_T$. Hence we shall write $(\Omega, \mathbb{F}, \mathbb{P})$, rather than $(\Omega, \mathcal{F}, \mathbb{F}, \mathbb{P})$, in what follows.

Definition A.2.1 A real-valued stochastic process $X = (X_t)_{t \in [0,T]}$ defined on $(\Omega, \mathbb{F}, \mathbb{P})$ is $\mathbb{F}$-*adapted* if for any $t \in [0, T]$ the random variable X_t is $\mathcal{F}_t$-measurable, i.e., for any $x \in \mathbb{R}$ the event $\{X_t \le x\}$ belongs to the σ-field $\mathcal{F}_t$.

Similarly, an $\mathbb{R}^k$-valued stochastic process $X = (X^1, \ldots, X^k)$ defined on $(\Omega, \mathbb{F}, \mathbb{P})$ is said to be $\mathbb{F}$-*adapted* if for any $t \in [0, T]$ the $\mathbb{R}^k$-valued random variable $X_t = (X_t^1, \ldots, X_t^k)$ is $\mathcal{F}_t$-measurable, i.e., for any choice of real numbers $x_1, \ldots, x_k \in \mathbb{R}$ the event

$$\{X_t^1 \le x_1, X_t^2 \le x_2, \ldots, X_t^k \le x_k\}$$

belongs to the σ-field $\mathcal{F}_t$.

We shall sometimes say that a process is adapted, rather than $\mathbb{F}$-adapted, if the filtration $\mathbb{F}$ is fixed so that no danger of confusion arises. Let us mention that any stochastic process X is adapted to its *natural filtration* (i.e., the filtration *generated by X*), which is defined as: $\mathcal{F}_t^X = \sigma(X_u \mid u \le t)$.

Any stochastic processes considered in the sequel is assumed to be a *càdlàg* process, that is, almost all sample paths are right-continuous functions with finite left-hand limits. We say the a stochastic process is *continuous* if almost all sample paths are continuous functions.

Definition A.2.2 We say that the processes X and Y defined on a common probability space are *indistinguishable* if the event $\{X_t = Y_t$ for $t \in [0, T]\}$ has probability 1. We identify any two processes that are indistinguishable.

From now on, we assume that the filtration $\mathbb{F}$ satisfies the *usual conditions*:
(i) $\mathbb{F}$ is $\mathbb{P}$-complete, i.e., any event $A \in \mathcal{F}_T$ such that $\mathbb{P}(A) = 0$ belongs to the σ-field $\mathcal{F}_0$, and thus it belongs to the σ-field $\mathcal{F}_t$ for any $t \in [0, T]$,
(ii) $\mathbb{F}$ is right-continuous, i.e., $\mathcal{F}_t = \bigcap_{u > t} \mathcal{F}_u$ for any $t \in [0, T)$.

If a given filtration $\mathbb{F}$ is not $\mathbb{P}$-complete and/or right-continuous, we need first to modify it as follows. First, we set $\tilde{\mathcal{F}}_t = \sigma(\mathcal{F}_t \cup \mathcal{N})$, where $\mathcal{N}$ is the class of all $\mathbb{P}$-null events from $\mathcal{F}_T$ (i.e, all events from $\mathcal{F}_T$ of probability zero). Second, for any $t \in [0, T)$ we define $\hat{\mathcal{F}}_t = \tilde{\mathcal{F}}_{t+}$, where $\tilde{\mathcal{F}}_{t+} = \bigcap_{\epsilon > 0} \tilde{\mathcal{F}}_{t+\epsilon}$. The filtration $\hat{\mathbb{F}}$ defined in that way is $\mathbb{P}$-complete and right-continuous; it is referred to as the $\mathbb{P}$-*augmentation* of $\mathbb{F}$.

A.3 Martingales

A martingale is an $\mathbb{F}$-adapted stochastic processes enjoying the property that the conditional expected value of its future value at time u, based on the information $\mathcal{F}_t$ available at time $t \le u$ coincides with its current value. The origins of this concept are related to the idea of a *fair game*, that is, a game in which the expected payoff matches the stake. Intuitively, a martingale describes the current value of our wealth when we make any bets associated with a fair game.

As already mentioned, we shall assume that all processes are defined on the finite interval $[0, T]$. For many results provided in what follows this assumption is not required, but to preserve consistency we will maintain it throughout this appendix. The formal definition of a martingale and closely related concepts of a super- and submartingale thus read as follows.

Definition A.3.1 A real-valued, $\mathbb{F}$-adapted process $M = (M_t)_{t \in [0,T]}$, defined on a filtered probability space $(\Omega, \mathbb{F}, \mathbb{P})$, is called an $\mathbb{F}$-*martingale* with respect to the filtration $\mathbb{F}$ if the following conditions are satisfied:
(i) M is integrable, that is, $\mathbb{E}_\mathbb{P}|M_t| < \infty$ for $t \in [0, T]$,
(ii) the following *martingale equality* holds, for any $0 \le t \le u \le T$,

$$\mathbb{E}_\mathbb{P}(M_u \mid \mathcal{F}_t) = M_t. \tag{A.10}$$

If (A.10) is replaced by the inequality $\mathbb{E}_\mathbb{P}(M_u \mid \mathcal{F}_t) \le M_t$ then M is said to be an $\mathbb{F}$-*supermartingale*. Finally, M is called an $\mathbb{F}$-*submartingale* if for any $0 \le t \le u \le T$ $\mathbb{E}_\mathbb{P}(M_u \mid \mathcal{F}_t) \ge M_t$. Note that the expected value of a martingale is constant, $\mathbb{E}_\mathbb{P}M_t = \mathbb{E}_\mathbb{P}M_0$ for $t \in [0, T]$. For a supermartingale, the expected value is a non-increasing function, that is, $\mathbb{E}_\mathbb{P}M_t \le \mathbb{E}_\mathbb{P}M_0$ for any $t \in [0, T]$. Finally, for a submartingale we have that $\mathbb{E}_\mathbb{P}M_t \ge \mathbb{E}_\mathbb{P}M_0$ for any $t \in [0, T]$.

The following result summarizes the basic properties of martingales. Its easy proof is left as an exercise.

Proposition A.3.1 *The following properties are valid:*
(i) *Let X be an $\mathcal{F}_T$-measurable and integrable random variable. Then the process $M_t = \mathbb{E}_\mathbb{P}(X \mid \mathcal{F}_t)$, $t \in [0, T]$, is a (uniformly integrable) $\mathbb{F}$-martingale.*
(ii) *Assume that M is either a super- or a submartingale with respect to $\mathbb{F}$. Then M is an $\mathbb{F}$-martingale if and only if the expected value of M is constant, or equivalently, whenever $\mathbb{E}_\mathbb{P}M_0 = \mathbb{E}_\mathbb{P}M_T$.*
(iii) *Let M be an $\mathbb{F}$-martingale and let $h : \mathbb{R} \to \mathbb{R}$ be a convex function. If the process $X_t = h(M_t)$, $t \in [0, T]$, is integrable then it is an $\mathbb{F}$-submartingale.*

A.4 Standard Brownian Motion

We start by stating the definition of the *standard Brownian motion* (also known as the *Wiener process*).

Definition A.4.1 A real-valued, $\mathbb{F}$-adapted process $W = (W_t)_{t \in [0,T]}$, with $W_0 = 0$, defined on a filtered probability space $(\Omega, \mathbb{F}, \mathbb{P})$, is called a *one-dimensional standard Brownian motion* with respect to the filtration $\mathbb{F}$ if:
(i) for any $0 \le t \le u \le T$, the increment $W_u - W_t$ is a random variable independent of the σ-field $\mathcal{F}_t$,
(ii) for any $0 \le t \le u \le T$, the probability distribution of $W_u - W_t$ is Gaussian with expected value 0 and variance $u - t$, that is, $W_u - W_t \stackrel{d}{=} N(0, u - t)$,
(iii) almost all sample paths of W are continuous functions, that is, the process W is sample-paths continuous.

If the filtration $\mathbb{F}$ is not given in advance, by a *standard Brownian motion* we mean a process W satisfying Definition A.4.1 with $\mathbb{F}$ being the natural filtration of W. More precisely, we assume that $\mathbb{F} = \mathbb{F}^W$, meaning that $\mathbb{F}$ is the $\mathbb{P}$-augmentation of the natural filtration $\sigma\{W_u \mid u \le t\}$ of the process W. In that case, condition (ii) in

Definition A.4.1 can be replaced by the requirement that W is a *process of independent increments*, that is, for any $n \in \mathbb{N}$ and any sequence $0 < t_1 < \ldots < t_n \le T$ the random variables $W_{t_1} - W_{t_0}, W_{t_2} - W_{t_1}, \ldots, W_{t_n} - W_{t_{n-1}}$ are independent.

We take here for granted the existence of a Brownian motion (see, for instance, Karatzas and Shreve (1998a)) and we focus on the most important properties of this process.

Lemma A.4.1 (i) *A Brownian motion W is a continuous $\mathbb{F}$-martingale.*
(ii) *A Brownian motion W has the Markov property with respect to $\mathbb{F}$, meaning that for any bounded Borel measurable function $g : \mathbb{R} \to \mathbb{R}$ and $0 \le t \le u \le T$*

$$\mathbb{E}_{\mathbb{P}}(g(W_u) \mid \mathcal{F}_t) = \mathbb{E}_{\mathbb{P}}(g(W_u) \mid W_t). \tag{A.11}$$

Proof. Let us prove the first part. It is clear that W is a continuous, $\mathbb{F}$-adapted process and $\mathbb{E}_{\mathbb{P}}|W_t| < \infty$ for any $t \in [0, T]$. Moreover the martingale property holds since we have, for any $0 \le t \le u \le T$,

$$\mathbb{E}_{\mathbb{P}}(W_u \mid \mathcal{F}_t) = \mathbb{E}_{\mathbb{P}}(W_u - W_t \mid \mathcal{F}_t) + \mathbb{E}_{\mathbb{P}}(W_t \mid \mathcal{F}_t) = W_t. \tag{A.12}$$

The last equality in formula (A.12) holds since condition (ii) implies that $\mathbb{E}_{\mathbb{P}}(W_u - W_t \mid \mathcal{F}_t) = \mathbb{E}_{\mathbb{P}}(W_u - W_t) = 0$. In fact, W is a *square-integrable martingale* as manifestly $\mathbb{E}_{\mathbb{P}}(W_t^2) = t < \infty$ for any $t \in [0, T]$.

For the second part, note that

$$\mathbb{E}_{\mathbb{P}}(g(W_u) \mid \mathcal{F}_t) = \mathbb{E}_{\mathbb{P}}(g(W_u - W_t + W_t) \mid \mathcal{F}_t) = \mathbb{E}_{\mathbb{P}}(g(\eta + \psi) \mid \mathcal{F}_t),$$

where $\psi = W_t$ is $\mathcal{F}_t$-measurable and $\eta = W_u - W_t$ is independent of $\mathcal{F}_t$. Hence, by part (v) in Lemma A.1.1, $\mathbb{E}_{\mathbb{P}}(g(W_u) \mid \mathcal{F}_t)$ is $\sigma(W_t)$-measurable. Consequently, by part (iii) in Lemma A.1.1, equality (A.11) holds. $\square$

A Brownian motion is an example of a process of *finite quadratic variation,* as made explicit by the following result.

Proposition A.4.1 *For every $0 \le u < t \le T$, and any sequence $(\pi_n)_{n\in\mathbb{N}}$ of finite partitions $\pi_n = \{u = t_0^n < t_1^n < \cdots < t_{m_n}^n = t\}$ of the interval $[u, t]$ satisfying $\lim_{n\to\infty} \delta(\pi_n) = 0$, where $\delta(\pi_n) \stackrel{\text{def}}{=} \max_{k=1,\ldots,m_n} (t_k^n - t_{k-1}^n)$, we have*

$$\lim_{n\to\infty} \sum_{k=1}^{m_n} (W_{t_k^n} - W_{t_{k-1}^n})^2 = t - u, \tag{A.13}$$

where the convergence in (A.13) is in $L^2(\Omega, \mathcal{F}_T, \mathbb{P})$.

Proof. Let us fix $0 \le u < t \le T$ and let us denote $\Delta_k^n = t_k^n - t_{k-1}^n$. Note that necessarily $\lim_{n\to\infty} m_n = \infty$. It suffices to check that $\lim_{n\to\infty} I_n = 0$, where we write

$$I_n = \lim_{n\to\infty} \mathbb{E}_{\mathbb{P}} \left(\sum_{k=1}^{m_n} \left((W_{t_k^n} - W_{t_{k-1}^n})^2 - \Delta_k^n \right) \right)^2.$$

Since the increments $W_{t_{j+1}^n} - W_{t_j^n}$ and $W_{t_{i+1}^n} - W_{t_i^n}$ are independent for $i \neq j$ and $\mathbb{E}_{\mathbb{P}}(W_{t_k^n} - W_{t_{k-1}^n})^2 = \Delta_k^n$, we obtain

$$I_n = \sum_{k=1}^{m_n} \left(\mathbb{E}_{\mathbb{P}}(W_{t_k^n} - W_{t_{k-1}^n})^4 - 2\Delta_k^n \, \mathbb{E}_{\mathbb{P}}(W_{t_k^n} - W_{t_{k-1}^n})^2 + (\Delta_k^n)^2 \right)$$

$$= 2 \sum_{k=1}^{m_n} (\Delta_k^n)^2 \leq 2(t - u) \max_{k=1,\dots,m_n} (t_k^n - t_{k-1}^n) = 2(t - u)\delta(\pi_n).$$

In the second equality we used the fact that the assumption that the increment $W_{t_k^n} - W_{t_{k-1}^n}$ has the Gaussian distribution $N(0, \Delta_k^n)$ yields

$$\mathbb{E}_{\mathbb{P}}(W_{t_k^n} - W_{t_{k-1}^n})^4 = 3(\Delta_k^n)^2 = 3(t_k^n - t_{k-1}^n)^2.$$

It is now clear that I_n tends to zero as n tends to infinity (so that $\delta(\pi_n)$ tends to zero). $\square$

It is worth mentioning that if the series $\sum_{n=1}^{\infty} \delta(\pi_n)$ is convergent then it can be deduced from the Borel-Cantelli lemma that the convergence in (A.13) is almost sure (that is, it holds with probability 1). Let $\langle W \rangle_t$ stand for the quadratic variation of W on $[0, t]$, so that $\langle W \rangle_t = t$ for any $t \in [0, T]$.

Lemma A.4.2 *The process $M = W^2 - \langle W \rangle$ is a continuous $\mathbb{F}$-martingale.*

Proof. It is easy to see that the process M is integrable and $\mathbb{F}$-adapted. Note also that $\mathbb{E}_{\mathbb{P}}(W_t W_u \mid \mathcal{F}_t) = W_t \, \mathbb{E}_{\mathbb{P}}(W_u \mid \mathcal{F}_t) = W_t^2$ for $0 \leq t \leq u \leq T$. Consequently

$$\mathbb{E}_{\mathbb{P}}(W_u^2 - W_t^2 \mid \mathcal{F}_t) = \mathbb{E}_{\mathbb{P}}((W_u - W_t)^2 \mid \mathcal{F}_t) = \mathbb{E}_{\mathbb{P}}(W_u - W_t)^2 = u - t,$$

and thus, for $0 \leq t \leq u \leq T$,

$$\mathbb{E}_{\mathbb{P}}(M_u \mid \mathcal{F}_t) = \mathbb{E}_{\mathbb{P}}(W_u^2 - u \mid \mathcal{F}_t) = W_t^2 - t = M_t. \qquad \square$$

Remarks. In view of the last result, the quadratic variation $\langle W \rangle$ can also be seen as the unique increasing and continuous process arising in the *Doob-Meyer decomposition* (see Proposition A.7.2) of continuous submartingale $X = W^2$.

Before stating an important corollary to Proposition A.4.1, let us recall that for any function $f : [0, T] \to \mathbb{R}$ the *variation of f on the interval $[t, u]$*, denoted as $\mathrm{var}_{[t,u]}(f)$, is defined as

$$\mathrm{var}_{[t,u]}(f) = \sup_{\pi \in \Pi_{[t,u]}} \sum_{k=1}^{n(\pi)} |f(t_k) - f(t_{k-1})|, \tag{A.14}$$

where $\Pi_{[t,u]}$ is the class of all finite partitions $\pi = \{t = t_0 < t_1 < \cdots < t_{n(\pi)} = u\}$ of the interval $[t, u]$. Note that $n(\pi)$ stands for the number of subintervals in a partition π.

Corollary A.4.1 *Sample paths of a Brownian motion are almost surely functions of infinite variation on any interval.*

Proof. The proof hinges on the observation that for any partition of $[t, u]$ we have

$$\sum_{k=1}^{n(\pi)} (f(t_k) - f(t_{k-1}))^2 \leq \max_{m=1,\ldots,n(\pi)} \big(f(t_m) - f(t_{m-1})\big) \sum_{k=1}^{n(\pi)} |f(t_k) - f(t_{k-1})|$$

$$\leq \max_{m=1,\ldots,n(\pi)} \big(f(t_m) - f(t_{m-1})\big) \, \mathrm{var}_{[t,u]}(f).$$

Since any continuous function is uniformly continuous on the compact interval $[t, u]$, it is not difficult to show that the last inequality would contradict Proposition A.4.1 if $\mathbb{P}\{\mathrm{var}_{[t,u]}(W) < +\infty\} > 0$. We conclude that for any $t < u$ we have $\mathrm{var}_{[t,u]}(W) = +\infty$ almost surely.   □

Remarks. Though in the proof of Proposition A.4.1 we used the fact that the increments of a Brownian motion have the Gaussian probability distribution, the property established in Corollary A.4.1 is by far more general, since it is enjoyed by any continuous local martingale. Specifically, any continuous local martingale with sample paths of finite variation is a constant process (see Proposition A.7.3).

The definition of a one-dimensional Brownian motion can be easily extended to the case of a d-dimensional process.

Definition A.4.2 An $\mathbb{R}^d$-valued, $\mathbb{F}$-adapted process $W = (W^1, \ldots, W^d)$, with $W_0 = 0$, defined on a filtered probability space $(\Omega, \mathbb{F}, \mathbb{P})$, is called a *$d$-dimensional standard Brownian motion* with respect to the filtration $\mathbb{F}$ if processes $W^1, \ldots, W^d$ are independent one-dimensional standard Brownian motions.

Lemma A.4.3 *Let $W = (W^1, \ldots, W^d)$ be a d-dimensional standard Brownian motion with respect to the filtration $\mathbb{F}$. Then for any $i \neq j$ the process $M = W^i W^j = (W_t^i W_t^j)_{t \in [0,T]}$ is a continuous square-integrable martingale.*

Proof. Let us fix $i \neq j$. It is easily seen that M is an $\mathbb{F}$-adapted and square-integrable process. Let us thus focus on the equality $\mathbb{E}_\mathbb{P}(M_u \mid \mathcal{F}_t) = M_t$ for any $0 \leq t \leq u \leq T$. We have

$$\mathbb{E}_\mathbb{P}(M_u \mid \mathcal{F}_t) = \mathbb{E}_\mathbb{P}(M_u - M_t \mid \mathcal{F}_t) + \mathbb{E}_\mathbb{P}(M_t \mid \mathcal{F}_t) = M_t \tag{A.15}$$

since

$$\mathbb{E}_\mathbb{P}(M_u - M_t \mid \mathcal{F}_t) = \mathbb{E}_\mathbb{P}(W_u^i W_u^j - W_t^i W_t^j \mid \mathcal{F}_t)$$

$$= W_t^i \, \mathbb{E}_\mathbb{P}(W_u^j - W_t^j \mid \mathcal{F}_t) + W_t^j \, \mathbb{E}_\mathbb{P}(W_u^i - W_t^i \mid \mathcal{F}_t)$$

$$+ \mathbb{E}_\mathbb{P}((W_u^i - W_t^i)(W_u^j - W_t^j) \mid \mathcal{F}_t)$$

$$= \mathbb{E}_\mathbb{P}((W_u^i - W_t^i)(W_u^j - W_t^j)) = \mathbb{E}_\mathbb{P}(W_u^i - W_t^i) \, \mathbb{E}_\mathbb{P}(W_u^j - W_t^j) = 0,$$

where we used, in particular, the independence of the increments $W_u^i - W_t^i$ and $W_u^j - W_t^j$ and their independence of $\mathcal{F}_t$.   □

Corollary A.4.2 *Assume that* $W = (W^1, \ldots, W^d)$ *is a d-dimensional standard Brownian motion with respect to the filtration* $\mathbb{F}$. *Then for any* $i, j = 1, \ldots, d$ *the real-valued process* $W_t^i W_t^j - \delta_{ij} t$ *is an* $\mathbb{F}$-*martingale, where* $\delta_{ij} = 0$ *for any* $i \neq j$ *and* $\delta_{ii} = 1$ *for* $i = 1, \ldots, d$.

Let us define the quadratic cross-variation $\langle W^i, W^j \rangle$ by setting

$$\langle W^i, W^j \rangle_t = \tfrac{1}{2} \left(\langle W^i + W^j, W^i + W^j \rangle_t - \langle W^i, W^i \rangle_t - \langle W^j, W^j \rangle_t \right).$$

It can be checked that $\langle W^i, W^j \rangle_t = \delta_{ij} t$ for $t \in [0, T]$.

A.5 Stopping Times and Martingales

The following concept will prove useful in what follows.

Definition A.5.1 A random variable $\tau : (\Omega, \mathbb{F}, \mathbb{P}) \rightarrow [0, T]$ is called a *stopping time* with respect to the filtration $\mathbb{F}$ (or briefly, an $\mathbb{F}$-*stopping time*) if, for every $t \in [0, T]$, the event $\{\tau \leq t\}$ belongs to the σ-field $\mathcal{F}_t$.

Recall that a process X is *progressively measurable* with respect to the filtration $\mathbb{F}$ if, for every t, the map $(u, \omega) \rightarrow X_u(\omega)$ from $[0, t] \times \Omega \rightarrow \mathbb{R}$ is $\mathcal{B}([0, t]) \otimes \mathcal{F}_t$-measurable. It is known that any $\mathbb{F}$-adapted process with right- or left-continuous sample paths is progressively measurable with respect to the filtration $\mathbb{F}$. For any progressively measurable process X and any stopping time τ the *stopped process* X^τ, which is defined by $X_t^\tau = X_{\tau \wedge t}$ for $t \in [0, T]$, is also progressively measurable. For any $\mathbb{F}$-martingale M and any $\mathbb{F}$-stopping time τ, the stopped process M^τ is known to be an $\mathbb{F}$-martingale. The following result furnishes a useful characterization of martingales.

Lemma A.5.1 *Let* $M = (M_t)_{t \in [0, T]}$ *be an* $\mathbb{F}$-*adapted and integrable stochastic process. Then* M *is an* $\mathbb{F}$-*martingale if and only if we have* $\mathbb{E}_{\mathbb{P}}(M_\tau) = \mathbb{E}_{\mathbb{P}}(M_0)$ *for any stopping time* τ *with values in* $[0, T]$.

Proof. It suffices to check that $M_t = \mathbb{E}_{\mathbb{P}}(M_T \mid \mathcal{F}_t)$ for every $t \in [0, T]$. By assumption, we have that $\mathbb{E}_{\mathbb{P}}(M_\tau) = \mathbb{E}_{\mathbb{P}}(M_T)$ for any stopping time τ with values in $[0, T]$ (this is true since $\tau = T$ is also a stopping time with values in $[0, T]$). Let us fix t and let us consider an event A belonging to $\mathcal{F}_t$. Then the random variable τ_A defined as follows

$$\tau_A = \begin{cases} t, & \text{if } \omega \in A, \\ T, & \text{if } \omega \notin A, \end{cases}$$

is a stopping time with values in $[0, T]$. The equality $\mathbb{E}_{\mathbb{P}}(M_{\tau_A}) = \mathbb{E}_{\mathbb{P}}(M_T)$ yields $\mathbb{E}_{\mathbb{P}}(\mathbb{1}_A M_t) = \mathbb{E}_{\mathbb{P}}(\mathbb{1}_A M_T)$. Since this equality holds for any event A in $\mathcal{F}_t$, by virtue of definition of conditional expectation, we obtain the equality $M_t = \mathbb{E}_{\mathbb{P}}(M_T \mid \mathcal{F}_t)$, which in turn implies that M is an $\mathbb{F}$-martingale. $\square$

It appears that the martingale property is too restrictive if we wish define the Itô integral for a large class of processes. Hence the following definition.

Definition A.5.2 A process M is said to be a *local martingale* with respect to $\mathbb{F}$ if there exists an increasing sequence $(\tau_n)_{n \in \mathbb{N}}$ of stopping times such that τ_n tends to T almost surely, and for every n the process M^n, given by the formula

$$M_t^n = \begin{cases} M_{t \wedge \tau_n(\omega)}(\omega), & \text{if } \tau_n(\omega) > 0, \\ 0, & \text{if } \tau_n(\omega) = 0, \end{cases}$$

is a uniformly integrable martingale. Any sequence $(\tau_n)_{n \in \mathbb{N}}$ with these properties is called the *reducing sequence* for a local martingale M.

Any martingale is also a local martingale. It is noteworthy that the class of (continuous) local martingales is essentially larger than the class of (continuous) martingales.

A.6 Itô Stochastic Integral

Since almost all sample paths of a Brownian motion have infinite variation on every open interval, the classical Lebesgue-Stieltjes integration theory cannot be applied to define an integral of a stochastic process with respect to a Brownian motion. We shall now describe briefly the Itô integration theory, which is the basis of the Itô stochastic calculus. Let W be a standard d-dimensional Brownian motion defined on a filtered probability space $(\Omega, \mathbb{F}, \mathbb{P})$. For simplicity, the horizon date $T > 0$ will be fixed throughout.

Itô integral for elementary processes. Let $\mathcal{K}$ stand for the space of d-dimensional *elementary processes,* that is, processes of the form

$$\gamma(t) = \gamma_{-1} \mathbb{1}_0 + \sum_{j=0}^{m-1} \gamma_j \mathbb{1}_{(t_j, t_{j+1}]}(t), \quad \forall t \in [0, T], \tag{A.16}$$

where $t_0 = 0 < t_1 < \cdots < t_m = T$, the $\mathbb{R}^d$-valued random variables γ_j, $j = 0, \ldots, m - 1$ are uniformly bounded and $\mathcal{F}_{t_j}$-measurable and the random variable γ_{-1} is $\mathcal{F}_0$-measurable. For any process $\gamma \in \mathcal{K}$, the *Itô stochastic integral* $\hat{I}_T(\gamma)$ with respect to W over the time interval $[0, T]$ is defined by the formula (the dot ' $\cdot$ ' stands hereafter for the Euclidean inner product in $\mathbb{R}^d$)

$$\hat{I}_T(\gamma) = \int_0^T \gamma_u \cdot dW_u \overset{\text{def}}{=} \sum_{j=0}^{m-1} \gamma_j \cdot (W_{t_{j+1}} - W_{t_j}). \tag{A.17}$$

Similarly, the Itô stochastic integral of γ with respect to W over any interval $[0, t]$, where $t \leq T$, is defined by setting

$$\hat{I}_t(\gamma) = \int_0^t \gamma_u \cdot dW_u \overset{\text{def}}{=} \hat{I}_T(\gamma \mathbb{1}_{[0,t)}) = \sum_{j=0}^{m-1} \gamma_j \cdot \left(W_{t_{j+1} \wedge t} - W_{t_j \wedge t} \right), \tag{A.18}$$

where $x \wedge y = \min\{x, y\}$. It is not difficult to check that for any process $\gamma \in \mathcal{K}$ the Itô integral $I_t(\gamma)$, $t \in [0, T]$, is a continuous $\mathbb{F}$-martingale; in particular, the equality $\mathbb{E}_{\mathbb{P}}(I_u(\gamma) \mid \mathcal{F}_t) = I_t(\gamma)$ holds for any $t \leq u \leq T$, or more explicitly

$$\mathbb{E}_{\mathbb{P}}\left(\int_0^u \gamma_s \cdot dW_s \,\Big|\, \mathcal{F}_t\right) = \int_0^t \gamma_s \cdot dW_s.$$

Moreover, the following properties are valid, for $t \in [0, T]$,

(i) (*linearity*) for any $a, b \in \mathbb{R}$ and any processes $\gamma, \eta \in \mathcal{K}$

$$\int_0^t (a\gamma_u + b\eta_u) \cdot dW_u = a \int_0^t \gamma_u \cdot dW_u + b \int_0^t \eta_u \cdot dW_u,$$

(ii) (*local property*) for any $\mathbb{F}$-stopping time τ and any process $\gamma \in \mathcal{K}$ we have

$$\int_0^t \gamma_u \mathbb{1}_{[0,\tau]}(u) \cdot dW_u = \int_0^t \gamma_u^\tau \cdot dW_u^\tau = I_t^\tau(\gamma).$$

Itô integral for processes from $\mathcal{L}_{\mathbb{P}}^2(W)$. In the next step, the definition of the Itô integral will be extended from the class $\mathcal{K}$ to the class $\mathcal{L}_{\mathbb{P}}^2(W)$ of all $\mathbb{F}$-progressively measurable processes γ defined on $(\Omega, \mathbb{F}, \mathbb{P})$ for which

$$\|\gamma\|_W^2 \stackrel{\text{def}}{=} \mathbb{E}_{\mathbb{P}}\left(\int_0^T |\gamma_u|^2 \, du\right) < \infty, \tag{A.19}$$

where $|\cdot|$ stands for the Euclidean norm in $\mathbb{R}^d$.

Lemma A.6.1 *The class $\mathcal{K}$ is a subset of $\mathcal{L}_{\mathbb{P}}^2(W)$ and for any $\gamma \in \mathcal{K}$*

$$\mathbb{E}_{\mathbb{P}}\left(\int_0^T \gamma_u \cdot dW_u\right)^2 = \|I_T(\gamma)\|_{L^2}^2 = \|\gamma\|_W^2 . \tag{A.20}$$

The space $\mathcal{L}_{\mathbb{P}}^2(W)$ equipped with the norm $\|\cdot\|_W$ is a complete normed linear space – that is, a Banach space. Moreover, the class $\mathcal{K}$ of elementary stochastic processes is a dense linear subspace of $\mathcal{L}_{\mathbb{P}}^2(W)$.

By virtue of Lemma A.6.1 and well known results from the classic analysis, the isometry $\hat{I}_T : (\mathcal{K}, \|\cdot\|_W) \to L^2(\Omega, \mathcal{F}_T, \mathbb{P})$ can be extended to the isometry $I_T : (\mathcal{L}_{\mathbb{P}}^2(W), \|\cdot\|_W) \to L^2(\Omega, \mathcal{F}_T, \mathbb{P})$. This leads to the following definition.

Definition A.6.1 For any process $\gamma \in \mathcal{L}_{\mathbb{P}}^2(W)$, the random variable $I_T(\gamma)$ is called the *Itô stochastic integral* of γ with respect to W over $[0, T]$, and it is denoted by $\int_0^T \gamma_u \cdot dW_u$.

In other words, for any $\gamma \in \mathcal{L}_{\mathbb{P}}^2(W)$ we set $I_T(\gamma) = \lim_{n \to \infty} I_T(\gamma)$ (with the limit in $L^2(\Omega, \mathcal{F}_T, \mathbb{P})$) for any sequence of processes $(\gamma^n)_{n \in \mathbb{N}}$ such that $\gamma^n \in \mathcal{K}$ for $n \in \mathbb{N}$ and

$$\lim_{n \to \infty} \mathbb{E}_{\mathbb{P}}\left(\int_0^T |\gamma_u - \gamma_u^n|^2 \, du\right) = \lim_{n \to \infty} \|\gamma - \gamma^n\|_W^2 = 0.$$

More generally, for every $\gamma \in \mathcal{L}^2_{\mathbb{P}}(W)$ and every $t \in [0, T]$, we set

$$I_t(\gamma) = \int_0^t \gamma_u \cdot dW_u \overset{\text{def}}{=} I_T(\gamma \mathbb{1}_{[0,t]}) \tag{A.21}$$

and thus we define the Itô stochastic integral $I(\gamma)$ as an $\mathbb{F}$-adapted stochastic process. The next result summarizes its most important properties.

Proposition A.6.1 *For any process* $\gamma \in \mathcal{L}^2_{\mathbb{P}}(W)$, *the Itô stochastic integral* $I(\gamma)$ *is a square-integrable continuous martingale on* $(\Omega, \mathbb{F}, \mathbb{P})$. *Moreover, the process*

$$(I_t(\gamma))^2 - \langle I(\gamma) \rangle_t, \quad \forall t \in [0, T], \tag{A.22}$$

is a continuous martingale on $(\Omega, \mathbb{F}, \mathbb{P})$, *where*

$$\langle I(\gamma) \rangle_t \overset{\text{def}}{=} \int_0^t |\gamma_u|^2 \, du, \quad \forall t \in [0, T]. \tag{A.23}$$

It is worth mentioning that the Itô integral is a linear map on $\mathcal{L}^2_{\mathbb{P}}(W)$ and it enjoys the local property with respect to $\mathbb{F}$-stopping times, that is, $I(\gamma \mathbb{1}_{[0,\tau]}) = I^\tau(\gamma)$ for any $\mathbb{F}$-stopping time τ and any process $\gamma \in \mathcal{L}^2_{\mathbb{P}}(W)$.

Itô integral for processes from $\mathcal{L}_{\mathbb{P}}(W)$. By applying the optional stopping technique (also known as a *localization*), we shall extend the definition of Itô stochastic integral to the class $\mathcal{L}_{\mathbb{P}}(W)$ of all progressively measurable processes γ for which

$$\mathbb{P}\left\{ \int_0^T |\gamma_u|^2 \, du < \infty \right\} = 1. \tag{A.24}$$

Let a process γ belong to $\mathcal{L}_{\mathbb{P}}(W)$. In order to define the Itô integral $I_T(\gamma)$, we first introduce an increasing sequence $(\tau_n)_{n \in \mathbb{N}}$ of $\mathbb{F}$-stopping times by setting

$$\tau_n = \inf\left\{ t \in [0, T] \,\Big|\, \int_0^t |\gamma_u|^2 \, du \geq n \right\},$$

where by convention $\inf \emptyset = T$. Since the process γ belongs to the class $\mathcal{L}_{\mathbb{P}}(W)$, it is not difficult to verify that $\lim_{n \to \infty} \tau_n = T$.

Furthermore, for any $n \in \mathbb{N}$ the process γ^n, which is given by the formula $\gamma^n_t = \gamma_t \mathbb{1}_{[0,\tau_n]}(t)$, $t \in [0, T]$, manifestly belongs to the class $\mathcal{L}^2_{\mathbb{P}}(W)$ and thus the Itô integral $I_T(\gamma^n)$ is well defined. Using the local property of the Itô integral, we obtain

$$I(\gamma^n) = I(\gamma \mathbb{1}_{[0,\tau_n]}) = \left(I(\gamma^n) \right)^{\tau_n}.$$

We may thus define the Itô integral $I_T(\gamma)$ by setting

$$I_T(\gamma) = \lim_{n \to \infty} I_T(\gamma^n).$$

We need, of course, to check that the right-hand side in the last equality is well defined and does not depend on the choice of the *localizing* (or *reducing*) sequence of $\mathbb{F}$-stopping times. The details of this construction are left to the reader (see, for instance, Karatzas and Shreve (1998a)).

Let us stress that the Itô integral of a process from the class $\mathcal{L}_\mathbb{P}(W)$ does not necessarily follow an $\mathbb{F}$-martingale. We have, however, the following result, which is a rather straightforward consequence of the definition of the Itô integral $I(\gamma)$ for a process γ belonging to the class $\mathcal{L}_\mathbb{P}(W)$.

Proposition A.6.2 *Assume that a stochastic process γ belongs to $\mathcal{L}_\mathbb{P}(W)$. Then the Itô stochastic integral $I(\gamma)$ follows a continuous local martingale with respect to the filtration $\mathbb{F}$.*

Remarks. It is clear that the space $\mathcal{L}_\mathbb{P}(W)$ of stochastic processes is invariant with respect to an equivalent change of probability measure on $(\Omega, \mathcal{F}_T)$. Specifically, the equality $\mathcal{L}_\mathbb{P}(W) = \mathcal{L}_{\tilde{\mathbb{P}}}(\tilde{W})$ holds whenever $\mathbb{P}$ and $\tilde{\mathbb{P}}$ are mutually equivalent probability measures on $(\Omega, \mathcal{F}_T)$ and the processes W and $\tilde{W}$ are Brownian motions under $\mathbb{P}$ and $\tilde{\mathbb{P}}$ respectively. Since we restrict ourselves to equivalent changes of probability measures, on a given reference probability space $(\Omega, \mathcal{F}_T, \mathbb{P})$, we shall write shortly $\mathcal{L}(W)$ instead of $\mathcal{L}_\mathbb{P}(W)$ in what follows. Hence a process γ is called *integrable with respect to W* if it belongs to the class $\mathcal{L}(W)$.

A.7 Continuous Local Martingales

It order to simplify the presentation, we assume that the σ-field $\mathcal{F}_0$ is trivial, that is, $\mathbb{P}(A) = 0$ or $\mathbb{P}(A) = 1$ for any $A \in \mathcal{F}_0$. This implies that the initial value M_0 can be interpreted as a real number. The foregoing result deals with non-negative (not necessarily continuous) local martingales.

Proposition A.7.1 *Any non-negative local martingale M with respect to $\mathbb{F}$ is also an $\mathbb{F}$-supermartingale. If, in addition, M the expected value of M is constant then M is an $\mathbb{F}$-martingale.*

Proof. Recall that if M is a local martingale with respect to $\mathbb{F}$ then there exists an increasing sequence $(\tau_n)_{n \in \mathbb{N}}$ of $\mathbb{F}$-stopping times such that

$$\mathbb{P}\{ \lim_{n \to \infty} \tau_n = T \} = 1$$

and for any $n \in \mathbb{N}$ the stopped process M^{τ_n} is an $\mathbb{F}$-martingale. This means that we have, for any fixed $n \in \mathbb{N}$ and arbitrary dates $0 \le t \le u \le T$,

$$M_t^{\tau_n} = \mathbb{E}_\mathbb{P}(M_u^{\tau_n} \mid \mathcal{F}_t).$$

Since the process M is non-negative, using the conditional form of Fatou's lemma (see Lemma A.1.2), we obtain, for $0 \le t \le u \le T$,

$$\mathbb{E}_{\mathbb{P}}(M_u \mid \mathcal{F}_t) = \mathbb{E}_{\mathbb{P}}(\liminf_{n\to\infty} M_u^{\tau_n} \mid \mathcal{F}_t) \le \liminf_{n\to\infty} \mathbb{E}_{\mathbb{P}}(M_u^{\tau_n} \mid \mathcal{F}_t)$$
$$= \liminf_{n\to\infty} M_t^{\tau_n} = M_t.$$

We conclude that M is an $\mathbb{F}$-supermartingale. In particular, M is an integrable process, that is, $\mathbb{E}_{\mathbb{P}} M_t \le \mathbb{E}_{\mathbb{P}} M_0 < \infty$ for $t \in [0, T]$.

Assume, in addition, that $\mathbb{E}_{\mathbb{P}} M_u = \mathbb{E}_{\mathbb{P}} M_t$. To show that M is an $\mathbb{F}$-martingale under this additional assumption, we shall argue by contradiction. Assume that M is not an $\mathbb{F}$-martingale, so that the supermartingale inequality is strict with positive probability, that is,

$$\mathbb{P}\{M_t > \mathbb{E}_{\mathbb{P}}(M_u \mid \mathcal{F}_t)\} > 0.$$

The last inequality implies that

$$\mathbb{E}_{\mathbb{P}} M_t > \mathbb{E}_{\mathbb{P}}(\mathbb{E}_{\mathbb{P}}(M_u \mid \mathcal{F}_t)) = \mathbb{E}_{\mathbb{P}} M_u$$

and this clearly contradicts our assumption that the expected value of M is constant. □

The assumption that M is non-negative can be replaced by the assumption that $M_t \ge \eta$ for some random variable η with $\mathbb{E}_{\mathbb{P}} \eta > -\infty$.

Corollary A.7.1 *Let M be a non-negative local martingale M with respect to $\mathbb{F}$. If $M_0 = 0$ then $M_t = 0$ for every $t \in [0, T]$.*

Proof. Assume that $M_{t_0} \ne 0$ for some $t_0 \in [0, T]$. Then $\mathbb{E}_{\mathbb{P}} M_{t_0} > 0 = \mathbb{E}_{\mathbb{P}} M_0$, which contradicts the property that M is a supermartingale. □

The following definition extends the concept of quadratic variation to the class of all continuous local martingales.

Definition A.7.1 Let M be a continuous local martingale with respect to $\mathbb{F}$. We denote by $\langle M \rangle$ the unique continuous, increasing and $\mathbb{F}$-adapted process with $\langle M \rangle_0 = 0$ such that the process $M^2 - \langle M \rangle$ is a continuous local martingale with respect to $\mathbb{F}$. The process $\langle M \rangle$ is termed the *quadratic variation* of M.

In view of Proposition A.6.1, it is clear that the notation introduced previously in formula (A.23) is consistent with this more general definition.

It is worth noting that the quadratic variation is invariant with respect to an $\mathcal{F}_0$-measurable shift of M; specifically, if M is a continuous local martingale and a continuous local martingale N satisfies $N = \psi + M$ for some $\mathcal{F}_0$-measurable random variable ψ then $\langle N \rangle = \langle M \rangle$. In particular, the quadratic variations of M and $M - M_0$ are identical, that is, $\langle M \rangle = \langle M - M_0 \rangle$.

The existence and uniqueness of the process $\langle M \rangle$ satisfying conditions of Definition A.7.1 is a non-trivial issue, but it can be deduced from the Doob-Meyer decomposition theorem. We state below without proof a suitable version of this classic result. Let us stress that the result given below is by no means the most general version of the Doob-Meyer decomposition theorem (see, for instance, Karatzas and Shreve (1998a)).

Proposition A.7.2 *Let X be a non-negative continuous submartingale. Then there exists a continuous increasing process A with $A_0 = 0$ such that the process $N = X - A$ is a continuous martingale. The processes N and A in the Doob-Meyer decomposition $X = N + A$ are unique up to indistinguishability of stochastic processes.*

It is common to refer to the process A as the *compensator* of a submartingale X.

In our case, we apply Proposition A.7.2 to the process $X = M^2$, which is (after localization) a bounded non-negative continuous submartingale. In that case, the compensator A of $X = M^2$ satisfies the definition of the quadratic variation $\langle M \rangle$.

Remarks. It is known that the quadratic variation $\langle M \rangle$ of a continuous local martingale M can also be interpreted as the following limit (cf. (A.13))

$$\langle M \rangle_t = \lim_{n \to \infty} \sum_{k=1}^{m_n} (M_{t_k^n} - M_{t_{k-1}^n})^2, \tag{A.25}$$

where the convergence in probability holds for any sequence $(\pi_n)_{n \in \mathbb{N}}$ of finite partitions $\pi_n = \{0 = t_0^n < t_1^n < \cdots < t_{m_n}^n = t\}$ of the interval $[0, t]$ satisfying $\lim_{n \to \infty} \delta(\pi_n) = 0$. Hence the name *quadratic variation* attributed to the process $\langle M \rangle$ is indeed justified.

Lemma A.7.1 *Let M be a continuous local martingale such that $\langle M \rangle_t = 0$ for $t \in [0, T]$. Then $M_t = M_0$ for every $t \in [0, T]$.*

Proof. We may and do assume, without loss of generality, that M is a continuous bounded martingale and $M_0 = 0$. Then $\mathbb{E}_{\mathbb{P}} M_t^2 = \mathbb{E}_{\mathbb{P}} \langle M \rangle_t = 0$ for any $t \in [0, T]$ and thus $M_t = 0$ for any $t \in [0, T]$. □

Definition A.7.2 *An $\mathbb{F}$-adapted stochastic process $X = (X_t)_{t \in [0,T]}$ is said to be of finite variation if almost all sample paths of X are functions of finite variation on $[0, T]$.*

More explicitly, a process X is of finite variation whenever (cf. (A.14))

$$\text{var}_{[0,T]}(X) \overset{\text{def}}{=} \sup_{\pi \in \Pi_{[0,T]}} \sum_{k=1}^{n(\pi)} |X_{t_k} - X_{t_{k-1}}| < \infty, \quad \mathbb{P}\text{-a.s.}, \tag{A.26}$$

where $\Pi_{[0,T]}$ is the class of all finite partitions $\pi = \{0 = t_0 < t_1 < \cdots < t_{n(\pi)} = T\}$ of the interval $[0, T]$.

Proposition A.7.3 *Let M be a continuous local martingale of finite variation. Then $M_t = M_0$ for every $t \in [0, T]$.*

Proof. It suffices to observe that the assumption that sample paths of M are continuous functions of finite variation implies that the quadratic variation of M vanishes. This follows from the observation that sample path of $\langle M \rangle$ can be obtained as the almost sure limit in (A.25) for some sequence of partitions. The proof is thus based on similar arguments as those already employed in the proof of Corollary A.4.1. □

The last result shows that continuous local martingales are never processes of finite variation. In particular,

Corollary A.7.2 *If $M_0 = 0$ and M is a continuous local martingales of finite variation then M vanishes (more precisely, it is indistinguishable from the null process).*

As in the special case of a Brownian motion, the stochastic integral of the form

$$\int_0^t \gamma_u \, dM_u$$

cannot be interpreted as the pathwise Lebesgue-Stieltjes integral. The theory of Itô integration with respect to continuous local martingales parallels the theory developed for the Brownian motion case. The only major modification in the one-dimensional case is that in the definition of spaces $\mathcal{L}_{\mathbb{P}}^2(M)$ and $\mathcal{L}_{\mathbb{P}}(M)$ the norm $\| \gamma \|_W^2$ should be replaced by its generalization:

$$\| \gamma \|_M^2 \overset{\text{def}}{=} \mathbb{E}_{\mathbb{P}} \Big(\int_0^T \gamma_u^2 \, d\langle M \rangle_u \Big).$$

A.8 Continuous Semimartingales

The concept of a continuous semimartingale is a natural extension of the notion of a continuous local martingale.

Definition A.8.1 A real-valued, continuous, $\mathbb{F}$-adapted process X is called a *(real-valued) continuous semimartingale* if it admits a decomposition

$$X_t = X_0 + M_t + A_t, \quad \forall t \in [0, T], \tag{A.27}$$

where X_0, M and A satisfy:
(i) X_0 is an $\mathcal{F}_0$-measurable random variable,
(ii) M is a continuous local martingale with $M_0 = 0$,
(iii) A is a continuous process whose almost all sample paths are of finite variation on the interval $[0, T]$ with $A_0 = 0$.

We denote by $\mathcal{S}^c(\mathbb{P})$ the class of all real-valued continuous semimartingales on the probability space $(\Omega, \mathbb{F}, \mathbb{P})$. It appears that the decomposition of X with the properties given in Definition A.8.1 is unique, up to indistinguishability of stochastic processes, as the following result shows.

Proposition A.8.1 *Let X be a continuous semimartingale with the decomposition $X = X_0 + M + A$ so that, in particular, $M_0 = A_0 = 0$. If X admits also a decomposition $X = X_0 + \tilde{M} + \tilde{A}$ for some continuous local martingale $\tilde{M}$ with $\tilde{M}_0 = 0$ and some continuous process $\tilde{A}$ of finite variation on $[0, T]$ with $\tilde{A}_0 = 0$ then $M_t = \tilde{M}_t$ and $A_t = \tilde{A}_t$ for $t \in [0, T]$.*

Proof. It is enough to observe that a difference of two continuous local martingales is also a continuous local martingales, and a difference of two continuous processes of finite variation also follows a continuous process of finite variation. In our case, we have $M - \tilde{M} = \tilde{A} - A$ and $M_0 - \tilde{M}_0 = \tilde{A}_0 - A_0 = 0$. Consequently, in view of Lemma A.7.3, we conclude that the equalities $M_t = \tilde{M}_t$ and $A_t = \tilde{A}_t$ hold for every $t \in [0, T]$. □

In view of the uniqueness of the decomposition $X = X_0 + M + A$ established in Proposition A.8.1, it is justified to call it the *canonical decomposition* of a continuous semimartingale X under $\mathbb{P}$. Since the continuous local martingale M in the canonical decomposition of a continuous semimartingale X, it is customary to denote it by X^c. The *quadratic variation* $\langle X \rangle$ of a continuous semimartingale X is then defined as as the quadratic variation of X^c, that is, $\langle X \rangle = \langle X^c \rangle$. Note that

$$\langle X \rangle_t = \lim_{n \to \infty} \sum_{k=1}^{m_n} (X_{t_k^n} - X_{t_{k-1}^n})^2 = \lim_{n \to \infty} \sum_{k=1}^{m_n} (X_{t_k^n}^c - X_{t_{k-1}^n}^c)^2$$

where the convergence in probability holds for any sequence $(\pi_n)_{n \in \mathbb{N}}$ of finite partitions $\pi_n = \{0 = t_0^n < t_1^n < \cdots < t_{m_n}^n = t\}$ of the interval $[0, t]$ satisfying $\lim_{n \to \infty} \delta(\pi_n) = 0$.

The next result is an immediate consequence of Corollary A.7.2.

Corollary A.8.1 *A continuous semimartingale is a continuous local martingale if and only if the process A it its canonical decomposition $X = X_0 + M + A$ vanishes, that is, is indistinguishable from the null process.*

If $\tilde{\mathbb{P}}$ is a probability measure equivalent to $\mathbb{P}$ on $(\Omega, \mathcal{F}_T)$ then X is also a continuous semimartingale under $\tilde{\mathbb{P}}$. However, its canonical decompositions under $\mathbb{P}$ and $\tilde{\mathbb{P}}$ are distinct, in general (cf. Theorem A.15.2).

The extension of the notion of a continuous semimartingale to a multidimensional setting is rather obvious. Specifically, we say that an $\mathbb{R}^d$-valued process $X = (X^1, \ldots, X^d)$ is a *d-dimensional continuous semimartingale* if each component $X^1, \ldots, X^d$ follows a real-valued continuous semimartingale. A d-dimensional Brownian motion $W = (W^1, \ldots, W^d)$ is a simple example of a d-dimensional continuous semimartingale.

Itô processes. Let W be a d-dimensional standard Brownian motion defined on a filtered probability space $(\Omega, \mathbb{F}, \mathbb{P})$. We shall now introduce a particular class of continuous semimartingales, referred to as *Itô processes*.

Definition A.8.2 An $\mathbb{F}$-adapted continuous process X is called an *Itô process* if it admits a representation

$$X_t = X_0 + \int_0^t \alpha_u \, du + \int_0^t \beta_u \cdot dW_u, \quad \forall \, t \in [0, T], \tag{A.28}$$

for some $\mathbb{F}$-adapted processes α and β that are defined on $(\Omega, \mathbb{F}, \mathbb{P})$ and satisfy suitable integrability conditions.

It is customary to represent the integral formula (A.28) using the differential notation as

$$dX_t = \alpha_t \, dt + \beta_t \cdot dW_t.$$

It is clear that the Itô process X follows a continuous semimartingale and formula (A.28) gives the canonical decomposition of X. In the present set-up, decomposition (A.28) is unique in the following sense: if X satisfies (A.28) and simultaneously we have that

$$X_t = X_0 + \int_0^t \tilde{\alpha}_u \, du + \int_0^t \tilde{\beta}_u \, dW_u, \quad \forall \, t \in [0, T],$$

for some $\mathbb{F}$-adapted processes $\tilde{\alpha}$ and $\tilde{\beta}$, then the following equalities hold, for every $t \in [0, T]$,

$$\int_0^t \tilde{\alpha}_u \, du = \int_0^t \alpha_u \, du, \quad \int_0^t \tilde{\beta}_u \, dW_u = \int_0^t \beta_u \, dW_u.$$

It follow immediately from Corollary A.8.1 that an Itô process X given by (A.28) is a continuous local martingale whenever it can be represented as

$$X_t = X_0 + \int_0^t \beta_u \cdot dW_u, \quad \forall \, t \in [0, T].$$

A.9 Itô's Lemma

In this section, we shall deal with the following problem: does the process $g(X_t)$ follow a semimartingale if X is a continuous semimartingale and g is a sufficiently regular function? It turns out that the class of continuous semimartingales is invariant with respect to compositions with C^2-functions (let us stress that more general results are also available).

One-dimensional case. Consider a real-valued function $g = g(x, t)$, where $x \in \mathbb{R}$ is the space variable and $t \in [0, T]$ is the time variable. It is evident that if X is a continuous semimartingale, and $g : \mathbb{R} \times [0, T] \to \mathbb{R}$ is a jointly continuous function, then the process $Y_t = g(X_t, t)$ is $\mathcal{F}_t$-adapted and has almost all sample paths continuous.

The next result, which is a special case of *Itô's lemma*, states that Y is a semimartingale provided that the function g is sufficiently smooth.

Proposition A.9.1 *Suppose that $g : \mathbb{R} \times [0, T] \to \mathbb{R}$ is a function of class $C^{2,1}(\mathbb{R} \times [0, T], \mathbb{R})$. Then for any Itô process X, the process $Y_t = g(X_t, t)$, $t \in [0, T]$, is an Itô process. Moreover, its canonical decomposition is given by the Itô formula*

$$dY_t = g_t(X_t, t) \, dt + g_x(X_t, t) \alpha_t \, dt + g_x(X_t, t) \beta_t \, dW_t + \tfrac{1}{2} g_{xx}(X_t, t) \beta_t^2 \, dt.$$

More generally, if $X = X_0 + M + A$ is a real-valued continuous semimartingale, and g is a function of class $C^{2,1}(\mathbb{R} \times [0, T], \mathbb{R})$ then the process $Y_t = g(X_t, t)$ follows a continuous semimartingale with the following canonical decomposition

$$dY_t = g_t(X_t, t) \, dt + g_x(X_t, t) \, dX_t + \tfrac{1}{2} g_{xx}(X_t, t) \, d\langle M \rangle_t. \tag{A.29}$$

Multidimensional case. Recall that a process $W = (W^1, \ldots, W^d)$ defined on a filtered probability space $(\Omega, \mathbb{F}, \mathbb{P})$ is called the d-*dimensional standard Brownian motion* if the components $W^1, \ldots, W^d$ are independent one-dimensional standard Brownian motions with respect to $\mathbb{F}$.

In this paragraph, W is assumed to be a d-dimensional standard Brownian motion. Let γ belong to $\mathcal{L}_W(\mathbb{P})$, that is, γ is a $\mathbb{R}^d$-valued $\mathbb{F}$-progressively measurable process satisfying

$$\mathbb{P}\left\{ \int_0^T |\gamma_u|^2 \, du < \infty \right\} = 1,$$

where $|\cdot|$ stands for the Euclidean norm in $\mathbb{R}^d$. Then the Itô stochastic integral of γ with respect to W is well defined and we have, for any $t \in [0, T]$,

$$I_t(\gamma) = \int_0^t \gamma_u \cdot dW_u = \sum_{i=1}^d \int_0^t \gamma_u^i \, dW_u^i.$$

Let $X = (X^1, \ldots, X^k)$ be a k-dimensional Itô process given as

$$X_t^i = X_0^i + \int_0^t \alpha_u^i \, du + \int_0^t \beta_u^i \cdot dW_u, \tag{A.30}$$

where α^i are real-valued processes and $\beta^i = (\beta^{i1}, \ldots, \beta^{id})$ are $\mathbb{R}^d$-valued processes for $i = 1, \ldots, k$. It is implicitly assumed in (A.30) that the processes $\alpha^i, \beta^i, i = 1, \ldots, k$ are integrable in a suitable sense.

Let us introduce notation for the *cross-variation* (or the *quadratic covariation*) of two continuous semimartingales. If $X^i = X_0^i + M^i + A^i$ are in $\mathcal{S}^c(\mathbb{P})$ for $i = 1, \ldots, k$ then we set $\langle X^i, X^j \rangle \stackrel{\text{def}}{=} \langle M^i, M^j \rangle$, where in turn $\langle M^i, M^j \rangle$ is given by the following *polarization equality*

$$\langle M^i, M^j \rangle \stackrel{\text{def}}{=} \tfrac{1}{2}\left(\langle M^i + M^j \rangle - \langle M^i \rangle - \langle M^j \rangle \right) = \tfrac{1}{4}\left(\langle M^i + M^j \rangle - \langle M^i - M^j \rangle \right).$$

It is easily seen that $\langle X^i, X^i \rangle = \langle X^i \rangle$. If X^i and X^j are the Itô processes given by (A.30) then it is easily seen that we have, for any $t \in [0, T]$,

$$\langle X^i, X^j \rangle_t = \int_0^t \beta_u^i \cdot \beta_u^j \, du = \int_0^t \sum_{l=1}^d \beta_u^{il} \beta_u^{jl} \, du$$

The following result extends Proposition A.9.1.

Proposition A.9.2 *Suppose that g is a function of class $C^2(\mathbb{R}^k, \mathbb{R})$. Then the following form of the Itô formula is valid*

$$dg(X_t) = \sum_{i=1}^k g_{x_i}(X_t)\alpha_t^i \, dt + \sum_{i=1}^k g_{x_i}(X_t)\beta_t^i \cdot dW_t + \frac{1}{2} \sum_{i,j=1}^k g_{x_i x_j}(X_t) \, \beta_t^i \cdot \beta_t^j \, dt.$$

More generally, if processes X^i are in $\mathcal{S}^c(\mathbb{P})$ for $i = 1, \ldots, k$ then

$$g(X_t) = g(X_0) + \sum_{i=1}^{k} \int_0^t g_{x_i}(X_u)\, dX_u^i + \frac{1}{2} \sum_{i,j=1}^{k} \int_0^t g_{x_i x_j}(X_u)\, d\langle X^i, X^j \rangle_u,$$

or equivalently

$$dg(X_t) = \sum_{i=1}^{k} g_{x_i}(X_t)\, dX_t^i + \frac{1}{2} \sum_{i,j=1}^{k} g_{x_i x_j}(X_t)\, d\langle X^i, X^j \rangle_t.$$

The *Itô integration by parts formula* is obtained by taking $g(x_1, x_2) = x_1 x_2$.

Corollary A.9.1 *Let X^1 and X^2 belong to $\mathcal{S}^c(\mathbb{P})$. Then*

$$X_t^1 X_t^2 = X_0^1 X_0^2 + \int_0^t X_u^1\, dX_u^2 + \int_0^t X_u^2\, dX_u^1 + \langle X^1, X^2 \rangle_t. \tag{A.31}$$

In particular, for any $X \in \mathcal{S}^c(\mathbb{P})$

$$X_t^2 = X_0^2 + 2 \int_0^t X_u\, dX_u + \langle X \rangle_t. \tag{A.32}$$

Itô-Tanaka-Meyer formula. The classic Itô formula is obtained under the assumption that the transformation is twice continuously differentiable in the space variable. It is noteworthy that in the one-dimensional case the preservation of the semimartingale property of X holds under a much weaker assumption that a function $g : \mathbb{R} \to \mathbb{R}$ is a difference of two convex functions.

Let X be a real-valued continuous semimartingale, with the canonical decomposition $X = X_0 + M + A$. For any fixed $a \in \mathbb{R}$, we denote by $L_t^a(X)$ the (right) semimartingale *local time* of X at the level a, that is, the process given explicitly by the formula

$$L_t^a(X) = |X_t - a| - |X_0 - a| - \int_0^t \operatorname{sgn}(X_u - a)\, dX_u$$

for every $t \in [0, T]$, where $\operatorname{sgn}(x) = 1$ for $x > 0$ and $\operatorname{sgn}(x) = -1$ for $x < 0$. By convention, we set $\operatorname{sgn}(0) = -1$ so that the function $\operatorname{sgn}(x - a)$ represents the left-hand side derivative of the function $|x - a|$. It is well known that the local time $L^a(X)$ of a continuous semimartingale X is an adapted process whose sample paths are almost all continuous, non-decreasing functions, and we have

$$L_t^a(X) = \int_0^t \mathbb{1}_{\{X_u = a\}}\, dL_u^a(X),$$

as well as

$$L_t^a(X) = \lim_{\epsilon \downarrow 0} \frac{1}{2\epsilon} \int_0^t \mathbb{1}_{\{|X_u - a| \le \epsilon\}}\, d\langle M \rangle_u.$$

In addition, for any bounded (or nonnegative) measurable function $g : \mathbb{R} \to \mathbb{R}$ the following *density of occupation time formula* holds:

$$\int_{\mathbb{R}} g(a) L_t^a(X) \, da = \int_0^t g(X_u) \, d\langle M \rangle_u.$$

For an arbitrary convex function $f : \mathbb{R} \to \mathbb{R}$ and a continuous semimartingale X, the following decomposition, referred to as the *Itô-Tanaka-Meyer formula*, is valid

$$f(X_t) = f(X_0) + \int_0^t f_l'(X_u) \, dX_u + \frac{1}{2} \int_{\mathbb{R}} L_t^a(X) \, \mu(da),$$

where f_l' is the left-hand-side derivative[2] of f and the measure $\mu = f''$ represents the second order derivative of f, in the sense of distributions. If the function f belongs to the class $C^2(\mathbb{R}, \mathbb{R})$, then $\mu(da) = f''(a) \, da$ for a nonnegative function f'', and the density of occupation time formula yields

$$\int_{\mathbb{R}} L_t^a(X) \, \mu(da) = \int_{\mathbb{R}} L_t^a(X) f''(a) \, da = \int_0^t f''(X_u) \, d\langle M \rangle_u.$$

Hence the Itô-Tanaka-Meyer formula reduces here to the classical Itô formula.

A.10 Lévy's Characterization Theorem

In some instances, it would be convenient to have the possibility of checking whether a given process is a standard Brownian motion by establishing its martingale property and by computing its quadratic variation. To this end, we may use the *martingale characterization* of a Brownian motion, due to Paul Lévy.

One-dimensional case. Recall that if W is a one-dimensional Brownian motion on a probability space $(\Omega, \mathbb{F}, \mathbb{P})$ then W is a continuous $\mathbb{F}$-martingale with $W_0 = 0$ and the process $W_t^2 - t$ is also a continuous $\mathbb{F}$-martingale.

Theorem A.10.1 *Let M be a continuous local martingale on a probability space $(\Omega, \mathbb{F}, \mathbb{P})$ such that $M_0 = 0$ and the process $M_t^2 - t$ is a continuous local martingale. Then M is a standard Brownian motion with respect to $\mathbb{F}$.*

The assumption that $M_t^2 - t$ is a continuous local martingale means that the quadratic variation of M satisfies $\langle M \rangle_t = t$ for every $t \in [0, T]$. Under the stronger assumption that M is a square-integrable continuous martingale, this property can be represented as follows

$$\mathbb{E}_{\mathbb{P}}(M_u^2 - M_t^2 \mid \mathcal{F}_t) = u - t, \quad \forall t \le u \le T, \tag{A.33}$$

or equivalently (in view of the martingale property of M)

$$\mathbb{E}_{\mathbb{P}}((M_u - M_t)^2 \mid \mathcal{F}_t) = u - t, \quad \forall t \le u \le T.$$

---

[2] Recall that the left-hand-side and right-hand-side derivatives of a convex function are non-decreasing functions.

Remarks. It should be stressed that the assumption of continuity of sample paths of M is essential in Theorem A.10.1 and thus it cannot be dropped. Indeed, if we take as M the *compensated Poisson process*, that is, the process $M = N_t - t$, where N is the Poisson process with unit intensity, then it can be easily verified that the process $M_t^2 - t$ is a martingale as well.

The proof of Theorem A.10.1 is omitted. Let us only mention that is based on the following lemma, which in turn can be proved by directly checking conditions of Definition A.4.1.

Lemma A.10.1 *A real-valued continuous $\mathbb{F}$-adapted process X defined on $(\Omega, \mathbb{F}, \mathbb{P})$ is a standard Brownian motion with respect to $\mathbb{F}$ if and only if for any $\lambda \in \mathbb{R}$ the process*

$$M_t^\lambda = \exp\left(\lambda X_t - \tfrac{1}{2}\lambda^2 t\right), \quad \forall t \in [0, T], \tag{A.34}$$

is a (local) martingale with respect to $\mathbb{F}$ and $M_0^\lambda = 1$.

The process M^λ is a special case of the *stochastic exponential* examined in some detail in Sect. A.13. It is easily seen that when W is a one-dimensional Brownian motion with respect to $\mathbb{F}$ then we have

$$\mathbb{E}_{\mathbb{P}}\left(e^{\lambda(W_u - W_t)} \,\Big|\, \mathcal{F}_t\right) = e^{\frac{1}{2}\lambda^2(u-t)}, \quad \forall t \le u \le T.$$

This implies that the process M^λ associated with W is an $\mathbb{F}$-martingale.

Multidimensional case. Theorem A.10.1 can be extended to a multidimensional setting. Let $\delta_{ij} = 1$ if $i = j$ and $\delta_{ij} = 0$ otherwise.

Theorem A.10.2 *Let $M = (M^1, \ldots, M^d)$ be a continuous $\mathbb{F}$-adapted process on a filtered probability space $(\Omega, \mathbb{F}, \mathbb{P})$ with the initial value $M_0 = (0, \ldots, 0)$. Then M is a d-dimensional standard Brownian motion if and only if the following conditions are satisfied:*
(i) for any $i = 1, \ldots, d$ the process M^i is an $\mathbb{F}$-martingale,
(ii) for any $i, j = 1, \ldots, d$ the process $M_t^i M_t^j - \delta_{ij} t$ is an $\mathbb{F}$-martingale.

The following variant of the multidimensional Lévy's characterization theorem will prove useful.

Theorem A.10.3 *Assume that $M = (M^1, \ldots, M^d)$ is an $\mathbb{F}$-adapted process on a probability space $(\Omega, \mathbb{F}, \mathbb{P})$ such that $M_0 = (0, \ldots, 0)$ and the components $M^1, \ldots, M^d$ are continuous local martingales with $\langle M^i, M^j \rangle_t = \delta_{ij} t$. Then M is a d-dimensional standard Brownian motion.*

A.11 Martingale Representation Property

In this section we shall assume that the filtration $\mathbb{F} = \mathbb{F}^W$ is the standard augmentation of the natural filtration $\sigma\{W_u \mid u \le t\}$ of a d-dimensional Brownian motion W.

In other words, we assume here that the underlying probability space is $(\Omega, \mathbb{F}^W, \mathbb{P})$. We will now state two closely related versions of the so-called *predictable representation property* of the Brownian filtration.

The first theorem is known as the *Itô representation theorem*. It states that any square-integrable and $\mathcal{F}_T^W$-measurable random variables admits a representation as the Itô integral of some stochastic process.

Theorem A.11.1 *For any random variable* $X \in L^2(\Omega, \mathcal{F}_T^W, \mathbb{P})$, *there exists a unique* $\mathbb{F}$-*predictable process* γ *from the class* $\mathcal{L}_{\mathbb{P}}^2(W)$ *such that the following equality is valid*

$$X = \mathbb{E}_{\mathbb{P}} X + \int_0^T \gamma_u \cdot dW_u. \tag{A.35}$$

The condition that γ belongs to $\mathcal{L}_{\mathbb{P}}^2(W)$ implies that the Itô integral $I(\gamma)$ is a square-integrable $\mathbb{F}$-martingale and thus we also have that, for every $t \in [0, T]$,

$$\mathbb{E}_{\mathbb{P}}(X \mid \mathcal{F}_t^W) = \mathbb{E}_{\mathbb{P}} X + \int_0^t \gamma_u \cdot dW_u. \tag{A.36}$$

Remarks. In Theorem A.11.1 we assert that γ is an $\mathbb{F}$-*predictable process* (for the definition see, for instance, Chapter 5 in Revuz and Yor (1999)). Let us only note that any left-continuous and $\mathbb{F}$-adapted process is $\mathbb{F}$-predictable. Also, if a process is $\mathbb{F}$-predictable then it is $\mathbb{F}$-progressively measurable, but the converse does not hold, in general.

The second result is commonly known as the *martingale representation theorem* for the Brownian filtration. Note that it is a rather straightforward consequence of Theorem A.11.1 (see, in particular, formula (A.36)).

Theorem A.11.2 *Let a process* $M = (M_t)_{t \in [0,T]}$ *be an* $\mathbb{F}^W$-*martingale such that* $\mathbb{E}_{\mathbb{P}} M_T^2 < \infty$. *Then there exists a unique* $\mathbb{F}$-*predictable process* γ *from the class* $\mathcal{L}_{\mathbb{P}}^2(W)$ *such that, for any* $t \in [0, T]$,

$$M_t = M_0 + \int_0^t \gamma_u \cdot dW_u. \tag{A.37}$$

Theorem A.11.2 can be extended to local martingales with respect to a Brownian filtration. As a consequence, we obtain the following notable result.

Corollary A.11.1 *Any local martingale with respect to the Brownian filtration* $\mathbb{F}^W$ *is necessarily a continuous process.*

Example A.11.1 Let $d = 1$ and let us first take $X = W_T^2$. Using the Itô formula of Proposition A.9.1 with $g(x, t) = x^2$ and $X = W$, we obtain

$$W_t^2 = \int_0^t 2W_u \, dW_u + t, \quad \forall t \in [0, T].$$

Since $\mathbb{E}_{\mathbb{P}} W_T^2 = T$, the equality (A.35) becomes

$$W_T^2 = \mathbb{E}_{\mathbb{P}} W_T^2 + \int_0^T 2W_u \, dW_u,$$

so that the process $\gamma_t = 2W_t$, $t \in [0, T]$, is independent of T.

A more interesting result is obtained for $X = W_T^3$. By applying the Itô formula to $g(x) = x^3$ and $X = W$, we obtain

$$W_t^3 = \int_0^t 3W_u^2 \, dW_u + \int_0^t 3W_u \, du, \quad \forall t \in [0, T].$$

Therefore

$$W_T^3 = \int_0^T 3(W_u^2 + (T - u)) \, dW_u. \tag{A.38}$$

since the Itô integration by parts formula (A.31) applied to $X_t^1 = W_t$ and $X_t^2 = T - t$ yields $\int_0^T W_u \, du = \int_0^T (T - u) \, dW_u$ (of course, here $\langle X^1, X^2 \rangle = 0$). Hence, for any fixed $T > 0$, we have $\gamma_t = 3W_t^2 + 3(T - t)$ for every $t \in [0, T]$.

Alternatively, by evaluating the conditional expectation, we obtain

$$\mathbb{E}_{\mathbb{P}}(W_T^3 \mid \mathcal{F}_t) = f(W_t, t), \quad \forall t \in [0, T],$$

where the function $f : [0, T] \times \mathbb{R} \to \mathbb{R}$ is given by $f(x, t) = x^3 + 3x(T - t)$. Note that f solves on $(0, T) \times \mathbb{R}$ the following partial differential equation

$$\frac{\partial f}{\partial t} + \frac{1}{2} \frac{\partial^2 f}{\partial x^2} = 0$$

with the terminal condition $f(x, T) = x^3$ for $x \in \mathbb{R}$. Hence, by applying the Itô formula to $g(x, t) = f(x, t)$ and $X = W$, we re-derive (A.38).

A.12 Stochastic Differential Equations

The theory of stochastic differential equations is covered by numerous textbooks and monographs. In this short section, we mainly deal with the special case of linear stochastic differential equations.

Let $\mu : \mathbb{R} \times [0, T] \to \mathbb{R}$ and $\sigma : \mathbb{R} \times [0, T] \to \mathbb{R}^d$ be some given functions and let W be a d-dimensional standard Brownian motion defined on a filtered probability space $(\Omega, \mathbb{F}, \mathbb{P})$.

Definition A.12.1 By a *solution* of the stochastic differential equation (SDE, for short)

$$dX_t = \mu(X_t, t) \, dt + \sigma(X_t, t) \cdot dW_t \tag{A.39}$$

with the initial condition X_0 given as an $\mathcal{F}_0$-measurable random variable, we mean an $\mathbb{R}^k$-valued, $\mathbb{F}$-adapted stochastic process X defined on the probability space $(\Omega, \mathbb{F}, \mathbb{P})$ and such that, for every $t \in [0, T]$,

$$X_t = X_0 + \int_0^t \mu(X_u, u) \, du + \int_0^t \sigma(X_u, u) \cdot dW_u. \tag{A.40}$$

It is clear from the definition above that a stochastic differential equation, usually represented in its differential form (A.39), is nothing else than the integral equation (A.40), in which the second integral is defined as the Itô stochastic integral. Definition A.12.1 implicitly assumes that both integrals in the right-hand side of (A.40) are well defined, that is,

$$\mathbb{P}\left(\int_0^T |\mu(X_u, u)| \, du < \infty \right) = 1$$

and the process γ given as $\gamma_t = \sigma(X_t, t)$, $t \in [0, T]$, belongs to the class $\mathcal{L}_{\mathbb{P}}(W)$. Note that any solution X to the SDE (A.39) is an Itô process, in the sense of Definition A.8.2.

The concept of uniqueness of solutions to an SDE can be defined in several alternative ways. For our purposes, the most convenient form is the so-called *pathwise uniqueness* of solutions. If X is a solution to (A.39), we shall informally say that X is driven by W.

Definition A.12.2 We say that the *pathwise uniqueness* of solutions to (A.39) holds is for any filtered probability space $(\Omega, \mathbb{F}, \mathbb{P})$, any d-dimensional standard Brownian motions W and $\tilde{W}$ defined on $(\Omega, \mathbb{F}, \mathbb{P})$, and any two solutions X and $\tilde{X}$ driven by W and $\tilde{W}$ respectively, the following implication is true

$$\mathbb{P}\{W_t = \tilde{W}_t \mid \forall t \in [0, T]\} = 1 \quad \Longrightarrow \quad \mathbb{P}\{X_t = \tilde{X}_t \mid \forall t \in [0, T]\} = 1.$$

The pathwise uniqueness of solutions (A.39) is sometimes referred to as the *strong uniqueness*. This is due to the fact that under pathwise uniqueness any solution to (A.39) is *strong*, meaning that it is adapted to the filtration generated by the driving Brownian motion W.

Linear stochastic differential equation. Let us first consider a time-homogeneous linear SDE of the form

$$dX_t = (aX_t + c) \, dt + (bX_t + f) \, dW_t, \tag{A.41}$$

where W is a one-dimensional Brownian motion and a, b, c and f are real numbers. It is clear that $\mu(x, t) = ax + c$ and $\sigma(x, t) = bx + f$ so that μ and σ are linear functions.

By setting $c = f = 0$, we obtain the SDE governing the stock price process in the Black-Scholes model (see Sect. 3.1). In the present notation, it becomes

$$dX_t = aX_t \, dt + bX_t \, dW_t = X_t(a \, dt + b \, dW_t). \tag{A.42}$$

Proposition A.12.1 *The unique solution of the SDE (A.42) is given by the expression*

$$X_t = X_0 \exp\left(bW_t + \left(a - \tfrac{1}{2}b^2\right)t \right). \tag{A.43}$$

Proof. See the proof of Proposition 3.1.1 in Sect. 3.1. □

The following result is useful in the case of Vasicek's model of the short-term rate (see Sect. 10.1).

Proposition A.12.2 *The unique solution of the SDE*

$$dX_t = (aX_t + c)\,dt + f\,dW_t \tag{A.44}$$

is given by the formula

$$X_t = X_0 e^{at} + \int_0^t c e^{a(t-u)}\,du + \int_0^t f e^{a(t-u)}\,dW_u. \tag{A.45}$$

Proof. See the proof of Proposition 10.1.2 in Sect. 10.1.   □

The definition of an SDE can also be extended to the case where the coefficients μ and σ explicitly depend on the time parameter t and ω. For instance, we may assume that the parameters (A.41) follow stochastic processes, so that it becomes

$$dX_t = (a_t X_t + c_t)\,dt + (b_t X_t + f_t) \cdot dW_t. \tag{A.46}$$

where W is a d-dimensional Brownian motion and the $\mathbb{F}$-adapted stochastic processes a, c are real-valued, whereas the processes b, f are $\mathbb{R}^d$-valued. Then we have the following extension of Proposition A.12.1.

Proposition A.12.3 *Let a and b be $\mathbb{F}$-adapted bounded processes. Then the unique solution of the SDE*

$$dX_t = X_t(a_t\,dt + b_t \cdot dW_t) \tag{A.47}$$

is given by the following generalization of formula (A.43)

$$X_t = X_0 \exp\left(\int_0^t b_u \cdot dW_u + \int_0^t \left(a_u - \tfrac{1}{2}|b_u|^2 \right) du \right). \tag{A.48}$$

Proof. To check that the process X given by (A.48) is a solution to (A.47), it suffices to differentiate the right-hand side in (A.48) using the Itô formula of Proposition A.9.2. The uniqueness of solutions to (A.47) can be proved directly or deduced from Theorem A.12.1.   □

The following result extends Proposition A.12.2.

Proposition A.12.4 *Let a, c and f be $\mathbb{F}$-adapted bounded processes. Then the unique solution of the linear SDE*

$$dX_t = (a_t X_t + c_t)\,dt + f_t \cdot dW_t \tag{A.49}$$

is given by the formula

$$X_t = \Phi_t \left(X_0 + \int_0^t \Phi_u^{-1} c_u\,du + \int_0^t \Phi_u^{-1} f_u \cdot dW_u \right) \tag{A.50}$$

where

$$\Phi_u = \exp\left(\int_0^u a_s\,ds \right).$$

Itô's existence and uniqueness theorem for SDEs. Let us end this section by stating the classic Itô's theorem, which provides sufficient conditions for the existence and uniqueness of solutions to the SDE (A.39).

Theorem A.12.1 *Let $\mu : \mathbb{R} \times [0, T] \to \mathbb{R}$ and $\sigma : \mathbb{R} \times [0, T] \to \mathbb{R}^d$ satisfy the following conditions:*
(i) the functions μ and σ are Lipschitz continuous with respect to the variable x, that is, there exists a constant $K_1 > 0$ such that, for any $x, y \in \mathbb{R}$ and $t \in \mathbb{R}_+$,

$$|\mu(t, x) - \mu(t, y)|^2 + |\sigma(t, x) - \sigma(t, y)|^2 \le K_1 |x - y|^2, \tag{A.51}$$

(ii) the functions μ and σ satisfy the linear growth condition: there exists a constant K_2 such that, for any $x \in \mathbb{R}$ and $t \in \mathbb{R}_+$,

$$|a(t, x)|^2 + |b(t, x)|^2 \le K_2(1 + |x|^2). \tag{A.52}$$

Then the SDE has the unique solution X.

For notational simplicity, in Theorem A.12.1 the symbol $|\cdot|$ is used to denote the Euclidean norms in both $\mathbb{R}$ and $\mathbb{R}^d$, but this slight abuse of notation should not be confusing.

Let us finally mention that an analogous result is valid in the multidimensional case, that is, when the k-dimensional process $X = (X^1, \ldots, X^k)$ is given as

$$dX_t = \mu(X_t, t) \, dt + \sigma(X_t, t) \, dW_t, \quad X_0 = x, \tag{A.53}$$

with the coefficients $\mu : \mathbb{R}^k \times [0, T] \to \mathbb{R}^k$ and $\sigma : \mathbb{R}^k \times [0, T] \to \mathbb{R}^{k \times d}$ satisfying the natural extensions of Itô's conditions (A.51) and (A.52).

A.13 Stochastic Exponential

Let W be a d-dimensional standard Brownian motion defined on a filtered probability space $(\Omega, \mathbb{F}, \mathbb{P})$. For an $\mathbb{R}^d$-valued process γ belonging to $\mathcal{L}(W)$, we define the real-valued $\mathbb{F}$-adapted process U by setting

$$U_t = I_t(\gamma) = \int_0^t \gamma_u \cdot dW_u, \quad \forall t \in [0, T]. \tag{A.54}$$

The process U defined in this way is, of course, a continuous local martingale under $\mathbb{P}$ with respect to $\mathbb{F}$.

Definition A.13.1 The *stochastic exponential* (also known as the *Doléans exponential*) of U is given by the formula, for $t \in [0, T]$,

$$\mathcal{E}_t(U) = \mathcal{E}_t\left(\int_0^{\cdot} \gamma_u \cdot dW_u\right) = \exp\left(\int_0^t \gamma_u \cdot dW_u - \frac{1}{2}\int_0^t |\gamma_u|^2 \, du\right),$$

that is, $\mathcal{E}_t(U) = \exp(U_t - \langle U \rangle_t / 2)$. More generally, for any continuous local martingale M we set $\mathcal{E}_t(M) = \exp(M_t - \langle M \rangle_t / 2)$ for $t \in [0, T]$.

Lemma A.13.1 *The stochastic exponential of U is the unique solution Z of the stochastic differential equation*

$$dZ_t = Z_t \, \gamma_t \cdot dW_t = Z_t \, dU_t \tag{A.55}$$

with the initial condition $Z_0 = 1$.

Proof. The assertion is a straightforward consequence of Proposition A.12.4.   □

Remarks. (i) It follows immediately from Lemma A.13.1 that

$$d\mathcal{E}_t(U) = \mathcal{E}_t(U) \, \gamma_t \cdot dW_t = \mathcal{E}_t(U) \, dU_t, \tag{A.56}$$

Note that $\mathcal{E}(U)$ is a strictly positive continuous local martingale under $\mathbb{P}$ and thus, in view of Proposition A.7.1, it follows a supermartingale with respect to $\mathbb{F}$. If, in addition,

$$\mathbb{E}_{\mathbb{P}}(\mathcal{E}_T(U)) = \mathbb{E}_{\mathbb{P}}(\mathcal{E}_0(U)) = 1,$$

so that the expected value $\mathbb{E}_{\mathbb{P}}\mathcal{E}_t(U)$, $t \in [0, T]$ is constant, then the process $\mathcal{E}(U)$ is a strictly positive continuous $\mathbb{F}$-martingale.

(ii) For any continuous local martingale M, the stochastic exponential $\mathcal{E}(M)$ is the unique solution Z of the stochastic differential equation

$$dZ_t = Z_t \, dM_t \tag{A.57}$$

with the initial condition $Z_0 = 1$.

A.14 Radon-Nikodým Density

Two probability measures $\tilde{\mathbb{P}}$ and $\mathbb{P}$ on $(\Omega, \mathcal{F}_T)$ are said to be *equivalent* if, for any event $A \in \mathcal{F}_T$, the equality $\mathbb{P}(A) = 0$ holds if and only if $\tilde{\mathbb{P}}(A) = 0$. In other words, $\tilde{\mathbb{P}}$ and $\mathbb{P}$ are equivalent on $(\Omega, \mathcal{F}_T)$ if they have the same set of null events in the σ-field $\mathcal{F}_T$. It is easily seen that if $\tilde{\mathbb{P}}$ and $\mathbb{P}$ are equivalent on $\mathcal{F}_T$ then they also enjoy this property on any σ-field $\mathcal{G} \subseteq \mathcal{F}_T$; in particular, $\tilde{\mathbb{P}}$ and $\mathbb{P}$ are equivalent on the σ-field $\mathcal{F}_t$, for any $t \in [0, T]$.

Definition A.14.1 The *Radon-Nikodým density* of $\tilde{\mathbb{P}}$ with respect to $\mathbb{P}$ is defined as the unique $\mathcal{F}_T$-measurable random variable η_T such that we have, for any event $A \in \mathcal{F}_T$,

$$\tilde{\mathbb{P}}(A) = \int_A \eta_T \, d\mathbb{P} = \mathbb{E}_{\mathbb{P}}(\eta_T \mathbb{1}_A). \tag{A.58}$$

The existence and uniqueness of the random variable η_T satisfying (A.58) is a consequence of the classic Radon-Nikodým theorem. Formula (A.58) implies that for any $\tilde{\mathbb{P}}$-integrable random variable ψ we have that $\mathbb{E}_{\tilde{\mathbb{P}}}\psi = \mathbb{E}_{\mathbb{P}}(\psi \eta_T)$. Note also that ψ is $\tilde{\mathbb{P}}$-integrable if and only if $\psi \eta_T$ is $\tilde{\mathbb{P}}$-integrable. Finally, it is easy to check that $\mathbb{P}\{\eta_T > 0\} = 1$ and $\mathbb{E}_{\mathbb{P}}\eta_T = \tilde{\mathbb{P}}(\Omega) = 1$.

Conversely, assume that η_T is any $\mathcal{F}_T$-measurable random variable such that the last two conditions hold. Then formula (A.58) defines a probability measure $\tilde{\mathbb{P}}$ on $(\Omega, \mathcal{F}_T)$, which is then equivalent to $\mathbb{P}$.

To emphasize the role of η_T as the link between the expectations with respect to $\tilde{\mathbb{P}}$ and $\mathbb{P}$, it is customary to use the short-hand notation

$$\eta_T = \frac{d\tilde{\mathbb{P}}}{d\mathbb{P}}.$$

Definition A.14.2 The *Radon-Nikodým density process* $\eta = (\eta_t)_{t \in [0,T]}$ of $\tilde{\mathbb{P}}$ with respect to $\mathbb{P}$ and a given filtration $\mathbb{F}$ is defined by setting, for $t \in [0, T]$,

$$\eta_t \stackrel{\text{def}}{=} \mathbb{E}_{\mathbb{P}}(\eta_T \mid \mathcal{F}_t) = \mathbb{E}_{\mathbb{P}}\left(\frac{d\tilde{\mathbb{P}}}{d\mathbb{P}} \,\Big|\, \mathcal{F}_t\right), \quad \forall\, t \in [0, T]. \tag{A.59}$$

It is clear that Radon-Nikodým density process η is a strictly positive martingale under $\mathbb{P}$. It is also uniformly integrable since clearly $\eta_t = \mathbb{E}_{\mathbb{P}}(\eta_T \mid \mathcal{F}_t)$. Finally, the random variable η_t is the Radon-Nikodým density of $\tilde{\mathbb{P}}$ with respect to $\mathbb{P}$ on $(\Omega, \mathcal{F}_t)$. The latter property is denoted as

$$\eta_t = \frac{d\tilde{\mathbb{P}}_{|\mathcal{F}_t}}{d\mathbb{P}_{|\mathcal{F}_t}} = \frac{d\tilde{\mathbb{P}}}{d\mathbb{P}}\Big|_{\mathcal{F}_t}. \tag{A.60}$$

Lemma A.14.1 *A stochastic process X is an $\mathbb{F}$-martingale under $\tilde{\mathbb{P}}$ if and only if the process $X\eta$ is an $\mathbb{F}$-martingale under $\mathbb{P}$.*

Proof. The proof relies on the application of the abstract Bayes formula (see Lemma A.1.4). Assume first that $X\eta$ is an $\mathbb{F}$-martingale under $\mathbb{P}$ so that equality $\mathbb{E}_{\mathbb{P}}(X_u\eta_u \mid \mathcal{F}_t) = X_t\eta_t$ holds for $0 \le t \le u \le T$. Then the Bayes formula yields, for $0 \le t \le u \le T$,

$$\mathbb{E}_{\tilde{\mathbb{P}}}(X_u \mid \mathcal{F}_t) = \frac{\mathbb{E}_{\mathbb{P}}(X_u\eta_T \mid \mathcal{F}_t)}{\mathbb{E}_{\mathbb{P}}(\eta_T \mid \mathcal{F}_t)} = \frac{\mathbb{E}_{\mathbb{P}}(X_u\mathbb{E}_{\mathbb{P}}(\eta_T \mid \mathcal{F}_u) \mid \mathcal{F}_t)}{\mathbb{E}_{\mathbb{P}}(\eta_T \mid \mathcal{F}_t)}$$
$$= \frac{\mathbb{E}_{\mathbb{P}}(X_u\eta_u \mid \mathcal{F}_t)}{\eta_t} = \frac{X_t\eta_t}{\eta_t} = X_t.$$

We conclude that X is an $\mathbb{F}$-martingale under $\mathbb{P}$. The proof of the converse implication goes along the same lines. $\quad\square$

A.15 Girsanov's Theorem

We are in the position to state the classic Girsanov theorem. It states that, under mild technical conditions, a Brownian motion with an absolutely continuous drift becomes a standard Brownian motion under an equivalent probability measure. Before proceeding to the general result, let us first consider a special case of a linear drift.

Proposition A.15.1 *Let W be a one-dimensional standard Brownian motion on a probability space $(\Omega, \mathbb{F}, \mathbb{P})$. For a real number $\gamma \in \mathbb{R}$, we define the process X by setting $\tilde{W}_t = W_t - \gamma t$ for $t \in [0, T]$. Let the probability measure $\tilde{\mathbb{P}}$, equivalent to $\mathbb{P}$ on $(\Omega, \mathcal{F}_T)$, be defined through the formula*

$$\frac{d\tilde{\mathbb{P}}}{d\mathbb{P}} = \exp\left(\gamma W_T - \tfrac{1}{2}\gamma^2 T\right) = \eta_T, \quad \mathbb{P}\text{-a.s.}$$

Then X is a standard Brownian motion on the probability space $(\Omega, \mathbb{F}, \tilde{\mathbb{P}})$.

Proof. To establish the proposition, it is enough to make use of Lemmas A.10.1 and A.14.1. In view of the former, it suffices to show that for any $\lambda \in \mathbb{R}$ the process

$$M_t^\lambda = \exp\left(\lambda \tilde{W}_t - \tfrac{1}{2}\lambda^2 t\right), \quad \forall\, t \in [0, T],$$

is an $\mathbb{F}$-martingale under $\mathbb{Q}$. Using the latter lemma, we see that it is enough to check that $M^\lambda \eta$ is an $\mathbb{F}$-martingale under $\mathbb{P}$. But clearly

$$M_t^\lambda \eta_t = \exp\left(\lambda(W_t - \gamma t) - \tfrac{1}{2}\lambda^2 t\right) \exp\left(\gamma W_t - \tfrac{1}{2}\gamma^2 t\right) = \exp\left(\alpha W_t - \tfrac{1}{2}\alpha^2 t\right)$$

where $\alpha = \lambda + \gamma$, and thus this process follows an $\mathbb{F}$-martingale under $\mathbb{P}$, by another application of Lemma A.10.1. We conclude that $\tilde{W}$ is a standard Brownian motion on $(\Omega, \mathbb{F}, \tilde{\mathbb{P}})$.   □

Proposition A.15.1 can be extended to the case of a stochastic drift term, as the following result shows.

Theorem A.15.1 *Let W be a standard d-dimensional Brownian motion on a filtered probability space $(\Omega, \mathbb{F}, \mathbb{P})$. Suppose that γ is an $\mathbb{R}^d$-valued $\mathbb{F}$-progressively measurable process such that*

$$\mathbb{E}_{\mathbb{P}}\left\{\mathcal{E}_T\left(\int_0^\cdot \gamma_u \cdot dW_u\right)\right\} = 1. \tag{A.61}$$

Define a probability measure $\tilde{\mathbb{P}}$ on $(\Omega, \mathcal{F}_T)$ equivalent to $\mathbb{P}$ by means of the Radon-Nikodým derivative

$$\frac{d\tilde{\mathbb{P}}}{d\mathbb{P}} = \mathcal{E}_T\left(\int_0^\cdot \gamma_u \cdot dW_u\right), \quad \mathbb{P}\text{-a.s.} \tag{A.62}$$

Then the process $\tilde{W}$ given by the formula

$$\tilde{W}_t = W_t - \int_0^t \gamma_u \, du, \quad \forall\, t \in [0, T], \tag{A.63}$$

follows a standard d-dimensional Brownian motion on the space $(\Omega, \mathbb{F}, \tilde{\mathbb{P}})$.

Proof. In view of Theorem A.10.2 and Lemma A.14.1, it suffices to check that the processes $\tilde{W}^i \eta$ and $(\tilde{W}^i_t \tilde{W}^j_t - \delta_{ij} t)\eta_t$ are continuous local martingales under $\mathbb{P}$. For the first process, an application of the Itô integration by parts formula gives

$$
\begin{aligned}
d(\tilde{W}^i_t \eta_t) &= \tilde{W}^i_t \, d\eta_t + \eta_t \, d\tilde{W}^i_t + d\langle \tilde{W}^i, \eta \rangle_t \\
&= \tilde{W}^i_t \eta_t \gamma_t \cdot dW_t + \eta_t \, dW^i_t - \eta_t \gamma^i_t \, dt + d\langle \tilde{W}^i, \eta \rangle_t \\
&= \tilde{W}^i_t \eta_t \gamma_t \cdot dW_t + \eta_t \, dW^i_t
\end{aligned}
$$

since $d\eta_t = \eta_t \gamma_t \cdot dW_t$ and thus $d\langle \tilde{W}^i, \eta \rangle_t = \gamma^i_t \eta_t \, dt$. We thus see that $\tilde{W}^i \eta$ is a continuous local martingale under $\mathbb{P}$.

Let us denote $\tilde{M}^{ij}_t = \tilde{W}^i_t \tilde{W}^j_t - \delta_{ij} t$. Then we have that (recall that $\langle W^i, W^j \rangle_t = 0$ for $i \neq j$ and $\langle W^i, W^i \rangle_t = t$ for $t \in [0, T]$)

$$
d\tilde{M}^{ij}_t = \tilde{W}^i_t \, d\tilde{W}^j_t + \tilde{W}^j_t \, d\tilde{W}^i_t,
$$

and this in turn implies that

$$
d\langle \tilde{M}^{ij}, \eta \rangle_t = \eta_t (\tilde{W}^i_t \gamma^j_t + \tilde{W}^j_t \gamma^i_t) \, dt.
$$

By applying once again the Itô integration by parts formula, we obtain

$$
\begin{aligned}
d(\tilde{M}^{ij}_t \eta_t) &= \tilde{M}^{ij}_t \, d\eta_t + \eta_t \, d\tilde{M}^{ij}_t + d\langle \tilde{M}^{ij}, \eta \rangle_t \\
&= \tilde{M}^{ij}_t \, d\eta_t + \eta_t (\tilde{W}^i_t \, d\tilde{W}^j_t + \tilde{W}^j_t \, d\tilde{W}^i_t) + \eta_t (\tilde{W}^i_t \gamma^j_t + \tilde{W}^j_t \gamma^i_t) \, dt \\
&= \tilde{M}^{ij}_t \eta_t \gamma_t \cdot dW_t + \eta_t (\tilde{W}^i_t \, dW^j_t + \tilde{W}^j_t \, dW^i_t)
\end{aligned}
$$

since $\tilde{W}^i_t = W^i_t - \gamma^i_t \, dt$. We conclude that the process $(\tilde{W}^i_t \tilde{W}^j_t - \delta_{ij} t)\eta_t$ follows a continuous local martingale under $\mathbb{P}$.   □

In order to make use of Girsanov's theorem for a particular process, we need, of course, to be able to verify condition (A.61). The next result furnishes a sufficient condition for the equality (A.61) to hold. The inequality (A.64) is commonly referred to as *Novikov's condition*.

Proposition A.15.2 *Assume that*

$$
\mathbb{E}_{\mathbb{P}}\left\{ \exp\left(\frac{1}{2} \int_0^T |\gamma_u|^2 \, du \right) \right\} < \infty. \tag{A.64}
$$

Then $\mathbb{E}_{\mathbb{P}}(\mathcal{E}_T(U)) = 1$, *where* U *is given by* (A.54). *Consequently, the process* $\mathcal{E}(U)$ *is a strictly positive continuous* $\mathbb{F}$-*martingale.*

In particular, if the process γ is uniformly bounded (that is, there exists a constant K such that $|\gamma_t| \leq K$ for $t \in [0, T]$) then Novikov's condition is satisfied and thus $\mathbb{E}_{\mathbb{P}}(\mathcal{E}_T(U)) = 1$.

A weaker, but also sufficient for the validity of (A.61), is the *Kazamaki condition*

$$\mathbb{E}_{\mathbb{P}}\left\{ \exp\left(\frac{1}{2} \int_0^t \gamma_u \cdot dW_u \right) \right\} < \infty, \quad \forall\, t \in [0, T]. \tag{A.65}$$

Let us comment briefly on relevant filtrations. Obviously, we always have that $\mathcal{F}_t^{\tilde{W}} \subseteq \mathcal{F}_t$ for any $t \in [0, T]$. The filtrations generated by W and $\tilde{W}$ do not coincide, in general. In the special case where the underlying filtration $\mathbb{F}$ is the $\mathbb{P}$-augmentation of the natural filtration of W, the inclusion $\mathcal{F}_t^{\tilde{W}} \subseteq \mathcal{F}_t^W$ holds for any $t \in [0, T]$, but the filtration generated by $\tilde{W}$ can be essentially smaller than $\mathbb{F}^W$.

The next well-known result shows that, if the underlying filtration is generated by a d-dimensional Brownian motion, the Radon-Nikodým density process of any probability measure equivalent to $\mathbb{P}$ has necessarily the form of the stochastic exponential (A.62) for some process γ.

Proposition A.15.3 *Assume that the filtration $\mathbb{F}$ is the usual augmentation of the natural filtration of W, that is, $\mathbb{F} = \mathbb{F}^W$. Then for any probability measure $\tilde{\mathbb{P}}$ on $(\Omega, \mathcal{F}_T)$ equivalent to $\mathbb{P}$ there exists an $\mathbb{F}^W$-progressively measurable, $\mathbb{R}^d$-valued process γ such that the Radon-Nikodým density of $\tilde{\mathbb{P}}$ with respect to $\mathbb{P}$ equals*

$$\frac{d\tilde{\mathbb{P}}}{d\mathbb{P}} = \mathcal{E}_T\left(\int_0^{\cdot} \gamma_u \cdot dW_u \right), \quad \mathbb{P}\text{-a.s.} \tag{A.66}$$

Proof. Let η_T be the Radon-Nikodým density of $\tilde{\mathbb{P}}$ with respect to $\mathbb{P}$ on $(\Omega, \mathcal{F}_T)$ (see (A.60)). It is clear that the process $\eta_t = \mathbb{E}_{\mathbb{P}}(\eta_T \mid \mathcal{F}_t)$ is an $\mathbb{F}$-martingale and $\eta_0 = 1$. Since the underlying filtration is a Brownian filtration, from remarks after martingale representation theorem A.11.2 we can deduce the existence of a process $\tilde{\gamma} \in \mathcal{L}_{\mathbb{P}}(W)$ such that

$$\eta_t = 1 + \int_0^t \tilde{\gamma}_u \cdot dW_u, \quad \forall\, t \in [0, T]. \tag{A.67}$$

Since $\mathbb{P}\{\eta_T > 0\} = 1$, we also have that $\mathbb{P}\{\eta_t > 0\} = 1$ for any $t \in [0, T]$, and thus, in view of continuity of η, which is apparent from representations (A.67), we obtain $\mathbb{P}\{\eta_t > 0 \mid \forall\, t \in [0, T]\} = 1$. Therefore, the process $\gamma_t = \tilde{\gamma}_t \eta_t^{-1}$ is well defined and we manifestly have

$$\eta_t = 1 + \int_0^t \tilde{\gamma}_u \cdot dW_u = 1 + \int_0^t \eta_u \gamma_u \cdot dW_u.$$

We conclude that the Radon-Nikodým density process is the unique solution to the SDE (cf. formula (A.55))

$$d\eta_t = \eta_t \gamma_t \cdot dW_t, \tag{A.68}$$

and thus, in view of Lemma A.13.1, it satisfies, for any $t \in [0, T]$,

$$\eta_t = \mathcal{E}_t\left(\int_0^{\cdot} \gamma_u \cdot dW_u \right).$$

This completes the proof of the proposition. □

Case of continuous semimartingales. We end this section by stating a generalization of Girsanov's theorem to the case of continuous semimartingales. Let $\tilde{\mathbb{P}}$ and $\mathbb{P}$ be two equivalent probability measures on a common filtered probability space. We assume, in addition, that the Radon-Nikodým density process η given by (A.59) is continuous (we know already that this holds if $\mathbb{F} = \mathbb{F}^W$ for some Brownian motion W).

Let us set, for any $t \in [0, T]$,

$$U_t = \int_0^t \eta_u^{-1}\, d\eta_u$$

so that U is a continuous local martingale and $\eta_t = \mathcal{E}_t(U)$ for $t \in [0, T]$.

Theorem A.15.2 *Let $\tilde{\mathbb{P}}$ be a probability measure equivalent to $\mathbb{P}$ on $(\Omega, \mathcal{F}_T)$ and such that the Radon-Nikodým density process η given by (A.59) is continuous. Then any continuous real-valued $\mathbb{P}$-semimartingale X is a continuous $\tilde{\mathbb{P}}$-semimartingale. If the canonical decomposition of X under $\mathbb{P}$ is $X = X_0 + M + A$ then its canonical decomposition under $\tilde{\mathbb{P}}$ is $X = X_0 + \tilde{M} + \tilde{A}$ where*

$$\tilde{M}_t = M_t - \int_0^t \eta_u^{-1}\, d\langle \eta, M\rangle_u = M_t - \langle U, M\rangle_t$$

and

$$\tilde{A}_t = A_t + \int_0^t \eta_u^{-1} d\langle \eta, M\rangle_u = A_t + \langle U, M\rangle_t.$$

In particular, X follows a local martingale under $\tilde{\mathbb{P}}$ if and only if the process $A + \langle U, M\rangle$ vanishes identically, that is, $A_t + \langle U, M\rangle_t = 0$ for every $t \in [0, T]$.

More generally, if $\eta = \mathcal{E}(U)$, where U is a local martingale under $\mathbb{P}$ (not necessarily with continuous sample paths), the last theorem remains valid under the assumption that the cross-variation $\langle U, M\rangle$ exists.

A.16 Martingale Measures

In this text, the main application of Girsanov's theorem is related to the issue of existence and uniqueness of a martingale measure for a given process.

Definition A.16.1 Let $X = (X_t)_{t\in[0,T]}$ be an $\mathbb{F}$-adapted process on a filtered probability space $(\Omega, \mathbb{F}, \mathbb{P})$. We say that a probability measure $\tilde{\mathbb{P}}$ equivalent to $\mathbb{P}$ on $(\Omega, \mathcal{F}_T)$ is a *martingale measure* for X whenever X is a local martingale under $\tilde{\mathbb{P}}$ with respect to $\mathbb{F}$.

Some authors prefer to make a clear terminological distinction between the concepts of a *martingale measure* and a *local martingale measure*. We decided to use only one term, and thus our *martingale measure* corresponds to what is sometimes called a *local martingale measure*.

Let us examine in detail the existence and uniqueness of a martingale measure for the unique solution X of the SDE (cf. equation (A.42))

$$dX_t = X_t(a\,dt + b\,dW_t) \qquad (A.69)$$

where a and $b \neq 0$ are real numbers and W is a one-dimensional standard Brownian motion. Note that X is an $\mathbb{F}$-martingale under $\mathbb{P}$ whenever $a = 0$.

From Proposition A.15.1, we know that for any real number γ the process $\tilde{W}_t = W_t - \gamma t$ is a one-dimensional standard Brownian motion on a probability space $(\Omega, \mathbb{F}, \tilde{\mathbb{P}})$, where $\tilde{\mathbb{P}}$ satisfies

$$\frac{d\tilde{\mathbb{P}}}{d\mathbb{P}} = \exp\left(\gamma W_T - \tfrac{1}{2}\gamma^2 T\right), \quad \mathbb{P}\text{-a.s.} \qquad (A.70)$$

By inserting the differential $dW_t = d\tilde{W}_t + \gamma\,dt$ into (A.69), we see that X satisfies under $\tilde{\mathbb{P}}$ the following SDE

$$dX_t = X_t((a + b\gamma)\,dt + b\,d\tilde{W}_t). \qquad (A.71)$$

Hence X is an $\mathbb{F}$-martingale under $\tilde{\mathbb{P}}$ provided that γ is chosen in such a way that the drift term in under (A.71) vanishes, that is, when $a + b\gamma = 0$. Of course, this means that $\gamma = -a/b$. This lead to the following result.

Proposition A.16.1 *Let W be a one-dimensional standard Brownian motion and let $\mathbb{F} = \mathbb{F}^W$ be generated by W. Then the unique martingale measure $\tilde{\mathbb{P}}$ for the solution X to the SDE (A.69) is given by (A.70) with $\gamma = -a/b$.*

Proof. We have already checked that $\tilde{\mathbb{P}}$ is a martingale measure for X. The uniqueness can be easily deduced from Proposition A.15.3.   □

The uniqueness in Proposition A.16.1 is no longer true when W is a d-dimensional standard Brownian motion for some $d > 1$, so that $b \in \mathbb{R}^d$. By applying Theorem A.15.1 and Proposition A.15.3, we conclude that a martingale measure for X corresponds to any $\mathbb{R}^d$-valued $\mathbb{F}$-progressively measurable process γ satisfying the equation $a + b \cdot \gamma_t = 0$, $t \in [0, T]$, and such that (A.61) is satisfied for γ. It is rather clear that there are infinitely many such processes, and thus uniqueness of a martingale measure for X fails to hold.

A.17 Feynman-Kac Formula

Let us first consider a particular example of a general set-up covered by Proposition A.17.2. If $f : \mathbb{R} \to \mathbb{R}$ is a bounded function and $g : \mathbb{R} \to \mathbb{R}$ is a function bounded from below then the unique bounded solution $v(t, x)$ of the partial differential equation (PDE)

$$\frac{\partial v}{\partial t}(t, x) = \frac{1}{2}\frac{\partial^2 v}{\partial x^2}(t, x) - g(x)v(t, x) \qquad (A.72)$$

with the initial condition $v(0, x) = f(x)$ can be represented by the following version of the *Feynman-Kac formula*

$$v(t, x) = \mathbb{E}_{\mathbb{P}}\Big(f(x + W_t) \exp\Big(- \int_0^t g(x + W_u)\, du \Big) \Big). \tag{A.73}$$

It is clear that this correspondence between expected values of certain functionals of a Brownian motion and solutions to partial differential equations allows us to compute some probabilistic quantities using the PDE approach and, conversely, to solve initial value problems for certain PDEs through purely probabilistic techniques (for instance, the Monte Carlo simulation of sample paths of a stochastic process).

One-dimensional case. Let us first examine the one-dimensional version of the Feynman-Kac formula.

Proposition A.17.1 *Let $\mu, \sigma : \mathbb{R} \to \mathbb{R}$ satisfy the Lipschitz condition (A.51) and the linear growth condition (A.52). Let $f : \mathbb{R} \to \mathbb{R}$ be a bounded function and $g : \mathbb{R} \to \mathbb{R}$ be a function bounded from below. Assume that $v : \mathbb{R} \times [0, T] \to \mathbb{R}$ is the unique bounded solution to the PDE*

$$\frac{\partial v}{\partial t}(t, x) = \frac{1}{2}\sigma^2(x)\frac{\partial^2 v}{\partial x^2}(t, x) + \mu(x)\frac{\partial v}{\partial x}(t, x) - g(x)u(t, x) \tag{A.74}$$

with the initial condition $v(0, x) = f(x)$. Then v has the following representation

$$v(t, x) = \mathbb{E}_{\mathbb{P}}\Big(f(X_t^x) \exp\Big(- \int_0^t g(X_u^x)\, du \Big) \Big), \tag{A.75}$$

where X^x is a solution to the SDE

$$dX_t^x = \mu(X_t^x)\, dt + \sigma(X_t^x)\, dW_t$$

with the initial condition $X_0^x = x$.

Remarks. The *infinitesimal generator* $\mathcal{A}$ of X is given by the formula

$$\mathcal{A}f(x) = \frac{1}{2}\sigma^2(x)\frac{\partial^2 f}{\partial x^2}(x) + \mu(x)\frac{\partial f}{\partial x}(x).$$

Then equation (A.74) can be rewritten as follows

$$\frac{\partial v}{\partial t} = \mathcal{A}v - gv.$$

In the proof of Proposition A.17.1 we will use the following lemma.

Lemma A.17.1 *Assume that the σ-field $\mathcal{F}_0$ is trivial. Let $M = (M_t)_{t \in [0,T]}$ be a local martingale with respect to $\mathbb{F}$ such that the random variable $M_T^* = \sup_{t \le T} |M_t|$ is integrable. Then M is an $\mathbb{F}$-martingale.*

Proof. Let $(\tau_n)_{n\in\mathbb{N}}$ be any reducing sequence of stopping times for M. For any $\in \mathbb{N}$ and $t \leq T$ the martingale property of the stopped process M^{τ_n} yields

$$\mathbb{E}_{\mathbb{P}}(M_{T\wedge\tau_n} \mid \mathcal{F}_t) = M_{t\wedge\tau_n}. \tag{A.76}$$

Since the random variable M_T^* dominates $M_{T\wedge\tau_n}$ and $X_{t\wedge\tau_n}$, using the dominated convergence theorem, we may pass to the limit in (A.76) and thus we obtain, for any $t \in [0, T]$, $\mathbb{E}_{\mathbb{P}}(M_T \mid \mathcal{F}_t) = M_t$, which is, of course, the desired martingale property of M.   □

Proof of Proposition A.17.1. The proof relies on finding a martingale $M = (M_s)_{s\in[0,t]}$ such that $\mathbb{E}_{\mathbb{P}}M_0 = v(t, x)$ and

$$\mathbb{E}_{\mathbb{P}}M_t = \mathbb{E}_{\mathbb{P}}\Big(f(X_t^x)\exp\Big(-\int_0^t g(X_u^x)\,du\Big)\Big).$$

Since for any martingale M we have that $\mathbb{E}_{\mathbb{P}}M_0 = \mathbb{E}_{\mathbb{P}}M_t$, this implies (A.75). Let us fix $t > 0$ and let us define the process $M = (M_s)_{s\in[0,t]}$ by the formula

$$M_s = v(t - s, X_s^x)\exp\Big(-\int_0^s g(X_u^x)\,du\Big). \tag{A.77}$$

Itô formula, when combined with the assumption that the function u solves (A.74), yield

$$dM_s = \exp\Big(-\int_0^s g(X_u^x)\,du\Big)\sigma(X_t^x)\frac{\partial u}{\partial x}(t - s, X_s^x)\,dW_s$$

and thus M is a local martingale. We assumed that u, g are bounded functions, so that $u(x) \leq L_1$ and $g(x) \leq L_2$ for some constants L_1 and L_2. From (A.77), we thus get

$$\sup_{u\leq t}|M_u| \leq L_1\exp(L_2 t) < \infty,$$

so that $\mathbb{E}_{\mathbb{P}}M_t^* < \infty$. It follows from Lemma A.17.1 that M is a martingale. Hence

$$v(t, x) = \mathbb{E}_{\mathbb{P}}M_0 = \mathbb{E}_{\mathbb{P}}M_t = \mathbb{E}_{\mathbb{P}}\Big(v(0, X_t^x)\exp\Big(\int_0^t g(X_u^x)\,du\Big)\Big)$$

$$= \mathbb{E}_{\mathbb{P}}\Big(f(X_t^x)\exp\Big(\int_0^t g(X_u^x)\,du\Big)\Big),$$

since obviously $v(0, X_t^x) = f(X_t^x)$. We conclude that (A.75) holds.   □

Multidimensional case. Let W be a d-dimensional standard Brownian motion on the underlying probability space $(\Omega, \mathbb{F}, \mathbb{P})$. We consider the k-dimensional diffusion process X given as

$$dX_t = \mu(X_t)\,dt + \sigma(X_t)\,dW_t, \quad X_0 = x, \tag{A.78}$$

where the coefficients $\mu : \mathbb{R}^k \to \mathbb{R}^k$ and $\sigma : \mathbb{R}^k \to \mathbb{R}^{k\times d}$ satisfy the Lipschitz condition (A.51) and the linear growth condition (A.52). By virtue of the multidimensional version of Theorem A.12.1, we conclude that the SDE (A.78) admits a unique solution. Then X is a continuous process and it enjoys the Markov property with respect to $\mathbb{F}$.

Let the d-dimensional matrix $a(x) = [a_{ij}(x)]$ be equal to $\sigma(x)\sigma^t(x)$, where $\sigma^t(x)$ is the transpose of the matrix $\sigma(x)$. We associate with the Markov process X its *infinitesimal generator* $\mathcal{A}$, which is given by the formula

$$\mathcal{A}f(x) = \sum_{i=1}^{k} \mu_i(x) \frac{\partial f}{\partial x_i}(x) + \frac{1}{2} \sum_{i,j=1}^{k} a_{ij}(x) \frac{\partial^2 f}{\partial x_i \partial x_j}(x),$$

for any function $f : \mathbb{R}^k \to \mathbb{R}$ belonging to a suitably defined class of functions, referred to as the domain $D(\mathcal{A})$ of the differential operator $\mathcal{A}$.

Typically, it is rather difficult to determine explicitly the domain $D(\mathcal{A})$ of the infinitesimal generator $\mathcal{A}$ of a diffusion process X. However, $\mathcal{A}$ is always well defined on the class of all functions with compact support (i.e., vanishing outside a bounded interval) belonging to the class $C^2(\mathbb{R}^k)$. Let us denote by $C_c^2(\mathbb{R}^k)$ the class of all such functions. Then we have $D(\mathcal{A}) \subseteq C_c^2(\mathbb{R}^k)$.

We have the following result, referred as the *Feynman-Kac formula*.

Proposition A.17.2 *Let* $g \in C(\mathbb{R}^k)$ *be a function bounded from below and let* f *belong to the class* $C_c^2(\mathbb{R}^k)$. *Define the function* $v : \mathbb{R}_+ \times \mathbb{R}^k \to \mathbb{R}$ *by the formula*

$$v(t, x) = \mathbb{E}_{\mathbb{P}}\left(f(X_t^x) \exp\left(-\int_0^t g(X_s^x)\, ds \right) \right).$$

Then v *satisfies the PDE*

$$\frac{\partial v}{\partial t} = \mathcal{A}v - gv \tag{A.79}$$

with the initial condition $v(0, x) = f(x)$ *for* $x \in \mathbb{R}^k$.

Under assumptions of Proposition A.17.2 we also have the following converse result.

Proposition A.17.3 *If* $w \in C^{1,2}(\mathbb{R} \times \mathbb{R}^k)$ *is bounded on* $[-K, K] \times \mathbb{R}^d$ *for any* $K > 0$ *(equivalently, the function* w *is bounded on* $A \times \mathbb{R}^d$ *for any compact subset* $A \subset \mathbb{R}$) *and* w *is a solution to the PDE* (A.79) *with the initial condition* $w(0, x) = f(x)$ *for* $x \in \mathbb{R}^k$ *then* $w(t, x) = v(t, x)$.

A.18 First Passage Times

In this section, we present the basic results concerning first passage times and probability distributions of some relevant functionals of a Brownian motion with drift. In this section, we find it convenient to assume that $T = \infty$, that is, we deal with a standard Brownian motion $W = (W_t)_{t \in \mathbb{R}_+}$ with respect to a reference filtration $\mathbb{F} = (\mathcal{F}_t)_{t \in \mathbb{R}_+}$. Let us denote by $\mathcal{F}_\infty$ the smallest σ-field containing $\mathcal{F}_t$ for all $t \in \mathbb{R}_+$.

Recall that a standard Brownian motion W has the Markov property with respect to the reference filtration $\mathbb{F}$ (see Lemma A.11), specifically, for any bounded Borel measurable function $g : \mathbb{R} \to \mathbb{R}$ and arbitrary $t, h \geq 0$ the following equality is valid

$$\mathbb{E}_{\mathbb{P}}(g(W_{t+h}) \mid \mathcal{F}_t) = \mathbb{E}_{\mathbb{P}}(g(W_{t+h}) \mid W_t). \qquad (A.80)$$

It appears that this property can be extended to some random times.

A random variable $\tau : (\Omega, \mathbb{F}, \mathbb{P}) \to \mathbb{R}_+ \cup \{\infty\}$ is called an $\mathbb{F}$-*stopping time* if, for every $t \in \mathbb{R}_+$, the event $\{\tau \leq t\}$ belongs to the σ-field $\mathcal{F}_t$.

For any $\mathbb{F}$-stopping time time τ we denote by $\mathcal{F}_\tau$ the σ-field of all event in $\mathcal{F}_\infty$ occurring prior to or at τ. Formally, the σ-field $\mathcal{F}_\tau$ is defined by the formula

$$\mathcal{F}_\tau = \{A \in \mathcal{F}_\infty \mid A \cap \{\tau \leq t\} \in \mathcal{F}_t \ \forall t \in \mathbb{R}_+\}.$$

The σ-field $\mathcal{F}_\tau$ should not be confused with the σ-field $\mathcal{F}(\tau)$ generated by the random variable τ. Recall that $\mathcal{F}(\tau) = \sigma\{\{\omega \in \Omega \mid \tau(\omega) \leq t\} \mid t \in \mathbb{R}_+\}$ and thus $\mathcal{F}(\tau) \subseteq \mathcal{F}_\tau$, but usually $\mathcal{F}_\tau$ and $\mathcal{F}(\tau)$ do not coincide.

For instance, if $\mathbb{F} = \mathbb{F}^X$ for some stochastic process X then $\mathcal{F}_\tau$ is the σ-field generated by the stopped process X^τ, that is, $\mathcal{F}_\tau = \sigma\{X_{\tau \wedge t} \mid t \in \mathbb{R}_+\}$.

A Brownian motion W enjoys also the following *strong Markov property* with respect to $\mathbb{F}$: for any bounded Borel measurable function $g : \mathbb{R} \to \mathbb{R}$, any stopping time τ such that $\mathbb{P}\{\tau < \infty\} = 1$, and any real number $t \geq 0$

$$\mathbb{E}_{\mathbb{P}}(g(W_{\tau+t}) \mid \mathcal{F}_\tau) = \mathbb{E}_{\mathbb{P}}(g(W_{\tau+t}) \mid W_\tau). \qquad (A.81)$$

It can be deduced from (A.81) that a Brownian motion W starts afresh after any finite stopping time τ, that is, the process $\tilde{W} = (\tilde{W}_t)_{t \in \mathbb{R}_+}$ given by the equality $\tilde{W}_t = W_{\tau+t} - W_\tau$ is a Brownian motion with respect to its natural filtration $\mathbb{F}^{\tilde{W}}$.

Given a one-dimensional standard Brownian motion W, let us denote by M_t^W and m_t^W the running maximum and minimum respectively. More explicitly, we set

$$M_t^W = \sup_{u \in [0,t]} W_u, \quad m_t^W = \inf_{u \in [0,t]} W_u.$$

It is well known that we have, for any $t > 0$,

$$\mathbb{P}\{M_t^W > 0\} = 1, \quad \mathbb{P}\{m_t^W < 0\} = 1. \qquad (A.82)$$

The following result, commonly referred to as the *reflection principle,* is a rather straightforward consequence of the strong Markov property of a standard Brownian motion and its symmetry (cf. Harrison (1985), Karatzas and Shreve (1998a) or Revuz and Yor (1999)).

Lemma A.18.1 *The formula*

$$\mathbb{P}\{W_t \leq x, \ M_t^W \geq y\} = \mathbb{P}\{W_t \geq 2y - x\} = \mathbb{P}\{W_t \leq x - 2y\} \qquad (A.83)$$

is valid for every $t > 0$, $y \geq 0$ and $x \leq y$.

Let the process Y follow a Brownian motion with the standard deviation σ and the drift v, specifically,

$$dY_t = v\,dt + \sigma\,dW_t, \quad Y_0 = y_0,$$

where W is a standard one-dimensional Brownian motion under $\mathbb{P}$ with respect to $\mathbb{F}$. Put another way

$$Y_t = y_0 + vt + \sigma W_t \tag{A.84}$$

for some constants $v \in \mathbb{R}$ and $\sigma > 0$. Let us note that Y inherits from a Brownian motion W the strong Markov property with respect to the reference filtration $\mathbb{F}$.

Probability distribution of the first passage time. Let τ stand for the *first passage time to zero* by the process Y, that is, the $\mathbb{F}$-stopping time τ given by

$$\tau = \inf\{\, t \geq 0 \mid Y_t = 0 \,\}.$$

It is well known that in an arbitrarily small time interval $[0, t]$ the sample path of the Brownian motion started at 0 passes through origin infinitely many times (see, for instance, Page 42 in Krylov (1995)). Using Girsanov's theorem and the strong Markov property of the Brownian motion, it is not difficult to show that first passage time by Y to zero coincides with the first crossing time by Y of the level 0, that is, with probability 1,

$$\tau = \inf\{\, t \geq 0 \mid Y_t \leq 0 \,\} = \inf\{\, t \geq 0 \mid Y_t < 0 \,\}.$$

Let us first prove the following well known result.

Lemma A.18.2 *Let W be a one-dimensional standard Brownian motion under $\mathbb{P}$ and let $X_t = vt + \sigma W_t$, $t \in \mathbb{R}_+$, where $\sigma > 0$ and $v \in \mathbb{R}$. Then for every $x > 0$ and any $s > 0$ we have*

$$\mathbb{P}\Big\{ \sup_{0 \leq u \leq s} X_u \leq x \Big\} = N\left(\frac{x - vs}{\sigma\sqrt{s}}\right) - e^{2v\sigma^{-2}x} N\left(\frac{-x - vs}{\sigma\sqrt{s}}\right) \tag{A.85}$$

and for every $x < 0$ and any $s > 0$

$$\mathbb{P}\Big\{ \inf_{0 \leq u \leq s} X_u \geq x \Big\} = N\left(\frac{-x + vs}{\sigma\sqrt{s}}\right) - e^{2v\sigma^{-2}x} N\left(\frac{x + vs}{\sigma\sqrt{s}}\right). \tag{A.86}$$

Proof. To derive the first equality, we will use Girsanov's theorem and the reflection principle for a Brownian motion. Assume first that $\sigma = 1$. Let $\mathbb{P}^*$ be the probability measure on $(\Omega, \mathcal{F}_s)$ given by

$$\frac{d\mathbb{P}^*}{d\mathbb{P}} = e^{-vW_s - \frac{v^2}{2}s}, \quad \mathbb{P}\text{-a.s.},$$

so that the process $W_t^* = X_t = W_t + vt$, $t \in [0, s]$, follows a standard Brownian motion under $\mathbb{P}^*$ and

$$\frac{d\mathbb{P}}{d\mathbb{P}^*} = e^{vW_s^* - \frac{v^2}{2}s}, \quad \mathbb{P}^*\text{-a.s.}$$

Moreover

$$\mathbb{P}\{\sup_{0\leq u\leq s} X_u > x,\ X_s \leq x\} = \mathbb{E}_{\mathbb{P}^*}\left(e^{\nu W_s^* - \frac{\nu^2}{2}s}\,\mathbb{1}_{\{\sup_{0\leq u\leq s} W_u^* > x,\ W_s^* \leq x\}}\right).$$

Let us define an $\mathbb{F}$-stopping time τ_x by setting

$$\tau_x = \inf\{t \geq 0 \mid W_t^* = x\},$$

and let us introduce an auxiliary process $\tilde{W}_t$, $t \in [0, s]$, defined by the formula

$$\tilde{W}_t = W_t^* \mathbb{1}_{\{\tau_x \geq t\}} + (2x - W_t^*)\mathbb{1}_{\{\tau_x < t\}}.$$

By virtue of the reflection principle, the process $\tilde{W}$ also follows a standard Brownian motion under $\mathbb{P}^*$. Moreover, we have

$$\{\sup_{0\leq u\leq s} \tilde{W}_u > x,\ \tilde{W}_s \leq x\} = \{W_s^* \geq x\} \subset \{\tau_x \leq s\}.$$

Let us denote

$$J = \mathbb{P}\{\sup_{0\leq u\leq s} (W_u + \nu u) \leq x\}.$$

Then we obtain

$$J = \mathbb{P}\{X_s \leq x\} - \mathbb{P}\{\sup_{0\leq u\leq s} X_u > x,\ X_s \leq x\}$$

$$= \mathbb{P}\{X_s \leq x\} - \mathbb{E}_{\mathbb{P}^*}\left(e^{\nu W_s^* - \frac{\nu^2}{2}s}\,\mathbb{1}_{\{\sup_{0\leq u\leq s} W_u^* > x,\ W_s^* \leq x\}}\right)$$

$$= \mathbb{P}\{W_s + \nu s \leq x\} - \mathbb{E}_{\mathbb{P}^*}\left(e^{\nu \tilde{W}_s - \frac{\nu^2}{2}s}\,\mathbb{1}_{\{\sup_{0\leq u\leq s} \tilde{W}_u > x,\ \tilde{W}_s \leq x\}}\right)$$

$$= \mathbb{P}\{X_s \leq x\} - \mathbb{E}_{\mathbb{P}^*}\left(e^{\nu(2x - W_s^*) - \frac{\nu^2}{2}s}\,\mathbb{1}_{\{W_s^* \geq x\}}\right)$$

$$= \mathbb{P}\{X_s \leq x\} - e^{2\nu x}\,\mathbb{E}_{\mathbb{P}^*}\left(e^{\nu W_s^* - \frac{\nu^2}{2}s}\,\mathbb{1}_{\{W_s^* \leq -x\}}\right)$$

$$= \mathbb{P}\{X_s \leq x\} - e^{2\nu x}\,\mathbb{P}\{W_s + \nu s \leq -x\}$$

$$= N\left(\frac{x - \nu s}{\sqrt{s}}\right) - e^{2\nu x} N\left(\frac{-x - \nu s}{\sqrt{s}}\right).$$

This ends the proof of the first equality for $\sigma = 1$. We have, for any $\sigma > 0$,

$$\mathbb{P}\{\sup_{0\leq u\leq s} (\sigma W_u + \nu u) \leq x\} = \mathbb{P}\{\sup_{0\leq u\leq s} (W_u + \nu\sigma^{-1}u) \leq x\sigma^{-1}\},$$

and this implies (A.85). Since $-W$ follows a standard Brownian motion under $\mathbb{P}$ we have, for any $x < 0$,

$$\mathbb{P}\{\inf_{0\leq u\leq s} (\sigma W_u + \nu u) \geq x\} = \mathbb{P}\{\sup_{0\leq u\leq s} (\sigma W_u - \nu u) \leq -x\},$$

and thus the second equality is a simple consequence of the first.   □

Proposition A.18.1 *Let Y be given by (A.84), where $v \in \mathbb{R}$, $\sigma > 0$ and W is a standard Brownian motion under $\mathbb{P}$. Then the random variable τ has an inverse Gaussian probability distribution under $\mathbb{P}$, that is, for $0 < s < \infty$,*

$$\mathbb{P}\{\tau \leq s\} = \mathbb{P}\{\tau < s\} = N(h_1(s)) + e^{-2v\sigma^{-2}y_0} N(h_2(s)), \qquad (A.87)$$

where N is the standard Gaussian cumulative distribution function and

$$h_1(s) = \frac{-y_0 - vs}{\sigma\sqrt{s}}, \quad h_2(s) = \frac{-y_0 + vs}{\sigma\sqrt{s}}.$$

Proof. Let us first observe that

$$\mathbb{P}\{\tau \geq s\} = \mathbb{P}\{\inf_{0 \leq u \leq s} Y_u \geq 0\} = \mathbb{P}\{\inf_{0 \leq u \leq s} X_u \geq -y_0\}, \qquad (A.88)$$

where $X_u = \sigma W_u + vu$. We know from Lemma A.18.2 that for every $x < 0$ we have

$$\mathbb{P}\{\inf_{0 \leq u \leq s} X_u \geq x\} = N\left(\frac{-x + vs}{\sigma\sqrt{s}}\right) - e^{2v\sigma^{-2}x} N\left(\frac{x + vs}{\sigma\sqrt{s}}\right).$$

When combined with (A.88), this yields (A.87).   □

The following corollary is a consequence of Proposition A.18.1 and the strong Markov property of the process Y with respect to the filtration $\mathbb{F}$.

Corollary A.18.1 *Under the assumptions of Proposition A.18.1 we have, for any $t < s$ on the event $\{\tau > t\}$,*

$$\mathbb{P}\{\tau \leq s \mid \mathcal{F}_t\} = N\left(\frac{-Y_t - v(s - t)}{\sigma\sqrt{s - t}}\right) + e^{-2v\sigma^{-2}Y_t} N\left(\frac{-Y_t + v(s - t)}{\sigma\sqrt{s - t}}\right).$$

Joint probability distribution of Y and τ. In the next step, we will examine the joint distribution of the process Y and its first passage time τ. For the process $X_t = vt + \sigma W_t$, we write

$$M_s^X = \sup_{u \in [0,s]} X_u, \quad m_s^X = \inf_{u \in [0,s]} X_u.$$

By the Girsanov theorem, the process $\sigma^{-1} X$ is a Brownian motion under an equivalent probability measure and thus we can deduce that, for any $s > 0$ (cf. (A.83))

$$\mathbb{P}\{M_s^X > 0\} = 1, \quad \mathbb{P}\{m_s^X < 0\} = 1.$$

Lemma A.18.3 *For any $s > 0$, the joint distribution of X_s and M_s^X is given by the formula*

$$\mathbb{P}\{X_s \leq x, M_s^X \geq y\} = e^{2vy\sigma^{-2}} \mathbb{P}\{X_s \geq 2y - x + 2vs\}, \qquad (A.89)$$

for every $x, y \in \mathbb{R}$ such that $y \geq 0$ and $x \leq y$.

Proof. Since

$$I = \mathbb{P}\{X_s \le x, \, M_s^X \ge y\} = \mathbb{P}\{X_s^\sigma \le x\sigma^{-1}, \, M_s^{X^\sigma} \ge y\sigma^{-1}\},$$

where $X_t^\sigma = W_t + vt\sigma^{-1}$, it is clear that we may and do assume, without loss of generality, that $\sigma = 1$.

We shall use similar arguments as in the proof of Lemma A.18.2. It follows from Girsanov's theorem that X is a standard Brownian motion under the probability measure $\mathbb{P}^*$ given on $(\Omega, \mathcal{F}_s)$ by the formula

$$\frac{d\mathbb{P}^*}{d\mathbb{P}} = e^{-vW_s - \frac{v^2}{2}s}, \qquad \mathbb{P}\text{-a.s.}$$

Therefore

$$\frac{d\mathbb{P}}{d\mathbb{P}^*} = e^{vW_s^* - \frac{v^2}{2}s}, \qquad \mathbb{P}\text{-a.s.},$$

where $W_t^* = X_t = W_t + vt$, $t \in [0, s]$, is a standard Brownian motion under $\mathbb{P}^*$. It is thus clear that

$$I = \mathbb{E}_{\mathbb{P}^*}\left(e^{vW_s^* - \frac{v^2}{2}s} \mathbb{1}_{\{X_s \le x, \, M_s^X \ge y\}}\right) = \mathbb{E}_{\mathbb{P}^*}\left(e^{vW_s^* - \frac{v^2}{2}s} \mathbb{1}_{\{W_s^* \le x, \, M_s^{W^*} \ge y\}}\right).$$

Since W^* is a standard Brownian motion under $\mathbb{P}^*$, an application of the reflection principle yields

$$I = \mathbb{E}_{\mathbb{P}^*}\left(e^{v(2y - W_s^*) - \frac{v^2}{2}s} \mathbb{1}_{\{2y - W_s^* \le x, \, M_s^{W^*} \ge y\}}\right)$$

$$= \mathbb{E}_{\mathbb{P}^*}\left(e^{v(2y - W_s^*) - \frac{v^2}{2}s} \mathbb{1}_{\{W_s^* \ge 2y - x\}}\right)$$

$$= e^{2vy} \mathbb{E}_{\mathbb{P}^*}\left(e^{-vW_s^* - \frac{v^2}{2}s} \mathbb{1}_{\{W_s^* \ge 2y - x\}}\right),$$

since under the present assumptions we have $2y - x \ge y$. Let us define an equivalent probability measure $\tilde{\mathbb{P}}$ on $(\Omega, \mathcal{F}_s)$ by setting

$$\frac{d\tilde{\mathbb{P}}}{d\mathbb{P}^*} = e^{-vW_s^* - \frac{v^2}{2}s}, \qquad \mathbb{P}^*\text{-a.s.}$$

Then we have that

$$I = e^{2vy} \mathbb{E}_{\mathbb{P}^*}\left(e^{-vW_s^* - \frac{v^2}{2}s} \mathbb{1}_{\{W_s^* \ge 2y - x\}}\right) = e^{2vy} \tilde{\mathbb{P}}\{W_s^* \ge 2y - x\}.$$

The process $\tilde{W}_t = W_t^* + vt$, $t \in [0, s]$, is a standard Brownian motion under $\tilde{\mathbb{P}}$ and we have

$$I = e^{2vy} \tilde{\mathbb{P}}\{\tilde{W}_s + vs \ge 2y - x + 2vs\}.$$

The asserted formula (A.89) follows easily from the last equality.   □

Let us observe that (a similar remark applies to any formula given below)

$$\mathbb{P}\{X_s \leq x, M_s^X \geq y\} = \mathbb{P}\{X_s < x, M_s^X > y\}.$$

The following result is a straightforward consequence of Lemma A.18.3.

Proposition A.18.2 *We have, for any* $x, y \in \mathbb{R}$ *satisfying* $y \geq 0$ *and* $x \leq y$,

$$\mathbb{P}\{X_s \leq x, M_s^X \geq y\} = e^{2vy\sigma^{-2}} N\left(\frac{x - 2y - vs}{\sigma\sqrt{s}}\right). \tag{A.90}$$

Hence for every $x, y \in \mathbb{R}$ *such that* $x \leq y$ *and* $y \geq 0$

$$\mathbb{P}\{X_s \leq x, M_s^X \leq y\} = N\left(\frac{x - vs}{\sigma\sqrt{s}}\right) - e^{2vy\sigma^{-2}} N\left(\frac{x - 2y - vs}{\sigma\sqrt{s}}\right). \tag{A.91}$$

Proof. For the first equality, note that

$$\mathbb{P}\{X_s \geq 2y - x + 2vs\} = \mathbb{P}\{-\sigma W_s \leq x - 2y - vs\} = N\left(\frac{x - 2y - vs}{\sigma\sqrt{s}}\right),$$

since $-\sigma W_t$ has Gaussian distribution with zero mean and variance $\sigma^2 t$. To establish (A.91), it is enough to observe that

$$\mathbb{P}\{X_s \leq x, M_s^X \leq y\} + \mathbb{P}\{X_s \leq x, M_s^X > y\} = \mathbb{P}\{X_s \leq x\}$$

and to apply (A.90). □

It is clear that, for every $y \geq 0$,

$$\mathbb{P}\{M_s^X \geq y\} = \mathbb{P}\{X_s \geq y\} + \mathbb{P}\{X_s \leq y, M_s^X \geq y\}$$

and thus

$$\mathbb{P}\{M_s^X \geq y\} = \mathbb{P}\{X_s \geq y\} + e^{2vy\sigma^{-2}} \mathbb{P}\{X_s \geq y + 2vs\}. \tag{A.92}$$

Hence

$$\mathbb{P}\{M_s^X \leq y\} = 1 - \mathbb{P}\{M_s^X \geq y\} = \mathbb{P}\{X_s \leq y\} - e^{2vy\sigma^{-2}} \mathbb{P}\{X_s \geq y + 2vs\}.$$

This leads to the following corollary.

Corollary A.18.2 *The following formula holds for every* $y \geq 0$

$$\mathbb{P}\{M_s^X \leq y\} = N\left(\frac{y - vs}{\sigma\sqrt{s}}\right) - e^{2vy\sigma^{-2}} N\left(\frac{-y - vs}{\sigma\sqrt{s}}\right). \tag{A.93}$$

Let us now focus on the distribution of the infimum of X. Observe that we have, for any $y \leq 0$,

$$\mathbb{P}\{\sup_{0 \leq u \leq s}(\sigma W_u - vu) \geq -y\} = \mathbb{P}\{\inf_{0 \leq u \leq s}(-\sigma W_u + vu) \leq y\} = \mathbb{P}\{\inf_{0 \leq u \leq s} X_u \leq y\},$$

where the last equality follows from the symmetry of the Brownian motion. Consequently we have, for every $y \leq 0$,

$$\mathbb{P}\{m_s^X \leq y\} = \mathbb{P}\{M_s^{\tilde{X}} \geq -y\},$$

where the process $\tilde{X}$ equals $\tilde{X}_t = \sigma W_t - vt$. The following result is thus not difficult to prove.

Proposition A.18.3 *For every $s > 0$, the joint distribution of (X_s, m_s^X) satisfies, for any $x, y \in \mathbb{R}$ such that $y \leq 0$ and $y \leq x$,*

$$\mathbb{P}\{ X_s \geq x, \ m_s^X \geq y\} = N \left(\frac{-x + vs}{\sigma \sqrt{s}} \right) - e^{2vy\sigma^{-2}} N \left(\frac{2y - x + vs}{\sigma \sqrt{s}} \right).$$

Corollary A.18.3 *The following formula is valid for every $y \leq 0$*

$$\mathbb{P}\{m_s^X \geq y\} = N \left(\frac{-y + vs}{\sigma \sqrt{s}} \right) - e^{2vy\sigma^{-2}} N \left(\frac{y + vs}{\sigma \sqrt{s}} \right).$$

Recall that $Y_t = y_0 + X_t$, where $X_t = vt + \sigma W_t$. We write

$$m_s^X = \inf_{u \in [0,s]} X_u, \quad m_s^Y = \inf_{u \in [0,s]} Y_u.$$

Corollary A.18.4 *For any $s > 0$ and $y \geq 0$ we have*

$$\mathbb{P}\{Y_s \geq y, \ \tau \geq s\} = N \left(\frac{-y + y_0 + vs}{\sigma \sqrt{s}} \right) - e^{-2v\sigma^{-2}y_0} N \left(\frac{-y - y_0 + vs}{\sigma \sqrt{s}} \right).$$

Proof. Since

$$\mathbb{P}\{Y_s \geq y, \ \tau \geq s\} = \mathbb{P}\{ Y_s \geq y, \ m_s^Y \geq 0\} = \mathbb{P}\{ X_s \geq y - y_0, \ m_s^X \geq -y_0\},$$

the formula is rather obvious. □

More generally, by applying the Markov property and time-homogeneity of Y we obtain the following result.

Lemma A.18.4 *Under the assumptions of Lemma A.18.1 we have, for any $t < s$ and $y \geq 0$ on the event $\{\tau > t\}$,*

$$\mathbb{P}\{Y_s \geq y, \ \tau \geq s \mid \mathcal{F}_t\} = N \left(\frac{-y + Y_t + v(s - t)}{\sigma \sqrt{s - t}} \right)$$
$$- e^{-2v\sigma^{-2}Y_t} N \left(\frac{-y - Y_t + v(s - t)}{\sigma \sqrt{s - t}} \right).$$

References

Aase, K.K. (1988) Contingent claim valuation when the security price is a combination of an Ito process and a random point process. *Stochastic Process. Appl.* 28, 185–220.

Aase, K.K., Øksendal, B. (1988) Admissible investment strategies in continuous trading. *Stochastic Process. Appl.* 30, 291–301.

Adams, K.J., van Deventer, D.R. (1994) Fitting yield curves and forward rate curves with maximum smoothness. *J. Fixed Income* 4(1), 52–62.

Adams, P.D., Wyatt, S.B. (1987) Biases in option prices: evidence from the foreign currency option market. *J. Bank. Finance* 11, 549–562.

Ahn, C.M. (1992) Option pricing when jump risk is systematic. *Math. Finance* 2, 299–308.

Ahn, H., Muni, A., Swindle, G. (1997) Misspecified asset price models and robust hedging strategies. *Appl. Math. Finance* 4, 21–36.

Ahn, H., Penaud, A., Wilmott, P. (1999) Various passport options and their valuation. *Appl. Math. Finance* 4, 275–292.

Aihara, S., Bagchi, A. (2000) Estimation of stochastic volatility in the Hull-White model. *Appl. Math. Finance* 7, 153–181.

Aït-Sahalia, Y. (1996a) Nonparametric pricing of interest rate derivative securities. *Econometrica* 64, 527–560.

Aït-Sahalia, Y. (1996b) Testing continuous-time models of the spot interest rate. *Rev. Finan. Stud.* 9, 385–426.

Aït-Sahalia, Y. (1998) Dynamic equilibrium and volatility in financial asset markets. *J. Econometrics* 84, 93–128.

Aït-Sahalia, Y. (2002) Telling from discrete data whether the underlying continuous-time model is a diffusion. *J. Finance* 57, 2075–2112.

Aït-Sahalia, Y. (2004) Disentangling diffusion from jumps. *J. Finan. Econom.* 74, 478–528.

Aït-Sahalia, Y., Lo, A.W. (1998) Nonparametric estimation of state-price densities implicit in financial asset prices. *J. Finance* 53, 499–547.

Aït-Sahalia, Y., Lo, A.W. (2000) Nonparametric risk management and implied risk aversion. *J. Econometrics* 94, 9–51.

Aït-Sahalia, Y., Mykland, P.A. (2003) The effects of random and discrete sampling when estimating continuous-time diffusions. *Econometrica* 71, 483–549.

Aït-Sahalia, Y., Mykland, P.A. (2004) Estimators of diffusions with randomly spaced discrete observations: a general theory. *Ann. Statist.* 32, 2186–2222.

Akahori, J. (1995) Some formulae for a new type of path-dependent option. *Ann. Appl. Probab.* 5, 383–388.

Alziary, B., Décamps, J.-P., Koehl, P.-F. (1997) A P.D.E. approach to Asian options: analytical and numerical evidence. *J. Bank. Finance* 21, 613–640.

Amin, K. (1991) On the computation of continuous time option prices using discrete approximations. *J. Finan. Quant. Anal.* 26, 477–496.

Amin, K. (1993) Jump diffusion option valuation in discrete time. *J. Finance* 48, 1833–1863.

Amin, K., Bodurtha, J. (1995) Discrete time valuation of American options with stochastic interest rates. *Rev. Finan. Stud.* 8, 193–234.

Amin, K., Jarrow, R. (1991) Pricing foreign currency options under stochastic interest rates. *J. Internat. Money Finance* 10, 310–329.

Amin, K., Jarrow, R. (1992) Pricing options on risky assets in a stochastic interest rate economy. *Math. Finance* 2, 217–237.

Amin, K., Khanna, A. (1994) Convergence of American option values from discrete-to continuous-time financial models. *Math. Finance* 4, 289–304.

Amin, K., Morton, A. (1994) Implied volatility functions in arbitrage-free term structure models. *J. Finan. Econom.* 35, 141–180.

Amin, K., Ng, V.K. (1997) Inferring future volatility from the information in implied volatility in Eurodollar options: a new approach. *Rev. Finan. Stud.* 10, 333–367.

Ammann, M. (1999) *Pricing Derivative Credit Risk.* Springer-Verlag, Berlin Heidelberg New York.

Andersen, L. (2000) A simple approach to the pricing of Bermudan swaptions in the multifactor LIBOR market model. *J. Comput. Finance* 3, 5–32.

Andersen, L., Andreasen, J. (2000a) Jump-diffusion processes: volatility smile fitting and numerical methods for option pricing. *Rev. Derivatives Res.* 4, 231–262.

Andersen, L., Andreasen, J. (2000b) Volatility skews and extensions of the Libor market model. *Appl. Math. Finance* 7, 1–32.

Andersen, L., Andreasen, J. (2001) Factor dependence of Bermudan swaptions: fact or fiction? *J. Finan. Econom.* 62, 3–37.

Andersen, L., Brotherton-Ratcliffe, R. (1997) The equity option volatility smile: an implicit finite-difference approach. *J. Comput. Finance* 1, 5–37.

Andersen, L., Brotherton-Ratcliffe, R. (2001) Extended Libor market models with stochastic volatility. Working paper, Gen Re Financial Products.

Andersen, L., Buffum, L. (2003) Calibration and implementation of convertible bond models. Working paper.

Andersen, L., Piterbarg, V. (2004) Moment explosions in stochastic volatility models. Working paper.

Andersen, L., Andreasen, J., Brotherton-Ratcliffe, R. (1998) The passport option. *J. Comput. Finance* 1, 15–36.

Andreasen, J., Dufresne, P.C., Shi, W. (1998) Applying the HJM-approach when volatility is stochastic. Working paper.

Ané, T., Geman, H. (2000) Order flow, transaction clock and normality of asset returns. *J. Finance* 55, 2259–2284.

Ang, J.S., Peterson, D.R. (1984) Empirical properties of the elasticity coefficient in the constant elasticity of variance model. *Finan. Rev.* 1, 372–380.

Ansel, J.-P., Stricker, C. (1992) Lois de martingale, densités et décomposition de Föllmer Schweizer. *Ann. Inst. H.Poincaré Probab. Statist.* 28, 375–392.

Ansel, J.-P., Stricker, C. (1993) Unicité et existence de la loi minimale. In: *Lecture Notes in Math. 1557.* Springer, Berlin Heidelberg New York, pp. 22–29.

Ansel, J.-P., Stricker, C. (1994) Couverture des actifs contingents. *Ann. Inst. H. Poincaré Probab. Statist.* 30, 303–315.

Arak, M., Goodman, L.S. (1987) Treasury bond futures: valuing the delivery options. *J. Futures Markets* 7, 269–286.

Arrow, K. (1964) The role of securities in the optimal allocation of risk-bearing. *Rev. Econom. Stud.* 31, 91–96.

Arrow, K. (1970) *Essays in the Theory of Risk Bearing.* North-Holland, London.

Artzner, P. (1997) On the numeraire portfolio. In: *Mathematics of Derivative Securities,* M.A.H. Dempster and S.R. Pliska, eds. Cambridge University Press, Cambridge, pp. 53–58.

Artzner, P., Delbaen, F. (1989) Term structure of interest rates: the martingale approach. *Adv. in Appl. Math.* 10, 95–129.

Artzner, P., Delbaen, F. (1992) Credit risk and prepayment option. *ASTIN Bull.* 22, 81–96.

Artzner, P., Delbaen, F. (1995) Default risk insurance and incomplete markets. *Math. Finance* 5, 187–195.

Artzner, P., Heath, D. (1995) Approximate completeness with multiple martingale measures. *Math. Finance* 5, 1–11.

d'Aspremont, A. (2003) Interest rate model calibration using semidefinite programming. *Appl. Math. Finance* 10, 183–213.

Athanassakos, G. (1996) On the application of the Black and Scholes formula to valuing bonds with embedded options: the case of extendible bonds. *Appl. Finan. Econom.* 6, 37–48.

Atlan, M. (2006) Localizing volatilities. Working paper.

Avellaneda, M., Parás, A. (1994) Dynamic hedging portfolios for derivative securities in the presence of large transaction costs. *Appl. Math. Finance* 1, 165–194.

Avellaneda, M., Parás, A. (1996) Managing the volatility risk of portfolios of derivative securities: the Lagrangian uncertain volatility model. *Appl. Math. Finance* 3, 21–52.

Avellaneda, M., Levy, A., Parás, A. (1995) Pricing and hedging derivative securities in markets with uncertain volatilities. *Appl. Math. Finance* 2, 73–88.

Avellaneda, M., Friedman, C., Holmes, R., Samperi, D. (1997) Calibrating volatility surfaces via relative-entropy minimization. *Appl. Math. Finance* 4, 37–64.

Ayache, E., Forsyth, P., Vetzal, K. (2003) Valuation of convertible bonds with credit risk. *Journal of Derivatives,* Fall 2003.

Ayache, E., Henrotte, P., Nassar, S., Wang, X. (2004) Can anyone solve the smile problem? *Wilmott* 78–96.

Babbs, S. (1990) A family of Itô process models for the term structure of interest rates. Working paper, University of Warwick.

Babbs, S., Webber, N.J. (1997) Term structure modelling under alternative official regimes. In: *Mathematics of Derivative Securities,* M.A.H. Dempster and S.R. Pliska, eds. Cambridge University Press, Cambridge, pp. 394–422.

Bachelier, L. (1900) Théorie de la spéculation. *Ann. Sci École Norm. Sup.* 17, 21–86. [English translation in: *The Random Character of Stock Market Prices,* P.H. Cootner, ed. MIT Press, Cambridge (Mass.) 1964, pp. 17–78.]

Back, K. (1991) Asset pricing for general processes. *J. Math. Econom.* 20, 371–395.

Back, K., Pliska, S.R. (1990) On the fundamental theorem of asset pricing with an infinite state space. *J. Math. Econom.* 20, 1–18.

Bajeux-Besnainou, I., Rochet, J.-C. (1996) Dynamic spanning: are options an appropriate instrument. *Math. Finance* 6, 1–16.

Bakshi, G.S., Cao, C., Chen, Z. (1997) Empirical performance of alternative option pricing models. *J. Finance* 52, 2003–2049.

Bakshi, G.S., Cao, C., Chen, Z. (2000) Pricing and hedging long term options. *J. Econometrics* 94, 277–318.

Ball, C.A., Roma, A. (1994) Stochastic volatility option pricing. *J. Finan. Quant. Anal.* 29, 589–607.

Ball, C.A., Torous, W.N. (1983a) A simplified jump process for common stock returns. *J. Finan. Quant. Anal.* 18, 53–65.

Ball, C.A., Torous, W.N. (1983b) Bond price dynamics and options. *J. Finan. Quant. Anal.* 18, 517–531.

Ball, C.A., Torous, W.N. (1984) The maximum likelihood estimation of security price volatility: theory, evidence and application to option pricing. *J. Business* 57, 97–112.

Ball, C.A., Torous, W.N. (1985) On jumps in common stock prices and their impact on call pricing. *J. Finance* 40, 155–173.

Ball, C.A., Torous, W.N. (1986) Futures options and the volatility of futures prices. *J. Finance* 41, 857–870.

Balland, P., Hughston, L.P. (2000) Markov market model consistent with cap smile. *Internat. J. Theor. Appl. Finance* 3, 161–181.

Bank, P., Föllmer, H. (2003) American options, multi-armed bandits, and optimal consumption plans: a unifying view. In: *Paris-Princeton Lectures on Mathematical Finance 2002*, R. Carmona et al., eds., Springer, Berlin Heidelberg New York, pp. 1–42.

Bardhan, I., Chao, X. (1993) Pricing options on securities with discontinuous returns. *Stochastic Process. Appl.* 48, 123–137.

Bardhan, I., Chao, X. (1995) Martingale analysis for assets with discontinuous returns. *Math. Oper. Res.* 20, 243–256.

Barles, G., Soner, M.H. (1998) Option pricing with transaction costs and a nonlinear Black-Scholes equation. *Finance Stochast.* 2, 369–397.

Barles, G., Burdeau, J., Romano, M., Samsœn, N. (1995) Critical stock price near expiration. *Math. Finance* 5, 77–95.

Barndorff-Nielsen, O.E. (1997) Normal inverse Gaussian distributions and stochastic volatility modelling. *Scand. J. Statist.* 5, 151–157.

Barndorff-Nielsen, O.E. (1998) Processes of normal inverse Gaussian type. *Finance Stochast.* 2, 41–68.

Barndorff-Nielsen, O.E., Rejman, A., Weron, A., Weron, R. (1996) Option pricing for asset returns described by generalized hyperbolic distributions: discrete versus continuous models. Working paper.

Barndorff-Nielsen, O.E., Nicolato, E., Shephard, N. (2002) Some recent advances in stochastic volatility modelling. *Quant. Finance* 2, 11–23.

Barone-Adesi, G., Elliott, R.J. (1991) Approximations for the values of American options. *Stochastic Anal. Appl.* 9, 115–131.

Barone-Adesi, G., Whaley, R.E. (1986) The valuation of American call options and the expected ex-dividend stock price decline. *J. Finan. Econom.* 17, 91–111.

Barone-Adesi, G., Whaley, R.E. (1987) Efficient analytic approximation of American option values. *J. Finance* 42, 301–320.

Barraquand, J., Pudet, T. (1996) Pricing of American path-dependent contingent claims. *Math. Finance* 6, 17–51.

Baxter, M., Rennie, A. (1996) *Financial Calculus. An Introduction to Derivative Pricing.* Cambridge University Press, Cambridge.

Baz, J., Chacko, G. (2004) *Financial Derivatives. Pricing, Applications and Mathematics.* Cambridge University Press, Cambridge.

Beaglehole, D., Tenney, M. (1991) General solutions of some interest rate contingent claim pricing equations. *J. Fixed Income* 1(2), 69–93.

Beckers, S. (1980) The constant elasticity of variance model and its implications for option pricing. *J. Finance* 35, 661–673.

Beckers, S. (1981) Standard deviations implied in option prices as predictors of future stock price variability. *J. Bank. Finance* 5, 363–381.

Beckers, S. (1983) Variances of security price returns based on high, low, and closing prices. *J. Business* 56, 97–112.

Bélanger, A., Shreve, S.E., Wong, D. (2004) A general framework for pricing credit risk. *Math. Finance* 14, 317–350.

Bellamy, N., Jeanblanc, M. (2000) Incompletenesss of markets driven by a mixed diffusion. *Finance Stochast.* 4, 209–222.

Benninga, S., Blume, M. (1985) On the optimality of portfolio insurance. *J. Finance* 40, 1341–1352.

Bensoussan, A. (1984) On the theory of option pricing. *Acta Appl. Math.* 2, 139–158.

Bensoussan, A., Elliott, R.J. (1995) Attainable claims in a Markov model. *Math. Finance* 5, 121–131.

Bensoussan, A., Lions, J.-L. (1978) *Applications des inéquations variationelles en contrôle stochastique.* Dunod, Paris.

Bensoussan, A., Crouhy, M., Galai, D. (1994) Stochastic equity volatility related to the leverage effect I: equity volatility behaviour. *Appl. Math. Finance* 1, 63–85.

Bensoussan, A., Crouhy, M., Galai, D. (1995) Stochastic equity volatility related to the leverage effect II: equity valuation of European equity options and warrants. *Appl. Math. Finance* 2, 43–59.

Berestycki, H., Busca, J., Florent, I. (2000) An inverse parabolic problem arising in finance. *C. R. Acad. Sci. Paris* 331, Série I, 965–969.

Berestycki, H., Busca, J., Florent, I. (2002) Asymptotics and calibration of local volatility models. *Quant. Finance* 2, 61–69.

Berestycki, H., Busca, J., Florent, I. (2004) Computing the implied volatility in stochastic volatility models. *Comm. Pure Appl. Math.* 57, 1352–1373.

Bergman, Y.Z. (1982) Pricing of contingent claims in perfect and imperfect markets. Doctoral dissertation, University of California, Berkeley.

Bergman, Y.Z. (1995) Option pricing with differential interest rates. *Rev. Finan. Stud.* 8, 475–500.

Bergman, Y.Z. (1998) General restrictions on prices of financial derivatives written on underlying diffusions. Working paper, The Hebrew University, Jerusalem.

Bergman, Y.Z., Grundy, D.B., Wiener, Z. (1996) General properties of option prices. *J. Finance* 51, 1573–1610.

Berle, S., Cakici, N. (1998) How to grow a smiling tree. *J. Fin. Engrg* 7, 127–146.

Bertsekas, D., Shreve, S. (1978) *Stochastic Optimal Control: The Discrete Time Case.* Academic Press, New York.

Bergenthum, J., Rüschendorf, L. (2006) Comparison of option prices in semimartingale models. *Finance Stochast.* 10, 222–249.

Bibby, B., Sørensen, M. (1997) A hyperbolic diffusion model for stock prices. *Finance Stochast.* 1, 25–41.

Bick, A. (1995) Quadratic-variation-based trading strategies. *Manag. Sci.* 41, 722–732.

Bick, A., Willinger, W. (1994) Dynamic spanning without probabilities. *Stochastic Process. Appl.* 50, 349–374.

Bielecki, T.R., Rutkowski, M. (2002) *Credit Risk: Modeling, Valuation and Hedging.* Springer, Berlin Heidelberg New York.

Bielecki, T.R., Jeanblanc, M., Rutkowski, M. (2004a) Hedging of defaultable claims. In: *Paris-Princeton Lectures on Mathematical Finance 2003*, R. Carmona et al., eds. Springer, Berlin Heidelberg New York, pp. 1–132.

Bielecki, T.R., Jeanblanc, M., Rutkowski, M. (2004b) Modelling and valuation of credit risk. In: *Stochastic Methods in Finance*, M. Frittelli and W. Runggaldier, eds., Springer, Berlin Heidelberg New York, pp. 27–126.

Bielecki, T.R., Jeanblanc, M., Rutkowski, M. (2005a) PDE approach to valuation and hedging of credit derivatives. *Quant. Finance* 5, 257–270.

Bielecki, T.R., Jeanblanc, M., Rutkowski, M. (2005b) Hedging of credit derivatives in models with totally unexpected default. In: *Stochastic Processes and Applications to Mathematical Finance*, J. Akahori et al., eds., World Scientific, Singapore, pp. 35–100.

Biger, N., Hull, J. (1983) The valuation of currency options. *Finan. Manag.* 12, 24–28.

Bingham, N.H., Kiesel, R. (1998) *Risk-Neutral Valuation. Pricing and Hedging of Financial Derivatives*. Springer, Berlin Heidelberg New York.

Bingham, N.H., Kiesel, R. (2002) Semi-parametric methods in finance: theoretical foundations. *Quant. Finance* 2, 241–250.

Bismut, J.M., Skalli, B. (1977) Temps d'arrêt optimal, théorie générale de processus et processus de Markov. *Z. Wahrsch. verw. Geb.* 39, 301–313.

Björk, T. (1995) On the term structure of discontinuous interest rates. *Survey Indust. Appl. Math.* 2, 626–657. [In Russian]

Björk, T. (1997) Interest rate theory. In: *Financial Mathematics, Bressanone, 1996*, W. Runggaldier, ed. *Lecture Notes in Math.* 1656, Springer, Berlin Heidelberg New York, pp. 53–122.

Björk, T. (1998) *Arbitrage Theory in Continuous Time*. Oxford University Press, Oxford.

Björk, T. (2004) On the geometry of interest rate models. In: *Paris-Princeton Lectures on Mathematical Finance 2003*, R. Carmona et al., eds. Springer, Berlin Heidelberg New York, pp. 133–215.

Björk, T., Christensen, B.J. (1999) Interest rate dynamics and consistent forward rate curves. *Math. Finance* 9, 323–348.

Björk, T., Gombani, A. (1999) Minimal realizations of interest rate models. *Finance Stochast.* 3, 413–432.

Björk, T., Landén, C. (2002) On the construction of finite dimensional realizations for nonlinear forward rate models. *Finance Stochast.* 6, 303–331.

Björk, T., Svensson, L. (2001) On the existence of finite dimensional realizations for nonlinear forward rate models. *Math. Finance* 11, 205–243.

Björk, T., Landén, C., Svensson, L. (2004) Finite dimensional markovian realizations for stochastic volatility forward rate models. *Proc. Royal Soc.* 460/2041, 53–84.

Björk, T., Di Masi, G., Kabanov, Y., Runggaldier, W. (1997a) Towards a general theory of bond market. *Finance Stochast.* 1, 141–174.

Björk, T., Kabanov, Y., Runggaldier, W. (1997b) Bond market structure in the presence of marked point processes. *Math. Finance* 7, 211–239.

Black, F. (1972) Capital market equilibrium with restricted borrowing. *J. Business* 45, 444–454.

Black, F. (1975) Fact and fantasy in the use of options. *Finan. Analysts J.* 31(4), 36–41, 61–72.

Black, F. (1976a) Studies of stock price volatility changes. In: *Proceedings of the 1976 Meetings of the American Statistical Association*, pp. 177–181.

Black, F. (1976b) The pricing of commodity contracts. *J. Finan. Econom.* 3, 167–179.

Black, F. (1986) Noise. *J. Finance* 41, 529–543.

Black, F. (1989) How we came up with the option formula. *J. Portfolio Manag.* 2, 4–8.

Black, F., Cox, J.C. (1976) Valuing corporate securities: some effects of bond indenture provisions. *J. Finance* 31, 351–367.

Black, F., Karasinski, P. (1991) Bond and option pricing when short rates are lognormal. *Finan. Analysts J.* 47(4), 52–59.

Black, F., Scholes, M. (1972) The valuation of option contracts and a test of market efficiency. *J. Finance* 27, 399–417.

Black, F., Scholes, M. (1973) The pricing of options and corporate liabilities. *J. Political Econom.* 81, 637–654.

Black, F., Jensen, M., Scholes, M. (1972) The capital asset pricing model: some empirical tests. In: *Studies in the Theory of Capital Markets*, M. Jensen, ed. Praeger, New York, pp. 79–121.

Black, F., Derman, E., Toy, W. (1990) A one-factor model of interest rates and its application to Treasury bond options. *Finan. Analysts J.* 46(1), 33–39.

Blattberg, R.C., Gonedes, N.J. (1974) A comparison of the stable and Student distributions as statistical models for stock prices. *J. Business* 47, 244–280.

Bliss, R.R. (1997) Testing term structure estimation models. *Adv. Futures Options Res.* 9, 197–231.

Bliss, R.R., Panigirtzoglou, N. (2002) Testing the stability of implied probability functions. *J. Bank. Finance* 26, 381–422.

Blomeyer, E.C. (1986) An analytic approximation for the American put price for options on stock with dividends. *J. Finan. Quant. Anal.* 21, 229–233.

Blomeyer, E.C., Johnson, H. (1988) An empirical examination of the pricing of American put options. *J. Finan. Quant. Anal.* 23, 13–22.

Bodie, Z., Taggart, R.A. (1978) Future investment opportunities and the value of call provision on a bond. *J. Finance* 33, 1187–1200.

Bodurtha, J.N., Courtadon, G.R. (1987) Tests of an American option pricing model on the foreign currency options market. *J. Finan. Quant. Anal.* 22, 153–168.

Bodurtha, J.N., Courtadon, G.R. (1995) Probabilities and values of early exercise: spot and futures foreign currency options. *J. Derivatives*, Fall, 57–70.

Bodurtha, J.N., Jermakyan, M. (1999) Nonparametric estimation of an implied volatility surface. *J. Comput. Finance* 2, 5–32.

Bollerslev, T. (1986) Generalized autoregressive conditional heteroskedasticity. *J. Econometrics* 31, 307–327.

Bollerslev, T., Chou, R., Kroner, K. (1992) ARCH modeling in finance: a review of the theory and empirical evidence. *J. Econometrics* 52, 5–59.

Boness, J. (1964) Elements of a theory of stock-option value. *J. Political Econom.* 12, 163–175.

Bookstaber, R., Clarke, R. (1984) Option portfolio strategies: measurement and evaluation. *J. Business* 57, 469–492.

Bookstaber, R., Clarke, R. (1985) Problems in evaluating the performance of portfolios with options. *Finan. Analysts J.* 41(1), 48–62.

Borodin, A., Salminen, P. (1996) *Handbook of Brownian Motion: Facts and Formulae*. Birkhäuser, Basel Boston Berlin.

Bouaziz, L., Briys, E., Crouhy, M. (1994) The pricing of forward-starting asian options. *J. Bank. Finance* 18, 823–839.

Bouleau, N., Lamberton, D. (1989) Residual risks and hedging strategies in Markovian markets. *Stochastic Process. Appl.* 33, 131–150.

Boyle, P.P. (1977) Options: a Monte Carlo approach. *J. Finan. Econom.* 4, 323–338.

Boyle, P.P. (1986) Option valuation using a three jump process. *Internat. Options J.* 3, 7–12.

Boyle, P.P. (1988) A lattice framework for option pricing with two state variables. *J. Finan. Quant. Anal.* 23, 1–12.

Boyle, P.P., Ananthanarayanan, A.L. (1977) The impact of variance estimation in option valuation models. *J. Finan. Econom.* 5, 375–387.

Boyle, P.P., Emanuel, D. (1980) Discretely adjusted option hedges. *J. Finan. Econom.* 8, 259–282.

Boyle, P.P., Tse, Y.K. (1990) An algorithm for computing values of options on the maximum or minimum of several assets. *J. Finan. Quant. Anal.* 25, 215–227.

Boyle, P.P., Evnine, J., Gibbs, S. (1989) Numerical evaluation of multivariate contingent claims. *Rev. Finan. Stud.* 2, 241–250.

Boyle, P.P., Broadie, M., Glasserman, P. (1997) Monte Carlo methods for security pricing. *J. Econom. Dynamics Control* 21, 1267–1321.

Brace, A. (1996) Dual swap and swaption formulae in the normal and lognormal models. Working paper, University of New South Wales.

Brace, A., Musiela, M. (1994) A multifactor Gauss Markov implementation of Heath, Jarrow, and Morton. *Math. Finance* 4, 259–283.

Brace, A., Musiela, M. (1997) Swap derivatives in a Gaussian HJM framework. In: *Mathematics of Derivative Securities*, M.A.H. Dempster and S.R. Pliska, eds. Cambridge University Press, Cambridge, pp. 336–368.

Brace, A., Womersley, R.S. (2000) Exact fit to the swaption volatility matrix using semidefinite programming. Working paper, University of New South Wales.

Brace, A., Gątarek, D., Musiela, M. (1997) The market model of interest rate dynamics. *Math. Finance* 7, 127–154.

Brace, A., Musiela, M., Schlögl, E. (1998) A simulation algorithm based on measure relationships in the lognormal market model. Working paper, University of New South Wales.

Brace, A., Dun, T., Barton, G. (2001a) Towards a central interest rate model. In: *Option Pricing, Interest Rates and Risk Management*, E. Jouini, J. Cvitanić and M. Musiela, eds. Cambridge University Press, Cambridge, pp. 278–313.

Brace, A., Goldys, B., Klebaner, F., Womersley, R. (2001b) Market model for stochastic implied volatility with application to the BGM model. Working paper, University of New South Wales.

Brace, A., Goldys, B., van de Hoek, J., Womersley, R. (2002) Markovian models in the stochastic implied volatility framework. Working paper, University of New South Wales.

Breeden, D. (1979) An intertemporal asset pricing model with stochastic consumption and investment opportunities. *J. Finan. Econom.* 7, 265–296.

Breeden, D., Gilkeson, J.H. (1993) A path-dependent approach to security valuation with application to interest rate contingent claims. *J. Bank. Finance* 21, 541–562.

Breeden, D., Litzenberger, R. (1978) Prices of state-contingent claims implicit in option prices. *J. Business* 51, 621–651.

Brennan, M.J. (1979) The pricing of contingent claims in discrete time models. *J. Finance* 34, 53–68.

Brennan, M.J., Schwartz, E.S. (1977a) The valuation of American put options. *J. Finance* 32, 449–462.

Brennan, M.J., Schwartz, E.S. (1977b) Convertible bonds: valuation and optimal strategies for call and conversion. *J. Finance* 32, 1699–1715.

Brennan, M.J., Schwartz, E.S. (1977c) Savings bonds, retractable bonds and callable bonds. *J. Finan. Econom.* 5, 67–88.

Brennan, M.J., Schwartz, E.S. (1978a) Finite-difference methods and jump processes arising in the pricing of contingent claims: a synthesis. *J. Finan. Quant. Anal.* 13, 461–474.

Brennan, M.J., Schwartz, E.S. (1978b) Corporate income taxes, valuation and the problem of optimal capital structure. *J. Business* 51, 103–114.

Brennan, M.J., Schwartz, E.S. (1979) A continuous-time approach to the pricing of bonds. *J. Bank. Finance* 3, 135–155.

Brennan, M.J., Schwartz, E.S. (1980a) Conditional predictions of bond prices and returns. *J. Finance* 35, 405–417.

Brennan, M.J., Schwartz, E.S. (1980b) Analyzing convertible bonds. *J. Finan. Quant. Anal.* 15, 907–929.

Brennan, M.J., Schwartz, E.S. (1982) An equilibrium model of bond pricing and a test of market efficiency. *J. Finan. Quant. Anal.* 17, 301–329.

Brennan, M.J., Schwartz, E.S. (1989) Portfolio insurance and financial market equilibrium. *J. Business* 62, 455–472.

Brenner, M., Galai, D. (1986) Implied interest rates. *J. Business* 59, 493–507.

Brenner, M., Galai, D. (1989) New financial instruments for hedging changes in volatility. *Finan. Analysts J.* 45(4), 61–65.

Brenner, M., Subrahmanyam, M.G. (1988) A simple formula to compute the implied standard deviation. *Finan. Analysts J.* 44(5), 80–83.

Brenner, M., Ou, E.Y., Zhang, J.E. (2000) Hedging volatility risk. Working paper.

Brigo, D., Mercurio, F. (2000) Option pricing impact of alternative continuous-time dynamics for discretely-observed stock prices. *Finance Stochast.* 4, 147–159.

Brigo, D., Mercurio, F. (2001a) *Interest Rate Models: Theory and Practice.* Springer, Berlin Heidelberg New York.

Brigo, D., Mercurio, F. (2001b) A deterministic-shift extension of analytically-tractable and time-homogeneous short-rate models. *Finance Stochast.* 5, 369–387.

Brigo, D., Mercurio, F. (2001c) Displaced and mixture diffusions for analytically tractable smile models. In: *Mathematical Finance – Bachelier Congress 2000*, H. Geman et al., eds. Springer, Berlin Heidelberg New York, pp. 151–174.

Brigo, D., Mercurio, F. (2002a) On stochastic differential equations with marginal laws evolving according to mixtures of densities. Working paper, Banca IMI.

Brigo, D., Mercurio, F. (2002b) Lognormal-mixture dynamics and calibration to market volatility smiles. *Internat. J. Theor. Appl. Finance* 5, 427–446.

Brigo, D., Mercurio, F. (2003) Analytical pricing of the smile in a forward LIBOR market model. *Quant. Finance* 3, 15–27.

Brigo, D., Mercurio, F., Sartorelli, G. (2003) Alternative asset-price dynamics and volatility smile. *Quant. Finance* 3, 173–183.

Britten-Jones, M., Neuberger, A. (2000) Option prices, implied price processes, and stochastic volatility. *J. Finance* 55, 839–866.

Briys, E., de Varenne, F. (1997) Valuing risky fixed rate debt: an extension. *J. Finan. Quant. Anal.* 32, 239–248.

Briys, E., Crouhy, M., Schöbel, R. (1991) The pricing of default-free interest rate cap, floor, and collar agreements. *J. Finance* 46, 1879–1892.

Briys, E., Bellalah, M., Mai, H.M., de Varenne, F. (1997) *Options, Futures and Exotic Derivatives.* J.Wiley, Chichester.

Broadie, M., Detemple, J. (1995) American capped call options on dividend-paying assets. *Rev. Finan. Stud.* 8, 161–191.

Broadie, M., Detemple, J. (1996) American option valuation: new bounds, approximations, and a comparison of existing methods. *Rev. Finan. Stud.* 9, 1211–1250.

Broadie, M., Detemple, J. (1997a) The valuation of American options on multiple assets. *Math. Finance* 7, 241–286.

Broadie, M., Detemple, J. (1997b) Recent advances in numerical methods for pricing derivative securities. In: *Numerical Methods in Finance*, L.C.G. Rogers D. Talay, eds. Cambridge University Press, Cambridge, pp. 43–66.

Broadie, M., Glasserman, P. (1997a) Pricing American-style securities using simulation. *J. Econom. Dynamics Control* 21, 1323–1352.

Broadie, M., Glasserman, P. (1997b) A stochastic mesh method for pricing high-dimensional American options. Working paper, Columbia University.

Broadie, M., Glasserman, P., Kou, S. (1997) A continuity correction for discrete barrier options. *Math. Finance* 7, 325–349.

Broadie, M., Cvitanić, J., Soner, H.M. (1998) Optimal replication of contingent claims under portfolio constraints. *Rev. Finan. Stud.* 11, 59–79.

Broadie, M., Glasserman, P., Kou, S. (1999) Connecting discrete and continuous path-dependent options. *Finance Stochast.* 3, 55–82.

Brody, D.C., Hughston, L.P. (2003) A coherent approach to interest rate modelling. Working paper, Imperial College and King's College London.

Brody, D.C., Hughston, L.P. (2004) Chaos and coherence: a new framework for interest rate modelling. *Proc. Roy. Soc. London* 460, 85–110.

Brown, H., Hobson, D., Rogers, L. (2001a) Robust hedging of options. *Appl. Math. Finance* 5, 17–43.

Brown, H., Hobson, D., Rogers, L. (2001b) Robust hedging of barrier options. *Math. Finance* 11, 285–314.

Brown, S.J., Dybvig, P.H. (1986) The empirical implications of the Cox, Ingersoll, Ross theory of the term structure of interest rates. *J. Finance* 41, 616–628.

Buchen, P., Kelly, M. (1996) The maximum entropy distribution of an asset inferred from option prices. *J. Finan. Quant. Anal.* 31, 143–159.

Buchen, P., Konstandatos, O. (2005) A new method of pricing lookback options. *Math. Finance* 15, 245–260.

Buff, R. (2002) *Uncertain Volatility Models - Theory and Applications.* Springer, Berlin Heidelberg New York.

Bühler, H. (2006) Expensive martingales. *Quant. Finance* 6, 207–218.

Bühler, W., Käsler, J. (1989) Konsistente Anleihenpreise und Optionen auf Anleihen. Working paper, University of Dortmund.

Bühler, W., Uhrig-Homburg, M., Walter, U., Weber, T. (1999) An empirical comparison of forward-rate and spot-rate models for valuing interest-rate options. *J. Finance* 54, 269–305.

Bühlmann, H., Delbaen, F., Embrechts, P., Shiryaev, A. (1996) No-arbitrage, change of measure and conditional Esscher transform in a semi-martingale model of stock price. *CWI Quarterly* 9, 291–317.

Bunch, D., Johnson, H.E. (1992) A simple and numerically efficient valuation method for American puts, using a modified Geske-Johnson approach. *J. Finance* 47, 809–816.

Burghardt, G., Hoskins, B. (1995) A question of bias. *Risk* 8(3), 63–70.

Butler, J.S., Schachter, B. (1986) Unbiased estimation of the Black/Scholes formula. *J. Finan. Econom.* 15, 341–357.

Büttler, H.-J., Waldvogel, J. (1996) Pricing callable bonds by means of Green's function. *Math. Finance* 6, 55–88.

Cakici, N., Chatterjee, S. (1991) Pricing stock index futures with stochastic interest rates. *J. Futures Markets* 11, 441–452.

Campbell, J.Y. (1986) A defense of traditional hypotheses about the term structure of interest rates. *J. Finance* 41, 183–193.

Campbell, J.Y., Lo, A.W., MacKinlay, A.C. (1997) *The Econometrics of Financial Markets.* Princeton University Press, Princeton (New Jersey).

Canina, L., Figlewski, S. (1992) The informational content of implied volatilities. *Rev. Finan. Stud.* 5, 659–682.

Carassus, L., Jouini, E. (1998) Investment and arbitrage opportunities with short sales constraints. *Math. Finance* 8, 169–178.

Carassus, L., Gobet, E., Temam, E. (2002) Closed formulae for super-replication prices with discrete time strategies. Working paper.

Carmona, R., Tehranchi, M. (2004) A characterization of hedging portfolios for interest rate contingent claims. *Ann. Appl. Probab.* 14, 1267–1294.

Carr, P. (1993) Deriving derivatives of derivative securities. Working paper, Cornell University.

Carr, P. (1995) Two extensions to barrier option valuation. *Appl. Math. Finance* 2, 173–209.

Carr, P., Chen, R.R. (1993) Valuing bond futures and the quality option. Working paper, Cornell University.

Carr, P., Faguet, D. (1994) Fast accurate valuation of American options. Working paper, Cornell University.

Carr, P., Jarrow, R. (1990) The stop-loss start-gain paradox and option valuation: a new decomposition into intrinsic and time value. *Rev. Finan. Stud.* 3, 469–492.

Carr, P., Lee, R. (2003) Robust valuation and hedging of derivatives on quadratic variation. Working paper, Courant Institute of Mathematical Sciences.

Carr, P., Madan, D. (1998a) Determining volatility surfaces and option values from an implied volatility smile. Working paper.

Carr, P., Madan, D. (1998b) Towards a theory of volatility trading. In: *Volatility: New Estimation Techniques for Pricing Derivatives*, R. Jarrow, ed. Risk Books, London, pp. 417–427.

Carr, P., Madan, D. (1999) Option valuation using the fast Fourier transform. *J. Comput. Finance* 2, 61–73.

Carr, P., Madan, D. (2004) A note on sufficient conditions for no arbitrage. Working paper, Courant Institute of Mathematical Sciences.

Carr, P., Jarrow, R., Myneni, R. (1992) Alternative characterizations of American put options. *Math. Finance* 2, 87–106.

Carr, P., Ellis, K., Gupta, V. (1998) Static hedging of exotic options. *J. Finance* 53, 1165–1190.

Carr, P., Geman, H., Madan, D., Yor, M. (2002a) The fine structure of asset returns: an empirical investigation. *J. Business* 75, 305–332.

Carr, P., Geman, H., Madan, D., Yor, M. (2002b) Self decomposability and option pricing. Working paper.

Carr, P., Geman, H., Madan, D., Yor, M. (2003) Stochastic volatility for Lévy processes. *Math. Finance* 13, 345–382.

Carr, P., Geman, H., Madan, D., Yor, M. (2005) Pricing options on realized variance. *Finance Stochast.* 9, 453–475.

Carverhill, A.P. (1994) When is the short rate Markovian? *Math. Finance* 4, 305–312.

Carverhill, A.P. (1995) A simplified exposition of the Heath, Jarrow and Morton model. *Stochastics Stochastics Rep.* 53, 227–240.

Carverhill, A.P., Clewlow, L.J. (1990) Flexible convolution. *Risk* 3(4), 25–29.

Chan, T. (1999) Pricing contingent claims on stocks driven by Lévy processes. *Ann. Appl. Probab.* 9, 504–528.

Chan, K.C., Karolyi, G.A., Longstaff, F.S., Sanders, A.B. (1992) The volatility of short-term interest rates: an empirical comparison of alternative models of the term structure of interest rates. *J. Finance* 47, 1209–1227.

Chance, D. (1989) *An Introduction to Options and Futures.* Dryden Press, Orlando.

Chance, D. (1990) Default risk and the duration of zero coupon bonds. *J. Finance* 45, 265–274.

Chatelain, M., Stricker, C. (1994) On componentwise and vector stochastic integration. *Math. Finance* 4, 57–65.

Chatelain, M., Stricker, C. (1995) Componentwise and vector stochastic integration with respect to certain multi-dimensional continuous local martingales. In: *Seminar on Stochastic Analysis, Random Fields and Applications*, E. Bolthausen, M. Dozzi and F. Russo, eds. Birkhäuser, Boston Basel Berlin, pp. 319–325.

Chen, L. (1996) *Interest Rate Dynamics, Derivatives Pricing, and Risk Management. Lecture Notes in Econom. and Math. Systems* 435. Springer, Berlin Heidelberg New York.

Chen, R., Scott, L. (1992) Pricing interest rate options in a two-factor Cox-Ingersoll-Ross model of the term structure. *Rev. Finan. Stud.* 5, 613–636.

Chen, R., Scott, L. (1995) Interest rate options in multifactor Cox-Ingersoll-Ross models of the term structure. *J. Derivatives*, 52–72.

Cheng, B.N., Rachev, S.T. (1995) Multivariate stable futures prices. *Math. Finance* 5, 133–153.

Cherny, A.S. (2003) General arbitrage pricing model: probability and possibility approaches. Working paper, Moscow State University.

Cherubini, U., Esposito, M. (1995) Options *in* and *on* interest rate futures contracts: results from martingale pricing theory. *Appl. Math. Finance* 2, 1–15.

Chesney, M., Scott, L. (1989) Pricing European currency options: a comparison of the modified Black-Scholes model and a random variance model. *J. Finan. Quant. Anal.* 24, 267–284.

Chesney, M., Elliott, R., Gibson, R. (1993a) Analytical solutions for the pricing of American bond and yield options. *Math. Finance* 3, 277–294.

Chesney, M., Elliott, R., Madan, D., Yang, H. (1993b) Diffusion coefficient estimation and asset pricing when risk premia and sensitivities are time varying. *Math. Finance* 3, 85–99.

Chesney, M., Geman, H., Jeanblanc-Picqué, M., Yor, M. (1997a) Some combinations of Asian, Parisian and barrier options. In: *Mathematics of Derivative Securities*, M.A.H. Dempster and S.R. Pliska, eds. Cambridge University Press, Cambridge, pp. 88–102.

Chesney, M., Jeanblanc-Picqué, M., Yor, M. (1997b) Brownian excursions and Parisian barrier options. *Adv. in Appl. Probab.* 29, 165–184.

Cheuk, T.H.F., Vorst, T.C.F. (1996) Complex barrier options. *J. Portfolio Manag.* 4, 8–22.

Cheyette, O. (1990) Pricing options on multiple assets. *Adv. in Futures Options Res.* 4, 69–81.

Chriss, N.A. (1996) *Black-Scholes and Beyond: Option Pricing Models*. Irwin Professional Publ.

Christie, A.A. (1982) The stochastic behavior of common stock variances: value, leverage and interest rate effects. *J. Finan. Econom.* 10, 407–432.

Christopeit, N., Musiela, M. (1994) On the existence and characterization of arbitrage-free measures in contingent claim valuation. *Stochastic Anal. Appl.* 12, 41–63.

Chung, Y.P. (1991) A transaction data test of stock index futures market efficiency and index arbitrage profitability. *J. Finance* 46, 1791–1809.

Clark, P.K. (1973) A subordinated stochastic process model with finite variance for speculative prices. *Econometrica* 41, 135–159.

Clewlow, L., Carverhill, A. (1994) On the simulation of contingent claim. *J. Derivatives*, 66–74.

Clewlow, L., Strickland, C. (1998) *Implementing Derivatives Models*. J.Wiley, Chichester.

Cohen, H. (1991) Testing pricing models for the Treasury bond futures contracts. Doctoral dissertation, Cornell University, Ithaca.

Cohen, H. (1995) Isolating the wild card option. *Math. Finance* 5, 155–165.

Colwell, D.B., Elliott, R.J. (1993) Discontinuous asset prices and non-attainable contingent claims. *Math. Finance* 3, 295–308.

Colwell, D.B., Elliott, R.J., Kopp, P.E. (1991) Martingale representation and hedging policies. *Stochastic Process. Appl.* 38, 335–345.

Constantinides, G.M., Bhattacharya, S., eds. (1987) *Frontiers of Financial Theory*. Rowman and Littlewood, Totowa, New Jersey.

Cont, R. (2002) Inverse problems in financial modeling: theoretical and numerical aspects of model calibration. Lecture notes, École Polytechnique.

Cont, R. (2006) Model uncertainty and its impact on pricing of derivative instruments. *Math. Finance* 16, 519–548.

Cont, R., da Fonseca, J. (2002) Dynamics of implied volatility surface. *Quant. Finance* 2, 45–60.

Cont, R., Tankov, P. (2002) Calibration of jump-diffusion option pricing models: a robust non-parametric approach. Working paper, École Polytechnique.

Cont, R., Tankov, P. (2003) *Financial Modelling with Jump Processes*. Chapman and Hall/CRC.

Conze, A., Viswanathan (1991) Path dependent options: the case of lookback options. *J. Finance* 46, 1893–1907.

Cooper, I., Martin, M. (1996) Default risk and derivative products. *Appl. Math. Finance* 3, 53–74.

Cootner, P., ed. (1964) *The Random Character of Stock Market Prices*. MIT Press, Cambridge (Mass.)

Cornell, B., French, K.R. (1983a) The pricing of stock index futures. *J. Futures Markets* 3, 1–14.

Cornell, B., French, K.R. (1983b) Taxes and the pricing of stock index futures. *J. Finance* 38, 675–694.

Cornell, B., Reinganum, M.R. (1981) Forward and futures prices: evidence from the foreign exchange markets. *J. Finance* 36, 1035–1045.

Corrado, C.J., Miller, T.W. (1996) A note on a simple, accurate formula to compute implied standard deviations. *J. Bank. Finance* 20, 595–603.

Cossin, D., Pirotte, H. (2000) *Advanced Credit Risk Analysis*. J. Wiley, Chichester.

Courtadon, G. (1982a) The pricing of options on default-free bonds. *J. Finan. Quant. Anal.* 17, 75–100.

Courtadon, G. (1982b) A more accurate finite-difference approximation for the valuation of options. *J. Finan. Quant. Anal.* 17, 697–703.

Cox, J.C. (1975) Notes on options pricing I: constant elasticity of variance diffusions. Working paper, Stanford University.

Cox, J.C., Huang, C.-F. (1989) Optimal consumption and portfolio policies when asset prices follow a diffusion process. *J. Econom. Theory* 49, 33–83.

Cox, J.C., Ross, S.A. (1975) The pricing of options for jump processes. Working Paper, University of Pennsylvania.

Cox, J.C., Ross, S.A. (1976a) A survey of some new results in financial options pricing theory. *J. Finance* 31, 382–402.

Cox, J.C., Ross, S.A. (1976b) The valuation of options for alternative stochastic processes. *J. Finan. Econom.* 3, 145–166.

Cox, J.C., Rubinstein, M. (1983) A survey of alternative option-pricing models. In: *Option Pricing, Theory and Applications*, M. Brenner, ed. Toronto, pp. 3–33.

Cox, J.C., Rubinstein, M. (1985) *Options Markets*. Prentice-Hall, Englewood Cliffs (New Jersey).

Cox, J.C., Ross, S.A., Rubinstein, M. (1979a) Option pricing: a simplified approach. *J. Finan. Econom.* 7, 229–263.

Cox, J.C., Ingersoll, J.E., Ross, S.A. (1981a) A re-examination of traditional hypotheses about the term structure of interest rates. *J. Finance* 36, 769–799.

Cox, J.C., Ingersoll, J.E., Ross, S.A. (1981b) The relation between forward prices and futures prices. *J. Finan. Econom.* 9, 321–346.

Cox, J.C., Ingersoll, J.E., Ross, S.A. (1985a) An intertemporal general equilibrium model of asset prices. *Econometrica* 53, 363–384.

Cox, J.C., Ingersoll, J.E., Ross, S.A. (1985b) A theory of the term structure of interest rates. *Econometrica* 53, 385–407.

Crépey, S. (2003) Calibration of the local volatility in the trinomial tree using Tikhonov regularization. *Inverse Problems* 19, 91–127.

Cutland, N.J., Kopp, P.E., Willinger, W. (1991) A nonstandard approach to option pricing. *Math. Finance* 1, 1–38.

Cutland, N.J., Kopp, P.E., Willinger, W. (1993a) A nonstandard treatment of options driven by Poisson processes. *Stochastic Process. Appl.* 42, 115–133.

Cutland, N.J., Kopp, P.E., Willinger, W. (1993b) From discrete to continuous financial models: new convergence results for option pricing. *Math. Finance* 3, 101–123.

Cvitanić, J. (1997) Optimal trading under constraints. In: *Financial Mathematics, Bressanone, 1996*, W. Runggaldier, ed. *Lecture Notes in Math.* 1656, Springer, Berlin Heidelberg New York, pp. 123–190.

Cvitanić, J., Karatzas, I. (1992) Convex duality in constrained portfolio optimization. *Ann. Appl. Probab.* 2, 767–818.

Cvitanić, J., Karatzas, I. (1993) Hedging contingent claims with constrained portfolios. *Ann. Appl. Probab.* 3, 652–681.

Cvitanić, J., Karatzas, I. (1996a) Hedging and portfolio optimization under transaction costs: a martingale approach. *Math. Finance* 6, 133–165.

Cvitanić, J., Karatzas, I. (1996b) Backward stochastic differential equations with reflection and Dynkin games. *Ann. Probab.* 24, 2024–2056.

Cvitanić, J., Ma, J. (1996) Hedging options for a large investor and forward-backward SDE's. *Ann. Appl. Probab.* 6, 370–398.

Cvitanić, J., Pham, H., Touzi, N. (1999a) A closed-form solution to the problem of super-replication under transaction costs. *Finance Stochast.* 3, 35–54.

Cvitanić, J., Pham, H., Touzi, N. (1999b) Super-replication in stochastic volatility models under portfolio constraints. *J. Appl. Probab.* 36.

Dai, Q., Singleton, K. (2000) Specification analysis of affine term structure models. *J. Finance* 55, 1943–1978.

Dalang, R.C., Morton, A., Willinger, W. (1990) Equivalent martingale measures and no-arbitrage in stochastic securities market model. *Stochastics Stochastics Rep.* 29, 185–201.

Dana, R.-A., Jeanblanc, M. (2003) *Financial Markets in Continuous Time*. Springer, Berlin Heidelberg New York.

Da Prato, G., Zabczyk, J. (1992) *Stochastic Equations in Infinite Dimensions*. Cambridge University Press, Cambridge.

Das, S. (1994) *Swaps and Financial Derivatives: The Global Reference to Products, Pricing, Applications and Markets*, 2nd ed. Law Book Co., Sydney.

Dassios, A. (1995) The distribution of the quantile of a Brownian motion with drift and the pricing of related path-dependent options. *Ann. Appl. Probab.* 5, 389–398.

Davis, M.H.A. (1988) Local time on the stock exchange. In: *Stochastic Calculus in Application*, J.R. Norris, ed. Longman Scientific and Technical, Harrow (Essex), pp. 4–28.

Davis, M.H.A. (1997) Option pricing in incomplete markets. In: *Mathematics of Derivative Securities*, M.A.H. Dempster and S.R. Pliska, eds. Cambridge University Press, Cambridge, pp. 227–254.

Davis, M.H.A. (1998) A note on the forward measure. *Finance Stochast.* 2, 19–28.

Davis, M.H.A. (2000) Mathematics of financial markets. In: *Mathematics Unlimited: 2001 and Beyond*, B. Engquist and W. Schmid, eds. Springer, Berlin Heidelberg New York.

Davis, M.H.A. (2004a) Complete-market models of stochastic volatility. *Proc. Royal Soc. London Ser. A* 460, 11–26.

Davis, M.H.A. (2004b) The range of traded option prices. Working paper, Imperial College London.

Davis, M.H.A., Clark, J.M.C. (1994) A note on super-replicating strategies. *Phil. Trans. Roy. Soc. London Ser. A* 347, 485–494.

Davis, M.H.A., Hobson, D.G. (2004) The range of traded option prices. Working paper.

Davis, M.H.A., Norman, A.R. (1990) Portfolio selection with transaction costs. *Math. Oper. Res.* 15, 676–713.

Davis, M.H.A., Panas, V.P. (1994) The writing price of a European contingent claim under proportional transaction costs. *Comp. Appl. Math.* 13, 115–157.

Davis, M.H.A., Mataix-Pastor, V. (2005) A note on the swap market model. Working paper, Imperial College, London.

Davis, M.H.A., Panas, V.P., Zariphopoulou, T. (1993) European option pricing with transaction costs. *SIAM J. Control Optim.* 31, 470–493.

Day, T., Lewis, C. (1988) The behavior of volatility implicit in the prices of stock index options. *J. Finan. Econom.* 22, 103–122.

Day, T., Lewis, C. (1992) Stock market volatility and the information content of stock index options. *J. Econometrics* 52, 267–287.

Décamps, J.-P., Rochet, J.-C. (1997) A variational approach for pricing options and corporate bonds. *Econom. Theory* 9, 557–569.

Deelstra, G., Delbaen, F. (1995a) Long-term returns in stochastic interest rate models. *Insurance Math. Econom.* 17, 163–169.

Deelstra, G., Delbaen, F. (1995b) Long-term returns in stochastic interest rate models: convergence in law. *Stochastics Stochastics Rep.* 55, 253–277.

De Jong, F., Driessen, J., Pelsser, A. (2001a) Libor market models versus swap market models for pricing interest rate derivatives: an empirical analysis. *European Finance Rev.* 5, 201–237.

De Jong, F., Driessen, J., Pelsser, A. (2001b) Estimation of the Libor market model: combining term structure data and option prices. Working paper.

Delbaen, F. (1992) Representing martingale measures when asset prices are continuous and bounded. *Math. Finance* 2, 107–130.

Delbaen, F. (1993) Consols in CIR model. *Math. Finance* 3, 125–134.

Delbaen, F., Haezendonck, J. (1989) A martingale approach to premium calculation principles in an arbitrage free market. *Insurance Math. Econom.* 8, 269–277.

Delbaen, F., Lorimier, S. (1992) Estimation of the yield curve and the forward rate curve starting from a finite number of observations. *Insurance Math. Econom.* 11, 259–269.

Delbaen, F., Schachermayer, W. (1994a) Arbitrage and free lunch with bounded risk for unbounded continuous processes. *Math. Finance* 4, 343–348.

Delbaen, F., Schachermayer, W. (1994b) A general version of the fundamental theorem of asset pricing. *Math. Ann.* 300, 463–520.

Delbaen, F., Schachermayer, W. (1995a) Arbitrage possibilities in Bessel processes and their relations to local martingales. *Probab. Theory Rel. Fields* 102, 357–366.

Delbaen, F., Schachermayer, W. (1995b) The no-arbitrage property under a change of numéraire. *Stochastics Stochastics Rep.* 53, 213–226.

Delbaen, F., Schachermayer, W. (1995c) The existence of absolutely continuous local martingale measures. *Ann. Appl. Probab.* 5, 926–945.

Delbaen, F., Schachermayer, W. (1996a) The variance-optimal martingale measure for continuous processes. *Bernoulli* 2, 81–105.

Delbaen, F., Schachermayer, W. (1996b) Attainable claims with p'th moments. *Ann. Inst. H.Poincaré Probab. Statist.* 32, 743–763.

Delbaen, F., Schachermayer, W. (1997a) The Banach space of workable contingent claims in arbitrage theory. *Ann. Inst. H.Poincaré Probab. Statist.* 33, 113–144.

Delbaen, F., Schachermayer, W. (1997b) The fundamental theorem of asset pricing for unbounded stochastic processes. *Math. Ann.* 312, 215–250.

Delbaen, F., Schachermayer, W. (1998) A simple counter-example to several problems in the theory of asset pricing. *Math. Finance* 8, 1–11.

Delbaen, F., Shirakawa, H. (1997) Squared Bessel processes and their applications to the square root interest rate model. Working paper.

Delbaen, F., Yor, M. (2002) Passport options. *Math. Finance* 12, 299–328.

Delbaen, F., Monat, P., Schachermayer, W., Schweizer, M., Stricker, C. (1997) Weighted norm inequalities and hedging in incomplete markets. *Finance Stochast.* 1, 181–227.

De Malherbe, K. (2002) Correlation analysis in the LIBOR and swap market model. *Internat. J. Theor. Appl. Finance* 4, 401–426.

Dempster, M.A.H., Hutton, J.P. (1997) Numerical valuation of cross-currency swaps and swaptions. In: *Mathematics of Derivative Securities*, M.A.H. Dempster and S.R. Pliska, eds. Cambridge University Press, Cambridge, pp. 473–503.

Dempster, M.A.H., Richard, D.G. (1999) Pricing exotic American options fitting the volatility smile. Working paper, University of Cambridge.

Derman, E. (1999) Regimes of volatility. *Risk* 12(4).

Derman, E., Kani, I. (1994) Riding on a smile. *Risk* 7(2), 32–39.

Derman, E., Kani, I. (1998) Stochastic implied trees: arbitrage pricing with stochastic term and strike structure of volatility. *Internat. J. Theor. Appl. Finance* 1, 61–110.

Derman, E., Karasinski, P., Wecker, J. (1990) Understanding guaranteed exchange-rate contracts in foreign stock investments. Working paper, Goldman Sachs.

Derman, E., Ergener, D., Kani, I. (1995) Static options replication. *J. Derivatives* 2(4), 78–95.

Derman, E., Kani, I., Chriss, N. (1996a) Implied trinomial trees and the volatility smile. *J. Derivatives* 3, 7–22.

Derman, E., Kani, I., Zou, J.Z. (1996b) The local volatility surface: unlocking the information in index option prices. *Finan. Analysts J.* 52(4), 25–36.

Döberlein, F., Schweizer, M., Stricker, C. (2000) Implied savings accounts are unique. *Finance Stochast.* 4, 431–442.

Dolinsky, Y., Kifer, Y. (2007) Hedging with risk for game options in discrete time. *Stochastics Stochastics Rep.* 79, 169–195.

Dothan, U. (1978) On the term structure of interest rates. *J. Finan. Econom.* 6, 59–69.

Douady, R. (1994) Options à limite et options à limite double. Working paper.

Dritschel, M., Protter, P. (2000) Complete markets with discontinuous security price. Working paper, Purdue University.

Duan, J.-C. (1995) The GARCH option pricing model. *Math. Finance* 5, 13–32.

Duan, J.-C. (1996) A unified theory of option pricing under stochastic volatility: from GARCH to diffusion. Working paper.

Duan, J.-C. (1997) Augmented GARCH(p, q) model and its diffusion limit. *J. Econometrics* 79, 97–127.

Dudenhausen, A., Schlögl, E., Schlögl, L. (1999) Robustness of Gaussian hedges and the hedging of fixed income derivatives. Working paper.

Duffee, G. (1996) On measuring credit risks of derivative instruments. *J. Bank. Finance* 20, 805–833.

Duffee, G. (1998) The relation between Treasury yields and corporate bond yield spreads. *J. Finance* 53, 2225–2242.

Duffie, D. (1988a) An extension of the Black-Scholes model of security valuation. *J. Econom. Theory* 46, 194–204.

Duffie, D. (1988b) *Security Markets: Stochastic Models*. Academic Press, Boston.

Duffie, D. (1989) *Futures Markets*. Prentice-Hall, Englewood Cliffs (New Jersey).

Duffie, D. (2001) *Dynamic Asset Pricing Theory*. 3rd ed. Princeton University Press, Princeton (New Jersey).

Duffie, D. (1998a) First-to-default valuation. Working paper, Stanford University.

Duffie, D. (1998b) Defaultable term structure models with fractional recovery of par. Working paper, Stanford University.

Duffie, D., Glynn, P. (1995) Efficient Monte Carlo estimation of security prices. *Ann. Appl. Probab.* 5, 897–905.

Duffie, D., Harrison, J.M. (1993) Arbitrage pricing of Russian options and perpetual lookback options. *Ann. Appl. Probab.* 3, 641–649.

Duffie, D., Huang, C.-F. (1985) Implementing Arrow-Debreu equilibria by continuous trading of few long-lived securities. *Econometrica* 53, 1337–1356.

Duffie, D., Huang, C.-F. (1986) Multiperiod security markets with differential information; martingales and resolution times. *J. Math. Econom.* 15, 283–303.

Duffie, D., Huang, M. (1996) Swap rates and credit quality. *J. Finance* 51, 921–949.

Duffie, D., Jackson, M. (1990) Optimal hedging and equilibrium in a dynamic futures market. *J. Econom. Dynamics Control* 14, 21–33.

Duffie, D., Kan, R. (1994) Multi-factor interest rate models. *Phil. Trans. Roy. Soc. London Ser.A* 347, 577–586.

Duffie, D., Kan, R. (1996) A yield-factor model of interest rates. *Math. Finance* 6, 379–406.

Duffie, D., Lando, D. (2001) The term structure of credit spreads with incomplete accounting information. *Econometrica* 69, 633–664.

Duffie, D., Protter, P. (1992) From discrete to continuous time finance: weak convergence of the financial gain process. *Math. Finance* 2, 1–15.

Duffie, D., Richardson, H.R. (1991) Mean-variance hedging in continuous time. *Ann. Appl. Probab.* 1, 1–15.

Duffie, D., Singleton, K. (1993) Simulated moments estimation of Markov models of asset prices. *Econometrica* 61, 929–952.

Duffie, D., Singleton, K. (1997) An econometric model of the term structure of interest-rate swap yields. *J. Finance* 52, 1287–1321.

Duffie, D., Singleton, K. (1998) Ratings-based term structures of credit spreads. Working paper, Stanford University.

Duffie, D., Singleton, K. (1999) Modeling term structures of defaultable bonds. *Rev. Finan. Stud.* 12, 687–720.

Duffie, D., Singleton, K. (2003) *Credit Risk: Pricing, Measurement and Management*. Princeton University Press, Princeton.

Duffie, D., Stanton, R. (1992) Pricing continuously resettled contingent claims. *J. Econom. Dynamics Control* 16, 561–573.

Duffie, D., Ma, J., Yong, J. (1995) Black's consol rate conjecture. *Ann. Appl. Probab.* 5, 356–382.

Duffie, D., Schroder, M., Skiadas, C. (1996) Recursive valuation of defaultable securities and the timing of resolution of uncertainty. *Ann. Appl. Probab.* 6, 1075–1090.

Duffie, D., Schroder, M., Skiadas, C. (1997) A term structure model with preferences for the timing of resolution of uncertainty. *Econom. Theory* 9, 3–22.

Duffie, D., Pan, J., Singleton, K. (2000) Transform analysis and asset pricing for affine jump-diffusions. *Econometrica* 68, 1343–1376.

Duffie, D., Filipović, D., Schachermayer, W. (2003) Affine processes and applications in finance. *Ann. Appl. Probab.* 13, 984–1053.

Dufresne, D. (1990) The distribution of a perpetuity, with applications to risk theory and pension funding. *Scand. Actuarial J.* 39–79.

Dufresne, D. (2001) The integrated square root process. Working paper, University of Melbourne.

Dumas, B., Fleming, J., Whaley, R. (1998) Implied volatility functions: empirical tests. *J. Finance* 53, 2059–2106.

Dun, T., Schlögl, E., Barton, G. (2000) Simulated swaption delta-hedging in the lognormal forward LIBOR model. Working paper, University of Sydney and University of Technology, Sydney.

Dupire, B. (1992) Arbitrage pricing with stochastic volatilities. *Journées Internationales de Finance, ESSEC-AFFI,* Paris, June 1992.

Dupire, B. (1993a) Pricing and hedging with smiles. *Journées Internationales de Finance, IGR-AFFI.* La Baule, June 1993.

Dupire, B. (1993b) Model art. *Risk* 6(10), 118–121.

Dupire, B. (1994) Pricing with a smile. *Risk* 7(1), 18–20.

Dupire, B. (1996) A unified theory of volatility. Working paper.

Dupire, B. (1997) Pricing and hedging with a smile. In: *Mathematics of Derivative Securities*, M.A.H. Dempster and S.R. Pliska, eds. Cambridge University Press, Cambridge, pp. 103–111.

Durrett, R. (1996) *Stochastic Calculus: A Practical Introduction.* CRC Press.

Durrleman, V. (2004) From implied to spot volatilites. Doctoral dissertation, Princeton University.

Dybvig, P.H. (1988) Bond and bond option pricing based on the current term structure. Working paper, Washington University.

Dybvig, P.H., Huang, C.-F. (1988) Non-negative wealth, absence of arbitrage, and feasible consumption plans. *Rev. Finan. Stud.* 1, 377–401.

Dybvig, P.H., Ingersoll, J.E., Ross, S.A. (1996) Long forward and zero-coupon rates can never fall. *J. Business* 69, 1–25.

Dynkin, E.B. (1969) Game variant of a problem on optimal stopping. *Soviet Math. Dokl.* 10, 270–274.

Eberlein, E. (1992) On modeling questions in security valuation. *Math. Finance* 2, 17–32.

Eberlein, E., Jacod, J. (1997) On the range of options prices. *Finance Stochast.* 1, 131–140.

Eberlein, E., Keller, U. (1995) Hyperbolic distributions in finance. *Bernoulli* 1, 281–299.

Eberlein, E., Keller, U., Prause, K. (1998) New insights into smile, mispricing and Vaue-at-Risk: the hyperbolic model. *J. Business* 71, 371–406.

Eberlein, E., Özkan, F. (2005) The Lévy LIBOR model. *Finance Stochast.* 9, 327–348.

Edwards, F.R., Ma, C.W. (1992) *Futures and Options.* McGraw-Hill, New York.

Ekström, E., Janson, S., Tysk, J. (2004) Superreplication of options on several underlying assets. *J. Appl. Probab.* 42, 27–38.

El Karoui, N. (1981) Les aspects probabilistes du contrôle stochastique. In: *Lecture Notes in Math. 876.* Springer, Berlin Heidelberg New York, pp. 73–238.

El Karoui, N., Chérif, T. (1993) Arbitrage entre deux marchés: application aux options quanto. *Journées Internationales de Finance, AFFI,* Tunis, June 1993.

El Karoui, N., Geman, H. (1994) A probabilistic approach to the valuation of floating rate notes with an application to interest rate swaps. *Adv. in Futures Options Res.* 7, 47–63.

El Karoui, N., Jeanblanc, M. (1990) Sur la robustesse de l'équation de Black-Scholes. International Conference in Finance, HEC.

El Karoui, N., Karatzas, I. (1991) A new approach to the Skorohod problem and its applications. *Stochastics Stochastics Rep.* 34, 57–82.

El Karoui, N., Lacoste, V. (1992) Multifactor models of the term structure of interest rates. Working paper, CERESSEC, Cergy Pontoise.

El Karoui, N., Quenez, M.C. (1995) Dynamic programming and pricing of contingent claims in an incomplete market. *SIAM J. Control Optim.* 33, 29–66.

El Karoui, N., Quenez, M.C. (1997) Nonlinear pricing theory and backward stochastic differential equations. In: *Financial Mathematics, Bressanone, 1996*, W. Runggaldier, *Lecture Notes in Math.* 1656, Springer, Berlin Heidelberg New York, pp. 191–246.

El Karoui, N., Rochet, J.C. (1989) A pricing formula for options on coupon bonds. Working paper, SDEES.

El Karoui, N., Saada, D. (1992) A review of the Ho and Lee model. *Journées Internationales de Finance, ESSEC-AFFI*, Paris, June 1992.

El Karoui, N., Lepage, C., Myneni, R., Roseau, N., Viswanathan, R. (1991) The valuation and hedging of contingent claims with Markovian interest rates. Working paper, Université Paris VI.

El Karoui, N., Myneni, R., Viswanathan, R. (1992) Arbitrage pricing and hedging of interest rate claims with state variables, theory and applications. Working paper, Stanford University and Université Paris VI.

El Karoui, N., Geman, H., Lacoste, V. (1995) On the role of state variables in interest rates models. Working paper, Université Paris VI and ESSEC.

El Karoui, N., Frachot, A., Geman, H. (1996) A note on the behaviour of long zero coupon rates in a no arbitrage framework. Working paper.

El Karoui, N., Peng, S., Quenez, M.C. (1997b) Backward stochastic differential equations in finance. *Math. Finance* 7, 1–72.

El Karoui, N., Jeanblanc-Picqué, M., Shreve, S. (1998) Robustness of the Black and Scholes formula. *Math. Finance* 8, 93–126.

Elliott, R.J. (1982) *Stochastic Calculus and Applications*. Springer, Berlin Heidelberg New York.

Elliott, R.J., Kopp, P.E. (1990) Option pricing and hedge portfolios for Poisson processes. *Stochastic Anal. Appl.* 8, 157–167.

Elliott, R.J., Kopp, P.E. (1999) *Mathematics of Financial Markets*. Springer, Berlin Heidelberg New York.

Elliott, R.J., Madan, D.B. (1998) A discrete time equivalent martingale measure. *Math. Finance* 8, 127–152.

Elliott, R.J., Jeanblanc, M., Yor, M. (2000) On models of default risk. *Math. Finance* 10, 179–195.

Elton, E.J., Gruber, M.J. (1995) *Modern Portfolio Theory and Investment Analysis*, 5th J.Wiley, New York.

Elton, E.J., Gruber, M.J., Roni, M. (1990) The structure of spot rates and immunization. *J. Finance* 45, 621–641.

Emanuel, D.C. (1983) Warrant valuation and exercise strategy. *J. Finan. Econom.* 12, 211–235.

Emanuel, D.C., Macbeth, J. (1982) Further results on the constant elasticity of variance call option pricing model. *J. Finan. Quant. Anal.* 17, 533–555.

Embrechts, P., Rogers, L.C.G., Yor, M. (1995) A proof of Dassios' representation for the α-quantile of Brownian motion with drift. *Ann. Appl. Probab.* 5, 757–767.

Embrechts, P., Klüppelberg, C., Mikosch, T. (1997) *Modelling Extremal Events for Insurance and Finance*. Springer, Berlin Heidelberg New York.

Engle, R. (1982) Autoregressive conditional heteroskedasticity with estimates of the variance of U.K. inflation. *Econometrica* 50, 987–1008.

Engle, R., Bollerslev, T. (1986) Modelling the persistence of conditional variances. *Econometric Rev.* 5, 1–50.

Engle, R., Mustafa, C. (1992) Implied ARCH models from options prices. *J. Econometrics* 52, 5–59.

Engle, R., Ng, V.K. (1993) Measuring and testing the impact of news on volatility. *J. Finance* 48, 1749–1779.

Evans, L.T., Keef, S.P., Okunev, J. (1994) Modelling real interest rates. *J. Bank. Finance* 18, 153–165.

Evnine, J., Rudd, A. (1985) Index options: the early evidence. *J. Finance* 40, 743–758.

Fajardo, J., Mordecki, E. (2006) Symmetry and duality in Lévy markets. *Quant. Finance* 6, 219–227.

Fama, E.F. (1965) The behaviour of stock market prices. *J. Business* 38, 34–105.

Fama, E.F. (1976) Forward rates as predictors of future spot rates. *J. Finan. Econom.* 3, 361–377.

Fama, E.F. (1981) Stock returns, real activity, inflation, and money. *Amer. Econom. Rev.* 71, 545–565.

Fama, E.F. (1984a) The information in the term structure. *J. Finan. Econom.* 13, 509–528.

Fama, E.F. (1984b) Term premiums in bond returns. *J. Finan. Econom.* 13, 529–546.

Fama, E.F. (1990) Stock returns, expected returns and real activity. *J. Finance* 45, 1089–1108.

Fama, E.F., French, K.R. (1992) The cross-section of expected stock returns. *J. Finance* 47, 427–465.

Fama, E.F., MacBeth, J. (1973) Risk, return and equilibrium: empirical tests. *J. Political Econom.* 81, 607–636.

Fama, E.F., Schwert, G.W. (1977) Asset returns and inflation. *J. Finan. Econom.* 4, 115–146.

Feiger, G., Jacquillat, B. (1979) Currency option bonds, puts and calls on spot exchange and the hedging of contingent foreign earnings. *J. Finance* 34, 1129–1139.

Figlewski, S. (1997) Forecasting volatility. *Financial Markets, Institutions and Instruments* 6(1), 1–88.

Figlewski, S., Green, T.C. (1999) Market risk and model risk for a financial institution writing options. *J. Finance* 54, 1465–1499.

Figlewski, S. (2002) Assessing the incremental value of option pricing theory relative to an informationally passive benchmark. *J. Derivatives*, Fall, 80–96.

Filipović, D. (1999) A note on the Nelson-Siegel family. *Math. Finance* 9, 349–359.

Filipović, D. (2000) Exponential-polynomial families and the term structure of interest rates. *Bernoulli* 6, 1–27.

Filipović, D. (2001a) *Consistency Problems for Heath-Jarrow-Morton Interest Rates Models*. Springer, Berlin Heidelberg New York.

Filipović, D. (2001b) A general characterization of one factor affine term structure models. *Finance Stochast.* 5, 389–412.

Filipović, D., Teichmann, J. (2002) On finite dimensional term structure models. Working paper.

Filipović, D., Teichmann, J. (2003) Existence of invariant manifolds for stochastic equations in infinite dimensions. *J. Funct. Anal.* 197, 398–4432.

Filipović, D., Teichmann, J. (2004) On the geometry of the term structure. *Proc. Roy. Soc. London A* 460, 129–167.

Flesaker, B. (1991) The relationship between forward and futures contracts: a comment. *J. Futures Markets* 11, 113–115.

Flesaker, B. (1993a) Testing the Heath-Jarrow-Morton/Ho-Lee model of interest rate contingent claims pricing. *J. Finan. Quant. Anal.* 28, 483–495.

Flesaker, B. (1993b) Arbitrage free pricing of interest rate futures and forward contracts. *J. Futures Markets* 13, 77–91.

Flesaker, B., Hughston, L. (1993) Contingent claim replication in continuous time with transaction costs. Working paper, Merrill Lynch.

Flesaker, B., Hughston, L. (1996a) Positive interest. *Risk* 9(1), 46–49.

Flesaker, B., Hughston, L. (1996b) Positive interest: foreign exchange. In: *Vasicek and Beyond*, L. Hughston, Risk Publications, London, pp. 351–367.

Flesaker, B., Hughston, L. (1997) Dynamic models of yield curve evolution. In: *Mathematics of Derivative Securities*, M.A.H. Dempster and S.R. Pliska, eds. Cambridge University Press, Cambridge, pp. 294–314.

Florio, S., Runggaldier, W.J. (1999) On hedging in finite security markets. *Appl. Math. Finance* 6, 159–176.

Föllmer, H., Kabanov, Yu.M. (1998) Optional decomposition and Lagrange multipliers. *Finance Stochast.* 2, 69–81.

Föllmer, H., Kramkov, D. (1997) Optional decomposition under constraints. *Probab. Theory Rel. Fields* 109, 1–25.

Föllmer, H., Leukert, P. (1999) Quantile hedging. *Finance Stochast.* 3, 251–273.

Föllmer, H., Leukert, P. (2000) Efficient hedging: cost versus shortfall risk. *Finance Stochast.* 4, 117–146.

Föllmer, H., Schied, A. (2000) *Stochastic Finance. An Introduction in Discrete Time.* De Gruyter.

Föllmer, H., Schweizer, M. (1989) Hedging by sequential regression: an introduction to the mathematics of option trading. *ASTIN Bull.* 18, 147–160.

Föllmer, H., Schweizer, M. (1991) Hedging of contingent claims under incomplete information. In: *Applied Stochastic Analysis*, M.H.A. Davis and R.J. Elliott, eds. Gordon and Breach, London New York, pp. 389–414.

Föllmer, H., Schweizer, M. (1993) A microeconomic approach to diffusion models for stock prices. *Math. Finance* 3, 1–23.

Föllmer, H., Sondermann, D. (1986) Hedging of non-redundant contingent claims. In: *Contributions to Mathematical Economics in Honor of Gérard Debreu*, W. Hildenbrand and A. Mas-Colell, eds. North Holland, Amsterdam, pp. 205–223.

Fouque, J.-P., Papanicolaou, G.C., Sircar, K.R. (2000a) *Derivatives in Financial Markets with Stochastic Volatility.* Cambridge University Press, Cambridge.

Fouque, J.-P., Papanicolaou, G.C., Sircar, K.R. (2000b) Mean-reverting stochastic volatility. *Internat. J. Theor. Appl. Finance* 3, 101–142.

Fournié, E., Lasry, J.-M., Lions, P.-L. (1997a) Some nonlinear methods to study far-from-the-money contingent claims. In: *Numerical Methods in Finance*, L.C.G. Rogers and D. Talay, eds. Cambridge University Press, Cambridge, pp. 115–145.

Fournié, E., Lasry, J.-M., Touzi, N. (1997b) Monte Carlo methods for stochastic volatility models. In: *Numerical Methods in Finance*, L.C.G. Rogers and D. Talay, eds. Cambridge University Press, Cambridge, pp. 146–164.

Fournié, E., Lasry, J.-M., Touzi, N. (1997c) Small noise expansion and importance sampling. *Asymptotic Anal.* 14, 361–376.

Fournié, E., Lasry, J.-M., Lebuchoux, J., Lions, P.-L., Touzi, N. (1999) An application of Malliavin calculus to Monte Carlo methods in finance. *Finance Stochast.* 3, 391–412.

Frachot, A. (1995) Factor models of domestic and foreign interest rates with stochastic volatilities. *Math. Finance* 5, 167–185.

Frachot, A. (1996) A reexamination of the uncovered interest rate parity hypothesis. *J. Internat. Money Finance* 15, 419–437.

French, K.R. (1980) Stock returns and the weekend effect. *J. Finan. Econom.* 8, 55–69.

French, K.R. (1983) A comparison of futures and forward prices. *J. Finan. Econom.* 12, 311–342.

French, K.R. (1984) The weekend effect on the distribution of stock prices: implication for option pricing. *J. Finan. Econom.* 13, 547–559.

French, K.R., Roll, R. (1986) Stock return variances: the arrival of information and the reaction of traders. *J. Finan. Econom.* 17, 5–26.

French, K.R., Schwert, G.W., Stambaugh, R.F. (1987) Expected stock returns and volatility. *J. Finan. Econom.* 19, 3–29.

Frey, R. (1996) Derivative asset analysis in models with level-dependent and stochastic volatility. *CWI Quarterly* 10, 1–34.

Frey, R. (1998) Perfect option hedging for a large trader. *Finance Stochast.* 2, 115–141.

Frey, R. (2000) Superreplication in stochastic volatility models and optimal stopping. *Finance Stochast.* 2, 161–187.

Frey, R., Sin, C.A. (1999) Bounds on European option prices under stochastic volatility. *Math. Finance* 9, 97–116.

Frey, R., Sommer, D. (1996) A systematic approach to pricing and hedging of international derivatives with interest rate risk. *Appl. Math. Finance* 3, 295–317.

Frey, R., Stremme, A. (1997) Market volatility and feedback effects from dynamic hedging. *Math. Finance* 7, 351–374.

Friedman, A. (1976) *Stochastic Differential Equations and Applications. Volume 2*, Academic Press.

Frittelli, M. (2000) The minimal entropy martingale measure and the valuation problem in incomplete markets. *Math. Finance* 10, 39–52.

Frittelli, M., Lakner, P. (1994) Arbitrage and free lunch in a general financial market model: the fundamental asset pricing theorem. In: *Mathematical Finance*, M.H.A. Davis, D. Duffie, W.H. Fleming and S.E. Shreve, eds. *Lecture Notes in Math.* Springer, Berlin Heidelberg New York.

Friz, P., Gatheral, J. (2005) Valuation of volatility derivatives as an inverse problem. *Quant. Finance* 5, 531–542.

Galai, D. (1977) Tests of market efficiency of the Chicago Board Options Exchange. *J. Business* 50, 167–197.

Galai, D. (1978) On the Boness and Black-Scholes models for valuation of call options. *J. Finan. Quant. Anal.* 13, 15–27.

Galai, D. (1983) The components of the return from hedging options against stocks. *J. Business* 56, 45–54.

Galluccio, S., Hunter, C. (2004a) The co-initial swap market model. *Notes by Banca Monte dei Paschi di Siena SpA* 33, 209-232.

Galluccio, S., Hunter, C. (2004b) Single and multicurrency co-initial swap market model. Working paper, BNP Paribas.

Galluccio, S., Le Cam, Y. (2004) Modelling hybrids with jumps and stochastic volatility. Working paper, BNP Paribas.

Galluccio, S., Huang, Z., Ly, J.-M., Scaillet, O. (2003a) Theory and calibration of swap market models. Forthcoming in *Math. Finance*.

Galluccio, S., Guiotto, P., Roncoroni, A. (2003b) Shape factors and cross-sectional risk. Working paper, BNP Paribas.

Garman, M.B. (1976a) A general theory of asset valuation under diffusion state model. Working paper, University of California, Berkeley.

Garman, M.B. (1976b) An algebra for evaluating hedge portfolios. *J. Finan. Econom.* 3, 403–427.

Garman, M.B. (1978) The pricing of supershares. *J. Finan. Econom.* 6, 3–10.

Garman, M.B. (1985a) Towards a semigroup pricing theory. *J. Finance* 40, 847–861.

Garman, M.B. (1985d) The duration of option portfolios. *J. Finan. Econom.* 14, 309–316.

Garman, M.B., Klass, M.J. (1980) On the estimation of security price volatilities from historical data. *J. Business* 53, 67–78.

Garman, M.B., Kohlhagen, S.W. (1983) Foreign currency option values. *J. Internat. Money Finance* 2, 231–237.

Gątarek, D. (1996) Pricing American swaptions in an approximate lognormal model. Working paper, University of New South Wales.

Gątarek, D. (2002) Calibration of the Libor market model: three prescriptions. Working paper.

Gątarek, D. (2003) LIBOR market models with stochastic volatility. Working paper.

Gatheral, J. (1999) The volatility skew: arbitrage constraints and asymptotic behaviour. Working paper, Merrill Lynch.

Gatheral, J. (2000) The structure of the implied volatility in the Heston model. Working paper, Merrill Lynch.

Gatheral, J. (2003) *Case Studies in Financial Modelling.* Course notes. Courant Institute of Mathematical Sciences.

Gatheral, J., Matytsin, A., Youssfi, C. (2000) Rational shapes of the volatility surface. Working paper, Merrill Lynch.

Gay, G.D., Manaster, S. (1984) The quality option implicit in futures contracts. *J. Finan. Econom.* 13, 353–370.

Gay, G.D., Manaster, S. (1986) Implicit delivery options and optimal delivery strategies for financial futures contracts. *J. Finan. Econom.* 16, 41–72.

Geman, H. (1989) The importance of the forward neutral probability in a stochastic approach of interest rates. Working paper, ESSEC.

Geman, H. (2002) Pure jump Lévy processes for asset price modelling. *J. Bank. Finance* 26, 1297–1316.

Geman, H., Ané, T. (1996) Stochastic subordination. *Risk* 9(9), 145–149.

Geman, H., Eydeland, A. (1995) Domino effect. *Risk* 8(4), 65–67.

Geman, H., Yor, M. (1992) Quelques relations entre processus de Bessel, options asiatiques, et fonctions confluentes hypergéométriques. *C.R. Acad Sci. Paris, Série I* 314, 471–474.

Geman, H., Yor, M. (1993) Bessel processes, Asian options and perpetuities. *Math. Finance* 3, 349–375.

Geman, H., Yor, M. (1996) Pricing and hedging double-barrier options: a probabilistic approach. *Math. Finance* 6, 365–378.

Geman, H., El Karoui, N., Rochet, J.C. (1995) Changes of numeraire, changes of probability measures and pricing of options. *J. Appl. Probab.* 32, 443–458.

Geman, H., Madan, D., Yor, M. (1998) Asset prices are Brownian motions only in business time. Working paper.

Geman, H., Madan, D., Yor, M. (2001a) Time changes for Lévy processes. *Math. Finance* 11, 79–96.

Geman, H., Madan, D., Pliska, S., Vorst, T., eds. (2001b) *Mathematical Finance. Bachelier Congress 2000.* Springer, Berlin Heidelberg New York.

Gentle, D. (1993) Basket weaving. *Risk* 6(6), 51–52.

Gerber, H.U., Shiu, E.S.W. (1994a) Option pricing by Esscher transforms. *Trans. Soc. Actuaries* 46, 51–92.

Gerber, H.U., Shiu, E.S.W. (1994b) Martingale approach to pricing perpetual American options. *ASTIN Bull.* 24, 195–220.

Gerber, H.U., Shiu, E.S.W. (1996a) Martingale approach to pricing perpetual American options on two stocks. *Math. Finance* 6, 303–322.

Gerber, H.U., Shiu, E.S.W. (1996b) Actuarial bridges to dynamic hedging and option pricing. *Insurance Math. Econom.* 18, 183–218.

Geske, R. (1977) The valuation of corporate liabilities as compound options. *J. Finan. Quant. Anal.* 12, 541–552.

Geske, R. (1978) Pricing of options with stochastic dividend yield. *J. Finance* 33, 617–625.

Geske, R. (1979a) The valuation of compound options. *J. Finan. Econom.* 7, 63–82.

Geske, R. (1979b) A note on an analytical valuation formula for unprotected American call options on stocks with known dividends. *J. Finan. Econom.* 7, 375–380.

Geske, R., Johnson, H.E. (1984) The American put option valued analytically. *J. Finance* 39, 1511–1524.

Geske, R., Roll, R. (1984) On valuing American call options with the Black-Scholes European formula. *J. Finance* 39, 443–455.

Geske, R., Shastri, K. (1985a) Valuation by approximation: a comparison of alternative option valuation techniques. *J. Finan. Quant. Anal.* 20, 45–71.

Geske, R., Shastri, K. (1985b) The early exercise of American puts. *J. Business Finance* 9, 207–219.

Ghysels, E., Gourieroux, C., Jasiak, J. (1995) Market time and asset price movements. Theory and estimation. Working paper.

Gibbons, M.R. (1982) Multivariate tests of financial models: a new approach. *J. Finan. Econom.* 10, 3–27.

Gibbons, M.R., Ferson, W. (1985) Testing asset pricing models with changing expectations and an unobservable market portfolio. *J. Finan. Econom.* 14, 217–236.

Gibbons, M.R., Ramaswamy, K. (1993) A test of the Cox, Ingersoll and Ross model of the term structure. *Rev. Finan. Stud.* 6, 619–658.

Gilster, J.E. (1990) The systematic risk of discretely rebalanced option hedges. *J. Finan. Quant. Anal.* 25, 507–516.

Gilster, J.E., Lee, W. (1984) The effects of transaction costs and different borrowing and lending rates on the option pricing model: a note. *J. Finance* 39, 1215–1221.

Glasserman, P. (2003) *Monte Carlo Methods in Financial Engineering*. Springer, Berlin Heidelberg New York.

Glasserman, P., Kou, S.G. (2003) The term structure of simple forward rates with jump risk. *Math. Finance* 3, 383–410.

Glasserman, P., Merener, N. (2003) Cap and swaption approximations in LIBOR market models with jumps. *J. Comput. Finance* 7, 1–11.

Glasserman, P., Zhao, X. (2000) Arbitrage-free discretization of lognormal forward Libor and swap rate model. *Finance Stochast.* 4, 35–68.

Glasserman, P., Heidelberger, P., Shahabuddin, P. (1999) Asymptotically optimal importance sampling and stratification for pricing path-dependent options. *Math. Finance* 9, 1117-1152.

Göing-Jaeschke, A., Yor, M. (1999) A survey and some generalizations of Bessel processes. Working paper, ETHZ and Université Pierre et Marie Curie.

Goldberg, L.R. (1998) Volatility of the short rate in the rational lognormal model. *Finance Stochast.* 2, 199–211.

Goldenberg, D.H. (1991) A unified method for pricing options on diffusion processes. *J. Finan. Econom.* 29, 3–34.

Goldman, B., Sosin, H., Gatto, M. (1979a) Path dependent options: "buy at the low, sell at the high". *J. Finance* 34, 1111–1128.

Goldman, B., Sosin, H., Shepp, L.A. (1979b) On contingent claims that insure ex-post optimal stock market timing. *J. Finance* 34, 401–414.

Goldys, B. (1997) A note on pricing interest rate derivatives when LIBOR rates are lognormal. *Finance Stochast.* 1, 345–352.

Goldys, B., Musiela, M. (1996) On partial differential equations related to term structure models. Working paper, University of New South Wales.

Goldys, B., Musiela, M. (2001) Infinite dimensional diffusions, Kolmogorov equations and interest rate models. In: *Option Pricing, Interest Rates and Risk Management*, E. Jouini, J. Cvitanić and M. Musiela, eds. Cambridge University Press, Cambridge, pp. 314–335.

Goldys, B., Musiela, M., Sondermann, D. (2000) Lognormality of rates and term structure models. *Stoch. Anal. Appl.* 18, 375–396.

Goodman, L.S., Ross, S., Schmidt, F. (1985) Are foreign currency options overvalued? The early experience of the Philadelphia Stock Exchange. *J. Futures Markets* 5, 349–359.

Gould, J.P., Galai, D. (1974) Transaction costs and the relationship between put and call prices. *J. Finan. Econom.* 1, 105–129.

Gouriéroux, C. (1997) *GARCH Models and Financial Applications*. Springer, Berlin Heidelberg New York.

Gouriéroux, C., Laurent, J.P., Pham, H. (1998) Mean-variance hedging and numéraire. *Math. Finance* 8, 179–200.

Gozzi, F., Vargiolu, T. (2002) Superreplication of European multiasset derivatives with bounded stochastic volatility. *Math. Oper. Res.* 55, 69–91.

Grabbe, J. Orlin (1983) The pricing of call and put options on foreign exchange. *J. Internat. Money Finance* 2, 239–253.

Grabbe, J. Orlin (1995) *International Financial Markets*. 3rd Prentice-Hall, Englewood Cliffs (New Jersey).

Grannan, E.R., Swindle, G.H. (1996) Minimizing transaction costs of option hedging strategies. *Math. Finance* 6, 341–364.

Greene, M.T., Fielitz, B.D. (1977) Long-term dependence in common stock returns. *J. Finan. Econom.* 4, 339–349.

Greene, M.T., Fielitz, B.D. (1979) The effect of long term dependence on risk-return models of common stocks. *Oper. Res.* 27, 944–951.

Greene, R.C., Jarrow, R. (1987) Spanning and completeness in markets with contingent claims. *JET* 41, 202–210.

Grinblatt, M., Jegadeesh, N. (1996) Relative pricing of Eurodollar futures and forward contracts. *J. Finance* 51, 1499–1522.

Grünbichler, A., Longstaff, F.A. (1996) Valuing futures and options on volatility. *J. Bank. Finance* 20, 985–1001.

Grundy, B.D. (1991) Option prices and the underlying asset's return distribution. *J. Finance* 46, 1045–1069.

Guo, C. (1998) Option pricing under heterogeneous expectations. *Finan. Rev.* 33, 81–92.

Gyöngy, I. (1986) Mimicking the one-dimensional marginal distributions of processes having an Itô differential. *Probab. Theory Rel. Fields* 71, 501–516.

Hagan, P.S., Woodward, D.E. (1999a) Equivalent Black volatilities. *Appl. Math. Finance* 6, 147–157.

Hagan, P.S., Woodward, D.E. (1999b) Markov interest rate models. *Appl. Math. Finance* 6, 223–260.

Hagan, P.S., Kumar, D., Lesniewski, A.S., Woodward, D.E. (2002) Managing smile risk. *Wilmott*, September, 84–108.

Hansen, A.T., Poulsen, R. (2000) A simple regime switching term structure model. *Finance Stochast.* 4, 371–389.

Hansen, L.P., Richard, S.F. (1987) The role of conditioning information in deducing testable restrictions implied by dynamic asset pricing models. *Econometrica* 55, 587–613.

Hansen, O.K. (1994) Theory of arbitrage-free term structure. *Econom. Finan. Computing*, Summer, 67–85.

Hansen, O.K., Myrup, K. (1994) The effect of the European currency turmoil on risk and prices of Danish bonds. *Econom. Finan. Computing*, Summer, 87–108.

Harrison, J.M. (1985) *Brownian Motion and Stochastic Flow Systems.* J. Wiley, New York.

Harrison, J.M., Kreps, D.M. (1979) Martingales and arbitrage in multiperiod securities markets. *J. Econom. Theory* 20, 381–408.

Harrison, J.M., Pliska, S.R. (1981) Martingales and stochastic integrals in the theory of continuous trading. *Stochastic Process. Appl.* 11, 215–260.

Harrison, J.M., Pliska, S.R. (1983) A stochastic calculus model of continuous trading: complete markets. *Stochastic Process. Appl.* 15, 313–316.

He, H. (1990) Convergence from discrete to continuous-time contingent claims prices. *Rev. Finan. Stud.* 3, 523–546.

He, H., Keirstead, W.P., Rebholz, J. (1998) Double lookbacks. *Math. Finance* 8, 201–228.

Heath, D.C., Jarrow, R.A. (1987) Arbitrage, continuous trading, and margin requirement. *J. Finance* 42, 1129–1142.

Heath, D.C., Jarrow, R.A. (1988) Ex-dividend stock price behavior and arbitrage opportunities. *J. Business* 61, 95–108.

Heath, D.C., Jarrow, R.A., Morton, A. (1990a) Bond pricing and the term structure of interest rates: a discrete time approximation. *J. Finan. Quant. Anal.* 25, 419–440.

Heath, D.C., Jarrow, R.A., Morton, A. (1990b) Contingent claim valuation with a random evolution of interest rates. *Rev. Futures Markets* 9, 54–76.

Heath, D.C., Jarrow, R.A., Morton, A. (1992a) Bond pricing and the term structure of interest rates: a new methodology for contingent claim valuation. *Econometrica* 60, 77–105.

Heath, D.C., Jarrow, R.A., Morton, A., Spindel, M. (1992b) Easier done than said. *Risk* 5(9), 77–80.

Hegde, S. (1988) An empirical analysis of implicit delivery options in Treasury bond futures contracts. *J. Bank. Finance* 12, 469–492.

Heitmann, F., Trautmann, S. (1995) Gaussian multi-factor interest rate models: theory, estimation, and implications for options pricing. Working paper, Johannes Gutenberg-University Mainz.

Henderson, V. (2005) Anaytical comparisons of option prices in stochastic volatility models. *Math. Finance* 15, 49–60.

Henderson, V., Hobson, D. (2000) Local time, coupling and the passport option. *Finance Stochast.* 4, 69–80.

Henderson, V., Hobson, D. (2001) Passport options with stochastic volatility. *Appl. Math. Finance* 8, 79–95.

Henderson, V., Hobson, D., Kentwell, G. (2002) A new class of commodity hedging strategies: a passport options approach. *Internat. J. Theor. Appl. Finance* 5, 255–278.

Henderson, V., Hobson, D., Kluge, T. (2003) Extending Figlewski's option pricing formula. Working paper.

Henrotte, P. (1991) Transaction costs and duplication strategies. Working paper, Stanford University and HEC.

Heston, S.L. (1993) A closed-form solution for options with stochastic volatility with applications to bond and currency options. *Rev. Finan. Stud.* 6, 327–343.

Heston, S.L., Nandi, S. (2000) A closed-form GARCH option valuation model. *Rev. Finan. Stud.* 13, 585–625.

Heynen, R. (1994) An empirical investigation of observed smile patterns. *Rev. Futures Markets* 13, 317–353.

Heynen, R.C., Kat, H.M. (1994) Partial barrier options. *J. Finan. Engrg* 2, 253–274.

Heynen, R.C., Kat, H.M. (1995) Lookback options with discrete and partial monitoring of the underlying price. *Appl. Math. Finance* 2, 273–284.

Heynen, R.C., Kemna, A.G.Z., Vorst, T.C.F. (1994) Analysis of the term structure of implied volatilities. *J. Finan. Quant. Anal.* 29, 31–56.

Ho, T.S.Y., Lee, S.-B. (1986) Term structure movements and pricing interest rate contingent claims. *J. Finance* 41, 1011–1029.

Ho, T.S.Y., Singer, R.F. (1982) Bond indenture provisions and the risk of corporate debt. *J. Finan. Econom.* 10, 375–406.

Ho, T.S.Y., Singer, R.F. (1984) The value of corporate debt with a sinking-fund provision. *J. Business* 57, 315–336.

Ho, T.S.Y., Stapleton, R.C., Subrahmanyam, M.G. (1997) The valuation of American options with stochastic interest rates: a generalization of the Geske-Johnson technique. *J. Finance* 52, 827–840.

Hobson, D.G. (1998a) Robust hedging of the lookback option. *Finance Stochast.* 2, 329–347.

Hobson, D.G. (1998b) Volatility misspecification, option pricing and superreplication via coupling. *Ann. Appl. Probab.* 8, 193–205.

Hobson, D.G., Rogers, L.C.G. (1998) Complete model with stochastic volatility. *Math. Finance* 8, 27–48.

Hodges, H. (1996) Arbitrage bounds on the implied volatility strike and term structures of European-style options. *J. Derivatives*, 23–35.

Hoffman, I.D. (1993) Lognormal processes in finance. Doctoral dissertation, University of New South Wales, Sydney.

Hofmann, N., Platen, E., Schweizer, M. (1992) Option pricing under incompleteness and stochastic volatility. *Math. Finance* 2, 153–187.

Hogan, M. (1993) Problems in certain two-factor term structure models. *Ann. Appl. Probab.* 3, 576–581.

Hogan, M., Weintraub, K. (1993) The log-normal interest rate model and Eurodollar futures. Working paper, Citibank, New York.

Hsu, D.A., Miller, R., Wichern, D. (1974) On the stable Paretian behavior of stock market prices. *J. Amer. Statist. Assoc.* 69, 108–113.

Huang, C.-F., Litzenberger, R.H. (1988) *Foundations for Financial Economics.* North-Holland, New York.

Huang, Y., Davison, M. (2002) Hedging options with mis-specified parameters. Working paper.

Huang, J., Subrahmanyam, M.G., Yu, G. (1996) Pricing and hedging American options: a recursive integration method. *Rev. Finan. Stud.* 9, 277–300.

Hubalek, F., Schachermayer, W. (1998) When does convergence of asset price processes imply convergence of option prices? *Math. Finance* 8, 385–403.

Hubalek, F., Klein, I., Teichman, J. (2002) A general proof of the Dybvig-Ingersoll-Ross theorem: long forward rates can never fall. *Math. Finance* 12, 447–451.

Huge, B., Lando, D. (1999) Swap pricing with two-sided default risk in a rating-based model. *European Finance Review* 3, 239–268.

Hughston, L. (2001) *The New Interest Rate Models.* Risk Books, London.

Hughston, L.P., Rafailidis, A. (2005) A chaotic approach to interest rate modelling. *Finance Stochast.* 9, 43–65.

Hui, C.H. (1996) One-touch double barrier binary option values. *Appl. Finan. Econom.* 6, 343–346.

Hull, J.C. (1994) *Introduction to Futures and Options Markets*. 2nd Prentice-Hall, Englewood Cliffs (New Jersey).

Hull, J.C. (1997) *Options, Futures, and Other Derivatives*. 3rd Prentice-Hall, Englewood Cliffs (New Jersey).

Hull, J.C., White, A. (1987a) The pricing of options on assets with stochastic volatilities. *J. Finance* 42, 281–300.

Hull, J.C., White, A. (1987b) Hedging the risks from writing foreign currency options. *J. Internat. Money Finance* 6, 131–152.

Hull, J.C., White, A. (1988a) The use of the control variate technique in option pricing. *J. Finan. Quant. Anal.* 23, 237–252.

Hull, J.C., White, A. (1988b) An analysis of the bias in option pricing caused by a stochastic volatility. *Adv. in Futures Options Res.* 3, 29–61.

Hull, J.C., White, A. (1988c) An overview of the pricing of contingent claims. *Canad. J. Admin. Sci.* 5, 55–61.

Hull, J.C., White, A. (1990a) Pricing interest-rate derivative securities. *Rev. Finan. Stud.* 3, 573–592.

Hull, J.C., White, A. (1990b) Valuing derivative securities using the explicit finite difference method. *J. Finan. Quant. Anal.* 25, 87–100.

Hull, J.C., White, A. (1993a) Bond option pricing based on a model for the evolution of bond prices. *Adv. in Futures Options Res.* 6, 1–13.

Hull, J.C., White, A. (1993b) One-factor interest rate models and the valuation of interest rate derivative securities. *J. Finan. Quant. Anal.* 28, 235–254.

Hull, J.C., White, A. (1993c) Efficient procedures for valuing European and American path-dependent options. *J. Derivatives*, Fall, 21–31.

Hull, J.C., White, A. (1994) The pricing of options on interest-rate caps and floors using the Hull-White model. *J. Financial Engrg* 2, 287–296.

Hull, J.C., White, A. (1995) The impact of default risk on the prices of options and other derivative securities. *J. Bank. Finance* 19, 299–322.

Hull, J.C., White, A. (2000) Forward rate volatilities, swap rate volatilities and the implementation of the LIBOR market model. *J. Fixed Income* 10, 46–62.

Hunt, P.J., Kennedy, J.E. (1996) On multi-currency interest rate models. Working paper, ABN-Amro Bank and University of Warwick.

Hunt, P.J., Kennedy, J.E. (1997) On convexity corrections. Working paper, ABN-Amro Bank and University of Warwick.

Hunt, P.J., Kennedy, J.E. (1998) Implied interest rate pricing model. *Finance Stochast.* 2, 275–293.

Hunt, P.J., Kennedy, J.E. (2000) *Financial Derivatives in Theory and Practice*. J.Wiley, Chichester New York.

Hunt, P.J., Kennedy, J.E., Scott, E.M. (1996) Terminal swap-rate models. Working paper, ABN-Amro Bank and University of Warwick.

Hunt, P.J., Kennedy, J.E., Pelsser, A. (2000) Markov-functional interest rate models. *Finance Stochast.* 4, 391–408.

Hurst, S.R., Platen, E., Rachev, S.T. (1997) Subordinated market index models: a comparison. *Financial Engineering and the Japanese Markets* 4, 97–124.

Hurst, S.R., Platen, E., Rachev, S.T. (1999) Option pricing for a logstable asset model. *Mathematical and Computer Modelling* 25, 105–119.

Huynh, C.B. (1994) Back to baskets. *Risk* 7(5), 59–61.

Hyer, T., Lipton, A., Pugachevsky, D. (1997) Passport to success. *Risk* 10, 127–131.

Ikeda, N., Watanabe, S. (1981) *Stochastic Differential Equations and Diffusion Processes.* North-Holland, Amsterdam (Kodansha, Tokyo).

Ingersoll, J. (1977) An examination of corporate call policies on convertible securities. *J. Finance* 32, 463–478.

Ingersoll, J.E., Jr. (1987) *Theory of Financial Decision Making.* Rowman and Littlefield, Totowa (New Jersey).

Jacka, S.D. (1991) Optimal stopping and the American put. *Math. Finance* 1, 1–14.

Jacka, S.D. (1992) A martingale representation result and an application to incomplete financial markets. *Math. Finance* 2, 239–250.

Jacka, S.D. (1993) Local times, optimal stopping and semimartingales. *Ann. Probab.* 21, 329–339.

Jäckel, P. (2002) *Monte Carlo Methods in Finance.* J. Wiley, Chichester.

Jäckel, P., Rebonato, R. (2003) The link between caplet and swaption volatilities in a Brace-Gatarek-Musiela/Jamshidian framework: approximate solutions and empirical evidence. *J. Comput. Finance* 6, 35–45.

Jackwerth, J.C. (2000) Option-implied risk-neutral distributions and implied binomial trees: a literature review. *J. Derivatives* 7, 66–82.

Jackwerth, J.C., Rubinstein, M. (1996) Recovering probability distributions from option prices. *J. Finance* 51, 1611–1631.

Jacod, J. (1979) *Calcul stochastique et problèmes de martingales. Lecture Notes in Math. 714.* Springer, Berlin Heidelberg New York.

Jacod, J., Shiryaev, A.N. (1998) Local martingales and the fundamental asset pricing theorems in the discrete-time case. *Finance Stochast.* 2, 259–273.

Jagannathan, R.K. (1984) Call options and the risk of underlying securities. *J. Finan. Econom.* 13, 425–434.

Jaillet, P., Lamberton, D., Lapeyre, B. (1990) Variational inequalities and the pricing of American options. *Acta Appl. Math.* 21, 263–289.

James, J., Webber, N. (2000) *Interest Rate Modelling.* J. Wiley, Chichester.

Jamshidian, F. (1987) Pricing of contingent claims in the one factor term structure model. Working paper, Merrill Lynch Capital Markets.

Jamshidian, F. (1988) The one-factor Gaussian interest rate model: theory and implementation. Working paper, Merrill Lynch Capital Markets.

Jamshidian, F. (1989a) An exact bond option pricing formula. *J. Finance* 44, 205–209.

Jamshidian, F. (1989b) The multifactor Gaussian interest rate model and implementation. Working paper, Merrill Lynch Capital Markets.

Jamshidian, F. (1991a) Bond and option evaluation in the Gaussian interest rate model. *Res. Finance* 9, 131–170.

Jamshidian, F. (1991b) Forward induction and construction of yield curve diffusion models. *J. Fixed Income* 1 (June), 62–74.

Jamshidian, F. (1992) An analysis of American options. *Rev. Futures Markets* 11, 72–80.

Jamshidian, F. (1993a) Option and futures evaluation with deterministic volatilities. *Math. Finance* 3, 149–159.

Jamshidian, F. (1993b) Price differentials. *Risk* 6(7), 48–51.

Jamshidian, F. (1994a) Corralling quantos. *Risk* 7(3), 71–75.

Jamshidian, F. (1994b) Hedging quantos, differential swaps and ratios. *Appl. Math. Finance* 1, 1–20.

Jamshidian, F. (1995) A simple class of square-root interest-rate models. *Appl. Math. Finance* 2, 61–72.

Jamshidian, F. (1996) Bond, futures and options evaluation in the quadratic interest rate model. *Appl. Math. Finance* 3, 93–115.

Jamshidian, F. (1997a) LIBOR and swap market models and measures. *Finance Stochast.* 1, 293–330.

Jamshidian, F. (1997b) A note on analytical valuation of double barrier options. Working paper, Sakura Global Capital.

Jamshidian, F. (1999) Libor market model with semimartingales. Working paper, NetAnalytic Limit

Janson, S., Tysk, J. (2003) Volatility time and properties of option proces. *Ann. Appl. Probab.* 13, 890–913.

Jarrow, R.A. (1987) The pricing of commodity options with stochastic interest rates. *Adv. in Futures Options Res.* 2, 19–45.

Jarrow, R.A. (1988) *Finance Theory*. Prentice-Hall, Englewood Cliffs (New Jersey).

Jarrow, R.A. (1994) Derivative security markets, market manipulation, and option pricing theory. *J. Finan. Quant. Anal.* 29, 241–261.

Jarrow, R.A. (1995) *Modelling Fixed Income Securities and Interest Rate Options*. McGraw-Hill, New York.

Jarrow, R.A., Madan, D. (1991) A characterization of complete markets on a Brownian filtration. *Math. Finance* 1, 31–43.

Jarrow, R.A., Madan, D. (1995) Option pricing using the term structure of interest rates to hedge systematic discontinuities in asset returns. *Math. Finance* 5, 311–336.

Jarrow, R.A., Madan, D. (1999) Hedging contingent claims on semimartingales. *Finance Stochast.* 3, 111–134.

Jarrow, R.A., O'Hara, M. (1989) Primes and scores: an essay on market imperfections. *J. Finance* 44, 1265–1287.

Jarrow, R.A., Oldfield, G.S. (1981a) Forward contracts and futures contracts. *J. Finan. Econom.* 9, 373–382.

Jarrow, R.A., Oldfield, G.S. (1981b) Forward options and futures options. *Adv. in Futures Options Res.* 3, 15–28.

Jarrow, R.A., Rudd, A. (1982) Approximate option valuation for arbitrary stochastic processes. *J. Finan. Econom.* 10, 347–369.

Jarrow, R.A., Rudd, A. (1983) *Option Pricing*. Dow Jones-Irwin, Homewood (Illinois).

Jarrow, R.A., Turnbull, S. (1994) Delta, gamma and bucket hedging of interest rate derivatives. *Appl. Math. Finance* 1, 21–48.

Jarrow, R.A., Turnbull, S. (1995) Pricing derivatives on financial securities subject to credit risk. *J. Finance* 50, 53–85.

Jarrow, R.A., Wiggins, J.B. (1989) Option pricing and implicit volatilities. *J. Econom. Surveys* 3, 59–81.

Jarrow, R.A., Yildirim, Y. (2003) Pricing Treasury inflation protected securities and related derivatives using an HJM model. *J. Finan. Quant. Anal.* 38, 337–359.

Jarrow, R.A., Lando, D., Turnbull, S. (1997) A Markov model for the term structure of credit risk spreads. *Rev. Finan. Stud.* 10, 481–523.

Jaschke, S.R. (1998) Arbitrage bounds for the term structure of interest rates. *Finance Stochast.* 2, 29–40.

Jeanblanc-Picqué, M., Pontier, M. (1990) Optimal portfolio for a small investor in a market with discontinuous prices. *Appl. Math. Optimization* 22, 287–310.

Jeanblanc, M., Rutkowski, M. (2002) Default risk and hazard processes. In: *Mathematical Finance – Bachelier Congress 2000*, H. Geman, D. Madan, S.R. Pliska and T. Vorst, eds. Springer, Berlin Heidelberg New York, pp. 281–312.

Jeanblanc, M., Yor, M., Chesney, M. (2006) *Mathematical Methods for Financial Markets.* Springer, Berlin Heidelberg New York.

Jeffrey, A. (1995) Single factor Heath-Jarrow-Morton term structure models based on Markov spot interest rate dynamics. *J. Finan. Quant. Anal.* 30, 619–642.

Jensen, B., Nielsen, J. (1991) The structure of binomial lattice models for bonds. Working paper, Institut for Finansiering, Copenhagen.

Ji, D.M., Yin, G. (1993) Weak convergence of term structure movements and the connection of prices and interest rates. *Stochastic Anal. Appl.* 11, 61–76.

Jiang, G.J. (1999) Stochastic volatility and jump-diffusion – implications on option pricing. *Internat. J. Theor. Appl. Finance* 2, 409–440.

Jin, Y., Glasserman, P. (2001) Equilibrium positive interest rates: a unified view. *Rev. Finan. Stud.* 14, 187–214.

Johnson, H.E. (1983) An analytic approximation to the American put price. *J. Finan. Quant. Anal.* 18, 141–148.

Johnson, H. (1987) Options on the maximum or the minimum of several assets. *J. Finan. Quant. Anal.* 22, 277–284.

Johnson, H., Shanno, D. (1987) Option pricing when the variance is changing. *J. Finan. Quant. Anal.* 22, 143–151.

Johnson, H., Stulz, R. (1987) The pricing of options with default risk. *J. Finance* 42, 267–280.

Johnson, L.L. (1960) The theory of hedging and speculation in commodity futures markets. *Rev. Econom. Stud.* 27, 139–151.

Jones, C., Milne, F. (1993) Tax arbitrage, existence of equilibrium, and bounded tax rebates. *Math. Finance* 3, 189–196.

Jones, C., Gautam, K., Lipson, M.L. (1994) Transactions, volumes and volatility. *Rev. Finan. Stud.* 7, 631–651.

Jones, E.P. (1984) Option arbitrage and strategy with large price changes. *J. Finan. Econom.* 13, 91–113.

Jonkhart, M.J.L. (1979) On the term structure of interest rates and the risk of default: an analytical approach. *J. Bank. Finance* 3, 253–262.

Jorion, P. (1995) Predicting volatility in the foreign exchange market. *J. Finance* 50, 507–528.

Joshi, M., Rebonato, R. (2003) A stochastic volatility displaced-diffusion extension of the LIBOR market model. *Quant. Finance* 3, 458–469.

Joshi, M., Theis, J. (2002) Bounding Bermudan swaptions in a swap-rate market model. *Quant. Finance* 2, 370–377.

Jouini, E. (1997) Market imperfections, equilibrium and arbitrage. In: *Financial Mathematics, Bressanone, 1996*, W. Runggaldier, ed. *Lecture Notes in Math.* 1656, Springer, Berlin Heidelberg New York, pp. 247–307.

Jouini, E., Kallal, H. (1995) Arbitrage in securities markets with short-sales constraints. *Math. Finance* 5, 197–232.

Kabanov, Yu.M., Kramkov, D.O. (1994a) No-arbitrage and equivalent martingale measures: an elementary proof of the Harrison-Pliska theorem. *Theory Probab. Appl.* 39, 523–527.

Kabanov, Yu.M., Kramkov, D.O. (1994b) Large financial markets: asymptotic arbitrage and contiguity theorem. *Theory Probab. Appl.* 39, 222–229.

Kabanov, Yu.M., Kramkov, D.O. (1998) Asymptotic arbitrage in large financial markets. *Finance Stochast.* 2, 143–172.

Kabanov, Yu.M., Safarian, M.M. (1997) On Leland's strategy of option pricing with transaction costs. *Finance Stochast.* 1, 239–250.

Kalay, A. (1982) The ex-dividend day behavior of stock prices: a re-examination of the clientele effect. *J. Finance* 37, 1059–1070.

Kalay, A. (1984) The ex-dividend day behavior of stock prices: a re-examination of the clientele effect: a reply. *J. Finance* 39, 557–561.

Kallsen, J., Kühn, C. (2004) Pricing derivatives of American and game type in incomplete markets. *Finance Stochast.* 8, 261–284.

Kallsen, J., Kühn, C. (2005) Convertible bonds: financial derivatives of game type. In: *Exotic Option Pricing and Advanced Lévy Models,* A. Kyprianou et al., eds. Wiley, pp. 277–288.

Kallsen, J., Taqqu, M. (1998) Option pricing in ARCH-type models. *Math. Finance* 8, 13–26.

Kani, I., Derman, E., Kamal, M. (1996) Trading and hedging local volatility. Working paper, Goldman Sachs.

Kaplanis, C. (1986) Options, taxes, and ex-dividend day behavior. *J. Finance* 41, 411–424.

Karatzas, I. (1988) On the pricing of American options. *Appl. Math. Optim.* 17, 37–60.

Karatzas, I. (1989) Optimization problems in the theory of continuous trading. *SIAM J. Control Optim.* 27, 1221–1259.

Karatzas, I. (1996) *Lectures on the Mathematics of Finance.* CRM Monograph Series, Vol. 8, American Mathematical Society, Providence (Rhode Island).

Karatzas, I., Kou, S.-G. (1996) On the pricing of contingent claims under constraints. *Ann. Appl. Probab.* 6, 321–369.

Karatzas, I., Kou, S.-G. (1998) Hedging American contingent claims with constrained portfolios. *Finance Stochast.* 2, 215–258.

Karatzas, I., Shreve, S. (1998a) *Brownian Motion and Stochastic Calculus.* 2nd ed. Springer, Berlin Heidelberg New York.

Karatzas, I., Shreve, S. (1998b) *Methods of Mathematical Finance.* Springer, Berlin Heidelberg New York.

Karatzas, I., Xue, X.-X. (1991) A note on utility maximization under partial observations. *Math. Finance* 1, 57–70.

Karatzas, I., Lehoczky, J.P., Shreve, S.E., Xu, G.-L. (1991) Martingale and duality methods for utility maximization in an incomplete market. *SIAM J. Control Optim.* 29, 702–730.

Kawai, A. (2001) Analytical and Monte Carlo swaption pricing under the forward swap measure. *J. Comput. Finance* 6(1), 101–111.

Kawai, A. (2003) A new approximate swaption formula in the LIBOR market model: an asymptotic expansion approach. *Appl. Math. Finance* 10, 49–74.

Keim, D.B., Stambaugh, R.F. (1986) Predicting returns in the stock and bond markets. *J. Finan. Econom.* 17, 357–390.

Kemna, A.G.Z., Vorst, T.C.F. (1990) A pricing method for options based on average asset values. *J. Bank. Finance* 14, 113–129.

Kennedy, D.P. (1994) The term structure of interest rates as a Gaussian random field. *Math. Finance* 4, 247–258.

Kennedy, D.P. (1997) Characterizing Gaussian models of the term structure of interest rates. *Math. Finance* 7, 107–118.

Kifer, Y. (1971) Optimal stopping in games with continuous time. *Theory Probab. Appl.* 16, 545–550.

Kifer, Y. (2000) Game options. *Finance Stochast.* 4, 443–463.

Kijima, M. (2002) Monotonicity and convexity of option prices revisited. *Math. Finance* 12, 411–425.

Kim, I.J. (1990) The analytic valuation of American options. *Rev. Finan. Stud.* 3, 547–572.

Kim, I.J., Ramaswamy, K., Sundaresan, S. (1993) Does default risk in coupons affect the valuation of corporate bonds? *Finan. Manag.* 22, 117–131.

Kind, P., Liptser, R.S., Runggaldier, W.J. (1991) Diffusion approximation in past dependent models and applications to option pricing. *Ann. Appl. Probab.* 1, 379–405.

Klebaner, F. (2002) Option price when the stock is a semimartingale. *Elect. Comm. in Probab.* 7, 79–83.

Klein, P. (1996) Pricing Black-Scholes options with correlated credit risk. *J. Bank. Finance* 20, 1211–1229.

Klein, I., Schachermayer, W. (1996a) Asymptotic arbitrage in non-complete large financial markets. *Theory Probab. Appl.* 41, 927–934.

Klein, I., Schachermayer, W. (1996b) A quantitative and a dual version of the Halmos-Savage theorem with applications to mathematical finance. *Ann. Probab.* 24, 867–881.

Kleinberg, N.L. (1995) A note on finite securities market models. *Stochastic Anal. Appl.* 13, 543–554.

Klemkosky, R.C., Lee, J.H. (1991) The intraday ex post and ex ante profitability of index arbitrage. *J. Futures Markets* 11, 291–311.

Klemkosky, R.C., Resnick, B.G. (1979) Put-call parity and market efficiency. *J. Finance* 34, 1141–1155.

Kloeden, P.E., Platen, E. (1995) *Numerical Solution of Stochastic Differential Equations.* 2nd ed. Springer, Berlin Heidelberg New York.

Kolb, R.W. (1991) *Understanding Futures Markets.* 3rd ed. Kolb Publishing, Miami.

Kon, S.J. (1984) Models of stock returns – a comparison. *J. Finance* 39, 147–165.

Korn, R. (1997) *Optimal Portfolios.* World Scientific, Singapore.

Kou, S. (2002) A jump-diffusion model for option pricing. *Manag. Science* 48, 1086–1101.

Kramkov, D. (1996) Optional decomposition of supermartingales and hedging contingent claims in incomplete security markets. *Probab. Theory Rel. Fields* 105, 459–479.

Kramkov, D.O., Mordecki, E. (1994) Integral option. *Theory Probab. Appl.* 39, 162–171.

Kramkov, D.O., Shiryaev, A.N. (1994) On the pricing of the "Russian option" for the symmetrical binomial model of (B, S)-market. *Theory Probab. Appl.* 39, 153–161.

Kreps, D. (1981) Arbitrage and equilibrium in economies with infinitely many commodities. *J. Math. Econom.* 8, 15–35.

Kruse, S., Nögel, U. (2005) On the pricing of forward starting options in Heston's model on stochastic volatility. *Finance Stochast.* 9, 233–250.

Krylov, N.V. (1995) *Introduction to the Theory of Diffusion Processes.* American Mathematical Society, Providence.

Kunitomo, N., Ikeda, M. (1992) Pricing options with curved boundaries. *Math. Finance* 2, 275–298.

Kusuoka, S. (1999) A remark on default risk models. *Adv. Math. Econom.* 1, 69–82.

Kwok, Y.K. (1998) *Mathematical Models of Financial Derivatives.* Springer, Berlin Heidelberg New York.

Lacoste, V. (1996) Wiener chaos: a new approach to option hedging. *Math. Finance* 6, 197–213.

Lagnado, R., Osher, S. (1997) A technique for calibrating derivative security pricing models: numerical solution of an inverse problem. *J. Comput. Finance* 1, 13–25.

Lakner, P. (1993) Martingale me asure for a class of right-continuous processes. *Math. Finance* 3, 43–53.

Lakonishok, J., Vermaelen, T. (1983) Tax reform and ex-dividend day behavior. *J. Finance* 38, 1157–1179.

Lakonishok, J., Vermaelen, T. (1986) Tax-induced trading around ex-dividend days. *J. Finan. Econom.* 16, 287–319.

Lamberton, D. (1993) Convergence of the critical price in the approximation of American options. *Math. Finance* 3, 179–190.

Lamberton, D. (1995) Critical price for an American option near maturity. In: *Seminar on Stochastic Analysis, Random Fields and Applications*, E. Bolthausen, M. Dozzi and F. Russo, eds. Birkhäuser, Boston Basel Berlin, pp. 353–358.

Lamberton, D., Lapeyre, B. (1993) Hedging index options with few assets. *Math. Finance 3*, 25–42.

Lamberton, D., Lapeyre, B. (1996) *Introduction to Stochastic Calculus Applied to Finance*. Chapman and Hall, London

Lamoureux, C., Lastrapes, W. (1993) Forecasting stock return variance: toward an understanding of stochastic implied volatilities. *Rev. Finan. Stud.* 6, 293–326.

Lando, D. (1994) On Cox processes and credit risky bonds. Working paper, University of Copenhagen.

Lando, D. (1995) On jump-diffusion option pricing from the viewpoint of semimartingale characteristics. *Surveys Appl. Indust. Math.* 2, 605–625.

Lando, D. (1998) On Cox processes and credit-risky securities. *Rev. Derivatives Res.* 2, 99–120.

Lando, D. (2004) *Credit Risk Modeling: Theory and Applications*. Princeton University Press, Princeton (New Jersey).

Langetieg, T.C. (1980) A multivariate model of the term structure. *J. Finance* 35, 71–97.

Lapeyre, B., Sulem, A., Talay, D. (2001) *Understanding Numerical Analysis for Option Pricing*. Cambridge University Press, Cambridge.

Latané, H., Rendleman, R.J. (1976) Standard deviations of stock price ratios implied in option prices. *J. Finance* 31, 369–381.

Lau, K.W., Kwok, Y.K. (2004) Anatomy of option features in convertible bonds. *J. Futures Markets* 24, 513–532.

Laurent, J.P., Pham, H. (1999) Dynamic programming and mean-variance hedging. *Finance Stochast.* 3, 83–110.

Lauterbach, B., Schultz, P. (1990) Pricing warrants: an empirical study of the Black-Scholes model and its alternatives. *J. Finance* 45, 1181–1209.

Leblanc, B. (1996) Une approche unifiée pour une forme exacte du prix d'une option dans les différents modèles à volatilité stochastique. *Stochastics Stochastics Rep.* 57, 1–35.

Leblanc, B., Scaillet, O. (1998) Path dependent options on yields in the affine term structure model. *Finance Stochast.* 2, 349–367.

Leblanc, B., Yor, M. (1996) Quelques applications des processus à accroissements indépendants en mathématiques financières. Working paper.

Leblanc, B., Yor, M. (1998) Lévy processes in finance: a remedy to the nonstationarity of continuous martingales. *Finance Stochast.* 2, 399–408.

Lee, R.W. (2001) Implied and local volatilities under stochastic volatilities. *Internat. J. Theor. Appl. Finance* 4, 45–89.

Lee, R.W. (2002) Implied volatility: statics, dynamics, and probabilistic interpretation. Working paper, Stanford University.

Lee, R.W. (2004a) The moment formula for implied volatility at extreme strikes. *Math. Finance* 14, 469–480.

Lee, R.W. (2004b) Option pricing by transform methods: extensions, unification, and error control. *J. Comput. Finance.* 7(3), 51–86.

Lehar, A., Scheicher, M., Schittenkopf, C. (2002) GARCH vs. stochastic volatility: option pricing and risk management. *J. Bank. Finance* 26, 323–345.

Lehoczky, J.P. (1997) Simulation methods for option pricing. In: *Mathematics of Derivative Securities*, M.A.H. Dempster and S.R. Pliska, eds. Cambridge University Press, Cambridge, pp. 528–544.

Leisen, D.P.J. (1998) Pricing the American put option: a detailed convergence analysis for binomial models. *J. Econom. Dynamics Control* 22, 1419–1444.

Leisen, D.P.J. (1999) The random-time binomial model. *J. Econom. Dynamics Control* 23, 1355–1386.

Leisen, D.P.J., Reimer, M. (1996) Binomial models for option valuation: examining and improving convergence. *Appl. Math. Finance* 3, 319–346.

Leland, H.E. (1980) Who should buy portfolio insurance? *J. Finance* 35, 581–594.

Leland, H.E. (1985) Option pricing and replication with transactions costs. *J. Finance* 40, 1283–1301.

Lepeltier, J.-P., Maingueneau, M. (1984) Le jeu de Dynkin en théorie générale sans l'hypothèse de Mokobodski. *Stochastics* 13, 25–44.

Lesne, J.P., Prigent, J.L. (2001) A general subordinated stochastic process for derivative pricing. *Internat. J. Theor. Appl. Finance* 4, 121–146.

Lesne, J.P., Prigent, J.L., Scaillet, O. (2000) A convergence of discrete time option pricing models under stochastic interest rates. *Finance Stochast.* 4, 81–93.

Levental, S., Skorohod, A.S. (1995) A necessary and sufficient condition for absence of arbitrage with tame portfolios. *Ann. Appl. Probab.* 5, 906–925.

Levental, S., Skorohod, A.S. (1997) On the possibility of hedging options in the presence of transactions costs. *Ann. Appl. Probab.* 7, 410–443.

Levy, A. (1989) A note on the relationship between forward and futures prices. *J. Futures Markets* 9, 171–173.

Levy, E. (1992) The valuation of average rate currency options. *J. Internat. Money Finance* 11, 474–491.

Levy, E., Turnbull, S. (1992) Average intelligence. *Risk* 5(2), 53–59.

Levy, H., Yoder, J.A. (1996) A stochastic dominance approach to evaluating alternative estimators of the variance for use in the Black-Scholes option pricing model. *Appl. Finan. Econom.* 6, 377–382.

Lewis, A. (2000) *Option Valuation under Stochastic Volatility: With Mathematica Code.* 2nd ed. Finance Press.

Lewis, A. (2001) A simple option formula for general jump-diffusion and other exponential Lévy processes. Working paper.

Li, A., Ritchken, P., Sankarasubramanian, L. (1995) Lattice models for pricing American interest rate claims. *J. Finance* 50, 719–737.

Lipton, A. (1999) Similarities via self-similarities. *Risk* 12(9), 101–105.

Lipton, A. (2001) *Mathematical Methods for Foreign Exchange. A Financial Engineer's Approach.* World Scientific, Singapore.

Litterman, R., Scheinkman, J. (1991) Common factors affecting bond returns. *J. Fixed Income* 1, 54–61.

Litzenberger, R., Ramaswamy, K. (1982) The effects of dividends on common stock prices. *J. Finance* 37, 429–443.

Lo, A.W., MacKinlay, A.C. (1988) Stock markets do not follow random walks: evidence from a simple specification test. *Rev. Finan. Stud.* 1, 41–66.

Lo, C.F., Yuen, P.H., Hui, C.H. (2000) Constant elasticity of variance option pricing model with time-dependent parameters. *Internat. J. Theor. Appl. Finance* 3, 661–674.

Longstaff, F.A. (1989) A nonlinear general equilibrium model of the term structure of interest rates. *J. Finan. Econom.* 23, 195–224.

Longstaff, F.A. (1990a) The valuation of options on yields. *J. Finan. Econom.* 26, 97–123.

Longstaff, F.A. (1990b) Time-varying term premia and traditional hypotheses about the term structure. *J. Finance* 45, 1307–1314.

Longstaff, F.A. (1993) The valuation of options on coupon bonds. *J. Bank. Finance* 17, 27–42.

Longstaff, F.A., Schwartz, E.S. (1992a) Interest rate volatility and the term structure: a two-factor general equilibrium model. *J. Finance* 47, 1259–1282.

Longstaff, F.A., Schwartz, E.S. (1992b) A two-factor interest rate model and contingent claims valuation. *J. Fixed Income* 2 (December), 16–23.

Longstaff, F.A., Schwartz, E.S. (1995) A simple approach to valuing risky fixed and floating rate debt. *J. Finance* 50, 789–819.

Longstaff, F.A., Santa-Clara, E., Schwartz, E.S. (2001a) Throwing away a billion dollars: the cost of suboptimal exercise in the swaptions market. *J. Finan. Econom.* 62, 39–66.

Longstaff, F.A., Santa-Clara, E., Schwartz, E.S. (2001b) The relative valuation of caps and swaptions: theory and empirical evidence. *J. Finance* 56, 2067–2109.

Lotz, C., Schlögl, L. (2000) Default risk in a market model. *J. Bank. Finance* 24, 301–327.

Lucic, V. (2004) Forward-start options in stochastic volatility models. Wilmott, May, 72–74.

Luenberger, D.G. (1984) *Introduction to Linear and Nonlinear Programming*. Addison-Wesley, Reading (Mass.)

Lyons, T.J. (1995) Uncertain volatility and risk-free synthesis of derivatives. *Appl. Math. Finance* 2, 117–133.

MacBeth, J.D., Merville, L.J. (1979) An empirical examination of the Black-Scholes call option pricing model. *J. Finance* 34, 1173–1186.

MacBeth, J.D., Merville, L.J. (1980) Tests of the Black-Scholes and Cox call option valuation models. *J. Finance* 35, 285–303.

McCulloch, J.H. (1971) Measuring the term structure of interest rates. *J. Business* 44, 19–31.

McCulloch, J.H. (1975) The tax-adjusted yield curve. *J. Finance* 30, 811–830.

McCulloch, J.H. (1993) A reexamination of traditional hypotheses about the term structure: a comment. *J. Finance* 48, 779–789.

McDonald, R., Siegel, D. (1984) Option pricing when the underlying asset earns a below-equilibrium rate of return: a note. *J. Finance* 39, 261–265.

McKean, H.P., Jr. (1965) Appendix: A free boundary problem for the heat equation arising from a problem in mathematical economics. *Indust. Manag. Rev.* 6, 32–39.

MacKinlay, A.C., Ramaswamy, K. (1988) Index futures arbitrage and the behavior of stock index futures prices. *Rev. Finan. Stud.* 1, 137–158.

MacMillan, L.W. (1986) Analytic approximation for the American put option. *Adv. in Futures Options Res.* 1, 119–139.

Madan, D.B., Milne, F. (1991) Option pricing with V.G. martingale components. *Math. Finance* 1, 39–55.

Madan, D.B., Milne, F. (1993) Contingent claims valued and hedged by pricing and investing in a basis. *Math. Finance* 4, 223–245.

Madan, D.B., Seneta, E. (1990) The variance gamma (V.G.) model for share market returns. *J. Business* 63, 511–524.

Madan, D.B., Unal, H. (1998) Pricing the risk of default. *Rev. Derivatives Res.* 2, 121–160.

Madan, D.B., Yor, M. (2002) Making Markov martingales meet marginals: with explicit constructions. *Bernoulli* 8, 509–536.

Madan, D.B., Milne, F., Shefrin, H. (1989) The multinomial option pricing model and its Brownian and Poisson limits. *Rev. Finan. Stud.* 2, 251–265.

Madsen, C. (1994a) The pricing of options on coupon bonds. Working paper, Realkredit Danmark, Copenhagen.

Madsen, C. (1994b) The pricing of interest rate contingent claims. Working paper, Realkredit Danmark, Copenhagen.

Maghsoodi, Y. (1996) Solution of the extended CIR term structure and bond option valuation. *Math. Finance* 6, 89–109.

Manaster, S., Koehler, G. (1982) The calculation of implied variances from the Black-Scholes model: a note. *J. Finance* 37, 227–230.

Mandelbrot, B. (1960) The Pareto-Lévy law and the distribution of income. *Internat. Econom. Rev.* 1, 79–106.

Mandelbrot, B. (1963) The variation of certain speculative prices. *J. Business* 36, 394–419.

Mandelbrot, B. (1967) The variation of some other speculative prices. *J. Business* 40, 393–413.

Marangio, L., Bernaschi, M., Ramponi, A. (2002) A review of techniques for the estimation of the term structure. *Internat. J. Theor. Appl. Finance* 5, 189–221.

Markowitz, H. (1952) Portfolio selection. *J. Finance* 7, 77–91.

Markowitz, H. (1987) *Mean-Variance Analysis in Portfolio Choice and Capital Markets*. Basil Blackwell, Cambridge, Mass.

Margrabe, W. (1978) The value of an option to exchange one asset for another. *J. Finance* 33, 177–186.

Marris, D. (1999) Financial option pricing and skewed volatility. Working paper.

Marsh, T.A., Rosenfeld, E.R. (1986) Non-trading, market marking, and estimates of stock price volatility. *J. Finan. Econom.* 15, 359–372.

Martini, C. (1999) Propagation of convexity by Markovian and martingalian semigroup. *Potential Anal.* 10, 133–175.

Matytsin, A. (1999) Modelling volatility and volatility derivatives. Working paper.

Melick, W.R., Thomas, C.P. (1997) Recovering an asset's implied PDF from option prices: an application to crude oil during the Gulf crisis. *J. Finan. Quant. Anal.* 32, 91–115.

Melino, A., Turnbull, S.M. (1990) Pricing foreign currency options with stochastic volatility. *J. Econometrics* 45, 239–265.

Mercurio, F. (2005) Pricing of inflation-indexed derivatives. *Quant. Finance* 5, 289–302.

Mercurio, F., Runggaldier, W.J. (1993) Option pricing for jump-diffusions: approximations and their interpretation. *Math. Finance* 3, 191–200.

Mercurio, F., Vorst, T.C.F. (1996) Option pricing with hedging at fixed trading dates. *Appl. Math. Finance* 3, 135–158.

Merrick, J.J. (1990) *Financial Futures Markets: Structure, Pricing and Practice*. Harper and Row (Ballinger), New York.

Merton, R.C. (1973) Theory of rational option pricing. *Bell J. Econom. Manag. Sci.* 4, 141–183.

Merton, R.C. (1974) On the pricing of corporate debt: the risk structure of interest rates. *J. Finance* 29, 449–470.

Merton, R.C. (1976) Option pricing when underlying stock returns are discontinuous. *J. Finan. Econom.* 3, 125–144.

Merton, R.C. (1980) On estimating the expected return on the market: an exploratory investigation. *J. Finan. Econom.* 8, 323–361.

Merton, R.C. (1990) *Continuous-Time Finance*. Basil Blackwell, Oxford.

Meulbroek, L. (1992) A comparison of forward and futures prices of an interest rate-sensitive financial assets. *J. Finance* 47, 381–396.

Mikkelsen, P. (2002) Cross-currency LIBOR market models. Working paper, The Aarhus School of Business.

Mikosch, T. (1999) *Elementary Stochastic Calculus with Finance in View*. World Scientific, Singapore.

Mills, T.C. (1993) *The Econometric Modelling of Financial Time Series*. Cambridge University Press, Cambridge.

Miltersen, K.R. (1994) An arbitrage theory of the term structure of interest rates. *Ann. Appl. Probab.* 4, 953–967.

Miltersen, K., Sandmann, K., Sondermann, D. (1997) Closed form solutions for term structure derivatives with log-normal interest rates. *J. Finance* 52, 409–430.

Mittnik, S., Rachev, S.T. (1993) Modeling asset returns with alternative stable distributions. *J. Econometric Rev.* 12, 261–330, 347–389.

Miura, R. (1992) A note on look-back options based on order statistics. *Hitotsubashi J. Commerce Manag.* 27, 15–28.

Miyahara, Y. (2006) GLP and MEMM pricing models and related problems. In: *Stochastic Processes and Applications to Mathematical Finance*, J. Akahori et al., eds., World Scientific, Singapore.

Modest, D.M. (1984) On the pricing of stock index futures. *J. Portfolio Manag.* 10, 51–57.

Monat, P., Stricker, C. (1993) Fermeture de $G_T(\Theta)$ et de $c + G_T(\Theta)$. Working paper, Université de Franche-Comté, Besançon.

Monat, P., Stricker, C. (1995) Föllmer-Schweizer decomposition and mean-variance hedging of general claims. *Ann. Probab.* 23, 605–628.

Moraleda, J.M., Pelsser, A. (1996) Forward versus spot interest-rate models of the term structure: an empirical comparison. *J. Derivatives* 7(3), 9–21.

Moraleda, J.M., Vorst, T.C.F. (1997) Pricing American interest rate claims with humped volatility models. *J. Bank. Finance* 21, 1131–1157.

Morse, J.N. (1988) Index futures and the implied volatility of options. *Rev. Futures Markets* 7, 324–333.

Morton, A.J. (1989) Arbitrage and martingales. Doctoral dissertation, Cornell University, Ithaca.

Mulinacci, S. (1996) An approximation of American option prices in a jump-diffusion model. *Stochastic Process. Appl.* 62, 1–17.

Mulinacci, S., Pratelli, M. (1998) Functional convergence of Snell envelopes: applications to American options approximations. *Finance Stochast.* 2, 311–327.

Müller, S. (1985) *Arbitrage Pricing of Contingent Claims. Lecture Notes in Econom. and Math. Systems 254.* Springer, Berlin Heidelberg New York.

Müller, S. (1989) On complete securities markets and the martingale property of securities prices. *Econom. Lett.* 31, 37–41.

Musiela, M. (1993) Stochastic PDEs and term structure models. *Journées Internationales de Finance, IGR-AFFI.* La Baule, June 1993.

Musiela, M. (1994) Nominal annual rates and lognormal volatility structure. Working paper, University of New South Wales.

Musiela, M. (1995) General framework for pricing derivative securities. *Stochastic Process. Appl.* 55, 227–251.

Musiela, M., Rutkowski, M. (1997) Continuous-time term structure models: forward measure approach. *Finance Stochast.* 1, 261–291.

Musiela, M., Sondermann, D. (1993) Different dynamical specifications of the term structure of interest rates and their implications. Working paper, University of Bonn.

Musiela, M., Zariphopoulou, T. (2004a) A valuation algorithm for indifference prices in incomplete markets. *Finance Stochast.* 8, 399–414.

Musiela, M., Zariphopoulou, T. (2004b) An example of indifference prices under exponential preferences. *Finance Stochast.* 8, 229–239.

Musiela, M., Turnbull, S.M., Wakeman, L.M. (1993) Interest rate risk management. *Rev. Futures Markets* 12, 221–261.

Myneni, R. (1992) The pricing of the American option. *Ann. Appl. Probab.* 2, 1–23.

Nachman, D. (1989) Spanning and completeness with options. *Rev. Finan. Stud.* 1, 311–328.

Naik, V., Lee, M. (1990) General equilibrium pricing of options on the market portfolio with discontinuous returns. *Rev. Finan. Stud.* 3, 493–521.

Natenberg, S. (1994) *Option Volatility and Pricing: Advanced Trading Strategies and Techniques*. Probus Publ. Co., Chicago (Illinois).

Ncube, M., Satchell, S. (1997) The statistical properties of the Black-Scholes option price. *Math. Finance* 7, 287–305.

Neftci, S.N. (1996) *An Introduction to the Mathematics of Financial Derivatives*. Academic Press, New York.

Nelson, D.B. (1990) ARCH models as diffusion approximations. *J. Econometrics* 45, 7–38.

Nelson, D.B. (1991) Conditional heteroskedasticity in asset returns: a new approach. *Econometrica* 59, 347–370.

Nelson, D., Ramaswamy, K. (1990) Simple binomial processes as diffusion approximations in financial models. *Rev. Finan. Stud.* 3, 393–430.

Nelson, C., Siegel, A. (1987) Parsimonious modelling of yield curves. *J. Business* 60, 473–489.

Neuberger, A. (1990a) Pricing swap options using the forward swap market. Working paper, London Business School.

Neuberger, A. (1990b) Volatility trading. Working paper, London Business School.

Ng, N. (1987) Detecting spot prices forecasts in futures prices using casuality tests. *Rev. Futures Markets* 6, 250–267.

Nielsen, J.Aa., Sandmann, K. (1996) The pricing of Asian options under stochastic interest rates. *Appl. Math. Finance* 3, 209–236.

Nielsen, L.T. (1999) *Pricing and Hedging of Derivative Securities*. Oxford University Press, Oxford.

Nyborg, K.G. (1996) The use and pricing of convertible bonds. *Appl. Math. Finance* 3, 167–190.

Omberg, E. (1987a) The valuation of American put options with exponential exercise policies. *Adv. in Futures Options Res.* 2, 117–142.

Omberg, E. (1987b) A note on the convergence of binomial pricing and compound option models. *J. Finance* 42, 463–470.

Omberg, E. (1988) Efficient discrete time jump process models in option pricing. *J. Finan. Quant. Anal.* 23, 161–174.

Øksendal, B. (2003) *Stochastic Differential Equations*. 6th edition. Springer, Berlin Heidelberg New York.

Pagès H. (1987) Optimal consumption and portfolio policies when markets are incomplete. Working paper, MIT.

Papanicolaou, G., Sircar, K.R. (1999) Mean-reverting stochastic volatility. *Internat. J. Theor. Appl. Finance* 3, 101–142.

Park, H.Y., Chen, A.H. (1985) Differences between futures and forward prices: a further investigation of the marking-to-market effects. *J. Futures Markets* 5, 77–88.

Park, H.Y., Sears, R.S. (1985) Estimating stock index futures volatility through the prices of their options. *J. Futures Markets* 5, 223–237.

Parkinson, M. (1977) Option pricing: the American put. *J. Business* 50, 21–36.

Parkinson, M. (1980) The extreme value method for estimating the variance of the rate of return. *J. Business* 53, 61–66.

Pearson, N.D., Sun, T.-S. (1994) Exploiting the conditional density in estimating the term structure: an application to the Cox, Ingersoll and Ross model. *J. Finance* 49, 1279–1304.

Pedersen, M.B. (1998) Calibrating Libor market models. Working paper.

Pelsser, A. (2000a) *Efficient Methods for Valuing Interest Rate Derivatives*. Springer, Berlin Heidelberg New York.

Pelsser, A. (2000b) Pricing double barrier options using Laplace transforms. *Finance Stochast.* 4, 95–104.

Pelsser, A., Pietersz, R. (2003) Risk managing Bermudan swaptions in the Libor BGM model. Working paper.

Pelsser, A., Pietersz, R., Regenmortel, M. (2002) Fast drift-approximated pricing in the BGM model. Working paper.

Pham, H., Touzi, N. (1996) Equilibrium state prices in a stochastic volatility model. *Math. Finance* 6, 215–236.

Pham, H., Rheinländer, T., Schweizer, M. (1998) Mean-variance hedging for continuous processes: new proofs and examples. *Finance Stochast.* 2, 173–198.

Pietersz, R. (2005) *Pricing Models for Bermudan-style Interest Rate Derivatives*. Erasmus University Rotterdam.

Pietersz, R., van Regenmortel, M. (2005) Generic market models. Working paper.

Piterbarg, V.V. (2003a) A stochastic volatility forward Libor model with a term structure of volatility smiles. Working paper, Bank of America.

Piterbarg, V.V. (2003b) Mixture of models: A simple recipe for a . . . hangover? Working paper, Bank of America.

Piterbarg, V.V. (2003c) A practitioner's guide to pricing and hedging callable Libor exotics in forward Libor models. Working paper.

Piterbarg, V.V. (2003d) Computing deltas of callable Libor exotics in a forward Libor model. Working paper.

Pitman, J.W., Yor, M. (1996) Quelques identités en loi pour les processus de Bessel. *Astérisque* 236, 249–276.

Platen, E. (1999) A short term interest rate model. *Finance Stochast.* 3, 215–225.

Platen, E. (2001) A benchmark model for financial markets. Working paper, University of Technology, Sydney.

Platen, E., Rebolledo, R. (1994) Pricing via anticipative stochastic calculus. *Ann. Appl. Probab.* 26, 1006–1021.

Platen, E., Schweizer, M. (1998) On feedback effects from hedging derivatives. *Math. Finance* 8, 67–84.

Pliska, S.R. (1986) A stochastic calculus model of continuous trading: optimal portfolios. *Math. Oper. Res.* 11, 371–382.

Pliska, S.R. (1997) *Introduction to Mathematical Finance: Discrete Time Models*. Blackwell Publishers, Oxford.

Polakoff, M.A., Dizz, F. (1992) The theoretical source of autocorrelation in forward and futures price relationships. *J. Futures Markets* 12, 459–473.

Popova, I., Ritchken, P., Woyczynski, W. (1995) Option pricing with fat tails: the case for Paretian-stable distributions. Working paper, Case Western Reserve University, Cleveland.

Poterba, J.M., Summers, L.H. (1988) Mean reversion in stock prices: evidence and implications. *J. Finan. Econom.* 22, 27–60.

Praetz, P.D. (1972) The distribution of share price changes. *J. Business* 45, 49–55.

Pratelli, M. (1996) Quelques résultats du calcul stochastique et leur application aux marchés financiers. *Astérisque* 236, 277–290.

Prigent, J.-L. (2003) *Weak Convergence of Financial Markets*. Springer, Berlin Heidelberg New York.

Prisman, E.Z. (1986) Valuation of risky assets in arbitrage free economies with frictions. *J. Finance* 41, 545–560.

Protter, P. (2003) *Stochastic Integration and Differential Equations*. 2nd ed. Springer, Berlin Heidelberg New York.

Pye, G. (1974) Gauging the default premium. *Finan. Analysts J.* 30(1), 49–52.

Rachev, S.T., Rüschendorf, L. (1994) Models for option prices. *Theory Probab. Appl.* 39, 120–152.

Rachev, S.T., Weron, A., Weron, R. (1996) CED model for asset returns and fractal market hypothesis. Working paper, University of California.

Rachev, S.T., Weron, A., Weron, R. (1997) Conditionally exponential dependence model for asset returns. *Appl. Math. Lett.* 10, 5–9.

Rady, S. (1997) Option pricing in the presence of natural boundaries and a quadratic diffusion term. *Finance Stochast.* 1, 331–344.

Rady, S., Sandmann, K. (1994) The direct approach to debt option pricing. *Rev. Futures Markets* 13, 461–514.

Ramaswamy, K., Sundaresan, S.M. (1985) The valuation of options on futures contracts. *J. Finance* 40, 1319–1340.

Ramaswamy, K., Sundaresan, S.M. (1986) The valuation of floating-rate instruments, theory and evidence. *J. Finan. Econom.* 17, 251–272.

Rebonato, R. (1998) *Interest Rate Option Models: Understanding, Analysing and Using Models for Exotic Interest-Rate Options.* J. Wiley, Chichester.

Rebonato, R. (1999a) On the simultaneous calibration of multifactor lognormal interest rate models to Black volatilities and to the correlation matrix. *J. Comput. Finance* 2, 5–27.

Rebonato, R. (1999b) On the pricing implications of the joint lognormal assumption for the swaption and cap markets. *J. Comput. Finance* 2, 57–76.

Rebonato, R. (2000) *Volatility and Correlation in the Pricing of Equity, FX and Interest-Rate Options.* J. Wiley, Chichester.

Rebonato, R. (2002) *Modern Pricing of Interest-Rate Derivatives: The Libor Market Model and Beyond.* Princeton University Press, Princeton.

Rebonato, R., Joshi, M. (2001) A joint empirical and theoretical investigation of the modes of deformation of swaption matrices: implications for model choice. *Internat. J. Theor. Appl. Finance* 5, 667–694.

Redhead, K. (1996) *Financial Derivatives: An Introduction to Futures, Forwards, Options and Swaps.* Prentice-Hall, Englewood Cliffs (New Jersey).

Reiner, E. (1992) Quanto mechanics. *Risk* 5(3), 59–63.

Renault, E., Touzi, N. (1996) Option hedging and implied volatilities in a stochastic volatility model. *Math. Finance* 6, 279–302.

Rendleman, R., Bartter, B. (1979) Two-state option pricing. *J. Finance* 34, 1093–1110.

Rendleman, R., Bartter, B. (1980) The pricing of options on debt securities. *J. Finan. Quant. Anal.* 15, 11–24.

Revuz, D., Yor, M. (1999) *Continuous Martingales and Brownian Motion.* 3rd ed. Springer, Berlin Heidelberg New York.

Rich, D. (1994) The mathematical foundations of barrier option pricing theory. *Adv. in Futures Options Res.* 7, 267–312.

Richard, S.F. (1978) An arbitrage model of the term structure of interest rates. *J. Finan. Econom.* 6, 33–57.

Richard, S.F., Sundaresan, M. (1981) A continuous time equilibrium model of forward and futures prices in a multigood economy. *J. Finan. Econom.* 9, 347–372.

Richardson, M., Smith, T. (1993) A test for multivariate normality in stock returns. *J. Business* 66, 295–321.

Ritchey, P. (1990) Call option valuation for discrete normal mixtures. *J. Finan. Res.* 13, 285–296.

Ritchken, P. (1987) *Options: Theory, Strategy and Applications.* Scott, Foresman and Co., Glenview (Illinois).

Ritchken, P. (1995) On pricing barrier options. *J. Derivatives*, 19–28.

Ritchken, P., Sankarasubramanian, L. (1995) Volatility structures of forward rates and the dynamics of the term structure. *Math. Finance* 5, 55–72.

Ritchken, P., Trevor, R. (1997) Pricing options under generalized GARCH and stochastic volatility processes. Working paper.

Roberts, G.O., Shortland, C.F. (1997) Pricing barrier options with time-dependent coefficients. *Math. Finance* 7, 95–105.

Rockafellar, R.T. (1970) *Convex Analysis*. Princeton University Press, Princeton.

Rogers, L.C.G. (1994) Equivalent martingale measures and no-arbitrage. *Stochastics Stochastics Rep.* 51, 41–49.

Rogers, L.C.G. (1995) Which model for term-structure of interest rates should one use? In: *IMA Vol.65: Mathematical Finance*, M.H.A. Davis et al., eds. Springer, Berlin Heidelberg New York, pp. 93–116.

Rogers, L.C.G. (1996) Gaussian errors. *Risk* 9(1), 42–45.

Rogers, L.C.G. (1997a) Arbitrage from fractional Brownian motion. *Math. Finance* 7, 95–105.

Rogers, L.C.G. (1997b) The potential approach to the term structure of interest rates and foreign exchange rates. *Math. Finance* 7, 157–176.

Rogers, L.C.G. (1998) Volatility estimation with price quanta. *Math. Finance* 8, 277–290.

Rogers, L.C.G., Satchell, S.E. (1991) Estimating variance from high, low and closing prices. *Ann. Appl. Probab.* 1, 504–512.

Rogers, L.C.G., Shi, Z. (1995) The value of an Asian option. *J. Appl. Prob.* 32, 1077–1088.

Rogers, L.C.G., Stapleton, E.J. (1998) Fast accurate binomial pricing. *Finance Stochast.* 2, 3–17.

Roll, R. (1977) An analytic valuation formula for unprotected American call options on stocks with known dividends. *J. Finan. Econom.* 5, 251–258.

Romagnoli, S., Vargiolu, T. (2000) Robustness of the Black-Scholes approach in the case of option on several assets. *Finance Stochast.* 4, 325–341.

Romano, M., Touzi, N. (1997) Contingent claims and market completeness in a stochastic volatility model. *Math. Finance* 7, 399–412.

Ross, S.A. (1976a) The arbitrage theory of capital asset pricing. *J. Econom. Theory* 13, 341–360.

Ross, S.A. (1976b) Options and efficiency. *Quart. J. Econom.* 90, 75–89.

Ross, S.A. (1978) A simple approach to the valuation of risky streams. *J. Business* 51, 453–475.

Rossi, A. (2002) The Britten-Jones and Neuberger smile-consistent with stochastic volatility option pricing model: a further analysis. *Internat. J. Theor. Appl. Finance* 5, 1–31.

Rouge, R., El Karoui, N. (2000) Pricing via utility maximization and entropy. *Math. Finance* 10, 259–277.

Rubinstein, M. (1976) The valuation of uncertain income streams and the pricing of options. *Bell J. Econom.* 7, 407–425.

Rubinstein, M. (1983) Displaced diffusion option pricing. *J. Finance* 38, 213–217.

Rubinstein, M. (1984) A simple formula for the expected rate of return of an option over a finite holding period. *J. Finance* 39, 1503–1509.

Rubinstein, M. (1985) Nonparametric tests of alternative option pricing models using all reported trades and quotes on the thirty most active CBOE option classes from August 23, 1976 through August 31, 1978. *J. Finance* 40, 455–480.

Rubinstein, M. (1991a) Somewhere over the rainbow. *Risk* 4(11), 61–63.

Rubinstein, M. (1991b) Exotic options. Working paper, University of California, Berkeley.

Rubinstein, M. (1994) Implied binomial trees. *J. Finance* 49, 771–818.

Rubinstein, M. (1998) Edgeworth binomial trees. *J. Derivatives* 7, 66–82.

Rubinstein, M., Reiner, E. (1991) Breaking down the barriers. *Risk* 4(8), 28–35.

Rutkowski, M. (1994) The early exercise premium representation of foreign market American options. *Math. Finance* 4, 313–325.

Rutkowski, M. (1996a) Valuation and hedging of contingent claims in the HJM model with deterministic volatilities. *Appl. Math. Finance* 3, 237–267.

Rutkowski, M. (1996b) Risk-minimizing hedging of contingent claims in incomplete models of financial markets. *Mat. Stos.* 39, 41–73.

Rutkowski, M. (1997) A note on the Flesaker-Hughston model of term structure of interest rates. *Appl. Math. Finance* 4, 151–163.

Rutkowski, M. (1998b) Dynamics of spot, forward, and futures Libor rates. *Internat. J. Theor. Appl. Finance* 1, 425–445.

Rutkowski, M. (1999a) Models of forward Libor and swap rates. *Appl. Math. Finance* 6, 1–32.

Rutkowski, M. (1999b) Self-financing trading strategies for sliding and rolling-horizon bonds. *Math. Finance* 9, 361–385.

Rutkowski, M. (2001) Modelling of forward Libor and swap rates. In: *Option Pricing, Interest Rates and Risk Management*, E. Jouini, J. Cvitanić and M. Musiela, eds. Cambridge University Press, Cambridge, pp. 336–395.

Ruttiens, A. (1990) Currency options on average exchange rates pricing and exposure management. *20th Annual Meeting of the Decision Science Institute,* New Orleans 1990.

Rydberg, T.H. (1997) A note on the existence of unique equivalent martingale measures in a Markovian setting. *Finance Stochast.* 1, 251–257.

Rydberg, T.H. (1999) Generalized hyperbolic diffusions with applications towards finance. *Math. Finance* 9, 183–201.

Sabanis, S. (2002) Stochastic volatility. *Internat. J. Theor. Appl. Finance* 5, 515–530.

Salopek, D. (1997) *American Put Options.* CRC Press.

Samperi, D. (2002) Calibrating a diffusion pricing model with uncertain volatility: regularization and stability. *Math. Finance* 12, 71–87.

Samuelson, P.A. (1965) Rational theory of warrant prices. *Indust. Manag. Rev.* 6, 13–31.

Samuelson, P.A. (1973) Mathematics of speculative prices. *SIAM Rev.* 15, 1–42.

Samuelson, P.A., Merton, C. (1969) A complete model of warrant pricing that maximizes utility. *Indust. Manag. Rev.* 10, 17–46.

Sandmann, K. (1993) The pricing of options with an uncertain interest rate: a discrete-time approach. *Math. Finance* 3, 201–216.

Sandmann, K., Sondermann, D. (1989) A term structure model and the pricing of interest rate options. Working paper, University of Bonn.

Sandmann, K., Sondermann, D. (1993a) A term structure model and the pricing of interest rate derivatives. *Rev. Futures Markets* 12, 391–423.

Sandmann, K., Sondermann, D. (1993b) On the stability of lognormal interest rate models. Working paper, University of Bonn.

Sandmann, K., Sondermann, D. (1997) A note on the stability of lognormal interest rate models and the pricing of Eurodollar futures. *Math. Finance* 7, 119–125.

Sandmann, K., Sondermann, D., Miltersen, K.R. (1995) Closed form term structure derivatives in a Heath-Jarrow-Morton model with log-normal annually compounded interest rates. In: *Proc. 7th Annual European Futures Research Symp. Bonn, 1994.* Chicago Board of Trade, pp. 145–165.

Santa Clara, P., Sornette, D. (2001) The dynamics of the forward interest rate curve as a stochastic string shock. *Rev. Finan. Stud.* 14, 149–185.

Sato, K. (1991) Self similar processes with independent increments. *Probab. Theory Rel. Fields* 89, 285–300.

Sato, K. (1999) *Lévy Processes and Infinitely Divisible Distributions.* Cambridge University Press, Cambridge.

Savine, A. (2002) A theory of volatility. In: *Mathematical Finance*, J. Yong, ed. World Scientific, Singapore, pp. 151–167.

Schachermayer, W. (1992) A Hilbert space proof of the fundamental theorem of asset pricing in finite discrete time. *Insurance Math. Econom.* 11, 249–257.

Schachermayer, W. (1993) A counter-example to several problems in the theory of asset pricing. *Math. Finance* 3, 217–229.

Schachermayer, W. (1994) Martingale measures for discrete time processes with infinite horizon. *Math. Finance* 4, 25–55.

Schachermayer, W., Teichmann, J. (2008) How close are the option pricing formulas of Bachelier and BlackMerton-Scholes. *Math. Finance* 18, 155–170.

Schaefer, S.M., Schwartz, E.S. (1984) A two-factor model of the term structure: an approximate analytical solution. *J. Finan. Quant. Anal.* 19, 413–424.

Schaefer, S.M., Schwartz, E.S. (1987) Time-dependent variance and the pricing of bond options. *J. Finance* 42, 1113–1128.

Scheinkman, J.A., LeBaron, B. (1989) Nonlinear dynamics and stock returns. *J. Business* 62, 311–337.

Schlögl, E. (2002) A multicurrency extension of the lognormal interest rate market models. *Finance Stochast.* 6, 173–196.

Schlögl, E., Schlögl, L. (1997) A tractable term structure model with endogenous interpolation and positive interest rates. Working paper, University of Bonn.

Schlögl, E., Schlögl, L. (2000) A square root interest rate model fitting discrete initial term structure data. *Appl. Math. Finance* 7, 183–209.

Schlögl, E., Sommer, D. (1994) On short rate processes and their implications for term structure movements. Working paper, University of Bonn.

Schlögl, E., Sommer, D. (1998) Factor models and the shape of the term structure. *J. Finan. Engrg* 7, 79–88.

Schmalensee, R., Trippi, R. (1978) Common stock volatility expectations implied by option premia. *J. Finance* 33, 129–147.

Schmidt, W.M. (1997) On a general class of one-factor models for the term structure of interest rates. *Finance Stochast.* 1, 3–24.

Schönbucher, P.J. (1998) Term structure modelling of defaultable bonds. *Rev. Derivatives Res.* 2, 161–192.

Schönbucher, P.J. (1999) A market model of stochastic implied volatility. *Phil. Trans. Royal Society* A 357/1758, 2071–2092.

Schönbucher, P.J. (2003) *Credit Derivatives Pricing Models*. J.Wiley, Chichester.

Schoenmakers, J., Coffey, B. (1999) Libor rates models, related derivatives and model calibration. Working paper.

Schoutens, W. (2003) *Lévy Processes in Finance: Pricing Financial Derivatives*. J.Wiley, Chichester.

Schroder, M. (1989) Computing the constant elasticity of variance option pricing formula. *J. Finance* 44, 211–219.

Schwartz, E.S. (1977) The valuation of warrants: implementing a new approach. *J. Finan. Econom.* 4, 79–93.

Schweizer, M. (1990) Risk-minimality and orthogonality of martingales. *Stochastics Stochastics Rep.* 30, 123–131.

Schweizer, M. (1991) Option hedging for semimartingales. *Stochastic Process. Appl.* 37, 339–363.

Schweizer, M. (1992a) Mean-variance hedging for general claims. *Ann. Appl. Probab.* 2, 171–179.

Schweizer, M. (1992b) Martingale densities for general asset prices. *J. Math. Econom.* 21, 363–378.

Schweizer, M. (1994a) A projection result for semimartingales. *Stochastics Stochastics Rep.* 50, 175–183.

Schweizer, M. (1994b) Risk-minimizing hedging strategies under restricted information. *Math. Finance* 4, 327–342.

Schweizer, M. (1994c) Approximating random variables by stochastic integrals. *Ann. Probab.* 22, 1536–1575.

Schweizer, M. (1995a) On the minimal martingale measure and the Föllmer-Schweizer decomposition. *Stochastic Anal. Appl.* 13, 573–599.

Schweizer, M. (1995b) Variance-optimal hedging in discrete-time. *Math. Oper. Res.* 20, 1–32.

Schweizer, M. (1996) Approximation pricing and the variance-optimal martingale measure. *Ann. Appl. Probab.* 24, 206–236.

Schweizer, M. (2001) A guided tour through quadratic hedging approaches. In: *Option Pricing, Interest Rates and Risk Management*, E. Jouini, J. Cvitanić and M. Musiela, eds. Cambridge University Press, pp. 509–537.

Scott, L.O. (1987) Option pricing when the variance changes randomly: theory, estimation, and an application. *J. Finan. Quant. Anal.* 22, 419–438.

Scott, L.O. (1991) Random-variance option pricing: empirical tests of the model and delta-sigma hedging. *Adv. in Futures Options Res.* 5, 113–135.

Scott, L.O. (1997) Pricing stock options in a jump-diffusion model with stochastic volatility and interest rates: applications of Fourier inversion method. *Math. Finance* 7, 413–426.

Selby, M., Hodges, S. (1987) On the evaluation of compound options. *Manag. Sci.* 33, 347–355.

Sethi, S.P. (1997) *Optimal Consumption and Investment with Bankruptcy.* Kluwer Academic Publishers, Dordrecht.

Seydel, P.J. (2002) *Tools for Computational Finance.* Springer, Berlin Heidelberg New York.

Sharpe, W. (1978) *Investments.* Prentice-Hall, Englewood Cliffs (New Jersey).

Shastri, K., Tandon, K. (1986) Valuation of foreign currency options: some empirical tests. *J. Finan. Quant. Anal.* 21, 145–160.

Shastri, K., Wethyavivorn, K. (1987) The valuation of currency options for alternate stochastic processes. *J. Finan. Res.* 10, 283–293.

Shea, G.S. (1984) Pitfalls in smoothing interest rate term structure data: equilibrium models and spline approximations. *J. Finan. Quant. Anal.* 19, 253–270.

Shea, G.S. (1985) Interest rate term structure estimation with exponential splines: a note. *J. Finance* 40, 319–325.

Sheedy, E.S., Trevor, R.G. (1996) Evaluating the performance of portfolios with options. Working paper, Macquarie University.

Sheedy, E.S., Trevor, R.G. (1999) Further analysis of portfolios with options. Working paper, Macquarie University.

Shephard, N. (1996) Statistical aspects of ARCH and stochastic volatility. In: *Time Series Models*, D. Cox, D. Hinkley and O. Barndorff-Nielsen, eds. Chapman and Hall, London, pp. 1–67.

Shepp, L.A., Shiryaev, A.N. (1993) The Russian option: reduced regret. *Ann. Appl. Probab.* 3, 631–640.

Shepp, L.A., Shiryaev, A.N. (1994) A new look at the "Russian option". *Theory Probab. Appl.* 39, 103–119.

Shiga, T., Watanabe, S. (1973) Bessel diffusions as one parameter family of diffusion processes. *Z. Wahrscheinlichkeitstheorie verw. Gebiete* 27, 37–46.

Shimko, D. (1993) Bounds of probability. *Risk* 6(4), 33–37.

Shirakawa, H. (1991) Interest rate option pricing with Poisson-Gaussian forward rate curve processes. *Math. Finance* 1, 77–94.

Shirakawa, H. (1999) Evaluation of yield spread for credit risk. *Adv. Math. Econ.* 1, 83–97.

Shiryaev, A.N. (1984) *Probability.* Springer, Berlin Heidelberg New York.

Shiryaev, A.N. (1994) On some basic concepts and some basic stochastic models used in finance. *Theory Probab. Appl.* 39, 1–13.

Shiryaev, A.N. (1999) *Essentials of Stochastic Finance: Facts, Models, Theory.* World Scientific, Singapore.

Shiryaev, A.N., Kabanov, Y.M., Kramkov, D.O., Melnikov, A.V. (1994a) Toward the theory of pricing of options of both European and American types. I. Discrete time. *Theory Probab. Appl.* 39, 14–60.

Shiryaev, A.N., Kabanov, Y.M., Kramkov, D.O., Melnikov, A.V. (1994b) Toward the theory of pricing of options of both European and American types. II. Continuous time. *Theory Probab. Appl.* 39, 61–102.

Shreve, S.E. (1991) A control theorist's view of asset pricing. In: *Applied Stochastic Analysis,* M.H.A. Davis and R.J. Elliott, eds. Gordon and Breach, New York, pp. 415–445.

Shreve, S.E. (2004) *Stochastic Calculus for Finance I. The Binomial Asset Pricing Model.* Springer, Berlin Heidelberg New York.

Shreve, S.E. (2005) *Stochastic Calculus for Finance II. Continuous-Time Model.* Springer, Berlin Heidelberg New York.

Shreve, S.E., Vecer, J. (2000) Options on a traded account: vacation calls, vacation puts and passport options. *Finance Stochast.* 4, 255–274.

Singleton, K., Umantsev, L. (2002) Pricing coupon-bond options and swaptions in affine term structure models. *Math. Finance* 12, 427–446.

Sidenius, J. (2000) LIBOR market models in practice. *J. Comput. Finance* 3(3), 5–26.

Sin, C. (1998) Complications with stochastic volatility models. *Adv. Appl. Probab.* 30, 256–268.

Sîrbu, M., Pikovsky, I., Shreve, S.E. (2004) Perpetual convertible bonds. *SIAM J. Control Optim.* 43, 58–85.

Sîrbu, M., Shreve, S.E. (2005) A two-person game for pricing convertible bonds. Working paper.

Sircar, K.R., Papanicolaou, G.C. (1998) General Black-Scholes models accounting for increased market volatility from hedging strategies. *Appl. Math. Finance* 5, 45–82.

Sircar, K.R., Papanicolaou, G.C. (1999) Stochastic volatility, smile and asymptotics. *Appl. Math. Finance* 6, 107–145.

Skiadopoulos, G. (2001) Volatility smile consistent option models: a survey. *Internat. J. Theor. Appl. Finance* 4, 403–437.

Skinner, D.J. (1989) Options markets and stock return volatility. *J. Finan. Econom.* 23, 61–78.

Smith, C.W., Jr. (1976) Option pricing: a review. *J. Finan. Econom.* 3, 3–51.

Soner, H.M., Shreve, S.E., Cvitanić, J. (1995) There is no nontrivial hedging portfolio for option pricing with transaction costs. *Ann. Appl. Probab.* 5, 327–355.

Stambaugh, R.F. (1988) The information in forward rates: implications for models of the term structure. *J. Finan. Econom.* 21, 41–70.

Stanton, R. (1997) A nonparametric model of term structure dynamics and the market price of interest rate risk. *J. Finance* 52, 1973–2002.

Stapleton, R.C., Subrahmanyam, A. (1984) The valuation of multivariate contingent claims in discrete time models. *J. Finance* 39, 207–228.

Stapleton, R.C., Subrahmanyam, A. (1999) The term structure of interest-rate futures prices. Working paper.

Steele, J.M. (2000) *Stochastic Calculus and Financial Applications*. Springer, Berlin Heidelberg New York.

Stein, E.M., Stein, J.C. (1991) Stock price distributions with stochastic volatility: an analytic approach. *Rev. Finan. Stud.* 4, 727–752.

Stigler, R.J. (1990) The term structure of interest rates. In: *Handbook of Monetary Economics, Vol.I*, B.M. Friedman and F.H. Hahn, eds. North-Holland, Amsterdam New York, pp. 627–722.

Stoll, H.R., Whaley, R.E. (1994) The dynamics of stock index and stock index futures returns. *J. Finan. Quant. Anal.* 25, 441–468.

Stricker, C. (1984) Integral representation in the theory of continuous trading. *Stochastics* 13, 249–257.

Stricker, C. (1990) Arbitrage et lois de martingale. *Ann. Inst. H.Poincaré Probab. Statist.* 26, 451–460.

Stulz, R.M. (1982) Options on the minimum or the maximum of two risky assets: analysis and applications. *J. Finan. Econom.* 10, 161–185.

Sundaresan, S. (1991) Futures prices on yields, forward prices, and implied forward prices from term structure. *J. Finan. Quant. Anal.* 26, 409–424.

Sundaresan, S. (1997) *Fixed Income Markets and Their Derivatives*. South-Western College Publ., Cincinnati (Ohio).

Sutcliffe, C.M.S. (1993) *Stock Index Futures*. Chapman & Hall, London.

Swidler, S., Diltz, J.D. (1992) Implied volatilities and transaction costs. *J. Finan. Quant. Anal.* 27, 437–447.

Szakmary, A., Ors, E., Kim, J.K., Davidson, W.N. (2003) The predictive power of implied volatility: evidence from 35 futures markets. *J. Bank. Finance* 27, 2151–2175.

Szatzschneider, W. (1998) Extended Cox, Ingersoll and Ross model. Working paper. Anahuac University.

Taksar, M., Klass, M.J., Assaf, D. (1988) A diffusion model for optimal portfolio selection in the presence of brokerage fees. *Math. Oper. Res.* 13, 277–294.

Tanudjaja, S. (1996) American option valuation in Gaussian HJM. Doctoral dissertation, University of New South Wales, Sydney.

Taqqu, M.S., Willinger, W. (1987) The analysis of finite security markets using martingales. *Adv. in Appl. Probab.* 19, 1–25.

Taylor, S.J. (1986) *Modelling Financial Time Series*. J.Wiley, Chichester.

Taylor, S.J. (1994) Modeling stochastic volatility: a review and comparative study. *Math. Finance* 4, 183–204.

Taylor, S.J., Xu, X. (1993) The magnitude of implied volatility smiles: theory and empirical evidence for exchange rates. *Rev. Futures Markets* 13, 355–380.

Tehranchi, M. (2005) A note on invariant measures for HJM models. *Finance Stochast.* 9, 389–398.

Tessitore, G., Zabczyk, J. (1996) Pricing options for multinomial models. *Bull. Pol. Acad. Sci.* 44, 363–380.

Titman, S., Torous, W. (1989) Valuing commercial mortgages: an empirical investigation of the contingent claims approach to pricing risky debt. *J. Finance* 44, 345–373.

Tompkins, R.G. (2001) Implied volatility surfaces: uncovering regularities for options on financial futures. *European J. Finance* 7, 198–230.

Toft, K.B. (1996) On the mean-variance tradeoff in option replication with transactions costs. *J. Finan. Quant. Anal.* 31, 233–263.

Torous, W.N. (1985) Differential taxation and the equilibrium structure of interest rates. *J. Business Finance* 9, 363–385.

Tucker, A.L., Peterson, D.R., Scott, E. (1988) Tests of the Black-Scholes and constant elasticity of variance currency option valuation model. *J. Finan. Res.* 11, 201–213.

Turnbull, S.M. (1994) Pricing and hedging diff swaps. *J. Finan. Engrg* 2, 297–333.

Turnbull, S.M., Milne, F. (1991) A simple approach to the pricing of interest rate options. *Rev. Finan. Stud.* 4, 87–120.

Turnbull, S.M., Wakeman, L.M. (1991) A quick algorithm for pricing European average options. *J. Finan. Quant. Anal.* 26, 377–389.

Uratani, T., Utsunomiya, M. (1999) Lattice calculation for forward LIBOR model. Working paper, Hosei University.

Van Moerbeke, P. (1976) On optimal stopping and free boundary problem. *Arch. Rational Mech. Anal.* 60, 101–148.

Vargiolu, T. (1999) Invariant measures for the Musiela equation with deterministic diffusion term. *Finance Stochast.* 3, 483–492.

Vasicek, O. (1977) An equilibrium characterisation of the term structure. *J. Finan. Econom.* 5, 177–188.

Vasicek, O., Fong, H.G. (1982) Term structure modeling using exponential splines. *J. Finance* 37, 339–348.

Veiga, C. (2004) Expanding further the universe of exotic options closed pricing formulas in the Black and Scholes framework. Working paper.

Vorst, T.C.F. (1992) Prices and hedge ratios of average exchange rate options. *Internat. Rev. Finan. Anal.* 1, 179–193.

West, K.D. (1988) Bubbles, fads and stock price volatility tests: a partial evaluation. *J. Finance* 43, 639–656.

Whaley, R.E. (1981) On the valuation of American call options on stocks with known dividends. *J. Finan. Econom.* 9, 207–211.

Whaley, R.E. (1982) Valuation of American call options on dividend-paying stocks: empirical tests. *J. Finan. Econom.* 10, 29–58.

Whaley, R.E. (1986) Valuation of American futures options: theory and empirical tests. *J. Finance* 41, 127–150.

Whaley, R.E. (1993) Derivatives on market volatility: hedging tools long overdue. *J. Derivatives*, Fall, 71–84.

Whalley, E., Wilmott, P. (1997) An asymptotic analysis of an optimal hedging model for option pricing with transaction costs. *Math. Finance* 7, 307–324.

Wiggins, J.B. (1987) Option values under stochastic volatility: theory and empirical estimates. *J. Finan. Econom.* 19, 351–372.

Williams, D. (1991) *Probability with Martingales*. Cambridge University Press, Cambridge.

Willinger, W., Taqqu, M.S. (1989) Pathwise stochastic integration and applications to the theory of continuous trading. *Stochastic Process. Appl.* 32, 253–280.

Willinger, W., Taqqu, M.S. (1991) Toward a convergence theory for continuous stochastic securities market models. *Math. Finance* 1, 55–99.

Willinger, W., Taqqu, M.S., Teverovsky, V. (1999) Stock market prices and long-range dependence. *Finance Stochast.* 3, 3–13.

Wilmott, P. (1999) *Derivatives: The Theory and Practice of Financial Engineering*. J.Wiley, Chichester New York.

Wilmott, P., Dewynne, J.N., Howison, S. (1993) *Option Pricing: Mathematical Models and Computations*. Oxford Financial Press, Oxford.

Wong, B., Heyde, C.C. (2002) Change of measure in stochastic volatility models. Working paper, University of New South Wales.

Wu, L. (2002) Fast at-the-money calibration of LIBOR market model through Lagrange multipliers. *J. Comput. Finance* 6, 33–45.

Wu, L., Zhang, F. (2002) LIBOR market model: from deterministic to stochastic volatility. Working paper.

Xu, X., Taylor, S.J. (1994) The term structure of volatility implied by foreign exchange options. *J. Finan. Quant. Anal.* 29, 57–74.

Yasuoka, T. (2001) Mathematical pseudo-completion of the BGM model. *Internat. J. Theor. Appl. Finance* 4, 375–401.

Yasuoka, T. (2005) A study of lognormal LIBOR forward rate model in connection with the HJM framework. Working paper.

Yor, M. (1992a) *Some Aspects of Brownian Motion. Part I*. Birkhäuser, Basel Boston Berlin.

Yor, M. (1992b) On some exponential functionals of Brownian motion. *Adv. in Appl. Probab.* 24, 509–531.

Yor, M. (1993a) On some exponential functionals of Bessel processes. *Math. Finance* 3, 229–239.

Yor, M. (1993b) From planar Brownian windings to Asian options. *Insurance Math. Econom.* 13, 23–34.

Yor, M. (1995) The distribution of Brownian quantiles. *J. Appl. Probab.* 32, 405–416.

Yor, M. (2001) *Functionals of Brownian Motion and Related Processes*. Springer, Berlin Heidelberg New York.

Zabczyk, J. (1996) *Chance and Decision. Stochastic Control in Discrete Time*. Scuola Normale Superiore, Pisa.

Zhang, X.L. (1997) Valuation of American option in jump-diffusion models. In: *Numerical Methods in Finance*, L.C.G. Rogers and D. Talay, eds. Cambridge University Press, Cambridge, pp. 93–114.

Zhu, Y., Avellaneda, M. (1998) A risk-neutral stochastic volatility model. *Internat. J. Theor. Appl. Finance* 1, 289–310.

Index

Stochastic Modelling and Applied Probability
formerly: Applications of Mathematics

Stochastic Modelling and Applied Probability
formerly: Applications of Mathematics

7272203R0

Made in the USA
Lexington, KY
07 November 2010